W9-BOE-940

COLLEGE PHYSICS

McGraw-Hill Book Company

New York St. Louis San Francisco
Düsseldorf Johannesburg Kuala Lumpur
London Mexico Montreal
New Delhi Panama Paris São Paulo
Singapore Sydney Tokyo Toronto

Robert L. Weber
Associate Professor of Physics
The Pennsylvania State University

Kenneth V. Manning
Professor Emeritus of Physics
The Pennsylvania State University

Marsh W. White
Professor Emeritus of Physics
The Pennsylvania State University

George A. Weygand
Professor of Physics
Bridgewater State College

Fifth Edition

COLLEGE PHYSICS

College Physics

3 4 5 6 7 8 9 0 VHVH 7 9 8 7

This book was set in Century Schoolbook by York Graphic Services, Inc. The editors were Jack L. Farnsworth and J. W. Maisel; the designer was Jo Jones; the production supervisor was Joe Campanella. New drawings were done by John Cordes, J & R Technical Services, Inc.
Von Hoffman Press, Inc., was printer and binder.

Library of Congress Cataloging in Publication Data
Main entry under title:

College physics.

First ed. by R. L. Weber, K. V. Manning, and M. W. White published under title: College technical physics; 2d–4th ed. by R. L. Weber, K. V. Manning, and M. W. White published under title: College physics.
Includes bibliographical references.
1. Physics. I. Weber, Robert L., date II. Weber, Robert L., date College technical physics.
QC21.2.C63 1974 530 73-13858
ISBN 0-07-068827-3

Contents

Preface

This fifth edition of "College Physics" is designed to further the objectives established by the preceding editions, that is, to provide a solid base in physics for those students who will go on to careers in such fields as medicine, engineering, biology, and earth science and for those students who simply seek to understand the physical nature of our environment. It is hoped through this book that the student will find evidence of the continuity which exists between all sciences and between all areas of our culture and recognize those concepts common to all sciences—the "ideas that travel."

The typical student who will use this book needs more than a nonmathematical, purely descriptive survey treatment of physics. On the other hand, a highly theoretical approach is not needed. For example, the reader may not be particularly interested in the derivation of physical equations per se. By blending a clear and careful presentation of fundamental concepts and methods of classical physics with a consideration of the relationship between science and our technologically oriented society, the authors feel that they have developed a book which is sufficiently sophisticated to challenge the brighter liberal art student of today who feels a need to see the relationship between physics and the "real" world. To this end, at appropriate places throughout the book, attention is given to contemporary developments in physics. For example, an expanded and updated treatment of space physics, in which the history as well as the physical principles are considered, is presented in one chapter as well as at several other points in the book. The physiological and psychological implications of supersonic flight and sonic booms are discussed. The special physical stresses and strains that man undergoes in the present-day environment in which he goes faster, flies higher, and is exposed to greater heat and pressure changes than ever before are explored.

The relationship of physics to other fields of science is stressed in this book through references to special topics of concern to the earth sciences, such as seismology, topics of concern to the chemical sciences, such as crystal formation and radioactivity, and topics of concern to the biological sciences, such as the effect of vibration on humans and echolocation by bats.

An attempt has been made to demonstrate the cause-and-effect relationship wherever possible. For example, discussion of force is presented before that of vectors and molecular theory before temperature and thermal expansion; moreover, the application or effect of an electrical discharge in a tube is presented in the form of an expanded discussion of electronics, which includes the nature of black-and-white television transmission and a consideration of other related electron-operated devices such as the electron microscope. Also, the application of the principles of sound to acoustical problems such as hi-fidelity and stereo reproduction are considered along with additional discussion of ultrasonics and supersonics.

The authors recognize and seek to extend the reader's interest in the physics of the atom and nucleus as interpreted by the theories of relativity and quantum mechanics. To this end, the section on Modern Physics has been expanded with new or revised sections on the Lorentz contraction, matter waves, the uncertainty principle, the laser, and cosmic rays.

To help attain the goal of a coherent understanding of classical and contemporary physics, we have revised somewhat the traditional compartments of physics and their sequence. We feel that interrelations are made apparent and fruitful by arranging the book into four major areas of physics, in each of which there is considerable unity in the concepts, principles, and methods of analysis used. These divisions are

Physics of particles
Wave physics
Physics of fields
Modern physics

It has been the experience of many of the physics professors who have offered suggestions about this revision that students find it easier to make the transition from the study of the motion of particles to a study of motion in a wave form than it is for them to embark directly on an investigation of electromagnetic fields after completing the study of Physics of Particles (mechanics). For this reason, in this edition the section on Wave Physics, which includes waves, sound, and optics, is presented before the Physics of Fields, electricity and magnetism. Some rearrangement in the order of chapters has also been made in the revision so that the story line would not be interrupted.

The authors acknowledge the wish of some professors to vary their course sequence. For this reason the book has been designed so that alternate sequences may be chosen. For example, an instructor who wishes to present the Physics of Fields following mechanics can readily do so by scheduling Chapters 32 to 44 to follow Chapter 18.

Perhaps more important than the rearrangement of topics in this edition has been a determined effort to break down communication barriers between the reader, the author, and the physics. Wherever possible, the authors have talked directly to the student. This often takes the form of helpful hints or notes provided at strategic points in the book. A special effort has been made to explain to the reader why the next step in the development of a concept is a logical one. Unnecessary scientific terms are avoided and careful definitions of unfamiliar terms and words are provided. A larger number of solved examples have been provided within the text to help the students see how the physical concepts are applied. To further facilitate communication, more than 150 new illustrations have been provided in this text, along with the revision of many previously existing illustrations.

The unusually extensive and varied questions and problems at the ends of the chapters have been revised, with more than 750 problems and an equal number of new discussion questions having been added. The discussion questions were designed to aid in developing and sharpening the reasoning ability of the student. The fact that some can be answered in various ways might often lead to spirited discussion in the classroom.

Mathematical and other complexities have not been introduced into the problems for a deliberate reason: emphasis has been placed instead on the illustration of physical principles. A few derivations and problems involving symbolic solutions have been included, however. Answers have been given to even-numbered problems; and a systematic plan for the solution of problems is included in the Appendix for the benefit of the students.

The mathematical background expected of students using this book includes a few basic trigonometric ideas, algebra, and the analytical ability associated with freshman-sophomore standing in science and engineering courses. No calculus appears in the body of the text, but use of the delta notation will permit many students to follow the calculus derivations included in the Appendix. This enables the experienced teacher to place more or less emphasis upon a mathematical approach to the topics considered. Vector notation has been introduced where it is appropriate and helpful in dealing with the physical concepts. Extensive use of vector representation has been avoided, particularly in cases in which understanding of the physical principles expressed in vector notation would require an extensive understanding of vector properties.

Electrical concepts are developed through a discussion of charges at rest and in motion, without dependence upon a study of fictitious isolated magnetic poles. Field concepts in electrostatics and magnetism are emphasized. Conservation laws are given special attention as the basis for some of the great generalizations of science. Particular emphasis is given to wave phenomena. In optics, the refraction of light at a spherical surface is introduced first, to lead naturally up to refraction at successive spherical surfaces in lenses.

The number of systems of units that must be considered by the student is minimized. In mechanics, use is made of the mks, absolute cgs, and British gravitational units.

A minimal use of the latter units is maintained to aid the student in adjusting comfortably to the quantitative expression of physical relationships without a forced and abrupt change into the metric system. The British gravitational system is used less and less toward the end of the book and seldom appears in latter sections, inasmuch as it is assumed that the reader will have by then obtained a facility with the use of the mks and cgs systems. In electricity, rationalized mksa units are generally used, with some mention of the cgs absolute systems in the few cases where these units are often employed in the literature. Various systems of units are tabulated and conversion tables are presented in the Appendix.

We have continued in this fifth edition the practice of adhering to the correct use of significant figures in all numerical data. Such attention to significant figures encourages the student to utilize the fact that physical data can be manipulated only with the precision of the experimentally known values.

Experience has shown that the summaries given at the end of each chapter are helpful to students. As outlines including all the important definitions, defining equations, units, laws, and principles, arranged in the order of topics in the text, they encourage the student to make a systematic review of the assigned material.

It is hoped that the references included in some chapters may be helpful to students and teachers in planning reports and will encourage students to develop a knowledgable avocational interest in some area of physics.

In the preparation of this revision of "College Physics" the suggestions of many people were most helpful. In particular, Dr. Richard Calusdian, Chairman of the Department of Physics at Bridgewater State College, offered much advice during the revision of the section on Modern Physics. Professor John Heller, a colleague, provided much of the illustration copy for this edition. Special thanks is extended to my wife, Beatrice Weygand, who typed, edited, and provided much of the inspiration for this fifth edition. We have gladly received and carefully considered suggestions and recommendations from some of the colleagues, teachers, and students who have used earlier editions of "College Physics" in nearly two hundred colleges and large universities. Further comments from users of this book are most welcome and freely invited. Finally, to the many people who assisted in the preparation of this fifth edition, sincere thanks is extended.

George A. Weygand

Wilhelm Konrad Röntgen, 1845–1923

Born in Lennep, Rhenish Prussia. Professor at Wurzburg and Munich. Awarded the 1901 Nobel Prize in Physics for his discovery of x-rays.

Introduction

The Nature of Physics

Man has an inquiring mind. The earliest history of man reveals his attempts to understand his environment—attempts made not just to satisfy his curiosity but also to assure his survival in a sometimes hostile world. Modern-day man still feels the need to understand the world around him, the cause-and-effect relationships which will enable him to predict events and, to a certain degree, control his environment. According to Jean Piaget[1] and others[2] who have studied the process of learning, so strong is this drive for comprehension that in the absence of acceptable explanations of phenomena, man will develop his own "informal science" based on his superficial observations and biases. From his moment of birth, man gathers informal science "facts" and organizes and applies them; but he eventually realizes that conclusions based upon these facts, while possibly satisfying immediate needs, are not acceptable in the end because of their acausal and often teleological nature. That is, man attempts to explain scientific phenomena by putting life into inanimate objects but then must turn to a more organized and disciplined approach based upon deductive and inductive reasoning—the scientific method.

Perhaps few readers of this book have escaped being exposed to, or perhaps even having been forced to learn by rote memory, the steps of the "scientific method," a procedure through which observations are made, hypotheses developed, experiments designed and conducted, and conclusions drawn about a clearly defined physical problem. While the mechanical memorizing of this procedure may be of little value to a person seeking to understand either the workings of science or his environment, the truth is that the process is significant in that it is the net result of man's search for a means to arrive at a valid interpretation of the natural phenomena which surround him.

The scientific method has been effective in that it has a unique capacity to be self-correcting. An observer starts with initial assumptions and with an initial point of view, and the process enables him, to some extent, to revise his point of view creatively. Through the scientific method, one tries to develop theory which is both self-consistent and consistent with all known experimental data. In this sense the process of science is a search for the eternal truth that man has been seeking since his creation.

[1]Jean Piaget, "The Child's Conception of Physical Causality," Humanities Press, New York, 1951.
[2]Robert Karplus, Beginning a Study in Elementary School Science, *American Journal of Physics,* January, 1962, p. 1.

Recently it has been asserted that the sciences and the humanities are producing two cultures which are not on speaking terms with one another. Another charge—that science is unsympathetic to the well-being of society—is based largely on the debatable use made by some scientists of the conclusions drawn and discoveries synthesized through the scientific method. In the last decade, however, the scientific community has become increasingly aware of its responsibility toward society, and many spokesmen for science have stated the urgency of defining science in humanistic terms. I. I. Rabi[1] notes that we have not been cautious enough of the meaning of science in our generation and offers the following definition of science: "Science is an adventure of the whole human race to learn to live in and perhaps to love the universe in which they are. To be a part of it is to understand, to understand oneself, to begin to feel that there is a capacity within man far beyond what he felt he had, of an infinite extension of human possibilities—not just on the material side. . . ." Rabi proposes that science be taught "with a certain historical understanding, with a certain philosophical understanding, with a social understanding and a human understanding."

Since this is a textbook on physics, one might well ask if this area of science can be defined in humanistic terms. Gerald Holton[2] provides an answer to this question by defining physics as a sequence of related ideas whose pursuit provides one with the cumulative effect of an ever higher vantage point and a more encompassing view of the workings of nature. Physics is neither an isolated, bloodless body of facts and theories with mere vocational usefulness, nor a glorious entertainment restricted to an elite of specialists. Rather, students of physics should realize that what has been achieved in physics has sooner or later influenced man's whole life. To be ignorant of physics may leave them unprepared for their own time. They can be neither participants nor even intelligent spectators in one of the great adventures. It is the philosophy expressed in this definition that has guided the authors in writing this book.

The word "science" has its derivation in the Latin word *scientia,* meaning "to know." Yet much controversy has arisen over the interpretation of the term "to know." Knowledge obtained through science should not be solely factual but should advance understanding of phenomena. It is of little value merely to know a large number of facts; it is most important to know what these facts mean. Perhaps this distinction between factual knowing and understanding can be illustrated more clearly by considering two words in the German language describing knowledge—*Kennen* and *Wissen*. *Kennen* means "to know" and *Wissen* means "to understand." It is the latter that is most important to a scientist, and it is this type of knowledge that the authors hope will result from a study of physics as presented in this book.

The practical fruits of science are increasingly more important to our material prosperity, the conveniences of life, and often to life itself. To some, an emphasis on gadgetry and on know-how confuses science with technology. Others call attention to the interplay and mutual dependence of science and technology; the invention of a device such as a microscope, radar, or laser has been of immediate benefit in "pure" science. While some discussion of the technological applications of science will be given in this book, the authors will emphasize the concepts and principles of physics, inasmuch as an understanding of them will enable a student to master any devices and applications he may later encounter.

Holton has referred to the "pyramidal structure of physics, so beautiful and almost unique to our field." Physics is found to have a unity the significance of which should not be lost in a concentration on the separate fields of study into which it is somewhat arbitrarily divided. Several areas of physics which historically were explored separately, such as electricity and magnetism,

[1] I. I. Rabi, Address given at the AAAS meeting of the Educational Policies Commission, 27, December, 1966, Washington, D.C.

[2] Gerald Holton, Paper given at APS-AAPT meeting, 1, February, 1967, New York.

have been found to be intimately related. In recent years the sweeping power of a few fundamental laws in physics has been recognized. Certain fundamental concepts, "ideas that travel," underlie all sciences and make arbitrary divisions between and within them unsound.

Physics has commonly been divided into mechanics, heat, sound, optics, electricity and magnetism, and modern physics. The authors have not followed this pattern, but have chosen rather to separate the book into four major areas of physics, each of which contains considerable unity in concepts, principles, and methods of analysis used. But we feel that a description of the six historical "compartments" does provide the student with an overview of the dimensions of physics, so we present it here.

1 *Mechanics* is the oldest and the most basic branch of physics. It deals with such ideas as inertia, motion, force, and energy. Mechanics includes the properties and laws of both solids and fluids, of point masses, and of continuous matter.
2 The subject of *heat* includes the principles of temperature measurement, that is, the effects of temperature on the properties of materials, heat flow, and thermodynamics—the study of transformations involving heat and work.
3 The study of *sound* is concerned with vibrations and waves and with their recording, transmission, and perception, as in music and speech.
4 *Optics* is concerned with the nature and propagation of light, including the refraction that occurs when light passes through prisms and lenses. Of importance also are discussions of the separation of white light into its constituent colors, the nature of spectra, and the wave aspects of light such as interference, diffraction, and polarization.
5 *Electricity and magnetism* deal with still other aspects of matter and space in which the key concepts are electric charge and current.
6 The fascinating portion of physics known as *modern physics* is the interpretation and extension of physics in light of key events which happened about 1900; the discovery of x-rays, radioactivity, and the electron, and the formulation of quantum theory and the theory of relativity. In this sense the term modern physics is not synonymous with "contemporary physics" but rather implies a viewpoint in contrast with that of pre-1900 "classical" physics. Contemporary physics is, of course, the work on the present frontiers of physics dealing with both experiment and theory. We seem to progress first in one aspect and then the other. There are attractive frontiers in all the areas of physics previously mentioned, including problems in quantum mechanics, low-temperature phenomena, unconventional sources of electric energy, coherent light radiation, and extension of atomic theory to the properties of the solid state.

1 MEASUREMENT

The scientific method is an effective approach to solving physical problems, providing that several conditions have been met. Two of the conditions are (1) that the observations made in the study be valid and relevant, and (2) that the conclusions drawn be expressed in such a manner that they may be communicated to others. Both conditions require that techniques of measurement be developed.

Care must be taken in the study of a physical problem to correctly analyze data obtained so that the important may be separated from the superficial. For example, if a red ball is falling from a tower, the color of the ball may be insignificant but the fact that it is falling is significant. Further, in science one must avoid being like the six blindmen of Indostan who each described an elephant from observation of a limited area of the animal. Detailed examination of a problem should be conducted with an aim toward permitting the making of overall generalizations and not for the sake of building up a quantity of unrelated data. Lord Kelvin is quoted as saying that "when you can measure what you are speaking about and express it in numbers, you know something about it; but when you cannot express it in numbers, your knowledge is of a meagre and unsatisfactory kind; it may be the beginning of knowledge, but you have scarcely, in your thoughts, advanced to the stage of science, whatever the matter may be." The late Professor William S. Franklin frequently said, "The most important thing for a young man to acquire from his first course in physics is an appreciation for precise ideas." On the other hand, a noted physicist recently has observed, "Show me a person who goes around measuring things to 0.1 mm and I'll show you a bore." Sir Arthur Eddington expressed this view earlier when he noted that "life would be stunted and narrow if we could feel no significance in the world around us beyond that which can be weighed and measured with the tools of the physicist or described by the metrical symbols of the mathematician." Actually both views are correct. Since most physical problems are solved in quantitative terms, precise measurements are important, but an observer must not become so engrossed in his measurements that he fails to see the larger implications of his data.

2 THE MEASURING PROCESS

Measuring anything means comparing it with a standard to determine its relationship to the standard. For example, we may be interested in knowing how an object's size or mass compares with a standard size or mass. To carry out accurate measurements, we have to establish a system of *standards* and a system of units in which to express the standards. In general, a *unit* (for example, 1 meter) is fixed by definition and is independent of such physical conditions as temperature. A standard is the physical embodiment of a unit, provided that certain conditions are met. For example, a standard meter bar has the length of 1 meter only when maintained under certain atmospheric conditions.

For convenience, it has been customary to select only a few standards to establish the units of specified quantities and to have the units of other quantities fixed by physical equations. Thus, so-called fundamental units have been established for the measurement of length, mass, time, temperature, and electric current, and other derived units may be obtained from this arbitrary set of fundamental units.

No physical measurement can be made "exactly." There is always some uncertainty in comparing an unknown with a standard or in estimating the final digit in reading a scale. While this final digit is significant in that it gives information about the quantity being measured, it is doubtful. In reading a physical measurement it is common practice to retain only one estimated or doubtful figure. It is suggested that the reader who is unfamiliar with these guidelines refer to the Appendix for a review of the proper use of significant figures. However, it may be helpful to note a few such accepted procedures here.

In computations involving measured quantities, the process is greatly simplified without any loss of accuracy if figures that are not significant are dropped. A typical example would be to find the area of a block 20.2*6* units long and 4.6*2* units wide.

$$\begin{array}{r} 20.2\mathit{6} \\ \times\ 4.6\mathit{2} \\ \hline \mathit{4052} \\ 1215\mathit{6} \\ 810\mathit{4} \\ \hline 93.\mathit{6012} \end{array}$$

In the product only one doubtful figure is retained and therefore the area is reported as $93.\bar{6}$ units squared. The short horizontal bar is placed over the 6 to indicate that the answer is significant to that figure. The answer is reported as having three significant figures. A useful approximation states that in multiplication and division, the result should have as many significant figures as does the least accurate of the factors entering into the procedure. In addition and subtraction problems, the process would be continued only as far as the first column that has a doubtful figure.

When insignificant figures are dropped, the last retained figure should remain unchanged if the first dropped is less than 5; thus 4.533 becomes

4.53. The last figure should be increased by 1 if the first figure dropped is greater than 5; 4.536 becomes $4.5\bar{4}$. If the dropped figure is exactly 5, the preceding figure is left unchanged if it is even but increased by 1 if it is odd; thus 4.535 becomes $4.5\bar{4}$, and 4.565 becomes $4.5\bar{6}$.

3 OUR PRESENT UNITS AND STANDARDS

For several years there has been a determined effort to get the United States to adopt a universally accepted decimal system called the *metric system* for its nonscientific as well as scientific measurements. Legislation has been presented to the Congress and scientific organizations have passed resolutions, but the conversion has not yet been made. The average citizen in this country still must live in a society with two standards of measurement, the so-called British system and the metric system. Although commonly used but not in the United Kingdom (which has adopted the metric system), the British gravitational system of measurement is not an easy system to work with, since there are no convenient or predictable ratios between units. For example, there are, for some nonscientific reason, 12 inches in 1 foot, 3 feet in a yard, and 5,280 feet in a mile. These units are largely based on nonreproducible standards and traditions. For example, it is said that King Henry I established the yard by measuring the distance between the tip of his finger and the tip of his nose. An inch is the length of three dry, round barley corns laid end to end, by pronouncement of King Edward II. Interestingly, the system of shoe sizes we use today is based upon that definition. The shoemakers of King Edward's era found that the longest foot of that day was 30 barley corns, or 13 inches, long. They called this size 13, and graded sizes downward by one barley corn to a size. The same basic system is used today, and the difference between shoe sizes is one-third of an inch—the length of a grain of barley. The mile comes from the Latin *mille,* or thousand, and was determined by the thousand double steps of the average Roman soldier. Edward Teller[1] has commented humorously upon the historical bases for our British system of measurement. In Noah's time carpenters had a measurement called the cubit. This was the length of the forearm from the tip of the middle finger to the elbow. Teller observes that, assuming several carpenters worked on the same project, it is a wonder that the Ark floated. The wittiness of Teller, a great nuclear physicist, is seen through a story he tells about the historical development of our commonly used Fahrenheit temperature scale. "There is a story that the erudite German, Gabriel Daniel Fahrenheit, once waited in Danzig until it had got as cold as he thought it could possibly get. Then, on that very cold day he stuck his thermometer out the window and that became *zero*. Then he put it under his arm. That became *100 degrees*. So the history of our system of temperature supposedly goes back to the fact that there was once, in a rather cold town, a rather hot guy!"

As noted above, the metric system is a decimal system, that is, units are a certain power of 10 larger or smaller than other units describing the same physical property. Table 1 indicates the prefixes used in the metric system and their relative sizes.

For this table to be most meaningful, it is recommended that the reader review the techniques of exponential notation. We should note here, however, that by rearranging the large and small numbers that occur in scientific measurement in such a way that there is only one digit to the left of the decimal point and expressing the number as a power of 10, approximations of the magnitude of answers may be obtained when using the data in mathematical operations.

Example Find the product of 125,000,000 and 0.0002.

By moving the decimal point in such a way

---

[1] Edward Teller, U.S. Scientists' Hidden Handicap, *This Week Magazine,* May 15, 1960, pp. 6–7.

Table 1

Multiples and submultiples	Prefixes	Symbols
$1\,000\,000\,000\,000 = 1 \times 10^{12}$	tera	T
$1\,000\,000\,000 = 1 \times 10^{9}$	giga	G
$1\,000\,000 = 1 \times 10^{6}$	mega	M
$1\,000 = 1 \times 10^{3}$	kilo	k
$100 = 1 \times 10^{2}$	hecto	h
$10 = 1 \times 10^{1}$	deka	da
$1 = 1 \times 10^{0}$		
$0.1 = 1 \times 10^{-1}$	deci	d
$0.01 = 1 \times 10^{-2}$	centi	c
$0.001 = 1 \times 10^{-3}$	milli	m
$0.000\,001 = 1 \times 10^{-6}$	micro	μ
$0.000\,000\,001 = 1 \times 10^{-9}$	nano	n
$0.000\,000\,000\,001 = 1 \times 10^{-12}$	pico	p
$0.000\,000\,000\,000\,001 = 1 \times 10^{-15}$	femto	f
$0.000\,000\,000\,000\,000\,001 = 1 \times 10^{-18}$	atto	a

as to leave one digit to the left of the decimal point we get 1.25×10^x and 2.0×10^y, where x and y are determined by the rules of mathematics. That is, by moving a decimal point to the left, the number becomes smaller by a factor of 10^x. Since this procedure is merely a shorthand method of expressing a quantity, care must be taken to ensure that the mathematical value of the quantity remain constant. Therefore, a movement of the decimal point one place to the left must be compensated for by multiplying the quantity by 10. If the decimal point is moved two steps to the left, the quantity must be multiplied by 10 and then by 10 again, or 10^2, and so on.

Therefore 125,000,000 becomes 1.25×10^8, since the decimal point was moved eight places to the left.

Conversely, a movement of the decimal point to the right increases the size of a quantity and must be compensated for by dividing the quantity by 10 for each place moved. Since $0.1 = \frac{1}{10} = 10^{-1}$ according to the rule of negative exponents, a one-place move to the right can be adjusted by multiplying the quantity by 10^{-1} and a two-place move to the right would necessitate multiplying the quantity by 10^{-1}, and then by 10^{-1} again, or 10^{-2}. Thus 0.0002 becomes 2×10^{-4}, since the decimal point was moved four places to the right.

The basic algebraic rules describing the treatment of exponents in multiplication and division apply here. Since

$$x^5 \times x^4 = x^{5+4} = x^9 \qquad 10^5 \times 10^4 = 10^9$$

$$\text{and} \qquad 10^5 \times 10^{-4} = 10^1$$

Also:

$$\frac{x^5}{x^4} = x^{5-4} = x^1 \qquad \frac{10^5}{10^4} = 10^{5-4} = 10^1$$

and

$$\frac{10^5}{10^{-4}} = 10^{5-(-4)} = 10^9$$

Therefore, the answer to the problem is

$$1.25 \times 10^8 \times 2 \times 10^{-4} = 2.5 \times 10^4$$

4 WORLD STANDARDS FOR MEASUREMENT

To be a functional system of measurement, the metric system must be based on standards which are not only dependable but accessible. The standard for length in the metric system, the *meter* (m), was originally designed to be one ten-millionth of the distance along the meridian from the equator, through Paris, to the North Pole. It was soon found that two meter bars could be compared with each other with greater precision than they could relate to the earth's quadrant. Accordingly, the International Commission of the Meter, meeting in 1872, resolved to take the meter in the Archives of Paris "as is" as the standard of length. Three years later many of the leading nations of the world signed the treaty of the meter. Thus began a procedure for coordinating the standards of measurement for the scientific world through an International Bureau of Weights and Measures and a General Conference on Weights and Measures.

Since the standard meter bar was stored in Paris, scientists around the world found it inaccessible, so that replicas were distributed to a limited number of bureaus of standards throughout the world. However, the need for a readily available standard of length was growing proportionately with the increased amount of scientific research. In 1960 the Eleventh General Conference on Weights and Measures redefined the meter in terms of a standard which was accessible to all who needed to work with a high degree of precision. To understand this new standard, it is necessary for us to look ahead to a discussion of the nature of light that is presented later in this book. It is shown there that light is transmitted in waves and that for each color light, there is a related wavelength called lambda (λ). For every chemical element excited so that it glows, for example, by igniting it or passing an electrical discharge through it, a predictable color or group of colors of light will be emitted. If this light is then passed through a glass prism, the components of this emitted light will be optically separated and, except in the case of white light, a series of lines called a line spectrum is formed. Each line in the spectrum has a specific wavelength. The number of spectral lines emitted by a burning element will never change, nor will the wavelength of each line. Since the wavelength of these spectral lines could be accurately determined by using a device such as an optical interferometer, which is readily available to most scientists, it was decided to base the standard of length upon the wavelength of a spectral reference line. The element krypton 86, a gas, produces a clearly defined spectral line in the orange-red end of the spectrum. The wavelength of this orange-red line produced when an electric discharge passes through the gas krypton 86 has been established as the world standard for length. It is found that 1,650,763.73 wavelengths of that light equals the length of the standard meter. Thus,

$$1 \text{ m} = 1{,}650{,}763.73 \text{ wavelengths}$$

This corresponds to a wavelength for the orange-red line in the krypton 86 spectrum of $\lambda = 6{,}057.802 \times 10^{-10}$ m (Figs. 1 and 2).

For convenience it should be noted that 1 meter is equivalent to 39.37 inches in the British system of measurement, according to legal definition in the United States.

The *mass* of an object is the property which indicates the amount of "matter" in the object as evidenced by its inertia. As will be seen later, inertia is the measure of resistance to change in motion. The standard for the unit of mass, the *kilogram* (kg), is a cylinder of platinum-iridium alloy, the International Prototype Kilogram, kept at the International Bureau of Weights and Measures in Sèvres, France. A duplicate in the custody of the National Bureau of Standards

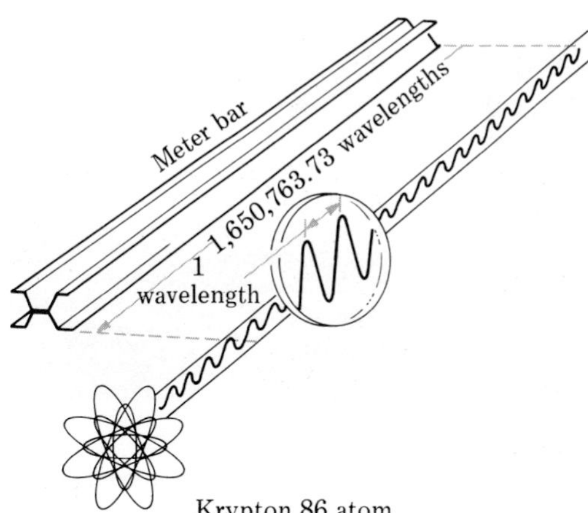

Figure 1
Definition of a standard meter. The meter is defined as 1,650,763.73 wavelengths in vacuum of the orange-red line of the spectrum of krypton 86.

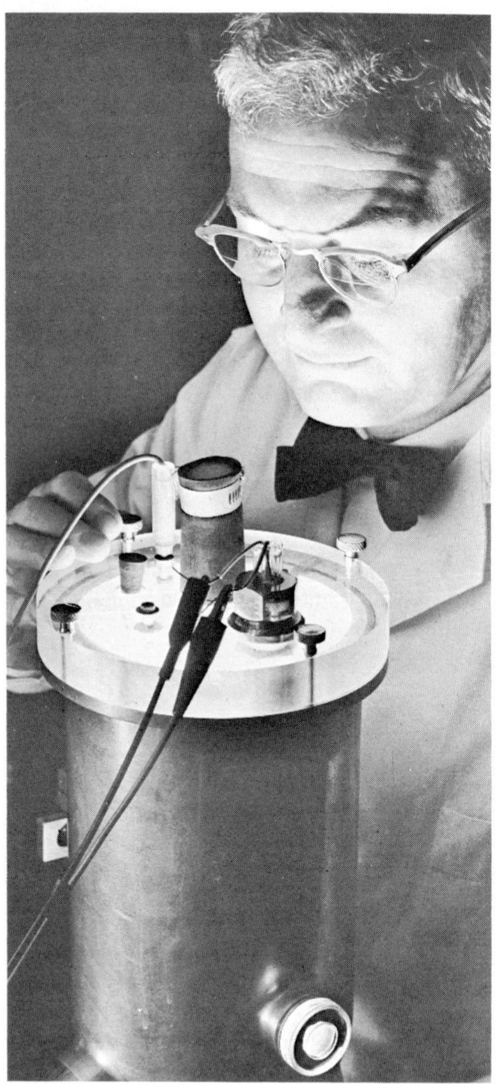

Figure 2
The wavelength of the orange-red light emitted by a krypton 86 lamp has been adopted as the international standard of length. Here a National Bureau of Standards scientist adjusts the device that holds the lamp.

serves as a mass standard for the United States (see Fig. 3). This is the only base unit still defined by an artifact. The kilogram is equal to 2.205 pounds, avoirdupois.

The *liter* (l) is not a defined standard but rather a derived unit of capacity or volume defined as the volume of 1 kilogram of water, at standard atmospheric pressure (a pressure capable of supporting 0.760 meter of mercury in a closed tube) and at the temperature of its maximum density, approximately 4° Celsius (C). The liter is slightly larger than the liquid quart in British units; 1 liter equals 1.057 liquid quarts.

Astronomical clocks have, without competition, been the foundation of the measurement of time since the dawn of history. For many years after the establishment of the General (International) Conference on Weights and Measures, no

Figure 3
The national standard of mass. Kilogram 20, a cylinder 39 mm in diameter and 39 mm high, with slightly rounded edges, made of an alloy containing 90 percent platinum and 10 percent iridium. It was furnished by the International Bureau of Weights and Measures in pursuance of the metric treaty of 1875.

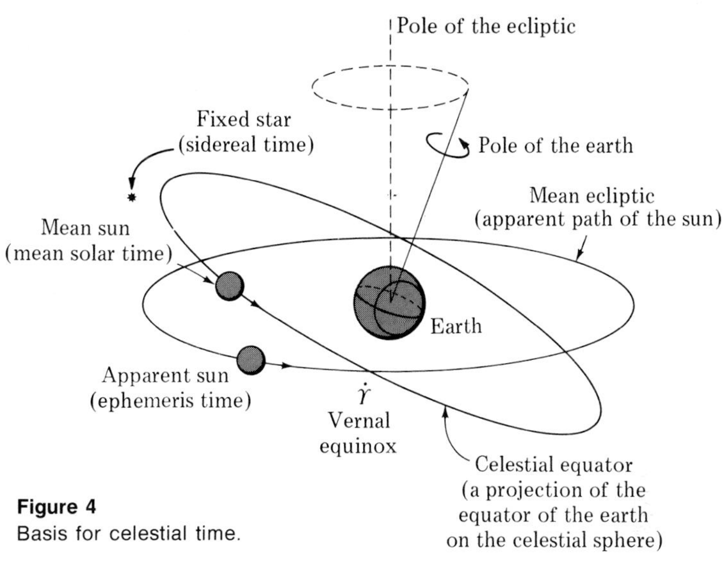

Figure 4
Basis for celestial time.

new standard for time was adopted. The ancient definition of the second as 1/86,400 part of a mean solar day was retained. *Mean solar time* is determined by the rotation of the earth about its axis with respect to the imaginary mean sun traveling along the celestial equator. (See Fig. 4.) Since the apparent solar day varies throughout the year owing to the eccentricity of the earth's orbit, astronomers kept track by observing star transits, in relation to which the earth's rotation is much more uniform. But precise astronomical observations revealed that this standard, called *sidereal time,* which is determined by measuring the rotation of the earth with respect to a fixed star, was not good enough. There is a gradual slowing down of the earth's rotational motion, as expected, from tidal friction; and there are also erratic fluctuations in its rotational speed. For this reason astronomers carry out their more precise calculations in *ephemeris time,* which is based on planetary motions (all of which are in substantial accord). In 1955 the International Astronomical Union recommended that the second be divorced from the mean solar second and be redefined in terms of the ephemeris second. In 1960 the General Conference on Weights and Measures, which now represents 39 nations, ratified a resolution of the International Conference on Weights and Measures (ICWM) which defined the *second* as 1/31,556,925.9747 of the tropical year 1900. (The tropical year is the time between successive arrivals of the apparent sun at the vernal equinox.)

Although the world had agreed upon a suitable astronomical standard for time, the search continued for a natural time unit similar to the krypton 86 standard for length, which would be available in all parts of the world. For some time it had been known that atoms emit or absorb radiation at a precise frequency if they are undisturbed. The problem was to measure this frequency as accurately as possible while disrupting the atoms as little as possible.

The idea of using some predictable vibrating object as a device to measure the passing of time is not new. Clocks which are run by a swinging pendulum are based on this principle. Also the household electric clock is another example because it uses the cycle of the alternating current as its "pendulum." The accuracy of the electric clock depends on the steadiness of the rate at which the current alternates. The 60-cycle rate produced by power-generating stations is steady enough for household use. In the search for higher precision, quartz-crystal clocks were developed for use in scientific research. When subjected to an alternating electric field, a quartz crystal tends to vibrate at its own specific, sharply defined rate. If the crystal is placed in an oscil-

Figure 5
This cesium atomic clock at the Boulder Laboratories of the National Bureau of Standards measures frequencies and time intervals to an accuracy equivalent to the loss of less than 1 s in 1,000 years. The liquid nitrogen being poured into a cold trap helps maintain a vacuum so that cesium atoms can be beamed through the device without being deflected by molecules of air.

lator circuit, its natural frequency will be imposed on the circuit. The resulting current can run a synchronous clock motor with a maximum error of one part in a billion (10^9). The quartz-crystal clock has some problems in that the frequency of the crystal varies as it ages and it is also affected by changes in temperature.

The first "atomic clock" made was based on the vibrations of an ammonia molecule,[1] which occur at a sharply defined frequency, 23,870 megahertz (mHz), and was constructed by scientists at the National Bureau of Standards in 1948. Eventually, an atomic clock of considerably greater precision was built by the National Bureau of Standards using cesium 133, a silvery metal which is liquid at room temperature. A series of three such cesium clocks have been built, each having an increasing degree of accuracy. The model built in 1965 has an accuracy of 5 parts in 10^{12}, which is equivalent to saying that there would be no more than 1-second variation in 6,000 years. It has been found that cesium clocks provide a more constant time reference than the earth and also that their basic internal period is very insensitive to external effects such as temperature. Consequently, the frequency of a cesium oscillator (Fig. 5) has been related to the defined *second* (s) by lengthy astronomical observations as being 9,192,631,770 hertz (1 hertz (Hz) = 1 cycle/s). At the Thirteenth General Conference on Weights and Measures, held October 13, 1967, a cesium 133 oscillator with suitable associated equipment was adopted as the laboratory standard for maintaining a scale of time. The frequency which the definition assigns to the cesium radiation was carefully chosen to make it impossible to distinguish the new second from the ephemeris second based on the earth's motion. This decision made it unnecessary to convert data obtained using the old standard for time. Currently the mean solar second differs from the ephemeris second by about -130 parts in 10^{10}, so that frequencies broadcast for the purpose of keeping mean solar time are offset from the cesium frequency by this amount. (This amount is not a constant but varies from year to year. For example, in 1960 it was -150 parts in 10^{10}.) In practice, time signals and frequencies broadcast by the National Bureau of Standards stations, the United States Navy, and several foreign countries are coordinated so that both the signals and the frequencies are mutually consistent.

As was the case for length, mass, and time, a unit has been established for electric current. The *ampere* (amp), the unit of electric current, is defined in terms of the attraction (Chap. 38) between two long parallel wires carrying the same current. If the wires are 1 meter apart and the current is adjusted until the measured force of attraction (measured in newtons (N), a unit of force) per unit of length between the wires, due to their magnetic field, is 2×10^{-7} newtons per meter, the current is defined to be 1 ampere. It should be noted that the ampere is defined in terms of force, which in turn is defined in terms of the fundamental units of length, mass, and time (Fig. 6).

For *temperature*, another fundamental quantity, a physical constant has been selected as the standard. While it will be necessary to defer dis-

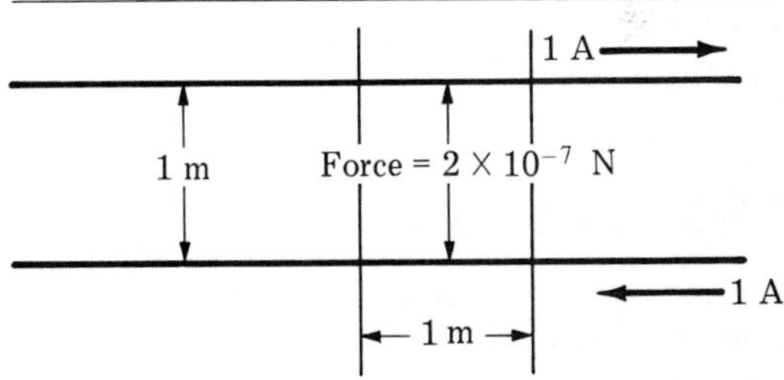

Figure 6
Definition of an ampere. An ampere is defined as the magnitude of the current that when flowing through each of two long parallel wires separated by one meter in free space results in a force between the two wires due to their magnetic fields of 2×10^{-7} newton for each meter of length.

[1] Harry M. Davis, "Radio Waves and Matter," *Scientific American,* September, 1948.

cussion of the implications of this accepted standard until later, for the purpose of illustrating the manner in which standards are set, the basis for the measurement of temperature is given here. The thermodynamic, or Kelvin, scale of temperature has its origin, or zero point, at absolute zero and has a reference point at the "triple point" of water, which is defined as an equilibrium point where three phases—solid, liquid, and saturated vapor—can exist together at 273.16° Kelvin. The triple point is defined as 0.01°C on the Celsius scale and is approximately 32.02°F on the Fahrenheit scale.

Modern standards are dynamic in nature. Accuracies are being continually increased. The range of conditions for which standards are required is continually expanding. Our concept of temperature, for example, has been expanded, and now measurements are needed from near 0°K in liquid helium to over 100,000°K in plasmas. Yet new standards must be related to and made rigorously consistent within the whole chain of measurement leading back to the national standards and to the international prototype standards. A trend in modern standardization is the move away from arbitrary prototype standards as the basis of measurement, toward a reliance on natural constants. If properly chosen, natural constants afford greater opportunity for increased accuracy and stability, and these constants are freely available to all scientists and all laboratories.

5
SYSTEMS OF UNITS

A complete set of units, both fundamental and derived, for all kinds of quantities is called a *system of units*. Several systems have been devised and are in common use. Each is named in terms of the fundamental units upon which it is based.

A system based on the fundamental units of the meter, the kilogram, the second, and the ampere is called the *mksa* system. The mksa system was adopted by the International Electrotechnical Commission to be put into universal use on January 1, 1940. Actual acceptance of the mksa system is progressing slowly. It did not abruptly displace other systems. Advantages of the mksa system are that it leads to the electrical units in practical use, it has some derived units that are of convenient size, and it has nomenclature that is simple in comparison with that of the cgs system.

The *cgs* system of units, formerly widely used in science, is based upon the centimeter as a unit of length, the gram as a unit of mass, and the second as a unit of time. Because of the decimal nature of the system (as is the case also with the mksa system), conversions within the system are relatively simple as compared with those in the *fps* system. Unfortunately, many of the derived units in the cgs system are inconveniently small.

A system of units used in commerce in the United States is based upon the foot as a unit of length, the pound as a unit of force, and the second as a unit of time. From the initials of these three units this system is called the fps system. The system is largely used in engineering, except in the field of electricity. The fps system is also called the British gravitational system.

Conversions between and within systems of units can be made by use of a very few conversion factors. Ideally, in mechanics, only three such factors are necessary, one for each of the fundamental quantities of mass, length, and time. (In other areas of physics, one recognizes other fundamental quantities, such as temperature and luminous intensity.) Some conversion factors are given in Table 2.

Table 2
EQUIVALENTS OF CERTAIN UNITS

1 meter (m) = 39.37 inches (in)
1 inch = 2.540 centimeters (cm)
1 liter (l) = 1.057 liquid quarts (qt)
1 kilogram (kg) weighs 2.205 pounds (lb), avoirdupois
453.6 grams (g) weigh 1 pound, avoirdupois

If all units are inserted into an equation, they can be handled as algebraic quantities; when they are handled in this manner, the correct final unit is obtained. This method, often called "cancelation of units," has an added advantage in that it may call attention to a factor that has been forgotten. The student should carefully develop this technique to avoid errors resulting from careless handling of units.

The system has its mathematical basis in the rules of multiplication. It is possible to multiply a number by 1 without changing its value.

Example Change 115 in to centimeters. Since 1 in = 2.54 cm,

$$\frac{1\text{ in}}{2.54\text{ cm}} \quad \text{or} \quad \frac{2.54\text{ cm}}{1\text{ in}} = 1$$

Therefore,

$$115\ \cancel{\text{in}} \times \frac{2.54\text{ cm}}{1\ \cancel{\text{in}}} = (115)(2.54)\text{ cm} = 292\text{ cm}$$

It should be noted that of the two possible fractions made from this equivalency, the fraction which was selected had its units presented in such a way that the canceled units were in the numerator and denominator, respectively. Also, it should be observed that only the units and not the quantities were canceled out.

Example Express 60 mi/h in feet per second.

$$60\frac{\cancel{\text{mi}}}{\cancel{\text{h}}} \times \frac{5{,}280\text{ ft}}{1\ \cancel{\text{mi}}} \times \frac{1\ \cancel{\text{h}}}{60\ \cancel{\text{min}}} \times \frac{1\ \cancel{\text{min}}}{60\text{ s}}$$

$$= \frac{60 \times 5{,}280\text{ ft}}{60 \times 60\text{ s}} = 88\text{ ft/s}$$

Many problems in motion require that miles per hour be changed to feet per second; therefore, it is helpful to know that

$$60\text{ mi/h} = 88\text{ ft/s}$$

Since it is necessary to be consistent in working with units when solving physical problems, measurements made of properties such as mass, length, or volume must be expressed in the same units before they can be treated mathematically. It is not possible, for example, to add 10 cm to 10 m without first changing one of these readings to the same unit as the other or to express both readings in some other agreed-upon unit.

Table 3

1 kilometer (km) = 1,000 m
1 hectometer (hm) = 100 m
1 dekameter (dam) = 10 m
1 meter = 1 m
1 decimeter (dm) = 0.1 m
1 centimeter (cm) = 0.01 m
1 millimeter (mm) = 0.001 m

Example The following readings were obtained when measuring the dimensions of a long, thin rectangular object 0.25 km long, 25 cm wide, and 25 mm thick. What is its volume in cubic meters (m^3)?

First, each unit must be changed to meters. This can be done if one recalls the relative sizes of the metric units of length as presented in Table 3.

It should be noted that

$$1\text{ km} = 10\text{ hm} = 100\text{ dam} = 1{,}000\text{ m} = 10{,}000\text{ dm}$$
$$= 100{,}000\text{ cm} = 1{,}000{,}000\text{ mm}$$

Thus, 0.25 km = 2.5 hm = 25 dam = 250 m

Similarly,

$$25\text{ cm} = 2.5\text{ dam} = 0.25\text{ m}$$

and 25 mm = 2.5 cm = 0.25 dam = 0.025 m

The volume of a rectangle = length × width × height. Therefore, the volume is 250 m ×

0.25 m × 0.025 m. Using powers of 10,

$$2.5 \times 10^2 \text{ m} \times 2.5 \times 10^{-1} \text{ m} \times 2.5 \times 10^{-2} \text{ m}$$

$$= (2.5)^3 \times 10^{-1} \text{ m}^3 = 15.6 \times 10^{-1} \text{ m}^3, \text{ or } 1.56 \text{ m}^3$$

SUMMARY

Science deals with concepts in seeking to extend the range of our experience and to reduce it to order. Physics deals with concepts which describe properties of matter, energy, space, and time.

Measurement in simple cases means comparing a thing with a standard to see how many times as big it is. In other cases, counting and statistical analysis may be a necessary part of measurement.

Three *fundamental quantities* are necessary in mechanics. These are commonly chosen as *length, mass,* and *time* or as *length, force,* and *time.*

For each fundamental quantity there is an arbitrarily chosen (but properly standardized) *fundamental unit.* The numerical value of a fundamental unit is fixed and preserved by defining it in terms of a prototype standard (the kilogram cylinder, for example) or in terms of a physical constant (such as the triple point of water). Other units based on combinations of the fundamental units are called *derived units.*

In the United States our units are defined in terms of the metric standards.

A complete set of units, both fundamental and derived, is called a *system of units.* Systems of units are named from the fundamental units used as a basis for the system, as mksa, fps, cgs. Only a few conversion factors are needed to convert all derived units from one system to another.

In measured and computed quantities *significant* figures only should be retained. Physicists usually express small and large quantities in powers of 10.

Mean solar time is determined by the rotation of the earth about its axis with respect to the imaginary mean sun traveling along the celestial equator. *Sidereal time* is determined by measuring the rotation of the earth with respect to a fixed star. *Ephemeris time* is based on planetary motions.

The *second* is defined as 1/31,556,925.9747 of the tropical year 1900.

References

Barber, Bernard: "Science and the Social Order," Collier Books, The Macmillan Company, New York, 1970.

Bitter, Francis: "Mathematic Aspects of Physics: An Introduction," Doubleday & Company, Inc., Garden City, N.Y., 1963.

Brown, Martin (ed.): "The Social Responsibility of the Scientist," The Free Press, New York, 1971.

Cline, Barbara Lovell: "Men Who Made a New Physics," Signet Books, New American Library, Inc., New York, 1969.

Deason, Hilary J. (ed.): "A Guide to Science Reading," Signet Books, New American Library, Inc., New York, 1963.

Karplus, Robert: "Physics and Man," W. A. Benjamin, Inc., New York, 1970.

Lyons, Harold: Atomic Clocks, *Scientific American,* February, 1957 (Reprint 225).

Parke, Nathan Grier, III: "Guide to the Literature of Mathematics and Physics," Dover Publications, Inc., New York, 1958.

Price, Derek J. de Solla: "Science since Babylon," Yale University Press, New Haven, Conn., 1961.

Questions

1 By what steps is a science developed?

2 What is meant by the scientific method?

3 Discuss the statement made by Jurgen Moltman, "If you don't expect the unexpected you will never find it."

4 What is meant by the term "informal science" as used by Robert Karplus?

5 What value may the study of physics have in character formation? Also in the attitude toward all problems of life?

6 Comment on the following quotation: "The supreme task of the physicist is to arrive at those universal elementary laws from which the cosmos can be built up by pure deduction. There is no logical path to those laws; only intuition, resting on sympathetic understanding of experience, can reach them" [Albert Einstein].

7 List five advantages and five disadvantages resulting from a total changeover to the metric system in this country.

8 What is the only base unit still defined by a scientific artifact?

9 What do you consider the most important characteristics of a standard used to fix the value of a unit?

10 What advantage is there in establishing the wavelength of the orange-red spectral line of krypton 86 as the world standard for the meter?

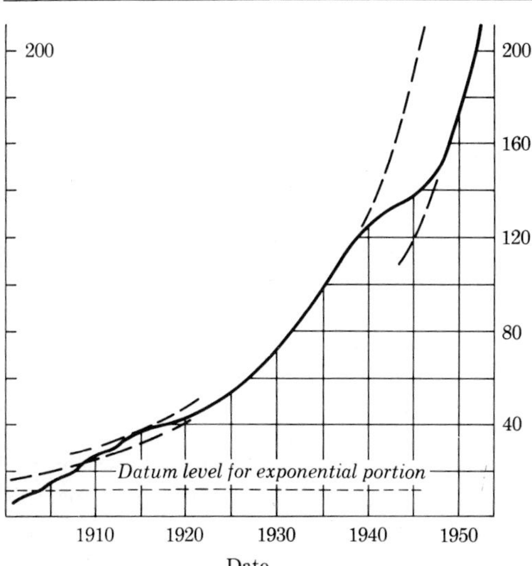

Figure 7
Total number of *Physics Abstracts,* in thousands, published since January 1, 1900. The full curve gives the total; the broken curve represents the exponential approximation. Parallel curves are drawn to enable the effect of the wars to be illustrated. *(From Price, "Science since Babylon," Yale University Press, New Haven, Conn., 1961.)*

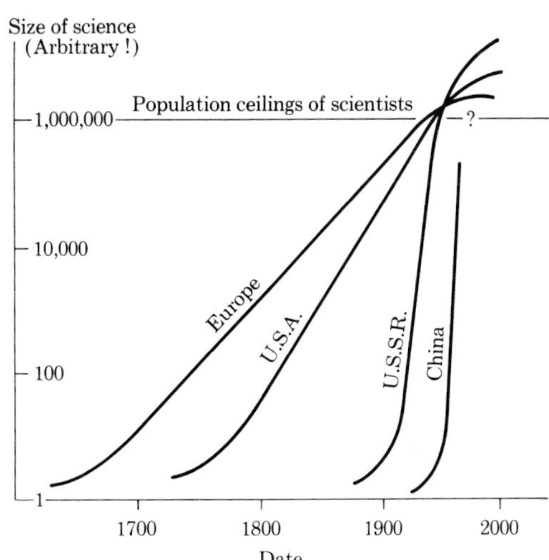

Figure 8
Schematic graph of the rise of science in various world regions. *(From Price, "Science since Babylon," Yale University Press, New Haven, Conn., 1961.)*

11 Figure 7 uses publications as one measure of the growth of physics. From it, what conclusions might you draw concerning (*a*) the effect of war on science, (*b*) the time required for physics to double its output, and (*c*) the problem of filing scientific information and making it available?

12 From Price's graph (Fig. 8) showing a projection of the "size" of science (measured in number of scientists, publications, or dollars), what speculation occurs to you concerning (*a*) the effect of the predicted merging of these curves on international cooperation or tension and (*b*) the value judgments that will have to be made in science when we can no longer increase dollars and manpower to investigate every problem?

13 What is "unscientific" about your answers to Questions 11 and 12?

14 Comment on the validity of the statement "Some 80 to 90 percent of all scientists that have ever been, are alive now."

Problems

1 Someone, possibly Henry VIII's hatchetman, has determined that the average human head weighs 14 lb. How many kilograms does the head weigh?

2 A certain brand of cigarettes is labeled 100's by their manufacturer. This refers to the fact that the length of each cigarette is 100 mm. How many inches long is each cigarette? *Ans.* 3.94 in.

3 A halfback covers 1.5 m per step while running at full speed. How many steps must he take when running the 100 yd from goal line to goal line on a football field?

4 What is the mass in grams of a US gallon of water? *Ans.* 3,770 g

5 Solve by powers of 10:

$$\frac{150{,}000 \times 0.0025 \times 20}{3{,}000{,}000 \times 0.015 \times 150}$$

6 The interior of the Skylab space station has a volume of 390 m^3. How many cubic feet is this volume? *Ans.* 14,000 ft^3.

7 Show that a car traveling 45 mi/h has a speed of about 66 ft/s.

8 Is it cheaper to buy gasoline at 36 cents/gal in Windsor, Ontario, or at 30 cents/gal in Detroit, Michigan? Assume that the tax is the same. (1 British imperial gal = volume of 10 lb of water at 62°F = 277.4 in^3 1 US fluid gal = 4 US fluid qt = 231.0 in^3.) *Ans.* Same cost.

9 Measure the thickness of 100 pages of this book. Calculate, and express to the proper number of significant figures, the thickness of the paper on which this page is printed.

10 The Apollo-Saturn V moon rocket measures 363 ft long. How many centimeters is this length? *Ans.* 11,050 cm.

11 A thin circular sheet of iron has a diameter of 14 cm. Find its area. If a square meter of the material has a mass of 0.30 kg, find the mass of the sheet.

12 Change 20 oz/ml to kilograms per gallon. *Ans.* 2.14×10^3 kg/gal.

13 The Empire State Building, 380 m tall, is observed by a man 3.22 km away. What is the angle between the ground and his line of sight to the top of the building?

14 The element mercury has a density of 13.6 g/cm^3. How much does a cubic foot of it weigh? *Ans.* 849 lb.

15 The speed v of radio waves is equal to the product of the wavelength λ and the frequency f. If $v = 3.00 \times 10^8$ m/s, find the wavelength of a radio wave from a transmitter broadcasting on a frequency of 760 kHz. Express the wavelength (*a*) in meters and (*b*) in feet.

16 Add the following and express the answer in centimeters: 1.5 km, 1.5 m, 1.5 mm, and 1.5 dm. *Ans.* 150,165.15 cm.

17 What is the distance from Philadelphia to New York, 90 mi, expressed in kilometers?

18 Bob Hayes set a record by running a 100-yd course as part of a four-man relay team in 1962 in 7.8 s. What was his average velocity in (*a*) miles per hour? (*b*) in kilometers per second? *Ans.* 26.22 mi/h; 1.18×10^{-2} km/s.

19 Five coins are placed on a balance and are found to have the following masses: 5.25 g, 4.95 g, 5.05 g, 4.90 g, 5.10 g. If the standard weight for these coins is 5.00 g, find the average percentage deviation of these readings.

20 Density in cgs units is expressed in grams per cubic centimeter and in mks units in kilograms per cubic meter. Convert a density of 7.45 g/cm^3 to mks units. *Ans.* 7.45×10^3 kg/m^3.

Give me a place on which to stand
and I shall move the earth.

Archimedes 215 B.C.

PART ONE

The Physics of Particles

Hendrik Antoon Lorentz, 1853–1928

Born in Arnhem, Holland. Professor at Leyden. In 1902 shared the Nobel Prize for Physics with his pupil Zeeman for their investigations of the effects of magnetism on the phenomena of radiation.

Pieter Zeeman, 1865–1943

Born in Zonnemaire, Holland. Lecturer at Leyden University, later director of the Physical Institute. Shared the 1902 Nobel Prize with Lorentz for their investigations of the effects of magnetism on the phenomena of radiation.

1

Forces Acting on an Object: Vectors

In the study of physics we are concerned with the exchange of energy between objects which make up our universe. The effect of an energy exchange may have great influence on our lives, but the fact that an energy exchange is taking place may not always be obvious to an observer. This is partly because energy may exist in a number of forms—as heat, light, and electric, mechanical, and nuclear energy, for example. We shall see later that mechanical energy is defined as the ability to do work. We usually think of this form of energy when we use the term, for it is easy to observe the effect of the transfer of the energy possessed by a body in motion when it reacts with another body. The effect of a moving car striking a parked car or a falling object landing on an object at rest on the ground is quite obvious. Since a goal of this book is to relate physics to the real-life world as well as to assist the student to build a knowledge of physics, we shall use this type of familiar event as a point of departure in the book. We shall consider the energy possessed by a body due to its state of motion or its position and then explore what happens when this energy is transferred to another object.

Before considering the exchange of mechanical energy, however, we must consider how this energy interacts between objects. The term "force" has been invented to describe the influence of an energy-bearing object on another object. Force is one of the most important physical concepts, but since it is based on evidence provided by our senses it is difficult to define satisfactorily. A force is arbitrarily defined as any influence capable of producing a change in the motion of an object. Forces can be expressed in the familiar unit pounds and also in the units of the cgs system, dynes, and the mksa system units, newtons. Both dynes and newtons will be defined later in this book. In order to analyze and predict the change resulting from the forces acting on a *particle* of matter or as far as we are concerned on a group of particles organized into *bodies,* it is necessary to develop certain mathematical and graphical tools.

Simon Stevin, a Flemish scientist who lived in the northern Netherlands from 1548 to 1620, when offering mathematical proof of the law of the inclined plane, introduced a technique which we use today: the triangle of forces method (the closed polygon of forces method) which revealed the additive properties of force. Through his work, forces were shown to be *vector* quantities. We shall use his discovery to study the effect of forces on a point or on a body.

If we consider an object which has a force, often described as a push or pull, acting on it, we can predict something about its resultant motion if we can represent the force graphically. For

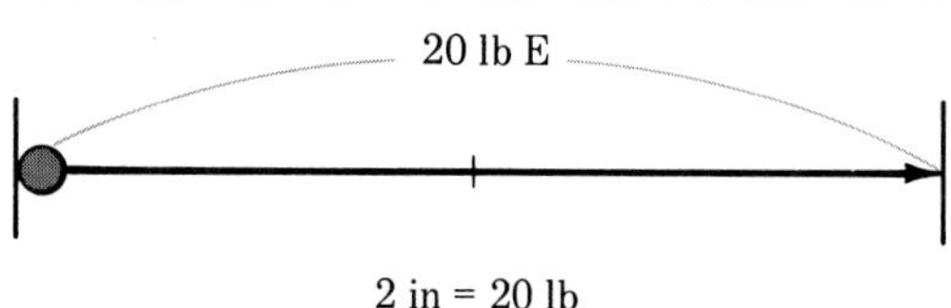

Figure 1-1
Representation of a vector quantity.

example, if a 20-lb force is applied to a block, we can represent the force by a scaled line with a sufficient length to represent the 20 lb. If 1 in of the line is set equal to 10 lb, then the line would be drawn 2 in long. We have a problem, however, in that whereas we know the magnitude of the force, we are unable to predict the direction in which the object will move unless we know the direction, or "sense," of the force. Therefore, we need to know whether the 20-lb force is acting easterly or northerly or any other compass direction from a reference direction. For example, if the 20-lb force acts easterly, we can represent the force with, let us say, a 2-in line with an arrowhead indicating the direction (easterly) of the force (Fig. 1-1). This scaled line with arrowhead is called a vector. The word "vector" comes from the Latin *vehere,* meaning *to carry,* which indicates that the force is being carried from one place to another—20 lb easterly in this case. If this force were the only force acting on the block, it would move in an easterly direction at a rate that could be predicted by using the appropriate equations of motion to be described later in the book.

It is sufficient for us to realize at this point that we can represent forces by vectors. It is now necessary to see how we can employ vectors, as Stevin did, to enable us to make predictions about the effect of forces on a particle.

1-1 VECTORS AND SCALARS

It is worth noting here that although we are concerned for the moment with forces acting on an object, there are other physical properties which can be represented as vector quantities. Other such vector quantities which will be discussed in detail later are displacement, velocity, acceleration, weight, momentum, torque, and electric field intensity. Once we have established procedures for solving force problems vectorially, these same procedures will enable us to solve problems involving these other vector quantities.

In addition, there are other physical quantities which can be described in detail by using just the magnitude of the quantity. These quantities which can be completely specified by a number and a unit are called *scalar* quantities. Typical scalar quantities are mass, volume, temperature, energy, length, and speed, all of which can be added by ordinary arithmetic: 3 s + 5 s = 8 s.

1-2 VECTORS

If two forces of 20-lb magnitude are applied to an object, one acting easterly and the other northerly, and these are the only two forces acting on the object, it is possible to find a third force which has the same or equivalent effect as the two forces acting on the object. This third force is called the *resultant* force of the two forces. If one wishes, this third force could be substituted for the two original forces.

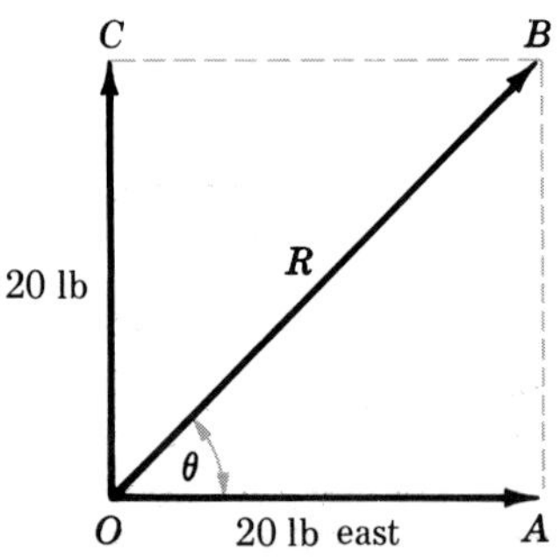

Figure 1-2
Parallelogram of forces diagram.

According to the work of Stevin it is possible to add these two forces vectorially. By constructing a parallelogram of forces, it is possible to show that the sum of vector **OA** and vector **OC** is equal to vector **OB** (Fig. 1-2). It should be noted that the vector **OB** is not equal to the sum of the absolute values of **OA** and **OC**, 20 lb + 20 lb $\neq$ 40 lb in this case, because of the vector nature of the forces. (*Note:* In this book boldface type is used to represent vector quantities and to portray their vector properties clearly.) That is, we have to consider not only the magnitude but the effect of direction in determining the resultant **OB.** A vector combines with another vector by *geometrical* addition to form a *resultant vector* which represents the combined effect of the quantities represented by the original vectors.

1-3 ADDITION OF VECTORS

In solving all vector problems, either graphical (geometrical) or mathematical procedures may be used. The graphical procedures involve carefully drawn diagrams showing the magnitude and direction of each vector. In solving the problem mentioned above, a scale is established, for example 1 in = 10 lb, and by constructing a parallelogram, the length and direction of the resultant **OB** is then determined with a ruler and a protractor. The magnitude of **OB** is thus obtained in terms of the established scale (Fig. 1-2).

Let $\quad$ 1 in = 10 lb

Therefore, $\quad$ **OA** = 2 in

OC = 2 in

and $\quad$ **OB** = 2.8 in

Therefore, $\quad$ **OB** = 28 lb

A protractor shows angle θ to b... ...e- of 45°

According to this graphi... sultant is a force of 28 lb... north of east.

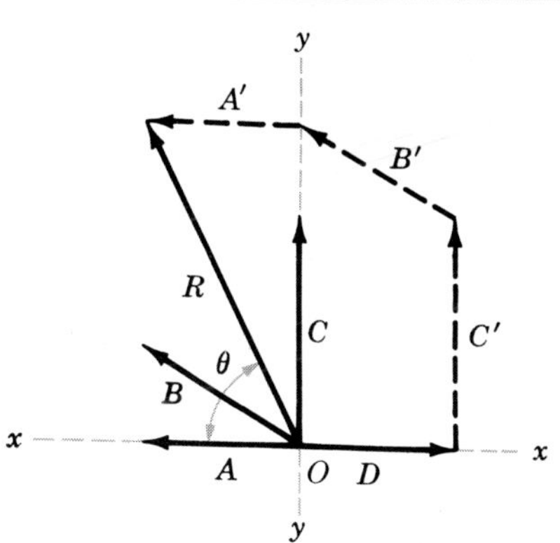

Figure 1-3
A vector polygon diagram.

If it were desirable to cancel out the effect of this resultant, as is often the case, a force equal in magnitude but opposite in direction would be applied. The force, called the *equilibrant,* would in this case be **OD,** 28 lb, 45°S of W.

Another graphical procedure for solving vector problems is the polygon-of-forces technique which is used when more than two forces are acting on an object.

Example What is the resultant of the following forces acting on an object at O? **A** = 6 lb W; **B** = 4 lb NW, **C** = 8 lb N, and **D** = 3 lb E.

First construct the vector diagram to scale (Fig. 1-3). Then, using any one vector as a ...se (**D** in this illustration), draw vector **C′** at ... end of ... arrowhead vector **D,** then draw vector **B′** a... end of vector of vector **C′**, and vector **A′**...tor, the resultant is **B′**. Since this is the ...drawn from the origin O determined by ...d, vector **A′**. The length and ...resultant can be found by the use to the la... a protractor. In this case the result... dir... force of 12 lb, about 60°N of W.

Should the last vector drawn come back to the origin, the figure is called a closed polygon and the resultant is zero because the forces canceled each other out. This is an example of linear equilibrium, a concept we shall encounter many times in the next few chapters of this book. When the vector sum of the forces acting on an object is equal to zero, the object is in equilibrium insofar as linear motion is concerned. This statement is called the *first condition for equilibrium*.

It should be noted also that had the equilibrant in the above example (12 lb, 60°S of E) been an original force acting on the object, the vector diagram would have been a closed polygon. In fact, the purpose of the equilibrant is to cancel out forces and to put the object in a state of equilibrium.

Graphical procedures have an advantage over mathematical procedures in that they may be used easily for multivector problems and for cases in which two forces are not acting at right angles to each other. Both types of problems require fairly lengthy mathematical procedures to solve analytically. The disadvantage to using graphical procedures is that the accuracy of the results depends on the care taken in drawing the vector diagram. Carelessly drawn diagrams will lead to unacceptable errors. For accurate solutions to problems, it is wise to be familiar with accepted mathematical techniques.

The mathematical, or analytical, method for determining the resultant **R** of any two vector quantities **A** and **B** employs rules of trigonometry. In the simplest case of two vectors acting at right angles to each other, use is made of the ratios of the three sides to each other.

The student should review the trigonometric ·les for determining the ratio of the sides in a righ· ·iangle (a triangle with a 90° angle). Summarizing these rules, it can be shown that any two nonparallel lines will intersect and form an angle θ. If the two lines are truncated by a line drawn at 90° to one of the lines (line DE or AB in Fig. 1-4), a right triangle will be formed. In solving physical problems, especially problems involving vector quantities, it is often helpful to know the ratio of the sides of a right triangle to each other. Direct measurement will show that the ratio of the side opposite the angle to the longest side in a right triangle, called the hypotenuse, will be the same in the example shown in Fig. 1-4 whether the ratio ED/OE or that of AB/OB is considered. For the given angle θ, the ratio will always be the same regardless of the size of the triangle. For convenience, the ratio of the side opposite the angle of reference, θ, to the hypotenuse is called the *sine* of the angle.

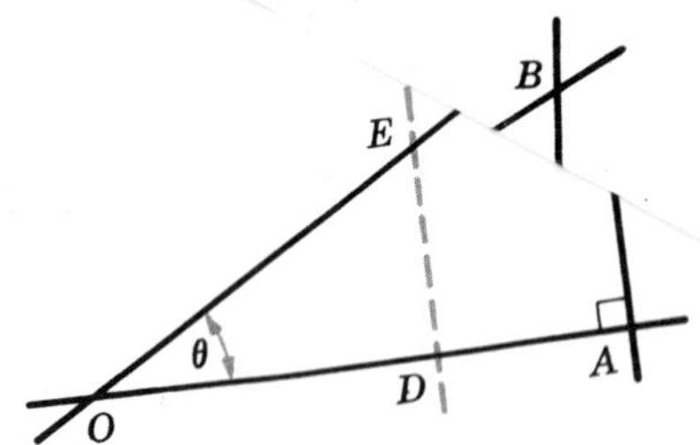

Figure 1-4
A right triangle.

In a similar manner, the ratio of the side adjacent to the angle to the hypotenuse can be shown to be constant for a given angle θ. The ratio of the side adjacent to the hypotenuse is called the *cosine* of the angle.

While other ratios can be set up between sides of a right triangle, the only other one of much value in this study of physics is the ratio of the side opposite the angle to the side adjacent to the angle. This, too, will be a constant for a given angle. The ratio of the side opposite to the side adjacent is called the *tangent*.

The length of any of the three sides in a right triangle can be determined, provided that the length of two of the sides are known, by the pythagorean theorem, which states that the square of the hypotenuse is equal to the sum of the squares of the other two sides:

$$(OB)^2 = (AB)^2 + (OA)^2$$

Example Find mathematically the resultant force when a 20-lb force east and a 20-lb force north act on an object (see Fig. 1-5). Since the figure is a parallelogram, $\mathbf{OC} = \mathbf{AB} = 20$ lb ·h.

$$\frac{\text{opposite}}{\text{·ent}} = \frac{20\text{ lb}}{20\text{ lb}} = 1.0$$

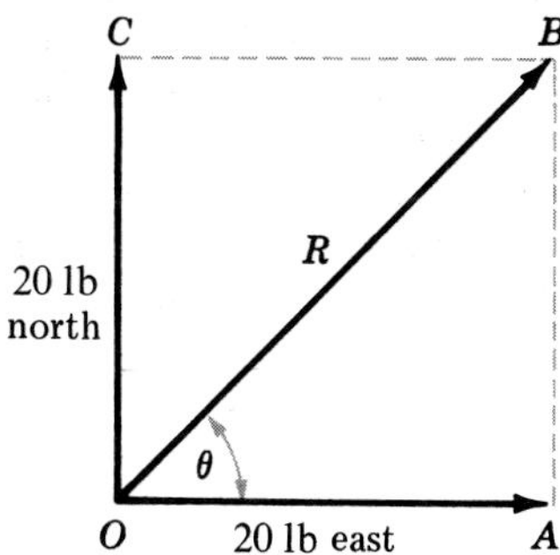

Figure 1-5
Parallelogram of forces diagram.

and $\theta = 45°$

Then,

$$\sin 45° = \frac{\mathbf{AB}}{\mathbf{OB}} \quad \text{and} \quad \mathbf{OB} = \mathbf{R} = \frac{\mathbf{AB}}{\sin 45°}$$

$$= \frac{20 \text{ lb}}{0.707} = 28 \text{ lb}$$

This is the same answer obtained when we solved the problem graphically earlier: 28 lb 45°N of E.

The trigonometric rules stated above hold only for right triangles.

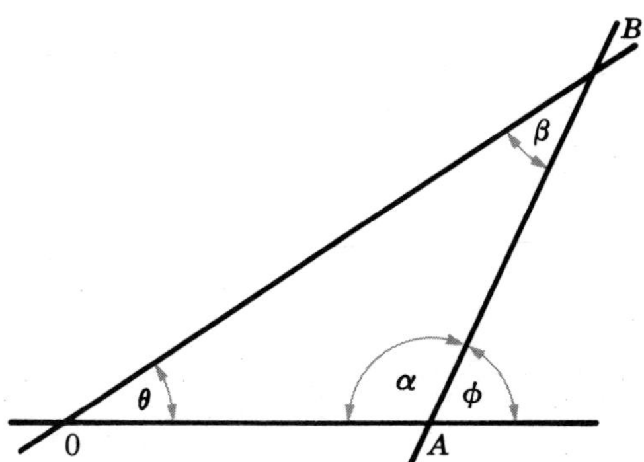

Figure 1-6
A triangle not containing a right angle.

In the case in which AB in Fig. 1-4 is not at right angles to OA, a new set of relationships exist (Fig. 1-6). The pythagorean theorem now expands into a new form, called the *law of cosines,* which appears as

$$(OB)^2 = (AB)^2 + (OA)^2 - 2(AB)(OA) \cos \alpha$$

Since the angle ϕ is often given in problems involving vectors, the law of cosines can appear as

$$(OB)^2 = (AB)^2 + (OA)^2 + 2(AB)(OA) \cos \phi$$

The change in sign in the last term of the equation is due to the rules establishing the sign values of the trigonometric functions as they appear in various quadrants (quarters of a 360° reference circle) (see Fig. 1-7).

Since angle ϕ in Fig. 1-6 is less than 90°, it

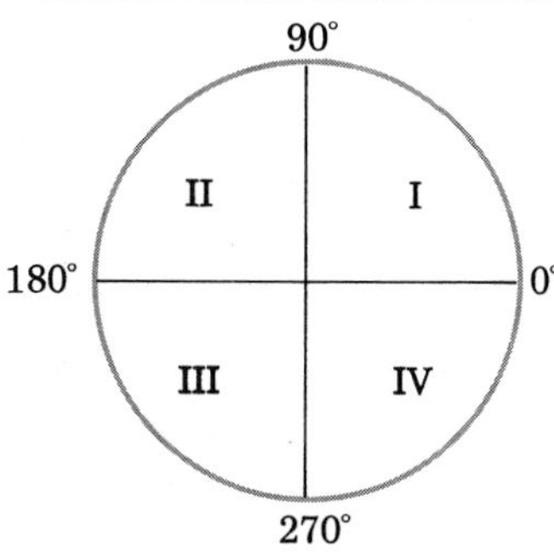

Figure 1-7*a*
Quadrants in a circle.

Quadrant	Angle	sin	cos	tan
I	0–90°	+	+	+
II	90–180°	+	–	–
III	180–270°	–	–	+
IV	270–360°	–	+	–

Figure 1-7*b*
Table showing sign of trigonometric functions for various quadrants.

follows the rules of signs for angles in quadrant I; and since angle α is between 90 and 180°, it follows the rules of signs for quadrant II. Therefore the cosine, which is − in quadrant II, is + in quadrant I; hence the sign change in the two forms of the law of cosines.

The law of cosines will enable us to find the value of the side OB, which is the resultant; but since we need also to determine the sense (direction) of this resultant, it is necessary to determine the angle θ. The angle can be found by using the *law of sines,* which relates the sides of a triangle to the angle opposite each side:

$$\frac{OB}{\sin \alpha} = \frac{AB}{\sin \theta} = \frac{OA}{\sin \beta}$$

If three of the four variables in two of these equivalencies are known, the fourth variable can be found.

In magnitude, the resultant of two vectors may be greater than, equal to, or less than either one of them, depending on the angle between them. In Fig. 1-8*a,* two vectors **A** and **B** of magnitude 2 and 3 units, respectively, are shown separately. In *b,* where the vectors are in the same direction, the magnitude of the resultant is merely their arithmetic sum. As the angle between the vectors increases, the resultant becomes less, as in *c, d,* and *e.* In *f,* the magnitude of the resultant is the difference of the magnitudes of the two vector quantities.

Example Two vectors of 8.0 units and 5.0 units make an angle of 60° with each other. What is their vector sum?

Using a convenient scale, draw vector **A** 8.0 units long (Fig. 1-9). Beginning at the end of **A,** draw **B** 5.0 units long so that there is a 60° angle between the direction of **A** and that of **B.** There is, of course, a 120° angle between the line seg-

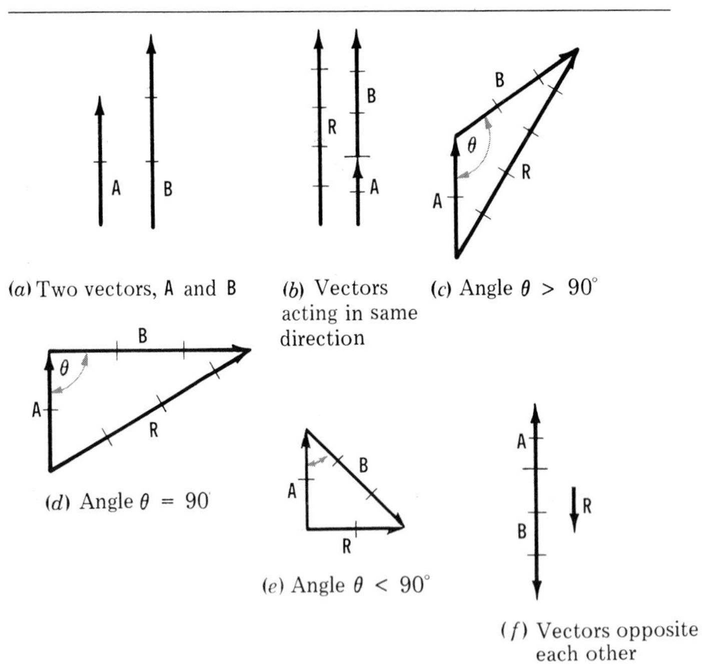

Figure 1-8
The resultant of two vectors depends on the angle between them.

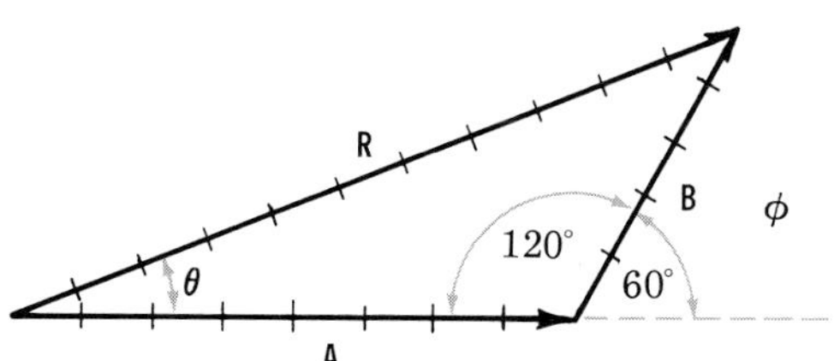

Figure 1-9
Addition of two vectors in given example.

ments. Complete the triangle by drawing the resultant vector **R** from *O* to the head of **B.** Measure the magnitude of **R,** in the scale selected, and the angle θ between **R** and the reference direction, say, that of **A.** It is found that **R** is 11+ units, at 20°+ to the direction of **A,** the 8.0-unit vector.

Applying the analytical method to obtain greater precision, we have

$$R^2 = A^2 + B^2 + 2AB \cos \phi$$
$$= 8.0^2 + 5.0^2 + (2)(8.0)(5.0)(0.50) = 129$$
$$R = 11.4 \text{ units}$$

The angle θ may be found from the sine law

$$\frac{R}{\sin 120°} = \frac{B}{\sin \theta}$$

$$\sin \theta = \frac{B \sin 120°}{R} = \frac{5.0}{11.4} \times 0.87 = 0.38$$

$$\theta = 22°$$

Hence **R** is 11.4 units at 22° to the direction of **A.**

It is sometimes confusing to find the trigonometric values of an angle greater than 90°. A table of rules for determining the trigonometric functions of obtuse angles is presented in the Appendix. It might prove helpful to remember that any trigonometric function of a given angle is numerically equal to the same function of its related angle. For example, the sin 120° = sin (180° − 120°) = sin 60°. Since sin 120° is for an angle in the second quadrant, the sign is positive. Also, the sine of the angle 258° = sin (180° + 78°) = sin 78°; but since the sin 258° is in the third quadrant, the sine has a negative value and sin 258° = −sin 78°.

1-4 COMPONENTS OF A VECTOR

There is frequently occasion to perform the operation inverse to geometric addition, namely, to determine a set of vectors whose effect when acting together is the same as that of a given vector. This process is called the resolution of a vector into its components. A set of components of a vector is a set of vectors whose resultant is the original vector.

A given vector may have any number of components. The most useful choice of components is usually that in which the *x* and *y* components are found, where *x* is the horizontal component and *y* is the vertical component, as in Fig. 1-10.

Consider the vector **A,** which makes an angle of 45° with the horizontal, as is shown in Fig. 1-10. In order to obtain a set of rectangular components of **A,** one of which shall be horizontal, draw a horizontal line through the tail of the vector **A.** Now from the head of **A,** draw a line perpendicular to the horizontal line. We see that

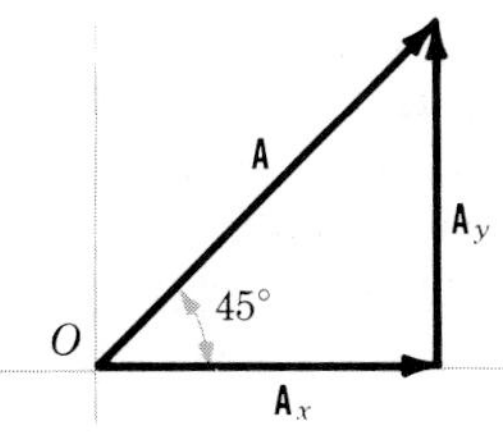

Figure 1-10
Vertical and horizontal components of a vector.

the vector **A** can be considered as the resultant of the vectors $\mathbf{A}_x$ and $\mathbf{A}_y$. The magnitudes of the horizontal and vertical components are $A \cos 45°$ and $A \sin 45°$, obtained by recalling that $\sin 45° = A_y/A$; $A_y = A \sin 45°$; and that $\cos 45° = A_x/A$; $A_x = A \cos 45°$. The directions of the arrowheads are important, for we are now considering that $\mathbf{A}_x$ has been added to $\mathbf{A}_y$ to give the resultant **A;** therefore, the arrows must follow head to tail along $\mathbf{A}_x$ and $\mathbf{A}_y$ so that **A** can properly be considered as a resultant drawn from the tail of the first arrow $\mathbf{A}_x$ to the head of the last arrow $\mathbf{A}_y$. This resolution into components now allows us to discard the vector **A** in our problem and keep only the two components $\mathbf{A}_x$ and $\mathbf{A}_y$. These two taken together are in every way equivalent to the single vector **A.**

What is the advantage of having two vectors to deal with where there was only one before? The advantage lies in the fact that several vectors making various odd angles with each other can be replaced by a set of vectors along one direction and a set in a direction at right angles to the first. The magnitudes of each of these two groups of vectors can then be summed up separately by ordinary arithmetic, thus reducing the problem to one of two vectors acting at right angles to each other.

1-5 COMPONENT METHOD OF ADDING VECTORS

To add a number of vectors **A, B, C,** and **D** (Fig. 1-11), where **A** = 10N, 70°S of E; **B** = 5N, E; **C** = 15N 45°N of E; and **D** = 15N 60°N of W, we proceed as follows. Place the vectors at the origin on a set of rectangular coordinates (x and y). Next resolve each vector into x and y components (Fig. 1-12).

Vector **A:**

$$\mathbf{A}_x = \mathbf{A} \cos 70° = 10\,(0.342) = +3.4 \text{ N}$$

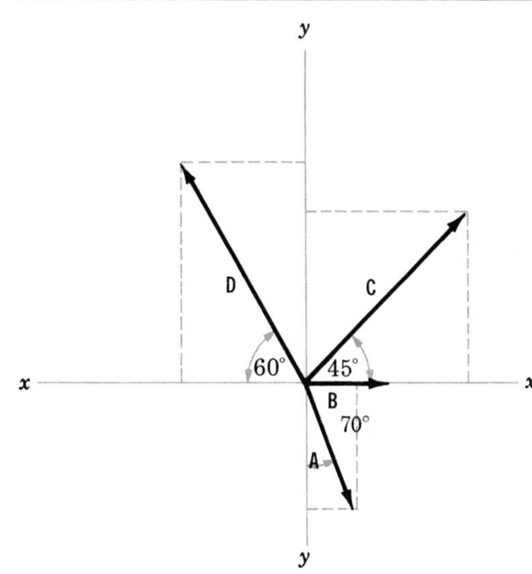

Figure 1-11
Addition of vectors by component method.

$$\mathbf{A}_y = \mathbf{A} \sin 70° = 10\,(0.940) = -9.4 \text{ N}$$

Vector **B:**

$$\mathbf{B}_y = 0 \qquad \mathbf{B}x = \mathbf{B} = +5 \text{ N}$$

Vector **C:**

$$\mathbf{C}_x = \mathbf{C} \cos 45° = 15\,(0.707) = +10.6 \text{ N}$$
$$\mathbf{C}_y = \mathbf{C} \sin 45° = 15\,(0.707) = +10.6 \text{ N}$$

Vector **D:**

$$\mathbf{D}_x = \mathbf{D} \cos 60° = 15\,(0.5) = -7.5 \text{ N}$$
$$\mathbf{D}_y = \mathbf{D} \sin 60° = 15\,(0.866) = +13.0 \text{ N}$$

Then enter in a tabular form and add the forces along the x axis, calling the sum Σx (the greek letter sigma (Σ) is used to express the sum), and then add the forces along the y axis, calling the sum Σy. Some of the components may be

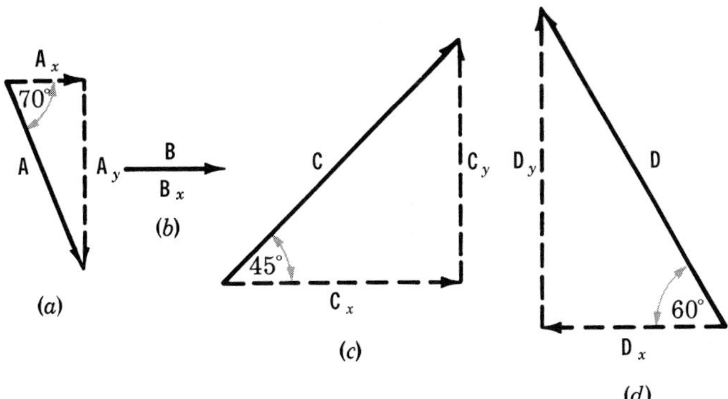

Figure 1-12
Resolution of vectors into x and y components.

negative (for example, $\mathbf{D}_x$ and $\mathbf{A}_y$). This is due to the sign convention which says that x to the right of the y axis is + and to the left is −; y above the x axis is + and y below the x axis is −. It should be noted that a vector has a zero rectangular component at right angles to itself. For instance, vector **B** has a zero y component.

We have now resolved the four vectors into two vectors acting 90° apart and can solve for the magnitude of the resultant by using the pythagorean theorem or by the addition of vectors (see Fig. 1-13). Using the latter gives

$$\tan\theta = \frac{14.2}{11.5} = 1.23$$

$$\theta = 51°\text{N of E} \qquad \text{and} \qquad \sin 51° = \frac{14.2}{R}$$

$$\mathbf{R} = \frac{14.2}{\sin 51°} = \frac{14.2}{0.777} = 18.3\text{N}$$

The resultant is an 18.3-N force acting 51°N of E.

Table 1
x AND y COMPONENTS OF VECTORS IN THE EXAMPLE

Vector	x Component, N	y Component, N
A	+ 3.4	− 9.4
B	+ 5.0	0
C	+ 10.6	+ 10.6
D	− 7.5	+ 13.0
	$\Sigma x = +11.5$	$\Sigma y = +14.2$

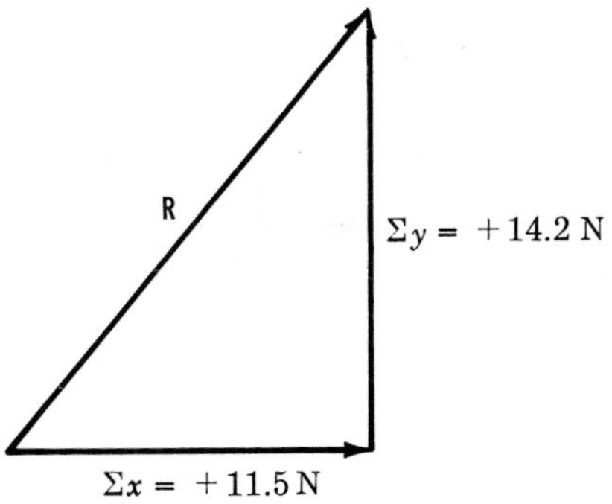

Figure 1-13
Vector diagram using the sums of the x and y components.

1-6 EQUILIBRIUM

Many important problems confronting the physicist and engineer involve several forces acting on a body under circumstances in which they pro-

Figure 1-14
After a parachute opens and falls a certain distance, it moves downward thereafter with uniform velocity. Such a system of balanced forces is in equilibrium.

duce no change in the motion of the body. The state in which there is no change in the motion of a body is called *equilibrium*. A body in equilibrium may be at rest but does not necessarily have to be at rest; it may be moving with uniform speed in a straight line or rotating uniformly around a fixed axis. An example of this situation occurs when a parachute reaches its terminal velocity when falling (Fig. 1-14). It continues to fall toward earth with a constant speed but, as we shall see later, no longer has a net force acting on it to cause it to accelerate.

In this chapter the discussion will be restricted to the action of forces that are in equilibrium and also that have lines of action which pass through the same point. Forces whose lines of action intersect at a common point are said to be concurrent.

We have seen earlier that a body in linear motion is in equilibrium if the vector sum of the forces acting upon it is zero. This statement is known as the *first condition for equilibrium*.

Several forces may be added by the use of any of the methods previously described. When the vectors are added by the graphical method, the vector sum is zero if the length of the arrow representing the resultant is zero. It was shown that this can occur only if the head of the last vector to be added comes back to touch the tail of the first vector. The vector sum is zero if the vector diagram is a closed polygon. This method is especially useful when there are three or more concurrent forces.

Example An object weighing 100 lb and suspended by a rope A (Fig. 1-15a) is pulled aside by the horizontal rope B and held so that rope A makes an angle of 30° with the vertical. Find the tensions in ropes A and B.

Consider the junction O as the body in equilibrium. It is acted upon by the three forces $\mathbf{W}$ (known), $\mathbf{F}_1$, and $\mathbf{F}_2$ (unknown). These forces are represented in Fig. 1-15b, with $\mathbf{W}$ scaled to represent the known weight and only the directions and not the magnitudes of $\mathbf{F}_1$ and $\mathbf{F}_2$ known. The junction O is in equilibrium, since it is not moving due to the influence of these three forces;

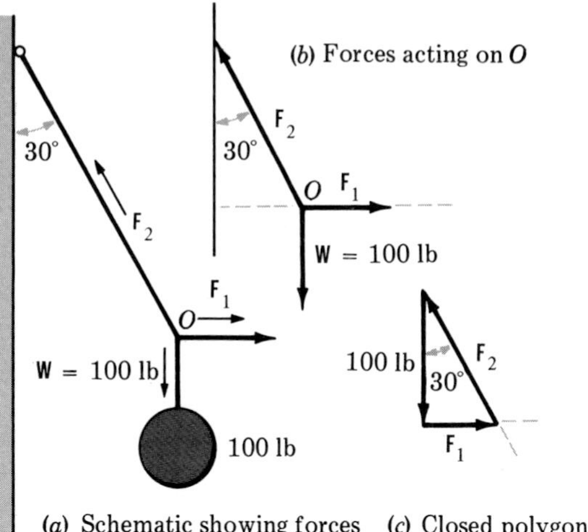

Figure 1-15
Finding an unknown force by the vector method.

their resultant must therefore be zero. The vectors representing **W**, $\mathbf{F}_1$, and $\mathbf{F}_2$ can be redrawn as in Fig. 1-15*c* so that they combine to form a closed triangle. Note the fact that each vector is drawn parallel to the force that it represents. It is also important to draw the vector diagram so that the forces considered are those which act upon the body that is in equilibrium.

In order to solve the vector triangle of Fig. 1-15*c*, it may be observed that

$$\frac{\mathbf{F}_1}{100 \text{ lb}} = \tan 30° = 0.58$$

so that $\mathbf{F}_1 = (100 \text{ lb})(0.58) = 58$ lb. To get $\mathbf{F}_2$, we can put

$$\frac{100 \text{ lb}}{\mathbf{F}_2} = \cos 30° = 0.866$$

Therefore,

$$\mathbf{F}_2 = \frac{100 \text{ lb}}{0.866} = 115 \text{ lb}$$

That is, in order to hold the system in the position of Fig. 1-15*a*, one must pull on the horizontal rope with a force of 58 lb. The tension of rope *A* is then 115 lb. The tension in the segment of rope directly supporting the weight is, of course, just 100 lb.

1-7 HINTS FOR SOLVING PROBLEMS INVOLVING A BODY IN EQUILIBRIUM

One of the first challenges that students face in a course in physics is that of learning to apply the proper techniques to solve physical problems. While there are many ways in which such problems can be approached, experience will show that certain techniques work best for particular problem-solving situations. The challenge is to recognize or identify the *type* of problem and then to apply the proper technique to solve it. The student must analyze each problem to determine what data are given and what data are sought. This analysis coupled with a knowledge of the formulas, either defined or derived, which relate to the problem provide the tools needed to solve all physical problems.

The authors of this book have made an attempt to assist the student in developing the techniques of problem solving by providing in several strategic positions in the book procedures which have proved to be quite effective in the solution of certain kinds of problems. It is strongly recommended that the student read these sections carefully.

This is the first of such sections, directed toward the solution of problems involving a body in equilibrium. In such problems, the following steps are recommended:

1 Draw a sketch to picture the apparatus, as in Fig. 1-15*a*. Include on the sketch the known data. Indicate by a suitable symbol, such as *O*, the point of concurrence of the several forces.
2 Make a figure, like Fig. 1-15*b*, to show the directions of all the forces acting through the

point of concurrence, as well as both the magnitudes and directions of the known forces, being sure that only forces acting on the body are included.

3 Draw a closed vector polygon, such as Fig. 1-15*c*, scaled to show both the magnitude and direction of each of the forces.

4 Finally, solve the vector problem by suitable mathematical methods.

Example By the method of components find the resultant and the equilibrant of a 7.0-lb horizontal force and a 12.0-lb force making an angle of 60° with the horizontal (Fig. 1-16).

The horizontal and vertical components of the 12.0-lb forces are

$$H = 12 \cos 60° = 6.0 \text{ lb} \qquad \text{and}$$
$$V = 12 \sin 60° = 10.4 \text{ lb}$$

The horizontal and vertical components of the 7.0-lb force are

$$H = 7.0 \text{ lb} \qquad \text{and} \qquad V = 0$$

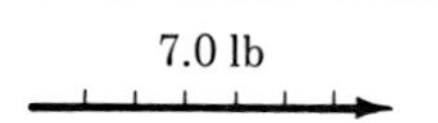

(*a*) Components of the 7-lb horizontal vector

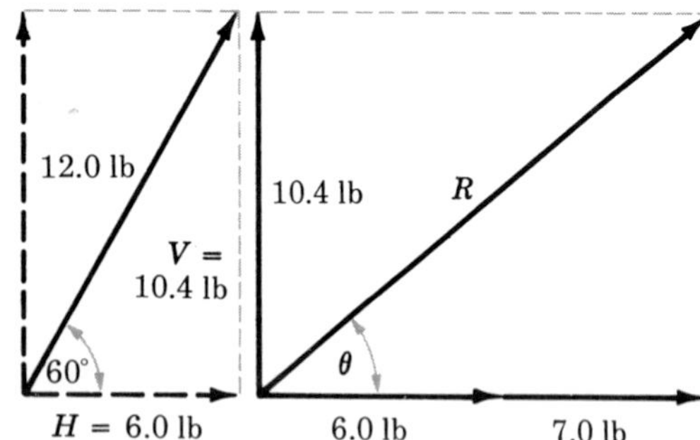

(*b*) Components of 12.0 lb vector at 60°

(*c*) Combined horizontal and vertical components of the two forces

Figure 1-16
Resultant forces by component method.

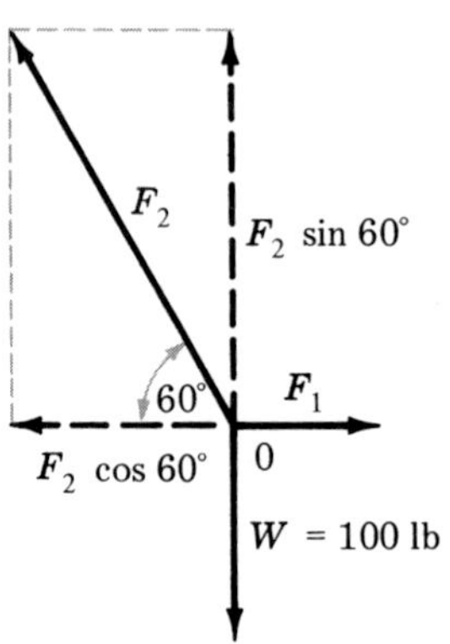

Figure 1-17
Solving the problem of Fig. 1-15 by the use of components.

We thus replace the original two forces by three forces, one vertical and two horizontal. Since the two horizontal forces are in the same direction, they may be added as ordinary numbers, giving a total horizontal force of 6.0 lb + 7.0 lb = 13.0 lb. The problem is now reduced to the simple one of adding two forces at right angles (Fig. 1-16*c*), giving the resultant

$$R = \sqrt{(10.4^2 + 13.0^2)} \text{ lb} = 16.6 \text{ lb}$$

The angle θ which R makes with the horizontal is given by $\tan \theta = 10.4/13.0 = 0.80$, so that $\theta = 38.7°$.

The equilibrant is equal in magnitude but opposite in direction to R. Hence it is a force of 16.6 lb at an angle of 218.7°.

Example An object weighing 100 lb and suspended by a rope A (Fig. 1-15) is pulled aside by the horizontal rope B and held so that rope A makes an angle of 30° with the vertical. Find the tensions in ropes A and B.

We have previously solved this problem by the straightforward method of adding the vectors to form a closed figure. The method is quite appropriate to simple cases, but for the sake of illustration, let us now solve the problem again by the more general method of components. In Fig. 1-17 are shown the same forces, separated for greater convenience of resolution. The horizontal and vertical components of the 100-lb force are, respectively, 0 and 100 lb down. The horizontal and vertical components of F_1 are, respectively, F_1 (to the right) and O. Although we do not yet know the numerical value of F_2, whatever it is, the horizontal and vertical components will certainly be $F_2 \cos 60°$ to the left and $F_2 \sin 60°$ up. We now have four forces, two vertical and two horizontal, whose vector sum must be zero to ensure equilibrium. In order that the resultant may be zero, the sum of the horizontal components and the sum of the vertical components must each be equal to zero. Therefore,

$$\Sigma F_V = F_2 \sin 60° - W = 0$$

$$F_2 \sin 60° = W \qquad 0.866\, F_2 = 100$$

$$F_2 = 115 \text{ lb}$$

and

$$\Sigma F_H = F_1 - F_2 \cos 60° = 0$$

$$F_1 = F_2 \cos 60°$$

$$F_1 = (115 \text{ lb})(0.5) = 58 \text{ lb}$$

Example A load of 100 lb is hung from the middle of a rope, which is stretched between two walls 30.0 ft apart (Fig. 1-18). Under the load the rope sags 4.0 ft in the middle. Find the tensions in sections A and B.

The midpoint of the rope is in equilibrium under the action of the three forces exerted on it by sections A and B of the rope and the 100-lb weight. A vector diagram of the forces appears in Fig. 1-19. The horizontal and vertical components of the 100-lb force downward are 0 and 100 lb, respectively. The horizontal and vertical components of F_1 are, respectively, $F_1 \cos\theta$ to the left and $F_1 \sin\theta$ upward. Similarly, the horizontal and vertical components of F_2 are, respectively, $F_2 \cos\theta$ to the right and $F_2 \sin\theta$ upward. In order for the resultant to be zero, the sum of the horizontal components and the sum of the vertical components must each be equal to zero. Therefore,

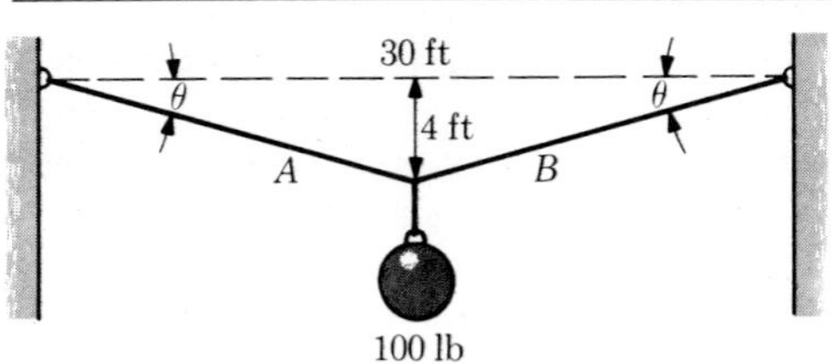

Figure 1-18
Finding the tension in a stretched rope.

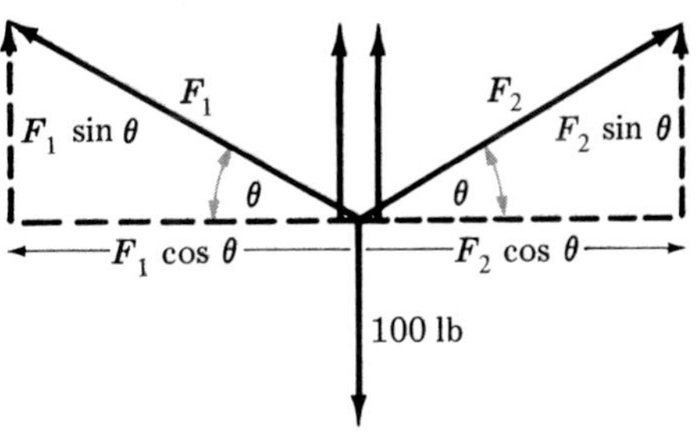

Figure 1-19
Horizontal and vertical components of the forces in a stretched rope.

$$\Sigma F_H = F_2 \cos\theta - F_1 \cos\theta = 0 \quad \text{horizontal} \quad (a)$$

$$\Sigma F_V = F_1 \sin\theta + F_2 \sin\theta - 100 \text{ lb} = 0 \quad \text{vertical} \quad (b)$$

Since these two equations involve three unknown quantities F_1, F_2, and θ, we cannot solve them completely without more information.

The value of $\sin\theta$ can be determined from the dimensions shown in Fig. 1-18:

$$\sin\theta = \frac{4.0 \text{ ft}}{A}$$

$$A = \sqrt{(15.0^2 + 4.0^2)\text{ ft}^2} = \sqrt{241 \text{ ft}^2} = 15.5 \text{ ft}$$

and

$$\sin\theta = \frac{4.0 \text{ ft}}{15.5 \text{ ft}} = 0.26$$

From Eq. (*a*), $F_1 = F_2$. Substituting in Eq. (*b*),

$$F_1 \sin\theta + F_1 \sin\theta - 100 \text{ lb} = 0$$
$$2F_1 \sin\theta = 100 \text{ lb}$$
$$2F_1(0.26) = 100 \text{ lb}$$
$$F_1 = \frac{100 \text{ lb}}{2(0.26)} = 1\bar{9}0 \text{ lb}$$

and

$$F_2 = 1\bar{9}0 \text{ lb}$$

Two things should be noticed about the problem just solved: (1) the value of a function of an angle in the vector diagram was needed in order to carry out the solution; and (2) the value of that function was determined from the geometry of the original problem.

Example Calculate the force needed to hold a 1,000-lb car on an inclined plane that makes an angle of 30° with the horizontal, if the force is to be parallel to the incline.

The forces on the car include (see Fig. 1-20) its weight **W** vertically downward, the force **B** parallel to the incline, and the force **A** exerted on the car by the inclined plane itself. The last force mentioned is perpendicular to the plane and is called the normal.

Since the car is in equilibrium under the action of the three forces, **A**, **B**, and **W**, a closed triangle can be formed with vectors representing them, as in Fig. 1-20*b*. In the vector diagram, $\mathbf{B} = \mathbf{W} \sin\theta$. Since θ is 30° and $\mathbf{W} = 1{,}000$ lb,

$$\mathbf{B} = (1{,}000 \text{ lb}) \sin 30°$$
$$= (1{,}000 \text{ lb})(0.500) = 500 \text{ lb}$$

The value of **A**, the perpendicular force exerted by the plane, can be found by observing that

$$\mathbf{A} = \mathbf{W} \cos 30° = (1{,}000 \text{ lb})(0.866) = 866 \text{ lb}$$

It should be noticed that **W** can be resolved into two components that are, respectively, parallel and perpendicular to the incline. These components are equal in magnitude and opposite in direction to **B** and **A**, respectively.

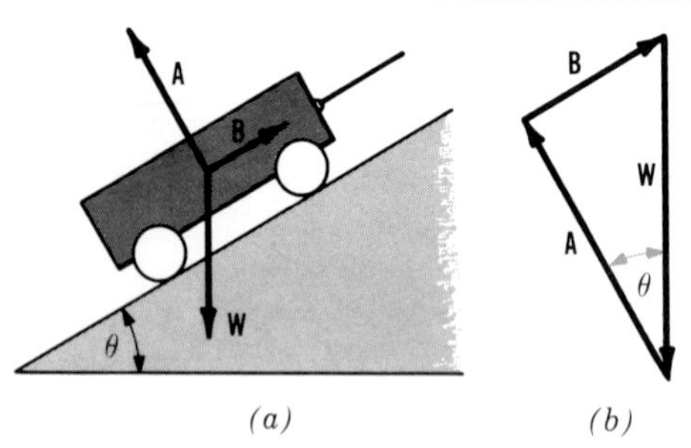

Figure 1-20
Finding the forces acting on a body on an incline.

1-8 DIFFERENCE OF TWO VECTORS

We shall see later in the study of relative velocity, acceleration, and certain other properties of matter which are vector in nature that the occasion arises when it is necessary to find the difference of two vector quantities of the same kind. The difference of two vectors is obtained by adding one vector to the negative of the other (the vector equal in magnitude and opposite in direction). As in arithmetic, $5 - 3 = 2$ may be written $5 + (-3) = 2$, so in the case of vectors we may understand the vector difference $\mathbf{A} - \mathbf{B} = \mathbf{C}$ as $\mathbf{A} + (-\mathbf{B}) = \mathbf{C}$. The vector $-\mathbf{B}$ is equal in magnitude and opposite in direction to **B**.

Note in Fig. 1-21 the direction of the difference vector **C**, and compare it with the vector which would be the sum of **A** and **B** (make your own diagram). The vector difference **C** is also frequently defined as the vector which must be added to **B** to give **A**. Does **C** satisfy this definition?

Example Two cars start from the same point but one travels north at 50 km/h and the other

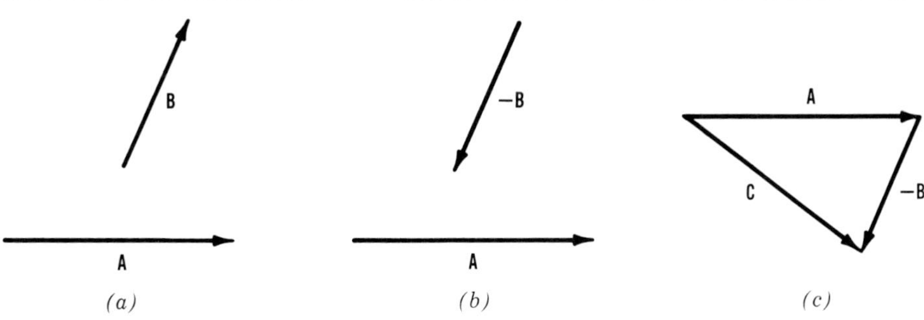

Figure 1-21
Difference of two vectors.

east at 30 km/h. What is the vector difference of these two velocities? What does it signify? (See Fig. 1-22).

$$\tan\theta = \frac{50}{30} = 1.67 \qquad \theta = 59°$$

$$\sin 59° = \frac{50}{C} \qquad C = \frac{50}{\sin 59°} = \frac{50}{0.8572}$$

$$C = 58.3 \text{ km/h}$$

Therefore, the vector difference is 58.3 km/h 59°S of E. This signifies the relative velocity of the two cars to each other.

1-9 THE PHYSICS OF SAILING

Lovers of the sport of sailing may claim that physicists are taking the fun out of sailing by attempting to analyze what happens when a sailboat is driven by the wind. Despite this objection, a sailboat in motion does provide an excellent example of forces acting on an object.

A question that is often asked is whether a sailboat can go faster than the wind. Let us consider a sailboat initially at rest with its stern (back) to the wind, Fig. 1-23*a*. If the sail is extended so that it makes a right angle with the boat, the boat will start to move, in a form of sailing known as running before the wind. Perhaps no type of sailing is more enjoyable to the

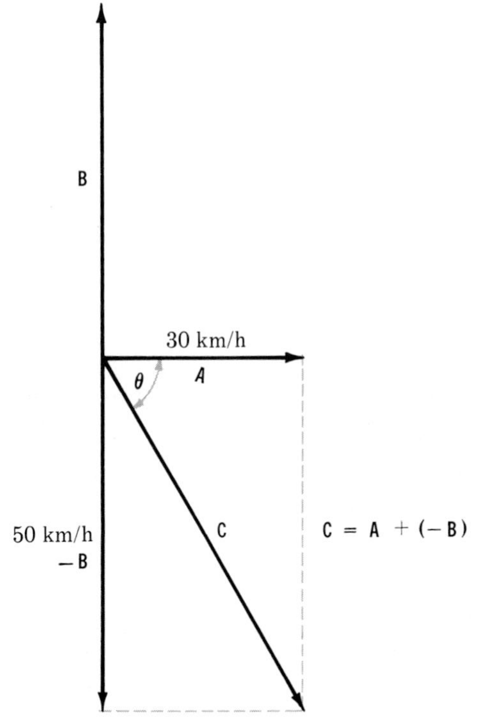

Figure 1-22
Finding the vector difference.

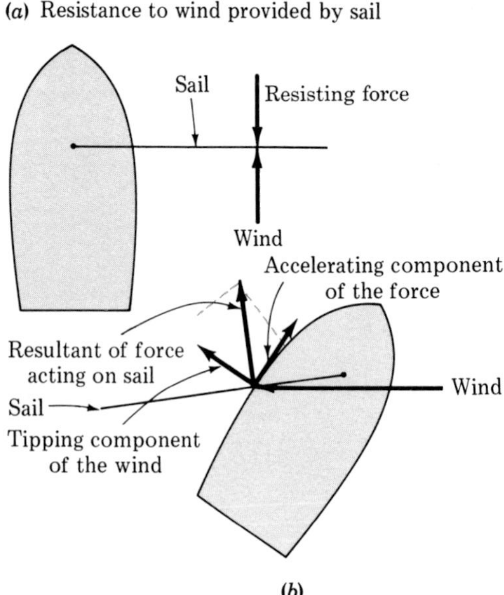

Figure 1-23
Forces acting on a sailboat.

weekend sailor than this. Could this be the configuration of the wind and sail to achieve the greatest speed? To answer this, let us analyze the forces that are acting on the sail. There is a force exerted on the sail by the wind and another resisting force provided by the sail due to the friction of the boat in the water and also initially due to the inertia of the boat. If the force of the wind is greater than the resisting force, the boat will accelerate according to Newton's second law of motion. This acceleration will continue until the boat starts moving as fast as the wind; then when the sail is moving at the same speed as the wind, the accelerating force disappears because the sail no longer provides a surface for the wind to push against. So, when running before the wind, the maximum velocity that can be attained by the boat is the velocity of the wind itself.

Suppose that we now turn the boat so that the sail makes an angle with the wind, as in Fig. 1-23*b*, where the boat is actually headed at some angle into the wind.[1]

Here, the resultant of the force of the wind on the sail can be represented as a force acting perpendicularly to the sail. It is then possible to resolve this resultant force into a force at right angles to the boat, a tipping force which has no value in moving the boat forward other than putting the boat on its side and cutting down the friction between the boat and the water (this type of sailing is called planing) and which must be compensated for by the keel or centerboard. It is also possible to resolve this resultant force into a force acting parallel to the boat. This is the accelerating component of the force of the wind and will cause the boat to move forward. It should be noted here that the sail does not move away from the wind as in the case of running before the wind. Therefore the force of the wind remains in contact with the sail and the boat keeps accelerating. We know that according to Newton's second law the boat should keep accelerating even after attaining the velocity of the wind. Further, as the boat travels faster, the force of impact between the sail and the wind is increased because the boat is actually heading somewhat into the wind. Hence the boat can go faster than the wind in this case.

Can the boat keep accelerating indefinitely? As we have seen in several other situations earlier in this chapter, the boat attains a terminal velocity for the same reason that the falling raindrop does. That is, at a certain velocity the forces opposing the motion, mainly water resistance on the hull of the boat, cancel the accelerating force due to the wind. In the design of boats used for racing, the hulls are carefully built to reduce this friction to a minimum. Anyone who has seen a catamaran-style sailboat, with its twin hull designed to keep friction down, moving with such great speed will recognize just how fast a sailboat can move when resistance is reduced. A further

[1] Although the sail in this case acts as an airfoil, much as the wing of an airplane which helps the boat to move forward, at this point in our discussion let us consider only the force due to the wind on the back of the sail.

example can be seen in iceboats, which have practically no hull resistance. These "boats" can attain a velocity several times faster than the wind. In answer to the question posed at the start of this section, sailboats can attain speeds of two to three times the speed of the wind. Furthermore, an analysis of forces that are acting will show that the maximum velocity will be attained when the sail makes a 45° angle into the wind.

SUMMARY

A *force* is arbitrarily defined as any influence capable of producing a change in the motion of an object.

Forces are usually expressed in *pounds* (in the fps system), *dynes* (in the cgs system), and *newtons* (in the mks system).

Quantities whose measurement is specified by *magnitude* and *direction* are called *vector quantities*. Those which have only magnitude are called *scalar quantities*.

A vector quantity may be represented graphically by a directed line segment, called a vector. A *vector* is the line whose length indicates to scale the magnitude of the vector quantity and whose direction indicates the direction of the quantity.

The *resultant* of two or more vector quantities is that single quantity of the same physical makeup that would produce the same result.

The resultant **R** of two vectors **M** and **N** having angle θ between their directions is conveniently found from the law of cosines and the law of sines,

$$R^2 = M^2 + N^2 + 2MN \cos \theta$$

$$\frac{M}{\sin \phi} = \frac{R}{\sin (180° - \theta)}$$

Vectors are conveniently added graphically by placing them head to tail and drawing the resultant from the origin to the head of the last vector, closing the polygon.

The *rectangular components* of a vector are its projections on a set of right-angle axes, for example, the horizontal and vertical axes.

$$R_x = R \cos \theta$$
$$R_y = R \sin \theta$$

The *component* method of adding vectors is to resolve each into its rectangular components, which are then added algebraically and the resultant found.

$$R^2 = R_x^{\ 2} + R_y^{\ 2}$$

A body is in *equilibrium* when there is no change in its motion.

When a body is in equilibrium, the vector sum of all the forces acting on it is zero. This is known as the *first condition for equilibrium*. When a body is in equilibrium, the force diagram is a closed polygon, or the sums of the rectangular components of all the forces must each equal zero: $\Sigma F_x = 0$ and $\Sigma F_y = 0$.

The *difference* of two vectors is obtained by adding one vector to the negative of the other (the vector equal in magnitude and reversed in sense).

Questions

1 What items must be stated to specify a vector quantity completely?

2 Give several examples of scalar quantities; of vector quantities.

3 Show that the maximum value that the resultant of two vectors can have is the arithmetic sum and that the minimum value is the arithmetic difference.

4 Show why the vector addition of a 5-N force E and a 5-N force N is not 10 N NE.

5 The sum of two vectors is at right angles to their difference. Show that the vectors are equal in magnitude.

6 It is found that the sum and the difference of two vectors have equal magnitudes. Prove that the two vectors are at a 90° angle to each other.

7 Are vectors necessary, or is the concept of a vector merely a convenience in expressing physical quantities?

8 Two forces of 50 N and 80 N act upon a body. What are the maximum and minimum possible values of the resultant forces?

9 What are the handicaps which one has in solving complicated vector problems by the parallelogram method?
10 Show by a series of vector polygons that it is immaterial which order is used in laying off the vectors end to end and that the resultant is always the same.
11 Show how the parallelogram method of solving vector problems can be simplified by the use of the polygon method.
12 An automobile is acted upon by the following forces: a horizontal force due to air resistance; the weight of the car; a force almost vertically upward on the front wheels; the force of the ground on the rear wheels. Draw a vector polygon to show these forces in equilibrium. What does this imply with respect to the velocity of the car?
13 When two vectors are drawn from a common origin in a vector parallelogram, what quantity does each diagonal of the parallelogram represent?
14 A wire is stretched horizontally between two supports (Fig. 1-18), with a load applied at the center. Assuming no stretching in the wire, plot a rough graph to show how the force in the wire would vary as the angle θ is varied from 0 to 90° by moving the supports.
15 Explain the difference between vector and algebraic sums.
16 Three forces of 12, 15, and 20 N are in equilibrium. If the 12-N force is directed horizontally to the right, what two configurations in a vertical plane may the other two forces have?
17 Give several examples of moving bodies which are in equilibrium.
18 Why does an airplane pilot prefer to take off and land into the wind? Explain by the use of a vector diagram.
19 In moving a sled over the snow is it better to pull the sled with a rope or to push on the sled with a pole? Explain by the use of vectors.
20 When telephone wires are covered with ice, is there more danger of their breaking when they are taut than when they sag? Why?

Problems

1 Resolve a force of 100 N acting at an angle of 37° with the horizontal into horizontal and vertical components.
2 An object weighing 40 N rests on an inclined plane making an angle of 30° with the horizontal. Resolve the force into components parallel to the plane and perpendicular (normal) to the plane.
Ans. Force parallel = 20 N, force perpendicular = 34.64 N.
3 Two forces of 10 N each are acting on a point. One force acts E and the other 70°N of E. Find the resultant by using the law of sines and cosines.
4 Resolve a force of 100 N into two components which lie on opposite sides of the force and each of which makes an angle of 30° with the force. *Ans.* Each is equal to 57.7 N.
5 Two airplanes start from the same point, one traveling 250 km E and the other 180 km N. What is the vector difference of these two displacements? What does it signify?
6 Three cities X, Y, and Z are connected by straight highways. X is 6 km from Y, Y is 4 km from Z, and X is 5 km from Z. Find the angle made by the highways XY and YZ.
Ans. 55.7°.
7 What is the angle between a 2.00-unit vector and a 3.00-unit vector so that their sum is 4.00 units?
8 What effect do the following forces have on a point: 100 N, 30°E of N; 200 N, 80°S of E; 150 N, 45°S of W; 175 N, 25°W of N; 50 N, due N. Solve graphically. *Ans.* 95.5 N, 4.5°S of W.
9 Solve Prob. 8 using the component method.
10 A 20-m-long rope attached at the top and the bottom of a flagpole is pulled 2 m away from the pole by a 100-N force acting at right angles to the pole at its midpoint. What is the tension on the segments of rope on each side of the 100-N force? *Ans.* 255 N.
11 An automobile which weighs 3,200 lb is on a road which rises 10 ft for each 100 ft of road. What force tends to move the car down the hill?
12 Four boxes each weighing 100 N are sus-

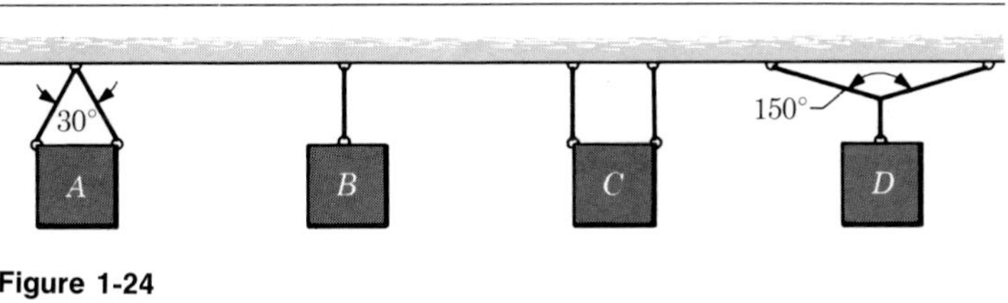

Figure 1-24

pended from a beam (Fig. 1-24). What is the tension in each of the wires?

Ans. 51.8 N; 100 N; 50 N; 193 N.

13 A boat which can travel at 10 m/s in still water attempts to reach a point directly across a river in which there is a current of 8 m/s. At what angle to the shore must the boat be steered to reach that point?

14 Two similar cylindrical polished bars weighing 5.00 N each lie next to one another in contact. A third similar bar is placed on the other two in the groove between them. Neglecting friction, what horizontal force on each lower bar is necessary to keep them together (Fig. 1-25)?

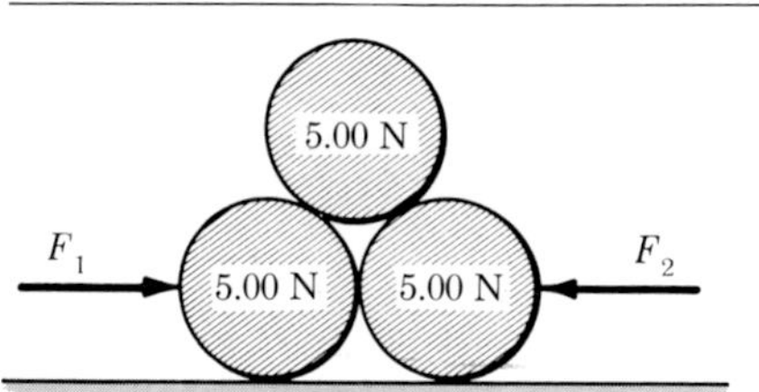

Figure 1-25

Ans. 1.44 N.

15 A train has a velocity of 150 km/h due east. A person on the train walks toward the front of the train with a velocity of 15 km/h relative to the train. What is this velocity relative to an observer on the ground?

16 Many presses and other tools make use of the toggle, which is a pair of bars jointed together so as to form a very obtuse angle between them (Fig. 1-26). When a 20-N force is exerted on the joint along the angle bisector, the opposite ends of the bars may exert very much larger forces on their restraints. If the angle at the joint is 170°, find the two horizontal forces necessary to hold the system in equilibrium. *Ans.* 114 N.

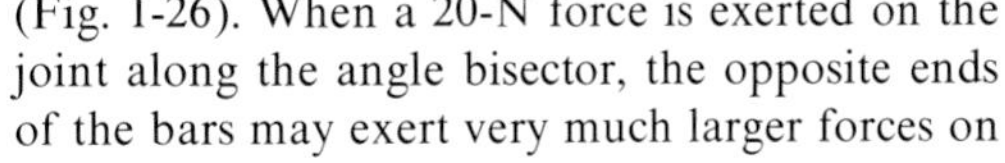

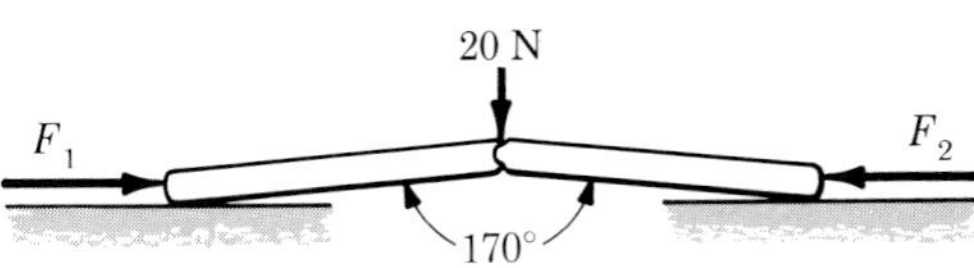

Figure 1-26

17 Add the following forces by the component method: 12.0 N at 30° to an x axis, 15.0 N at 140°, and 8.0 N at 290°.

18 An iron sphere weighs 10 N and rests in a V-shaped trough whose sides form an angle of 60°. What is the normal force exerted by the sphere on each side of the trough?

Ans. 10 N.

19 As a flag is hoisted up the mast at 15 ft/s, a ship goes south at 22 ft/s, and the tide moves east at 4.0 ft/s. What is the speed of the flag relative to the earth?

20 A boat travels 10.0 km/h in still water. If it is headed 60°S of W in a current that moves at 12.0 km/h due East, what is the resultant velocity of the boat? *Ans.* 11.1 km/h at 51°S of E.

21 When a sail is hoisted, the rope or halyard passes over a pulley and down the mast, where it is often wrapped around a cleat at the bottom. To effect the final tightening, the halyard is grasped at the middle, if possible, and pulled away from the mast, after which the slack so gained is taken up at the cleat. If the halyard is 20 m from pulley to cleat and is pulled at the middle, so that it is 1.0 m out from the mast, with a force of 20 N, what is the tension in the halyard?

22 A person who is 35 ft east of you runs north at 16 ft/s. At what angle north of east would you throw a ball at 60 ft/s groundspeed in order to hit him? *Ans.* 15.4°.

23 A rope 10 m long is stretched between a tree and a car. A man pulls with a force of 10 N at right angles to and at the middle point of the rope, and moves this point 0.5 m. Assuming no stretching of the rope, what is the tension in the rope at the final position?

24 A man walks westward on a boat with a speed of 4.0 km/h; the ship's propeller drives it 15 km/h northwest; tide and wind drive the ship 5.0 km/h south. What is the actual speed of the man relative to the earth? What is the direction of his velocity? *Ans.* 16 km/h at 21°N of W.

25 A particle in a cyclotron travels in a circle of 2,000-m radius. What is the difference in the magnitude of its displacement (chord) and its distance of travel (arc) in one-eighth of a circle?

26 Three men pull on ropes attached to the top of a heavy object which is level with the ground. Man *A* is 6 ft tall, stands 6 ft away, 45°N of E from the center of the object and exerts a force of 72 lb. Man *B* is 5 ft 6 in tall, is 6 ft away, 60°N of W and pulls with a force of 60 lb. Man *C* is 5 ft tall, also stands 6 ft away, 30°S of E and pulls with a force of 80 lb. Assuming the ropes to be attached to their shoulders which are $\frac{2}{3}$ of their height from the ground, what is the horizontal resultant of these forces on the object?
Ans. 93.5 lb, 33°N of E.

Henri Antoine Becquerel, 1852–1908

Born in Paris. Professor at the Paris Polytechnic School. Awarded, with the Curies, the 1903 Nobel Prize for Physics for his discovery of spontaneous radioactivity.

Pierre Curie, 1859–1906

Born in Paris. Professor at the Municipal School of Industrial Physics and Chemistry. Pierre and Marie Curie shared the 1903 prize with Henri Becquerel for their work on the phenomena of the radiation discovered by Becquerel.

Marie Sklodovska Curie, 1867–1934

Born in Warsaw. Director of applied physics in the University of Paris. Marie Curie shared the 1903 Nobel Prize for Physics with Pierre Curie and Henri Becquerel and was awarded the 1911 Nobel Prize for Chemistry for her discovery of radium and polonium.

2

Velocity and Acceleration

In Chap. 1 it was shown that forces acting on an object can cancel each other out and cause the object to be in equilibrium. In this chapter we will see what happens to an object if the forces do not cancel each other out, that is, if there is a net unbalanced force acting upon the object. As an illustration of this, common sense will tell us that if two men push on opposite ends of an object with equal force, the object will not move. However, if one man stops pushing, the force still being applied by the other man will cause the object to move in the direction of the remaining force. In the next chapter we will see just how this unbalanced force is related to the movement of the object. For the present, however, we shall be concerned with the description of the resultant movement of objects due to the influence of unbalanced forces.

Just as the nature of the forces acting on an object can vary, the type of motion resulting from these forces can vary. While some motion in nature may be simple in form, most motion is complex. In fact, we sometimes are endangered by the motion of objects around us, especially if that motion is erratic and uncontrolled as we observe it in a flooded river, a hurricane, or a runaway automobile. On the other hand, controlled motion can be of service to man. It is necessary to study the motions of objects if we are to understand the behavior of objects in motion and learn to control them. While it might be more exciting to begin our study by observing motions such as that of a satellite orbiting the earth or a rocket traveling to the moon, this type of motion will be considered only after we have studied some simpler cases. It is surprising to most students of physics to find that even the most complicated motions can be analyzed and represented in terms of a few elementary types, when these simple types of motion are thoroughly understood. In fact, it will be shown that all complex motion is merely a combination of two or more simple motions.

2-1 SPEED AND VELOCITY

The simplest kind of motion that an object can have is a uniform motion in a straight line. By uniform motion is meant that in every second the body moves the *same distance* in the *same direction* as it did in every other second. Every part of the body moves in exactly the same way. An object moving in this manner is moving with constant *velocity*. It should be carefully noted that the term constant velocity implies not only *constant speed,* that is, no change in the rate of movement, but *unchanging direction* as well.

The *speed* of a moving body is the distance

it moves per unit of time. If the speed is uniform, the object moves equal distances in each successive unit of time. Whether or not the speed is constant, the *average speed* is the distance the body moves, its displacement in space (which is defined as its change in position specified by a length and a direction), divided by the time required for the motion.

$$\text{Average speed} = \frac{\text{distance traveled}}{\text{time elapsed}}$$

$$\bar{v} = \frac{s}{t} \tag{1a}$$

where s is the distance traversed, $\bar{v}$ the average speed, and t the time. In this book, a bar placed over a symbol, as with $\bar{v}$, indicates that this is an average. The mks unit of speed is the meter per second (m/s); the fps unit is the foot per second (ft/s); and many of the other units are common, such as the mile per hour (mi/h), centimeter per second (cm/s), knot, etc. Equation (1*a*) may be put in the following form:

$$s = \bar{v}t \tag{1}$$

If the speed is constant, its value is, of course, identical with the average speed. If, for example, an automobile travels 200 km in 4 h, its average speed is 50 km/h. In 6 h it would travel 300 km.

As we saw earlier, speed is a scalar quantity, and hence the concept of speed does not involve the idea of direction. A body moving with constant speed may move in a straight line or in a circle or in any one of an infinite variety of paths so long as the distance moved in any unit of time is the same as that moved in another equal unit of time.

The concept of velocity includes the idea of direction as well as magnitude and velocity and is therefore a vector quantity. A car moving at constant speed along a winding road has a changing velocity because the direction of motion is changing. *Velocity,* a vector quantity, is defined as the time rate of change of displacement. The definition of *average velocity* is given by

$$\text{Average velocity} = \frac{\text{displacement}}{\text{time elapsed}}$$

$$\bar{\mathbf{v}} = \frac{\mathbf{s}}{t} \tag{1b}$$

Constant velocity is a particular case of constant speed. Not only does the distance traveled per unit time remain the same, but the direction as well does not change. An automobile that travels for 1 h at a constant velocity of 20 mi/h N reaches a place 20 mi N of its original position. If, on the other hand, it travels around a racetrack at a constant speed of 20 mi/h, it may traverse the same distance without any final displacement. At one instant its velocity may be 20 mi/h E; at another, 20 mi/h S.

The statement "An automobile is moving with a *velocity* of 20 mi/h" is incorrect because it is incomplete, inasmuch as the direction of motion must be stated in order to specify a velocity. For this reason one should always use the word *speed* when the direction of the motion is not specified or when the direction is changing.

Example An automobile traveled by a circuitous path for a time of 3.0 h and covered a total distance of 180 mi. (*a*) What was the average speed in feet per second? (*b*) If at the end of the trip the car was exactly 60 mi N of its starting point, what was the average velocity?

(*a*) The average speed was

$$\bar{v} = \frac{s}{t} = \frac{180\ \text{mi}}{3.0\ \text{h}} = 60\ \text{mi/h} = 88\ \text{ft/s}$$

and

(*b*) The average velocity was

$$\bar{\mathbf{v}} = \frac{\mathbf{s}}{t} = \frac{60\ \text{mi N}}{3.0\ \text{h}} = 20\ \text{mi/h N}$$

2-2 INSTANTANEOUS VELOCITY

We are frequently interested in knowing the speed and velocity of a body at a particular instant and not merely the average value over a considerable time interval. For this purpose it is convenient to consider the ratio $\Delta s/\Delta t$, where Δs is the change of displacement which the body has during the small time interval Δt. If Δt is made smaller and smaller, approaching zero but never reaching it, Δs becomes equally infinitestimal and the *instantaneous velocity* v is the limit of this ratio. From the calculus, this fact is written

$$\mathbf{v} = \lim_{\Delta t \to 0} \frac{\Delta \mathbf{s}}{\Delta t} \qquad (1c)$$

Thus instantaneous velocity is the time rate of change of displacement.

Instantaneous speed is similarly defined as the time rate of change of distance, which is the rate of change of the magnitude of the displacement; hence the instantaneous speed of a particle is the magnitude of its instantaneous velocity. Unless otherwise stated, the terms speed and velocity refer to the instantaneous values.

Suppose that we consider the case of a body which is traveling in a straight line with a uniformly increasing speed. A curve of displacement against time for such a body is shown in Fig. 2-1. By definition, the slope of the curve at a point P is determined from a line drawn tangent to the curve at P as the change in the displacement (Δs), the ordinate, divided by the change in the time interval (Δt), the abscissa. Thus the velocity of a body at the point P_1 is equal to the slope of the line drawn tangent to the curve at P_1. If the velocity were constant, the slope would be the same at every point and the curve would be a straight line. From the shape of the curve in Fig. 2-1 it is evident that the velocity of the body in question is not constant, since the slope is steadily increasing, as shown by the slope at the second reference point, P_2.

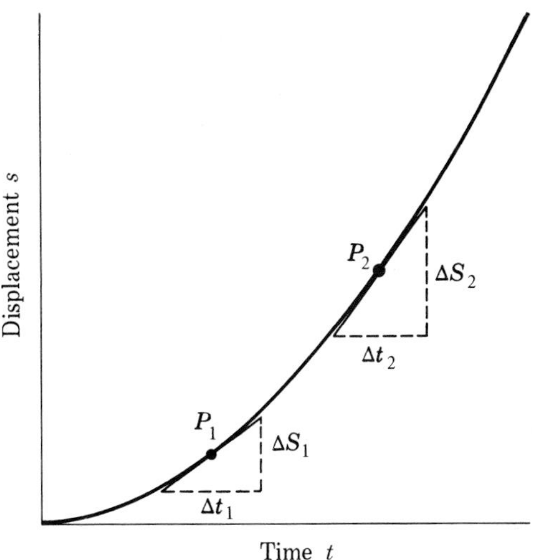

Figure 2-1
Displacement–time curve.

2-3 ACCELERATED MOTION

Objects seldom move with constant velocity. In almost all cases the velocity of an object is continually changing in magnitude, direction, or both. Motion in which the velocity is changing is called *accelerated motion*. The time rate at which the velocity changes is called the *acceleration*.

The velocity of a body may be changed by changing the speed, the direction, or both speed and direction. If the direction of the acceleration is parallel to the direction of motion, only the speed changes; if the acceleration is at right angles to the direction of motion, only the direction changes. Acceleration in any other direction produces changes in both speed and direction.

The average acceleration of a body is defined by

$$\text{Average acceleration} = \frac{\text{change in velocity}}{\text{time elapsed}}$$

$$\bar{\mathbf{a}} = \frac{\Delta \mathbf{v}}{\Delta t} = \frac{\mathbf{v}_1 - \mathbf{v}_0}{t} \qquad (2a)$$

where $\Delta\mathbf{v}$ is the vector difference of $\mathbf{v}_1$ and $\mathbf{v}_0$ as defined in Sec. 1-9, $\mathbf{v}_0$ is the velocity when $t = 0$, t is the time elapsed, and $\mathbf{v}_1$ is the velocity at time t.

Since units of acceleration are obtained by dividing a unit of velocity by a unit of time, it may be seen that the mks unit of acceleration is the meter per second per second. Recall from the law of multiplication of fractions that $(\mathrm{m/s})/(\mathrm{s}/1) = \mathrm{m/s} \times 1/\mathrm{s} = \mathrm{m/s^2}$. Similarly, the fps unit is the foot per second per second ($\mathrm{ft/s^2}$), and the cgs unit is the centimeter per second per second ($\mathrm{cm/s^2}$).

Although we use the word speed to describe the magnitude of a velocity, there is no corresponding word for the magnitude of an acceleration. Hence the term acceleration is used to denote either the vector quantity or its magnitude.

Example An automobile accelerates at a constant rate from 15 mi/h to 45 mi/h in 10 s while traveling in a straight line. What is the average acceleration?

The magnitude of the average acceleration, or the rate of change of speed in this case, is the change in speed divided by the time in which it took place, or

$$\bar{a} = \frac{45\ \mathrm{mi/h} - 15\ \mathrm{m/h}}{10\ \mathrm{s} - 0} = \frac{30\ \mathrm{mi/h}}{10\ \mathrm{s}}$$

$$= 3.0\ (\mathrm{mi/h})/\mathrm{s}$$

indicating that the speed increases 3.0 mi/h during each second. Since

$$30\ \mathrm{mi/h} = 30\ \cancel{\mathrm{mi/h}}\ \frac{88\ \mathrm{ft/s}}{60\ \cancel{\mathrm{mi/h}}} = 44\ \mathrm{ft/s}$$

the average acceleration can be written also as

$$\bar{a} = \frac{44\ \mathrm{ft/s}}{10\ \mathrm{s}} = 4.4\ \mathrm{ft/s^2}$$

This statement means simply that the speed increases 4.4 ft/s during each second, or 4.4 $\mathrm{ft/s^2}$.

Example A proton in a cyclotron changes in velocity from 30 km/s N to 40 km/s E in 20 microseconds (μs). (1 $\mu\mathrm{s} = 10^{-6}$ s.) What is the average acceleration during this time?

The vector character of the velocities in this case is important. The vector difference of the velocities is found as in Fig. 2-2.

The magnitude of the difference of velocities represented by $|\mathbf{v}_1 - \mathbf{v}_0|$ can be obtained by the pythagorean theorem in this case.

$$|\mathbf{v}_1 - \mathbf{v}_0| = \sqrt{(40\ \mathrm{km/s})^2 + (30\ \mathrm{km/s})^2}$$

$$= 50\ \mathrm{km/s}$$

$$\frac{30\ \mathrm{km/s}}{50\ \mathrm{km/s}} = \sin 37^\circ$$

$$\mathbf{a} = \frac{\mathbf{v}_1 - \mathbf{v}_0}{t} = \frac{50\ \mathrm{km/s\ at\ 37^\circ S\ of\ E}}{20 \times 10^{-6}\ \mathrm{s}}$$

$$= 2.5\ \mathrm{km/s^2\ at\ 37^\circ S\ of\ E}$$

As in the case of velocity, we are frequently concerned with instantaneous values of acceleration. *Instantaneous acceleration* is defined by

$$\mathbf{a} = \lim_{(\Delta t \to 0)} \frac{\Delta \mathbf{v}}{\Delta t} \qquad (2b)$$

Hereafter the term acceleration will be used to refer to the instantaneous and not the average value, unless otherwise stated.

2-4 UNIFORMLY ACCELERATED MOTION

The simplest type of accelerated motion is *uniformly accelerated motion,* defined as motion in a straight line in which the direction is always the same and the speed changes at a constant rate. It is important to note that the vector difference of velocities becomes simply the algebraic difference, and one may work with simple algebraic equations. One direction along the line of motion

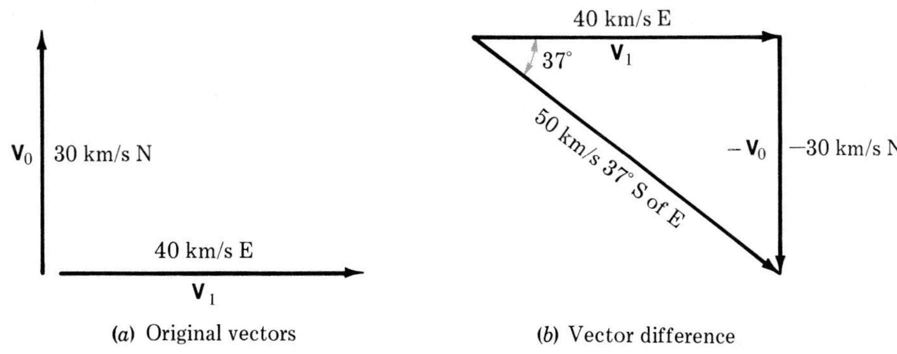

Figure 2-2
Vector differences of two velocities.

is called positive; the opposite is called negative. In this type of motion, the average value of the acceleration is the same as the constant instantaneous value and Eq. (2*a*) becomes for this case

$$a = \frac{v_1 - v_0}{t}$$

or

$$v_1 - v_0 = at \tag{2}$$

The velocity of a body in uniformly accelerated motion changes steadily, with equal changes of velocity in equal intervals of time, as shown in Fig. 2-3, where the slope is constant at all points. The slope taken between P_1 and P_2 and then between P_3 and P_4 will be the same since the slope is $\Delta v/\Delta t$, which equals the acceleration—in this case constant. This would not be true in general for a nonuniformly accelerated motion, which would be represented by a curved graph line. The slope of the tangent to such a curve at any point would represent the instantaneous acceleration. For the case of uniform acceleration,

$$\bar{v} = \frac{v_0 + v_1 + v_2 \cdots + v_n}{n + 1}$$

Since we usually refer to an initial and a final velocity in solving problems of motion,

$$\bar{v} = \frac{v_0 + v_1}{2} \tag{3}$$

The distance covered at this average velocity is given by $s = \bar{v}t$.

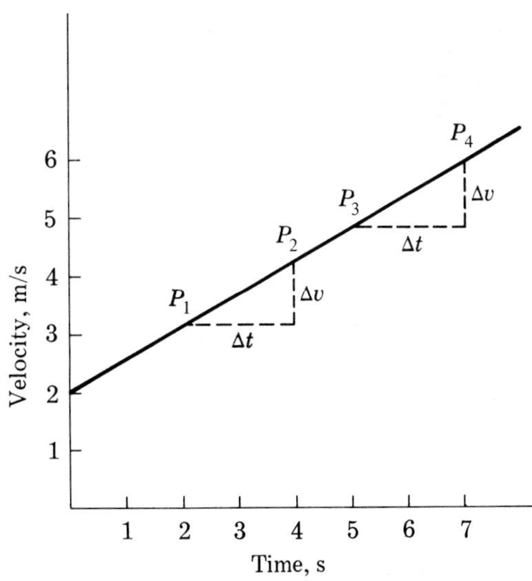

Figure 2-3
Velocity in uniformly accelerated motion.

Example How far does an automobile move while its speed increases uniformly from 15 mi/h to 45 mi/h in 10 s?

$$\bar{v} = \tfrac{1}{2}(15 + 45)\ \text{mi/h} = 30\ \text{mi/h} = 44\ \text{ft/s}$$

$$s = \bar{v}t = (44\ \text{ft/s})(10\ \text{s}) = 440\ \text{ft}$$

Three equations for uniformly accelerated motion have now been considered:

$$s = \bar{v}t \qquad (1)$$

$$v_1 - v_0 = at \qquad (2)$$

$$\bar{v} = \frac{v_0 + v_1}{2} \qquad (3)$$

These three equations have been defined by setting certain conditions and limitations. It is possible to combine these fundamental equations in such a manner that two other very useful equations are obtained.

It is important for the student to realize that the equations used throughout this study of physics have all been either defined or have been derived from defined equations. That is, the equations do not appear by magic but rather are based on agreed-upon conditions and restrictions. In the eyes of many physicists, a well thought out and executed derivation of a working equation is a thing of beauty. While the authors of this book share this opinion to a degree, they are also aware that the study of physics as presented in this text could become encumbered by an inordinately large number of derivations. Consequently, only a few key derivations are given throughout the book. The derivations presented below are typical of the techniques used in establishing working equations.

Example Derive the equation $s = v_0t + \frac{1}{2}at^2$.

Since $s = \bar{v}t$ and $\bar{v} = \dfrac{(v_0 + v_1)}{2}$

then, $s = \left(\dfrac{v_0 + v_1}{2}\right)t$ and $v_0 + v_1 = \dfrac{2s}{t}$

also $a = \dfrac{v_1 - v_0}{t}$ and $v_0 = v_1 - at$

but, $v_1 = v_0 + at$ and $v_1 = \dfrac{2s}{t} - v_0$

Therefore,

$$v_0 + at = \frac{2s}{t} - v_0 \quad \text{or} \quad 2v_0 = \frac{2s}{t} - at$$

Multiplying both sides by t and then dividing by 2, we obtain

$$2v_0t = 2s - at^2 \quad \text{and} \quad v_0t = s - \tfrac{1}{2}at^2$$

Rearranging these terms, we get

$$s = v_0t + \tfrac{1}{2}at^2 \qquad (4)$$

Example Derive the equation $2as = v_1^2 - v_0^2$. Since

$$a = \frac{v_1 - v_0}{t} \quad \text{and} \quad t = \frac{s}{\bar{v}} = s/[(v_0 + v_1)/2] = \frac{2s}{v_0 + v_1}$$

then, $a = v_1 - v_0/[2s/(v_0 + v_1)]$

Therefore,

$$2as = (v_1 - v_0)(v_1 + v_0)$$

and

$$2as = v_1^2 - v_0^2 \qquad (5)$$

2-5 HINTS FOR SOLVING PROBLEMS INVOLVING UNIFORMLY ACCELERATED MOTION

In the five working equations given above for solving problems in uniformly accelerated motion, there are six variables: $\bar{v}$, v_0, v_1, a, s, and

t. To solve problems of this type, the student should first determine which of these variables are given and which are not. While not all six variables will be involved in each problem, such an analysis will permit a judgment to be made about the nature of the problem and which of the five equations will provide the best route to obtain the answer.

Since the equations are solved by algebraic addition, it is important to choose a direction to call positive and apply it consistently to distance, speed, and acceleration when values are being inserted in the equations.

Since acceleration is generally expressed in feet per second squared and time is given in seconds in the fps system, it is usually necessary to express velocity in feet per second. Therefore, it is wise to change any velocity given in miles per hour to feet per second. It was shown earlier that 60 mi/h = 88 ft/s. This equivalency can be used as a conversion factor.

Example A car accelerates uniformly from a standstill to 60 mi/h in 4.0 s. What is its acceleration? How far does it travel during this time interval?

$$v_0 = 0 \qquad v_1 = 60 \text{ mi/h} = 88 \text{ ft/s}$$

$$t = 4.0 \text{ s} \qquad a = ? \qquad s = ?$$

From Eq. (2),

$$a = \frac{v_1 - v_0}{t} = \frac{88 \text{ ft/s} - 0}{4.0} = 22 \text{ ft/s}^2$$

From Eq. (4),

$$s = v_0 t + \tfrac{1}{2} a t^2 = 0 + \tfrac{1}{2}(22 \text{ ft/s}^2)(4.0 \text{ s})^2$$

$$= 1\bar{8}0 \text{ ft}$$

or from Eq. (1),

$$s = \bar{v}t = \frac{(v_0 + v_1)\,t}{2} = \frac{(0 + 88 \text{ ft/s})(4.0 \text{ s})}{2}$$

$$= 1\bar{8}0 \text{ ft}$$

Example An airplane lands on a carrier deck at 150 mi/h and is brought to a stop uniformly, by an arresting device, in 500 ft. Find the acceleration and the time required to stop.

$$v_0 = 150 \text{ mi/h} = 220 \text{ ft/s} \qquad v_1 = 0$$

$$s = 500 \text{ ft} \qquad a = ? \qquad t = ?$$

From Eq. (5),

$$2as = v_1^2 - v_0^2$$

$$2a\,(500 \text{ ft}) = 0 - (220 \text{ ft/s})^2$$

$$a = \frac{-(220 \text{ ft/s})^2}{2\,(500 \text{ ft})} = -48.4 \text{ ft/s}^2$$

(*Note:* The negative sign on the acceleration indicates that the plane is slowing down.) From Eq. (2),

$$t = \frac{v_1 - v_0}{a} = \frac{0 - 220 \text{ ft/s}}{-48.4 \text{ ft/s}^2} = 4.55 \text{ s}$$

2-6 FREELY FALLING BODIES: ACCELERATION DUE TO GRAVITY

The most common example of uniformly accelerated motion is the motion of a body falling freely, i.e., a body which is falling under the action of its weight alone. If a stone is dropped, it falls to the earth. If air resistance is negligible, the stone is uniformly accelerated.

The acceleration of freely falling bodies is so important and so frequently used that it is customary to represent it by the special symbol g. At sea level and 45° latitude, g has a value of 32.17 ft/s², or 980.6 cm/s², or 9.806 m/s². For some purposes it is sufficiently accurate to use $g = 32$ ft/s², or 980 cm/s², or 9.80 m/s².

The value of g is not quite the same at all places on the earth. We shall see later that the weight of a body depends upon its distance from the center of the earth. It will also be seen that the acceleration of a freely falling body depends upon this distance. At a given latitude the value is greater at sea level than at higher altitude. At sea level, the value is greater near the poles than at the equator. Locally there may be small variations because of irregularities in the layers of rock beneath the surface. Such local variations are the basis of one type of prospecting for oil.

Since a freely falling body is uniformly accelerated, the five equations already developed for that type of motion may be applied when air resistance is neglected.

Example A boy lets go of a tree limb and falls freely. What is his speed at the end of 0.50 s? How far does he fall during this time?

$v_0 = 0 \qquad v_1 = ? \qquad s = ? \qquad a = +32\ \text{ft/s}^2$

$t = 0.50\ \text{s}$

It is convenient in this case to call downward quantities positive. From Eq. (2),

$$v_1 = v_0 + at = 0 + (32\ \text{ft/s}^2)(0.50\ \text{s}) = 16\ \text{ft/s}$$

From Eq. (4),

$$s = v_0 t + \tfrac{1}{2}at^2 = 0\,\tfrac{1}{2}(32\ \text{ft/s}^2)(0.50\ \text{s})^2 = 4.0\ \text{ft}$$

Table 1 and Fig. 2-4 show the speed at the end of time t and the distance fallen during time t for a body that starts from rest and falls freely.

When an object is thrown with initial speed v_0, instead of falling from rest, the first term of Eq. (4) is no longer zero. If it is thrown downward, both v_0 and a have the same direction and hence are given the same algebraic sign. If the object is thrown upward, however, v_0 is directed upward while a is directed downward, and it is convenient to call upward positive and downward negative.

Table 1

Time t, s	Speed, ft s, at end of time t	Distance, ft, fallen in time t
1	32	16
2	64	64
3	96	144
4	128	256

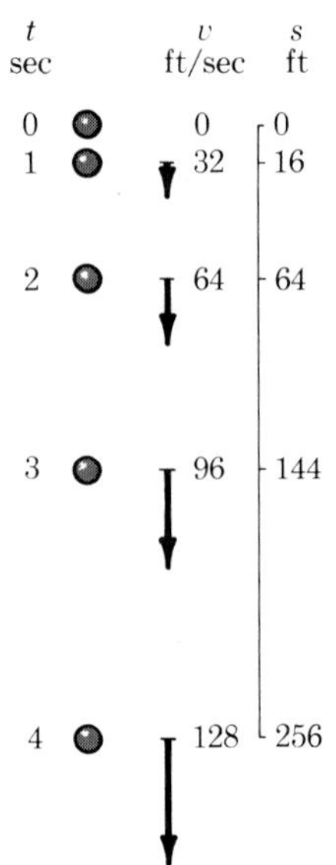

Figure 2-4
Position and speed of a body falling freely from rest after successive intervals of time.

Example A ball is thrown upward with an initial speed of 80 ft/s. (*a*) How high does it go? (*b*) What is its speed at the end of 3.0 s? (*c*) How high is it at that time?

(*a*) Since both upward and downward quantities are involved, upward will be called positive. At the highest point the ball stops, and hence at that point $v_1 = 0$.

$v_0 = 80\ \text{ft/s} \qquad v_1 = 0 \qquad a = g$

$g = -32\ \text{ft/s}^2 \qquad s = ? \qquad t = ?$

From Eq. (5),

$$2as = v_1^2 - v_0^2$$

$$2(-32\ \text{ft/s}^2)s_1 = 0 - (80\ \text{ft/s})^2$$

$$s_1 = \frac{-(80\ \text{ft/s})^2}{2(-32\ \text{ft/s}^2)} = 100\ \text{ft}$$

From Eq. (2),

$$\begin{aligned} v_1 &= v_0 + at \\ &= 80\ \text{ft/s} + (-32\ \text{ft/s}^2)(3.0\ \text{s}) \\ &= 80\ \text{ft/s} - 96\ \text{ft/s} = -16\ \text{ft/s} \end{aligned}$$

From Eq. (4),

$$\begin{aligned} s_2 &= v_0 t + \tfrac{1}{2}at^2 \\ &= (80\ \text{ft/s})(3.0\ \text{s}) + \tfrac{1}{2}(-32\ \text{ft/s}^2)(3.0\ \text{s})^2 \\ &= 240\ \text{ft} - 144\ \text{ft} = 96\ \text{ft} \end{aligned}$$

or, from Eq. (1)

$$\begin{aligned} s_2 &= \bar{v}t = \frac{v_1 + v_0}{2}t \\ &= \frac{-16\ \text{ft/s} + 80\ \text{ft/s}}{2}3.0\ \text{s} = 96\ \text{ft} \end{aligned}$$

Note that s is the magnitude of the displacement, not the total distance traveled. If the ball returns to the starting point or goes on past it, s will be zero or negative, respectively.

2-7 TERMINAL VELOCITY

In the preceding discussion we assumed that there is no air resistance. In the actual motion of every falling body this is far from true. The frictional resistance of the air depends upon the speed of the moving object. The resistance is quite small for the first one or two seconds, but as the speed of fall increases the resistance becomes large enough to reduce appreciably the net downward force on the body and the acceleration decreases. After some time of uninterrupted fall, the body is moving so rapidly that the drag of the air is as great as the weight of the body, so that there is no acceleration. The body has then reached its *terminal speed,* a speed that it cannot exceed in falling from rest. As we saw earlier in our discussion of equilibrium, a body falling at terminal speed has no unbalanced forces acting on it, so the conditions for equilibrium have been met and the body ceases to accelerate.

Very small objects, such as dust particles and water droplets, and objects of very low density and large surface, such as feathers, have very low terminal speeds; hence they fall only small distances before losing most of their acceleration. The effect of air friction on falling bodies can be shown by a classic experiment called the "guinea and feather tube" demonstration in which a coin (the guinea) and a feather are enclosed in a long tube. When the tube filled with air is inverted, the coin falls much faster than the feather. If the air is pumped out and the tube is again inverted, the coin and feather fall together. The same striking result can be shown by dropping a book and an unfolded sheet of paper. If the paper is then crumpled and the experiment repeated, the book and the paper will arrive at the ground at the same time. Try it and see for yourself.

A man jumping from a plane reaches a terminal speed of about 120 mi/h if he delays opening his parachute. When the parachute is opened, the terminal speed is reduced because of the increased air resistance to about 14 mi/h, which is about equal to the speed gained in jumping from a height of 7 ft. A large parachute encounters more air resistance than a small one and hence causes slower descent. A plane in a vertical dive without the use of its motor can attain a speed of about 400 mi/h before reaching terminal speed.

The fact that falling bodies do reach a terminal speed is of much importance to us because if such objects did not behave in this manner, life would be far different in this world. For example, using Eqs. (1) through (5), it can be shown that a raindrop falling without friction from an altitude of 32,000 ft would acquire a velocity given by

$$v_0 = 0 \qquad v_1 = ? \qquad a = +32 \text{ ft/s}^2$$
$$s = 32{,}000 \text{ ft} \qquad t = ?$$

From Eq. (5),

$$2as = v_1^2 - v_0^2$$

and

$$2as = v_1^2 \qquad \text{or} \qquad v_1 = \sqrt{2as}$$
$$v_1 = \sqrt{2 \times 32 \text{ ft/s}^2 \times 32{,}000 \text{ ft}}$$
$$= \sqrt{2.05 \times 10^6 \text{ ft}^2\text{/s}^2}$$
$$= 1{,}430 \text{ ft/s, or } 955 \text{ mi/h}$$

The impact of the raindrops falling at that rate on a person would be great enough to cause physical harm and actually knock him down. Happily, in the motion of small light objects such as raindrops, air resistance does exert the necessary retarding force to counterbalance the gravitational force.

2-8 RELATIVE VELOCITY

If car A going 50 km/h overtakes car B going 40 km/h, the first car has a relative velocity of 10 km/h with respect to the second. It is this relative velocity which the driver uses to judge the time needed to pass the slower car. The relative velocity of one body with respect to another is the vector difference of the two velocities. The order of the velocities in the difference is important. Generally some common reference point is taken in determining relative velocity. In the case illustrated, the velocity of car A with respect to the ground can be designated $\mathbf{v}_{AG}$ and that of the ground to car B be called $\mathbf{v}_{GB}$. Then by the so-called "domino rule,"

$$\mathbf{v}_{AG} + \mathbf{v}_{GB} = \mathbf{v}_{AB}$$

which is the velocity of A relative to B, the relative velocity. It should be noted that the velocity of the ground with respect to car B is negative. (Imagine B standing still, then G is moving to the left, or negatively.) Since vector quantities can be added algebraically as long as they are acting in a rectilinear fashion (straight line), we can say that

$$\mathbf{v}_{AB} = \mathbf{v}_A + \mathbf{v}_B'$$

let $\mathbf{v}_B' = -\mathbf{v}_B$

then, $$\mathbf{v}_{AB} = \mathbf{v}_A - (-\mathbf{v}_B') = \mathbf{v}_A - \mathbf{v}_B \qquad (6)$$

Thus far the velocities of bodies have been treated as if they were absolute quantities; but in reality *all* velocities are relative. The $\mathbf{v}_A$ above was designated $\mathbf{v}_{AG}$, that is, velocity with respect to the ground. But the ground is not always used as the frame of reference. The velocity of a comet, for example, may be calculated relative to the center of the earth or to the center of the sun. Between 1881 and 1887, A. A. Michelson and E. W. Morley performed a historic experiment to determine the velocity of the earth relative to the supposed "fixed ether," the medium postulated

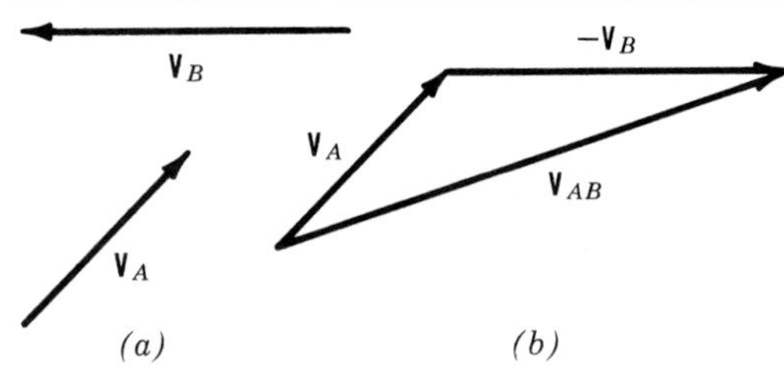

Figure 2-5
Relative velocity.

to be that through which light waves are propagated through all space. (This experiment is discussed in detail later in the book.) If the expected success had been achieved, this ether could have been used as a kind of absolute frame of reference for at least the velocities encountered in astronomy. To the astonishment of the scientific world the result was a failure to find such a relative velocity, a result which has been confirmed by many increasingly accurate experiments up to the present day. In the years following the experiment many explanations were offered for the negative result, but none that were consistent with all the known facts. Our present interpretation results from a proposal in 1905 by Albert Einstein that the speed of light in empty space is the same for all observers, regardless of their motions. This statement is, of course, contrary to the meaning of Eq. (6). Einstein showed that his proposal led to the conclusion that the speed of light c is the limit to the possible speed of any body relative to any observer and that Eq. (6) is valid only for speeds small compared with the speed of light in a vacuum, namely, $c = 186{,}000$ mi/s, or 3.00×10^8 m/s. For two bodies A and B moving in the same or opposite directions at speeds comparable with the speed of light, their relative speed v_{AB} is given by

$$v_{AB} = \frac{v_A - v_B}{1 - v_A v_B/c^2} \tag{7}$$

The basis of this equation is discussed in Chap. 45 along with other consequences of the special theory of relativity.

Example Two electrons A and B have speeds of $0.90c$ and $0.80c$, respectively. Find their relative speeds (a) if they are moving in the same direction and (b) if they are moving in opposite directions

$$v_{AB} = \frac{v_A - v_B}{1 - v_A v_B/c^2}$$

$$(a)\quad v_{AB} = \frac{0.90c - 0.80c}{1 - 0.72c^2/c^2} = \frac{0.10c}{0.28} = 0.36c$$

$$(b)\quad v_{AB} = \frac{0.90c - (-0.80c)}{1 - (-0.72c^2/c^2)} = \frac{1.70c}{1.72} = 0.99c$$

If Eq. (6) had been used, the relative speeds would have been computed to be $0.10c$ and $1.70c$, respectively.

2-9 PROJECTILE MOTION

An object launched into space without motive power of its own, which travels freely under the action of gravity and air resistance alone, is called a *projectile,* or ballistic missile. While a rocket is self-powered for a small part of its flight, it becomes a projectile when its fuel is shut off and travels thereafter in the same way as a bullet. For flights short enough so that the curvature of the earth may be neglected, the motion of a projectile is one of constant downward acceleration, but it differs from the uniformly accelerated motion already discussed in that the direction of the acceleration is seldom the same as that of the initial velocity. Hence the velocity is continually changing in both magnitude and direction.

It is convenient in studying such projectile motion to consider it as made up of two components, one vertical and the other horizontal. Since the gravitational force is vertically downward, it produces an acceleration only in that direction, leaving the horizontal component of the velocity unchanged if air resistance is neglected. The complex motion of the projectile reduces to two simple motions, constant horizontal velocity and uniformly accelerated vertical motion.

Suppose that we ask ourselves how a stone will move if it is thrown horizontally at a speed of 50 ft/s. Neglecting air resistance, the stone will travel with a constant horizontal speed of 50 ft/s until it strikes something. At the same time it will execute the uniformly accelerated motion of an object falling freely from rest; that is, beginning with a vertical speed of zero, it will acquire downward speed at the rate of 32 ft/s in each second. It will fall 16 ft during the first second, 48 ft during the next, 80 ft during the third, and

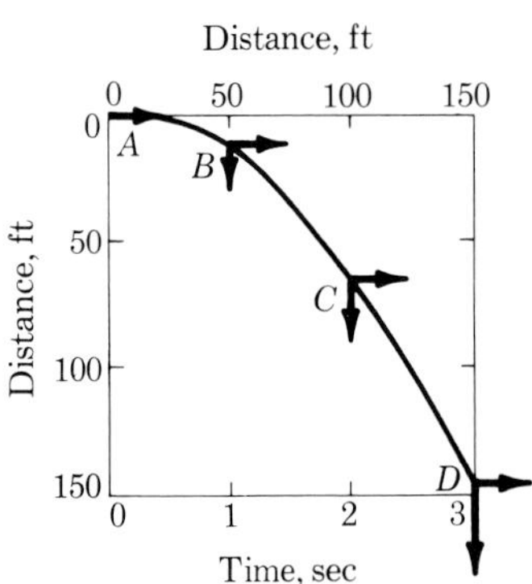

Figure 2-6
Path of stone thrown horizontally with a speed of 50 ft/s.

so on, just as if it had no horizontal motion. Its progress during the first three seconds is illustrated in Fig. 2-6. At A the stone has no vertical speed; at B (after 1 s) its vertical speed is 32 ft/s; at C, 64 ft/s; and at D, 96 ft/s. The curved line $ABCD$ in Fig. 2-6 is the path that the stone follows, and the arrows at A, B, C, and D represent the velocities at those places. Note that the horizontal arrows are all the same length, indicating the constant horizontal speed, while the vertical arrows increase in length, to indicate the increasing vertical speed. The vertical arrow at C is twice as long as that at B, while that at D is three times as long. The resultant velocity of the projectile is at each point tangent to the curve $ABCD$. It is constantly changing both in magnitude and in direction.

No matter what may be the initial direction of motion of the projectile, its motion can be resolved into horizontal and vertical parts that are independent of each other. This important, but often hard to accept, fact that the horizontal and vertical components are independent of each other can be illustrated in several ways.

If a rifle is held horizontally and at the instant a bullet is fired from the rifle another bullet is dropped from the same height, both bullets will hit the ground, assuming it is level, at the same time even though the fired bullet may land hundreds of yards away. The reason for this can be understood if one realizes that the only force causing both bullets to fall is the force due to gravity. The fact that the fired bullet is moving horizontally does not influence the rate of fall. The horizontal and vertical motions are independent of each other.

Another illustration of this independence of horizontal and vertical motion of an object can be seen by considering the situation in which a ball is thrown vertically upward by someone sitting in a convertible car with its roof down, first when the car is at rest and then when the car is moving with constant velocity. An analysis of both cases will show that, ignoring wind resistance, both balls would be caught by the person in the car. The first situation where the car is at rest should prove no mystery, but perhaps an explanation for the second ball's being caught is needed. It should be noted that both the ball and the car have the same constant horizontal velocity; and since the vertical motion of the ball would be the same whether the car was moving or not, the ball thrown upward should return to the person throwing it in spite of the fact that it is moving.

Suppose that a projectile is fired with a speed and direction such as represented by vector $\mathbf{v}_0$ in Fig. 2-7. Consider the components of velocity $\mathbf{v}_0$ along the x (horizontal) and y (vertical) axes.

An object that had the *simultaneous* horizontal and vertical speeds represented by $v_0 \cos \theta_0$ and $v_0 \sin \theta_0$ would follow exactly the same path, in the initial direction of $\mathbf{v}_0$. In discussing the motion of the projectile, one may use either the whole speed in the direction of $\mathbf{v}$ or the horizontal and vertical parts. The latter viewpoint simplifies the problem.

The horizontal acceleration a_x is zero, if air resistance is negligible. Hence the horizontal component of the velocity v_x remains constant, and at any time t

$$v_x = v_0 \cos \theta_0 \tag{8}$$

The vertical acceleration of the projectile is in

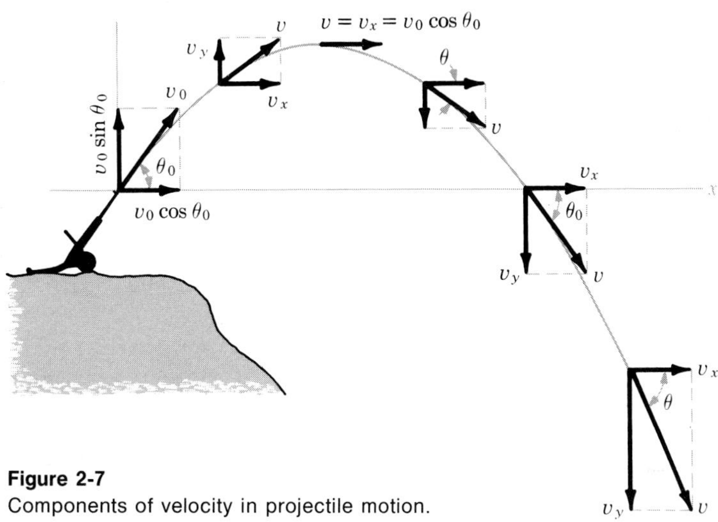

Figure 2-7
Components of velocity in projectile motion.

the negative (downward) direction along the y axis; thus the vertical component of velocity at any time t is

$$v_y = v_0 \sin \theta_0 - gt \tag{9}$$

The magnitude of the velocity at any instant is $v = \sqrt{v_x^2 + v_y^2}$, and the angle this resultant velocity makes with the horizontal can be found from

$$\tan \theta = \frac{v_y}{v_x} \tag{10}$$

In projectile motion one frequently wishes to determine the height to which the projectile rises, the time of flight, and the horizontal range. The first two may be obtained by the use of Eqs. (5) and (2), while the range is determined by multiplying v_x by the time of flight.

Example A projectile is thrown with a speed of 100 ft/s in a direction 30° above the horizontal. Find the height to which it rises, the time of flight, and the horizontal range.

Initially,

$$v_y = v \sin \theta = (100 \text{ ft/s})(\sin 30°) = 50.0 \text{ ft/s}$$

$$v_x = v \cos \theta = (100 \text{ ft/s})(\cos 30°) = 86.6 \text{ ft/s}$$

Using Eq. (5) as applied to the vertical motion,

$$v_1^2 - v_0^2 = 2as$$

$$0 - (50 \text{ ft/s})^2 = 2(-32 \text{ ft/s}^2)s$$

$$s = \frac{2{,}500 \text{ ft}^2/\text{s}^2}{64 \text{ ft/s}^2} = 39 \text{ ft}$$

The time required to reach the highest point is, from Eq. (2),

$$v_1 - v_0 = at$$

$$0 - 50 \text{ ft/s} = (-32 \text{ ft/s}^2)t$$

$$t = \frac{50 \text{ ft/s}}{32 \text{ ft/s}^2} = 1.6 \text{ s}$$

Assuming that the surface above which the projectile moves is horizontal, an equal time will be required for the projectile to return to the surface. This can be seen by realizing that v_0 (downward) is now zero at the highest point. Also, $s = 39$ ft, $a = +32 \text{ ft/s}^2$ and v_1 and t are unknown.

Using $s = v_0t + \frac{1}{2}at^2 = \frac{1}{2}at^2$,

$$t = \sqrt{\frac{2s}{a}} = \sqrt{\frac{2 \times 39 \text{ ft}}{32 \text{ ft/s}^2}} = 1.6 \text{ s}$$

Hence the time elapsed before the projectile strikes the surface is

$$t' = 2t = 2 \times 1.6 \text{ s} = 3.2 \text{ s}$$

During all this time the projectile travels horizontally with a uniform speed of 86.6 ft/s. The horizontal range R is therefore

$$R = v_x \times \text{time in air}$$
$$= v_xt' = (86.6 \text{ ft/s})(3.2 \text{ s}) = 2\bar{8}0 \text{ ft}$$

If the surface above which the projectile moves is not level, the time of flight will be increased or decreased depending upon whether the striking point is below or above the firing point. The range is correspondingly increased or decreased.

The motion of any projectile, with air resistance neglected, may be treated in this manner no matter what may be the initial speed and angle of projection. The initial velocity is resolved into vertical and horizontal components and the two are considered separately.

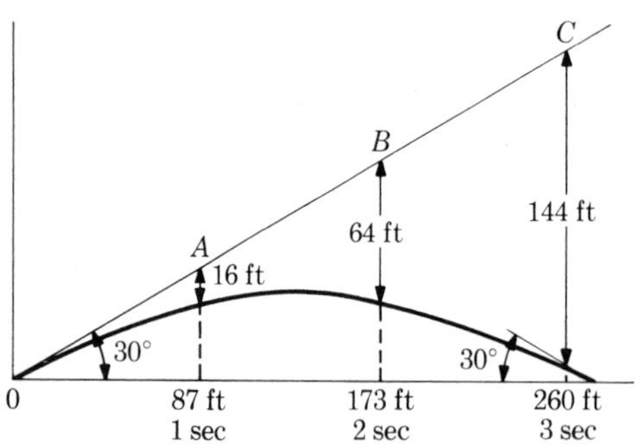

Figure 2-8
Path of a projectile fired at an angle of 30° above the horizontal with an initial speed of 100 ft/s. The projectile strikes with a speed equal to the original speed and at an angle of 30° to the horizontal.

In Fig. 2-8 we note that the path may be found by considering a uniform motion in the initial direction OC and finding the distance the projectile has fallen from this path at each instant. In 1 s under the action of gravity, the projectile falls 16 ft; hence at the end of 1 s it is 16 ft below A; in 2 s it falls 64 ft and hence is 64 ft below B, and so on.

2-10 MONKEY AND HUNTER

A classic problem, the monkey and hunter, illustrates the projectile-type problem. A hunter aims a rifle at a monkey sitting at the top of a 100-ft-tall tree which is 200 ft away from the hunter. If the monkey drops from the tree the instant it sees the flash of the rifle being fired, will the bullet hit the monkey? If it does, at what elevation will the monkey be when hit? Assume the muzzle velocity of the bullet to be 500 ft/s (see Fig. 2-9).

The angle θ is determined by recalling that

$$\tan\theta = \frac{\text{elevation of tree}}{\text{range}} = \frac{100 \text{ ft}}{200 \text{ ft}} = 0.5$$

Therefore, $\theta = 26.6°$. The horizontal and vertical components of the velocity of the bullet are

$$v_y = v \sin\theta = 500 \sin 26.6° = 224 \text{ ft/s}$$
$$v_x = v \cos\theta = 500 \cos 26.6° = 447.5 \text{ ft/s}$$

The bullet will take $t = s/v_x = 200 \text{ ft}/(447.5 \text{ ft/s}) = 0.447 \text{ s}$ to reach the tree. The bullet will have an elevation at 0.477 s of

$$s = v_yt + \tfrac{1}{2}at^2 = (224 \text{ ft/s})(0.447 \text{ s}) + \tfrac{1}{2}(-32 \text{ ft/s}^2)(0.477 \text{ s})^2$$
$$= 100 \text{ ft} - 3.2 \text{ ft} = 96.8 \text{ ft}$$

The monkey at the end of 0.477 s will have fallen

$$s = \tfrac{1}{2}at^2 = \tfrac{1}{2}(32 \text{ ft/s}^2)(0.44 \text{ s})^2 = 3.2 \text{ ft}$$

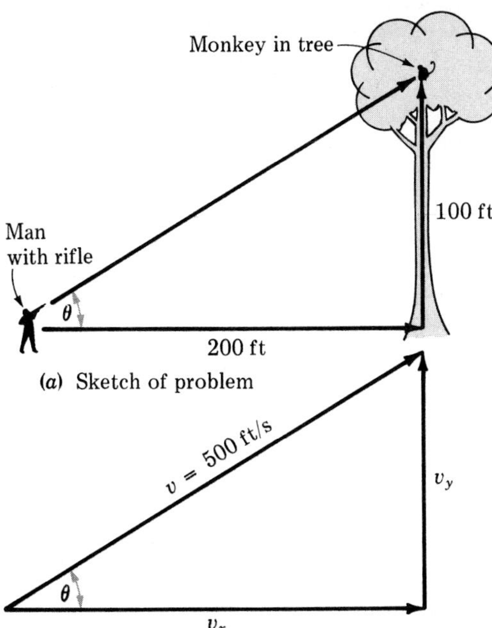

Figure 2-9
Sketches for monkey and hunter problem.

and will be 96.8 ft high. Therefore, the elevation of the monkey and the bullet is 96.8 ft and the monkey will have been hit.

2-11 RANGE AND ANGLE OF ELEVATION

The range of a projectile depends upon the angle at which it is fired as well as upon the initial speed. If the angle is small, the horizontal velocity is relatively high but the time of flight is so short that the range is small. On the other hand, if the angle of projection is very large, the time of flight is long but the horizontal velocity is small. In the absence of air resistance the maximum horizontal range is attained when the angle of elevation is 45°.

When the range of a projectile is of the order of 100 mi or more, account must be taken of the fact that the *direction* of the acceleration due to gravity does not remain constant but changes significantly, always pointing toward the center of the earth. The trajectory in such a case, even in the absence of air resistance, departs from the parabolic and follows an elliptical path. The discussion of the motion of a missile or satellite for long trajectories is postponed to Chap. 9.

2-12 AIR RESISTANCE

So far in the discussion of the motion of projectiles the resistance of the air has been neglected. However, for high-speed projectiles this resistance is no small factor. It introduces a force which opposes the motion, a force which varies both in magnitude and in direction and hence produces variable acceleration in addition to the constant acceleration we have assumed previously. This resistance reduces the height of flight, the range of the projectile, and the speed of the projectile when it strikes its target. A representation of these effects is given in Fig. 2-10.

SUMMARY

Displacement is a change in position, specified by a length and a direction.

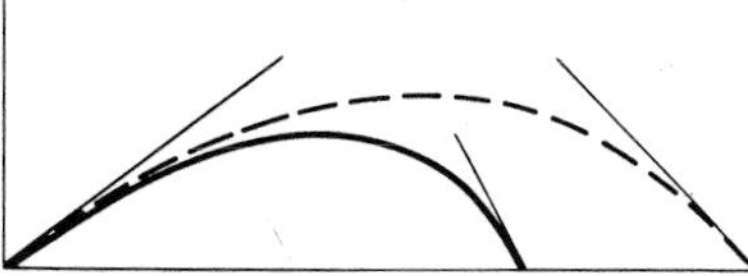

Figure 2-10
Path of a projectile. The dotted curve represents the path that would be followed if there were no air resistance, while the solid line is an actual path. The maximum height, range, and striking speed are decreased, while the striking angle is increased.

Speed is distance per unit time, and *velocity* is displacement per unit time. Their *average* values are defined by the equations

$$\bar{v} = \frac{s}{t} \qquad \bar{\mathbf{v}} = \frac{\mathbf{s}}{t}$$

Instantaneous velocity is the time rate of change of displacement

$$\mathbf{v} = \lim_{\Delta t \to 0} \frac{\Delta \mathbf{s}}{\Delta t}$$

Velocity is a vector quantity; therefore a statement of velocity must specify the direction as well as the speed, for example, 25 km/h E, 30 cm/s SW.

Acceleration is the change of velocity per unit time. Average acceleration is defined by

$$\bar{\mathbf{a}} = \frac{\mathbf{v}_1 - \mathbf{v}_0}{t}$$

Instantaneous acceleration is the time rate of change of velocity

$$\mathbf{a} = \lim_{\Delta t \to 0} \frac{\Delta \mathbf{v}}{\Delta t}$$

Uniformly accelerated motion is defined as motion in a straight line in which the direction is always the same and the speed changes at a constant rate.

The equations of uniformly accelerated motion for the particular case in which the direction of motion remains fixed and the speed changes uniformly are

$$s = \bar{v}t \qquad (1)$$

$$v_1 - v_0 = at \qquad (2)$$

$$\bar{v} = \frac{v_0 + v_1}{t} \qquad (3)$$

$$s = v_0 t + \tfrac{1}{2}at^2 \qquad (4)$$

$$2as = v_1^2 - v_0^2 \qquad (5)$$

A freely falling body is one that is acted on by no forces of appreciable magnitude other than its weight.

The acceleration of a freely falling body at sea level and 45° latitude is approximately 32 ft/s^2, or 980 cm/s^2, or 9.80 m/s^2.

The *terminal speed* of a falling object is the vertical speed at which the force of air resistance is just sufficient to balance its weight.

The relative velocity of one body with respect to a second body is the velocity of the first minus the velocity of the second, vector subtraction being used.

These relations are approximations to relativistic equations for velocities and lose accuracy near the speed of light.

A *projectile* is an object which is given an initial velocity and which is then allowed to move under the action of gravity.

In projectile motion the vertical and horizontal motions may be treated separately. If air resistance is neglected, the horizontal motion is uniform, while the vertical motion is uniformly accelerated. Under these conditions the path is parabolic.

The *range* of a projectile depends upon its initial speed and the angle of projection. If air resistance is negligible, maximum range is attained with an angle of 45°.

Air resistance decreases the speed, the maximum height, and the range of a projectile.

Questions

1 A satellite travels with constant speed at a fixed elevation above the earth. Show why the velocity is not constant. What is the direction of the acceleration?

2 Show by the use of graphs why the average speed of an object is $\frac{1}{2}(v_0 + v_1)$ only for the case of uniform acceleration and is not true for variable acceleration.

3 A man on a moving flatcar throws a ball toward his companion on the other end of the car. Describe the velocity of the ball (*a*) relative to the companion and (*b*) relative to the earth, when the car is moving (i) forward and (ii) backward.

4 What is the average speed of a car which goes at 40 km/h for 20 km and at 60 km/h for 20 km?

5 Show by a vector diagram how much the smokestack on a moving train caboose would have to be inclined in order for a vertically falling raindrop to pass through the stack without hitting the sides.

6 Can a body have a velocity without an acceleration? Can it have an acceleration with zero velocity? Give examples.

7 Cite an example to show that it is possible for an object to have an acceleration without its speed changing.

8 In a famous paradox of the Greek philosopher Zeno, Achilles, a fast runner, was proved by means of the following argument to be unable to overtake and pass a tortoise: Achilles would first have to arrive at the place where the tortoise started, by which time the tortoise would have moved on. Then he would have to arrive at this second place, by which time the tortoise would have moved to a new place, etc. Is there anything wrong with the proof?

9 A marble rolls with negligible friction down an inclined plane. Show by means of a vector polygon how the acceleration parallel to the plane may be expressed in terms of the geometrical dimensions of the plane. Galileo referred to this experiment as "diluting gravity." Show why this is an appropriate designation.

10 Sketch rough curves to illustrate the velocity as a function of time for the following cases: (*a*) a baseball thrown vertically upward, starting from the instant in which it leaves the thrower's hand and continuing until the ball strikes the ground; (*b*) an elevator on a complete upward trip.

11 State two reasons why the value of g is different at various places on the earth. What would one expect about this value on the moon? on the sun?

12 If a body falls from a great height, can its speed reach a maximum value and thereafter decrease? Explain.

13 Discuss the statement "If a projectile is fired at more than 18,000 mi/h horizontally, it won't come down."

14 Describe the flying maneuver used to simulate a condition of "weightlessness" for the training of astronauts.

15 Why are the rear sights of a long-range rifle adjustable?

16 Show clearly how the concepts expressed in Newton's laws of motion apply to the motion of a projectile.

17 Describe the effect that an increase in the angle of elevation has on the range of a projectile for various angles of elevation.

18 Derive an expression for the speed of a projectile at time t after it is fired with velocity v at an angle of elevation θ.

19 A man stands in the center of a flatcar moving with a uniform speed of 40 km/h. He throws a baseball into the air with a speed of 40 km/h. Compare the path of the ball as viewed by the man with that as viewed by an observer on the ground for the following cases: (*a*) ball thrown vertically upward, (*b*) ball thrown forward horizontally, and (*c*) ball thrown backward horizontally.

20 Discuss the factors that would affect the acceleration of an Atlas rocket as it rises during the 70 s that the fuel burns.

21 Derive an equation for the range of a projectile as a function of the angle of its initial velocity above the horizontal, neglecting air resistance. (*Hint:* Write equations for x and y as functions of t and θ; place y equal to zero; eliminate t between the equations; use the trigonometric identity for the sine of 2θ.)

22 Derive an equation for y as a function of x for a projectile, treating v_0 and θ as known, and eliminating the time t.

23 What would be the appearance of a speed-time curve if the falling body were so light that the effect of air friction could not be neglected?

Problems

1 A motorist has to travel 3.50 km in a city where his average speed should not exceed

25 km/h. If he increases his average speed to 40 km/h, how much time will he gain in his journey?

2 A runner A can run the mile race in 4.25 min. Another runner B requires 4.55 min to run this distance. If they start out together and maintain their normal speeds, how far apart will they be at the finish of the race? *Ans.* 348 ft.

3 A train starts from rest and at the end of 90 s has a speed of 30 km/h. What is its acceleration?

4 A car going 50 mi/h overtakes and passes another car moving at 45 mi/h. What length of road is required for the operation? Assume that each car is 15 ft long and that there is a 60-ft space between them before and after passing. Taking into account the approach of a car from the opposite direction at 50 mi/h, what clear length of road is required?

Ans. 1,500 ft; 3,000 ft.

5 An object falls from a plane flying horizontally at an altitude of 40,000 ft at 500 mi/h. How long will it take to hit the ground?

6 The initial speed of a car having excellent brakes is 30 mi/h (44 ft/s). When the brakes are applied, it stops in 2.0 s. Find the acceleration.

Ans. -22 ft/s^2.

7 A bullet accelerates from rest to a speed of 600 m/s while traveling the 0.600-m length of a rifle barrel. What is its average acceleration? (Treat this as a case of uniform acceleration.) How many times as great as the acceleration of gravity is this?

8 A body slides down a frictionless plane and during the third second after starting from rest it travels 19.4 m. What is the angle of inclination of the plane? *Ans.* 53.1°.

9 A body slides down a frictionless incline 10.0 m long. If the incline makes an angle of 30.0° with the horizontal, calculate (a) the time of descent, (b) the speed with which it reaches the bottom, and (c) the distance traversed during the second second after it starts from rest.

10 An object is thrown downward from a 150-ft cliff with a velocity of 45 mi/h so that it makes an angle of 30° with the horizontal. (a) How fast will it be going vertically when it hits the ground? (b) How long will it take to reach the ground? (c) How far will it fall from the base of the cliff?

Ans. (a) 103.4 ft/s; (b) 2.2 s; (c) 125.8 ft.

11 An elevator is ascending with an upward acceleration of 4.0 ft/s^2. At the instant its upward speed is 8.0 ft/s a bolt drops from the top of the cage 9.0 ft from the floor. Find the time until the bolt strikes the floor and the distance it has fallen.

12 A rocket ship is on its way to the moon at 15,600 mi/h when it is alerted that a second ship is observed 1,000 mi away (viewed at an angle 30° from the line of motion of the rocket ship). The second ship is traveling at 9,000 mi/h and will cross the path of the first ship at right angles. Are they on a collision path? If so how much time does the command pilot of the first rocket ship have to take evasive action?

Ans. Yes, they will collide; 3.33 min.

13 A car travels with a constant speed of 30 km/h for 15 min. It then quickly speeds up to 50 km/h and maintains this velocity for 30 min. Neglecting the time needed to accelerate, what is the average velocity for the whole period?

14 An automobile has a speed of 60 mi/h. When the brakes are applied, it slows to 15 mi/h in 4 s. What is its acceleration? How far does it travel during the fourth second?

Ans. $a = -16.5$ ft/s^2; $s = 30$ ft.

15 A sport car starting from rest can attain a speed of 60 mi/h in 8.0 s. A runner can do a 100-yd dash in 9.8 s. Assume that the runner is moving with uniform speed and that the car starts at the instant he passes it. How far will both travel until the car overtakes the runner?

16 A stone falls from a railroad overpass which is 36 ft high into the path of a train which is approaching the overpass with uniform speed. If the stone falls when the train is 50 ft away from the overpass and the stone hits the ground just as the train arrives at that spot, how fast is the train moving? *Ans.* 22.8 mi/h.

17 Two tall buildings are 60 m apart. With what speed must a ball be thrown horizontally from a window 150 m above the ground in one building so that it will enter a window 15 m from the ground in the other?

18 How high will a body rise that is projected vertically upward with a speed of 100 ft/s? How long will it take for the body to reach its maximum height? *Ans.* 156 ft; 3.1 s.

19 An arrow is shot vertically upward with a speed of 288 ft/s, and 3.00 s later another is shot up at a speed of 240 ft/s. Will they meet? If so, where?

20 A balloon which is ascending at the rate of 12 m/s is 80 m above the ground when a stone is dropped. How long a time is required for the stone to reach the ground? *Ans.* 5.4 s.

21 A cannon is fired with a muzzle velocity of 300 m/s at an angle of 60°. What is its range?

22 A stone is dropped from a high altitude, and 3.00 s later another is projected vertically downward with a speed of 150 ft/s. When and where will the second overtake the first?

Ans. 5.70 s; 520 ft.

23 A baseball is batted into the air and caught at a point 100 m distant horizontally in 4 s. If air resistance is neglected, what is its maximum height in meters above the ground?

24 A bomb is dropped from an airplane traveling horizontally with a speed of 300 mi/h. If the airplane is 10,000 ft above the ground, how far from the target must it be released? Neglect air friction. *Ans.* 2.09 mi.

25 A bomb is dropped from an airplane 1,500 m above the ground when the plane is moving horizontally at the rate of 160 km/h. Where should the plane be with respect to the target when the bomb is dropped if a hit is to be made?

26 Two distant nebulae are observed to be moving away from the earth at 0.150 and 0.250 times the speed of light c. If they and the earth are in nearly the same straight line, how fast would one nebula appear to be moving, as seen from the other? Determine their relative speed by the classical equation first, then by the relativistic equation. *Ans.* $0.100c$; $0.104c$.

27 A ball thrown by a boy in the street is caught 2.0 s later by another boy on the porch of a house 15.0 m away and 5.0 m above the street level. What was the speed of the ball and the angle above the horizontal at which it was thrown?

28 A missile is fired with a launch velocity of 15,000 ft/s at a target 1,200 mi away. At what angle must it be fired to hit the target? How long after it is fired will the target be hit? (Assume that the accelerating power is cut off the instant it leaves the ground.) *Ans.* 32°, 496 s.

29 A projectile is fired at an angle of 30° above the horizontal from the top of a cliff 600 ft high. The initial speed of the projectile is 2,000 ft/s. How far will the projectile move horizontally before it hits the level ground at the base of the cliff?

30 At what speed must a rifle bullet be fired so that it hits a monkey when the monkey is 180 ft high and at the instant of firing drops from the top of a 200-ft-high tree, 350 ft away?

Ans. 364 ft/s.

Sir Joseph John Thomson, 1856–1940

Born in Cheetham Hall, near Manchester. Professor at the Royal Institute for Natural Philosophy in London, later master of Trinity College, Cambridge. Discoverer of isotopy. Awarded the 1906 Nobel Prize for Physics for his theoretical and experimental investigations of the passage of electricity through gases.

John William Strutt, Third Baron Rayleigh, 1842–1919

Born in Essex, England. Chairman of the Davy-Faraday Research Laboratory in London. Awarded the 1904 Nobel Prize for Physics for his investigations on the density of the more important gases, and for his discovery of argon, one of the results of those investigations.

Philipp Lenard, 1862–1947

Born in Pozony, Hungary. Professor of experimental physics at Kiel, later at Heidelberg. Received the 1905 Nobel Prize for Physics for his work on cathode rays.

3

Force and Motion

We have considered the conditions under which there is no change in motion of a particle. If the net force on a particle is zero, that is, there is no unbalanced external force, its motion does not change. We have studied the motion of bodies that are moving with uniform acceleration, but we have not inquired about the forces that produce such acceleration. We shall now seek to analyze the relationship between resultant forces and the accelerations they produce.

When a body is at rest, we know from experience that it will remain at rest unless something is done to change that state. We walk without fear in front of a standing locomotive because we know that it will not suddenly move. A heavy box on the floor will stay in place unless it is pushed or pulled. We must exert a force upon it to change its motion, that is, to give it an acceleration.

We readily accept the fact that no body can be set in motion without having a force act upon it. It may not be so easy to accept the equally true fact that a body in motion cannot change its motion unless a resultant force acts on it. We seldom if ever observe a body that has no force acting on it.

A box resting on the floor has more than one force acting on it, but they do not produce a change in motion. A rather large horizontal force must be exerted to start the box moving, and it stops quickly when the force is removed. If the box is mounted on wheels, a smaller force is required to start it and it continues to move longer. If more care is taken to reduce the friction, it becomes easier to start the box and it continues to move more readily. We are finally led to the conclusion that if the friction of the floor could be entirely removed, *any* horizontal force could start the box moving and once started it would continue to move indefinitely unless a force were exerted to stop it. The property of a body by virtue of which a net force is required to change its motion is called *inertia*.

3-1 THE LAW OF INERTIA: NEWTON'S FIRST LAW

The conclusion which has been reached regarding the need of a force to change the motion of a body was stated by Sir Isaac Newton (1642–1727). *There is no change in the motion of a body unless an unbalanced external force is acting upon it.* If the body is at rest, it will continue at rest. If it is in motion, it will continue in motion with constant speed *in a straight line* unless there is a net force acting. This law of inertia is usually called *Newton's first law of motion*. It is the

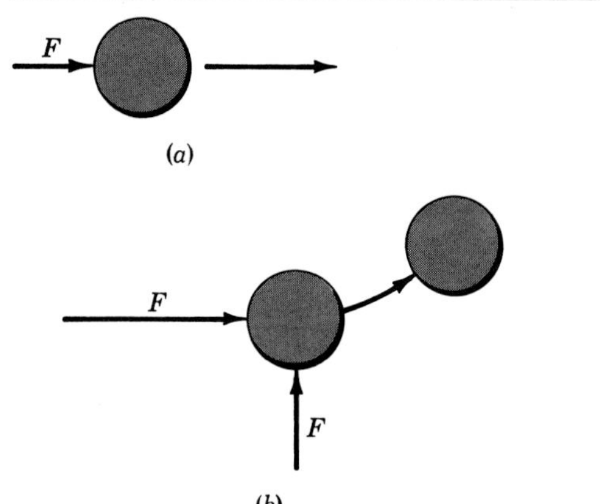

Figure 3-1
The effect of an external force on moving objects.

purely negative statement that no acceleration will occur without a net force to cause that change.

It should be noted that this law states that in order to change either the rate of motion or the direction in which an object moves, a force is required. Further, the direction of this applied force is important because, as Fig. 3-1 shows, the force in *a* acting in the direction in which the object was moving will change only the rate of motion, while the force in *b* which is applied at right angles to the original motion will change the direction only. Since velocity is a vector quantity defined in terms of magnitude and direction, either force can be considered as causing a change in *velocity* of the object. Case *b* will become more important when we consider uniform circular motion and central acceleration later in this book.

There are many examples of the first law of motion, but perhaps the student would appreciate knowing the specific illustrations of this law that Newton himself gave in his book the "Principia" in 1687. He observed that

1 Projectiles continue in their motion until retarded by air resistance and pulled down by gravity

2 A turning wheel does not stop rotating as long as it is not retarded by air resistance

3 Planets and comets maintain both their progressive and their revolving motion longer in spaces that offer less resistance

3-2 FORCE AND ACCELERATION: NEWTON'S SECOND LAW

Let us consider an experiment in which we have a spring and several identical blocks of metal. Suppose that one metal block is placed on a horizontal frictionless surface. The spring is attached to the metal block and stretched a small known distance while the block is held (Fig. 3-2), exerting a force F_1 on the metal block. If the block is released and the stretch of the spring is kept constant by pulling it along, the block is accelerated. The acceleration a_1 of the block may be determined by measuring the time required to move a known distance s starting from rest and then by using the equation $s = v_0t + \frac{1}{2}at^2$. Repeat the experiment with the spring stretched twice as much to give a force $F_2 = 2F_1$, and again with three times the initial stretch to give $F_3 = 3F_1$. The accelerations will be found to be directly proportional to the net force F applied and in the direction of the net force.

$$F \propto a$$

or

$$\frac{F_1}{a_1} = \frac{F_2}{a_2} = \frac{F_3}{a_3} = \text{a const} \tag{1}$$

This constant ratio of the net force to the acceleration produced is a measure of the inertia of the body being accelerated and is thus the mass of the body.

Suppose that we repeat our experiment with spring and metal blocks, but this time we shall keep the stretch of the spring constant and apply the same force F to one block, two blocks, three blocks, etc., and measure the resultant acceler-

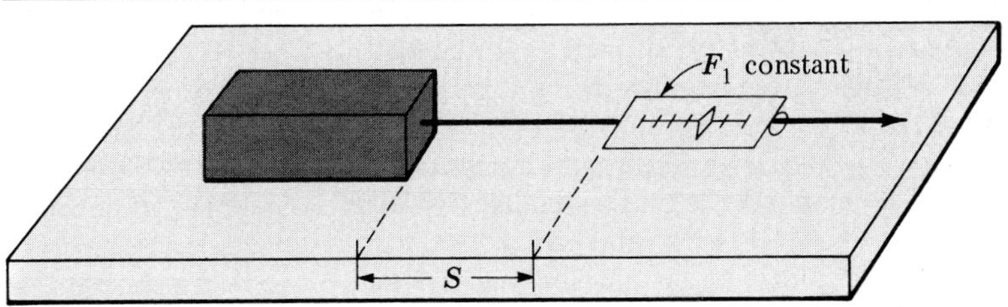

Figure 3-2
Force and acceleration apparatus.

ations. We find that the acceleration is smaller when the number of blocks is increased. For two blocks the acceleration is one-half that for one block, for three blocks one-third, etc. If m_1 is the mass of one block,

$$\frac{F}{a_1} = m_1$$

$$\frac{F}{a_2} = m_2 = 2m_1 \qquad (2)$$

$$\frac{F}{a_3} = m_3 = 3m_1$$

etc.

The acceleration of a body is directly proportional to the net force and inversely proportional to the mass being accelerated. The generalization of this statement is Newton's second law of motion. *Whenever a net (resultant) force acts on a body, it produces an acceleration in the direction of the resultant force that is directly proportional to the resultant force and inversely proportional to the mass of the body.*

According to the Newton's second law, the following proportions may be written:

$$a \propto F \qquad \text{and} \qquad a \propto \frac{1}{m}$$

These proportions may be combined as

$$a \propto \frac{F}{m}$$

and written as the equation

$$F = kma \qquad (3)$$

In Eq. (3) any unit of force, any unit of mass, and any unit of acceleration can be used, provided that the proper value is assigned to the constant k. In general, a different value of k would have to be assigned for each combination chosen. For simplicity, it is customary to use a system of units for which k has a value of 1.

It is important to recognize the significance of proportionality constants, as k above. In science many physical relationships are observed as proportions, or ratios. While these ratios help to give an understanding of the relationship of two properties, it is often desirable to express this relationship in quantitative terms by means of some statement of equality, an equation. A typical example can be seen if we consider the geometric ratio of the area of a circle to its radius. The area of a circle is proportional to the square of the radius:

$$A \propto r^2$$

While this can be proved to be true, we are unable to use this proportion in that form to find the area of a circle. By selecting a proportionality constant, however, we are able to change the proportion to an equation. For the particular case of a circle where the proportionality constant equals 3.14+, or π, the following equation holds:

$$A = \pi r^2$$

For Newton's second law, the proportionality constant k is customarily defined as 1. The value of k depends on units in which force, mass, and acceleration are presented. We shall see many situations in physics where proportionality constants are employed, e.g., Coulomb's law and Newton's universal law of gravitation.

Since the utilization of Newton's second law of motion depends greatly upon the units which are used, we will defer consideration of Newton's third law until we have examined in some detail the choice of units available to us.

3-3 SYSTEMS OF UNITS

Earlier we observed that three fundamental quantities are required to set up a system of units in mechanics. The choice of these fundamental quantities is rather arbitrary, and the kinds of units set up depend upon the choice. Commonly, the fundamental quantities are length, mass, and time. Another system of fundamental quantities utilizes length, force, and time.

Mass is independent of the place at which observation is made, and hence a system of units based on length, mass, and time is called an *absolute* system. In the alternative choice, length, force, and time, the force commonly chosen is a gravitational force, or weight; and hence the system of units is called a *gravitational* system.

3-4 ABSOLUTE SYSTEMS OF UNITS

In an absolute system of units a fundamental unit is arbitrarily assigned to each of the fundamental quantities length, mass, and time as described in the Introduction. Three such absolute systems are commonly set up: two metric, *mks* and *cgs,* and one British.

The mks system is that for which the fundamental units selected are the meter, the kilogram, and the second, the initials of these units forming the name of the system. In this system the unit of acceleration is the meter per second per second. From the kilogram unit of mass and meter per second per second as a unit of acceleration we can derive a unit of force that will make k of Eq. (3) unity. This unit of force is called the newton. A *newton* (N) is the force that will give to a mass of one kilogram an acceleration of one meter per second per second (m/sec^2).

When we use a set of units, such as the mks system, in which one unit is defined in such a way as to make k unity, Eq. (3) reduces to

$$\mathbf{F} = m\mathbf{a} \tag{4}$$

Equation (4) can be used only *when a consistent set of units is employed.*

In the cgs system the centimeter, the gram, and the second are used as starting units. The acceleration is measured in centimeters per second per second. As before, we can derive a unit of force that makes k of Eq. (3) unity. The *dyne* (dyn) is the net force that will give to a mass of one gram (g) an acceleration of one centimeter per second per second (cm/s^2) (Fig. 3-3). A mosquito weighs approximately 1 dyn.

The cgs system was used for many years as the principal metric system. The mks system was adopted by an international conference to become effective in 1940. Since that time it has increasingly replaced the cgs system. An advantage of the mks system is that it leads to the "practical" electrical units in common use.

A British absolute system is based upon the foot, the pound, and the second. In it the pound is used as a unit of mass, and as before a unit of force is defined to make k of Eq. (3) unity. The *poundal* is the force that will give to a mass of one pound an acceleration of one foot per second per second (ft/s^2). We shall not use this system of units in this book.

3-5 GRAVITATIONAL SYSTEMS OF UNITS

In a gravitational system of units the fundamental unit of force is defined in terms of the pull of the earth upon an arbitrarily chosen body. In a

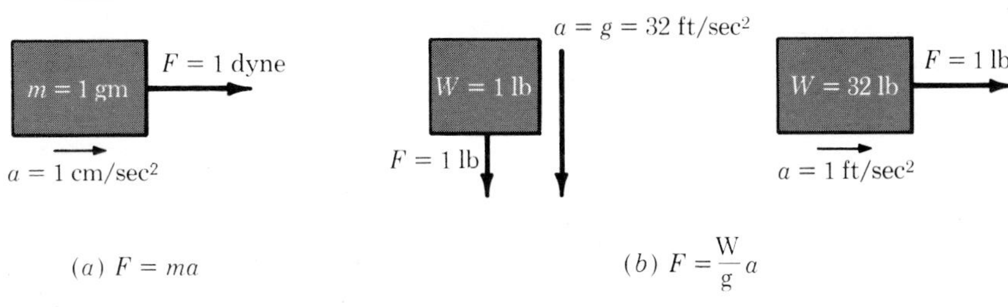

(a) $F = ma$ (b) $F = \frac{W}{g} a$

Figure 3-3

metric gravitational system the gram is used as a unit of force. The *gram force* is defined as one-thousandth the pull of the earth upon a standard kilogram at a place where g has a value of 980.665 cm/s^2. In this gravitational system the unit of mass is derived from Eq. (3) in such a manner as to make k unity. No name has been assigned to this gravitational unit of mass. It is the mass to which a gram force would give an acceleration of 1 cm/s^2. This system of units will not be used in this book.

In the British gravitational system, fps, the fundamental unit of force, the pound (lb), is 1/2.2046 the force with which the earth pulls on a standard kilogram at a place where g is 32.17398 ft/s^2. In this system we define a unit of mass, the slug, from Eq. (3). The *slug* is the mass to which a force of one pound will give an acceleration of one foot per second per second.

3-6 CHOICE OF UNITS TO BE USED

We have outlined five systems of units each of which is consistent, logical, and suitable for use in Eq. (4). It is unfortunate that in the different sets of units the same word is used to designate a unit of mass in one set but a unit of force in another. Therefore throughout the mechanics section of this book we shall omit completely reference to the *British absolute* system and to the *cgs gravitational system*. Whenever the term *gram* or *kilogram* is used, it will refer to *mass*. Whenever the term *pound* is used, it will refer to *force*. Summarizing, we shall limit ourselves to the mks, cgs (abs), and British gravitational (fps) systems.

3-7 WEIGHT: RELATION BETWEEN MASS AND WEIGHT

The *weight* of a body at any point in space may be defined as the resultant gravitational force acting on the body, due to all other bodies in space. When this definition is used, the body has weight at all points in space except at those very special points at which the resultant gravitational force is zero. For a body near the surface of the earth the gravitational forces due to outside bodies are almost negligible in comparison with that due to the earth; the only gravitational force that we need consider is that due to the earth, and this force is weight. Because the earth is rotating, observations of the gravitational force give a result that is slightly less than the gravitational force, as we shall see later. Since this effect is small, we shall ignore it for the moment.

Some prefer to define weight as the reaction of a measuring instrument to the gravitational force. According to this definition the weight depends upon the conditions under which the measurement is made. For example, a body falling freely is accelerated by the resultant gravitational force, and the reaction of the measuring instrument as it and the body fall freely is zero. Or in a satellite, the gravitational force produces

Table 1

System	F	=	m	a
mks	N	=	kg	m/s^2
cgs (abs)	dyn	=	g	cm/s^2
British (grav)	lb	=	slug	ft/s^2

the central acceleration necessary to hold both satellite and occupant in the orbit. Hence there is no reaction of the capsule on the occupant. This condition of zero reaction is often referred to as *weightlessness*. The sensations of an observer under these conditions are the same as those in a condition of true weightlessness as defined above.

When a body falls freely, the only force acting on it is its weight. This net force produces the acceleration g observed in freely falling bodies. From Eq. (3) we obtain

$$F = kma$$

$$W = kmg$$

If we use units that are consistent with Eq. (4), the value of k is unity and

$$W = mg \tag{5}$$

or

$$m = \frac{W}{g} \tag{6}$$

Since weight is a force, we use Eq. (5) to express weights in newtons or in dynes when we use the absolute mks or cgs units, respectively. In the British gravitational system we commonly express the mass in *slugs* from Eq. (6). In any case, where we use units that are consistent with Eq. (4), we can always substitute W/g for m and mg for W.

Example Find the mass of an object that weighs 320 lb.

$$m = \frac{W}{g}$$

$$m = \frac{320 \text{ lb}}{32 \text{ ft/s}^2} = 10 \text{ slugs}$$

Table 1 lists consistent sets of mechanical units.

3-8 HINTS FOR THE SOLUTION OF PROBLEMS USING NEWTON'S SECOND LAW

In applying the second law of motion to the solution of a problem, much difficulty can be avoided by following a definite procedure. The most common source of difficulty is the failure to recognize that the F of the second law always refers to the *resultant*, or *unbalanced*, force acting on a body and the m refers to the entire mass of that same body. In solving any problem involving force and motion, the following steps are recommended:

1 Make a sketch showing the conditions of the problem. Indicate on it dimensions or other data given in the problem.
2 Select for consideration the *one body whose motion is to be studied.* Construct a force vector diagram. On this vector diagram, represent by vectors *all* the forces acting *on* the body that has been selected. If any forces are unknown, represent them also by vectors, and label them as unknown quantities.
3 From the vector diagram, find the *resultant* force acting on the body. This resultant is the F of Eq. (4).
4 Find the unknown quantity (a, F, or m) from the relation $F = ma$. If the weight of the body is given, compute m from $m = W/g$. If the problem asks for a distance, velocity, or time, apply the equations of accelerated motion (Chap. 2) as required.

Example A 50-kg block rests at the top of a smooth plane whose length is 2.00 m and whose height is 0.50 m. How long will it take for the block to slide to the bottom of the plane when released?

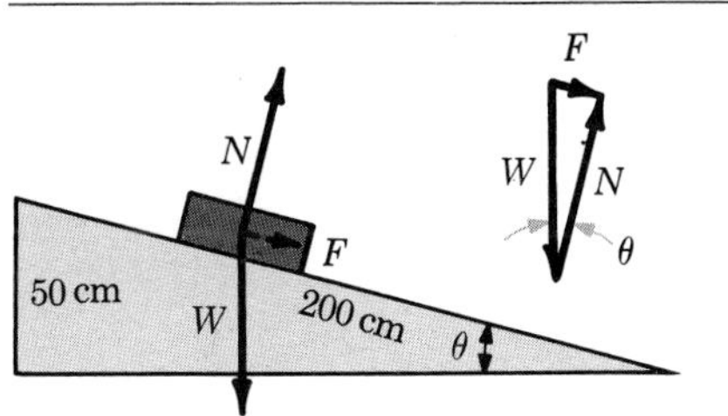

Figure 3-4
Inclined plane.

As indicated in Fig. 3-4, the forces acting on the block are the weight W downward and the force N of the plane against the block. This force N, called the normal, is perpendicular to the plane. The resultant (unbalanced) force F is parallel to the plane. The force triangle and the space triangle are similar, and the angle θ is common to the two triangles. From the force triangle

$$F = W \sin \theta$$

From the space triangle

$$\sin \theta = \frac{0.50 \text{ m}}{2.00 \text{ m}} = 0.25$$

$$F = mg \sin \theta = (50 \text{ kg})(9.8 \text{ m/s}^2)(0.25)$$
$$= 1.2 \times 10^2 \text{ N}$$

From Eq. (4),

$$a = \frac{F}{m} = \frac{1.2 \times 10^2 \text{ N}}{50 \text{ kg}} = 2.4 \text{ m/s}^2$$

Since the block starts from rest, the time of descent is determined from $s = \frac{1}{2}at^2$.

$$t = \sqrt{\frac{2s}{a}} = \sqrt{\frac{2 \times 2.00 \text{ m}}{2.4 \text{ m/s}^2}} = 1.3\text{s}$$

Example A 60.0-lb block rests on a smooth plane inclined at an angle of 20° with the horizontal (Fig. 3-5). The block is pulled up the plane with a force of 30.0 lb parallel to the plane. What is its acceleration?

Here three forces are acting on the block. Its weight W is 60 lb downward. The force of the plane on the block is a thrust N normal to the plane. There is a pull P parallel to the plane. Addition of these vectors by the polygon rule shows an unbalanced force F acting on the block parallel to the plane.

The weight of the block may be resolved into components of 60.0 lb × cos 20° normal to the plane and 60.0 lb × sin 20° parallel to the plane. The normal component is balanced by the force N. Hence the unbalanced force F parallel to the plane and directed up the plane is

$$F = 30.0 \text{ lb} - 60.0 \text{ lb} \times \sin 20°$$
$$= 30.0 \text{ lb} - (60.0 \times 0.342) \text{ lb} = 9.5 \text{ lb}$$

$$m = \frac{W}{g} = \frac{60.0 \text{ lb}}{32 \text{ ft/s}^2} = 1.87 \text{ slugs}$$

From Eq. (4),

$$a = \frac{F}{m} = \frac{9.5 \text{ lb}}{1.87 \text{ slugs}} = 5.1 \text{ ft/s}^2$$

Note that if the angle were 30°, the component of the weight down the plane would be equal to the force up the plane and there would be no unbalanced force acting on the block. Hence it would not be accelerated. If the angle were greater than 30°, the block would be accelerated down the plane.

Example A 2.0-ton elevator is supported by a cable that can safely support 6,$\bar{4}$00 lb. What is the shortest distance in which the elevator can be brought to a stop when it is descending with a speed of 4.0 ft/s?

The maximum net force acting on the elevator (Fig. 3-6) is

$$6{,}\bar{4}00 \text{ lb} - 4{,}\bar{0}00 \text{ lb} = 2{,}\bar{4}00 \text{ lb} \qquad \text{upward}$$

The mass being accelerated is

$$m = \frac{W}{g} = \frac{4{,}\bar{0}00 \text{ lb}}{32 \text{ ft/s}^2} = 125 \text{ slugs}$$

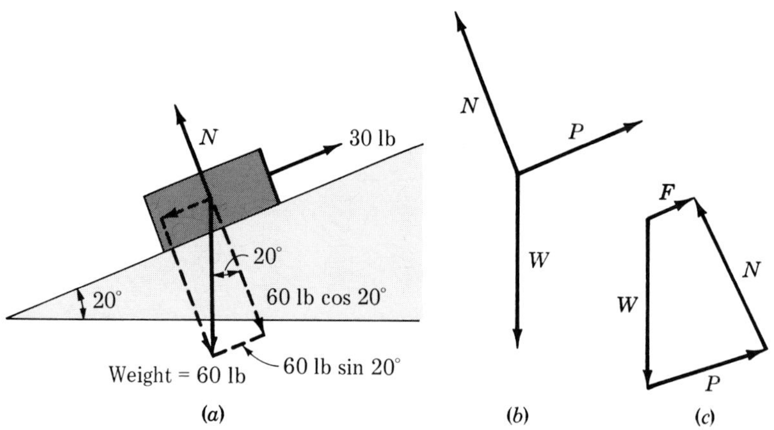

Figure 3-5
Determination of resultant force acting on block.

From Eq. (4),

$$a = \frac{F}{m} = \frac{2{,}400 \text{ lb}}{125 \text{ slugs}} = 19 \text{ ft/s}^2 \qquad \text{upward}$$

Since the initial velocity is downward, the upward acceleration will be considered negative.

The time required to stop the elevator is

$$t = \frac{v_1 - v_0}{a} = \frac{0 - 4.0 \text{ ft/s}}{-19 \text{ ft/s}^2} = 0.21 \text{ s}$$

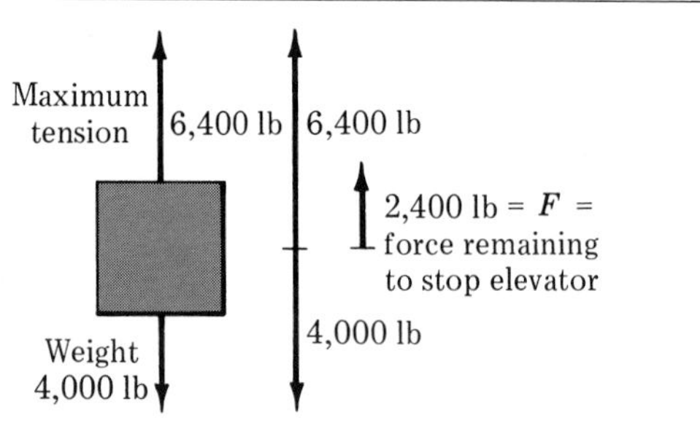

Figure 3-6
Finding unbalanced force from vector addition of forces.

In this time the elevator will have covered a distance

$$s = \bar{v}t = \frac{4.0 \text{ ft/s} + 0}{2}(0.21 \text{ s}) = 0.42 \text{ ft}$$

Example A 5.0-kg block is placed on a smooth horizontal surface (Fig. 3-7). A horizontal cord attached to the block passes over a light

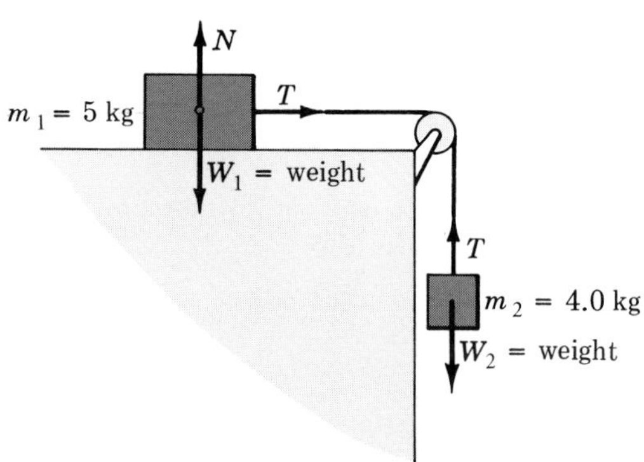

Figure 3-7
Block accelerated by a falling mass.

frictionless pulley and is attached to a 4.0-kg body. Find the acceleration and the tension in the cord when the system is released.

Consider first the 5.0-kg block. The forces acting on this body are its weight W_1 downward, the reaction N of the plane upward, and the tension T of the cord to the right. Since there is no vertical acceleration, $N = W_1$ and the resultant force is T. Thus, from Eq. (4),

$$T = m_1 a$$

The forces on the 4.0-kg body are its weight W_2 downward and the tension T upward. The net downward force is $W_2 - T$ and

$$W_2 - T = m_2 a$$

Since the two bodies move together, the accelerations are equal in magnitude although the directions are not the same. If we add the two equations by solving them simultaneously, we obtain

$$\begin{aligned} T &= m_1 a \\ W_2 - T &= m_2 a \\ \hline W_2 &= m_1 a + m_2 a \\ W_2 &= (m_1 + m_2) a \end{aligned}$$

But $\quad W_2 = m_2 g = 4.0 \text{ kg} \times 9.8 \text{ m/s}^2 = 39 \text{ N}$
Therefore,

$$39 \text{ N} = (5 \text{ kg} + 4 \text{ kg})a$$

$$a = \frac{39 \text{ N}}{9 \text{ kg}} = 4.4 \text{ m/s}^2$$

Then, substituting in $T = m_1 a$,

$$T = (5 \text{ kg})(4.4 \text{ m/s}^2) = 22.0 \text{ N}$$

Example Two bodies having masses $m_1 = 30$ g and $m_2 = 40$ g are attached to the ends of a string of negligible mass and suspended from a light frictionless pulley as shown in Fig. 3-8. Find the accelerations of the bodies and the tension in the string.

Consider the body of mass m_1. Two external forces act on it, the weight $m_1 g$ downward and the upward pull T of the string. The resultant force on this body is $T - m_1 g$ upward. From Eq. (4) we may write

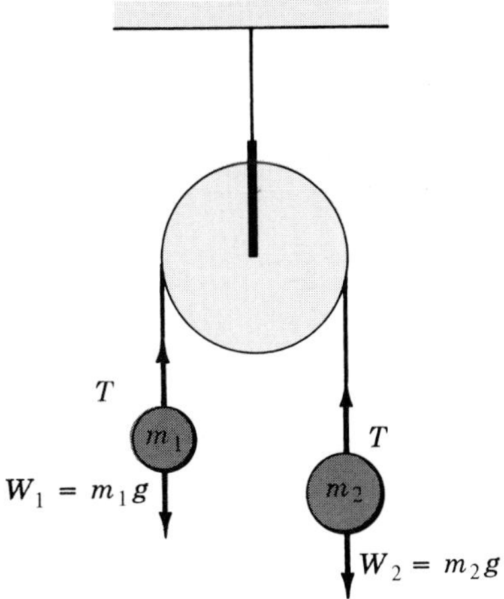

Figure 3-8
Unequal masses suspended from a pulley.

$$T - m_1 g = m_1 a$$

where a is the upward acceleration of this body.

Now consider the body of mass m_2. The forces acting on this body are its weight $m_2 g$ downward and the tension T upward. The resultant force is $m_2 g - T$ downward and from Eq. (4)

$$m_2 g - T = m_2 a$$

where a is the downward acceleration of this body. Since the two bodies move together, the accelerations are equal in magnitude but opposite in direction.

$$m_1 = 30 \text{ g} = 0.030 \text{ kg}$$

$$m_2 = 40 \text{ g} = 0.040 \text{ kg}$$

$$T - (0.030 \text{ kg})(9.8 \text{ m/s}^2) = (0.030 \text{ kg})a$$

and

$$(0.040 \text{ kg})(9.8 \text{ m/s}^2) - T = (0.040 \text{ kg})a$$

When the two equations are added by solving simultaneously, we obtain

$$\begin{aligned} T - 0.294 \text{ N} &= (0.030 \text{ kg})a \\ -T + 0.392 \text{ N} &= (0.040 \text{ kg})a \\ \hline +0.098 \text{ N} &= (0.070 \text{ kg})a \end{aligned}$$

$$a = \frac{0.098 \text{ N}}{0.070 \text{ kg}} = 1.40 \text{ m/s}^2$$

Then

$$\begin{aligned} T &= 0.294 \text{ N} + 0.030 \text{ kg } (1.40 \text{ m/s}^2) \\ &= (0.294 + 0.042) \text{ N} \\ &= 0.336 \text{ N} \end{aligned}$$

3-9 VARIABLE FORCE: VARIABLE ACCELERATION

In our discussion thus far we have assumed that the forces considered have been constant in magnitude and in direction. In general, forces may vary in any manner. If there is any such variation, Newton's second law applies at each instant and the instantaneous value of the acceleration is proportional to the net force at that instant and in the direction of that force.

The mass of a body does not necessarily remain constant during its motion. As a rocket consumes fuel and expels the exhaust gas to achieve the thrust, the mass continually decreases. Other bodies may pick up material as they move, as in the case of growing raindrops.

The principle of relativity (Chap. 45) states that even though no material is added, the mass of a particle increases with speed according to the relation

$$m = \frac{m_0}{\sqrt{1 - v^2/c^2}} \qquad (7)$$

where m_0 is called the *rest mass*, m is the mass at speed v, and c is the speed of light. This change of mass with speed is negligible for speeds that are small compared with the speed of light and hence is important only for the small particles that can be accelerated to speeds approaching that of light.

Example What would be the mass of an object having a rest mass of 1.0 g if it moves at 0.8 the speed of light?

$$m = \frac{m_0}{\sqrt{1 - (v^2/c^2)}} = \frac{1.0 \text{ g}}{\sqrt{1 - [(0.8c)^2/c^2]}}$$

Then

$$\begin{aligned} m &= \frac{1.0 \text{ g}}{\sqrt{(c^2 - 0.64c^2)/c^2}} = \frac{1.0 \text{ g}}{\sqrt{1 - 0.64}} \\ &= \frac{1.0}{\sqrt{0.36}} \\ &= \frac{1.0}{0.6} = 1.67 \text{ g} \end{aligned}$$

3-10 REACTING FORCES: NEWTON'S THIRD LAW

For every force that acts on one body there is a second force equal in magnitude but opposite in direction that acts upon another body. These forces are often referred to as *acting* and *reacting* forces. Here the term *acting force* means the force that one body exerts on a second body, while *reacting force* means the force that the second body exerts on the first. There can be no force unless the mutual interaction of two bodies is involved (or fields arising from two different sources). It should be remembered that acting and reacting forces, though equal in magnitude and opposite in direction, can never neutralize each other for they always act on *different* objects. In order for two forces to neutralize each other, they must act on the *same* object.

When a baseball bat strikes a ball, it exerts

a force on the ball while the two are in contact. During the same time the ball exerts a force of the same magnitude but opposite in direction on the bat. A freely falling body is accelerated by the net force with which the earth attracts the body. The earth in turn is accelerated by the opposite reacting force the body exerts on the earth. Because of the great mass of the earth this acceleration is too small to be observed. In throwing a light object, one has the feeling that he cannot put much effort into the throw, for he cannot exert any more force on the object thrown than that object exerts in reaction against his hand. This reacting force is proportional to the mass of the object ($F \propto m$) and to the acceleration ($F \propto a$). The thrower's arm must be accelerated along with the object thrown; hence the larger part of the effort exerted in throwing a light object is expended in "throwing" one's arm.

When one steps from a small boat to the shore, he observes that the boat is pushed away as he steps. The force he exerts on the boat is responsible for its motion, while the force of reaction, exerted by the boat on him, is responsible for his motion toward the shore. The two forces are equal in magnitude and opposite in direction, while the accelerations which they produce (in boat and passenger, respectively) are inversely proportional to the masses of the objects on which they act. Thus a large boat will experience only a small acceleration when one steps from it to shore.

A book lying on a table is attracted by the earth. At the same time it attracts the earth, so that they would be accelerated toward each other if the table were not between them. Hence, each exerts a force on the table, and, in reaction, the table exerts an outward force on each of them, keeping them apart. It is interesting to note that the table exerts outward forces on the book and the earth by virtue of being slightly compressed by the pair of inward forces, which they exert on it.

Another example of the third law in action is the manner in which a rocket flies. While this will be treated in detail in the chapter on Space Physics, a simple analogy may prove helpful here. No one has missed seeing what happens when the nozzle of an inflated balloon is released. While the motion quickly becomes quite random due to the aerodynamics of the balloon, the first reaction of the balloon is to move away from the direction of the escaping air, an action-reaction phenomenon. A force diagram of the pressure on a balloon (Fig. 3-9) reveals that when the balloon is inflated, the internal pressure due to the forces acting on the inner surface cancel each other out. This must be true or else the balloon would not be at equilibrium and would be moving around. Once the nozzle is released, the equilibrium is destroyed; and where every internal force had been counterbalanced by another internal force (F_1 and F_2, for example), an unbalanced force is now present because F_2 has escaped out the nozzle and F_1, behaving according to the second law, causes an acceleration in the direction of F_1. This is the same principle upon which a rocket works. Gases produced by a rocket engine are allowed to escape, much as the air from the balloon, and the unbalanced force in the chamber then causes the resulting acceleration.

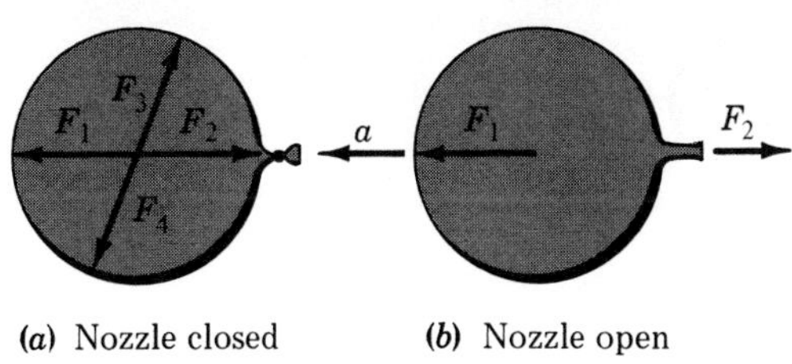

(a) Nozzle closed (b) Nozzle open

Figure 3-9
Forces inside an inflated balloon.

Example A 0.96-lb ball A and a 1.28-lb ball B are connected by a stretched spring of negligible mass as shown in Fig. 3-10. When the two balls are released simultaneously, the initial acceleration of B is 5.0 ft/s² westward. What is the initial acceleration of A?

$$m_A = \frac{W_A}{g} = \frac{0.96 \text{ lb}}{32 \text{ ft/s}^2} = 0.030 \text{ slug}$$

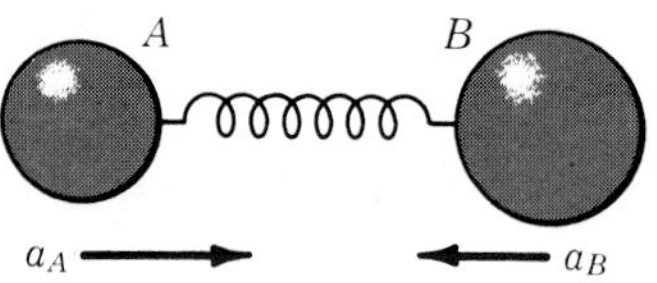

Figure 3-10
Acceleration inversely proportional to mass.

$$m_B = \frac{W_B}{g} = \frac{1.28\ \text{lb}}{32\ \text{ft/s}^2} = 0.040\ \text{slug}$$

Since

$$F_A = -F_B$$

$$m_A a_A = (m_B)(-a_B)$$

$$a_A = \frac{(-a_B)(m_B)}{m_A} = -5.0\ \text{ft/s}^2 \frac{0.040\ \text{slug}}{0.030\ \text{slug}}$$

$$= -6.7\ \text{ft/s}^2$$

Since the westward acceleration of B was taken as positive, the negative sign for the acceleration of A indicates that its acceleration is eastward.

3-11 UNIVERSAL GRAVITATION

In addition to the three laws of motion, Newton formulated a law of great importance in mechanics, the law of *universal gravitation: Every particle in the universe attracts every other particle with a force that is directly proportional to the product of the masses of the two particles and inversely proportional to the square of the distance between their centers of mass.* This relation may be expressed symbolically by the equation

$$F = \frac{Gm_1m_2}{s^2} \tag{8}$$

where F is the force of attraction, m_1 and m_2 are the respective masses of the two particles, s is the distance between their centers, and G is a constant called the *gravitational constant.* The value of G depends upon the system of units used in Eq. (8). If the force is expressed in newtons, the mass in kilograms, and the distance in meters, G has the value $6.670 \times 10^{-11}\ \text{N}\cdot\text{m}^2/\text{kg}^2$.

For extended bodies having a larger volume and consisting of many particles, the gravitational attraction is given in magnitude and direction by the vector sum of the attractions by the individual particles. This sum can be readily calculated for many regular bodies. Some of the results are very useful. For any point outside a uniform sphere the result is the same as if the whole mass were concentrated at the center of the sphere. Thus two such spheres attract as if their masses were concentrated at their centers. Inside a uniform spherical shell the resultant force is zero. We may use this property to examine the gravitational force inside any uniform solid sphere or one that is made up of any number of concentric shells each of which is uniform. Consider the sphere of Fig. 3-11. A body embedded in the sphere at a radius r experiences gravitational forces due to the shell between the radii r and R. The resultant force

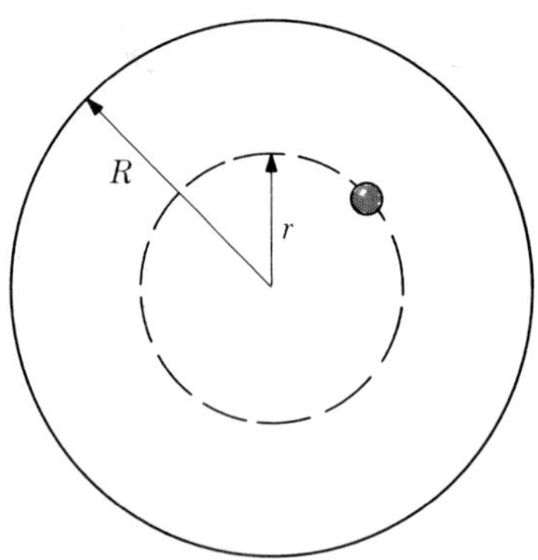

Figure 3-11
A body embedded in a solid sphere. The outer shell where the radius is greater than r produces a resultant gravitational force of zero. The resultant gravitational force on the body is that due to the part of the sphere inside the radius r.

due to this shell is zero. The body also experiences a gravitational force due to the sphere of radius r to which it is external. For the gravitational force on this body we use in Eq. (8) the mass of the inner sphere of radius r, and the distance of the body from the center of the sphere: the radius r. From these considerations we may conclude that the gravitational force is a maximum at the surface of a uniform sphere. Inside the sphere the effective mass decreases as the cube of the radius, while the distance decreases only as the square of that same distance. Outside the sphere the attracting mass does not change, but the distance increases. To a first approximation the earth may be considered as made up of many spherical shells, and hence it approximates the behavior just described. Near the surface local variations cause g to change both in magnitude and in direction.

Newton checked his law of gravitation by calculations based upon the orbit of the moon. With the approximate data at his disposal he still found reasonable agreement between his calculations and observations.

Newton's law of universal gravitation was not experimentally proved until more than a century after it was first published. While few physicists had any doubts about the truth of his hypothesis, the law was proposed in an age in which direct experimental evidence was required to prove its validity. It remained for Henry Cavendish (1731–1810) to measure in the laboratory the force of attraction between two masses. This was a formidable task because the force between any two masses which could be conveniently handled in the laboratory was extremely small and required great skill to measure. Since the force between the balls he used would amount to only about 1/50,000,000 of their weight, it was necessary to guard against any external factor such as changes in temperature and air currents. Figure 3-12 illustrates the essential apparatus Cavendish used. A 6-ft-long wooden arm was suspended by a thin wire 40 in long. Two small (2-in-diameter) balls were suspended at the ends of the arms. Two larger balls (8 in diameter) were positioned one in front of one small ball and the other in back of the other small ball. By use of a telescope the minute deflection of the rod bearing the small balls was measured and from the interpretation of these data the validity of the law was established and the value of G was determined.

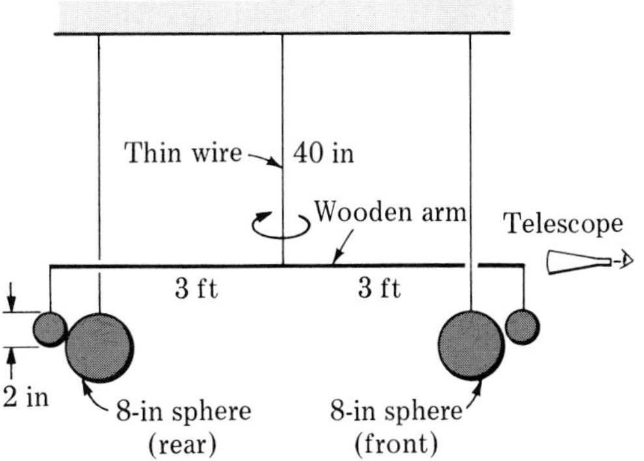

Figure 3-12
Cavendish apparatus.

Example Two lead balls whose masses are 5.20 kg and 0.250 kg are placed with their centers 50.0 cm apart. With what force do they attract each other?

From Eq. (8),

$$F = G\frac{m_1 m_2}{s^2}$$

$$= 6.670 \times 10^{-11}\ \mathrm{N \cdot m^2/kg^2} \times \frac{5.20\ \mathrm{kg} \times 0.250\ \mathrm{kg}}{(0.500\ \mathrm{m})^2}$$

$$= 3.47 \times 10^{-10}\ \mathrm{N}$$

3-12 MEASUREMENT OF MASS

We observe that we now have two equations that involve mass. We have defined mass from the property of inertia that is inherent in all material bodies. We have observed a second property common to all material bodies, namely, that they show a gravitational attraction for each other. It is not immediately apparent that the property of inertia is involved in gravitation. We may examine the relationship between inertia and gravitation. Suppose that we compare the masses m_1 and m_2 of two bodies by observing the accelerations a_1 and a_2 produced by equal forces. From Eq. (4),

$$\frac{m_1}{m_2} = \frac{a_2}{a_1}$$

Or, if we apply forces necessary to produce equal accelerations,

$$\frac{F_1}{F_2} = \frac{m_1}{m_2}$$

Let us consider the gravitational force when each of these bodies is attracted by the earth. The gravitational forces W_1 and W_2 produce the same acceleration in each body. Hence

$$\frac{W_1}{W_2} = \frac{m_1}{m_2}$$

That is, the gravitational force is proportional to the mass measured by the inertial property. Thus masses may be compared by observing inertial effects or by observing gravitational effects, and the results are identical. It was precisely this proportionality between gravitational force and inertia that Newton used in accounting for planetary motion.

A classic problem in physics is to calculate the mass of the earth, or any other planet using the universal law of gravity.

Example At the surface of the earth $g = 9.806\ \text{m/s}^2$. Assuming the earth to be a sphere of radius 6.371×10^6 m, compute the mass of the earth.

Since

$$F = m_1 a \qquad \text{and} \qquad W = m_1 g$$

and

$$F = G\frac{m_1 m_2}{s^2} = \text{weight}$$

then

$$m_1 g = G\frac{m_1 m_2}{s^2}$$

Let m_1 be the mass of any object on the earth's surface and m_2 be the mass of the earth. Since m_1 appears on both sides of the equation it cancels out. Therefore,

$$g = G\frac{m_2}{s^2}$$

or

$$m_E = \frac{g\,s^2}{G} = \frac{(9.806\ \text{m/s}^2)\,(6.371 \times 10^6\ \text{m})^2}{6.670 \times 10^{-11}\ \text{N}\cdot\text{m}^2/\text{kg}^2}$$
$$= 5.967 \times 10^{24}\ \text{kg}$$

This compares fairly well with the "best" estimate of the mass of the earth which is 5.975×10^{24} kg. Similarly, the mass of the moon or of any planet can be approximated using certain assumptions and conditions.

Example Compute the mass of the moon.

Since the moon is about one-fourth as large as the earth, $s \approx (6.371 \times 10^6\ \text{m})/4$ (the value we shall use is 1.74×10^6 m). Also it is commonly said that a person weighs one-sixth as much on the moon as he does on earth. Since in the equation $W = mg$, m is constant, then the weight and g are proportional. Therefore g on the moon must be approximately equal to $(9.806\ \text{m/s}^2)/6$ or $1.62\ \text{m/s}^2$. Using the equation,

$$m_{\text{moon}} = \frac{g\,s^2}{G} = \frac{(1.62\ \text{m/s}^2)(1.74 \times 10^6\ \text{m})^2}{6.670 \times 10^{-11}\ \text{N}\cdot\text{m}^2/\text{kg}^2}$$

$$m_{\text{moon}} = 7.39 \times 10^{22}\ \text{kg}$$

This compares with the value calculated by more sophisticated means for the mass of the moon, 7.349×10^{22} kg.

SUMMARY

The property of a body by virtue of which a resultant force is required to change its motion is called *inertia. Mass* is a numerical measure of inertia.

The relation between forces and the motions produced by them was described by Newton in three laws of motion. They are:

1 A body at rest remains at rest, and a body in motion continues to move at constant speed in a straight line unless it is acted upon by an external, unbalanced force.
2 An unbalanced force acting on a body produces an acceleration in the direction of the net force, an acceleration that is directly proportional to the unbalanced force and inversely proportional to the mass of the body.
3 For every force that one body exerts on a second body there is a force equal in magnitude but opposite in direction that the second body exerts upon the first body.

The relation expressed in Newton's second law may be expressed in equation form as

$$F = kma$$

where F, m, and a can be in any units, provided that the proper value is assigned to k, or

$$F = ma$$

where one can use *only* those consistent sets of units in which one of the units is defined in such a manner as to make $k = 1$.

A *newton* is defined as the force that will impart to a 1-kg mass an acceleration of 1 m/s^2.

The *dyne* is defined as the force that will impart to a 1-g mass an acceleration of 1 cm/s^2.

The *slug* is the mass to which a force of 1 lb will give an acceleration of 1 ft/s^2.

When the speed of a body approaches the speed of light, the mass of the body increases according to the relation

$$m = \frac{m_0}{\sqrt{1 - v^2/c^2}}$$

The law of universal gravitation expresses the fact that every particle attracts every other particle with a force directly proportional to the product of their masses and inversely proportional to the square of the distance between their centers of mass. In equation form

$$F = G\frac{m_1 m_2}{s^2}$$

The *weight* of a body at any point in space may be defined as the resultant gravitational force acting on the body due to all other bodies in space.

Questions

1 Motion and rest are referred to as being relative terms. What does this mean?
2 Is the weight of a body the same thing as its mass? Discuss briefly. Is the weight of a body constant at all places on the earth? Is the mass? Explain your answers.
3 If a man standing on a scale grabs his shoelaces and pulls up on them, will the scale reading be affected? If so, how?
4 According to biblical account, David killed the giant Goliath by using a sling. What scientific principle did he use to sling the stone?
5 Consider an object on a horizontal frictionless plane, acted upon by a single horizontal force.

- **a** If the mass is 1 g and the force 1 dyn, the acceleration is ______.
- **b** If the mass is 1 g and the force 5 dyn, the acceleration is ______.

c If the mass is 5 g and the force 10 dyn, the acceleration is ______.
d If the weight is 32 lb and the force 1 lb, the acceleration is ______.
e If the weight is 320 lb and the force 20 lb, the acceleration is ______.
f If the weight is 500 lb and the force 10 lb, the acceleration is ______.
g If the mass is 10.0 g and the force 9,800 dyn, the acceleration is ______.
h If the mass is 6.0 kg and the force is 2.0 N, the acceleration is ______.
i If the mass is 1.0 kg and the force is 9.8 N, the acceleration is ______.

6 Explain why an empty train starts more quickly than a loaded train.

7 Explain how a baseball pitcher's windup enables him to throw the ball with greater speed than he otherwise could throw it.

8 Which is greater, the attraction of the earth for a pound of lead, or the attraction of the pound of lead for the earth?

9 Why would an object have greater weight at the North Pole than at the Equator?

10 If an elevator supported by a cable is stopped quickly, it may oscillate up and down. Explain.

11 If bullets are fired from a plane in its direction of flight, will the velocity of the plane be affected? Explain.

12 If you found yourself on perfectly smooth ice in the center of a pond so that there was no friction, how could you get off the pond?

13 If acting and reacting forces are "equal and opposite," why can they never balance or cancel?

14 Describe how a stone which is being whirled in a circular path on the end of a string will move if the string breaks. Explain your answer.

15 The distance of sea level from the center of the earth is 3,963.34 mi at the Equator and decreases to 3,949.99 mi at the poles. Suggest an experiment by which this information about the shape of the earth might be obtained. In view of it, what is meant by vertical? by horizontal? What basis is there for the statement sometimes made that the Mississippi River flows uphill?

16 Approximate values of g in various places are shown in the following table.

Location	Latitude, deg	Elevation, m	g, m/s^2
St. Michael, Alaska	63.5	1	9.822
Denver, Colorado	39.7	1,638	9.796
Portland, Oregon	45.5	8	9.806
Key West, Florida	24.6	1	9.790
Canal Zone	8.9	6	9.782

A single object is taken to each of these stations. Compare the mass at the various places. Compare the weights.

17 If a ball is thrown vertically upward by a person on a train which has a speed of 80 km/h, the ball will be caught by the person if his hand has not changed position. Explain why this happens.

18 What is the meaning of weightlessness as applied to objects in space? Is there any place where weight is zero? If so, where?

19 Derive an equation that expresses the total downward force on a light pulley over which a pair of masses M_1 and M_2 are suspended by a light cord.

20 Describe how a satellite which orbits the earth is constantly being accelerated toward the earth.

Problems

1 If a force of 2 N is applied to a 0.5-kg mass, what acceleration should result?

2 An unbalanced force of 50 N acts on an object weighing 100 N. What acceleration is produced? *Ans.* 4.9 m/s^2.

3 How much does an object having a mass of 10 kg weigh? How much mass does a 64-lb object have?

4 A rope is attached to a 100-lb object and is pulled upward with a force of 150 lb. What is the upward acceleration of the object? *Ans.* 16 ft/s^2.

5 An airplane in taking off from a field makes a run of 2,300 ft and leaves the ground in 15.0 s from the start. (a) What is its acceleration, assumed constant? (*b*) With what speed does it leave the ground?

6 A 10-g rifle bullet acquires a speed of 400 m/s in traversing a barrel 50 cm long. Find the average acceleration and accelerating force.
Ans. 1.6×10^7 cm/s²; 1.6×10^8 dyn.

7 An electron ($m = 9.11 \times 10^{-31}$ kg) in empty space experiences an upward electric force equal to 25 times its weight. What is its acceleration?

8 Calculate the accelerating force needed to change the speed of a 20-lb object from 18 ft/s to 50 ft/s in a distance of 40 ft. *Ans.* 17 lb.

9 A plumb bob hangs from the roof of a railway coach. What angle will the plumb line make with the vertical when the train is accelerating 2.3 m/s²?

10 A 1.50-ton automobile crashed into a wall at a speed of 10 mi/h. The car moved 5.00 in before being brought to rest. What was the average force exerted on the wall by the car?
Ans. 24,$\bar{3}$00 lb

11 What pull must a locomotive exert on a 12,000-ton train to attain a speed of 60 mi/h in 5.0 min? Assume uniform acceleration, and assume that 30 percent of the applied force is used against friction.

12 A 1,000-g block on a smooth table is connected to a 500-g piece of lead by a light cord that passes over a small pulley at the end of the table. (*a*) What is the acceleration of the system? (*b*) What is the tension in the cord?
Ans. 327 cm/s²; 3.27×10^5 dyn.

13 A rocket has a mass of 2.00×10^4 kg of which half is fuel. Assume that the fuel is consumed at a constant rate as the rocket is fired and that there is a constant thrust of 5.0×10^6 N. Neglecting air resistance and any possible variation of g, compute (*a*) the initial acceleration and (*b*) the acceleration just as the last fuel is used.

14 An object whose mass is 12.0 kg is acted upon by two forces that are in opposite directions: one of 540 N, the other of 1,260 N. What is the acceleration produced? How is the direction of the acceleration related to the forces?
Ans. 60 m/s².

15 What force, applied parallel to the plane, is necessary so that a 100-lb object will slide down the frictionless plane with an acceleration of 8.0 ft/s², if the plane makes an angle of 30° with the horizontal?

16 An object of mass 8.00 kg is pulled up an inclined plane, making an angle of 30° with the horizontal, by a cord which passes over a pulley at the top of the plane and is fastened to a 10.0-kg mass. Neglecting friction, find the acceleration and the tension in the string.
Ans. 327 cm/s²; 6.52×10^6 dyn.

17 A 2.00-lb hammer traveling at a speed of 15.0 ft/s strikes a nail and drives it 0.75 in into a block of wood. Assume that the resisting force in the wood is constant. Find (*a*) the acceleration of the hammer and (*b*) the constant force.

18 An elevator and its load weigh 1,600 lb. Find the tension in the supporting cable when the elevator, originally moving downward at 20 ft/s, is brought to rest with constant acceleration in a distance of 50 ft. *Ans.* 1,800 lb.

19 The reaction time of the average motorist is 0.70 s. Assume that by means of the brakes, a retarding force equal to three-fourths the weight of the car can be applied to the car. If the car is traveling 60 mi/h, find (*a*) the acceleration when the brakes are applied and (*b*) the distance the car travels after the motorist receives the stopping signal.

20 A 100-lb box slides down a frictionless skid inclined at an angle of 60° with the horizontal. Find (*a*) the accelerating force, (*b*) the time required to travel the first 20 ft, and (*c*) the time required to travel the next 20 ft.
Ans. 87 lb; 1.2 s; 0.49 s.

21 What force, applied parallel to the plane, is necessary to move a 16.0-kg object up a frictionless plane with a uniform acceleration of 2.00 m/s², if the plane makes an angle of 60° with the horizontal?

22 A 200-lb man stands in an elevator. What force does the floor exert on him when the elevator is (*a*) stationary; (*b*) accelerating upward 16.0 ft/s²; (*c*) moving upward at constant speed; and (*d*) moving upward but decelerating at 12.0 ft/s²? *Ans.* 200 lb; 300 lb; 200 lb; 125 lb.

23 One side of a double-inclined plane somewhat like that in Fig. 3-13 makes an angle of 30°

with the horizontal; the other makes an angle of 60°, so that there is a 90° angle at the top. A 6.0-kg mass and a 2.0-kg mass are attached to the ends of a string which passes over a pulley at the top of the smooth double plane with the 2.0-kg mass on the steeper side. Find the acceleration and the tension in the cord when the system is released.

24 In Fig. 3-13 the blocks A and B of masses $m_A = 2m_B$ are on frictionless planes. Find the magnitude and direction of the acceleration of each block.

Ans. $a_A = 0.54\ \text{m/s}^2$ down; $a_B = 0.54\ \text{m/s}^2$ up.

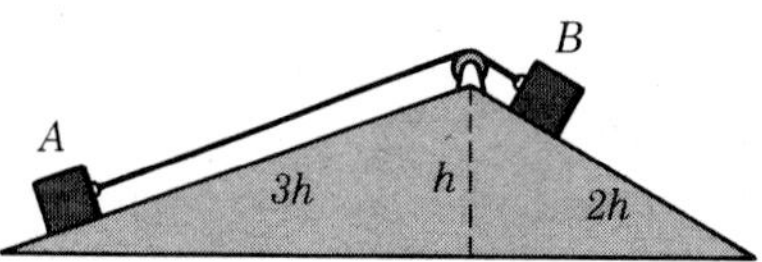

Figure 3-13

25 An elevator, which weighs 8.0 tons with its load, is descending with a speed of 900 ft/min. If the load on the cables must not exceed 14 tons, what is the shortest distance in which the elevator can be stopped?

26 A spring balance fastened to the roof of a moving elevator car indicates 90 lb as the weight of a 120-lb body. (*a*) What is the magnitude and direction of the acceleration of the elevator? (*b*) Can one determine from these data the direction in which the elevator is moving?

Ans. 8.0 ft/s^2 down; no.

27 In the system shown in Fig. 3-14 find the acceleration of each of the bodies at the instant the system is in the configuration pictured. Will the acceleration remain the same as the motion progresses? If not, what will be the manner in which it changes?

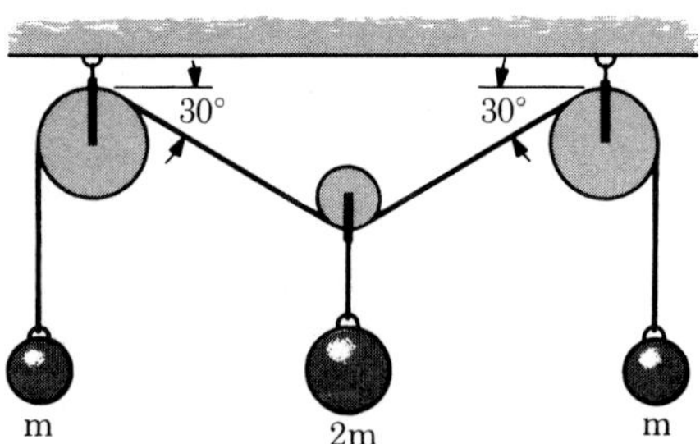

Figure 3-14

28 A light frictionless pulley carries a light cord to which is attached at one end a 48-lb weight and at the other a 64-lb weight. The weights are suddenly released. Find the acceleration and the tension in the cord. *Ans.* 4.6 ft/s^2; 55 lb.

29 In the arrangement of Fig. 3-15 express the acceleration of each block in terms of the three masses and g. Find the acceleration of each block if $m_2 = 3m_1 = 2m_3$. Assume the plane to be frictionless and neglect the mass of the pulleys.

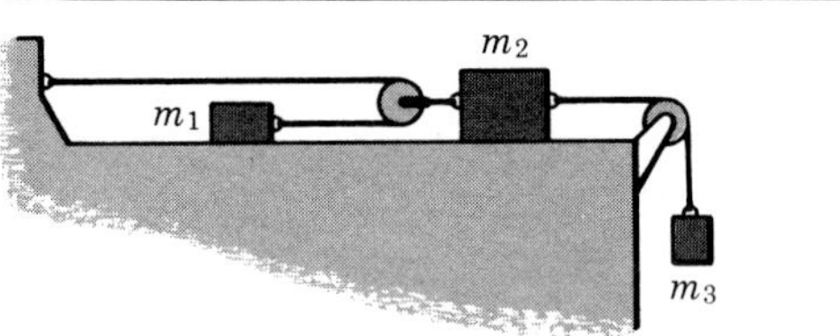

Figure 3-15

30 Two objects of mass 500 g each are fastened together by a cord and suspended over a frictionless pulley at the top of a double-inclined plane. One side of the plane makes an angle of 45° with the horizontal, and the other side makes an angle of 30°, so that there is an angle of 105° at the top of the plane where the pulley is attached. Calculate the acceleration of the system and the force in the cord.

Ans. 102 cm/s^2; 2.96×10^5 dyn.

31 An electron has a rest mass of 9.11×10^{-31} kg. Find its mass when its speed is $0.50c$, $0.90c$, and $0.99c$.

32 What average force is necessary to accelerate a proton, rest mass 1.67×10^{-27} kg from rest to one-tenth the speed of light ($c = 3.0 \times 10^8$ m/s) in a distance of 3.0 cm? What force would be required to produce this same acceleration if the initial speed were 0.50*c*?

Ans. 2.5×10^{-11} N; 2.9×10^{-11} N.

33 The mass of the earth is approximately 5.98×10^{24} kg and that of the moon is 0.0123 times as great. The mean distance between them is 3.84×10^5 km. (*a*) Compute the gravitational force of attraction for each other. (*b*) Find the acceleration of each that is produced by this force.

34 What is the acceleration due to gravity on the surface of the moon if the mass of the moon is 0.0127 that of the earth and the radius of the moon is 0.25 that of the earth? *Ans.* 6.5 ft/s^2.

Albert Abraham Michelson, 1852–1931

Born in Streino, Prussia. Director of the physics department, University of Chicago. The 1907 Nobel Prize for Physics was conferred on him for his optical instruments of precision and the spectroscopic and meteorologic investigations which he carried out by means of them.

4

Work, Energy, and Power

A significant difference between our civilization and that of the ancients is our extensive utilization of energy from sources other than the muscles of men and animals. Many of the early advances in physics were made by men who were trying to understand and control sources of energy and apply them to men's tasks. As the study of physics has advanced, energy has continued to be a principal concern, playing such a crucial role that physics has been called the "science of energy and its transformations."

4-1 WORK

The term work, commonly used in connection with widely different activities, is restricted in physics to cases in which there is a force and a displacement along the line of the force. In this technical sense of the word work, a pier does no work in supporting a bridge, and a man does no work if he merely holds up a suitcase, though he may experience muscular fatigue. But in lifting the suitcase to a rack he would perform work. When a force **F** moves through a displacement **s** and the directions of these two vectors are not the same, the work $\mathcal{W}$ is defined as the product of the magnitude of the average force $\bar{\mathbf{F}}$ and the displacement $s \cos \theta$ in the direction of the force. (See Fig. 4-1.)

$$\mathcal{W} = (\bar{\mathbf{F}} \cos \theta)(\mathbf{s}) \qquad (1)$$

where θ is the angle between the direction of the force and that of the displacement, and the force $\bar{\mathbf{F}}$ is averaged over the displacement. In the special case where the force is constant and has the same direction as the displacement, $\theta = 0°$, $\cos 0° = 1$, and the work done by the force is the product of the constant force and the distance.

Although work is the product of two vector quantities, force and displacement, it is itself a *scalar* quantity. When two vector quantities are multiplied so as to produce a scalar quantity, this process is called *scalar multiplication* of vectors.

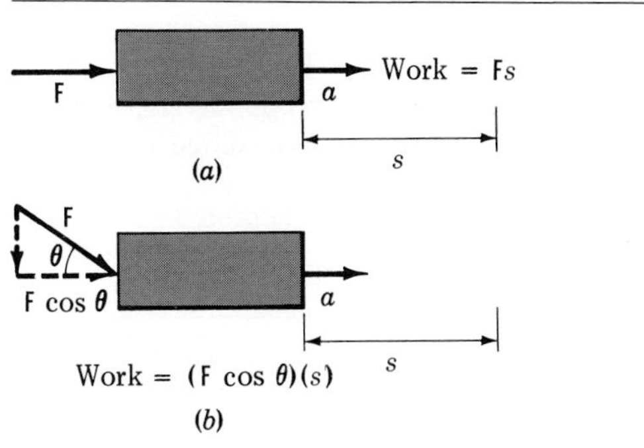

Figure 4-1
Work done on a block.

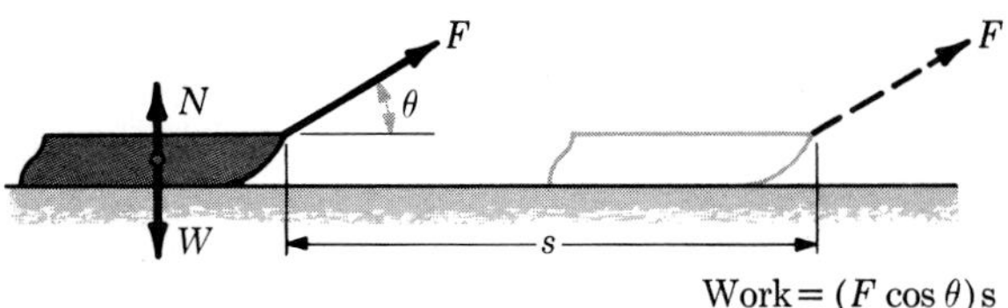

Figure 4-2
Work done in dragging a sled.

This procedure is usually denoted by the use of a dot between the two vectors to indicate scalar multiplication; thus

$$\mathcal{W} = \mathbf{F} \cdot \mathbf{s} \tag{2}$$

The magnitude of this scalar product is $Fs \cos \theta$, as appears in Fig. 4-2. Another kind of product of two vector quantities is the vector product described in Chap. 5.

4-2 UNITS

In the mks system the unit of work is the *newton-meter,* the work done by a force of one newton exerted through a distance of one meter when the force is parallel to the displacement. The newton-meter is called a *joule,* after the British physicist James Prescott Joule (1818–1889), whose experiments contributed heavily to the acceptance of the relationship between heat and work. In the cgs system the unit of work is the *erg,* which is the work done by one dyne exerted over a distance of one centimeter. An erg is 10^{-7} J. In the British system the unit of work is the *foot-pound,* the work done by a force of one pound acting over a displacement of one foot in the direction of the force. Another unit of work, used especially in electrical measurements, is the *kilowatthour* (Sec. 4-16).

Example A box is pushed without acceleration 5.0 m along a horizontal floor against a frictional force of 180 N. How much work is done?

$$\mathcal{W} = Fs = (180\text{ N})(5.0\text{ m})$$
$$= 900\text{ N}\cdot\text{m} = 900\text{ J}$$

Example What work is performed in dragging a sled 50 ft horizontally without acceleration when the force of 60 lb is transmitted by a rope making an angle of 30° with the ground (Fig. 4-2)?

The component of the force in the direction of the displacement is $F \cos 30°$.

$$\mathcal{W} = (F \cos 30°)s$$
$$= 60\text{ lb} \times 0.866 \times 50\text{ ft} = 2.6 \times 10^3\text{ ft}\cdot\text{lb}$$

4-3 ENERGY: THE ABILITY TO DO WORK

That property of a body or system of bodies by virtue of which work can be performed is called *energy* (a scalar quantity). Energy can exist in many forms and can be transformed from one

Table 1
UNITS TO DESCRIBE WORK

System	Work	=	Force	×	Displacement	
mks	joule (J)	=	newton		meter	(N·m)
cgs (abs)	erg	=	dyn	×	cm	(dyn·cm)
British (grav)	foot-pound	=	pound	×	ft	(ft·lb)

form to another. The energy possessed by an object by virtue of its motion is called *kinetic energy,* or energy of motion. Energy of position, or configuration, is called *potential energy*. When work is done on a body in the absence of frictional forces, the work done is equal to the sum of the increase in kinetic energy and the increase in potential energy. The units in which energy is expressed are the same as the units for work.

Many problems in mechanics can be solved by the laws of motion discussed in Chap. 3. Given certain information about the initial status of an object and the forces to which it is subjected, we can predict its position and velocity at any future time. In some situations, however, as in the description of the motion of a pendulum, direct application of the laws of motion would require complicated calculations of forces and accelerations in order to obtain a relatively simple result. Consideration of the potential and kinetic energy involved and the relation between work and energy simplifies the solution of many problems in mechanics. Moreover, the concept of energy leads to the principle of the conservation of energy, which unifies a wide range of phenomena in the physical sciences.

4-4 POTENTIAL ENERGY

The energy which bodies possess by virtue of their positions, configurations, or internal mechanisms is called *potential energy* E_p. Important forms of this type of energy are electrical, elastic, chemical, and nuclear potential energy. The most common form of potential energy is *gravitational* potential energy. Since the earth attracts every body, work is required to lift the body to a higher level. When a brick is carried to the top of a building, the work done on the brick (weight of brick times vertical distance) represents energy that can be recovered. By virtue of its position at the top of the building the brick possesses more ability to do work than it had when it was at ground level. It has increased its potential energy. The work done on the brick, and hence the potential energy gained, is the product of the weight W and the height h to which it is raised. This increase in potential energy is given by

$$E_p = Wh = mgh \tag{3}$$

If W is in newtons, m is in kilograms, and h is in meters, E_p is given in joules. If W is in pounds, m is in slugs, and h is in feet, E_p is given in foot-pounds. If W is in dynes, m is in grams, and h is in centimeters, E_p is given in ergs.

The gravitational potential energy is expressed relative to a specified arbitrary reference level. This reference level may be any point that is agreed upon by those concerned. For example, the arbitrary reference for zero gravitational potential energy may be chosen as sea level, or floor level, or "at infinity," a point so far from the earth as to be effectively outside the earth's gravitational field.

Example A 40-lb stone is hoisted to the top of a building 100 ft high. How much does its potential energy increase?

Friction being neglected, the increase in potential energy is just the amount of work done in lifting the stone, so that

$$E_p = Fs = (40\text{ lb})(100\text{ ft}) = 4.0 \times 10^3\text{ ft}\cdot\text{lb}$$

Example A 40-lb stone is carried up a ramp, along a path making a 30° angle to the horizontal, to the top of a building 100 ft high. How much work is done? (Neglect friction.)

The distance traversed is now 200 ft, as shown in Fig. 4-3*a*. The force exerted on the stone is equal in magnitude to its weight. The angle between the force and the displacement is 60°.

$$\begin{aligned}\mathcal{W} &= Fs\cos\theta = (40\text{ lb})(200\text{ ft})(0.500)\\ &= 4.0 \times 10^3\text{ ft}\cdot\text{lb}\end{aligned}$$

Notice that the work is the same as for the previous example (again neglecting friction). If the stone is rolled up on rollers by a force parallel

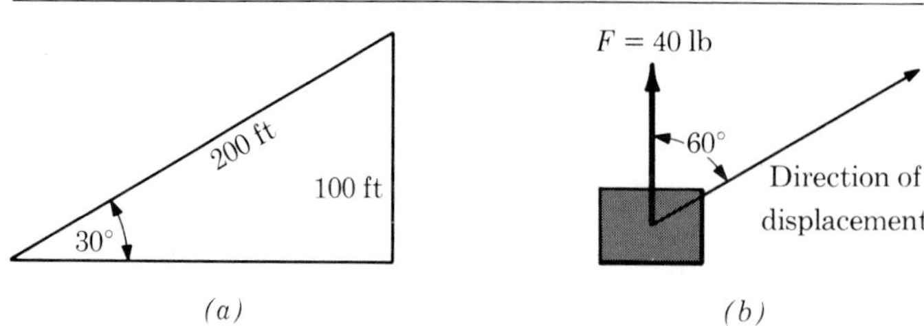

Figure 4-3
Work done when force is not in the direction of the displacement.

to the ramp, angle θ would be zero but F would be only 20 lb.

$$\mathcal{W} = Fs\cos\theta = (20\text{ lb})(200\text{ ft})(1)$$
$$= 4.0 \times 10^3\text{ ft}\cdot\text{lb}$$

In these cases a momentary force somewhat greater than that discussed is needed to set the body in motion. The work thus done on the body at the beginning is recovered when the body comes to rest.

In all three modes of (frictionless) transport, the work done equals the gain of E_p given by Eq. (3). In Eq. (3) we have assumed that when we elevate an object a distance h which is small compared with the radius of the earth, the gravitational force acting on that object remains constant. For any system in which the force is not constant the gain in potential energy is the product of the *average* force and the distance moved in its direction. If an object such as a rocket is lifted to a height of several times the earth's radius, the gravitational force is far from constant, since this force varies inversely as the square of the distance of the object from the center of the earth. This is shown in Fig. 4-4 for a body which weighs 1,000 lb at the surface of the earth. If such a body is lifted to a high altitude, the work can be obtained by considering the operation as a large number of bits of work, each performed with a different average force. The area of each rectangle in Fig. 4-5 represents the product of a *force* and a *distance*, hence a quantity of work. The total work done, and hence the total potential energy acquired by the body, is represented approximately by the total area of the rectangles and accurately by the area under the curve.

It is shown in Chap. 9 that the potential energy of a body at high altitude with respect to the surface of the earth is given by

$$E_p = GMm\left(\frac{1}{R} - \frac{1}{r}\right) \tag{4}$$

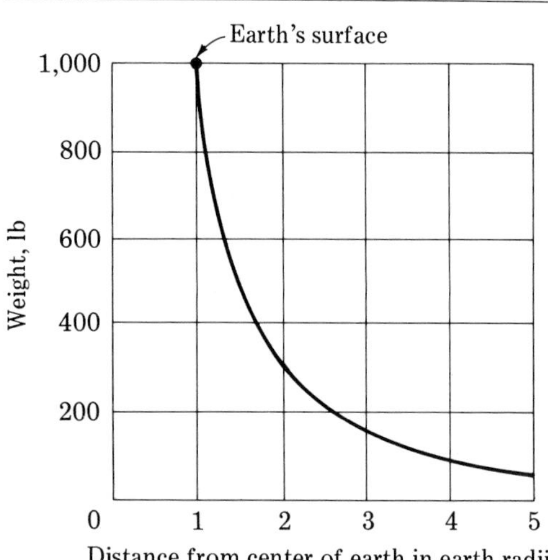

Figure 4-4
Variation of weight with distance from the center of the earth for a body of mass 31.1 slugs.

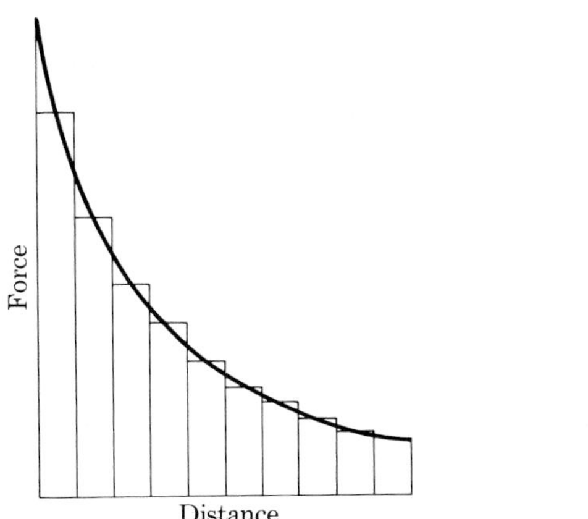

Figure 4-5
Work done by a varying force.

where G is the constant of universal gravitation, M is the mass of the earth, m is the mass of the body, R is the radius of the earth, and r is the distance of the body from the center of the earth (note that r is not the altitude above the surface of the earth).

Example A 200-kg satellite is lifted to an orbit of 2.20×10^4 mi radius. How much additional potential energy does it acquire relative to the surface of the earth?

The solution is found from Eq. (4), where

$$R = 6.37 \times 10^6 \text{ m}$$

$$r = 2.20 \times 10^4 \text{ mi} = 3.54 \times 10^7 \text{ m}$$

$$M = 5.98 \times 10^{24} \text{ kg}$$

$$m = 200 \text{ kg}$$

$$E_p = (6.67 \times 10^{-11} \text{ m}^3/\text{kg-s}^2) \times (5.98 \times 10^{24} \text{ kg})(200 \text{ kg}) \times \left(\frac{1}{6.37 \times 10^6 \text{ m}} - \frac{1}{3.54 \times 10^7 \text{ m}}\right)$$

$$= 1.03 \times 10^{10} \text{ J}$$

This is about equal to the work needed to lift an object weighing 3,800 tons to a height of 1,000 ft above the earth.

4-5 KINETIC ENERGY

In addition to energy of position or state, objects may possess energy due to their motions. A car or bullet in motion, a stream of water, or a revolving flywheel possesses kinetic energy. The kinetic energy of a moving object can be measured by the amount of work it will do if brought to rest or by the amount of work originally needed to impart the velocity to it, in circumstances where the work cannot also go into potential energy.

Consider a body with an initial speed v_0 on which a steady unbalanced force F acts as it moves a distance s. The body gains speed at a rate given by $a = F/m$ until it reaches a final speed v_1. The work done on the body by the unbalanced force that accelerated it appears as a change in its kinetic energy. Since $F = ma$, multiplying by s gives $Fs = mas$ and

$$\Delta(E_k) = Fs = mas \tag{5}$$

From Eq. (5), Chap. 2,

$$2as = v_1^2 - v_0^2$$

or

$$as = \tfrac{1}{2}(v_1^2 - v_0^2)$$

$$\Delta(E_k) = \tfrac{1}{2} m\,(v_1^2 - v_0^2) = \tfrac{1}{2} mv_1^2 - \tfrac{1}{2} mv_0^2$$

If the body was initially at rest, $v_0 = 0$ and the gain in kinetic energy is the final kinetic energy. Thus the kinetic energy of a body at any instant is

$$E_k = \tfrac{1}{2} mv^2 \tag{6}$$

If m is in kilograms and v in meters per second, Eq. (6) gives the kinetic energy in joules (newton-meters). If m is expressed in slugs and v in feet per second, Eq. (6) gives the kinetic energy in foot-pounds. If m is in grams and v in centimeters per second, Eq. (6) gives the kinetic energy in ergs.

Although a steady force has been assumed here, the result is independent of the particular manner in which a body attains its velocity.

Example What is the kinetic energy of a 3,000-lb automobile which is moving at 30 mi/h (44 ft/s)?

$$m = \frac{W}{g} = \frac{3{,}000 \text{ lb}}{32 \text{ ft/s}^2} = 94 \text{ slugs}$$

$$E_k = \tfrac{1}{2}mv^2 = \tfrac{1}{2} \times 94 \text{ slugs} \times (44 \text{ ft/s})^2$$
$$= 9.1 \times 10^4 \text{ ft·lb}$$

When an accelerating force is applied to a body, the work done by that force produces a change in the kinetic energy of the body. If a resultant force F acts to start a body in motion or to stop one initially in motion,

$$Fs = \tfrac{1}{2}mv^2 \qquad (7)$$

Example What average force is necessary to stop a bullet of mass 20 g and speed 250 m/s as it penetrates wood to a distance of 12 cm?

The work done by the retarding force is equal to the initial kinetic energy of the bullet

$$Fs = \tfrac{1}{2}mv^2$$

$$F \times 0.12 \text{ m} = \tfrac{1}{2}(0.020 \text{ kg})(250 \text{ m/s})^2$$

$$F = 5.2 \times 10^3 \text{ N}$$

This force is nearly 30,000 times the weight of the bullet.

The initial kinetic energy, $\frac{1}{2}mv^2 = 6\overline{2}0$ J, is largely wasted in heat and in work done in deforming the bullet.

4-6 FRICTION

Whenever an object moves while in contact with another object, frictional forces oppose the relative motion. These forces are caused by the adhesion of one surface to the other and by the interlocking of the irregularities of the rubbing surfaces. The force of frictional resistance depends upon the properties of the surfaces and upon the force keeping the surfaces in contact.

The effects of friction are often undesirable. Friction increases the work necessary to operate machinery, it causes wear, and it generates heat, which often does additional damage. To reduce this waste of energy, friction is minimized by the use of wheels, bearings, rollers, and lubricants. Automobiles and airplanes are streamlined in order to decrease air friction, which is large at high speeds.

On the other hand, friction is desirable in many cases. Nails and screws hold boards together by means of friction. Power may be transmitted from a motor to a machine by means of a clutch or a friction belt. In walking, driving a car, striking a match, tying shoes, or sewing fabric together we find friction a useful force. Sand is placed on rails in front of the drive wheels of locomotives, cinders are scattered on icy streets, chains are attached to the wheels of automobiles, and special materials are developed for use in brakes—all for the purpose of increasing friction where it is desirable. Frictional forces are important in determining the path of a space vehicle and the heating produced when the vehicle reenters the earth's atmosphere.

4-7 SLIDING FRICTION

When we slide a box across a floor, we find that we must continue to apply a steady horizontal force to cause the box to slide uniformly over the horizontal surface. We conclude that there is a force, parallel to the surfaces in contact, opposing the motion. This opposing force is called *friction*. If the applied force is just equal to the frictional force, the body will continue to move uniformly; if the applied force is greater than the frictional force, the body will be accelerated.

The friction between solids sliding over one another is due to several causes acting at once. If the surfaces are very rough, an interlocking process gives rise to large forces; but as one or both of the surfaces are made progressively smoother, it is found that the friction diminishes at first, then becomes fairly constant. Thus the friction between polished steel blocks remains appreciable and even increases with further smoothing. For the surfaces of the same metal in contact it is found that microscopic welds are formed and broken; and even for dissimilar metals strong adhesive bonds are formed. The presence or absence of oxide coatings and water films is significant. All these processes cause vibrations which set up waves through the materials and produce heating. Despite the complexity and variety of the processes involved for various materials, certain simple observations have been found to apply to nearly all cases of sliding friction to within a few percent accuracy.

The observations regarding sliding friction are these:

1 The frictional force is parallel to the surfaces sliding over one another.
2 The frictional force is proportional to the force which is normal (perpendicular) to the surfaces and which presses them together.
3 The frictional force is roughly independent of the area of the surface of contact.

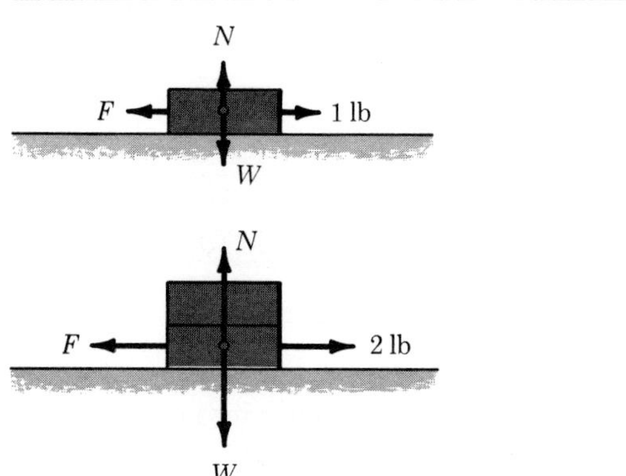

Figure 4-6
The frictional force is directly proportional to the normal force pressing the two surfaces together.

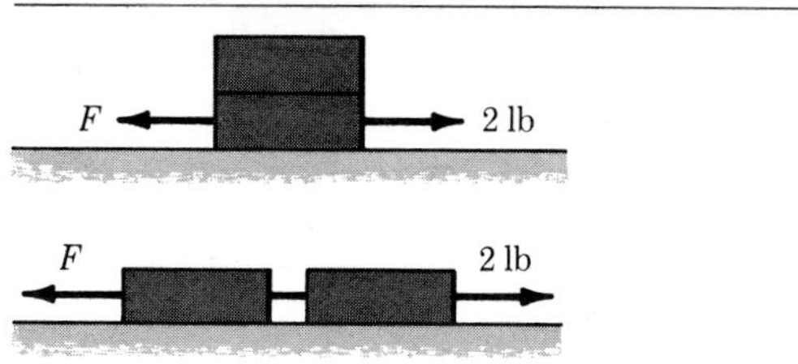

Figure 4-7
The frictional force is independent of the area of the surface of contact.

4 The frictional force is roughly independent of the speed of sliding provided that the resulting heat does not alter the condition of the surfaces.
5 The frictional force depends upon the nature of the substances in contact and the condition of the surfaces (i.e., on polish, roughness, grain, and wetness).

Sliding friction is sometimes called *kinetic* friction. These observations may be illustrated by simple experiments. By the use of a spring balance to pull a brick uniformly on a tabletop (Fig. 4-6), the frictional force may be found to be 1 lb, whereas a force of 2 lb is required to maintain the motion if a second brick is placed on top of the first to increase the normal force. If the second brick is tied behind the first (Fig. 4-7), the frictional force is still 2 lb, showing its independence of area of contact. If the bricks are placed on glass or metal surfaces, one finds that friction depends upon the nature of the surfaces in contact.

4-8 COEFFICIENT OF KINETIC FRICTION

When one body is in uniform motion on another body, the ratio of the frictional force to the perpendicular force pressing the two surfaces together is called the *coefficient of kinetic friction.*

$$\mu_k = \frac{F}{N} \tag{8}$$

Here μ (mu) is the coefficient of friction, F the frictional force, and N the normal or perpendicular force.

Example A 65-lb horizontal force is sufficient to draw a 1,200-lb sled on level, well-packed snow at uniform speed. What is the value of the coefficient of friction?

$$\mu_k = \frac{F}{N} = \frac{65 \text{ lb}}{1{,}200 \text{ lb}} = 0.054$$

The frictional force is proportional to the normal force, which must include all normal components of forces pressing the surfaces together. Only in very special cases is the normal force the weight of the body.

Example A 1,200-lb sled is pulled along a horizontal surface at uniform speed by means of a rope that makes an angle of 30° above the horizontal (Fig. 4-8). If the tension in the rope is 100 lb, what is the coefficient of friction?

The frictional force (parallel to the surface) is

$$F = 100 \text{ lb} \times \cos 30° = 100 \text{ lb} \times 0.866 = 86.6 \text{ lb}$$

The normal force N is the weight of the sled downward minus the vertical component of the tension upward.

$$N = 1{,}200 \text{ lb} - 100 \text{ lb} \times \sin 30° = 1{,}150 \text{ lb}$$

$$\mu_k = \frac{F}{N} = \frac{86.6 \text{ lb}}{1{,}150 \text{ lb}} = 0.0753$$

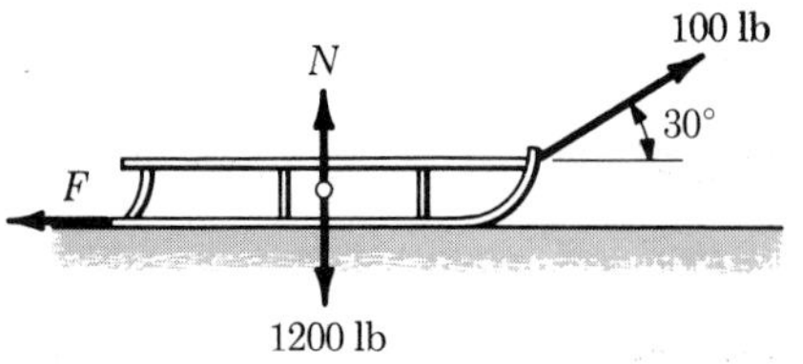

Figure 4-8
Forces acting on a sled.

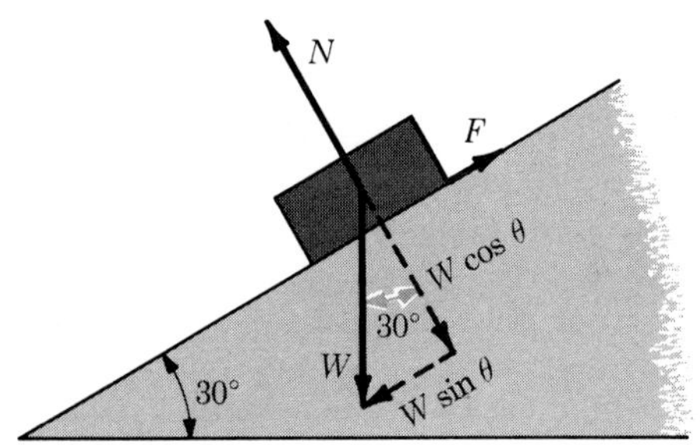

Figure 4-9
Forces acting on a body on an inclined plane.

Example A 50-kg box is placed on an inclined plane making an angle of 30° with the horizontal (Fig. 4-9). If the coefficient of kinetic friction is 0.30, find the resultant force on the box.

The weight of the box is a force acting vertically downward. This force may be separated into components parallel (W sin 30°) and perpendicular (W cos 30°) to the plane. A frictional force (parallel to the plane) acts up the plane.

$$\begin{aligned} N &= W \cos 30° = 50 \text{ kg} \times 9.8 \text{ m/s}^2 \times 0.87 \\ &= 4\bar{3}0 \text{ N} \end{aligned}$$

The frictional force is given by

$$F = \mu_k N = 0.30 \times 4\bar{3}0 \text{ N} = 1\bar{3}0 \text{ N}$$

The resultant force is

$$\begin{aligned} R &= W \sin 30° - F \\ &= 50 \text{ kg} \times 9.8 \text{ m/s}^2 \times 0.50 - 1\bar{3}0 \text{ N} \\ &= 1\bar{2}0 \text{ N down the plane} \end{aligned}$$

When two surfaces are lubricated, friction is reduced by the substitution of the internal friction of the lubricant for the friction between the original surfaces. The ratio F/N is then not a simple constant but depends upon the properties of the lubricant and the area and relative speed of the moving surfaces.

Table 1
COEFFICIENTS OF FRICTION

Material	μ_s	μ_k
Steel on steel	0.15	0.09
Metal on metal, lubricated	0.03	0.03
Leather on oak	0.4	0.3
Rubber tire on dry concrete road	1.0	0.7
Rubber tire on wet concrete road	0.7	0.5

4-9 STATIC FRICTION

When a body at rest on a horizontal surface is pushed gently sideward, it does not move, because there is a frictional force just equal to the sideward push. If the push is increased, the frictional force increases until a *limiting friction* is reached. If the side push exceeds the limiting friction, the body is accelerated. When there is no relative motion between the two surfaces in contact, the friction is called *static friction* and the frictional force can have any value from zero up to the limiting value.

For limiting friction (but not for all static friction) the same laws apply as for sliding friction except, of course, that referring to the speed of sliding. The *coefficient of static (limiting) friction* is the ratio of the limiting frictional force to the normal force.

$$\mu_s = \frac{F_{max}}{N} \tag{9}$$

For any two surfaces the coefficient of static friction μ_s is somewhat greater than the coefficient of sliding or kinetic friction μ_k.

4-10 LIMITING ANGLE (ANGLE OF REPOSE)

The coefficient of static friction may be found without measurement of forces by the following simple and convenient method. Let the plane *AC*

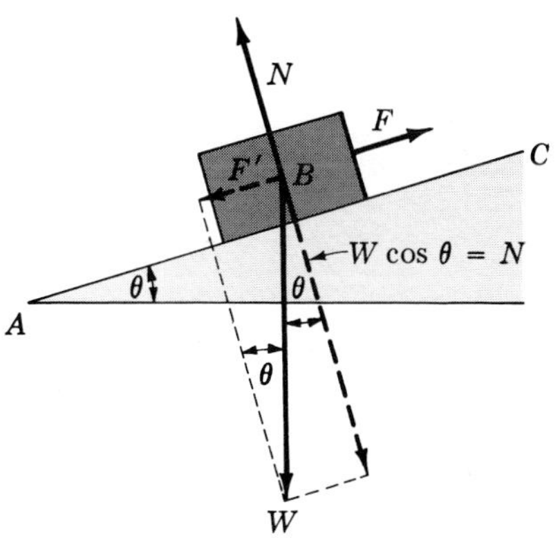

Figure 4-10
Limiting angle for friction.

(Fig. 4-10) be tilted upward (gradually) until the body *B* just begins to slide down the incline. Consider the weight W of the body to be resolved into components $W \cos \theta$ and $F' = W \sin \theta$, respectively perpendicular and parallel to the incline. The component $W \cos \theta = N$ presses the two surfaces together, the component F' is directed down the plane, and the frictional force F is directed up the plane. Just before sliding occurs, there is no acceleration, and F' is balanced by the friction F. From Fig. 4-10,

$$\frac{F'}{N} = \frac{F}{N} = \frac{W \sin \theta}{W \cos \theta} = \tan \theta$$

and hence the coefficient of friction is given by

$$\mu_s = \tan \theta$$

4-11 ROLLING FRICTION

Rolling friction is the resistance to motion caused chiefly by the deformation produced where a wheel or cylinder pushes against the surface on which it rolls. The deformation of an automobile tire in contact with the pavement is readily visible. Even in the case of a steel wheel rolling on

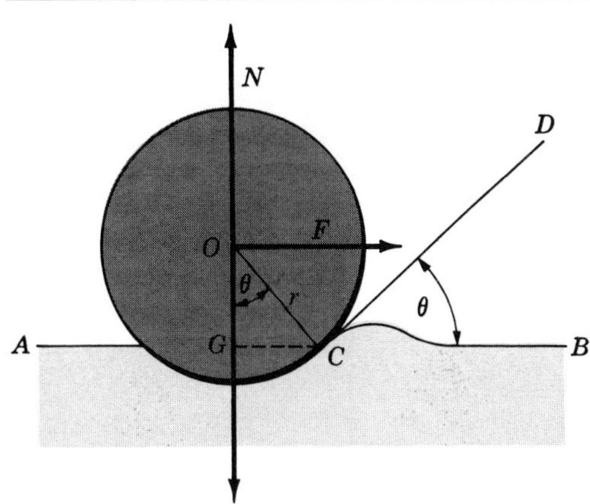

Figure 4-11
Rolling friction.

a steel rail there is some deformation of the two surfaces. A force F (Fig. 4-11) is required to roll the wheel on the horizontal rail because the surfaces are continually distorted as illustrated by the hill along line CD. The deformations of the two surfaces produce internal friction in the two bodies. This topic will be explored again when we study the elastic properties of solids. The force of rolling friction varies inversely as the radius of the roller, and it is less the more rigid the surfaces.

Rolling friction is ordinarily much smaller than sliding friction. Sliding friction on the axle of a wheel is replaced by rolling friction through the use of roller or ball bearings.

4-12 FLUID FRICTION

The friction encountered by solid objects in passing through fluids and the frictional forces set up within liquids and gases in motion are examples of fluid friction. The laws of fluid friction differ greatly from those of sliding and rolling friction. The amount of frictional resistance encountered by an object moving through a fluid depends on the size, shape, and speed of the moving object, as well as on the nature of the fluid itself. We have already seen fluid friction in action when we discussed the terminal speed of a body falling through air (a fluid).

4-13 STOPPING DISTANCE

The fact that the kinetic energy of a moving object is proportional to the square of its speed has an important bearing upon the problem of stopping an automobile. Doubling the speed of the car quadruples the amount of work that must be done by the brakes in making a quick stop.

A consideration of the equation

$$v_1^2 - v_0^2 = 2as$$

shows that, if $v_1 = 0$ (indicating a stop), $s = -v_0^2/2a$, so that the distance in which an automobile can be stopped is likewise proportional to the square of the speed, on the assumption of a constant negative acceleration. Actually, however, the deceleration accomplished by the brakes is smaller at high speed because of the effect of heat upon the brake linings, so that the increase in stopping distance with speed is even more rapid than is indicated by theoretical considerations.

Example In what distance can a 3,000-lb automobile be stopped from a speed of 30 mi/h (44 ft/s) if the coefficient of friction between tires and roadway is 0.70?

The retarding force furnished by the roadway can be no greater than

$$Fs = \mu N = (0.70)(3{,}000 \text{ lb}) = 2{,}\bar{1}00 \text{ lb}$$

Since the work done by this force is equal to the kinetic energy of the car, the stopping distance can be found by substituting in Eq. (7).

$$Fs = \tfrac{1}{2} mv^2$$

$$m = \frac{W}{g} = \frac{3{,}000 \text{ lb}}{32 \text{ ft/s}^2} = 94 \text{ slugs}$$

$$s = \frac{\frac{1}{2}mv^2}{F} = \frac{94 \text{ slugs } (44 \text{ ft/s})^2}{2 \times 2{,}\bar{1}00 \text{ lb}} = 43 \text{ ft}$$

4-14 CONSERVATION OF ENERGY

Energy is given to a body or system of bodies when work is done upon it. In this process there is merely a transfer of energy from one body to another. *In such transfer no energy is created or destroyed; it merely changes from one form to another.* This statement is known as the *law of conservation of energy.* It is true that in most processes some of the energy becomes unavailable. Work done against friction is converted into heat energy in such a form that it can seldom be used. Thus, although the energy is not destroyed, it is wasted as far as its usefulness in the process is concerned.

It was mentioned in Chap. 3 that if a body is accelerated to a speed approaching the speed of light its mass increases appreciably. As a consequence of this fact Eq. (6) for kinetic energy ceases to be accurate for such high-speed bodies, since this equation was derived on the assumption that the mass remains constant. Taking the mass variation into account consists not merely in substituting the relativistic mass in Eq. (6) but in deriving the equation from the beginning with mass variable. This yields for the kinetic energy of a high-speed particle

$$E_k = mc^2 - m_0c^2 \qquad (10)$$

where m is the mass at the high speed, m_0 is the mass when the body is at rest, and c is the speed of light. The quantity m_0c^2 is called the *rest energy* of the body, and the quantity mc^2 is called its *total energy*. This equation may be written $E_k = (m - m_0)c^2 = \Delta mc^2$ so as to associate an increase of kinetic energy with an increase of mass Δm. Einstein's prediction of the equivalence of mass and energy is readily verifiable in nuclear reactions (Chap. 49). In certain reactions a particle may receive energy of the order of 10^{-12} J at the expense of a decrease in the mass of the reactants which is measurable with a mass spectrometer. In ordinary chemical reactions the energy released per molecule is so much smaller ($\approx 10^{-18}$ J) that the corresponding changes in mass are not apparent. All other forms of energy may be similarly associated with mass changes, which are in most cases unmeasurably small.

4-15 TRANSFORMATIONS OF KINETIC AND POTENTIAL ENERGY

Very frequently in mechanical systems at low speeds there is an interchange of kinetic and potential energies. If a ball is held at the top of a building, it possesses potential energy. When it is released and falls, the kinetic energy increases as the potential energy decreases. The sum of E_k and E_p remains constant and equal to the potential energy at the top if no energy is lost against air resistance.

Example A 3,000-lb automobile at rest at the top of an incline 30 ft high and 300 ft long is released and rolls down the hill. What is its speed at the bottom of the incline if the average retarding force due to friction is 200 lb? (Fig. 4-12.)

The potential energy at the top of the hill is available to do work against the retarding force F and to supply kinetic energy.

$$Wh = Fs + \tfrac{1}{2}mv^2$$

$$m = \frac{W}{g} = \frac{3{,}000 \text{ lb}}{32 \text{ ft/s}^2} = 94 \text{ slugs}$$

$$3{,}000 \text{ lb} \times 30 \text{ ft} = 200 \text{ lb} \times 300 \text{ ft} + \tfrac{1}{2} \times 94 \text{ slugs} \times v^2$$

$$9.0 \times 10^4 \text{ ft}\cdot\text{lb} - 6.0 \times 10^4 \text{ ft}\cdot\text{lb} = \tfrac{1}{2} \times 94 \text{ slugs} \times v^2$$

$$v^2 = \frac{3.0 \times 10^4 \text{ ft}\cdot\text{lb}}{47 \text{ slugs}} = 6\bar{4}0 \text{ ft}^2/\text{s}^2$$

$$v = 25 \text{ ft/s}$$

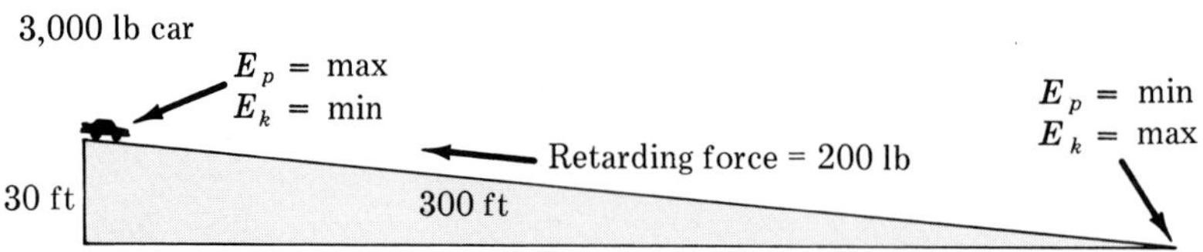

Figure 4-12
Figure for example in Sec. 4-15.

The motion of a pendulum furnishes another simple example of energy transformations. A small ball (Fig. 4-13) of mass m is suspended from a fixed point P by a string of length l. When the ball is pulled aside from O to position B, it is raised a distance h and hence given potential energy mgh. When the ball is released, it moves toward its lowest point and its energy while remaining constant changes from potential to kinetic, the sum of the two forms always being equal to mgh. At point O all the energy will be kinetic. The ball will have a speed v obtained from

$$mgh = \tfrac{1}{2}mv^2 \qquad (11)$$

or

$$v = \sqrt{2gh} \qquad (12)$$

This is the *speed* it would have acquired if it had fallen freely through a vertical distance h. However, the *velocity* is directed toward the left at O. Under the constraint of the string the ball will continue to move along the arc BOA, gaining potential energy at the expense of its kinetic energy as it approaches A. If no energy is lost to its surroundings, the ball will reach point A at a height h above its lowest position. It will then retrace its path AOB, and the motion will be repeated.

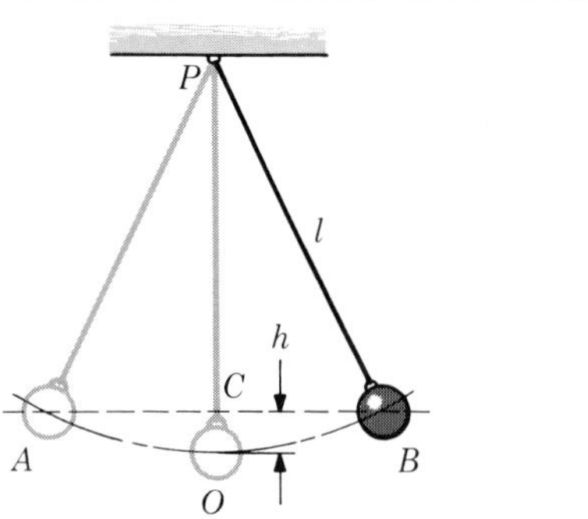

Figure 4-13
Pendulum; an example of energy transformations.

Example The bob of a pendulum has its rest point 1.00 m below the support. The bob is pulled aside until the string makes an angle of 15° with the vertical. Upon release, with what speed does the bob swing past its rest point?

From the geometry of Fig. 4-13,

$$h = l - \overline{PC} = l - l\cos 15° = l(1 - \cos 15°)$$

$$= (1.00\ \text{m})(1.000 - 0.966) = 0.034\ \text{m}$$

$$v = \sqrt{2gh} = \sqrt{2 \times 9.80\ \text{m/s}^2 \times 0.034\ \text{m}}$$

$$= 0.82\ \text{m/s}$$

4-16 POWER

In science and technology the word *power* is restricted to mean the *time rate of doing work*. The average power is the work performed divided by the time required for the performance. In measuring power P, both the *work* $\mathcal{W}$ and the *elapsed time* t must be measured

$$\bar{P} = \frac{\mathcal{W}}{t} \qquad (13)$$

The same work is done when a 500-lb steel girder is lifted to the top of a 100-ft building in 2 min as is done when it is lifted in 10 min. However, the average power required is five times as great in the first case as in the second, for the power needed to do the work varies inversely as the time.

Enormous amounts of power are used in many events that last only very short times, as in the case of lightning. In some thermonuclear experiments moderate amounts of energy are released in a few millionths of a second to produce tremendous amounts of power. For example, in a plasma experiment called Zeus, energy from a capacitor bank is released in a brief pulse with a power comparable with that consumed normally in the entire United States.

Since work is frequently done in a continuous fashion (or energy is transported in a continuous stream), another expression for power (work/time) is useful; thus

$$P = \frac{\mathbf{F} \cdot \Delta \mathbf{s}}{\Delta t} = \mathbf{F} \cdot \left(\frac{\Delta \mathbf{s}}{\Delta t}\right)$$

since $\Delta \mathbf{s}/\Delta t$ represents the velocity of the body on which the force is applied,

$$P = \mathbf{F} \cdot \mathbf{v} \qquad (14a)$$

or in scalar terms

$$P = Fv \cos \theta \qquad (14b)$$

Table 2
UNITS OF POWER

1 watt (w) = 1 newton-meter per second (N·m/s) = 1 joule per second (J/s) = 10^7 ergs per second (erg/s)
1 horsepower (hp) = 550 foot-pounds per second (ft·lb/s) = 33,000 foot-pounds per minute (ft·lb/min)
1 horsepower = 746 watts
1 kilowatt = 1,000 watts = 1.34 horsepower

where θ is the angle between the force and the velocity.

The mks unit of power is the joule per second, called the *watt;* the kilowatt (1,000 W) is also commonly used. The cgs unit, the *erg per second,* is inconveniently small. In the British system the *foot-pound per second* and the *horsepower* are common units of power. The latter was defined by James Watt in connection with his engineering studies on the steam engine. One horsepower is defined as 550 ft·lb/s.

Since work is the product of power and time, any power unit multiplied by a time unit may be used as a unit of work. Commonly used units of work formed in this manner are the watt-second (joule), the watthour, the kilowatthour, and the horsepowerhour.

Example By the use of a pulley a man raises a load of 50 kg to a height of 15 m in 65 s. Find the average power required.

$$\bar{P} = \frac{\mathcal{W}}{t} = \frac{Fs}{t} = \frac{50 \text{ kg} \times 9.8 \text{ m/s}^2 \times 15 \text{ m}}{65 \text{ s}}$$

$$= 113 \text{ W}$$

Example What power is needed to move a 3,000-lb car up an 8.0° incline with a constant speed of 50 mi/h against a frictional force of 80 lb?

By vector solution the downhill component of the car's weight is

$$F_D = 3{,}000 \text{ lb} \times \sin 8.0° = 4\bar{2}0 \text{ lb}$$

The total force needed is

$$F = 4\bar{2}0 \text{ lb} + 80 \text{ lb} = 5\bar{0}0 \text{ lb}$$

The power needed is given by

$$P = Fv = 5\bar{0}0 \text{ lb} \times 50 \text{ mi/h} \times \frac{5{,}280 \text{ ft/mi}}{3{,}600 \text{ s/h}}$$

$$= 3.7 \times 10^4 \text{ ft·lb/s}$$

$$= \frac{3.7 \times 10^4 \text{ ft·lb/s}}{550 \text{ (ft·lb/s)/hp}} = 68 \text{ hp}$$

4-17 SIMPLE MACHINES

A machine is a device for applying energy to do work in a way suitable for a given purpose. No machine can create energy. To do work, a machine must receive energy from some source, and the maximum work it does cannot exceed the energy it receives.

Machines may receive energy in different forms, such as mechanical energy, heat, electric energy, or chemical energy. We are here considering only machines that employ mechanical energy and do work against mechanical forces. In the so-called "simple machines," the energy is supplied by a *single applied force* and the machine does useful work against a *single resisting force.* The frictional force which every machine encounters in action and which causes some waste of energy will be neglected for simplicity in treating some of the simple machines. Most machines, no matter how complex, are combinations of two or more simple machines.

There are really two major classes of simple machines, the lever and the inclined plane. However, these usually have been modified into more specialized simple machines so that it may be considered that there are the following six simple machines: the lever, the pulley, the wheel and axle, the inclined plane, the screw, and the wedge.

4-18 ACTUAL MECHANICAL ADVANTAGE

The utility of a machine is chiefly that it enables a person to perform some desirable work by changing the amount, the direction, or the point of application of the force. The ratio of the output force F_o exerted by the machine on a load to the input force F_i exerted by the operator on the machine is defined as the *actual mechanical advantage* (AMA) of the machine.

$$\text{AMA} = \frac{F_o}{F_i}$$

For example, if a machine is available that enables a person to lift 500 lb by applying a force of 25 lb, its actual mechanical advantage is 500 lb/25 lb = 20. For most machines the AMA is greater than unity. A machine that is designed to increase the force has an AMA greater than 1; for example, a bench vise, a crowbar, or a block and tackle. A machine designed to increase speed has an AMA less than 1; for example, a catapult, a fly casting rod, the gears in a hand-operated beater, or the chain drive of a bicycle. A machine designed to simply change the direction of the applied force has an AMA of 1. An example of this would be a machine consisting of a rope which is attached to an object to be lifted and then passed over a beam, or a single fixed pulley. With the force being exerted downward on the rope, the resultant force would be of the same magnitude but upward.

4-19 IDEAL MECHANICAL ADVANTAGE

In any machine, because of the effects of friction, the useful work done *by* the machine is always less than the work done *on* the machine. The input work done by the applied force F_i is measured by the product of F_i and the distance s_i through which it acts. The output work is measured by the product of the output force F_o and the distance s_o through which it acts. Hence

$$F_o s_o < F_i s_i$$

If we divide each member of the inequality by $F_i s_o$, we obtain

$$\frac{F_o}{F_i} < \frac{s_i}{s_o}$$

that is, the ratio of the forces F_o/F_i is less than the ratio of the distances s_i/s_o for any machine. If the effects of friction are very small, the value of the output work approaches that of the input work, or the value of F_o/F_i becomes nearly that of s_i/s_o. The *ideal mechanical advantage* (IMA) is defined as the ratio of the distance s_i through which the input force acts to the distance s_o through which the output force acts.

$$\text{IMA} = \frac{s_i}{s_o}$$

Since the forces move these distances in equal times, the ratio s_i/s_o is also frequently called the *velocity ratio*. In a "frictionless" machine the inequality of the ratio of the forces to the ratio of the distances would become an equality.

Each of the six simple machines has a predictable ideal mechanical advantage based on the physical structure of the machine. For example, in the lever, which is a long rigid rod designed to balance on some balancing edge, a fulcrum, the IMA may be predicted if we know the ratio of the length of the effort (input) arm to the length of the resistance (output) arm. IMA = length of effort arm/length of resistance. In reality although one of the most efficient machines, the lever is not rigid, so it will bend; and there may be some friction at the fulcrum; and, further, the rod may not be of uniform density—all factors which make the machine less efficient. Yet the use of the rule of thumb for determining the IMA of machines does permit the making of helpful approximations.

We saw earlier that some machines have mechanical advantages less than 1. Using the rule-of-thumb method we can show that the human forearm is a lever and can make some prediction about its IMA. Figure 4-14 shows that the biceps muscle is attached a short distance in front of the elbow on the forearm. The length of the effort arm is the distance from the elbow to the point of contact of the biceps with the forearm. This distance is small compared with the length of the resistance arm, the distance from the elbow to the hand.

Therefore, since

$$\text{IMA} = \frac{\text{length of effort arm}}{\text{length of resistance arm}}$$

the IMA < 1

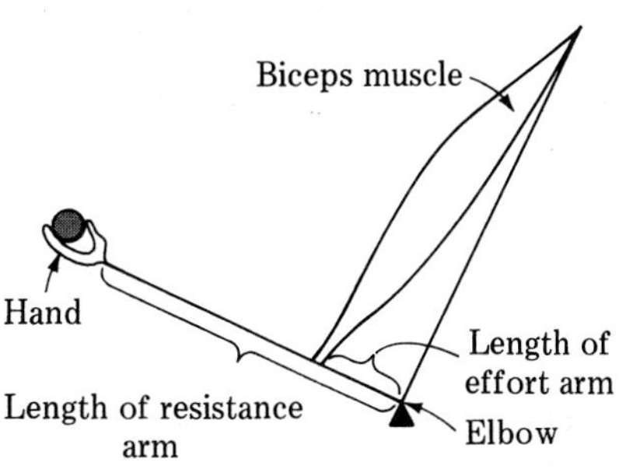

Figure 4-14
The human forearm as a simple machine.

The human forearm is thus designed to give mobility and speed to the hand and not lifting power. The biceps, contrary to popular belief, especially by those who work at developing large biceps muscle, is not designed to give strength to the arm. Other muscles with other mechanical arrangements, such as the triceps in back of the arm, will provide the necessary lifting strength.

Example A pulley system is used to lift a 1,000-lb block of stone a distance of 10 ft by the application of a force of 150 lb for a distance of 80 ft. Find the actual mechanical advantage and the ideal mechanical advantage.

$$\text{AMA} = \frac{F_o}{F_i} = \frac{1{,}000 \text{ lb}}{150 \text{ lb}} = 6.67$$

$$\text{IMA} = \frac{s_i}{s_o} = \frac{80 \text{ ft}}{10 \text{ ft}} = 8.0$$

4-20 EFFICIENCY

Because of frictional losses and other losses in all moving machinery, the useful work done by a machine is less than the energy supplied to it. From the principle of conservation of energy,

Energy input = energy output + energy wasted

assuming no energy is stored in the machine.

The efficiency of a machine is defined as the ratio of its output work to its input work. This ratio is always less than 1, and is usually multiplied by 100 percent and expressed in percent. A machine has a high efficiency if a large part of the energy supplied to it is expended by the machine on its load and only a small part wasted. The efficiency (eff) may be as high as 98 percent for a large electric generator and will be less than 50 percent for a screw jack.

$$\text{Eff} = \frac{\text{output work}}{\text{input work}} = \frac{F_o s_o}{F_i s_i}$$

Also since

$$\frac{F_o s_o}{F_i s_i} = \frac{F_o/F_i}{s_i/s_o}$$

$$\text{Eff} = \frac{\text{AMA}}{\text{IMA}}$$

Note that the work input times the efficiency is equal to the work output

$$(\text{Eff})(F_i s_i) = F_o s_o$$

Example What is the efficiency of the pulley system (described in the previous example) which lifts a 1,000-lb block of stone a distance of 10 ft by the application of a force of 150 lb for a distance of 80 ft?

$$\text{Eff} = \frac{F_o s_o}{F_i s_i} = \frac{(1{,}000\text{ lb})(10\text{ ft})}{(150\text{ lb})(80\text{ ft})} = 0.83 = 83\%$$

Also

$$\text{Eff} = \frac{\text{AMA}}{\text{IMA}} = \frac{6.67}{8.0} = 0.83 = 83\%$$

To calculate the mechanical advantage of a machine, one can imagine it to have carried out a chosen motion. Expressions are written separately for the input distance s_i and the output distance s_0. The ratio of these is the ideal mechanical advantage.

Example A 60-lb sled is pulled up an inclined plane 50 ft long and 30 ft high by a boy who exerts a force of 45 lb parallel to the plane. (*a*) What is the efficiency of this plane? (*b*) What force would the boy have to exert if the plane were frictionless?

$$\text{Eff} = \frac{\mathcal{W}_{\text{out}}}{E_{\text{in}}} = \frac{60\text{ lb} \times 30\text{ ft}}{45\text{ lb} \times 50\text{ ft}} = 0.80 = 80\%$$

$$F_{\text{ideal}} = F_{\text{actual}} \times \text{Eff} = 45\text{ lb} \times 0.80 = 36\text{ lb}$$

An alternative solution for this force is

$$F = W \sin\theta = 60\text{ lb} \times \frac{30}{50} = 36\text{ lb}$$

SUMMARY

Work is the product of force and the displacement in the direction of the force:

$$\mathcal{W} = Fs\cos\theta$$

Energy is the capacity for doing work.

The *foot-pound* is the work done by a force of 1 lb exerted through a distance of 1 ft.

The *joule* (*newton-meter*) is the work done by a force of 1 N exerted through a distance of 1 m. A joule is 10^7 ergs.

The *erg* is the work done by a force of 1 dyn exerted through a distance of 1 cm.

Energy is that property of a body or physical system of bodies by virtue of which work can be done.

Potential energy is energy of position or configuration. For gravitational potential energy,

$$E_p = Wh = mgh$$

Kinetic energy is energy of motion.

$$E_k = \tfrac{1}{2}mv^2$$

For transformations between work and kinetic energy,

$$Fs = \frac{1}{2}mv^2$$

Energy can be neither created nor destroyed, only transformed. This is the *principle of the conservation of energy.*

Frictional force F is proportional to the normal force N pressing the two surfaces together and is directed parallel to these surfaces.

$$F = \mu N$$

The coefficient of friction μ is defined as the ratio of the frictional force to the normal force.

$$\mu_k = \frac{F}{N} \quad \text{and} \quad \mu_s = \frac{F_{max}}{N}$$

The coefficient of static friction μ_s may be calculated from the limiting angle of repose, θ.

Rolling friction is the resistance to motion caused chiefly by the deformation produced where a wheel or cylinder pushes against the surface on which it rolls.

Fluid friction is resistance encountered by solids passing through fluids and the friction set up with liquids and gases.

Power is the time rate of doing work.

$$\bar{P} = \frac{\mathcal{W}}{t} = Fv \cos \theta$$

A *horsepower* is 550 ft·lb/s.

A *watt* is 1 J/s.

One *horsepower* is equivalent to 746 W.

A *machine* is a device for applying energy to do work in a way suitable for a given purpose.

In a *simple machine* the energy is supplied by a single applied force and the machine does useful work against a single resisting force.

The *actual mechanical advantage* of a machine is the ratio of the output force to the input force:

$$\text{AMA} = \frac{F_o}{F_i}$$

The *ideal mechanical advantage* is the ratio of the distance s_i through which the input force acts to the distance s_o through which the output force acts:

$$\text{IMA} = \frac{s_i}{s_o}$$

The *efficiency* of a machine is the ratio of the output work to the input energy.

Questions

1 A man is supported on a ladder which is leaning against a wall. Is the wall doing any work on the ladder?

2 If you try for 20 min to move a heavy object but cannot make it move, have you done any work on the object?

3 A man rowing his boat upstream is just able to hold his position with respect to the shore. Is he doing work? If so, on what?

4 Account for the energy possessed by the water stored above a dam.

5 Which performs work: the hammer or the nail? the powder or the bullet? the catcher or the baseball? the baseball or the bat? Explain your answers.

6 Two boys are throwing and catching a ball on a moving train. Does the kinetic energy of the ball at any instant depend on the speed of the train? Explain.

7 Show that during the motion of a simple pendulum the work done by the tension in the string is zero.

8 Show that when a body of mass m is dropped from a height h, the sum of its kinetic and potential energies at any instant is constant and equals mgh.

9 Trace the changes in the energy of a roller-coaster car.

10 A 90-kg man is seated in an automobile moving at a speed of 80 km/h. What is his kinetic energy relative to his fellow passengers in the car? Does he have this same kinetic energy relative to the ground?

11 If increasing the speed of a body increases its mass, does heating a gas cause a mass increase?
12 How can one calculate the minimum speed with which a projectile would have to be fired vertically in order to escape from the earth?
13 Name several types of mechanisms in which friction is essential for proper operation.
14 What becomes of the energy expended against friction?
15 Why should one take short steps rather than long ones when walking on ice?
16 An automobile is moving along a concrete road with the same speed as that of a streetcar alongside it. Which vehicle can stop in the shorter distance? (The coefficient of friction for rubber on concrete is 0.7, for steel on steel 0.2)
17 A body rests on a rough horizontal plane. Show that no force, however great, applied toward the plane at an angle with the normal less than the limiting angle for friction, can push the body along the plane.
18 Can a moving automobile be stopped in shorter distance (*a*) by applying the brakes to "lock" the wheels or (*b*) by applying a braking force just short of that which causes the tires to slip? Explain.
19 Distinguish carefully between doing work and the exerting force. Give examples.
20 Show that no work is done on a body that moves with constant speed in a circle.
21 In a tug-of-war, team *A* is slowly giving ground to team *B*. Is any work being done? If so, by what force?
22 Does the use of a lever increase one's power?
23 What kind of machine would you select if you desired one having a mechanical advantage of 2? of 500 or more? Which machine would be likely to have the greater efficiency if both machines were as mechanically perfect as it is possible to make them?
24 Why does a road wind up a steep hill instead of going directly up the slope?
25 How does the IMA of an inclined plane vary with the angle θ of the plane; that is, what trigonometric function of θ gives the value of IMA?
26 Describe several machines in which the load moves at a greater speed than the applied force. Which is greater in each case, the applied force or the force exerted by the machine?
27 If a block rests on a plane and the plane is gradually tilted until the block just begins to slip, the angle of inclination of the plane is called the angle of limiting repose. Show why this angle θ is given by $\tan \theta = \mu_s$.

Problems

1 A loaded cart has a total mass of 227 kg. If a 312-N force acting at an angle of 30° to the ground is applied, how much work is done in moving the cart 15 m?
2 A force of 50 N exerted over a distance of 3.0 m causes a block having a mass of 12.75 kg to move 1.0 m. Find the AMA; the IMA; the efficiency. *Ans.* 2.5, 3.0, 83 percent.
3 A safe weighing 10 tons is to be loaded on a truck 5.0 ft high by means of planks 20 ft long. If it requires 350 lb to overcome friction on the skids, find the least force necessary to move the safe.
4 A man weighing 450 N sits on a platform suspended from a movable pulley and raises himself by a rope passing over a fixed pulley. Assuming the ropes parallel, what force does he exert? (Neglect the weight of the platform.) *Ans.* 150 N.
5 A box weighing 150 lb is moved across a horizontal floor by dragging it by means of a rope attached to the front end. If the rope makes an angle of 30° upward with the floor and if the coefficient of friction between box and floor is 0.400, find the force exerted by the man pulling the rope.
6 A 1,000-N piano is moved 20 m across a floor by a horizontal force of 350 N. Find the coefficient of friction. What happens to the energy expended? *Ans.* 0.35.
7 A 350-g block of wood on a horizontal plane is fastened to a cord passing over a frictionless pulley and attached to a 265-g load. The coefficient of kinetic friction between block and plane is 0.45. (*a*) Determine the acceleration of the system after it is set in motion. (*b*) What is the force in the cord?

8 A sled weighing 100 lb reaches the foot of a hill with a speed of 40 ft/s. The coefficient of kinetic friction between the sled and the horizontal surface of the ice at the foot of the hill is 0.030. How far will the sled travel on the ice?

Ans. $8\overline{3}0$ ft.

9 What average net thrust must a 17-ton airplane have to reach an altitude of 5,000 ft and a speed of 600 mi/h at an airline distance of 10 mi from its starting point?

10 An automobile traveling at 50 mi/h on a level road is stopped by sliding the wheels. If the coefficient of kinetic friction between tires and road is 0.75, what is the minimum distance in which the car can be brought to rest?

Ans. 112 ft.

11 A panelboard sheet hangs in a vertical position while gripped between the jaws of a clamp. One side has a coefficient of static friction of 0.30 with the clamp and the other a coefficient of 0.45. How much horizontal force must each face of the clamp exert if the sheet weighs 150 N? How much vertical force will each face of the clamp then exert?

12 A 3,200-lb car starts to roll from the top of a 400-ft hill which is 500 ft long. How fast is it going when it reaches the bottom? *Ans.* 160 ft/s.

13 A man lifts a 100-N can of oil by pressing his two hands against the smooth sides, toward each other. If the coefficient of static friction is 0.30, what force must he apply with each hand?

14 A worker can supply a maximum pull of 150 lb on a cart which has a handle set at an angle of 40° above the horizontal. What is the coefficient of friction if he can just move a total load of 1,500 lb? *Ans.* 0.082.

15 A 4.45×10^4 N truck is traveling at 100 km/h. If the coefficient of friction between the tires and the roadway is 0.50, what is the minimum distance the truck will go before stopping?

16 A roller coaster which weighs 640 lb starts from rest at point A and begins to coast down the track. If the frictional force is 5.0 lb, how fast will the coaster be going at point B, which is 100 ft down the track and 30 ft below point A, and at point C, which is 50 ft farther along the track and 20 ft higher than point B?

Ans. 43.3 ft/s at B; 23.8 ft/s at C.

17 A prony brake is a device often used to measure the power of an engine. In a small motor the brake may be a strap passing halfway around the pulley of the motor, with the ends of the strap supported by spring balances (Fig. 4-15). In one case a pulley had a diameter of 45.2 cm, and the balances indicated a net difference in force of 0.525 N, when the motor speed was 500 r/min. The power input was 16.7 W. What was the efficiency?

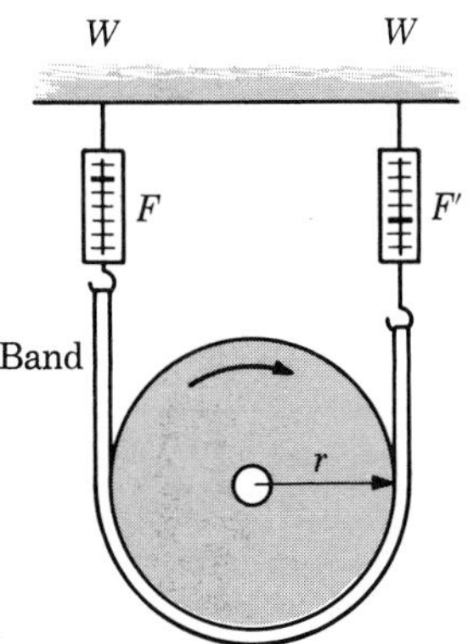

Figure 4-15
A prony, or band, brake.

18 A rocket rises to a height of 20.0 mi, 200 mi, 2,000 mi, and 20,000 mi above the surface of the earth on a flight toward the moon. If it weighs 3,000 lb at the surface of the earth (at a radius of 4,000 mi), what is its potential energy at each of these altitudes? (For which cases must the variation of gravity with altitude be taken into account?) $G = 3.44 \times 10^{-8}$ ft³/(slug)(s²); $M_e = 4.10 \times 10^{23}$ slugs.

Ans. Taking $Ep = 0$ at earth's surface: 3.17×10^8 ft·lb; 3.00×10^9 ft·lb; 2.08×10^{10} ft·lb; 5.25×10^{10} ft·lb.

19 A proton is accelerated to a kinetic energy of 1.60×10^{-9} J in a modern synchrotron. Find its mass increase over its rest mass of 1.67×10^{-27} kg and its relative mass gain.

20 A loaded sled weighing 1,250 lb is given a speed of 25.0 mi/h while moving a distance of 140 ft from rest on a horizontal ice surface. If the coefficient of friction is 0.105, what constant force, applied horizontally, would be necessary to produce this motion? *Ans.* 320 lb.

21 An electron whose mass is 9.11×10^{-31} kg is moving at 3.00×10^7 m/s. What constant force is required to bring it to rest in 1.5×10^{-9} cm, as might occur at the target of an x-ray tube?

22 The coefficient of friction between a block and the surface on which it slides is 0.18, and the surface is inclined at an angle of 15° with the horizontal. If the block weighs 300 lb, what force is needed to drag it up the incline at uniform speed? What force is required to let it slide down the plane at uniform speed?

Ans. 130 lb; 26 lb.

23 A 100-lb box slides down a skid 20.0 ft long and inclined at an angle of 60° to the horizontal. At the bottom of the skid the box slides along a level surface of equal roughness. If the coefficient of kinetic friction for the surfaces is 0.100, how far will the box travel, and how long a time will elapse after it reaches the level surface before it comes to rest?

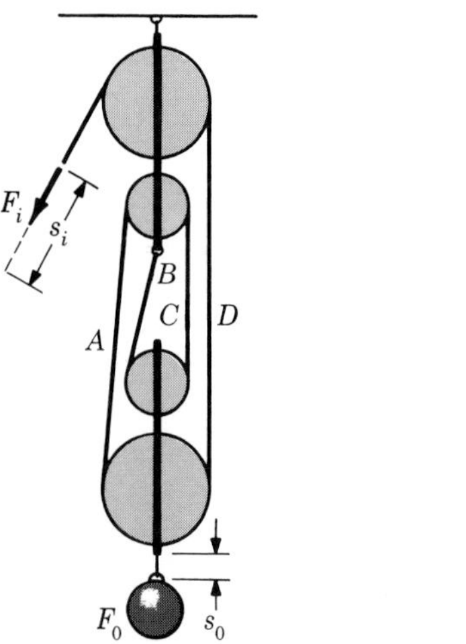

Figure 4-16
Block and tackle.

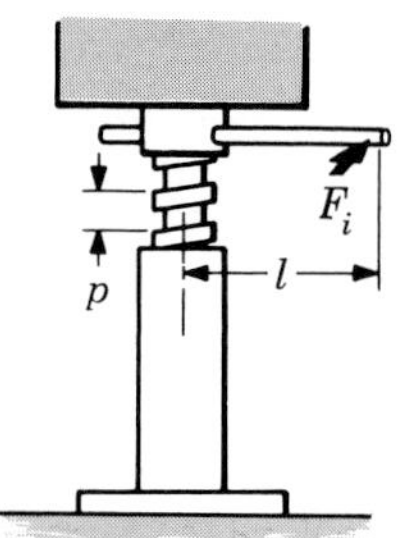

Figure 4-17
Screw jack.

24 A coin lying on a meterstick starts to slide when one end of the stick is raised 36.4 cm above the horizontal. (*a*) What is the limiting angle for friction? (*b*) What is the coefficient of starting friction? *Ans.* 21°; 0.38.

25 In the pulley system shown in Fig. 4-16 how far must the applied force move if the load moves 1.00 ft? If the load weighs 200 lb and the efficiency of the system is 90 percent, what must be the applied force?

26 Seven pulleys, three movable and four fixed, are connected by a single cord. What resistance will a 100-N effort move with this machine?

Ans. 700 N.

27 The screw jack of Fig. 4-17 has a pitch p of 0.25 in, a lever arm 18 in long, and an efficiency of 40 percent. How far must the force, applied at the end of the lever arm, move in one turn? What is the ratio of the load to the applied force (called the actual mechanical advantage)?

28 A pulley system has a mechanical advantage of 5 and is used to lift a load of 1,000 lb. If the effort moves 10 ft, how far does the load move? If this work was done in 15 s, what horsepower was developed? If the machine was 80 percent efficient, how much effort would be needed to lift the load? *Ans.* 2 ft; 0.243 hp, 250 lb.

29 For the compound machine of Fig. 4-18, the radii of the wheel and axle are 18 in and 6.0 in,

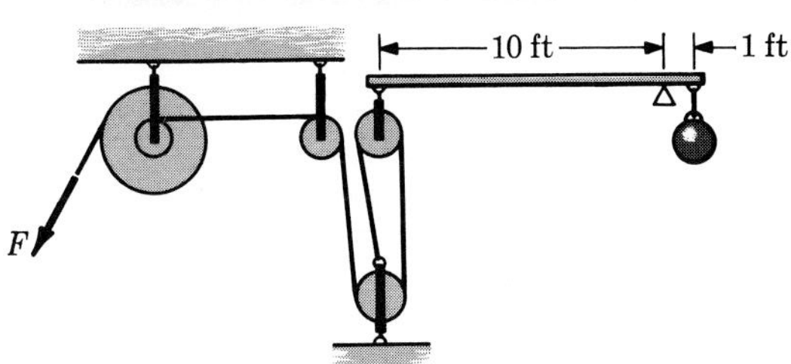

Figure 4-18
A compound machine.

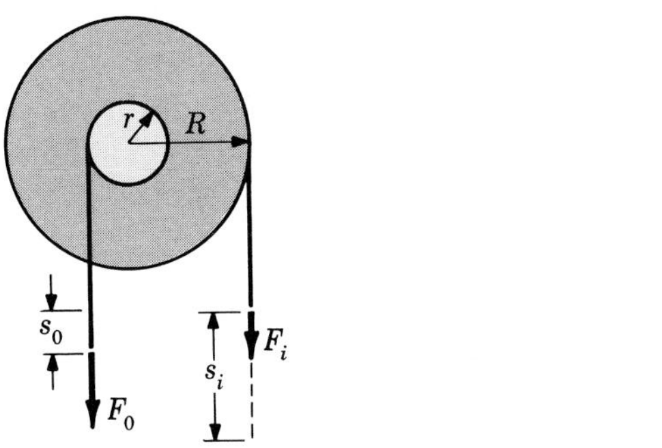

Figure 4-19
Wheel and axle.

respectively. If the load is lifted 2.00 in, through what distance does the applied force move? What would be the ratio of the load to the applied force if the efficiency were 100 percent?

30 A man raises a 500-N stone by means of a lever 5.0 m long. If the fulcrum is 0.65 m from the end that is in contact with the stone, what is the ideal mechanical advantage? *Ans.* 6.7.

31 The radius of a wheel is 1.8 ft, and that of the axle is 2.5 in (Fig. 4-19). What force, neglecting friction, must be applied at the rim of the wheel in order to lift a load of 800 lb which is attached to a cable wound around the axle?

32 Using a wheel and axle, a force of 30 lb applied to the rim of a wheel can lift a load of 240 lb. The diameters of the wheel and axle are 3 ft and 4 in, respectively. Determine the efficiency of the machine. *Ans.* 89 percent.

33 A load of 400 N is lifted by means of a screw whose pitch is 5.0 mm. The length of the lever is 20 cm, and the force applied is 5.0 N. Determine the efficiency of the machine.

34 Out of a total of 300 ft · lb of work put into a machine, an amount equal to 75 ft·lb is lost in overcoming friction. What is the efficiency of this machine? *Ans.* 75 percent.

35 A motor operates at its rated load of 10 hp for 8.0 h a day. Its efficiency is 87 percent. What is the daily cost of operation if electric energy costs 5 cents/kWh?

Gabriel Lippmann, 1845–1921

Born in Hollerich, Luxembourg. Director of the physical laboratory at the Sorbonne, Paris. Awarded the 1908 Nobel Prize for Physics for his method of photographic reproduction of colors, based upon the phenomenon of interference.

5

Torque

When a body is in equilibrium, it is either at rest or moving uniformly. Forces may act either to change the linear motion of a body or to change the rotation of the body, or both. If all the forces acting upon the body intersect at a common point and their vector sum is zero, creating a closed polygon, they have no tendency to change either translation or rotation.

Since most bodies are acted upon by forces that do not act through a single common point, we must consider the effect of each force in changing the rotation, as well as its effect in changing the linear motion of the body. The same force applied at different places or in different directions produces greatly different rotational effects. Therefore, in studying equilibrium, we must consider the place at which a force is applied as well as its magnitude and direction.

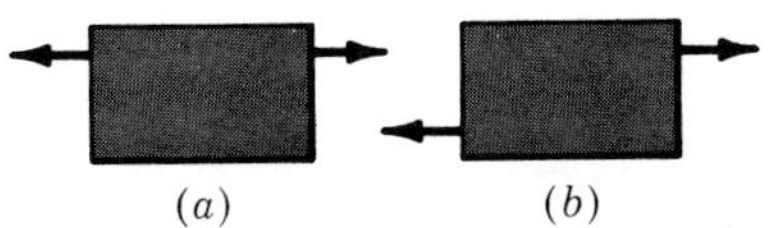

Figure 5-1
Forces which are equal in magnitude and opposite in direction produce equilibrium when they have a common line of action (*a*) but do not produce equilibrium when they do not have the same line of action (*b*).

5-1 CONDITIONS FOR EQUILIBRIUM

Consider an arrangement in which two opposing forces equal in magnitude act on a block as in Fig. 5-1*a*. It is obvious that if the block is originally at rest it will remain so under the action of these two forces. We say, as before, that the vector sum of the forces is zero.

Now suppose that the forces are applied as in Fig. 5-1*b*. The vector sum of the forces is again zero; yet it is plain that under the action of these two forces, the block will rotate. In fact, when the vector sum of the applied forces is equal to zero, we can be sure only that the body as a whole will have no change in its *linear* motion; we cannot be sure that there will be no change in its *rotary* motion. Hence complete equilibrium is not assured. In addition to the first condition for equilibrium previously stated (Chap. 1), a second condition is necessary, a condition eliminating the possibility of a change in rotational motion. The example of Fig. 5-1*b* indicates that this second condition is concerned with the *placement* of the forces, as well as their magnitudes and directions.

In order to study the factors that determine the effectiveness of a force in changing rotational motion, consider the familiar problem of turning a heavy wheel by pulling on a spoke (Fig. 5-2).

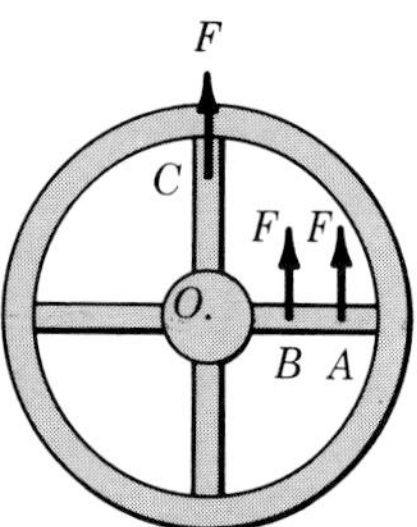

Figure 5-2
An unbalanced force produces rotational acceleration if its line of action does not pass through the axis of rotation.

It is a common experience that the wheel can be set into motion more easily and quickly by applying a force F perpendicular to a spoke at some point A far from the axis than by applying the same force at a point B nearer the axis. The effect of a given force upon the rotational motion of a body is greater the farther the *line of action* of the force is from the axis of rotation. The distance of the line of action from the axis is measured perpendicular to the line of action of the force. It is not merely the distance from the axis to the point of application of the force. If the same force F of Fig. 5-2 is applied at C rather than at A, where $OC = OA$, there is no effect upon the rotation of the wheel, since the line of action of F passes through the axis and it merely pulls the wheel as a whole upward. Though the magnitude of the force, its direction, and the distance of its point of application from the axis are the same in the two examples, rotation is affected when the force is applied at A but it is not affected when the force is applied at C.

5-2 MOMENT ARM

The factor that determines the effect of a given force upon rotational motion is the *perpendicular distance from the axis of rotation to the line of action of the force.* This distance is called the *moment arm* of the force. In Fig. 5-3, the moment arm of the force F is indicated by the dotted line $s = OP$. The line of action of the force is a mere geometrical construction and may be extended indefinitely either way in order to intersect the perpendicular OP. It has nothing to do with the length of the force vector. The force F in Fig. 5-2 produces no change in the rotation when it is applied at C, because its line of action passes through the axis of rotation and its moment arm is therefore zero. The same force applied at A has a moment arm OA and, therefore, tends to change the rotation.

5-3 TORQUE

For a fixed moment arm, the greater the force the greater the effect upon rotational motion. The two quantities, force and moment arm, are of equal importance. They can be combined into a single quantity, *torque* (also called *moment of force*), which measures the effectiveness of the force in changing rotation about the chosen axis. Torque will be represented by the symbol L.

The torque (moment of force) about a chosen axis is *the product of the force and its moment arm.*

$$L = sF \qquad (1)$$

where s is the perpendicular distance from the

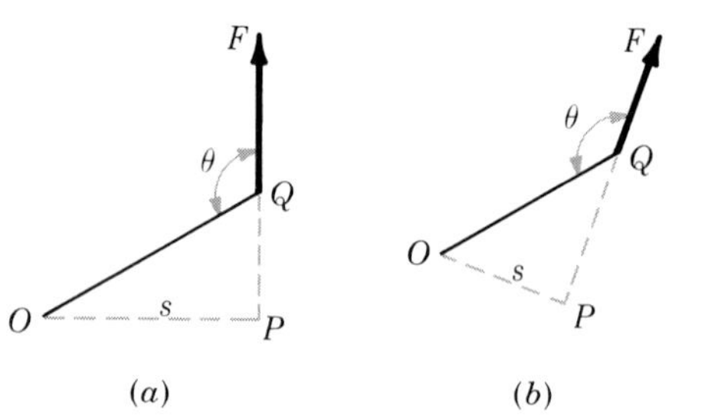

Figure 5-3
Measurements of moment arm.

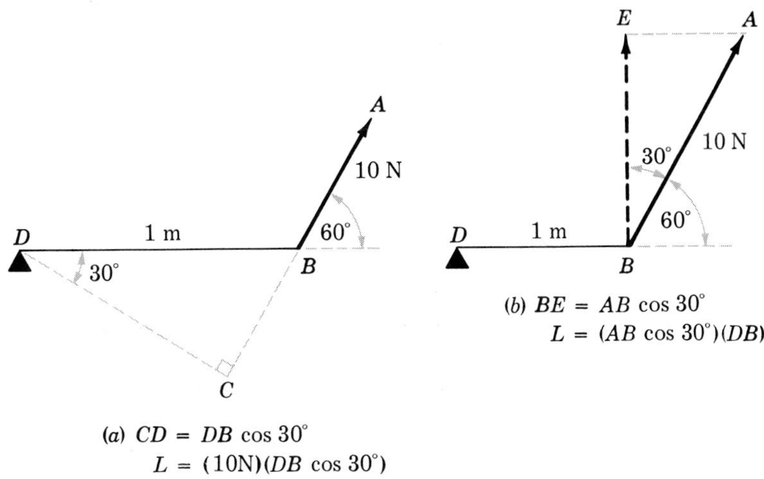

Figure 5-4
Finding the torques by two procedures.

axis to the line of action of the force. An axis must always be selected about which torques are to be measured. The value of the torque produced by a given force depends, of course, upon the axis chosen. The selection of an axis is quite arbitrary; it need not be any actual axle or fulcrum. In many cases, however, a wise selection of the axis about which torques are to be calculated greatly simplifies a problem, because it reduces to zero the torque due to a force whose magnitude or direction is unknown.

Example Find the torque created by a 10-N force acting 60°N of E at a distance of 1 m from the axis of rotation of a lever.

Two approaches can be used. First, find the perpendicular distance from the fulcrum to the line of action of the force by extending the line of action AB and then dropping a perpendicular to that line (Fig. 5-4a). Then from the trigonometry of a right triangle,

$$\cos 30° = \frac{CD}{DB}$$

$$CD = DB \cos 30° = (1\text{ m})(0.866) = 0.866\text{ m}$$

Therefore,

$$L = (0.866\text{ m})(10\text{ N}) = 8.66\text{ m}\cdot\text{N}$$

A second procedure involves taking the vertical component of the 10-N force. (Fig. 5-4b). From the figure it can be seen that

$$\cos 30° = \frac{BE}{10\text{ N}}$$

$$BE = (10\text{ N})(\cos 30°) = (10\text{ N})(0.866)$$

$$= 8.66\text{ N}$$

Therefore,

$$L = (1\text{ m})(0.866\text{ N}) = 8.66\text{ m}\cdot\text{N}$$

Since torque is a product of a force and a distance, its unit is a force unit times a distance unit, such as the pound-foot, the usual unit in the British system. An mks unit of torque is the meter-newton. The cgs unit of torque is the centimeter-dyne. Because the units here formed are products of force units and distance units, as are also the units of work, the order of the units is here interchanged to call attention to the fact that in torque, force and distance are at right angles. Any similar combination of force and distance units makes a suitable unit for torque.

5-4 VECTOR NATURE OF TORQUE

In our discussion of the addition of vectors we established the requirement that the vectors that were being added be similar in nature. That is, force vectors were added to force vectors and velocity vectors to velocity vectors. We will see, however, that it is possible to multiply vectors of different kinds. We should be aware that since vectors have direction as well as magnitude, vector multiplication does not follow exactly the algebraic rules that the multiplication of scalars follow. There are three types of vector-multiplication situations that may occur: (1) the multiplication of a vector by a scalar, (2) the multiplication of two vectors in such a manner as to produce a scalar, and (3) the multiplication of two vectors so that another vector is produced.

The multiplication of a scalar, **k,** by a vector, **A,** produces another vector which has the same direction but a magnitude of **kA.** For example, $2 \times \mathbf{A} = 2\mathbf{A}$.

When we multiply two vector quantities, it is necessary to distinguish between the scalar (or dot) product and the vector (or cross) product. The *scalar product* is the product of the *magnitude* of one vector by the *magnitude of the component* of the other vector quantity in the direction of the first. (See Fig. 5-5*a*.)

To find the scalar product of vector quantities **A** and **B** in Figure 5-5*a*, the component of **B** along **A** is found to be $B \cos \theta$. Therefore, $\mathbf{A} \cdot \mathbf{B} = AB \cos \theta$. Since A is only the magnitude of **A** and B is the magnitude of **B** and the cosine θ is a pure number, this product is a scalar quantity.

It is seen that any relation involving the cosine of an included angle may be written in terms of the scalar product. For example, we have seen that the mechanical work $\mathcal{W}$ done by a force **F** which makes an angle θ with the displacement **s** is $\mathcal{W} = Fs \cos \theta$ or, in vector notation, $\mathcal{W} = \mathbf{F} \cdot \mathbf{s}$. Other scalar products appearing in a study of physics are electric power and electric potential, gravitational potential energy, and electromagnetic energy density.

The vector product of two vectors **A** and **B** is written as $\mathbf{A} \times \mathbf{B}$ and produces another vector **C.** $\mathbf{A} \times \mathbf{B} = \mathbf{C}$. The magnitude of **C** is found to be $AB \sin \theta$. The proof of this can be found in several mathematics books.[1] The direction of **C**

[1] H. Margeneau and G. Murphy, "The Mathematics of Physics and Chemistry," D. Van Nostrand Company, Inc., New York, 1957, p. 143.

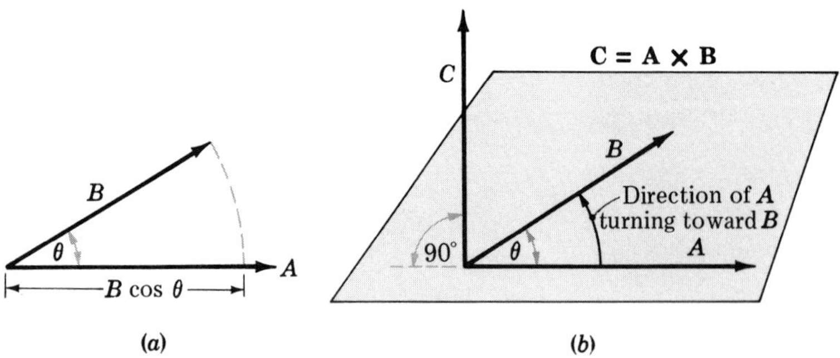

Figure 5-5
(*a*) Scalar (dot) product of two vectors. (*b*) Vector (cross) product of two vector quantities.

is perpendicular to the plane of **A** and **B.** (See Fig. 5-5*b*.)

The direction of **C** is determined by the so-called "right-hand rule," actually the rule of a right-hand screw. If the fingers of the right hand are wrapped about an imaginary axis perpendicular to the plane of **A** and **B,** the fingers indicating the direction in which **A** will be turned into **B** through an angle of less than 180°, and the thumb kept erect, the direction of the thumb gives the direction of the vector product **A × B.** From this sign convention, it follows that the vector product **B × A** has the same magnitude as **A × B** but has opposite direction, **A × B = −B × A.**

Vector products are often encountered in physics, for example, in angular momentum, the force on a moving charge in a magnetic field, and the topic we are presently considering, torque.

In Fig. 5-6 we represent the two vectors involved in a torque, the force **F** and the length vector **r** from the axis to the point of application of **F.** The vector product is

$$\mathbf{L} = \mathbf{r} \times \mathbf{F} \qquad (2)$$

which is represented by a vector of magnitude $rF \sin \theta$ perpendicular to the plane of r and F and directed into the paper. We observe that $r \sin \theta = s$ is the moment arm. The consequences of the vector nature of torque will be discussed in more detail later in connection with rotary motion. For the present we shall confine our attention to cases in which all the forces act in the same plane. For these cases the axes, and therefore the torques, are parallel, and only the algebraic signs of the torques need be considered.

The algebraic sign of such torques is determined by consideration of the direction of the rotation the torque tends to produce. For example, the torques in Fig. 5-3 tend to produce counterclockwise accelerations about O, while the torque in Fig. 5-6 tends to produce a clockwise acceleration. One may refer to these torques as positive and negative, respectively. Note that a given force may produce a counterclockwise torque about another axis. The direction of a torque is not known from the direction of the force alone.

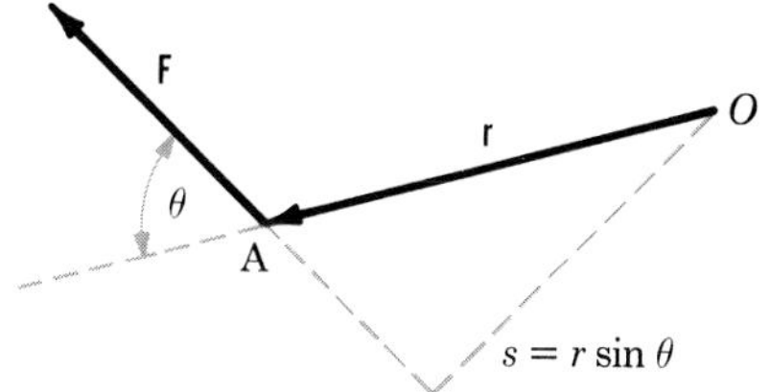

Figure 5-6
A clockwise torque **r × F**. The vector representing the torque is perpendicular to the plane of the paper and directed into the paper. The magnitude of the torque is $rF \sin \theta$, where $s = r \sin \theta$ is the moment arm.

Example A light horizontal bar is 4.0 m long. A 3.0-N force acts vertically upward on it 1.0 m from the right-hand end. Find the torque about each end.

Since the force is perpendicular to the bar, the moment arms are measured along the bar.

About the right-hand end,

$$L_r = 1.0 \text{ m} \times 3.0 \text{ N} = 3.0 \text{ m}\cdot\text{N} \qquad \text{clockwise}$$

About the left-hand end,

$$L_l = 3.0 \text{ m} \times 3.0 \text{ N} = 9.0 \text{ m}\cdot\text{N} \qquad \text{counterclockwise}$$

The torques produced by this single force about the two axes differ in both magnitude and direction.

5-5 CONCURRENT AND NONCONCURRENT FORCES

Concurrent forces are those whose lines of action intersect in a common point (Fig. 5-7). If an axis passing through this point is selected, the torque

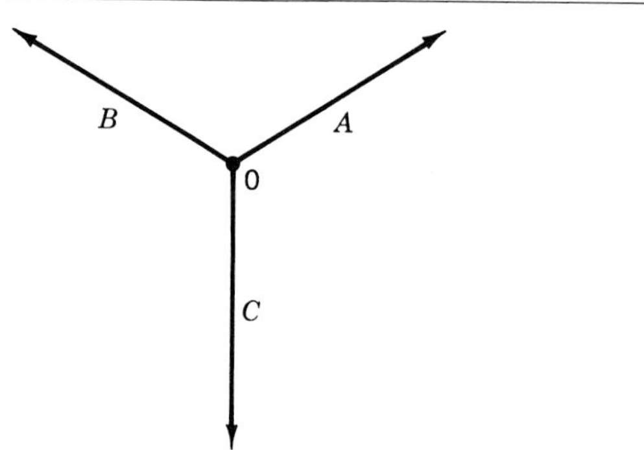

Figure 5-7
Illustration of concurrent forces.

produced by each force of such a set is zero. If any other axis is chosen in place of that through the common point, the sum of the torques will not, in general, be zero. For the special case in which the resultant of the concurrent forces is zero, the sum of the torques about any axis is zero. Hence a consideration of torque is not necessary in the study of a set of *concurrent* forces in equilibrium.

For a set of nonconcurrent forces, there exists no single axis about which no torque is produced by any of the forces. Therefore, in studying a set of nonconcurrent forces in equilibrium, it is essential to take into account the relation existing among the torques produced by such a set of forces. This relation is expressed in the second condition for equilibrium.

5-6 THE TWO CONDITIONS FOR EQUILIBRIUM

We have previously considered (Chap. 1) the condition necessary for equilibrium under the action of concurrent forces, namely, that the vector sum of all the forces acting shall be equal to zero. This condition must also be fulfilled when the forces are not concurrent. This first condition may be expressed by the statement that the sum of the components in any three perpendicular directions shall be zero.

$$\Sigma F_x = 0 \tag{3}$$

$$\Sigma F_y = 0 \tag{4}$$

$$\Sigma F_z = 0 \tag{5}$$

or

$$\Sigma \mathbf{F} = 0 \tag{6}$$

For an object to be in equilibrium under the action of a set of forces, the sum of the torques (about any axis) acting upon the body must be zero. This statement is known as the *second condition for equilibrium.* It may be represented by the equation

$$\Sigma \mathbf{L} = 0 \tag{7}$$

In the first and second conditions we have a complete system for solving problems involving bodies in static equilibrium. These same conditions are useful in certain problems involving uniform motion. If the first condition is satisfied, the vector sum of the forces is zero and no translational acceleration is produced. If the second condition is satisfied, the vector sum of the torques is zero and there is no rotational acceleration. This means not that there is no motion but only that the forces applied to the body produce no *change* in its motion. While in equilibrium, the body may have a uniform motion including both translation and rotation.

5-7 CENTER OF GRAVITY; CENTER OF MASS

The most common force acting upon a body is its weight. For every body, no matter how irregular its shape, there exists a point such that the entire weight may be considered as concentrated at that point. This point is called the *center of gravity* of the body. This point must be that for which the sum of the torques about horizontal axes through the center of gravity produced by the weights of the particles that make up the body must be equal to zero. If x_c and y_c (Fig. 5-8) are coordinates of the center of gravity, the y component L_y of torques (about an axis parallel to the y axis) is

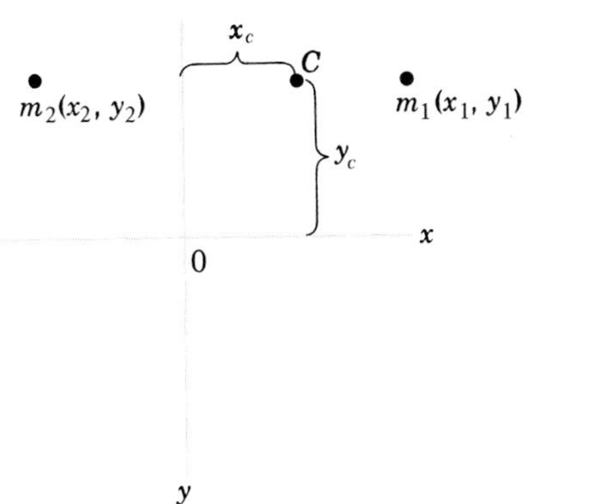

Figure 5-8
Coordinates of center of gravity.

$$L_y = (x_1 - x_c)m_1 g + (x_2 - x_c)m_2 g + \cdots$$
$$= \Sigma(x - x_c)mg = 0 \qquad (8)$$

where the x's are the x coordinates of the particles. Similarly, for the x components of the torque.

$$L_x = (y_1 - y_c)m_1 g + (y_2 - y_c)m_2 g + \cdots$$
$$= \Sigma(y - y_c)mg = 0 \qquad (9)$$

This procedure could be extended to a third dimension (z) to find the torque about an axis parallel to the z axis, L_z.

The center of gravity may be either within or outside the body. If a single force equal to the weight of the body and acting vertically upward could be applied at the center of gravity, it would support the body in equilibrium, no matter how the body might be tipped about the center of gravity.

The center of mass is the point about which the product of the mass and the moment arm sum up to zero. We may then write

$$(x_1 - x_c)m_1 + (x_2 - x_c)m_2 + \cdots$$
$$= \Sigma(x - x_c)m = 0 \quad (10)$$

and similar equations for y and z.

If, over the region occupied by the body, the acceleration due to gravity is everywhere the same in magnitude and direction, Eq. (10) is obtained by dividing Eqs. (8) and (9) by g and *the center of gravity and the center of mass are coincident.*

A knowledge of the position of the center of gravity is very useful in problems of equilibrium, for that is the point of application of the vector representing the weight. It is *never* necessary and *seldom* convenient to break the weight up into parts when considering the effect of weight in a torque equation.

Example A uniform bar 9.0 ft long and weighing 5.0 lb is supported by a fulcrum 3.0 ft from the left end as in Fig. 5-9. If a 12-lb load is hung from the left end, what downward pull at the right end is necessary to hold the bar in equilibrium? With what force does the fulcrum push up against the bar?

Consider the bar as an object in equilibrium. The first step is to indicate clearly all the forces that act on it. Since the bar is uniform, its center of gravity is at its midpoint and hence the weight of the bar, 5.0 lb, can be considered to be concentrated at its middle. A 12-lb force acts downward at the left end of the bar, a force R acts upward at the fulcrum, and there is an unknown downward force F at the right end.

The first condition for equilibrium indicates that the vector sum of the forces applied to the bar is zero, or that

$$R - 12\text{ lb} - 5.0\text{ lb} - F = 0$$

Without further information we certainly can-

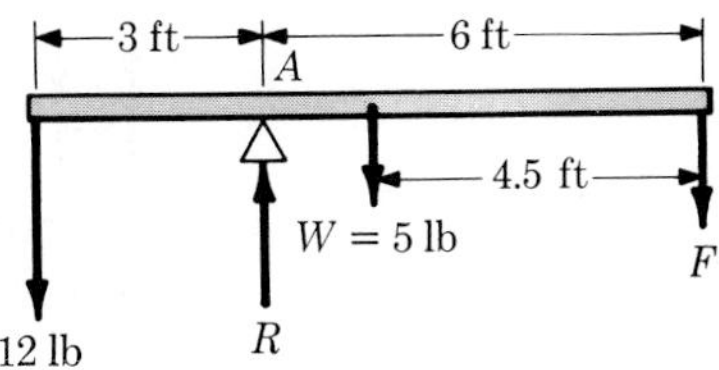

Figure 5-9
The forces acting on a lever.

not solve this equation, since it has two unknown quantities in it, R and F. Let us set it aside for a moment and employ the second condition for equilibrium, calculating the torques about some axis and equating their algebraic sum to zero.

The first thing we must do is select an axis from which to measure moment arms. We shall choose an axis through the point A about which to calculate all the torques. Beginning at the left end of the bar, we have (12 lb)(3.0 ft) = 36 lb·ft of torque, counterclockwise about A. Next, we see that the force R produces no torque, since its line of action passes through the point A. (Is it clear now why we decided to take A as an axis?) Third, the torque produced by the weight W of the bar is (5.0 lb)(1.5 ft) = 7.5 lb·ft, clockwise. Finally, F produces a torque F(6.0 ft), clockwise.

Taking the counterclockwise torque as positive and clockwise torque as negative and equating the algebraic sum of all the torques to zero, we write

$$\Sigma L = (12 \text{ lb})(3.0 \text{ ft}) + R(0) - (5.0 \text{ lb})(1.5 \text{ ft}) - F(6.0 \text{ ft}) = 0$$

$$36 \text{ lb}\cdot\text{ft} + 0 - 7.5 \text{ lb}\cdot\text{ft} - F(6.0 \text{ ft}) = 0$$

$$F(6.0 \text{ ft}) = 28.5 \text{ lb}\cdot\text{ft}$$

$$F = 4.8 \text{ lb}$$

Substituting this value in the equation obtained from the first condition for equilibrium, we find $R - 12 \text{ lb} - 5.0 \text{ lb} - 4.8 \text{ lb} = 0$, or $R = 22$ lb.

Another technique for finding the center of gravity of an object is illustrated in the following example.

Example A thin sheet of metal 1.0 ft square has a weight of 10.0 lb. If 1.0-lb, 2.0-lb, 4.0-lb, and 3.0-lb weights are hung on different corners (Fig. 5-10), where would a fulcrum have to be placed to balance the sheet and the weights?

By looking at the sheet from side AC, one can see the force diagram indicated in Fig. 5-10b. To balance this sheet so that the vertical forces add to zero, a force P must be acting upward at some specific location x feet from A. Also since

$$F_y = 0 \qquad P = 7.0 + 10.0 + 3.0 = 20.0 \text{ lb}$$

By arbitrarily taking a fulcrum at A, the following torques are observed. Since counterclockwise

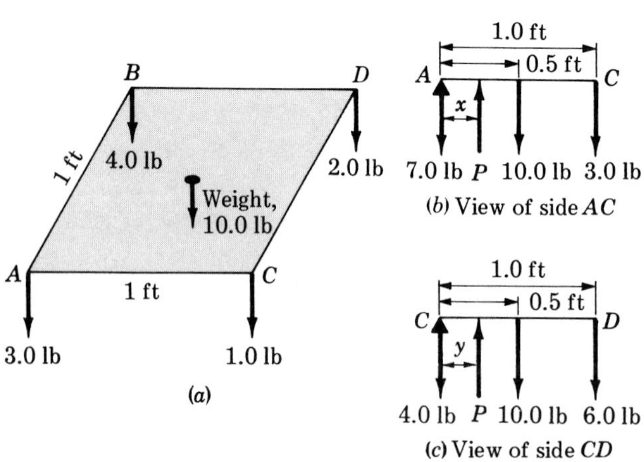

Figure 5-10
Illustration for center-of-gravity problem.

torques equal clockwise torques,

$$Px = (10.0 \text{ lb})(0.5 \text{ ft}) + (3.0 \text{ lb})(1 \text{ ft})$$

$$20x = 5 + 3$$

$$x = \frac{8}{20} = 0.4 \text{ ft}$$

Therefore, the fulcrum will be placed on a line 0.4 ft to the right of side *AB*.

Viewing the sheet from side *CD*, we see that

$$Py = (10 \text{ lb})(0.5 \text{ ft}) + (6.0 \text{ lb})(1.0 \text{ ft})$$

$$20y = 5 + 6 = 11$$

$$y = \frac{11}{20} = 0.55 \text{ ft}$$

Therefore, the fulcrum will be placed 0.55 ft above side *AC*.

This point 0.4 ft to the right of *AB* and 0.55 ft above the side *AC* is the center of gravity of the system consisting of the metal sheet and the attached weights.

Example A rope (Fig. 5-11) helps to support a uniform 200-lb beam, 20 ft long, one end of which is hinged at the wall and the other end of which supports a 1.0-ton load. The rope makes an angle of 50° with the wall and the beam makes an angle of 60° with the wall. Determine the tension in the rope.

Since all the known forces act on the 20-ft beam, let us consider it as the object in equilibrium. In addition to the 200-lb and 2,000-lb forces straight down, there are the pull of the rope on the beam and the force *F* which the hinge exerts on the beam at the wall. Let us not make the mistake of assuming that the force at the hinge is straight up or straight along the beam. A little thought will convince us that the hinge must be pushing both up and out on the beam. The exact direction of this force, as well as its magnitude,

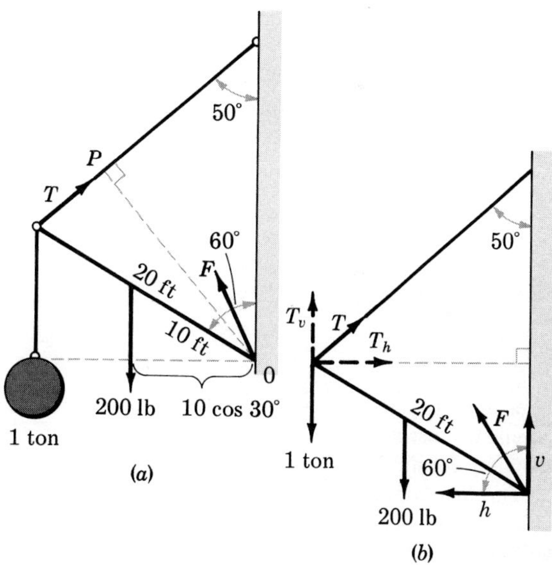

Figure 5-11
The forces acting on a beam.

is unknown. The second condition for equilibrium is an excellent tool to employ in such a situation, for if we use an axis through the point 0 as the axis about which to take torques, the unknown force at the hinge has zero moment arm and, therefore, causes zero torque. The remarkable result is that we can determine the tension T in the rope without knowing either the magnitude or the direction of the force at 0.

The torques about an axis through 0 are

The counterclockwise torque due to the weight of the boom:

$$(200 \text{ lb})(10 \text{ ft} \cos 30°)$$
$$= (200 \text{ lb})(10 \text{ ft})(0.866) = 1{,}730 \text{ lb·ft}$$

The counterclockwise torque due to the load:

$$(2{,}000 \text{ lb})(20 \text{ ft} \cos 30°)$$
$$= (2{,}000 \text{ lb})(20 \text{ ft})(0.866) = 34{,}\bar{6}00 \text{ lb·ft}$$

The clockwise torque due to the tension:

The moment arm of T is OP which can be found to be $20 \cos 20° = 20(0.94) = 18.\bar{8}$ ft.

Therefore, $T(18.8 \text{ ft})$ or $18.8T$; and, since $\Sigma L = 0$,

$$18.8T + 34{,}600 + 1{,}730 = 0$$
$$T = \frac{36{,}400}{18.8} = 1{,}9\bar{3}0 \text{ lb}$$

The problem of finding the magnitude and direction of the force at the hinge is another excellent illustration of the application of conditions for equilibrium in torque problems.

The first condition for equilibrium ($\Sigma F = 0$) gives us our tools for this problem. The technique will be to convert all known forces so that these components act either horizontally or vertically.

The load and the weight are already acting vertically downward. We need now to find T_h and T_v, the horizontal and vertical components of the tension. By extending T_h across the diagram, we can see that a right triangle is formed and

$$T_h = 1{,}930 \text{ lb} \cos 40° = 1{,}930 \text{ lb} \times 0.7660 = +1{,}480 \text{ lb}$$

and

$$T_v = 1{,}930 \text{ lb} \sin 40° = 1{,}930 \text{ lb} \times 0.6430 = +1{,}240 \text{ lb}$$

Also the force F has a vertical v and a horizontal h component. We will solve for these. Calling forces up + and forces down − and the forces right + and left −, we have

$$\Sigma F_v = -\text{load} - \text{weight of boom} + T_v + v = 0$$
$$-2{,}000 \text{ lb} - 200 \text{ lb} + 1{,}240 \text{ lb} = -v$$
$$v = 960 \text{ lb}$$
$$\Sigma F_h = +T_h - h = 0$$
$$h = T_h = 1{,}480 \text{ lb}$$

We now have the vertical and horizontal components of F and by trigonometry, we see that ϕ, the angle that F makes with the horizontal, can be found because

$$\tan \phi = \frac{960 \text{ lb}}{1{,}480 \text{ lb}} = 0.65 \qquad \text{and} \qquad \phi = 33°$$

Therefore, $F = (960 \text{ lb})(\sin 33°) = 523$ lb. It should be noted that F is not acting up the beam but rather 3° above it.

Example A 30-ft ladder weighing 100 lb having its center of mass one-third of the way up from the bottom rests against a smooth wall so that it makes an angle of 60° with the ground. If the coefficient of friction between the ground and the ladder is 0.4, how high can a 150-lb man go before the ladder slips? (Fig. 5-12.)

Write the torque equations around point A as an axis.

Clockwise torque = counterclockwise torque

$$(100 \text{ lb})(10 \text{ ft} \cos 60°) + (150 \text{ lb})(x) = F(BC)$$

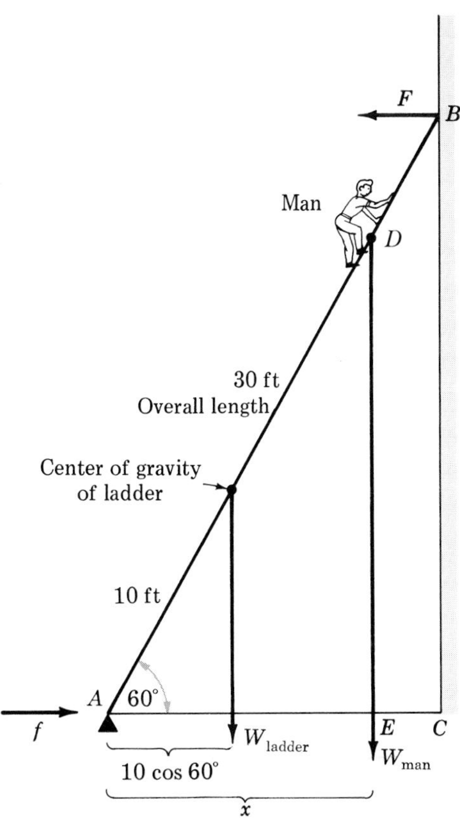

Figure 5-12
Illustration of ladder problem.

where $BC = (30 \text{ ft})(\sin 60°) = 26$ ft and F is the force exerted by the wall on the ladder.

F can be found by using the first condition for equilibrium, $\Sigma F_x = 0$. Therefore, f, the frictional force between the ground and the ladder, equals F. Since $\mu = f/N$; $f = \mu N = 0.4$ (weight of ladder + weight of man), and $f = (0.4)(250) = 100 \text{ lb} = F$.

Therefore,

$$(100 \text{ lb})(5 \text{ ft}) + (150 \text{ lb})(x) = (100 \text{ lb})(26 \text{ ft})$$

$$(150 \text{ lb})(x) = 2600 \text{ lb}\cdot\text{ft} - 500 \text{ lb}\cdot\text{ft}$$

$$x = \frac{2100 \text{ lb}\cdot\text{ft}}{150 \text{ lb}} = 14 \text{ ft}$$

But x is the perpendicular distance (AE) to the line of action of the man's weight. We need to find its projection on the ladder (AD).

$$AD = (x)/(\cos 60°)$$

$$= (14)/(\cos 60°) = 28 \text{ ft}$$

Therefore, the man can climb up to 28 ft before the ladder slips.

5-8 HINTS FOR SOLVING TORQUE PROBLEMS

The technique used in removing the unknown force from the beam problem by taking torques about the hinge as an axis is a standard device in statics. The student should always be on the lookout for the opportunity to sidestep (temporarily) a troublesome unknown force by selecting an axis of torques that lies on the line of action of the unknown force he wishes to avoid.

5-9 COUPLES

In general, the application of one or more forces to an object results in both translational and rotational acceleration. An exception to this result is the case in which a single force is applied along a line passing through the center of gravity of the object, in which case there is no rotational acceleration. Another special case is the one in which two forces equal in magnitude, opposite in direction, and not in the same line are applied to the object as in Fig. 5-1*b*. In this case there is no translational acceleration, but there is a net torque, and hence a rotational acceleration. *A pair of forces equal in magnitude, opposite in direction, and not in the same line is called a couple.* The torque produced by a couple is independent of the position of the axis and is equal to the product of one of the forces and the perpendicular distance between them.

Consider the torque produced by the couple

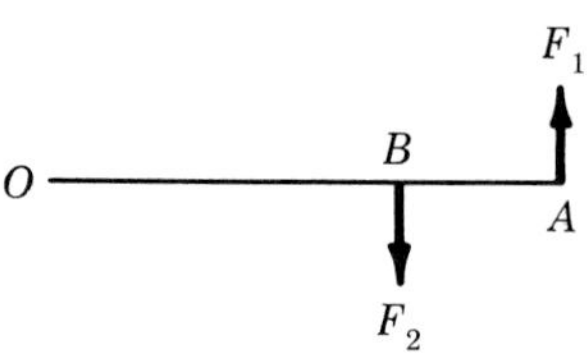

Figure 5-13
Forces constituting a couple.

shown in Fig. 5-13. About the axis through 0, the torque produced by F_1 is $(\overline{OA})F_1$, and that produced by F_2 is $-(\overline{OB})F_2$. Since $F_1 = F_2 = F$, the total torque is

$$(\overline{OA})F - (\overline{OB})F = (\overline{OA} - \overline{OB})F = (\overline{AB})F$$

This verifies the statement that the torque produced by a couple is the product of one (either) of the forces and the perpendicular distance between them, a product independent of the location of the axis. A couple cannot be balanced by a single force, but only by the application of another couple equal in magnitude and opposite in direction.

SUMMARY

The motion of a body is determined by the *placement* of the forces acting on it, as well as the magnitude and direction of the forces.

The *moment arm* of a force is the perpendicular distance from the axis to the line of action of the force.

The *torque* produced by a force is equal to the product of the force and its moment arm,

$$L = sF$$

or

$$\mathbf{L} = \mathbf{r} \times \mathbf{F}$$

Torque is measured in meter-newtons, centimeter-dynes, pound-feet, or any similar combination of force and distance units.

Torque is a vector quantity.

The *scalar product,* also called the *dot product,* is the product obtained by multiplying the *magnitude* of one vector by the *magnitude of the component* of the other vector in the direction of the first.

The *vector product,* or *cross product,* of two vectors produces another vector having a magnitude equal to the product of the magnitude of the two vectors times the sine of the angle between them and having a direction perpendicular to the plane of the two vectors, the sense of which is determined by the right-hand rule.

For an object to be in equilibrium, it is necessary (1) that the vector sum of the forces applied to it be zero and (2) that the vector sum of the torques acting on it be zero.

$$\Sigma\mathbf{F} = 0$$

$$\Sigma\mathbf{L} = 0$$

The *center of gravity* of a body is the point at which its weight may be considered as acting.

The *center of mass* of a body or system is the point about which the product of the mass and moment arm sum up to zero.

A *couple* consists of two forces equal in magnitude, opposite in direction, and not in the same line. The torque produced by the couple is equal to the magnitude of one (either) of the forces times the perpendicular distance between them.

Questions

1 How may the torque of a given force be increased?

2 What two conditions must be satisfied for an object to be in equilibrium?

3 Explain why the force that a wall exerts on a ladder leaning against it must equal the force that the ladder exerts on the wall.

4 Explain why in a ladder problem the force exerted on the ladder by the wall is considered to be equal and opposite in direction to the horizontal component of the force exerted by the ground on the base of the ladder.

5 Does the center of gravity of an object have to be within the geometric confines of the object? Explain your answer.

6 Explain why a person trying to push a stalled automobile must lean forward to get the best results.

7 Show that if the resultant of a set of concurrent forces acting on a point is zero, the sum of the torques about any point due to these forces must also be zero.

8 A man wishes to check the weight of a purchase but has available only a spring balance which can read only to about one-third of the presumed weight of the object. Show how he can perform an accurate weighing with the spring balance and a yardstick.

9 Show how one might locate the center of gravity of a thin, uniform sheet of metal by using only a plumb bob.

10 Where is the center of gravity of a doughnut? Explain your reasoning.

11 In order to assist a horse pull a wagon out of a rut, where on the wheel could you most effectively apply a force. Why?

12 Three unequal forces act upon a body at a point so that the body is in equilibrium. If the magnitudes of two of the forces are doubled, how must the third force be changed to preserve equilibrium? Use diagrams to support your answer.

13 How might one use the property of center of gravity to locate the geographic center of a state?

14 Discuss the stability of a body in equilibrium. How is stability related to potential energy? Give illustrations.

15 Archimedes is reputed to have said, "Give me a place to stand and I can lift the earth." Comment on this statement. What are some of the auxiliary devices that he would have needed for this cosmic experiment?

16 Show that a couple cannot be balanced except by another couple.

Problems

1 A rope is wound around a shaft 10.0 cm in diameter. If a pull of 500 N is exerted on the rope, what torque is imparted to the shaft?

2 A bar 2.0 m long makes an angle of 30° with the horizontal. A vertical force of 40 N is applied 0.4 m from the upper end. Calculate the torque due to this force about each end.

Ans. $5\bar{5}$ m·N; $1\bar{4}$ m·N.

3 A uniform board 20.0 cm × 10.0 cm has a mass of 200 g. Masses of 50.0 and 80.0 g are attached at two corners at the ends of one of the longer sides. Locate the center of gravity.

4 A rigid rod resting on a fulcrum has the following 20-lb forces acting on it. *A* is attached 2 ft to the left of the fulcrum acting vertically upward; *B* is 4 ft to the left of the fulcrum, acting 30°S of E; *C* is 6 ft to the left of the fulcrum, acting 45°N of E; *D* is attached 2 ft to the right of the fulcrum, acting 60°N of W; *E* is acting vertically downward; 4 ft to the right of the fulcrum; and *F* is 6 ft to the right of the fulcrum and acting 50°N of W. Is the system in equilibrium? If not, where can a 20-lb force be placed to put it in equilibrium? *Ans.* No; 1.9 ft to cause a counterclockwise moment.

5 A uniform steel rod with a linear mass of 2 g/m is bent in the shape of a letter F. The vertical rod is 2 m long, the cap is 1 m long, and the smaller arm, attached 1.5 m from the bottom, is 0.5 m long. Find the center of mass of the object.

6 A letter E has the following dimensions: its base is 3 ft long, the middle arm is 1 ft long, and the cap is 2 ft long. The vertical arm is 4 ft long and the middle arm is attached 1 ft from the top. If the linear weight is 20 lb per linear foot, find the center of mass. *Ans.* 0.7 ft to the right of the vertical arm; 1.9 ft above the base.

7 A uniform sheet of metal 6 ft square weighs 50 lb. If the following weights are hung on each corner of the square in a consecutive manner: 20 lb at *A*, 60 lb at *B*, 80 lb at *C*, and 40 lb at *D*, find the center of gravity of the system.

8 The legs of a wheelbarrow are 3.0 horizontal ft from the axle. When the wheelbarrow is unloaded, a force of 12 lb applied to the handles 4.0 horizontal ft from the axle is needed to raise the legs from the ground. If a 120-lb box is placed in the wheelbarrow with its center of gravity 1.5 ft from the axle, what force must be applied to the handles to raise the legs from the ground?

Ans. 57 lb.

9 A painter stands on a horizontal uniform scaffold hung by its ends from two vertical ropes

A and *B*, 6 m apart. The scaffold weighs 200 N. The tension in *A* is 750 N and that in *B* is 250 N. (*a*) What is the weight of the painter? (*b*) How far from *A* is he standing?

10 A uniform pole 20 ft long and weighing 80 lb is supported by a boy 2.0 ft from end *A* and a man 5.0 ft from end *B*. At what point should a load of 100 lb be placed so that the man will support twice as much as the boy?

Ans. 8.8 ft from *B*.

11 A uniform 30-lb beam 10 ft long is carried by two men *A* and *B*, one at each end of the beam. (*a*) If *A* exerts a force of 25 lb, where must a load of 50 lb be placed on the beam? (*b*) What force does *B* exert?

12 A bricklayer who weighs 200 lb stands 4.0 ft from one end of a uniform 20-ft scaffold. A 60-lb pile of bricks has its center of gravity 7.0 ft from the same end. The scaffold weighs 90 lb. If the scaffold is supported at the two ends, find the force on each support. *Ans.* 106 lb, 244 lb.

13 A nonuniform bar weighs 400 N and is 4 m long. When it is supported by a fulcrum at its midpoint, a load of 80 N must be supplied at the small end to hold the bar in a horizontal position. Where is the center of gravity?

14 A crane boom which weighs 200 lb and which is 10 ft long is supported so that it makes an angle of 50° with the supporting wall and so that the rope makes an angle of 70° with the wall. If a load of 700 lb is hung on the end of the boom, what tension occurs in the rope and what thrust does the support exert on the lower end of the boom? *Ans.* Tension = 707 lb; thrust = 930 lb, 45°N of E.

15 The boom of a wall crane 4 m long is held at right angles to the wall by a tie that is attached to the wall 3.0 m above the foot of the boom. If the load lifted is 4.5×10^4 N, find the tension in the tie and the compressional force in the boom. Neglect the weight of the boom.

16 A crane boom 30 ft long, weighing 200 lb, and having its center of gravity 10 ft from the bottom is hinged at the lower end and makes an angle of 30° with the vertical. It is held in position by a horizontal cable fastened at the upper end. The boom supports a 1,000-lb load at its upper end. Find the tension in the cable, the horizontal and vertical components of the thrust at the hinge, and the resultant thrust there.

Ans. $6\bar{2}0$ lb; $6\bar{2}0$ lb; $1{,}\bar{2}00$ lb; $1{,}\bar{3}00$ lb, 63° above the horizontal.

17 A uniform ladder 8 m long and weighing 350 N rests against a smooth vertical wall at an angle of 30° to the wall. A 700-N man stands 6 m up from the bottom of the ladder. Find the horizontal force necessary at the base to keep the ladder from slipping.

18 A uniform ladder 20.0 ft long weighing 30.0 lb rests on horizontal ground and leans at an angle of 60° with the horizontal against a smooth vertical wall. How far up the ladder may a 160-lb man go before the ladder slips if the coefficient of friction between ladder and ground is 0.433. *Ans.* 15.9 ft.

19 A 150-lb man is able to climb up 15 ft on a 20-ft ladder before it slips. What must the coefficient of friction be if the ladder is arranged so that its base makes an angle of 50° with the ground? Assume the ladder to be uniform and weighing 100 lb.

20 A stone having a mass of 10 kg is supported in the middle of a rope, which makes an angle of 150° at its point of contact with the stone. Find the tension in the rope. *Ans.* 190 N.

21 A uniform rod 30 ft long and weighing 200 lb is pivoted at its upper end. It is drawn aside so that it makes an angle of 60° with the vertical by means of a rope fastened at the lower end and making an angle of 90° with the rod. Find (*a*) the tension in the rope, (*b*) the horizontal force at the pivot, (*c*) the vertical force at the pivot, and (*d*) the resultant force at the pivot.

22 A 20-lb farm gate is supported by two hinges 4.0 ft apart. The gate is 12.0 ft long. Its weight is entirely supported by the upper hinge. If the gate is of uniform construction, what forces are exerted (*a*) at the upper hinge and (*b*) at the lower hinge? *Ans.* 36 lb, 56° from vertical; 30 lb, horizontally.

Carl-Ferdinand Braun, 1850–1918

Born in Fulda, Germany. Director of the Physical Institute at Strasbourg. Shared the 1909 Nobel Prize for Physics with Marconi for their development of wireless telegraphy.

Guglielmo Marconi, 1874–1937

Born in Bologna, Italy. Founder of the Marconi Wireless Telegraph Company (1897). Marconi and Braun received the 1909 Nobel Prize for Physics for their development of wireless telegraphy.

6

Rotation of Rigid Bodies

Thus far we have considered linear motion: bodies in equilibrium where there is no change in the motion and bodies that undergo linear acceleration when acted upon by resultant linear forces. We have noted that a resultant torque may cause rotation. At any time the motion of a body may consist of translation, or rotation, or a combination of translation and rotation. Let us examine first the ways in which rotary or angular motion can be described. We shall then see how the action of a torque in changing angular motion can be expressed by relations similar to Newton's laws for translational motion.

6-1 ROTARY MOTION; ANGULAR SPEED

As a disk turns about its axis, not all points move with the same linear speed; as the disk makes one rotation, a point at the edge must move farther than one near the axis and the points move these different distances in the same time. In Fig. 6-1 the point A has greater speed than B, and B has greater speed than C.

If we consider the line ABC rather than the points, we notice that the line as a whole turns about the axis. In a certain interval of time it will turn through an *angle* shown by the shaded area. The angle turned through per unit time is called

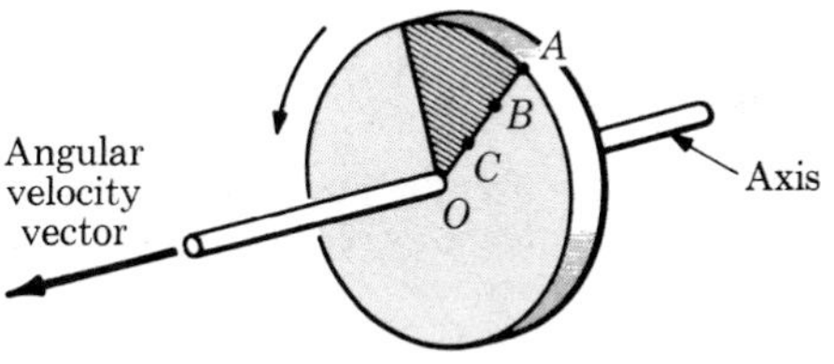

Figure 6-1
Rotation about axis O. The angular speed is the same for all parts of the disk, but the linear speed increases as the radius increases.

the *average angular speed,*

$$\bar{\omega} = \frac{\theta}{t} \tag{1a}$$

where $\bar{\omega}$ (omega) is the average angular speed and θ is the angle turned through in time t. The angle may be expressed in degrees, in revolutions ($1\ \text{r} = 360°$), or in radians. The angular speed is then expressed in degrees per second, revolutions per second, or in radians per second.

The radian is a unit of angular measure which is based on the relative dimensions of a sector of a circle. In radian measure, there is a very simple relation between angular motion and the linear motion of the points. In Fig. 6-2, the ratio

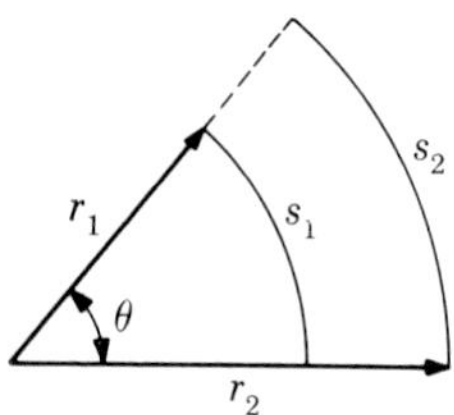

Figure 6-2
The ratio of arc to radius is a measure of the angle.

of length of arc to radius is the same for both concentric arcs. This ratio is used as a measure of the angle

$$\theta = \frac{s}{r}$$

where s is the length of the arc and r is the radius. The unit of angle in this system is the *radian* (rad), *which is the angle whose length of arc is equal to the radius*. The length of the circumference is $2\pi r$. Hence

$$1 \text{ r} = 360° = \frac{2\pi r}{r} = 2\pi \text{ rad}$$

$$1 \text{ rad} = \frac{360°}{2\pi} = 57.3° \text{ approx}$$

If the time interval of Eq. (1*a*) is taken shorter and shorter, the average angular speed approaches the instantaneous value ω.

$$\omega = \lim_{\Delta t \to 0} \frac{\Delta\theta}{\Delta t} \tag{1b}$$

The distinction between speed and velocity, used in describing linear motion, applies also to angular motion. An angular velocity is represented by a vector along the axis of rotation pointing in the direction of the thumb of a right hand which grasps the axis, with the fingers encircling the axis in the direction of rotation. The general vector equations for rotary motion are beyond the scope of this discussion, and we shall here restrict ourselves to scalar properties.

6-2 ANGULAR ACCELERATION

As in the case of linear motion, angular motion may be uniform or accelerated. *Angular acceleration* α (alpha) is the time rate of change of angular velocity

$$\bar{\alpha} = \frac{\omega_1 - \omega_0}{t} \tag{2a}$$

where ω_0 is the initial and ω_1 the final angular velocity. The change in angular velocity may be a change in magnitude, or a change in direction, or both.

In studying uniformly accelerated angular motion about a fixed axis, we treat it in a manner similar to that in which uniformly accelerated linear motion was handled in Chap. 2. For solving problems in uniformly accelerated rotary motion the following five equations are useful:

$$\theta = \bar{\omega}t \tag{1}$$

$$\omega_1 - \omega_0 = \alpha t \tag{2}$$

$$\bar{\omega} = \tfrac{1}{2}(\omega_0 + \omega_1) \tag{3}$$

$$\theta = \omega_0 t + \tfrac{1}{2}\alpha t^2 \tag{4}$$

$$\omega_1^2 - \omega_0^2 = 2\alpha\theta \tag{5}$$

Note that these equations become identical with Eqs. (1) to (5) of Chap. 2 if s is substituted for θ, v for ω, and a for α. These equations hold whatever the angular measure may be, as long as the same measure is used throughout a single problem. As in the case of linear motion, Eq. (1) holds for any type of angular motion, while the other four are true only for uniformly accelerated angular motion.

When radian measure is used, there is a very simple relationship between angular and linear motions. These relationships are given by the equations

$$s = \theta r \tag{6}$$

$$\bar{v} = \frac{\Delta s}{\Delta t} = r\frac{\Delta \theta}{\Delta t} = r\bar{\omega} \tag{7}$$

$$\bar{a} = \frac{\Delta v}{\Delta t} = r\frac{\Delta \omega}{\Delta t} = r\bar{\alpha} \tag{8}$$

Example A flywheel rotating at 200 r/min slows down at a constant rate of 2.00 rad/s². What time is required to stop the flywheel, and how many revolutions does it make in the process? *Hint:* It is common practice, and often necessary, to change revolutions per minute to radians per second in solving this kind of problem.

$$\omega_0 = 200 \text{ r/min} = 200(2\pi) \text{ rad/min}$$
$$= \frac{200(2\pi)}{60} \text{ rad/s}$$

From Eq. (2), $\omega_1 - \omega_0 = \alpha t$,

$$0 - \frac{400\pi}{60} \text{ rad/s} = (-2.00 \text{ rad/s}^2)t$$

$$t = 10.5 \text{ s}$$

From Eq. (5), $\omega_1^2 - \omega_0^2 = 2a\theta$,

$$0 - \left(\frac{400\pi}{60} \text{ rad/s}\right)^2 = 2(-2.00 \text{ rad/s}^2)\theta$$

$$\theta = 110 \text{ rad} = \frac{110}{2\pi} \text{ r} = 17.5 \text{ r}$$

6-3 ROTATIONAL INERTIA (MOMENT OF INERTIA)

It has been found that a force is necessary to change the motion of a body, i.e., to produce an acceleration. A greater force is required to give an acceleration to a body of large mass than to cause the same acceleration in a smaller one. If the rotation of a body about an axis is to be changed, a *torque* about that axis must be applied. The angular acceleration produced by a given torque depends not only upon the mass of the rotating body but also upon the *distribution of mass* with respect to the axis. In Fig. 6-3, a bar with adjustable masses m_1 and m_2 is supported on an axle. If a string is wrapped around the axle and an object of weight W is hung on the string, the axle and rod will rotate. The rate of gain in speed of rotation will be much greater when m_1 and m_2 are near the axle, as shown by the dots, than when they are near the ends of the rod. The mass is not changed by this shift, but the distribution of mass is altered, and the *rotational inertia* is changed.

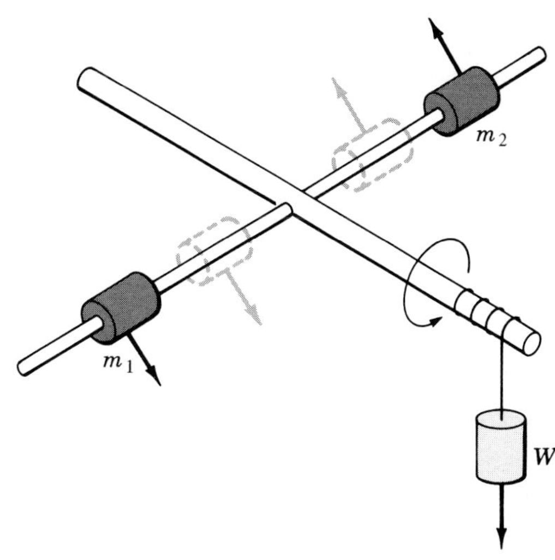

Figure 6-3
Angular acceleration produced by a torque depends upon the distribution of mass.

If a small body of mass m is guided in a circular path at a distance r from an axis O (Fig. 6-4), a tangential force will give the body an acceleration $a = F/m$ along the circumference. The torque of the force F about the axis O is $L = rF$, from which we obtain

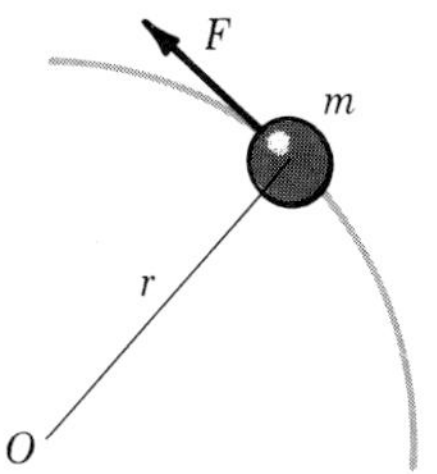

Figure 6-4
Particle accelerated by a tangential force has an angular acceleration about an axis.

$$L = mra = mr^2\alpha \tag{9}$$

The torque produces an angular acceleration proportional to the torque. The product mr^2 is characteristic of the configuration of the particle with respect to the chosen axis. The *rotational inertia I* (also called *moment of inertia*) of a particle about the chosen axis is the product of the mass m of the particle and the square of the radius r

$$I = mr^2 \tag{10}$$

In an extended body, each particle of matter in the body contributes to the rotational inertia an amount mr^2. The rotational inertia I of the body is the sum of the contributions of the individual elements

$$I = m_1r_1^2 + m_2r_2^2 + m_3r_3^2 + \cdots$$

or $$I = \Sigma\, mr^2 \tag{11}$$

The unit of rotational inertia is made up as a composite unit. In the mks system we have the kilogram-meter². In the cgs system the corresponding unit is the gram-centimeter². In the British gravitational system, where the weight is in pounds, we get the mass in slugs from W/g, and the unit for I is the slug-foot².

For many regular bodies the rotational inertia can be computed without difficulty and expressed in terms of the total mass of the body and the dimensions of the body. Note that the value of the rotational inertia depends upon the position of the axis chosen.

The rotational inertias of some regular bodies are given in Table 1. The pictures show the position of the axis. A sample calculation of one of these expressions is found in the Appendix.

Example What is the rotational inertia about an axis through the center of a 25-kg solid sphere whose diameter is 0.30 m?

From Table 1,

$$I = \tfrac{2}{5}mR^2 = \tfrac{2}{5}(25\text{ kg})(0.15\text{ m})^2 = 0.22\text{ kg}\cdot\text{m}^2$$

Example What is the rotational inertia of a 50-lb cylindrical flywheel whose diameter is 16 in?

For a cylinder about its axis,

$$I = \tfrac{1}{2}mR^2$$

$$m = \frac{W}{g} = \frac{50\text{ lb}}{32\text{ ft/s}^2} = 1.6\text{ slugs}$$

$$R = 8.0\text{ in} = \tfrac{2}{3}\text{ ft}$$

$$I = \tfrac{1}{2}(1.6\text{ slugs})(\tfrac{2}{3}\text{ ft})^2 = 0.35\text{ slug}\cdot\text{ft}^2$$

6-4 PARALLEL-AXIS THEOREM

It is frequently desirable to be able to compute the rotational inertia of a body about an axis other than its common geometrical axis. This is relatively simple if the axes are parallel. We make use of the fact that the rotational inertia about an axis that is distant s from the center of gravity is the sum of the rotational inertia about an axis through the center of gravity and the rotational inertia of the body considered concentrated at the center of gravity about the new axis. That is, we must add ms^2 to the rotational inertia about the axis through the center of gravity.

Example A solid cylinder of radius R rolls on a flat surface. Find the rotational inertia I_s

TABLE 1
ROTATIONAL INERTIA OF REGULAR BODIES ABOUT AXES INDICATED

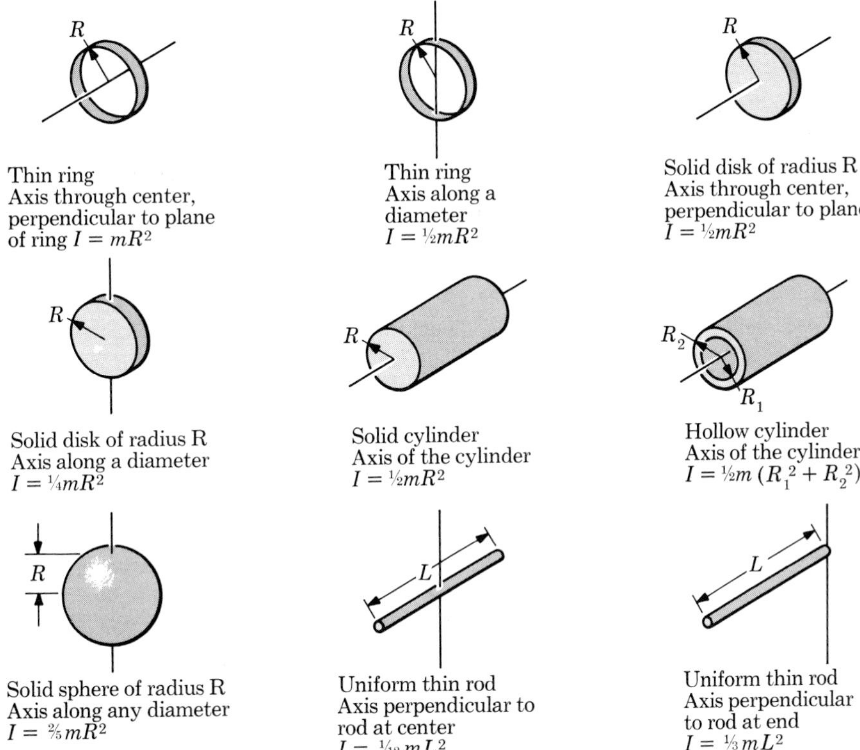

Thin ring Axis through center, perpendicular to plane of ring $I = mR^2$	Thin ring Axis along a diameter $I = \frac{1}{2}mR^2$	Solid disk of radius R Axis through center, perpendicular to plane $I = \frac{1}{2}mR^2$
Solid disk of radius R Axis along a diameter $I = \frac{1}{4}mR^2$	Solid cylinder Axis of the cylinder $I = \frac{1}{2}mR^2$	Hollow cylinder Axis of the cylinder $I = \frac{1}{2}m\,(R_1^2 + R_2^2)$
Solid sphere of radius R Axis along any diameter $I = \frac{2}{5}mR^2$	Uniform thin rod Axis perpendicular to rod at center $I = \frac{1}{12}mL^2$	Uniform thin rod Axis perpendicular to rod at end $I = \frac{1}{3}mL^2$

of the cylinder about its line of contact with the surface (see Fig. 6-5).

From the parallel-axis theorem and Table 1,

$$I_s = I_{cg} + ms^2 = \tfrac{1}{2}mR^2 + ms^2$$
$$= \tfrac{1}{2}mR^2 + mR^2 = \tfrac{3}{2}mR^2$$

since in this case $s = R$.

6-5 NEWTON'S LAWS FOR ANGULAR MOTION

Newton's laws for rotary motion are very similar to those for linear motion. The first law applies to a condition of equilibrium. *A body does not change its angular velocity unless it is acted upon by an external, unbalanced torque.* A body at rest does not begin to rotate without a torque to cause it to do so. Neither does a body that is rotating

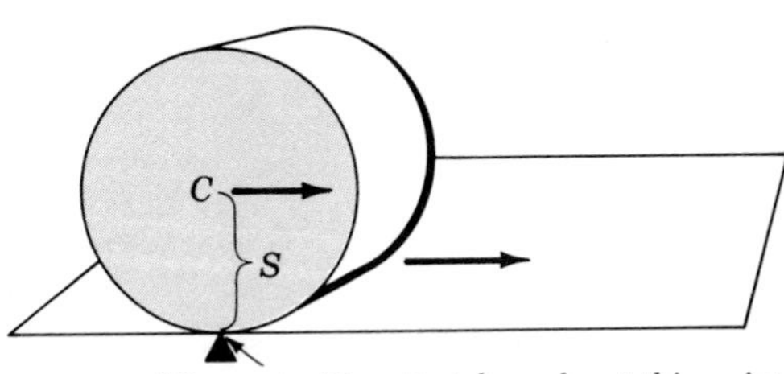

Figure 6-5
Parallel-axis diagram.

change its rotation or change its axis unless a torque acts. A rotating wheel would continue to rotate forever if it were not stopped by a torque such as that due to friction.

An unbalanced torque about an axis produces an angular acceleration, about that axis, which is directly proportional to the torque and inversely proportional to the rotational inertia of the body about that axis. In the form of an equation this becomes

$$L = I\alpha \tag{12}$$

where L is the unbalanced torque, I is the rotational inertia, and α is the angular acceleration. Torque must always be referred to some axis as are also moment of inertia and angular acceleration. In Eq. (12) we must be careful to use the same axis for all three quantities. As in the case of the force equation for linear motion, we must be careful to use a consistent set of units in Eq. (12). The angular acceleration must be expressed in radians per second per second.

Since torque and angular acceleration are both vector quantities, Eq. (12) may be written as a vector equation

$$\mathbf{L} = I\boldsymbol{\alpha} \tag{13}$$

The vector representing $\boldsymbol{\alpha}$ is in the same direction as that representing **L.**

The following sets of units apply to Eq. (12):

L	I	α
m·N	kg·m^2	rad/s^2
cm·dyn	g·cm^2	rad/s^2
lb·ft	slug·ft^2	rad/s^2

Example A wheel and its axle (Fig. 6-6) have a total mass of 50.0 kg. The axle has a diameter of 10.0 cm. A 2.50-kg ball is fastened to a cord wrapped around the axle. When started from rest, the ball moves downward a distance of 2.80 m in 7.00 s. Find the rotational inertia of the rotating system. Neglect friction.

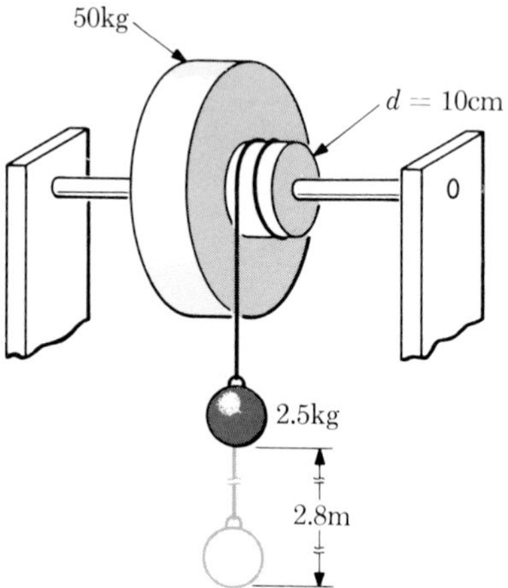

Figure 6-6
Angular acceleration of a wheel.

The linear acceleration of the ball is

$$a = \frac{2s}{t^2} = \frac{2(2.80 \text{ m})}{(7.00 \text{ s})^2} = 0.114 \text{ m/s}^2$$

The unbalanced force acting on the ball is the difference between its weight and the unknown tension T in the cord. From Newton's second law

$$\begin{aligned} W &= mg \\ &= (2.50 \text{ kg})(9.80 \text{ m/s}^2) \\ &= 24.5 \text{ N} \\ 24.5 \text{ newtons} - T &= (2.50 \text{ kg})(0.114 \text{ m/s}^2) \\ T &= 24.2 \text{ N} \end{aligned}$$

The torque L on the rotating system is

$$\begin{aligned} L = rT &= (0.0500 \text{ m})(24.2 \text{ N}) \\ &= 1.21 \text{ m·N} \end{aligned}$$

The angular acceleration is

$$\alpha = \frac{a}{r} = \frac{0.114 \text{ m/s}^2}{0.0500 \text{ m}} = 2.28 \text{ rad/s}^2$$

From Eq. (12), the rotational inertia of the wheel and its axle is

$$I = \frac{L}{\alpha} = \frac{1.21\ \text{m}\cdot\text{N}}{2.28\ \text{rad/s}^2} = 0.532\ \text{kg}\cdot\text{m}^2$$

Example A flywheel, in the form of a uniform disk 4.0 ft in diameter, weighs 600 lb. What will be its angular acceleration if it is acted upon by a net torque of 225 lb·ft

$$m = \frac{W}{g} = \frac{600\ \text{lb}}{32\ \text{ft/s}^2} = 18.8\ \text{slugs}$$

$$I = \tfrac{1}{2}mR^2 = \tfrac{1}{2}(18.8\ \text{slugs})(2.0\ \text{ft})^2$$

$$= 38\ \text{slug}\cdot\text{ft}^2$$

$$L = I\alpha$$

$$225\ \text{lb-ft} = (38\ \text{slug}\cdot\text{ft}^2)\alpha$$

$$\alpha = 5.9\ \text{rad/s}^2$$

Note: *In radian measure the angle is a ratio of two lengths and hence is a pure number. The unit "radian" therefore does not always appear in the algebraic handling of units.*

Example If the disk of the preceding example is rotating at 1,200 r/min, what torque is required to stop it in 3.0 min?

From Eq. (2),

$$\omega_1 - \omega_0 = \alpha t$$

$$\omega_0 = 1{,}200\ \text{r/min} = 20\ \text{r/s}$$

$$= 40\pi\ \text{rad/s}$$

$$t = 3.0\ \text{min} = 180\ \text{s}$$

$$0 - 40\pi\ \text{rad/s} = \alpha(180\ \text{s})$$

$$\alpha = -\frac{40\pi}{180}\ \text{rad/s}^2$$

$$L = I\alpha$$

$$= (38\ \text{slug}\cdot\text{ft}^2)\left(-\frac{40\pi}{180}\ \text{rad/s}^2\right)$$

$$= -26\ \text{lb}\cdot\text{ft}$$

The negative sign is consistent with a retarding torque.

For every torque applied to one body, there is a torque equal in magnitude and opposite in direction applied to another body. If a motor applies a torque to a shaft, the shaft applies an opposite torque to the motor. If the motor is not securely fastened to its base, it may turn in a direction opposite to that of the shaft. If an airplane engine exerts a torque to turn the propeller clockwise, the airplane experiences a torque tending to turn it counterclockwise and this torque must be compensated by the thrust of the air on the wings.

6-6 WORK, POWER, ENERGY

If a constant torque L turns a body through an angle θ, work is done by the torque. The torque may be the result of a force F acting at a distance r from the axis. The work done by the force is

$$\mathcal{W} = sF \qquad (14)$$

but the distance s in the direction of the force is the arc $s = r\theta$, and $L = rF$,

$$\mathcal{W} = r\theta F = L\theta \qquad (15)$$

Since power is work per unit time,

$$P = \frac{\Delta\mathcal{W}}{\Delta t} = L\frac{\Delta\theta}{\Delta t} = L\omega = L(2\pi n) \qquad (16)$$

where n is the number of rotations per unit time.

The expression for the kinetic energy of rotation of a body can be derived from Eqs. (5), (12), and (15) in the same manner that Eq. (6), Chap. 4, was derived to express the linear kinetic energy. For pure rotation, starting from rest, the work done by a constant torque is equal to the gain in kinetic energy

$$E_k = L\theta$$

but $\quad L = I\alpha$

and $$2\alpha\theta = \omega_1^2 - \omega_0^2 = \omega_1^2 - 0$$

$$\theta = \frac{\omega_1^2}{2\alpha}$$

$$E_k = I\alpha\frac{\omega_1^2}{2\alpha}$$

$$E_k = \tfrac{1}{2}I\omega_1^2 \quad \text{or} \quad \tfrac{1}{2}I\omega^2 \tag{17}$$

Frequently a body has simultaneous linear and angular motions. For example, the wheel of an automobile rotates about its axle, but the axle advances along the road. It is usually easier to deal with the kinetic energy of such a body if we consider the two parts: (1) that due to translation of the center of mass ($\frac{1}{2}mv^2$) and (2) that due to rotation about an axis through the center of mass ($\frac{1}{2}I\omega^2$)

$$E_k = \tfrac{1}{2}mv^2 + \tfrac{1}{2}I\omega^2 \tag{18}$$

Example What is the kinetic energy of a 3.0-kg ball whose diameter is 15 cm, if it rolls across a level surface with a speed of 2.0 m/s?

$$E_k = \tfrac{1}{2}mv^2 + \tfrac{1}{2}I\omega^2$$

$$\omega = \frac{v}{R} = \frac{2.0\ \text{m/s}}{0.075\ \text{m}} = 27\ \text{rad/s}$$

$$I = \tfrac{2}{5}mR^2 = \tfrac{2}{5}(3.0\ \text{kg})(0.075\ \text{m})^2$$

$$= 6.8 \times 10^{-3}\ \text{kg}\cdot\text{m}^2$$

$$E_k = \tfrac{1}{2}(3.0\ \text{kg})(2.0\ \text{m/s})^2 + \tfrac{1}{2}(6.8 \times 10^{-3}\ \text{kg}\cdot\text{m}^2)(27\ \text{rad/s})^2 = 8.5\ \text{J}$$

When energy is supplied to a body so that it is divided between energy of translation and energy of rotation, the way in which the energy is divided is determined by the distribution of mass. If two cylinders of equal mass, one being solid and the other hollow, roll down an incline, the solid cylinder will roll faster. Its rotational inertia is less than that of the hollow cylinder, and hence the kinetic energy of rotation is smaller than that of the hollow cylinder; but its kinetic energy of translation is greater than that of the hollow cylinder. Hence the solid cylinder has a greater speed.

Example A solid cylinder 30 cm in diameter at the top of an incline 2.0 m high is released and rolls down the incline without loss of energy due to friction. Find its linear and angular speeds at the bottom.

The potential energy of the cylinder at the top of the incline is converted into kinetic energy of translation and rotation as the cylinder rolls down. At the bottom of the incline all the potential energy lost has been converted into kinetic energy

$$\Delta E_p = \Delta E_k$$

$$mgh = \tfrac{1}{2}mv^2 + \tfrac{1}{2}I\omega^2$$

but $$\omega = \frac{v}{R} \quad \text{and} \quad I = \tfrac{1}{2}mR^2$$

Then

$$mgh = \tfrac{1}{2}mv^2 + \tfrac{1}{2}(\tfrac{1}{2}mR^2)\left(\frac{v}{R}\right)^2$$

$$= \tfrac{1}{2}mv^2 + \tfrac{1}{4}mv^2 = \tfrac{3}{4}mv^2$$

$$v^2 = \tfrac{4}{3}gh$$

$$v = \sqrt{\tfrac{4}{3}gh}$$

$$= \sqrt{\tfrac{4}{3}(9.8\ \text{m/s}^2)(2.0\ \text{m})} = 5.1\ \text{m/s}$$

$$\omega = \frac{v}{R} = \frac{5.1\ \text{m/s}}{0.15\ \text{m}} = 34\ \text{rad/s}$$

Note that the linear speed does not depend upon the size or the mass of the cylinder. Its dependence on the distribution of mass can be seen by working the same example but this time for a ring of the same dimensions.

As above,

$$mgh = \tfrac{1}{2}mv^2 + \tfrac{1}{2}I\omega^2$$

but $$I_{\text{ring}} = mr^2 \quad \text{and} \quad v = r\omega$$

Then, $mgh = \frac{1}{2}mv^2 + \frac{1}{2}mv^2$

and $gh = v^2$

$$v = \sqrt{gh} = \sqrt{9.8\ \text{m/s}^2 \times 2.0\ \text{m}}$$

$$= 4.43\ \text{m/s}$$

$$\omega = \frac{v}{r} = \frac{4.43\ \text{m/s}}{0.15\ \text{m}} = 29.6\ \text{rad/s}$$

If the cylinder and the ring were to be released at the same time from the same position, the cylinder would reach the bottom of the incline first. Suppose a 30-cm sphere were to roll down the incline. Would it beat the cylinder, the ring, or both? Try it and find out.

6-7 COMPARISON OF LINEAR AND ANGULAR MOTIONS

In our discussion of motions and forces we have found the equations of angular motion to be quite similar to those of linear motion. They can be obtained directly from the equations of linear motion if the following substitutions are made: θ for s, ω for v, α for a, L for F, I for m. In Table 2 is listed a set of corresponding equations.

SUMMARY

For a rotating body the *average angular speed* is the angle turned through per unit time by a line that passes through the axis of rotation

$$\bar{\omega} = \frac{\theta}{t}$$

Angular distance, in radians, is the ratio of the length of arc to its radius.

A *radian* is the angle whose length of arc is equal to the radius.

Average angular acceleration is the time rate of change of angular velocity

$$\bar{\alpha} = \frac{\omega_2 - \omega_1}{t}$$

Equations of uniformly accelerated angular motion are similar to those for linear motion, with angle substituted for distance, angular speed for linear speed, and angular acceleration for linear acceleration.

The *rotational inertia* (moment of inertia) of a body about a given axis is the sum of the products of the mass and square of the radius for

Table 2
CORRESPONDING EQUATIONS IN LINEAR AND ANGULAR MOTION

	Linear	Angular
Velocity	$\bar{v} = \frac{s}{t}$	$\bar{\omega} = \frac{\theta}{t}$
Acceleration	$\bar{a} = \frac{v_1 - v_0}{t}$	$\bar{\alpha} = \frac{\omega_1 - \omega_0}{t}$
Uniformly accelerated motion	$v_1 - v_0 = at$ $s = v_0t + \frac{1}{2}at^2$ $v_1^2 - v_0^2 = 2as$	$\omega_1 - \omega_0 = \alpha t$ $\theta = \omega_0 t + \frac{1}{2}\alpha t^2$ $\omega_1^2 - \omega_0^2 = 2\alpha\theta$
Newton's second law	$F = ma$	$L = I\alpha$
Work	$\mathcal{W} = Fs$	$\mathcal{W} = L\theta$
Power	$P = Fv$	$P = L\omega$
Kinetic energy	$E_k = \frac{1}{2}mv^2$	$E_k = \frac{1}{2}I\omega^2$

each particle of the body

$$I = \Sigma mr^2$$

For angular motion *Newton's laws* may be stated:

1 A body does not change its angular velocity unless it is acted upon by an external, unbalanced torque.
2 An unbalanced torque about an axis produces an angular acceleration about that axis, which is directly proportional to the torque and inversely proportional to the moment of inertia of the body about that axis:

$$L = I\alpha$$

3 For every torque applied to one body there is a torque equal in magnitude and opposite in direction applied to another body.

In angular motion the *work* done by a torque L in turning through an angle θ is

$$\mathcal{W} = L\theta$$

The *power* supplied by a torque is

$$P = L\omega$$

Kinetic energy of rotation is given by the equation

$$E_k = \tfrac{1}{2}I\omega^2$$

For a rolling or spinning body the total kinetic energy, both translational and rotational, is

$$E_k = \tfrac{1}{2}mv^2 + \tfrac{1}{2}I\omega^2$$

In the last five equations the angles must be expressed in radian measure.

Questions

1 What is the geometric definition of a radian?
2 Why doesn't the radian unit always appear in the algebraic handling of the units of angular motion?
3 What is the purpose of a flywheel? Describe the necessary distribution of mass in a flywheel so that this objective will be attained.
4 Describe how you can use a simple experiment involving moment of inertia to determine whether an egg is raw or hard-boiled.
5 A bicycle wheel is supported by its axle on two inclined rods. It is allowed to roll down first with the axle free to turn on its bearings and second with the cones tightened so that the wheel must turn with the axle. In which case does it reach the bottom of the incline quicker? Explain.
6 Define and state the name of the common mks unit of each of the following: average angular speed, instantaneous angular speed, average angular acceleration, torque, and moment of inertia.
7 Distinguish between angular speed and angular velocity.
8 From the definition of moment of inertia, $I = L/\alpha$, show that the moment of inertia of a particle of mass m, distant r from the axis, is mr^2.
9 From the units of torque and angular acceleration, show that the mks unit of moment of inertia is the kilogram-meter^2.
10 What portion of the total kinetic energy of a rolling solid disk is energy of translation and what portion is energy of rotation?
11 A solid cylinder rolls down an inclined plane. Its time of descent is noted. Then a hole is bored along the axis of the cylinder. When it is again allowed to roll down the incline, will it require more, less, or the same time to reach the bottom? Explain.
12 Show that when a hoop rolls down an incline half the kinetic energy is rotational and half translational.
13 A solid cylinder, a hollow cylinder, and a solid sphere roll down an incline starting simultaneously. In what order do they reach the bottom of the incline?
14 A hollow cylinder and rectangular block are placed at the top of an incline. When they are released, the cylinder rolls down the incline without loss of energy while the block slides down

with half its energy being used to do work against friction. Do they reach the bottom at the same time? If not, which arrives first? Explain.

15 A disk whose rotational inertia is I is given an angular acceleration by wrapping a string around the disk and passing the string over a pulley to a mass m. Derive an equation relating the angular acceleration α of the disk to the rotational inertia I, the radius r of the disk, and the mass m of the body attached to the string. Neglect friction.

Problems

1 A shaft 15.0 cm in diameter is to be turned on a lathe with a surface linear speed of 180 m/min. What is its angular speed?

2 A pulley 18.0 in in diameter makes 300 r/min. What is the linear speed of the belt if there is no slippage? The belt passes over a second pulley. What must be the diameter of the second pulley if its shaft turns at a rate of 400 r/min?
Ans. 1,410 ft/min; 13.5 in.

3 Assuming the orbit to be, to a first approximation, circular, calculate the angular speed of the earth in its orbit around the sun.

4 What is the angular speed of each of the three hands of an electric clock?
Ans. 0.105 rad/s; 1.74×10^{-3} rad/s; 1.45×10^{-4} rad/s.

5 An earth satellite in a circular orbit 300 mi above the surface of the earth makes a complete revolution in 96 min. Assuming the earth to be a sphere of radius 4,000 mi, calculate the angular and linear speed of the satellite.

6 The moon is 384,000 km from the earth and makes 1 r in 27.3 d. Find its angular and linear speed in its orbit.
Ans. 2.66×10^{-6} rad/s; 1.02×10^{3} m/sec.

7 A wheel has its speed increased from 120 to 240 r/min in 20 s. (*a*) What is the angular acceleration? (*b*) How many revolutions of the wheel are required?

8 A flywheel revolving at 400 r/min slows down with a deceleration of 4.00 rad/s^2. How long will it take the wheel to come to rest and how many revolutions will it make in doing so?
Ans. 10.4 s; 34.8 r.

9 What constant torque must be applied to a 200-lb cylindrical flywheel having a radius of 2.0 ft in order to increase the angular speed by 1,800 r/min in 15 s?

10 A uniform circular disk 3.0 ft in diameter weighs 960 lb. What is its moment of inertia about its usual axis? *Ans.* 34 slug·ft^2.

11 A cylinder 12.0 cm in diameter having a mass of 3.00 kg rests on a horizontal plane. Compute the moment of inertia of the cylinder about an axis along the line of contact with the plane.

12 A 20-lb bowling ball 8 in in diameter rolls along the ground at 20 ft/s. Find the moment of inertia about its line of contact with the ground.
Ans. 0.0973 slug·ft^2.

13 A solid cylinder of mass 2.00 kg and radius 8.00 cm is mounted with its axis horizontal. A thread is wound around the circumference of the cylinder and a 200-g mass is hung on the end of the thread and released. Neglecting friction, find the angular acceleration of the cylinder and the linear acceleration of the mass.

14 A 400-lb flywheel has a radius of 2.0 ft. (*a*) What constant torque is required to bring the wheel from rest to a speed of 120 r/min in 30 s? (*b*) How much work is done in this interval?
Ans. 21 lb·ft; 4,000 lb·ft.

15 A solid cylinder of mass 300 g and radius 1.5 cm starts from rest and rolls down a plane 1,470 cm long inclined at 30° to the horizontal. How long will it take to descend if there is no loss of energy due to friction? What will be its energy of rotation at the bottom?

16 What is the total kinetic energy possessed by a 20-lb bowling ball 8 in in diameter rolling along the alley without slipping at 20 ft/s?
Ans. 175 slug·ft^2/s^2.

17 A 50-g solid sphere is rolling without slipping along a horizontal surface at a speed of 800 cm/s. It comes to a point where the surface rises 30° above the horizontal. Neglect energy losses due to friction. (*a*) What is the total energy of the rolling sphere? (*b*) To what vertical height will it roll up the plane?

18 A flywheel in the form of a disk 4.0 ft in

diameter weighs 500 lb. What will its angular acceleration be if it is acted upon by a net torque of 300 lb·ft? *Ans.* 9.6 rad/s^2.

19 If the disk in Prob. 18 is rotating at 1,500 r/min, what torque is required to stop it in 3.0 min?

20 A uniform disk of 12.0-in diameter and 50.0-lb weight is mounted on an axle having a diameter of 1.00 in. A string is wrapped around the axle, and a constant force of 1.00 lb is exerted on it. What speed will the wheel acquire in 3.00 min? *Ans.* 38.5 rad/s.

21 The extremity of the hour hand of a clock travels one-eighteenth as fast as the extremity of the minute hand. If the minute hand is 15.24 cm long, how long is the hour hand?

22 (*a*) What is the constant torque which must be applied to a flywheel weighing 400 lb and having an effective radius of 2.00 ft if starting from rest and moving with uniform angular acceleration, it develops an angular speed of 1,800 r/min in 10.0 s? (*b*) If the shaft on which the pulley is mounted has a radius of 6.00 in and there is a tangential frictional force of 20.0 lb, how much must be the total torque?

Ans. 942 lb·ft; 952 lb·ft.

23 A steel ball rolls down an incline 200 cm long, making an angle of 4.0° with the horizontal. It requires 3.00 s for the ball to reach the bottom of the incline after starting from rest at the top. Calculate the value of g for this location.

24 A wheel of an automobile traveling 30.0 mi/h has an external radius of 14.0 in and weighs 80.0 lb. Assuming the effective radius to be 10.0 in, find (*a*) the kinetic energy of translation, (*b*) the kinetic energy of rotation, and (*c*) the total kinetic energy of the wheel.

Ans. 2,420 lb·ft; 1,2$\bar{3}$0 lb·ft; 3,680 lb·ft.

25 In the arrangement shown in Fig. 6-7, the solid disk and the pulley have the same radii, and the disk, pulley, and bob have equal masses. The plane has a slope of 30.0°; the disk rolls on the incline without slipping or loss of energy. Find the acceleration of the hanging block.

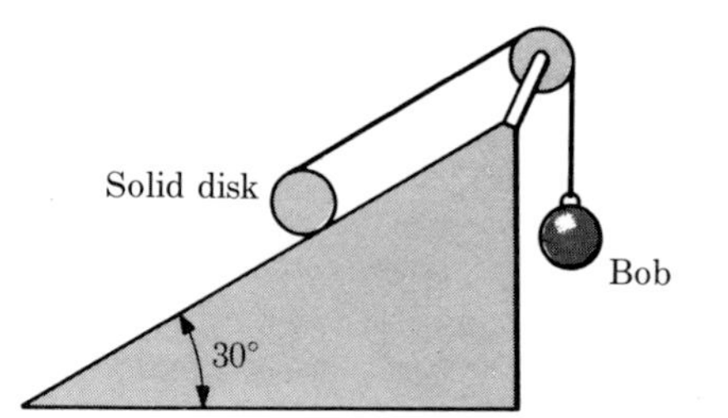

Figure 6-7

26 A 25-lb solid sphere is at the top of an incline 5.00 ft high and 13.0 ft long. What linear speed does the sphere acquire in rolling down the slope? *Ans.* 15.0 ft/s.

27 A spool of thread with an inner radius r_1 and an outer radius r_2 has a moment of inertia I_0 about the center of mass. The spool is pulled by a thread, as shown in Fig. 6-8, with a constant force F. Find the linear acceleration of the spool if $\theta = 0°$. (Assume that the spool rolls without slipping.)

28 The rotational inertia of an automobile wheel is found by hanging it on a knife-edge near the rim and allowing it to swing as a pendulum. About this axis, which is 15.0 in from the axis through the center of the wheel, the rotational inertia is 1.50 slug·ft^2. If the wheel weighs 20.0 lb, what is the rotational inertia about the axis through the center? *Ans.* 0.52 slug·ft^2.

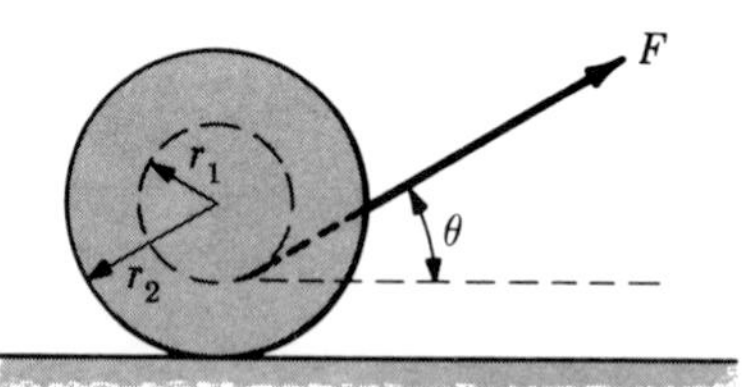

Figure 6-8

Johannes Diderik van der Waals, 1837–1923

Born in Leyden. Professor at Amsterdam University. Received the 1910 Nobel Prize for Physics for his work concerning the equations of state of gases and liquids.

7

Momentum

In collisions between objects in motion the forces involved may be extremely large. When the bat strikes a baseball, both ball and bat are greatly distorted while they are in contact. If a heavy, fast-moving truck strikes a house, the force may be sufficient to move the building from its foundation.

While the force may be very great during the impact, the large force acts for only a very short time. The force is not constant during the contact, varying between wide limits. The way in which the force varies during the collision depends upon the elastic properties of each of the bodies involved as well as upon their speeds. Because of the complicated manner in which the forces vary, it is usually convenient to study impact problems from the standpoint of momentum.

The laws of linear and angular momentum apply equally to the motions of atomic and subatomic particles as well as to larger bodies. We shall see later that the application of these laws has enabled us to deduce some of the properties of these particles.

7-1 IMPULSE AND MOMENTUM

When a body is acted upon by a resultant force the body is accelerated in accordance with Newton's second law of motion

$$\mathbf{F} = m\mathbf{a} = m\frac{\Delta \mathbf{v}}{\Delta t} = \frac{\Delta(m\mathbf{v})}{\Delta t} \tag{1}$$

$$\mathbf{F}\,\Delta t = m\,\Delta \mathbf{v} = m\mathbf{v}_1 - m\mathbf{v}_0 \tag{2}$$

The change in velocity of the body depends upon the mass of the body, upon the force F that acts, and upon the length of time Δt for which the force acts. The product of a force and the time during which it acts is called *impulse*. The impulse produces the change in motion shown in Eq. (2). In this equation it is implied that the mass m is constant and that there is simply a change in velocity, but this is not necessarily the case. *The product of the mass* m *of a body and the velocity* **v** *of the body is called its momentum* **p**.

$$\mathbf{p} = m\mathbf{v} \tag{3}$$

Every material particle has momentum when it is in motion. The impulse produces a change in the momentum.

Momentum is a *vector quantity,* its direction being the direction of the velocity. The units of momentum are made up from those of mass and velocity, for example, the kilogram-meter per second, gram-centimeter per second, and slug-foot per second.

When the term *momentum* is applied to an extended body, the velocity used is the velocity of the center of mass. To find the momentum of

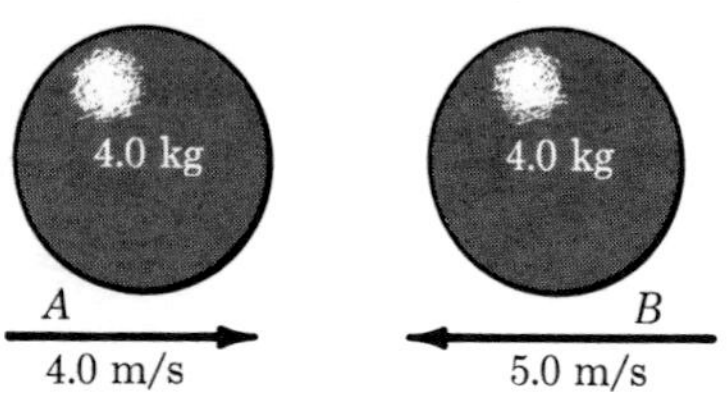

Figure 7-1
Two balls of equal masses having velocities that are unequal in magnitude and opposite in direction.

a system of two or more bodies, we must add the momentums of the individual bodies vectorially. Consider two 4.0-kg balls moving toward each other with unequal speeds of 4.0 m/s and 5.0 m/s as shown in Fig. 7-1. The momentum of A is

$$p_A = (4.0 \text{ kg})(4.0 \text{ m/s}) = 16 \text{ kg-m/s}$$

to the right, while that of B is

$$p_B = (4.0 \text{ kg})(5.0 \text{ m/s}) = 20 \text{ kg}\cdot\text{m/s}$$

to the left. The vector sum of the two momentums and hence the momentum of the *system* is 4.0 kg·m/s to the left.

7-2 CONSERVATION OF MOMENTUM

From consideration of Eq. (2) we see that the total impulse is equal to the change of momentum produced,

$$\mathbf{F}\,\Delta t = \Delta\mathbf{p} \tag{4}$$

or

$$\mathbf{F} = \frac{\Delta\mathbf{p}}{\Delta t} \tag{5}$$

The force continues to produce a change in momentum as long as it lasts. We may then make an alternative statement of Newton's second law of motion: *The rate of change of momentum of a body is proportional to the resultant force acting on the body.*

During the time when the momentum of a body or system of bodies is changing, the force may be constant or it may be changing in any manner. Under these conditions the total impulse is the sum of all the infinitesimal impulses ($\Sigma F\Delta t$) during the time interval as the force changes, or the force F of Eq. (4) represents the average force during the time interval Δt.

If there is no net external force acting upon a system of bodies, the momentum of the system does not change. This statement is a simple expression of *the law of conservation of momentum.* Frequently bodies that make up a system exert forces upon each other. These are *internal* in that, while they change the momentum of individual parts of the system, they do not change the momentum of the system as a whole. For example, when a missile explodes while in flight, the forces that arise due to the explosion are internal. The center of mass of the fragments continues to follow the path previously established. We assume that no new external forces arise. The use of the law of conservation of momentum simplifies the description of the mechanical behavior of common objects. In the motion of one particle or in every reaction between two or more particles the law of conservation of momentum applies.

If a net external force acts upon a system of bodies, the momentum of the system is changed, but in the process, some other set of bodies must gain (or lose) an amount of momentum equal to that lost (or gained) by the system. In every process where velocity is changed the momentum lost by one body or set of bodies is equal to that gained by another body or set of bodies.

$$\text{Momentum lost} = \text{momentum gained} \tag{6}$$

Let us consider further the balls shown in Fig. 7-1. If they continue to move toward each other, they will collide and in the collision each will exert a force on the other. The momentum of the system of two balls is 4.0 kg·m/s to the left before the impact. By the law of conservation of momentum it must be the same after the impact.

If the balls are elastic, they will rebound and the conservation law requires that the speeds of recoil shall be such that the total momentum shall remain unchanged.

The recoil of a gun is an example of conservation of momentum. The momentum of gun and bullet is zero before the explosion. The bullet gains forward momentum, and hence the gun must gain an equal backward momentum so that the sum will remain zero.

Example A 4.0-g bullet is fired from a 5.0-kg gun with a speed of 600 m/s. What is the speed of recoil of the gun?

The momentum of the gun is equal in magnitude but opposite in direction to that of the bullet:

$$m_1 v_1 + m_2 v_2 = 0$$

$$v_1 = -\frac{m_2}{m_1} v_2$$

$$v_1 = -\frac{0.0040 \text{ kg}}{5.0 \text{ kg}} (600 \text{ m/s})$$

$$= -0.48 \text{ m/s}$$

In the firing of the gun, forces are exerted, one on the gun and the other on the projectile. These forces, however, are *internal;* i.e., they are within the system of the gun and bullet that we considered. If we consider the bullet alone, the force becomes an external force and causes a change in momentum of the bullet, but in accordance with Newton's third law, a force equal in magnitude but opposite in direction acts on the gun, giving it a momentum equal in magnitude to that given to the bullet but opposite in direction.

The operation of jet engines and rockets depends upon conservation of momentum. Gases expelled at high speed by the engine require large forces to give them the high momentums involved. The reaction force on the engine supplies the motive force for the vehicle.

Molecules, atoms, electrons, and the multitude of subatomic particles interact in collisions in which there is conservation of momentum. Even in the collisions of photons and electrons in which the photons are scattered and the electrons recoil, conservation of momentum is involved.

Example A 6.6×10^4 N car traveling with a speed of 30 km/h strikes an obstruction and is brought to rest in 0.10 s. What is the average force on the car?

From Eq. (2),

$$F \Delta t = mv_1 - mv_0$$

$$m = \frac{W}{g} = \frac{6.6 \times 10^4 \text{ N}}{9.8 \text{ m/s}^2} = 6.8 \times 10^3 \text{ kg}$$

$$v_0 = 30 \text{ km/h} = 8.33 \text{ m/s}$$

$$F = \frac{m(v_1 - v_0)}{\Delta t} = \frac{6.8 \times 10^3 \text{ kg}(0 - 8.33 \text{ m/s})}{0.10 \text{ s}}$$

$$= -5.66 \times 10^5 \text{ N}$$

7-3 ELASTIC AND INELASTIC COLLISIONS

In every collision or interaction, momentum is conserved; i.e., the total momentum before the collision is equal to the total momentum after the collision. Energy is conserved in the collision, but the type of energy usually changes. In any ordinary collision the total kinetic energy never increases and usually decreases as a result of the collision,

$$(E_k)_2 \lesseqgtr (E_k)_1 \tag{7}$$

The exceptions to this general statement occur when there is a release of new energy as a result of the collision. If one is unwary enough to hit a dynamite cap with a hammer, the release of the energy of the cap results in a considerable increase in kinetic energy of the remaining parts of the system. A collision of a particle with an atomic nucleus frequently results in release of energy, with resulting high kinetic energy of the product particles.

In the collision of two bodies A and B having

masses of m_A and m_B, initial velocities of v_{1A} and v_{1B}, and final velocities v_{2A} and v_{2B}, the momentum equation gives

$$m_A v_{1A} + m_B v_{1B} = m_A v_{2A} + m_B v_{2B} \qquad \text{vector sum} \qquad (8)$$

From kinetic-energy considerations,

$$\tfrac{1}{2} m_A v_{2A}{}^2 + \tfrac{1}{2} m_B v_{2B}{}^2 \leqq \tfrac{1}{2} m_A v_{1A}{}^2 + \tfrac{1}{2} m_B v_{1B}{}^2 \qquad \text{scalar sum} \qquad (9)$$

Collisions in which the kinetic energy is conserved are said to be *elastic,* such as two steel balls; all others are *inelastic,* such as putty colliding with some other object. Rearranging Eqs. (8) and (9), we obtain

$$m_A(v_{2A} - v_{1A}) = m_B(v_{1B} - v_{2B}) \qquad (10)$$

and $$m_A(v_{2A}{}^2 - v_{1A}{}^2) \leqq m_B(v_{1B}{}^2 - v_{2B}{}^2) \qquad (11)$$

If we divide Eq. (11) by Eq. (10), we have

$$\frac{v_{2A}{}^2 - v_{1A}{}^2}{v_{2A} - v_{1A}} \leqq \frac{v_{1B}{}^2 - v_{2B}{}^2}{v_{1B} - v_{2B}} \qquad (12)$$

or $$v_{2A} + v_{1A} \leqq v_{1B} + v_{2B}$$

and $$v_{2A} - v_{2B} \leqq v_{1B} - v_{1A} \qquad (13)$$

For an elastic collision,

$$v_{2A} - v_{2B} = v_{1B} - v_{1A} \qquad (14)$$

The relative velocity after collision, $v_{2A} - v_{2B}$, is less than or equal to the negative of $v_{1A} - v_{1B}$, the relative velocity before collision. The negative ratio of the relative velocity after collision to the relative velocity before collision is called the *coefficient of restitution, e.*

$$e = -\frac{v_{2A} - v_{2B}}{v_{1A} - v_{1B}} = \frac{v_{2A} - v_{2B}}{v_{1B} - v_{1A}} \qquad (15)$$

If the collision is perfectly elastic, $e = 1$. If the collision is completely inelastic, $e = 0$ and in this case the two colliding bodies adhere and move as one body after collision.

Example A 2.0-kg ball B traveling with a speed of 22 m/s overtakes a 4.0-kg ball A traveling in the same direction as the first, with a speed of 10 m/s. If the coefficient of restitution is 0.80, find the speeds of the two balls after the collision. The velocity to the right will be considered +.

$$e(v_{1B} - v_{1A}) = v_{2A} - v_{2B}$$

$$0.80\,(22\text{ m/s} - 10\text{ m/s}) = v_{2A} - v_{2B}$$

$$v_{2A} - v_{2B} = 9.6\text{ m/s}$$

$$m_A v_{1A} + m_B v_{1B} = m_A v_{2A} + m_B v_{2B}$$

$$(4.0\text{ kg})(10\text{ m/s}) + (2.0\text{ kg})(22\text{ m/s}) = (4.0\text{ kg})(v_{2A}) + (2.0\text{ kg})(v_{2B})$$

$$(2.0\text{ kg})(v_{2A}) + (1.0\text{ kg})(v_{2B}) = 42\text{ kg}\cdot\text{m/s}$$

or $$2.0v_{2A} + v_{2B} = 42\text{ m/s}$$

and $$v_{2A} - v_{2B} = 9.6\text{ m/s}$$

$$3.0v_{2A} = 51.6\text{ m/s}$$

Final velocity of A:

$$v_{2A} = 17.2\text{ m/s}$$

Substituting,

$$17.2\text{ m/s} - v_{2B} = 9.6\text{ m/s}$$

Final velocity of B:

$$v_{2B} = 7.6\text{ m/s}$$

Therefore, both balls continue to move in the initial direction but with their speeds changed.

7-4 EXPERIMENTAL VERIFICATION OF THE LAW OF CONSERVATION OF MOMENTUM

Experimental verification of the law of conservation of momentum depends upon the ability to measure velocities before and after impact. Where all the velocities have the same direction, measurement of speed is sufficient.

In a typical experiment a steel sphere A is rolled down an incline (a grooved track) and the point of impact with the floor is observed (Fig. 7-2). Then it is rolled down from the same height but this time collides with a smaller steel sphere B resting at the end of the incline. The two spheres then fall off the end of the incline to the ground, and the two points of impact are observed.

The object of the experiment is to determine whether the momentum of A before impact is equal to the sum of the momentum of A after impact and the momentum of B. Knowing the mass of A and B and the vertical distance from the end of the incline to the ground S_v and by measuring the horizontal range S_h for two impacts of A with the ground (one without collision and one with collision) and for the one impact of B with the ground, the momentums can be computed.

The velocity of B as it reaches the bottom of the incline can be found by using the following equations. Since

$$S_v = \tfrac{1}{2}gt^2 \qquad \text{and} \qquad t = \sqrt{\frac{2S_v}{g}}$$

$$\text{and} \qquad S_h = vt \qquad \text{or} \qquad v = \frac{S_h}{t}$$

assuming the horizontal velocity v to be constant, then,

$$v = \frac{S_h}{\sqrt{2S_v/g}}$$

Therefore, the velocities of A, with and without collision, and for B can be determined by knowing only the range and the height, assuming g to be a constant.

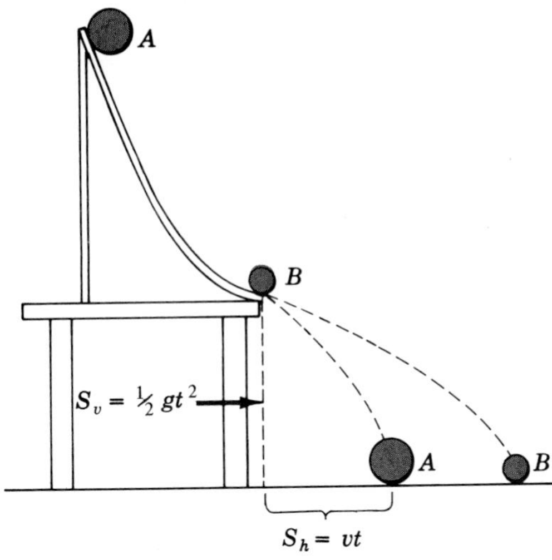

Figure 7-2
Conservation-of-momentum demonstration.

$$\underset{\text{(without collision)}}{\text{mass}_A \times \text{vel}_A} = \underset{\text{(with collision)}}{\text{mass}_A \times \text{vel}_A} + \text{mass}_B \times \text{vel}_B$$

or

$$\underset{\text{(without collision)}}{\text{mom}_A} = \underset{\text{(with collision)}}{\text{mom } A} + \text{mom } B$$

7-5 COLLISIONS OF BODIES NOT MOVING ALONG THE SAME LINE

Thus far we have considered only those collisions in which each of the bodies moves along the same line. If the motions are not thus simply related but lie in one plane, we must consider two components of momentum at right angles to each other and the corresponding components of velocity. From the law of conservation of momentum we may write two equations, one for each component, since the momentum in each of these directions must be conserved separately. The kinetic-energy relation gives one more equation. These three equations are not sufficient to determine the four quantities necessary to specify the motion of the two bodies, i.e., the two speeds and the two directions. If the direction of motion of one of the bodies after collision is known, the problem can be solved. Many types of collisions are studied in cloud-chamber photographs. These tracks enable us to determine directions, often in three dimensions, from which the mechanics of the collisions can be inferred.

7-6 A VARIATION OF A CLASSIC TWO-BODY-COLLISION EXPERIMENT

Collision experiments are frequently used in physics laboratories to investigate two fundamental principles—the conservation of momentum and the conservation of energy. Analysis of tracks made by the collisions of subatomic particles in photographic emulsions, cloud chambers, and bubble chambers suggest that these principles hold for small- as well as large-scale interactions. This observation has led to the successful use of large-body-collision events as models for the study of small-scale collisions. The experiment described here is a typical attempt to construct a macroscopic analog of a microscopic event. Efforts to enhance the understanding of physical phenomena through the construction of models is a widely accepted technique. It is the intent of the authors to propose the following variation of a two-body-collision experiment to serve as a model to further the student's comprehension of the conservation of momentum and energy.

It can be shown mathematically and approximated experimentally by the technique proposed below that an elastic collision between a large body in motion and an infinitely small body at rest will result in the latter attaining a velocity roughly equal to twice that of the large body at the instant of impact. An example of this phenomenon would be the striking of a golf ball by a golf club. The ball should leave the tee at twice the speed that the club head possesses at the instant it struck the ball. The mathematical proof of this statement is presented here in detail for the convenience of the reader.

Let M = mass of large body
m = mass of small body
V = initial velocity of large body
v = initial velocity of small body (zero)
V' = final velocity of large body
v' = final velocity of small body

The momentum and kinetic energy of the system before and after collision are

$$MV_{\text{club}} + mv_{\text{ball}} = MV'_{\text{club}} + mv'_{\text{ball}} \qquad \text{where } mv = 0 \quad (a)$$

$$\tfrac{1}{2}MV^2_{\text{club}} + \tfrac{1}{2}mv^2 = \tfrac{1}{2}MV'^2 + \tfrac{1}{2}mv'^2 \qquad \text{where } \tfrac{1}{2}mv^2 = 0 \quad (b)$$

From (a),

$$V' = \frac{(MV - mv')}{M} \quad \text{and} \quad V' = V - \frac{m}{M}v'$$

Substituting into (b) and solving for v',

$$\tfrac{1}{2}MV^2 = \tfrac{1}{2}M\left(V - \frac{m}{M}v'\right)^2 + \tfrac{1}{2}mv'^2$$

$$\tfrac{1}{2}MV^2 = \tfrac{1}{2}MV^2 - mVv' + \frac{m^2v'^2}{2M} + \tfrac{1}{2}mv'^2$$

Reducing,

$$0 = -Vv' + \frac{1}{2}\frac{m}{M}v'^2 + \frac{1}{2}v'^2$$

Combining terms,

$$V = \left(\frac{m}{2M} + \frac{1}{2}\right)v' \quad \text{and} \quad v' = \frac{V}{(m/2M) + \frac{1}{2}}$$

Since $m/2M$ is small, $v' = V/\frac{1}{2}$ and

$$v' = 2V \qquad (c)$$

To prove this experimentally, a hammer pivoted at its handle end serves as the large mass and a coin (a dime) as the small mass at rest. The hammer is suspended so that it hangs freely with the center of its head just touching the top of a lab table. The hammer is drawn back through an angle ϕ and released so that it strikes the coin resting in front of the hammer and extending slightly over the edge of the table. The angle ϕ is held constant by the use of a stand mounted on the floor as a control and to restrict the backward swing of the hammer (Fig. 7-3).

The velocity of the hammer at the point of collision can be approximated by considering the hammer as a pendulum of mass M. At the starting point it has a potential energy Mgh which is changed to kinetic energy equal to $\frac{1}{2}MV^2$ at the lowest position in its swing. Assuming that all potential energy is changed to kinetic energy, the hammer's velocity at this point is

$$\frac{1}{2}MV^2 = Mgh$$

$$V = \sqrt{2gh} = \text{velocity of hammer}$$

The velocity of the coin can be determined by observation of the average distance it is moved by the impact and by direct measurement of the length of time the coin is in motion by using a stopwatch or by strobe photography in which the flashes come at a predetermined interval. Then, since $\bar{v} = s/t$ and $(v' + v'')/2 = \bar{v}$, where $v'' = 0$, $v' = 2s/t$. An analysis of data will verify that $v' = 2V$.

It should be noted that many variables enter into this determination and certain assumptions and limitations are made. We should keep in mind that the purpose of "model building" in science is to permit generalizations to be made about phenomena so that more precise experimentation can be carried on later. In the illustration above, the hammer is nonuniform in density and geometric design, an error is introduced if the angle ϕ becomes too large, the mass of the coin is considered to be infinitesimally small, and skill is required to time the motion of the coin, yet even the beginning physicist can reproduce the experiment and find that the coin will leave the point of impact with the hammer at approximately twice the speed that the hammer approached the point of impact with the coin.

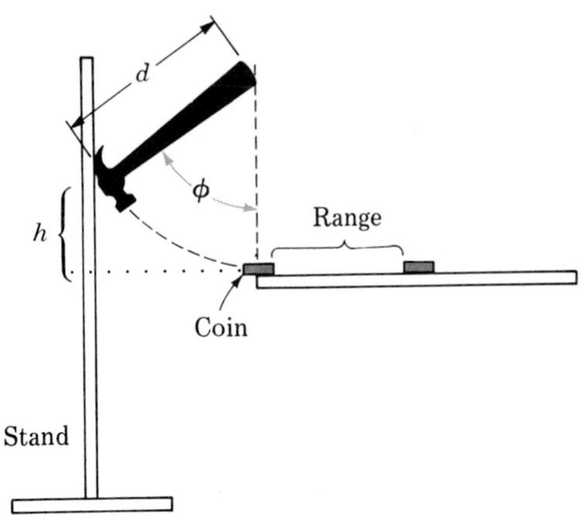

Figure 7-3
Experimental setup for hammer-and-coin problem.

7-7 COLLISION CROSS SECTION

Let us consider two particles whose lines of motion are parallel but displaced a distance s from each other. The two particles will "collide" if the force that they exert upon each other in the approach is great enough to cause measurable deviation in at least one of the paths. Physical "contact" is not necessary in a collision. For example, electrically charged particles exert appreciable forces at considerable distances. Astronomical bodies exert gravitational forces that cause changes in initial paths at very great distances. Whatever the nature of the forces involved, momentum is conserved in the collision. For these collisions the paths of the two bodies change gradually rather than abruptly from a point, as illustrated in Fig. 7-4. Around the center of mass of each particle we may draw a circle whose plane is perpendicular to the direction of motion and of such radius that if the initial line of motion of another particle passes within that area, there would be sufficient force to indicate a collision.

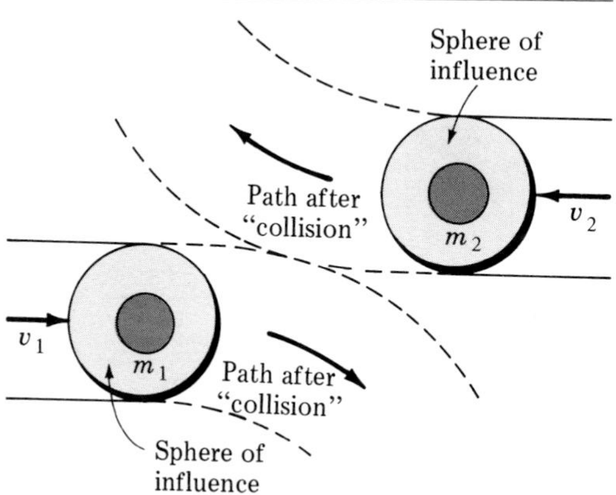

Figure 7-4
Two objects in "collision." Two objects, m_1 and m_2, whose lines of motion are parallel but separated by a distance s, interact in a "collision" that deflects each. Momentum is conserved in the collision.

The area of the circle, actually a sphere of influence, would represent a *collision cross section* for the particle. *The collision cross section of a particle usually depends upon the energy of the approaching particle,* being smaller for a high-energy particle than for a lower-energy particle. For a given force the deflection of the high-energy particle is less than that for a lower-energy particle; hence a closer approach is required for the high-energy particle to produce a measurable deflection.

In nuclear physics we can learn things about the forces involved by measuring collision cross sections experimentally (Fig. 7-5). In such experiments it is not possible to make measurements for individual collisions, but data from many impacts may be analyzed statistically.

7-8 ANGULAR MOMENTUM

Let us consider a particle of mass m that is moving with velocity $\mathbf{v}$. It has momentum $\mathbf{p} = m\mathbf{v}$. With reference to an axis through an origin O (Fig. 7-6) the particle has an *angular momentum* $\mathbf{H}$ defined by the vector equation

$$\mathbf{H} = \mathbf{r} \times \mathbf{p} = \mathbf{r} \times m\mathbf{v} \tag{16}$$

Since angular momentum is the product of linear momentum and a moment arm, it is also called *moment of momentum*. The magnitude of H is given by

$$H = rp \sin \theta = mrv \sin \theta \tag{17}$$

where θ is the angle between r and p. *The direction of* H *is perpendicular to the plane determined by* r *and* p *in the sense described by the right-hand rule given in Chap.* 5.

The units in which angular momentum is expressed can be determined by examination of Eq. (17). In the mks system mass is in kilograms, distance in meters, and velocity in meters per second. Hence the angular momentum is in kilogram-meters^2 per second. Similar analysis gives a cgs unit of gram-centimeters^2 per second and a British unit of slug-feet^2 per second.

In the rotation of extended rigid bodies angular momentum about the axis of rotation is an important property. Each particle of the body has an angular momentum, at a certain instant, given by Eq. (17). Since the velocity at every instant is perpendicular to r, $\sin \theta = 1$ and

$$H = mrv = mr^2\omega \tag{18}$$

In a rigid body all the particles have the same angular velocity, and the total angular momentum of the body is

$$H = (\Sigma mr^2)\omega = I\omega \tag{19}$$

In vector form Eq. (19) becomes

$$\mathbf{H} = I\boldsymbol{\omega} \tag{20}$$

where I is the rotational inertia of the body.

Figure 7-5
Collision between an electrically charged particle and the nucleus of an atom. The picture is a track in a "cloud chamber." Momentum is conserved in the collision.

Example In the Bohr-atom model an electron of mass 9.11×10^{-31} kg revolves in a circular orbit about the nucleus. It completes an orbit of radius 0.53×10^{-10} m in 1.51×10^{-16} s. What is the angular momentum of the electron in this orbit?

$$H = mrv = mr^2\omega$$

$$= (9.11 \times 10^{-31}\ \text{kg}) \frac{2\pi\ \text{rad}}{1.51 \times 10^{-16}\ \text{s}} (0.53 \times 10^{-10}\ \text{m})^2$$

$$= 1.06 \times 10^{-34}\ \text{kg} \cdot \text{m}^2/\text{s}$$

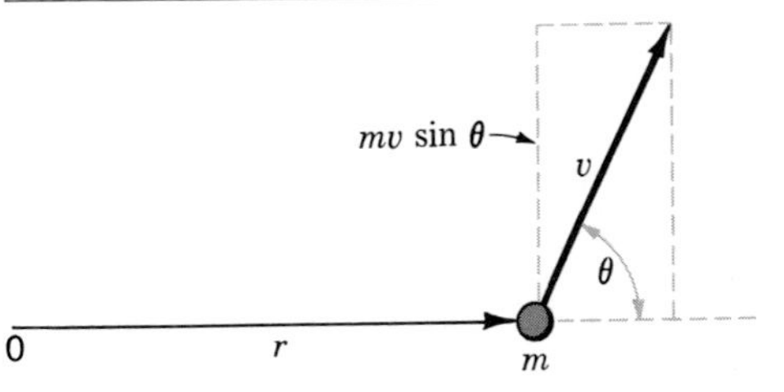

Figure 7-6
Particle of mass m, velocity $\mathbf{v}$, and momentum $\mathbf{p} = m\mathbf{v}$ has an angular momentum $\text{H} = \mathbf{r} \times \mathbf{p} = m\mathbf{r} \times \mathbf{v}$ about an axis through O.

7-9 CONSERVATION OF ANGULAR MOMENTUM

In Chap. 6 we discussed Newton's laws of motion as extended to rotary motion of a rigid body. We may restate the laws in terms of angular momentum: *If there is no resultant external torque acting upon a body, the angular momentum of the body remains unchanged.* This statement is *the law of conservation of angular momentum.* When a rigid body is set into rotation, in the absence of a net external torque, it will maintain its initial angular velocity in both magnitude and direction. Hence the rotating body tends to maintain the same plane of rotation. The rotation of the wheels helps maintain the balance of a bicycle or motorcycle. The barrel of a gun is rifled to cause the bullet to spin so that it will not "tumble."

If the distribution of mass of a rotating body is changed as it rotates, the angular velocity must also change to maintain the same angular momentum. Suppose that a man stands on a stool that is free to rotate with little friction (Fig. 7-7). If he is set in rotation with his arms outstretched, he will rotate at a constant rate. If he lowers his arms, his rotational inertia is decreased and his rate of rotation increases. A figure skater changes the rate of rotation by changing the distance of arms and legs from the axis of rotation. A diver achieves fast or slow rotation by doubling the body or extending the parts away from the axis.

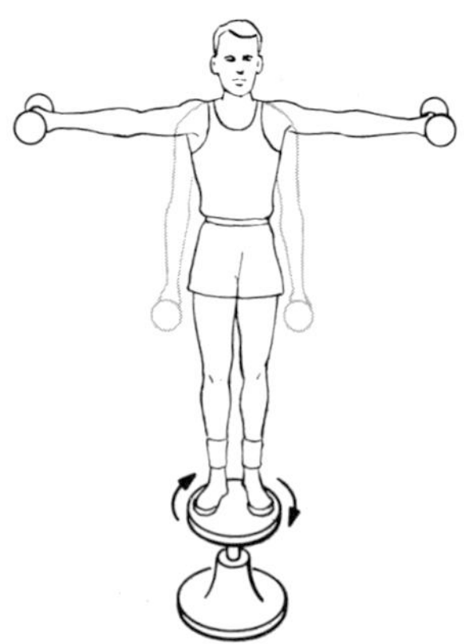

Figure 7-7
Conservation of angular momentum.

Example A figure skater holds two heavy objects in his hands. Assume that he is turning at 1 r/s with the objects held at arm's length (30 in from the axis of rotation). If he were to pull the objects in by folding his arms until the objects were 10 in from the axis of rotation, how fast will he be turning?

The initial angular velocity of the objects is 1 r/s $= 2\pi$ rad/s $= \omega_1$, and $v_1 = r_1\omega_1$. Therefore,

$$v_1 = \frac{2\pi r}{t_1}$$

The angular momentum H for each object is $r_1 m v_1$ and the combined angular momentum is $2r_1 m v_1$; and substituting,

$$H_1 = 2r_1 m \frac{2\pi r_1}{t_1}$$

$$= \frac{4\pi r_1^2 m}{t_1}$$

When the objects are drawn in toward the axis of rotation,

$$H_2 = \frac{4\pi r_2^2 m}{t_1}$$

According to the law of conservation of angular momentum $H_1 = H_2$. Therefore,

$$\frac{4\pi r_1^2 m}{t_1} = \frac{4\pi r_2^2 m}{t_2}$$

and

$$\frac{r_1^2}{t_1} = \frac{r_2^2}{t_2}$$

$$t_2 = \frac{r_2^2 t_1}{r_1^2} = \frac{(10\text{ in})^2}{(30\text{ in})^2}(1\text{ s})$$

$$= \tfrac{1}{9}\text{ s}$$

So we see the skater will be making 1 r in $\frac{1}{9}$ s or 9 r/s.

The second law of motion may be stated: *When a net external torque acts on a body, the time rate of change of angular momentum is proportional to the net torque and is in the direction of the net torque.*

$$\mathbf{L} = \frac{\Delta \mathbf{H}}{\Delta t} = \frac{\Delta(I\boldsymbol{\omega})}{\Delta t} = I\frac{\Delta \boldsymbol{\omega}}{\Delta t} \qquad (21)$$

$$\mathbf{L} = I\boldsymbol{\alpha} \qquad (22)$$

where α is the angular acceleration. From Eq. (21)

$$\mathbf{L}\,\Delta t = \Delta \mathbf{H} = I\,\Delta \boldsymbol{\omega} \qquad (23)$$

The product of torque and time is the *angular impulse.* Angular impulse is equal to the change in angular momentum.

7-10 VECTOR PROPERTIES IN ANGULAR MOTION

Angular velocity, angular acceleration, angular momentum, and torque are all vector quantities. The vector representing the vector quantity is parallel to the axis of rotation in the sense given by the right-hand rule described in Chap. 5. In Fig. 7-8 the angular velocity of the rotating disk is represented in magnitude and direction by the vector $\boldsymbol{\omega}$ parallel to the axis AB. If a torque is applied to the disk in such a manner that the axis of the torque is the same as the axis of rotation of the disk, the direction of the torque will be parallel to AB and the resultant angular acceleration will also be parallel to AB. Hence such a torque will produce a change only in the *magnitude* of the angular velocity.

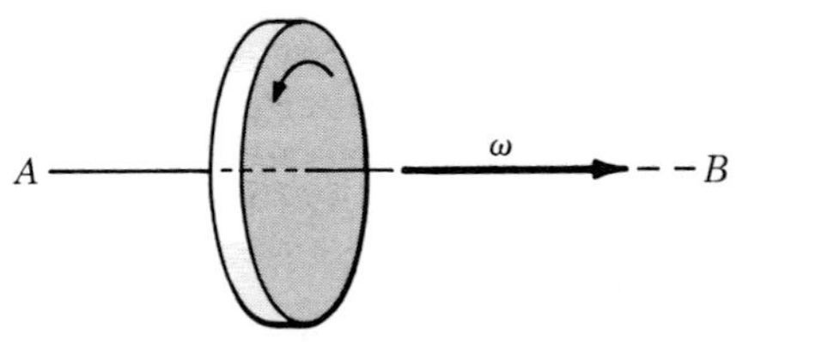

Figure 7-8
Angular velocity represented by a vector ω parallel to AB, the axis of spin.

7-11 PRECESSION

If a torque is applied to the disk by two equal vertical forces F at the ends of the axle (Fig. 7-9), the axis of the torque is a horizontal line such as CD and the torque is represented by the vector $\mathbf{L}$. The angular acceleration produced is in the direction CD. When the disk is viewed from above, its motion is described by the vectors drawn in Fig. 7-10, $\boldsymbol{\omega}$ representing the original angular velocity of the disk and $\boldsymbol{\alpha}$ representing the angular acceleration produced by the torque. Since the angular acceleration is at right angles to the angular velocity, no change in angular speed is produced but only a change in direction. That is, the axis of rotation changes its direction, rotating in a counterclockwise sense. Note that the direction of motion of the end of the axle is at right angles to the original direction of the axle and also at right angles to the direction of

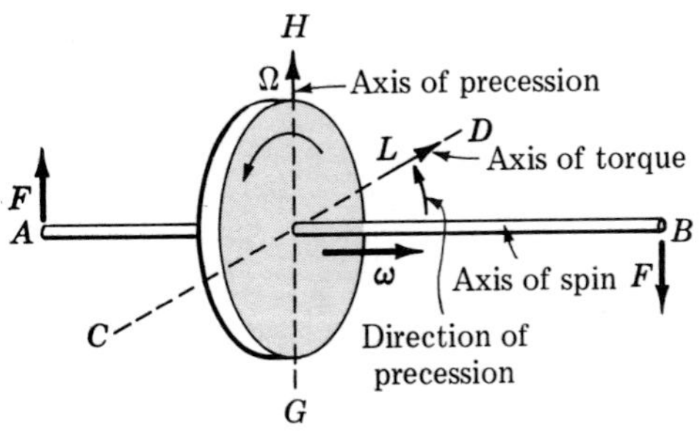

Figure 7-9
Torque L acting on a rotating disk, due to forces F.

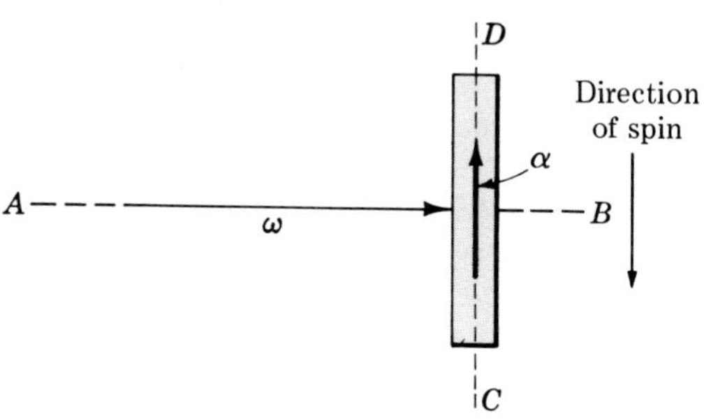

Figure 7-10
Vectors representing angular velocity ω and angular acceleration α for the disk of Fig. 7-9. View is of spinning disk from above.

the *force* applied. The direction of motion of the end of the axle is in the direction of the *torque* applied. Such shift in the axis of spin of a rotating body is called *precession*. The vertical axis GH about which the turning of the spin axis takes place is called the *axis of precession*. In Fig. 7-11 are shown the three axes involved in the motion. Each axis is perpendicular to the other two. The line AB is the axis of spin along which is drawn the vector representing the angular velocity $\boldsymbol{\omega}$ of the rotating body; CD is the axis of torque along which is drawn the vector representing **L**; and GH is the axis of precession along which is drawn a vector representing the angular velocity of precession $\boldsymbol{\Omega}$.

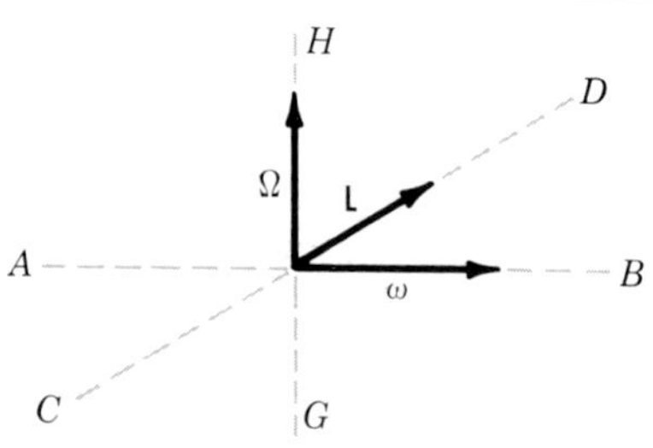

Figure 7-11
Three axes describing precession: axis of spin *AB*, axis of torque *CD*, and axis of precession *GH*.

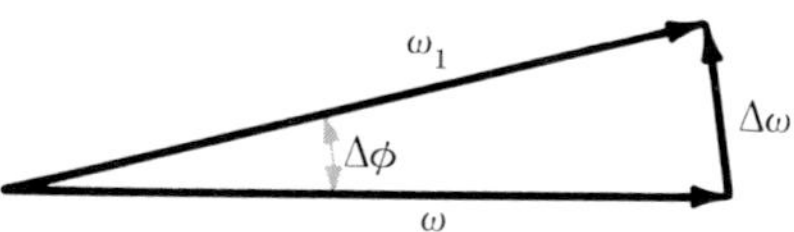

Figure 7-12
Vector diagram for precession.

The rate of precession, i.e., the angular speed about the axis of precession, depends upon the angular momentum of the rotating body about the axis of spin and the torque applied. In Fig. 7-12 $\boldsymbol{\omega}$ represents the original angular velocity, $\boldsymbol{\omega}_1$ represents the angular velocity a short time Δt later, and $\Delta\boldsymbol{\omega}$ represents the change in angular velocity. The change in angular velocity is a change in direction only, since the torque is at right angles to the initial angular velocity. If $\Delta\phi$ is small

$$\Delta\omega = \omega\,\Delta\phi \qquad \text{approx}$$

As $\Delta\phi$ approaches zero, the approximation becomes better and in the limit the equation becomes exact.

From Eq. (23),

$$L(\Delta t) = \Delta(I\omega) = I(\Delta\omega) = I\omega(\Delta\phi)$$

$$L = I\omega\,\frac{\Delta\phi}{\Delta t}$$

But $\Delta\phi/\Delta t = \Omega$, the angular velocity of precession. Then

$$L = I\omega\Omega \tag{24}$$

or

$$\Omega = \frac{L}{I\omega} \tag{25}$$

The rate of precession is directly proportional to the torque applied and inversely proportional to the angular momentum of the rotating body about the axis of spin.

Example A disk with its axle has a weight of 4.0 lb and a moment of inertia of 0.0040

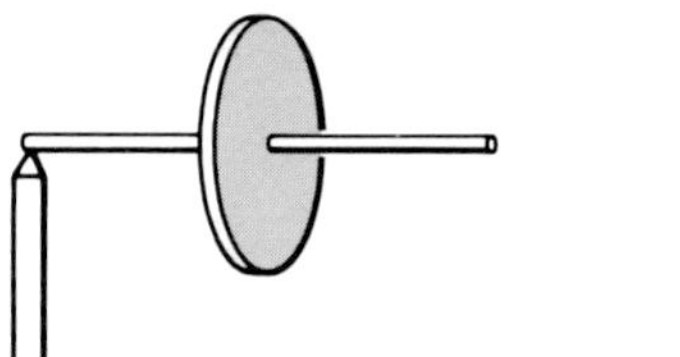

Figure 7-13
A rotating disk supported at one end precesses.

slug·ft². It is supported at one end of the axle (Fig. 7-13), the support being 3.0 in from the center of gravity. When the disk is turning at the rate of 1,800/min, what is the rate of precession?

The weight, considered as concentrated at the center of gravity, produces a torque

$$L = Ws = (4.0 \text{ lb})(0.25 \text{ ft})$$

$$\omega = 1{,}800 \text{ r/min} = 30 \text{ r/s}$$

$$= 2\pi \times 30 \text{ rad/s}$$

$$\Omega = \frac{L}{I\omega} = \frac{(4.0 \text{ lb})(0.25 \text{ ft})}{(0.0040 \text{ slug}\cdot\text{ft}^2)(2\pi \times 30 \text{ rad/s})}$$

$$= 1.3 \text{ rad/s}$$

In the foregoing discussion it is assumed that the rate of change of angular momentum is small compared with the original angular momentum, in which case the change is one of direction only. If the precession is prevented by restraints on the rotation of the axle, the torque set up by the restraint turns the axle downward so that the disk behaves just as would a nonrotating disk. If the torque on the rotating disk is too large, the motion becomes suddenly unstable and the above description no longer applies.

7-12 GYROSCOPES

A gyroscope is a rotating body that is so mounted as to be free to turn about any of three mutually perpendicular axes. Such a mounting is shown in Fig. 7-14. If the wheel spins with high angular speed about axis 1, the base may be turned in any manner without transmitting a torque, except for frictional torque, to the rotating wheel, which will therefore maintain its axis of rotation unchanged as the support is tilted in any manner so long as the wheel rotates rapidly. Since the angular momentum depends upon the moment of inertia and upon the angular velocity, a heavy wheel rotating at high speed would have a large angular momentum and correspondingly great stability. If a torque is applied perpendicular to the axis of spin, there will be precession of the axis, as previously described.

The two principal characteristics of the behavior of gyroscopes are (1) stability of the axis and (2) precession. Both these characteristics are employed in the many applications of gyroscopes. In those applications which require stability great care must be taken in mounting the gyroscope wheel so that as little torque as possible is transmitted to the axis. In this class of application are the gyropilot, gyrohorizon indicator, directional gyro, and to some extent the gyrocompass. The latter, however, is so constructed that when it is in any position except that with its axis parallel to the axis of the earth, there will be a torque

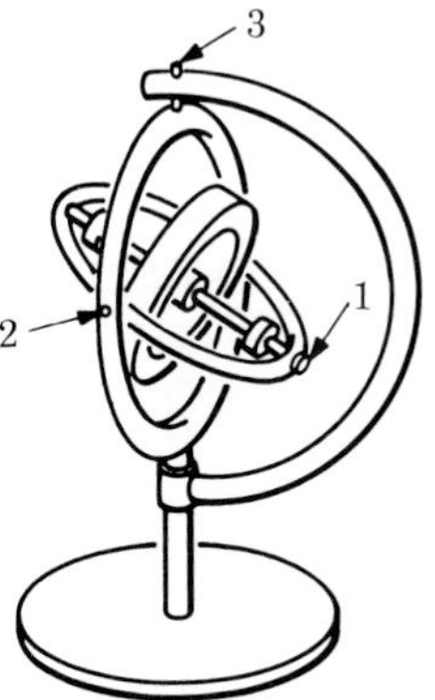

Figure 7-14
A gyroscope has three axes of freedom.

which will cause a precession into that position.

For rotating bodies that show gyroscopic action, the greater the angular momentum, the more marked will be the effect. Some of these effects are useful, others harmful. As a car turns a corner, the gyroscopic action of the wheels produces a torque tending to overturn the car. If the flywheel rotates counterclockwise as one looks forward in a car, the force on the front wheels decreases when the car turns to the right but increases when the car turns to the left.

The gyrostabilizer may be used to reduce the roll of a boat by exerting a torque opposite to the roll. The gyroscope wheel, spun at high speed by a motor, is mounted with its axis vertical in such a manner that the axis may be tilted forward or backward but not sideways. Assume that the spin of the gyroscope is counterclockwise as viewed from above. When the boat rolls, say, to the right, a control gyro closes contacts of a motor which tilts the axis forward. There results a torque opposing the roll to the right. Similarly if the roll is to the left, the motor tilts the axis backward, supplying a torque again opposing the roll.

Inertial guidance systems make use of the properties of rotating bodies to maintain a selected orientation or to turn the system by properly applied torques.

SUMMARY

Momentum is the product of the mass and velocity of a body. It is a vector quantity.

$$\mathbf{p} = m\mathbf{v}$$

Common units of momentum are the kilogram-meter per second, gram-centimeter per second, and slug-foot per second.

Impulse is the product of a force and the time during which it acts. Impulse is equal to the change in momentum.

$$F\,\Delta t = p_2 - p_1 = mv_1 - mv_0$$

Some units of impulse are the newton-second, the dyne-second, and the pound-second.

The *law of conservation of momentum* states that the momentum of a body or system of bodies does not change unless a resultant external force acts upon it.

An elastic collision is one in which kinetic energy is conserved, as well as momentum. In an inelastic collision, momentum is conserved but kinetic energy is not. The *coefficient of restitution* is the negative ratio of the relative velocity after a collision to the relative velocity before collision. Its value is unity for a perfectly elastic collision, 0 for a completely inelastic collision, and between 0 and 1 for all others.

Angular momentum is the vector product of the displacement from the axis and linear momentum.

$$\mathbf{H} = \mathbf{r} \times \mathbf{p} = m\mathbf{r} \times \mathbf{v} = I\omega$$

$$H = mvr \sin\theta$$

The *law of conservation of angular momentum* states that the angular momentum of a rotating body remains unchanged unless it is acted upon by a resultant external torque.

Angular momentum is a *vector quantity,* as are also angular velocity, angular acceleration, and torque. The direction of the vectors representing these quantities is parallel to the axis in the sense given by the right-hand rule.

Precession is the change in direction of the axis of spin under the action of a torque.

A gyroscope exhibits the properties of precession and stability of axis. These properties result in many useful applications.

Questions

1 Explain why momentum is considered to be a vector quantity.

2 Why do we say momentum must be conserved? Give examples of physical events or devices depending upon the conservation of momentum.

3 Why does a gun appear to have a greater kick when fired with the butt held loosely against the shoulder than when held tightly?

4 Explain how the term "conservation" applies (*a*) to energy and (*b*) to momentum.

5 During high windstorms such as tornadoes, bits of straw have been seen embedded in telephone posts. Explain why this happens.

6 When one billiard ball strikes a second in such a manner that their centers of gravity are not in the line of motion of the first ball, their paths after collision do not lie on the same line. Draw a vector diagram to represent the momentums before and after such a collision if the angle between the paths is 100°.

7 The historical development of the 45-caliber pistol is traced back to the time of a native uprising in the Philippine Islands. The standard army 38-caliber pistol was not effective in stopping the charges of machete-swinging natives. Why do you suppose the change of weapons was made?

8 The Russian spaceships have returned to earth by landing on solid land, whereas the United States spaceships have landed in the ocean. What are the advantages of both systems for returning space vehicles?

9 Prove that if the coefficient of restitution is 1, kinetic energy is conserved in collision.

10 How may the speed of a rifle bullet be measured with simple apparatus?

11 Suggest some probable reasons for the difference between observed values of momentum before and after impact.

12 Some automobiles are now being built with hydraulic pistons attached to the front and rear bumpers of the cars. Explain how these pistons may make driving a car more safe.

13 A tank truck which is half-filled with a liquid is considered to be more difficult to drive on icy roads than a completely filled truck. Why do you suppose this might be so?

14 How may a high diver turning somersaults in the air arrange on the way down to strike the water head first? What physical principle does he make use of?

15 A stone is dropped in the center of a deep vertical mine shaft. Will the stone continue in the center of the shaft or will it strike the side? If it strikes the side, will it be on the north, south, east, or west side? Consider the linear speed of rotation of the earth and assume ideal conditions.

16 Show that the fractional loss in kinetic energy before and after impact of a ballistic pendulum of mass m struck inelastically by a moving ball of mass M is given by $m/(M + m)$. Into what form of energy is the lost kinetic energy transformed?

17 On what physical principle or law is based the statement that the horizontal component of the velocity of a projectile remains constant?

18 A projectile is fired due south in the Northern Hemisphere. When it strikes the ground will it be east or will it be west of the north-south line along which it started? Consider the speed of rotation of the earth. If the projectile is fired north, on which side of the line will it strike? On which side of an east-west line will it strike if fired east? if fired west?

19 What happens to the momentum of a meteorite if it enters the earth's atmosphere? What happens to its kinetic energy? What happens to its center of gravity if the meteorite explodes above the earth?

20 Why is a rifle barrel "rifled"?

21 An airplane propeller rotates counterclockwise as viewed from the pilot's seat. What effect does its gyroscopic action have when the plane is turning toward the right? when the plane is diving?

22 What is the gyroscopic action of the front wheels of a car traveling at high speed when one turns toward the right?

Problems

1 What is the momentum of a 100-kg shell if its speed is 1,500 m/s?

2 What is the momentum of a 1.5-N baseball which is dropped from the top of the 170-m Washington Monument? *Ans.* 8.84 kg·m/s.

3 What is the momentum of a 4.45×10^4 N truck when traveling at the rate of 100 km/h?

4 A projectile has a momentum of 1.6×10^4 kg·m/s and a speed of 200 m/s. What is its mass? What is its weight? *Ans.* 80 kg; 784 N.

5 A 5.0-oz baseball arrives at the bat with a speed of 160 ft/s. It remains in contact with the bat for 0.020 s and leaves with the direction of its motion reversed and at a speed of 280 ft/s. (*a*) What impulse does the bat impart to the ball? (*b*) What is the value of the average force exerted by the bat on the ball?

6 At what speed does a gun recoil if it weighs 44.5 N and fires a 1.0-g bullet at 2,500 m/s? *Ans.* 0.55 m/s.

7 Two balls *A* and *B*, weighing 49 N each, approach each other with speeds of 20 m/s and 30 m/s, respectively. On the assumption of a perfectly elastic collision, what will be their speeds after collision?

8 Two perfectly elastic balls, weighing 6.0 lb and 4.0 lb, approach each other with speeds of 20 ft/s and 35 ft/s, respectively. What will be their speeds after they collide? *Ans.* −24 ft/s; 31 ft/s.

9 A 5.00-lb ball and a 10.0-lb ball have speeds of 10.0 ft/s and −14.0 ft/s, respectively, as they approach each other. Find their speeds after collision if the coefficient of restitution is 0.800.

10 A 10-ton truck traveling east at 60 mi/h collides head-on with a 1-ton auto heading west at 45 mi/h. Assuming the coefficient of restitution to be 70 percent, find the velocities of the truck and the car after the collision.
Ans. Truck: 43.8 mi/h, east; car: 117 mi/h, east.

11 Two lead spheres *A* and *B* of masses 25 g and 75 g, respectively, are hung on strings so that they just touch when the strings are vertical. *A* is drawn aside until it has risen 15 cm. When it is released, it makes a perfectly inelastic collision with *B*. What is the speed of *A* at the instant of collision? With what speed do *A* and *B* move off together?

12 Two balls of mass 30.0 g and 90.0 g are supported by strings 100 cm long. The larger ball is pulled aside until its center of gravity has been raised 5.00 cm and then released. Assuming the collision to be perfectly elastic, find the velocity of each ball after the collision. *Ans.* 148 cm/s; 49.4 cm/s.

13 A $5\frac{1}{4}$-oz baseball is thrown so that it is captured by the 11.7-lb block of a ballistic pendulum. The block is displaced so that its center of gravity is raised 2.7 in. With what speed was the ball pitched?

14 Two objects *A* and *B* having masses of 0.15 kg and 0.10 kg approach each other at 4.0 m/s and 3.0 m/s, respectively. (*a*) If the collision is perfectly inelastic, what are the post-collision speeds of the two objects? (*b*) If the coefficient of restitution is 0.60, what are their post-collision speeds?
Ans. (*a*) Both move at 1.2 m/s in direction of object *A*. (*b*) *B* is moving at 3.7 m/s and *A* at 0.48 m/s, both having reversed their directions.

15 A 0.250-kg ball is fired from a spring gun into a 12.0-kg block of a ballistic pendulum. The block is displaced so that its center of gravity is raised 6.00 cm. (*a*) Find the speed of the ball as it left the gun. (*b*) Find the loss in kinetic energy during the collision.

16 A 30-ft sloop moving at 6 mi/h and weighing 6,400 lb collides with a dock piling when landing and comes to a stop after moving the piling 2 ft. What is the force on the boat? *Ans.* 3,850 lb.

17 The rotor of an electric motor has a moment of inertia of 25 slug·ft^2. If it is rotating at a rate of 1,200 r/min, what is its angular momentum?

18 The force applied to an object increases uniformly with time at the rate of 60 lb/s for 0.20 s. What is the average force during this period of time. What is its impulse? *Ans.* 6.0 lb; 1.2 lb·s.

19 What torque is required to change the speed of the rotor of an electric motor from 600 r/min to 1,200 r/min in 2.0 s if the rotor has a moment of inertia of 25 slug·ft^2.

20 A 10-kg axe strikes a log with a force of 500 N and comes to rest in 0.05 s. Find the impulse. How fast did the axe approach the log? *Ans.* 25.0 N·s; 2.5 m/s.

21 Consider the earth as a uniform sphere of mass 5.98×10^{24} kg, revolving about the sun in 365 d in an approximate circle of radius 1.50×10^8 km. Find the magnitude of its angular momentum in this motion.

22 A fire hose with a 1.5-in-diameter nozzle directs a stream of water horizontally with a speed of 80 ft/s against a vertical wall. What force is exerted on the wall, assuming the water moves parallel to the wall after striking it? Water has a density of 62.4 lb/ft^3. *Ans.* 150 lb.

23 An electron has a mass of 9.11×10^{-31} kg. It revolves about a nucleus in a circular orbit of radius 0.529×10^{-10} m at a speed of 2.2×10^6 m/s. Find the magnitude of its linear and angular momentum in this motion.

24 A gyroscope wheel weighs 7.0 lb and has an effective radius of 4.0 in. It spins with its axis horizontal at a rate of 3,000 r/min, clockwise as viewed from the pivot. The gyroscope is supported by a pivot near one end of the axle 6.0 in from the center of gravity (Fig. 7-13). What is the angular velocity of precession? When viewed from above is the precession clockwise or is it counterclockwise? *Ans.* 0.46 rad/s.

Wilhelm Wien, 1864–1928

Born in Geffken, East Prussia. Röntgen's successor at Würzburg and Munich Universities. In 1911 Wien was awarded the Nobel Prize for Physics for his discoveries regarding the laws governing the radiation of heat.

8

Uniform Circular Motion

Uniform motion along a straight line seems "natural." We accept this type of motion and usually seek no cause-and-effect relationship for it. However, if there is a change in the direction of the motion, we recognize that some disturbing force is in action. Just as a force is required to change the speed of an object, so must a force act to cause a change in the *direction* of the motion. Whenever the net force on a body acts in a direction other than the original direction of motion, it changes the direction of the motion. Such acceleration is very common, for it is present whenever a car turns a corner, an airplane changes its direction, a wheel turns, a planet moves in its orbit around the sun, or an electron moves in its path around an atomic nucleus. The simplest type of motion in which the direction changes is uniform circular motion in which there is no change in speed but only a change in direction.

8-1 CENTRAL ACCELERATION

When an object is moving in a circular path with constant speed, there must be a constant force acting at right angles to the motion of the object. Since velocity depends on direction as well as magnitude, its velocity is continually changing. The acceleration produced by this force results in a change in *direction* but no change in speed. Therefore the acceleration must always be at right angles to the motion, since any component in the direction of the motion would produce a change in speed. The acceleration is always directed toward the center of the circle in which the body moves. It is constant in magnitude but continually changing direction. In Fig. 8-1 a body is moving with uniform speed v and constant angular speed ω in a circular path with a radius r. The velocities of the body at the points A and B are, respectively, v_A and v_B, equal in magnitude but differing in direction by a small angle $\Delta\theta$. In the vector triangle, Δv represents the change in velocity in the time Δt required for the body to move from A to B.

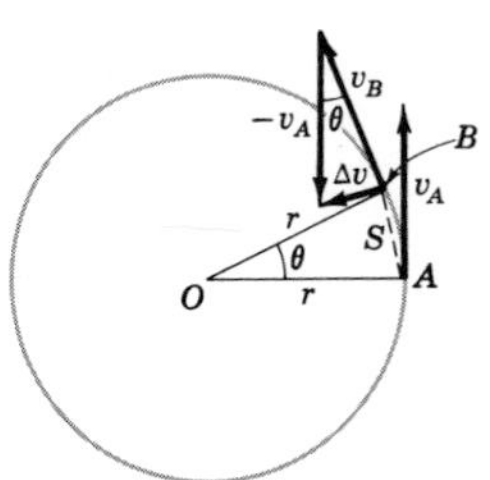

Figure 8-1
Changing velocity in uniform circular motion.

The vector triangle formed by determining the vector difference of v_A and v_B and the triangle formed by drawing radii to reference points A and B are similar, that is, the corresponding sides are proportional in length and the included angle θ is the same for both cases. Therefore, $\Delta v/v = s/r$, where v is the average velocity at A and B. The arc AB is the actual distance that the object moves in a given time t. From $v = s/t$, this distance is equal to vt. It can be shown that for a small angle θ, the length of the chord s is approximately equal to the length of the arc AB, or vt. In the limit as Δt approaches zero we can consider vt to equal s. Also it can be seen that the direction of Δv becomes more and more nearly perpendicular to that of v.

Substituting in the above equation, $\Delta v/v = vt/r$ and $v/t = v^2/r$. Since acceleration is defined as a change of velocity in a given time, $a = \Delta v/t = v^2/r$. In the limit as Δt approaches zero, the instantaneous acceleration is found to be directed toward the center of the circle and is called the central acceleration a_c. Therefore,

$$a_c = \lim_{\Delta t \to 0} \frac{\Delta v}{\Delta t} = \omega^2 r = \frac{v^2}{r} \qquad (1)$$

Since $v = r\omega$, we were able to substitute to get $a_c = r^2\omega^2/r = \omega^2 r$. This equation states that the acceleration increases as the speed is increased and, for a given speed, is greater for a shorter radius. The acceleration is at right angles to the velocity and hence is directed toward the center of the circle.

In Eq. (1) the angular speed ω must be expressed in radians per second, and the units of a then depend upon the units in which r and v are expressed.

Example What is the acceleration of a point on the rim of a flywheel 0.90 m in diameter, turning at the rate of 1,200 r/min?

$$\omega = 1{,}200 \text{ r/min} = 20 \text{ r/s}$$

$$= 20 \times 2\pi \text{ rad/s}$$

$$r = 0.45 \text{ m}$$

$$a = \omega^2 r = (20 \times 2\pi \text{ rad/s})^2(0.45 \text{ m})$$

$$= 7{,}\bar{1}00 \text{ m/s}^2$$

Example A train whose speed is 100 km/h rounds a curve whose radius of curvature is 150 m. What is its acceleration?

$$v = 100 \text{ km/h} = 27.8 \text{ m/s}$$

$$a_c = \frac{v^2}{r} = \frac{(27.8 \text{ m/s})^2}{150 \text{ m}} = 5.15 \text{ m/s}^2$$

8-2 CENTRIPETAL FORCE

According to Newton's laws of motion any object that experiences an acceleration is acted upon by an unbalanced force, a force which is proportional to the acceleration and in the direction of the acceleration. The net force that produces the central acceleration is called *centripetal force* and is directed toward the center of the circular path. Every body that moves in a circular path does so under the action of a centripetal force. A body moving with uniform speed in a circle is not in equilibrium.

From Newton's second law the magnitude of the centripetal force is given by

$$F_c = ma = m\frac{v^2}{r} = m\omega^2 r \qquad (2)$$

where m is the mass of the moving object, v is its linear speed, r is the radius of the circular path, and ω is the angular speed. If the mass is in kilograms, the radius in meters, and the speed in meters per second, the force is in newtons. If m is in grams, v in centimeters per second, and r in centimeters, F_c is in dynes. If m is in slugs, v in feet per second, and r in feet, F_c is in pounds.

Example A 1.44×10^4 N car traveling with a speed of 100 km/h rounds a curve whose radius is 150 m. Find the necessary centripetal force.

$$m = \frac{W}{g} = \frac{1.44 \times 10^4 \text{ N}}{9.8 \text{ m/s}^2} = 1.47 \times 10^3 \text{ kg}$$

$$v = 100 \text{ km/h} = 27.8 \text{ m/s}$$

$$F_c = m\frac{v^2}{r} = (1.47 \times 10^3 \text{ kg})\frac{(27.8 \text{ m/s})^2}{150 \text{ m}}$$

$$= 7.5\bar{7}0 \times 10^3 \text{ N}$$

An inspection of Eq. (2) discloses that the centripetal force necessary to pull a body into a circular path is directly proportional to the square of the speed at which the body moves and inversely proportional to the radius of the circular path. Suppose, for example, that a 4.0-kg object is held in a circular path by a string 1.2 m long. If the object moves at a constant speed of 0.80 m/s,

$$F_c = \frac{mv^2}{r} = \frac{(4.0 \text{ kg})(0.80 \text{ m/s})^2}{1.2 \text{ m}}$$

$$= 2.1 \text{ N}$$

If the speed is doubled, with the radius kept constant, F_c becomes four times as great, or 8.4 N. If, instead, the radius is decreased from 1.2 m to 0.60 m, with the speed maintained at 0.80 m/s, F_c increases to 4.2 N. If at any instant the string breaks, eliminating the centripetal force, the object will retain the velocity it has at the instant the string breaks and will travel at constant speed along a line tangent to the circle, according to Newton's first law. The paths taken by sparks from a grinding wheel are an illustration of this fact. This action is illustrated in Fig. 8-2.

A body can travel at uniform speed in a circular path only when the resultant of the forces acting on the body is constant in magnitude and always directed toward the center of the circle.

8-3 EFFECT OF ROTATION OF THE EARTH ON ACCELERATION DUE TO GRAVITY

In our discussion of the acceleration due to gravitational forces at the surface of the earth (Sec. 3-11) we neglected the rotation of the earth. Be-

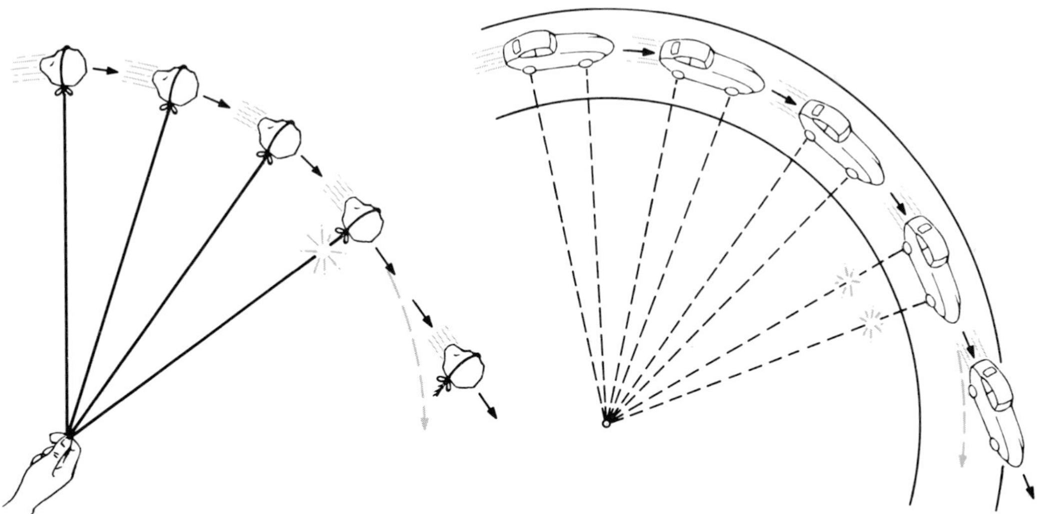

Figure 8-2
If the string breaks, the rock flies off. If friction "breaks," the car skids off.

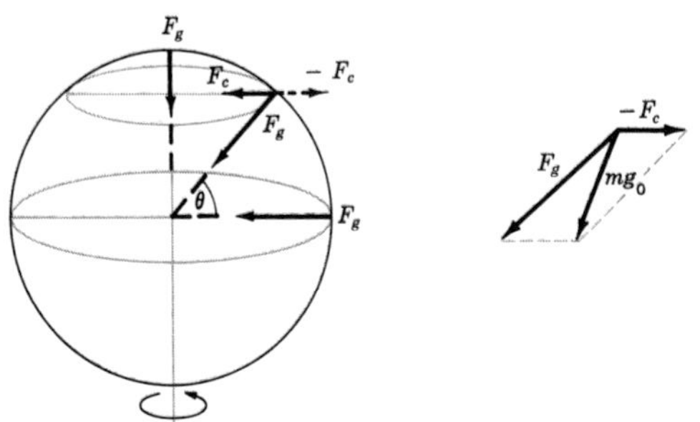

Figure 8-3
Gravitational and centripetal forces on a rotating earth.

cause of this rotation a part of the gravitational force is needed to produce the centripetal force necessary to hold a body in its circular path as the earth rotates. Consider the diagram of the earth in Fig. 8-3. At the equator the center of the circle of revolution is the center of the earth and

$$F_g = \frac{GM_e m}{R_e^2} = mg_0 + ma_c \qquad (3)$$

where g_0 is the observed acceleration due to gravity at the equator and a_c is the central acceleration in the circular path. The measured value of the acceleration due to gravity at the equator would then be less than the acceleration produced by the gravitational force by an amount a_c, where

$$a_c = \omega^2 r$$

For the earth,

$$\omega = \frac{2\pi \text{ rad}}{24 \text{ h}} = \frac{\pi}{12 \times 3{,}600} \text{ rad/s}$$

$$= 7.3 \times 10^{-5} \text{ rad/s}$$

$$r = R_e = 6.4 \times 10^6 \text{ m}$$

Then

$$a_c = (7.3 \times 10^{-5} \text{ rad/s})^2 (6.4 \times 10^6 \text{ m})$$

$$= 3.4 \times 10^{-2} \text{ m/s}^2$$

This represents a decrease in g of only 3.4 parts in 980, or less than 0.4 percent.

At any latitude θ the centripetal force F_c is not directed toward the center of the earth.

$$\mathbf{F}_g = m\mathbf{g}_0 + \mathbf{F}_c$$

The observed force mg_0 is the vector difference of the two forces

$$m\mathbf{g}_0 = \mathbf{F}_g - \mathbf{F}_c \qquad (4)$$

The vector difference, which is the observed weight, differs in both magnitude and direction from $\mathbf{F}_g$, but in both respects the differences are small.

At the geographical poles of the earth the central force is zero, and the observed value of g is the true gravitational value.

8-4 CENTRIPETAL FORCE DOES NO WORK

Work has been defined as the product of force and the displacement in the direction of the force. Since centripetal force acts at right angles to the direction of motion, there is no displacement in the direction of the centripetal force and it accomplishes no work. Aside from the work done against friction, which has been neglected, no energy is expended on or by an object while it is moving at constant speed in a horizontal circular path. This conclusion is consistent with the observation that if the speed is constant, the kinetic energy of the body is also constant.

8-5 ACTION AND REACTION

Newton's third law expresses the fact that for every force that is exerted on one body there is a second force equal in magnitude but opposite in direction acting on a second body. When an object not free to move is acted upon by an external force, it is pushed or pulled out of its natural shape. As a consequence it exerts an *elastic* reaction in an attempt to resume its normal shape. On the other hand, the action of a force upon a free object results in an acceleration. The object exerts an inertial reacting force upon the agent of the accelerating force.

The elastic reacting force of a stretched body is equal in magnitude to the stretching force but opposite in direction. So also the inertial reacting force of an accelerated body is equal in magnitude to the accelerating force but opposite in direction. It should be remembered, however, that a force of reaction is exerted *by* the reacting object, not on it.

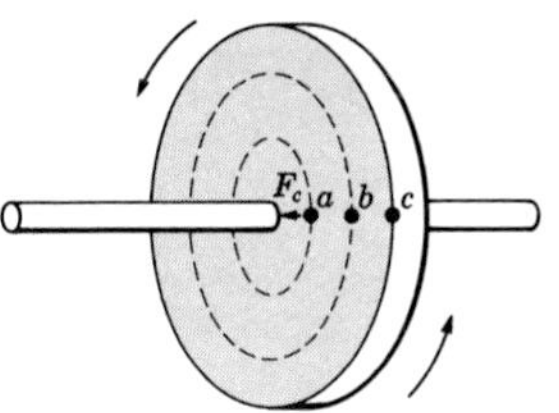

Figure 8-4
Centripetal force acting on rotating flywheel.

8-6 CENTRIFUGAL REACTION

It should be noted that the often-used term *centrifugal force* is a misnomer. To a person in a car that is traveling around a curve the outward movement of his body is obvious. If one really analyzes what is happening, however, it will become apparent that the observed effect is due to the person's body attempting to continue on in a straight line in keeping with Newton's first law of motion while the car is deviating from this straight line. The momentum of the person's body increases with the rate of motion and in general is directly proportional to the centripetal force. In general usage, the centrifugal force is considered to be equal in magnitude and opposite in direction to the centripetal force. Care should be taken though to use the term correctly. An illustration may prove helpful here. If a ball is whirling around at the end of a string which you are holding in your hand, the *ball* is undergoing a centripetal force (inward), but according to Newton's third law of motion (action-reaction) the ball exerts an outward force on your *hand,* a centrifugal outward force.

As the speed of a heavy solid wheel, a flywheel, increases, the force needed to hold the parts of the wheel in circular motion increases with the square of the speed, as indicated by Eq. (2). Finally the cohesive forces between the molecules are no longer sufficient, and the wheel disintegrates, the parts flying off along tangent lines like mud from an automobile tire. The inward-directed force is greatest near the center of the wheel, at a in Fig. 8-4, for each ring must supply the force required to accelerate all rings (b and c) farther from the axis.

When a container full of liquid is being whirled at a uniform rate, the pail exerts an inward force on the liquid sufficient to keep it in circular motion (Fig. 8-5). The bottom of the pail presses on the layer of liquid next to it; that layer in turn exerts a force on the next; and so on. In each layer the pressure (force per unit area) must be the same all over the layer or the liquid will not remain in the layer. If the liquid is of uniform density (mass per unit volume), each element of volume of mass m in a given layer will experience an inward force $m(v^2/r)$ just great enough to maintain it in that layer and there will be no motion of the liquid from one layer to another.

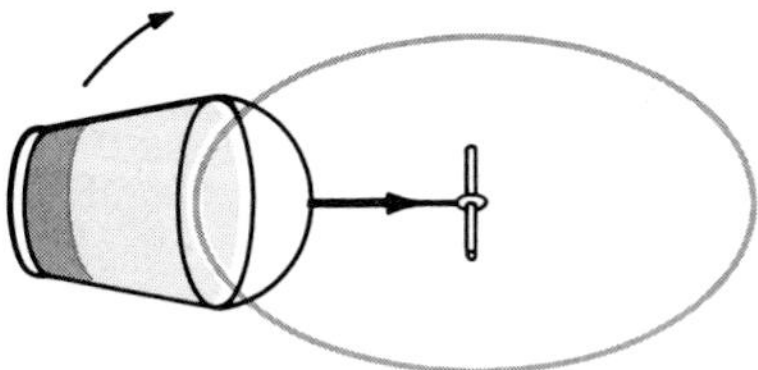

Figure 8-5
Centripetal force on a liquid. The principle of the centrifuge.

If, however, the layer is made up of a mixture of particles of different densities, the force required to maintain a given element of volume in the layer will depend upon the density of the liquid in that element. Since the inward force is the same on *all* the elements in a single layer, there will be a motion between the layers. For those parts which are less dense than the average, the central force is greater than that necessary to hold them in the layer; hence they are forced inward. For the parts more dense than the average the force is insufficient to hold them in the circular path, and they will move to a layer farther out. As rotation continues, the parts of the mixture will be separated, with the least dense nearest the axis and the most dense farthest from the axis. This behavior is utilized in the centrifuge, a device for separating liquids of different densities. Types of centrifuge are commonly used to separate mixtures of liquids or mixtures of solids in liquids. Very high speed centrifuges may be used to separate gases of different densities. The ultracentrifuge, designed by J. W. Beams, operates at angular speeds greater than 10^6 r/s and may produce centripetal accelerations higher than 10^9 *g*. By the use of such a device it is possible to separate materials whose densities are very nearly equal, such as those composed of different isotopes of a given substance.

8-7 THE CONICAL PENDULUM

A *conical pendulum* consists of a body of mass m attached to a flexible string, the body revolving in a horizontal circle with uniform circular motion about a vertical axis (Fig. 8-6). The two forces acting on the body are its weight mg vertically downward and the tension T in the string directed along the string. The resultant of these forces is the centripetal force directed toward the center of the horizontal circle. For different speeds of revolution the angle θ must change so that this condition is satisfied.

Using the closed polygon of forces method,

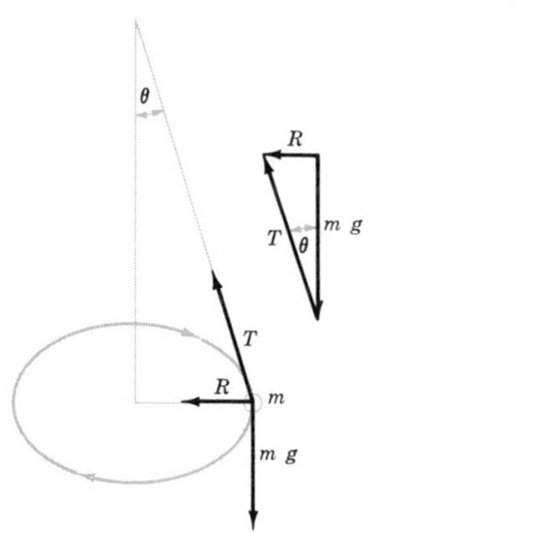

Figure 8-6
A conical pendulum.

since m is in equilibrium, it can be shown that R is equal to

$$\tan\theta = \frac{R}{mg}$$

$$R = mg\tan\theta$$

An adaptation of the conical pendulum is the centrifugal governor, in which the string is replaced by rigid arms that are hinged. As the mass

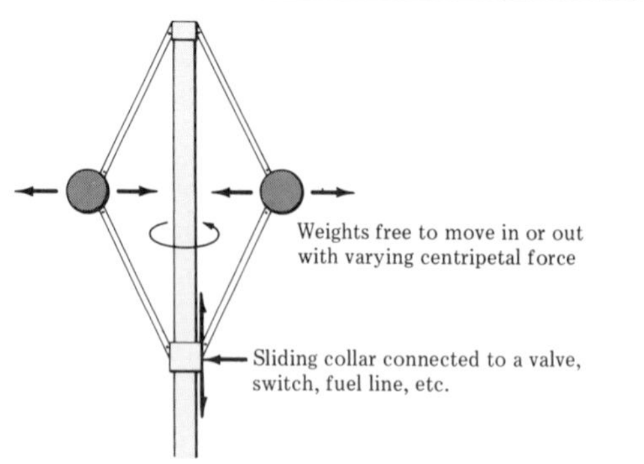

Figure 8-7
Centrifugal governor.

moves in or out, a valve mechanism is opened or closed. In this case the forces do not all act through a single point.

8-8 BANKING OF CURVES

A runner, in going around a curve, leans inward to obtain the centripetal force that causes him to turn (Fig. 8-8). The roadway must exert an upward force sufficient to sustain his weight, while at the same time it must supply a horizontal centripetal force. If the roadway is flat, the horizontal force is entirely frictional. In that case the frictional force may not be large enough to cause a sharp turn when the surface of the roadway is smooth.

If the roadway is tilted from the horizontal, a part of the horizontal force is still supplied by friction but the remainder is a result of the reaction of the surface. If the angle of banking is properly selected, the force the roadway exerts is perpendicular to its surface and no frictional force is necessary.

For this ideal case, as shown in Fig. 8-8, the thrust F of the roadway is perpendicular to the surface AC. The weight W of the runner is directed vertically downward. The resultant of these two forces F and W is the force F_c, the horizontal centripetal force. In the force triangle the angle θ is the angle of bank of the roadway,

$$\tan\theta = \frac{F_c}{W} = \frac{mv^2/r}{mg} = \frac{v^2}{rg} \qquad (5)$$

Equation (5) indicates that since the angle of banking depends upon the speed, the curve can be ideally banked for only one speed. At any other speed the force of friction must be depended upon to prevent slipping. The banking of highway curves, by reducing the lateral force of friction on the tires, greatly reduces wear in addition to contributing to safety.

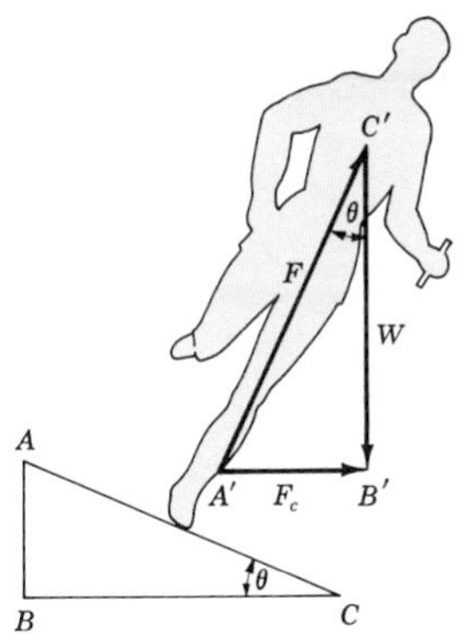

Figure 8-8
Advantage of banking curves.

Example A curve on a highway forms an arc whose radius is 150 ft. If the roadbed is 30 ft wide and its outer edge is 4.0 ft higher than the inner edge, for what speed is it ideally banked?

The tangent of the angle of bank is the ratio of the difference in elevation of the two edges and the width of the road.

$$\tan\theta = 4.0\text{ ft}/30\text{ ft}$$

$$= \frac{v^2}{rg}$$

hence

$$v^2 = gr\tan\theta$$

so that

$$v = \sqrt{\frac{(32\text{ ft/s}^2)(150\text{ ft})(4.0\text{ ft})}{30\text{ ft}}} = 25\text{ ft/s}$$

Example An unbanked curve has a radius of 80.0 m. What is the maximum speed at which a car can make the turn if the coefficient of static friction μ_s is 0.81?

When a curve is not banked, the centripetal force must be supplied by friction between the wheels and the roadway. Since the normal force is the weight,

$$F_c = \mu_s mg = m\frac{v^2}{r}$$

whence

$$v^2 = \mu_s gr = (0.81)(9.80\ \text{m/s}^2)(80.0\ \text{m})$$

$$v = 25\ \text{m/s}$$

8-9 CURVILINEAR MOTION

Frequently the net force acting upon a body is neither parallel to the direction of its motion nor at right angles to that direction. In this case neither the speed nor the direction remains constant. Such motion may be readily studied by considering two components of the acceleration, one parallel to the original direction of motion, the other perpendicular to that direction.

One of the most common of such motions is planetary motion, in which the force on the moving body is inversely proportional to the square of the radius and always directed toward a fixed point. The body travels in an ellipse, the fixed point being at one focus. The speed is greatest when the moving body is near the focus, perigee, less when it is farther away, apogee. This motion is called planetary motion because the planets move in this manner in their journeys around the sun. The gravitational forces acting are inversely proportional to the square of the radius.

Artificial satellites are subject to the inverse-square attraction of the earth and their motion is very nearly planetary motion. The path could be circular if the final speed and direction of firing were exactly right to give that path. If the direction of firing is not horizontal, or if the speed is above or below that necessary for the circular path, an elliptical path will result. The plane of the orbit does not generally remain fixed with respect to the surface of the earth, because of the fact that the satellite is fired from a rotating earth and the rotation continues after the firing. Planetary motion is discussed in more detail in Chap. 9 when Kepler's laws are presented.

Electrified particles show an inverse-square law of attraction. Hence the path of such a particle, such as an electron as it revolves about the nucleus of a simple atom, would be similar to that of a planet around the sun. Its path would be approximately circular or elliptical. In his first picture of the hydrogen atom, Bohr assumed circular motion of the electrons (Chap. 47).

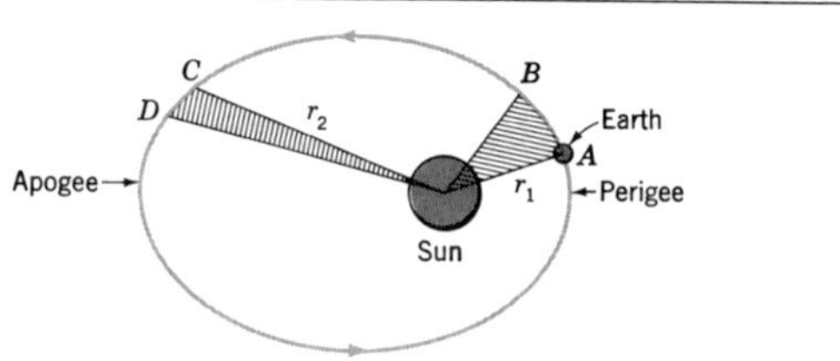

Figure 8-9
Elliptical path of earth around sun showing points of perigee (closest) and apogee (farthest) distance to sun. Shaded sections are equal in area and represent area "swept out" during a given time t.

8-10 PUTTING IT ALL TOGETHER

Earlier in the book we referred to a description of the "pyramidal structure of physics, so beautiful and almost unique to our field." Let us now consider a problem, parts of which we have seen in several chapters of the book. Now we shall put the "parts together," to build the pyramid, to solve the problem.

A cart rolls down a frictionless inclined plane and gains just enough speed to stay on a circular loop at the end of the incline. How high up the plane must the cart start and what is the relationship between that height h and the diameter of the loop, d? Our study of the conservation of energy gives us our first clue. The potential energy at the height h equals mgh, and assuming all of this energy is converted to kinetic energy at the bottom of the incline, $\frac{1}{2}mv^2$, the speed of the cart at the bottom will be

$$v = \sqrt{2gh}$$

Our second clue is provided by our knowledge of centripetal force and weight. At the top of the loop, the centripetal force equals the gravitational force on the cart. That is, assuming the radius of the loop to be r,

$$F_c = \frac{mv^2}{r} = mg$$

and $\quad v^2 = gr \quad$ or $\quad v = \sqrt{gr}$

The total energy at the top of the loop is the sum of the potential, $mg(2r)$, and the kinetic energy, $\frac{1}{2}mv^2$ or $\frac{1}{2}mgr$, which equals the total energy at the starting point h.

$mgh = 2mgr + \frac{1}{2}mgr \quad$ and $\quad h = 2\frac{1}{2}r = 1\frac{1}{4}d$

Therefore, whatever diameter of the loop is chosen the cart must start at an elevation equal to $1\frac{1}{4}$ times this diameter.

SUMMARY

In *uniform circular motion* (1) the speed v is constant; (2) the direction of the motion is continually and uniformly changing; (3) the acceleration a_c is constant in magnitude and is directed toward the center of the circular path. The magnitude of the *central acceleration* is given by

$$a_c = \frac{v^2}{r} = \omega^2 r$$

where v is the linear speed, r is the radius, and ω is the angular speed.

The *centripetal force,* the inward force that causes the central acceleration, is given by

$$F_c = m\frac{v^2}{r} = m\omega^2 r$$

The *centrifugal reaction* is the outward force exerted by the moving object on the agent of its centripetal force. The magnitude of the centrifugal reaction is equal to that of the centripetal force.

The proper banking of a curve to eliminate the necessity for a sidewise frictional force is given by the relation

$$\frac{v^2}{gr} = \tan\theta$$

Often in curvilinear motion the accelerating force is neither parallel nor perpendicular to the direction of motion. The acceleration produces change in both speed and direction.

Questions

1 Derive the expression for the central acceleration of a particle in uniform circular motion.
2 It is said that no work is done by a centripetal force. Explain why this is true.
3 In data used for computing centripetal force which would be more serious: a 1 percent error in observing the time or a 1 percent error in measuring the radius? Why?
4 Show that the units of v^2/r are those of acceleration.
5 Show that the acceleration of a body traveling in uniform circular motion is always toward the center.
6 Could a horizontal axis of rotation be used satisfactorily in an experiment on uniform circular motion? Explain.
7 Upon what principle does the centrifuge work? Give examples of devices employing this principle.
8 A person clings to a rotating merry-go-round to keep from being thrown from it. What kind of force do his muscles exert on his body?
9 Discuss the statement that it is not helpful to use the idea of centrifugal force, since it is not a real force.
10 A person standing at the center of a large, rotating turntable shoots an arrow at two paper targets, also on the turntable, one mounted some distance in back of the other. When the turntable

is stopped, the position of the holes in the targets indicate that the arrow followed a curved path. Explain how and why this happened.

11 Astronauts leaving the earth's surface are said to experience between 6 and 10 g's when the rocket begins to accelerate. Explain.

12 A candle is mounted in each of two beakers fastened upright at the ends of a stick. If the candles are lit, taking care to have the flames below the lip of the beakers, and the stick rotated horizontally in a circular path, the flames will bend in toward the center of rotation. Explain.

13 What do you expect would happen to the gravitational acceleration as you descend into a deep vertical mine shaft? Explain your answer.

14 A satellite orbiting the earth is said to "fall" around the earth. Prove that this is true.

15 How could you determine the mass of the earth? How could you find out how much the sun weighs?

16 A body at the end of a string moves in a vertical circle. If the string always pulls toward the center of the circle, can the speed of the body be constant? Discuss any variation and the reasons for it.

Problems

1 What is the least speed at which an airplane can execute a loop of 120-m radius so that there will be no tendency for the pilot to fall out at the highest point?

2 An airplane performs a loop-the-loop maintaining a constant speed of 180 mi/h. What is the maximum radius of the loop? If the pilot weighs 150 lb, what will his apparent weight be at the bottom of the loop? *Ans.* 2.18×10^3 ft; 300 lb.

3 An aviator loops-the-loop in a circle 120 m in diameter. If he is traveling 192 km/h, how many g's does he experience?

4 A 440-N boy swings on a 3.0-m-long swing. If his horizontal speed at the lowest point is 3.0 m/s, what total force must the ropes holding the swing be able to withstand? *Ans.* 575 N.

5 Assuming the earth to be a sphere 13,000 km in diameter, how much is the acceleration due to gravity changed by the rotation of the earth (*a*) at the equator, (*b*) at 40° latitude, and (*c*) at the pole? Is this change an increase or a decrease in the value of g?

6 Compute the minimum speed that a pail of water must have in order to swing without splashing in a vertical circle of radius 3.8 ft.

Ans. 11 ft/s.

7 What must be the speed of a satellite to stay in a circular orbit 1,600 km above the surface of the earth?

8 A swing is 5 m long. If a person swings high enough to cause it to loop-the-loop, how fast is he going when he goes over the top?

Ans. 7 m/s.

9 If the coefficient of friction between tires and roadway is 0.50, what is the smallest radius at which a car can turn on a horizontal road when its speed is 48 km/h?

10 A 3,200-lb car is driven over a circular-shaped knoll in a country road. If the knoll has a radius of 75 ft, how fast can the car pass over the bump without leaving the ground?

Ans. 49 ft/s, or 33.5 mi/h.

11 An amusement device has a mast with cross-arms extending 6 m from the center at the top. A car is suspended from the end of the crossarm by a rope 9 m long. Find the angular speed in radians per second and in revolutions per minute that will cause the rope to make an angle of 30° with the vertical.

12 A 1,500-lb car rounds a curve with a radius of 100 ft. If the coefficient of friction is 0.5, how fast can the car go before skidding?

Ans. 40 ft/s, or 27.3 mi/h.

13 A 9.0-kg ball is suspended from a hook in the ceiling by a string 1.0 m long. It is set into motion in a horizontal circle with a speed such that the string maintains an angle of 30° with the horizontal. Calculate the speed of the ball and the tension in the string.

14 A 100-lb boy is standing on a merry-go-round platform 10 ft from the center. The platform is turning at the rate of 4.0 r/min. Find the boy's linear speed, his radial acceleration, the frictional force needed to prevent him from slipping off the platform, and the coefficient of static friction if he is on the verge of slipping at this speed. *Ans.* 4.2 ft/s; 1.8 ft/s^2, 5.6 lb; 0.056.

15 A 35.6-N body swings in a horizontal circle at the end of a string 0.6 m long at a rate of 72 r/min. Find the tension in the string and the angle that the string makes with the horizontal.

16 An 8-lb ball attached to the end of a 10-ft-long string is rotating as a conical pendulum. If it rotates at 30 r/min, what angle will the string make with the vertical? *Ans.* 71°.

17 A 1.44×10^4 N automobile is moving with a constant speed of 6 km/h on a curve of 30-m radius. (*a*) What is its acceleration? (*b*) What is the centripetal force on the automobile? (*c*) What supplies this force?

18 Each metal stud on a snow tire weighs 0.1 oz. What centripetal force acts on each stud embedded in the 24-in-diameter tire when the car is traveling 60 mi/h? *Ans.* $1.5\overline{1}0$ lb.

19 A ball having a mass of 2.27 kg is swung at the end of a cord in a vertical circle of radius 0.6 m at the rate of 2.0 r/s. What is the tension in the cord when the ball is (*a*) at the level of the center, (*b*) at the bottom, and (*c*) at the top?

20 A 0.1-oz fly is trying desperately to hold on to a 4-in steel ball which is spinning at 100 r/min. If he is on the "equator" of this ball, how much force must the suction pads on his feet exert to prevent him from falling off the ball? *Ans.* 3.4×10^{-3} lb.

21 The governor of an engine has arms that are 30 cm long and stand at an angle of 30° with the vertical when the governor is in constant rotation. Find the ang
of the shaft c

22 A 4-ft-lo
porting a 32-l
to it. If the ro
path, what is
attain as it rot

23 A 1.44 ×
of 120 m radius
at an angle of
frictional force
mum coefficien

24 The designers of an expressway wish to have automobiles round a curve at 70 mi/h. If the roadway is not banked and the coefficient of friction between the tires and the road is 0.3, what must the radius of the curve be to permit a 4,800-lb car to negotiate the turn safely? *Ans.* 1,110 ft.

25 A car whose wheels are 54 in apart laterally and whose center of gravity is 18 in above the road rounds a curve of 200 ft radius. Assuming no slipping of the wheels on the road, find the greatest speed at which the car can round the curve without tipping over.

26 A ball rolls down an inclined track and around a vertical loop 40.0 cm in diameter that is built into the track. How high above the lowest point in the loop must the ball be released in order that it will just go over the loop if half the energy of the ball is expended in work against friction? *Ans.* 100 cm.

Nils Gustaf Dalen, 1869–1937

Born in Stenstorp, Sweden. Engineer and inventor. Dalen was awarded the 1912 Nobel Prize for Physics for his invention of the automatic regulators that can be used in conjunction with gas accumulators for lighting lighthouses and light buoys.

Heike Kamerlingh Onnes, 1853–1926

Born in Groningen, Holland. Founder of the Cryogenic Laboratory at Leyden. Awarded the 1913 Nobel Prize for Physics in recognition of his investigations into the properties of matter at low temperatures, which led, among other things, to the production of liquid helium.

9

Aerospace Physics

We first throw a little something into the skies, then a little more, then a shipload of instruments—then ourselves.

Fritz Zwicky
California Institute of Technology

Man's curiosity has served as a constant impelling force, pushing him to search out the unknown and leading him to invent the tools and the techniques to explore it. As man developed the technological skills which permitted him to reach space, it was inconceivable that he should not explore it. At this point in time it is impossible to foretell what man will gain from the exploration of space. The rewards are just now starting to appear, but possibly the most important return from space exploration will be the vast addition to man's knowledge about the universe in which he lives.

Up to this point in the book we have examined some of the basic laws of mechanics and have attempted to show how important these laws are in understanding and, to some degree, controlling our environment. Perhaps the application of these laws of motion is nowhere more dramatically illustrated than in the exciting frontier of space research. In an effort to provide the reader with information that will enhance his understanding of the great adventure in space, this chapter is devoted to space physics, a special application of the laws of physics.

In this chapter we will first trace the history of man's journey into space. By the term *space* we mean our environment in the solar system (only about 10^{-24} of the volume occupied by our galaxy). We shall look at some of the questions about our universe that we hope our exploration of space will answer. Finally, we shall discuss how this exploration may be helped by our understanding of the principles of physics.

Obviously, all areas of physics are heavily utilized in the space research program. Communication, heat, biophysical, and geophysical problems are all of concern to physicists. Since many of the concepts relating to these problems have not yet been considered in our study, only the areas of physics which are related to motion in space will be discussed here and other aspects of aerospace physics will be deferred until later.

9-1 THE HISTORY OF SPACE FLIGHT[1]

Although the launching of the first artificial satellite, Sputnik I, on October 4, 1957, was a dramatic

[1]Several NASA publications contain excellent reports on the history and goals of man's exploration in space, especially the pamphlet Space . . . the New Frontier edited by James Dean, which proved helpful in preparing this section of this book.

and much-publicized event, we should not look upon that accomplishment as the beginning of man's attempt to conquer space. Someone once observed that no great discovery or event ever stands alone, that it is always preceded by other discoveries or events which have paved the way for it. This is especially true in space research, since its inception can be traced back to the earliest days of man's existence when he first began to wonder and hypothesize about the luminous objects apparently moving across the heavens. Indeed, the written record of man's concept of the universe and his dream of leaving the earth and traveling to a distant world can be traced back many centuries through his writings. The record goes back to the second century B.C. when a part of "Cicero's Republic" entitled "Scipio's Dream" appeared which presented the concept of the earth as being an insignificant part of a vast universe containing "stars which we never see from earth." Perhaps the first written evidence of man's urge to fly to the moon was provided by Lucian of Greece in the second century A.D., in a story he wrote of a lunar flight entitled "Vera Historia."

Whereas man may have maintained his interest in space flight, no further written evidence of this interest appeared for many centuries until the advent of the scientific renaissance. This period saw the emergence of men such as Copernicus (1473–1543), who devoted all his activities to evolving a picture of the universe with the sun in a central position; Tycho Brahe (1546–1601), who made such precise observations of the movement of the planets and stars; Kepler (1571–1630), who interpreted Brahe's data and from it derived mathematical relationships about the movements of celestial bodies; Galileo (1564–1642), who was one of the first to use the telescope to observe the sky; and Newton (1642–1727), who established the definite interrelationship of earthly and heavenly physics, between mechanics and theoretical astronomy. Under the influence of these men, people once again began to dream of journeying through space, and the literature revealed these dreams. Over the centuries writers such as Jules Verne, Edgar Allan Poe, and H. G. Wells (among many other writers) captured the imagination of their readers by their exciting stories of space travel. Perhaps the most famous of these works is Jules Verne's "From the Earth to the Moon" which was published in 1865 and in which the concept of "weightlessness" was a main feature. A lesser known but fascinating novel published in 1869 entitled "The Brick Moon" by Edward Everett Hale (of "Man Without a Country" fame) presented for the first time not only the concept of placing a man-made satellite into orbit but also the possibility of weather satellites, manned orbital laboratories (a project to be completed in the 1970s), and communication and navigation satellites.

As much as man dreamed about space travel, it could not come to fruition until the mechanical skills and the necessary spacecraft were developed. The development of rockets parallels quite closely the literary interest in space travel with alternating periods of activity and inactivity. This history has been traced in considerable detail in Table 1, wherein the quite natural relationship between missilery and space flight becomes evident. The first evidence of man's creation of a "rocket" engine can be traced back to the second century A.D. when Hero of Alexandria used jets of steam issuing from a rotatable, mounted globe (an aeolipyle) to turn it. The historical record then disappears until about 1,000 years later when rockets reappeared which had been developed to the point where they could be used as weapons of war. In 1232 A.D. the Chinese in an effort to drive back attacking Mongols used gunpowder to propel "arrows of flying fire," something equivalent to our present-day skyrockets. By the middle of the thirteenth and early fourteenth century, the rockets had spread to Europe where they were used in the third Venetian-Genovese war. Historical evidence shows that a rocket destroyed a tower in the vital battle for the Isle of Chiozza during that war. By the late 1700s a fairly advanced type of rocket had been developed in India. In the war against the British in 1792, troops of Tipu, the Sultan of Mysore, used such rockets with considerable effectiveness.

From what has been noted above, the first

development of rockets quite obviously was directed along military lines. Their potential as weapons was even further improved when Sir William Congreve of Great Britain developed a solid rocket propellant which considerably increased the range of rockets. These rockets were used extensively in the Napoleonic Wars and in the War of 1812. American history notes that in the latter war the British used rockets in their attack on Fort McHenry in Baltimore, a fact commemorated in the line "and the rockets' red glare" in the American national anthem.

Not all rocket development was war-oriented. For example, the Congreve lifesaving rocket was developed in 1838 in Britain, a rocket which was used to shoot a line out to grounded ships in order to transfer people over a rescue device called a "breeches buoy."

Only limited development took place in rocketry for most of the next century because rockets fueled by black powder had attained their maximum capacity. In 1919 Robert H. Goddard, a professor of physics at Clark University, prepared a report for the Smithsonian Institution entitled A Method of Reaching Extreme Altitudes, in which he discussed the possibility of shooting a rocket to the moon and exploding a load of powder on its surface. At the same time Goddard concluded that a liquid-fueled rocket would overcome some of the difficulties encountered with the pellets of powder he had used to power his rockets. In 1926, Goddard launched the first liquid-fueled rocket at Auburn, Massachusetts. Although the rocket flew only 184 ft, it was the breakthrough needed. As Table 1 indicates, developments came rapidly from that point in history. The German development of the buzz bombs in World War II and the development of the intercontinental ballistic missile program in this country in 1946 were forerunners to the launching of the first man-made satellite in 1957, to the putting of man in space in 1961, and to the successful culmination of one of history's greatest efforts, landing a man on the moon in 1969.

What is next in space research? Should man's landing on the moon be the end of space exploration? There are some who say it should be and give reasons to support their views. They feel that economic consideration must be given to any future space programs. The cost of the aerospace program founded under the Space Act of 1958 which created the National Aeronautics and Space Administration (NASA) is a very large factor in this nation's economy. While it cost Goddard only a few thousand dollars to launch his rocket, a single launching of the Atlas Agena rocket cost $7.5 million and the total cost of sending Apollo 16 to the moon, the United States' fourth manned landing there, was $445 million. For the early years of the 1970s, the average budget provided to NASA was between $600 and $700 million. The total Space Research and Technology budget for that period was estimated to have created a tax burden of $65 per family per year. Others argue that we should order our priorities and that space research should come only after some of our earthly problems are solved.

On the other hand, those who feel that the rewards from space research will be worth the expenditure of resources, time, and energy note that since such research draws broadly from all fields of science and engineering, space technology holds the promise of uncovering many new benefits for mankind. To this end, as an aid in identifying and disseminating new processes, materials, and equipment which can improve life on earth, NASA has established an Office of Technology Utilization.

One such benefit already realized is provided by satellites now in orbit which are equipped with television cameras, infrared sensors, and meteorological instruments which make possible better short- and long-range weather predictions. This predictive capacity has a direct effect on millions of people, for early warnings of impending tornadoes, floods, blizzards, and hurricanes enable communities to make the necessary preparations to save lives and property.

It is estimated that there are 20,000 surface craft at all times on the Atlantic Ocean. Hundreds of aircraft are in the skies over the world. It is vital to these navigators and pilots to know exactly where they are. Navigational satellites are

Table 1

MILESTONES IN MISSILERY AND SPACE FLIGHT[1]

c. 200	Hero of Alexandria uses the reacting force of escaping steam to propel an experimental device.
c. 1200	Chinese use gunpowder to propel "arrows of flying fire," equivalent to present-day sky rockets.
c. 1780	Advanced type of rocket developed in India.
1792	Troops of Tipu, Sultan of Mysore, use rockets against British in second Mysore War.
c. 1800	Sir William Congreve of Great Britain improves rocket propellant to provide considerable increase in range.
1812	British use rockets in attack on Fort McHenry (Baltimore).
c. 1830	William Hale, an American, increases stability of rockets by adding nozzle vanes.
1846	Mexican War sees first use of rocket weapons by United States in a war. Lifesaving rockets developed by English and German inventors.
1913	Ramjet proposed and patented in France.
c. 1915	World War I sees advent of guided missile to supplant aimed rockets.
1926	Robert H. Goddard, professor of physics at Clark University, fires first successful liquid-fuel rocket.
c. 1930	Germans experiment with the pulse jet, used to power the Nazi V-1 buzz bomb of World War II.
1931	Germany uses liquid rocket fuel.
1932	Captain Walter Dornberger undertakes development of liquid-fuel rocket weapons for the German army.
1936	German Peenemunde Project is organized, to develop war rockets.
1941	United States starts work on controllable rocket weapons.
1942	American Razon missile, controllable in both azimuth and range, is developed.
1944	United States government awards first contract for research and development of guided missile to General Electric Company.
c. 1945	Germany uses V-1 buzz bomb, V-2, and other rocket missiles in World War II. United States uses "Weary Willie" unmanned bombers.
1946	Work is started in the United States on an intercontinental ballistic missile program, the MX-774.
1949	First flight of a missile beyond earth's atmosphere is made at White Sands, New Mexico
c. 1952	United States long-range missile program is stimulated by Atomic Energy Commission warhead developments.
1954	United States starts ICBM program; USAF awards contracts to Convair, North American Aviation, and General Electric.
1957	First artificial earth satellites—Sputniks I and II—launched by rocket (October 4 and November 3).
1958	Explorer I, first United States satellite, launched (January 31).
1958	Vanguard I, first "permanent" satellite, launched by the United States (March 17).
1958	Pioneer I, first lunar probe, launched by the United States (October 11).
1958	Project Score (Atlas) launched broadcasting a human voice from outer space for the first time (December 18).
1959	Russia launches Lunik, first satellite to orbit around sun (January 2).
1959	Pioneer IV launched, first United States satellite to orbit sun (March 3).
1959	Russia launches first space vehicle to land on moon (September 12).
1959	Russia launches first satellite to orbit moon (October 4).
1960	United States recovers first space vehicle from orbit (August 11).
1961	Manned orbital flight achieved in Soviet Vostok satellite (April 12).
1962	Project Mercury succeeds in manned orbital flight (February 20).
1962	Telstar satellite relays first transatlantic television programs (July 10).

Table 1 (Continued)

1962	Two Russian astronauts put in related orbits (August 13).
1962	Mariner II launched to encounter Venus (August 27).
1962	Mariner II passed within 22,000 mi of Venus, reporting data on temperature, cloud cover, magnetic field, particles, and radiation dosage encountered throughout voyage (December 14).
1963	First long flight is made by an American (Cooper), 34 h 20 min (May 15).
c. 1965	First man goes outside spacecraft in orbit in 10-min space walk (Leonov) (March 18).
1966	First docking in space achieved by Armstrong and Scott in United States Gemini 8 (March 16).
1967	Heaviest manned spacecraft (USSR Soyuz I) is launched. Crashed killing Komaruv (April 23).
1968	First manned voyage around the moon is made by Borman, Lovell, and Anders in United States, Apollo 8 (December 21).
1969	Two spacecraft, USSR Soyuz IV and V, are launched the same day and rendezvous in space, with two astronauts transferring to the other spacecraft (January 15).
1969	Docking with the lunar module in space by United States Apollo 9 (March 3).
1969	Man (Stafford, Cernan, and Young in United States Apollo 10) descends to within 9 mi of the moon's surface (May 18).
1969	Man lands on the Moon. Armstrong and Aldrin land in the lunar module and Collins remains in the command module of United States Apollo 11 (July 16).
1969	Three spacecraft (Soyuz VI, VII, and VIII) are launched on consecutive days with seven men aboard. This was a step toward a manned space platform (October 11, 12, and 13).
1969	Man achieves second landing on the moon (United States Apollo 12). Retrieve parts of Surveyor 3, an unmanned spacecraft that landed on the moon in 1967 (November 14).
1970	Third manned lunar attempt aborted after 56 h due to loss of pressure in liquid oxygen (United States Apollo 13, April 11).
1971	First of three global communications satellites, United States Intelsat IV F-2, launched in 1971 to form part of a satellite communications system operated by the multination International Communications Satellite Consortium (January 25).
1971	The world's first unmanned space station, USSR Salyut I, launched for scientific exploration (April 19). Later, Soyuz 10 (April 22) and Soyuz 11 (June 6) dock with Salyut I.
1971	Fourth manned lunar landing achieved United States Apollo 15). First mission to use an electrically powered car, the lunar surface rover. Explored moon 3 d (July 26).
1972	United States Apollo 16 lands on the rugged upland region of the moon in the area of Descartes (March 17).
1972	U.S. Apollo 17, manned moon-landing mission, marked the end of the flight phase of lunar exploration (December 7).
1973	Project Skylab, a multifaceted, long-term project, begun with the first-stage launching successful (May 1973). The beginning of a series of earth-oriented investigations that will extend through the mid-1970s, designed to aid man in predicting and controlling his environment.

[1]Adapted from the table Milestones in Missilery appearing in R. L. Weber, "Physics for Teachers: a Modern Review," McGraw-Hill Book Company, New York, 1964, pp. 4 and 5.

already stationed in space which enable them to pinpoint their location at any time of day or night in all kinds of weather.

Satellites have opened a new era in global communication. Echo, Telstar, and Relay have augmented current facilities as well as made possible global telecasts and other type of worldwide communication not previously available through any other system.

Astronomers have been hindered in their observations by the earth's atmosphere, which blocks out and distorts electromagnetic radiation from space. By placing telescopes and other astronomical equipment in satellites orbiting above the atmosphere (at 20 mi one is above 99 percent of the atmosphere), information about stars and galaxies hitherto unavailable has become accessible.

Satellites also are currently assisting in determining exact distances and locations and precise shapes of land and sea areas on earth.

Further, the study of aerospace medicine promises benefits in the treatment of heart and blood illnesses. Studies have been made on human behavior and performance under conditions of great stress, emotion, and fatigue. Sometimes there is a spin-off benefit. For example, a derivative of hydrazine (isoniazid), developed as a liquid space propellant, has been found to be useful in treating tuberculosis and certain mental illnesses. The space industry has developed the skills necessary to produce reliable and accurate miniature parts, such as valves, which may be used someday to replace worn-out human organs.

One of the world's great problems is hunger. Scientists and dietitians are working on the problem of space feeding and nutrition. The information gained from this research will have great influence on future food and agricultural processes. This involves the growth of synthetic and new foods and the compressing of large numbers of calories into pill-sized packages. This research also involves new methods of food growth and storage, a major concern in underprivileged nations that will benefit from the discoveries made in this area.

While man is looking out, he is also looking

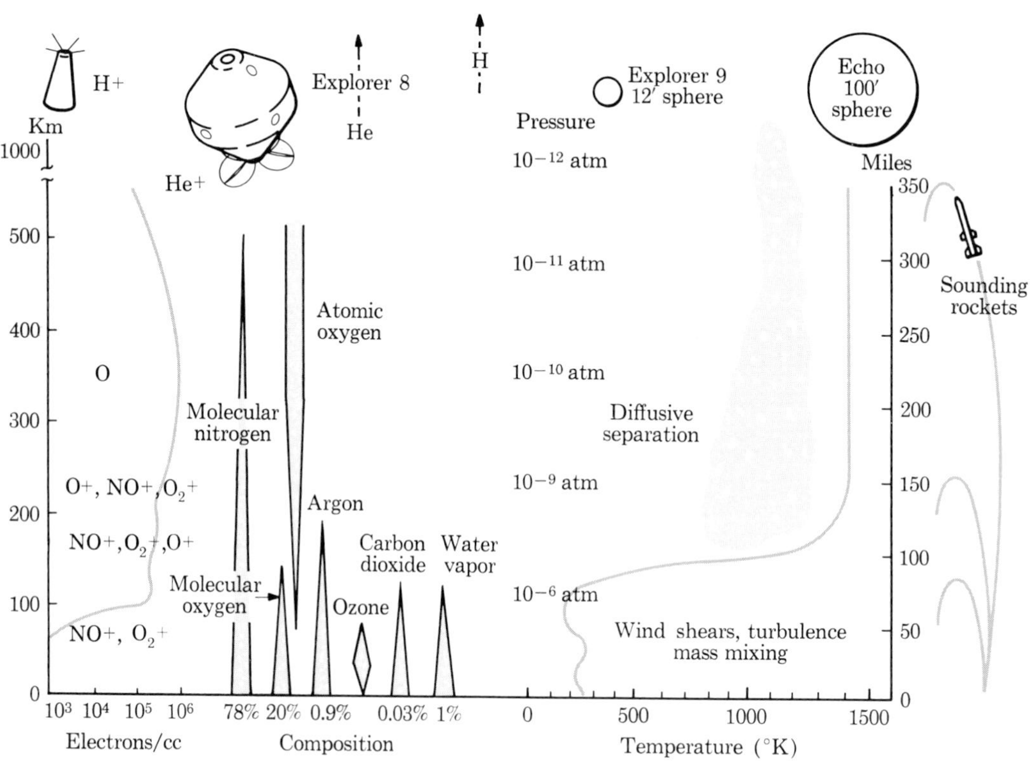

Figure 9-1
Characteristics of the earth's atmosphere. (*NASA.*)

in. One of the major space goals for the 1970s is to launch and support a permanent space platform eventually to be manned by 50 to 100 men (project Skylab is the forerunner of these large space stations), to serve as a scientific research laboratory. The development of the space station fits in well with NASA's current policy of utilizing its technological knowledge for the solution of problems closer to earth. From such a station, in addition to providing further improvement in weather forecasting, man can locate schools of fish, differentiate between diseased and healthy crops, locate mineral deposits, detect the dumping of manufacturing wastes into inland streams, measure soil fertility, and predict crop yields on a worldwide basis.

A better understanding of the composition, pressure, temperature, and turbulence of the earth's atmosphere was obtained with the aid of sounding rockets (Fig. 9-1). However, much is still to be learned about our environment. As late as 1972, the Apollo 16 moon mission provided new information which changed our understanding of our atmosphere. Photographs taken from the moon by an ultraviolet camera showed that there are three dense atmospheric rings around earth, not two as formerly believed. It was found that earth had an extra ring of oxygen and nitrogen gases. This ring had been hidden from astronomers on earth by the other two dense gaseous layers.

Only a small but significant step has been taken in the exploration of our solar system. It is the long-range goal of those participating in the space research program to learn more about the planets Mars and Venus and also Mercury, Jupiter, and the other more distant planets.

9–2 ESCAPING FROM THE EARTH: TYPES OF ORBITS

When man attempts to leave the earth, there are certain mathematical and physical laws which must be obeyed before this can be accomplished. The next few sections of the book are designed to illustrate some of the basic "rules of the game" of space flight.

When a satellite revolves around a central body, the satellite follows a path known mathematically as a conic curve. These curves may be visualized by taking plane slices of a solid circular cone (Fig. 9-2). A body "bound" by a central body, such as a planet in the solar system, follows an orbit which is elliptical. A nonrecurring comet passing through the solar system would follow a hyperbolic path relative to the sun. The other conic curves, the circle and the parabola, are unique orbit paths. The circle is a special case of an elliptical orbit. The parabolic path is the borderline case between an elliptical (binding) orbit and a hyperbolic (escaping) orbit. Both the circular and parabolic orbits are unstable, for any slight disturbance would cause the body to enter either an elliptical or a hyperbolic orbit.

Study of the special case of a circular orbit is justified by the simplicity with which certain relationships of general importance can be illustrated for the circular path. In what immediately follows we shall assume that (1) the gravitational force due to the main body varies only with the radial distance from the center of that body; (2) the satellite body has negligible mass compared with the main body, and so the center of mass of the system is practically at the center of the main body (Fig. 9-3); and (3) both bodies are perfectly spherical.

Actually, the earth is not strictly spherical in shape. Its slight asymmetry causes perturbations of the orbits of close-by satellites. The plane of the orbit is gradually rotated by the unsymmetrical gravitational pull. This precession of an orbit may be an advantage, for it permits the satellite to "see" more of the earth's surface.

9-3 ENERGY FOR A CIRCULAR ORBIT

The potential energy of a small body at a distance r from a very large body may be defined as the work which would be done by an external agent in bringing the small body from a very great

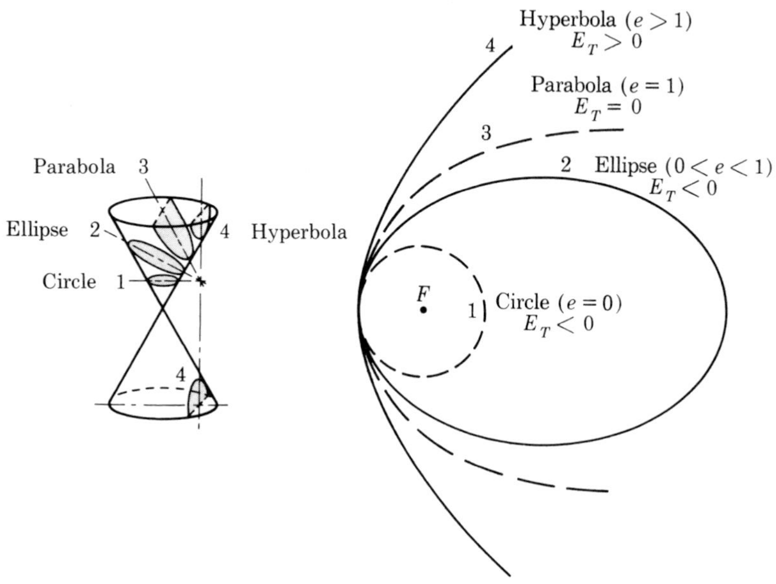

Figure 9-2
Basic orbits related to conic sections.

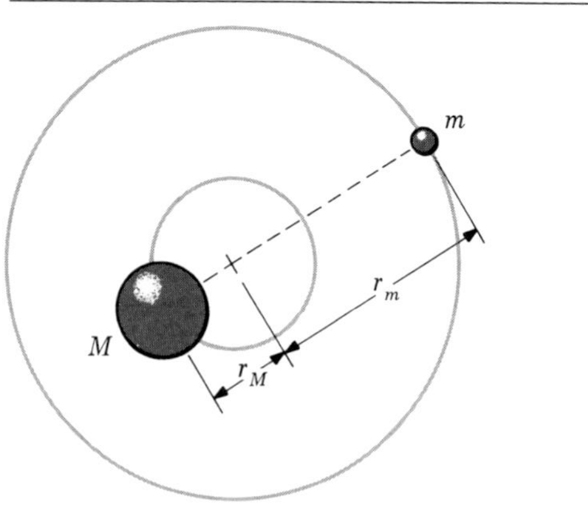

Figure 9-3
Rotation of two-body system about its center of mass.

distance (infinity) to a distance r from the large body. Here we arbitrarily assign the value zero to potential energy when the small body lies outside the influence of the larger body.

Consider now the work done by an agent in moving a small body in the opposite sense, away from the surface of the earth. Imagine the distance from R to r (Fig. 9-4) to be divided into small equal intervals so that over each interval the gravitational force F_G will be practically constant. Then we can easily calculate the work done in each interval and add these contributions to get the total. At the surface, $F_G = GMm/R^2$. At the top of the first interval, F_G is GMm/r_1^2. Since these values are nearly the same, we can use GMm/Rr_1 for the average force in the first interval. The work done in the first interval is as follows.

Since $\mathcal{W}_1 = F_G(\Delta r)$ then,

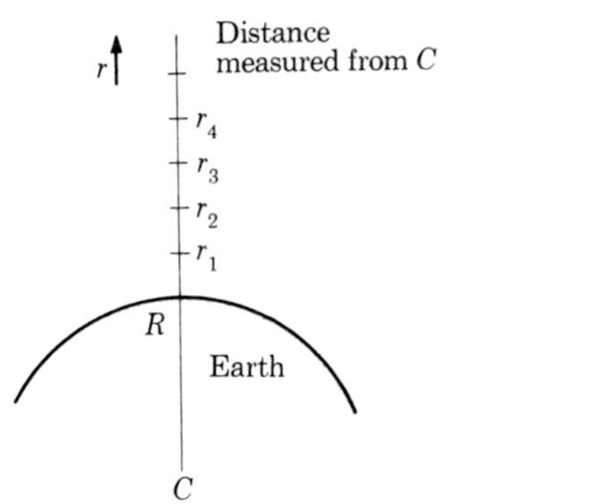

Figure 9-4
Calculation of gravitational potential energy.

$$\mathcal{W}_1 = F_G(r_1 - R) = \frac{GMm}{Rr_1}(r_1 - R)$$
$$= GMm\left(\frac{1}{R} - \frac{1}{r_1}\right) \qquad (1)$$

Likewise the work done in the second interval is

$$\mathcal{W}_2 = \frac{GMm}{r_1 r_2}(r_2 - r_1) = GMm\left(\frac{1}{r_1} - \frac{1}{r_2}\right) \qquad (2)$$

In the third interval,

$$\mathcal{W}_3 = GMm\left(\frac{1}{r_2} - \frac{1}{r_3}\right) \qquad (3)$$

If we add these three expressions, the intermediate values r_1 and r_2 cancel out and the work done in the first three intervals can be expressed in terms of the values of r at the ends, R and r_3; thus

$$\mathcal{W} = GMm\left(\frac{1}{R} - \frac{1}{r}\right) \qquad (4)$$

is the general expression for the work required to move a mass m against the earth's gravitational field out to a distance r.

Returning now to the definition of the gravitational potential energy E_p of mass m when at distance r from the earth's center, we have

$$E_p = -\mathcal{W}_{r\to\infty} = -GMm\left(\frac{1}{r} - \frac{1}{\infty}\right)$$
$$= -\frac{GMm}{r} \qquad (5)$$

Thus the potential energy of a body in orbit is always negative. This is a consequence of the fact that the force between the bodies is one of attraction and of our arbitrary choice in taking the potential energy of the system to be zero when the two bodies are separated by infinite distance.

The kinetic energy of a moving body is $\frac{1}{2}mv^2$. For circular motion, $v = \omega r$ may be substituted to obtain

$$E_k = \tfrac{1}{2}m\omega^2 r^2 \qquad (6)$$

Since the gravitational force supplies the centripetal force,

$$F_G = \frac{GmM}{r^2} = mr\omega^2 \qquad (7)$$

and we obtain

$$GM = \omega^2 r^3 \qquad (8)$$

The kinetic energy is then

$$E_k = \frac{1}{2}\frac{GmM}{r} \qquad (9)$$

for a body of mass m in circular orbit at radius r from a central body of mass M. The kinetic energy is always positive. This expresses the ability of the moving body to do work in bringing its energy to zero.

The total energy E of a body is the sum of its kinetic and potential energies. So, for a circular orbit, we have

$$E = E_k + E_p = \frac{GmM}{2r} - \frac{GmM}{r}$$

$$= \frac{GmM}{2r} - \frac{2GmM}{2r} = -\frac{GmM}{2r} \quad (10)$$

The total energy is negative for a body in orbit. This indicates that the body is held in the system. Positive work must be done to free the body from the gravitational force.

Figure 9-5 shows that as the radius of the circular path about the central body increases, the potential energy increases (toward zero) and the kinetic energy decreases (as the angular speed becomes smaller).

A *potential-well model* may be constructed by plotting a graph of the potential energy E_p which a body of mass m would have at various distances from the center of the earth (Fig. 9-6). When the body is infinitely far from earth, $E_p = 0$. As the body is brought closer and closer to the earth, work is done on it by the earth's field and the potential energy of the body acquires a larger and larger negative value. Thus on the surface of earth we live in a gravitational well thousands of miles deep. To reach the moon or another planet we must climb out of this well onto the plane marked "gravitational free space" in Fig. 9-6.

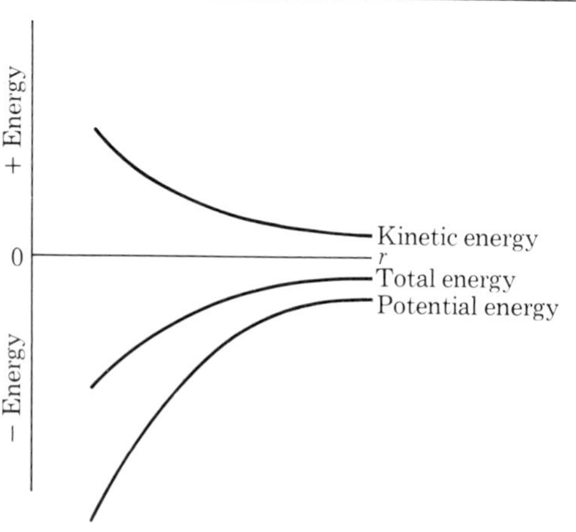

Figure 9-5
Energy in a circular orbit.

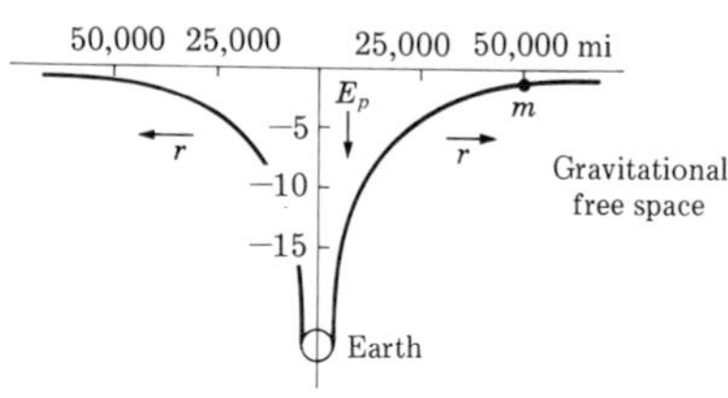

Figure 9-6
Gravitational potential energy of mass m, showing "well" analogy for earth's field.

Example A 90-kg space probe is shot from earth with an initial speed of 12 km/s. Can the probe attain a circular orbit about the earth?

If the total energy is negative, the probe is bound and may go into orbit. If the total energy is positive, the probe will escape from the earth's attraction. At take-off

$$E_p = \frac{-GMm}{R} = -6.67 \times 10^{-11}\ \text{m/kg}\cdot\text{s}^2 \times \frac{(5.97 \times 10^{24}\ \text{kg})(90\ \text{kg})}{6.37 \times 10^6\ \text{m}}$$

$$= -5.62 \times 10^9\ \text{J}$$

$$E_k = \tfrac{1}{2}mv^2 = \tfrac{1}{2}(90\ \text{kg})(12 \times 10^3\ \text{m/s})^2$$

$$= 6.49 \times 10^9\ \text{J}$$

$$E = E_p + E_k = 0.87 \times 10^9\ \text{J}$$

With atmospheric effects neglected, the total energy of the probe in space will be the same as that at take-off. Since the total energy relative to the earth is positive, the probe will not become a satellite of the earth.

Example What is the minimum take-off speed which would enable a space vehicle to escape from the earth?

By neglecting atmospheric effects and assuming a stationary earth, the smallest take-off speed needed for escape is determined from the requirement that the vehicle's kinetic energy be equal to the negative of its potential energy at take-off

$$\tfrac{1}{2}mv^2 = \frac{GMm}{R}$$

This gives

$$v^2 = \frac{2GM}{R}$$

$$= \frac{2(6.67 \times 10^{-11}\ \mathrm{N \cdot m^2/kg^2})(5.97 \times 10^{24}\ \mathrm{kg})}{6.37 \times 10^6\ \mathrm{m}}$$

$$= 1.23 \times 10^8\ \mathrm{m^2/s^2}$$

$$v = 11.2\ \mathrm{km/s} = 6.96\ \mathrm{mi/s}$$

$$= 2.50 \times 10^4\ \mathrm{mi/h}$$

Example At what speed would a projectile have to leave a space platform, horizontally, 300 mi above the earth in order to enter a circular orbit around the earth, i.e., to "fall continuously" around the earth?

For a circular orbit at a distance h beyond the radius R of the earth, the speed would have to have a value to make the required centripetal force just equal to the gravitational force at that height.

$$F = \frac{mv^2}{R+h} = \frac{GMm}{(R+h)^2}$$

$$v^2 = \frac{GM}{R+h}$$

$$= \frac{(6.67 \times 10^{-11}\ \mathrm{N \cdot m^2/kg^2})(5.97 \times 10^{24}\ \mathrm{kg})}{(3{,}960 + 300)\ \mathrm{mi}} \times \frac{1\ \mathrm{mi}}{1{,}609\ \mathrm{m}} = 58.1 \times 10^6\ \mathrm{m^2/s^2}$$

$$v = 7.60\ \mathrm{km/s} = 4.75\ \mathrm{mi/s}$$

$$= 1.70 \times 10^4\ \mathrm{mi/h}$$

9-4 CONICS

Some earth satellites are in nearly circular orbits and can be described to a good approximation by the equations just developed for the circular orbit. But we shall now examine the more general paths suggested by the conic sections of Fig. 9-2.

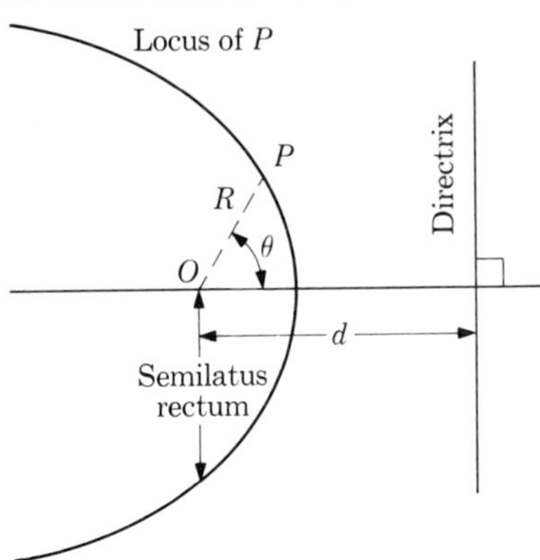

Figure 9-7
Definition of a conic.

A conic is defined to be the locus of a point P relative to a line called the directrix and a fixed point O (Fig. 9-7), which is given by the equation

$$\frac{R}{d - R\cos\theta} = e = \text{const} \tag{11}$$

where R and d are distances defined in Fig. 9-7. The constant e is known as the eccentricity of the curve and indicates its shape:

1	Circle	$e = 0$
2	Ellipse	$0 < e < 1$
3	Parabola	$e = 1$
4	Hyperbola	$e > 1$
5	Straight line	$e \to \infty$

9-5 THE ELLIPSE

The conic called an *ellipse* is the locus of a point such that the sum of its distances from two fixed points (called the *foci*) is constant. In Fig. 9-8 $F'P + PF = 2a$, the length of the major axis. The equation for an ellipse, in rectangular coor-

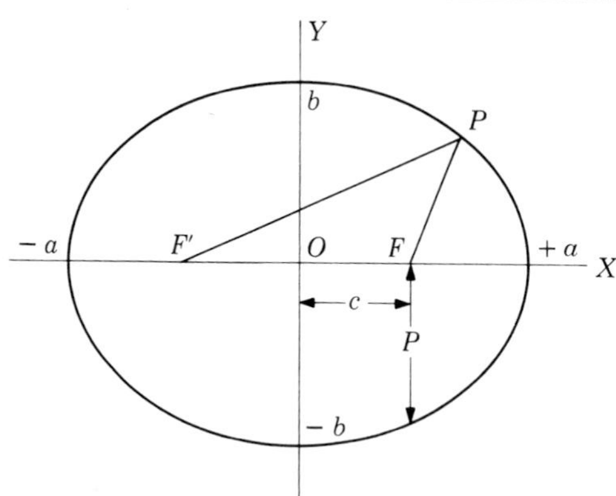

Figure 9-8
The ellipse.

dinates, follows directly from the definition,

$$\frac{x^2}{a^2} + \frac{y^2}{b^2} = 1 \tag{12}$$

where $b^2 = a^2 - c^2$.

The equation for the ellipse in polar coordinates has the form $r = p/(1 + e \cos \theta)$, where p is the semilatus rectum (Fig. 9-7) and where eccentricity e is positive but less than 1. This equation assumes the origin to be at one focus; Eq. (12) assumes the origin to be at the center of the ellipse. The eccentricity e is the ratio of c to a (Fig. 9-8).

9-6 KEPLER'S LAWS

A space vehicle when not under power is governed by the laws which determine the motions of stars, planets, and comets. Johannes Kepler (1571–1630) used inductive reasoning to formulate laws to fit the astronomical observations and calculations given him by his patron Tycho Brahe. Newton in his "Principia Mathematica" (1687) showed that the kind of planetary motion described by Kepler's laws can be deduced from the universal law of gravitation. Kepler's description of planetary motion may be stated as follows:

First law: The planets move in ellipses having a common focus situated at the sun (Fig. 9-9).

Second law: The line which joins a planet to the sun sweeps over equal areas in equal intervals of time (Fig. 9-9).

Third law: The squares of the periods of the planets are proportional to the cubes of their mean distances from the sun expressed in astronomical units (au), where 1 au equals the average distance between the sun and the earth, 92,900,000 mi. (Mean distance refers to the semimajor axis a, Fig. 9-8.)

To illustrate Kepler's third law, we may equate the mean gravitational attraction of the sun to the planet's average centripetal attraction,

$$\frac{GmM_s}{r^2} = m\omega^2 r = \frac{4\pi^2 mr}{T^2} \tag{13}$$

where G is the constant of gravitation, m the mass of the planet, M_s the mass of the sun, and r the average of the nearest and farthest distances from sun to planet. A relation for Kepler's third law follows:

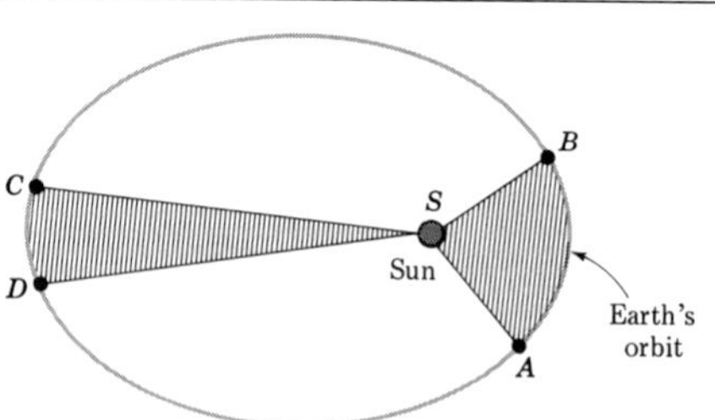

Figure 9-9
Orbit of earth around sun. In time t the area swept out when the earth moves from A to B equals the area swept out when it moves from C to D. Thus its orbital velocity is greatest between A and B.

$$\frac{r^3}{T^2} = \frac{GM_s}{4\pi^2} = \text{const} \qquad (14)$$

Example Assume that the satellite Echo has a mass of 75.0 kg and an orbital period of 117 min. Calculate the semimajor axis a.

For an ellipse (Fig. 9-8), $a = \frac{1}{2}(r_{max} + r_{min})$. So Eq. (14) becomes

$$T^2 = \frac{4\pi^2 a^3}{GM_e}$$

$$a^3 = \frac{T^2 GM_e}{4\pi^2}$$

$$= (117 \times 60\ \text{s})^2 \times (6.67 \times 10^{-11}\ \text{N}\cdot\text{m}^2/\text{kg}^2) \times (5.97 \times 10^{24}\ \text{kg})/4\pi^2$$

$$= 4.98 \times 10^{20}\ \text{m}^3$$

$$a = 7.94 \times 10^3\ \text{km} = 4.93 \times 10^3\ \text{mi}$$

Note: It should be observed that the solution does not depend upon the mass of the satellite.

9-7 COORDINATE SYSTEMS

To define an orbit in space or a position in an orbit, we need a reference system. Since all positions are relative, the common practice is to relate a position in space to a convenient coordinate system that moves with the observer.

To define the position of an earth satellite in the solar system and to describe its path, one needs to know the period of the satellite and the constants which fix the position and shape of the orbit:

The *period* is the time for a satellite to make one revolution around the earth.

Perigee is the position of closest approach to the center of the earth. *Apogee* is the position of the satellite farthest from the earth (Fig. 9-10a).

The *eccentricity* describes the elongation of the orbit as the ratio of c to a (Fig. 9-10a).

The *angle of inclination* i of the orbit is the angle between the plane of the orbit and the equatorial plane.

The complete description of a planetary orbit

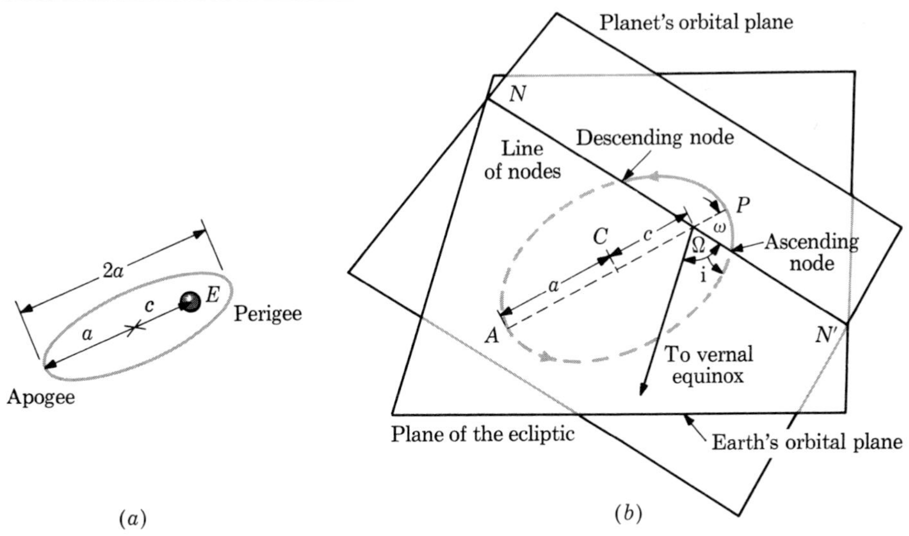

Figure 9-10
(a) Elliptic orbit; (b) the constants (elements) which define an orbit.

also requires specification of the two angles shown in Fig. 9-10*b*, ω, called the angle of ascending nodes, and Ω, called the angle of perigee.

9-8 ROCKET PROPULSION

For extremely long ranges, rockets are more practical than missiles propelled from guns. A rocket has been defined as "any machine that propels itself by ejecting material brought along for the purpose." A machine gun mounted on a car (Fig. 9-11*a*) might come under this broad definition, for when bullets are fired, the reaction forces may set the car in motion. Each bullet receives momentum mv toward the right as it leaves the gun. If n bullets are fired in time t, the average force exerted by the gun on these bullets is $F = nmv/t$, toward the right. The reaction is a force nmv/t in the opposite direction, acting on the gun. This reaction force accelerates the gun and car toward the left.

A rocket is an internal-combustion engine that carries its own supply of oxygen. Therefore it does not require air but can operate in empty space. When the burning gases are expelled with a velocity v relative to the rocket (Fig. 9-11*b*), discharging mass at a constant rate $\Delta m/\Delta t$, the change of momentum per second of the material passing from the rocket into the exhaust jet is $v\ \Delta m/\Delta t$. The magnitude of the reaction force, the thrust on the rocket, is

$$F = v\frac{\Delta m}{\Delta t} \qquad (15)$$

Example A 3.60×10^4 kg rocket rises vertically from rest. It ejects gas at an effective velocity of 1,800 m/s at a mass rate of 580 kg/s for 40 s before the fuel is expended. Determine the upward acceleration of the rocket at times $t = 0$, 20, and 40 s.

The constant upward force is $F = v\ \Delta m/\Delta t = 1{,}800$ m/s (580 kg/s) $= 10.44 \times 10^5$ N. This upward force is opposed by the weight of the rocket, which is 3.53×10^5 N at take-off, 2.39×10^5 N after 20 s, and 1.25×10^5 N after 40 s. Hence the net upward forces at times $t = 0$, 20, and 40 s, respectively, are

$$F_0 = 6.91 \times 10^5 \text{ N}$$

$$F_{20} = 8.05 \times 10^5 \text{ N}$$

$$F_{40} = 9.19 \times 10^5 \text{ N}$$

The acceleration of the rocket at any instant is given by $a = (F_{\text{net}}/W)\ g$, so that, at the three times considered,

$$a_0 = 1.96\ g \qquad a_{20} = 2.97\ g \qquad a_{40} = 7.35\ g$$

In this example we have neglected air friction and the variation of g with altitude.

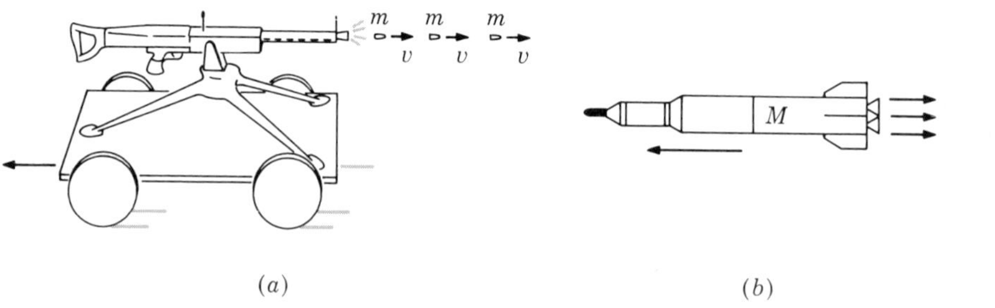

Figure 9-11
Rocket propulsion: reaction force from ejected matter.

9-9 BURNOUT VELOCITY

The range and accuracy of a rocket depend on the velocity v_b attained by the time the fuel has burned out or is cut off. A simplified case will be considered. Suppose that a rocket is launched vertically from the earth and continues upward in a straight line, encountering negligible atmospheric resistance. The total rate of change of momentum is equal to the resultant force F acting on the rocket,

$$F = m\frac{\Delta v}{\Delta t} + \overline{v}_e\frac{\Delta m}{\Delta t} \tag{16}$$

Here m is the mass of the rocket at any instant (a variable) and $\overline{v}_e$ is the average exhaust velocity of the gases ejected from the rocket motor, in the direction opposite to the velocity v of the rocket. In our simplified case, F includes only the gravitational force acting on the rocket, and $F = mg$. Equation (16) may be solved (Appendix A-6) for the burnout velocity v_b,

$$v_b = \overline{v}_e \ln\frac{m_0}{m_b} - \overline{g}t_b \tag{17}$$

where m_0 is the mass of the rocket at take-off, m_b is its mass at burnout at time t_b later, and $\overline{g}$ is the average value of the gravitational acceleration.

9-10 MULTIPLE-STAGE ROCKETS

In a single-stage rocket the propulsion energy must be used to accelerate the entire empty mass of the rocket even after most of that mass is no longer useful. This severely limits the speed attainable. In fact with present fuels a single-stage rocket cannot achieve the speeds of the order of 25,000 ft/s and greater required to place a satellite in orbit or to escape the earth's gravitational field.

A multiple-stage rocket is made up of a number of independent sections each equipped with a propulsion system and a portion of the total propellant load. After the first (booster) stage has lifted the entire rocket and has reached its burnout velocity, its empty mass is dropped from the rocket. A second (sustainer) stage carrying the payload is then fired and continues to accelerate the now lightened missile to the appropriate final velocity. Of course more than two stages can be and are used, but design and operational difficulties become more numerous as stages are added.

9-11 POWER SOURCES FOR ROCKETS

Today's rockets utilize chemical reactions, the burning of a solid or liquid fuel (Fig. 9-12*a* and *b*). Other types are being developed or have been suggested more or less speculatively. The use of a nuclear reactor to heat and expel a fuel is a possibility. Or energy could be obtained from isotope decay. Several electrical systems have been proposed. Ion propulsion (Fig. 9-12*c*) might be achieved by forming gas ions, accelerating them in an electric field, then expelling them to obtain a reaction thrust. Arc heating (Fig. 9-12*d*) would use electrical energy to heat and expel a fuel. Magnetoplasma propulsion (Fig. 9-12*e*) proposes to ionize a gas at high temperature, then use a magnetic field to accelerate the ions out of the rocket to obtain a reaction force. Each of these electrical methods requires a formidable amount of electric power, which implies an unwanted increase in overall mass.

The performance of a rocket engine is conveniently described by its *specific impulse*. This is the thrust produced divided by the weight of propellant consumed per second,

$$I_s = \frac{F}{\Delta W/\Delta t} \tag{18}$$

The unit of specific impulse is the second. In Fig. 9-13, the classes of rockets just described are compared. The various electrical types are seen to have a better specific impulse than conventional chemical rockets. But chemical rockets can

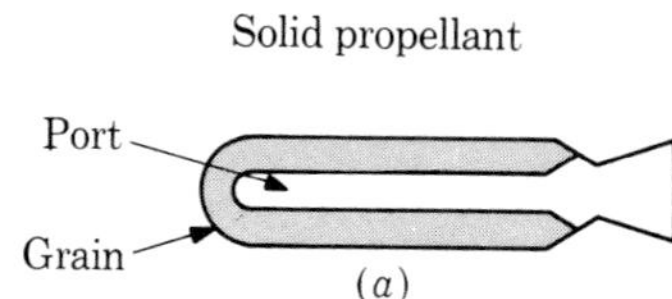

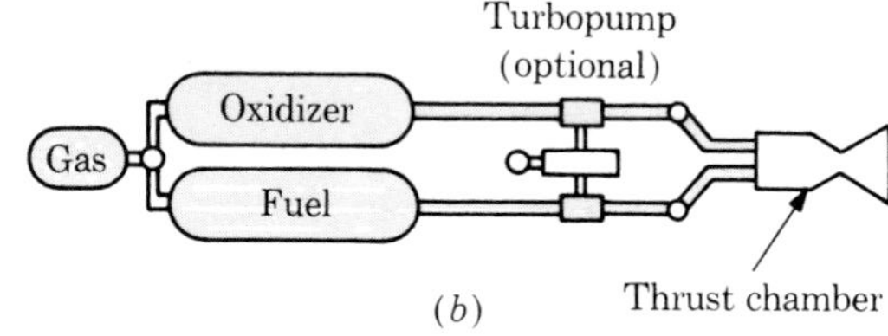

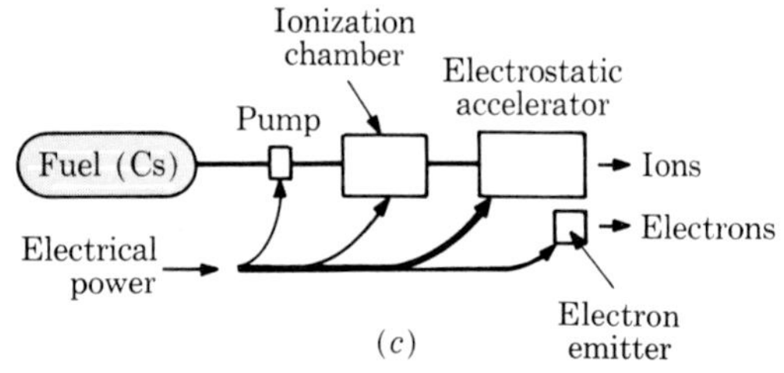

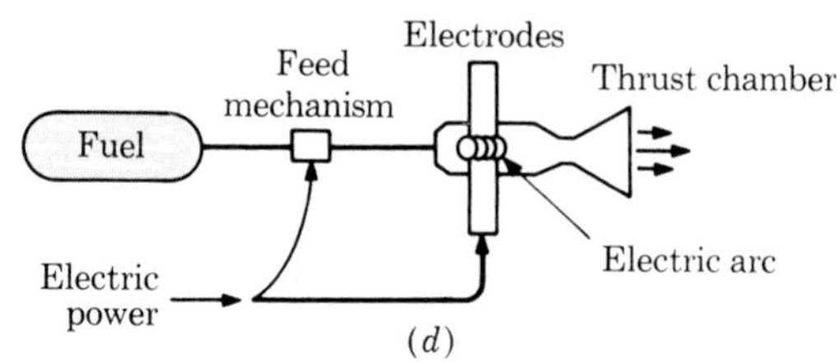

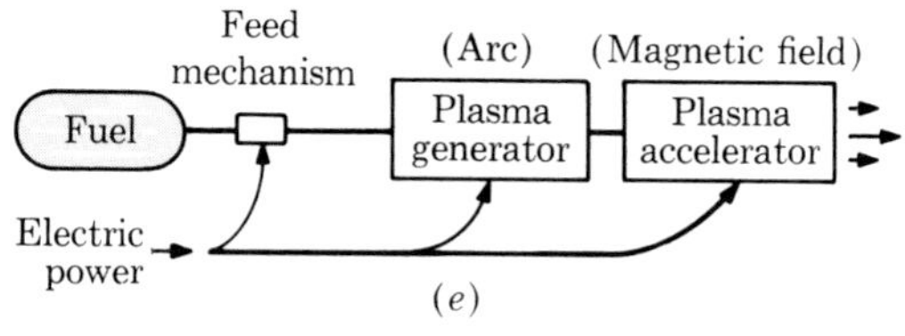

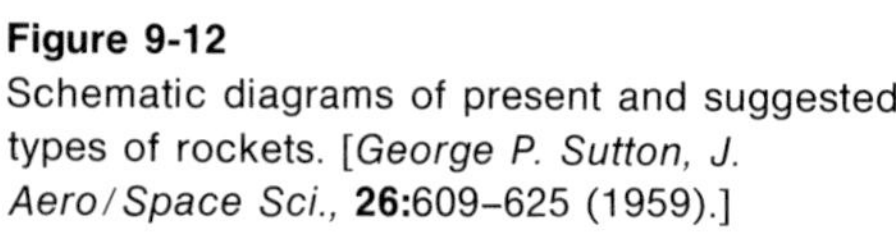

Figure 9-12
Schematic diagrams of present and suggested types of rockets. [*George P. Sutton, J. Aero/Space Sci.,* **26:**609–625 (1959).]

now be built of much lighter weight, and hence they have the advantage in thrust-to-weight ratio, which determines the rocket's acceleration.

In the future it may be feasible to use chemical boosters in initial stages at take-off, with an ion or nuclear rocket to maneuver in space where the accelerations required are small and where long life and high specific impulse are advantageous.

SUMMARY

Aerospace physics, possibly the most dramatic illustration of the application of the laws of motion, refers to the study of the space surrounding the earth and of the universe, using the tools developed through physics and other related sciences.

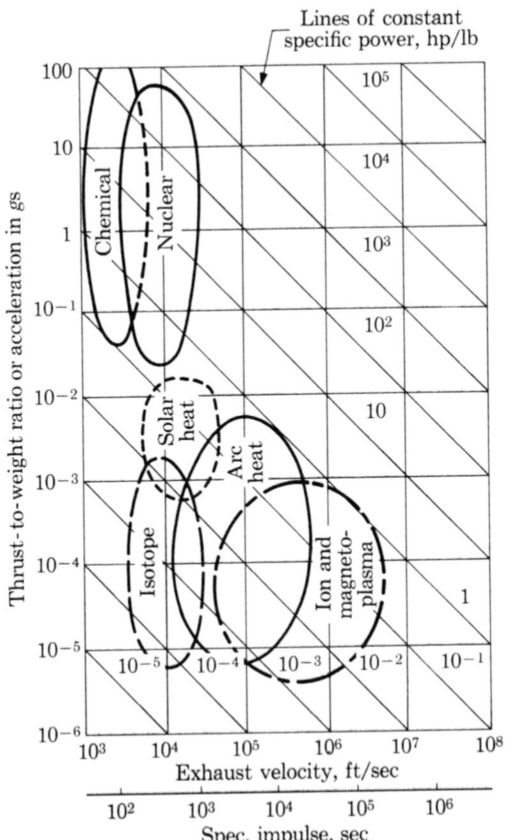

Figure 9-13
Acceleration and specific power for various rockets. [*George P. Sutton, J. Aero/Space Sci.*, **26**:609–625 (1959).]

Space is defined as man's environment in the solar system, which is itself small in the order of magnitudes in the universe. The solar system occupies only 10^{-24} of the volume of man's galaxy, the Milky Way.

Man has long had the dream of flying to the moon: the first written record can be traced back to the second century in Greece. His goal was achieved with the first landing on the moon in 1969.

Earth satellites, planets, and comets follow paths which can be described mathematically by conic sections.

A body held in a circular orbit by the attraction of a central body has negative potential energy which is greater in magnitude than the kinetic energy; hence the total energy of the system is negative,

$$E = E_p + E_k = -\frac{GmM}{2r}$$

To escape from earth, a missile must have a velocity at take-off which makes its initial kinetic energy at least equal to the negative of its potential energy.

Kepler's laws, which describe the motion of stars, planets, and comets, also govern a space vehicle when not under power:

First law: The planets move in ellipses having a common focus situated at the sun.

Second law: The line which joins a planet to the sun sweeps over equal areas in equal intervals of time.

Third law: The squares of the periods of the planets are proportional to the cubes of their mean distances from the sun.

A satellite orbit is defined by six constants. The *perigee, apogee,* and *eccentricity* describe the orbit as observed from earth. The orientation of the orbit in the solar system requires in addition the specification of three orbital angles.

The *thrust* which accelerates a rocket is the reaction to the force which expels its propellant particles, $F = v\,\Delta m/\Delta t$.

The *mass ratio* m_0/m_b of a rocket is important in determining the velocity attained at burnout.

The *specific impulse* of a rocket is the thrust produced divided by the weight of propellant consumed per second.

References

Adams, Carsbie C.: "Space Flight," McGraw-Hill Book Company, New York, 1958.

———, Wernher von Braun, and Frederick I.

Ordway: "Careers in Astronautics and Rocketry," McGraw-Hill Book Company, New York, 1962.
Berman, Arthur I.: "Astronautics: Fundamentals of Dynamical Astronomy and Space Flight," John Wiley & Sons, Inc., New York, 1961.
Lundquist, C. A.: "Space Science," McGraw-Hill Book Company, New York, 1966.
NASA: Objectives and Goals in Space Science and Applications, A Report of the Office of Technology Utilization, NASA, Washington, D.C., 1968.
———: Space . . . the New Frontier, a publication supervised by James Dean, NASA, Washington, D.C., 1964.
———: Space Resources for Teachers: Biology, a publication prepared at the University of California, Berkeley, for NASA, Washington, D.C., 1969.
———: Space Resources for Teachers: Chemistry, a publication prepared at Ball State University for NASA, Washington, D.C., 1971.
———: Space Mathematics, A Resource for Teachers, a publication prepared at Duke University for NASA, Washington, D.C., 1972.
Newton, Isaac: "Mathematical Principles," trans. by A. Motte in 1729, rev. by F. Cajori, University of California Press, Berkeley, Calif., 1934.
Rea, D. G.: Whither lunar and planetary exploration in the 1970's? *Science,* November 28, 1969.
Weber, Robert L.: "Physics for Teachers: A Modern Review," McGraw-Hill Book Company, New York, 1964.

Questions

1 What is meant by the term space?

2 Give examples of some specific problems of space travel that physics has been called upon to solve.

3 Compare the length of the flight of Robert Goddard's first liquid-fuel rocket in 1926 to the length of the manned flight to the moon in 1969.

4 Discuss the pros and cons of future space exploration. List some of the fringe benefits attained through space research to this point.

5 Comment on the remark, "Space stations will be obsolete when they are feasible."

6 Will prolonged "weightlessness" affect the muscular system of an animal? Give illustrations to support your answer.

7 Show the similarity between the "fall" of the moon and fall of a pendulum bob.

8 What does Kepler's second law say about the duration of winter in the Southern Hemisphere (which occurs in July when the earth is farthest from the sun) as compared with winter in the Northern Hemisphere?

9 At what point in its trajectory does a projectile have its minimum speed?

10 Discuss the various possible trajectories of earth satellites and the relationship of each to the escape velocity.

11 Why is the upper (dotted) path in Fig. 9-14 not a possible orbit for an earth satellite?

12 Show that if frictional forces cause a satellite to lose total energy, it will move into an orbit closer to the earth with an actual increase in speed.

13 After a certain satellite was put into orbit, it was stated that the satellite would not return to earth but would burn up on its descent. Why should this occur, since the satellite did not burn up on ascent?

14 How can rocket action be demonstrated with a toy balloon?

15 Verify the statement: "Near the surface of the earth, gravity robs a vertically rising rocket

Figure 9-14

of about 32.2 km/h in speed each second, or about 3,860 km/h for each 2 min of acceleration."

16 If in a satellite a cork is immersed in a bottle of water, will the cork rise? Explain.

17 An astronaut in spaceship *A* in orbit about the earth wishes to achieve a rendezvous with spaceship *B*, which he sees far ahead in approximately the same orbit. How does he maneuver? (If the speed of spaceship *A* is increased, it will no longer be in its original orbit.)

18 Show by an equation the relationship between thrust, mass of propellant, and exhaust velocity for chemical rockets.

19 If it becomes possible to convert the energy of nuclear fusion in a plasma directly into electric energy, without the usual rotating generator, would this make ion propulsion of rockets more feasible?

20 List the elements and compounds present in the latest concept of the earth's atmosphere.

21 How would the fact that Jupiter has a higher escape velocity than the moon's affect the nature of the atmosphere on their surfaces?

22 The earth satellite Explorer 3 had a highly eccentric orbit with perigee at a height of 109 mi. At this point the speed was 2.76×10^4 ft/s in a direction perpendicular to the radius to the center of the earth. Show that this speed is too great for a circular orbit at the radius of 4,109 mi. Hence the satellite described an elliptical orbit. Its apogee was at the height 1,630 mi. Show that the speed at apogee was too small for a circular orbit at radius 5,630 mi.

23 A rocket is launched horizontally and continues in a path parallel to the earth. Show that if the atmosphere offers negligible resistance and the rocket is initially at rest, the burnout velocity becomes $v_b = v_e \ln (m_0/m_b)$, where the exhaust velocity v_e is constant.

Problems

1 An object weighs 445 N on earth and is placed 80.5 km above the surface of the earth? (Radius of earth = 6.38×10^3 km.) What is its weight at that altitude?

2 An object weighing 100 lb on the surface of the moon is moved up to the following altitudes: 50 mi, 100 mi, 200 mi, 500 mi, and 1,080 mi. What would its physical weight be at these altitudes?
Ans. 92 lb, 84 lb, 71 lb, 47 lb, 25 lb.

3 If the mass of the moon is one-eightieth the mass of the earth and its diameter is one-fourth that of the earth, what is the acceleration due to gravity at the surface of the moon? How far will a 2.0-kg mass fall in 1.0 s on the moon?

4 Calculate the mean distance of Mercury to the sun if its period of revolution is 0.241 year.
Ans. 0.39 au, 3.6×10^7 mi, or 5.8×10^7 km.

5 The periods of revolution of the planets Mercury, Venus, Mars, and Jupiter are, respectively, 0.241, 0.617, 1.88, and 11.9 years. Find their mean distances from the sun, expressed in astronomical units (1 au = distance from sun to earth).

6 Calculate the period of Jupiter if the distance from it to the sun is 7.78×10^8 km.
Ans. 11.86 years.

7 A point on the earth's Equator is carried about 1,610 km/h by the rotation of the earth. Jupiter has an equatorial diameter 11 times that of the earth and a day of 10 h. Calculate the speed of a point on the equator of Jupiter.
Ans. 2.64×10^4 mi/h.

8 Identify the conics represented by the following eccentricities: $e = 0$; $0 < e < 1$; $e = 1$; $e > 1$.
Ans. circle; ellipse; parabola; hyperbola.

9 Compute the eccentricity of the orbit of Sputnik I which had a perigee of 212.5 km and an apogee of 938.6 km. *Ans.* 0.052.

10 If a rocket attains a speed of 966 km/h by the time it reaches 305 m, how many times *g* is its acceleration? *Ans.* 12 *g*.

11 Compute the circular orbital velocity of a spaceship at an altitude of 160 km above the surface of the earth.

12 What would the necessary circular orbital velocity be for a satellite at an altitude of 500 mi? When 1,000 mi high?
Ans. 16,700 mi/h; 15,700 mi/h.

13 What velocity would be needed to attain a circular orbit 96.6 km above the surface of the moon? (Mass of the moon is 0.012 times the mass of the earth.)

14 A spacecraft in orbit above the earth has apogee and perigee altitudes of 1,000 mi and 500 mi, respectively. Find the respective velocities. *Ans.* 15,400 mi/h, 17,100 mi/h.

15 A rocket whose thrust is 2.70×10^4 lb weighs initially 2.20×10^4 lb, of which 80 percent is fuel. Assuming constant thrust, find the initial acceleration and the acceleration just before burnout. Neglect air resistance and variation of g.

16 A rocket has a gross weight of about 64,000 lb at lift-off. Its engine can produce 96,000 lb of thrust. If the exhaust velocity of the gases from the rocket is 7,500 ft/s, at what rate is the mass ejected? What is the acceleration of the rocket at launch and after 100 s of flight?
Ans. 12.8 slugs/s; 16 ft/s^2 or $\frac{1}{2}g$; 101 ft/s^2 or 3.17 g.

17 A fueled rocket of mass 9.1×10^3 kg ejects hot gases at a speed of 1.3×10^3 m/s and at a mass rate of 150 kg/s. (*a*) What is the thrust exerted on the rocket? (*b*) If burnout takes place in 20 s, what is the rocket speed at burnout, assuming vertical launching? (*c*) What is the specific impulse?

Max von Laue, 1879–1960

Born in Phaffendorf on the Rhine. Professor at Zurich, Frankfurt am Main, and Berlin. Awarded the 1914 Nobel Prize for Physics for his discovery of the diffraction of Röntgen rays in crystals.

10

Elastic Properties of Solids and Liquids

From early times, man has designed and built structures of great size and beauty. Only in relatively modern times has the design of structures and machines become a science as well as an art. Galileo applied the principles of equilibrium to determine the internal stresses in a cantilever beam. Hooke, Young, and others devised ways of measuring significant elastic properties of materials. With such data on the strength of materials, engineers can design with confidence structures which are economical in their use of material and vehicles which are economical in their power requirements. The physicist finds it a challenge to account for observed elastic properties in terms of atomic arrangements and interatomic forces. Elastic properties are important in the theory of vibration and sound.

Man's adventures in space have shown that the unusual conditions encountered there sometimes produce unexpected and even disastrous results. The effect of extreme temperature changes, exposure to radiation, and vibrational stress which occur in space travel make a study of the structural properties of matter increasingly important.

10-1 ELASTICITY

As used in physics, elasticity means the property by which an object changes its shape and size under the action of opposing forces and recovers its original configuration when the forces are removed. The recovery is practically perfect for many kinds of materials, provided that the distorting forces are not too great. If the distorting forces are too large, the body does not recover completely its original configuration when the forces are removed but acquires a permanent set or permanent deformation. It is then said that the deformation exceeded the *elastic limit*.

Materials like dough, lead, and putty for which the elastic limit is very small are called *inelastic*, or *plastic*, materials. Steel is a highly elastic material; it returns closely to original dimensions even after being subjected to relatively large forces. From this point of view, rubber is not highly elastic, even though it may be readily stretched.

10-2 HOOKE'S LAW

The English physicist Robert Hooke (1635–1703), a great experimentalist who was the curator of experiments of the Royal Society, showed experimentally that the deformation of an elastic body is directly proportional to the applied force, provided that the elastic limit is not exceeded.

An experiment may be performed with a helical spring of initial length l_0 suspended from a rigid support (Fig. 10-1*a*). Weights are added successively to the spring, and the corresponding elongations y are noted. It is found that the de-

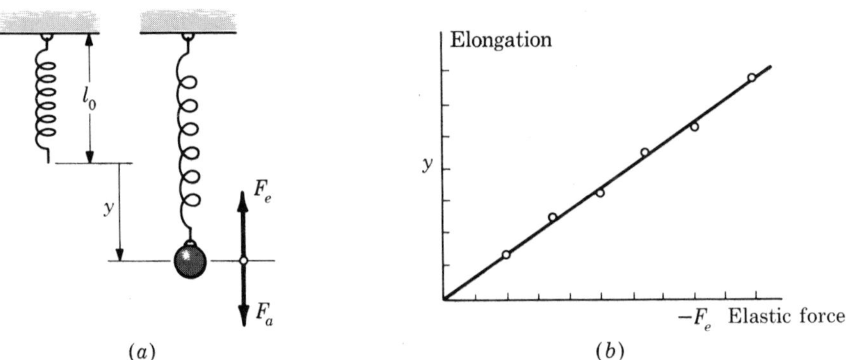

Figure 10-1
(*a*) The weight of the ball elongates the spring. The elastic force F_e is the reaction of the spring on the ball. (*b*) The deformation y is proportional to the applied force F_a.

forming force F_a is proportional to the lengthening of the spring, $F_a \propto y$ or $F_a = ky$, and the elastic force F_e exerted by the spring in the direction *opposite* to the stretching is

$$F_e = -ky \tag{1}$$

The constant k is called the *force constant* of the spring; k is the force per unit displacement—that is, how much force is required to stretch the spring one unit of length.

The change in the elastic potential energy of the spring is the work done by the applied force during slow elongation, or the negative of the work done by the elastic force F_e. As the spring is extended (Fig. 10-1), the restoring force changes linearly from 0 to $-ky$. The average value of the force, averaged with respect to dis-

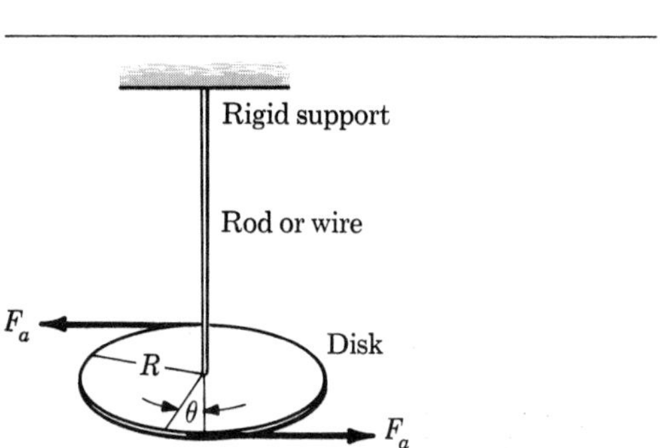

Figure 10-2
Torsional elasticity.

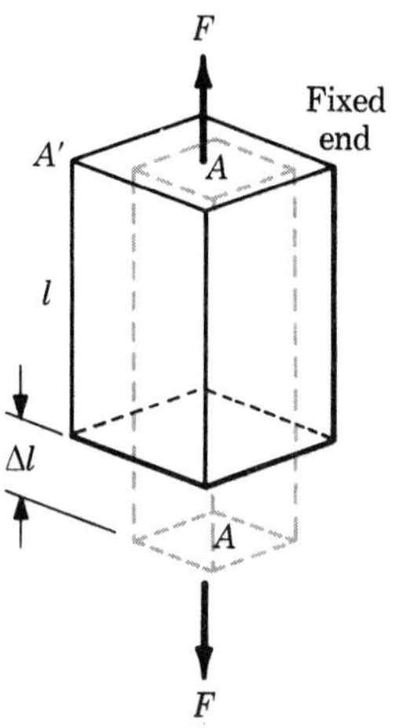

Figure 10-3
Longitudinal stress on a bar, with deformations exaggerated.

tance, is $-\frac{1}{2}ky$. Since work equals a force $(-\frac{1}{2}ky)$ moving through a distance (y), the restoring force on the spring does work equal to $(-\frac{1}{2}ky)(y) = -\frac{1}{2}ky^2$. The negative of this quantity is by definition the potential energy of the elongated spring relative to its initial condition,

$$E_p = \tfrac{1}{2}ky^2 \qquad (2)$$

To investigate the validity of Hooke's law for a deformation due to a twisting or a torque, one may use a rod clamped at its upper end (Fig. 10-2) and to the lower end of which is applied a torque $L_a = 2RF_a$. Experiment shows that the angular displacement θ is directly proportional to the applied torque L_a;

$$L_a = c\theta \qquad \text{or} \qquad L_e = -c\theta \qquad (3)$$

where the elastic reaction L_e of the wire is the negative of L_a and the torsion constant c is the torque per unit angular displacement. The elastic potential energy stored in the twisted rod is

$$E_p = \tfrac{1}{2}c\theta^2 \qquad (4)$$

10-3 STRESS AND STRAIN

From measurements on particular bodies, we should like to be able to generalize and describe the elastic properties of the materials of which the bodies are made. The concepts of stress and strain enable us to state Hooke's law in such a general form. Stress is related to the force producing a deformation. Strain is related to the amount of the deformation.

If stretching forces F are applied to the ends of a long elastic rod of square cross section and length l (Fig. 10-3), the rod becomes longer and thinner and assumes the shape shown (somewhat exaggerated) by the dotted lines. The longitudinal *stress* is defined as the force per unit cross-sectional area, $\Delta F/A$. For a uniform rod, the stress is the same at all cross sections. The longitudinal *strain* is defined as the change in length per unit length, $\Delta l/l$. Strain is the fractional deformation, expressed as a number without units if l and Δl are measured in the same unit.

10-4 YOUNG'S MODULUS

Hooke's law may now be generalized in the case of stretched bars or wires: stress is proportional to strain, the constant of proportionality being characteristic of the material

$$\frac{\text{Stress}}{\text{Strain}} = \text{const} \qquad \frac{\Delta F/A}{\Delta l/l} = Y \qquad (5)$$

The ratio Y of the tensile stress to the tensile strain $\Delta l/l$ is called *Young's modulus*. Values of Y for several common materials are given in Table 1. The same values apply to compression and to tension, for moderate deformations. Note that the physical dimensions of Y are those of force per unit area.

Although stretching a rubber band does increase the restoring force, the stress and strain do not vary in a direct proportion; hence Young's modulus for rubber is not a constant. Moreover, a stretched rubber band does not return immediately to its original length when the deforming force is removed. This failure of an object to regain its original size and shape as soon as the deforming force is removed is called *elastic lag*, or *hysteresis* (a lagging behind).

We shall see other types of hysteresis when we study electromagnetism and the optics of the eye where residual vision occurs.

In listing a single value of Y for a material we imply that we are considering only isotropic materials, materials whose physical properties are independent of direction. An anisotropic material, such as wood, has different compressive properties in the different directions of applying compressive stress. Three different Young's moduli are used to describe the elastic behavior of an anisotropic material. Single crystals of most substances are anisotropic, having different elastic, optical, and electrical properties along differ-

Table 1
ELASTIC MODULI (Approximate Values)

Material	Young's modulus Y		Shear modulus S		Bulk modulus B	
	10^{10} N/m^2	10^6 lb/in^2	10^{10} N/m^2	10^6 lb/in^2	10^{10} N/m^2	10^6 lb/in^2
Aluminum	7.0	10	2.6	3.8	7.7	11
Brass (60% Cu)	10	14	3.5	5.1	11	16
Copper	13	19	4.8	7.0	14	20
Glass	6.0	8.7	3.1	4.5	3.7	5.4
Ice (−2°C)	0.28	0.41				
Iron, wrought	20	29	8.0	12	17	25
Lead	1.6	2.3	0.56	0.81	4.6	6.7
Nickel	20	29	7.9	11	16	23
Steel	20	29	8.4	12	17	25
Tungsten	40	58	15	22		

ent axes. But when such crystals are packed together with random orientations, as in a piece of metal, the metal as a whole ordinarily is isotropic in its properties.

Example A steel bar, 6 m long and of rectangular cross section 5.0 × 2.5 cm, supports a load of 1.78 × 10^6 N. How much is the bar stretched?

From Eq. (5),

$$\Delta l = \frac{\Delta F l}{YA}$$

$$\Delta F = 1.78 \times 10^6 \text{ N}$$

$$A = 5 \text{ cm} \times 2.5 \text{ cm} = 12.5 \text{ cm}^2 = 1.25 \times 10^{-3} \text{ m}^2$$

$$\Delta l = \frac{(1.78 \times 10^6 \text{ N})(6 \text{ m})}{(20 \times 10^{10} \text{ N/m}^2)(1.25 \times 10^{-3} \text{ m}^2)} = 4.272 \times 10^{-2} \text{ m}$$

Table 2
BULK MODULI FOR LIQUIDS (15°C)

Liquid	Bulk modulus B	
	10^{10} N/m^2	10^6 lb/in^2
Ethyl alcohol	0.110	0.16
Glycerin	0.40	0.58
Mercury	2.8	4.0
Petroleum	0.14	0.20
Water	0.21	0.31

10-5 VOLUME ELASTICITY; BULK MODULUS

Consider a cube on which forces are applied normal to each surface and uniformly distributed over the surface. The resultant force on each face is indicated by the vectors of Fig. 10-4. The cube will be in equilibrium under the action of these forces, but its volume will be reduced. The *volume stress* is the normal force per unit area. The *volume strain* ($\Delta V/V$) is the change in volume per unit volume. The ratio of the volume stress to the volume strain is called the coefficient of volume elasticity, or *bulk modulus*,

$$\frac{\text{Volume stress}}{\text{Volume strain}} = \text{const} \qquad \frac{\Delta F/A}{\Delta V/V} = B \quad (6)$$

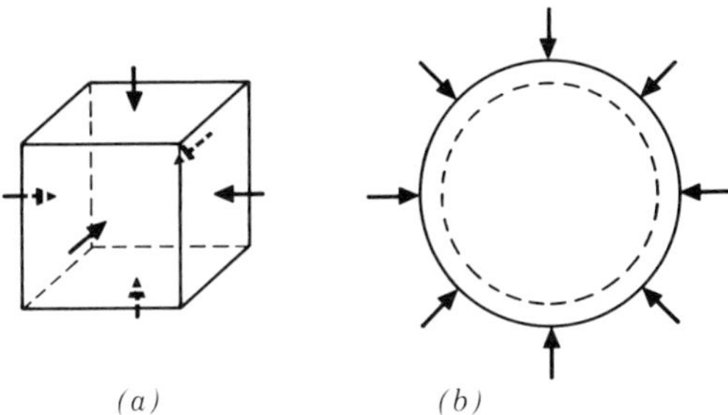

Figure 10-4
Volume stress. Normal forces on (*a*) the faces of a cube and (*b*) the surface of a sphere.

This type of deformation, which involves only volume changes, applies to liquids as well as to solids. The elastic moduli of liquids and solids (Tables 1 and 2) are large numbers, expressing the familiar fact that large forces are needed to produce even minute changes in volume. Gases are more easily compressed and have correspondingly smaller coefficients. The *compressibility* of a material is the reciprocal of its bulk modulus.

Example Find the weight density of water at a pressure of 4,000 lb/in², taking the weight density at normal atmospheric pressure as 62.4 lb/ft³.

From Eq. (6) and Table 2,

$$-\frac{\Delta V}{V} = \frac{\Delta F/A}{B}$$

$$= \frac{(4{,}000 - 15)\ \text{lb/in}^2}{144\ \text{in}^2/\text{ft}^2} 3.22 \times 10^{-6}\ \text{in}^2/\text{lb}$$

$$= 8.91 \times 10^{-5}$$

Since weight density is mg/V, the fractional decrease in volume just calculated will result in a fractional increase in density $\Delta D/D = 8.91 \times 10^{-5}$ and $D = [62.4 + 8.91 \times 10^{-5}(62.4)]\text{lb/ft}^3 = 62.406\ \text{lb/ft}^3$. Water has such a small compressibility that there is not an appreciable increase in density.

10-6 ELASTICITY OF SHEAR

A third type of elasticity concerns changes in shape. This is called *elasticity of shear*. As an illustration of shearing strain, consider a cube of the paper in a large book (Fig. 10-5), fixed at its lower face and acted upon by a tangential force F at its upper face. This force causes the consecutive horizontal layers of the cube to be slightly displaced, or sheared, relative to one another. A line such as BD or CE in the cube is rotated through an angle φ. The shearing strain is defined as the angle φ (expressed in radians). Usually this angle is small and may be approximated by $\varphi = BB'/BD$. The shearing stress is the ratio of the tangential force F to the area of the face $BCGH$ over which it is applied. The ratio shearing stress divided by shearing strain is the *shear modulus,* or coefficient of rigidity, n,

$$\frac{\text{Shearing stress}}{\text{Shearing strain}} = \text{const} \qquad \frac{\Delta F/A}{\varphi} = n \tag{7}$$

A wire or rod when twisted undergoes shear strain, as may be visualized by drawing a line along its length before twisting (Fig. 10-6). Delicate electrical meters often use a fine wire clamped at one end to support a moving coil. When there is a current in the coil, it reacts with an external magnetic field to produce a torque

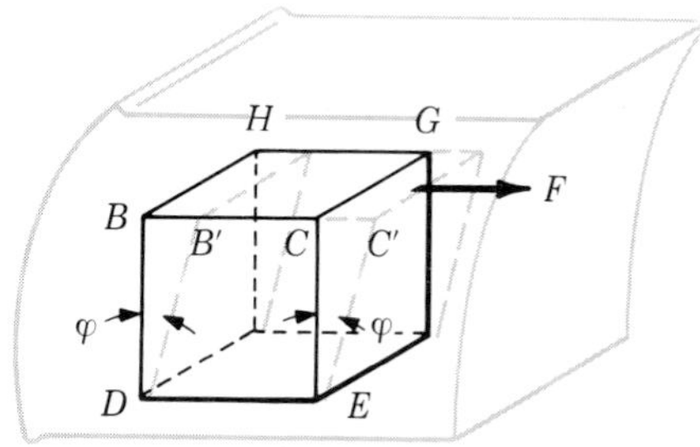

Figure 10-5
Shearing deformation. The cube has been sheared through angle φ by a force F.

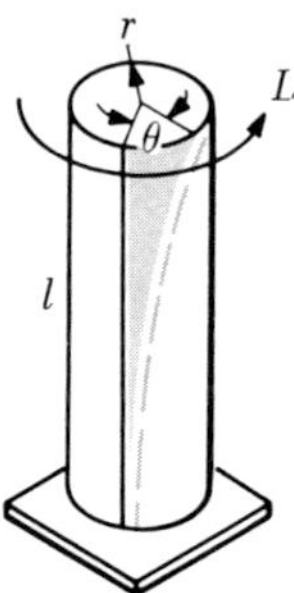

Figure 10-6
Shear θ of a cylinder. The dotted line was drawn as a straight line on the surface of the cylinder before it was twisted.

L_a. The coil turns, twisting the fiber until at equilibrium the elastic restoring torque L_e balances the magnetic torque. This elastic restoring torque is proportional to the deflection,

$$L_e = \frac{\pi r^4 n}{2l} \theta = k\theta \tag{8}$$

where k is constant for a given supporting fiber.

10-7 TYPES OF STRESS-STRAIN RELATIONS

The ways in which samples of different materials are deformed by various loads are illustrated in Fig. 10-7. For each load the tensile strain is calculated as the ratio of the elongation to the original length. This ratio is plotted against the tensile stress, and a curve is drawn through the points so obtained.

Generally the stress-strain diagram shows an initial range in which the sample obeys Hooke's law, i.e., Young's modulus is a constant; the specimen returns to its original length when the stress is removed. The sample will support stresses in excess of the proportional limit (elastic limit), but when unloaded it is found to have acquired a permanent set. The proportional limit is the greatest stress a material can sustain without departure from a linear stress-strain relation. (A few materials such as carbon steel show a sudden yielding without increase in stress, Fig. 10-7*b*.) For many materials which have a gradual knee in the stress-strain curve, yield strength is specified as the stress accompanying a small (say, 0.2 percent) permanent deformation which is considered not to have impaired the usefulness of the material.

If the applied stress is increased slowly, the sample will finally break. The maximum stress applied in rupturing the sample is called the *ultimate strength*. Although the ultimate strength of the sample lies far up on its strain-stress curve, it cannot safely be expected to carry such loads in structures. Axles and other parts of machines which are subject to repeated stresses are never loaded beyond the elastic limit. Ultimate strength is defined in relation to the maximum resistance to tensile, compressive, or shearing forces and is stated in terms of the stress which produces fracture or the stress which produces some specified deformation. For brittle materials the ultimate strength is the breaking stress.

The stress-strain diagram is the basis for evaluating a number of mechanical properties: the strength, deformation, and energy characteristics of materials. As an example of the latter, consider a material which shows marked hysteresis in regaining its original shape as an applied stress is removed (Fig. 10-8). In one cycle of a varying stress, energy proportional to the area of the hysteresis loop is absorbed by the material and converted into heat. Such a material might be suitable to place under a piece of machinery to minimize the vibration transmitted to the floor.

Whenever a machine part is subjected to repeated stresses over long periods of time, the internal structure of the material is changed. Each time the stress is applied, the molecules and crystals realign. Each time the stress is removed, this alignment retains some permanent set. As this process continues, certain regions are weakened, particularly around areas where there are microscopic cracks on the surface. This loss of strength

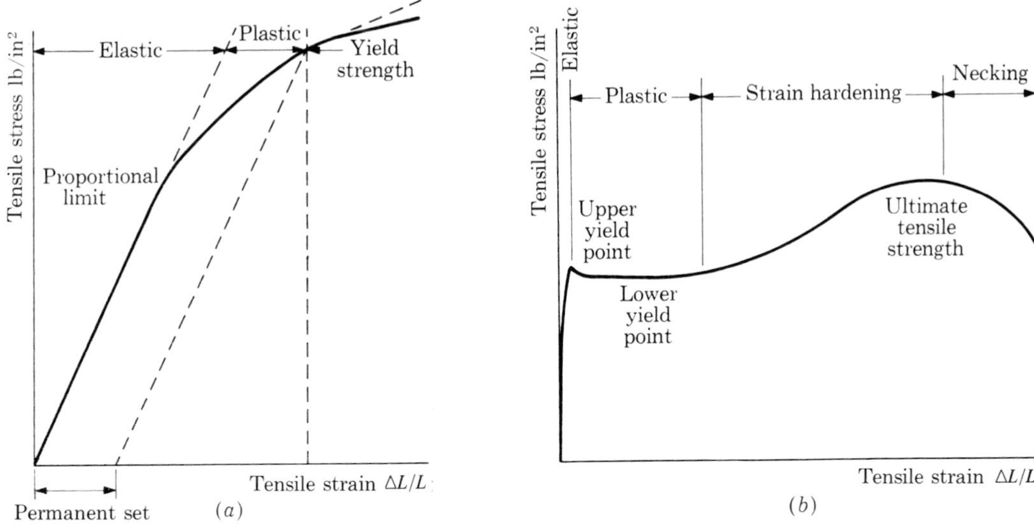

Figure 10-7
Stress-strain diagrams. (*a*) An aluminum alloy; (*b*) a soft steel.

because of repeated stresses is known as *fatigue*. Since failure due to fatigue occurs much sooner if flaws are present originally than it does in a perfect part, it is important to detect such flaws before a part is installed. In many plants, x-rays are used to detect hidden flaws in the components of complicated machines such as airplanes and rockets. Recommended procedures for making numerous tests have been developed cooperatively by industry and such organizations as the American Society for Testing and Materials and the National Bureau of Standards.

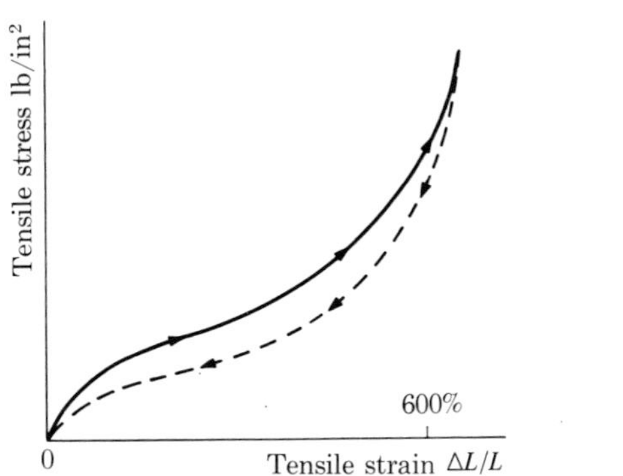

Figure 10-8
Elastic hysteresis shown in stress-strain curve for vulcanized rubber.

Weakening due to fatigue can be demonstrated by bending a piece of copper wire several times. Whereas it was impossible to break the wire at first, the wire breaks very easily after repeated bendings.

10-8 ROLLING FRICTION

We saw earlier that rolling friction was caused by a deformation of the surface in front of the rolling wheel. To be more precise, elastic hysteresis accounts for some kinds of rolling friction. In the case of a pneumatic tire, for which the road surface may deform very little but the tire may flatten considerably, the rolling friction is due primarily to the elastic hysteresis of the rubber. Inflating the tire to a higher pressure reduces the flexing of the rubber. This reduces not only the forces due to the deformation of the rubber but also the area of flattening.

10-9 THERMAL STRESSES

When a structure such as a bridge is put together, the design must take into account changes in shape due to changes in temperature. If such provision is not made, tremendous forces develop that may shatter parts of the structure. Anyone who has ever seen a concrete pavement shattered by these forces on a hot day realizes the violence of such a phenomenon.

Once again the special stress put on an object in space should be noted. The temperature differential on the moon is as much as 330°C and may occur in only a few hours. The sunlit side of the moon during the 14 d of sunlight during the lunar month reaches 130°C, while the shaded side of the moon can reach as low as −200°C during the sunless period.

A more spectacular example of thermal stress was observed by NASA in its experiments with the Fire I spaceship. It was found that the temperature of reentry reaches 20,000°F in the compressed-air area known as the gas cap in front of the vehicle. This temperature is many times higher than can be contained in almost any of the earth's furnaces. The heating rate is very high and the "heat pulse" lasts for about 1 min.

When selecting material for missions in space, the nature of the strain resulting from thermal stress must be carefully predicted.

10-10 RELATIONS AMONG ELASTIC CONSTANTS

Tension tests of metals are made on suitably machined bars in a testing machine which elongates the specimen and measures the resisting force. The true stress is computed on the basis of the reduced sectional area which accompanies an applied force. The ratio of the contraction in diameter $\Delta d/d$ to the extension in length $\Delta l/l$ (Fig. 10-3) is a constant, called *Poisson's ratio,* σ. For many metals the value of σ is about 0.3. We have defined four basic elastic moduli for a material: Young's modulus $Y = (\Delta F/A)/(\Delta l/l)$, the bulk modulus $B = (\Delta F/A)/(\Delta V/V)$, the shear modulus $n = (\Delta F/A)/\varphi$, and Poisson's ratio $\sigma = (\Delta d/d)/(\Delta l/l)$. One might expect the existence of a relationship among these parameters describing the elastic properties of a material. More advanced treatments of mechanics show that there is a definite relationship expressible as

$$B = \frac{Y}{3(1 - 2\sigma)} \qquad n = \frac{Y}{2(1 + \sigma)}$$

$$\sigma = \frac{Y}{2n} - 1 \tag{9}$$

10-11 ELASTIC BEHAVIOR AND ATOMIC STRUCTURE

The atoms of a crystal may be thought of as tiny spheres in thermal agitation about equilibrium points in some regular structure, such as the cubic lattice of Fig. 10-9*a*. These points lie about 0.2 to 0.4 nm apart (1 nanometer = 10^{-9} m), as determined from the way the atoms diffract x-rays. Each atom consists of a nucleus of diameter about 10^{-6} nm, which includes most of the mass of the atom and carries a positive electric charge. Electrons, very much lighter and negatively charged, orbit around each nucleus, to a distance of 0.1 to 0.2 nm from the nucleus. If one tries to push the atoms closer than their equilibrium distance, so that the electronic orbits of neighboring atoms begin to overlap, strong forces of repulsion come into play.

A helpful representation of interatomic forces in a crystal is the picture (Fig. 10-9*b*) of springs having equilibrium length connecting neighboring atoms. To compress the solid, one would have to exert a force to compress the springs. To expand the solid, one would have to exert a force

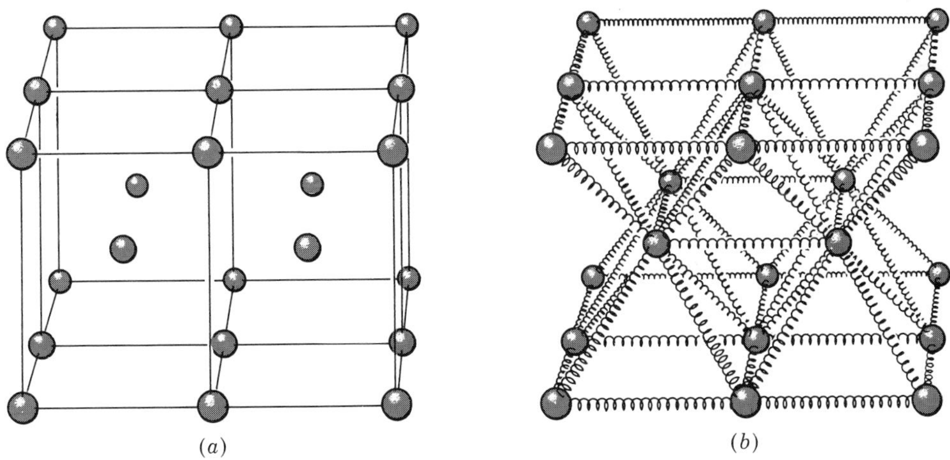

Figure 10-9
(*a*) Cubic arrangement of atoms in a crystal as inferred from x-ray data, and (*b*) the springlike forces between neighboring atoms.

to stretch the springs. The existence of internal forces implies movement of the atoms of the solid away from their equilibrium positions, and hence distortion of the body. Because of the cross bracing, the model of Fig. 10-9*b* will resist shear, but not so strongly as it resists extension or compression. Many solids require only about one-half to one-third the force per unit area to effect the same relative motion of the atoms in shear as is required for compression or extension (Table 1).

The vibration of the atoms is not linear harmonic motion. The increased vibration which accompanies a rise in temperature displaces the equilibrium positions of the atoms, evident in the expansion of the material. The increased agitation with rise in temperature may result in a shearing motion that breaks down the crystal structure. The solid melts and changes to a liquid. But interatomic forces tending to prevent expansion or contraction are still sufficiently great so that the liquid has a definite volume, but not definite shape. The liquid is pictured as having amorphous structure; it has a definite density, the average separation between atoms is definite, but there is not a unique position for each atom.

10-12 SOME FURTHER PROPERTIES OF MATTER

The suitability of a material for a certain application may depend on properties other than the mechanical ones just discussed. Among those are flammability, moisture absorption, electrical conductivity, and thermal conductivity. For most structural purposes, however, materials are selected chiefly on the basis of their cost and their elastic properties.

Materials possess several characteristics that are closely related to the elastic properties. Among these are ductility, malleability, compressibility, and hardness.

The *ductility* of a material is the property that represents its adaptability for being drawn into wire. *Malleability* is the property of a material by virtue of which it may be hammered or rolled into a desired shape. In the processes of drawing or rolling, stresses are applied that are much above the elastic limit so that a "flow" of the material occurs. For many materials the elastic limit is greatly reduced by raising the tempera-

Table 3
SCALE OF HARDNESS

Diamond	10	Apatite	5*
Ruby	9	Fluorite	4*
Topaz	8	Calcite	3*
Quartz	7	Gypsum	2†
Orthoclase	6	Talc	1†

*A knife will scratch.
†A fingernail will scratch.

ture; hence processes requiring flow are commonly carried on at high temperature.

The *hardness* of a mineral is determined from its ability to scratch other materials. Geologists, and the astronauts who landed on the moon, used such a procedure based on the Mohs' scale presented in Table 3. Diamonds, being the hardest substance known, are hence rated highest on the scale. It should be noted that a knife will scratch a mineral falling between levels 5 and 6 on the scale and a fingernail will scratch a mineral rated between levels 2 and 3. The property of hardness of engineering materials is now commonly measured by either the *Brinell number* or the *Rockwell number,* which are based on two somewhat different test procedures.

The Brinell number is the ratio of load (in kilograms) on a sphere used to indent the material, to the spherical area (in square millimeters) of the indentation. The standard indentor is a hardened steel ball of 10 mm diameter and the usual load is 3,000 kg, although 500-kg loads are used in testing some softer nonferrous materials.

In the Rockwell test, hardness is measured by the depth of penetration of a spherical-tipped conical indentor under certain specified conditions. The Rockwell number is read directly from a scale, lower numbers corresponding to soft materials, which suffer deeper penetration.

Hardness is not a fundamental property of materials but a composite one dependent on the elastic moduli, the elastic limit, the hardening produced by "working" a metal, etc. Empirical relations are used to determine other properties from the easily measured hardness, but all such schemes are of doubtful or limited validity.

The useful life of a structure or a machine part depends quite as much on its surface properties as on its bulk properties. A glass fiber 0.003 mm in diameter may have a tensile strength 30 times that of a fiber 1 mm in diameter, owing to the greater freedom from surface flaws in the smaller fiber. The surface layers of metal tools, glass plates, and rubber tires may be treated so as to introduce compressive stresses in the surface, thereby greatly increasing the durability of the part.

One difficulty encountered when attempting to measure the elastic properties of a material is that of providing a uniform or typical sample. If examined under sufficient magnification, no material is found to be uniform (homogeneous). Rock, brick, and concrete have a structure that can readily be seen. Elastic constants for such materials should not be taken for samples that are not large compared with the size of the unit structure. Resistance to crushing varies from 800 to 3,800 lb/in^2 for concrete, while that of granite varies from 9,700 to 34,000 lb/in^2.

Equally as important as correct sampling is the choice of a testing procedure which permits measurement of a given property of the sample under the same conditions as those under which the material will be used. It may be convenient to measure the ultimate strength of a steel by subjecting a small polished cylinder to steadily increasing tension until rupture occurs. But such a test tells little about the possible failure of a machine part made of the same steel when that part has an unpolished surface and when it is subjected to rapidly recurring loads or to twisting and bending. A valid test must duplicate the conditions of actual use.

SUMMARY

Elasticity is that property of a body which enables it to resist deformation and to recover after removal of the deforming force.

If the distorting forces acting on a body are

too large, the body does not recover completely its original configuration when the forces are removed. The deformation is said to have exceeded the *elastic limit*. The smallest stress that produces a permanent deformation is known as the elastic limit.

Materials for which the elastic limit is very small are called *inelastic* or *plastic* materials, such as dough, lead, and putty.

Hooke's law expresses the fact that within the limits of elasticity, stress is proportional to strain or the elongation s is proportional to the force F,

$$F = ks$$

Tensile stress is the ratio of the force to the cross-sectional area.

Tensile strain is the ratio of the increase in length to the original length.

A *modulus of elasticity* is found by dividing the stress by the corresponding strain.

Young's modulus is the ratio of tensile stress to tensile strain,

$$Y = \frac{F/A}{\Delta L/L}$$

The failure of an object to regain its original size and shape as soon as the deforming force is removed is called *elastic lag*, or *hysteresis*.

The *coefficient of volume elasticity*, or *bulk modulus*, is the ratio of volume stress to volume strain,

$$B = \frac{F/A}{\Delta V/V}$$

Compressibility is the reciprocal of the bulk modulus.

The *shear modulus*, or *coefficient of rigidity*, is the ratio of shearing stress to shearing strain,

$$n = \frac{F/A}{\varphi}$$

Rolling friction is caused by a deformation of the surface in front of the wheel and also the elastic hysteresis of a rubber tire on the wheel.

Materials possess several characteristics closely related to the elastic properties. Among these are *ductility, malleability, compressibility,* and *hardness*.

Questions

1 Assuming perfect elasticity show that Young's modulus for a material is numerically equal to the force that would be necessary to stretch a rod of unit cross section to double its original length.

2 What is the purpose of the steel in a horizontal reinforced concrete beam? in a vertical column? Does concrete need reinforcement more under compressive or more under tensile stresses? Why?

3 Which is the more elastic, rubber or steel? air or water?

4 What kind of elasticity is utilized in a suspension bridge? an automobile tire? an automobile drive shaft? a coil spring? a water lift pump? rubber heels?

5 Which will introduce more uncertainty in the determination of Y, an uncertainty of 1.0 mm in the 50.0-cm length of the wire or 0.00010 cm in an elongation of 0.0112 cm?

6 In what way do the numerical magnitudes of (*a*) strain, (*b*) stress, and (*c*) modulus of elasticity depend upon the units of force and length?

7 Reduce a modulus of elasticity of 19×10^{11} dyn/cm^2 to lb/in^2.

8 Two 12.0-m wires of the same material have diameters whose ratio is n. How much more will the smaller wire be stretched under a given load?

9 An elevator is suspended by a heavy steel cable. If this cable were replaced by two steel cables each having the same length as the original one but half its diameter, how would the amount of stretch in the pair of thin cables compare with that of the original cable?

10 An identical force is used to stretch two springs having the same length but different degrees of stiffness. Is more work done on the stiffer or the weaker spring?

11 Can one use a slender wire in the laboratory to estimate the load capacity of a large cable on a bridge? Explain.

12 A certain force is required to break a piece of cord. What force is required to break a cord made of the same material which is (*a*) twice as long and (*b*) twice as large in diameter and the same length?

13 From Table 1, which material would be preferable for the spiral spring of a spring balance? Why?

Problems

1 A force of 10 N stretches a spring 50 cm. Find the force constant of the spring. If you wish to stretch the spring an additional 25 cm, what total force will be required?

2 From the values in Table 1, calculate Y for aluminum and for steel in dynes per square centimeter.

Ans. 6.9×10^{11} dyn/cm^2; 20×10^{11} dyn/cm^2.

3 From the values in Table 1, calculate Y for aluminum and for steel in newtons per square meter.

4 How much will a steel wire 20 ft long and 0.22 in in diameter stretch when a load of 200 lb is hung on it? *Ans.* 0.043 in.

5 Compute the elongations of the aluminum wire of 40 mil (0.040 in) diameter and the 60-mil copper wire in the arrangement of Fig. 10-10.

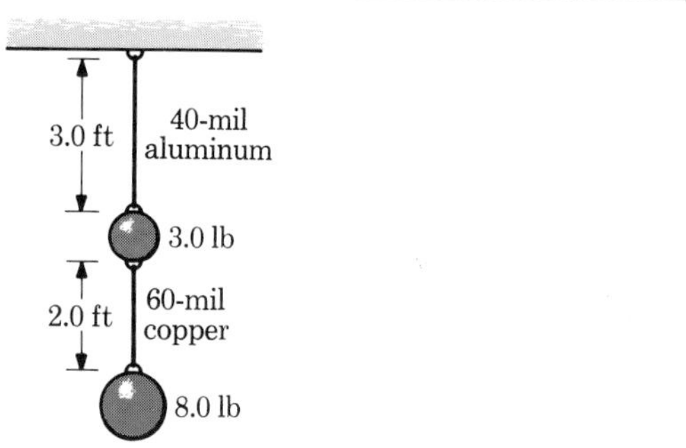

Figure 10-10

6 How much will an annealed steel rod 100 ft long and 0.0400 in^2 in cross section be stretched by a force of 1,000 lb? *Ans.* 1.03 in.

7 A hollow metal post is 10.0 ft long, and the cross-sectional area of the metal is 2.5 in^2. When a load of 25.0 tons is applied on top of it, the length decreases 0.0100 ft. Compute (*a*) the stress, (*b*) the strain, and (*c*) Young's modulus.

8 A wire 1,000 in long and 0.010 in^2 in cross section is stretched 4.0 in by a force of 2,000 lb. What are (*a*) the stretching stress, (*b*) the stretching strain, and (*c*) Young's modulus?

Ans. 2.0×10^5 lb/in^2; 0.00040; 5.0×10^7 lb/in^2.

9 A steel wire 2.0 m long and a copper wire 1.0 m long, each 0.5 cm^2 in cross section, are fastened together end to end and are then subjected to a tension of 10,000 N. Calculate the elongation of each wire.

10 A steel wire 100 cm long having a cross-sectional area of 0.025 cm^2 is stretched a distance of 0.30 cm. What is the stretching force?

Ans. 340 lb or 1.5×10^8 dyn.

11 Many high-voltage electrical cables have a solid steel core to support the aluminum wires that carry most of the current. Assume that the steel is 0.50 in in diameter, that each of the 12 aluminum wires has a diameter of 0.13 in, and that the strain is the same in the steel and the aluminum. If the total tension is 1.0 ton, what is the tension sustained by the steel?

12 A load of 9.0 tons is imposed on a vertical steel support 18 ft high having a cross-sectional area of 3.0 in^2. How much is the column shortened by the load? *Ans.* 0.045 in.

13 Could a steel piano wire 1.0 m long be stretched 8.0 mm without exceeding its elastic limit, which is about 1.20×10^5 lb/in^2 (or 8.26×10^8 N/m^2)?

14 A steel wire 8.0 ft long has a cross section of 0.050 in^2. When a stretching force of 1,600 lb is applied, the wire increases 0.106 in in length. (*a*) What is the stress in the wire? (*b*) What is Young's modulus for the wire?

Ans. 32,000 lb/in^2; 2.9×10^7 lb/in^2.

15 Young's modulus for the tendon in a man's leg is 1.6×10^8 N/m^2. If the tendon is 10 cm long and 0.45 cm in diameter, how much will it be stretched by a force of 10 N?

16 A steel wire $\frac{1}{2}$ mi long hangs vertically in a deep well. How much does it stretch under its own weight? The density of steel is 7.85 g/cm^3. (Suggestion: Compute the average elongation per unit length at the middle of the wire.)
Ans. 0.408 ft.

17 Fibers of spun glass have been found capable of sustaining unusually large stresses. Calculate the breaking stress of a fiber 0.00035 in in diameter, which broke under a load of 0.385 oz.

18 To maintain 200 in^3 of water at a reduction of 1 percent in volume requires a force per unit area of 3,400 lb/in^2. What is the bulk modulus of the water? *Ans.* 3.4×10^5 lb/in^2.

19 A 3,628-kg freight car moving along a horizontal railroad spur track at 7.2 km/h strikes a bumper whose coil springs experience a maximum compression of 30 cm in stopping the car. Calculate the elastic potential energy of the springs at the instant when they are compressed 15 cm.

20 If the density of seawater is 1.03 g/cm^3 at the surface, what is its density at a depth where the pressure is 10^9 dyn/cm^2? *Ans.* 1.08 g/cm^3.

21 A 5-kg projectile is fired horizontally from an 1,800-kg gun at a muzzle speed of 360 m/s. The initial recoil energy of the gun is completely transformed into potential energy of a spring. What should be the force constant of the spring to limit the recoil to 0.60 m?

22 A 4.0-ft-square steel plate 1.0 in thick is supported vertically with its lower edge fixed rigidly. Shear stress is applied, and the upper edge is observed to move parallel to the lower edge through 0.020 in. Find the shear strain.
Ans. 0.00042.

23 At the surface of the ocean the normal force per unit area on a body is 15 lb/in^2. At a depth of 10,000 ft below the surface it is 64×10^4 lb/ft^2. By what fraction is the volume of an aluminum sphere reduced as it is lowered from the surface to a 10,000-ft depth?

24 A 12-in cubical block of sponge has two parallel and opposite forces of 2.5 lb each applied to opposite faces. If the angle of shear is 0.020 rad, calculate the relative displacement and the shear modulus. *Ans.* 0.24 in; 0.87 lb/in^2.

25 When a 2.4-kg block is attached to the end of a spring hanging vertically, the spring experiences an elongation of 5.0 cm. What is the potential energy of the stretched spring?

26 A 2.0-in cube of gelatin has its upper surface displaced $\frac{1}{4}$ in by a tangential force of 1.0 oz. What is the shear modulus of gelatin?
Ans. 0.125 lb/in^2.

Sir William H. Bragg, 1862–1942

Born in Westward, Cumberland. Professor at Leeds University, later director of the Davy-Faraday Research Laboratory. Shared the 1915 Nobel Prize for Physics with his son W. L. Bragg for their contribution to the study of crystal structure by means of x-rays.

Sir William L. Bragg, 1890–1971

Born in Adelaide, South Australia. Cavendish Professor at Cambridge since 1938. Shared the 1915 Nobel Prize for Physics with his father W. H. Bragg for their contribution to the study of crystal structure by means of x-rays.

11

Vibratory Motion

Three types of motion have been treated in the earlier chapters. The simplest is that of an object in equilibrium, a motion consisting of constant speed and unchanging direction. The second type of motion, which is produced by the action of a constant force parallel to the direction of motion, is that in which the direction is constant and the speed increases uniformly. Projectile motion was discussed as a combination of these two simple types of motion. The third type of motion discussed is uniform circular motion, that produced by a (centripetal) force of constant magnitude directed inward along the radius of the circular path of the moving object.

It is clear that the sum of the forces we commonly observe acting on an object is not always zero, nor constant in magnitude and direction, nor constant in magnitude and in a rotating direction; consequently, the motions commonly observed are not always uniformly rectilinear, uniformly accelerated, uniformly circular, nor even combinations of the three. In general, the forces acting on a body vary in both magnitude and direction, resulting in complicated types of nonuniformly accelerated motion, which cannot be investigated in an elementary physics course. However, there is one common and important type of nonuniformly accelerated motion that can be analyzed rather simply. This motion is called periodic motion.

11-1 PERIODIC MOTION

A type of motion that is particularly important in mechanics is the to-and-fro, or vibrating, motion of objects stretched or bent from their normal positions and then released. Such an object moves back and forth along a fixed path, repeating over and over a fixed series of motions and returning to each position and velocity after a definite period of time. Such motion is called *periodic* motion, or *harmonic* motion. This type of motion is produced by varying forces, and hence the body experiences varying accelerations. While many periodic motions are quite complicated, they can usually be studied as combinations of relatively simple types of vibration. It is fortunate that the simple vibrations, though produced by varying forces, can be analyzed rather easily and completely by elementary methods.

11-2 SIMPLE HARMONIC MOTION

When an elastic spring is stretched by a force, the amount of the force required is proportional to the stretch. Suppose that an object of mass m (Fig. 11-1) hanging at the end of a spiral spring is pulled down a distance s below the equilibrium

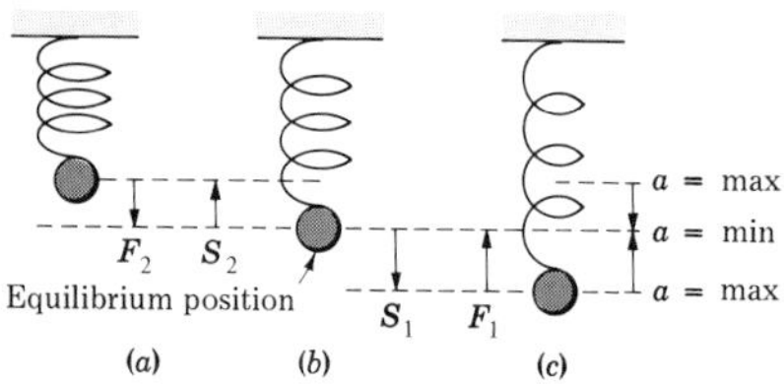

Figure 11-1
An object supported by a spring vibrates with simple harmonic motion.

position. The spring exerts a restoring force on the object, tending to pull it back toward its original position. This force is proportional to the displacement **s** but opposite in direction to the displacement.

$$\mathbf{F} = -k\mathbf{s} \qquad (1)$$

When the object is released, the restoring force produces an acceleration that is proportional to **F** and inversely proportional to the mass m being accelerated,

$$\mathbf{a} = \frac{\mathbf{F}}{m} = -\frac{k}{m}\mathbf{s} = -K\mathbf{s} \qquad (2)$$

Hence the acceleration is proportional to the displacement but opposite in direction.

As the object moves toward its equilibrium position, its speed increases but the force, and consequently the acceleration, decreases until it becomes zero when the object reaches the initial position. Because of its inertia the object continues past the equilibrium position, but at once a retarding force comes into being which increases until the object reaches a highest position, where it stops and begins its return trip. At all times during this motion the net force, and hence the acceleration, is proportional to the displacement and directed toward the equilibrium position. *The type of vibratory motion in which the acceleration is proportional to the displacement and always directed toward the equilibrium position is called simple harmonic motion* (SHM). This motion is always motion along a straight line, the acceleration and velocity constantly changing as the vibrating body moves through its series of positions. The direct proportionality of acceleration and displacement distinguishes simple harmonic motion from all other types of vibratory motion.

Very few vibrating bodies execute motion that is strictly simple harmonic, but many vibrate with a motion that is so nearly simple harmonic that it can be treated as such without appreciable error. Suppose that a steel ball is mounted on a flat spring that is clamped in a vise as in Fig. 11-2. Pull the ball sideways, bending the spring, and you will observe a restoring force that tends to move the ball back toward its initial position. This force increases as the ball is pulled farther away from its original position. The motion of the ball is only approximately SHM, since it moves along the arc of a circle instead of along a straight line and the direction of the force is tangent to the circle rather than toward the initial position. However, if the displacement is small, the departure from SHM is so slight that no great error is introduced by assuming that the motion is simple harmonic. The motion of a pendulum is also approximately SHM.

There are many illustrations of simple harmonic motion in physics that can be given to point out the importance of this form of motion. A few examples are the motion of a vibrating

Figure 11-2
A ball and spring in approximate simple harmonic motion.

tuning fork, and the motion of the particles of a medium through which a wave is traveling.

11-3 PERIOD, FREQUENCY, AND AMPLITUDE

The *period* T of a vibratory motion is the time required for a complete to-and-fro motion, or oscillation. In a complete oscillation the vibrating body moves from the equilibrium position to one end of the path, back to the equilibrium position, to the other end of the path, and back to the equilibrium position ready to repeat the cycle.

The *frequency* f of the vibratory motion is the number of complete oscillations per unit time. The frequency is the reciprocal of the period: $f = 1/T$.

The *amplitude* of a vibratory motion is the maximum displacement from the equilibrium position.

11-4 THE CIRCLE OF REFERENCE

When a body moves with uniform speed in a circle, the projection of this motion on a diameter is simple harmonic motion. In Fig. 11-3, the body

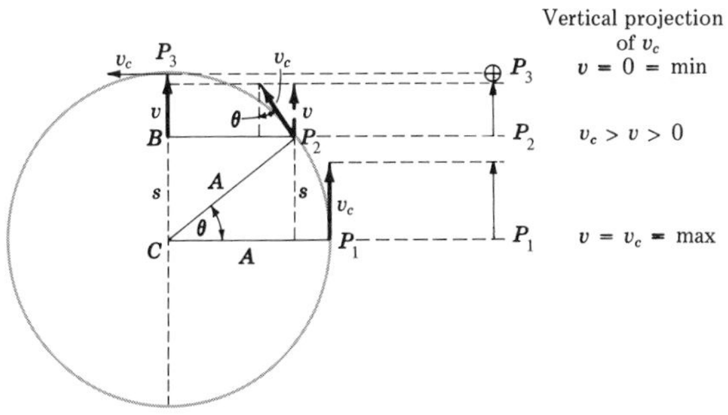

Figure 11-3
Circle of reference for analyzing SHM, showing velocity.

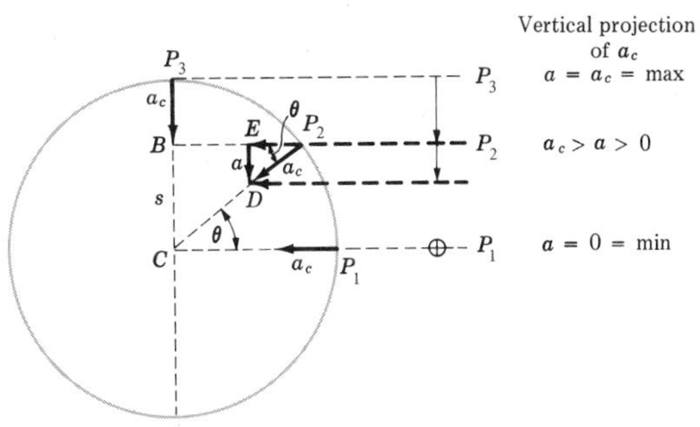

Figure 11-4
Acceleration from the reference circle.

P is moving with uniform speed v_c and uniform angular speed ω in a circular path. The projection B moves up and down along the vertical diameter. Assume that the time is assigned a zero value when the body B passes through the equilibrium position C, moving upward. At any later time t the rotating body P will have moved through an angle $\theta = \omega t$. At this instant the projection B has a displacement

$$s = A \sin \theta$$

Since

$$\omega = \frac{\theta}{t}$$

$$s = A \sin \omega t \qquad (3)$$

If a graph were made plotting various positions of s at different stages of SHM, the curve formed would be a sine curve. (See Fig. 11-5.)

At every position of the vibrating body, the velocity of the body is the component, taken parallel to the chosen diameter, of the velocity in the reference circle; similarly, the acceleration in the vibration is the component, taken parallel to the diameter, of the acceleration in the circular path. Consider Fig. 11-4. The velocity v in the vibration is

$$v = v_c \cos \theta$$

Since

$$v = r\omega = A\omega \qquad \text{and} \qquad \theta = \omega t$$

$$v = A\omega \cos \omega t \tag{4}$$

If the velocity was plotted on a graph as above it would form a cosine curve. At the same instant the acceleration of the particle at P is a_c directed toward the center of the circle. The component a of the central acceleration of particle P parallel to the chosen diameter is the acceleration of B. From Fig. 11-4,

$$a = -a_c \sin \theta$$

Since

$$\sin \theta = \frac{S}{A}$$

$$a = -a_c \frac{S}{A} \tag{5}$$

But as we saw in Chap. 8, $a_c = A\omega^2$ and $\theta = \omega t$; therefore,

$$a = -A\omega^2 \sin \omega t = -\omega^2 s \tag{6}$$

A plot of the acceleration would produce a negative sine curve (Fig. 11-5). Thus the acceleration of B is proportional to the displacement but is opposite in direction. Hence the motion of B is SHM, and the projection of uniform circular motion upon a diameter is SHM.

This example of simple harmonic motion is very useful in studying SHM, since it can be used to determine relationships among velocity, acceleration, period, frequency, and amplitude. The circle used here is commonly called the *reference* circle.

In terms of the reference circle the period of the SHM is the same as the time of one revolution in the reference circle. The amplitude in SHM is the same as the radius of the reference circle, and the frequency is the number of revolutions per unit time in the reference circle. For every SHM a reference circle can be set up from these relationships.

11-5 ACCELERATION AND SPEED IN SHM

At the position of greatest displacement, i.e., at the end points of the motion, the vibrating object comes momentarily to a stop. It should be noticed that at the instant when its speed is zero the object is acted upon by the maximum restoring force, so that the acceleration is greatest when the speed is zero. The restoring force (and therefore the acceleration) decreases as the object moves toward the equilibrium position. At the equilibrium

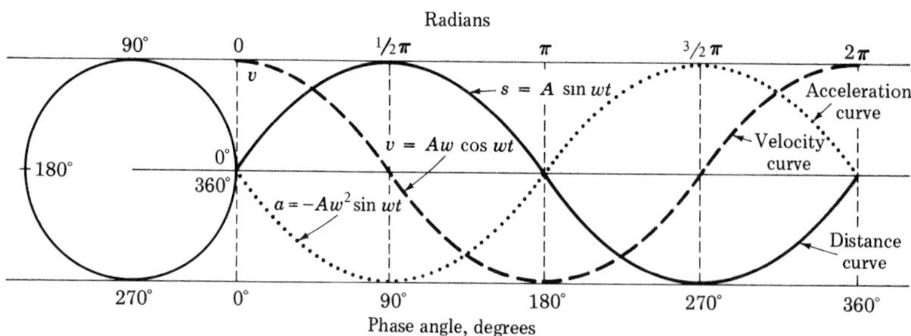

Figure 11-5
Graphs of displacement *s* (solid line), velocity *v* (dashed line), and acceleration *a* (dotted line) in simple harmonic motion. The curves indicate the relative phases of these three quantities.

position the acceleration is zero, and the speed is the greatest. The direction of the acceleration reverses as the object passes through the equilibrium position, increasing as the displacement increases, and reaching a maximum at the other extreme of displacement.

In Fig. 11-5 are plotted on the same set of axes graphs of s, v, and a. We have seen from Eqs. (3), (4), and (5) that the displacement is a sine curve, the velocity is a cosine curve, and the acceleration is a negative sine curve. We see that these three quantities do not reach corresponding parts of their values at the same time. The angle ωt expresses the relationship. This angle is called the *phase* angle. Since the cosine curve is the same shape as the sine curve but displaced 90°, we see that the velocity is 90° out of phase with the displacement. The acceleration is 180° out of phase with the displacement.

11-6 PERIOD AND FREQUENCY IN SHM

The frequency f of the SHM is the same as the frequency, or number of revolutions per unit time, in the reference circle. The angular velocity ω is

$$\omega = 2\pi f = \frac{2\pi}{T} \tag{7}$$

From Eq. (6),

$$a = -\omega^2 s = -\frac{4\pi^2}{T^2}s \tag{8}$$

and

$$T^2 = -4\pi^2\frac{s}{a}$$

$$T = 2\pi\sqrt{-\frac{s}{a}} \tag{9}$$

The period can be expressed in terms of the force constant of the spring (or other agency) that supplies the restoring force. From Eq. (2),

$$-\frac{s}{a} = \frac{m}{k}$$

If we substitute this value of $-s/a$ in Eq. (9), we obtain

$$T = 2\pi\sqrt{\frac{m}{k}} \tag{10}$$

If the mass m is expressed in grams and the force constant k in dynes per centimeter, Eq. (10) gives the period in seconds. If the mass is expressed in kilograms and the force constant in newtons per meter, Eq. (10) gives the period in seconds. In the British system the mass is computed in slugs from W/g, and the force constant is expressed in pounds per foot. The resulting period is in seconds.

Equation (10) expresses the fact that the period in SHM depends upon only two factors, the mass of the vibrating body and the force constant of the spring (or other agent). It should be noted that the *period is independent of the amplitude.*

Reference to Eq. (10) will show that if the object is replaced by another whose mass is four times as great, the period will be doubled. If, instead, the spring is replaced by another four times as stiff, the period is halved.

Example A 5.0-N ball is fastened to the end of a flat spring (Fig. 11-2). A force of 2.0 N is sufficient to pull the ball 6.0 cm to one side. Find the force constant and the period of vibration.

$$k = \frac{F}{s} = \frac{2.0\ \text{N}}{0.06\ \text{m}} = 33.3\ \text{N/m}$$

$$m = \frac{W}{g} = \frac{5.0\ \text{N}}{9.8\ \text{m/s}^2} = 0.51\ \text{kg}$$

$$T = 2\pi\sqrt{\frac{m}{k}} = 2\pi\sqrt{\frac{0.51\ \text{kg}}{33.3\ \text{N/m}}} = 0.78\ \text{s}$$

11-7 ENERGY IN SIMPLE HARMONIC MOTION

When there is a displacement of a body that is subject to elastic forces, work must be done on the body and elastic potential energy is stored up.

When the force is proportional to the displacement, as in simple harmonic motion, the work done, and hence the potential energy, is the scalar product of the average applied force $\bar{\mathbf{F}}$ and the displacement **s**.

As we saw in Chap. 10, the force applied changes linearly from 0 to $-ks$, and on the average is equal to $-\frac{1}{2}ks$. The negative of the work done in stretching a distance s was defined as the potential energy. Therefore,

$$E_p = \bar{\mathbf{F}} \cdot \mathbf{s} = (\tfrac{1}{2}ks)s = \tfrac{1}{2}ks^2 \qquad (11)$$

The potential energy is maximum for the greatest displacement, the amplitude A:

$$(E_p)_{max} = \tfrac{1}{2}kA^2 \qquad (12)$$

This maximum potential energy is the energy available for the vibration. When the body is released, the restoring force accelerates the body, creating kinetic energy at the expense of potential energy. If conditions are such that there is no dissipation of energy, at every instant during the vibration the sum of the kinetic energy and the potential energy is constant and equal to the initial energy given to the system.

$$E_p + E_k = \text{const} = \tfrac{1}{2}kA^2 \qquad (13)$$

or

$$\tfrac{1}{2}ks^2 + \tfrac{1}{2}mv^2 = \tfrac{1}{2}kA^2 \qquad (14)$$

In the case of a vibrating spring, such as that in Fig. 11-1, where the spring constant is known, the right-hand term of Eq. (14) may be computed directly. In many cases of simple harmonic motion, observations of mass, frequency, and amplitude enable us to compute the maximum kinetic energy $(\frac{1}{2}mv^2)_{max}$ and from that result find an effective force constant.

If there is dissipation of energy during the vibration, the amplitude will decrease as the motion continues and the vibration is said to be a *damped* vibration.

Example A 2.0-kg body vibrates in SHM with an amplitude of 3.0 cm and a period of 5.0 s. Find the speed, the acceleration, the kinetic energy, and the potential energy (*a*) at the midpoint of the vibration, (*b*) at the end of the path, and (*c*) at a point 2.0 cm from the midpoint. (*d*) Find the force on the body at the point 2.0 cm from the midpoint.

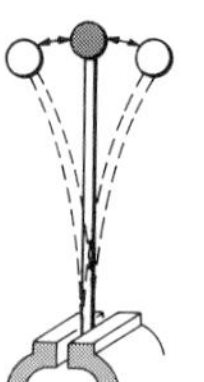

Figure 11-6
A ball and spring in approximate simple harmonic motion.

Since the amplitude is 3.0 cm = 0.030 m, the radius of the reference circle is 0.030 m. The speed v_c of the particle in the reference circle is the circumference divided by the time for one cycle:

$$v_c = \frac{2\pi A}{T} = \frac{2\pi(0.030\ \text{m})}{5.0\ \text{s}} = 3.8 \times 10^{-2}\ \text{m/s}$$

(*a*) At the midpoint of the path the particle in the reference circle is moving parallel to the chosen diameter, and therefore the velocity of the vibrating body is the same as the velocity in the reference circle, shown as the vertical projection on the y axis of Fig. 11-3.

$$v_1 = v_c = 3.8 \times 10^{-2}\ \text{m/s}$$

At the midpoint the acceleration in the reference circle is perpendicular to the diameter, and hence the component parallel to the path of the vibrating body is zero. Therefore, at the midpoint

$$a_1 = 0$$

The kinetic energy at the midpoint is

$$(E_k)_1 = \tfrac{1}{2}mv_1^2 = \tfrac{1}{2}(2.0\ \text{kg})(3.8 \times 10^{-2}\ \text{m/s})^2$$
$$= 1.4 \times 10^{-3}\ \text{J}$$

The potential energy at the midpoint is

$$(E_p)_1 = \tfrac{1}{2}ks^2 = 0 \qquad \text{since } s = 0$$

(*b*) At the end point, the velocity in the reference circle is perpendicular to the path of vibration and hence has no component in the direction of that path.

$$v_2 = 0$$

At the end point the acceleration in the reference circle is the same as the acceleration in the vibration

$$a_2 = a_c = \frac{v_c^2}{A} = \frac{(3.8 \times 10^{-2}\ \text{m/s})^2}{0.030\ \text{m}}$$
$$= 4.8 \times 10^{-2}\ \text{m/s}^2$$

At the end point the speed is zero, and therefore the kinetic energy is zero.

$$(E_k)_2 = 0$$

At the end point the potential energy is equal to the kinetic energy that the body had at the midpoint.

$$(E_p)_2 = \tfrac{1}{2}mv_1^2 = 1.4 \times 10^{-3}\ \text{J}$$

(*c*) At the point 2.0 cm from the midpoint, the velocity in the path of vibration is the component v of the velocity v_c in the reference circle (Fig. 11-3). From the geometry of Fig. 11-3,

$$v_3 = v_c \cos\theta$$
$$\cos\theta = \frac{BP}{CP} = \frac{\sqrt{(0.030\ \text{m})^2 - (0.020\ \text{m})^2}}{0.030\ \text{m}} = 0.75$$
$$v_3 = (3.8 \times 10^{-2}\ \text{m/s})(0.75)$$
$$= 2.8 \times 10^{-2}\ \text{m/s}$$

The acceleration at the point 2.0 cm from the midpoint may be found from the proportionality of acceleration and displacement,

$$\frac{a}{a_c} = \frac{-s}{r}$$
$$a_3 = \frac{-0.020\ \text{m}}{0.030\ \text{m}}(4.8 \times 10^{-2}\ \text{m/s}^2)$$
$$= -3.2 \times 10^{-2}\ \text{m/s}^2$$

The kinetic energy at this point is

$$(E_k)_3 = \tfrac{1}{2}mv_3^2 = \tfrac{1}{2}(2.0\ \text{kg})(2.8 \times 10^{-2}\ \text{m/s})^2$$
$$= 7.9 \times 10^{-4}\ \text{J}$$

The potential energy at this point may be found from the fact that the sum of the potential and kinetic energy is constant. We have found this total energy as

$$(E_k)_1 = (E_p)_2 = 14 \times 10^{-4}\ \text{J}$$

Therefore

$$(E_p)_3 = (E_p)_2 - (E_k)_3$$
$$= 14 \times 10^{-4}\ \text{J} - 7.9 \times 10^{-4}\ \text{J}$$
$$= 6.1 \times 10^{-4}\ \text{J}$$

(*d*) From Newton's second law, the force on the body at the point 2.0 cm from the midpoint is

$$F = ma = (2.0\ \text{kg})(-3.2 \times 10^{-2}\ \text{m/s}^2)$$
$$= -6.4 \times 10^{-2}\ \text{N}$$

The negative sign signifies that the force is opposite in direction to the displacement.

11-8 THE SIMPLE PENDULUM

One of the most common of approximate simple harmonic motions is the motion of a pendulum. A pendulum consisting of a small relatively heavy bob at the end of a very light string is called a *simple pendulum*. If such a pendulum is displaced as shown in Fig. 11-7, the weight mg of the bob

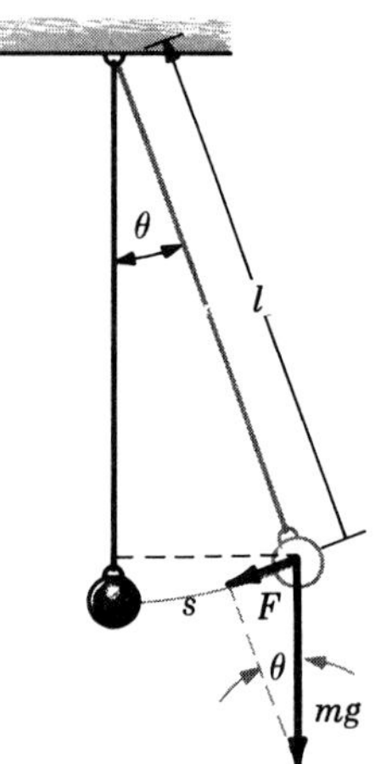

Figure 11-7
A simple pendulum.

supplies a restoring force

$$F = -mg \sin \theta$$

The displacement along the arc is $s = l\theta$. Hence, the force is proportional to $\sin \theta$, while the displacement is proportional to θ. The restoring force is not proportional to the displacement. Moreover, the restoring force is not directed toward the equilibrium position, but rather it is tangent to the arc. Thus the motion is not simple harmonic. However, if θ is small, we may replace $\sin \theta$ by θ (in radians) without serious error. To this degree of approximation the motion may be considered as simple harmonic motion. Then

$$F = ma = -mg\theta = -\frac{mgs}{l}$$

and

$$-\frac{s}{a} = \frac{l}{g}$$

From Eq. (9),

$$T = 2\pi\sqrt{\frac{l}{g}} \tag{15}$$

Equation (15) shows that to the degree of approximation involved the period depends only upon the length of the pendulum and the acceleration due to gravity. The period is independent of the mass of the bob. The period is also independent of the amplitude when the amplitudes are small. If the amplitude is large, the period is greater than that for small amplitudes. For an angle of 10° the error is about 0.2 percent; for 30°, about 1.7 percent.

11-9 THE COMPOUND (PHYSICAL) PENDULUM

Certain assumptions were made in our discussion of the simple pendulum. Among other things, we assumed that the cord supporting the mass at the end of the pendulum was massless and did not enter into the computation of the period of the pendulum.

There are many cases where this is not true, that is, the supporting arm may be massive enough that it cannot be ignored. A long, thin rod supported as a pendulum is one example of a physical pendulum, as is the pendulum of a "grandfather's" clock. The shape of the pendulum does not have to be uniform. However, a knowledge of the distribution of mass in such a pendulum is important since its center of gravity must be known.

It can be shown that the period of a compound pendulum can be expressed by using the following equation:

$$T = 2\pi\sqrt{\frac{I}{mgh}} \tag{16}$$

where I = moment of inertia about an axis at the point of suspension of the pendulum
m = mass of the pendulum
h = distance of the center of gravity from the point of suspension

Since it is sometimes desirable to relate the period of a simple pendulum to that of a compound pendulum, it can be seen from the equations for finding the period of both types of pendulum that in order for both to have the same period

$$T_{(\text{simple})} = T_{(\text{compound})}$$

$$2\pi\sqrt{\frac{l}{g}} = 2\pi\sqrt{\frac{I}{mgh}}$$

and $$l = \frac{I}{mh} \tag{17}$$

Therefore, a compound pendulum will have the same period as a simple pendulum having a length l equal to I/mh. This length is the length of an equivalent simple pendulum.

Those who have attempted to hit a baseball with a bat realize that if a certain point on the bat makes contact with the ball, the ball will travel farthest. This is called the "sweet" part of the bat. Actually, this point is known as the *center of percussion* or the *center of oscillation* of the bat. Any compound pendulum vibrates as if its mass were concentrated at one point, the center of percussion, at a distance l from the point of suspension where $l = I/mh$. A blow delivered to a compound pendulum at this point will cause it to rotate about the point of suspension smoothly. If struck at any other point, the pendulum will tend to quiver.

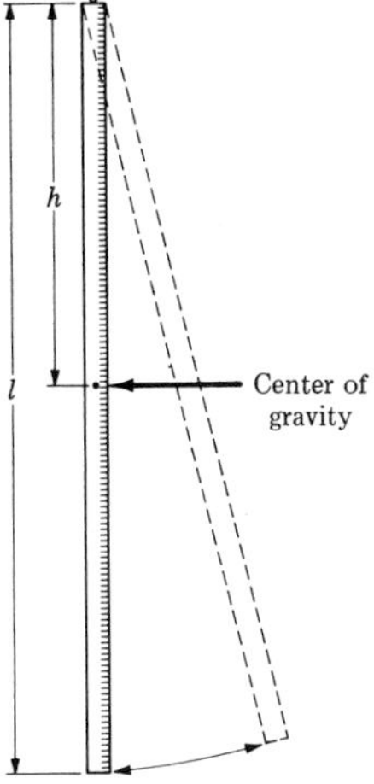

Figure 11-8
A compound pendulum in the form of a meterstick.

Example A meterstick is hung by one end and allowed to vibrate as a compound pendulum. (*a*) Find its period. (*b*) Find the length of a simple pendulum that would have the same period. (*c*) Find the center of oscillation.

(*a*) Assuming the meterstick to be a rod,

$$Icg_{(\text{rod})} = \tfrac{1}{12}ml^2$$

Then, $Is = Icg + mh^2$ and $Is = \frac{1}{12}m(2h^2) + mh^2$, where h is the distance between the point of suspension and the center of gravity; 0.5 m in this case, and $l = 2h$. Therefore,

$$T = 2\pi\sqrt{\frac{4/3\ mh^2}{mgh}}$$

and $$T = 2\pi\sqrt{\frac{4/3\ h}{g}} = 6.28\sqrt{\frac{4/3(0.5)\text{ m}}{9.8\text{ m/s}^2}}$$

$$= 1.64\text{ s}$$

(*b*) The length of an equivalent simple pendulum is

$$l = \frac{I}{mh} = \frac{4/3\ mh^2}{mh} = 4/3\ h$$

$$= \left(\frac{4}{3}\right)\left(\frac{1}{2}\right)\text{m} = 0.67\text{ m}$$

(*c*) Since the center of percussion is located 0.67 m from the point of suspension, and the point of suspension is at the end of the meterstick, the center of percussion is located 0.67 m down from the top of the meterstick pendulum.

11-10 SIMPLE ANGULAR HARMONIC MOTION

If a heavy cylinder is supported at the end of a thin rod (torsion pendulum) and twisted through an angle θ (Fig. 11-9) about an axis along the

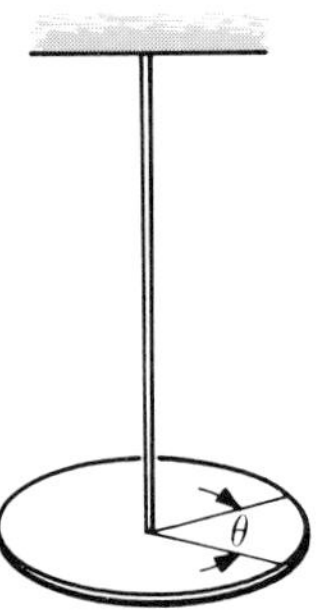

Figure 11-9
A torsion pendulum. Simple angular harmonic motion.

rod, the rod supplies a restoring torque proportional to the angle of twist.

$$L = -\kappa\theta \tag{18}$$

where the negative sign is introduced because L and θ are always opposite in sign. The constant κ is called the *moment of torsion* of the rod and depends upon the length, diameter, and material of the rod. This constant is of considerable importance in the design of instruments in which the sensitivity depends upon the twist of a wire or fiber.

When the pendulum is released, the restoring torque produces an angular acceleration proportional to the angular displacement. The motion produced is *simple angular harmonic motion*. The period depends upon the moment of torsion of the support and upon the moment of inertia of the oscillating system. By analogy to Eq. (10) we can write the equation

$$T = 2\pi\sqrt{\frac{I}{\kappa}} \tag{19}$$

where K is the moment of torsion of the supporting fiber and I is the moment of inertia of the vibrating system about an axis along the supporting fiber.

Example A solid cylinder of mass 5.0 kg and radius 6.0 cm is suspended by a vertical wire as a torsion pendulum. The axis of the cylinder is along the line of the wire. The period of vibration is 4.0 s. Find the moment of torsion of the wire.

$$I = \tfrac{1}{2}mR^2 = \tfrac{1}{2}(5.0\text{ kg})(0.060\text{ m})^2$$
$$= 9.0 \times 10^{-3}\text{ kg}\cdot\text{m}^2$$

From Eq. (19),

$$\kappa = \frac{4\pi^2 I}{T^2} = \frac{4\pi^2 \times 9.0 \times 10^{-3}\text{ kg}\cdot\text{m}^2}{(4.0\text{ s})^2}$$
$$= 2.2 \times 10^{-2}\text{ m}\cdot\text{N/rad}$$

11-11 RESONANCE

Suppose that the natural frequency of vibration of the system represented in Fig. 11-6 is 10 vib/s. Now imagine that, beginning with the system at rest, we apply to it a to-and-fro force, say, 25 times per second. In a short time this force will set the system to vibrating regularly 25 times a second, but with a very small amplitude, for the ball and spring are trying to vibrate at their natural rate of 10 vib/s. During part of the time, therefore, the system is, so to speak, "fighting back" against the driving force, whose frequency is 25 vib/s. We call the motion of the system in this case *forced vibration*.

Now suppose that the alternation of the driving force is gradually slowed down from 25 vib/s to 10 vib/s, the natural frequency of the system, so that the alternations of the driving force come just as the system is ready to receive them. When this happens, the amplitude of vibration becomes very large, building up until the energy supplied by the driving force is just sufficient to overcome friction. Under these conditions the system is said to be in *resonance* with the driving force.

A small driving force of proper frequency can build up a very large amplitude of motion in a system capable of vibration. We have all heard car rattles that appear only at certain speeds, or

Figure 11-10
Dangerous resonance. Excessive vibration caused collapse of the bridge.

vibrations set up in dishes, table lamps, cupboards, and the like, by musical sounds of particular frequency. A motor running in the basement will often set certain pieces of furniture into vibration. This problem of resonant vibrations may become particularly important with heavy machinery. The problem is to find the part that is vibrating in resonance with the machinery and to change its natural frequency by changing its mass or its binding force (force constant).

An example of the tremendous forces that can be built up by resonance is shown in Fig. 11-10, a photograph of the Tacoma Narrows bridge—nicknamed "Galloping Gerty." This bridge was constructed in such a manner that the central span resonated until the resonance became so great that it eventually caused the bridge to collapse. The hazards of such uncontrolled resonance were vividly demonstrated in this incident.

A common example of resonance is furnished by radio circuits. Each radio station transmits its signal by means of electromagnetic waves of a particular, assigned carrier frequency. The station selector, or dial, on a radio receiver is a device designed to put the radio in resonance with the transmitter's frequency to permit the reception of its signal. When a person tunes his receiver he is in effect altering what corresponds to the spring constant of a mechanical system. By thus changing the natural frequency of the radio receiver, he can bring the circuit into a state of resonance with the sending station.

A further example of resonance will be seen in the study of sound. A proper frequency sound wave falling on a piece of matter can cause it to resonate by amplifying the natural frequency of the motion of the particles in that matter. This important aspect of resonance produces much of the beautiful quality of sounds coming from such instruments as the violin and cello.

11–12 THE PHYSIOLOGICAL EFFECTS OF VIBRATION

We are subjecting humans today to greater physiological stresses than ever before in history. Man is going faster and higher, accelerating and decelerating more rapidly, and undergoing greater temperature changes than ever before. While much of this exposure has been in connection with space research, aquanauts experimenting with life under the seas, engineers developing more rapid means of transportation, and scientists involved in medical research, i.e., the use of ultrasonics as a form of bloodless surgery and in the development of radio-therapy techniques have all undergone unusual physiological hazards. Man also is exposed to vibrational hazards, therefore not only must we be concerned with the stress and strain experienced by inanimate objects due to vibrations, but we must also be aware of the effect of vibrational strains on animate objects.

During the lift-off and boost phases of space flight and also during reentry into the earth's atmosphere, astronauts experience considerable vibration. Such vibration consists of forces of various directions, magnitudes, and frequencies and of a periodic and oscillatory nature. A goal of dynamics research is to reduce these vibrations to levels that avoid damage to the vehicle. For manned vehicles, there is the additional requirement of further curbing vibration so that it does not interfere with crucial human activity.

Several studies conducted by NASA[1] and others have attempted to assess the damage done to humans by vibration. Studies of vibration tolerances of humans at constant acceleration have shown that certain frequencies of vibrations are more damaging than others and also indicate variations in human vibration tolerance. Work done at the Naval Medical Acceleration Laboratory in Johnsville, Pennsylvania,[2] showed that a human exposed to a 20- to 25-cycle-per-second (Hz) vibration evidenced nausea, internal bleeding, and cramps. Other symptoms of humans exposed to vibration at frequencies from 1 to 20 Hz have been described. These include head pain, blurred vision, and a painful lump in the throat. Most alarming is the observation that the "resonant frequency" for the human body is between 6 and 8 Hz, a relatively low and common rate of oscillation. At 6 to 8 Hz, the jaw resonates so as to make it impossible to speak. Heart and lung displacement may cause chest pain. Abdominal pain and pelvic pain have been found to be associated with distortion and stretching of internal structures. Postexperimental weariness has also been found to be associated with exposure to vibration. Evidence of this can be seen in the general weariness that a person feels after a long ride in an auto, even though the roads have been smooth.

It will be important for man to be able to continue to make quick decisions and responses under conditions of much noise and vibration. Therefore a basic knowledge of man's ability to perform under vibrational stress is important. It appears that considerable physiological investigation remains to be done in this field.

---

[1] Ferdinand S. Ruth (Ed.), Space Resources for Teachers, Biology, NASA, Washington, D.C., 1969.

[2] James D. Hardy (Ed.), "Physiological Problems in Space Exploration," Charles C Thomas, Publisher, Springfield, Ill., 1964.

SUMMARY

Periodic motion is that motion in which a body moves back and forth over a fixed path, repeating over and over a fixed series of motions and returning to each position and velocity after a definite interval of time.

Simple harmonic motion is that type of vibratory motion in which the acceleration is proportional to the displacement and is always directed toward the position of equilibrium.

$$a = -\frac{k}{m}s = -Ks$$

Simple harmonic motion is always motion along a straight line. Many vibrations that are not strictly simple harmonic are very close approximations and may be treated as such without serious error.

The projection on a diameter of the motion of a point that moves at constant speed on the "circle of reference" describes simple harmonic motion. The displacement, velocity, and acceleration in the SHM are related to the reference circle by the following equations:

$$s = A \sin \omega t$$

$$v = A\omega \cos \omega t$$

$$a = -A\omega^2 \sin \omega t$$

where t is the time elapsed after the vibrating body passes through the equilibrium position and ω is the angular velocity in the reference circle.

The *period* of a vibratory motion is the time required for one complete oscillation.

$$T = 2\pi\sqrt{\frac{m}{k}}$$

The *frequency* is the number of complete oscillations per second.

$$f = \frac{1}{T}$$

The *amplitude* of the motion is the maximum displacement from the equilibrium position. The radius of the reference circle is equal to the amplitude.

In undamped SHM the sum of kinetic and potential energy is constant and equal to the initial energy supplied to the vibrating system.

A *simple pendulum* is one which consists of a concentrated bob supported by a very light string. Its motion is approximately SHM, and the period is given by the equation

$$T = 2\pi\sqrt{\frac{l}{g}}$$

A *compound (physical) pendulum* has a support which is not massless and therefore enters into the determination of the period which is given by the equation

$$T = 2\pi\sqrt{\frac{I}{mgh}}$$

Each compound pendulum can be related to a simple pendulum so that both have the same period by making the length of the simple pendulum l equal to I/mh of the compound pendulum.

$$l = \frac{I}{mh}$$

A *torsion pendulum* vibrates with simple angular harmonic motion. Its period of oscillation is given by

$$T = 2\pi\sqrt{\frac{I}{\kappa}}$$

The *moment of torsion* κ is the ratio of the torque to the angle of twist produced by that torque. It depends upon the length, diameter, and material of the rod or wire.

Resonance occurs when a periodic driving force is impressed upon a system whose natural frequency of vibration is the same as that of the driving force. When this happens, the amplitude of vibration builds up until the energy supplied by the driving force is just sufficient to overcome friction in the system.

The physiological effects of vibration on man can be severe. Man has a low (6 to 8 Hz) resonant frequency, a frequency that causes physiological damage and fatigue.

Questions

1 Define SHM (simple harmonic motion).
2 What are the major differences between uniformly accelerated motion and periodic motion?
3 Define the force constant of a spring. State its defining equation and the cgs unit.
4 What is the relationship between the restoring force acting on a stretched spring and the displacement of the spring? Between the acceleration produced and the displacement?
5 Why are approximate simple harmonic motions common in nature? Why are true simple harmonic motions extremely rare?
6 Give illustrations other than those listed in the chapter of objects which normally move in the form of simple harmonic motion.
7 Describe clearly how the motion of the piston in the cylinder of a steam locomotive differs from simple harmonic motion.
8 Show by a diagram the relationship of a circle of reference to simple harmonic motion.
9 State the general equation for the period of a body executing SHM. State the equation for the period of a vibrating spring.
10 Show why the amplitude of vibration does not appear in equations for the period of various kinds of SHM.
11 Show that in SHM the acceleration is zero when the velocity is greatest and that the velocity is zero when the acceleration is greatest.
12 How does the period of a vertically oscillating spring vary with each of the following factors: mass of bob; amplitude of vibration; force constant of the spring; acceleration due to gravity?
13 Devise a method for the measurement of g from observations made upon a vertically oscillating spring.
14 Within a solid sphere of uniform density the gravitational force on an object varies directly with the first power of the distance from the center. Assuming that the earth were such a sphere and a hole could be drilled completely through it along a diameter, what would happen to an object dropped into the hole?
15 Under what conditions does the addition of two simple harmonic motions produce a resultant that is simple harmonic?
16 What would the effect be on the period of a simple pendulum if the pendulum was moved from sea level to the top of a mountain? to the moon? to the sun? Explain.
17 Explain how a simple pendulum might be used to assist in geophysical exploration for locating oil.
18 A simple pendulum has a period of 2.00 s at sea level and 45° latitude. What will be the effect qualitatively on the period if the pendulum is at sea level (*a*) at the equator? (*b*) at latitude 60°? What will be the effect of taking it to elevation 5,000 ft at latitude 45°?
19 Note the major differences between a simple pendulum and a compound or physical pendulum.
20 What is meant by the center of percussion of an object? Where might this point be on a hammer?
21 Why do marching men break step when crossing a light bridge?
22 It has been claimed that the late tenor Enrico Caruso could actually shatter a crystal goblet by singing a certain note. If this were possible, explain how and why this might have happened.
23 Describe several common phenomena in which resonance is an important factor.

Problems

1 A spring has a force constant of 100 N/m. It oscillates at a frequency of 40 cycles/min when an object is attached to it. What is the mass of the object?

2 What is the force constant of a spring that is stretched 10.0 cm by a force of 50.0 N? What is the period of vibration of a 100-N body if it is suspended by this spring?

Ans. 5.0×10^2 N/m; 8.97×10^{-1} s.

3 A 1,000-g cage is suspended by a spiral spring. When a 200-g bird sits in the cage, the cage is pulled 0.50 cm below its position when empty. Find the period of vibration of the cage (*a*) when empty and (*b*) when the bird is inside.

4 A 16-kg mass is hung on a spring. When an additional 1-kg mass is added, the spring stretches 0.80 m. If the spring and the 16-kg mass is stretched 2 m and then released, what is the period of vibration? *Ans.* 7.2 s.

5 The drive wheels of a locomotive whose piston has a stroke of 2.00 ft make 185 r/min. Assuming that the piston moves with SHM, find the speed of the piston relative to the cylinder head at the instant when the piston is at the center of its stroke.

6 A 10-lb block of iron is caused to vibrate with SHM by means of a spring. If the amplitude of vibration is 12 in and the time of a complete vibration is 0.60 s, find the maximum kinetic energy of the block. *Ans.* 17 ft·lb.

7 A spring is stretched 25.0 cm by a load of 200 g. A 300-g object is attached to the spring and displaced 10.0 cm from the equilibrium position. Find the potential energy of the system in this position. From consideration of energy find the speed of the object when it is 5.00 cm from the equilibrium position.

8 A 10.0-lb object vibrates in SHM with an amplitude of 10.0 in and a period of 5.0 s. What is the radius of its related circle of reference? What is the speed of the object at its midpoint? At the end of its path? What is the acceleration at the midpoint and at the end of its path? What is its frequency, the number of revolutions per second in the reference circle?

Ans. 10.0 in; 12.56 in/s; 0; 0; 15.8 in/s^2; 0.20 vib/s.

9 A 4.0-kg body is caused to vibrate with SHM by means of a spring. If the amplitude is 30 cm and the time of a complete vibration is 0.60 s, find (*a*) the maximum speed, (*b*) the maximum kinetic energy, (*c*) the minimum kinetic energy, and (*d*) the force constant of the spring.

10 What is the period of a vibrating object which has an acceleration of 8.0 m/s^2 when its displacement is 1.0 m? *Ans.* 2.2 s.

11 An 8.0-kg body performs SHM of amplitude 30 cm. The restoring force is 60 N when the displacement is 30 cm. Find (*a*) the period, (*b*) the acceleration when the displacement is 12 cm, (*c*) the maximum speed, and (*d*) the kinetic and the potential energy when the displacement is 12 cm.

12 A body moves with SHM of an amplitude 24 cm and a period of 1.2 s. (*a*) Find the speed of the object when it is at its midposition and when 24 cm away. (*b*) What is the magnitude of the acceleration in each case?

Ans. 130 cm/s; 0; 0; 660 cm/s^2.

13 A 10.0-kg body vibrates in SHM with a period of 4.0 s and an amplitude of 10.0 cm. Find the maximum speed and the maximum acceleration. Find the speed and acceleration when the body is one-sixth period from the equilibrium position. Find the net force on the vibrating body at the latter position.

14 A 4.0-kg mass is attached to the end of a flat spring which is pulled 0.080 m to one side by a force of 10.0 N. Find the force constant, the period of vibration, and the frequency of the vibration. *Ans.* 125 N/m; 1.125 s; 0.89 vib/s.

15 A horizontal platform moves up and down, executing simple harmonic motion with an amplitude of 6.0 in and a period of 0.50 s. (*a*) Calculate the speed of the platform when the displacement is 5.2 in from the equilibrium position. (*b*) Calculate the maximum value of the acceleration. (*c*) If a block is placed on top of a similar platform and they move up and down together executing simple harmonic motion with an amplitude of 6.0 in, what is the maximum frequency that the motion can have so that the block will remain in contact with the platform continuously?

16 A body having simple harmonic motion of amplitude 5.0 cm has a speed of 50 cm/s when

its displacement is 3.0 cm. What is its period?
Ans. 0.50 s.

17 A simple pendulum is used to determine the value of g. When the length of the pendulum is 98.45 cm, the period is measured to be 1.990 s. Find the value of g.

18 A simple pendulum is 1.00 m long. What is its period? *Ans.* 2 s.

19 A simple pendulum was accurately adjusted to have a period of 2.00 s. The supporting fiber broke and was shortened 2.00 in. Find the change in period, assuming $g = 32.2$ ft/s^2.

20 A body of mass 60.0 g is moving with a uniform angular speed in a vertical circle of radius 10.0 cm at the rate of 2.0 r/s. (*a*) What is the magnitude and direction of the centripetal force 0.00625 s after the body passes a horizontal diameter going in the upward direction? (*b*) What is the velocity of a companion particle, executing SHM on a horizontal diameter of the circle?
Ans. 9.51×10^4 dyn; 890 cm/s.

21 At a certain place a simple pendulum 100 cm long makes 250 complete vibrations in 8.38 min. What is the length of a simple seconds pendulum at that place?

22 A yardstick is hung by one end and allowed to vibrate as a compound pendulum. What is its period? What is the length of an equivalent simple pendulum? What is its center of percussion?
Ans. 1.56 s; 24 in; 24 in from top.

23 A hoop 4 ft in diameter is hung by a point on its rim and vibrates as a compound pendulum. What is the length of an equivalent simple pendulum?

24 A yardstick is hung on a nail 6 in from one end and allowed to swing about the nail in the form of a compound pendulum. What is its period? What is the length of an equivalent simple pendulum? What is its center of percussion?
Ans. 1.47 s; 21 in below the nail or 27 in from the top; 27 in from the top.

25 A thin rod 2 m long is suspended about an axis at one end and swings as a compound pendulum. What is its period? What is the length of an equivalent simple pendulum? What is its center of percussion.

26 A solid cylinder, whose weight is 16.0 lb and radius is 9.0 in, is supported along the axis by a wire 2.0 ft long. The cylinder is twisted through an angle of 120° by a torque of 4.0 ft·lb. Find the moment of torsion of the wire and the period of the pendulum when released.
Ans. 1.9 ft·lb/rad; 1.7 s.

27 A watch has a balance wheel which moves with an angular acceleration of 41 rad/s^2 when it is displaced 15° from its equilibrium position. What is its frequency?

28 A 200-g sphere of radius 12.0 cm is supported by a wire as a torsion pendulum. The frequency of the pendulum is 0.250 vib/s. Find the moment of torsion of the wire and the energy of the system when it is displaced 12.5° from its equilibrium position.
Ans. 2.85×10^4 cm·dyn/rad; 677 ergs.

29 A torsion pendulum begins moving with an angular acceleration of 15 rad/s^2 when its displacement is 90°. What is the frequency of the pendulum?

30 A watch has a balance wheel which moves with an angular acceleration of 25 rad/s^2 when it is displaced 45° from its equilibrium position. What is its frequency? *Ans.* 0.90 vib/s.

Charles Glover Barkla, 1877–1944

Born in Widnes, Lancashire. Professor at the University of Edinburgh. Awarded the 1917 Nobel Prize for Physics for his discovery of the secondary x-radiation characteristic of elements. This revealed the number of electrons in an atom and which elements were still unknown. H. G. J. Moseley would have shared the award but for his death at Gallipoli.

12

Fluids at Rest

According to the kinetic molecular theory, which we shall study in detail later, all matter, whether solid, liquid, or gas, consists of molecules that are in motion when above a zero-activity reference temperature called absolute zero. Further, these molecules attract each other in varying degrees. In solids, the molecules are relatively close to each other. That is, the *mean free path,* the average distance a molecule moves before colliding with another molecule, is generally smaller in a solid than in a liquid or a gas. In fact the molecular motion in a solid can be described as being vibrational in nature in that the molecules tend to stay localized and remain in one region. In a solid, the forces of attraction are great enough to hold the molecules in a regular pattern and thus maintain a definite volume and shape.

In a liquid, the molecules are, on the average, farther apart. This can be explained kinetically if we consider that there is a direct relationship between the energy possessed and the rate of motion of the particles in a piece of matter. As energy is pumped into matter (i.e., in the form of heat energy), the molecules begin to move faster. These molecules have mass; and since, as we saw earlier, an object in motion possesses momentum, the magnitude of which can be found by taking the product of its mass and its velocity, an increase in the rate of motion of these molecules results in an increase in momentum. Due to this increased momentum, the molecules colliding with their neighbors clear out a larger area for themselves and the mean free path increases. This is in keeping with the commonly observed property that matter normally expands when heated and contracts when cooled. The net result of this increased motion in a liquid is that the attractive forces are smaller and the type of motion changes from vibrational to what may be called translational. That is, the molecules in a liquid can move from place to place within the substance. Therefore, while the liquid maintains a definite volume, it assumes the shape of its container.

In a gas the distances between molecules are large compared with their size. This can be explained by continuing the above analogy. If additional heat energy is pumped into the substance, the molecules continue to increase in velocity and momentum and create a larger space for themselves, increasing their mean free path. The forces between these molecules are very small and as a result the type of molecular motion in a gas is random. A gas therefore has neither shape nor volume of its own but assumes those of its container. Liquids and gases are frequently grouped together as *fluids,* since they flow readily and do not resist shearing stresses. We shall first direct our study to the physics of fluids at rest and then to fluids in motion.

The class into which a substance falls depends upon the physical conditions surrounding it at the time of observation. Under varying conditions a single substance may be observed in any one of the three states. We are all familiar with water in three phases: ice, water, and vapor. Other substances such as iron and most other metals, which are not familiar in the liquid and gaseous phases, nevertheless exist in those phases if the temperature is sufficiently high. Those substances which are commonly observed as gases can all be liquefied and solidified if the temperature is lowered far enough and the pressure is made great enough.

12-1 DENSITY

One of the properties characteristic of every material is its density. We observe that a small piece of one material may be heavier than a much larger piece of another material. The *mass per unit volume* of a substance is called its *density*

$$\rho = \frac{m}{V} \tag{1}$$

Units of density are determined by dividing the chosen unit of mass by the unit of volume, as kilogram per cubic meter, gram per cubic centimeter, or slug per cubic foot.

It is sometimes helpful to use another quantity called *weight-density,* or weight per unit volume:

$$D = \frac{W}{V} \tag{2}$$

Since $W = mg$, we have a simple relation between density and weight-density:

$$D = \rho g \tag{3}$$

Weight-density is commonly used when we are concerned with effects depending upon force, while density is used when mass is to be considered. Values of density for some substances are given in Table 1.

Solids and liquids are only slightly compressed by even large stresses; hence their densities are almost constant under usual conditions. Gases are readily compressed; hence it is necessary to state the conditions under which the densities are measured.

12-2 SPECIFIC GRAVITY; RELATIVE DENSITY

The *specific gravity,* or *relative density,* of a substance is the ratio of its density to that of some standard substance. The standard usually chosen is water at the temperature of its maximum density, 4°C (39.2F). Thus, if ρ is the density of the substance and ρ_w the density of water, the relative density ρ_r of the substance is

$$\rho_r = \frac{\rho}{\rho_w} \tag{4}$$

and also

$$\rho_r = \frac{D}{D_w} \tag{4a}$$

Since each of the two densities has the same unit, their quotient is dimensionless and has no units. Relative density is often more convenient to tabulate than density, the values of which are different in the various systems of units. One may easily compute density from relative density by the use of Eq. (4):

$$\rho = (\rho_r)(\rho_w)$$

The units of density thus obtained will be those of the system in which the density of water is expressed.

Since the density of water in the cgs system is 1 g/cm^3, densities in that system are *numerically* equal to the relative density.

12-3 HYDROSTATIC PRESSURE

When a fluid is confined in a container, the fluid exerts a force on every part of the surface of the container that the fluid touches. Since a fluid

Table 1
RELATIVE DENSITY, DENSITY, AND WEIGHT-DENSITY

Substance	Relative density, ρ_r	Density		Weight-density, lb/ft^3
		kg/m^3	g/cm^3	
Solids:				
Aluminum	2.70	2,700	2.70	169
Brass	8.44–8.70	8,440–8,700	8.44–8.70	527–543
Carbon, graphite	2.25	2,250	2.25	141
Copper	8.89	8,890	8.89	555
Germanium	5.46	5,460	5.46	342
Glass	2.4–2.8	2,400–2,800	2.4–2.8	160–170
Gold	19.3	19,300	19.3	1,204
Ice	0.917	917	0.917	57.2
Iron, wrought	7.85	7,850	7.85	490
Lead	11.34	11,340	11.3	705
Wood, oak	0.8	800	0.8	50
Silicon	2.42	2,420	2.42	151
Silver	10.5	10,500	10.5	655
Steel	7.8	7,800	7.8	487
Tungsten	19.3	19,300	19.3	1,204
Zinc	7.1	7,100	7.1	443
Uranium	18.7	18,700	18.7	1,170
Liquids:				
Alcohol (ethyl) at 20°C	0.79	790	0.79	49
Ether	0.74	740	0.74	46
Gasoline	0.68	680	0.68	42
Mercury	13.595	13,595	13.595	850
Water, at 4°C	1.000	1,000	1.000	62.4
Water, at 20°C	0.998	998	0.998	62.3
Gases, 0°C and 76 cm Hg:				
Air	1.293×10^{-3}	1.293	1.293×10^{-3}	0.0807
Carbon dioxide	1.997×10^{-3}	1.997	1.997×10^{-3}	0.1246
Hydrogen	0.090×10^{-3}	0.090	0.090×10^{-3}	0.0058
Helium	0.178×10^{-3}	0.178	0.178×10^{-3}	0.0111
Nitrogen	1.251×10^{-3}	1.251	1.251×10^{-3}	0.0781
Oxygen	1.429×10^{-3}	1.429	1.429×10^{-3}	0.0892

cannot support a tangential force without moving, it follows that in a fluid at rest, the force on the walls of the container is always perpendicular to the containing surface. The *normal force per unit area* is called *pressure*. In symbols the average pressure $\overline{P}$ is

$$\overline{P} = \frac{F}{A} \tag{5}$$

The direction of the force resulting from the pressure is determined by the orientation of the surface, and therefore pressure acts as a scalar quantity. Since the force may not be uniformly distributed over a surface, Eq. (5) represents an average pressure over the area. In any small area ΔA where there is a normal force ΔF, the average pressure is $\overline{P} = \Delta F/\Delta A$. At any point on the surface the pressure is

$$P = \lim_{\Delta A \to 0} \frac{\Delta F}{\Delta A} \tag{6}$$

A unit of pressure is obtained from any force unit divided by an area unit. Pressures are commonly expressed in newtons per square meter, dynes per square centimeter, or pounds per square inch. Sometimes pressures are expressed in terms of certain commonly observed pressures as, for example, an *atmosphere*, representing a pressure equal to that exerted by the air under standard conditions, or a *centimeter of mercury*, representing a pressure equal to that exerted by a column of mercury one centimeter high.

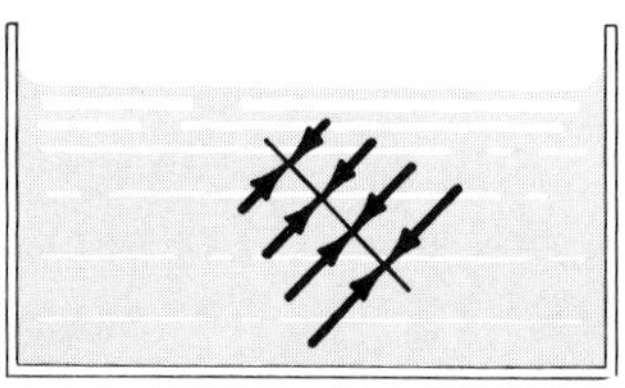

Figure 12-1
Forces on a plane inserted into a liquid.

The concept of pressure is particularly useful in discussing the properties of liquids and gases. The force exerted by a liquid on a plane surface immersed in the liquid at rest is always normal to the surface and is given by the product of the average pressure and the area of the surface. In Fig. 12-1 a plane inserted into a liquid experiences forces from each side perpendicular to the plane.

12-4 PRESSURE DUE TO THE WEIGHT OF A LIQUID

The atoms and molecules of which a liquid is composed are attracted to the earth in accordance with Newton's law of universal gravitation. Hence liquids collect at the bottoms of containers, and the upper layers exert forces on those underneath.

The pressure at a point in a liquid means the force per unit area of a surface placed at the point in question. Imagine a horizontal area A (Fig. 12-2) which is a distance h below the surface of the liquid. Because of its weight, the column of liquid directly above the area exerts a force F downward on the area equal to the weight W of the liquid in the column. The liquid is relatively incompressible; hence ρ is constant. The weight of the column is the volume times the weight of unit volume:

$$F = W = hA\rho g$$

The pressure P_l due to the liquid is

$$P_l = \frac{F}{A} = \frac{hA\rho g}{A}$$

$$P_l = h\rho g \tag{7}$$

Example Find the pressure due to a column of mercury 74.0 cm high.

$$P_l = h\rho g = (0.740\ \text{m})(1.36 \times 10^4\ \text{kg/m}^3)(9.80\ \text{m/s}^2)$$

$$= 9.86 \times 10^4\ \text{N/m}^2$$

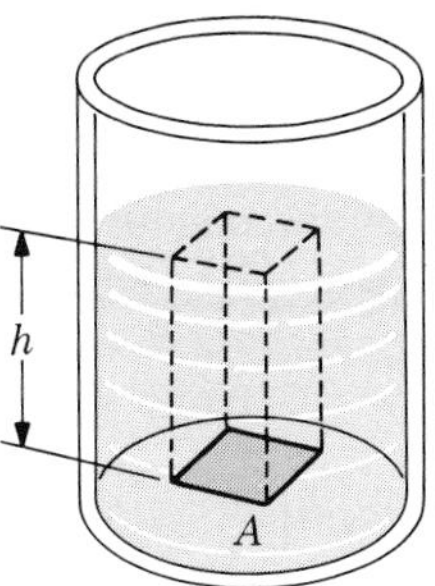

Figure 12-2
Pressure in a liquid.

In Eq. (7) the pressure is that due to the liquid alone. If there is a pressure on the surface of the liquid, this pressure must be added to that due to the liquid to find the pressure at a given level. The pressure at any level in the liquid is

$$P = P_{\text{surface}} + P_{\text{liquid}}$$

$$P = P_s + h\rho g \qquad (8)$$

where P_s is the pressure at the surface of the liquid, which is simply the atmospheric pressure in the case of a liquid in an open container.

Example A rectangular tank 6.0 × 8.0 ft is filled with gasoline to a depth of 8.0 ft. The pressure at the surface of the gasoline is 14.7 lb/in². Find the pressure at the bottom of the tank and the force exerted on the bottom.

From Table 1,

$$\rho g = D = 42 \text{ lb/ft}^3$$

$$P_s = (14.7 \text{ lb/in}^2)(144 \text{ in}^2/\text{ft}^2) = 2{,}1\bar{2}0 \text{ lb/ft}^2$$

$$P = P_s + h\rho g = 2{,}1\bar{2}0 \text{ lb/ft}^2 + (8.0 \text{ ft})(42 \text{ lb/ft}^3)$$

$$= 2.5 \times 10^3 \text{ lb/ft}^2$$

$$F = PA = (2.5 \times 10^3 \text{ lb/ft}^2)(6.0 \text{ ft})(8.0 \text{ ft})$$

$$= 1.2 \times 10^5 \text{ lb}$$

12-5 A LIQUID SEEKS ITS OWN LEVEL

If tubes of various shapes and sizes are connected to a common reservoir as in Fig. 12-3, a liquid poured into them will rise to the same level in all the tubes. This occurs because the pressure within a liquid is directly proportional to its depth below the free surface. The pressure is the same within a liquid at rest at a common level.

In Fig. 12-3 the base of each liquid container has the same area. From our discussion above, it can be shown that the forces on each base must be equal.

Since $P = F/A$,

$$F = PA \qquad \text{and} \qquad P = h\rho g$$

where h = distance from the base to the surface

ρ = density of the liquid

g = gravitational acceleration

All three variables, h, ρ, and g, are constant in each tube. Since A, the area of each base, is kept constant, the right-hand side of the equation $F = PA$ is a constant and the force on each base must be the same for each cylinder. Further, there appears to be a paradox here in that the weight of the liquid is obviously different in each container and since liquid pressure depends upon the weight of the fluid above a reference point, it would seem that the pressure should be greatest at the base of the largest container. This can be answered if we look at the shape of container C

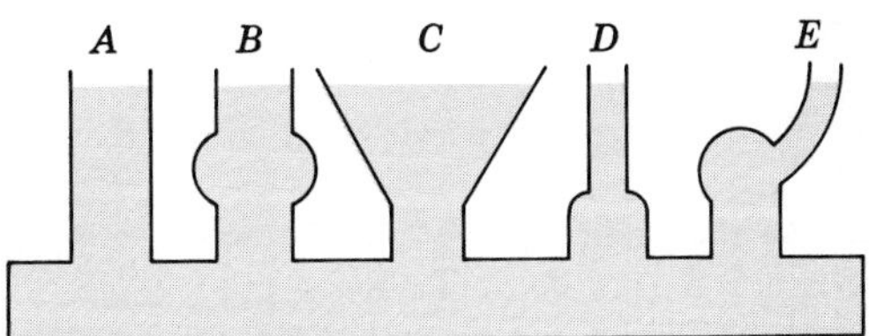

Figure 12-3
Liquid seeks its own level.

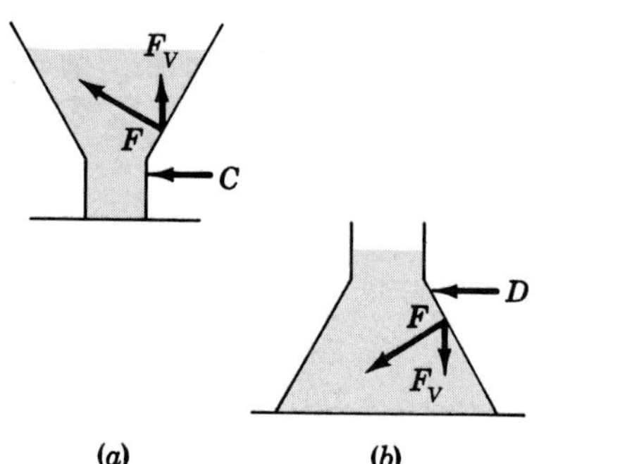

Figure 12-4
Forces acting on a liquid in various-shaped containers.

in the figure (Fig. 12-4*a*). The sloping sides of the container exert a force on the liquid. Since this force has an upward component, this component of the force supports some of the weight of the liquid. The converse of this is true in the case of container *D* which is narrower at the top than at the base (Fig. 12-4*b*). In *D*, the force exerted by the cylinder on the liquid is at some downward angle and the vertically downward component of that force, added to the weight of the liquid, will add up to a total force equal to the total force at the base of *C*; hence the pressures are the same. This can be shown for each container in the apparatus in Fig. 12-3.

12-6 PRESSURE IN LIQUIDS AT REST

The following general statements apply to the pressure in a liquid at rest:

1 Pressure exists at every point within the liquid.
2 As indicated by Eq. (7), the pressure P_l is proportional to the depth below the surface.
3 At any point in a liquid the magnitude of the force (due to pressure) exerted on a surface is the same no matter what the orientation of the surface is. If this statement were not true, there would be a net force in one direction and the liquid would be set in motion.
4 The pressure is the same at all points at the same level within a single liquid.
5 The force on the surfaces of the container due to the pressure is everywhere perpendicular to the surfaces of the container.
6 The force on the bottom of a container is the pressure at that level times the area of the bottom. The force may be greater than, equal to, or less than the weight of the liquid in the container. Why?

Example In a U tube (Fig. 12-5) the right-hand arm is filled with mercury, while the other arm is filled with a liquid of unknown density, the levels being as shown in the diagram. Find the density of the unknown liquid.

At the level of separation the pressure is the same in the two liquids, $P_1 = P_2$. At that level the pressure in the mercury is

$$P_1 = P_s + h_1\rho_1 g$$

and the pressure in the unknown liquid is

$$P_2 = P_s + h_2\rho_2 g$$

$$P_1 = P_2$$

$$h_1\rho_1 g = h_2\rho_2 g$$

$$\rho_2 = \frac{h_1\rho_1}{h_2} = \frac{(2.0 \text{ cm})(13.6 \text{ g/cm}^3)}{14 \text{ cm}}$$

$$= 1.9 \text{ g/cm}^3$$

12-7 PRESSURE IN GASES

When a gas is confined in a container, it exerts forces perpendicular to the walls of the container because shearing stress cannot exist in the gas. Here we do not have a simple pressure-depth relation. The pressure is primarily due to the motion of the molecules as they bombard the walls. For a small container the pressure is everywhere the same within the container.

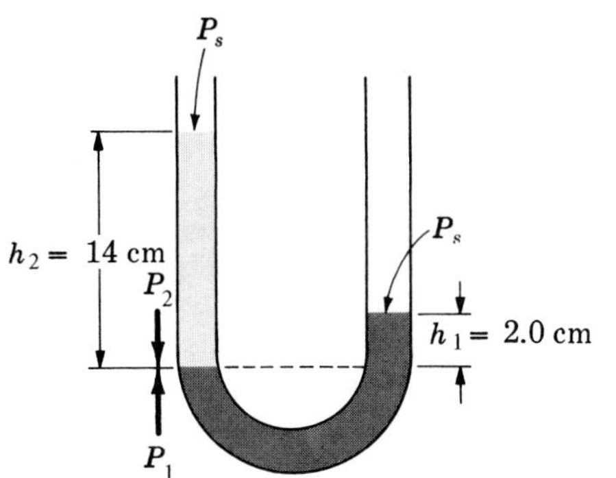

Figure 12-5
Columns of unequal heights may produce equal pressures.

Our most common gas is air, a mixture of several gases but principally nitrogen and oxygen. Since the air is always present, we seldom notice the forces that air exerts unless these forces become so great that they produce inconvenience, discomfort, or destruction.

It is a common expression to characterize something "as light as air," but air is hardly light. Air is attracted by the earth as is every other substance, and the total weight of the air is tremendous, roughly 6×10^{15} tons. This huge weight is always pressing on the surface of the earth, but since these radial forces exist over the entire surface of the earth, the resultant force is zero.

At the surface of the earth we observe a pressure due to the weight of the air. This pressure under standard conditions is about 14.7 lb/in^2, or 1.01×10^5 N/m^2. As a result of this pressure very large forces are exerted on even moderately large areas. On an ordinary window, which measures, say, 3.0 by 6.0 ft, the force is (14.7 lb/in^2)(36 in)(72 in) = $3\bar{8},000$ lb = 19 tons. Fortunately this large force is normally balanced by another force equal in magnitude but opposite in direction on the other side of the window, for no ordinary window would of itself be able to withstand so great a force. If, during a tornado, the exterior pressure suddenly falls, the greater interior pressure may actually cause a house to explode.

If a container such as an ordinary tin can is closed tightly and air pumped out, it soon collapses because of the greater force on the outside. This action is used in certain types of conveyors. A spout is inserted into grain or other loose material, air is removed from the spout by means of a blower, and the outside air pushes the material up the spout.

At higher levels atmospheric pressure is less than at the surface of the earth. Atmospheric pressure at sea level is 14.7 lb/in^2, but this decreases until at an elevation of 6 mi it is only 4 lb/in^2 and at 10 mi it is 2 lb/in^2. The decrease is not proportional to the height, however, since the density of the air decreases as the altitude increases. If the air were all at the same temperature and at rest, the decrease in pressure would be a simple exponential change. Under actual conditions it is much more complicated.

Atmospheric pressure varies from time to time and from place to place. It can be measured by using a manometer as shown in Fig. 12-6. *Standard atmospheric pressure* is defined as that equivalent to the pressure due to a column of mercury 76.00 cm long. It is sometimes confusing to a person unfamiliar with reading barometers to find that atmospheric pressure readings are given in centimeters of mercury or inches of mercury. Since pressure has been defined as a force per

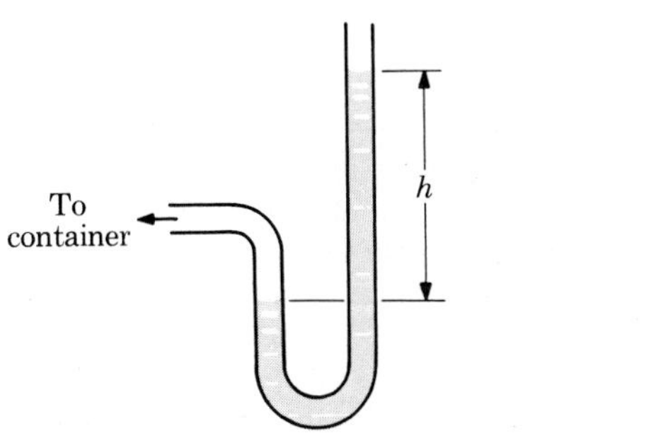

Figure 12-6
A simple U-tube manometer.

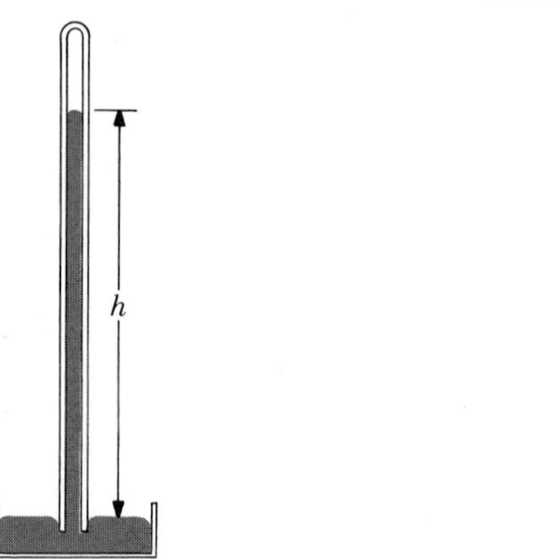

Figure 12-7
Principle of the mercury barometer.

unit area, $P = F/A$, one would expect the units of barometric readings to be in force per area units. However, if one observes a mercurial barometer (Fig. 12-7), it will be observed that what is being read is the height of a column of mercury which has a volume of $\pi r^2 h$. Since the πr^2 (the cross-sectional area of a given barometer) stays constant, the only variable is the height h. Hence the relationship between the height and the weight of the column of mercury to the atmospheric pressure can be expressed directly.

$$1 \text{ atm} = 76.00 \text{ cm Hg} = 29.92 \text{ in Hg}$$
$$= 101{,}300 \text{ N/m}^2 = 14.70 \text{ lb/in}^2$$

The mercurial barometer has certain limitations which prevent it from being used in many situations. For example, the barometer can be used only when it is mounted perpendicularly to the ground. The aneroid barometer is a more adaptable and portable form of barometer. This barometer consists of an evacuated, thin-walled, flat, collapsible metal cylinder, or "can" (Fig. 12-8). As the pressure increases, the can collapses to a greater degree; and as the pressure decreases, the can expands. An arm fastened to the surface of the can is attached through a variety of gear mechanisms to an indicator on a dial which has been calibrated to show readings in atmospheric pressure. Such a device can be used in any position and has found use as an altitude indicator (altimeter) in aircraft.

The barometer provides a record of increasing and decreasing atmospheric pressure. In the study of meteorology it is found that weather patterns can be predicted by observing changes in barometric pressure readings. A falling barometer, indicating a pressure drop, foretells of inclement weather; whereas a rising barometer, indicating an increase in atmospheric pressure, is a prediction of good or fair weather.

This statement of fact has caused much confusion for it (as in the case of fluid pressure at the bottom of odd-shaped reservoirs, described in Sec. 12-5) seems to be a paradox. A falling barometer indicates the approach of inclement, moist weather. Common sense "tells" us that wet air "weighs" more than dry air, so the advent of wet air should not cause a drop in atmospheric pressure but rather a rise in pressure, since the pressure in a fluid (air) is proportional to the weight of the fluid above. What is the answer, then? The fact that has been overlooked in this misconception is that the water in the air is in the form of a vapor or gas. While water in the liquid state is denser than air, water in the gaseous (vapor) state is less dense than air, so that the addition of water to the atmosphere actually

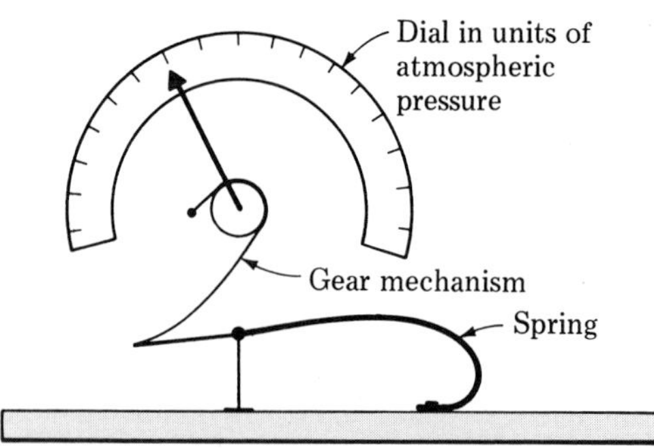

Figure 12-8
Sketch of an aneroid barometer.

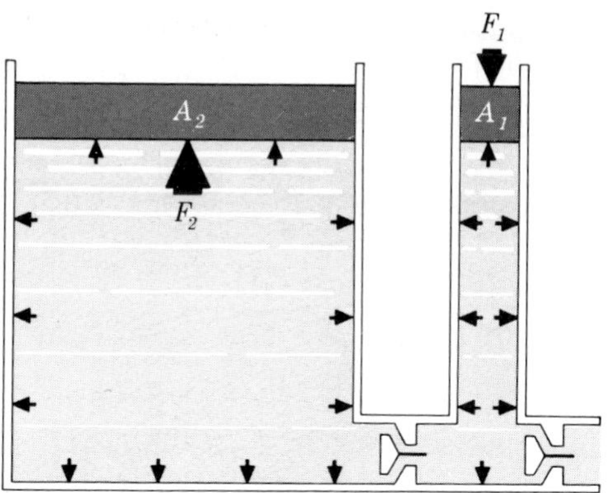

Figure 12-9
Hydraulic press.

has a diluting effect. That is, a liter of dry air is more dense (1.29 g/l) than a liter of moist air (1.18 g/l). Hence, the prediction of inclement weather when the atmosphere is "light" and fair weather when the atmosphere is "heavy."

12-9 EXTERNAL PRESSURE; PASCAL'S LAW

If any external pressure is applied to a confined fluid, the pressure will be increased at every point in the fluid by the amount of the external pressure. This is called *Pascal's law* after the French philosopher Blaise Pascal (1623–1662), who first clearly expressed it. In Eq. (8) we included an external pressure P_s which is most commonly atmospheric pressure. The fact that an external pressure applied to a liquid at rest increases the pressure at all points in the liquid by the amount of the external pressure has an important application in a machine called the *hydraulic press.* Small forces exerted on this machine cause very large forces exerted by the machine. In Fig. 12-9, the small force F_1 is exerted on a small area A_1. This increases the pressure in the liquid under the piston by an amount P. The force that this increase of pressure will cause on the large piston will be $F_2 = PA_2$, since th
under both pistons is the sa
Simply by changing the r
force F_2 may be made as la
big piston to carry. Larger p
transfer of liquid and are co
in action.

Example In a hydraulic press the small cylinder has a diameter of 8.0 cm, while the large piston has a diameter of 20.0 cm. If a force of 500 N is applied to the small piston, what is the force on the large piston, neglecting friction?

Since the pressure is increased the same amount at both pistons,

$$P_2 = P_1$$

$$\frac{F_2}{A_2} = \frac{F_1}{A_1}$$

$$F_2 = \frac{A_2}{A_1} F_1 = \frac{\pi(10\text{ cm})^2}{\pi(4\text{ cm})^2} 500\text{ N} = 3.12 \times 10^3\text{ N}$$

12-10 BOYLE'S LAW

Consider a container such as that shown in Fig. 12-10, one wall of which is a movable piston. In its initial position (*a*) there is a pressure P_1 when the volume is V_1. If, however, the piston is pressed down until it is in the new position shown in (*b*), the volume has been decreased to V_2, while the pressure has been increased to P_2. If this

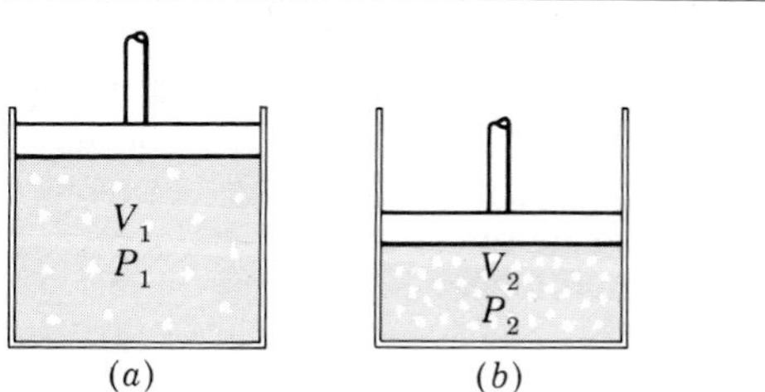

(*a*) (*b*)

Figure 12-10
P-V relation; temperature constant.

change takes place so slowly that there is no change in temperature, the volume occupied by the gas is inversely proportional to the pressure

$$V \propto \frac{1}{P} \tag{9}$$

or the product of pressure and volume is always the same, i.e.,

$$P_1V_1 = P_2V_2 \tag{10}$$

This relationship is expressed in *Boyle's law*, named after the Irish physicist Robert Boyle (1627–1691), who first stated it. It may be written as follows: If the temperature of a confined gas does not change, the product of the pressure and volume is constant. In symbols

$$PV = k \tag{11}$$

Example The volume of a gas under standard atmospheric pressure (76.0 cm Hg) is 200 in³. What is the volume when the pressure is 80 cm Hg if the temperature is unchanged?

From Eq. (10),

$$P_1V_1 = P_2V_2$$

$$76 \text{ cm} \times 200 \text{ in}^3 = 80 \text{ cm} \times V_2$$

$$V_2 = \tfrac{76}{80}\, 200 \text{ in}^3 = 1\bar{9}0 \text{ in}^3$$

Note: Boyle's law is approximately true over considerable ranges of temperature and pressure. There are, however, conditions under which it cannot be applied. If the temperature is near that at which the gas will liquefy, there will be large deviations from the simple law. The change in volume is greater than that predicted by the law. Also, if the pressure becomes very great, the deviation from Boyle's law is large, in this case the change in volume being less than that predicted by the law.

Gauge pressures cannot be used in Boyle's law. Whenever the pressure indicated is a gauge pressure, atmospheric pressure must be added before using it in the law.

Example An automobile tire whose volume is 1,500 in³ is found to have a pressure of 20.0 lb/in² when read on the tire gauge. How much air (at standard pressure) must be forced in to bring the pressure to 35.0 lb/in²?

The 1,500 in³ of air in the tire at 20 lb/in² is compressed into a smaller volume at 35.0 lb/in².

$$P_1V_1 = P_2V_2$$

$$P_1 = 20.0 \text{ lb/in}^2 + 14.7 \text{ lb/in}^2 = 34.7 \text{ lb/in}^2$$

$$P_2 = 35.0 \text{ lb/in}^2 + 14.7 \text{ lb/in}^2 = 49.7 \text{ lb/in}^2$$

$$(34.7 \text{ lb/in}^2)(1{,}500 \text{ in}^3) = (49.7 \text{ lb/in}^2)V_2$$

$$V_2 = 1{,}0\bar{5}0 \text{ in}^3$$

The volume of air added to the tire is

$$1{,}5\bar{0}0 \text{ in}^3 - 1{,}0\bar{5}0 \text{ in}^3 = 4\bar{5}0 \text{ in}^3$$

when its gauge pressure is 35.0 lb/in².

The volume at atmospheric pressure will be found from Boyle's law,

$$14.7 \text{ lb/in}^2 \times V = 49.7 \text{ lb/in}^2 \times 4\bar{5}0 \text{ in}^3$$

$$V = \frac{49.7}{14.7}\, 4\bar{5}0 \text{ in}^3 = 1{,}\bar{5}00 \text{ in}^3$$

12-11 EXPERIMENTAL PROOF OF BOYLE'S LAW

To study Boyle's law by a simple experiment, a quantity of mercury is poured into a J-shaped glass tube which has the short arm sealed off. While the procedure may vary, the tube shown in Fig. 12-11 has been adjusted, by tipping, until the mercury in the closed tube is higher than that in the open tube.

As additional mercury is poured into the tube, the volume of the trapped air in the closed tube becomes smaller due to the increased pressure caused by the added mercury. If it can be shown that the decrease in volume is proportional to the increased pressure, then Boyle's law would seem to hold.

The "added" pressure p on the trapped air is proportional to the difference in the heights of

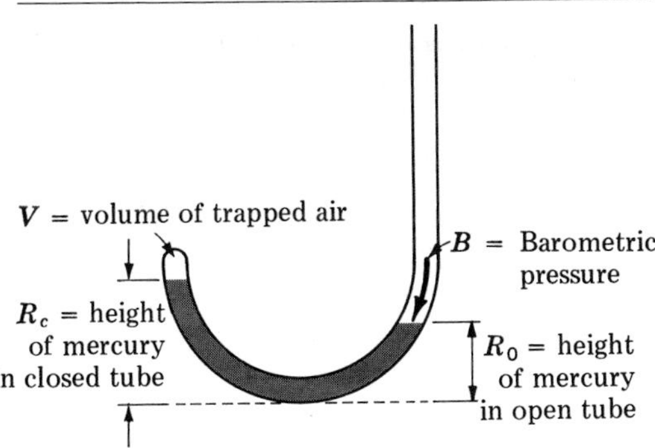

Figure 12-11
J tube used in experimental proof of Boyle's law.

the mercury in the open and the closed tubes, $R_o - R_c$. The actual pressure on the air is the barometric pressure B plus the added pressure p, or $B + p$.

Since Boyle's law states that

$$PV = k \text{ (a constant)}$$

$$(B + p)(V) = k$$

and $$[B + (R_o - R_c)]\,V = k$$

Example The mercury level in the closed end of a J tube is 90 mm and in the open tube is 18 mm. The volume of the trapped air is found to be 110 cm³. If mercury is added and the new level in the closed tube is 94 mm and the new level in the open tube is 45.5 mm and the volume of the trapped air is 106 cm³, does this agree with Boyle's law? Assume the atmospheric pressure to be 764 mm of mercury.

At the start of the experiment,

$$\begin{aligned}&[B + (R_o - R_c)]\,V\\ &= [764 \text{ mm} + (18 \text{ mm} - 90 \text{ mm})]\,(110 \text{ cm}^3)\\ &= (764 \text{ mm} - 72 \text{ mm})(110 \text{ cm}^3)\\ &= 76{,}200 \text{ mm Hg} \cdot \text{cm}^3\end{aligned}$$

After additional mercury is added,

$$\begin{aligned}&[B + (R_o - R_c)]\,V\\ &= [764 \text{ mm} + (45.5 \text{ mm} - 94 \text{ mm})]\,(106 \text{ cm}^3)\\ &= (764 \text{ mm} - 48.5 \text{ mm})(110 \text{ cm}^3)\\ &= 76{,}000 \text{ mm Hg} \cdot \text{cm}^3\end{aligned}$$

Therefore,

$$76{,}200 \text{ mm Hg} \cdot \text{cm}^3 \approx 76{,}000 \text{ mm Hg} \cdot \text{cm}^3$$
$$P_1V_1 \approx P_2V_2 \approx k$$

It should be noted that in the above example we simply attempted to show the relative sizes of the product of the pressure and the volume of the confined gas. To be more exact, we should change the height readings (692 and 715.5 mm) to pressure units. This can be done by multiplying each reading of height by the density of mercury, 13.6 g/cm³, to change the height reading to grams per centimeter², and then multiplying by 980 cm/s² to obtain the weight per unit area in dynes per centimeter², an acceptable unit for pressure.

$$\text{Pressure} = \text{height}_{\text{Hg}} \text{ (cm)} \times 13.6 \text{ g/cm}^3 \times 980 \text{ cm/s}^2 = X \text{ dyn/cm}^2$$

12-12 BUOYANCY; ARCHIMEDES' PRINCIPLE

Everyday observation has shown us that when an object is lowered into water it apparently loses weight and indeed may even float on the water. Evidently a liquid exerts an upward, buoyant force upon a body placed in it. Archimedes (287–212 B.C.), a Greek mathematician and inventor, recognized and stated the fact that *a body wholly or partly submerged in a fluid experiences an upward force equal to the weight of the fluid displaced.*

Archimedes' principle can readily be verified experimentally. One can deduce this principle from a consideration of Fig. 12-12. Consider a block of rectangular cross section A, immersed in a liquid of weight-density ρg. On the vertical faces, the liquid exerts horizontal forces, which are balanced on all sides. On the top face it exerts a downward force $h_1\rho gA$ and on the bottom face an upward force $h_2\rho gA$. The net upward force on the block is

$$h_2\rho Ag - h_1\rho Ag = hA\rho g$$

which is just the weight (volume hA times weight-density ρg) of the liquid displaced by the block.

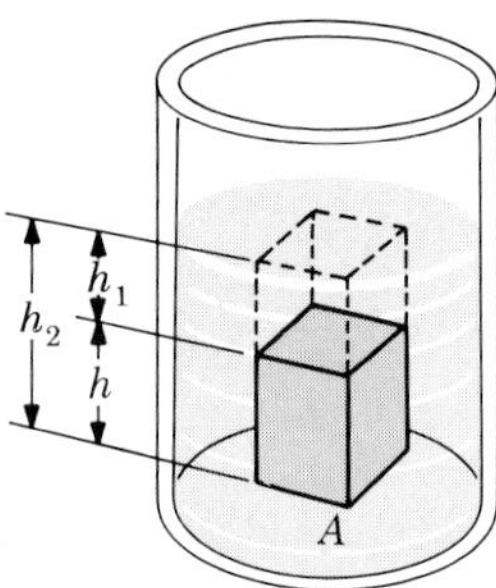

Figure 12-12
The upward force on the bottom of the block is greater than the downward force on the top.

In order to float, an object must be less dense than the fluid in which it is immersed. An object which has a specific gravity (density relative to water) of less than 1.0 will float on the surface when placed in water. That is, the effect of the force of gravity on that object is not sufficient to push aside or displace the volume of water necessary for the object to be totally submerged. For an object having a weight-density of 60 lb/ft^3 and a volume of 1 ft^3 to be submerged in water, it must push aside a cubic foot of water. But since fresh water has a weight-density of 62.4 lb/ft^3, it requires a force of 62.4 lb to displace a cubic foot of water. Since the only force acting downward on the object is its weight of 60 lb (we assume that it has not been dropped from some height or else it would momentarily submerge), this is not sufficient to push aside the cubic foot of water and, therefore, the object floats.

An object having a specific gravity greater than 1.0 will have the necessary force to push aside its volume of fluid and will sink in water, since a cubic foot of it weighs more than a cubic foot of water. An object having a specific gravity of exactly 1.0 will float, but it may float at any level within the water.

Normally the human body will float when the lungs are filled with air, but even under these conditions some people will not float.

Flotation depends upon the density of an object and the density of the human body varies considerably from person to person. Contrary to popular belief, obese people are not necessarily good floaters. The determining factor is the skeletal structure or frame of the person. A large-boned person will probably be a "sinker" and a small-boned person will be a "floater." You can tell which of these you are by simply attempting a prone float, lying motionless, face down, in a pool and determining how much of your body sinks. Being a sinker can have its drawbacks. For example, consider the plight of the sinker in the problem given below.

Example A man who has a volume of 3.5 ft^3 and a weight of 250 lb wishes to travel across the ocean on a ship. In order to obtain insurance he must promise to wear a life preserver at all times. If the life preserver weighs 10 lb, what is the minimum volume that the life preserver can have so that he will be a good insurance risk?

Since salt water has a weight-density of 64 lb/ft^3, the combined density of the man and the preserver must be less than that of the water.

Density of the man + preserver

$$= \frac{\text{weight of man} + \text{weight of preserver}}{\text{volume of man} + \text{volume of preserver}}$$

$$D = \frac{250 \text{ lb} + 10 \text{ lb}}{3.5 \text{ ft}^3 + V\,(\text{preserver})} = 64 \text{ lb/ft}^3 \text{ (max)}$$

Therefore,

$$260 \text{ lb} = 64 \text{ lb/ft}^3\,(3.5 \text{ ft}^3 + V)$$

$$260 \text{ lb} = 224 \text{ lb} + (64 \text{ lb/ft}^3)(V)$$

$$V = \frac{36 \text{ lb}}{64 \text{ lb/ft}^3} = 0.56 \text{ ft}^3$$

The life preserver must have a minimum volume of 0.56 ft^3 to make the man a floater.

The buoyant effect of air or other gas is by no means negligible. A balloon is supported as it floats in air by a buoyant force equal to the weight of air displaced. At the surface it may displace a weight of air greater than the combined weight of bag, gas contained in the bag, and load. If this is the case, the balloon is accelerated upward. As the balloon rises, the external pressure decreases and the balloon expands; but at the same time the density of air becomes less so that the buoyancy per unit volume is decreased. The balloon will rise until the weight of air displaced in the new position is equal to the total weight of balloon and contents.

The buoyant force of air on solids is less important because it is only a small fraction of the weight of the solid. However, in accurate weighing buoyancy cannot be neglected.

Example Brass weights are used in weighing an aluminum cylinder whose approximate mass is 89 g. What error is introduced if the buoyant effect of air ($\rho = 0.0013\ \text{g/cm}^3$) is neglected?

$$\rho_{\text{brass}} = 8.9\ \text{g/cm}^3$$

$$\rho_{\text{Al}} = 2.7\ \text{g/cm}^3$$

$$V = \frac{m}{\rho}$$

$$V_B = \frac{89\ \text{g}}{8.9\ \text{g/cm}^3} = 10\ \text{cm}^3$$

$$V_{\text{Al}} = \frac{89\ \text{g}}{2.7\ \text{g/cm}^3} = 33\ \text{cm}^3$$

The difference in volume V of air displaced on the two pans of the balance is

$$V = V_{\text{Al}} - V_B = 33\ \text{cm}^3 - 10\ \text{cm}^3 = 23\ \text{cm}^3$$

Hence, the mass error

$$m = V\rho = 23\ \text{cm}^3 \times 0.0013\ \text{g/cm}^3 = 0.030\ \text{g}$$

The error introduced is only a small fraction of the total mass, but in many experiments where accuracy is important an error of 0.030 g in 89 gm is too great to allow.

Example A balloon is to operate at a level where the weight-density of air is 0.060 lb/ft³. How much load can it support if it has a volume of 800 ft³ at that level and is filled with hydrogen, $D = 0.0050\ \text{lb/ft}^3$? The weight W_b of the bag is 30 lb.

$$W = VD$$

Weight of air displaced:

$$W_a = 800\ \text{ft}^3 \times 0.060\ \text{lb/ft}^3 = 48\ \text{lb}$$

Weight of hydrogen:

$$W_H = 800\ \text{ft}^3 \times 0.0050\ \text{lb/ft}^3 = 4\ \text{lb}$$

The load supported:

$$L = W_a - W_H - W_b = 48\ \text{lb} - 4\ \text{lb} - 30\ \text{lb}$$

$$= 14\ \text{lb}$$

12-13 VOLUME ELASTICITY OF GASES

If the pressure on a confined gas is increased, the volume decreases in accordance with Boyle's law. When the pressure is returned to its original value, the gas returns to its original volume. The gas has the property of volume elasticity.

If the temperature remains the same during compression and expansion, from Eq. (6), Chap. 10, the bulk modulus is

$$B = \frac{\text{stress}}{\text{strain}} = \frac{\Delta P}{\Delta V/V} = V\frac{\Delta P}{\Delta V} \qquad (12)$$

A minus sign may be introduced because an increase in P causes a decrease in V.

If the pressure is increased by a small amount ΔP to $P + \Delta P$, the volume is decreased by a small amount ΔV to $V - \Delta V$. From Boyle's law,

$$PV = (P + \Delta P)(V - \Delta V)$$
$$= PV + V\Delta P - P\Delta V - \Delta P\Delta V$$

Since ΔP and ΔV are small, their product can be neglected and

$$PV = PV + V\Delta P - P\Delta V$$

$$V\Delta P = P\Delta V$$

$$\frac{\Delta P}{\Delta V} = \frac{P}{V}$$

By substitution in Eq. (12),

$$B = V\frac{\Delta P}{\Delta V} = V\frac{P}{V}$$

$$= P$$

Thus at constant temperature the bulk modulus of a gas is equal to its pressure.

12-14 SURFACE PHENOMENA

Suppose that a straight piece of wire hangs in a horizontal position from a spring attached to the midpoint of the wire. The spring is stretched by the weight of the wire. If now the wire is dipped into a liquid that wets the wire and is then pulled up just above the surface as in Fig. 12-13, it will be found that the spring is stretched more than it was before. Evidently the film of liquid that extends over the wire exerts a force F on the wire. If the length of the wire is increased, F is found to increase proportionately. Thus the force appears to be proportional to the length of the film. From Fig. 12-13 we see that there is a film on each side of the wire. We might assume that the length of the film is twice the length of the wire. We may test this assumption by using first a ring of wire in which there will be a film on each side of the wire and then a flat disk of the same radius. For the disk there is only a single film around the edge. If we perform these experiments, we find that the force due to the film is twice as great for the ring as for the disk, which confirms our assumption. If several liquids are used, one after the other, the force is found to depend upon the liquid used. The force also depends upon the gas above the liquid.

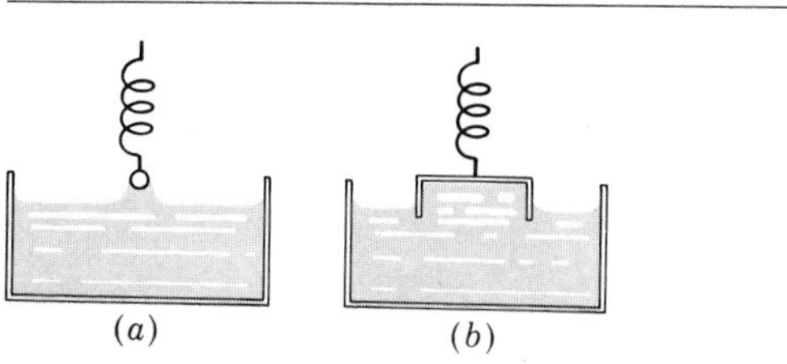

Figure 12-13
Surface film.

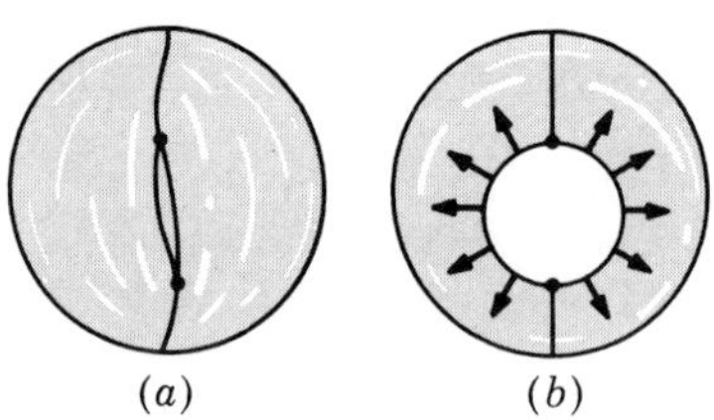

Figure 12-14
Force of surface film at right angles to boundary.

If a thread with a small loop is tied across a wire frame (Fig. 12-14*a*) and the frame is dipped into soap solution, the film will cover the whole frame and the loop may be of any shape. If, however, the film is broken within the loop, the loop immediately becomes circular in shape as in Fig. 12-14*b*.

12-15 SURFACE TENSION

These experiments suggest that the surface film exerts a force perpendicular to any line in the surface. Such a force must be a result of molecular forces. If the surface is flat, the force across the line in one direction is just equal in magnitude to that in the opposite direction and there is equilibrium (Fig. 12-15). However, if the film is broken on one side (Fig. 12-14*b*) or if the surface is so shaped that the forces are not in

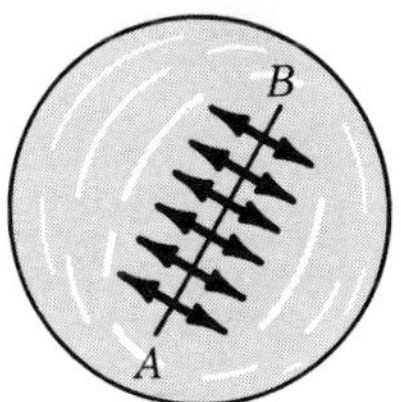

Figure 12-15
Forces across an imaginary line in a surface of a liquid.

opposite directions, the surface forces do not themselves produce equilibrium.

The force per unit length across such a line in the surface is called the *surface tension*

$$T = \frac{F}{L} \tag{13}$$

where F is the force across the line of length L. The unit of surface tension may be any force unit divided by any length unit. Since the forces are small, it is customary to use the force in dynes and the length in centimeters so that T is given in dynes per centimeter. Surface tension is the cause of several phenomena that would not be expected of liquids at rest. A steel needle has a density so high that the buoyant force of water upon it is not sufficient to support its weight. However, if it is oiled slightly so that the water will not wet it and it is laid very carefully upon the surface so that the film is not broken, it will float (Fig. 12-16). Note that the surface is depressed under the needle so that the film is able to exert an upward force.

Table 2
SURFACE TENSION FOR PURE LIQUIDS IN CONTACT WITH AIR

Liquid	Temperature, °C	T, dyn/cm
Benzene	20	27.6
	50	24.7
Carbon tetrachloride	20	26.8
Ethyl alcohol	20	22.3
	50	19.8
Methyl alcohol	20	22.6
	50	20.1
Mercury	20	465
Water	10	74.2
	20	72.8
	50	67.9
Liquid lead	400	445
Liquid tin	400	520

Example The maximum force, in addition to the weight, required to pull a wire frame (Fig. 12-13) 5.00 cm long from the surface of water at temperature 20°C is 728 dyn. Calculate the surface tension of water.

There is a film on either side of the wire and hence the length of the film is twice the length of the wire.

$$T = \frac{F}{L} = \frac{F}{2l} = \frac{728 \text{ dyn}}{2 \times 5.00 \text{ cm}} = 72.8 \text{ dyn/cm}$$

Consider a rectangular wire frame (Fig. 12-17) one side of which is movable. If the frame is dipped into soap solution so that there is a film

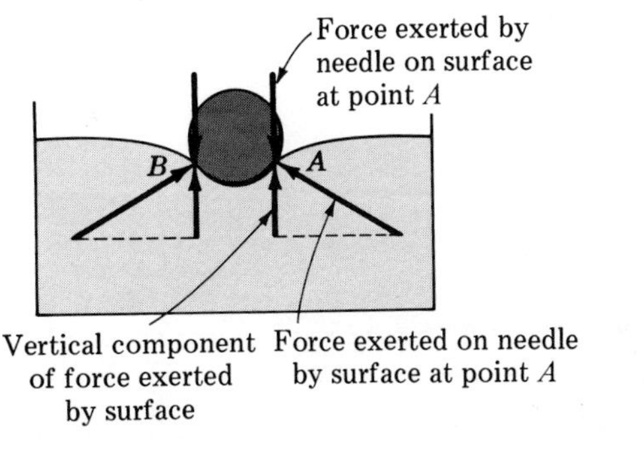

Figure 12-16
The force due to surface tension is sufficient to support a needle.

on the inside of the rectangle, there will be a force on the movable side so as to decrease the area of the loop. To hold the movable side in equilibrium, a force F must be applied. Since there is a film on either side of the wire,

$$F = LT = 2lT \tag{14}$$

where l is the length of the movable wire. If the movable wire is pulled slowly a distance s so as to increase the area, the force F remains constant as the area of the film increases. In this process work is done. The work done in moving the wire is

$$\mathcal{W} = Fs = 2lsT \tag{15}$$

But the increase in area A of the double film (front and back) is $2ls$.

Thus $$\mathcal{W} = AT$$

or $$T = \frac{\mathcal{W}}{A} \tag{16}$$

Hence the surface tension may also be regarded as the work done per unit area in increasing the area of the film. The surface tension can therefore be expressed in ergs per square centimeter. This unit is equivalent to a dyne per centimeter.

The surface under tension will always tend to contract until its area is the smallest possible for

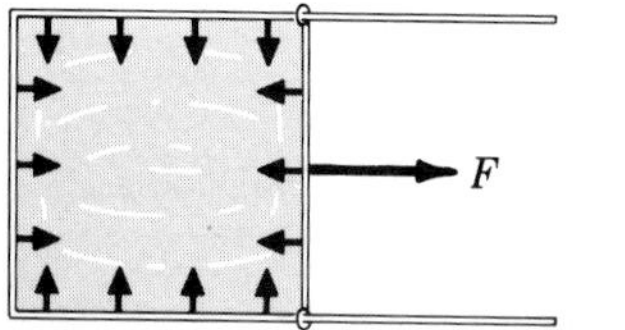

Figure 12-17
Force per unit length and work per unit area in a film.

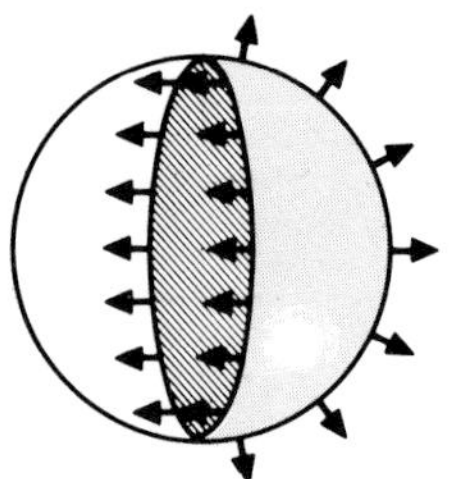

Figure 12-18
Forces that produce pressure within a droplet.

the conditions present. The film of Fig. 12-17 tends to contract until its area becomes zero. A droplet takes the shape that will make its surface area least, that is, a spherical shape if the drop is undisturbed by its weight.

12-16 PRESSURE DUE TO SURFACE TENSION

Within a liquid droplet the pressure is greater than the pressure outside. Its spherical shape is one indication of the pressure difference. Consider the droplet of Fig. 12-18. The excess pressure ΔP inside the droplet causes a force on each element of the surface perpendicular to that surface element. The resultant of these forces is a force perpendicular to a cross section of the droplet and given by the excess pressure times the area of the cross section,

$$F = \Delta P \times \pi r^2$$

This force is supplied by the film pulling across the perimeter of this cross section,

$$F = 2\pi r T$$

Thus $$\Delta P \times \pi r^2 = 2\pi r T$$

or $$\Delta P = \frac{2T}{r} \tag{17}$$

12-17 CONTACT OF LIQUIDS, SOLIDS, AND GASES

At the boundary of the surface of a liquid in a vessel there are three substances in contact. We may consider that there is a surface effect at each surface of separation and that there is a force parallel to each such surface. In addition there is a force of adhesion between the molecules of the liquid and those of the solid. This adhesive force is perpendicular to the wall of the container. The point where the three substances are in contact is in equilibrium under the action of four forces, and the liquid surface will set itself so that this condition is realized. The liquid surface near this point of contact will, in general, be curved. Two such combinations are shown in Fig. 12-19. In Fig. 12-19*a* the adhesive force is relatively large, the liquid wets the solid, and the liquid surface is concave upward. In Fig. 12-19*b* the adhesive force is relatively smaller, the liquid does not wet the solid, and the liquid surface is concave downward. If a line is drawn tangent to the surface, it makes an angle θ with the wall. This angle is called the *angle of contact*. When the liquid wets the solid the angle of contact will be less than 90°; when the liquid does not wet the solid the angle of contact is greater than 90°. Certain angles of contact are given in Table 3.

Table 3
ANGLES OF CONTACT BETWEEN PURE LIQUIDS AND CLEAN SOLIDS

Solid	Liquid	θ, deg
Glass	Water	0
Glass	Ethyl alcohol	0
Glass	Mercury	128
Paraffin	Water	107

12-18 CAPILLARITY

It is a common observation that water rises in fine glass tubes. If a string lies over the edge of a pan of water, the water will rise in the string over the top and down the outside. These effects are caused by surface tension.

In the tube shown in Fig. 12-20, the liquid wets the tube making an angle of contact θ. The liquid rises in the tube until the force due to

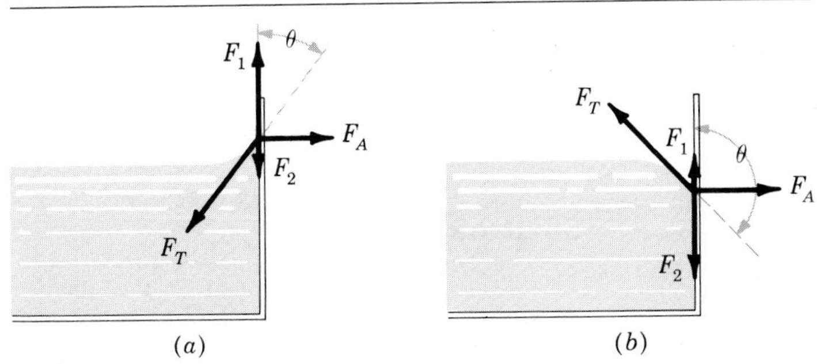

Figure 12-19
Angle of contact θ. The forces F_1, F_2, and F_T are each parallel to one of the surfaces of contact. F_A represents the adhesive force. F_1, F_2, and F_T represent the surface tensions of the solid-vapor, solid-liquid, and liquid-vapor, respectively.

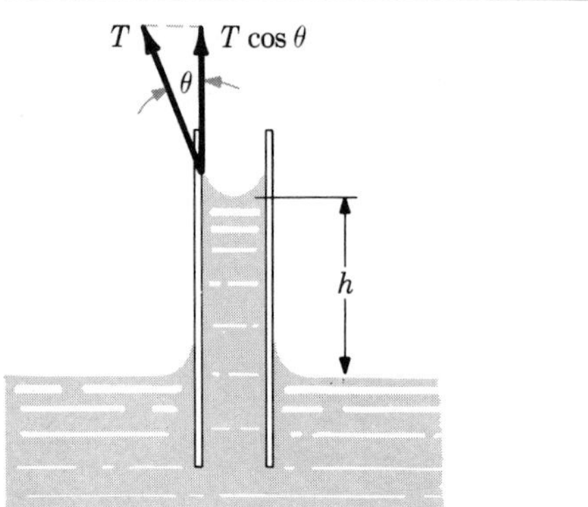

Figure 12-20
Rise of liquid in a capillary tube.

surface tension is equal to the weight of the liquid lifted up.

$$F = W$$

The upward force F is the product of the length of the film, i.e., the circumference $2\pi r$ of the tube and the upward component $T \cos \theta$ of the surface tension.

$$F = 2\pi r T \cos \theta \qquad (18)$$

The weight of liquid lifted is the product of the volume $\pi r^2 h$ and the weight per unit volume ρg.

Since $F = W$,

$$W = \pi r^2 h \rho g \qquad (19)$$

$$2\pi r T \cos \theta = \pi r^2 h \rho g$$

$$h = \frac{2T \cos \theta}{r \rho g} \qquad (20)$$

We note that if the angle of contact is greater than 90°, the cosine is negative and the liquid goes down in the tube. That is, if the liquid does not wet the tube, it descends.

Example The tube of a mercury barometer is 3.0 mm in diameter. What error is introduced into the reading because of surface tension?

For mercury:

$$\theta = 128° \qquad \cos \theta = -0.616$$

$$h = \frac{2T \cos \theta}{r \rho g}$$

$$= \frac{2(465 \text{ dyn/cm})(-0.616)}{0.15 \text{ cm } (13.6 \text{ g/cm}^3)(980 \text{ cm/s}^2)}$$

$$= -0.286 \text{ cm} = -2.86 \text{ mm}$$

The negative sign indicates that the reading is lower than the correct reading.

The curved surface of the liquid in the capillary tube indicates that the pressure inside the liquid is not the same as that outside. The difference in pressure is

$$\Delta P = \frac{F}{A} = \frac{2\pi r T \cos \theta}{\pi r^2} = \frac{2T \cos \theta}{r} \qquad (21)$$

Also $\Delta P = h \rho g \qquad (22)$

Hence

$$h \rho g = \frac{2T \cos \theta}{r}$$

$$h = \frac{2T \cos \theta}{r \rho g} \qquad (23)$$

Equation (23) is identical with Eq. (20).

At the surface of separation between two liquids, there is also a tension similar to surface tension. It is called *interfacial tension.*

SUMMARY

The *mean free path* is defined as the average distance a molecule moves before colliding with another molecule in that piece of matter.

The *mean free path* is generally smallest in solids, larger in liquids, and largest in gases. The type of molecular motion in a substance above absolute zero is generally vibrational in solids, translational in liquids, and random in gases.

Liquids and gases are frequently grouped together as fluids.

Density of a substance is mass per unit volume.

$$\rho = \frac{m}{V}$$

Weight-density is weight per unit volume.

$$D = \frac{W}{V}$$

Specific gravity, or *relative density,* of a substance is the ratio of the density of the substance to the density of water.

$$\rho_r = \frac{\rho}{\rho_w}$$

Pressure is normal force per unit area. The average pressure is

$$\overline{P} = \frac{F}{A}$$

The pressure at a point is given by

$$P = \lim_{\Delta A \to 0} \frac{\Delta F}{\Delta A}$$

At a depth h below the surface of a liquid, the pressure due to the weight of the liquid is

$$P = h\rho g$$

Pressure in an unconfined gas is due to the weight of the gas, but there is not the simple relation between pressure and depth that there is for liquids, because the density does not remain constant. Pressure in a gas may be measured by means of a manometer or by a barometer. Standard barometric pressure is 76.00 cm Hg at sea level and 45° latitude.

Water vapor in the air dilutes the air and causes it to weigh less than dry air. Hence, a falling barometric pressure reading indicates the presence of moist air and the advent of inclement weather.

Pascal's law states that an external pressure applied to a confined fluid increases the pressure at every point in the fluid by an amount equal to the external pressure.

Boyle's law states that if the temperature of a confined gas is unchanged, the product of the pressure and volume is constant.

$$PV = k$$

Archimedes' principle states that a body wholly or partly submerged in a fluid is buoyed up by a force equal to the weight of the fluid displaced.

At the surface of a liquid there is a *surface tension*. It is expressed as the force per unit length of the surface film,

$$T = \frac{F}{L}$$

or as the work done per unit area in increasing the area of the film,

$$T = \frac{\mathcal{W}}{A}$$

Surface tension is usually expressed in dynes per centimeter or ergs per square centimeter.

Liquids rise or are depressed in fine tubes. The height to which they rise is given by the expression

$$h = \frac{2T\cos\theta}{r\rho g}$$

Questions

1 Present evidence that would support the theory that molecules are in motion in a gas; in a liquid; and in a solid. How do the motions of molecules differ in these phases?

2 What reasons have you for believing in the existence of molecules?

3 Water is most dense at 4°C; below this temperature water expands and becomes less dense. What do you suppose is peculiar about water that causes it to behave in this manner?

4 What effect does a rise in temperature have on the density of a solid; on the density of most liquids; on the density of water at 0°C?

5 In stating that the pressure at a point in a liquid is proportional to the depth, what is assumed regarding the density?

6 Explain how you could determine the relative density of a liquid by weighing a solid in it.

7 An ice cube floats in a glass of water filled to the brim. Will the water overflow when the ice melts?

8 Devise a procedure to determine the specific gravity of a block of cork.

9 A test often conducted on wines and liquors is to draw samples from them into a hydrometer. What does the hydrometer measure?

10 In some regions after a heavy rain, water backs up into basements of houses. When valves are inserted to prevent the flow into the house, the force on the floor is sometimes sufficient to break the floor. One household adviser recommended that the valve be replaced by a pipe at the drain high enough to reach above the water level. Comment on the effect of this device on the force that is exerted on the floor.

11 An open-tube manometer consists of a U tube open to the air at one end and connected to a pressure chamber at the other. It can be filled with mercury, water, or an oil of specific gravity 0.60. Discuss the advantages and disadvantages of each. Under what circumstances would each be useful?

12 A body is immersed in a liquid in such a manner that it is closely in contact with the bottom and there is no liquid beneath the body. Is there a buoyant force on the body? Explain.

13 If you take a deep breath when floating in water, will you float with more or less of your body out of water? Explain.

14 Does a ship wrecked in midocean sink to the bottom or does it remain suspended at some great depth? Justify your opinion.

15 The palm of our hand has an area of approximately 20 in^2. When we hold our hand out with the palm upward, we are supporting a column of air about 60 mi high which presses down on the hand with a pressure of almost 300 lb. Why doesn't our hand collapse under this pressure?

16 As you ride in an elevator your ears may become "blocked" and by yawning you are able to relieve the situation. Explain why this blocking occurs and how the yawn cures it.

17 What does a falling barometer foretell? Explain why this indicator is fairly reliable.

18 If an object has a specific gravity of 1, what can you say about its ability to float?

19 A can full of water is suspended from a spring balance. Will the reading of the balance change (*a*) if a block of cork is placed in the water and (*b*) if a piece of lead is suspended in the water? Explain.

20 A free balloon can be arranged to float at a constant elevation in the air, but this is not possible for a submarine in water. Show why this is the case.

21 Show clearly why the constant k in the Boyle's-law equation $PV = k$ is not a pure number but has the dimensions of work.

22 Why is the work of digging a tunnel under a river done in a high-pressure chamber? What determines how high the pressure should be in the chamber? What danger is involved if the pressure becomes too low? Too high?

23 Cite the physical principle involved in the operation of (*a*) a hydrometer, (*b*) hydraulic brakes, (*c*) an aspirator, (*d*) a mercury barometer, and (*e*) an airship.

24 What does Archimedes' principle really tell us? Give illustrations of how it may be applied.

25 Explain how Archimedes' principle may be used to measure the specific gravity of an object lighter than air.

26 Explain the buoyant force exerted by liquids on submerged bodies. Would a denser liquid exert a more buoyant force than a less dense liquid? Explain.

27 Explain why a piece of iron may float in mercury but sink in water.

28 Boats are often described by their "displacement." What information is being given when we say a boat has a displacement of 2,600 lb?

29 When bits of camphor gum are dropped on water, they move about erratically. Explain.

30 Show that the pressure due to surface tension in a bubble is given by $P = 4T/r$.

31 If two soap bubbles of radii 1.0 and 2.0 cm could be joined by a tube without bursting, what would happen? Why?

32 A needle can be made to float on the surface of water. If even a drop of detergent is added to the water, the needle will sink immediately. Explain.

33 What physical principles are applied when candle wax is removed from clothing by covering the spot with blotting paper and then passing a hot iron over the paper?

34 Water will not run from the upper end of a short capillary tube the lower end of which dips below the surface of water, but sap runs from the end of a maple branch that has been cut off. Explain.

Problems

1 An irregular gold nugget is found, but no balance is available to determine its mass. Water is poured into a graduate, and when the nugget is dropped into the water the reading increases by 3.75 cm^3. What is the mass of the nugget? Specific gravity of gold = 19.3.

2 The densest element, osmium, has a specific gravity of 22.5. (*a*) What is the weight of a cubic inch of osmium? (*b*) What is the volume of 150 lb of this metal? *Ans.* 0.813 lb/in^3; 185 in^3.

3 A liter of milk has a mass of 1,032 g. It contains 4.0 percent butterfat by volume, and the specific gravity of the butterfat is 0.865. What is the density of the fat-free skimmed milk?

4 Uranium has a specific gravity of 18.7. (*a*) What is the weight of a cubic inch? (*b*) How large a volume could a man carry if he can lift a load of 200 lb? *Ans.* 0.675 lb; 296 in^3.

5 A box whose base is 1.0 m square has a mass of 100 kg. What is the pressure beneath the box?

6 A vertical force of 4.0 oz pushes a phonograph needle against the record surface. If the point of the needle has an area of 0.0010 in^2, find the pressure in pounds per square inch.

Ans. 2.5×10^2 lb/in^2.

7 To what pressure can a 500-N man raise the air pressure in an automobile tire if he uses a simple pump without levers and the area of cross section of the piston is 20 cm^2?

8 Calculate the approximate pressure in pounds per square inch sustained by submarines at a 200-ft depth in the ocean; by Professor Beebe's bathysphere at a depth of 1,426 ft; by anything in the deepest part of the ocean, 35,000 ft. Assume density of salt water to be 64.4 lb/ft^3.

Ans. 90 lb/in^2; 640 lb/in^2; 15,600 lb/in^2.

9 If the casing of an oil well 150 m deep is full of water, what is the pressure at the bottom due to the weight of water?

10 The barometric pressure is 30.0 in of mercury. Express this in pounds per square foot and in pounds per square inch and millimeters of mercury.

Ans. 2,120 lb/ft^2; 14.7 lb/in^2; 762 mm.

11 A vertical U tube is partly filled with mercury and a solution of unknown specific gravity is poured into one arm. The surface of separation of the liquids is at 6.35 cm on the scale, the free surface of the mercury at 8.46 cm, and the free surface of the solution at 24.84 cm. Find the specific gravity of the solution.

12 A swimming tank 50 ft long and 20 ft wide has a sloping floor so that the water is 4.0 ft deep at one end and 7.0 ft deep at the other. Find the total force due to the water on the bottom and that on each end.

Ans. 340,000 lb; 10,000 lb; 31,000 lb.

13 In constructing a concrete wall of a basement, a form is built 15.0 ft high and 30.0 ft long. There is a space of 6.0 in between the form and the earthen bank. After a heavy rain the 6.0-in space fills with water to a depth of 8.0 ft. What is the force on the form?

14 Find the force tending to crush an outside hatch of a submarine 2.00 ft^2 in area when the boat is 100 ft below the surface of the sea. Specific gravity of sea water is 1.03. *Ans.* 12,900 lb.

15 An irregularly shaped piece of metal of density 7.80 g/cm^3 weighs 429 g when completely immersed in nitrobenzene, which has a density of 1.20 g/cm^3. Find the volume of the piece of metal.

16 When a polar bear jumps on an iceberg, he

notices that his 420-lb weight is just sufficient to sink the iceberg. What is the weight of the iceberg? *Ans.* 3,500 lb.

17 A rectangular scow 150 ft long and 15.0 ft wide weighs 50,000 lb. Find the depth of freshwater required to float it.

18 A barge has vertical sides and a flat bottom of 320 ft^2 area. When partly filled, the barge is immersed in freshwater to a depth of 2.00 ft. Upon removal of the load the barge rises 16.0 in. (*a*) What is the approximate weight of the barge? (*b*) What vertical displacement will result if a 10-ton truck is loaded on the barge?

Ans. 13,300 lb; 1.0 ft.

19 A solid glass sphere has a weight of 21.0 lb in air and an apparent weight of 14.0 lb when immersed in turpentine. When immersed in water its apparent weight is 11.6 lb. Calculate the specific gravity of the glass and that of the turpentine. Find the density of each.

20 A 2.0-lb iron ball is supported by a wire and immersed in oil of specific gravity 0.80. What is the tension in the wire? The specific gravity of iron is 7.8. *Ans.* 1.8 lb.

21 Water from city mains with a pressure of 75 lb/in^2 is used to operate a hydraulic lift in a garage. If the piston area is 1.50 ft^2 and the efficiency is 90 percent, what is the maximum load that can be raised?

22 A stone of specific gravity 2.50 starts from rest and sinks in a freshwater lake. Allowing for a 25 percent frictional force, calculate the distance the stone sinks in 3.00 s. *Ans.* 1,980 cm.

23 What is the net lifting ability of a balloon containing 400 m^3 of hydrogen, the mass of the balloon without gas being 250 kg, if the density of air is 1,200 g/m^3?

24 A 165-lb man floats with nearly all his body below the surface of a lake. What is his volume?

Ans. 2.65 ft^3.

25 The volume of an air bubble increases tenfold in rising from the bottom of a lake to its surface. If the height of the barometer is 762.0 mm, and if the temperature of the air bubble is constant, what is the depth of the lake?

26 What volume of lead of specific gravity 11.3 must be placed on top of a 20.0-g block of cork of specific gravity 0.240 to cause the cork to be barely submerged in water? *Ans.* 5.61 mm^3.

27 A tube 120 cm long, closed at one end, is half-filled with mercury and then inverted into a mercury trough in such a way that no air escapes from the tube. If the barometric pressure is 75 cm, what is the height of the mercury column inside the tube?

28 A diver and his suit weigh 200 lb. It requires 30.0 lb of lead to sink him in freshwater. If the specific gravity of lead is 11.3, what is the volume of the diver and his suit? *Ans.* 3.96 ft^3.

29 The volume of a tire is 1,500 in^3 when the pressure is 30 lb/in^2 above atmospheric pressure. (*a*) What volume will this air occupy at atmospheric pressure? Assume that atmospheric pressure is 15 lb/in^2. (*b*) How much air will come out of the tire when the valve is removed?

30 A stone weighs 30.0 lb in air and 21.0 lb in water. What is its (*a*) specific gravity, (*b*) weight-density, and (*c*) volume?

Ans. 3.33; 208 lb/ft^3, 0.144 ft^3.

Max Planck, 1858–1947

Born in Kiel. Succeeded Kirchhoff at the University of Berlin. Received the 1918 Nobel Prize for Physics for his contribution to the development of physics by his discovery of the element of action (quantum theory).

Johannes Stark, 1874–1957

Born in Schickenhof, Bavaria. Professor at Göttingen and Würzburg. Awarded the 1919 Nobel Prize for Physics for his discovery of the Doppler effect in canal rays and the separation of spectral lines in an electric field.

13

Fluids in Motion

The harnessing of waterpower and the building of efficient steam turbines require knowledge of the behavior of fluids in motion. The designing of streamlined cars, trains, and airplanes is based on the study of problems involving fluids in motion—particularly air in motion. An understanding of the fundamental principles of flight and the operation of certain aircraft instruments follows from a logical extension of the principles of mechanics to fluids in motion.

We have seen that the properties of fluids at rest can be described by the simple concepts of pressure and density, by Archimedes' principle of buoyancy, and by Pascal's law of the transmission of pressure. When fluids are in motion, new properties become apparent. In predicting what happens we cannot always rely on our previous experience or intuition. But careful consideration shows that the phenomena of fluids in motion can be described in terms of the familiar principles of mechanics. The term "fluid" is applicable to both liquids and gases and is used in considering their common properties.

13-1 FLUID FLOW

Knowledge of the laws that govern fluid flow is important in providing for the distribution of water, gas, and oil in pipelines and the efficient transmission of energy in hydraulic machines. The rate of flow of a liquid through a pipe or channel is usually measured as the volume that passes a certain cross section per unit time, as cubic feet per second, liters per second, etc. If the average speed of the liquid at section S in Fig. 13-1 is v, the distance l through which the stream moves in time t is vt. This may be regarded as the length of an imaginary cylinder that has passed section S in time t. If A is the area of cross section, then the volume of the cylindrical section is $V = Al = Avt$ and the rate of flow of the liquid is given by

$$\frac{V}{t} = \frac{Avt}{t} = Av \qquad (1)$$

In a fluid at rest the pressure is the same at all points at the same elevation. This is no longer true if the fluid is moving. When water flows in

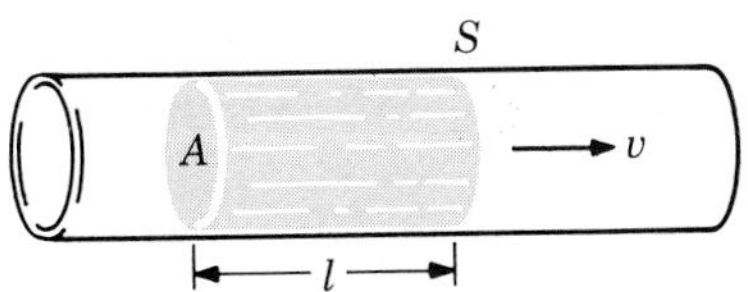

Figure 13-1
Rate of flow of liquid through a pipe.

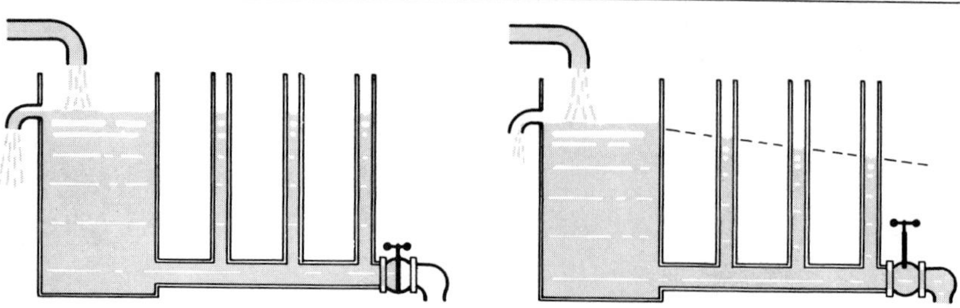

Figure 13-2
Friction causes a fall in pressure along a tube in which a liquid flows.

a uniform horizontal pipe, there is a fall in pressure along the pipe in the direction of flow. The reason for this fall in pressure is that force is required to maintain motion against friction. If the liquid is being accelerated, additional force is required.

When the valve of Fig. 13-2 is closed, water rises to the same level in each vertical tube. When the valve is opened slightly to permit a small rate of flow, the water level falls in each tube, indicating a progressive decrease of pressure along the pipe. The pressure drop is proportional to the rate of flow. Frictional effects are very important when water is distributed in city mains or when petroleum is transported long distances in pipelines. Pumping stations must be placed at intervals along such lines to maintain the flow.

13-2 STREAMLINES AND TUBES OF FLOW

We shall first consider steady flow. A fluid flowing in a pipe (Fig. 13-3) will have a certain velocity v_1 at a, a velocity v_2 at b, and so on. If, as time goes on, the velocity of whatever fluid particle happens to be at a is still v_1, that at b is still v_2, etc., then the flow is said to be *steady* and the line *abc* which represents the path followed by a particle is called a *streamline*. It represents the fixed path followed by an orderly procession of particles. In streamline flow, all particles passing through a also pass through b and c. This is not the case in turbulent flow. When the flow is unsteady or turbulent, there are eddies and whirlpools in the motion and the paths of the particles are continually changing.

It is convenient to imagine the whole region in which flow occurs to be divided into tubes. A *tube of flow* is a tube that follows in form the streamlines on its surface. It may be thought of as made up of a bundle of lines of flow. The fluid in a tube of flow remains in that tube. It is assumed that all particles passing a given cross section in a tube of flow have the same velocity. In regions where the streamlines are crowded together the speed is increased (Fig. 13-4), since from Eq. (1) $A_a v_a = A_b v_b$.

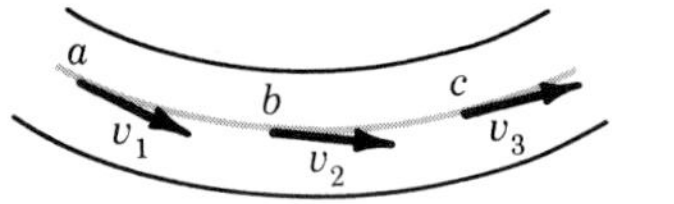

Figure 13-3
Steady flow. Path *abc* is a streamline.

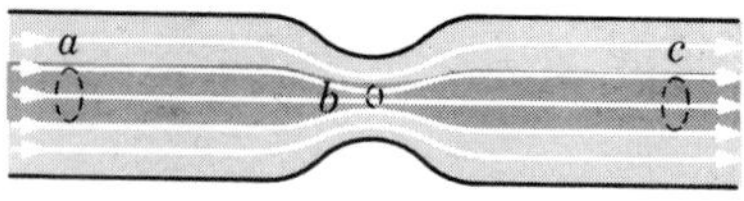

Figure 13-4
Steady flow. Volume *abc* is a tube of flow.

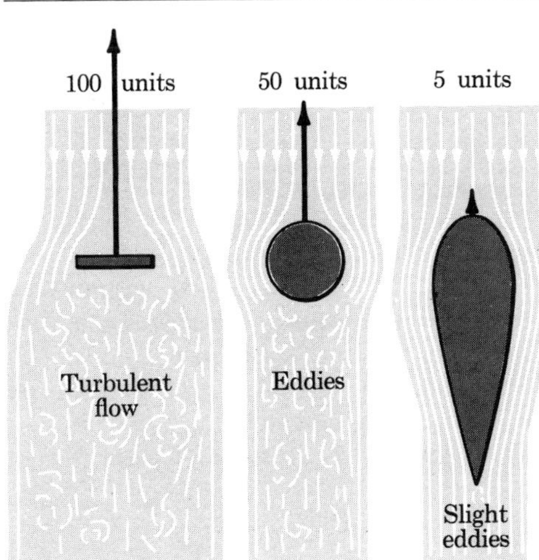

Figure 13-5
Comparative drag.

When a body moves through a fluid, turbulence increases the resistance offered by the fluid to the moving body. The streamlining of a car or a plane is intended to permit a steady flow of air past its surfaces so as to reduce the resistance or drag which turbulent flow would produce. Figure 13-5 shows a disk, a sphere, and a streamlined object of equal cross sections. They are placed in streams of fluid moving with the same speeds. The forces required to hold the objects stationary in the streams are indicated by the vectors. Note the reduction in force obtained by streamlining the object to reduce turbulence. A well-designed airplane may have a drag equivalent to that of a flat plate only 20 in square, moving broadside.

13-3 FLOW THROUGH A CONSTRICTION

When a liquid is flowing through a pipe of varying cross-sectional area (Fig. 13-6), there can be no accumulation between a and b, provided that the liquid is incompressible. Hence the mass of liquid passing through the cross section A_1 with speed v_1 must equal the mass passing in the same time t through cross section A_2 with speed v_2,

$$A_1 v_1 \rho t = A_2 v_2 \rho t \tag{2}$$

where ρ is the density of the liquid.

Two important consequences are immediately apparent from this equation. Since $A_1 v_1 = A_2 v_2$, it follows that the speed of flow in a pipe is greater in those regions where there is a constriction in the cross-sectional area A.

Furthermore, if as in Fig. 13-6 the speed is greater at b than at a, the liquid experiences an acceleration between a and b. This requires an accelerating force. This accelerating force can be present only if the pressure at a is greater than the pressure at b. We conclude that, in the steady flow of a liquid, the pressure is least where the speed is greatest.

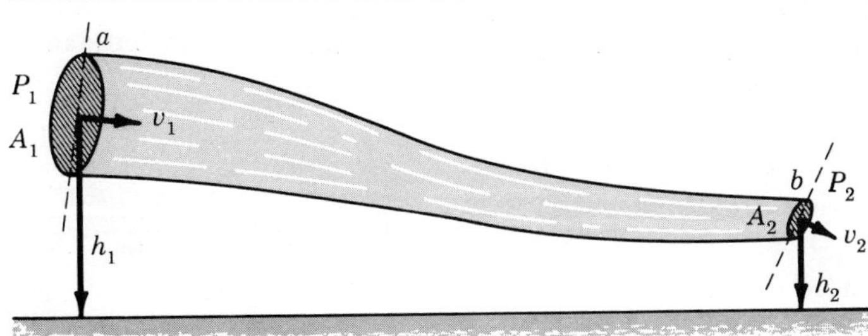

Figure 13-6
Liquid flow in a tube.

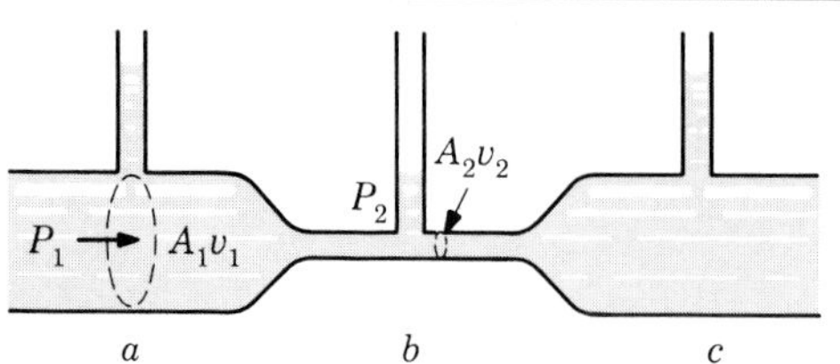

Figure 13-7
Flow through a constriction. Decrease in pressure accompanies increase in speed.

In Fig. 13-7 manometers connected to various locations show different pressures at an area of constriction. When water flows through a pipe that has a constriction, the water necessarily speeds up as it enters the narrow part of the tube, and there is a corresponding drop in pressure.

13-4 BERNOULLI'S THEOREM

The preceding description of the steady flow of a liquid can be put in more definite form by an application of the principle of conservation of energy to the incompressible liquid flowing between the two planes a and b of Fig. 13-6. In any time t, the volume V that flows through a is the same as that which flows through b. Since the pressure is different at the two ends, work is done on the liquid of an amount $P_1V - P_2V$ [since work $= Fs = (PA)(vt) = PV$]. The work done is equal to the change in energy (both potential and kinetic).

$$P_1V - P_2V = (mgh_2 - mgh_1) + (\tfrac{1}{2}mv_2^2 - \tfrac{1}{2}mv_1^2) \quad (3)$$

Since $V = m/\rho$,

$$P_1\frac{m}{\rho} - P_2\frac{m}{\rho} = (mgh_2 - mgh_1) + (\tfrac{1}{2}mv_2^2 - \tfrac{1}{2}mv_1^2)$$

Simplifying and rearranging terms to group initial terms on one side of the equation and final terms on the other gives

$$\frac{P_1}{\rho g} + h_1 + \frac{v_1^2}{2g} = \frac{P_2}{\rho g} + h_2 + \frac{v_2^2}{2g} \quad (4)$$

Each term in this form of Bernoulli's equation has the dimensions of a length. In the pressure-depth relation the depth h is frequently called the *head*. In analogy then, each term of Eq. (4) is called a head: $P/\rho g$, the pressure head; $v^2/2g$, the velocity head; and h, the elevation head.

Multiplication by ρg in Eq. (4) gives

$$P_1 + h_1\rho g + \frac{v_1^2\rho}{2} = P_2 + h_2\rho g + \frac{v_2^2\rho}{2} \quad (4a)$$

and one may state Bernoulli's theorem thus: *At any two points along a streamline in an ideal fluid in steady flow, the sum of the pressure, the potential energy per unit volume, and the kinetic energy per unit volume have the same value.*

Although Bernoulli's theorem is rigorously correct only for incompressible, nonviscous liquids, it is often applied to ordinary liquids with sufficient accuracy for many engineering purposes.

Example Water flows at the rate of 300 ft^3/min through an inclined pipe (Fig. 13-6). At a, where the diameter is 12 in, the pressure is 15 lb/in^2. What is the pressure at b, where the diameter is 6.0 in and the center of the pipe is 2.0 ft lower than at a?

$$A_1v_1 = A_2v_2 = \frac{300\ \text{ft}^3/\text{min}}{60\ \text{s/min}} = 5.0\ \text{ft}^3/\text{s}$$

$$\frac{A_1}{A_2} = \frac{v_2}{v_1} = \frac{\pi(6.0\ \text{in})^2}{\pi(3.0\ \text{in})^2} = 4.0$$

$$v_1 = \frac{5.0\ \text{ft}^3/\text{s}}{\pi(\frac{1}{2}\ \text{ft})^2} = 6.4\ \text{ft/s}$$

$$v_2 = 4v_1 = 26\ \text{ft/s}$$

$$P_1 = (15\ \text{lb/in}^2)(144\ \text{in}^2/\text{ft}^2) = 2,\bar{2}00\ \text{lb/ft}^2$$

For water, $D = 62.4$ lb/ft^3, and therefore $\rho = 1.94$ slugs/ft^3. From Eq. (4),

$$\frac{P_1}{\rho g} + h_1 + \frac{v_1^2}{2g} = \frac{P_2}{\rho g} + h_2 + \frac{v_2^2}{2g}$$

$$\begin{aligned} P_2 &= P_1 + \rho g(h_1 - h_2) + \frac{\rho}{2}(v_1^2 - v_2^2) \\ &= 2{,}\bar{2}00 \text{ lb/ft}^2 + (62.4 \text{ lb/ft}^3)(2.0 \text{ ft}) \\ &\quad + \frac{1.94 \text{ slugs/ft}^3}{2}[(6.4 - 26)^2 \text{ ft}^2/\text{s}^2] \\ &= 2{,}\bar{2}00 \text{ lb/ft}^2 + 1\bar{2}0 \text{ lb/ft}^2 - 6\bar{2}0 \text{ lb/ft}^2 \\ &= 1{,}\bar{7}00 \text{ lb/ft}^2 = 12 \text{ lb/in}^2 \end{aligned}$$

13-5 EFFECT OF FRICTION ON FLOW

When friction is present, some of the energy of a flowing stream is converted into heat. There is a gradual reduction in pressure along the direction of flow (Fig. 13-8). By analogy with the other terms in Eq. (4), the pressure reduction due to friction is treated as a *friction head.* Its effect at the outlet is to reduce the velocity head and the rate of discharge.

13-6 VISCOSITY

Viscosity is that property of a fluid, its internal friction, which causes it to resist flow. Consider a layer of liquid in a shallow pan on the surface of which a flat plate A is placed (Fig. 13-9). A force F is required to maintain the plate at a constant speed v with respect to the other surface B.

On the surface of each solid, A and B, there will be a layer of liquid that adheres to the solid and has zero speed with respect to it. The next layer of liquid moves slowly over the first, the third layer moves slowly over the second, and so on. This distribution of speeds results in a continual deformation of the liquid. The portion of the liquid C, which is cubical in shape at one instant, is rhomboidal a moment later at R.

If the thickness s of the fluid between the surfaces is increased, application of the same force produces a greater speed: $v \propto s$. If the area A of the plate is increased, there is a corresponding decrease in speed: $v \propto 1/A$. An increase in the force produces a proportional increase in speed: $v \propto F$. Thus the speed is proportional to F, to s, and also to $1/A$. Hence we can write the equation

$$v = \frac{Fs}{\eta A}$$

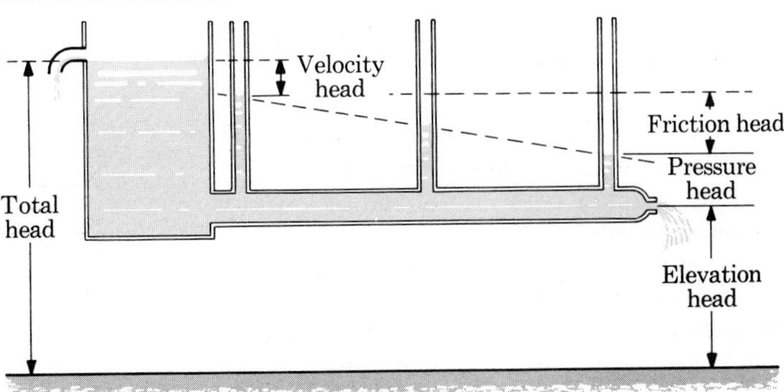

Figure 13-8
As the friction head increases, the pressure head decreases.

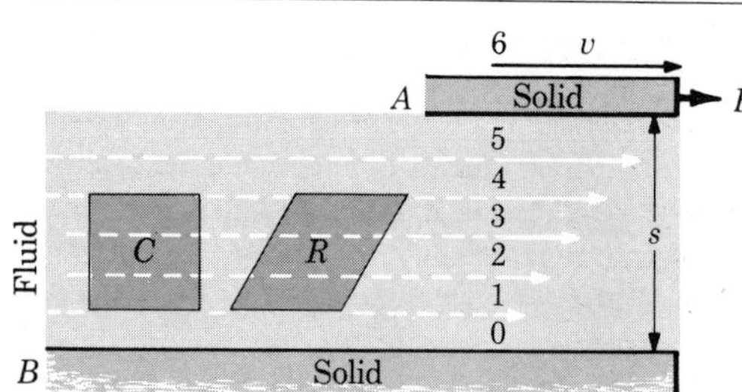

Figure 13-9
Fluid friction.

or

$$\eta = \frac{Fs}{Av} \qquad (5)$$

where the factor η (eta) depends only on the fluid and its temperature. This equation defines the *coefficient of viscosity* η of the liquid.

The absolute cgs unit of the coefficient of viscosity is called the *poise* (P) after Poiseuille, a pioneer experimenter in this field.

From Eq. (5),

$$\eta = \frac{Fs}{Av}$$

$$1\ \text{P} = \frac{1\ \text{dyn} \times 1\ \text{cm}}{1\ \text{cm}^2 \times 1\ \text{cm/s}}$$

$$= 1\ \text{dyn}\cdot\text{s/cm}^2$$

A coefficient of viscosity of 1 P is one that requires a tangential force of 1 dyn for each square centimeter of surface to maintain a relative velocity of 1 cm/s between two planes separated by a layer of the fluid 1 cm thick.

The centipoise ($\frac{1}{100}$ P) is the unit of viscosity commonly used.

In setting up this definition of coefficient of viscosity, we have chosen a case in which the application of forces to a fluid produces differences of speed between adjacent thin layers (laminae) within the fluid as indicated in Fig. 13-9. It is this laminar flow which is characterized by the proportionality between force and speed, which is expressed by Eq. (5). In the case of *turbulent* flow the lines of flow are complex and changing, and Eq. (5) is not applicable.

The viscosity of liquids decreases with increase in temperature. A liquid that flows at low temperature as slowly as the proverbial molasses in January may pour freely at higher temperature. Lubricating oil may fail to form a protective film at low temperatures. Hence, in starting a car on a cold day, it is wise to allow the engine to idle for a time until the oil is warmed. The viscosities of gases, unlike those of liquids, increase with increase in temperature.

13-7 NATURE OF VISCOSITY

The internal friction of liquids is attributed to the cohesive forces between molecules. Our knowledge of the details of this mechanism is far from complete, but the picture of liquid viscosity just described does explain qualitatively the experimentally observed facts. On this basis, for example, one would expect a decrease in the viscosity of a liquid with increase in temperature. This is observed experimentally.

In the case of gases, whose molecules have relatively large separations, cohesive forces are small, and some other mechanism must be sought for internal friction. One supposes that in the laminar flow of a gas there is a continual migration of molecules from one layer to another. Molecules diffuse from a fast-moving layer to a

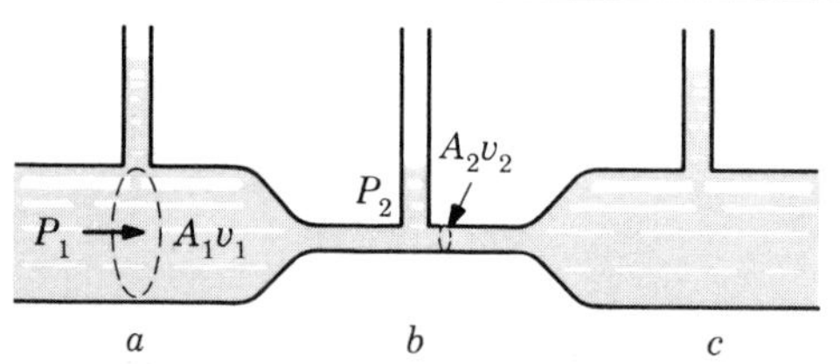

Figure 13-10
Flow through a constriction. Decrease in pressure accompanies increase in speed.

slower-moving layer (5 to 4, Fig. 13-9) and from the slower-moving layer to the faster. Thus each layer exerts a drag on the other. The amount of this drag depends on the mass of the molecules and their speeds and hence on the product *mv*. This description of gas viscosity accounts for the fact that an increase in temperature, which increases molecular speeds, results in an increase in the viscosity of a gas.

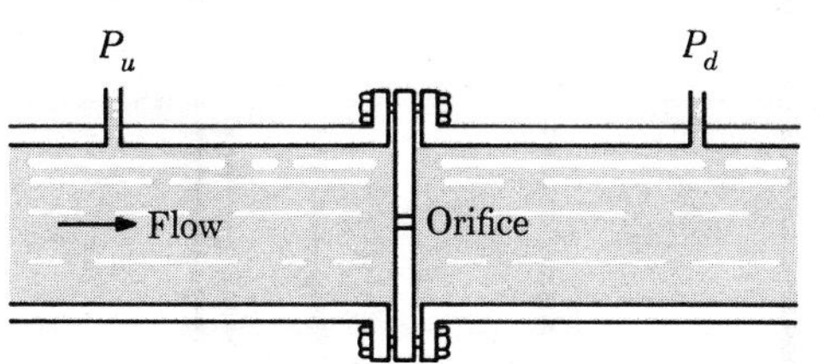

Figure 13-11
An orifice-type flow gauge.

13-8 PRESSURE AND SPEED

When water flows through a pipe that has a constriction (Fig. 13-10), the water necessarily speeds up as it enters the constriction, and as we have seen there is a decrease in pressure between *a* and *b*. Consider the case of a horizontal pipe. Applying Bernoulli's theorem, we get

$$\frac{P_1}{\rho} + \frac{v_1^2}{2} = \frac{P_2}{\rho} + \frac{v_2^2}{2} \tag{6}$$

Combining Eq. (6) with Eq. (2) gives

$$P_1 - P_2 = \frac{\rho}{2}(v_2^2 - v_1^2)$$

$$= \frac{\rho v_1^2}{2}\left(\frac{A_1^2}{A_2^2} - 1\right) \tag{7}$$

Equation (7) may be used to determine the rate of flow of the liquid, i.e., the volume per second of liquid passing *a*.

A tube similar to that of Fig. 13-10 having a constricted throat section between larger diameter inlet and outlet sections is called a *venturi tube*. A meter that utilizes such a tube and is calibrated by the relation of Eq. (7) is called a *venturi flowmeter*.

Another important application of Bernoulli's principle is the orifice-type flow gauge. A polished disk with an accurately sized hole is placed in the stream whose flow rate is to be measured (Fig. 13-11). With a manometer or other gauge the pressures are measured on the upstream and downstream sides of the orifice. From these measurements and the density and viscosity of the fluid its rate of flow is computed.

13-9 AIRFOILS

Imagine that the walls of a venturi tube are moved apart as suggested in Fig. 13-13. The result will be as shown at the right in Fig. 13-13. The nearby streamline follows the curved surface closely; at increasing distances above the surface the streamlines are less curved, and at a distance equal to four times the chord length the curvature

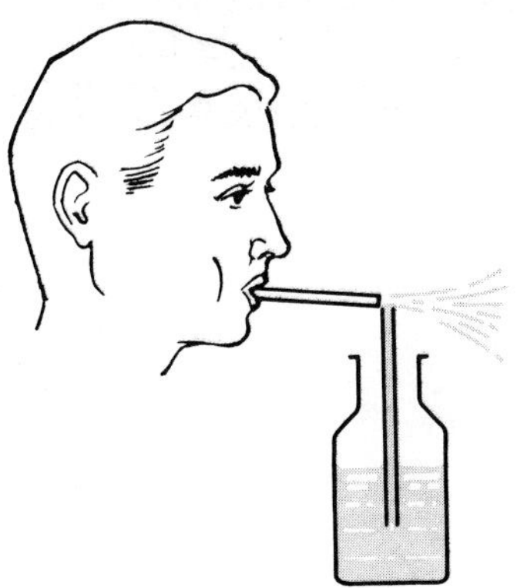

Figure 13-12
Bernoulli's principle applied to a spray jet.

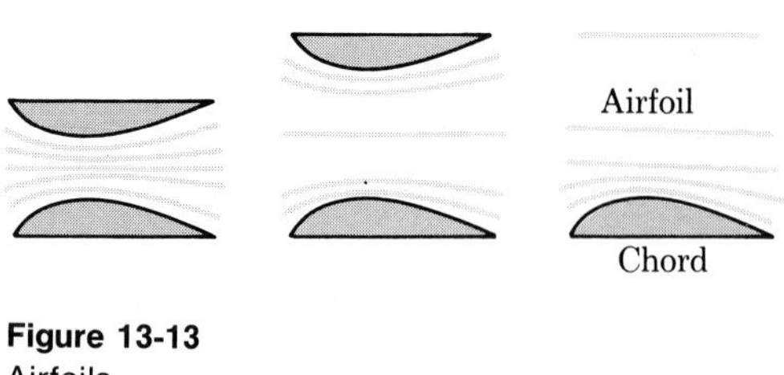

Figure 13-13
Airfoils.

is negligible. The increase in speed of the air on the upper curved surface results in a decreased pressure at that surface. Because the pressure below the section is greater than that above it, the section experiences a lift. Any such surface designed to obtain reacting force from the air through which it moves is called an *airfoil*. In this sense airplane wings, ailerons, rudders, and propellers are all airfoils.

When an airfoil is inclined upward a few degrees with respect to the wind direction, air will be deflected from the lower surface and the reacting force will produce a pressure at the lower surface greater than atmospheric by an amount indicated by the arrows in Fig. 13-14. At the upper surface the pressure is less than atmospheric, according to Bernoulli's principle. Both effects, but chiefly the second, constitute the lifting force L. Vectors representing the weight W of the wing, the drag D due to air friction, the propeller thrust, and the lift add to form a closed polygon (zero resultant) when the wing is moving with constant velocity.

13-10 DISCHARGE FROM AN ORIFICE

When liquid escapes from a small sharp-edged orifice in a vessel (Fig. 13-15), the outgoing liquid gains kinetic energy at the expense of the potential energy of the remaining liquid. In the absence of friction the kinetic energy, if changed back to potential energy, should be sufficient to raise the escaping liquid to the level of the surface in the vessel.

Liquid of mass m in leaving the orifice with speed v has kinetic energy $\frac{1}{2}mv^2$. The potential energy of this same liquid when at the upper surface of the liquid is mgh. Equating these, we get

$$\tfrac{1}{2}mv^2 = mgh \qquad \text{or} \qquad v^2 = 2gh \tag{8}$$

This relation is known as *Torricelli's theorem*.

Since the pressure P at a depth h below the liquid surface is ρgh, Eq. (8) can be expressed as

$$v^2 = \frac{2P}{\rho} \tag{9}$$

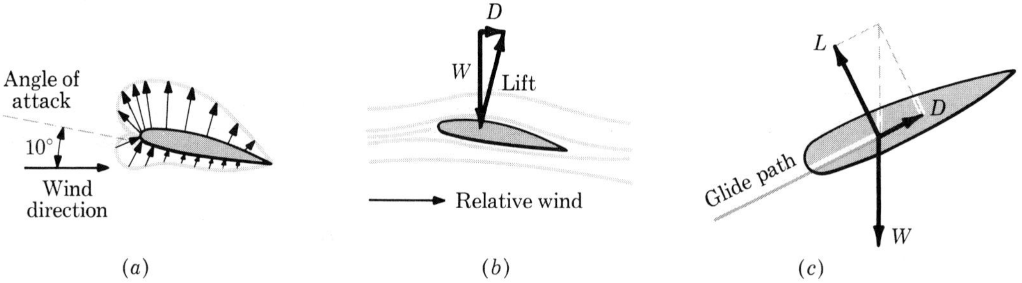

Figure 13-14
Lift produced by air flow past an airfoil.

This expression is likewise applicable to the escape of a gas at pressure P.

These results are only approximate. During the flow, the streamlines from all parts of the vessel crowd together at the orifice, and the stream contracts somewhat in cross section outside the orifice, owing to the inertia of the liquid as it follows the curved streamlines. Because of the converging of the streamlines as they approach the orifice, the cross section of the stream continues to diminish for a short distance outside the tank. For a sharp-edged circular opening, the area of the vena contracta is about 65 percent as great as the area of the orifice. Contraction can be eliminated by the use of an orifice whose walls curve to fit the streamlines (Fig. 13-15*b*). The effect of frictional resistance is to decrease the rate of discharge to about 0.6 of that given by Eqs. (8) and (9).

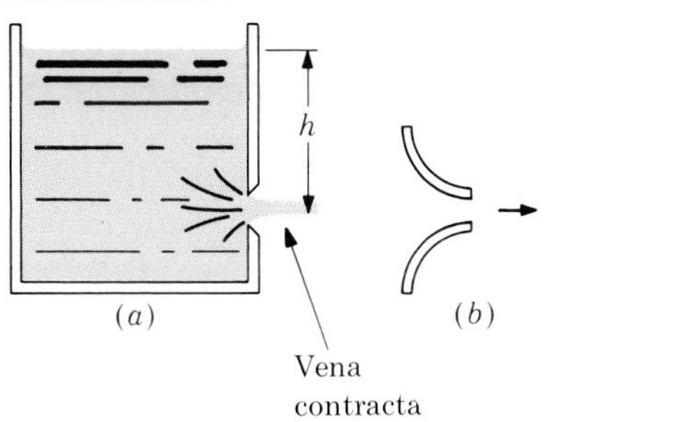

Figure 13-15
Discharge from an orifice.

Example A tank containing water has an orifice of 8.0 cm² in one vertical side, 3.0 m below the free surface level in the tank. (*a*) Find the speed of discharge, assuming that there is no wasted energy. (*b*) Assuming that the cross section of the stream contracts to 0.64 of the area of the circular orifice, find the flow.

From Eq. (8),

$$v = \sqrt{2gh}$$

$$= \sqrt{2 \times 9.8\ \text{m/s}^2 \times 3.0\ \text{m}}$$

$$= 7.66\ \text{m/s}$$

$$\text{Rate of flow} = (\text{area of cross section})(\text{speed})$$

$$= (0.64)(8.0\ \text{cm}^2)(7.66\ \text{m/s})$$

$$= (0.64)\left(\frac{8.0}{10^4}\ \text{m}^2\right)(7.66 \times 60\ \text{m/min})$$

$$= 0.235\ \text{m}^3/\text{min}$$

Example Life *can* be beautiful! You and a movie starlet are shipwrecked on a South Sea Island. By chance, she finds a tin can of square cross section of 40 × 40 cm and height 70 cm lying on the beach. You decide to make a shower out of it by punching 20 holes, each of 0.10 cm² area, into the bottom of the can. If she likes to stay in the shower for 5.0 min, does the can have sufficient capacity? Ignore the vena contracta effect.

Let us consider that the 0.1-cm²-area holes are combined to make one large hole 20 times as large in area. That is, 2.0 cm² in area. The water that comes out will generate a cylinder of water which has a volume of As, where s equals the length of the cylinder and A the cross-sectional area.

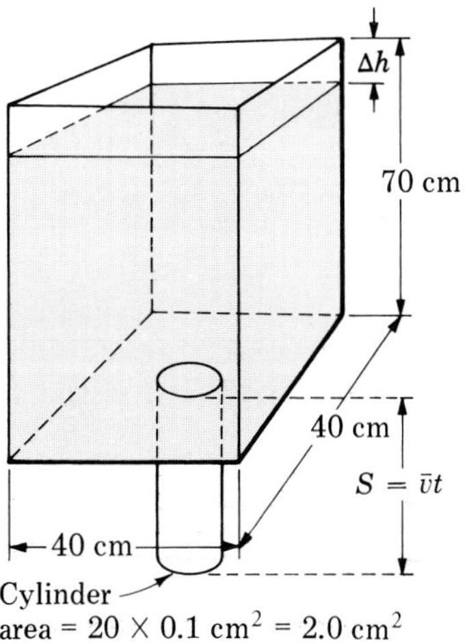

Figure 13-16
Shower for the starlet problem.

The speed with which the water leaves the hole is given by $v = \sqrt{2gh}$. It should be noted that the speed varies with the height of the liquid in a similar way that the speed of a body falling freely varies with the distance fallen. Let v_1 equal the speed when $h = 70$ cm and v_2 equal the speed when $h = 0$; $v_2 = 0$, and

$$\bar{v} = \frac{v_1 + v_2}{2} = \frac{\sqrt{2gh}}{2}$$

since the change in speed is constant.[1]

The total volume lost from the tank in time t is the area $\times$ height $= (1600 \text{ cm}^2)(\text{H})$, where H is the total height of the tank. The volume leaving the bottom in the same time t is $a\bar{v}t$, where a is the area of the column of water generated out the bottom of the tank. The two volumes must be equal, therefore

$$(1600 \text{ cm}^2)(\text{H}) = a\bar{v}t$$

$$(1600 \text{ cm}^2)(\text{H}) = \frac{2.0 \text{ cm}^2(\sqrt{2g\text{H}})(t)}{2}$$

$$= (\sqrt{2g\text{H}})(t)$$

when $\text{H} = 70$ cm, $t = (1600 \text{ cm}^2 \times 70 \text{ cm})/\sqrt{(1{,}960 \text{ cm/s}^2)(70 \text{ cm})} = 303$ s or 5 min and 3 s. Therefore, the starlet will be able to take her 5-min shower.

Note: The starlet problem provides an excellent illustration of the application of the calculus to solve physics problems. For those students who are familiar with the calculus, the problem has been solved here by the use of that form of mathematics. Let $h =$ depth of water at time t. The water which flows out generates a cylinder of height ($v\ dt$) centimeter and having a base of 2.0 cm². The volume of this cylinder is $2.0\ v\ dt$. Since $v = \sqrt{2gh}$,

$$\text{vol}_{\text{cyl}} = 2.0\ \sqrt{2gh}\ dt$$

Let $-dh$ represent the corresponding drop in the surface level. Then, the loss in volume equals $-A\ dh$ or $-(40)(40)\ dh$ or $-1{,}600dh$. Then, since the volume of the cylinder equals the loss of volume in the can,

$$2\ \sqrt{2gh}\ dt = -1{,}600\ dh$$

and
$$dt = \left(\frac{-1{,}600}{2}\right)\left(\frac{dh}{\sqrt{2gh}}\right) = \frac{-800\ dh}{\sqrt{1960\ h}}$$

$$= \frac{-800}{44.2}\frac{dh}{\sqrt{h}} = -18.1\frac{dh}{\sqrt{h}}$$

Integrating the expression,

$$\int dt = \int -18.1\frac{dh}{\sqrt{h}}$$

we get

$$t = (-18.1)(2)\sqrt{h} + C$$

$$= -36.2\sqrt{h} + C$$

At time $t = 0$, $h = 70$ cm; and substituting in the equation,

$$0 = -36.2\sqrt{70} + C$$

$$C = +303$$

Thus,
$$t = -36.2\sqrt{h} + 303$$

When the can is empty, $h = 0$ and $t = 303$ s or 5 min plus 3 s. So we see that the calculus computations show that the starlet will have sufficient time to take her shower.

13-11 TURBINES

In a turbine the direction of flow of water, steam, or burning gases is changed by blades or buckets on a wheel, and the force resulting from the change in momentum is used to rotate the wheel. For efficient operation turbine blades should be so formed that the impelling fluid flows smoothly over them. Also, the fluid should be discharged from the turbine with as small a speed as possible. Figure 13-17 shows how this is accomplished in

[1] If a graph is constructed in which the speed of the water coming out the bottom of the tank is plotted versus the time, the result is a straight line, that is, the acceleration is constant.

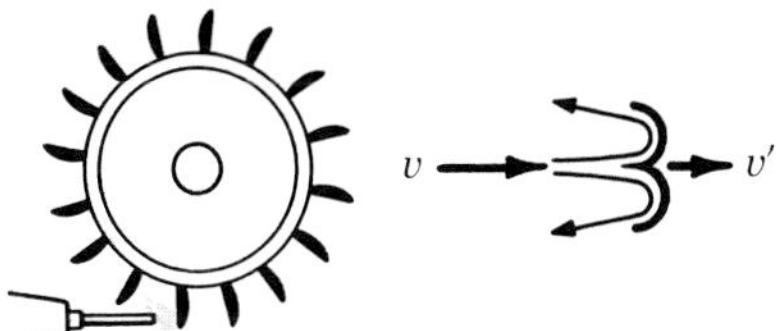

Figure 13-17
Action in a Pelton wheel.

a Pelton wheel. If the velocities of the jet and bucket are v and v', respectively, and if v' is half of v, then the velocity of the outgoing water relative to ground will be zero and the whole of the kinetic energy of the water in the jet is converted into useful work. In practice, 70 to 90 percent of this ideal work may be obtained. Figure 13-18 shows a steam turbine in which the direction of flow of high-speed steam is repeatedly changed by passage past alternate stator and rotor blades until much of its energy has been delivered to the rotor.

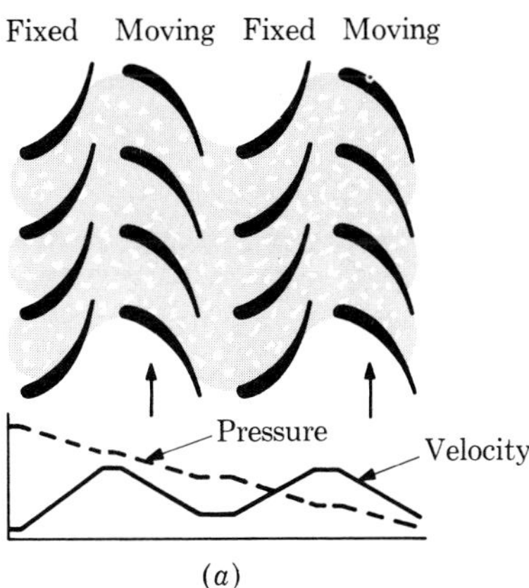

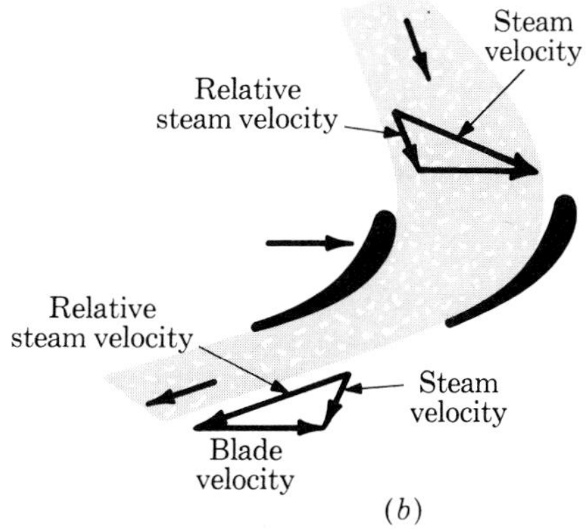

Figure 13-18
The steam turbine. (*a*) In a reaction turbine steam expands in alternate rings of fixed and moving blades that act as jets. The diagram shows two stages. Steam pressure falls in both fixed and moving blades. As the steam expands, the direction of its motion is changed and the reaction rotates the moving blades. Steam velocity referred to the earth increases in the fixed blades but decreases in the moving blades. (*b*) In the vector diagram, the horizontal arrows represent the velocity of the blades relative to earth. The flow lines are those which would be seen by an observer on the moving blades. The blades are shaped so that the velocity of the steam *relative to the moving blades* is parallel to the blade surfaces on entering and on leaving.

SUMMARY

The rate of flow of a liquid through a pipe is usually measured as the volume that passes a certain cross section per unit time.

$$\frac{V}{t} = Av$$

In the steady flow of a liquid the pressure is least where the speed is greatest.

The steady flow of a liquid can be described by *Bernoulli's theorem* in the statement that the sum of the pressure head, the elevation head, and the velocity head remains constant.

$$\frac{P_1}{\rho g} + h_1 + \frac{v_1^2}{2g} = \frac{P_2}{\rho g} + h_2 + \frac{v_2^2}{2g}$$

Bernoulli's principle is applied in a venturi flowmeter to measure the rate of flow of a fluid.

Fluid friction is measured in terms of the

coefficient of viscosity η, which is defined as the ratio of the tangential force per unit area of surface (F/A) to the velocity gradient (v/s) between two planes of fluid in laminar flow.

$$\eta = \frac{F/A}{v/s} = \frac{Fs}{Av}$$

An airfoil is any surface designed to obtain a reacting force from the air through which it moves.

The speed with which a liquid escapes from a vessel through an orifice is given by *Torricelli's theorem*

$$v = \sqrt{2gh}$$

In a turbine the change in momentum of a fluid as it is deflected by blades causes the shaft to rotate.

Questions

1 When an object, such as a pearl, is dropped into a dense liquid it quickly acquires a uniform, nonaccelerated motion downward. Why does this happen?

2 Why does the flow of water from a faucet decrease when someone opens another valve in the same building?

3 Why is a water tower generally placed on the top of a hill? If the hill is 50 m high, can it supply water to the top floor of a 100-m-tall office building? If not, how does water get to that floor?

4 Why will marbles not fall vertically in a cylinder of water?

5 How is air drawn into a gas burner?

6 If a person is standing near a fast-moving railroad train, is there danger that he will fall toward it? Explain.

7 Air is blown between two Ping-Pong balls suspended independently so that they are very close to each other; describe and explain what happens.

8 A Ping-Pong ball will remain suspended in the jet of air from a vacuum cleaner which is directed vertically. Explain with the aid of diagrams why this happens.

9 A playing card is placed upon a spool of thread so that it covers the hole. If air is blown through the spool hole from the opposite end, the card will not blow off the spool, even though the card may be facing downward toward the ground. Explain why this happens.

10 How does the speed of flow of liquid from a small hole in a tank depend upon the depth of the hole below the liquid surface?

11 A jet of water strikes a surface and the water may be assumed to run off with a negligible speed. Derive an expression for the force exerted by the jet.

12 Show by diagram the nature of the streams of water coming from holes punched into the sides of a container filled with water if the holes are $\frac{1}{4}$, $\frac{1}{2}$, and $\frac{3}{4}$ of the way up the sides of the container.

13 A TV commercial shows a hole being punched in the side of a can of a certain brand of antifreeze, and after a period the flow from the hole stops. What do you think causes this to happen?

14 It requires much greater gas consumption for an automobile to travel at high speeds than at low speeds due to a law describing the behavior of an object as it moves through a fluid. What is this law? Explain it.

15 New high-speed aircraft have wings which are very thin and almost flat. What do you think gives them sufficient lift to fly?

16 Slow-flying airplanes have very curved upper surfaces to their wings. Why is this necessary?

17 How will an increase in altitude affect the difference in pressure on the upper and lower surfaces of an airplane wing? Justify your answer.

18 A stretched elastic band is wrapped around a Ping-Pong ball. The ball is snapped forward by the elastic band, which also causes the top of the ball to spin in the forward direction. Will the ball "rise" or "drop"?

19 Draw a diagram to show what makes a baseball curve when thrown in a certain manner.

20 Explain how you would pitch a ball for an incurve; for a drop.

21 Describe how a spray gun or atomizer operates.

22 A tube can be used to siphon liquid from a container. Discuss how a siphon operates.

23 Two tall cylinders are filled with liquid to the same height. One contains water, the other mercury. In which cylinder is the pressure greater at a given depth? If small holes are opened in the wall halfway up the cylinders, from which will the liquid emerge at higher speed? Which jet will travel farther before striking the tabletop?

24 Explain why a layer of oil helps two flat metal objects to slide over each other with relative ease.

25 Two plates of glass are ground to a high degree and put in contact with each other. Will they be easy or difficult to separate? Why?

26 Water when falling from an opening in the bottom of a container will remain in a "solid" column for a while and then will break up into droplets. Why does the column break into the droplets?

27 In the situation referred to in Question 26, list as many factors as possible which would cause the length of the water column to vary before "beading" up.

Problems

1 A liquid flows through a tube having a diameter of 15.0 m. If its average speed is 15.0 m/min, find its rate of flow.

2 How much higher than a faucet must the surface of the water in a reservoir be if the pressure at the faucet is 60 lb/in^2? *Ans.* 138.5 ft.

3 Liquid flows in a pipe of inside diameter 1.5 in at an average speed of 3.5 ft/s. What is the rate of flow in gallons per minute? (1 gal = 231 in^3.)

4 Water flows from a pipe of 2.0 cm internal diameter at the rate of 8.0 l/min. What is the speed of the water in the pipe?

Ans. 42 cm/s.

5 What is the kinetic energy of each cubic meter of water in a stream that is moving with a speed of 20 m/s?

6 It has been estimated that a man's heart is about 33 cm below his brain. Considering the density of human blood to be 1.1×10^3 kg/m^3, what pressure must the heart produce to supply blood to the brain? *Ans.* 3.59×10^3 N/m^2.

7 Liquid of specific gravity 0.90 flows in a horizontal tube 6.0 cm in diameter. In a section where the tube is constricted to 4.2 cm in diameter, the liquid pressure is less than that in the main tube by 16,000 dyn/cm^2. Calculate the speed of the liquid in the tube. *Ans.* 110 cm/s.

8 Water flows normally through a 12-in pipe at the rate of 80 gal/min. At a point in the pipe an 8-in-diameter constriction exists. Find the velocity in feet per second at this constriction. (1 gal = 231 in^3 and 1 ft^3 = 1,728 in^3.)

Ans. 0.51 ft/s.

9 How much work is done in forcing 50 ft^3 of water through a $\frac{1}{2}$-in pipe if the difference in pressure at the two ends of the pipe is 15 lb/in^2?

10 A horizontal pipe has a diameter of 12 in and a constriction having a diameter of 8 in. If the pressure within the pipe is 30 lb/in^2 and the water flows at 16 ft/s, find the velocity and pressure at the constriction. *Ans.* 36 ft/s; 23.0 lb/in^2.

11 A horizontal pipe of cross section 4.0 in^2 has a constriction of cross section 1.0 in^2. Gasoline (weight-density 42 lb/ft^3) flows with a speed of 6.0 ft/s in the large pipe, where the pressure is 10.0 lb/in^2. Find (*a*) the speed and (*b*) the pressure in the constriction.

12 Find the terminal velocity of a copper pellet 4 mm in diameter falling in a beaker of glycerin. Density of copper is 8.9 g/cm^3; density of glycerin is 1.3 g/cm^3. Viscosity of glycerin is about 8.3 P. Hint: use the equation, $v_T = (219)(r^2g/n)(\rho - \rho')$. *Ans.* 8 cm/s.

13 Water flows steadily through a pipe at the rate of 64 ft^3/min. A pressure gauge placed on a section of the pipe where the diameter is 4.0 in reads 16 lb/in^2. Determine the pressure in a section of the pipe where the diameter is constricted to 2.0 in.

14 Calculate the maximum speed with which water at atmospheric pressure can flow past an obstacle without breaking into turbulent flow. Use Bernoulli's principle and neglect viscous drag. *Ans.* 14 m/s.

15 Water flowing at 1.0 ft/s in a pipe passes into a constriction whose area is one-tenth the normal pipe area. What is the decrease in water pressure in the constriction? The weight-density of water is 62.4 lb/ft^3.

16 In a wind-tunnel experiment the pressure at the upper surface of an airfoil is found to be 12.95 lb/in^2, while the pressure at the lower surface is 13.05 lb/in^2. What is the lifting force of a wing of this design if it has a span of 24 ft and a width of 5.0 ft? *Ans.* 1,700 lb.

17 An airplane weighing 800 lb has a wing area of 120 ft^2. What difference in pressure on the two sides of the wing is required to sustain the plane in level flight?

18 Water flows through a 2-in opening in the side of a tank. If the opening is 8 ft below the surface of the water, what volume of water would escape from the tank per minute?

Ans. 29.7 ft^3/min.

19 Water in a storage tank stands 4.0 m above the level of a valve in the side of the tank. (*a*) With what speed will water come out of the valve if friction is negligible? (*b*) To what height will this water rise if the opening in the outlet tube is directed upward?

20 A horizontal stream of water leaves an opening in the side of a tank. If the opening is 1 m above the ground and the stream hits the ground 2 m away, what is the speed of the water as it leaves the tank and what gauge pressure pushes on it? *Ans.* 4.4 m/s; 9.85×10^3 N/m^2.

21 A stream of water escapes from a hole 2.0 m above the base of a standpipe and strikes the ground at a horizontal distance of 15.0 m. How far below the surface of the water is the hole?

22 Water flows through an opening at the bottom of a tank at a rate of 4 ft^3/min. The water in the tank is 24 ft deep. If 16 lb/in^2 added pressure is applied to the surface, at what rate does the water leave the tank?

Ans. 6.39 ft^3/min.

23 A 1.0-kg pail will hold 10.0 kg of water. Water from a faucet fills the pail in 12.0 s. At the instant when the pail is half full, the scales on which it is supported read 6.5 kg. What is the velocity of the flowing water at that instant, on the assumption that no splashing takes place?

24 Water is being discharged through a circular orifice having a diameter of 5 cm. What is the speed of discharge and the rate of flow if the orifice is 3 m below the surface of the water?

Ans. 7.66 m/s; 15×10^{-3} m^3/s.

25 Castle Geyser at Yellowstone shoots a spire of water 76.0 m into the air. By how much must the pressure at its base exceed atmospheric pressure?

26 What is the velocity with which water escapes through an opening in the side of a tank if the pressure at that point is 2,000 N/m^2 above the pressure of the atmosphere; and if the pressure is 30 lb/in^2 above atmospheric pressure?

Ans. 2 m/s; 66.6 ft/s.

27 It is desired to refuel an airplane at the rate of 50 gal/min (1.0 ft^3 = 7.5 gal). The fuel line is a 3.0-in hose connected to a pump 3.0 ft above ground; a 2.0-in nozzle delivers the gasoline (specific gravity 0.72) to the airplane 15.0 ft above the ground. Find (*a*) the speed of the fuel at the nozzle, (*b*) the speed of the fuel in the line near the pump, and (*c*) the pressure in the line near the pump.

28 The pressure at an opening in the side of a tank of water is 40 lb/in^2 greater than atmospheric pressure. What will the velocity of the water be as it leaves the opening?

Ans. 77 ft/s.

29 A tank is filled to a depth of 3.0 m with salt water having specific gravity 1.10. If there is a small hole in the side of the tank 0.5 m above its bottom, with what speed will water flow out?

30 Water is flowing through a pipe at 120 ft^3/min. A gauge located where the pipe has a diameter of 4 in reads 50 lb/in^2. What will a second gauge placed where the pipe is 3 in in diameter read if both gauges are at the same horizontal level? *Ans.* 42.4 lb/in^2.

31 A turbine is driven by water which falls from a height of 60 ft. If the flow rate is 480 ft^3/min, determine the maximum power that can be produced.

32 If the second gauge in Prob. 30 is placed 5 ft

lower than the first gauge, what pressure will be read? *Ans.* 44.6 lb/in^2.

33 The speed of a vacuum pump, the volume of gas it will extract from a container per unit time, is nearly independent of pressure over a wide range of pressures. When a pump with a speed of 10 l/min is connected to a closed system containing air at an initial pressure of 76.0 cm Hg, the pressure is observed to drop to 38.0 cm Hg in 6.00 s. Calculate the approximate time required to reduce the pressure to 0.100 mm Hg.

Charles Edouard Guillaume, 1861–1938

Born in Fleurier, Switzerland. Director of the International Bureau of Weights and Measures, Sèvres. Awarded the 1920 Nobel Prize for Physics for his discovery of the anomalies of nickel-steel alloys (including Invar) and their importance in the physics of precision.

Albert Einstein, 1879–1955

Born in Ulm, Würtemberg. Director of the Kaiser Wilhelm Institute for Physics, Berlin. Life member of the Institute for Advanced Study, Princeton, from 1933. Awarded the 1921 Nobel Prize for Physics for his attainments in mathematical physics and especially for his discovery of the law of the photoelectric effect.

14

Molecular Theory of Matter

In our discussion of the properties of liquids and those of gases we assumed that the fluids are composed of small particles called *molecules* and that liquids and gases differ in the spacing and forces between the molecules. These assumptions are in accordance with present ideas of the structure of matter.

Most of the conclusions that we reach regarding the molecular nature of matter are the result of indirect observations, since the molecules are much too small to see and in all except a few experiments the effects of individual molecules are too small to observe. From observation of the effects of *large groups* of molecules, we make up a picture or theory of the arrangement and behavior of the molecules. We then *test by experiment* the results predicted by the theory. Observation can be made only on large numbers of molecules at one time or on the average effects of individual molecules. Hence only *average* effects are subject to observation. In short, we observe macroscopic events and attempt to draw conclusions about microscopic events.

14-1 BUILDING A MODEL OF THE MOLECULAR NATURE OF MATTER

In building a model to understand the molecular nature of matter, we start with a set of assumptions. In model building the assumptions made or limits established are considered valid, though they may not seem to be immediately so, if the results predicted are verified by experiment. The fundamental assumptions we make in building this model include the following:

1 We assume that all matter is composed of distinct particles called molecules. In the case of a pure substance, these molecules are alike in all respects. The *chemical* properties, such as the ability for atoms of various elements to combine or interact with one another, are determined by the character of the molecules. The *physical* properties depend upon the forces the molecules exert on each other and the relation of these forces to the distances between molecules. Also, the motion of the molecules must be taken into account.

2 In a gas, we assume that the molecules obey Newton's laws of motion. Except in very special circumstances, we assume that collisions between molecules are perfectly elastic, i.e., that both momentum and kinetic energy are unchanged (for pairs of molecules in collision).

3 It is assumed further that above a certain reference level of internal energy of the body, the molecules are continually in motion. The average speed is dependent upon the temperature, a property which determines the direction of flow of heat energy between an object and its surroundings. When the temperature is constant, the average speed is constant. If one looks at a solid, it is not possible to see the molecules in motion; hence it may be difficult to accept this basic assumption. It is not surprising that we are not able to observe this motion when it is realized that these molecules are going very rapidly, are traveling very short distances, and are reversing or changing directions so quickly that any motion would blur out of our vision. For example, it has been determined that a molecule of hydrogen gas at 0°C has an average speed of 170,000 cm/s, travels 1.6×10^{-5} cm between collisions (mean free path) and is involved in 10 billion (1×10^{10}) collisions per second. Similarly, oxygen, which is more dense and moves more slowly, still has an average speed at 0°C of 43,000 cm/s, a mean free path of 0.9×10^{-5} cm and undergoes 4.5 billion (4.5×10^{9}) collisions per second.

As impossible as it is to see this motion, these movements can be amplified to discernible levels by molecular "pumping," which is the basis upon which masers and lasers work.

The model we build must be based upon our observations. From observations made of the physical properties of solids, liquids, and gases, we can state some conclusions about the forces which molecules exert upon each other. In order to pull a solid apart, it is necessary to exert rather large forces of tension. Hence, there must be forces of attraction between the molecules when they are at the distances of separation existing in solids. Much smaller forces are required to separate parts of liquids, and in gases the molecules if left to themselves separate without apparent external forces. These facts indicate that as the distance between molecules increases, the force of attraction decreases rapidly. On the other hand, we find that very large forces of compression are necessary to reduce the volume of a solid or a liquid. We must therefore conclude that when the distance between the molecules is very small, there are forces of repulsion, and these forces increase very rapidly as the distance between the molecules decreases. The forces must vary in a manner something like that shown in Fig. 14-1. As the molecules approach each other from distances which are large compared with their own dimensions, the forces exerted on one another are at first extremely small; then on closer approach a force of attraction develops

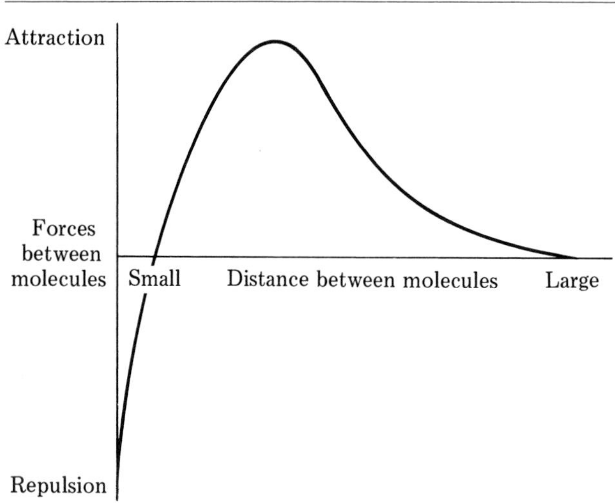

Figure 14-1
Variation of attractive force with distance between molecules.

which reaches a maximum value, decreases to zero, and finally becomes a repulsive force which increases very rapidly as the distance is further reduced.

14-2 PRESSURE IN A GAS

If our picture of the molecular nature of matter is correct, it must lead to the laws which are observed in the behavior of gases. Using the assumptions stated above, we will build a model called the "billiard-ball model," so named because the molecules are considered to be perfectly elastic spheres.

Consider the gas contained in a cubical box the length of each edge being l (Fig. 14-2).

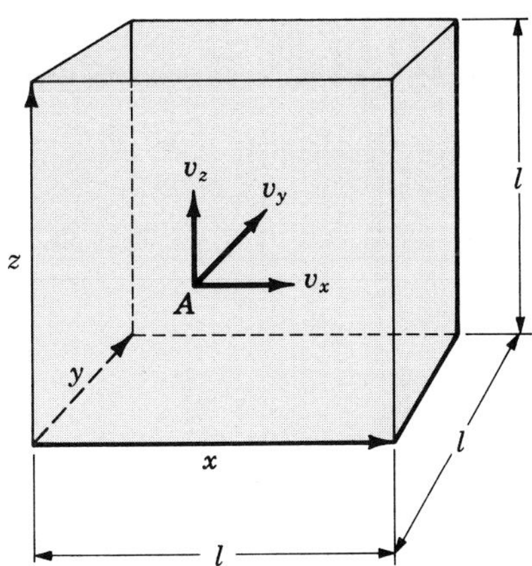

Figure 14-2
Pressure in a gas is related to molecular motion.

1 There are n molecules of a gas in a unit volume. The velocity of each can be divided into three rectangular components. If we consider one molecule,

$$\mathbf{v} = \mathbf{v}_x + \mathbf{v}_y + \mathbf{v}_z$$

or

$$\mathbf{v}^2 = \mathbf{v}_x^{\ 2} + \mathbf{v}_y^{\ 2} + \mathbf{v}_z^{\ 2} \qquad (1)$$

2 Consider only the x (horizontal) component of the motion. When the molecule strikes the face yz it rebounds with a speed equal to $-v_x$. Therefore,

$$\text{Change in speed} = v_x - (-v_x) = 2v_x$$

and

$$\text{Change in momentum} = mv_x - (-mv_x) = 2\,mv_x$$

3 The distance between collisions with face yz is $2l$ and the time between collisions with this face equals distance divided by speed:

$$t = 2\frac{l}{v_x}$$

and the period is $1/t = v_x/2l$ collisions per unit time.

4 The change in momentum per unit time at face yz is

$$\frac{\Delta\rho}{\Delta t} = \frac{2\,mv_x}{2l/v_x} = \frac{2\,mv_x^{\ 2}}{2l} = \frac{mv_x^{\ 2}}{l}$$

5 But, since impulse equals $F\Delta t = \Delta\rho$, $F = \Delta\rho/\Delta t$; therefore,

$$F = \frac{mv_x^{\ 2}}{l} \qquad (2)$$

and this is the force one molecule exerts on face yz.

6 Since there are n molecules per unit volume and the total volume is l^3, the total number of molecules is nl^3.

7 Thus, the average total force on the yz face is

$$F_{yz} = nl^3\left(m\bar{v}^2\frac{x}{l}\right) = nl^2\,m\bar{v}_x^{\ 2} \qquad (3)$$

where $\bar{v}_x^{\ 2}$ = the average of the squares of the x components of the speeds.

8 The average kinetic energy of a molecule equals

$$\tfrac{1}{2}m\bar{v}^2 = \tfrac{1}{2}m\bar{v}_x^{\ 2} + \tfrac{1}{2}m\bar{v}_y^{\ 2} + \tfrac{1}{2}m\bar{v}_z^{\ 2}$$

and

$$m\overline{v}_x{}^2 = m\overline{v}_y{}^2 = m\overline{v}_z{}^2 = \tfrac{1}{3}\, m\overline{v}^2 \qquad (4)$$

where $\overline{v}^2$ is the root-mean-square (rms) speed, the square root of the average square of the speeds.

9 Then, from step 7, the force on this side is

$$F_{yz} = nl^2\,(\tfrac{1}{3}\, m\overline{v}^2)$$

$$F_{yz} = \tfrac{1}{3}\, nl^2\, m\overline{v}^2 \qquad (5)$$

10 But since $P = F/A$, the pressure on the side equals

$$P = \frac{\tfrac{1}{3}\, nl^2\, m\overline{v}^2}{l^2} = \tfrac{1}{3}\, nm\overline{v}^2 \qquad (6)$$

11 In terms of average kinetic energy, we may rewrite Eq. (6)

$$P = \tfrac{1}{3}\, n\,(\tfrac{2}{2})\, m\overline{v}^2 = \tfrac{2}{3}\, n\,(\tfrac{1}{2}\, m\overline{v}^2) \qquad (7)$$

12 In a volume V of gas, the number of molecules is $N = Vn$, where n is the number of molecules in a unit volume. Then, multiplying both sides of the equation in step 11 by V,

$$PV = \tfrac{2}{3}\, nV\,(\tfrac{1}{2}\, m\overline{v}^2)$$

but $n = N/V$; thus,

$$PV = \tfrac{2}{3}\, N\,(\tfrac{1}{2}\, m\overline{v}^2) \qquad (8)$$

At a given temperature, the kinetic energy $(\tfrac{1}{2}\, m\overline{v}^2)$ is constant and since N is constant, the right-hand side of the equation is constant. Therefore, $PV = k$ (when t is constant), which is Boyle's law.

Example Compute the rms speed of the molecules of oxygen at 76.0 cm Hg pressure and 0°C, at which temperature and pressure the density of oxygen is 0.00143 g/cm³ (= 1.43 kg/m³).

From Eq. (6),

$$P = \tfrac{1}{3}\, nm\overline{v^2}$$

Since $nm = \rho$,

$$P = \tfrac{1}{3}\, \rho\overline{v^2} \qquad \text{or} \qquad \overline{v^2} = \frac{3P}{\rho}$$

$$\begin{aligned} P &= h\rho g \\ &= 0.760\ \text{m} \times 13.6 \times 10^3\ \text{kg/m}^3 \times 9.8\ \text{m/s}^2 \\ &= 1.013 \times 10^5\ \text{N/m}^2 \end{aligned}$$

$$\begin{aligned} \overline{v^2} &= \frac{3P}{\rho} = \frac{3 \times 1.013 \times 10^5\ \text{N/m}^2}{1.43\ \text{kg/m}^3} \\ &= 2.12 \times 10^5\ \text{m}^2/\text{s}^2 \\ v &= 461\ \text{m/s} \end{aligned}$$

In Chap. 15 we will develop the general gas law in the form

$$PV = nRT \qquad (9)$$

where n is the number of moles, of gas, which is the mass of the gas divided by the molecular mass, R is the universal gas constant, and T is the Kelvin temperature. While these terms will be defined later, we have sufficient knowledge of constants and temperature to make the following important point meaningful. If we compare Eq. (8) with Eq. (9), we obtain

$$\tfrac{2}{3}\, N\,(\tfrac{1}{2}\, m\overline{v^2}) = nRT \qquad (10)$$

Let N_0 represent Avogadro's number, the number of molecules in a mole of gas. Then $n = N/N_0$. Hence

$$\tfrac{2}{3}\,(\tfrac{1}{2}\, m\overline{v^2}) = \frac{R}{N_0}\, T \qquad (11)$$

$$\tfrac{1}{2}\, m\overline{v^2} = \frac{3}{2}\,\frac{R}{N_0}\, T = \tfrac{3}{2}\, kT \qquad (12)$$

The average kinetic energy per molecule is proportional to the Kelvin temperature of the gas. The universal constant k of Eq. (12) is called the *Boltzmann constant* and is related to the universal gas constant and Avogadro's number by

$$k = \frac{R}{N_0} \qquad (13)$$

We may compute its value by using the values of $R = 8.31$ J/(mol)(°K) and $N_0 = 6.023 \times 10^{23}$ molecules/mol,

$$k = \frac{8.31 \text{ J/(mol)(°K)}}{6.023 \times 10^{23} \text{ molecules/mol}}$$
$$= 1.38 \times 10^{-23} \text{ J/(molecule)(°K)}$$

From Eq. (12) we predict that at any one temperature T the average kinetic energy per molecule is the same for all gases. This is in accordance with the experimental observations that the molar heat capacity is nearly the same for all monatomic gases. For polyatomic gases the energies of rotation and vibration must be considered. It follows that at a single temperature the rms speed of the molecules of gases is inversely proportional to the square root of the mass of the molecules.

$$\tfrac{1}{2} m_1 \overline{v_1^2} = \tfrac{1}{2} m_2 \overline{v_2^2}$$

$$\frac{v_{1,\text{rms}}}{v_{2,\text{rms}}} = \sqrt{\frac{m_2}{m_1}} \qquad T \text{ const} \qquad (14)$$

14-3 MEAN FREE PATH

In our development of the model for the behavior of gases we did not take into account collisions between molecules as they move from one surface to the other. There are, of course, many such collisions, but since we have assumed all collisions to be perfectly elastic, no energy is lost and the average energies are unchanged.

We saw earlier (p. 217) that the average length of path traversed by a molecule between two successive collisions is called the *mean free path*. The mean free path depends upon the number n of molecules per unit volume, the diameter d of a molecule, and the relative speed of the molecules. We may compute an expression for the mean free path.

1 Assume all molecules are standing still except one which moves with a velocity v_r.

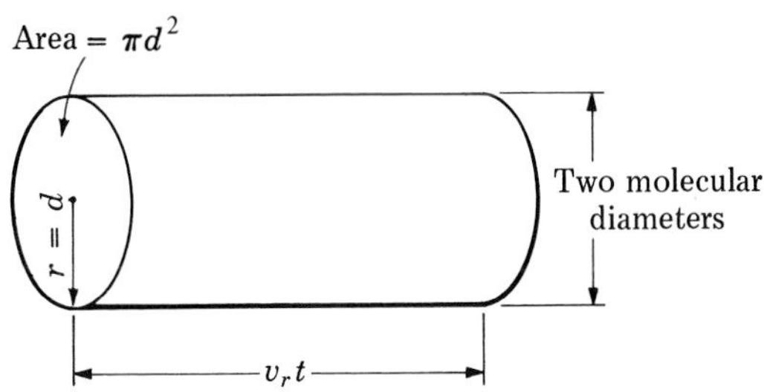

Figure 14-3
Cylinder swept out by molecule in t seconds.

2 If no collisions occur, this molecule would move a distance of $v_r t$ in t seconds, since $v_r = s/t$.
3 But the molecule will strike every molecule whose *center* will come within a distance of one molecular diameter of its own center.
4 In t seconds it will strike every molecule whose center should fall within a cylinder of length equal to $v_r t$ and cross-sectional area of πd^2, where d is a molecular diameter (Fig. 14-3).
5 The volume of this cylinder (volume $= \pi r^2 h$) is

$$\text{Volume} = \pi d^2 v_r t \qquad \text{cm}^3$$

6 The number of molecular centers in it are $N = Vn$. Substituting for V,

$$N = (\pi d^2 v_r t)(n)$$

7 The molecule will strike

$$\frac{N}{t} = \pi d^2\, n v_r \qquad \text{molecules/s}$$

which is the frequency of collision, f.
8 Since the molecule travels with a speed v_r, the distance it will travel between collisions can be determined as below. By definition the frequency f is the reciprocal of the period, $f = 1/t$. Also,

$$v = \frac{s}{t} \qquad \text{and} \qquad t = \frac{s}{v}$$

Thus,

$$f = \frac{1}{(s/v)} = \frac{v}{s}$$

and the distance between collision $s = v/f$ or

$$s = \frac{v_r}{(\pi d^2 \, n v_r)} = \frac{1}{\pi d^2 \, n} = \overline{l}$$

where l is the mean free path.

9 Since none of the molecules is at rest, the speed v_r is not the speed of the single molecule. It can be shown that on the average, $v/v_r = 1/\sqrt{2}$ and hence

$$l = \frac{1}{\pi \sqrt{2} \, n d^2} \qquad (15)$$

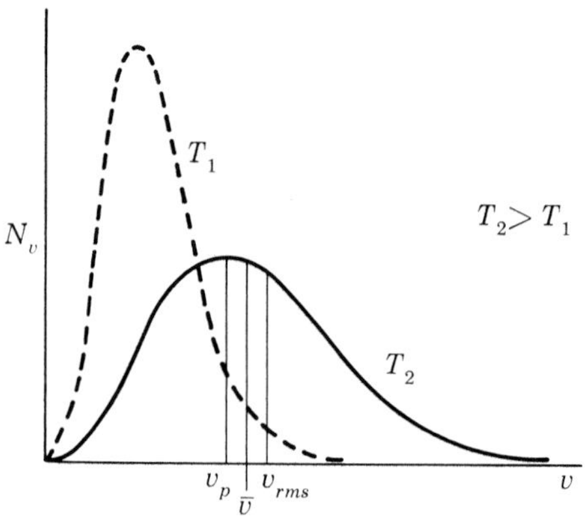

Figure 14-4
Maxwellian distribution of molecular speeds. v_p is the most probable speed, $\bar{v}$ is the average speed, and $\bar{v}_{\mathrm{rms}}$ is the root-mean-square speed.

14-4 DISTRIBUTION OF MOLECULAR SPEEDS

The individual molecules of a gas may have speeds ranging in value between zero and infinity. The distribution of speeds of the molecules depends upon the temperature. James Clerk-Maxwell worked out an expression for this distribution. If $N_v \, \Delta v$ represents the number of molecules with speeds between v and $v + \Delta v$,

$$N_v \, \Delta v = 4\pi \, N \left(\frac{m}{2\pi k T}\right)^{3/2} v e^{-(mv^2/2kT)} \, \Delta v \qquad (16)^1$$

If we plot N_v against v, we get a distribution curve that for a single gas is dependent simply upon the temperature (Fig. 14-4). The value of v for the maximum ordinate of the curve is the most probable speed v_p. The curve is not symmetrical about the most probable speed. The average speed $\bar{v}$ is somewhat higher than the most probable speed; the root-mean-square speed v_{rms} is somewhat greater than the average speed.

[1] Leigh Page, "Introduction to Theoretical Physics," 3d ed., Chap. 9, D. Van Nostrand Company, Inc., Princeton, N.J., 1952.

14-5 COHESION AND ADHESION

The force with which like molecules attract each other is called *cohesion*. Within a solid the forces of cohesion are large, since the molecules are close together. If the solid is broken, it is difficult to force the molecules close enough together for these forces to become large again. When an iron bar is broken, the pieces may be fitted together but they will not cohere. However, by heating the iron until it softens and pounding it with a hammer, the molecules may again be brought close enough to cohere and the bar is welded. Local heating until the metal flows also serves to bring the molecules close enough for cohesion. Two ordinary plates of glass or two blocks of steel, when brought together, show little evidence of cohesion, because they touch at only a few points. It is possible, however, to grind these surfaces flat enough so that when they are brought into contact the cohesive forces become very large. The highest-grade gauge blocks will thus cohere and can be pulled apart only with considerable difficulty.

The existence of cohesive forces between water molecules can be dramatically illustrated by a simple demonstration. A strip of paper is folded like an accordion and one end is brought close to a water surface. As long as there is no water on the paper it will not be noticeably attracted to the water. On the other hand, if the same folded paper has one end of the paper wet and then this end is brought near the water, the wet end will move forcefully into the water. See Fig. 14-5.

At the surface between two different substances or in a mixture, unlike molecules attract each other. The attraction of unlike molecules is called *adhesion*. Glue adheres to wood; solder adheres to brass. Adhesive forces may be greater than cohesive forces; in fact, the relative magnitude of the cohesive and adhesive forces determines the nature of the surface between two substances in contact.

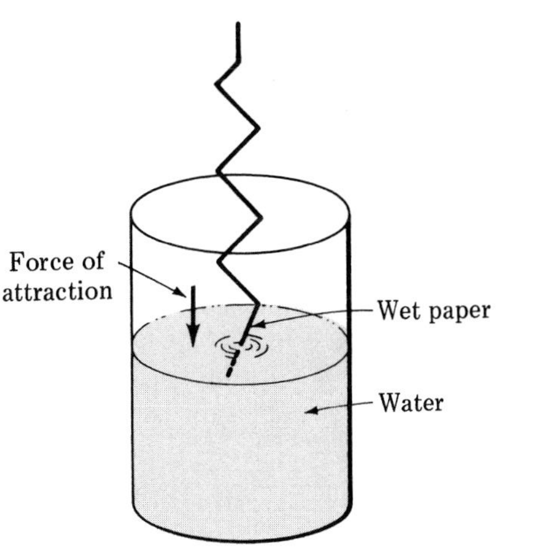

Figure 14-5
A folded paper will be drawn to the water if the lower end of the paper is wet.

14-6 DIFFUSION

If a bottle of ammonia is opened in one corner of a closed room, the odor is soon apparent in all parts of the room, even though there are no air currents. The ammonia molecules reach the observer because of their own motion. The molecules in the air of the room are relatively far apart. As the ammonia molecules move, they pass between the molecules of the air, with occasional collisions. Some of the molecules reach every part of the enclosure in a short time. The process of one substance mixing with another because of molecular motion is called *diffusion*. If the gas is confined in a small container and the pressure is reduced, diffusion takes place more rapidly, for the gas molecules are farther apart and collisions are less frequent.

Diffusion occurs in liquids as well as in gases. This may be shown by dropping a few copper sulfate crystals into a jar of water. As the crystals dissolve, the blue color appears first at the bottom and slowly spreads upward. After a period of some weeks, all the liquid will be colored. Since the copper sulfate solution is denser than water, the spreading cannot be caused by convection currents, as it would be if the crystals were suspended near the top. If this were done, the mixing would be quite rapid.

Diffusion of gases or liquids through solids is also common. A light gas such as hydrogen passes quite readily through porous earthenware. Such a porous jar, arranged as at *A* in Fig. 14-6 with

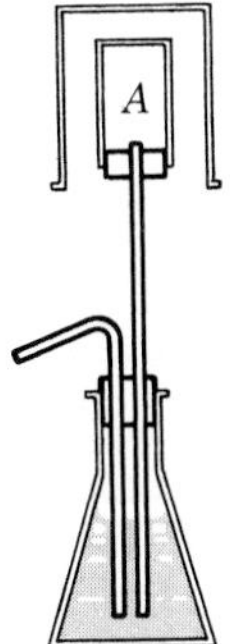

Figure 14-6
Diffusion of gas through a porous cup.

a tube extending into a flask of water, is filled with air. When a container of hydrogen is lowered over the porous jar, the hydrogen diffuses into the jar faster than air diffuses out. Hence the pressure is increased inside the system, and water is forced out.

While it is not as common an occurrence, solids will diffuse into other solids. Spacecraft have many electronic devices which have dissimilar metals that are in direct contact. Since the signals sent by these devices could be altered if a significant amount of diffusion took place, care must be taken in selecting materials for such missions.

14-7 OSMOSIS

Some substances, such as parchment or vegetable materials, have the property of allowing some molecules to diffuse through but not others. Suppose that a piece of parchment is fastened over a thistle tube and some sugar syrup placed inside (Fig. 14-7). When this is immersed in water, the water molecules diffuse in through the parchment but the larger sugar molecules cannot diffuse out. This process is called *osmosis*. The liquid rises in the tube until the pressure, due to the liquid, is sufficient to stop the diffusion or the membrane breaks. The pressure set up by this one-way diffusion is called *osmotic pressure*.

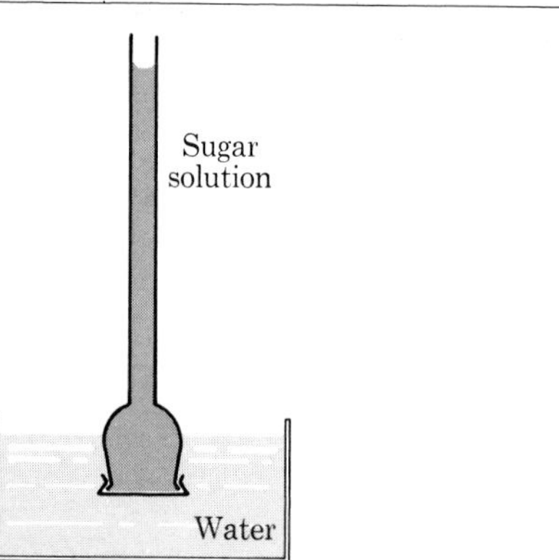

Figure 14-7
Osmosis.

Osmosis is an important process in animal and vegetable life, since it determines the flow of liquids through membranes. When fruit or vegetables are placed in fresh water, the skin becomes stretched and may even burst because water passes in through the skin. If, however, the fruit or vegetable is placed in a sugar or salt solution, liquid may pass out through the skin, with resultant shriveling.

14-8 BROWNIAN MOVEMENT

Individual molecules are too small to be visible, but it is possible to observe the effects of collisions of molecules with small particles, which can be seen with the aid of a microscope. If some gamboge, a gum, is dissolved in alcohol and a small amount poured into a dish of water, a very fine suspension is formed. When this is viewed by means of a microscope, the particles are seen to be in very erratic motion. This motion is called *Brownian motion,* after the man who first observed it (1827). The motion is due to collisions of molecules with the particles. Each impact imparts a motion to the particle, and since it is struck first from one direction and then another, the motion is quite erratic. Brownian motion constitutes one of the most direct pieces of evidence for the belief in the motion of molecules.

14-9 MOLECULAR ACTIVITY AND CRYSTAL FORMATION

In Chap. 10 we observed some of the elastic properties of matter. We can now show that many of the physical properties of matter, i.e., flexibility and strength, can be explained by the kinetic molecular theory. As a demonstration of this, a razor blade is heated in the flame of a burner until it becomes incandescent and then is cooled rapidly by immersion in cold water (quenched). A second blade then is heated similarly but is cooled slowly by being allowed to cool at room temperature. When both have cooled, the difference in the physical properties of the two blades from their original condition and from each other is striking. According to the kinetic molecular theory, an addition of heat energy to the blades produces increased molecular activity. Indeed, if we have a hot enough source of heat energy we can add sufficient energy to the blade to cause the steel to liquefy and to become molten due to the destruction of bonds. In this illustration, we supply enough heat energy to disrupt the molecular patterns. Upon cooling, the steel blades return to a more stable condition and the crystal patterns are reestablished, however, with a marked difference from the original pattern. The blade which cools off rapidly becomes extremely brittle and when a stress is applied to it, it reacts like an eggshell. The blade which has cooled slowly, on the other hand, becomes quite flexible and pliable and reacts like a strip of soft lead. An analogy can be drawn to illustrate the effect of rapid cooling on molecular arrangement in matter. At the rim of volcanic craters a substance called obsidian (volcanic glass) is often found. This dark, crystalline material has a translucent property due to the fact that the crystals formed in it are extremely small. If this material had cooled more slowly, it would have lost this translucent property. There is a direct relationship between the rate of cooling of a substance and the size of crystals formed. Rapid cooling produces tiny crystals and slow cooling produces larger crystals. Figure 14-8 illustrates the different crystal patterns formed when the razor blades have been cooled as above.

It will be observed in Fig. 14-8*b* that the large number of small crystals arrange themselves in such a way that it is possible to locate many "fracture lines," only three of which are shown, along which cleavage can take place. This accounts for the brittle nature of the rapidly cooled razor blades. Figure 14-8*c* illustrates the long crystal patterns established with slow cooling. Few fracture lines occur in this case; and when a stress is applied to the blade, the crystals slide over one another rather than fracture, giving the blade its flexibility. Another example of this type of crystal is found in graphite, which is used in "lead" pencils because of the ease with which this crystalline carbon rubs off onto a paper.

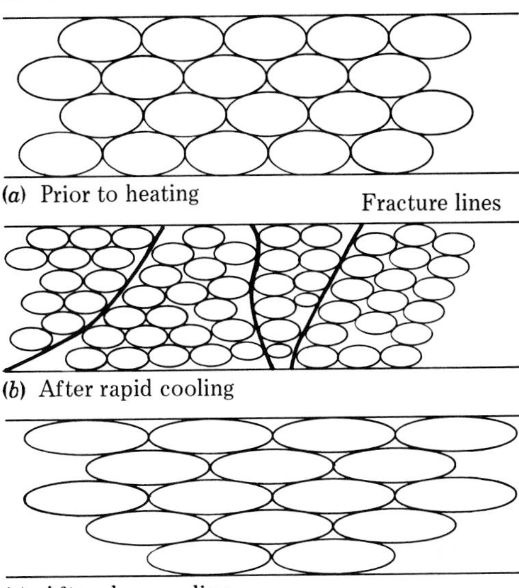

Figure 14-8
Molecular patterns formed when matter is cooled at different rates.

The technique used to preserve food, called "deep freezing," depends upon this property of crystal formation. All living tissue contains a large percentage of water which, as is commonly known, expands upon freezing. Bacterial action leading to the spoiling of food depends to a large degree upon damaged cellular walls and torn tissue for the bacteria to make an entry. If the tissue is allowed to freeze slowly, the ice crystals formed will be large and much cell damage will be done and spoilage will result. However, if the tissue (plant or animal) is frozen quickly, the crystals of ice will be tiny and very little damage will be done to the cell walls, bacterial decomposition will be reduced to a very low level, and the tissue can be preserved for a long period. A classic example of this prolonged preservation of tissue by rapid freezing is the remains of the prehistoric beasts in Siberia called woolly mammoths that were covered by ice after having fallen into a crevasse and lay in the tundra for many centuries. These beasts were frozen so quickly that they were almost perfectly preserved.

SUMMARY

The *molecular theory* of matter is based upon a set of assumptions. Predictions made upon the basis of these assumptions are tested experimentally to determine whether the theory is tenable.

In building a model to understand a physical principle, the assumptions made or limits established are considered valid if the results predicted are verified by experiment.

The fundamental assumptions made in the molecular theory are that all matter is made up of molecules which are in constant motion. In a gas, Newton's laws of motion are assumed to hold. In collisions between molecules, the laws of conservation of momentum and conservation of kinetic energy hold. The molecules exert forces upon each other, which depend upon the distance between them.

The pressure of a gas is given by the expression

$$P = \frac{1}{3} n m \overline{v^2}$$

The *Boltzmann constant* k is the ratio of the universal gas constant R to Avogadro's number N_0, $k = R/N_0$.

The average kinetic energy per molecule is proportional to the Kelvin temperature of the gas.

$$\frac{1}{2} m \overline{v^2} = \frac{3}{2} kT$$

The mean free path $\bar{l}$ of a molecule is the average distance it travels between successive collisions.

$$\bar{l} = \frac{1}{\pi \sqrt{2}\, n d^2}$$

where d is the diameter of the molecule and n is the number of molecules per unit volume.

In a Maxwellian distribution of speeds the average speed $\bar{v}$ is greater than the most probable speed v_p, and the rms speed v_{rms} is greater than the average.

Cohesion is the attraction between like molecules, while *adhesion* is the attraction between unlike molecules.

Diffusion takes place because the molecules of a gas or liquid move between the molecules of the substance that is being penetrated.

Osmosis is one-way diffusion through a semipermeable membrane.

Brownian movements are motions of small particles that are bombarded by molecules.

There is a direct relationship between the rate of cooling of a substance and the size of crystals formed. Rapid cooling produces tiny crystals and slow cooling produces larger crystals.

Questions

1 All scientific "models" built must have certain characteristics if the model is to be considered valid. List some of these characteristics.

2 Why can't you see molecules in motion?

3 Is absolute zero an imaginary, theoretical point? Explain.

4 What is meant by molecular "pumping"? Give examples of instruments depending upon this principle.

5 Is it true that the closer two objects get, the greater the attractive force between them? Are there any limitations to this statement?

6 Describe any observable physical effect resulting from an increase in the mean free path of the molecules in an object.

7 Show that the kinetic energy per mole (gram-molecular mass) is the same for all gases at the same temperature.

8 Define, point out the significant differences, and give examples of cohesion and adhesion.

9 In terms of molecular forces, explain how glue works.

10 Explain by the use of the kinetic theory why the pressure of a gas on its container rises when the volume is reduced. Is the surface area of a container always and necessarily reduced when the volume is reduced? If not, would the pressure of a gas in the container necessarily rise with a reduction of volume?

11 Explain by the use of kinetic theory why the pressure of a gas increases as the square of the velocity of the molecules.

12 A tank of gas contains both light and heavy molecules. Do they all have the same speed; the same kinetic energy?

13 Why is it that you can immerse your hand in mercury and your hand will be dry? How and why does this differ from the effect of putting your hand in water?

14 Will water evaporate more rapidly from a shallow pan than from a tall narrow pan having the same volume? Explain.

15 On a cold day, rubbing the hands together not only makes them feel warmer, they are warmer. Why is this the case?

16 Why do packages of frozen foods have printed on them the warning, "Do not refreeze"?

17 Describe the main differences, from a molecular properties viewpoint, between soldering and welding two pieces of metal together.

18 Can you relate the process of osmosis to the thirst one has after consuming too many salted peanuts or, for that matter, too much alcohol?

19 Can you explain by means of the kinetic theory the drop in pressure that occurs when air flowing through a tube encounters a constricted part of the tube (Bernoulli effect)? Is the pressure on the side wall of the tube the same as that on a surface facing the stream?

Problems

1 The mass of a hydrogen atom is about 1.66×10^{-24} g. How many atoms are present in 1 g of hydrogen?

2 How many molecules are there in 1 mg of ethane? The molecular mass of ethane is 30 g/mol. *Ans.* 2.0×10^{19} molecules.

3 Copper has atomic mass 63.6 and specific gravity 8.9. What is the average volume per atom?

4 There are 6.02×10^{23} molecules in a mass of any gas numerically equal to its molecular mass. How many molecules are there in 1.00 cm^3 of nitrogen under standard conditions? *Ans.* 2.68×10^{19}.

5 From the data of Table 1, Chap. 12, compute the rms speed at 0°C of the molecules of (*a*) carbon dioxide, (*b*) hydrogen, and (*c*) nitrogen.

6 Calculate the rms speed of a hydrogen molecule at 20°C and 70.0 cm Hg when the density of hydrogen at 0°C is 0.089 g/l. *Ans.* 1.9×10^5 cm/s.

7 Oxygen is confined in a container at 0°C and normal atmospheric pressure. The temperature of the gas is then increased until the pressure is doubled. Neglect any change in volume of the container. Find the rms speed of the molecules at this temperature.

8 Compute the rms speed of oxygen molecules at 0°C and 76.0 cm Hg when the density is 1.429 g/l. *Ans.* 4.6×10^4 cm/s.

9 The rms speed of helium atoms under standard conditions is 1.30×10^3 m/s. What is the density of helium under these conditions?

10 Calculate the rms speed of methane molecules at a temperature of −120°C. Methane has a molecular mass of 16 g/mol. *Ans.* 4.7×10^4 cm/s.

11 Under normal conditions the average distance a hydrogen molecule travels between collisions is 1.83×10^{-5} cm. Compute (*a*) the time between collisions and (*b*) the frequency of collisions.

12 What is the velocity of a molecule of methane gas at 37°C. Molecular mass of methane is 16 g/mol. *Ans.* 695 m/s.

13 If the mean distance between collisions of carbon dioxide molecules under standard conditions is 6.29×10^{-4} cm, what is the time between collisions?

14 Find the temperature at which the rms velocity of an oxygen molecule equals that of a hydrogen molecule which is at a temperature of 300°K. *Ans.* 4800°K.

15 The conduction of heat by a gas virtually stops when the mean free path becomes as large as the dimensions of the container. At what pressure will the mean free path of nitrogen molecules become as great as the 5.0-mm spacing of the walls of a vacuum bottle? Assume that the diameter of a nitrogen molecule is 1.9×10^{-10} m and that the temperature is 300°K.

16 A 1.00-l flask contains 2.68×10^{22} molecules of oxygen, each of mass 5.31×10^{-23} g and having an rms speed of 4.61×10^{4} cm/s. Compute the pressure in the flask, and reduce it to atmospheres. *Ans.* 1.01×10^{5} N/m^2; 1.0 atm.

17 If the average distance between collisions of CO_2 molecules at 1 atm pressure and 0°C is 6.29×10^{-6} m, what is the molecular diameter?

18 Does a molecule of nitrogen at 27°C travel faster than sound at that temperature? Compare these speeds. (Speed of sound at 27°C = 1,140 ft/s.) *Ans.* Nitrogen = 1,700 ft/s; 560 ft/s faster than sound.

19 If all the molecules in 1.00 cm^3 of nitrogen at standard conditions were lined up in contact (1.9×10^{-10} m from center to center), how long a line would they make?

20 Twenty grams of oxygen at 27°C possess a certain amount of translational kinetic energy. Find this kinetic energy. Molecular mass of oxygen = 16 g. *Ans.* 2,340 J.

21 In a certain electron microscope, electrons travel 1.0 m from the electron gun to the screen. To avoid scattering electrons by residual molecules of nitrogen in the vacuum chamber, below what pressure would you recommend operating the microscope? The diameter of a nitrogen atom is about 1.9×10^{-10} m.

22 Find the kinetic energy of the molecules in 1 g of ammonia gas at 27°C. Molecular mass of ammonia = 17 g/mol. *Ans.* 220 J.

23 What is the mean free path in centimeters of the average molecule in a mole of ammonia gas at standard pressure? Molecular mass of ammonia is 17 g/mol and its molecular diameter is 3×10^{-8} cm.

Niels Bohr, 1885–1962

Born in Copenhagen, Denmark. Director, Institute for Theoretical Physics, Copenhagen. The 1922 Nobel Prize for Physics was conferred on Bohr for the value of his study of the structure of atoms and of the radiation emanating from them.

15

Temperature; Thermal Expansion

As was the case in the study of motion, before we can study the nature of heat energy it is necessary to develop units of measurement and appropriate measuring techniques with which to express this form of energy. The *temperature* of an object is defined as the property that determines the transfer of heat energy to or from other objects. Of two objects having different temperatures, the one that transfers heat to the other is said to be at a higher temperature. Our subjective idea of temperature is obtained from the sensation of warmth or cold which we experience upon touching an object. If a stove is very hot, we can sense this even without touching it. Under some conditions our temperature sense is an unreliable guide. For example, if the hand has been in hot water, tepid water will feel cold, whereas if the hand has been in cold water, the same tepid water will feel warm.

A group of psychologists studying the behavior of man uncovered evidence that illustrates the variable tolerance humans have to heat energy. A study of the length of cigarette butts thrown away by individuals varied considerably among members of the group observed, but each individual developed a quite predictable throwaway length. It was found that the almost-automatic rejections of the cigarettes came when the smoke from the cigarettes, which is cooled by the unburned tobacco, became hotter than the tongue could comfortably tolerate. The different lengths indicate that man has varying degrees of sensitivity to heat energy and provides some evidence that man must depend upon something other than his senses to determine the quantity of heat energy possessed by an object.

In order to specify temperatures on a numerical scale, we find that we must measure temperature indirectly in terms of the change of some physical property such as length, pressure, or electrical resistance. It turns out that a temperature scale based upon one material and property differs somewhat from scales based upon other choices of thermometric materials and properties. This raises an interesting question. How do we define for scientific measurements a scale of temperature which is "absolute" in the sense that it is independent of the peculiarities of any one material and in the sense that its zero has physical significance?

In this chapter we shall see how we can define an absolute scale of temperature which has a well-defined (though unattainable) zero and no upper limit. In Chap. 18 we shall discuss this concept of temperature in the macroscopic viewpoint of thermodynamics and in the microscopic viewpoint of the atomic ("kinetic") theory of heat.

15-1 THERMAL EQUILIBRIUM

Consider an object A, such as a small iron disk, which feels warm to the hand and an identical iron disk, object B, which feels cold. If these objects are placed in contact with each other, after a sufficient time each will produce the same temperature sensation when touched. They are said to be in *thermal equilibrium* when there is no net transfer of heat from A to B or from B to A.

If four thin, square sheets consisting of metal, marble, wood, and glass, respectively, are mounted on a flat surface and allowed to stand at room temperature, they will all seem to be at different temperatures when touched by a person. However, if each square has a thermometer embedded in it, the temperature will read the same in each square. How does one explain this phenomenon? We first have to understand that *any inanimate object without a built-in heating system will eventually attain the temperature of the environment,* that is, it will reach an equilibrium temperature. Strange as it may at first seem, ice which normally freezes at 0°C can be at a lower temperature, say −10°C, if it is stored in a freezer having a temperature of −10°C. If the squares are at the same temperature as the room, why do they seem to vary in temperature? Another property of matter which we shall observe again later in our study of heat is that certain objects have different ability to transfer heat away from or to another object. Metal, for example, generally has a good ability to transfer heat energy; hence it is called a good thermal conductor. Our hand, which is warmer than a metal object at normal room temperature, will have its heat conducted rapidly away by the metal in an attempt to reach an equilibrium temperature. This rapid loss of heat by the hand gives us the sensation that the metal is "cold." Wood, which is a poor conductor of heat energy, does not remove the heat as rapidly from the hand and appears to be

Table 1
COMMON TEMPERATURE-MEASURING METHODS

Method	Thermometric property and substance	Approximate range, °C
Mercury-in-glass thermometer	Length of mercury thread (depends on expansion of mercury less expansion of glass)	−38 to 350
Alcohol-in-glass thermometer	Length of alcohol column	−80 to 100
Bimetallic thermometer	Relative expansion of two different metals	−40 to 500
Thermocouple	Emf generated when dissimilar metals form circuit with junctions at different temperatures	−250 to 500
Resistance thermometer	Resistance of a platinum wire	−272 to 1600
Optical pyrometer	Monochromatic brightness of a tungsten lamp	800 up
Total-radiation pyrometer	Electric power generated by radiation in all wavelengths received from test body	100 up
Constant-volume gas thermometer	Pressure of a confined mass of gas	−269 to 1600
Speed of sound	Thermodynamic relation between speed of sound and gas temperature	No limits

"warmer" than the metal. The glass and the marble also have varying capacities to transfer heat and will appear to be at different temperatures.

In making temperature measurements it is important that the thermometer be in thermal equilibrium with the body whose temperature it is intended to indicate. This requirement of equilibrium can be satisfied easily in measuring the temperature of water in a calorimeter, but not so easily in measuring the temperature of hot gases expelled from a jet engine.

15-2 MEASUREMENT OF TEMPERATURE

There are many possible kinds of thermometers, since almost all the properties of material objects (except mass) change as the temperature changes. A thermometer is specified by choosing a particular thermometric substance and a particular thermometric property of that substance. A thermometric property of matter is a property that varies predictably with an increase or decrease in temperature. It could be the change of pressure in a gas thermometer, a change in electromotive force (emf) in a thermocouple, or the change in height of a liquid in a liquid-in-glass thermometer. It is assumed that there is a one-to-one relationship between the measured values of that property and the temperature. Some practical choices are listed in Table 1.

In principle, the choice of the equation which relates temperature T to the measured value x of a thermometric property is arbitrary. It is simplest to choose the linear function

$$T_x = cx \tag{1}$$

where c is a constant to be defined. Any two temperatures are in the same ratio as the corresponding values of x

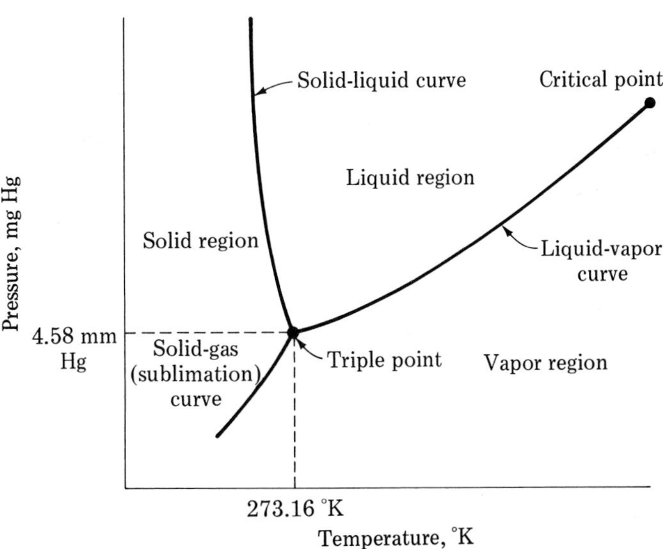

Figure 15-1
Triple-point diagram for water. Note that all three states, solids, liquids, and vapors, exist at this point.

$$\frac{T_1}{T_2} = \frac{x_1}{x_2} \tag{2}$$

The value of c is determined if we specify a standard fixed point at which all thermometers must give the same reading, T_0, corresponding to the value x_0 of a thermometric property. The fixed point chosen is the triple point of pure water, the temperature at which liquid water, ice, and water vapor coexist in equilibrium (Fig. 15-1). The water-vapor pressure then is 4.58 mm Hg. The temperature of this reference point is arbitrarily set at 273.16°K. This value was adopted by the Tenth General Conference on Weights and Measures in 1954, in Paris. The ice point, still used for many practical calibrations, is 273.15°K. By this definition, when the thermometric property has the value x, the temperature T is

$$T_x = 273.16°\text{K}\left(\frac{x}{x_0}\right) \tag{3}$$

where the Kelvin degree is the unit temperature interval.

Obviously the arbitrary choices of thermometric substances and properties allow for the possibility of many different kinds of thermometers whose temperature indications, unfortunately, are found to be mutually inconsistent by as much as several degrees. In order to have a definite and reproducible temperature scale for scientific use, we must be more specific about how we are to use Eq. (3). A constant-volume gas thermometer provides a feasible means of defining a reproducible, absolute temperature scale.

15-3 IDEAL-GAS TEMPERATURE

A constant-volume gas thermometer uses the pressure of a gas confined at constant volume as the thermometric property. A simple version is shown in Fig. 15-2. Gas is confined in the bulb

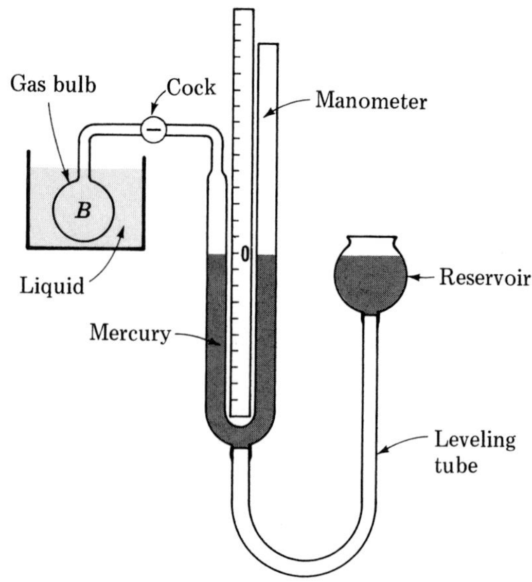

Figure 15-2
A gas thermometer.

B, which is connected to a U-tube manometer M in which the mercury level may be adjusted by raising or lowering the reservoir R.

With the bulb surrounded by an ice bath (more precisely, a bath at the triple point), the mercury levels may be adjusted to zero on the scale. The pressure of the gas P_0 is then just atmospheric pressure. If the temperature of the bulb is raised to T, both the volume and the pressure of the gas will increase, as indicated by the change in levels of the mercury. By raising the reservoir, the level of mercury in the left-hand column may be restored to the index 0. The gas then occupies its original volume. Its pressure P can be found by measuring the difference in levels of the mercury columns and adding to it the atmospheric pressure. Then the temperature T is calculated from

$$T_P = 273.16°\text{K}\left(\frac{P}{P_0}\right) \qquad V \text{ const} \tag{4}$$

Now suppose that we compare temperature indications obtained from a thermometer employing the same gas at successively different initial pressures P_0. First, when the bulb is surrounded by water at the triple point, let us admit dry air until the pressure P_0 in the bulb has some definite value, say, 1 atm. Then surround the bulb with steam condensing at 1 atm pressure. With the volume of the gas kept constant, measure the gas pressure P_s, and calculate the steam-point temperature $(T_P)_s$ from Eq. (4). Next, remove some air from the bulb so that the pressure at the triple point is, say, 0.5 atm. Again place the bulb in the steam bath, and calculate its temperature, using Eq. (4), from $(T_P)_s = 273.16°\text{K}\ (P/0.5\ \text{atm})$. Make similar additional measurements of $(T_P)_s$, reducing the amount of gas in the bulb each time. If the values so found for the steam temperature are plotted on a graph of temperature as a function of filling pressure (P_0), they fall on a straight line which may be extrapolated to intersect the T axis when $P_0 = 0$.

If this procedure is carried out now with different gases, different lines are obtained, showing that the temperature indication of a constant-volume gas thermometer depends on the gas used, at ordinary reference pressures. But as the reference pressure P_0 is decreased, the temperature readings of a gas thermometer using different gases approach the same value (Fig. 15-3). Hence the temperature indication extrapolated to $P_0 = 0$ is independent of the peculiarities of any gas and depends only on the general properties of gases. We shall define an *ideal-gas temperature* by the relation

$$T = 273.16°\text{K} \lim_{P_0 \to 0} \frac{P}{P_0} \qquad V \text{ const} \qquad (5)$$

We shall choose a constant-volume gas thermometer as our standard and the scientific temperature scale as defined by Eq. (5).

There are limitations to the temperature scale thus defined. First, it implies that, in measuring even the lowest temperatures, the thermometric substance remains a gas. Actually, even helium

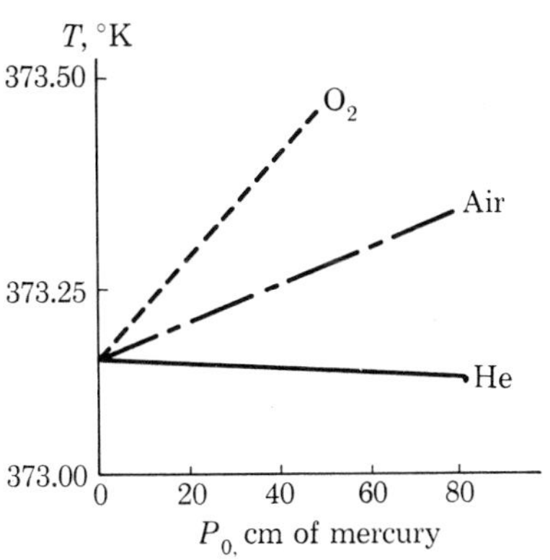

Figure 15-3
The steam-point temperature T as calculated with a gas thermometer using different gases at different initial pressures P_0.

liquefies at about 4°K, and its vapor pressure cannot be used to indicate temperatures below about 1°K. Second, at high temperatures we encounter difficulty with our assumption that temperatures are measured by using a gas in *equilibrium* with the test body. At temperatures above 1800°K, even quartz and platinum bulbs fail as containers of the gas. And, at still higher temperatures, processes are far from being in equilibrium. So the gas-thermometer scale applies directly to only a limited region in the broad range of temperatures now of importance in science (Fig. 15-4). Our concept of temperature is actually under continuing revision, particularly in the regions of extreme temperatures.

In 1848, William Thomson (Lord Kelvin) showed that it is possible to construct a temperature scale in terms of the efficiency of an ideal heat engine, such that the scale in all regions is independent of any thermometric substance. Anticipating Chap. 18, we shall state that temperatures on this thermodynamic temperature scale, called the *Kelvin scale,* and temperatures measured on the ideal-gas scale are identical. This justifies our writing °K after an ideal-gas temperature, as we have done.

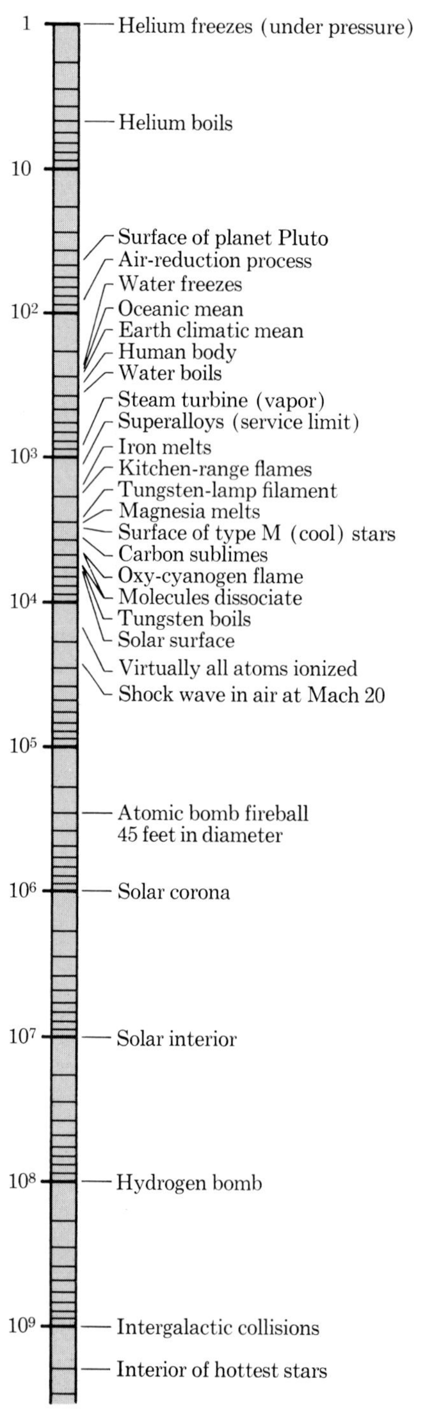

15-4 CELSIUS AND FAHRENHEIT SCALES

The Ninth General Conference on Weights and Measures (1948) decided that the name *centigrade* should be abandoned and *Celsius* used instead, to honor the Swedish astronomer Anders Celsius, who in 1742 invented a scale on which zero referred to the boiling point of water and 100 to the temperature of melting ice. A year later, Linne and Stromer reversed this designation and established the centigrade scale.

The Celsius temperature scale uses a degree (unit of temperature) of the same size as the degree on the Kelvin (or ideal-gas) scale. The Celsius temperature t, which corresponds to a temperature T on the Kelvin scale, is given by

$$t = T - 273.15^\circ \tag{6}$$

The triple point of water corresponds to 0.01°C. Zero on the Celsius scale is at the ice point, the temperature at which ice and air-saturated water at 1 atm pressure are in equilibrium. The Celsius temperature at which steam condenses at 1 atm pressure is $t_s = 373.15^\circ - 273.15^\circ = 100.00^\circ\text{C}$.

It is no accident that temperatures of ice point and steam point have the convenient values 0 and 100°C. We have described a logical path rather than the historic sequence in defining temperature scales. Actually, the Celsius (centigrade) scale was first defined in terms of these *two* fixed points, to which the assignment of the numbers 0 and 100 served to fix the *size* of the Celsius degree. Later in defining the ideal-gas scale (or the Kelvin scale) on the basis of a *single* fixed point, the numerical constant in Eq. (4) was so chosen as to preserve (very closely) the same size of degree on the Kelvin scale as on the Celsius scale. This is an example of the customary proce-

Figure 15-4
Range of temperature, °K. (*Adapted from a chart by F. J. Dyson in Scientific American, September 1954, p. 58.*)

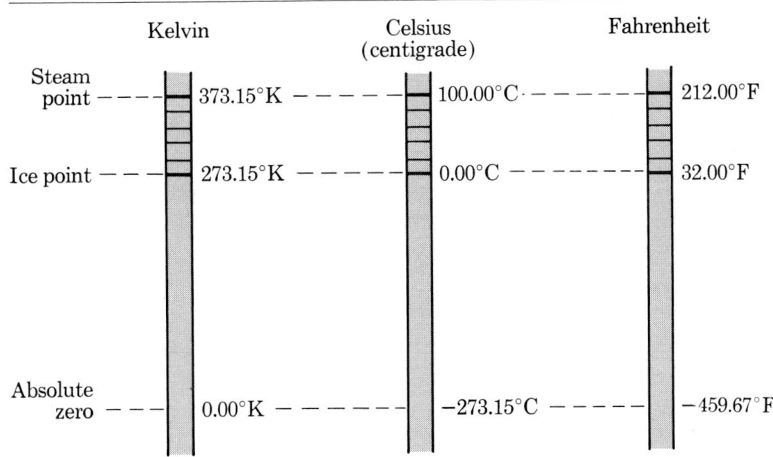

Figure 15-5
Reference points on various temperature scales. (At atmospheric pressure, or 760 mm Hg.)

dure, mentioned in the introductory chapter, that when improvements are made in the definitions of units or standards the new and the old are made to be consistent at least to the accuracy within which measurements could be made on the old system.

The Fahrenheit scale, used in many English-speaking countries, assigns the value 32°F to the ice point and 212°F to the steam point. The Fahrenheit degree is thus five-ninths as large as one Celsius degree. Conversions from one scale to the other may be made by means of a simple proportion, suggested by Fig. 15-5,

$$\frac{\text{C} - 0^\circ}{\text{F} - 32^\circ} = \frac{100^\circ - 0^\circ}{212^\circ - 32^\circ}$$

This equation may be solved for either C or F to give

$$\text{C} = \tfrac{5}{9}(\text{F} - 32^\circ) \tag{7}$$

$$\text{F} = \tfrac{9}{5}\text{C} + 32^\circ \tag{8}$$

Example A Fahrenheit thermometer indicates a temperature of 14°F. What is the corresponding reading on the Celsius scale?

The temperature of 14°F is 32° − 14° = 18 F° below the freezing point of water. A temperature interval of 18 F° is equivalent to $\frac{5}{9}(18^\circ) = 10$ C°. Hence the reading on the Celsius scale is 10° below the freezing point of water, or 0° − 10° = −10°C.

Example A comfortable room temperature is 72°F. What is this temperature, expressed in degrees Celsius?

$$\text{C} = \tfrac{5}{9}(72^\circ - 32^\circ) = \tfrac{5}{9}(40^\circ) = 22^\circ\text{C}$$

From the manner in which the standard interval is subdivided in the two scales (Fig. 15-5), it is evident that 100 C° = 180 F°. Dividing each side of this equation by 180 F° and 100 C° in turn, we obtain

$$1\ \text{F}^\circ = \tfrac{5}{9}\text{C}^\circ \qquad \text{and} \qquad 1\ \text{C}^\circ = \tfrac{9}{5}\text{F}^\circ \tag{9}$$

These conversion factors may be used for the convenient transfer of temperatures from one scale to the other.

When expressing *a temperature,* we write after the number one of the common symbols, °C, °F, etc.; here, when expressing *a difference of temperature* (temperature interval), we have used the symbols C° and F° for the words *Celsius degrees* and *Fahrenheit degrees.*

Example Over a period of 50 years, the extremes of temperature in New York differed by 116 F°. Express this range in Celsius degrees.

$$116\ \text{F}^\circ = \tfrac{5}{9}(116 - 32)\ \text{C}^\circ = 46.7\ \text{C}^\circ$$

15-5 INTERNATIONAL TEMPERATURE SCALE

Once the constant-volume gas thermometer has been used, in the method suggested by Eq. (5), to determine the Kelvin temperatures of selected, reproducible reference points, then these fixed points can be used to calibrate other temperature-measuring devices more convenient to use in everyday measurements. An International Practical Temperature Scale (IPTS) was adopted in 1927 (and revised in 1948) for the practical calibration of scientific and industrial instruments. The IPTS specifies (1) the values assigned to fixed points (Table 2), (2) the standard instruments (thermocouple, resistance thermometer, and optical pyrometer) to be calibrated at those fixed points, and (3) the equations to be used in calculating temperatures from the indications of the instruments. The IPTS was originally planned to agree with the thermodynamic scale as closely as possible with the techniques available in 1927. It was expected that experimental techniques would be refined. Hence it was anticipated that the IPTS would eventually need to be revised and also extended to apply to temperatures below the oxygen point.

15-6 LINEAR EXPANSION

Two frequently observed effects of temperature changes are change in size and change of state of materials. Change of state will be discussed in Chap. 16. Both types of change can be interpreted in the atomic model suggested in Sec. 10-11. We shall consider here changes of size which occur without change of state.

Atoms of a solid are held together in a regular array by electrical forces represented by the springs in Fig. 15-6. At any temperature the atoms are in vibration, with frequency roughly 10^{13} vib/s and amplitude roughly 10^{-9} cm. These vibrations are not simple-harmonic; as the temperature rises, the increase in amplitude of atomic vibrations results in a shifting apart of the posi-

Table 2

PRIMARY FIXED POINTS OF THE INTERNATIONAL PRACTICAL TEMPERATURE SCALE

Substance	Equilibrium point	Value adopted, 1948	
		°C	°K
Oxygen	Normal boiling point	−182.970	90.18
Ice	Normal melting point	0.000	273.15
Water	Normal boiling point	100.000	373.15
Sulfur	Normal boiling point	444.60	717.75
Silver	Normal melting point	960.8	1233.95
Gold	Normal melting point	1063.0	1336.15

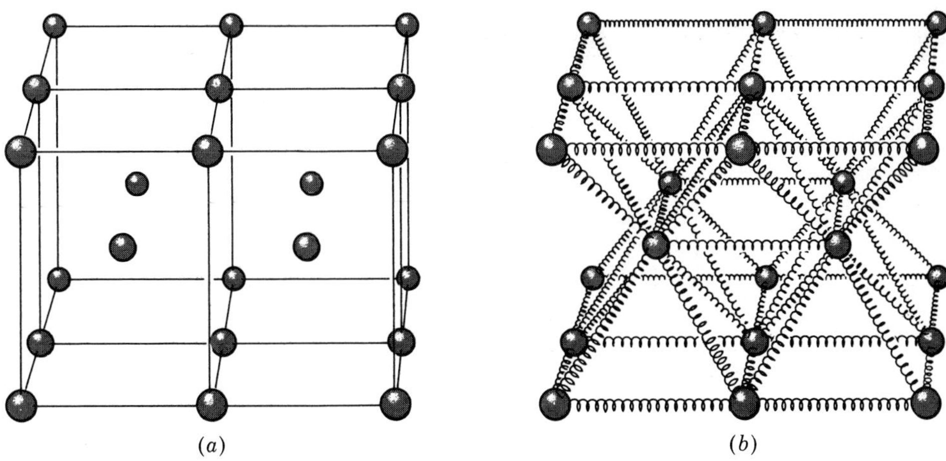

Figure 15-6
(*a*) Cubic arrangement of atoms in a crystal as inferred from x-ray data, and (*b*) the springlike forces between neighboring atoms.

tions of equilibrium of individual atoms. This produces an expansion of all the linear dimensions of a solid body, and thus an increase in volume. Liquids and gases have no shapes of their own, and therefore only volume expansion has meaning. For solids, we are primarily concerned with linear expansion.

Linear expansivity (or the *coefficient of linear expansion*) is defined as the change in length per unit length per degree rise in temperature,

$$\bar{\alpha}_0 = \frac{L_t - L_0}{L_0(t - t_0)} \tag{10}$$

where $\bar{\alpha}_0$ is the average linear expansivity for the region about temperature t_0 and L_0 and L_t are the initial and final lengths, respectively. Logically, when using Eq. (10) for calculations, one should be sure that α and L_0 are based on the same reference temperature, usually 0°C. However, for most metals the numerical value of α does not change rapidly with temperature in the range of ordinary use.

Measurements of the change in length and the initial length are expressed in the same unit of length; hence the value of $\bar{\alpha}$ will be independent of the length unit used, but it will depend on the unit used to measure the temperature interval. The value of the expansivity must be specified as *per Celsius degree* or *per Fahrenheit degree,* as the case may be. If we let ΔL represent the change in length of a bar, then

$$\Delta L \approx \alpha L_0 \, \Delta t \tag{11}$$

The final length will be

$$L_t \approx L_0 + \Delta L = L_0 + \alpha L_0 \, \Delta t$$
$$= L_0(1 + \alpha \, \Delta t) \tag{12}$$

Example Find the percentage difference between α_0 and α_{50} for aluminum.

To find a general relation between the linear expansivity α_0 based on 0°C and α_t based on some other temperature t, we can apply Eqs. (10) and (11),

$$\alpha_0 = \frac{L_t - L_0}{L_0(t - 0)} \qquad \text{or} \qquad L_t = L_0(1 + \alpha_0 t)$$

Table 3
LINEAR EXPANSIVITIES (AVERAGE, IN RANGE 0 TO 100°C)

Material	$\bar{\alpha}$, per C°	$\bar{\alpha}$, per F°
Aluminum	23×10^{-6}	13×10^{-6}
Brass	19×10^{-6}	11×10^{-6}
Copper	17×10^{-6}	9.3×10^{-6}
Germanium	6.0×10^{-6}	3.3×10^{-6}
Glass, ordinary	9×10^{-6}	5×10^{-6}
Glass, Pyrex	3.3×10^{-6}	1.8×10^{-6}
Invar (nickel-steel alloy)	0.9×10^{-6}	0.5×10^{-6}
Iron	12×10^{-6}	6.6×10^{-6}
Platinum	9.0×10^{-6}	5.0×10^{-6}
Fused quartz	0.5×10^{-6}	0.27×10^{-6}
Silicon	2.4×10^{-6}	1.3×10^{-6}
Steel	11×10^{-6}	6.1×10^{-6}
Tungsten	4.4×10^{-6}	2.5×10^{-6}
Uranium	15×10^{-6}	8.2×10^{-6}
Wood, along grain	$(3 \text{ to } 6) \times 10^{-6}$	$(2 \text{ to } 4) \times 10^{-6}$
Wood, across grain	$(35 \text{ to } 60) \times 10^{-6}$	$(20 \text{ to } 35) \times 10^{-6}$

and

$$\alpha_t = \frac{L_t - L_0}{L_t(t - 0)} \qquad \text{or} \qquad L_0 = L_t(1 - \alpha_t t)$$

By substituting the value of L_0 from the last equation into the equation above it, we find

$$\alpha_t = \frac{\alpha_0}{1 + \alpha_0 t}$$

Then, if $\alpha_0 = 23 \times 10^{-6}/\text{C}°$ (Table 3),

$$\alpha_t = \frac{23 \times 10^{-6}/\text{C}°}{1.00115}$$

Hence α_{50} is about 0.1 percent smaller than α_0.

Example A copper bar is 8.0 ft long at 68°F and has an expansivity $\alpha_{68} = 9.3 \times 10^{-6}/\text{F}°$. What is its increase in length when heated to 110°F?

$$\begin{aligned} \Delta L &= L\alpha\, \Delta t \\ &= (8.0 \text{ ft})(9.3 \times 10^{-6}/\text{F}°)(110° - 68°)\text{F} \\ &= 0.0031 \text{ ft} \end{aligned}$$

Example A steel plug has a diameter of 10.000 cm at 30.0°C. At what temperature will the diameter be 9.986 cm?

$$\Delta L = L\alpha_{30}\, \Delta t$$

From Table 3,

$$\alpha_{30} = \frac{11 \times 10^{-6}/\text{C}°}{1 + \alpha_0(30\ \text{C}°)}$$

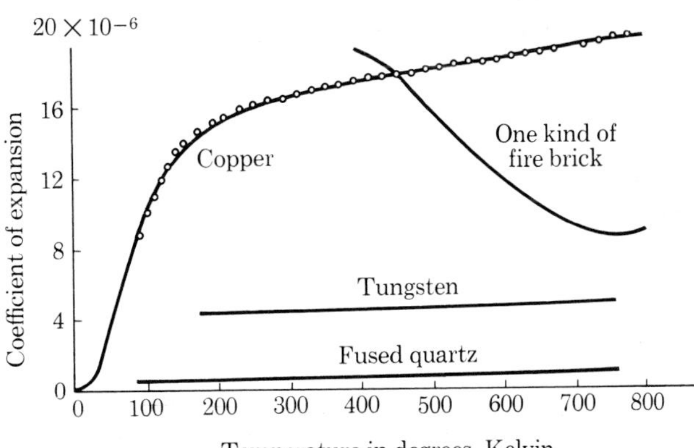

Figure 15-7
Variation of linear expansivity (per C°) for several materials.

$$\Delta t = \frac{9.986\text{ cm} - 10.000\text{ cm}}{(10.000\text{ cm})(11 \times 10^{-6}/\text{C}°)} = -1\bar{3}0\text{ C}°$$

Hence the required temperature $t = 30.0°\text{C} - 1\bar{3}0°\text{C} = -100°\text{C}$.

Precise measurements show that, for a given value of $\bar{\alpha}$, Eqs. (11) and (12) are inadequate to describe the expansion of a material over all ranges of temperature (Fig. 15-7). More exact expressions are obtained by introducing terms with higher powers in the right-hand member of Eq. (12).

When the temperature of a solid is raised, it expands in all directions. Certain crystals are found to have different expansivities along different axes. However, many of the common materials have the *same properties in all* directions. The latter are called *isotropic* substances.

Thermal expansion may be large enough to require special expansion joints in bridges, buildings, and pavements. Thermal expansion makes possible the shrink fitting of collars on shafts. Sometimes, as in constructing a furnace or sealing metal electrodes through a glass bulb, it is necessary to choose materials whose expansivities closely match over the temperature range to be used so that thermal strains do not cause failure. This would be especially hazardous in space flights when the temperature differential is great.

In the thermal expansion of an isotropic solid, the distance between any two points, such as a and b in Fig. 15-8, increases in the ratio α per degree rise in temperature.

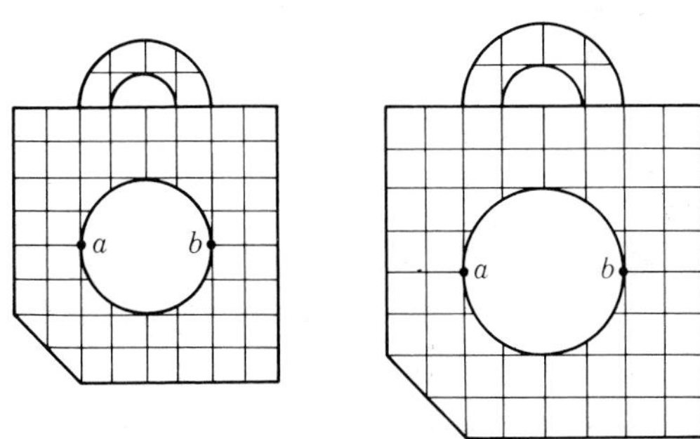

Figure 15-8
Thermal expansion, like photographic enlargement, increases every linear dimension by the same proportion. (The exaggerated expansion shown here would correspond to a temperature rise of about 25,000 C° for a constant α.)

A classic demonstration of uniform thermal expansion is given by the ball-and-ring device. A metallic ball is designed to just fit through a metal ring of the same material when at the same temperature. When the ball alone is heated, it will not pass through the ring due to the expansion of the ball. When the ring alone is heated, the hole does not get smaller as one might expect, but rather it expands according to the theory expressed above and the ball can still fit through the ring.

A bimetallic strip, used in some thermometers and thermostats (Fig. 15-9), is made by welding together side by side two metal strips each of length l_0 and thickness d so that their ends coincide at temperature T_0. One metal has linear expansivity α_1, the other α_2, where $\alpha_1 \neq \alpha_2$. When the bimetallic strip is heated to a temperature greater than t_0, one strip becomes longer than the other, causing the bimetallic element to bend in a circular arc. This deformation can move a pointer over a thermometer scale or actuate electrical contacts in a thermoregulator.

Example Show that the coefficient of area expansion, $\Delta A/A_0\,\Delta t$, of an isotropic solid is twice its linear expansivity α.

Consider a rectangular sheet of the solid material (Fig. 15-10). When the temperature increases by Δt, a increases by $\Delta a = \alpha a\,\Delta t$ and b

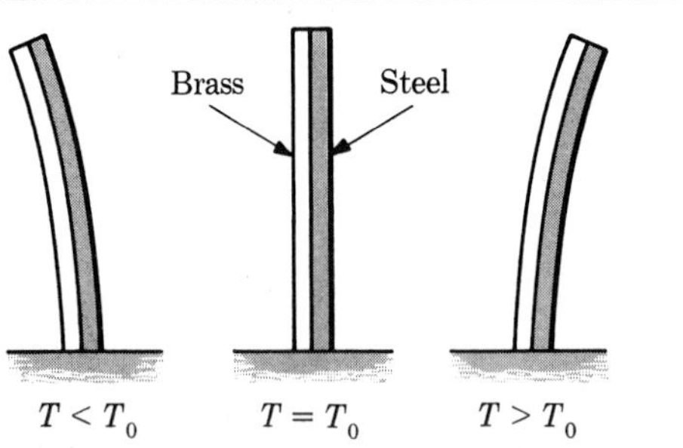

Figure 15-9
A bimetallic element for temperature measurement or control.

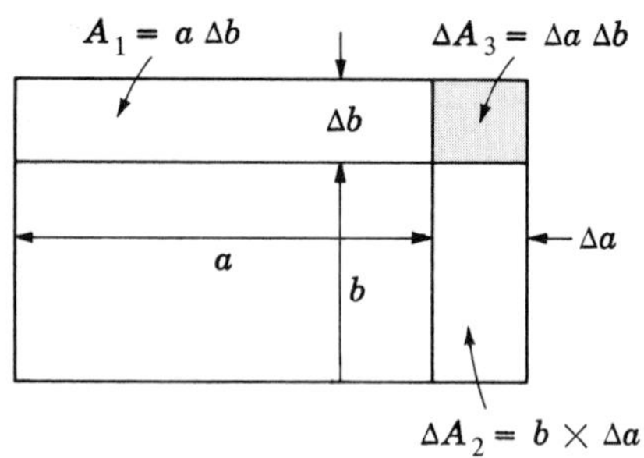

Figure 15-10
Increase of area in thermal expansion.

increases by $\Delta b = \alpha b\,\Delta t$. From Fig. 15-10, the increase in area $\Delta A = \Delta A_1 + \Delta A_2 + \Delta A_3$:

$$\begin{aligned}\Delta A &= a\,\Delta b + b\,\Delta a + \Delta a \cdot \Delta b \\ &= a\alpha b\,\Delta t + b\alpha a\,\Delta t + \alpha^2 ab(\Delta t)^2 \\ &= \alpha ab\,\Delta t(2 + \alpha\,\Delta t)\end{aligned}$$

Since $\alpha \approx 10^{-5}$ per C°, from Table 3, the product $\alpha\,\Delta t$ for practical temperatures is small in comparison with 2 and may be neglected. Hence

$$\Delta A \approx 2\alpha A\,\Delta t$$

In this approximation we have neglected the area of the smallest rectangle $\Delta a \cdot \Delta b$ in Fig. 15-10.

15-7 VOLUME EXPANSION

The *volume expansivity* for a material is the change in volume per unit volume per degree rise in temperature,

$$\bar{\beta} = \frac{V_t - V_0}{V_0\,\Delta t} = \frac{\Delta V}{V_0\,\Delta t} \tag{13}$$

where $\bar{\beta}$ is the average volume expansivity, V_t is the volume at temperature t, V_0 the volume at 0°C, and Δt the temperature change.

Table 4
VOLUME EXPANSIVITIES OF LIQUIDS

Liquid	β, per C°	β, per F°
Alcohol, ethyl	1.0×10^{-3}	6.1×10^{-4}
Mercury	1.8×10^{-4}	1.0×10^{-4}
Water (15–100°)	3.7×10^{-4}	2.0×10^{-4}

There is a simple approximate relation between the linear expansivity for an isotropic solid and the corresponding volume expansivity

$$\beta \approx 3\alpha \tag{14}$$

This relation can be derived by consideration of the expansion of a cube of the material: $V_t = V_0 (1 + \beta \, \Delta t) = L_0^3(1 + \alpha \, \Delta t)^3$. If the right-hand side of this equation is expanded and terms in α^2 and α^3 are neglected, Eq. (14) follows.

15-8 THE GAS LAWS

We have found that pressure, volume, and temperature of a gas are related. In Chap. 12 we considered the relation between pressure P and volume V when the temperature remains constant, namely, that the product of pressure and volume is constant.

$$PV = k \tag{15}$$

This relation is known as *Boyle's law.*

We have now found that there are relationships between temperature and pressure if the volume remains unchanged and between temperature and volume if the pressure is constant. These relations are expressed simply if an *absolute* (Kelvin) temperature scale is used.

From Eq. (4),

$$\frac{P}{T} = k_1 \qquad \text{at constant volume} \tag{16}$$

A similar relation for a constant-pressure gas thermometer gives

$$\frac{V}{T} = k_2 \qquad \text{at constant pressure} \tag{17}$$

These two equations express *Charles's laws: If the*

Table 5
VOLUME EXPANSIVITIES OF GASES

Substance	Temperature, °C	Pressure, cm of Hg	Volume expansivity, per C°	Pressure coefficient, per C°
Air	0–100	0.6		376.66×10^{-5}
	0–100	25		365.80×10^{-5}
	0–100	76	367.1×10^{-5}	366.50×10^{-5}
	0–100	100	367.28×10^{-5}	367.44×10^{-5}
	0–100	2,000		388.66×10^{-5}
Hydrogen	0–100	76.4	366.00×10^{-5}	365.04×10^{-5}
Carbon dioxide	0–100	100	374.10×10^{-5}	372.48×10^{-5}
Sulfur dioxide	0–100	76	390.3×10^{-5}	384.5×10^{-5}

volume of a confined gas is constant, the pressure is directly proportional to the absolute temperature, or

$$\frac{P_1}{T_1} = \frac{P_2}{T_2} \tag{18}$$

If the pressure of a confined gas is unchanged, the volume is directly proportional to the absolute temperature or

$$\frac{V_1}{T_1} = \frac{V_2}{T_2} \tag{19}$$

15-9 THE GENERAL GAS LAW

The three gas laws just stated can be combined into a single law that applies to changes in conditions of a gas whether the factors change one at a time or simultaneously.

$$\frac{P_1V_1}{T_1} = \frac{P_2V_2}{T_2} = k$$

assuming that the mass is kept constant. This relationship can be written in a simpler form if we express the mass in *kilomoles, n,* of the gas. A mole (mol) is defined as one molecular mass of the substance expressed in grams. From chemistry we find that the molecular mass of oxygen is 32; thus 1 mol of oxygen has a mass of 32 g. Also, 1 mol of hydrogen, which has a molecular mass of 2, consists of 2 g. A kilomole is simply 1,000 mol of a gas.

By definition, then, $n = m/W$; where n is the number of kilomoles, m is the mass of the gas expressed in grams, and W is the atomic mass of the gas. To find n for a particular gas, we divide the mass of the gas by the atomic or molecular mass.

By experiment it has been found that when the constant k in the combined equation has been measured for various kinds of gases,

$$\frac{PV}{T} = k = \frac{m}{W}R \quad \text{and} \quad \frac{PV}{T} = nR$$

where R is a constant called the *universal gas constant* which can be used for all low-density gases. Therefore, P, V, T, and n can be related to each other by what is known as the *general gas law:*

$$PV = nRT \tag{20}$$

The value of R depends upon the units selected for pressure and volume in the equation. When n is in kilomoles and T in degrees Kelvin,

$$R = 8{,}317 \text{ J/(kmol)(K°)} \tag{21}$$

In calculations it is often convenient to express the quantities in Eq. (20) in other than mks units. For example, P may be given in atmospheres and V in liters. Equivalent values of the gas constant R expressed in different units may be useful for reference:

$$\begin{aligned} R &= 0.08205 \text{ l·atm/(mol)(K°)} \\ &= 8.317 \times 10^7 \text{ ergs/(mol)(K°)} \\ &= 1.987 \text{ cal/(mol)(K°)} \end{aligned} \tag{22}$$

If we introduce the restrictions stated in the special laws, the general gas law reduces to the simpler expressions. If mass and temperature are constant, Eq. (20) reduces to $PV = k$ (Boyle's law). If m and V are constant, $P/T = k_1$, while if P and m are constant, $V/T = k_2$. These are Charles's laws.

In using the general gas law it may be convenient to write it in other forms. If we divide both sides of Eq. (20) by mT and express the result in terms of initial and final conditions, we have

$$\frac{P_1V_1}{m_1T_1} = \frac{P_2V_2}{m_2T_2} \tag{23}$$

or, since the density $\rho = m/V$,

$$\frac{P_1}{\rho_1T_1} = \frac{P_2}{\rho_2T_2} \tag{24}$$

The general gas law can be used to study any changes in the conditions of the gas so long as *absolute* temperatures and *complete* (absolute) pressures are used. Gauge pressures cannot be used in the gas law.

Example A 5,000-cm³ container holds 4.90 g of a gas when the pressure is 75.0 cm Hg and the temperature is 50°C. What will be the pressure if 6.00 g of this gas is confined in a 2,000-cm³ container at 0°C?

From Eq. (23),

$$\frac{P_1V_1}{m_1T_1} = \frac{P_2V_2}{m_2T_2}$$

$$\frac{P_1 \times 2{,}000 \text{ cm}^3}{6.00 \text{ g} \times 273°\text{K}} = \frac{75.0 \text{ cm Hg} \times 5{,}000 \text{ cm}^3}{4.90 \text{ g} \times 323°\text{K}}$$

$$P_1 = 194 \text{ cm Hg}$$

Example The weight-density of air at 32°F and 29.92 in Hg pressure is 0.081 lb/ft³. What is its weight-density at an altitude where the pressure is 13.73 in Hg and the temperature is −40°F?

From Eq. (24),

$$\frac{P_1}{D_1T_1} = \frac{P_2}{D_2T_2}$$

$$T_1 = 32°\text{F} = 273°\text{K}$$

$$T_2 = -40°\text{F} = 233°\text{K}$$

$$\frac{29.92 \text{ in Hg}}{0.081 \text{ lb/ft}^3 \times 273°\text{K}} = \frac{13.73 \text{ in Hg}}{D_2 \times 233°\text{K}}$$

$$D_2 = 0.044 \text{ lb/ft}^3$$

Example Air at pressure 14.7 lb/in² is pumped into a tank whose volume is 42.5 ft³. What volume of air must be pumped in to make the gauge read 55.3 lb/in² if the temperature is raised from 70 to 117°F in the process?

$$P_1 = 14.7 \text{ lb/in}^2$$

$$P_2 = 14.7 \text{ lb/in}^2 + 55.3 \text{ lb/in}^2 = 70.0 \text{ lb/in}^2$$

$$T_1 = 70°\text{F} = 294°\text{K}$$

$$T_2 = 117°\text{F} = 320°\text{K}$$

Since m is constant,

$$\frac{P_1V_1}{T_1} = \frac{P_2V_2}{T_2}$$

$$\frac{14.7 \text{ lb/in}^2 \times V_1}{294°\text{K}} = \frac{70.0 \text{ lb/in}^2 \times 42.5 \text{ ft}^3}{320°\text{K}}$$

$$V_1 = 199 \text{ ft}^3$$

Since 42.5 ft³ of air was in the tank at the beginning, the volume the added air would occupy at atmospheric pressure is

$$V = 199 \text{ ft}^3 - 42.5 \text{ ft}^3 = 157 \text{ ft}^3$$

SUMMARY

The *temperature* of an object is that property which determines the direction of flow of heat between it and its surroundings.

Our subjective idea of temperature is obtained from the sensation of warmth or cold that we experience on touching an object.

Thermal equilibrium exists when there is no net transfer of heat between two objects.

Any inanimate object without a built-in heating system will eventually attain the temperature of the environment.

A thermometer scale is established by choosing a simple relation between a measurable physical property and temperature, the zero of the scale being fixed by assigning a numerical value to an easily reproducible temperature (triple point of water).

A thermometric property of matter is a property that varies predictably with an increase or decrease in heat energy.

Conversions between Celsius and Fahrenheit scale readings are made by the relations

$$\text{F} = \tfrac{9}{5}\text{C} + 32°$$

$$\text{C} = \tfrac{5}{9}(\text{F} - 32°)$$

The International Practical Temperature Scale specifies fixed points, instruments, and calculations to be used in the practical measurement of temperature, to give values that are reproducible and internationally acceptable.

The *expansivity* is the fractional change (in length or in volume) per degree change in temperature. The units, per C° or per F°, must be expressed.

$$\alpha = \frac{l_t - l_0}{l_0\,\Delta t} \qquad \beta = \frac{V_t - V_0}{V_0\,\Delta t}$$

The expansion of a material is equal to the product of the expansivity, the original size (length or volume), and the temperature change

$$\Delta l = \alpha l_0\,\Delta t \qquad \text{and} \qquad \Delta V = \beta V_0\,\Delta t$$

The pressure coefficients of expansion and the volume expansivities of all gases are approximately equal to $\frac{1}{273}$ per C°.

The general gas equation is

$$PV = nRT \qquad \text{or} \qquad \frac{P_1V_1}{m_1T_1} = \frac{P_2V_2}{m_2T_2}$$

References

Conant, James B. (Ed.): "Harvard Case Histories in Experimental Science," vol. I, Harvard University Press, Cambridge, Mass., 1957.

Herzfeld, Charles M.: "Temperature: Its Measurement and Control in Science and Industry," Reinhold Publishing Corporation, New York, 1962.

Hoyle, Fred: Ultrahigh Temperatures, *Scientific American,* September 1954, pp. 144–154.

MacDonald, D. K. C.: "Near Zero, the Physics of Low Temperature," Science Study Series, Doubleday & Company, Inc., Garden City, N.Y., 1962.

Zemansky, Mark W.: The Use and Misuse of the Word "Heat" in Physics Teaching, *The Physics Teacher,* September 1970, p. 294.

Questions

1 Give illustrations supporting the statement that our body's temperature sense is not reliable in determining the temperature of objects.

2 Explain what is meant by a thermometric property and give illustrations of objects possessing this property.

3 Explain what the term *isotropic* means and give examples of isotropic substances.

4 List several methods of measuring temperatures, and mention the factors that limit the range of each. What means are available for measuring the highest attainable temperatures?

5 What factors must be taken into account in the design of a sensitive thermometer?

6 Mercury is commonly used in thermometers. Give three advantages of using mercury in thermometers. State one disadvantage.

7 Why might alcohol be a good choice for the indicator in a thermometer used to measure very low temperatures?

8 Can water be used as a thermometric fluid in a thermometer? What disadvantages do you see resulting from using water in this manner?

9 Outline carefully the logical steps in the definition of a temperature scale.

10 Describe a gas thermometer and explain how it operates.

11 Clinical thermometers are used to measure a person's body temperature. How do they differ from ordinary thermometers?

12 Suggest a procedure by which a constant record of the body temperature of an astronaut can be obtained.

13 How do you account for the fact that solids and liquids generally expand when heated?

14 Explain why the water in a lake freezes at the top and not at the bottom.

15 If in constructing a thermometer, you wish to use a liquid which would give you the greatest space between gradations on the thermometer, which of the following liquids would you use: mercury, alcohol, or water? Why?

16 Define the coefficient of linear expansion. What are the common metric and British units for it?

17 Describe a method of determining the coefficient of linear expansion for a metal rod.

18 Would the linear expansivity be different if the inch were used as the unit of length rather than the centimeter? If the Fahrenheit scale were used rather than the Celsius scale? Explain.

19 An observation made on the expansion of a liquid contained in a glass bulb or tube does not give the true expansion of the liquid. Explain.

20 Suggest several practical uses of the differential expansion of two different materials. In what cases is differential expansion undesirable?

21 Describe how a bimetallic rod behaves when heated and how it can be used to operate some mechanical device. Give examples.

22 In the early manufacture of light bulbs, platinum wire was sealed through the glass. By reference to Table 3 show why this was feasible.

23 A steel plug fits into a hole in a brass plate. What will be the effect on the closeness of fit if the plug alone is heated? If the plate alone is heated? If both are heated equally?

24 To open a tightly capped jar, the cap is held under a faucet through which hot water is running. Why would this help to loosen the cap?

25 From Table 3 find the relative coefficients of linear expansion for Pyrex and regular glass. Do these data suggest why Pyrex glass can withstand larger and more rapid changes of temperature than ordinary glass? Explain.

26 Explain why the mercury column first descends and then rises when a mercury-in-glass thermometer is plunged into hot water.

27 Show that the coefficient of volume expansion is three times that of the coefficient of linear expansion.

28 If a hole is bored in a thin steel plate and the plate is then heated, will the hole become larger or smaller? Explain your answer.

Problems

1 Change $-15°F$ to degrees Kelvin.

2 Change $303°K$ to degrees Celsius and °F.

Ans. 30°C; 86°F.

3 On a temperature scale proposed by Anders Celsius in 1742, the 0 reading referred to the boiling point of water and the 100 reading to the temperature of melting ice. Convert a temperature of 68°F to this Celsius scale.

4 Liquid oxygen freezes at $-218.4°C$ and boils at $-183.0°C$. Express these temperatures in terms of the Fahrenheit scale.

Ans. $-361.1°F$; $-297.4°F$.

5 Convert $-14°C$, 20°C, and 60°C to Fahrenheit readings. Convert 98°F, $-13°F$, and 536°F to Celsius readings.

6 Man has been able to produce temperatures as low as $-459°F$. Express this temperature in degrees Celsius and degrees Kelvin.

Ans. $-272.8°C$; 0.2°K.

7 What is the approximate temperature of a healthy person in degrees Celsius?

8 Calculate the temperature at which the readings of a Fahrenheit and a Celsius thermometer are the same. *Ans.* $-40°$.

9 On a hypothetical temperature scale X, the ice point is assigned the reading 40° and the steam point 160°. For another scale Y, the assigned values are $-20°$ and 180°, respectively. Convert a reading of 20°X to the Y scale.

10 Which is colder, Dry Ice (solid CO_2) subliming at $-78.5°F$ or ethyl alcohol freezing at $-179°F$? *Ans.* Ethyl alcohol is colder ($-117°C$).

11 The brass scale of a mercury barometer gives the correct length of the mercury column at 0°C. If the barometer reads 740 mm at 25°C, what is the reading when corrected for expansion of the brass scale and for the expansion of the mercury (which decreases its density)?

12 A 2-m-long aluminum pipe at 27°C is heated until it is 2.0024 m at 77°C. What is the coefficient of linear expansion of aluminum?

Ans. $2.4 \times 10^{-5}/°C$.

13 A wheel is 0.915 m in circumference. An iron tire measures 0.912 m around its inner face. How much must the temperature of the tire be raised in order that it may just slip onto the wheel?

14 An aluminum piston having a diameter of 3.000 in at 0°C is designed to fit into a steel cylinder with a clearance of 0.005 in (difference

in radii) at 0°C. What is the maximum temperature at which the piston can operate?
Ans. 278°C.

15 The ends of a 2.00-ft steel rod having a cross section of 1.00 in^2 are clamped to rigid supports when the rod is at 212°F. If the rod is allowed to cool to 92°F without shortening, what will be the tension in the rod?

16 The lower end of a vertical steam pipe 50 ft long is supported rigidly by a hanger attached to the basement ceiling. When the pipe is at 40°F, a steam radiator attached rigidly to the upper end of the pipe rests on the attic floor. Find the distance the radiator is lifted off the floor when the iron steam pipe is at 220°F. *Ans.* 0.72 in.

17 A vertical cylindrical glass tube 100 cm long is closed at its lower end. To what height must it be filled with mercury in order that the volume of the tube above the mercury shall remain constant as the temperature changes?

18 The hot-water heating system of a building contains 60.0 ft^3 of water in steel pipes. How much water will overflow into the expansion tank if the system is filled at 40°C and then is heated to 80°C? *Ans.* 0.82 ft^3.

19 A thin brass sheet at 10°C has the same surface area as a thin steel sheet at 20°C. At what common temperature, if any, will the two sheets have the same area?

20 A 20-mi marathon course is measured off by using a steel tape measure correct at 0°C. If the temperature is 25°C when the course is laid, will the runners run a greater or lesser distance than they are scheduled to run? How many miles difference will there be?
Ans. Too far; 0.0056 mi.

21 A thin-walled aluminum hoop is to be swung from a pivot at its circumference during a change in temperature. Calculate the percentage change in its moment of inertia per centigrade degree rise in temperature, at a temperature of 20°C.

22 A metal bar is 52.302 cm long at 18.0°C and 52.359 cm long at 98.0°C. What is the linear expansivity of the metal? *Ans.* 1.36×10^{-5}/°C.

23 A Pyrex-glass graduated cylinder calibrated at 0°C holds 200.0 cm^3 of water at 0°C. What is the apparent volume of the liquid as read on the scale when both the water and the cylinder are heated to 50°C?

24 The coefficient of linear expansion of steel is 1×10^{-5}/°C. How much expansion should engineers anticipate in a 2,000-ft steel bridge if it undergoes a change in temperature from 0°C to 30°C? *Ans.* 0.6 ft.

25 A 50-gal steel drum is filled with gasoline when the temperature is 50°F. How much of the gasoline will overflow when the temperature becomes 110°F? The volume expansivity of gasoline is 0.00096 per C°.

26 An iron rod and a zinc rod have lengths of 25.55 cm and 25.50 cm, respectively, at 0°C. At what temperature will the rods have the same lengths? The coefficients of expansion of iron and zinc are 0.000010 and 0.000030 per °C, respectively. *Ans.* 98°C.

27 A tire gauge registers 32 lb/in^2 when air in a certain tire is at 10°C. After running for a time, the tire heats up to 30°C. Calculate the pressure that would be registered then by the gauge.

28 The spherical space satellite Explorer V was made of aluminum and had a diameter of 36.0 cm at 0°C. It cooled off to −35°C when it reached its orbital height. What was its change in volume? *Ans.* Decreased by 56.5 cm^3.

29 Explorer V had copper antennas that were each 25 cm long. If they changed in temperature from 0°C to −35°C, what change in length resulted?

30 A petroleum sample has a specific gravity of 0.847 at 20°C. If the coefficient of volume expansion of the petroleum is 0.000899 per °C, what is the specific gravity at 70°C? *Ans.* 0.810.

31 Compare the densities of the air at the bottom and the top of a mine shaft when the temperatures and pressures are, respectively, 5°C and 770 mm Hg at the bottom and 20°C and 760 mm Hg at the top.

32 A gas having a volume of 100 ft^3 at 27°C is expanded to 120 ft^3 by being heated at constant pressure. To what temperature has it been heated to have this new volume? *Ans.* 87°C.

33 The pressure of the gas in a gas thermometer

is 76.5 cm Hg at 27.0°C. By how much will the pressure increase when the temperature rises 1.0 C°?

34 A gas thermometer contains 100 cm^3 of air at 20°C. At what temperature will the volume of the air be 110 cm^3. *Ans.* 49.3°C.

35 An airtight cylinder 300 cm long is divided into two parts by a freely moving, airtight, heat-insulating piston. When the temperatures in the two chambers are equal to 27°C, the piston is located 100 cm from one end of the cylinder. How far will the piston move if the gas in the smaller part of the cylinder is heated to 74°C and the temperature in the larger section remains constant? Assume that the ideal-gas laws hold.

36 The volume of a gas held at constant pressure increases from 4.0 cm^3 at 0°C to 5.0 cm^3 at 100°C. At what temperature will this gas occupy a volume of 4.6 cm^3. *Ans.* 37°C.

37 A liter of alcohol at 0°C is heated to 40°C. What will its new volume be?

Robert Andrews Millikan, 1868–1953

Born at Morrison, Illinois. Director, Norman Bridge Laboratory of Physics, California Institute of Technology. Presented the 1923 Nobel Prize for Physics for his work on the elementary electric charge and on the photoelectric effect.

Karl Manne Georg Siegbahn, 1886–

Born in Örebro, Sweden. Professor at the University of Uppsala, later director of the Research Institute for Physics, Stockholm. The 1924 Nobel Prize for Physics was conferred on Siegbahn for his research and discoveries in x-ray spectroscopy.

16

Heat Phenomena

During the last two centuries, from the speculations of the caloric theory and from the experiments, in various countries, by Carnot, Colding, Rumford, Mayer, Helmholtz, and Joule, there arose the concept that heat is a form of kinetic energy, associated with the random agitation of molecules. Most of the energy of practical importance to us comes from the sun, and heat is involved in some stage in the conversion of this energy to do useful work. The production of heat is often necessary to maintain life and in the working of metals, glass, and other materials. In some cases prompt dissipation of unwanted heat is important, as during a space vehicle's reentry into the atmosphere or during the braking of a car on a long downgrade. So the laws governing heat transfer are important.

We can use the term *internal energy* to designate the total energy content of a system. The internal energy of a system of particles may be either *ordered* or *disordered,* or perhaps both. When the internal energy is ordered it can be observed as potential or kinetic energy. For example, a spring with a load attached to it which is oscillating in simple harmonic motion is a system in which the internal energy is ordered, that is, the kinetic and the potential energies of the masses and the spring are identifiable. When the internal energy is disordered, no large manifestations of kinetic or potential energy are observed. To illustrate this, let us consider gases expanding due to combustion in a cylinder of an engine (Fig. 16-1). Only a small fraction of the energy created in the burning of the hot gas will be used to drive the piston downward and to do useful work. This is due to the fact that only a few of the randomly moving particles in the cylinder will strike the piston in a direction that will cause it to move.

The internal energy of an *ideal gas* is energy arising from the random, disorganized, or *dis-*

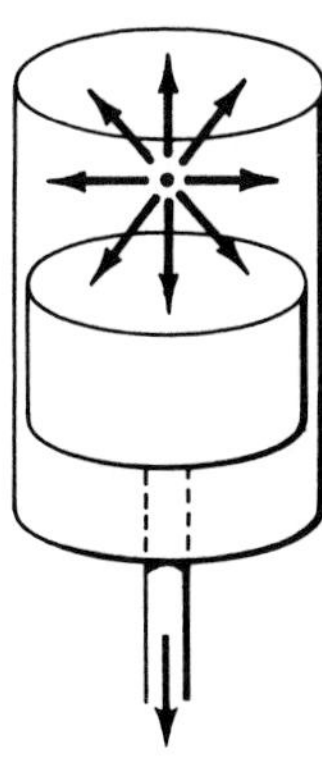

Figure 16-1
Disordered motion of particles due to an exploding gas in a cylinder.

ordered motion of many particles. The disordered internal energy of a system is called *thermal energy*.

On the view that heat is disordered energy, one might expect that the specification of heat requires two numbers: one to measure the quantity of energy, the other to measure the degree of disorganization of the energy in a system (the disorder). We shall look further at the energy aspect in this chapter and then, in Chap. 17, examine the mathematical concept of *entropy,* to measure the amount of random molecular energy which is available for conversion to useful work.

16-1 MEANING OF HEAT

Generally, when two bodies isolated from heat sources at different temperatures are placed together and allowed to remain indefinitely in contact, they attain a final temperature which is somewhere between the two original temperatures. This was explained, up to the start of the nineteenth century, by assuming the existence of a weightless, invisible fluid called *caloric* which was produced when a substance burned and which could be conducted from one body to another. The caloric theory served surprisingly well. But, in time, evidence that heat could not be a substance was given by Benjamin Thompson (1753–1814), an American who became Count Rumford of Bavaria. In one of his many enterprises, Rumford supervised the boring of cannon. This operation produced heat, enough to boil water placed in the bore. The apparent production of caloric was explained according to the prevailing theory by supposing that when matter was finely divided (as by the boring tool) it lost some of its ability to retain caloric. The caloric liberated from the metal shavings caused the water to boil.

Rumford surrounded a cannon by a wood box in which he placed water. The rise in temperature of this water was a measure of the heat produced during boring. Rumford showed that even a dull tool which did not cut the metal was apparently an inexhaustible source of caloric as long as mechanical work was done to rotate the tool. Writing of this experiment (1798), he ruled out possible caloric interpretations and concluded "It appears to me to be extremely difficult, if not quite impossible, to form any distinct idea of anything capable of being excited and communicated in the manner the Heat was excited and communicated in these experiments, except it be MOTION."

The idea that *mechanical work* was responsible for the creation of *heat* was stated in clearer terms by Mayer, Joule, Helmholtz, and Colding. Each measured the heat produced by the performance of definite amounts of mechanical work. In each case the quotient of the energy expended and the heat produced was found to be a constant. This experience was generalized and stated as a consequence of the law of conservation of energy: *the total quantity of energy in a closed system (heat being a form of energy) is constant.* This principle is called the *first law of thermodynamics.*

16-2 QUANTITY OF HEAT

When a new area of physics is being investigated, units are defined in terms of the experimental procedures. These units may later have to be related to more general units, or it may even be desirable to replace the tentative units by those of more general applicability. Heat was originally measured by noting the rise in temperature of a measured quantity of water which absorbed the heat. The *calorie* was defined as the quantity of heat required to raise the temperature of one gram of water through one degree Celsius. The *British thermal unit* (Btu) was defined as the quantity of heat required to raise the temperature of one pound mass of water[1] through one Fahrenheit degree. One Btu is equivalent to approximately two hundred and fifty-two calories. The

[1] Instead of defining thermal units in terms of unit mass (1 slug), engineering practice uses the pound mass (equal to 1/32.174 slug) in heat measurements.

kilocalorie (kcal), or kilogram-calorie,[1] is the amount of heat necessary to raise the temperature of one kilogram of water one Celsius degree.

Since the amount of heat required to raise the temperature of unit mass of water one degree is not quite constant throughout the temperature range, the definitions of the Btu and calorie given above may be regarded as the average values in the region from freezing to boiling. When greater precision of definition is desired, the exact temperature range is specified: the 15° calorie is the heat required to change the temperature of 1 g of water from 14.5 to 15.5°C.

Refined versions of Rumford's experiment, especially those performed by Joule, established the number of units of work equivalent to 1 unit of heat, called the *mechanical equivalent of heat*. Thus if work $\mathcal{W}$ is converted into heat Q, $\mathcal{W} \propto Q$; $\mathcal{W} = kQ$; $\mathcal{W}/Q = k = J$

$$\mathcal{W} = JQ \tag{1}$$

where J (after Joule) is a constant, independent of the magnitude of $\mathcal{W}$ or Q, but whose value depends on the units in which $\mathcal{W}$ and Q are expressed. Once it is established that heat is a form of energy, the joule, the unit used to measure energy in mechanics and electricity, could replace the somewhat vaguely defined water-temperature units of heat, the calorie and the Btu. So, while these units are still used, they are now defined by international agreement as certain multiples of the joule,

$$1 \text{ cal} = \tfrac{1}{860} \text{ W}\cdot\text{h} = 4.18605 \text{ J} \tag{2}$$

The Btu is defined through the relation 1 cal/(g)(C°) = 1 Btu/(lb_m)(F°), giving

$$1 \text{ Btu} = 778.26 \text{ ft}\cdot\text{lb} = 251.996 \text{ cal} \tag{3}$$

[1]Nutritionists in discussing food metabolism use the term *Calorie* (often capitalized to distinguish it from the "small calorie") for what we have defined as the kilocalorie. Their statement that there are 63 Cal in a slice of white bread means that about 63 kcal will be liberated when the dried sample of bread is completely burned in an atmosphere of oxygen in a calorimeter.

The values of J, the mechanical equivalent of heat, are now accepted as 4.18605 J/cal and 778.26 ft·lb/Btu.

Example If a 10-kg piece of copper falls 100 m, how much heat might be produced? How much could its temperature be changed by falling? Specific heat of copper is about 0.1 cal/(g)(C°) or 0.1 kcal/(kg)(C°). (See Sec. 16-3.)

The energy given up in falling is

$$E_p = mgh = 10 \text{ kg} \times 9.8 \text{ m/s}^2 \times 100 \text{ m}$$
$$= 9{,}800 \text{ J}$$

Since 1 cal = 4.186 J, the heat produced is

$$Q = 9{,}800 \text{ J} \frac{1 \text{ cal}}{4.186 \text{ J}}$$
$$= 2{,}340 \text{ cal} \qquad \text{or } 2.34 \text{ kcal}$$

and

$$Q = mc\,\Delta t$$

$$\Delta t = \frac{Q}{mc} = \frac{2.34 \text{ kcal}}{10 \text{ kg} \times 0.1 \text{ kcal/(kg)(C°)}}$$

Therefore, $\Delta t = 2.34$ C°.

16-3 SPECIFIC HEAT

The heat needed to change the temperature of a unit mass of a substance one degree is characteristic of that substance. The specific heat[1] of a substance is defined as the heat per unit mass per degree change in temperature,

$$c = \frac{Q}{m\,\Delta t} \tag{4}$$

where c is the specific heat, Q the heat added to

[1]Some authors call this quantity the *thermal capacity* of the substance and define specific heat as the ratio of the thermal capacity of the substance to that of water.

Table 1
SPECIFIC HEATS

Substance	Cal/(g)(C°), or Btu/(lb)(F°)	J/(kg)(C°)
Alcohol, ethyl	0.60	2.5×10^3
Aluminum	0.217	0.909×10^3
Brass	0.090	0.38×10^3
Carbon, graphite	0.160	0.65×10^3
Carbon, diamond	0.120	0.49×10^3
Copper	0.093	0.39×10^3
Ethylene glycol, Prestone	0.528	2.21×10^3
Glass, soda	0.16	0.65×10^3
Gold	0.0316	0.130×10^3
Hydrogen (at 15°C, constant pressure)	3.389	14.2×10^3
Ice	0.51	2.14×10^3
Iron	0.115	0.481×10^3
Lead	0.031	0.13×10^3
Mercury	0.033	0.14×10^3
Silver	0.056	0.23×10^3
Tungsten	0.034	0.14×10^3
Water (by definition)	1.00	4.19×10^3
Water vapor (100°C, constant pressure)	0.482	2.02×10^3
Zinc	0.093	0.39×10^3

the material of mass m, and Δt the rise in temperature.

In the British system the specific heat is expressed in Btu per pound mass (= 1/32.174 slug) per Fahrenheit degree; and in the metric system, in calories per gram per Celsius degree. Because of the way the Btu and the calorie are defined, the specific heat of a substance in metric units is the same numerically as that expressed in the British system. For example, the specific heat of salt, which is 0.204 Btu/(lb_m)(F°) is also 0.204 cal/(g)(C°). Water has a specific heat much larger than that of most common materials. Specific heat is not strictly a constant; it varies somewhat with temperature.

Knowing the specific heat c of a material, one can calculate the heat Q necessary to change the temperature of a mass m from an initial value t_i to a final value t_f from the relation

$$Q = mc(t_f - t_i) \tag{5}$$

Example How much heat is necessary to raise the temperature of 0.80 kg of ethyl alcohol from 15.0°C to its boiling point, 78.3°C?

$$Q = (0.80\ \text{kg})[2.5 \times 10^3\ \text{J/(kg)(C°)}] \times (78.3°\text{C} - 15.0°\text{C})$$

$$= 1.3 \times 10^5\ \text{J}$$

Alternatively,

$$Q = (800\ \text{g})[0.60\ \text{cal}/(\text{g})(\text{C}^\circ)] \times (78.3^\circ\text{C} - 15.0^\circ\text{C})$$

$$= 3.0 \times 10^4\ \text{cal}$$

The experimental determination of the specific heat of a metal by the *method of mixtures* consists essentially in adding a known mass of metal at a known high temperature to a known mass of water at a known low temperature and determining the equilibrium temperature that results. The heat absorbed by the water and containing vessel can be equated to the heat given up by the hot metal. From this equation the unknown specific heat can be computed.

$$\text{Heat lost} = \text{heat gained} \tag{6}$$

The heat lost by the warm object is $m_x c_x \Delta t_x$, where m_x is the mass of the object, c_x the specific heat of the object, and Δt_x the change in its temperature. The heat gained by the container and water will be $m_c c_c \Delta t_c + m_w c_w \Delta t_w$, where m_c and c_c are the mass and specific heat of the container and m_w and c_w are the mass and specific heat of the water in the container. The temperature change Δt_c refers to the container, and Δt_w is the change in the temperature of the water. To minimize the exchange of heat with the surroundings, a double-walled vessel (Fig. 16-2) is ordinarily used in calorimetric experiments. Such a container is called a *calorimeter*.

Example When 2.00 lb of brass at 212°F is dropped into 5.00 lb of water at 35.0°F, the resulting temperature is 41.2°F. Find the specific heat of brass.

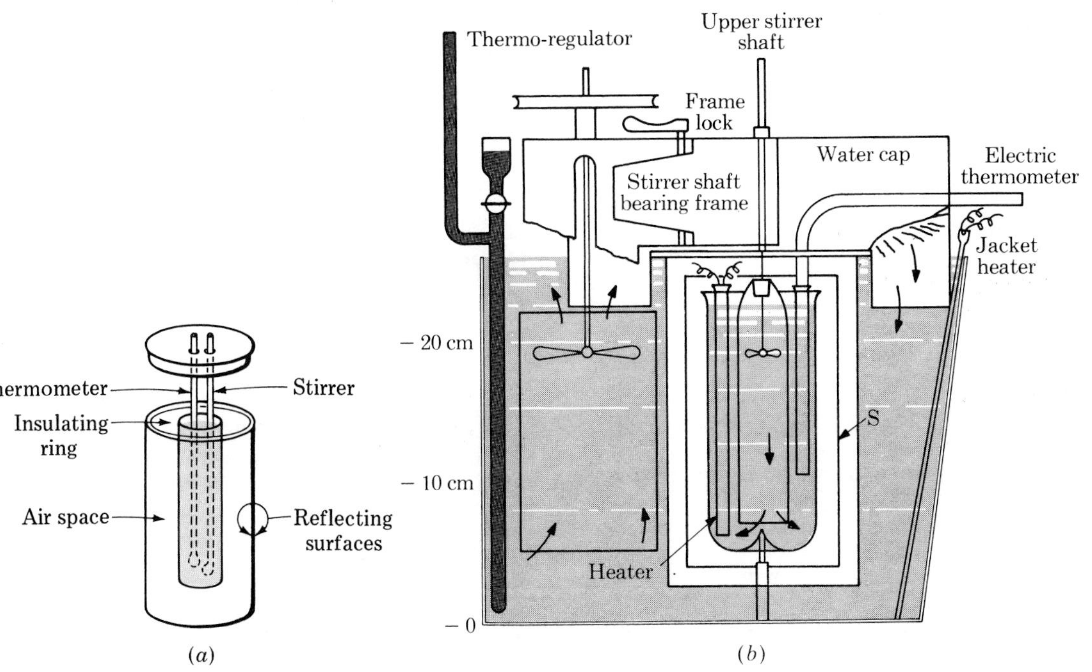

Figure 16-2
(*a*) A simple calorimeter. (*b*) A precision calorimeter, with external water bath to compensate for heat losses.

Heat lost by brass = heat gained by water

$$m_B c_B \, \Delta t_B = m_w c_w \, \Delta t_w$$

$$(2.00 \text{ lb})c_B(212°\text{F} - 41.2°\text{F}) = (5.00 \text{ lb}) \times [1.00 \text{ Btu/(lb)(F°)}](41.2°\text{F} - 35.0°\text{F})$$

$$c_B = \frac{(5.00 \text{ lb})[1.00 \text{ Btu/(lb)(F°)}](41.2°\text{F} - 35.0°\text{F})}{(2.00 \text{ lb})(212°\text{F} - 41.2°\text{F})}$$

$$= 0.091 \text{ Btu/(lb)(F°)}$$

Example If 80 g of iron shot at 100.0°C is dropped into 200 g of water at 20.0°C contained in an iron vessel of mass 50 g, find the resulting temperature.

In this mixture heat is lost by the shot, and heat is gained by the water and its container.

Heat lost by shot = heat gained by water + heat gained by vessel

$$(80 \text{ g})[0.12 \text{ cal/(g)(C°)}](100.0°\text{C} - t) = (200 \text{ g})[1.00 \text{ cal/(g)(C°)}](t - 20.0°\text{C}) + (50 \text{ g})[0.12 \text{ cal/(g)(C°)}](t - 20.0°\text{C})$$

$$t = 24°\text{C}$$

16-4 MOLAR HEAT CAPACITY

The molar heat capacity C is defined by using the gram mole as the unit of mass. One gram mole is a number of grams of a substance numerically equal to its molecular mass M. The number of moles n equals mass in grams divided by molecular mass, $n = m/M$. Equation (4) becomes

$$C = \frac{Q}{n \, \Delta t} = Mc \qquad (7)$$

The molar heat capacity of water is about 18 cal/(mol)(C°).

Empirical rules regarding heat capacities served as a guide in the early determination of atomic masses and of molecular formulas. According to the rule of Dulong and Petit, for the heavier solid elementary substances (with atomic masses above 35) the product of the specific heat and the atomic mass is approximately constant, with a value 6.2 cal/(g)(C°). To estimate the atomic mass of bismuth, Bi, we may divide the measured specific heat of Bi, 0.0294 cal/(g)(C°) (Table 2), into 6.2 to obtain 211 as a rough value of the atomic mass. According to Kopp's rule, the molar heat capacity of a solid compound is the sum of its atomic specific heats. The value 6.2 is used for all atoms except the lightest ones, for which the values used are H 2.5, Br 3.0, B 2.5, C 2.0, N 3.0, O 4.0, and F 5.0. This rule predicts for $CaCO_3$ a molar heat capacity of 20.2. The experimental value is 20.8 cal/(mol)(C°). Nowadays, x-ray diffraction data provide definite values of the atomic numbers and definite evidence for the molecular formulas of solid compounds.

Variation of molar heat capacity with temperature provides important information on the energy of the particles that constitute matter, but quantum mechanics is required for the interpretation of this information. The measurement of heat capacities of chemical compounds, metals, alloys, and superconducting metals, especially at low temperatures, while often difficult, is an active and important area of contemporary physics.

16-5 CHANGE OF PHASE

Not all the heat that an object receives necessarily raises its temperature. Surprisingly large amounts of energy are needed to do the work of separating the molecules when solids change to liquids and liquids change to vapors. Water will serve as a familiar example. The solid phase of water is ice. Ice has a specific heat of about 0.5 cal/(g)(C°). Water has a specific heat of 1 cal/(g)(C°).

To raise the temperature of 1 g of ice from −1 to 0°C requires $\frac{1}{2}$ cal of heat energy. To raise the temperature of 1 g of water in the liquid phase from 0 to 1°C requires 1 cal. To melt 1 g of ice requires 80 cal, although the temperature does not change while this large amount of heat is being added. The heat per unit mass needed to change a substance from the solid to the liquid

Table 2
AVERAGE SPECIFIC HEATS AND MOLAR HEAT CAPACITIES OF SOME METALS

Metal	Temperature range, °C	$\bar{c}$, cal/(gm)(C°)	M, gm/mole	$\bar{C} = M\bar{c}$, cal/(mole)(C°)
Aluminum	20–100	0.217	27.0	5.86
Bismuth	20–100	0.0294	209	6.2
Copper	15–100	0.093	63.5	5.90
Iron	20–100	0.115	55.9	6.43
Lead	20–100	0.031	207	6.42
Mercury	20–100	0.033	201	6.64
Nickel	20–100	0.110	58.7	6.45
Silver	15–100	0.056	108	6.05

state at its melting temperature is called the *heat of fusion*, L_f.

$$\text{Heat of fusion} = \frac{\text{heat needed}}{\text{mass}}$$

$$L_f = \frac{Q}{m} \tag{8}$$

The heat of fusion is expressed in Btu per pound or in calories per gram. The heat of fusion of ice is about 144 Btu/lb, or 80 cal/g. Whereas specific heats are numerically the same in British and metric units, the numerical value of a heat of fusion in the metric system is $\frac{5}{9}$ that in the British system.

Once a gram of ice is melted, 100 cal is required to raise the temperature of the gram of water from the melting point to the boiling point. As we continue to add heat at the boiling point, the temperature remains the same until the liquid is changed entirely to vapor. The steps by which a gram of ice is heated through fusion and vaporization are shown to scale in Fig. 16-3. The amount of heat per unit mass necessary to change a liquid from the liquid to the vapor phase without changing the temperature is called the *heat of vaporization*, L_v.

$$L_v = \frac{Q}{m} \tag{9}$$

The heat of vaporization is usually measured at the normal boiling point of a liquid. But occasionally L_v is measured under other conditions of temperature and pressure, and its value depends on those conditions. For water the heat of vaporization L_v is approximately 540 cal/g, or 970 Btu/lb. This is over five times as much energy as is needed to heat water from the melting to the boiling point. Where this energy goes in a

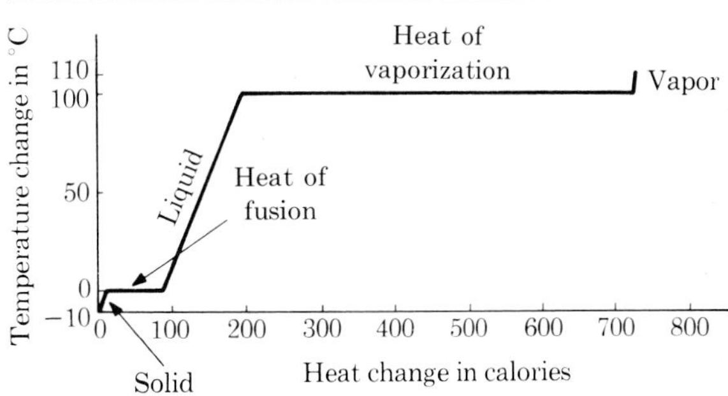

Figure 16-3
Heat required to change 1 g of ice at −10° to steam at 110°C.

change of phase may be better understood if we think of the liquid as made up of a myriad of molecules packed closely but rather irregularly, compared with the neat arrangement in the crystals that make up the solid. One gram of water occupies approximately 1 cm^3 of space as a liquid. The same amount of water (and therefore the same number of molecules) in the vapor state at 1 atm of pressure and a temperature of 100°C fills 1,671 cm^3 instead of 1 cm^3. The work to vaporize the water has been done in performing external work against the atmosphere and in separating the molecules to much larger distances than in the liquid phase.

Example How much heat is required to change 50 g of ice at −15°C to steam at 150°C?

The student may find it helpful to construct a diagram as shown below to solve problems of this nature.

The heat gained by the ice will be the heat to melt it (it is at 0°C when put into the calorimeter), plus the heat to warm it to the final temperature once it is all melted. This is

$$Q_g = m_i L_i + m_i c_w (t_f - 0)$$

where Q_g represents heat gained by the mass m_i of melting ice whose heat of fusion L_i is to be measured, c_w is the specific heat of the water which was ice before it melted, and t_f is the final temperature. The heat lost by the calorimeter and the water in it will be

$$Q_l = m_c c_c \, \Delta t_c + m_w c_w \, \Delta t_w$$

where the symbols have meanings analogous to those following Eq. (6).

State	Temperature	Calories required
Gas, $c = 0.5$ cal/(g)(C°)	150°C	$Q = mc\,\Delta t =$ (50 g)[(0.5 cal/(g)(C°)](50°C) = 1,250 cal
Boiling point	100°C	(H of V) = (540 cal)(50 g) = 27,000 cal
Liquid, $c = 1.0$ cal/(g)(C°)		$Q = mc\,\Delta t =$ (50 g)[(1.0 cal/(g)(C°)](100°C) = 5,000 cal
Freezing point	0°C	(H of F) = (80 cal)(50 g) = 4,000 cal
Solid, $c = 0.5$ cal/(g)(C°)	−15°C	$Q = mc\,\Delta t =$ (50 g)[(0.5 cal/(g)(C°)](15°C) = 375 cal
		Total calories = 37,625 cal

Heats of fusion and vaporization, like specific heats, are determined by calorimetric experiments. The only change needed in Eq. (6) is the addition of a term giving the amount of heat required to change the phase.

Example When 150 g of ice at 0°C is mixed with 300 g of water at 50.0°C, the resulting temperature is 6.7°C. Calculate the heat of fusion of ice.

Heat lost by water
= (300 g)[1.00 cal/(g)(C°)](50.0°C − 6.7°C)
= 13,$\bar{0}$00 cal

$$\text{Heat to melt ice} = (150\text{ g})L_i$$

Heat to raise temperature of ice water to final temperature
= (150 g)[1.00 cal/(g)(C°)](6.7°C − 0°C)
= 1$\bar{0}$00 cal

$$\text{Heat lost} = \text{heat gained}$$
$$13{,}\bar{0}00 \text{ cal} = (150 \text{ g})L_i + 1{,}\bar{0}00 \text{ cal}$$
$$L_i = 80.0 \text{ cal/g}$$

Any method of mixtures problem may be solved by choosing at will some standard state and then calculating the heat that would be gained or lost by each material in going from its initial condition to the standard state.

Example If 150 g of ice at 0°C is added to 200 g of water in a 100-g aluminum cup at 30°C, what is the resulting temperature?

Here let us choose water at 0°C as the standard state. Next compare the heat gained by the ice in going to the standard state with the heat lost by the water and cup in cooling to 0°C.

Heat gained by ice in melting
$= (150 \text{ g})(80.0 \text{ cal/g}) = 12{,}\bar{0}00 \text{ cal}$

Heat lost by cup
$= (100 \text{ g})[0.217 \text{ cal/(g)(C°)}](30°\text{C} - 0°\text{C})$
$= 6\bar{5}0 \text{ cal}$

Heat lost by water
$= (200 \text{ g})[1.0 \text{ cal/(g)(C°)}](30°\text{C} - 0°\text{C})$
$= 6\bar{0}00 \text{ cal}$

Since the melting of *all* the ice would require more heat ($12{,}\bar{0}00$ cal) than that available from the water and cup ($6\bar{6}00$ cal), not all the ice will melt. The final temperature will then be 0°C. The amount of ice remaining is given by

$$m = \frac{(12{,}\bar{0}00 - 6{,}\bar{6}00) \text{ cal}}{80.0 \text{ cal/g}} = 67 \text{ g}$$

16-6 VAPORIZATION

Vaporization is the change of a substance into the state of a vapor or a gas. Vaporization may occur in three ways. When a liquid is confined in a closed vessel, an equilibrium state is reached: molecules are continually leaving the surface of the liquid and forming a vapor; the vapor molecules are continually striking the liquid surface and entering it again. When the number of molecules leaving the liquid surface equals the number returning in the same time, the vapor is said to be *saturated*. The pressure of the saturated vapor is called the *vapor pressure;* it is characteristic of the substance and the temperature but independent of the volume of the vapor.

Evaporation is the vaporization of a liquid at its surface only. This proceeds without visible disturbance. Liquids left in uncovered dishes generally disappear in time by evaporation. The rate of evaporation depends on the material and such factors as the temperature, the surface area, amount of ventilation, and the pressure exerted on the surface. Evaporation is a cooling process, since it is the more energetic molecules that are able to escape from the liquid. Rapid evaporation may cool the remaining liquid enough to freeze it.

Boiling is the vaporization of a liquid in bubbles in the body of the liquid as well as at the free surface. It is accompanied by agitation of the liquid as the bubbles rise, expand, and burst. The boiling point of a liquid is the temperature at which its vapor pressure is equal to the pressure exerted on the liquid. The temperature at which a liquid boils at standard pressure, 76.0 cm Hg, is called its *normal boiling point.*

Sublimation is the changing of a solid directly into a vapor without passing through a liquid phase. Clothes hung to dry on a cold day may freeze and then dry. Solid carbon dioxide sublimes without wetting its container; hence its common name "Dry Ice." The odor of solid camphor and naphthalene (moth balls) is evidence that they sublime at room temperature.

16-7 THE EFFECT OF PRESSURE ON A CHANGE OF PHASE

It is a familiar fact that at high altitudes where the atmospheric pressure is reduced, water boils at temperatures lower than at sea level. On the other hand, if the pressure on the liquid surface

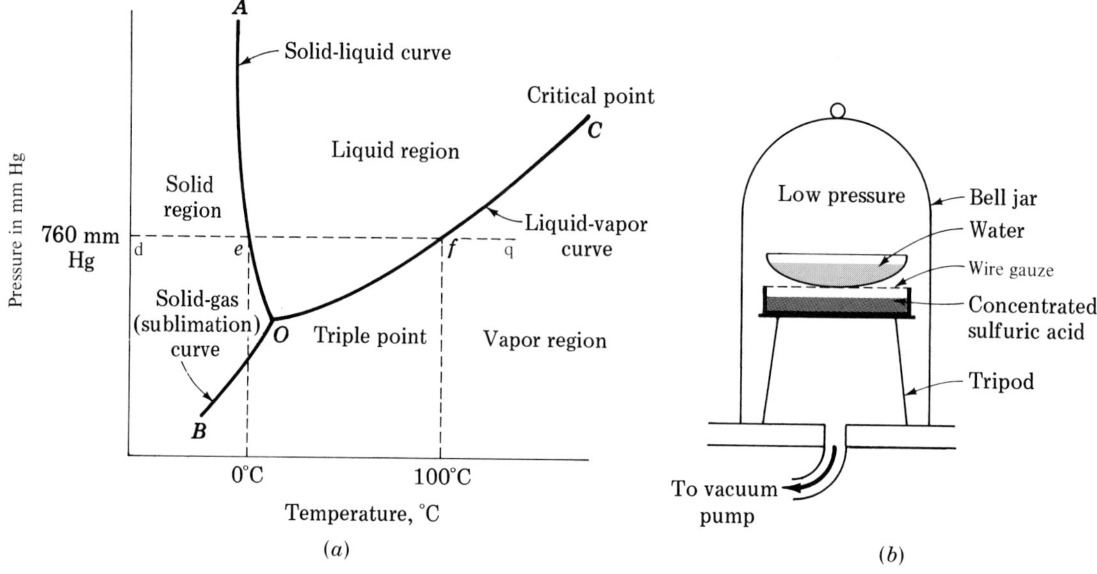

Figure 16-4
(*a*) Triple-point diagram for water showing melting and boiling points at standard barometric pressure. (*b*) Apparatus to demonstrate triple point of water.

is increased in a steam boiler or in a pressure cooker, boiling occurs at temperatures higher than the normal boiling point. Boiling can be brought about either by increasing the temperature of a liquid until its vapor pressure is equal to the pressure on the liquid or by reducing the pressure on the liquid to the value of the saturated vapor pressure. The vaporization curve of Fig. 16-4*a* represents conditions of equilibrium between the liquid and vapor phases of water.

The vaporization of water results in a large increase in the volume occupied by the molecules. An increase in pressure therefore raises the boiling point. Water can be boiled under reduced pressure at temperatures less than 100°C, as suggested by curve *OC* in Fig. 16-4*a*.

The freezing of a liquid is accompanied by a change of volume, although this change is much smaller than that which occurs on vaporization. For the few substances which, like water, expand on freezing, an increase in pressure lowers the freezing point. The line *OA* in Fig. 16-4*a* is the fusion curve for water. Each point on it represents a pressure and a temperature at which ice and water can exist together in equilibrium, the relative amounts of each remaining the same if no heat is added or removed.

16-8 TRIPLE POINT

Three cases of equilibrium have been discussed, those between liquid and vapor, between solid and liquid, and between solid and vapor. In each case the temperature of equilibrium depends on the pressure. This dependence can be represented conveniently in an equilibrium graph in which the coordinates of the points plotted represent temperatures and the corresponding pressures for equilibrium between two states of the substance.

In Fig. 16-4*a*, the equilibrium curve *OC* shows

the raising of the boiling point of the substance with increase of pressure. Curve *BO* shows the raising of the sublimation point with increase of pressure. Curve *OA* is typical of a substance which, like water, expands on freezing and shows for such a substance the lowering of the melting point with increase of pressure.

It can be shown that for any substance the three equilibrium curves intersect at a common point, *O* in Fig. 16-4*a*, called the *triple point,* at which conditions of temperature and pressure the three phases, solid, liquid, and saturated vapor, can exist together in equilibrium. For water the triple point is at 0.0075°C and 4.62 mm Hg pressure. If these conditions are changed, one of the phases will disappear. If the temperature is raised, for example, the ice will melt and the state of the system (water and saturated vapor) will be represented by a point on curve *OC*.

Consider a solid substance kept under constant pressure while its temperature is gradually increased. At each temperature its state will be represented by a point on the line *dg* in Fig. 16-4*a*. The substance will remain solid from *d* to *e*. At *e*, which represents the melting temperature, the solid substance can exist in equilibrium with the liquid. From *e* to *f* the substance will be in the form of liquid. At *f* the liquid and saturated vapor can exist together. This is the boiling point for the substance at the pressure chosen. At higher temperatures, along *fg*, the vapor will be superheated and no longer saturated.

In Figure 16-4*b* a laboratory setup used to demonstrate the triple point of water is sketched. An evaporating dish containing water is supported on a wire gauze (or clay triangle) over some concentrated sulfuric acid in a bell jar which is connected to a vacuum pump. The technique requires that a fairly low pressure be created in the bell jar. The concentrated sulfuric acid, an excellent dehydrating agent, is used to absorb some of the water vapor in the bell jar to help the pump reduce the pressure. As the water evaporates, the remaining water becomes cooler, for it is the more energetic molecules which escape to the vapor phase. As the pressure continues to drop, the water boils but the heat energy is used up so rapidly that the surface of the water in the evaporating dish is cooled below its freezing point and freezes, even though the water below the surface is still boiling. Therefore, in the bell jar we have water appearing as a solid, a liquid, and a gas (water vapor) at the same time.

16-9 CRITICAL POINT

An instructive experiment can be performed, with suitable precautions, using a strong glass tube in which liquid carbon dioxide and carbon dioxide vapor have been sealed. As the tube is slowly heated, the meniscus (liquid-vapor boundary) can be observed to remain practically stationary, becoming flatter as the temperature increases. At 31°C the meniscus vanishes. The density of the liquid has decreased with rising temperature and the density of the vapor has increased, until they become equal at 31°C. The temperature at which the liquid and vapor densities of a substance become equal is called its *critical temperature.* The pressure of the saturated vapor at this temperature is called the *critical pressure.*

A vapor kept above its critical temperature will never form a meniscus; i.e., it will never liquefy no matter how greatly it is compressed. Above the critical temperature and under high pressure the vapor may become as dense and as incompressible as the liquid at lower temperatures. The vaporization curve *OC* of Fig. 16-4 or Fig. 16-5 *terminates* at the critical point *C*. For water the critical point is at 374°C and 218 atm pressure.

When a substance in the gaseous phase is below its critical temperature, it is called a *vapor;* when above that temperature, it is called a *gas.* Only gases with densities low compared with the density of the condensed phase obey the gas laws. Otherwise there are marked departures from the gas laws. For saturated vapors the pressure is independent of volume.

In the two-dimensional phase diagrams, possible equilibrium states are represented by points on the curves (Figs. 16-4*a* and 16-5). In Fig. 16-6, the three variables, pressure, volume, and tem-

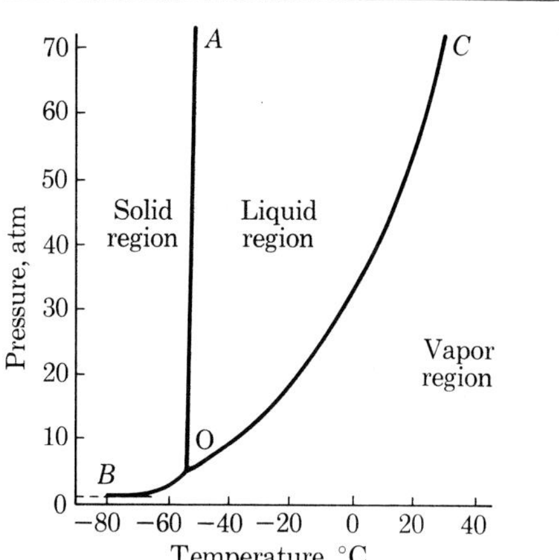

Figure 16-5
Triple-point diagram for CO_2.

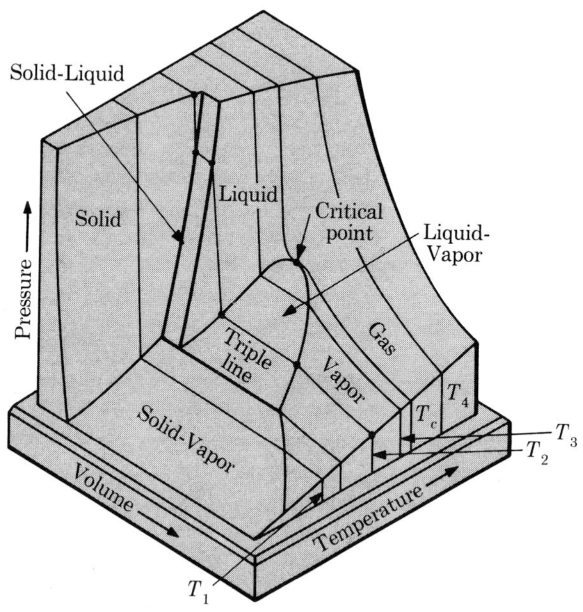

Figure 16-6
Thermodynamic model: *P-V-T* surface for a substance that expands on melting. (Not drawn to scale.)

perature, are plotted along mutually perpendicular axes. Every possible equilibrium state between two phases corresponds to a point on this surface. The triple point of Figs. 16-4*a* and 16-5 becomes a triple line on the *P-V-T* surface and represents conditions under which the three phases, solid, liquid, and vapor, can coexist in equilibrium. Dimensions are not constructed to scale in the model shown in Fig. 16-6. A model for a substance which, like water, contracts on melting would show the sloping band labeled "solid-liquid" to be concave toward the right.

16-10 HUMIDITY

At all times water is present in the atmosphere in one or more of its physical forms—solid, liquid, and vapor. The invisible vapor is always present in amounts that vary over a wide range, while water drops (rain or cloud) or ice crystals (snow or cloud) are usually present.

If a shallow pan of water is allowed to stand uncovered in a large room, the liquid will soon disappear, although the water will still be present as invisible vapor. If a similar pan of water is placed in a small enclosure, evaporation begins as before but after a time stops or becomes very slow and droplets begin to condense on the walls of the enclosure. The air is said to be *saturated.* When this condition has been reached, the addition of more water vapor merely results in the condensation of an equal amount. The amount of water vapor required for saturation depends upon the temperature (Table 5 in Appendix A-10); the higher the temperature, the greater the amount of water vapor required to produce saturation. If the air is not saturated, it can be made so either by adding more water vapor or by reducing the temperature until that already present will produce saturation. The temperature to which the air must be cooled, at constant pressure, to produce saturation is called the *dew point.* If a glass of water collects moisture on the outside, its temperature is below the dew point.

When the temperature of the air is reduced

to the dew point, condensation takes place if there are present nuclei on which droplets may form. These may be tiny salt crystals, smoke particles, or other particles that readily take up water. In the open air such particles are almost always present. In a closed space where such particles are not present, the temperature may be reduced below the dew point without consequent condensation. The air is then said to be *supersaturated.*

In a mixture of gases, such as air, the pressure exerted by the gas is the sum of the partial pressures exerted by the individual gases. The portion of the atmospheric pressure due to water vapor is called its *vapor pressure*. When the air is saturated, the pressure exerted by the water vapor is the *saturated vapor pressure*.

The mass of water vapor per unit volume of air is called the *absolute humidity*. It is commonly expressed in grains per cubic foot or in grams per cubic meter. *Specific humidity* is the mass of water vapor per unit mass of air and is expressed in grams per kilogram, grains per pound, etc. Specific humidity is more useful since it remains constant when pressure and temperature change, while the absolute humidity varies because of the change in volume of the air involved.

Relative humidity is defined as the ratio of the actual vapor pressure to the saturated vapor pressure at that temperature. It is commonly expressed as a percentage. At the dew point the relative humidity is 100 percent. From a knowledge of the temperature and dew point the relative humidity can be readily determined by the use of a table of vapor pressures.

Example In a weather report the temperature is given as 68°F and the dew point as 50°F. What is the relative humidity?

To use Table 2, we must change the temperatures to the Celsius scale.

$$C = \tfrac{5}{9}(F - 32°)$$

$$C_1 = \tfrac{5}{9}(68° - 32°) = \tfrac{5}{9}(36°) = 20°C$$

$$C_2 = \tfrac{5}{9}(50° - 32°) = \tfrac{5}{9}(18°) = 10°C$$

From the table we find the vapor pressures,

$$P_1 = 17.6 \text{ mm Hg}$$

$$= \text{pressure of saturated vapor}$$

$$P_2 = 9.2 \text{ mm Hg} = \text{actual vapor pressure}$$

$$\text{Relative humidity} = \frac{P_2}{P_1} = \frac{9.2 \text{ mm Hg}}{17.6 \text{ mm Hg}}$$

$$= 0.52 = 52\%$$

Whenever the temperature of the air is reduced to the dew point, condensation occurs. When the dew point is above the freezing point, water droplets are formed; when it is below, ice crystals are formed. The formation of dew, frost, clouds, and fog are examples of this process. The cooling may be caused by contact with a cold surface, by mixing with cold air, or by expansion in rising air.

16-11 CLOUD CHAMBER; BUBBLE CHAMBER

Change of phase is used in two devices which enable physicists to see or to photograph nuclear events: these devices are the cloud chamber and the bubble chamber.

To produce a cloud it is necessary that certain conditions exist. In the atmosphere, clouds form when the air is moist, when "seeds," or nuclei, around which water vapor can condense into droplets are available, when there is low-pressure, and when initially warm air is cooled. While low pressure and high temperature are important, attempts at artificial rain-making depend primarily upon the availability of moist air. If this condition exists, clouds are seeded with some material such as sodium iodide crystals that will serve as the nucleus for the formation of the droplets which, in turn, will produce heavy enough drops to fall as precipitation.

In 1897, C. T. R. Wilson observed that in air supersaturated with water vapor, *ions* would serve as the nuclei droplets, or condensation centers, forming on the ions. Later (1912) he applied this discovery to make an instrument, called a cloud chamber, for studying the paths of particles or rays that have the capability of ionizing the atoms of the gas through which they pass. The type of cloud chamber Wilson built causes the gas to expand and produces a low pressure. Thus a supersaturated vapor, a seed (the ions), and a temporary low pressure are available to produce the cloud. Although warm air is not present, this expansion-type cloud chamber works effectively. A simple form of this cloud chamber is shown in Fig. 16-7. It consists of a piston, diaphragm, or bulb above which there is a chamber in which the air is saturated with water vapor. If the piston or bulb is suddenly moved to increase the volume, the air expands, loses energy in performing work, cools, and becomes super-saturated with water vapor. If there are ions present, formed by the passage of an ionizing particle or radiation, water droplets form on the ions and mark the path. When brightly illuminated, these droplets can be seen or photographed. Many important discoveries in nuclear physics have been made with the aid of the Wilson cloud chamber.

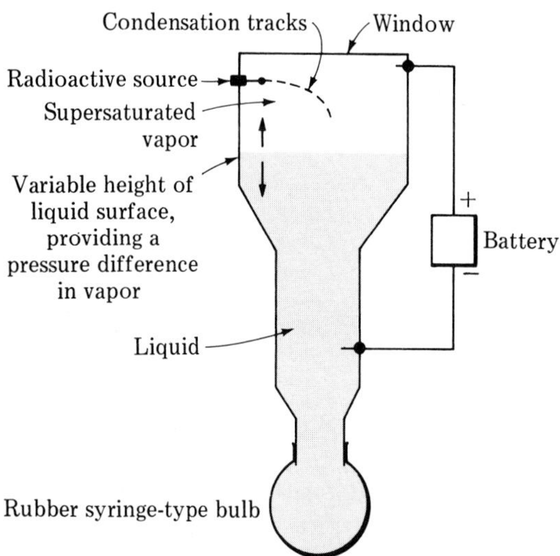

Figure 16-7
Demonstration-type expansion cloud chamber. The battery, not essential, sweeps ions out of the chamber to clear it for the next event.

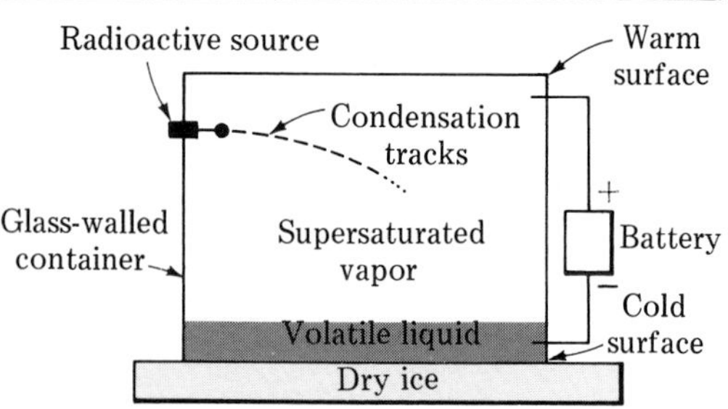

Figure 16-8
Diffusion cloud chamber.

Another form of the cloud chamber, the diffusion type, depends upon the existence of a temperature differential between the top and the bottom of the chamber. In this type, no attempt to create a low pressure is made. The chamber, which is supersaturated with either water or some other volatile liquid, such as alcohol, is placed on a very cold surface (Dry Ice) and then when ions are formed in the chamber, condensation tracks are produced in a manner similar to that in the expansion-type cloud chamber (Fig. 16-8).

In 1952, D. A. Glaser at the University of Michigan, developed a procedure which used a superheated liquid to display the tracks of ionizing particles, the liquid being analogous to the use of a supersaturated vapor in the cloud chambers. The instrument used in this procedure is called a *bubble chamber* because the tracks consist of a stream of closely spaced bubbles which correspond to the droplets formed in the cloud chamber. In practice, the superheated liquid (propane or hydrogen) is kept in the chamber under pressure at a temperature above its normal boiling point. When the pressure is suddenly reduced (corresponding to expansion in the cloud chamber), the liquid is superheated and vapor bubbles form at local hot spots where an energetic particle has passed through the liquid. The track of a high-energy particle is made visible as a trail of tiny bubbles and may be photographed. The density of the liquid is much greater than the density of vapor in a cloud chamber, which increases the probability that high-energy particles will interact with nuclei in the liquid. Also, the products of the interaction will be slowed down by the liquid and often stopped, within one of the larger bubble chambers, so that the entire event can be analyzed.

SUMMARY

Heat is a form of energy, associated with the kinetic energy of the random motion of large numbers of molecules.

The internal energy of a system of particles may be either *ordered* or *disordered,* or perhaps both. The internal energy of an *ideal gas* is energy arising from the random or disordered motion of many particles. The disordered internal energy of a system is called *thermal energy.*

The most commonly used units of heat are the calorie, the British thermal unit, and the joule.

The *calorie* is the amount of heat required to change the temperature of 1 g of water 1 C°.

The *British thermal unit* is the amount of heat required to change the temperature of 1 lb of water 1 F°.

The *first law of thermodynamics* states that the total energy in a closed system is constant. If heat Q enters a system and the system performs external work $\mathcal{W}$, then the change in the internal energy ΔE_i is given by $\Delta E_i = Q - \mathcal{W}$.

The specific heat of water varies so slightly with temperature that for most purposes it can be assumed constant [1 cal/(g)(C°)] between 0 and 100°C.

The heat lost or gained by a body when the temperature changes is given by the equation

$$Q = mc\,\Delta t$$

In a *calorimeter* the heat lost by the hot bodies is equal to the heat gained by the cold bodies.

The *heat of fusion* is the heat per unit mass required to change a substance from solid to liquid at its melting point.

The heat of fusion of ice is approximately 80 cal/g, or 144 Btu/lb.

The *heat of vaporization* is the heat per unit mass required to change a substance from liquid to vapor.

The heat of vaporization of water at its normal boiling point is approximately 540 cal/g, or 970 Btu/lb. It depends on the temperature at which vaporization takes place.

The *boiling point* of a liquid is the temperature at which its vapor pressure is equal to the pressure exerted on the liquid.

The boiling point of a liquid is raised by an increase in pressure.

The *freezing point* of water and of the few other materials which expand on freezing is lowered by an increase in pressure. The freezing point of a substance which contracts on freezing is raised by an increase in pressure.

The *triple point* is the condition of pressure and temperature at which the three phases can coexist in equilibrium.

The *critical point* is the condition of pressure and temperature at which a liquid and its vapor are indistinguishable. The critical temperature is the highest temperature at which a gas can be liquefied by pressure alone.

Absolute humidity is the mass of water vapor per unit volume of air. *Specific humidity* is the mass of water vapor per unit mass of air.

Relative humidity is defined as the ratio of the actual vapor pressure to the saturated vapor pressure at that temperature.

The *cloud chamber* and the *bubble chamber* are two instruments which may be used to study nuclear events and which depend on a phase change to operate.

The *dew point* is the temperature to which the air must be cooled, at constant pressure, to produce saturation.

References

Dyson, Freeman J.: What Is Heat? *Scientific American,* September 1954, pp. 58–63.

Mott-Smith, Morton: "Heat and Its Workings," Dover Publications, Inc., New York, 1962.

Nelkon, M.: "Heat, A Textbook for Advanced Level and Intermediate Students," Blackie & Son, Ltd., Glasgow, 1955.

Questions

1 Describe the difference between and give examples of ordered and disordered energy.

2 Discuss the term caloric as it was used in explaining the nature of heat.

3 State and discuss the implications of the first law of thermodynamics.

4 What is the distinction between quantity of heat and temperature?

5 What is the experimental evidence for considering heat to be a form of energy?

6 What variables determine the specific heat of an object?

7 State the metric and British units for specific heat, heat of fusion, and heat of vaporization.

8 It has been noted by environmental biologists that the high specific heat of water is important for the survival of living organisms. Why is this so?

9 If 10 cal of heat raises the temperature of object A 2.0°C while 10 cal raises the temperature of object B 1.0°C, and object A has a mass of 30 g and object B has a mass of 40 g, what can be said about the specific heats of objects A and B?

10 Which produces a more severe burn, boiling water or steam? Why?

11 Which would be liable to explode with greater violence, a high-pressure steam boiler or a tank containing air at the same pressure? Why?

12 Define and point out the differences between evaporation, boiling, and sublimation.

13 A human body is said to be cooling off when it perspires. Explain the cooling effect of perspiring.

14 A person with a high fever is often given an alcohol rub. How and why does this prove beneficial?

15 Explain why a tire pump gets hot when it is used to inflate an object.

16 If a low barometric pressure is an indicator of inclement weather and a high barometric pressure an indicator of fair weather, explain how it can be that "wet" air (inclement weather) does not weigh more than "dry" air (fair weather)—or does it? Explain.

17 At how low a temperature can water be made to boil?

18 A tall vertical pipe is closed at the bottom, filled with water, and heated at the bottom by a bunsen burner. Water and steam erupt from the tube as from a geyser. Explain.

19 One sometimes places a tub of water in a fruit storage room to keep the temperature above 30°F during a cold night. Explain.

20 In what sense is freezing a heating process? How does the heat thus produced protect plants and temper the climate?

21 What is meant by the "method of mixtures" in heat-exchange situations?

22 The triple-point diagram provides an explanation of why a snowball can be made from loose snow. Can you explain from an examination of this diagram why this happens?

23 Name three substances which undergo sublimation under normal conditions.

24 What is the effect of decreased pressure on the boiling point and the freezing point of water?

25 What is meant by the critical point on a triple-point diagram?

26 Differentiate between relative, absolute, and specific humidity.

27 Explain why the dew point is not a fixed temperature in the sense that the freezing point of water is.

28 Discuss the significance of the often-heard comment, "It is not the heat, but the humidity that causes discomfort."

29 Is it possible for the relative humidity to be greater than 100 percent? Explain your answer.

30 Assume that to determine the temperature of a bunsen flame in the laboratory you heat an iron washer in the flame and then drop the washer into water. (*a*) List the measurements you would make. (*b*) What physical constants would you look up before calculating? (*c*) Assign symbols to items in *a* and *b*, and write the equation from which you would calculate the temperature *t*. (*d*) Would you expect the temperature determination to be accurate? Explain, briefly.

Problems

1 How much heat is required to change the temperature of 200 g of heat (specific heat 0.030 cal/(g)(°C) from 10°C to 60°C?

2 When 25 g of ice at −20°C have absorbed 2,500 cal of heat, what is the temperature of the resulting water? *Ans.* 10°C.

3 If 1,000 l of air at 27°C and pressure of 1.0 atm has a mass of 1.115 kg and a specific heat at constant pressure of 1.0×10^3 J/(kg)(C°), how much heat is required to raise the temperature of this gas from 27 to 177°C at constant pressure?

4 How much heat is required to raise the temperature of 1.5 lb of water in an 8.0-oz aluminum vessel from 48°F to the boiling point, assuming no loss of heat to the surroundings? *Ans.* $\bar{2}60$ Btu.

5 A 15-kg automobile battery when fully charged can provide about 3.0×10^6 J of electric energy. (*a*) How much would the temperature of the battery rise if all this energy were converted into heat and no heat escaped? The average specific heat of the battery materials is 0.25 cal/(g)(C°). (*b*) If the energy were instead fully utilized to propel the battery upward, how high could it be lifted?

6 When hot coffee is poured into a cup, how much heat is lost from the coffee to the 200-g cup [specific heat 0.20 cal/(g)(C°)] in raising its temperature from 70 to 180°F? *Ans.* 2,440 cal.

7 Calculate the amount of energy required to heat the air in a house $30 \times 50 \times 40$ ft from 10 to 70°F. The density of air is about 0.080 lb/ft^3, and its specific heat at constant pressure is about 0.24 Btu/(lb)(F°). Discuss the assumptions made in your calculation.

8 A 10-kg object falls 200 m. How much heat does the object gain when it lands? ($J =$ 4.186 J/cal.) How many Btu's would this be equivalent to? *Ans.* 4.68×10^3 cal; 18.5 Btu.

9 If all the energy of falling water is converted into heat, what is the difference in the temperature of the water at the top of a 78-ft waterfall and the temperature of the water at the bottom of the fall? (778 ft·lb = 1 Btu.)

10 There are 100 g of water in a brass calorimeter of mass 200 g. It is found that 590 cal are required to raise the temperature of water and container 5.0°C. What is the specific heat of brass? *Ans.* 0.090 cal/(g)(°C).

11 When 200 g of aluminum at 100°C is dropped into an aluminum calorimeter of mass 120 g and containing 150 g of kerosene at 15°C, the mixture reaches a temperature of 50°C. What is the specific heat of kerosene?

12 A tank of 10.0 ft^3 capacity is half filled with glycerin. The weight density of the glycerin is 78.6 lb/ft^3 and its specific heat is 0.57 Btu/(lb)(°F). (*a*) How much heat is needed to raise the temperature of the glycerin from 40 to 70°F? (*b*) If heat is supplied to the glycerin at the rate of 750 Btu/min, at what rate does its temperature rise? *Ans.* 6.7×10^3 Btu; 3.3 F°/min.

13 A calorimeter contains 0.66 kg of turpentine at 10.6°C. When 0.147 kg of alcohol at 75.0°C is added, the temperature rises to 25.2°C. The specific heat of turpentine is 1.95×10^3 J/(kg)(C°), and the calorimeter is thermally equivalent to 30 g of water. Find the specific heat of the alcohol.

14 A 3.0-lb lead ball at 160°F is dropped into 5.0 lb of oil [specific heat 0.60 Btu/(lb)(°F)] at 70°F contained in a 4.0-lb copper vessel. Assuming heat exchanges to be restricted to this system, what temperature will be reached finally by the mixture? *Ans.* 72°F.

15 A 30-g piece of ice at −20°C is dropped into a 25.0-g calorimeter of specific heat 840 J/(kg)(C°), containing 100 g of water at 35.0°C. The final equilibrium temperature, corrected for thermal leakage, is found to be 7.2°C. What is the specific heat of the ice?

16 What is the final temperature if 200 g of copper with a specific heat of 0.1 cal/(g)(°C) at a temperature of 80°C is added to 200 g of water at 20°C that is held in a 50-g aluminum cup having a specific heat of 0.22 cal/(g)(°C).

Ans. 25.2°C.

17 Find the resulting temperature when 80 g of iron shot at 100.0°C are dropped into 200 g of water at 20.0°C contained in an iron vessel of mass 50 g [specific heat 0.12 cal/(g)(°C)].

18 A 150-g iron ball at 95°C is dropped into a cavity in a block of ice. The cavity is then found to contain 21.0 g of water. Calculate the heat of fusion of ice. *Ans.* 79.4 cal/g.

19 How many centimeters of rainfall at 10°C are required to just melt a layer of ice 0.635 cm thick?

20 A substance has the following properties; boiling point 120°C; freezing point −20°C; specific heat as a gas, liquid and solid 0.4 cal/(g)(°C), 1.5 cal/(g)(°C), 1.0 cal/(g)(°C), respectively, heat of fusion 200 cal/g and the heat of vaporization 600 cal/g. How much heat would be given up by 15 g of this substance at 180°C changing to −40°C? *Ans.* 15,810 cal.

21 When a 1.445-g sample of coal is completely burned in a bomb calorimeter, the temperature of the 2,510 g of water is raised from 74.85 to 82.65°F. (*a*) How much heat is given to the water as a result of the combustion of the coal? (*b*) Assuming that the figure 2,510 g includes the water equivalent of the metal parts of the calorimeter, what is the heat of combustion of the coal sample?

22 How much heat must be removed by the refrigerator coils from a 0.50-lb aluminum tray containing 3.0 lb of water at 70°F to freeze all the water, and then to cool the ice to 10°F?

Ans. $5\bar{8}0$ Btu.

23 How many joules are needed to change 20.0 g of water at 25°C to steam at 150°C? Assume the specific heat of steam to be 2.02×10^3 J/(kg)(C°).

24 How many Btu's are needed to change 50 lb of ice at 12°F to steam at 232°F?

Ans. $6\bar{6}$,000 Btu.

25 A 2.0-lb aluminum pail contains 20 lb of water at 70°F. What mass of ice at the melting point must be placed in the pail to cool the water to 50°F, assuming the heat exchanges to be limited to pail and its contents?

26 Calculate the heat of fusion of ice according to the following laboratory data obtained by a not-too-precise student. Mass of the calorimeter cup 200 g, specific heat of the cup 0.11 cal/(g)(°C), mass of the water in the cup 400 g, initial temperature of water 50°C, mass of ice 50 g, initial temperature of ice 0°C, final temperature 30°C. *Ans.* 139 cal/g.

27 Water is heated in a boiler from 100 to 284°F where, under a pressure of 52.4 lb/in^2, it boils. The heat of vaporization of water at 284°F is 511.5 cal/g, or 920.7 Btu/lb. How much heat is required to raise the temperature and to evaporate 500 gal of water?

28 Calculate the heat of vaporization of water from the following data obtained in a lab by the same student in Prob. 26. Mass of the calorimeter cup 200 g, specific heat of the cup 0.22 cal/(g)(°C), mass of the water 200 g, initial temperature of water 25°C, final temperature of water 55°C, and mass of steam condensed 12 g. Assume initial temperature of the steam to be 100°C.

Ans. 565 cal/g.

29 In which case does the air hold more water vapor: (*a*) temperature 32°F, dew point 32°F, or (*b*) temperature 80°F, dew point 50°F? What is the relative humidity in each case?

30 In a room where the temperature is 24°C, an experiment shows the dew point to be 12°C. What is the relative humidity?

Ans. 47 percent.

31 The relative humidity in a certain room is 60 percent at 20°C. (*a*) Calculate the relative humidity if the temperature drops to 15°C. (*b*) What is the dew point at that temperature?

32 The pressure of saturated water vapor at 23.0 and 12.0°C is equal to 21.0 and 10.5 mm of mercury, respectively. What is the relative humidity when the temperature is 23.0°C, the dew point is 12.0°C, and the height of mercury in the barometer is 30 in? *Ans.* 50 percent.

33 Outdoor air at 30°C and 90 percent relative humidity is drawn into an air-conditioning system, where it is first chilled to 12°C in a water spray, and then allowed to warm to 22°C without the addition of moisture. (*a*) What is the relative humidity of the conditioned air? (*b*) How much moisture is removed from each cubic meter of air treated?

34 What is the value of the relative humidity in a room where the temperature is 26°C and in which dew begins to form on the outside of a pitcher of iced water at a temperature of 10°C?

Ans. 37 percent.

35 A small drop of water is noticed on top of the mercury column of a barometer at 25°C. (*a*) What correction must be applied to the barometer reading? (*b*) Suppose that 1.0×10^{-5} g of water had been trapped in the 5.0-cm^3 volume above the mercury, what would the correction have been? (*c*) How much would the correction have to be changed for a 1° rise in temperature from 25°C, in each case?

36 An airtight room contains 65 m^3 of air at a temperature of 22°C, 760 mm of mercury pressure, and 20 percent relative humidity. (*a*) What is the dew point in the room? (*b*) How much water must be added to the air of the room to raise the relative humidity to 50 percent? (*c*) If all the water vapor in the air of the room were removed, what would be the pressure exerted by the air, if the temperature remained at 22°C?

Ans. −2°C; $3\bar{8}0$ g; 756 mm of mercury.

James Franck, 1882–1964

Born in Hamburg. Director of the Physical Institute, University of Göttingen. Professor of physical chemistry, University of Chicago. The 1925 Nobel Prize for Physics was shared by Franck and Hertz for their discovery of the laws governing the collision of an electron and an atom.

Gustav Hertz, 1887–

Born in Hamburg. Head of the Physical Institute at Halle University. Shared the 1925 Nobel Prize for Physics with Franck for their discovery of the laws governing the collision of an electron and an atom.

17

Heat Transfer

Heat is one of the most common forms of energy. Every object that we see or feel possesses heat energy. Heat is continually being transferred from one body to another.

We are often concerned with heat and its flow. Sometimes we want to get it from one place to another, from a furnace to the rooms of a house or from an automobile engine to the surrounding air. In other cases we want to prevent the flow of heat, as from a heated room to a cold exterior or from a welder's tongs to his hand. In the first problem we are confronted with the fact that there are no perfect conductors of heat. The second problem, that of heat storage, is complicated by the fact that there are no perfect heat insulators, so that one cannot confine heat. In order to utilize heat to the best advantage, it is necessary to know the laws that govern heat transfer.

17-1 HEAT FLOW

Heat is always being transferred in one way or another, wherever there is any difference in temperature. Just as water will run downhill, always flowing to the lowest possible level, so heat, if left to itself, flows down the temperature hill, always warming the cold objects at the expense of the warmer ones. The rate at which heat flows depends on the steepness of the temperature hill as well as on the properties of the materials through which it has to flow. The difference of temperature per unit distance is called the *temperature gradient,* in analogy to the idea of steepness of grade, which determines the rate of flow of water.

17-2 TYPES OF HEAT TRANSFER

There are three ways in which heat is transferred. Since heat itself is the energy of molecular activity, the simplest mode of transfer of heat, called *conduction,* is the direct communication of molecular disturbance through a substance by means of the collisions of neighboring molecules. Metals contain so-called free electrons which make them good conductors of electricity; these electrons also contribute to the conduction of heat, so metals have high thermal conductivities (Table 1).

Convection is the transfer of heat from one place to another by actual motion of the hot material. Heat transfer is accomplished also by a combination of *radiation* and *absorption*. In radiation, thermal energy is transformed into radiant energy, similar in nature to light. In fact, a part of such radiant energy is light. While in the form of radiation, the energy may travel a

Table 1
THERMAL CONDUCTIVITIES (near 20°C)

Substance	Cal / (cm²)(s)(C°/cm)	Btu / (ft²)(h)(F°/in)
Silver	0.990	2870
Copper	0.918	2660
Aluminum	0.504	1460
Steel	0.11	320
Oak, across grain	0.048	140
Concrete	0.0041	12.0
Glass	0.0025	7.2
Brick	0.0017	5.0
Water	0.0014	4.1
Hydrogen	0.00038	1.1
Corkboard	0.00010	0.30
Glass wool	0.00009	0.27
Air	0.000053	0.15

tremendous distance before being absorbed or changed back into heat. For example, energy radiated from the surface of the sun is converted into heat at the surface of the earth only 8 min later.

17-3 CONDUCTION

The amount of heat that flows through any body by conduction depends upon the time of flow, the area through which it flows, the temperature gradient, and the kind of material. Stated as an equation,

$$Q = kAt\frac{\Delta T}{\Delta L} \quad (1)$$

where k is called the *thermal conductivity* of the material, A is the area measured at right angles to the direction of the flow of heat, t is the time the flow continues, and $\Delta T/\Delta L$ is the average temperature gradient. The symbol ΔT represents the difference in temperature between two parallel surfaces distant ΔL apart (Fig. 17-1).

In the British system these quantities are usually measured in the following units: Q in Btu, A in square feet, t in hours, ΔT in Fahrenheit

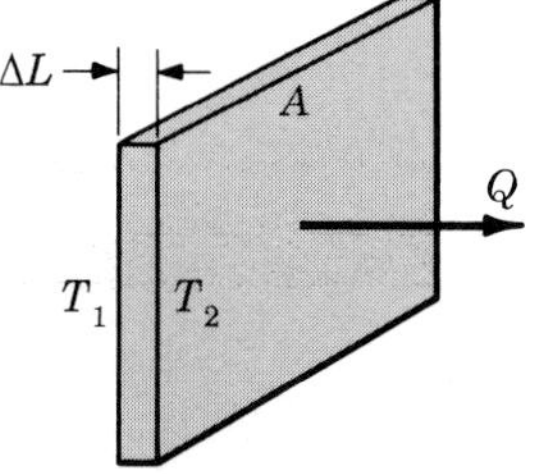

Figure 17-1
Heat conduction through a thin plate.

degrees, and ΔL in inches. The thermal conductivity k is then expressed in Btu/(ft^2)(hr)(F°/in). The corresponding unit of k in the metric system is cal/(cm^2)(s)(C°/cm).

Example A 3.0-in wall of fire brick, $k =$ 8.0 Btu/(ft^2)(h)(F°/in), has one surface at 335°F, the other at 80°F. Find (*a*) the temperature gradient in the brick and (*b*) the heat conducted through an area of 1.0 ft^2 in 1.0 d.

(*a*) Temperature gradient $\Delta T/\Delta L$

$$= \frac{335°\text{F} - 80°\text{F}}{3.0 \text{ in}} = 85 \text{ F°/in}$$

(*b*) Heat transferred $Q = \dfrac{kA(T_1 - T_2)t}{L}$

$$= \frac{8.0 \text{ Btu} \times 1 \text{ ft}^2 \times 255 \text{ F°} \times 24 \text{ h}}{(\text{ft}^2)(\text{h})(\text{F°/in})\ 3.0 \text{ in}}$$

$$= 1\bar{6},000 \text{ Btu}$$

Example Heat is conducted through a compound wall composed of parallel layers of two different conductivities, 0.32 and 0.14 cal/(cm^2)(s)(C°/cm), and of thicknesses 3.6 and 4.2 cm, respectively (Fig. 17-2). The temperatures of the outer faces of the wall are 96 and 8.0°C. Find (*a*) the temperature T_i of the interface and (*b*) the temperature gradient in each section of the wall.

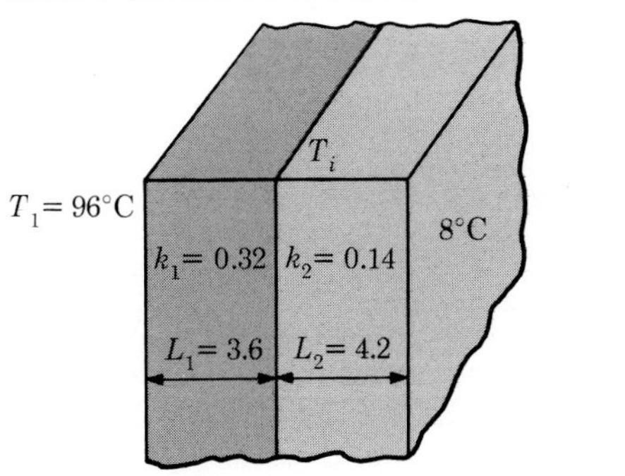

Figure 17-2
Heat conduction through a compound wall.

For steady flow, the same quantity of heat per unit area passes through the first layer per unit time as passes through the second layer.

$$\frac{1}{A}\frac{\Delta Q}{\Delta t} = \frac{k_1(T_1 - T_i)}{L_1} = \frac{k_2(T_i - T_2)}{L_2}$$

or
$$\frac{k_1 L_2}{k_2 L_1}(T_1 - T_i) = T_i - T_2$$

$$\frac{0.32}{0.14} \times \frac{4.2}{3.6}(96° - T_i) = T_i - 8.0°$$

$$T_i = 72°\text{C}$$

Temperature gradient in first layer

$$= \frac{(96 - 72)\text{C°}}{3.6 \text{ cm}} = 6.7 \text{ C°/cm}$$

Temperature gradient in second layer

$$= \frac{(72 - 8)\text{C°}}{4.2 \text{ cm}} = 15 \text{ C°/cm}$$

The general equation for steady, one-dimensional flow of heat may be written

$$\frac{\Delta Q}{\Delta t} = -kA\frac{\Delta T}{\Delta s} \qquad (2)$$

The minus sign is introduced because the temperature decreases in the direction of heat flow.

There are large differences in the thermal conductivities of various materials. Gases have very low conductivities. Liquids also are, in general, quite poor conductors. The conductivities of solids vary over a wide range, from the very low values for asbestos fiber or brick to the relatively high values for most metals. Fibrous materials such as hair felt or asbestos are very poor conductors (or good insulators) when dry; if they become wet, they conduct heat rather well. One of the difficult problems in using such materials for insulation is to keep them dry.

17-4 CONVECTION

The transfer of heat by convective circulation in a liquid or a gas is associated with pressure differences, most commonly brought about by local changes in density. A rise in temperature is accompanied by a decrease in density in most fluids. Hence if heat is applied at the bottom of a sample, the less dense fluid at the bottom is continually displaced by fluid of greater density from above. Heat transfer accompanies this motion, called *natural,* or *free, convection*. The primary circulation of the earth's atmosphere is an example of natural convection, as is also the circulation of water in an ordinary hot-water heating system in a house. Sometimes the pressure differences are produced mechanically by a pump, or blower, in which case the heat transfer is said to occur by *forced convection*. In all cases of convection, heat is transferred into or out of the fluid stream somewhere in its path.

Since convection is a very effective method of heat transfer, it must be considered in designing a system of insulation. If large air spaces are left within the walls of a house, convection currents are set up readily and much heat is lost. If, however, the air spaces are broken up into small, isolated regions, no major convection currents are possible and little heat is lost by this method. For this reason the insulating material used in the walls of a refrigerator or in a house is a porous material—cork, rock wool, glass wool, or other materials of like nature. Not only are they poor conductors in themselves, but they leave many small air spaces, which are very poor conductors and at the same time are so small that no effective convection currents can be set up.

The primary circulation of the earth's atmosphere is an important example of convection. A wind is a convection current resulting from unequal heating of the atmosphere. On a smooth, stationary earth warm air would rise near the equator (a region of low barometric pressure) and flow at high altitude toward the poles. Cool air would descend at the poles (regions of high barometric pressure) and flow toward the equator. Actually this simple equator-to-pole cycle is broken into several cycles (Fig. 17-3), chiefly because of the rotation of the earth and the interaction of cold air masses from the poles with the warm air masses.

17-5 EFFECT OF SURFACE FILMS ON HEAT TRANSFER

When a cool fluid flows along a heated wall, heat is transferred from the wall to the fluid. Langmuir (1912) dealt with this situation by assuming that heat is transferred by conduction through a thin

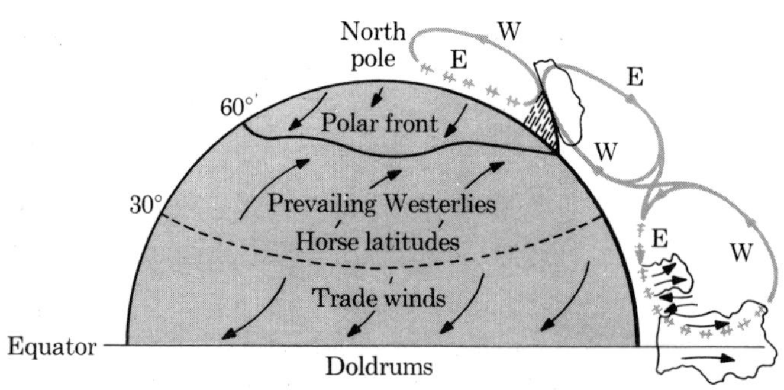

Figure 17-3
General circulation of the atmosphere on a uniform earth.

layer of fluid close to the wall, comprising roughly the laminar-flow region, to a region in which the ambient temperature is maintained by turbulent flow. A convective heat transfer of this type is treated approximately in the same manner as heat conduction, using the relation

$$\frac{\Delta Q}{\Delta t} = Ah\,\Delta T \qquad (3)$$

where ΔT is the temperature difference between the wall and the fluid at some distance from it and h is called a *film coefficient,* or *convection coefficient*. It represents the conductivity of the fluid divided by the effective thickness of the conducting layer.

The conducting layer is assumed to extend from the wall to the region where the temperature of the main body of fluid is reached, on the assumption that the temperature gradient across the conducting layer is constant and is equal to that actually existing at the wall. The film coefficient is found experimentally to depend on many factors: the nature of the fluid-wall surface (including curvature and orientation); the speed, density, viscosity, specific heat, and thermal conductivity of the fluid; and the occurrence of evaporation or condensation at the fluid-wall surface.

In the steady-flow passage of heat through a wall separating two fluids (say, water, steel, air, as in Fig. 17-4), the heat passes through the three layers in succession, at the same rate,

$$\frac{\Delta Q}{\Delta t} = k_1\frac{\Delta T_1}{L_1}A = k_2\frac{\Delta T_2}{L_2}A = k_3\frac{\Delta T_3}{L_3}A$$

where subscripts 1 and 3 refer to the fluid films and 2 to the wall. If ΔT is the total temperature difference across the three layers, the resulting heat flow is

$$\frac{\Delta Q}{\Delta t} = \frac{A\,\Delta T}{1/h_1 + L_2/k_2 + 1/h_3} \qquad (4)$$

If there were no films present, Eq. (4) would reduce to the simpler form of Eq. (1).

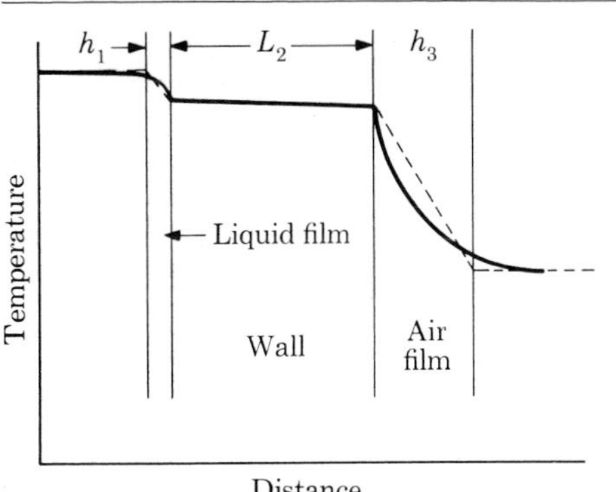

Figure 17-4
Temperature drop across a good conductor and the nonconvecting films on either side of it (solid line). Dotted line shows temperature drop based on effective thicknesses of fluid films.

Example Compute the heat conducted per hour through a plate-glass window of area 15 ft² and thickness $\frac{1}{4}$ in, when the inside temperature is 70°F and the outside temperature is 20°F. The thermal conductivity of the glass is 5.8 Btu/(h)(ft²)(F°/in), and the film coefficient for air is about 2 Btu/(h)(ft²)(F°).

From Eq. (4),

$$\frac{\Delta Q}{\Delta t} = \frac{15\text{ ft}^2(70-20)\text{F}^\circ}{\left(\frac{1}{2}+\frac{\frac{1}{4}}{5.8}+\frac{1}{2}\right)\frac{(\text{h})(\text{ft}^2)(\text{F}^\circ)}{\text{Btu}}}$$

$$= 7\bar{2}0\text{ Btu/h}$$

Note that if the air films were neglected, we should get for the rate of heat conduction the unrealistic figure $1\bar{7}{,}000$ Btu/h.

17-6 RADIATION

The transfer of heat by radiation does not require a material medium for the process. Energy traverses the space between the sun and the earth, and when it is absorbed it becomes heat energy.

Energy emitted by the heated filament of an electric lamp traverses the space between the filament and the glass, even though there is no gas in the bulb. Energy of this nature is emitted by all bodies. A body which absorbs this radiant energy converts the energy into heat, with a resulting increase in the random motion of its molecules.

All bodies emit radiant energy. Radiant energy travels out from a hot stove until it encounters some object, where, in general, it is partly reflected, partly absorbed, and partly transmitted. In every way it behaves in the same manner as light, except that it does not produce the sensation of sight. Heat radiation comprises wavelengths longer than those of the visible spectrum.

There are large differences in the transparency of various substances to heat radiations. Certain materials, such as hard rubber, nickel oxide, special glasses, or a deep-black solution of iodine in carbon disulfide, which are opaque to light, are almost perfectly transparent to heat radiation. Ordinary window glass, quite transparent optically, absorbs heat radiations. The glass roof of a greenhouse is transparent to the visible and near-infrared radiations received from the sun and admits them. Their energy is converted into heat when they are absorbed by the objects inside the greenhouse. These objects become warmer and themselves radiate energy. But since their temperatures are not high, the heat radiation they emit is not identical with the radiation that entered. Glass does not transmit this heat radiation, and hence the energy radiated by the bodies in the greenhouse cannot get out. A greenhouse thus acts as an energy trap, and since heat losses by radiation and convection are largely prevented, the temperature inside is greater than the temperature outside when the greenhouse receives direct sunlight. This is called the "greenhouse effect" and is the principle upon which solar-heated houses operate.

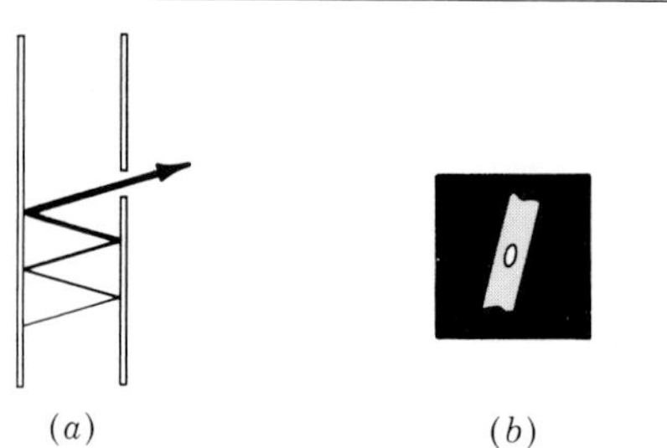

Figure 17-5
A blackbody radiator: an incandescent tungsten tube with a small hole in its wall. (*a*) Radiation emerging from the hole has had many internal reflections; its characteristics depend on temperature only. (*b*) The hole appears brighter than the surrounding incandescent tungsten.

17-7 THE IDEAL RADIATOR

Since white objects reflect and black objects absorb energy, an *ideal blackbody* is defined as one which absorbs all the radiation which falls upon it. No perfect blackbody is known, but a surface coated with lampblack is a good approximation. A small hole in the wall of a metal tube appears darker than the surrounding surface, because almost all the light entering the hole is absorbed. Thus a cavity having only a small aperture is almost a blackbody in the technical sense of absorbing all radiation incident upon it (Fig. 17-5).

Objects whose surfaces are in such condition that they are good absorbers of radiation are also good radiators. If two silver coins, one of which has been blackened, are placed in a small furnace, the blackened surface will absorb more radiation than the bright surface. The blackened surface will also radiate faster than the polished surface if the two are held at the same temperature. A blackbody is the ideal radiator.

Another illustration of the effect of radiant energy on a body is the *radiometer*. This device has four vanes mounted at right angles to each other, one side of each vane being white and the other side being black; all of the white sides face the same direction. The vanes are mounted on a low-friction vertical axle and placed in a glass bulb which has been partially evacuated of air.

When exposed to radiation, i.e., heat energy, the vanes begin to turn, the black vanes turning

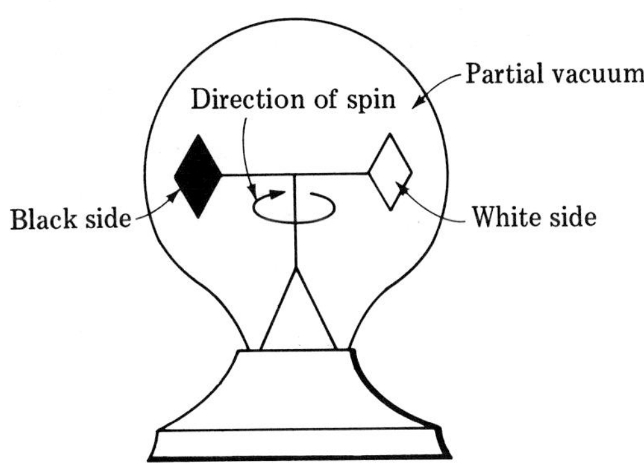

Figure 17-6
Radiometer.

away from the heat source. This can be explained by recalling that the black vane absorbs energy, and the particles of gas remaining in the glass bulb which come in contact with the black vane absorb heat from it and then move away from the black vane with increased momentum; as they do, they set up a Newton's third-law situation. As the particles leave the warmer (black) vane, they exert a reacting force on the vane and cause it to move (action-reaction). This reaction multiplied many times in the glass bulb causes the radiometer to spin with the black side of the vane "backing up." The radiometer can be used as a sensing device for radiant energy.

17-8 THEORY OF EXCHANGES

The rate at which a body radiates energy depends only on its temperature and the nature of its surface. The rate of radiation increases very rapidly as the temperature rises. But all objects radiate. Prevost stated in the form of a theorem the idea that all objects continuously radiate heat to their surroundings and receive heat from their surroundings. When the temperature of a body remains constant, it is receiving heat at the same rate as that at which it is radiating. A piece of metal left in the sun rises in temperature until it loses heat at the same rate at which it absorbs heat. A piece of ice radiates energy less rapidly than one's hand held near it and thus seems cold, while a heated iron radiates energy faster than the hand and thus seems warm.

A thermos bottle (Fig. 17-7) illustrates how the principles of heat transfer may be used to decrease the amount of heat flowing into (or out of) a container. It consists of two bottles, one inside the other, touching each other only at the neck. The space between the two bottles is evacuated, and the surfaces are silvered. Transfers by conduction are minimized by using a very small area of a poorly conducting material; those due to convection are lessened by removing the air. The transfer by radiation is made small, because the polished silver acts as a poor emitter for one surface and a poor absorber for the other.

17-9 RATE OF RADIATION

The law that expresses the total energy of all wavelengths radiated from a blackbody at a given temperature was stated originally by Stefan on the basis of careful measurements of the energy radiated from a blackbody cavity. Subsequently Boltzmann derived the same law from thermodynamic theory. The rate P at which energy is radiated by a blackbody is proportional to the area of the body and to the fourth power of its absolute temperature,

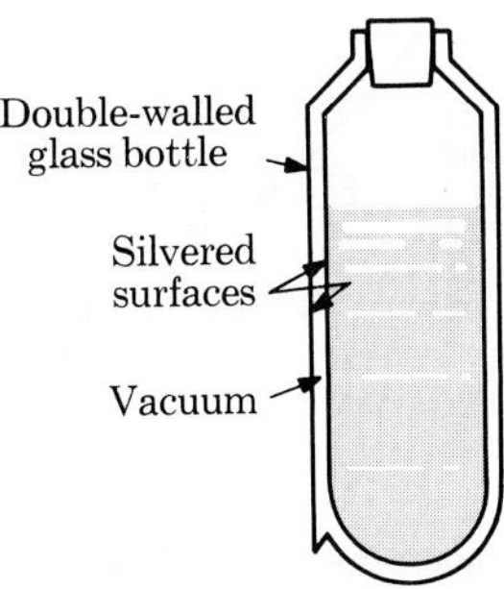

Figure 17-7
Dewar, or thermos, flask.

$$P = \sigma A T^4 \qquad (5)$$

The relation expressed by Eq. (5) is called the *Stefan-Boltzmann law*. If the power P is in watts, the area A in square centimeters, and the absolute temperature T in degrees Kelvin, the constant σ has the value 5.70×10^{-12} W/(cm^2)(K°)^4. It is important to note that a nonblackbody will radiate at a smaller rate.

Particularly when radiation is used to determine the temperature of a body, it is often necessary to know how the radiation from the body, such as a tungsten filament or a steelmaking furnace, compares with the radiation from the ideal blackbody at the same temperature. The *total emissivity* e of a surface is defined as the ratio of the radiancy of that surface to the radiancy of a blackbody at the same temperature. By radiancy R is meant the time rate per unit surface area at which energy is radiated into the hemisphere beyond the surface. The units commonly used for R are watts per square centimeter. The *spectral emissivity* e_λ of a surface is defined as the ratio of the spectral radiancy of that surface to that of a blackbody at the same temperature. By spectral radiancy R_λ is meant the rate per unit area per unit wavelength (λ) interval at which energy is radiated. The units commonly used for R_λ are watts per square centimeter per micron. In Fig. 17-8, the ratio of the area under the solid line to that under the dotted line is the total emissivity of tungsten, which is seen to be about 0.25 at 2000°K. The spectral emissivity at wavelength 1.0 μ is obtained from the ratio of ordinates as $e_\lambda = \frac{10}{30} = 0.33$. The value of e_λ is seen to be different at other wavelengths.

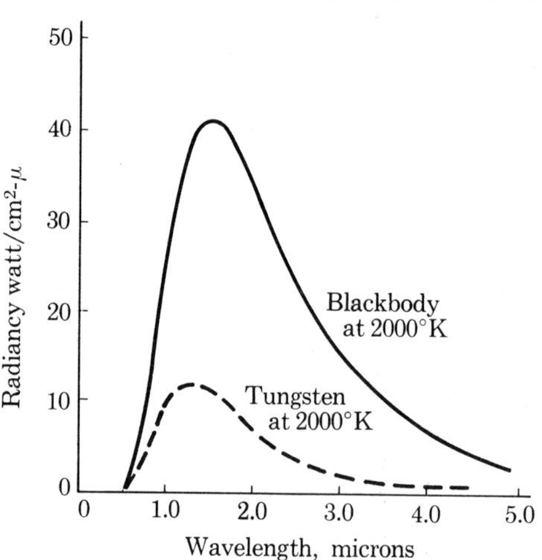

Figure 17-8
Radiancy of tungsten compared with that of a blackbody at 2000°K. One micron (1 μ) = 10^{-6} m = 10^4 A.

A blackbody at temperature T radiates energy and also receives energy from its surroundings at temperature T_0. According to Prevost's theorem of heat exchanges, the *net* rate at which the body loses heat by radiation is the difference between the rate at which it loses energy and the rate at which it receives energy,

$$P = \sigma A (T^4 - T_0^{\,4}) \qquad (6)$$

For the case of small temperature differences, the rate of cooling, due to conduction, convection, and radiation combined, is proportional to the difference in temperature,

$$P = bA(T - T_0) \qquad (7)$$

This is *Newton's law of cooling*. It is a valid approximation in the transfer of heat from a radiator to a room, the loss of heat through the walls of a room, or the cooling of a cup of coffee on the table.

Example What power is radiated from a tungsten filament 20 cm long and 0.010 mm in diameter when the filament is kept at 2500°K in an evacuated bulb? The tungsten radiates at 30 percent of the rate of a blackbody at the same temperature. Neglect conduction losses.

$$A = \pi(0.0010 \text{ cm})(20 \text{ cm}) = 0.0628 \text{ cm}^2$$

$$T^4 = (2500\text{°K})^4 = 39.1 \times 10^{12} (\text{°K})^4$$

From Eq. (5),

$$P = 0.30\sigma AT^4$$
$$= (0.30)[5.70 \times 10^{-12}\ \text{W}/(\text{cm}^2)(\text{K}^\circ)^4]$$
$$\times (0.0628\ \text{cm}^2)[39.1 \times 10^{12}(\text{K}^\circ)^4] = 4.2\ \text{W}$$

17-10 PLANCK'S QUANTUM THEORY OF RADIATION

Since a knowledge of Planck's quantum theory is fundamental to an understanding of modern physics, the student is urged to study this section with care.

The radiation from a blackbody cavity radiator can be analyzed with a grating (Chap. 30), and the radiancy in each wavelength region can be measured with sensitive detectors of radiation. When these data are plotted, for different temperatures of the blackbody, curves like those in Fig. 17-9 are obtained. Two features are notable. The area under a curve, representing total power radiated, is proportional to the fourth power of the temperature of the source (Stefan-Boltzmann law). Also, as the temperature is increased the wavelength λ_{max} of maximum radiancy shifts toward higher frequency (or shorter wavelength) in accordance with the relation

$$\lambda_{max}T = \text{const} = 2897.2\ \mu\text{K}^\circ \qquad (8)$$

which is known as *Wien's displacement law.*

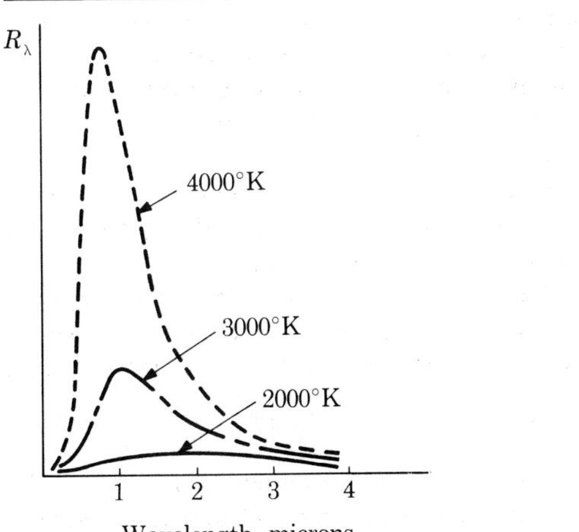

Figure 17-9
Spectral distribution of blackbody radiation.

Physicists tried to find an equation which would fit the experimental curves (Fig. 17-9) and to devise a theory of blackbody radiation. In 1896, Wilhelm Wien suggested an empirical equation, still used in calculating temperatures of incandescent bodies,

$$R_\lambda = c_1\lambda^{-5}e^{-c_2/\lambda T} \qquad (9)$$

where constants c_1 and c_2 were determined by fitting a curve to the experimental data. But attempts to construct a satisfactory theory of radiation within the framework of nineteenth century physics failed.

In 1901, Max Planck suggested a theory of radiation which predicted results fully in accordance with the experimental data by introducing a radical concept, the "quantum" of energy, which has profoundly influenced physics ever since. Working a decade before there was a detailed model of the atom, Planck built his theory of radiation on the following assumptions: The walls of a cavity radiator are made up of tiny electric oscillators. In the process of radiation and absorption, an oscillator (atom) loses energy in the form of an electromagnetic wave and later has its energy replenished by absorbing a wave or by thermal agitation. But the energy of an oscillator can have only certain definite values. When the oscillator spontaneously changes from one energy level to another, lower energy level, the difference in energy is radiated instantaneously in a single packet or "quantum" hf, where h is a universal constant which has been named *Planck's constant*. The value of h at present accepted is 6.62×10^{-34} J·s. Thus the frequency f of radiation is proportional to the energy given its quanta.

Planck proceeded to calculate the spectral-energy distribution in blackbody radiation by considering the immense number of oscillators in a cavity as some spontaneously made energy jumps downward from higher energy levels to lower energy levels, emitting quanta. He borrowed a mathematical treatment used successfully by Maxwell and Boltzmann in developing the classical kinetic theory of gases. Planck assumed that the probability that an oscillator has energy u is $e^{-u/kT}$, where k is the Boltzmann constant (Chap. 14). Of course, in Planck's theory, u can have only certain values, integral multiples of hf, whereas there is no such quantum restriction on the kinetic energy of a gas molecule. By working out the relative number of oscillators having each allowed energy level, for a given temperature, and then by considering the radiation that results when this statistically large number of oscillators undergoes spontaneous changes in energy levels, Planck arrived at an expression for the spectral radiancy of a blackbody,

$$R_\lambda = \frac{2\pi hc^2\lambda^{-5}}{e^{hf/kT}-1} = \frac{2\pi h}{c^3}\frac{f^5}{e^{hf/kT}-1} \qquad (10)$$

which is in accordance with experimental data. Note that Planck's law differs in form from Wien's law, Eq. (9), only in the term -1 in the denominator. This is sufficient to make the significant improvement in satisfying experimental data. But of course the great significance of Planck's work was the introduction of the concept of a quantum of energy and a theory which replaced the empirical constants c_1 and c_2 of Eq. (9) by certain fundamental constants of nature: the speed of light c, the Boltzmann constant k, and Planck's constant h. This topic will reappear in our discussion of modern physics.

SUMMARY

Heat is the most common form of energy.

The three ways in which heat may be transferred from one place to another are *conduction, convection,* and *radiation absorption.*

Conduction is heat transfer from molecule to molecule through a body or through bodies in contact.

Temperature gradient is temperature difference per unit distance along the direction of heat flow. Its units may be centigrade degrees per centimeter, Fahrenheit degrees per inch, etc.

Thermal conductivity k is a quantity that expresses how well a substance conducts heat. It may have units of calories per square centimeter per second for a gradient of 1 C°/cm or Btu per square foot per hour for a gradient of 1 F°/in.

$$k = \frac{Q}{At(\Delta T/\Delta L)}$$

Convection is heat transfer due to motion of matter.

The passage of heat through a surface film is described by the relation

$$\frac{\Delta Q}{\Delta t} = Ah\,\Delta T$$

where the film coefficient h represents the conductivity of the fluid divided by the effective thickness of the conducting layer.

In the process of *radiation* energy is transferred, without the aid of a material medium, from one body to another where, upon *absorption,* it again becomes energy of thermal motion. Radiant energy travels as an electromagnetic wave.

Good absorbers of radiation are also good radiators. A *blackbody* is a perfect absorber of radiation and an ideal radiator.

The total energy radiated per unit time by a blackbody is proportional to the fourth power of its absolute temperature (Stefan-Boltzmann law),

$$P = \sigma AT^4$$

Planck's theory of radiation introduced the concept that energy is radiated and absorbed only in multiples of a *quantum* of energy, *hf.* This led to an expression for the spectral radiancy of a blackbody, called *Planck's law,*

$$R_\lambda = \frac{2\pi hc^2\lambda^{-5}}{e^{hf/kT}-1} = \frac{2\pi h}{c^3}\frac{f^5}{e^{hf/kT}-1}$$

References

Barker, M. E.: Warm Clothes, *Scientific American,* March 1951, pp. 55–60.

Jakob, Max, and George A. Hawkins: "Elements of Heat Transfer," John Wiley & Sons, Inc., New York, 1958.

McAdams, William H.: "Heat Transmission," 3d ed., McGraw-Hill Book Company, New York, 1954.

Questions

1 Discuss the statement that "every object that we see or feel possesses heat energy."

2 Suits which were developed for lunar astronauts had to withstand extreme heat and cold yet were only a few millimeters thick. What special properties must the material in these suits possess?

3 A light fluffy quilt can be more efficient than a closely knit blanket in keeping a person warm. Explain why this is true.

4 What kinds of bodies are good insulators? Why?

5 An interesting demonstration consists of boiling water in a paper cup supported over a direct flame from a laboratory gas burner. The water can be made to boil, but the paper cup will not burn. Why does this happen?

6 A metal plate, a wooden plate, a paper plate, and a marble plate after having been in the same room for a period of time are touched by your hand. Each one feels to be at a different temperature. Give reasons to prove that in spite of your observations, they must be at the same temperature. Rank the order of apparent "coldness" among these plates and give supporting evidence for your ranking.

7 Explain why a moistened finger may freeze quickly to a piece of metal on a cold day but not to a piece of wood.

8 A metal rod and a wooden rod are attached to each other end to end so as to make one long rod. If a piece of paper is wrapped around the junction of the wood and the metal and then a flame is directed momentarily at the paper, the section of the paper in contact with the metal rod is not affected but the section of the paper in contact with the wood is scorched. Explain why this happens.

9 Can you suggest an apparatus to demonstrate the differences in thermal conductivities of various metals which gives results that do not depend also on differences in the specific heats?

10 What is the role of molecular action in convection and in conduction?

11 It has been stated that the laws of nature require that extremes or excesses be moderated. This can be shown to be true in the case of osmosis where a greater concentration of a liquid on one side of a membrane than on the other results in a flow of liquid through the membrane from the more dense side; also in the diffusion of gases from a more concentrated to a less concentrated area. Show that this principle holds for heat. What is this principle called in heat-transfer situations?

12 There are three ways in which heat is transferred. Name these and give an example of each.

13 In order to get a chimney in a fireplace to work properly it is sometimes necessary to ignite a roll of paper and hold it up the chimney for a few seconds. What does this accomplish?

14 Does warm air over a fire rise, or is it pushed up? Explain.

15 Since at any given time one side of the moon is very hot (130°C) and the other side very cold (−200°C), if the moon had an atmosphere, what climatic problems would this have presented for astronauts on the moon?

16 Explain how solar houses consisting of glass roofs can be heated by the sun's rays.

17 To insulate a house, a substance such as rock wool is used to fill hollow exterior walls. Why would this be a better insulator than air alone?

18 What is the function of the inert gas in modern tungsten filament lamps?

19 A test tube filled with water is inclined so that the flame of a bunsen burner reaches only the upper part of the tube. It is found that the lower part of the tube can be held in the hand painlessly indefinitely. If, however, the positions of hand and flame are interchanged, the tube cannot be held long. What do you conclude from these observations?

20 Conduits for hot-air heating systems are frequently made of bright sheet metal. The addition of a layer of asbestos paper on the conduit may actually increase the loss of heat through the surface. Explain.

21 Explain how a thermos flask minimizes energy losses from convection, conduction, and radiation.

22 Describe a means for detecting radiant energy.

23 Heat energy reaches the earth from the sun (93,000,000 mi away). Explain how this energy is transmitted. Is it necessary that some medium be present for this energy to reach the earth?

Problems

1 One side of an iron plate 4 cm thick and having a cross section of 5,000 cm^2 is at 150°C and the other side is at 120°C. Thermal conductivity of iron is 0.115 cal/(cm^2)(s)(°C/cm). How much heat is transmitted per second?

2 How much heat is conducted in 1.0 h through an iron plate 2.0 cm thick and 1,000 cm^2 in area, the temperatures of the two sides being kept at 0 and 20°C? *Ans.* 4.0×10^6 cal.

3 A pond is covered by a sheet of ice 2.0 cm thick [thermal conductivity 168 J/(m^2)(s)(C°/cm)]. The temperature of the lower surface of the ice is 0°C, and that of the upper surface is −10°C. At what rate is heat conducted through each square meter of the ice? What is the direction of this flow of energy?

4 A certain window glass, 30 × 36 in, is $\frac{1}{8}$ in thick. One side has a uniform temperature of 70°F, and the second face a temperature of 10°F. What is the temperature gradient?

Ans. 480°F/in.

5 How much heat is conducted through a sheet of plate glass, $k = 0.0024$ cal/(cm^2)(s)(C°/cm), which is 2.0 × 3.0 m and 5.0 mm thick, when the temperatures of the surfaces are 20 and −10°C? Why is considerably less heat transmitted through a window glass of these dimensions when room temperature is 20°C and the outdoor temperature is −10°C?

6 A nickel plate 0.8 cm thick has a temperature difference of 64°C between its faces. It transmits 200 kcal/h through an area of 10 cm^2. Calculate the thermal conductivity of nickel in cgs units.

Ans. 0.07 cal/[(cm^2)(s)(°C/cm)].

7 The value of a thermal conductivity is known in cal/(cm^2)(s)(C°/cm). By what factor should this value be multiplied to express it in (*a*) Btu/(ft^2)(h)(F°/in) and (*b*) kcal/(m^2)(s)(C°/m)?

8 A concrete foundation for a house 50 × 30 ft is 4 in thick. The ground in midwinter cools the outside of this foundation to 20°F. What is the temperature on the inside surface? Thermal conductivity of concrete is 12.0 Btu/(ft^2)(h)(F°/in). The heat loss is 1.8×10^5 Btu/hr. 1 Btu = 778 ft·lb = 252 cal = 4.186 J. *Ans.* 60°F.

9 A copper kettle, the bottom of which has an area 0.20 ft^2 and thickness 0.062 in, is placed over a gas flame. Assuming that the average temperature of the outer surface of the copper is 300°F and that the water in the kettle is at its normal boiling point, how much heat is conducted through the bottom in 1.0 min? Comment on the reasonableness of the assumptions and your numerical answer.

10 If the thermal conductivity of oak is 1.02 Btu/(ft^2)(h)(°F/in), how much heat will pass in 24 h through a door 3.0 × 7.0 ft whose thickness is 1.5 in, when the inside and outside temperatures are 72 and 10°F, respectively?

Ans. 2.1×10^4 Btu.

11 A copper rod whose diameter is 2.0 cm and length 50 cm has one end in boiling water and the other end in a jacket cooled by flowing water which enters at 10°C. The thermal conductivity of the copper is 0.102 kcal/(m^2)(s)(C°/m). If 0.20 kg of water flows through the jacket in 6.0 min, by how much does the temperature of this water increase?

12 When one end of the copper rod 30 cm long and 8.0 mm in diameter is kept in boiling water at 100°C and the other end is kept in ice, it is found that 1.2 g of ice melts per minute. What is the thermal conductivity of the rod?

Ans. 0.96 cal/(cm^2)(s)(°C/cm).

13 The thermal insulation of a woolen glove may be regarded as being essentially a layer of quiescent air 3.0 cm thick, of conductivity 5.7×10^{-6} kcal/(m^2)(s)(C°/min). How much heat does a person lose per minute from his hand, of area 200 cm^2 and skin temperature 35°C, on a winter day at −5°C?

14 A container used to store ice is made of wood 5.0 cm thick has an effective area of $1\bar{2}{,}000$ cm^2 and thermal conductivity of 0.00027 cal/(cm^2)(s)(C°/cm). How much ice inside the box would be melted each 24-h day if the outside temperature is 25° and the temperature inside the box is 5°C? *Ans.* 14 kg.

15 A ship has a steel hull 2.54 cm thick whose thermal conductivity is 0.011 kcal/(m^2)(s)(C°/m). A layer of insulating material 5.0 cm thick whose thermal conductivity in these units is 10×10^{-6} lines the hull. When the temperature inside the ship is 25.0°C and the temperature of the surrounding water is 5.0°C, what is the temperature on the inside surface of the steel hull? What is the difference in temperature between the inside and outside of the insulating layer? How many calories are transmitted through each square meter of the hull per minute?

16 One end of a 40-cm metal rod 2.0 cm^2 in cross section is in a steam bath while the other end is embedded in ice. It is observed that 13.3 g of ice melts in 15 s from the heat conducted to it by the rod. What is the thermal conductivity of the metal rod?

Ans. 2.24 cal/(cm^2)(s)(°C/cm); time = 15 min.

17 The temperature directly beneath a 3.0-in-concrete road is 5°F, and the air temperature is 20°F. (*a*) Calculate the steady heat flow per square foot through the concrete. What is the direction of this flow? The thermal conductivity of concrete is 6.0 Btu/(ft^2)(h)(F°/in). (*b*) How thick a layer of ice on the concrete would reduce the steady heat flow to 90 percent of its original value? The thermal conductivity of ice is 12 Btu/(ft^2)(F°/in).

18 What will be the rise in temperature in 30 min of a block of copper of 500-g mass if it is joined to a cylindrical copper rod 20 cm long and 3.0 mm in diameter when there is maintained a temperature difference of 80°C between the ends of the rod? The thermal conductivity of copper is 1.02 cal/(cm^2)(s)(°C/cm). Neglect heat losses. *Ans.* 11.3°C.

19 An insulating wall consists of 4.0 cm of material of thermal conductivity k_1 and a second layer 8.0 cm thick whose conductivity k_2 is equal to $4k_1$. The innermost and outermost surfaces are kept at −10 and 76°C, respectively. Calculate the temperature T_x at the interface.

20 A composite wall consists of three layers each 1 in thick whose thermal conductivities are 0.010, 0.020, and 0.030 Btu/(ft^2)(s)(°F/in), respectively. What conductivity should a single layer of material 3.0 in thick have to transmit the same heat flow for the same temperature difference? *Ans.* 0.016 Btu/(s)(ft^2)(°F/in).

21 Outdoor snow having a density of 0.100 g/cm^3 is coated with soot to make it a perfect absorber. If sunshine delivers 85 kcal/(m^2)(s), what depth of snow would be melted in 1 h?

22 Heat is conducted through a slab composed of parallel layers of two different conductivities, 0.0050 and 0.0025 cal/(cm^2)(s)(°C/cm) and thicknesses 0.36 and 0.48 cm, respectively. The temperatures of the outer faces of the slab are 96 and 8°C. Find (*a*) the temperature of the interface, and (*b*) the temperature gradient in each material. *Ans.* 72°C; 67°C/cm; $1\bar{3}0$°C/cm.

23 Calculate the power in watts radiated from a filament of an incandescent lamp at 2000°K if the surface area is 5.0×10^{-1} cm^2 and its emissivity is 0.85.

24 A picture window has seventy-two 6 × 8-in panes of glass. The glass is $\frac{1}{4}$ in thick and the thermal conductivity of glass is 5.8 Btu/(ft^2)(h)(F°/in) and the film coefficient for air is 2 Btu/(ft^2)(h)(F°). Compute the heat loss if the outside temperature is 27°F and the inside temperature is 72°F. *Ans.* 1035 Btu/hr.

25 At what rate does the sun lose energy by radiation? The temperature of the sun is about 6000°K, and its radius is 6.95×10^5 km.

26 The temperature of the sun is about 6000°K, and when the sun is directly overhead its radiation provides about 1.4×10^3 W/m^2 at the earth's surface. Calculate the decrease in this rate of heat transfer if the sun's temperature should drop 500 C°.

Ans. 0.59 cal/(cm^2)(min), or 29 percent.

27 What fraction of the heat reaching the earth from the sun would reach the earth if the distance between them were twice the present distance?

28 A can filled with water and containing an electric heating element rises to a temperature 45 C° above its surroundings when the power supplied to the heater is 120 W. The can and water are equivalent to 1.35 kg of water. What will be the initial rate of cooling when the heater is turned off? *Ans.* 0.021 C°/s.

29 Calculate the radiation, in watts per square centimeter, from a block of copper at 200 and at 1000°C. The oxidized copper surface radiates at 0.60 the rate of a blackbody.

30 How many watts will be radiated from a spherical blackbody 15.0 cm in diameter at a temperature of 800°C? *Ans.* 5.4 kW.

Jean Baptiste Perrin, 1870–1942

Born in Lille. Professor at the University of Paris. Given the 1926 Nobel Prize for Physics for his work on discontinuity in the structure of matter, and in particular for his discovery of the equilibrium of sedimentation.

18

Thermodynamics

Thermodynamics deals with the conversion of *mechanical energy* into *thermal energy* and the reverse process, the *conversion of heat into work.* The first and second laws of thermodynamics summarize experience that shows that in such processes (1) energy is conserved and (2) the process can never be fully reversed. Starting from these very general laws, *thermodynamics* allows us to derive practical relations about a particular equilibrium system that are independent of detailed assumptions about atomic structure or the exact mechanism by which energy is exchanged.

The related study, *statistical mechanics,* deals with the motions of a large number of individual particles of a substance and how these motions lead to the observable, large-scale properties, such as pressure and temperature of a gas. Statistical mechanics is more detailed and somewhat more complicated than thermodynamics. Like thermodynamics, it is limited to the treatment of systems in equilibrium.

Kinetic theory seeks to describe the rates of atomic and molecular processes by fairly direct means, by using general principles of mechanics. This approach is superior to statistical mechanics and to thermodynamics in just two respects: it makes use of only well-known elementary methods, and it can handle problems (such as reaction rates) relating to systems not in equilibrium.

In this chapter, our concern with the broad field of thermodynamics will be limited chiefly to these questions: How is the conversion of heat into mechanical work and the reverse procedure described in typical processes involving an ideal gas? How can thermodynamic principles be applied to establish an absolute temperature scale? How is the efficiency of heat engines (including refrigerators) related to temperature? Since we have identified heat as kinetic energy associated with the random (thermal) motions of molecules, how can we define a quantity (entropy) which measures "randomness"? Does thermodynamics offer any guidance in the development of unconventional energy sources for future use?

18-1 THERMODYNAMIC PROCESSES

A system is said to be in *thermodynamic equilibrium* when it is in a state of mechanical, thermal, and chemical equilibrium. There is no unbalanced force in the interior of the system or between the system and its environment. All parts of the system are at the same temperature, and this is the same as the temperature of the environment. The net rate of any chemical reaction or change of internal structure is zero. When a system is in thermodynamic equilibrium, its condition can be specified by giving the values of only a few quantities (such as pressure, volume, temperature, and quantity of a particular substance) called *variables of state*. By a *thermodynamic process*, or a *change of state*, is meant any change (however small) that involves changes in these variables. A change of *phase*, however, refers to a more fundamental change of physical form, as from a liquid to a vapor or from one crystalline form to another.

Any process that can be made to go in the reverse direction by an infinitesimal change in the conditions is called a *reversible process*. No actual change is *fully* reversible, but many processes when carried out slowly are *practically* reversible.

The slow compression of a spring is practically a reversible process. If the compressing force is slightly decreased, the spring expands and performs work equal to the work done in compressing it. The slow evaporation of a substance in an insulated container is practically reversible, for if the temperature is slightly lowered, condensation can be made to occur, returning energy to the heater until both it and the substance are in their original condition. The slow compression of a gas can be altered to expansion by a slight decrease in the force applied to the piston; hence this process, too, is reversible.

Any process that is not reversible is irreversible. All changes which occur suddenly or which involve friction or electrical resistance are inherently irreversible. An explosion is a highly irreversible change. Another type of irreversible process is represented by the following experiment. Let a flask containing air at atmospheric pressure be connected through a tube and stopcock with a second flask, which has been evacuated. If, now, the stopcock is opened, air rushes into the evacuated flask until the pressures in the two flasks become equal. No external work is done by the gas in expanding under these conditions; yet the gas cannot be restored to its original container and condition without energy from an outside source. The process is irreversible.

A *cycle* is a succession of changes that ends with the return of the body or system to its initial state. A reversible cycle is a cycle all of whose changes are reversible.

18-2 FIRST LAW OF THERMODYNAMICS

As we saw earlier, when heat is added to a substance, there is an increase in the *internal energy*. This increase in internal energy is manifested by a rise in temperature, an increase in pressure, or a change in phase. If at the same time the substance is allowed to perform external work, by expanding, for example, the heat Q required will be the heat necessary to change the internal energy of the substance from U_1 in the first state to U_2 in the second state plus the heat equivalent of the external work $\mathcal{W}$ performed,

$$Q = (U_2 - U_1) + \mathcal{W} \tag{1}$$

In applying Eq. (1), all quantities must be expressed in the same units; Q is positive for heat entering the system, and $\mathcal{W}$ is positive for energy leaving the system as external work. An arbitrary value may be assigned to the internal energy U in some standard reference state; its value in any other state is then defined, since $Q - \mathcal{W}$ is the same for all processes connecting the states.

The *first law of thermodynamics* states that *when heat is transformed into any other form of energy, or when other forms of energy are converted into heat, the total amount of energy* (*heat plus other forms*) *is constant*. The first law is extended to apply to interchanges of all forms of energy and thus extended becomes the law of the conservation of energy. This means that kinetic energy, gravitational potential energy, heat, electric and magnetic energy, and the energy of chemical reaction are in principle convertible into one another, without loss or gain of energy. Although the law of conservation of energy cannot be proved directly, it is in accord with a wide range of experience, and the many scientific conclusions based on this principle have been confirmed by experiment.

18-3 SPECIFIC HEATS OF A GAS

The heat necessary to raise the temperature of a gas depends on how the gas is confined. If the gas is held at constant volume, as indicated in Fig. 18-1*a*, the heat received is converted entirely into internal energy, in the form of molecular kinetic energy, thus raising the temperature. The heat per unit mass per degree required to raise the temperature of the gas under these conditions is called the *specific heat at constant volume,* c_v.

When the gas is confined in a cylinder under a piston that maintains constant pressure (Fig. 18-1*b*), the gas will expand on being heated. It does work in moving the piston. Hence heat must be supplied to change the internal energy of the gas and to perform external work. The heat per unit mass per degree required to raise the temperature of the gas under constant pressure is called the *specific heat at constant pressure,* c_p. Since the change of internal energy is the same in both cases, the specific heat at constant pressure c_p is greater than the specific heat at constant volume c_v, because external work is also performed when the gas expands at constant pressure. For air and for diatomic gases such as hydrogen, nitrogen, and oxygen, the ratio of the specific heats $\gamma = c_p/c_v$ is 1.40. For other gases the value of γ lies between 1.00 and 1.67.

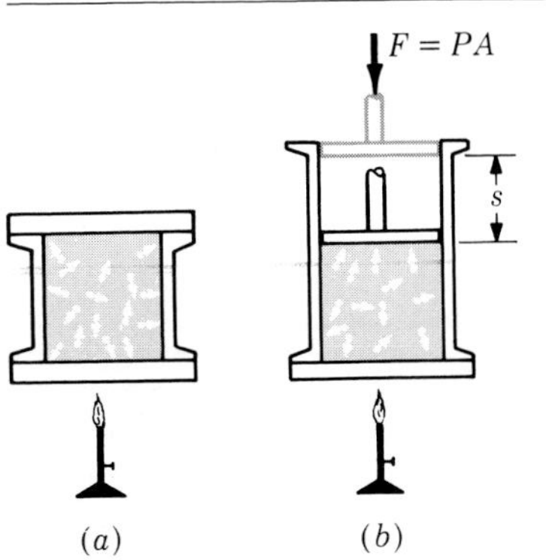

Figure 18-1
Specific heat of a gas (*a*) at constant volume and (*b*) at constant pressure.

The work done by a gas in moving a piston a distance s against a constant pressure P (Fig. 18-1*b*) is

$$\mathcal{W} = Fs = PAs$$

But As is the change in volume ΔV of the gas. Hence the work done by a gas expanding at constant pressure is

$$\mathcal{W} = P\,\Delta V \qquad (2)$$

From Eqs. (1) and (2) and the general gas law, at constant pressure,

$$\Delta Q_p = \Delta U + nR\,\Delta T$$

It is convenient to define a molar specific heat C (Sec. 16-4) as the heat required per mole per degree change in temperature. Then

$$C_p = \frac{1}{n}\frac{\Delta Q_p}{\Delta T} = \frac{1}{n}\frac{\Delta U}{\Delta T} + \frac{nR}{n}\frac{\Delta T}{\Delta T}$$

$$C_v = \frac{1}{n}\frac{\Delta Q_v}{\Delta T} = \frac{1}{n}\frac{\Delta U}{\Delta T}$$

Hence $C_p = C_v + R$ (3)

18-4 WORK DONE BY EXPANDING GAS

As an example of the transformation of heat into work, consider a process in which a gas is heated and caused to do external work as it expands from state 1 to state 2 (Fig. 18-2). The expansion can be regarded as made up of many infinitesimal

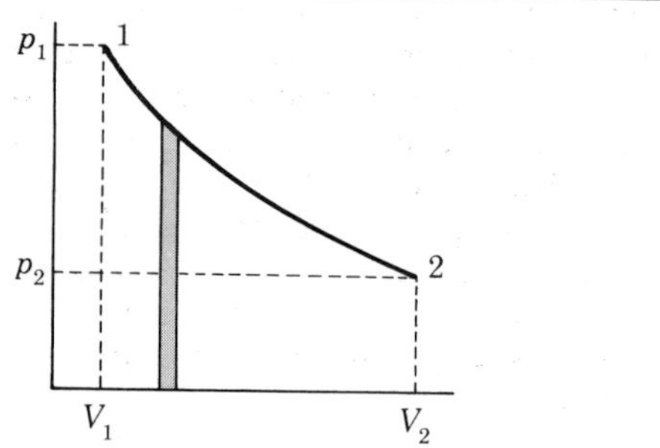

Figure 18-2
The area under a pressure-volume curve represents work.

expansions, such as the one shown as shaded, in any one of which the pressure can be assumed constant. The work $\Delta\mathcal{W}$ in an infinitesimal expansion is the product of $P\,\Delta V$, from Eq. (2). The work done during the entire expansion is the sum of the areas of all the vertical elements under the curve,

$$\mathcal{W} = \sum_{V_1}^{V_2} P\,\Delta V \tag{4}$$

The results for three important types of expansion will be mentioned.

Isothermal expansion occurs when the gas expands without change of temperature. For a gas expanding isothermally, the ideal-gas law becomes

$$PV = nRT = \text{const} \tag{5}$$

In this case $\Delta\mathcal{W} = P\,\Delta V = nRT_1 \dfrac{\Delta V}{V}$ (6)

One may use the property of logarithms[1] $\Delta(\ln x) \approx \Delta x/x$ to write Eq. (6) as

$$\Delta\mathcal{W} = nRT_1\,\Delta(\ln V) \tag{7}$$

The work done when n moles of ideal gas expand isothermally at temperature T_1 from volume V_1 to volume V_2 is

$$\mathcal{W} = nRT_1 \ln\frac{V_2}{V_1} = 2.303\,nRT_1 \log\frac{V_2}{V_1} \tag{8}$$

An equal amount of heat is transferred to keep the temperature constant.

Example Two moles of an ideal gas are compressed slowly and isothermally from a volume of 4.0 to 1.0 ft³, at a temperature of 300°K. How much work is done?

From Eq. (8)

$$\mathcal{W} = 2.30(2.0\text{ mol})[8.317\text{ J/(mol)(K°)}] \times (300\text{°K}) \log\tfrac{1}{4}$$
$$= -6,\bar{9}00\text{ J}$$

The negative sign indicates that this work was done *on* the gas by an external agent.

Adiabatic expansion occurs without the addition or withdrawal of heat. In Fig. 18-3 a small change in state is represented by increments ΔV and ΔP in the variables V and P. Here ΔV and

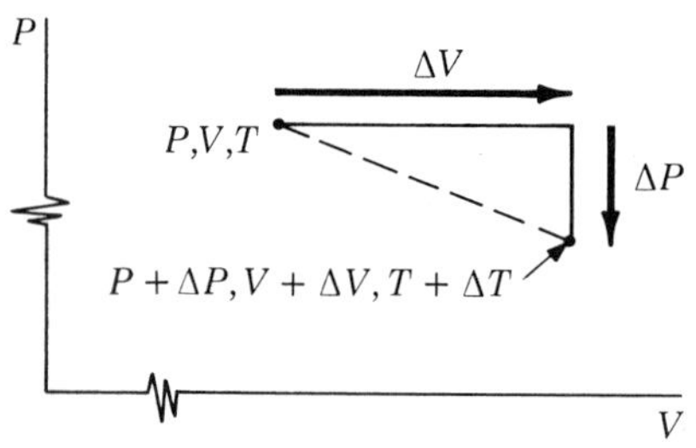

Figure 18-3
Small change of state represented on a *P-V* diagram.

[1] Interesting noncalculus derivations of this and other thermodynamic relations may be found in J. S. Marshall and E. R. Pounder, "Physics," pp. 292–300, The Macmillan Company, New York, 1957.

ΔP may be either positive or negative; they are small in comparison with V and P, respectively. The change can be considered arbitrarily as a step at constant pressure, followed by a step at constant volume. For the first step, the equation for an ideal gas gives

$$\frac{\Delta V}{V} = \frac{\Delta T_p}{T} \tag{9}$$

where ΔT_p refers to the temperature increment during a change at constant pressure. The heat transferred is

$$\Delta Q_p = nC_p \, \Delta T_p \tag{10}$$

Likewise during the second step

$$\frac{\Delta P}{P} = \frac{\Delta T_v}{T + \Delta T_p} \tag{11}$$

If we assume that ΔT_p is negligible as a term added to T, this relation becomes

$$\frac{\Delta P}{P} \approx \frac{\Delta T_v}{T} \tag{12}$$

The heat transferred is

$$\Delta Q_v = nC_v \, \Delta T_v \tag{13}$$

The total heat transferred in the two-step change of state is

$$\Delta Q = \Delta Q_p + \Delta Q_v = n(C_p \, \Delta T_p + C_v \, \Delta T_v) \tag{14}$$

or

$$\frac{\Delta Q}{T} = n\left(C_p \frac{\Delta T_p}{T} + C_v \frac{\Delta T_v}{T}\right) \tag{15}$$

From Eqs. (9) and (12),

$$\frac{\Delta Q}{nT} = C_p \frac{\Delta V}{V} + C_v \frac{\Delta P}{P} \tag{16}$$

Now, for an adiabatic change, $\Delta Q = 0$. If we let $\gamma = C_p/C_v$ and divide Eq. (16) by C_v, we have

$$\frac{\Delta P}{P} + \gamma \frac{\Delta V}{V} = 0 \tag{17}$$

We again use the approximation $\Delta(\ln x) \approx \Delta x/x$, which becomes exact as Δx approaches zero, to write Eq. (17) in the form

$$\ln P + \gamma \ln V = \text{const} \tag{18}$$

or

$$PV^{\gamma} = \text{const} \tag{19}$$

This is the relation we sought for the pressure and volume of a gas undergoing an adiabatic change. Equation (17) further shows that the ratio of the slope of an adiabatic curve through any point to the slope of the isothermal curve through the same point is equal to the ratio γ of the specific heats.

The work done by a gas expanding adiabatically is

$$\mathcal{W} = mc_v(T_1 - T_2) \tag{20}$$

The performance of external work is at the expense of a decrease in internal energy, accompanied by a decrease in the temperature of the gas. Since this expression involves the difference between the initial temperature T_1 of the gas and its final temperature T_2, the temperatures need not be expressed on an absolute scale. The units commonly used for c_v will give $\mathcal{W}$ in calories or in Btu, but the work is still represented by the area under the pressure-volume curve and can be converted into other units (say, joules) by using the proper value for J, the mechanical equivalent of heat.

Expansion at constant pressure is represented by a horizontal line in Fig. 18-3, and the work done is represented by the area under that line.

$$\mathcal{W} = P(V_2 - V_1) \tag{21}$$

Example Air which occupies 5.0 ft^3 at 15 lb/in^2 gauge pressure is expanded isothermally to atmospheric pressure and then cooled at constant pressure until it reaches its initial volume. Compute the work done by the gas.

The volume V_B of the gas after expansion is obtained from Boyle's law

$$V_B = \frac{P_A V_A}{P_B} = \frac{(15 + 14.7)(144)\ \text{lb/ft}^2(5.0\ \text{ft}^3)}{(14.7 \times 144)\ \text{lb/ft}^2}$$

$$= 10\ \text{ft}^3$$

The work done during the isothermal expansion is, from Eq. (8),

$$\mathcal{W}_{AB} = PV \ln \frac{V_B}{V_A}$$

$$= (14.7 \times 144)\ \text{lb/ft}^2 \times 10\ \text{ft}^3 \ln \frac{10}{5}$$

$$= (14.7 \times 144)\ \text{lb/ft}^2 \times 10\ \text{ft}^3 \times 2.303 \log 2$$

$$= 21{,}200\ \text{ft}\cdot\text{lb} \times 2.303\ (0.301)$$

$$= 14{,}700\ \text{ft}\cdot\text{lb} = 1\bar{5}{,}000\ \text{ft}\cdot\text{lb}$$

The work done during change in volume at constant pressure is, from Eq. (21),

$$\mathcal{W}_{BC} = P_B(V_B - V_C)$$

$$= (14.7 \times 144)\ \text{lb/ft}^2(10 - 5)\ \text{ft}^3$$

$$= 1\bar{1}{,}000\ \text{ft}\cdot\text{lb}$$

Hence the net work done by the gas is

$$\mathcal{W}_{AC} = 1\bar{5}{,}000 - 1\bar{1}{,}000 = \bar{4}{,}000\ \text{ft}\cdot\text{lb}$$

18-5 THROTTLING PROCESS

James Joule in collaboration with William Thomson (Lord Kelvin) investigated the forces between the molecules of a gas by means of a throttled expansion of the gas through a porous plug. Consider a gas forced by a pump to circulate as shown in Fig. 18-4. Follow a sample of gas, shown shaded, between imaginary pistons in Fig. 18-4*b*. As the gas passes through the plug or throttling valve, work equal to P_1V_1 is done on the chosen sample by the gas behind it on the high-pressure side. As it passes through the valve, the gas does work equal to P_2V_2 on the gas ahead of it. The expansion takes place in a thermally insulated chamber. Hence the first law of thermodynamics gives

$$Q = (U_2 - U_1) + \mathcal{W} = 0$$

But the net work is $\mathcal{W} = P_2V_2 - P_1V_1$; hence

$$\text{Energy}_{\text{first state}} + \text{Work}_{\text{first state}}$$
$$= \text{Energy}_{\text{second state}} + \text{Work}_{\text{second state}}$$

$$U_1 + P_1V_1 = U_2 + P_2V_2 = \text{const} \qquad (22)$$

This result applies to all fluids and is of great importance in steam engineering and in refrigeration. *The quantity $U + PV$ is called the enthalpy.* The enthalpy per unit mass is tabulated for steam and for many refrigerants. The throt-

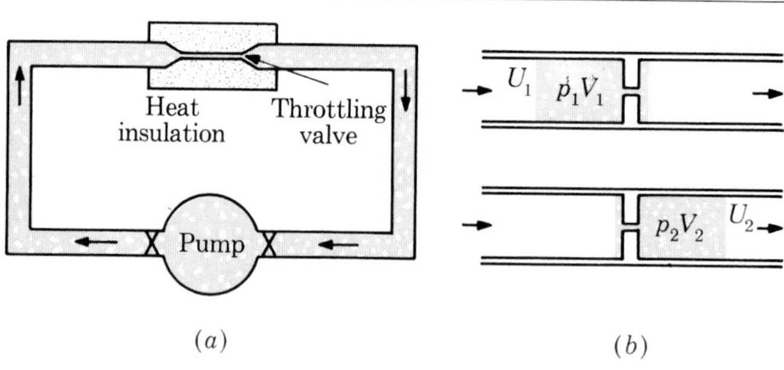

Figure 18-4
Throttling process.

tling process is the important one in the action of a refrigerator; it is the process that causes the drop in temperature needed for refrigeration. Liquids undergo a drop in temperature and partial vaporization in a throttling process. However, a Joule-Thomson expansion of a gas may result in either cooling or heating, depending on the gas used and the initial temperature and pressure. Cooling is readily interpreted as due to the fact that work is done against the attractive forces between molecules. Heating is associated largely with changes that occur in PV during flow through the valve. The work P_1V_1 done on the gas about to enter the valve may be greater than the work P_2V_2 done by this gas on emerging from the valve.

18-6 THE CARNOT CYCLE

Carnot made important thermodynamic studies using an ideal, reversible heat engine, which operated through a sequence of isothermal and adiabatic steps now known as a *Carnot cycle*. The engine may be a cylinder (Fig. 18-5) fitted with a piston and filled with any substance that expands with rising temperature and decreasing pressure. We shall consider the working substance to be an ideal gas. The initial state of this working substance is represented by point A in Fig. 18-6. As a first step, the cylinder is placed in contact with a reservoir of heat and the gas allowed to expand at constant temperature T_1 taking in heat Q_1, this change being represented by the isothermal curve AB. Next, the cylinder is insulated and expansion allowed to continue. This expansion is adiabatic, along BC, and the temperature drops from T_1 to T_2. The cylinder is then placed on a heat reservoir at temperature T_2, and the gas is compressed isothermally, along CD, the heat of compression Q_2 being transferred to the low temperature reservoir. Finally the cylinder is again insulated and the compression continued adiabatically, along DA, the gas now being heated to its original temperature T_1.

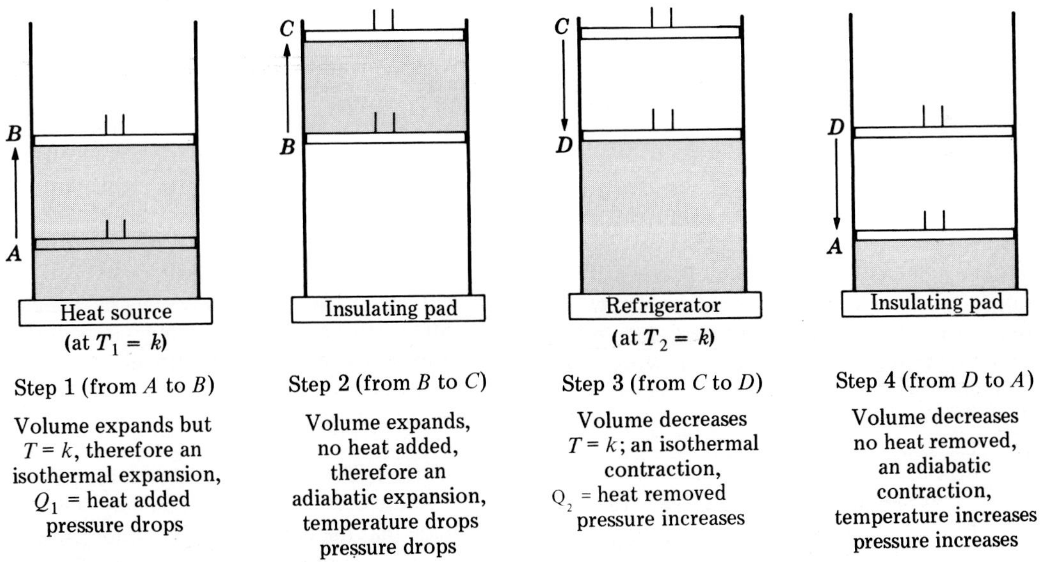

Figure 18-5
Equipment and steps for operation of a Carnot cycle, related to Fig. 18-6.

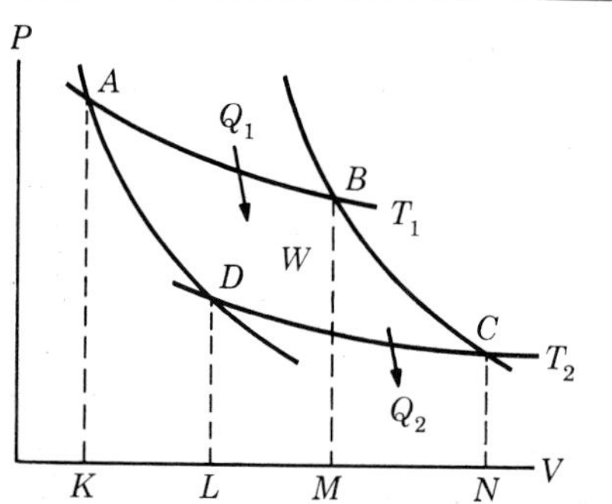

Figure 18-6
Carnot cycle.

Let us examine the significance of the Carnot-cycle diagram (Fig. 18-6).

The area *ABMKA* represents the energy Q_1 taken from the hot source and converted into work $\mathcal{W}$ in moving the piston.

The area *BCNMB* represents the energy taken from the working substance in cooling it from T_1 to T_2.

The area *CNLDC* represents the energy taken from the external machinery and given as heat Q_2 to the cool reservoir.

The area *ADLKA* represents the energy taken from the external machinery and given to the working substance, heating it from T_2 and T_1.

The work done *by* the gas in expanding from *A* to *B* to *C* equals the area under the upper curves, *ABCNKA*. The work done *on* the gas during compression from *C* to *D* to *A* equals the smaller area under the lower curves, *CDAKNC*. Area *ABCDA* = the net useful work done by the heat engine during one cycle. This work is the difference between Q_1, the energy received from the hot reservoir, and Q_2, the energy given to the cool reservoir.

$$\text{Eff} = \frac{\text{output work}}{\text{input heat}} = \frac{\text{area } ABCDA}{\text{area } ABMKA}$$

$$= \frac{Q_1 - Q_2}{Q_1} \tag{23}$$

We have thought of the working substance as an ideal gas, as an aid in visualizing the Carnot process. However, the results [Eq. (23)] are independent of the nature of the working substance.

18-7 SECOND LAW OF THERMODYNAMICS

In each of the heat-engine cycles described, we have seen that not all the heat supplied to the engine is converted into useful work; some heat is always rejected to some outside reservoir. This is true of all heat engines and leads to an important generalization known as *the second law of thermodynamics: It is impossible for an engine unaided by external energy to transfer heat from one body to another at a higher temperature.*

Heat of its own accord will always flow from high temperature to low temperature. It is impossible to utilize the immense amount of heat in the ocean, for example, to run an engine unless there can be found a reservoir at a lower temperature into which the engine can discharge heat.

Whenever heat is transferred from low temperature to higher temperature, an expenditure of energy is required from some external source. This takes place in a refrigerator in which electrical energy is used to pump heat from the cool interior to the warmer room.

It follows from the second law of thermodynamics that a Carnot (reversible) heat engine has a greater efficiency than any other heat engine operating between the same temperature limits. If we assumed the contrary, then the more efficient engine could be used to run the reversible engine as a heat pump and we would have a self-acting pair transferring heat from low to high temperature. But this is contrary to experience. As a result of this type of reasoning we may conclude that the efficiency of a Carnot (reversible) engine represents the maximum efficiency any heat engine can have with the given temperatures.

18-8 THE ABSOLUTE, OR KELVIN, TEMPERATURE SCALE

In each of the methods for measuring temperature that were discussed in earlier chapters, the results depend on the particular thermometric substance used. Most of the phenomena of heat depend on the properties of the particular substances involved. The efficiency of an ideal heat engine is unique in that it does not depend on the working substance or the particular mechanical device used.

Lord Kelvin (1827–1907) recognized the possibility of using an ideal heat-engine cycle to define a temperature scale which would be "absolute" in the sense that it did not depend on the thermometric substance. He suggested that temperatures on the absolute scale be defined from the relation

$$\text{Eff} = \frac{Q_1 - Q_2}{Q_1} = \frac{T_1 - T_2}{T_1} \qquad (24)$$

giving

$$\frac{Q_2}{Q_1} = \frac{T_2}{T_1} \qquad (25)$$

This equation states that any two temperatures are in the same ratio as the heat quantities absorbed and ejected in a Carnot cycle operated between those two temperatures.

Consider a Carnot cycle operating between two fixed temperatures, say, the boiling point of water (373°K) and the freezing point (273°K). There will be a certain area representing useful work (Fig. 18-7). We can define the temperature (323°K) midway between the boiling point and the freezing point as the temperature at which a Carnot engine operating between the boiling point and the midpoint does the same work, represented by area *A*, as the work represented by area *B*, done by a Carnot engine operating between the midpoint and the ice point. Obviously the interval so defined can be subdivided in the same manner, and the scale can be extended to higher or lower temperatures. The temperature at which a Carnot engine ejects no heat would be the zero on the absolute temperature scale. Absolute zero may be thought of as the sink temperature T_2 of a Carnot engine operating at an efficiency of 100 percent.

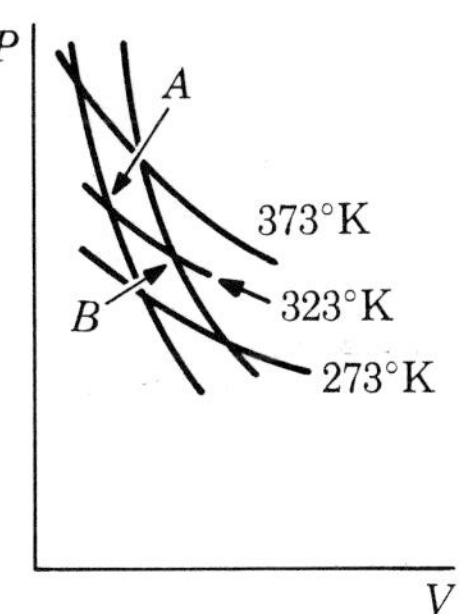

Figure 18-7
Definition of absolute temperature from a Carnot cycle.

It is generally accepted as a law of nature that although one may closely approach 0°K it is impossible actually to reach the zero of temperature. This statement of the unattainability of absolute zero is one version of *the third law of thermodynamics*.

The Kelvin, or thermodynamic, temperature scale is the same as the absolute scale determined by a perfect gas. Temperatures on this absolute scale are determined in practice with a gas thermometer, the readings being corrected for the deviation of the particular gas from the ideal-gas law as calculated from certain other experiments.

18-9 THE EFFICIENCY OF HEAT ENGINES

The ideal, or thermodynamic, efficiency of a heat engine is defined from Eq. (24),

$$\text{Ideal eff} = \frac{T_1 - T_2}{T_1} = 1 - \frac{T_2}{T_1} \qquad (26)$$

Owing to heat losses and friction, no actual engine ever attains the efficiency defined by Eq.

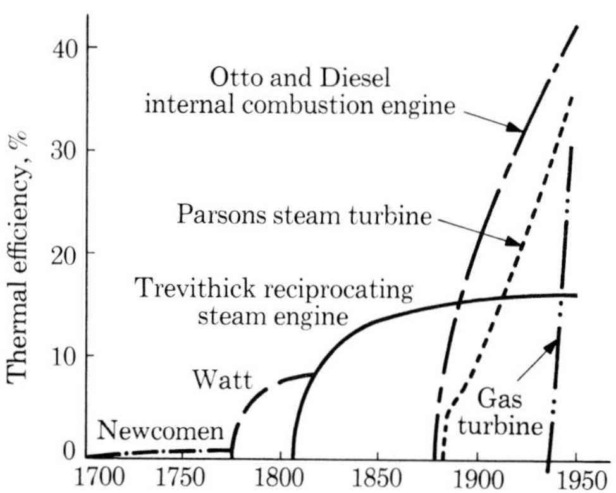

Figure 18-8
Improvement in heat-engine efficiencies. [*Adapted from a paper by M. W. Thring, J. Inst. Fuel,* **27**:*401–407 (1954)*.]

(26). The ideal efficiency remains as an upper limit to the efficiency of any heat engine.

To improve the efficiency of a practical engine it is evident from Eq. (26) that one should use a high input temperature and a low exhaust temperature. The use of cooling water in a condenser may reduce T_2 from about 100°C to about 40°C. Large steam turbines may have an input temperature T_1 as high as 600°C and gas turbines somewhat higher. A practical limit is imposed by changes in the thermal characteristics of high-temperature steam and by the endurance of turbine blades. The improvement in heat-engine efficiencies over the years is illustrated in Fig. 18-8.

Example A simple steam engine receives steam from the boiler at 180°C (about 150 lb/in^2 gauge pressure) and exhausts directly into the air at 100°C. What is the upper limit of its efficiency?

$$\text{Ideal eff} = \frac{(180 + 273)°\text{K} - (100 + 273)°\text{K}}{(180 + 273)°\text{K}}$$

$$= 0.176 = 17.6\%$$

18-10 GASOLINE AND DIESEL ENGINES

In the common gasoline engine four processes take place in each cycle. For computation of the efficiency, these processes can be represented approximately by the Otto cycle (Fig. 18-9*a*). Starting with the piston at the top of its stroke, *a*, an explosive mixture of air and gasoline vapor is compressed adiabatically in the cylinder to point *b*, heated by combustion to point *c*, allowed to expand adiabatically with performance of exter-

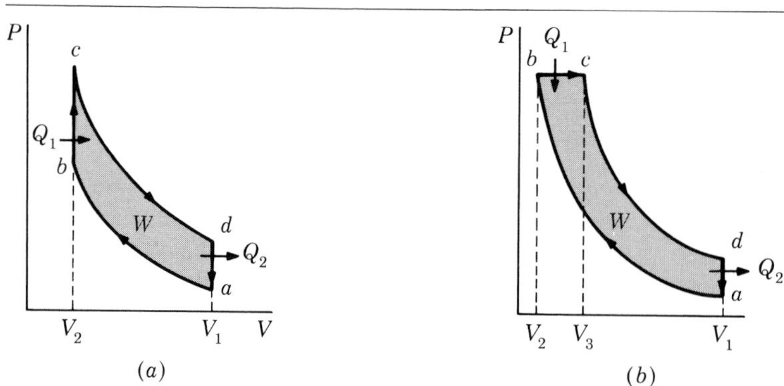

Figure 18-9
P-V diagrams: (*a*) Otto cycle for gasoline engine; (*b*) diesel cycle.

nal work to point d, and cooled at constant volume to point a, during the exhaust stroke. The ratio V_1/V_2 is called the *compression ratio* and is about 8 in automobile engines.

The work output is represented by the shaded area of Fig. 18-9a. The heat input is Q_1, and the exhaust heat is Q_2. By assuming the mixture in the cylinder to behave like an ideal gas, the heat input and the work output can be computed in terms of the compression ratio; then

$$\text{Eff} = \frac{W}{Q_1} = 1 - \frac{1}{(V_1/V_2)^{\gamma-1}} \qquad (27)$$

The efficiency computed for $V_1/V_2 = 8$, and $\gamma = 1.4$ is about 56 percent. The attainable efficiency is less, because of loss of heat to the cylinder walls, friction, turbulence, etc.

An idealized cycle for a diesel engine is shown in Fig. 18-9b. Starting at a, air is compressed adiabatically to b, at which point fuel oil is injected and starts to burn at approximately constant pressure, b to c. The remainder of the power stroke is an adiabatic expansion to d. During the exhaust stroke the gas cools at constant volume to a. Because there is no fuel in the cylinder during compression, the compression ratio V_1/V_2 can be larger than in a gasoline engine; it is usually about 15. The ratio V_1/V_3 may be about 5. Using these values, one could compute the diesel efficiency to be about the same as that for a gasoline engine.

18-11 REFRIGERATOR AND HEAT PUMP

A refrigerator may be thought of as a heat engine operated in reverse. The refrigerator takes heat Q_2 per cycle from a *cold* reservoir, the compressor supplies mechanical work *input* $\mathcal{W}$, and a *larger* quantity of heat Q_1 is delivered to a *hot* reservoir. Food and ice cubes may comprise the cold reservoir, an electric motor supplies the input work, and the air in the kitchen receives heat as the hot reservoir. By the sign convention used for Eq. (1), Q_1, Q_2, and $\mathcal{W}$ are all negative quantities. The coefficient of performance (or heating-energy ratio) of a refrigerator expresses its effectiveness in removing heat (Q_2) for the least expenditure of mechanical work $\mathcal{W}$,

$$\text{Coefficient of performance} = \frac{Q_2}{\mathcal{W}} = \frac{Q_2}{Q_1 - Q_2} \qquad (28)$$

The term *heat pump* is applied to a heat engine operated in reverse in situations where the useful output is regarded as the heat Q_1 transferred to the warm reservoir. Thus a heat pump might be used in winter to transfer heat from a pond to the interior of a house. No violation of the second law of thermodynamics occurs, for work $\mathcal{W}$ has to be supplied externally, by a motor. The heat engine can operate in the other direction in summer to cool the house.

The coefficient of performance of a heat pump is defined as

$$\text{Coefficient of performance} = \frac{Q_1}{\mathcal{W}} = \frac{Q_1}{Q_1 - Q_2} \qquad (29)$$

18-12 ENTROPY

The entropy of a body is a quantity which depends on the quantity of heat in the body and on its temperature, which, when multiplied by any lower temperature, gives the unavailable energy, or unavoidable waste, when mechanical work is derived from the heat energy of the body in a process which terminates at the lower temperature. The following two viewpoints illustrate different aspects of the utility of the concept of entropy: First, consider the term $\Delta Q/T$ in Eq. (16). It is found analytically that the sum $\Sigma_1^2\, \Delta Q/T$ between two states has a definite value, independent of the path. It has hence been found useful to define entropy S by a statement about its increment.

$$\Delta S = \frac{\Delta Q}{T} \qquad \text{for a reversible process} \qquad (30)$$

Entropy is one of four properties, P, V, T, and S, any two of which can specify the state of the gas. For example, in place of the P-V diagram to represent the Carnot cycle (Fig. 18-6), one could plot a T-S diagram. During an adiabatic transformation, $\Delta Q = 0$, and so $\Delta S = 0$. During an isothermal change $\Delta T = 0$. So the adiabatic steps are represented by straight lines parallel to the T axis; the isothermal steps are represented by straight lines parallel to the S axis. The Carnot diagram becomes a rectangle when plotted in terms of T and S.

To compute an entropy change in an irreversible process, we devise any reversible process connecting the given initial and final states and calculate $S_2 - S_1$ for the reversible process.

Example An ideal gas in a flask of volume v_1 expands into an evacuated flask to occupy a final volume v_2. What is the change in entropy if the temperature is held constant by a water bath?

Consider this irreversible expansion to be replaced by a reversible isothermal expansion of the gas. Then, from the first law of thermodynamics and the equation of state for the gas,

$$\Delta Q = \Delta \mathcal{W} = P\,\Delta V = nRT\frac{\Delta V}{V}$$

When the volume increases from V_1 to V_2

$$S_2 - S_1 = \sum \frac{\Delta Q}{T} = nR \sum \frac{\Delta V}{V}$$

$$= nR(\ln V_2 - \ln V_1)$$

$$= nR \ln \frac{V_2}{V_1} \qquad (31)$$

A second approach to the concept of entropy is available in the interpretation of the second law of thermodynamics in terms of the statistics of a large number of particles. Statistical mechanics defines entropy S by the relation

$$S = k \ln w \qquad (32)$$

where k is Boltzmann's constant and w is the probability that the system exists in the state it is in relative to all possible states it could be in. (Here *state* is defined in terms of the motions of all the individual particles, not in terms of the macroscopic variables P, V, T, and n, as earlier in this chapter.)

Consider the change in the entropy of an ideal gas during isothermal expansion. The temperature and the number of molecules remain unchanged. The probability w_1 that a single molecule may be found in a region having a volume V is proportional to V,

$$w_1 = cV \qquad (33)$$

where c is a constant. The probability of finding N molecules simultaneously in the volume V is the product of the individual probabilities, or the N-fold product of w_1. The probability of a state of N molecules in a volume V is

$$w = w_1{}^N = (cV)^N \qquad (34)$$

By combining Eqs. (32) and (34) we have

$$S = kN(\ln c + \ln V) \qquad (35)$$

Therefore the change in entropy between a state of volume V_1 and a state of volume V_2 (with temperature and number of molecules remaining constant) is

$$S_2 - S_1 = kN(\ln c + \ln V_2) - kN(\ln c + \ln V_1)$$

$$= kN \ln \frac{V_2}{V_1} = \frac{RN}{N_0} \ln \frac{V_2}{V_1}$$

$$= nR \ln \frac{V_2}{V_1} \qquad (36)$$

where N_0 is Avogadro's number and R is the gas constant per mole. This result, based on viewing entropy in terms of statistical mechanics and probability, is seen to be in accord with the result of Eq. (31), based on the purely thermodynamic definition of entropy.

An interesting modern application of entropy is to information theory. Ignorance (of alternatives) is regarded as essentially the same as disorder and entropy is used to describe both.

18-13 UNCONVENTIONAL ENERGY SOURCES

The gradual depletion of familiar sources of energy, notably coal, oil, and gas, has led physicists and engineers to seek other sources of energy, or, more precisely, other systems of energy conversion. Consideration of some possible sources of energy is so speculative that they are referred to as *esoteric* sources. The term *unconventional* is

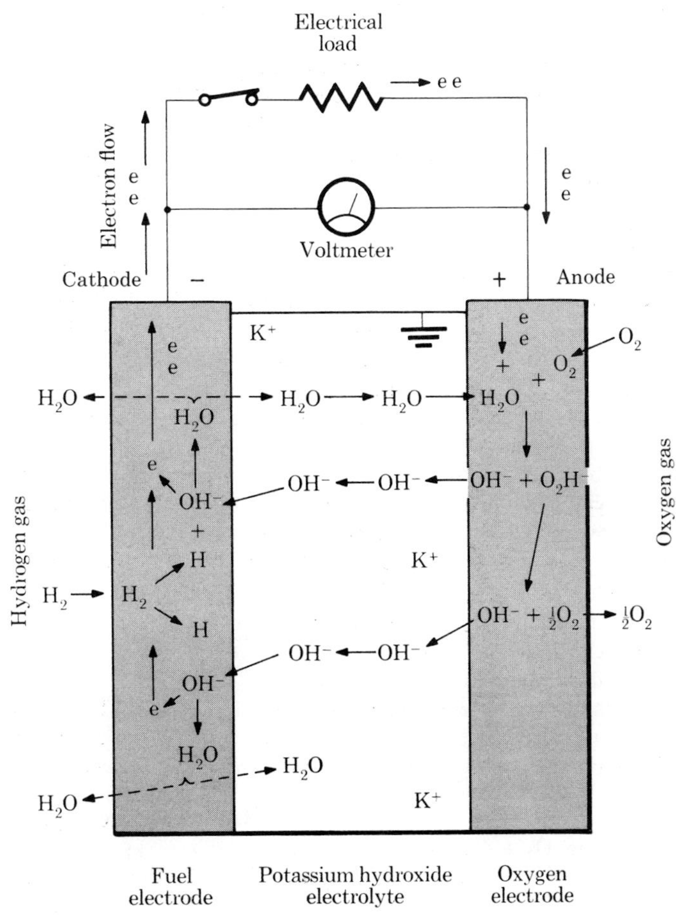

Figure 18-10
Hydrogen-oxygen fuel cell.

reserved for those untapped sources about which enough is understood today so that one may reasonably predict that engineering refinements will soon make of them practical energy sources. Some which will probably become increasingly important in our economy are: nuclear reactors; thermoelectric, thermionic, and magnetohydrodynamic generators; solar cells; and fuel cells.

Thermodynamic considerations are having a large influence on the development of each of these devices. The fuel cell looks particularly promising for thermodynamics. A fuel cell is a continuous-feed electrochemical device in which the chemical energy of reaction of a fuel and air (oxygen) is converted directly and usefully into electrical energy (Fig. 18-10). A fuel cell differs from a battery in that (1) its electrolyte remains unchanged and (2) it can operate continuously as long as an external supply of fuel and air is available.

It is an attractive feature of a fuel cell that its efficiency is *not* subject to the limitation derived for heat engines [Eq. (24)], for the energy being converted never deteriorates into the random motion of heat! Of the continuous-flow devices, the fuel cell is uniquely direct in its conversion of chemical energy into electrical energy. Partly because of this incentive, fuel cells are probably the most highly developed of the unconventional energy-conversion methods. Under favorable conditions, efficiencies of 80 and even 90 percent have been reported with hydrogen fuel.

In the cell pictured, anode and cathode are joined externally by a metallic circuit through which flow valence electrons from the fuel. Anode and cathode are connected internally by an electrolyte through which ions flow to complete the circuit. The electrode reactions are

Anode	$2H_2 \rightarrow 4H^+ + 4e^-$
Cathode	$O_2 + 4H^+ + 4e^- \rightarrow 2H_2O$
Overall	$O_2 + 2H_2 \rightarrow 2H_2O$

The electron does useful work in passing from anode to cathode in the external circuit.

SUMMARY

A system is said to be in *thermodynamic equilibrium* when it is in a state of mechanical, thermal, and chemical equilibrium.

By a *thermodynamic process,* or *change of state,* is meant any change that involves changes in the variables (such as pressure, volume, and temperature) needed to specify an equilibrium state.

A *reversible* process is one that is in equilibrium at each instant.

The *first law of thermodynamics* states that when heat is transformed into any other form of energy the total amount of energy is constant.

$$Q = (U_2 - U_1) + \mathcal{W}$$

The specific heat of a gas at constant pressure is greater than its specific heat at constant volume, $C_p = C_v + R$.

An isothermal process is one occurring at constant temperature. For isothermal expansion of an ideal gas, $PV = \text{const}$. The work done in isothermal expansion of an ideal gas is

$$\mathcal{W} = PV \ln \frac{V_2}{V_1}$$

An *adiabatic* process is one in which there is no exchange of heat with the surroundings. For adiabatic expansion of an ideal gas, $PV^\gamma = \text{const}$. The work done in adiabatic expansion of an ideal gas is

$$\mathcal{W} = mc_v(T_1 - T_2) = nC_v(T_1 - T_2)$$

The *second law of thermodynamics* states that a heat engine cannot transfer heat from a body to another at higher temperature unless external energy is supplied to the engine.

Work can be obtained from a source of heat energy only by a process that transfers some of the heat to a reservoir at a temperature lower than that of the source.

The *Kelvin,* or thermodynamic, temperature scale is independent of the thermometric substance, being based on the efficiency of an ideal heat engine.

The maximum efficiency of a heat engine supplied with heat at temperature T_1 and delivering heat to a reservoir at temperature T_2 is

$$\text{Eff} = \frac{T_1 - T_2}{T_1}$$

The coefficient of performance of a heat pump is defined as

$$\frac{\text{Heat given hot reservoir}}{\text{Mechanical work input}} = \frac{Q_1}{\mathcal{W}} = \frac{Q_1}{Q_1 - Q_2}$$

The change in entropy for a reversible process occurring at temperature T is

$$\Delta S = \frac{\Delta Q}{T}$$

The efficiency of a device such as a fuel cell in which energy is transformed without being converted into heat is not limited by the equation expressing the thermodynamic efficiency of a heat engine.

References

Carnot, Sadi: "Reflections on the Motive Power of Fire," E. Mendoza (ed.), Dover Publications, Inc., New York, 1960. Contains other papers on the second law of thermodynamics by E. Clapeyron and R. Clausius.

Morse, Philip M.: "Thermal Physics," W. A. Benjamin, Inc., New York, 1962.

Planck, Max: "Treatise on Thermodynamics," transl. A. Ogg, Dover Publications, Inc., New York, 1922.

Sandfort, John F.: "Heat Engines," Anchor Book, Doubleday & Company, Inc., New York, 1962.

Wilks, J.: "The Third Law of Thermodynamics," Oxford University Press, New York, 1961.

Zemansky, Mark W.: "Temperatures Very Low and Very High," Momentum Book, D. Van Nostrand Company, Inc., New York, 1964.

Questions

1 What three forms of equilibria must be satisfied for a system to be in thermodynamic equilibrium? Explain.

2 What is the difference between a change of state and a change of phase?

3 Define and give several examples of irreversible processes.

4 When is the conduction of heat a reversible process? When an irreversible process?

5 Distinguish between the two often-confused terms *entropy* and *enthalpy*.

6 The gas CO_2 escaping from a fire extinguisher puts out a fire because CO_2 does not support combustion and also because of its cooling effect. Why would recently released CO_2 have this latter effect?

7 Why does air released through a tire valve feel cool to the hand?

8 Trace the successive transformations by which sunlight is changed into the energy of an electric lamp.

9 It is said that 3,000 kcal in food is required every day for an average adult. What becomes of this energy?

10 Heat engines have very low efficiencies. Explain why this is the case.

11 Discuss the reasoning which leads to the idea of an absolute temperature. Explain the steps needed in defining an absolute scale. Can there be more than one absolute scale of temperature?

12 Explain how very low temperatures can be obtained with compressed gases.

13 Of what practical value is liquid air? Liquid helium?

14 How hot can a given object be heated? How cold can a given object be cooled?

15 "Heat goes downhill" is another way of expressing what law of thermodynamics? Explain what the expression means.

16 Two samples of a gas initially at the same temperature and pressure are compressed from a volume V to a volume $\frac{1}{2}V$, one isothermally, the other adiabatically. In which sample is the pressure greater?

17 Is there a principle of conservation of entropy? Consider a high-speed bullet which has ordered kinetic energy ($\frac{1}{2}mv^2$) striking a steel plate and producing disordered energy.

18 Show that when a substance of mass m which has a constant specific heat c is heated from T_1 to T_2 the change in entropy of $S_2 - S_1 = mc \ln(T_2/T_1)$.

19 Show that the efficiency e of a reversible heat engine is related to the coefficient of performance k of the refrigerator, obtained by running the engine backward, by the equation $ke = T_2/T_1$.

Problems

1 An engine used to pump water raises 1,000 lb of water 100 ft. In doing this, 1 lb of fuel having a heat of combustion of 1,000 Btu/lb was burned. What percentage of the heat produced was transformed into useful work?

2 A boiler and engine deliver 10 hp and use 30 lb of coal per hour. The heat value of coal is 15,000 Btu/lb. What percentage of the heat is transformed to work? *Ans.* 5.7 percent.

3 A 5.0-lb lead ball of specific heat 0.032 Btu/(lb)(F°) is thrown downward from a 50-ft building with an initial vertical speed of 20 ft/s. If half its energy at the instant of impact with the ground is converted to heat and absorbed by the ball, what will be its rise of temperature?

4 A 250-g copper calorimeter contains 550 g of oil. The oil is stirred by a 300-g steel paddle. A couple (torque) of 6.0×10^6 cm·dyn is applied to rotate the paddle. Calculate the rise in temperature produced after 1,000 r. The specific heats are as follows: copper, 0.092; oil, 0.511; and steel, 0.114 in cal/(g)(C°). *Ans.* 2.7°C.

5 At what speed must a lead bullet at 20°C strike an iron target in order that the heat produced on impact be just sufficient to melt the bullet, assuming no heat is lost to the surroundings? Lead has a specific heat of 0.032 cal/(g)(C°), its melting point is 327°C, and the heat of fusion is 5.4 cal/g.

6 How much heat in Btu/h is generated due to friction and electric resistance by a motor which loses 0.25 hp to these effects? *Ans.* 636 Btu/h.

7 What quantity of butter (6,000 cal/g) would supply the energy needed for a 160-lb man to ascend to the top of Mount Washington, elevation 6,288 ft?

8 A 120-lb wheel whose radius of gyration is 1 ft revolves at 480 r/min. Calculate the heat produced when the wheel is brought to rest by friction. *Ans.* 6.1 Btu.

9 A 200-ton train has its speed reduced from 40 to 30 mi/h in 0.50 min. If the whole of the work done against the frictional resistance of the brakes is converted into heat, find the heat developed.

10 A motor designed to agitate a 20-gal tank of water is capable of producing 0.5 hp. Assuming that the work done by the motor goes into heating the water, how much time will be required to raise the temperature of the water 20 F°? (Water = 8.34 lb/gal). *Ans.* 157 min.

11 How much heat (expressed in Btu) is produced in stopping by friction a 112-lb flywheel rotating 1.0 r/s if the mass is concentrated in the rim of mean radius 2.0 ft?

12 A simple steam engine has a piston area of 72 in² and an 18-in stroke. The average gauge pressure at the piston is 30 lb/in². The piston makes 180 power strokes per minute. What is the indicated horsepower of this engine? *Ans.* 18 hp.

13 How many tons of coal per hour are consumed by a locomotive working at the rate of $3,\bar{0}00$ hp if the heat of combustion of the coal is $12,\bar{0}00$ Btu/lb and the overall efficiency is 9.0 percent?

14 Determine the theoretical limiting efficiency of a steam engine when steam enters at 400°C and leaves the cylinder at 105°C. *Ans.* 43.8 percent.

15 A steam engine operates at an overall efficiency of 15 percent. How much heat must be supplied per hour in order to develop 3.0 hp?

16 A 10,000-ton ship is raised 16 ft in the locks of a canal. (*a*) What is the thermal equivalent (in Btu) of the work done? (*b*) How much coal (heating value of $12,\bar{0}00$ Btu/lb) would be required to produce this energy?

Ans. 4.1×10^5 Btu; 34 lb.

17 A certain ideal gas has $\gamma = 1.67$. (*a*) Compute the molar specific heats C_p and C_v. A 0.70-m^3 sample of this gas initially at pressure 4.5×10^4 N/m^2 is compressed adiabatically to a volume of 0.50 m^3. Find (*b*) the final pressure and (*c*) the ratio of final to original temperature.

18 A Carnot engine works between 450 and 50°C. What is its efficiency? *Ans.* 59.4 percent.

19 The weight-density of steam at 212°F is 0.0373 lb/ft^3. What is the heat equivalent of expanding water (weight-density 60 lb/ft^3 at 212°F) into steam against the force due to atmospheric pressure? What fraction of the total heat of vaporization does this represent?

20 A certain quantity of air at 76 cm Hg pressure is compressed adiabatically to two-thirds its initial volume. Calculate the final pressure.

Ans. 130 cm Hg.

21 A quantity of air at a pressure of 76 cm Hg is suddenly compressed to half its volume. (*a*) Calculate the new pressure. (*b*) What would the pressure be if the change were isothermal?

22 A steam generator was found to use 2.0 lb of coal ($12,\bar{0}00$ Btu/lb) for every horsepower-hour supplied by the engines. What proportion of the heat obtained from the coal was turned into work? What became of the rest of the heat? *Ans.* 11 percent.

23 Assume that you have an ideal gas for which $\gamma = 1.50$, initially at 1.0 atm pressure. The gas is compressed to one-half its original volume. What is the final pressure if the compression is (*a*) isothermal and (*b*) adiabatic?

24 What work is required to compress 320 g of oxygen isothermally at 37°C so that its final volume is one-tenth its original volume. (Molecular mass of oxygen is 32.) *Ans.* 59,400 J.

25 The temperature of steam from the boiler of an engine of 10 hp is 390°F and the condenser temperature is 176°F. How much heat must leave the boiler per hour if the efficiency is 20 percent of that of a reversible engine working between the same limits of temperature?

26 What is the thermodynamic efficiency of a steam engine which operates with a boiler temperature of 177°C and a condenser temperature of 77°C? *Ans.* 22 percent.

27 An ideal-gas engine operates in a Carnot cycle between 227 and 127°C. It absorbs 5.0×10^4 cal at the higher temperature. What amount of work (joules) is the engine theoretically capable of performing?

28 To 200 g of water at 20°C is added 80 g of lead shot [$c_p = 0.035$ cal/(g)(C°)] initially at 100°C. When the system has reached equilibrium, what change has occurred in its entropy?

Ans. 0.67 cal/K°.

29 In a mechanical refrigerator the low-temperature coils are at a temperature of −37°C, and the compressed gas in the condenser has a temperature of 62°C. What is the theoretical maximum efficiency?

30 Consider a mechanical refrigerator as a heat engine transferring heat from the cooling coils at −10°C to the room whose temperature is 20°C. How many kilowatthours of electric energy are needed to form 4.0 lb of ice at −10°C from water at 20°C? *Ans.* 0.023 kWh.

31 A Carnot engine has an efficiency of 25 percent when its low-temperature reservoir is at 25°C. It is desired to increase the efficiency to 35 percent. (*a*) To what temperature should the low-temperature reservoir be decreased if the temperature of the high-temperature reservoir remains constant? (*b*) To what temperature should the high-temperature reservoir be raised if the temperature of the low-temperature reservoir remains constant?

32 Calculate the amount of work done by a gas expanding from an initial volume of 6 l at 10 atm of pressure to 24 l. Assume the temperature is kept constant. *Ans.* 8.4×10^3 J.

33 Consider an ideal, reversible heat engine which transfers heat from a room at 17°C to the outdoors at −5°C. (*a*) What is the maximum efficiency of this engine? (*b*) If the heat engine is reversed and is used to pump heat from outdoors into the room, energy being supplied by an electric motor, how many joules of heat will be delivered to the room for the expenditure of 1.0 J of electric energy?

34 A volume of a gas at 20°C expands adiabatically until its volume is doubled. Find the resultant temperature of the gas. (The ratio of the specific heats of the gas is 1.4.) *Ans.* −56.7°C.

35 A brass rod [$k = 1.0$ J/(cm)(s)(C°)] extends from a heat reservoir at 127°C to a heat reservoir at 27°C. What change in entropy occurs in the process of conduction of 300 J of heat through the rod?

An ocean traveler has even more vividly the impression that the ocean is made of waves than that it is made of water.

A. S. Eddington

PART TWO

Wave Physics

Charles Thomson Rees Wilson, 1869–1959

Born in Glencorse, near Edinburgh. Professor at Cambridge. Shared the 1927 Nobel Prize for Physics for his discovery of a method of rendering discernible the paths of electrically charged particles by the condensation of vapor.

Arthur Holly Compton, 1892–1962

Born in Wooster, Ohio. Professor of Physics, University of Chicago, and later chancellor, Washington University. The 1927 Nobel Prize for Physics was awarded jointly to Compton and Wilson, to the former for his discovery of the Compton effect, which confirmed the quantum theory of radiation and assigned momentum as well as energy to light quanta.

19

Wave Motion

In all our activities the availability of suitable energy is a continual problem. The conversion of energy into the form that is useful for a particular purpose and the transportation of this energy from the place at which it is converted to the place at which it is needed are problems of immediate concern.

Developing means of communication has always been a challenge to man. Let us consider the situation where two men are on opposite sides of a lake. If one of the men wishes to get the attention of the other man, there are several means by which this can be done. For example, one can shout to the other, provided he has sufficient lung power. Another method would be for one of the men to shine a flashlight (assuming he has one). Still another possibility would be to throw a stone, or shoot an arrow to get the other person's attention. One other method would be to create a wave disturbance in the lake near one shore which, on a perfectly calm day, would carry across to the other side. Perhaps you can come up with many more ideas—signal flags, explosions, mirrors, etc. However, if we were to analyze each of the methods suggested above, we would find that they all fit into two large categories of communicative procedures. We are attempting to carry our message either through the motion of waves or through the motion of matter (particles). For example, shouting is a process depending upon a wave to carry it. Shining a flashlight, using a mirror, or creating an explosion (unless bits of exploded matter reached the other side of the lake) all depend upon waves for the transmission of their message. Creating ripples on the water is also a procedure in which wave motion is used to carry a signal, in this case that something or someone has imparted energy to the water. Waves provide a mechanism for the transfer of energy from one point to another without the physical transfer of material between the points. The other category, the transmission of a message by matter, is illustrated by the stone being thrown or the arrow being shot. Here a piece of matter is actually moving from one location to another as the message carrier. If we analyze the suggested procedures further we find that they are all procedures in which the transmission of energy is involved. It could be light energy, the transfer of mechanical energy (potential energy to kinetic energy), the release of chemical energy (the explosion), or a combination of these plus still other forms of energy. In short, each suggested procedure depends upon energy being carried from one location to another.

We do not have to limit ourselves, of course, to the simple illustrations of energy transmission listed above. For example, the transfer of electric energy by wire is another procedure involving particle transmission. Charged particles (electrons) upon entering a wire create a movement of electrons already present in the wire which by their motion supply energy to any point in the circuit. When a temperature difference is maintained between ends of a metal bar, energy is conducted through the bar by the collisions between the particles in the bar as they are agitated in the metal. Air currents set up by unequal heating of the surface of the earth transmit energy that can be harnessed at any place the wind blows. The ocean tides are another illustration of the transmission of energy by a fluid body.

In this section of the book we will be looking specifically at waves as a means of transmitting energy. Energy from the sun reaches us in large amounts. By far the greatest part of this comes as a wave motion. Radiant energy traverses the space between sun and earth with neither particles nor bodily motion of a fluid. Sounds reach our ears by means of waves in the air around us. Radio signals are waves that are used to transmit music or speech. Waves in strings or air columns are used in producing musical sounds. All waves have certain properties in common that will now be considered.

19-1 WAVES

If a stone is dropped into a quiet pool of water, a disturbance is created where the rock enters the liquid. However, the disturbance is not confined to that place alone but spreads out so that it eventually reaches all parts of the pool.

When the stone enters the water, it sets into motion the particles of water with which it comes in contact. These particles set into motion neighboring particles. They in turn produce similar motion in others, and so on, until the disturbance reaches particles at the edge of the pool. In all this disturbance no particle moves far from its initial position. Only the disturbance moves through the water. This can be seen by observing a wave passing through a long coiled spring. As the wave moves through the spring, each coil is momentarily displaced horizontally but then returns to its own position as it loses its energy. This behavior is characteristic of all wave motions. The particles move over short paths about their initial positions, and as a result a wave moves through the medium. *A wave is a disturbance that moves through a medium* in such a manner that at any point the displacement (or other quantity that varies) is a function of the time, while at any instant the displacement is a function of the position of the point. *The medium as a whole does not progress in the direction of motion of the wave.*

The motion of the wave through the medium is a result of the action of the successive parts of the medium on each other. Hence such a wave can travel only in an *elastic* medium. If the particles were entirely independent of each other, no waves could pass through.

For example, if the coils in the spring which were displaced by the wave did not attempt by exerting lateral forces to take along adjacent coils with them as they were displaced; and if the adjacent coils did not attempt and finally succeed in pulling back the displaced coils, no wave would form and we would simply have a disintegration of the medium, the spring.

There are several types of waves, their classification being made in accordance with the motion of the local part of the medium with respect to the direction of propagation. The most common types are *transverse* waves and *longitudinal* waves, but other types are frequently observed, usually as combinations of transverse and longitudinal.

19-2 TRANSVERSE WAVES

Consider a long string that is stretched by a tension T. If a small portion of the string is given a sudden lateral displacement (Fig. 19-1*a*), the displaced portion of the string will exert lateral forces tending to displace adjacent parts of the string and at the same time the displaced part will undergo forces tending to return it to the undisplaced position. The result is that a *pulse* travels out in each direction from the original displaced part of the string. After a short time t_1 the two pulses will appear as in Fig. 19-1*b* and still later as in Fig. 19-1*c*. The speed at which

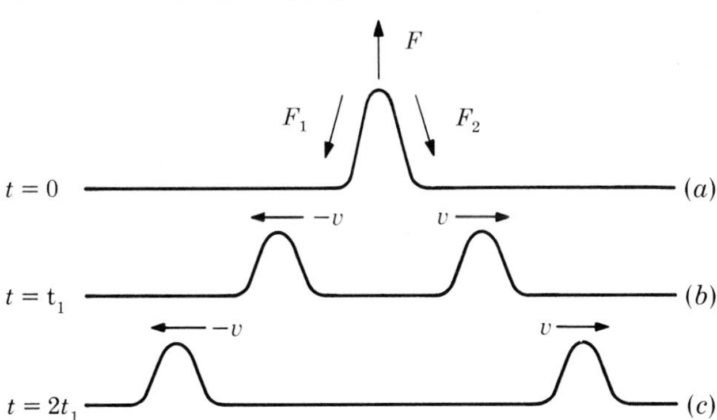

Figure 19-1
A pulse travels along a string in both directions from the source.

a pulse travels along the string is characteristic of the condition of the string.

Let us look more closely at the reasons why an elastic medium, such as a string, when displaced vertically, produces a pulse which travels laterally along the medium. At present we will limit our discussion to a wave pulse traveling down a cord. Several conditions must exist before such a transverse wave pulse is created. First, there must be forces acting along the string which cause the displacement as the pulse passes. Second, the string must have elasticity, that is, it does not tear apart under the stress created by the passing wave nor on the other extreme be so rigid that it will not yield to the wave pulse. Third, the string must have inertia so that its motion does not stop when it returns to its equilibrium position but goes on beyond it, producing a form of simple harmonic motion. This oscillatory motion will stop only if there is some mechanism for removing the energy in the string. The forces acting along the cord (F_1 through F_6 on Fig. 19-2*a*), in attempting to stretch and to contract the string, provide just such a mechanism. The segment of the string nearer the source passes its energy to the string segment adjacent to it by doing work on it (stretching the segment, for example). Once it has given up this energy, the crest formed by the wave pulse collapses and the string attempts to return to its original position, but due to the inertia of the string, it overshoots the equilibrium position and creates a downward crest which is then passed on to the next string segment in line.

This process has been presented graphically in Fig. 19-2*a*. In the sketch, the upward distorting force which had acted on string segment *AB* has just been counterbalanced by the forces F_1 and F_2 acting along the string and the upward movement stops. Since F_1 and F_2 are not acting directly opposite to each other, there is a resulting downward force on that segment, F_{net} (Fig. 19-2*b*). This force acts to accelerate it downward according to Newton's second law of motion. In a similar manner the downward force (or upward in the last half of the cycle, as in segment *EF*) can be shown for any segment of the string. It should be noted that in segment *CD*, the sum of the forces F_3 and F_4 are directly opposite and cancel out; hence there is no force and no acceleration at that point (an observation noted in the discussion of the circle of reference).

We can also illustrate that energy is transmitted along the string by the pulse if we observe that every force acting along the string at any point (force F_5, for example, in Fig. 19-2*a*) has a vertical component. For F_5, the vertical compo-

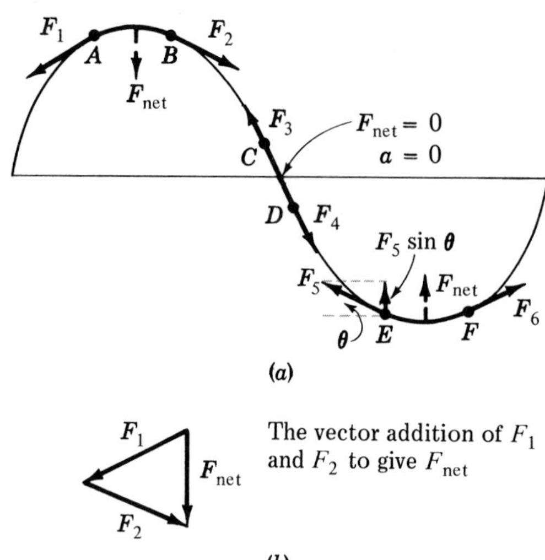

Figure 19-2
(*a*) Forces acting along a string due to a wave pulse. (*b*) The vector addition of F_1 and F_2 to give F_{net}.

nent is $F_5 \sin \theta$, where θ is the angle F_5 makes with the horizontal. This force is acting upward and will move a particle at point E upward with a velocity equal to $\bar{v} = s/t$. Since the particle at E moves up a distance $\bar{v}t$, an amount of work is done on the point by the adjacent segment of the string equal to $W = Fs = (F_5 \sin \theta)(\bar{v}t)$. Therefore work, or energy, is done by each particle in the string on the adjacent particle. The string passes along the energy transmitted to it by the wave in this manner and eventually resumes an equilibrium position. In summary, energy is always transmitted along a string by a wave pulse moving toward the right in such a way that any point to the right of a reference point has work done on it by the part to the left.

Let us consider the forces acting on a small section of the pulse. Since we are interested in the relative speed that exists between pulse and string, let us think of the string moving over the top portion of a stationary hump with the relative speed v. Consider the top portion as a circular arc of length $r\,\Delta\theta$ (Fig. 19-3). The net downward force on this portion of the string is the sum of the vertical components of F_1 and F_2.

$$F_{net} = F_1 \sin \frac{\Delta\theta}{2} + F_2 \sin \frac{\Delta\theta}{2}$$

Since $F_1 = F_2 = T$, where T = tension in cord,

$$F_{net} = 2T \sin \frac{\Delta\theta}{2}$$

When θ is small, $\sin \Delta\theta/2$ is approximately equal to $\Delta\theta/2$. Therefore,

$$F_{net} = 2T \frac{\Delta\theta}{2} = T\,\Delta\theta \qquad (1)$$

The central acceleration in the circular arc that is this portion of the string is

$$a = \frac{v^2}{r} \qquad (2)$$

and

$$F = ma = \frac{mv^2}{r} \qquad (3)$$

Let μ be the mass per unit length of the string, $\mu = m/l$, since

$$l = r\,\Delta\theta \qquad \text{and} \qquad m = \mu r(\Delta\theta)$$

Then,

$$F = T\,\Delta\theta = \frac{\mu r\,\Delta\theta\, v^2}{r} \qquad (4)$$

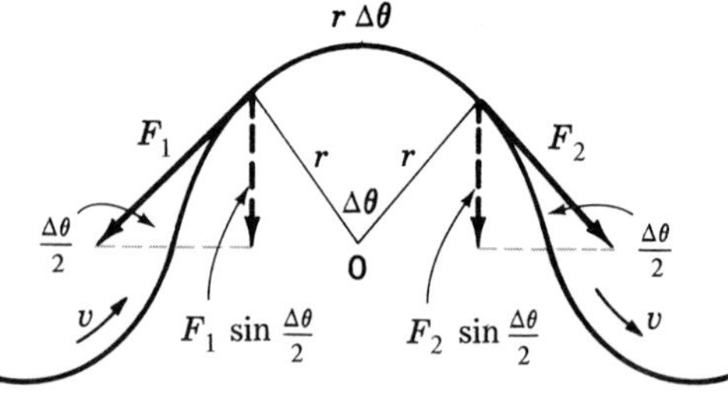

Figure 19-3
Forces acting on a section of the string in a pulse.

Reducing,

$$v^2 = \frac{T}{\mu}$$

$$v = \sqrt{\frac{T}{\mu}} \qquad (5)$$

Each pulse will travel along the string with constant speed until it reaches the end of the string. The assumption that the angle $\Delta\theta$ is small means that Eq. (5) holds strictly only for transverse pulses that are small, *but the pulse may be of any shape.*

Example A string 4.0 m long has a mass of 3.0 g. One end of the string is fastened to a stop, and the other end hangs over a pulley with a 2.0-kg mass attached. What is the speed of a transverse wave in this string?

$$T = Mg = 2.0 \text{ kg} \times 9.8 \text{ m/s}^2 = 19.6 \text{ N}$$

$$\mu = \frac{m}{l} = \frac{0.0030 \text{ kg}}{4.0 \text{ m}} = 7.5 \times 10^{-4} \text{ kg/m}$$

$$v = \sqrt{\frac{T}{\mu}} = \sqrt{\frac{19.6 \text{ kg}\cdot\text{m/s}^2}{7.5 \times 10^{-4} \text{ kg/m}}}$$

$$= \sqrt{2.6 \times 10^4 \text{ m}^2/\text{s}^2} = 1\bar{6}0 \text{ m/s}$$

A harmonic wave will travel along a string if one end is displaced with simple harmonic motion. In Fig. 19-4 the end of a string at $x = 0$ is made to vibrate with amplitude A and frequency f. A wave travels along the string with speed v given by Eq. (5). Each particle of the string vibrates in SHM with a common frequency, but each successive particle has a later phase of vibration. Between $x = 0$ and $x = \lambda$, a distance of one wavelength, the successive particles have gone through all possible phase differences and finally return to the initial phase. The distance between any two adjacent particles that are in the same phase is called the *wavelength* λ. As the simple harmonic vibration continues at the $x = 0$ end of the string, waves continue to travel out to the right along the string. The individual particles vibrate in directions at right angles to the

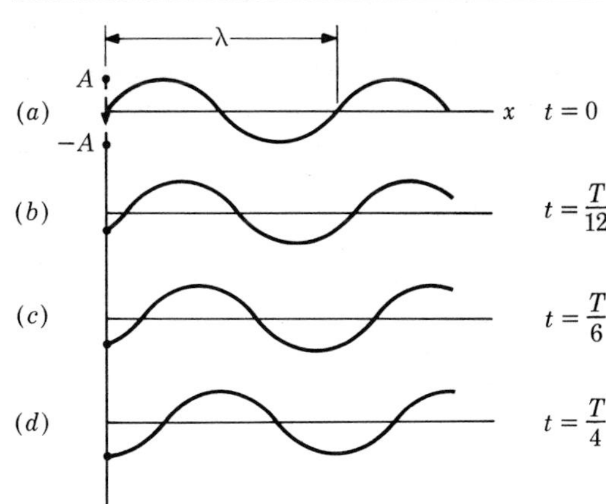

Figure 19-4
Successive positions of a wave traveling along a string.

direction in which the wave travels, and hence the wave is called a *transverse* wave (*trans* = carry; *verse* = across). Since the transverse wave requires that there be a shearing force in the medium, transverse waves can be propagated only in those mediums which will support a shearing stress.

There is a very simple relationship between the speed v of a wave, the frequency f of the wave, and the wavelength λ. The distance λ is the distance the wave travels in the time T of one complete vibration, i.e., the period since

$$\text{Average speed} = \frac{\text{distance}}{\text{time}}$$

$$v = \frac{\lambda}{T}$$

or since $1/T = f$,

$$v = f\lambda \qquad (6)$$

We may express a mathematical relation in the wave motion from consideration of the waves of Fig. 19-4. Let us take the time $t = 0$ at the instant the vibrating particle at $x = 0$ is at zero displacement and moving downward. At this instant the shape of the string will be given by expressing the displacement of each particle of the string in terms of the phase angle. If y is the vertical displacement of a given particle about its median line, and A is the maximum displacement, it can be shown by referring to the circle of reference (Fig. 19-5) that the $\sin\theta = y/r$, and since at the maximum displacement $r = A$, then $y = A\sin\theta$. Since $\theta = \omega t$ and $\omega = 2\pi$ rad/T, $\theta = (2\pi/T)t$. Also,

$$v = f\lambda = \frac{\lambda}{T} \quad \text{and} \quad \frac{v}{\lambda} = \frac{1}{T}$$

Since the wave moves a distance x in a given time t, $v = x/t$, and $x = vt$. Thus,

$$\theta = \frac{2\pi vt}{\lambda} = \frac{2\pi x}{\lambda}$$

So

$$y = A\sin\frac{2\pi x}{\lambda} \tag{7}$$

If we consider some point farther down the string, when x is not zero, it is later in phase than the adjacent part toward the origin of the curve. To find the displacement y at some phase angle θ,

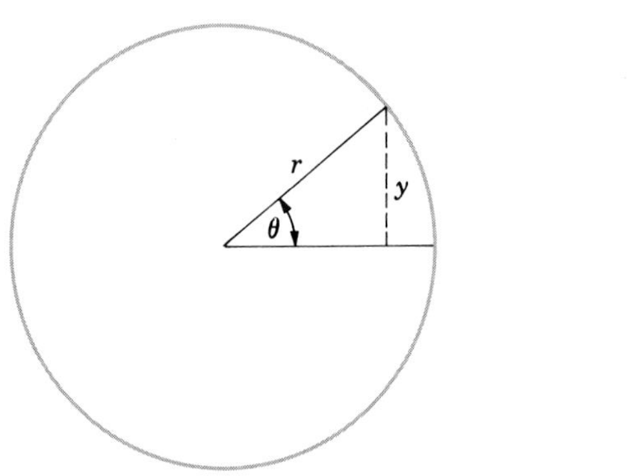

Figure 19-5
Circle of reference.

$$y = A\sin\left(\frac{2\pi x}{\lambda}\right) - (A\sin\theta)$$

$$= A\sin\left(\frac{2\pi x}{\lambda} - \theta\right)$$

and

$$= A\sin\left(\frac{2\pi x}{\lambda} - \frac{2\pi t}{T}\right) \tag{8}$$

Equation (8) represents a harmonic wave of amplitude A traveling in the positive x direction.

At a time one-twelfth of a period later (Fig. 19-4b), $t = T/12$, and the displacement is given by

$$y = A\sin\left(\frac{2\pi x}{\lambda} - \frac{2\pi T}{12T}\right) = A\sin\left(\frac{2\pi x}{\lambda} - \frac{\pi}{6}\right)$$

and at $t = T/6$ (Fig. 19-4c),

$$y = A\sin\left(\frac{2\pi x}{\lambda} - \frac{\pi}{3}\right)$$

The wave equation can be written in several alternative forms if we recall that $f = 1/T$ and that $v = f\lambda$

$$y = A\sin\left(\frac{2\pi x}{\lambda} - 2\pi ft\right) \tag{9}$$

or

$$y = A\sin\frac{2\pi}{\lambda}(x - vt) \tag{10}$$

Since a wave traveling the negative x direction merely reverses the direction of the velocity, v changes to $-v$ and Eq. (10) changes to

$$y = A\sin\frac{2\pi}{\lambda}(x + vt) \tag{11}$$

The wave equations here are expressed in terms of the sine function. The wave can be represented equally well by a cosine function. This change merely represents a different choice of initial position and time.

Example A wave is represented by the equation $y = 0.20\sin 0.40\pi(x - 60t)$, where all distances are measured in centimeters and time in seconds. Find: (a) the amplitude, (b) the wavelength, (c) the speed, and (d) the frequency of the wave. (e) What is the displacement at $x = 5.5$ cm and $t = 0.020$ sec?

From Eq. (10),

$$y = A \sin \frac{2\pi}{\lambda}(x - vt)$$

By comparison,

(a) $A = 0.20$ cm

(b) $\frac{2\pi}{\lambda} = 0.40\pi$

$\lambda = \frac{2}{0.40}\ \text{cm} = 5.0\ \text{cm}$

(c) $v = 60$ cm/s

(d) $f = \frac{v}{\lambda} = \frac{60\ \text{cm/s}}{5.0\ \text{cm}} = 12/\text{s}$

(e)
$$\begin{aligned} y &= (0.20\ \text{cm}) \sin 0.40\pi(5.5 - 60 \times 0.020) \\ &= (0.20\ \text{cm}) \sin 0.40\pi(5.5 - 1.2) \\ &= (0.20\ \text{cm}) \sin (0.40 \times 4.3\pi) \\ &= (0.20\ \text{cm}) \sin 1.72\pi \\ &= (0.20\ \text{cm})(-0.77) = -0.15\ \text{cm} \end{aligned}$$

19-3 LONGITUDINAL WAVES

In *longitudinal waves* the vibration of the individual particles is parallel to the direction the wave travels. Consider a stretched spring. In Fig. 19-6a the lines represent the positions of the coils of a spring when the spring is undisturbed. Suppose that the left-hand end of the spring is suddenly compressed (Fig. 19-6b) by moving the end to the right. These coils will then exert forces on the

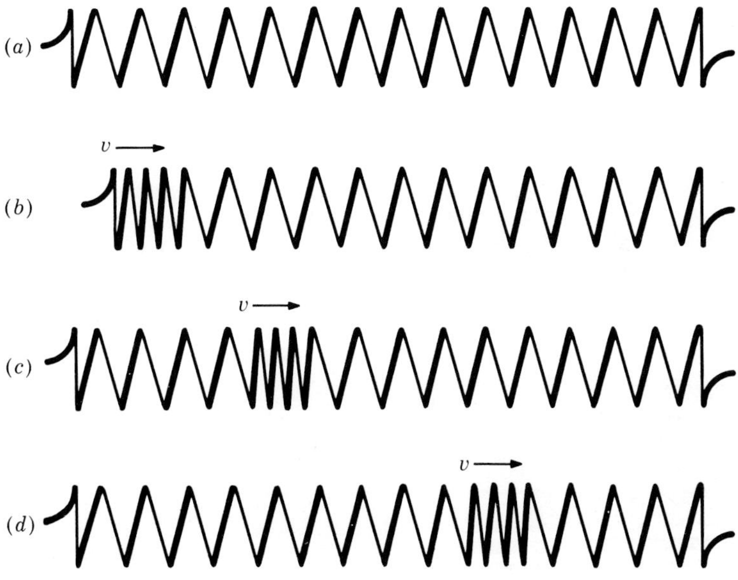

Figure 19-6
Compressional pulse in a spring. (a) The spring is undisturbed; (b) the pulse is started by moving the coils to the right; (c) and (d) the pulse has continued to move along the spring.

adjacent coils, causing the compression to travel along the spring as a compressional pulse. The speed of the pulse will depend upon the elastic constant of the spring and the mass per unit length of the spring. No part of the spring moves very far from its equilibrium position, but the pulse continues to travel along the spring.

A harmonic longitudinal wave can be set up in the spring if the end is moved with simple harmonic motion in a line parallel to the spring. The spring is compressed when the end is to the right of its normal position but expanded when the end is to the left of normal. The result is that, as the end is moved in SHM, a harmonic succession of compressions and following expansions will move along the spring.

Suppose that we replace the spring by a long tube of air with a piston at the left-hand end. Set the piston into SHM near the end of the tube. Harmonic compressions and rarefactions will travel along the tube. As the driving vibration continues, a steady condition will be set up, with the wave traveling to the right as illustrated in Fig. 19-7. The lines close together represent the compressions and those far apart the rarefactions.

Longitudinal waves do not require shearing stress and hence may be transmitted through any elastic medium. The derivation of the relation between the speed of the longitudinal waves and the properties of the medium is somewhat involved, and we shall give only the end result. For a fluid or a solid that is essentially one-dimensional, such as a thin rod,

$$v = \sqrt{\frac{E}{\rho}} \tag{12}$$

where E is the appropriate elastic modulus and ρ is the density. For a rod, E is Young's modulus; for a fluid, E is the bulk modulus B.

Equations (8) to (10) can be used to represent a longitudinal wave when the displacement is measured along the direction of travel of the wave.

Example Find the speed of a compressional wave in an iron rod whose specific gravity is 7.7 and whose Young's modulus is 27.5×10^6 lb/in².

$$v = \sqrt{\frac{E}{\rho}}$$

$$E = 27.5 \times 10^6 \text{ lb/in}^2 = 27.5 \times 10^6 \times 144 \text{ lb/ft}^2$$

$$\rho = \frac{D}{g} = \frac{7.7 \times 62.4 \text{ lb/ft}^3}{32 \text{ ft/s}^2} = 15.0 \text{ slugs/ft}^3$$

$$v = \sqrt{\frac{27.5 \times 10^6 \times 144 \text{ lb/ft}^2}{15.0 \text{ slugs/ft}^3}}$$

$$= 1.6 \times 10^4 \text{ ft/s}$$

19-4 WATER WAVES

In a liquid the motion of the particles may be neither purely transverse nor purely longitudinal but a combination of the two. In the latter case

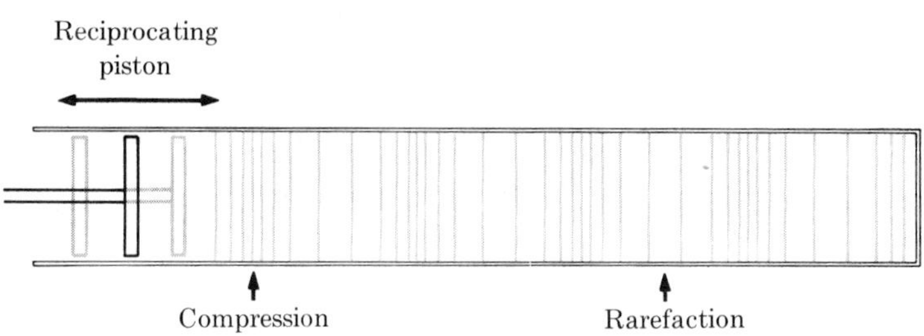

Figure 19-7
Compressions and rarefactions in a compressional wave in air.

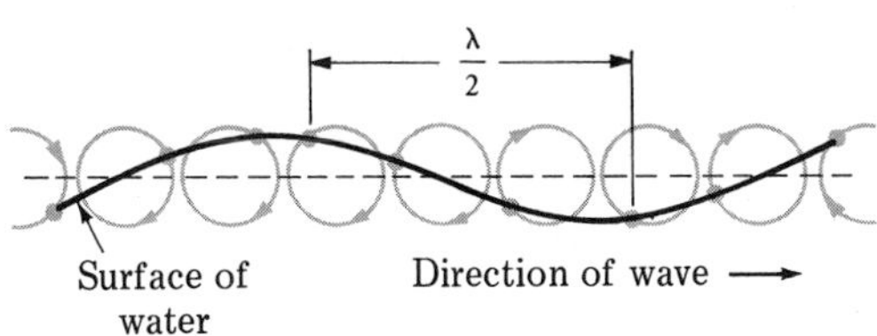

Figure 19-8
Particle motion in a water wave.

the path followed by the particles is either a circle or an ellipse. In Fig. 19-8 are shown the paths of several selected particles, their positions, and the wave shape. Figure 19-9 shows that a cork floating on water traces a circular path as a water wave passes it. The general rule for determining the shape, circular or elliptical, of the path the particles follow in a water wave is that the particles move in circular paths if the wavelength is as small or smaller than the depth of the water. As the water becomes deeper, the path becomes elliptical and at the lower levels of deep water, the wave motion is entirely longitudinal.

19-5 WAVE PROPERTIES

In observing waves, we note several of their characteristics. Waves travel with a definite speed through a uniform medium. Also, if we watch a single spot, we see that waves pass that spot at regular intervals of time. In the wave motion such factors as wave speed, frequency, phase, wavelength, and amplitude must be considered. Some of these properties we have discussed already. Let us recapitulate. The speed of a wave is the distance it advances per unit time. We have given expressions for the speed of two types of wave. We have noted that all the particles of the medium in which there is a wave vibrate about their respective positions of equilibrium in the same manner, but they reach corresponding positions in their paths at different times. These relative positions represent the phase of the motion. In Eq. (8) the relative phase of the particles is expressed by the angle that varies for position (x) and time (t). The number of waves that pass a point per unit time is the *frequency* f of the wave motion. The time required for a single wave to pass is called the *period* T of the wave motion. The *wavelength* λ is the distance between two adjacent particles that are in the same phase. The *amplitude* A of the wave is the maximum displacement of the particle from its equilibrium position. Waves can be described in terms of quantities other than displacement, and then the amplitude represents the maximum departure of the chosen quantity from normal. A *wave front* is a surface that passes through all points in the wave that are in the same phase. In a medium in which the speed is the same in all directions, the wave front is perpendicular to the direction that the wave travels.

An important point to note is that if the wave passes from one medium into another, the speed changes. In this process the frequency remains the same, but the wavelength will change in proportion to the speed; if v increases, λ also increases.

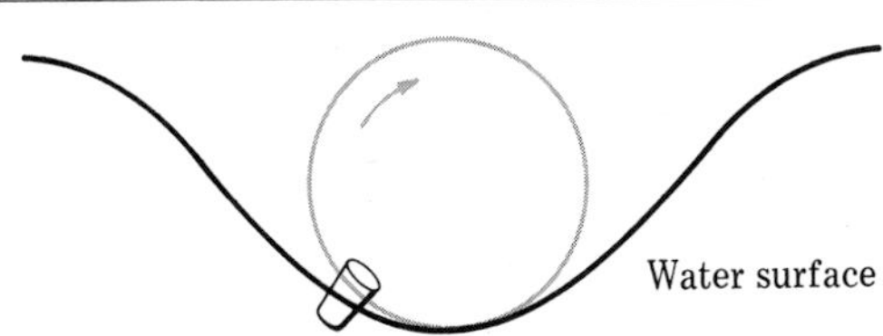

Figure 19-9
Path traced by a cork as a water wave passes.

Example A compressional wave of frequency 250 per second is set up in an iron rod and passes from the rod into air. The speed of the wave is 1.6×10^4 ft/s in iron and 1.1×10^3 ft/s in air. Find the wavelength in each material.

In iron:

$$\lambda = \frac{v}{f} = \frac{1.6 \times 10^4 \text{ ft/s}}{250/\text{s}} = 64 \text{ ft}$$

In air:

$$\lambda = \frac{v}{f} = \frac{1.1 \times 10^3 \text{ ft/s}}{250/\text{s}} = 4.4 \text{ ft}$$

19-6 GENERAL WAVE MOTION

We have developed a wave equation by considering the displacement of a string as a transverse wave passes along the string. This wave could also be described by specifying the velocity of each particle as the wave progresses. We remember from our study of simple harmonic motion that the displacement and velocity are 90° out of phase. Hence also the velocity wave and the displacement wave are 90° out of phase, and if the displacement wave is a sine wave, the velocity wave is a cosine wave. Any property of a medium that varies harmonically as the wave passes may be used to describe the wave. Such properties include displacement, velocity, pressure, density, and electric and magnetic field intensity. A string confines the wave that is propagated along it to one direction (or the reverse direction). Other waves may travel in any or all directions from the source. Equations (8) to (10) can be rewritten to represent the wave that travels in any one direction in a uniform medium. Let ϕ represent the property that varies in the medium. Then

$$\phi = A \sin 2\pi\left(\frac{x}{\lambda} - \frac{t}{T}\right)$$

$$= A \sin\frac{2\pi}{\lambda}(x - vt) \tag{13}$$

When the wave is confined to a single direction and there is no dissipation of energy, the amplitude remains constant. If the waves travel in all directions from the source, the amplitude decreases as the wave gets farther from the source. The decrease in amplitude occurs because the energy is spread over successively larger areas.

19-7 TRANSMISSION OF ENERGY

In all traveling waves energy travels through the medium in the direction in which the wave travels. Each particle of the medium has energy of vibration and passes energy on to succeeding particles.

In simple harmonic motion, where there is no damping, that is, loss of amplitude due to an external or internal force such as friction, the energy of the vibrating particle changes from kinetic energy to potential energy and back, with the total energy constant. We may get this constant energy E by finding the maximum kinetic energy.

$$E = \tfrac{1}{2}m\,(v_{\text{max}})^2$$

Since $v = s/t$ = circumference of circle of reference per period,

$$E = \tfrac{1}{2}m\left(\frac{s}{t}\right)^2 = \tfrac{1}{2}m\left(\frac{2\pi A}{T}\right)^2$$

$$= \tfrac{1}{2}m\,(2\pi f A)^2 = 2\pi^2 m f^2 A^2 \tag{14}$$

where A is the amplitude of vibration, T is the period, f is the frequency, and m is the mass of the vibrating particle.

As a wave passes through a medium, the energy per unit volume in the medium is the energy per particle times the number n of particles per unit volume,

$$\frac{E}{V} = (n)(2\pi^2 m f^2 A^2) \tag{15}$$

Since $\rho = m/V$ and $n = 1/V$, nm is the density ρ of the medium; hence $\rho = nm$.

$$\frac{E}{V} = 2\pi^2 \rho f^2 A^2 \tag{16}$$

The transfer of energy per unit time per unit area perpendicular to the direction of motion of the wave is called the *intensity* I of the wave. The energy that travels through such an area per unit time is that contained in a cylinder of unit cross section and of length numerically equal to the speed v of the wave. From Eq. (16),

$$I = 2\pi^2 v\rho f^2 A^2 \tag{17}$$

We see that the intensity is directly proportional to the square of the amplitude and to the square of the frequency. By doubling the amplitude of a wave or doubling the frequency, we get four times the energy.

Example The speed of a certain compressional wave in air at standard temperature and pressure is 330 m/s. A point source of frequency 300 per second radiates energy uniformly in all directions at the rate of 5.00 W. What is the intensity of the wave at a distance of 20.0 m from the source? What is the amplitude of the wave there?

At any concentric spherical surface the energy from a point source is spread over an area $4\pi r^2$. The intensity I of a wave at a given point in space was defined earlier to be the average power crossing a small area A, divided by the area. Also recalling that power = work/time = energy/time,

$$I = \frac{E}{tA} = \frac{P}{A} = \frac{5.00 \text{ W}}{4\pi \times (20.0 \text{ m})^2}$$
$$= 0.99 \times 10^{-3} \text{ W/m}^2$$

From Table 1, Chap. 12, for air $\rho = 1.29$ g/l = 1.29 kg/m^3. From Eq. (17),

$$A^2 = \frac{I}{2\pi^2 v\rho f^2}$$
$$= \frac{0.99 \times 10^{-3} \text{ W/m}^2}{2\pi^2(330 \text{ m/s})(1.29 \text{ kg/m}^3)(300/\text{s})^2}$$
$$= 1.32 \times 10^{-12} \text{ m}^2$$
$$A = 1.15 \times 10^{-6} \text{ m} = 1.15 \times 10^{-4} \text{ cm}$$

When a wave travels out in a uniform medium from a point source, the energy at some instant is passing through the surface of a sphere. At a later instant the same energy is passing through a larger spherical surface. The amount of energy per unit area per second, the intensity, is less at the second surface than at the first. Since the total energy per unit time is the same at the two surfaces, the intensity is inversely proportional to the area $4\pi r^2$ of the surface.

$$I = \frac{k_1}{4\pi r^2} = \frac{k_2}{r^2} \tag{18}$$

Equation (18) expresses the fact that the intensity of a wave from a small source varies inversely as the square of the distance from the source when the wave diverges uniformly in all directions.

If in place of a point source we have a line source, the energy is spread over a cylindrical surface. Again the energy spreads over successively larger surfaces, and the intensity is inversely proportional to the area $2\pi rl$ of the cylindrical surface.

$$I = \frac{k_3}{2\pi rl} = \frac{k_4}{r} \tag{19}$$

For cylindrical divergence, the intensity is inversely proportional to the first power of the distance.

If a plane source is large, i.e., large compared with the distance from the source, the surface over which the energy spreads is a plane and the areas of successive planes are the same. In this case the intensity is independent of the distance.

As a wave passes through any material medium, energy is absorbed by the medium, usually converted into heat. Thus the energy passing through each surface, and hence the intensity, decreases faster than is expected from the change in area alone. The decrease in intensity due to absorption is called *damping,* and a wave whose intensity decreases for this reason is called a *damped* wave.

19-8 SUPERPOSITION OF WAVES

When two or more waves exist simultaneously in the same medium, each wave travels through the medium as though the other were not present. In the sense of propagation through the medium neither wave affects the other. However, at any point which two waves of the same kind reach simultaneously, the medium will have a displacement that is the *sum of the displacements of the individual waves.* Here "displacement" refers to the departure from normal of that property of the medium which varies as the wave passes through the medium. If the property is a vector quantity, the sum is a vector sum. Let ϕ_1 and ϕ_2 represent the displacements of the individual waves. Then at every point in the medium and at each instant of time the resultant displacement is

$$\phi = \phi_1 + \phi_2 \tag{20}$$

For any wave in a material medium such as the transverse wave in a string or the longitudinal wave in a solid or fluid, Eq. (20) has the same restriction to small amplitude that holds in Eq. (8) or Eq. (13). For electromagnetic waves in empty space Eq. (20) holds strictly.

Example Two sources B and C that vibrate in the same phase radiate waves represented by the equations

$$y_B = (0.50 \text{ cm}) \sin 0.20\pi(x - 100t)$$

$$y_C = (0.30 \text{ cm}) \sin 0.40\pi(x - 100t)$$

Find the amplitude of the combined wave at a point D that is 25 cm from B and 15 cm from C.

From the equations

$$\lambda_B = 10 \text{ cm} \qquad \text{and} \qquad \lambda_C = 5.0 \text{ cm}$$

We can expand the equations above using the relation $\sin(a - b) = \sin a \cos b - \cos a \sin b$,

$$y_B = (0.50 \text{ cm})(\sin 0.20\pi x \cos 20\pi t - \cos 0.20\pi x \sin 20\pi t)$$

$$y_C = (0.30 \text{ cm})(\sin 0.40\pi x \cos 40\pi t - \cos 0.40\pi x \sin 40\pi t)$$

At the given point D,

$$\begin{aligned} y_B &= (0.50 \text{ cm})(\sin 0.20\pi \times 25 \cos 20\pi t - \cos 0.20\pi \times 25 \sin 20\pi t) \\ &= (0.50 \text{ cm})(\sin 5\pi \cos 20\pi t - \cos 5\pi \sin 20\pi t) \\ &= (0.50 \text{ cm})(0 - \cos 5\pi \sin 20\pi t) \\ &= (0.50 \text{ cm}) \sin 20\pi t = (0.50 \text{ cm}) \sin 2\pi t \\ y_C &= (0.30 \text{ cm})(\sin 6\pi \cos 40\pi t - \cos 6\pi \sin 40\pi t) \\ &= -(0.30 \text{ cm}) \sin 40\pi t = -(0.30 \text{ cm}) \sin 2\pi t \\ y &= y_B + y_C = (0.50 \text{ cm} - 0.30 \text{ cm}) \sin 2\pi t \end{aligned}$$

The amplitude is the maximum value of y,

$$A = y_{\max} = 0.50 \text{ cm} - 0.30 \text{ cm} = 0.20 \text{ cm}$$

In the example above the wave from B travels 2.5 wavelengths to D, while the wave from C travels 3.0 wavelengths to D. The waves are a half wave out of phase at D, and we see that the amplitude is the difference between the individual amplitudes. At a point where the waves arrive in phase, the amplitude would be the sum of the individual amplitudes. At points of other phase differences, the amplitudes will be between these two values.

Let us consider the waves that travel out from two sources S_1 and S_2 which are vibrating in phase with each other. In Fig. 19-10 the waves propagated from the two sources are represented by concentric arcs that are assumed to be one-half wavelength apart. Every point on an arc is in the same phase since they are equidistant from the source. Hence the arcs represent wave fronts. At every point at which these wave fronts from the two sources cross the waves will be in phase, or 180° out of phase, and hence the amplitude will be the sum of the individual amplitudes.

The solid and broken lines correspond to the

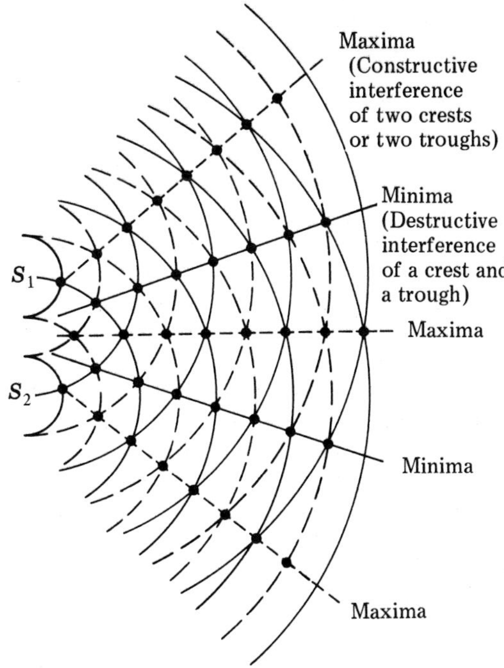

Figure 19-10
Interference patterns from two sources showing points of maxima and minima. Dotted lines represent troughs and solid lines crests of the waves.

crests and troughs of water waves. Since sound waves are longitudinal, we can draw an analogy between crests and areas of maximum compression and between the troughs and areas of maximum rarefaction. The amplitude at points where two crests or two troughs cross will be greatest and will be least where a crest and a trough cross. In Fig. 19-9 such maxima occur at all points equidistant from S_1 and S_2 and also at all points whose distances from the two sources differ by one, two, three, or any whole number of wavelengths. Some of these points lie on the heavy solid lines. At other points, indicated by the dotted lines, the two waves arrive a half wave out of phase, and the resulting amplitude is the difference between the individual amplitudes. The combining of two (or more) waves by superposition is known as *interference*. When two waves arrive at a point in phase with each other, the amplitudes add and the interference is said to be *constructive*. If they arrive a half wave (180°) out of phase, the resultant amplitude is the difference of the two amplitudes and the interference is said to be *destructive*. Interference is characteristic of wave motion, and its appearance in any phenomenon testifies to the wave nature of that phenomenon.

In any interference pattern the amplitude varies in a regular manner. Since the intensity of the wave is proportional to the square of the amplitude, we see that the rate of flow of energy is

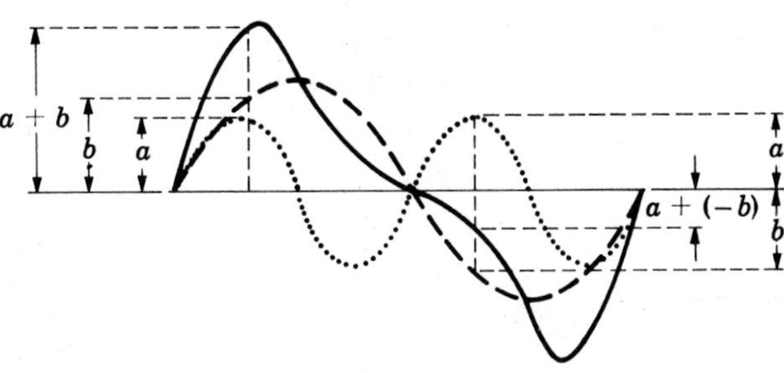

Figure 19-11
Two simple waves combined to give a complex wave (solid line).

not distributed uniformly as in a single wave. In the interference pattern energy is diverted from the regions of destructive interference and appears at the regions of constructive interference.

So far we have considered superposition of waves of the same frequency. The principle applies equally well to waves of different frequency. Two such waves traveling in the same direction will combine their displacements at any place and time. One such example is illustrated in Fig. 19-11. Two simple sine waves (dotted lines), one of which has twice the frequency and about half the amplitude of the other, combine to give a complex wave represented by the solid line. Note that at every point the displacement in the solid line is the sum of the displacements of the dotted lines.

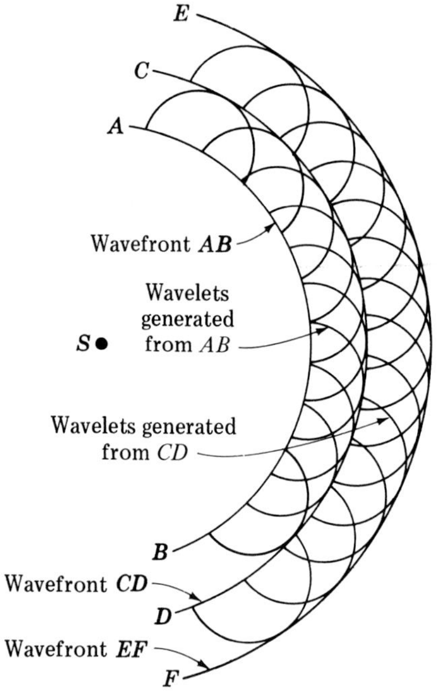

Figure 19-12
The generation of wave fronts by Huygens' construction.

19-9 HUYGENS' PRINCIPLE

As a wave progresses through any medium, a relation exists between successive wave fronts. Christian Huygens (1629–1695) devised a method of construction by which a known wave front may be used to find a succeeding wave front. Huygens' principle may be stated as follows: *Every point on a wave front may be considered as a new source of disturbance, sending wavelets in forward directions.* The new wave front is the envelope of the wavelets. In Fig. 19-12 there is shown a simple Huygens' construction. The arcs *AB*, *CD*, and *DE* represent wave fronts that have come from the source *S*. For several points on each wave front, arcs are drawn, each with a radius equal to the distance the wave will travel in a time t. In this simple case, in which it is assumed that the wave speed is everywhere the same, the radii are all equal. The new wave fronts are drawn tangent to the wavelets.

19-10 REFRACTION AND DISPERSION

So far we have considered only waves which have the same speed in all parts of the medium and for which the speed is the same for all frequencies. If the speed of the waves is not the same in all parts of the medium, the wave usually does not travel in straight lines. The deviation from linear propagation due to this change in speed is called *refraction*. Most waves undergo refraction at times.

For some types of wave the speed of the wave is dependent upon the frequency. This property is known as *dispersion*. Where waves of different frequency travel simultaneously, dispersion may cause separation of the waves due to refraction. If the two waves of Fig. 19-11 traveled at different speeds, the resultant wave shape would change as the waves proceed. A pulse, such as that of Fig. 19-1, may be represented as the resultant of many simple waves. If these component waves travel with different speeds, the shape of the pulse will change as it proceeds. A pulse does not maintain the same shape in a dispersive medium.

SUMMARY

A *wave* is a disturbance that travels through a medium.

Energy may be transmitted by waves.

In *transverse* waves the particles of the medium vibrate in paths perpendicular to the direction the wave moves. In *longitudinal* waves the paths in which the particles vibrate are parallel to the direction the wave travels.

In the case of a vibrating string, energy is transmitted along the string by a wave pulse moving toward the right so that any point to the right of a reference point has work done on it by the part to the left.

In a liquid, the motion of the particle may be neither purely transverse nor purely longitudinal but a combination of the two.

The *speed* of a wave is the distance it moves per unit time. The speed depends upon the kind of wave and the properties of the medium. For a transverse wave in a string

$$v = \sqrt{\frac{T}{\mu}}$$

For a longitudinal or compressional wave

$$v = \sqrt{\frac{E}{\rho}}$$

Frequency f is the number of waves per unit time that pass a point.

Period T is the time required for one wave to pass the point in question.

$$f = \frac{1}{T}$$

Wavelength λ is the distance between two adjacent particles that are in the same phase.

Two particles are in the same *phase* if they have the same displacement and are moving in the same direction.

For harmonic waves the displacement at time t and position x is given by

$$y = A \sin\left(\frac{2\pi x}{\lambda} - \frac{2\pi t}{T}\right) = A \sin\left(\frac{2\pi x}{\lambda} - 2\pi f t\right)$$

$$= A \sin\frac{2\pi}{\lambda}(x - vt)$$

Speed v, frequency f, and wavelength λ are related by the equation

$$v = f\lambda$$

The intensity of a wave is the energy transferred per unit time per unit area through a surface perpendicular to the direction of motion of the wave. The intensity is proportional to the square of the amplitude.

$$I = 2\pi^2 v\rho f^2 A^2$$

When waves travel uniformly in all directions without absorption from a small source, the intensity varies inversely as the square of the distance.

Two or more waves that travel through the same medium are each propagated as though the others were not present. In the region in which the waves arrive simultaneously, the displacement at each point and at each time is the sum of the individual displacements. This *superposition* of the waves is referred to as *interference*. Where the waves arrive *in phase*, the interference is called *constructive* and the resultant *amplitude* is the sum of the individual amplitudes. Where the waves arrive out of phase by a half wave (180°), the interference is *destructive* and the resultant amplitude is the difference of the individual amplitudes.

Huygens' principle states that each point on a wave front may be considered as a new source of disturbance sending wavelets in forward directions. It can be used to predict the propagation of waves.

Waves travel in straight lines within a medium if the speed of the wave is the same in all parts of the medium and in all directions. If the speed changes, there is usually a change in direction of the wave. The change of direction of a wave due to changes of speed is called *refraction*.

Dispersion occurs if the speed of the waves is dependent upon the frequency.

Questions

1 Describe the primary difference between longitudinal and transverse waves.

2 What conditions must exist for a transverse wave pulse to be created in a medium?

3 Describe and show by a diagram how and why a pulse can be created in a string.

4 Describe a simple experiment that illustrates longitudinal waves; transverse waves. Of what does each consist?

5 Show that each small piece of a string through which a wave pulse is passing is acted on by a Hooke's law force.

6 Draw a diagram representing some form of wave motion, and indicate five characteristics of a wave. Define each.

7 Could a wave motion be set up in which the parts of the medium vibrate with angular simple harmonic motion? If so, describe such a wave.

8 Write the equation that expresses the displacement of a particle executing SHM as a harmonic function of time.

9 Show that the equation $v = \sqrt{T/(m/l)}$ does give the appropriate units for velocity.

10 String A has one and one-half times the mass/unit length of string B, but is stretched only two-thirds as tightly as B. Compare the wave velocities in the two strings.

11 In discussing transverse and longitudinal waves, it was assumed that the particles vibrate in simple harmonic motion. Could either type of wave exist if the vibratory motion were not simple harmonic?

12 If a single disturbance sends out both transverse and longitudinal waves that travel with known speeds in the medium, how can the distance to the point of disturbance be determined?

13 We consider two speeds in a study of transverse waves in a string, the speed of the wave along the string and the transverse (across) motion of the particles of the string. Are they related to each other at all?

14 Can the speed of a wave traveling down a stretched string ever be equal to the maximum vertical speed of a particle in the string? If so, under what conditions?

15 Does increasing the tension in a vibrating string increase the number of nodes? Does it increase the frequency?

16 What possible changes could occur to the velocity, frequency, and wavelength of a wave as it goes from one medium to another?

17 What is the effect of the speed of a wave in a string if the tension is doubled and the mass/unit length is cut in half?

18 A rubber cord about 5.0 cm long is held between two fixed points and plucked and the pitch is observed. If this rubber cord is now stretched two, three, and four times its original length and then plucked, compare the changes noted. Explain why this behaves differently from a violin string in which you increase the length of the vibrating portion of the string.

19 Draw a diagram showing two transverse waves (*a*) having the same wavelength but amplitudes in the ratio of 2 : 1, (*b*) with the same amplitudes but wavelengths in the ratio of 1:2, and (*c*) with the same amplitude and the same wavelength but differing in phase by 90°.

20 When two waves interfere, does one change the progress of the other?

21 Show that $y = \sin\theta = \sin(2\pi x/\lambda)$.

22 Show how the intensity of a spherical wave moving from a point source varies with the distance from the source. Show how the amplitude varies.

Problems

1 Two identical strings hang from a support and are stretched by masses of 1 and 2 kg, respectively. Determine the ratio of the velocities for transverse waves moving along the strings.

2 A 50-ft rope that weighs 2.0 lb is stretched by a force of 150 lb. A wave is started down the rope by plucking it. What is the speed of the wave? *Ans.* 346 ft/s.

3 A piano string is 80 cm long and weighs 5.0×10^{-2} N. If it is stretched by a force of 500 N, what is the speed of the wave set up when the hammer strikes the string?

4 A 16-lb wire cable 100 ft long is stretched between two poles under tension of 500 lb. If the cable is struck at one end, how long will it take for the wave to travel to the far end and return? *Ans.* 0.63 s.

5 From the speed of a compressional wave in water, 1,450 m/s at 20°C, compute the adiabatic bulk modulus for water. Compare it with the isothermal value listed earlier in the book.

6 Assume that Young's modulus for silver is 7.75×10^{10} N/m^2. If it has a density of 1.05×10^3 kg/m^3, how fast does sound travel through the silver? *Ans.* 2.67×10^3 m/s.

7 Find the speed of a compressional wave in an aluminum rod.

8 The speed of a compressional wave in silver, specific gravity 10.5, is 2,610 m/s. Compute Y for silver. *Ans.* 7.16×10^{10} N/m^2.

9 Two waves whose frequencies are 20.0 and 30.0 per second travel out from a common point. How will they differ in phase at the end of 0.75 s?

10 Two waves whose frequencies are 500 and 511 per second travel out from a common point. Find their difference in phase after 1.40 s. *Ans.* 144°.

11 The speed of sound in water at 0°C is 1,346 m/s. Compute the adiabatic bulk modulus for water.

12 Copper has a density of 9 g/cm^3 and a bulk modulus of 1.2×10^{11} N/m^2. What is the speed of a wave through it? *Ans.* 3.6×10^3 m/s.

13 Calculate Young's modulus for steel in pounds per square inch if it has a density of 480 lb/ft^3 and if the speed of sound in it is 16,400 ft/s.

14 What must be the stress in a stretched steel wire for the speed of longitudinal waves to be equal to 100 times the speed of transverse waves? *Ans.* 2.9×10^3 lb/in^2.

15 Calculate the speed of a transverse wave in a string having a mass of 10 g and a length of 50 cm if the tension in the string is 0.05 N.

16 A string 1 m long and having a mass of 9.7×10^{-1} g is attached to one end of a tuning fork vibrating at 60 Hz. What tension must be applied to the string to cause it to vibrate in four segments? *Ans.* 8.73×10^4 dyn.

17 A long string, 8.0 m of which has a mass of 0.50 kg, is subjected to a tension of 30 N. One end is attached to a vibrator whose frequency of simple harmonic vibration is 6.0 per second. At $t = 0$ the source is at its maximum displacement of 12 cm. (*a*) Write the equation for the wave traveling away from the source along the string. (*b*) What is the displacement of the string at the source when $t = \frac{1}{36}$ s? (*c*) What is the displacement at a distance of 0.61 m from the source when $t = \frac{1}{36}$ s?

18 The equation of a wave moving through a cord is $y = 2 \sin [2\pi(2x - 100t)]$. Find the magnitude of the amplitude, wavelength, frequency, and velocity. (Assume y and x are in centimeters.) *Ans.* 2 cm; 0.5 cm; 100 Hz; 50 cm/s.

19 The equation of a transverse wave in a rope is given by $y = 10 \sin [\pi (0.010x - 2.00t)]$, where x and y are expressed in centimeters and t in seconds. Find the amplitude, frequency, speed, and wavelength of the wave.

20 A wave is represented by the equation $y = 0.025 \cos (3.14x - 62.8t)$, where distances are in meters and time is in seconds. Find the amplitude, the speed, the wavelength, and the frequency of the wave. Find the displacement at time $t = 0.10$ s at a point $x = 0.50$ m. *Ans.* 0.025 m; 20 m/s; 2.0 m; 10/s; zero.

21 What is the wavelength in water of a compressional wave whose frequency is 400 per second? The speed of the compressional wave in water is 1,450 m/s.

22 What is the velocity of a wave along a cord having a linear mass of 4 g/cm if the cord is stretched to have a tension of 9×10^4 dyn? *Ans.* 150 cm/s.

23 Water waves pass a reference point with a velocity of 15 mi/h. If the distance between wave crests is 11 ft, what is the frequency of the waves?

24 Two sources of waves are located at points A and B 20 cm apart. The waves radiated are represented by equation $y_A = 3.0 \cos [2\pi(0.40x - 100t)]$ and $y_B = 5.0 \cos [2\pi(0.5x - 100t)]$, where x and y are in centimeters and t is in seconds. Find (*a*) the amplitude of each wave, (*b*) the wavelength of each wave, (*c*) the frequency of each wave, and (*d*) the amplitude of the combined waves at a point that is 30 cm from A and 35 cm from B. *Ans.* 30 cm, 5.0 cm; 2.5 cm; 2.0 cm; 100/s, 100/s; 2.0 cm.

25 The adiabatic bulk modulus of water is 2.14×10^{10} dyn/cm^2. Find the speed of a compressional wave in water. What is the wavelength in water of a compressional wave whose frequency is 400 per second?

26 What is the wavelength in air under standard conditions of a compressional wave whose frequency is 250 per second? Assume that the bulk modulus for air is 1.40×10^6 dyn/cm^2 and the density of air is 1.29 g/l. *Ans.* 132 cm.

27 A compressional wave with frequency 300 per second travels through air at standard temperature and pressure, with a speed of 331 m/s. What is the amplitude of vibration when the intensity of the wave is 1.0×10^{-10} W/cm^2?

28 A string which weighs 1.0 oz for 16 ft is attached to a vibrator whose constant frequency is 80 per second. How long must the string be in order for it to vibrate in two segments when the stretching force is 16.0 lb? *Ans.* 4.5 ft.

29 A wave has a frequency of 13 per second and an amplitude of 2.0 in. Assuming the motion of the particles of the medium to be simple harmonic, find the displacement and the speed of the particle 0.33 s after it is in the equilibrium position.

30 If the intensity of a compressional wave is 16 erg/cm$^2\cdot$s at a distance of 10 m from a small source, what is the intensity at a distance of 100 m? In solving this problem, what assumptions have you made that would not be valid if the wave were produced in a closed space? *Ans.* 0.16 erg/cm$^2\cdot$s.

31 Two waves each of frequency 540 per second travel at a speed of 330 m/s. If the sources are in phase, what is the phase difference of the waves at a point that is 4.40 m from one source and 4.00 m from the other?

Sir Owen Willans Richardson, 1879–1959

Born in Dewsbury, Yorkshire. Director of the physical laboratory at King's College, London. Awarded the 1928 Nobel Prize for Physics for his work on thermionic phenomena and for discovery of the law which bears his name.

Louis Victor (Prince) de Broglie, 1892–

Born at Dieppe. Professor at the Poincaré Institute of the Sorbonne. The 1929 Nobel Prize for Physics was conferred on de Broglie for his discovery of the wave character of electrons.

20

Stationary Waves

In the previous chapter we have considered waves that travel through a medium without boundaries. That is, we have considered only those things which occur in the wave before it reaches a boundary. Most waves do reach a terminal point or boundary, and changes occur there. In particular, there is almost always reflection of the wave at the boundary of the medium, and hence a wave travels back into the first medium in a manner that is largely determined by boundary conditions.

When waves are reflected at a boundary, they travel back into the initial medium and there the two waves are superposed. Such superposition must always be considered when waves travel in a bounded medium.

20-1 REFLECTION OF WAVES

Consider a pulse similar to that in Fig. 19-1 which travels toward a boundary of a medium. If the medium were not bounded, the pulse would continue to move to the right. In reality the wave is reflected at the boundary and after reflection travels to the left in the original medium. The way the pulse is reflected depends upon the conditions imposed by the boundary. In Fig. 20-1, the boundary is a *fixed boundary*, i.e., the property of the pulse that varies in the wave (displacement, velocity, etc.) is kept constant. Forces arise at the fixed boundary such as to maintain unchanged the property that varies in the traveling wave. These forces are identical with those that would arise if a second wave (theoretical or virtual) were traveling to the left across the boundary to produce the reflected pulse. In Fig. 20-1, the virtual pulse is represented by dotted lines. It is of such form that the oncoming pulse and the reflected pulse always cancel each other. In Fig. 20-1*b*, the real pulse has just reached the boundary, traveling to the right; the virtual pulse (dotted) has also just reached the boundary, traveling to the left. As the two cross the boundary, they will at every instant cancel each other and the displacement at the boundary will always be zero, thus satisfying the *boundary condition* at the fixed boundary. The reflected pulse that came up as a hump reflects as a depression. We may make the general statement that a wave reflected at a fixed boundary will be reflected with an abrupt change of phase of 180°. Figure 20-1*d* shows the reflected wave traveling to the left. A string tied at one end serves as an example of this kind of reflection. A transverse pulse sent along the string will exhibit a phase change of 180° on reflection at the fixed end.

Another way of looking at the reflection of a wave from a fixed boundary is as follows. When

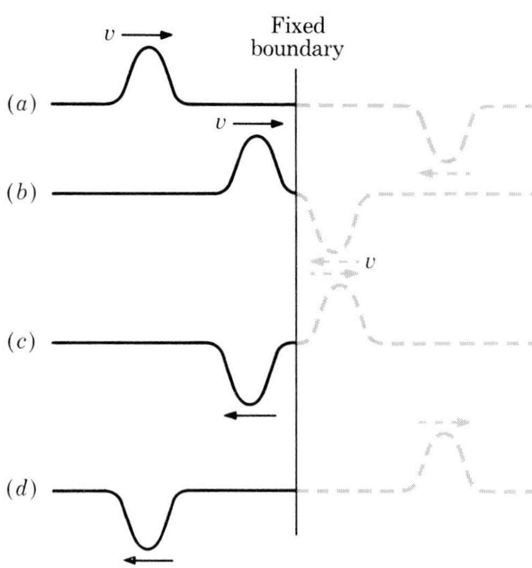

Figure 20-1
Reflection of a pulse at a fixed boundary with 180° change of phase.

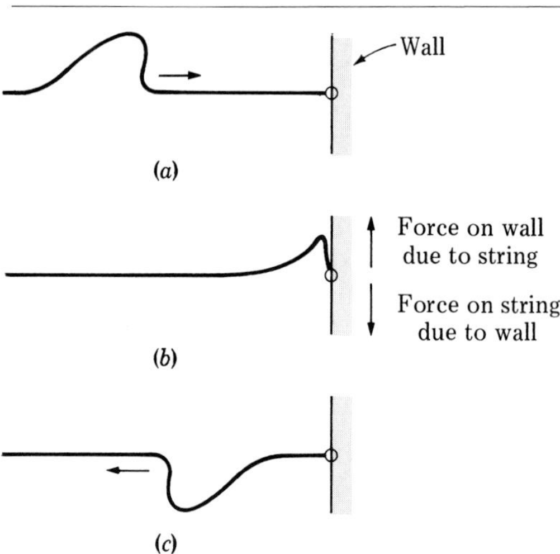

Figure 20-2
Reflection of a wave pulse at a fixed end.

the wave pulse (which we must remember is carrying a certain amount of energy) traveling in a string encounters a fixed end, a wall for example, the string exerts a transverse (upward or downward) force on the wall as shown in Fig. 20-2. Since the wall is massive, it resists the force exerted on it by the string and exerts an equal force, but opposite in direction, downward (or upward as the case may be) on the string with the result that the string is forced downward (or upward) with a displacement equal in magnitude but opposite in direction and in an inverted phase.

Reflection will also occur at a free boundary (complete freedom of change), but the different boundary conditions result in addition of incident and reflected waves at the boundary to produce a doubled rise and fall of the wave property. Figure 20-3*a* represents the approaching pulse traveling to the right and the virtual pulse to the left. In Fig. 20-3*b* the pulse has just reached the boundary. In Fig. 20-3*c* the incident and reflected pulses have combined at the free boundary. In Fig. 20-3*d* the reflected pulse just leaves the boundary, and in Fig. 20-3*e* the reflected pulse is traveling to the left at a later time. At a free boundary the wave is reflected without change of phase.

As with the case of wave reflection from a fixed end, we can analyze what happens to a wave pulse in a string when it reaches a free end of the string from a consideration of the exchange of energy that occurs. When the wave pulse comes to the free end of the string, the energy which it is carrying must be conserved. Since there is nothing at the free end to receive the energy, the string must keep it. The free end of the string is therefore moved upward until it stops at a height greater than the amplitude, a result of the kinetic energy within the pulse changing to potential energy. The free, elevated string then starts to fall and the potential energy is changed back to kinetic energy which is passed on to the string (much as the energy was given to the string at the other end when the wave pulse was first created). A wave pulse is formed traveling in the opposite direction, but in an erect rather than inverted orientation, as shown in Fig. 20-4*c*.

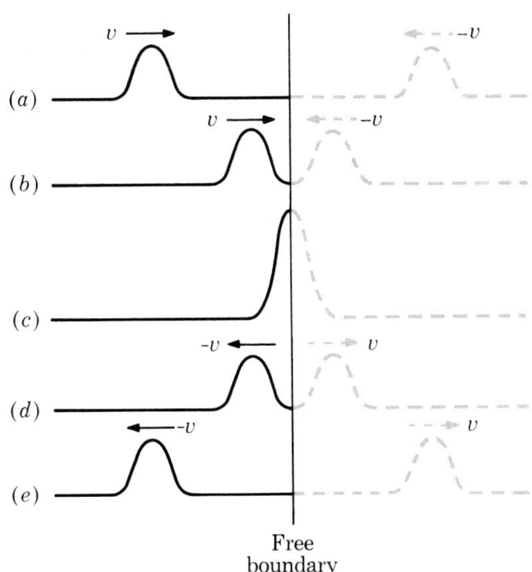

Figure 20-3
Reflection of a pulse at a free boundary with no change of phase.

One of the important uses of these reflection properties is in the study of the waves that travel between the boundaries of a vibrating body.

20-2 STATIONARY WAVES

If two sinusoidal waves of the same amplitude and frequency travel in opposite directions through a medium, the two waves will be superposed in such a manner that *stationary waves* (also called *standing waves*) will result.

Consider the two waves ϕ_1 (dashed lines) and ϕ_2 (dotted lines) shown in Fig. 20-5*a* and *b*. The waves have equal amplitudes and frequencies and travel with speed v in opposite directions. We shall consider the time $t = 0$ at the instant the two waves are coincident, that is, have the same phase angle (Fig. 20-5*c*). The resultant wave ϕ at this instant is found by adding the ordinates at every point. The resultant wave is shown by the solid line in Fig. 20-5*c*.

Figure 20-5*d* shows the waves an eighth period later ($t = T/8$). The wave ϕ_1 has moved an eighth wavelength to the right, and ϕ_2 has moved an eighth wavelength to the left. The resultant wave, however, has not moved at all. Its maximum, minimum, and zero points are in exactly the same positions as in Fig. 20-5*c*, but the displacement at each point is reduced. The waves in Fig. 20-5*e* have moved another eighth period ($t = T/4$), ϕ_1 to the right, ϕ_2 to the left. At this instant the resultant displacement is everywhere zero. The motions of the two individual waves continue in Fig. 20-5*f* and *g*, with the resultant wave form remaining stationary. In Fig. 20-5*g*, each wave has moved a half wavelength since the time $t = 0$, and the waves are again coincident, but 180° out of phase with the waves of Fig. 20-5*c*.

The resultant stationary wave has several important characteristics. We observe that there are several points, marked *N*, that are never displaced. These points are called *nodes*. We observe that the nodes are spaced at intervals of a half wavelength. Every point between nodes vibrates but the amplitude of vibration differs from point

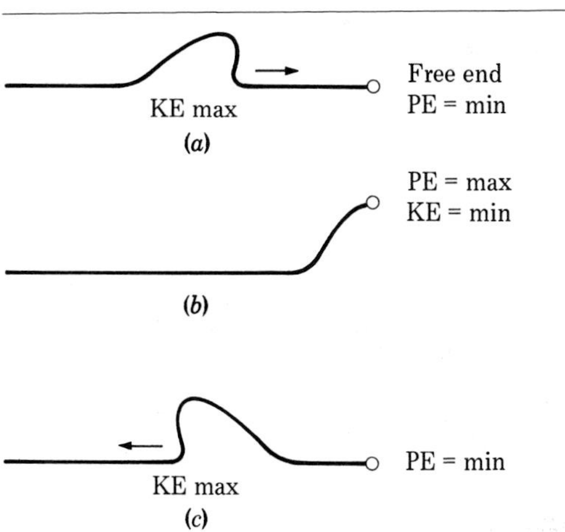

Figure 20-4
Reflection of a wave pulse at a free end.

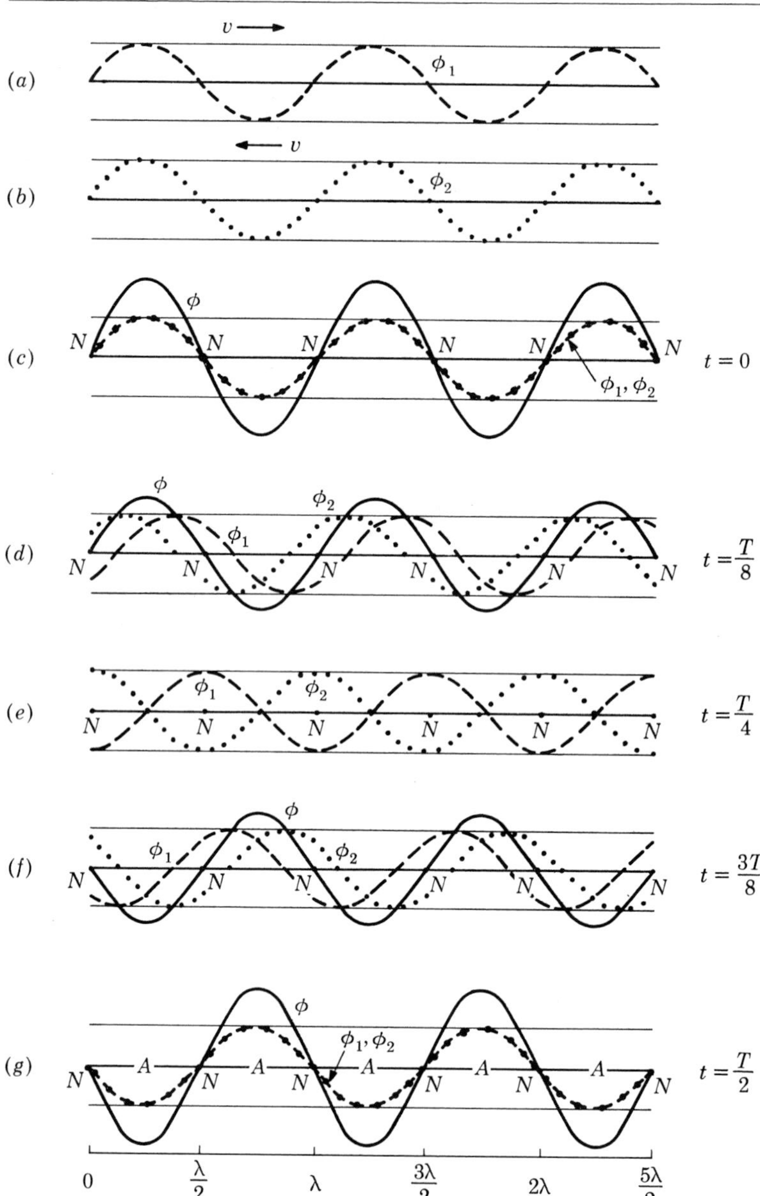

Figure 20-5
Two traveling waves moving in opposite directions combine to produce a stationary wave.

to point. The points of maximum amplitude lie halfway between the nodes. The points of maximum vibration are called *antinodes*. Their positions in Fig. 20-5 are marked by A. The amplitude of the stationary wave is the maximum amplitude of vibration and is the sum of the amplitudes of the individual waves.

We note also that, in each segment between two nodes of the stationary wave, the vibrations are all in phase with each other. In adjacent segments the vibrations are 180° out of phase. This phase relation is illustrated in Fig. 20-6. At an instant when the displacement is everywhere zero (as in Fig. 20-5*e*), the velocities in the vibration are represented by the arrows. Within a segment the arrows are all in the same direction but different in length.

Within a stationary wave there is no flow of energy through the medium. The component waves traveling in opposite directions have equal amplitudes, and hence the energy transfer in one direction by one wave is equal to the energy transfer in the opposite direction by the other wave. There is energy of vibration within a vibrating segment, but this energy is not transferred across a node and "stands," or is stationary.

The equation that can be used to represent a stationary wave may be obtained from the equations of the individual traveling waves. From Eq. (13), Chap. 19,

$$\phi_1 = A \sin\left(\frac{2\pi x}{\lambda} - 2\pi ft\right) \tag{1}$$

and

$$\phi_2 = A \sin\left(\frac{2\pi x}{\lambda} + 2\pi ft\right) \tag{2}$$

From the equation for the sine of the sum and difference of two angles these equations may be expanded to the form

$$\phi_1 = A\left(\sin\frac{2\pi x}{\lambda}\cos 2\pi ft - \cos\frac{2\pi x}{\lambda}\sin 2\pi ft\right) \tag{3}$$

$$\phi_2 = A\left(\sin\frac{2\pi x}{\lambda}\cos 2\pi ft + \cos\frac{2\pi x}{\lambda}\sin 2\pi ft\right) \tag{4}$$

At every point in the stationary wave

$$\phi = \phi_1 + \phi_2 \tag{5}$$

The form of this sum depends upon the conditions that prevail at the point $x = 0$. For the stationary wave illustrated in Fig. 20-5, the point $x = 0$ is taken at a point where the waves always cancel, a node. For this case the sum is given by

$$\phi = \phi_1 + \phi_2 = 2A \sin\frac{2\pi x}{\lambda}\cos 2\pi ft \tag{6}$$

If the reference origin is taken at an antinode, the stationary wave may be given by

$$\phi = 2A \cos\frac{2\pi x}{\lambda}\sin 2\pi ft \tag{7}$$

The amplitude of the vibration at every point in the stationary wave is given by the factor $2A \sin (2\pi x/\lambda)$ [or $2A \cos (2\pi x/\lambda)$], and the maximum amplitude, at the antinode, is $2A$.

Most stationary waves arise because of reflections at boundaries of the medium. These reflections result in two waves traveling in opposite

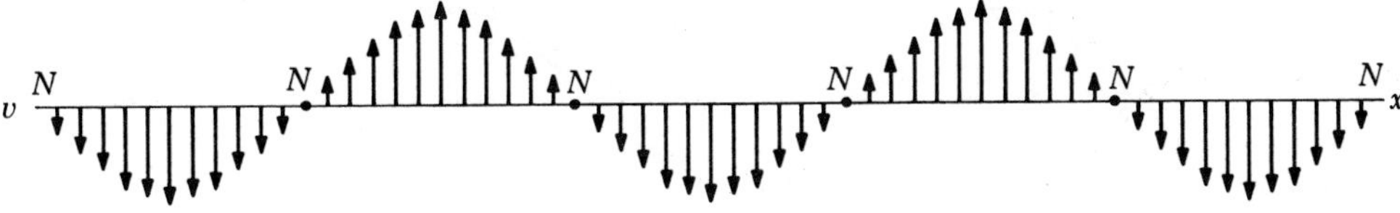

Figure 20-6
Particle velocities in a stationary wave.

directions. The boundary conditions at these ends determine the characteristics of the stationary wave. Let us assume that the origin is at a fixed boundary. Since it is fixed, no vibration takes place there and Eq. (6) represents the stationary wave. We may find the positions of nodes and antinodes by observing the points for which the amplitude is zero or maximum, respectively. The nodes occur at points for which $\sin(2\pi x/\lambda) = 0$, that is, at $x = 0$, $x = \lambda/2$, $x = 2\lambda/2$, $x = 3\lambda/2$, etc. In general, nodes occur at $x = N\lambda/2$, where N is any integer. The nodes are at intervals of a half wavelength.

Antinodes occur at those points for which $\sin(2\pi x/\lambda) = 1$, that is, at $x = \lambda/4$, $x = 3\lambda/4$, $x = 5\lambda/4$, etc. In general, antinodes occur at any odd quarter wavelength. We see that the antinodes are also spaced a half wavelength apart and a quarter wavelength from nodes.

If the boundary is free, we may use Eq. (7) in like manner to show that the antinodes appear at $x = 0$, $\lambda/2$, λ, etc., and that the nodes appear at the odd quarter wavelengths.

When the medium is bounded at both ends by either perfectly fixed or perfectly free conditions, the ends will be nodes or antinodes, respectively. This requires that there be a relationship between the length of the vibrating medium and the wavelength of the waves. If both ends are nodes, or both antinodes, the distance between

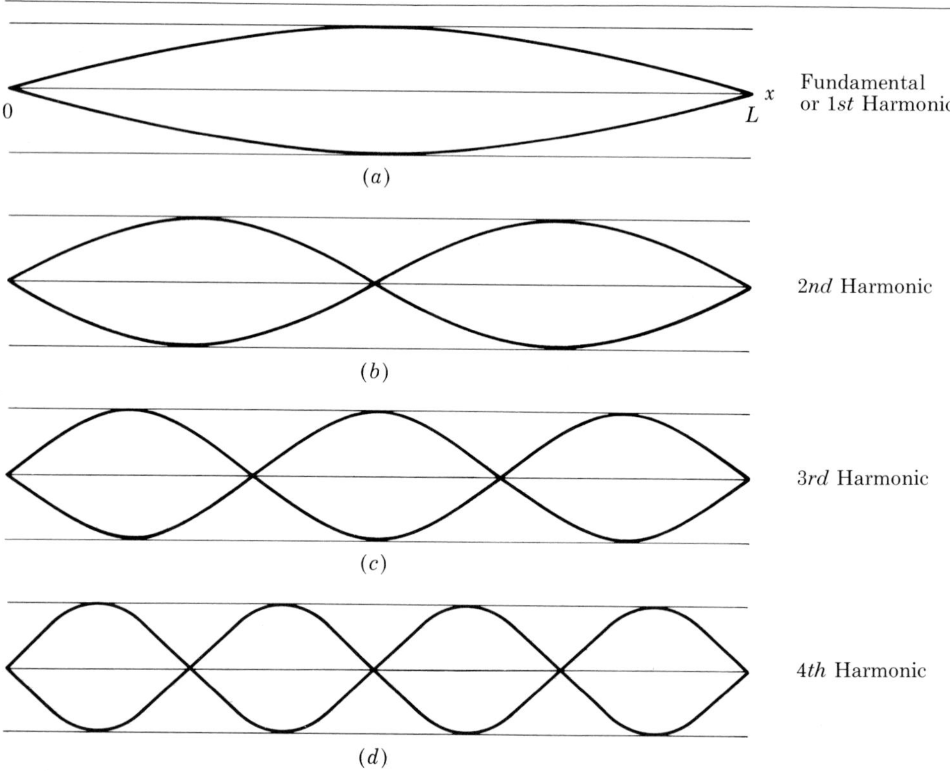

Figure 20-7
Four modes of vibration in a stretched string.

the ends must be an integral number of half wavelengths. If one end is a node and the other is an antinode, the length must be an odd number of quarter wavelengths.

20-3 MODES OF VIBRATION

Consider a string of length L stretched between two fixed supports. When a wave is set up in the string, the wave will travel in both directions and will be reflected at each end. For a stationary wave to exist in the string, each end, $x = 0$ and $x = L$, must be a node. From the conditions stated above at $x = L$, $L = N\lambda/2$, and the string must be an integral number of half wavelengths. Since the speed v of the waves is dependent only upon the tension and mass per unit length of the string, waves of all permitted frequencies have the same speed. It follows that the string may vibrate with only those frequencies for which the length of the string is an integral number of half wavelengths.

Figure 20-7 shows four of the possible modes of vibration. In (*a*) the length $L = \lambda/2$, and $\lambda = 2L$. The frequency for this mode is found from $v = f\lambda$; $f_1 = v/\lambda = v/2L$. In (*b*), $L = 2\lambda/2$, and $f_2 = 2(v/2L)$; in (*c*), $L = 3\lambda/2$, and $f_3 = 3(v/2L)$; in (*d*), $L = 4\lambda/2$, and $f_4 = 4(v/2L)$. The string may vibrate with a lowest frequency f_1 called the *fundamental* frequency, and any integral multiple of the lowest frequency. These possible modes of vibration whose frequencies are all multiples of a lowest frequency are called *harmonic* frequencies, or simply *harmonics*. The fundamental is the first harmonic, and each of the higher frequencies is named by the integer used to express its frequency in terms of the fundamental, for example (*d*) is the fourth harmonic.

The string may be forced to vibrate with frequencies other than these natural harmonic frequencies, but for this situation no stationary pattern is set up and the amplitude is small and inconstant.

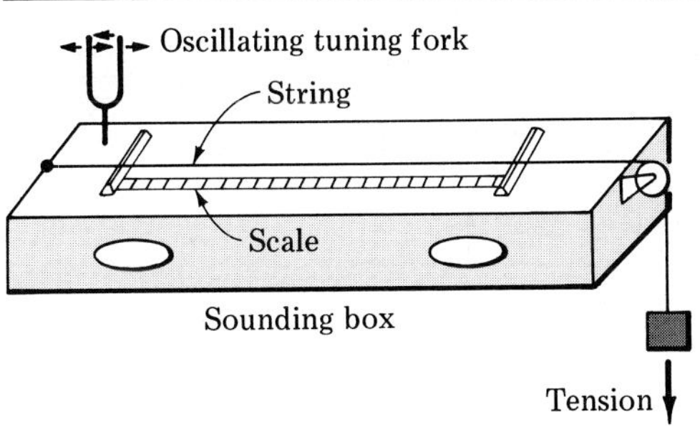

Figure 20-8
Sonometer.

Example A flexible wire 80 cm long has a mass of 0.40 g. It is stretched across stops on a sonometer that are 50 cm apart by a force of 500 N. Find the frequencies with which the wire may vibrate. (See Fig. 20-8.)

$$v = \sqrt{\frac{T}{\mu}} = \sqrt{\frac{500\ \text{N}}{0.40 \times 10^{-3}\ \text{kg}/0.80\ \text{m}}}$$
$$= \sqrt{10^6\ \text{m}^2/\text{s}^2} = 1{,}000\ \text{m/s}$$

For the fundamental mode of vibration

$$L = \frac{\lambda}{2} = 0.50\ \text{m}$$

or $$\lambda = 2L = 2 \times 0.50\ \text{m}$$

$$f = \frac{v}{\lambda} = \frac{1{,}000\ \text{m/s}}{1.00\ \text{m}} = 1{,}000\ \text{vib/s}$$

The other possible frequencies are the integral multiples of 1,000 vib/s, that is, 2,000, 3,000, 4,000, . . . , vib/s.

The modes of vibration of an air column that is open at one end and closed at the other are illustrated in Fig. 20-9. When a compressional wave is set up in the tube, the displacement must be zero (node) at the closed end but the displacement is free (antinode) at the open end. All modes of vibration are possible for which the open end is a displacement antinode and the closed end is a displacement node. Analysis similar to that for the string shows that only odd harmonics are possible. This topic will be explored further in our study of sound.

If the medium is one in which the waves travel in two dimensions, there are usually lines of no vibration, or *nodal lines*. An interesting example of such vibration is the Chladni plate.

In the beginning of the nineteenth century, Ernst Chladni (1756–1827) made a study of vibrations in cords, rods, membranes, and plates. His name is best known in connection with the figures formed in sand on a vibrating plate by nodal lines. Typically, a thin plate is clamped at its center and sand is sprinkled over it. A violin bow is drawn over the edge causing the plate to vibrate in two dimensions, setting up nodal lines. The nodal lines appear at the stationary points, where the sand heaps up. Antinodes appear in regions of maximum vibration and the sand bounces away from these areas and into the nodal areas. Many patterns of vibration are possible depending on the manner in which the plate is supported and on the means of excitation. In the case of a drumhead, the rim serves as a nodal line and all other nodal lines form concentric circles around the center of the drumhead, which is normally an area of maximum vibration, an antinode (Fig. 20-10*a*). Other plates and disks can vibrate in a rather complicated fashion. Figure 20-10*b* shows a typical pattern of Chladni's figures produced by a rectangular plate vibrating in this manner.

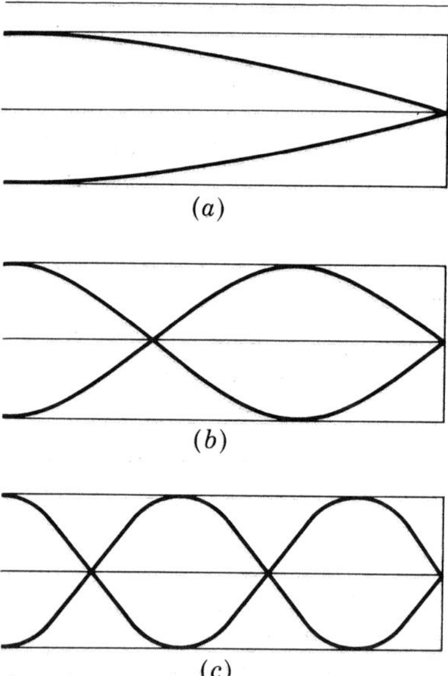

Figure 20-9
Modes of vibration in an air column open at one end and closed at the other.

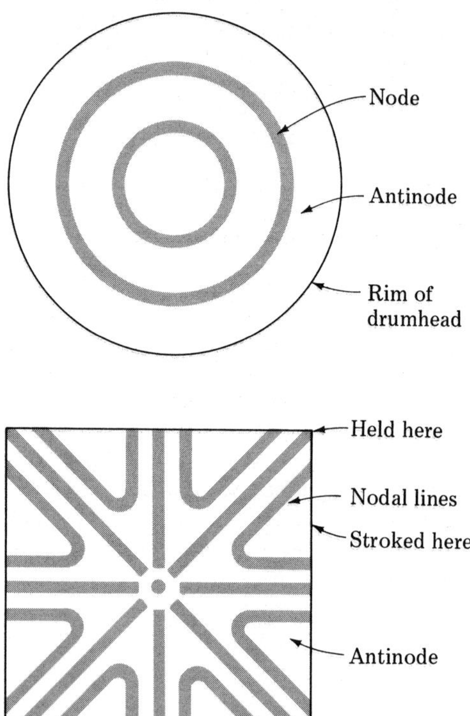

Figure 20-10
(*a*) Chladni's sand figures on a vibrating drumhead. (*b*) Chladni's sand figures for a vibrating plate held at a corner and stroked with a bow.

SUMMARY

A *stationary* wave is produced when two waves of equal amplitude and frequency travel in opposite directions in a medium.

In a stationary wave there are points called *nodes* at which the amplitude is always zero. At other points, called *antinodes*, the amplitude is a maximum and equal to the sum of the amplitudes of the individual waves.

Nodes appear at intervals of a half wavelength. Antinodes are also spaced a half wavelength apart. Antinodes are midway between nodes.

There is no flow of energy through the medium in a stationary wave.

A stationary wave is represented by an equation of the form

$$\phi = 2A \sin \frac{2\pi x}{\lambda} \cos 2\pi ft$$

or

$$\phi = 2A \cos \frac{2\pi x}{\lambda} \sin 2\pi ft$$

Bodies can vibrate with frequencies that are determined by the conditions at the two boundaries. A stretched string can vibrate with a fundamental frequency $f = v/2L$ and all its integral multiples.

An air column open at one end and closed at the other can vibrate in the modes of the odd harmonics.

In those media in which waves can travel in two dimensions, *nodal lines,* lines of no vibration, are usually formed.

Questions

1 What are the boundary conditions which determine the nature of a reflected wave? Give illustrations of the effect of each on a wave.

2 How are standing waves produced?

3 Illustrate the difference between a sinusoidal and a cosinusoidal wave and point out the physical significance of this difference.

4 Show by referring to a circle of reference that a node occurs when the acceleration of the particle in a vibrating cord is zero and the velocity is at a maximum.

5 What happens to a wave motion when it comes to the boundary of a medium? Distinguish between the effects at a free and at a fixed boundary.

6 Is there a transfer of energy through the medium when a stationary wave is produced in it? Explain.

7 If a vibrating string had several knots tied at irregular intervals along its length, could standing waves be set up in it? Why?

8 Two wave pulses traveling in opposite directions completely cancel each other out as they pass. What becomes of the energy possessed by the wave pulses when this happens?

9 Discuss the reflections of a compressional wave at the open end of a gas-filled tube. Consider the various properties of the wave, such as displacement, velocity, and pressure.

10 Discuss the reflections at the closed end of a gas-filled tube from the same considerations as in Question 1.

11 Sketch the fundamental wave and the second and third harmonics for (*a*) a vibrating string, (*b*) a vibrating rod clamped at one end, and (*c*) a vibrating rod clamped at the center.

12 Assuming that the solid wall between two rooms will not transmit a sound wave, is it possible for sound produced in one room to be heard in the other? If so, what wave processes are involved?

13 When a new phenomenon is observed in which there is a transfer of energy, there may be a question about whether it is a particle or a wave phenomenon. What experiments could be used as tests?

14 Can more than one set of standing waves be present simultaneously within a single medium?

15 How might one locate the positions of nodes and antinodes in a vibrating body?

16 How might fine dust on the body of a violin be used to show how the violin is vibrating?

17 What would one have to do to prevent a standing wave being set up in a cord by a traveling wave?

18 Write the equation for a wave whose speed is 20 m/s, whose wavelength is 2.0 m, and whose amplitude is 0.5 m. This wave reflects perfectly upon itself at a fixed boundary so as to travel backward through the oncoming waves. Describe the result by a diagram, with all dimensions labeled. Write the equation of the reflected wave.

Problems

1 A wire 1 m long has a mass of 0.50 g. If it is stretched between two clamps 1 m apart by a force of 1,000 N, find four frequencies with which the wire may vibrate.

2 What tension would be required to create a standing wave with four segments in a string 100 cm long weighing 0.50 g if it is attached to a vibrator with a frequency of 100 Hz?
Ans. 1.25×10^5 dyn.

3 A stationary wave in a medium is given by the equation $y = 8 \cos (\pi/6)x \; \sin (\pi/4)t$, where x and y are in meters and t is in seconds. The two boundaries of the medium are at $x = 0$ and $x = 9.0$ m. Find (*a*) the maximum amplitude of the vibration, (*b*) the wavelength, (*c*) the speed of the wave, and (*d*) the frequency. Show for each end whether it is a node or an antinode.

4 A stationary wave is represented by the equation $y = 2A \sin (2\pi x/\lambda) \cos (2\pi t/T)$. What in the equation represents the amplitude of vibration at various points? Let one end of the medium be at $x = 0$, the other at $x = 1$. What must be the relation between l and λ in order that there may be stationary waves in this medium? Use the equation to show the positions (values of x) of the possible nodes. *Ans.* $2A \sin (2\pi x/\lambda)$; $l = \lambda/N$; 0; $l/2$; l; $3l/2$. . . .

5 If a spring 2.0 m long vibrates with nodes at the ends and a single antinode in the middle, what is the wavelength of the standing wave?

6 A cord which is attached at one end is caused to vibrate by a vibrator attached at the other end. If the vibrator has a frequency of 10 Hz, and the velocity of the wave is 10 m/s, where will the first two nodes and the first two antinodes be from the attached end? *Ans.* Nodes: at end, 0.5 m; antinodes: 0.25 m, 0.75 m.

7 An oscillator at $x = 0$ which vibrates with amplitude of 0.30 cm and period 2.0 s radiates a wave of the same amplitude and of wavelength 100 cm in the positive x direction. (*a*) Write the equation of this wave. (*b*) Find the speed of the wave. A reflector acts as a fixed boundary at $x = 125$ cm. (*c*) Write the equation of the reflected wave. (*d*) Is a stationary wave produced? If so, write the equation representing it.

8 An experiment was performed with a stationary wave in a wire vibrating at 120 Hz. The wire had a mass of 25.0 g and a length of 200 cm. The tension on the wire was adjusted by means of weights, and the length of the vibrating segment could also be adjusted. (*a*) What was the wave speed when a 3.0-kg mass was suspended from the wire? (*b*) What was the length of the wire segment vibrating in the fundamental mode at the tension in *a*? *Ans.* 48.5 m/s; 0.202 m.

9 A string 4.0 ft long is attached to the prong of a tuning fork that vibrates at right angles to the string with a constant frequency of 50 per second. The weight of the string is 0.064 lb. What stretching force is necessary for the string to vibrate in four segments? What would be the effect of turning the fork so that it vibrates parallel to the length of the string?

10 A Kundt's tube, consisting of a horizontal, long glass tube open at both ends, is used to measure the speed of sound in a steel rod. The rod is clamped at its midpoint and one end of the rod protrudes into the glass tube along the bottom of which a thin, level layer of cork dust is distributed. When the rod is stroked, it vibrates longitudinally and emits its fundamental tone causing the cork dust to arrange itself into evenly distributed heaps along the length of the tube. If the rod is 120 cm long and the distance between the heaps of cork dust is 8 cm, calculate the speed of sound in the rod. Assume the speed of sound in air to be 340 m/s. *Ans.* 5,100 m/s.

11 A string which weighs 1.0 oz for 16 ft is attached to a transverse vibrator whose constant frequency is 80 per second. How long must the string be in order for it to vibrate in two segments when the stretching force is 16.0 lb?

12 In the Kundt's tube described in Prob. 10, the distance between heaps of cork dust was 8 cm when the steel rod was caused to vibrate longitudinally. If the ends of the glass tube are sealed and the air replaced by a gas, and the experiment repeated, the distance between heaps is observed to be 10 cm. What is the velocity of sound in the gas? Assume the speed of sound in air to be 340 m/s. *Ans.* 425 m/s.

13 What is the tension in a 3.0-m cord, whose mass is 0.15 kg, if a transverse wave is observed to travel in the cord with a speed of 16 m/s?

14 A brass rod ($Y = 9.0 \times 10^{11}$ dyn/cm^2 and density 8.6 g/cm^3) is clamped in the middle so that there is a node at that point and antinodes at the ends. If the rod is 90 cm long, what is the frequency of the compressional vibrations in the rod? *Ans.* 1.8×10^3 Hz.

15 What must be the length of an organ pipe, closed at one end, in order to have a fundamental frequency of 16 Hz?

16 The air column in a tube closed at one end has a fundamental frequency of 340 Hz. What is the frequency of the harmonic produced when a hole is opened 8.33 cm from the closed end? The speed of sound at room temperature is 340 m/s. *Ans.* 1020 Hz.

Sir Chandrasekhara Venkata Raman, 1888–1970

Born in Trichinopoly, South India. Professor at the University of Calcutta. Awarded the 1930 Nobel Prize for Physics for the discovery of the Raman effect: that radiation is scattered by various substances with a change in frequency, the change being characteristic of the scattering atoms or molecules.

Werner Heisenberg, 1901–

Born in Duisberg, Rhenish Prussia. Professor at Leipzig. Received the 1932 Nobel Prize for Physics for his creation of the quantum mechanics whose application has led, among other things, to the discovery of the allotropic forms of hydrogen.

21

Sound Waves

Great science nobly labors
To increase the people's joys,
But every new invention
Seems to add another noise.

A. P. Herbert

One of the most commonly observed types of wave is that called a sound wave. By means of a sound wave, minute amounts of energy are carried to our ears and stimulate the sensation of sound. Usually the medium that transmits these waves to the ear is the air that surrounds us. The fact that the waves are transmitted by air and other gases gives an immediate clue as to the kind of wave that comes to the ear, for we have observed (Chap. 19) that a gas can transmit only longitudinal (compressional) waves.

Three elements are necessary for the production of a sensation of sound. The process must be started by a *vibrating source* which in its vibration supplies energy to the surrounding medium. The *medium* transmits this energy from the source to the receiver by means of a sound wave. When the wave arrives at a *receiver,* energy is transferred to the receiver. If the receiver is the ear, a sensation of sound may be produced.

A classic problem is stated as follows: if a tree falls in a primeval forest will there be any sound? The falling tree provides a vibrating object, air must be present or the forest would not have grown, so we have the medium necessary to carry the vibrations. Since it is a primeval forest we can assume that man, who would normally have served as the receiver, is not present. Therefore the problem seems to be solved, for one of the three necessary elements is missing. However, there is a perplexing dimension to the problem. If a graph of the intensity of a sound was plotted against distance from the source of the sound, it would be observed that the intensity approaches zero but never reaches it. There are some who point out that the plot shows that a sound never completely dies. Theoretically, if man were to position himself in front of the wave with a sensitive enough listening device at any later time, the sound of the tree would be "heard" and, therefore, the falling tree *would* have made a sound. This argument does point out the need to carefully define sound.

21-1 NATURE OF SOUND

A clue to the answer to the problem presented above is given by the observation that sound may be defined psychologically or physically. In the first aspect it is said to be the *sensation* produced when the proper disturbance comes to the ear. From the second viewpoint, sound is the *stimulus* capable of producing the sensation of sound.

Whether an observer is necessary for sound to exist depends entirely upon which of the two definitions is used. For our purposes we shall use the second, or physical, definition.

As an aid in visualizing the production of sound, let us consider a flat, thin, metal rod which is caused to vibrate in air. By using a stroboscopic light (to "stop" the motion) we can clearly see the left and right motion of the rod as it vibrates. As the rod moves to the right (Fig. 21-1*a*) it pushes together the air molecules which are directly in front of the rod. We have seen earlier that a general rule of wave transmission is that the particles making up a vibrating medium (air in this case) do not move far from their original position but simply pass on the energy acquired from the pulse moving through the medium to adjacent particles. These particles of air will, in turn, compress the air beyond it, and so on. In this manner a compressional (longitudinal) wave pulse is created. The compression that is started by the rod moving to the right will thus travel away from the source into the surrounding medium. During the reverse cycle, the flat rod moves to the left drawing particles adjacent to it along with it as it moves, causing a rarefaction to occur to the right of the rod (Fig. 21-1*b*). As in the case of the compression, the rarefaction travels out from the source, each particle moving to the left to fill the "hole" created by the rarefaction, with the effect being a movement of the hole, or rarefaction, to the right. If the rod is moved back and forth at a regular interval, a succession of compressions and rarefactions will travel out from the rod. It should be noted that a similar wave pattern is established to the left of the rod. Also if a spherical object, such as a balloon, were caused to alternately expand and contract, the rarefactions and compressions would appear alternately as concentric circles around the balloon. Such a regular succession of disturbances traveling out from a source constitutes a wave motion. The compression and the following rarefaction make up a compressional wave.

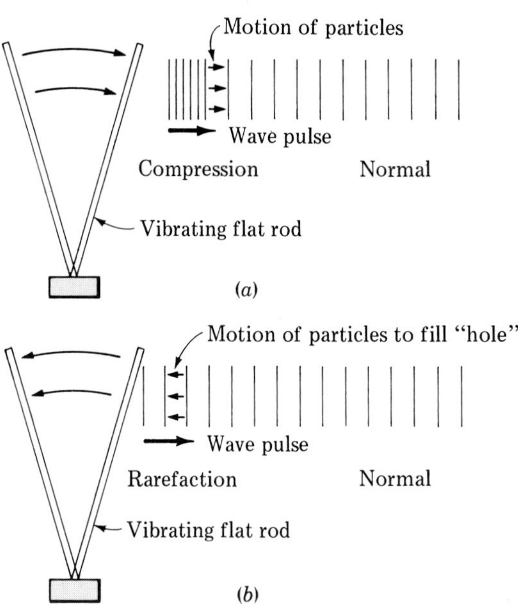

Figure 21-1
(*a*) Compression formed as rod moves to the right. (*b*) Rarefaction formed as rod moves to left.

If the back-and-forth motion of the rod is made rapid enough, an observer in the neighborhood will be able to hear a sound as the disturbance reaches his ear, provided that the vibrating rod has a frequency of between 20 and 20,000 Hz (vib/s). This is the approximate range of human hearing. These compressional waves are able to cause the sensation of hearing and are referred to as *sound waves*.

21-2 VIBRATING SOURCES

Any body that has the properties of inertia and elasticity may be set into vibration. Most, but not all, bodies may vibrate in more than one manner, and for each of these modes of vibration there is an associated frequency. In each of these modes of vibration a set of standing waves is set up in the body. Such standing waves may be transverse or longitudinal, depending upon the shape and characteristics of the vibrating body.

One common source of sound is the vibrating string which we discussed in Chap. 20. When the string is plucked, or bowed, or struck, transverse waves travel in both directions along the string and are reflected at the ends. As a result the string vibrates in any of its natural harmonic frequencies. In Chap. 20 we found these characteristic frequencies to be

$$f = N\frac{v}{\lambda} = N\frac{v}{2L} \tag{1}$$

where N is any integer, L is the length of the string, and v is the speed of the transverse wave in the string.

Example A string 80 cm long has a mass of 6.4×10^{-2} g and is stretched by a force of 96 N. What is the frequency of the fundamental vibration?

$$v = \sqrt{\frac{F}{\mu}} = \sqrt{\frac{96 \text{ N}}{6.4 \times 10^{-5} \text{ kg}/0.80 \text{ m}}}$$

$$= 1.1 \times 10^3 \text{ m/s}$$

$$f = \frac{v}{\lambda} = \frac{v}{2L} = \frac{1.1 \times 10^3 \text{ m/s}}{2 \times 0.80 \text{ m}}$$

$$= 6.8 \times 10^2 \text{ vib/s}$$

In practice a string usually vibrates with many of its characteristic frequencies simultaneously. Hence the wave that is sent out is not a simple sine wave but a complex wave involving many frequencies. The relative amplitudes of the various modes of vibration depend upon the stiffness of the string, the point at which it is set into vibration, and how it is excited. For example, if the string is plucked at a point that is a node for one of the modes of vibration, that mode is suppressed, or very much weakened, while the modes that have an antinode at the point of excitation will be relatively strengthened.

Another common source of sound is a vibrating or resonating air column. We shall consider the topic of resonance later in this chapter. Such a column may be open at one end and closed at the other, as discussed in Chap. 20, or it may be open at both ends. Organ pipes and wind instruments are common examples of vibrating air columns. We have observed that the air column which is closed at one end and open at the other must have a displacement node at the closed end and a displacement antinode at the open end. We have found that for this case the fundamental frequency is given by

$$f = \frac{v}{\lambda} = \frac{v}{4L} \tag{2}$$

where v is the speed of the compressional wave in the air column and L is the effective length of the air column. The possible frequencies of vibration are this lowest frequency and the *odd* multiples of the lowest frequency.

An approach that may be used to determine the proper length of an air column in a tube having one end closed so that it will vibrate at a particular frequency is illustrated in the following example.

Example (*a*) What is the shortest length of a column of air, closed at one end, so that a sound of 256 vib/s, or Hz, can cause it to vibrate? (*b*) What would be the next shortest length? Assume the speed of sound in air to be 331 m/s.

Since, $v = f\lambda$ $\quad \lambda = \dfrac{v}{f} = \dfrac{331 \text{ m/s}}{256 \text{ vib/s}}$

Therefore, $\lambda = 1.30$ m/vib

Since the column can vibrate only when a node occurs at the closed end and an antinode at the open end, Fig. 21-2*a* shows this can occur when the column is $\frac{1}{4}\lambda$ long. Therefore the tube must be $1.30/4$ m $= 0.325$ m long.

(*b*) The next configuration in which the tube will vibrate (resonate) will be as shown in Fig. 21-2*b*; hence, the length will be $\frac{3}{4}\lambda$ or $\frac{3}{4}$ (1.30 m) or 0.975 m long.

For an air column that is open at both ends,

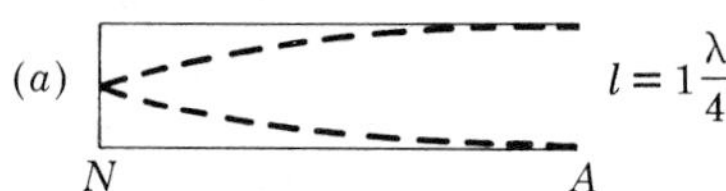

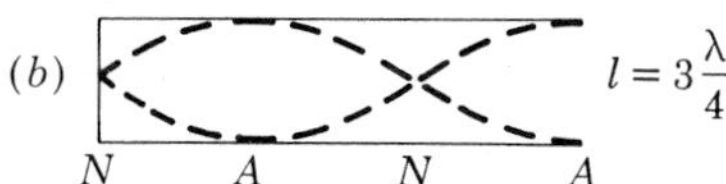

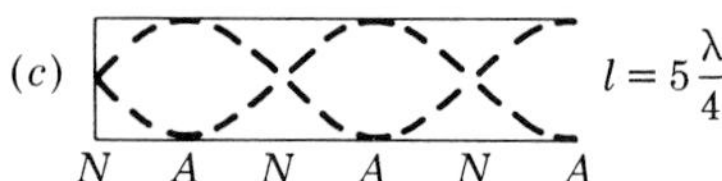

Figure 21-2
Vibration of air column in a tube closed at one end.

both ends are displacement antinodes, and the lowest frequency is that for which the length of the column is a half wavelength. The equation for the possible frequencies is the same as Eq. (1), except that v is the speed of the compressional wave in the gas. The arrangement of nodes and antinodes in an open-ended tube is shown in Fig. 21-3. In this figure the dotted lines represent graphs of the horizontal displacement of the particles of the gas.

As in the case of a vibrating air column in a tube with one end closed, we can determine appropriate tube lengths for the case where the tube has both ends open.

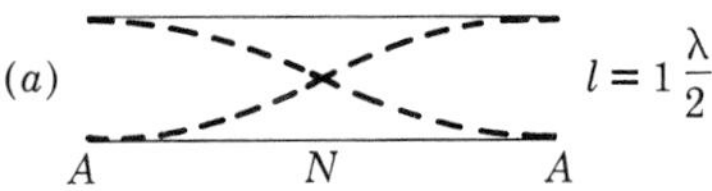

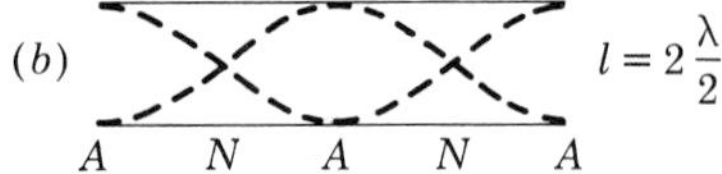

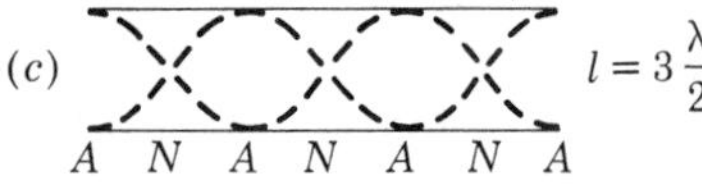

Figure 21-3
Vibration of air column in an open tube.

Example (*a*) What is the shortest length of a column of air in a tube which is open at both ends that would vibrate at 256 vib/s, or Hz?

(*b*) What would be the next shortest length? Assume the speed of sound to be 331 m/s.

Since, $v = f\lambda$ and $\lambda = \frac{v}{f} = \frac{331 \text{ m/s}}{256 \text{ vib/s}}$

$$\lambda = 1.30 \text{ m/vib}$$

Since an open tube will vibrate or resonate when an antinode is at both ends as shown in Fig. 21-3*a*, the distance between two antinodes will be equal to $\frac{1}{2}\lambda$ or $\frac{1}{2}(1.30 \text{ m})$ or 0.65 m.

(*c*) The next configuration at which the open-ended tube will vibrate is shown in Fig. 21-3*b*. The distance between the two antinodes in that case will be $(2)(\lambda/2)$, or λ or 1.30 m.

Here and in Chap. 20 we have discussed the compressional wave in terms of displacement of particles. The wave can be described as well in terms of the departure of *pressure* from normal pressure. This description is equally valid, and, in many cases, it is very useful. We must observe, however, that the variation of pressure in the air column is quite different from the variation of displacement. At an open end the pressure is atmospheric pressure, and it does not change. Therefore the open end is a pressure node. On the other hand, at the closed end the pressure changes most rapidly, and the closed end is a pressure antinode. It is left to the student to analyze this condition to show that this viewpoint makes no change in the possible frequencies.

Many other types of vibrating body are capable of radiating sound waves. Whether the vibration of the body is transverse as in the string or longitudinal as in the air column is immaterial so long as the vibration can set the surrounding medium into longitudinal vibration. Rods may vibrate transversely or longitudinally. Plates, bells, diaphragms, and many other bodies send

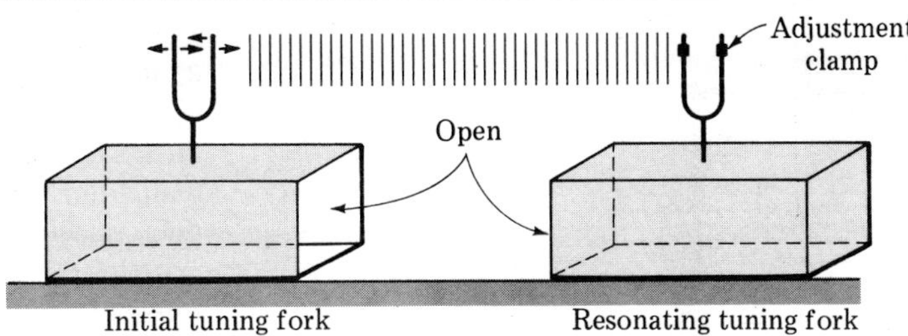

Figure 21-4
Two tuning forks to demonstrate resonance.

out sound waves when they vibrate. Many vibrating bodies are not at all efficient in sending out sound waves into the surrounding medium. A string stretched between two immovable posts would transmit very little energy to the surrounding air. Many primary vibrators require coupling to another body so that, when the second body is caused to vibrate, it is able to transmit energy more efficiently.

21-3 FORCED VIBRATION AND RESONANCE

Whenever a vibrating body is coupled to a second body in such a manner that energy can be transferred, the second body is made to vibrate with a frequency equal to that of the original vibrator. Such a vibration is called a *forced vibration*. If the base of a vibrating tuning fork is set against a tabletop, the tabletop is forced to vibrate. This combination radiates energy faster than the fork could alone. Similarly, a vibrating string is inefficient in transferring energy to surrounding air unless it is coupled to some sounding board.

Whenever the coupled body has a *natural frequency* of vibration equal to that of the source, there is a condition of *resonance* (Chap. 11). Under this condition the vibrator releases more energy per unit time, and the sound is greatly reinforced. Hence the external power supplied to a resounding system must be increased, otherwise its vibrations will be quickly damped.

The reinforcement of sound by resonance with its accompanying release of large amounts of energy has many useful and many obnoxious consequences. The resonance of the air column in an organ pipe amplifies the otherwise almost inaudible sound of the vibrating air jet. Resonance of a radio loudspeaker to certain frequencies would produce an objectionable distortion of speech or music.

A typical demonstration of resonance is given by using two tuning forks mounted on sounding boxes to amplify the sound, with one of the tuning forks being adjustable to provide a variable frequency (Fig. 21-4). When the forks have been adjusted so that they have the same frequency, one of the forks is caused to vibrate. This sound wave then falls upon the other tuning fork which starts to resonate, emitting an audible sound wave. By manually stopping the vibration of the first tuning fork and then releasing it, the process will be reversed and the second tuning fork will transmit its sound waves to the first which will then begin to resonate. Thus, sound waves can be "bounced" back and forth between the tuning forks.

21-4 TRANSMITTING MEDIUM

The motion of a vibrating source sets up waves in a surrounding elastic medium. Whether the vibration of the source is transverse or longitudinal, the wave set up in the surrounding medium is a compressional wave whose frequency is the

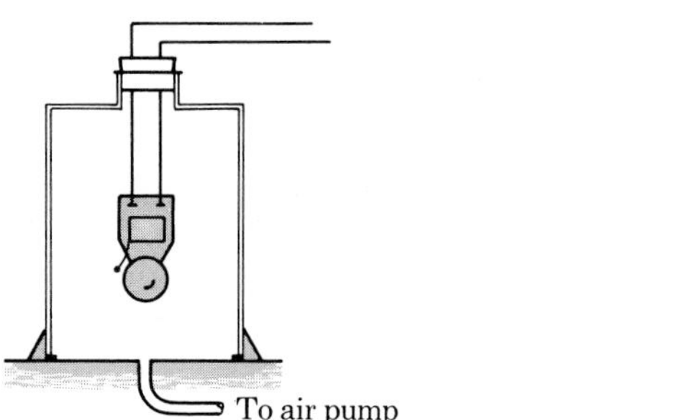

Figure 21-5
Sound is not transmitted through a vacuum.

same as the frequency of vibration of the source. If the frequency is within the range to which the ear is sensitive, the wave is a sound wave.

Since a sound wave involves compression and expansion of some material, sound can be transmitted only through a material medium having mass and elasticity. No sound can be transmitted through a vacuum. This fact can be demonstrated experimentally by mounting an electric bell under a bell jar and pumping the air out while the bell is ringing (Fig. 21-5). As the air is removed, the sound becomes fainter and fainter until it finally ceases but it again becomes audible if the air is allowed to reenter the jar.

Since there is no atmospheric gas on the moon, it is not possible for astronauts to talk directly to one another even if they could remove their helmets. An interesting problem would be to design methods by which they could be in voice contact should their electronic communication fail. One possible solution would be to touch helmets together because the voice-caused vibrations of the air in the helmet of one astronaut would strike his helmet and cause it to vibrate. The helmet vibrations could then be passed to the helmet of the second astronaut by direct contact and then be reconverted to sound waves in the air of his helmet. Can you think of other procedures that they might use to communicate?

Sound waves will travel through any elastic material. In general, they travel through solids with greater ease than they do through liquids, and through liquids with greater ease than they do through gases. There are many illustrations of these differences. For example, if one end of a meterstick is held against your ear and you lightly scratch the other end of the stick, you will hear the scratching quite clearly although it is barely audible through the air. This is the basis upon which the stethoscope operates as well as the tin-can telephones you might have made when a child by stretching a string passing through a hole punched in the bottom of each of two tin cans or paper cups (Fig. 21-6*a*). (If you have missed this experience in your childhood, try it. You may be surprised.) We are all familiar with sounds being transmitted through the closed windows, walls, and floors of a building. The sound of an approaching train may be heard by waves carried through the rails as well as by those transmitted through the air. Caution is advised, however, lest you try to hear the sound being carried by the rail where there is a curve in the track. The earth can be used as a carrier of sound waves, even for listening for approaching buffalo, as the cartoon in Fig. 21-6*b* suggests.

In a more serious vein, the transmission of waves through the earth are important in the study of earthquakes. When an earthquake occurs due to a shifting of the earth's crust, mechanical waves are produced. There are body waves, waves that travel through the earth, and there are surface waves produced. Two kinds of body waves have been identified. The first is similar to a sound wave in that it is a pressure wave vibrating along its direction of travel and producing a compressional wave; it is called the *P wave*. The *P* waves are the fastest of the seismic waves produced in an earthquake, traveling between $3\frac{1}{2}$ and $8\frac{1}{2}$ mi/s (depending upon the density of the material through which they are passing) and occur about 2 to 3 s apart. The second type of body wave is a transverse, or shear, wave and is called the *S wave*. S waves move more slowly than the P waves, usually about 60 percent as fast, and they occur 1 to 15 s apart. Further, while P waves will travel through fluids, S waves will not. The surface waves, called L waves, travel more slowly

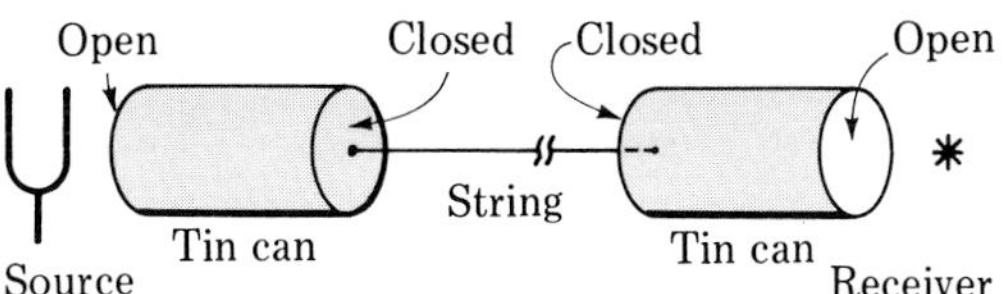

Redeye **By Gordon Bess**

Figure 21-6
(*a*) A "tin-can" telephone. (*b*) The transmission of sound through a solid.

than either P or S waves and are restricted to the surface. They are analogous to water waves in that the particles in the paths of L waves are moved in complex orbits. L waves travel at a fairly uniform speed, about $2\frac{1}{2}$ mi/s, and have a period of anywhere from 10 s to 1 min. Once they are set in motion, L waves continue to circle the earth and may last for several days after a major earthquake.

The P and the S waves are most useful in locating the point of origin (epicenter) of an earthquake. When the two waves arrive at a seismographic station, the lag of the S waves behind the P waves is a clue to the distance between the station and the earthquake. If the times of arrival of P and S waves at several such stations around the earth are available, the location of the earthquake can be determined graphically by drawing a circle with the estimated distance from each station as a radius and the location of the station as the circle's center. The epicenter lies at the common point of overlapping of these circles. In addition to providing information on earthquakes, P and S waves have also played a major role in determining the nature of the interior of our earth due to their varying ability to pass through materials of different densities and state.

As noted above, sound waves travel efficiently through liquids. If you have submerged beneath the water while swimming, you may have noticed how clearly you can hear the outboard motors of boats even though they are a great distance away. In fact, submarines are detected by the underwater sound waves produced by their propellers. The procedure for underwater detection and navigation called sonar (*so*und *na*vigation and *r*anging) is based upon the emission of a pulsed signal by the tracking ship. The wave travels through water at a predictable speed, strikes an object, and is reflected back to the source. This is the same technique that is employed in radar (*ra*dio *d*etection *a*nd *r*anging), except that sonar generally occurs underwater. In liquids and solids the alternate compressions and rarefactions are transmitted in the same manner as they are in air.

21-5 SPEED OF SOUND

If one watches the firing of a gun at a considerable distance, he will see the smoke of the discharge before he hears the report. This delay represents the time required for the sound to travel from the gun to the observer (the light reaches him almost instantaneously). The speed of sound may be found directly by measuring the time required for the waves to travel a measured distance. The speed of sound varies greatly with the material through which it travels. Table 1 shows values for the speed of sound in several common substances.

Table 1
SPEED OF SOUND AT 0°C (32°F) IN VARIOUS MEDIUMS

Medium	ft/s	m/s
Air	1,087	331.5
Hydrogen	4,167	1,270
Carbon dioxide	846	258.0
Water	4,757	1,450
Iron	16,730	5,100
Glass	18,050	5,500

Since sound waves are conducted through gases, we must conclude that the waves are longitudinal (compressional). The compressions and rarefactions of such a compressional wave are illustrated in Fig. 21-7. In the upper part of the figure the spacing of the lines indicates the relative pressures in the medium, above normal pressure where the lines are close together, and below normal pressure where the lines are farther apart. The lower half of Fig. 21-7 shows a graph of pressure as a function of distance. The zero line represents the normal pressure of the undisturbed medium, while the ordinates of the curve show the departure of the pressure from normal at various points.

In Chap. 19 we discussed the speed of a compressional wave in terms of an appropriate elastic modulus E and density ρ of the medium. From Eq. (12), Chap. 19,

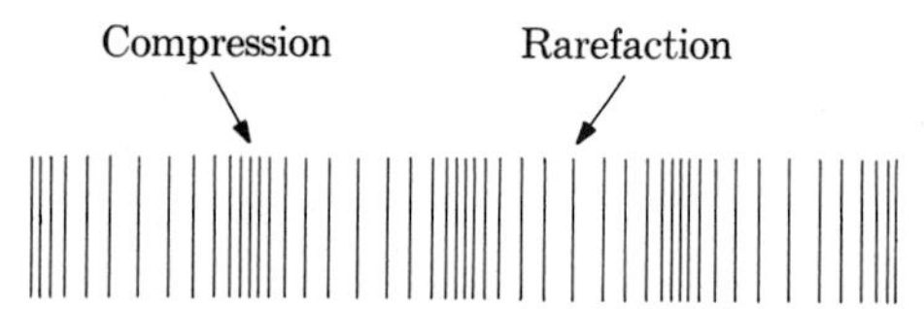

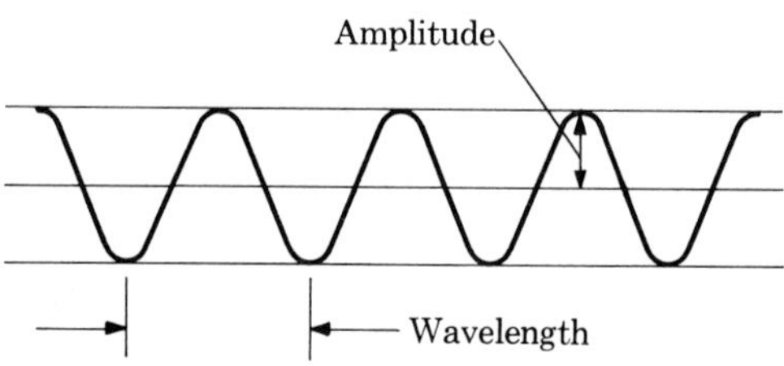

Figure 21-7
Representation of a sound wave.

$$v = \sqrt{\frac{E}{\rho}} \tag{3}$$

Equation (3) represents the speed of propagation in a fluid or in a solid wire or rod whose diameter is small compared with its length. It is not valid for an extended solid where shearing stresses occur as the compressional wave passes through. The units appropriate for Eq. (3) are any consistent set. In the mks system E is expressed in newtons per square meter, ρ in kilograms per cubic meter, and v in meters per second. In the cgs system E is in dynes per square centimeter, ρ in grams per cubic centimeter, and v in centimeters per second. In the customary British system E is in pounds per square foot, ρ in slugs per cubic foot, and v in feet per second.

For solid rods the appropriate elastic modulus is Young's modulus since only one dimension is involved. Equation (3) then becomes

$$v = \sqrt{\frac{Y}{\rho}} \tag{4}$$

Example Compute the speed of sound in the steel rails of a railroad track. The weight-density of steel is 490 lb/ft^3, and Young's modulus for steel is 29×10^6 lb/in^2.

$$v = \sqrt{\frac{Y}{\rho}} = \sqrt{\frac{Yg}{D}}$$

$$= \sqrt{\frac{(29 \times 10^6 \times 144 \text{ lb/ft}^2)(32 \text{ ft/s}^2)}{490 \text{ lb/ft}^3}}$$

$$= 1.7 \times 10^4 \text{ ft/s}$$

For an extended solid, the speed of the longitudinal wave depends upon the bulk modulus B and the shear modulus n

$$v = \sqrt{\frac{B + \frac{4}{3}n}{\rho}} \tag{5}$$

In a fluid the shear modulus is zero, and hence the speed depends only upon the bulk modulus and the density

$$v = \sqrt{\frac{B}{\rho}} \tag{6}$$

We have observed (Chap. 12) that in a gas at constant temperature the bulk modulus is the pressure of the gas. However, when a sound wave passes through the gas, the compressions and rarefactions are so rapid that the heat is not conducted from one part of the medium to another. Hence the changes are practically adiabatic, and the appropriate modulus of elasticity is the *adiabatic bulk modulus*. In Chap. 16 we observed that the adiabatic bulk modulus is γP, where γ is the ratio of the specific heat at constant pressure to the specific heat at constant volume. For air and other diatomic gases $\gamma = 1.40$. Thus, for a gas

$$v = \sqrt{\frac{\gamma P}{\rho}} \tag{7}$$

For an ideal gas, from Eq. (20), Chap. 15,

$$PV = nRT \tag{8}$$

where n is the number of moles of gas, R is the universal gas constant, and T is the absolute temperature. The mass m of the gas is nM, where M is the molecular mass of the particular gas. Then

$$PV = \frac{m}{M} RT \tag{9}$$

$$\frac{P}{\rho} = \frac{P}{m/V} = \frac{RT}{M} \tag{10}$$

Thus, from Eq. (7),

$$v = \sqrt{\frac{\gamma RT}{M}} \tag{11}$$

From Eq. (11) we see that the speed of a sound wave in a gas is independent of the pressure but is proportional to the square root of the absolute temperature.

Example Find the theoretical speed of sound in hydrogen at 0°C. For a diatomic gas $\gamma = 1.40$, and for hydrogen $M = 2.016$ g/mole.

$$v = \sqrt{\frac{\gamma RT}{M}}$$

$$= \sqrt{\frac{1.40[8.317 \text{ J/(mol)(K°)}](273\text{°K})}{2.016 \times 10^{-3} \text{ kg/mol}}}$$

$$= \sqrt{\frac{1.40 \times 8.317 \times 273 \text{ J}}{2.016 \times 10^{-3} \text{ kg}}}$$

$$= 1.26 \times 10^3 \text{ m/s}$$

It is often acceptable to use an approximation for the change in the speed of sound with a change in temperature. The speed of sound at 0°C is 1,087 ft/s or 331 m/s. An increase in temperature of 1°C will cause an increase of 2 ft/s or 0.61 m/s and a similar decrease will result from a 1°C drop. That is, sound in air at 20°C has a speed of (1,087 + 40) ft/s = 1,127 ft/s or (331 + 12.2) m/s = 343.2 m/s.

21-6 REFRACTION OF SOUND

Have you ever noticed that it is occasionally possible to hear people who are talking in a boat on a body of water at a greater distance away from you than if they were the same distance over land? This is due to a property of sound called refraction. In a uniform medium at rest sound travels with constant speed in all directions. If, however, the medium is not uniform, the sound will not spread out uniformly but the direction of travel changes because the speed is greater in one part of the medium. The bending of sound due to change of speed is called *refraction*.

The nonuniform spreading of sound in the open air is an example of this effect. If the air were at rest and at a uniform temperature throughout, sound would travel uniformly in all directions. Rarely, if ever, does this occur, for the air is seldom at rest, and almost never is the temperature uniform. On a clear summer day the surface of the earth is heated, and the air immediately adjacent to the surface has a much higher temperature than do the layers above. Since the speed of sound increases as the temperature rises, the sound travels faster near the surface than it does at higher levels. As a result of this difference in speed the wave is bent away from the surface, as shown in Fig. 21-8*a*. To an observer on the surface, sound does not seem to travel very far on such a day since it is deflected away from him.

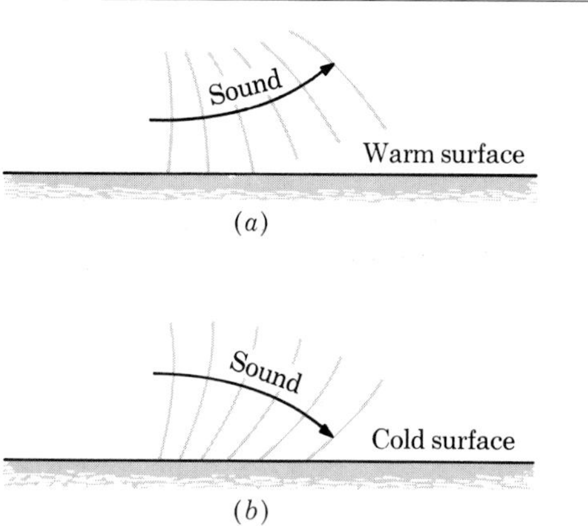

Figure 21-8
Refraction of sound, due to temperature difference.

On a clear night the ground cools more rapidly than the air above; hence the layer of air adjacent to the ground may become cooler than that at a higher level. As a result of this condition sound travels faster at the higher level than at the lower level and consequently is bent downward, as shown in Fig. 21-8*b*. Since the sound comes down to the surface, it seems to travel greater distances than at other times.

Sound waves over water are bent for similar reasons. In summer, the water is cooler than the air above it. Since sound travels faster at higher temperatures, the sound adjacent to the cold water moves more slowly than that some distance above the water, and the waves curve downward toward the surface of the water. In winter, however, the water is warmer than the air and the reverse process occurs. The air adjacent to the water is warmer than that of the layers at some distance above the water. Therefore, the waves refract upward with the effect that sound does not travel as far as when the water was cooler than air.

Wind is also a factor in refraction of sound. In discussing the speed of sound in air, we assume that the air is stationary. If the air is moving, sound travels through the moving medium with its usual speed *relative to the air* but its speed relative to the ground is increased or decreased by the amount of the speed of the air, depending upon whether the air is moving in the same direction as the sound or in the opposite direction. If the air speed is different at various levels, the direction of travel of sound is changed, as shown in Fig. 21-9. Friction causes the wind speed to be lower at the surface than at a higher level; hence sound traveling against the wind is bent upward and leaves the surface, while that traveling with the wind is bent downward. As a result, the observer on the surface reports that sound travels farther with the wind than against the wind.

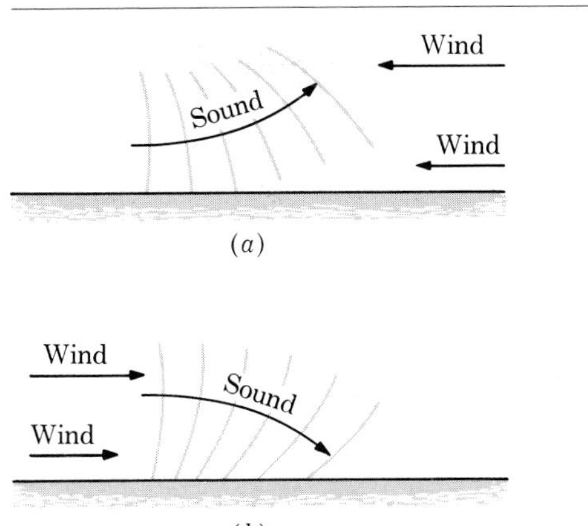

Figure 21-9
Refraction of sound, due to wind.

Combinations of the two phenomena just discussed may cause some effects that seem very peculiar. Sound may carry over a mountain and be heard on the other side, while similar sounds are not transmitted in the opposite direction. Frequently sound "skips" a region; i.e., it is audible near the source and also at a considerable distance, but at intermediate distances it is not audible. Such an effect is quite troublesome in the operation of such devices as foghorns. Refraction effects increase the difficulties in locating airplanes, guns, or submarines by means of sound waves.

21-7 REFLECTION OF SOUND WAVES

Sound waves are reflected from surfaces such as walls, mountains, clouds, or the ground. A sound is seldom heard without accompanying reflections, especially inside a building, where the walls and furniture supply the reflecting surfaces. The "rolling" of thunder is largely due to successive reflections from clouds and land surfaces.

The ear is able to distinguish two sounds as separate only if they reach it at least 0.1 s apart; otherwise, they blend in the hearing mechanism to give the impression of a single sound. If a sound of short duration is reflected back to the observer after 0.1 s or more, he hears it as a repetition of the original sound, an *echo*. In order that an echo may occur, the reflecting surface must be at least 55 ft away, since sound, traveling at a speed of 1,100 ft/s, will go the 110 ft from the observer to the reflector and back in 0.1 s.

Example A siren emits a sound which returns to the source in 4 s. How far away is the object which caused the echo from the source? Assume the temperature to be 30°C.

Sounds travels at

$$v = (331 + 18)\ \text{m/s}$$

or

$$v = 349\ \text{m/s}$$

Since

$$\bar{v} = \frac{s}{t}$$

$$s = \bar{v}t = 349\ \text{m/s} \times 4\ \text{s}$$

$$= 1{,}4\bar{0}0\ \text{m}$$

But this is the distance the sound traveled to the object and back. Therefore the distance is $1{,}4\bar{0}0$ m, or $7\bar{0}0$ m.

It was noted in Sec. 21-4 that use is made of the reflection of sound waves in the determination of ocean depths. A sound pulse is sent out under water from a ship. After being reflected from the sea bottom, the sound is detected by an underwater receiver also mounted on the ship, and the time interval is recorded by a special device. If the elapsed time and the speed of sound in water are known, the depth of the sea at that point can be computed. Measurements may thus be made almost continuously as the ship moves along.

Nature has endowed the bat with the capacity to locate objects by ultrasonic echoes, a form of natural sonar. Research has shown that the sonar system used by bats is billions of times more efficient and sensitive than man-made radar and sonar. Studies were conducted of the inaudible signals emitted by the big brown bat (*Eptesicus fuscus*) when it was cruising and when pursuing its prey. These signals were electronically reproduced in such a manner that data could be obtained directly about their frequency and wavelength. It was found that bats can vary the frequency of the pulsed sound waves to "home" in on its prey. When cruising, each pulse is ten- to fifteen-thousandths of a second long, but during pursuit the pulses are shortened to less than a thousandth of a second and may occur as often as 200 pulses per second. According to Griffin,[1] within each pulse of sound the frequency changes from 25,000 to 50,000 Hz. It should be noted that this frequency range is well above man's range of hearing (around 16,000 to 20,000 Hz). It was observed that as the pitch changes, the wavelength changes from about 6 to 12 mm, which is just about the size range of the insects which serve as food for the bats. It is believed that fluctuations in the frequency may be used to sweep the target as it moves erratically ahead of the pursuing bats. While the range of detection of bats is short (2 m) compared with radar (80,000 to 150,000 m) and sonar (2,500 m), it is much more sensitive in that it can detect targets having diameters of 0.01 m, whereas radar and sonar can detect targets which have a minimum diameter between 3 and 5 m. The relative size of these natural and man-made ranging systems is interesting. The bat's system has an average weight of 0.012 kg, radar a weight range of 90 to 12,000 kg, and sonar weighs on the average 450 kg. Another indication of the supersensitivity of the sonar of bats is that the echoes return to the bat only 1/2,000 as loud as emitted, yet the bat can distinguish these signals from all other background noise.

Sound waves may be reflected from curved surfaces for the purpose of making more energy travel in a desired direction, thus making the sound more readily audible at a distance. The curved sounding board placed behind a speaker in an auditorium throws forward some of the sound waves that otherwise would spread in various directions and be lost to the audience. In the same way, a horn may be used to collect sound waves and convey their energy to an ear or other detector.

[1] Griffin, Donald R., More about bat "Radar," *Scientific American*, July 1958, pp. 40–44.

21-8 INTERFERENCE OF WAVES; BEATS

Whenever two wave motions pass through a single region at the same time, the motion of the particles in the medium will be the result of the combined disturbances of the two sets of waves. The effects due to the combined action of the two sets of waves are known in general as *interference* and are important in all types of wave motion.

If a shrill whistle is blown continuously in a room whose walls are good reflectors of sound, an observer moving about the room will notice that the sound is exceptionally loud at certain points and unusually faint at others. At places where a compression of the reflected wave arrives at the same time as a compression of the direct wave, their effects add together and the sound is loud; at other places where a rarefaction of one wave arrives with a compression of the other, their effects partly or wholly cancel and the sound is faint.

Contrasted with the phenomenon of interference in space, we may have two sets of sound waves of slightly *different frequency* sent through the air at the same time. An observer will note a regular swelling and fading of the sound, a phenomenon called *beats*. Since the compressions and rarefactions are spaced farther apart in one set of waves than in the other, at one instant two compressions arrive together at the ear of the observer and the sound is loud. At a later time the compression of one wave arrives with the rarefaction of the other, and the sound is faint.

At a single point the time variation of pressure in the two waves may be written

$$p_1 = P \sin 2\pi f_1 t$$

and

$$p_2 = P \sin 2\pi f_2 t \qquad (12)$$

For simplicity the pressure amplitude P is assumed to be the same for the two waves. At the chosen point the waves are superposed, and the pressure at each instant is

$$p = p_1 + p_2 = P(\sin 2\pi f_1 t + \sin 2\pi f_2 t) \quad (13)$$

From the trigonometric identity

$$\sin a + \sin b = 2 \cos \tfrac{1}{2}(a - b) \sin \tfrac{1}{2}(a + b)$$

$$p = 2P \cos 2\pi \frac{f_1 - f_2}{2} t \sin 2\pi \frac{f_1 + f_2}{2} t \quad (14)$$

This may be considered as a vibration of the average frequency, $\frac{1}{2}(f_1 + f_2)$, in which the amplitude is $2P \cos 2\pi \frac{1}{2}(f_1 - f_2)t$. The amplitude of vibration varies with a frequency $\frac{1}{2}(f_1 - f_2)$. The amplitude will be a maximum and the sound will be loudest when the cosine has a value of either 1 or -1. Since these values appear twice during a cycle, the frequency of beats is twice the frequency of vibration, that is, $2 \times \frac{1}{2}(f_1 - f_2) = f_1 - f_2$, the difference of the frequencies. Thus, in Fig. 21-10, two sets of waves of frequencies 10 vib/s and 12 vib/s combine and give a resultant wave that fluctuates in amplitude 12 to 10, or 2 times per second.

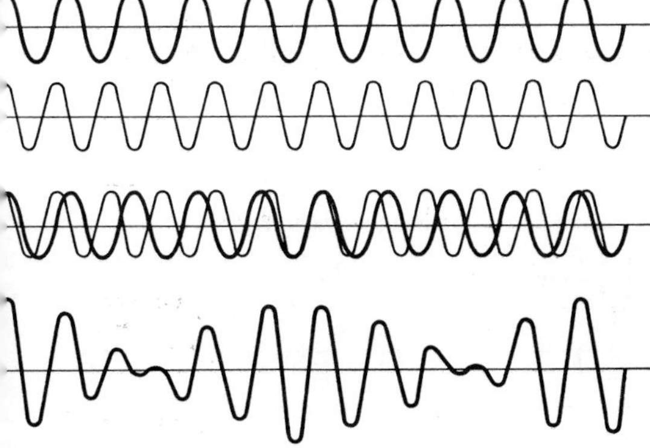

Figure 21-10
Two waves of different frequency combined to cause beats.

When the difference in frequency of the two waves is small, the variation in intensity is readily observed by listening to it. As the difference increases beyond 8 or 10 per second, it becomes increasingly difficult to distinguish them as separate. If the difference frequency reaches the audible range, a *beat note* may be heard. The ability to hear this beat note is due largely to lack of linearity in the response of the ear.

The two tuning forks shown in Fig. 21-4 can be used to demonstrate beats. The tuning forks are adjusted so that they have slightly different frequencies. Then, if *both* tuning forks are caused to vibrate, very noticeable beat waves will be produced where destructive and constructive interference occur alternately.

21-9 THE DOPPLER EFFECT

So far in our discussion of waves we have assumed that the source is at rest, the observer is at rest, and the medium in which the wave travels is at rest. At certain times we may have any one, or two, or all three of them moving. The general result of any or all of these motions is that the observer notes an apparent change in frequency from the frequency of vibration of the source.

Case A The source is moving toward the receiver; the medium and the observer are stationary.

Let V equal the speed of the wave in the medium, v_s equal the speed of the source (+ when in the same direction as the wave is being propagated, − when in the opposite direction), and f_s equal the frequency of the source.

In time t the *wave* from source S moves a distance equal to Vt toward the observer O (Fig. 21-11) to point b. During the same time t, the *source* moves a distance $v_s t$ to point a, and during this time the source has emitted a number of waves equal to $f_s t$. These waves are all crowded

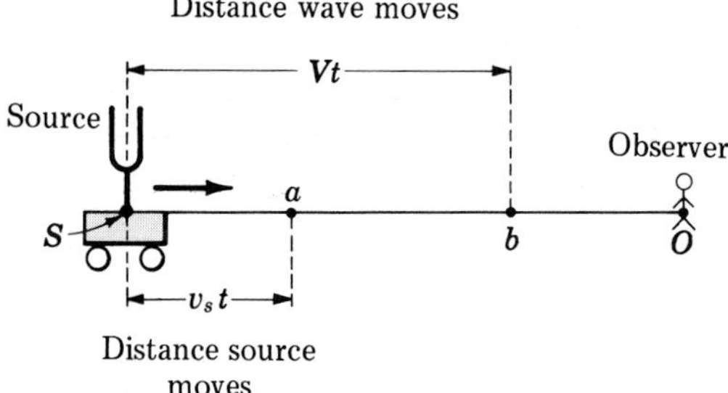

Figure 21-11
Doppler effect; moving source.

into the distance ab. Therefore, the new distance between the waves (the new wavelength λ') has become

$$\lambda' = \frac{\text{distance } ab}{\text{number of waves}} = \frac{Vt - v_s t}{f_s t}$$

and $$\lambda' = \frac{V - v_s}{f_s} \tag{15}$$

To the observer the speed of the wave coming to him is V, and the wavelength is λ'. The frequency f_o of the waves coming to him is then

$$V = f_o \lambda' \qquad \text{and} \qquad f_o = \frac{V}{\lambda'}$$

$$f_o = \frac{V}{(V - v_s)/f_s} = f_s\left(\frac{V}{V - v_s}\right) \tag{16}$$

where f_s is the observed frequency when the source is moving toward the observer.

It can be shown that when the *source* moves *away* from a stationary observer,

$$f_o = f_s\left(\frac{V}{V + v_s}\right)$$

In general, then, when the source moves,

$$f_o = f_s\left(\frac{V}{V \mp v_s}\right)$$

where $-$ is used for approach toward and $+$ for movement away from the observer.

Case B The observer is moving away from a stationary source with speed v_o and with the medium stationary.

If the observer is moving away from the stationary source, the wavelength of the waves is not changed but the number that reaches him in a given time t is decreased. If the observer remained at rest at O (Fig. 21-12), he would receive in time t all the waves in the distance Vt, that is, a number of waves equal to $f_s t$ or, since $V = f_s\lambda$, Vt/λ. If during the time t, he moves a distance $v_o t$ to c, he receives only the waves between b and O, that is $Vt/\lambda - v_o t/\lambda$ or $[(V - v_o)t]/\lambda$ and his apparent frequency is

$$f_o = \frac{(V - v_o)t/\lambda}{t} = \frac{V - v_o}{\lambda} = \frac{V - v_o}{V/f_s}$$

and $$f_o = f_s\left(\frac{V - v_o}{V}\right) \tag{17}$$

As above, a general equation can be presented for the case when the observer is moving and the source is stationary. That is, $f_o = f_s[(V \pm v_o)/V]$, where $+$ is used for an approach to the source and $-$ is used for motion away from the source.

Case C Both the source and the observer are moving, with the medium stationary.

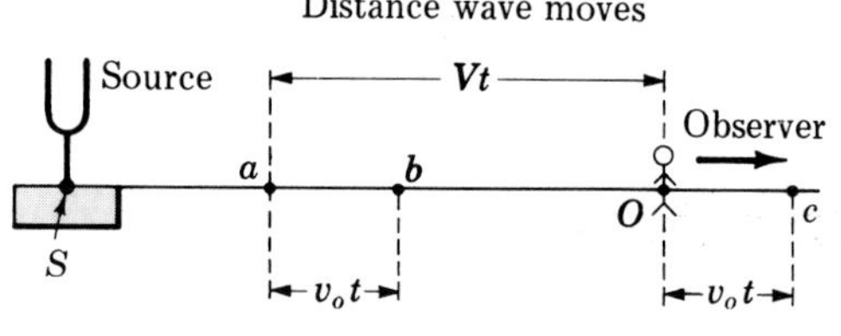

Figure 21-12
Doppler effect; moving observer.

If both source and observer are moving, both changes occur and

$$f_o = \frac{V}{(V - v_s)} \frac{(V - v_o)}{V} f_s = \frac{V - v_o}{V - v_s} f_s \quad (18)$$

We note that when the source and the observer are moving with the same speed in the same direction there is no shift in frequency.

Case D Equation (18) represents conditions when the medium is at rest or when all speeds are measured relative to the medium. If the medium is moving relative to the ground and all speeds are measured relative to the ground, an added term, the speed of the medium, appears. Then

$$f_o = \frac{V + v_m - v_o}{V + v_m - v_s} f_s \quad (19)$$

In the foregoing discussion it was assumed that all the motions were parallel to the line joining source and observer. If the motions are in other directions, we must use components of the velocities in the direction of this line.

The Doppler effect can be observed for all kinds of waves so long as the speed of the source is small compared with the wave speed. In light waves the frequency is independent of the speed of the medium, since in relativity the speed of light is considered to be the same in all frames of reference. The observed shift toward the red (lower frequencies) in the spectra of some stars has been interpreted to indicate that the stars are moving away from observers on the earth and gives support to the "expanding-universe" concept. That is, that some 10 to 12 billion years ago an event occurred which gave our universe its form and shape. A tremendous mass of primary matter, such as electrons, protons, and neutrons, was drawn together in a process called the "big squeeze." During this process temperatures as high as a billion degrees occurred. This was followed by a rapid expansion, an explosion on the nature of a thermonuclear fusion explosion, a "big-bang" event. Huge fragments were shot out into space. It is proposed that these fragments became galaxies, such as our Milky Way and the Crab Nebula (Andromeda). We know from our experience in space travel that once an object is put into motion in space, there is little effort needed to keep it moving. Further, observation of spectral lines coming from stars in other galaxies reveal that there is a shift toward the red end of the spectrum. This could only occur when a light source is moving away from the observer, as would be the case if the universe is expanding. While there is need for further study, it appears at present that the concept of an expanding universe is valid as opposed to the so-called steady-state theory which states that new hydrogen is "formed" in space and eventually forms into new galaxies, maintaining a perpetual universe. If a source of light, a star, moves toward the receiver, it causes a shift of spectral lines to the blue-violet end of the spectrum. It is possible to observe double stars, which alternately give evidence of a red shift and then a blue shift. This is explained as being a case where a double star has an orbit edgewise to the earth, and as these stars revolve they sometimes move toward earth and sometimes away from earth.

21-10 SONIC BOOMS

When the speed of the source approaches the wave speed, the wave front is distorted and the simple analysis above does not hold. If the speed of the source is the same as the wave speed, the energy piles up in directions making small angles with the direction of motion. This effect accounts for the "sonic boom" associated with airplanes as their speeds pass the speed of sound; about 762 mi/h at sea level and 664 mi/h at 35,000 ft.

When the speed of the source is greater than the phase speed of the wave, the Doppler principle no longer has meaning. A *shock wave* is produced that is unlike an ordinary wave. This condition is illustrated in Fig. 21-13. The source moves from S to S' during time t. From each of several positions along this path circles are drawn to represent the distance the waves have traveled from the point at which they are emitted

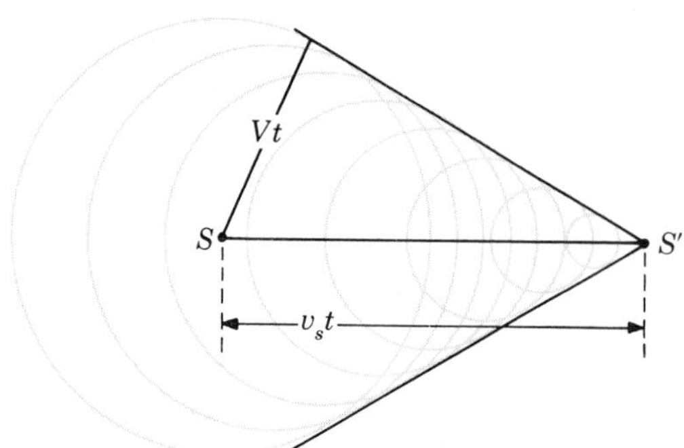

Figure 21-13
Shock wave.

by the end of the time t. If V is the wave speed and v_s is the source speed, the first circle is drawn with radius Vt to represent the wave front of the wave emitted at S at the instant the source reaches S'. The successive circles have centers at the points that the source has reached at $t_1 = 0.1t$, $t_2 = 0.2t$, $t_3 = 0.3t$, etc. The radius of each circle is the distance the wave has traveled between the time that it was emitted and t, that is, $V(t - t_1)$, $V(t - t_2)$, etc. The circles are successively smaller in radius until finally at time t at the point S' the radius is zero. The heavy line that is the envelope of the circles is the shock wave. The shock wave created when a high-speed missile travels through air approximates this shape, as does the bow wave of a boat.

Sonic booms are of much interest today since they not only pose a physical but also a psychological problem to man. We saw earlier that sound is a pressure wave. A sonic-boom shock wave, since it is a highly concentrated sound wave, is also a pressure wave. It is this wave of increased pressure that produces a sound like a thunderclap or an explosion as it hits your eardrum. When the shock waves (there are two waves, one formed by the nose and one by the tail of a plane) strike an object, they pass over and around it pressing on it from all sides. The two waves result in two explosive sounds, or booms, observed as such when the shock wave hits, lasting about 0.3 s (Fig. 21-14).

It has been noted by those studying sonic booms that the pressure variations they cause are not large when compared with others experienced in daily living. These pressure changes are less than would be experienced in a 15-m drop or rise in an elevator and considerably less than that experienced when you dive into a swimming pool. Therefore, harmful physiological effects are not likely.

The evidence is mixed regarding possible adaptation of humans to sonic booms. There is a quite definite trend observed toward building up a tolerance to sonic booms in individuals. However, some psychologists express concern over this and note that human beings can adapt to sonic booms only at the risk of lowering their sensitivity to proper alarm stimuli. Further, there are certain individuals who are hypersensitive to transient noises and will find adaptation difficult.

There are several misconceptions regarding the ability of the shock wave to cause structural damage. While the pressure in this narrow, 1/10,000-in-thick shock wave is 10 times as great as the strongest thunderclap, the pressures are no greater than 5 lb/ft². This is much less than the estimated 70 lb/ft² needed to cause structural damage. However, many complaints (mostly

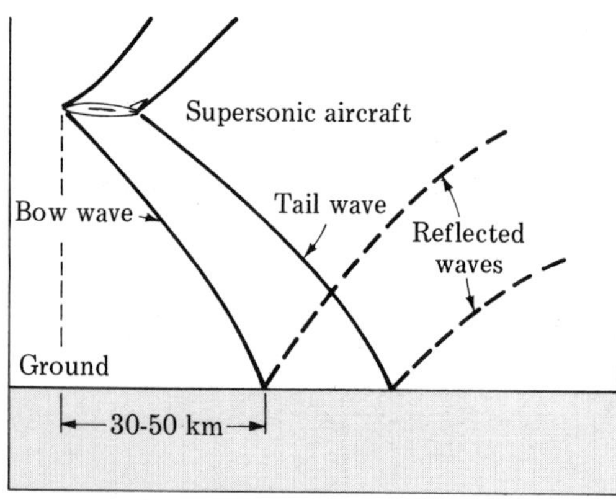

Figure 21-14
Characteristics of sonic waves created by a supersonic aircraft.

about superficial damage) are lodged about structural damage having been caused by the shock wave. Many of these reported incidents have been found to be attributable to weaknesses existing in a building prior to the sonic boom. The shock wave served merely as a triggering influence in some cases, as could truck traffic, windstorms, and heavy falling objects, under the right conditions. However, even though sonic booms serve only as a trigger to such damage, the existence of any unpredictable hazard in one's home is certainly unwelcome.

Much study has been done and will be done on limiting sonic booms. The controversy over large-scale introduction of supersonic transports (SST) will not be resolved until advanced-design aircraft that minimize the effects of sonic booms are developed.

21-11 ULTRASONICS VS. SUPERSONICS

In this chapter we have used the often-confused terms *ultrasonic* and *supersonic*. Since these terms are so important in our present-day society, let us clearly differentiate between the two.

We saw in the last section that the term supersonic refers to an object traveling with a velocity greater than that of sound. The problems encountered when one breaks the sound barrier were listed there.

The second term, ultrasonics, was mentioned in our discussion of echolocation by bats. This term refers to the sound produced by an object vibrating at a frequency higher than the human ear can hear. This frequency can run from 20,000 Hz to any desired frequency, but normally within a range of 20,000 to 100,000 Hz. However, one ultrasonic device has been developed that vibrates at 25 billion Hz.

The value of ultrasonics, especially in the fields of medicine and in various industries, is great and promises to become even greater. An ultrasonic wave is a pressure wave which has an extremely short wavelength because of its high frequency. This can be seen from the relationship $v = f\lambda$, where v is a constant and f and λ are inversely proportional to each other. The ultrasonic wavelength may even be shorter than that of visible light (10^{-6} m). In fact, the 25-billion-Hz wave mentioned above has a wavelength of 10^{-8} m, comparable to x-ray wavelength (10^{-8} to 10^{-11} m). It can be shown that a wave is affected only by an object which is larger than its wavelength. Sound waves are not substantially interfered with by objects in their paths which have diameters of a few inches. We shall see that light waves also are able to pass "through" objects which are small compared with their wavelength. X-rays, which have an even shorter wavelength, can penetrate solid matter because the average diameter of a molecule is smaller (10^{-10} m) than the x-ray wavelength. Accordingly, there is a direct relationship between the depth of penetration and the wavelength of the wave falling on an object. Herein lies the great value of an ultrasonic wave. By our increasing the frequency of sound to this range, the wavelength becomes proportionately smaller and eventually becomes small enough to have great penetrating power. Whereas light rays and x-rays are electromagnetic waves, ultrasonic waves are pressure waves. The effect of a pressure wave falling upon a molecule can be striking. For example, it is possible to have ultrasonic welding where an ultrasonic wave is directed upon two objects which are in contact with each other. The molecular agitation caused by the wave falling on the objects produces sufficient thermal energy to cause the objects to melt. Then, when the ultrasonic vibration is stopped, the objects solidify and are fused together. This is especially valuable in the "welding" together of plastic devices.

The penetrating power of ultrasonic waves make them valuable in medicine for diagnostic work and bloodless surgery. In diagnostic work, the technique is one of transmitting an ultrasonic signal through a patient and then by an analysis of the transmission, or perhaps reflection or refraction, of the signal, organs and growths, such as cysts and tumors, can be located. The use of ultrasonic waves for surgery is still very much experimental but holds great promise for man especially in the control and removal of tumors. In this process, very high frequency ultrasonic waves are directed on the growth from two sources. The sources are arranged so that the waves they emit strike the target in such a manner that constructive interference occurs at that point and an increase of pressure results there. This pressure wave is sufficient to cause the molecular bonding to break down and the tissue to disintegrate. To date, experiments in bloodless surgery by ultrasonics have been limited to laboratory animals. In one typical experiment a very small portion of an animal's brain, in a difficult-to-reach area, which controlled the opening and closing of the iris of the eye, was dissolved by ultrasonic surgery. This procedure, if successful in man, will permit the removal of a brain tumor without the necessity of surgically penetrating the skull, always a hazardous operation. Much more will have to be known about side effects of ultrasonics, however, before it can be used on humans in this manner.

A less dramatic application of ultrasonics is, however, quite widely used. Ultrasonic waves are employed to "clean" material in ultrasonic cleaners. In this process, operating with frequencies of around 30,000 Hz, the wave is produced in a liquid which causes many small bubbles to form. These bubbles implode with the release of individual pressure waves which strike any object submerged in the liquid. This process, called *cavitation*, actually blasts dirt and surface deposits off such things as jewelry, laboratory, glassware, and watches (of course, water is not used for this last example). The ultrasonic cleaner also has a sterilizing effect in that it kills many bacteria while it cleans. Such ultrasonic cleaners are effective and inexpensive and can be purchased for home use.

The potential application of ultrasonics is too great to allow the presentation of more than a short list of the functions that can be performed by ultrasonic irradiation. In chemistry, it is used for emulsification, for the dislocation of crystalline materials, and for molecular relaxation. In physics it is used in viscometry and in flaw detection.

SUMMARY

Sound is a disturbance of the type capable of being detected by the ear. The disturbance is produced by the *vibration* of some material body.

Vibrations may be set up in bodies that have the properties of inertia and elasticity. Those frequencies of vibration are possible for which stationary waves can be set up within the body. Most vibrating bodies vibrate in a complex manner with many of the characteristic modes of vibration. These usually consist of a fundamental mode of lowest frequency and several multiples of this frequency that are called *harmonics*. The fundamental is the first harmonic.

Forced vibrations are set up in an elastic body when it is coupled to a vibrating body. If a natural frequency of the coupled body is the same as that of the vibrator, there is *resonance*. For resonance there is a rapid transfer of energy, and a resultant louder sound.

Sound is transmitted through air and other media, solid or fluid, in the form of *compressional* (longitudinal) waves.

The speed of sound in any medium depends upon the elastic constants of the medium and upon its density. For a fluid or a thin rod the speed is given by

$$v = \sqrt{\frac{E}{\rho}}$$

The speed of sound in a gas depends upon the temperature of the gas but is independent of the pressure of the gas or the frequency of the wave.

$$v = \sqrt{\frac{\gamma P}{\rho}} = \sqrt{\frac{\gamma RT}{M}}$$

Although sound waves travel through all elastic materials, they travel through solids with greater ease than through liquids and through liquids with greater ease than they do through gases.

The transmission of waves through the earth give information about earthquakes. A *P wave* is a seismic wave that travels through the body of the earth. The *S wave* is a seismic wave that travels through the body of the earth but more slowly than a P wave and will not travel through fluids. *L waves* are surface seismic waves that travel more slowly than either of the other two waves.

In air at ordinary temperature the speed of sound is approximately 331 m/s, or 1,100 ft/s.

A sound wave may be *refracted* if the speed of the wave is not the same in all parts of the medium or if the parts of the medium are moving. A wave may also be refracted as it passes from one medium to another.

An *echo* occurs when reflected sound waves return to the observer 0.1 s or more after the original wave reaches him, so that a distinct repetition of the original sound is perceived.

The procedure for underwater detection and navigation, called *sonar,* is based on the emission from and return to a tracking ship of a pulsed signal. Bats use a similar technique called *echolocation* to navigate.

Beats occur when two sources of slightly different frequencies are sounded at the same time. The number of beats per second is equal to the difference in frequency of the two waves.

The apparent frequency of a source of sound is changed if there is relative motion between the source and the observer. This statement is called *Doppler's principle.* The relation between the apparent frequency and the frequency of the source is

$$f_o = \frac{V - v_o}{V - v_s} f_s$$

The Doppler effect observed when light falls on the earth from a receding star supports the expanding-universe concept of the origin of our universe.

A *sonic-boom* shock wave is a highly concentrated pressure wave produced when an object flies faster than the speed of sound.

The term *supersonic* refers to an object traveling faster than sound. *Ultrasonic* refers to the sound produced by an object which is vibrating at a frequency higher than a human can hear. Ultrasonic waves can be very penetrating due to their short wavelength.

Questions

1 Distinguish between the physical and the psychological definitions of sound.

2 Would a wave created on the surface of a body of water ever totally disappear? Discuss.

3 What is one fundamental difference between a sound wave and a light wave?

4 What indications are there that standing waves exist in a vibrating body?

5 Is the wave in a vibrating string a sound wave? If not, what kind of wave is it?

6 In order to get the air in a bottle to resonate when air is blown across its mouth, what conditions must exist for the vibrating air column in the bottle?

7 Suggest practical uses that can be made of the fact that sound waves travel through solids with greater ease than through liquids or gases.

8 By placing an inverted cup against a wooden door and then placing the ear against the cup, sounds on the other side of the door can be plainly heard. Explain why this happens.

9 What happens to the pitch of a "soda-straw flute" if the length of the straw is shortened? Explain.

10 While water is poured into a tall bottle, why does the pitch of the sound rise as the level of water approaches the top?

11 Is it possible for an object which is vibrating transversely to produce a sound wave? Explain your answer.

12 Explain the vibration in an organ pipe. What kind of wave is there within the pipe?

13 A sound produced by a tuning fork can be greatly amplified if the base of the vibrating fork is held against the top of a table. Explain why this happens.

14 What fundamental condition must exist for an object to resonate when a sound wave falls upon it?

15 Explain why sound will not travel through

a vacuum. Give an illustration other than the bell in a vacuum to show this to be true.

16 Why does sound travel faster in solids than in gases?

17 (*a*) If a person inhales hydrogen and then speaks, how will the characteristics of his voice be changed? (*b*) How would the situation be changed if carbon dioxide were used?

18 Why does the speed of a sound wave in a gas change with temperature?

19 Give some examples of the role that resonance plays in musical instruments. Under what circumstances should resonance be avoided in these instruments?

20 How are beats useful in tuning a musical instrument?

21 Describe three types of seismic waves and their characteristics.

22 What is the explanation given by bat experts for the varying of the frequency of the sound emitted by a bat pursuing a flying insect?

23 A great debate has raged for years concerning the advisability of building supersonic transport airplanes. Discuss some of the pros and cons of SST aircraft.

24 What is meant by the big-bang theory as described in this chapter. What evidence is there to support this theory?

25 How can energy of a sound wave be brought to a focus?

26 In Statuary Hall in the Capitol in Washington, a person standing a few feet from the wall can hear the whispering of another person who stands facing the wall at the corresponding point on the opposite side 15 m away. At points between, the sound is not heard. Explain.

27 A Kundt's tube, consisting of a metal rod clamped at its middle with one arm protruding into a horizontally mounted glass tube in which a thin uniform layer of cork dust rests, is put into compressional vibration by stroking the rod with a resined cloth. The end of the rod in the tube has a disk mounted at right angles to the glass tube. Describe how the speed of sound in the metal rod is measured by using this apparatus.

28 The intensity of the sound from a small source diminishes as the square of the distance from the source. How does absorption in the medium affect this relation?

29 Justify the statement that the speed of sound in air increases about 0.61 m/s for each centigrade degree rise in temperature from 0°C.

30 Is it easier or more difficult for an airplane to break the sound barrier at a high or at a low altitude? Explain.

Problems

Assume temperature to be 20°C unless otherwise stated.

1 What is the normal range of frequencies which the human ear can hear? What are the corresponding wavelengths for the minimum and maximum frequencies in air?

2 The sound of a gun is heard by an observer 6.0 s after the flash of the gun is seen. Calculate the distance from gun to observer. The temperature is 20°C. *Ans.* $20\bar{6}0$ m.

3 Thunder was heard 2.0 s after the lightning. If the temperature was 24°C, how far away was the lightning?

4 A string 2.0 m long has a mass of 1.2×10^{-4} kg and is stretched by a force of 100 N. What is the frequency of the fundamental vibration? *Ans.* 3.23×10^2 vib/s, or Hz.

5 A string 36 in long subjected to a tension of 30 lb vibrates 500 times per second. What will be the rate of vibration of the string if the tension is reduced to 15 lb?

6 (*a*) What is the shortest length of a column of air, closed at one end, so that a sound of 420 Hz can cause it to vibrate? (*b*) What would be the next shortest length? Assume the speed of sound to be 1,100 ft/s.

Ans. (*a*) 0.655 ft; (*b*) 1.96 ft.

7 Calculate the tension in a stretched cord that is vibrating with a frequency whose third overtone is equal to the frequency of the second overtone produced when the tension is 54 N.

8 Calculate the tension in a stretched string for which the frequency of the fundamental vibration equals the frequency of the second overtone when the tension is 4.96×10^5 dyn. *Ans.* 44.6 N.

9 (*a*) What is the shortest length of a column of air in a tube which is open at both ends that would vibrate at 420 Hz? (*b*) What would be the next shortest length? Assume the speed of sound to be 340 m/s.

10 (*a*) In order for a closed tube to emit a sound having a frequency of 256 Hz if the speed of sound is 340 m/s, what minimum length must the tube have. (*b*) If it were an open tube?

Ans. (*a*) 0.332 m; (*b*) 0.664 m.

11 Aluminum has a specific gravity of 2.6 and a Young's modulus of 7.8×10^{11} dyn/cm^2. An aluminum rod 90 cm long is clamped at its midpoint and set into longitudinal vibration by stroking it with a resined cloth. Find (*a*) the speed of the compressional wave in the rod, (*b*) the wavelength of this wave in the rod, (*c*) the frequency of the sound produced, and (*d*) the wavelength of the sound wave in air.

12 If you wish to send a message by banging out a code on steel railroad tracks to a person 2 mi away, how long will it take the message to get to that person? The weight density of steel is 490 lb/ft^3 and Young's modulus for steel is 29×10^6 lb/in^2. *Ans.* 0.621 s.

13 (*a*) Compute the frequency of the tone produced when air is blown through the holes of a rotating disk if there are 20 holes and the disk is turning at the rate of 1,800 r/min. (*b*) What is the wavelength of this sound in air when the speed of sound is 350 m/s?

14 A tuning fork vibrates 200 times per second and sends out a compressional wave that travels with a speed of 340 m/s. Find (*a*) the period and (*b*) the wavelength. *Ans.* 5.0×10^{-3} s; 1.7 m.

15 A stone is dropped from a cliff into a lake 30 m below. How much later will the impact be heard?

16 An object is dropped into a deep well which has a layer of water in its bottom. If the well is 500 ft deep, how long after the object is dropped will the splash be heard? Assume the temperature in the well to be 20°C. *Ans.* 6.04 s.

17 An airplane flies east through clouds at 400 km/h at a height of 2,440 m. At a certain instant the sound of the airplane appears to an observer on the ground to come from a point directly overhead. What is the approximate position of the airplane?

18 What is the theoretical speed of sound in oxygen at 0°C. For a diatomic gas $\gamma = 1.40$, and for oxygen $M = 32.00$ g/mol.

Ans. 3.17×10^2 m/s.

19 The density of oxygen is 16 times that of hydrogen. For both, $\gamma = 1.40$. If the speed of sound is 317.5 m/s in oxygen at 0°C, what is the speed in hydrogen?

20 A sonar device on a submarine sends out a signal and receives an echo 5 s later. Assuming the speed of sound in water to be 1,450 m/s, how far away is the object that is reflecting the signal?

Ans. 3,625 m.

21 A tuning fork of frequency 800 Hz is held over a resonance tube open at the upper end and closed at the lower end by water whose height can be varied. Resonance is observed when the distances from the open end of the tube to the water are 9.75 cm, 31.25 cm, and 52.75 cm. Calculate the speed of sound in air under these conditions.

22 In foggy weather a lighthouse sends sound signals simultaneously under water (0°C) and through the air (10°C). A vessel is 1,000 m from the lighthouse. How much later does one signal arrive than the other? *Ans.* 2.27 s.

23 A whistle closed at one end has a fundamental frequency of 160 Hz. What is the frequency of the first possible overtone?

24 What is the frequency of the tone emitted by a chime if it produces resonance in a tube 8.0 in long when the tube is closed at one end?

Ans. $4\bar{1}0$ Hz.

25 The first overtone of an open organ pipe has the same frequency as the first overtone of a closed pipe 3.6 m in length. What is the length of the open pipe?

26 When a sound wave enters a medium of different acoustical density, its speed changes but the frequency remains constant. What will be the change in wavelength when sound of frequency 1,000 per second passes from air to carbon dioxide? *Ans.* 0.241 ft.

27 A depth-measuring device emits a signal of frequency 36,000 Hz in water. The impulse is reflected from the ocean bed and returned to the device 0.60 s after the signal is emitted. (*a*) What is the depth of the water? (*b*) What is the wavelength of the wave in water? (*c*) What is the frequency of the wave in air? (*d*) What is the wavelength of the wave in air?

28 In a typical experiment performed with a Kundt's tube, described in Question 27, the following data were obtained: temperature, 20°C; length of steel rod, 125 cm; distance between nodes of cork dust, 8.00 cm. What is the frequency of the note emitted? What is the Young's modulus of the steel if its density is 7.85 g/cm^3?
Ans. 2,150 Hz; 22.5×10^{11} dyn/cm^2.

29 How many beats are heard when two tuning forks having frequencies of 300 and 305 Hz, respectively, are sounded together?

30 A tuning fork is held over a resonance tube and resonance occurs when the surface of the water is 10.00 cm below the fork. It next occurs when the water is 26.00 cm below the fork. If the velocity of sound is 345 m/s, calculate the frequency of the fork. *Ans.* 1,080 Hz.

31 An experimenter connects two rubber tubes to a box containing an electrically driven tuning fork and holds the other ends of the tubes to his ear. One tube is gradually made longer than the other, and when the difference in length is 7.0 in, the sound he perceives is a minimum. What is the frequency of the fork? (Use $v = 1{,}100$ ft/s.)

32 A pipe open at one end is closed at the other by a movable plunger. A vibrating rod is held near the open end. The air column resonates when the plunger is in any of several positions spaced 11.0 cm apart. The speed of sound in air is 333 m/s at the temperature of observation. What is the frequency of vibration of the rod?
Ans. $1{,}5\bar{1}0$ Hz.

33 A locomotive approaching a crossing at a speed of 80 mi/h sounds a whistle of frequency 440 when 1.00 mi from the crossing. There is no wind, and the speed of sound in the air is 0.200 mi/s. (*a*) What frequency is heard by an observer at the crossing? (*b*) What frequency is heard by a second observer 0.60 mi from the crossing on the straight road which crosses the railroad at right angles?

34 A source of sound moves away from an observer toward a smooth reflecting wall at a speed of 10 ft/s. If the source has a frequency of 500 Hz, what is the apparent frequency (*a*) of the sound waves coming directly to the observer and (*b*) of the waves bouncing off the wall and returning to the observer. (*c*) Would any beats be heard; if so, how many beats per second? Assume the speed of sound to be 1,100 ft/s.
Ans. 495 Hz; 505 Hz; 10 beats per second.

35 As a speedboat approaches a vertical cliff, the pilot notices that the sound of his own boat whistle reflected from the cliff changes in pitch from la ($A = 440$ Hz) to ti ($B = 495$ Hz). What is the speed of the boat?

Paul Adrien Maurice Dirac, 1902–

Born in Bristol. Professor of mathematics at Cambridge. Shared the 1933 Nobel Prize for Physics with Schrödinger for the creation of wave mechanics.

Erwin Schrödinger, 1887–1961

Born in Wien. Succeeded Planck as professor at the University of Berlin. Shared the 1933 Nobel Prize for Physics with Dirac for the creation of wave mechanics, a theory of the interaction of elementary particles.

22

Acoustics

We have been concerned with the mechanical aspects of sound production in our discussion so far, but we are also interested in the relationship of sound to hearing. The science of *acoustics* ties together the production and transmission of sound to our sense of hearing.

The hearing mechanism is able to receive compressional waves within the range to which it is sensitive and to convert the stimulus into a sensation of hearing. The ear is able to analyze the waves that come to it, distinguishing between two or more sounds that arrive simultaneously when the sounds differ in one or more of the characteristics, *pitch, quality* (timbre), and *loudness*. Each of these characteristics is closely associated with the *physical* characteristics of the sound wave that comes to the ear. Pitch is primarily associated with *frequency,* quality with the *complexity* of the wave, and loudness with the *rate* at which *energy* is transmitted to the ear. These are the principal associations between characteristics of sound and those of the waves. However, each of the three sound characteristics depends to a limited extent upon the other two physical characteristics not primarily related to it.

Not all compressional waves are able to excite the sensation of hearing. Some waves may not transmit sufficient energy to excite the sensation, although this threshold is remarkably low. Others may transmit too much energy, and the sensation becomes one of pain rather than one of sound. Further, if the frequency is too low, no sensation of sound is produced but one can feel the changes in pressure. Compressional waves whose frequency is less than that to which the ear is sensitive are called *infrasonic* waves. On the other hand, if the frequency is too high, no sensation of sound is produced. It is possible to detect these high-frequency waves by means other than the ear, and extensive studies of them have been made. As we saw earlier, these waves are called *ultrasonic* waves.

22-1 MUSICAL TONES

The sensation that we describe as a *musical tone* is produced by a *regular succession* of compressions and the following rarefactions that come to the ear. If the vibrations of the source are regularly spaced, that source will produce a musical tone; it has a fixed frequency. Other vibrating sources may not maintain a single frequency, and hence they do not produce musical tones, even though they cause a sensation of sound. If a jet of air is directed toward the outer, evenly spaced holes in the disk shown in Fig. 22-1, a regular succession of puffs of air will come through the holes as the disk is rotated. A sound wave is produced that will give rise to a musical tone. If the air jet is directed toward the inner circle, where the holes are unevenly spaced, the puffs will be irregular and an *unpitched sound* is pro-

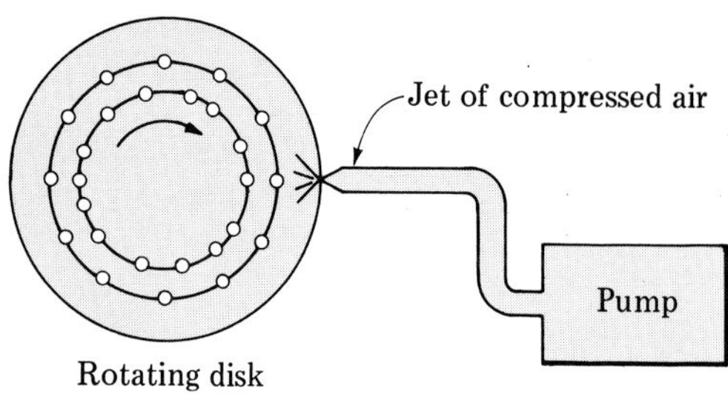

Figure 22-1
A siren disk.

duced. Other sounds cannot be classed as musical tones, because they are of such short duration that the ear is unable to distinguish a regular succession of pulses. Such sounds as that produced by a rifle shot or a sharp blow on a table top with a ruler fall in this class. There is no sharp dividing line between such sounds and musical tones. When a single small piece of wood is dropped on a table, there is little evidence of tone, but when a succession of sticks of varying lengths is dropped one after the other, the ear readily detects the variation in sound.

A *noise* may be defined as *any undesired sound.* On the basis of this definition, any sound can be a noise to some observer whether or not it is a musical tone.

22-2 PITCH AND FREQUENCY

Pitch is the characteristic of sound by which the ear assigns it a place in a musical scale. The ear assigns such a place to each musical tone but not to other sounds. When a stretched string is plucked, sound is produced that causes a given pitch sensation. If the tension of the string is increased, the pitch becomes higher. We have found that increasing the tension in a string increases its frequency of vibration. The principal physical characteristic associated with pitch is the *frequency* of the sound wave.

The range of frequency to which the human ear is sensitive varies considerably with the individual. For the average normal ear, it is from about 20 to 20,000 Hz. These limits are determined by testing a large number of seemingly normal individuals. The upper limit decreases, in general, as the age of the individual increases.

The satisfactory reproduction of speech and music does not require a range of frequencies as great as that to which the ear is sensitive. To have satisfactory fidelity of reproduction a range from 100to 8,000 Hz is required for speech and from 40 to 14,000 Hz for orchestral music.

The frequency range of most sound-reproducing systems, such as radio, telephone, and phonograph, is considerably less than that of the hearing range of the ear. A good radio transmitter and receiver in the broadcast band may cover the range from 40 to 8,000 Hz. This limited range allows it to reproduce speech faithfully, but it does detract from the quality of orchestral music. If the frequency range is further restricted, the quality of reproduction is correspondingly reduced.

Perfect fidelity of reproduction would require that the sound wave emitted by the reproducing system be exactly similar to that of the original source, both in frequencies present and in relative amplitudes of the various frequencies. High-fidelity reproduction attempts to approximate this condition.

Although pitch is associated principally with frequency, other factors also influence the sensation. Increase in intensity raises the pitch of a high-frequency tone but lowers the pitch of a low-frequency tone. The pitch of a complex tone depends upon its overtone structure; in some cases the pitch corresponds to a frequency that is not present. The difference tones discussed in Chap. 21 are examples. Tones of very short duration have a lower pitch than those of longer duration, for the same frequency.

22-3 HIGH-FIDELITY REPRODUCTION

Relatively recent developments in the recording and reproduction of sound has resulted in high-fidelity reproduction by which the full audio range is reproduced with little distortion. This became possible when electronic engineers successfully fused the art of amplification of minute fluctuating electric currents with the modern understanding of acoustics and music. As is the case with standard (AM) radios, the old 78-rpm records reproduce a frequency range of 50 to 7,500 Hz, while the hi-fi records ($33\frac{1}{3}$ rpm) reproduce a frequency range of 30 to 15,000 Hz and FM radios a range of 20 to 20,000 Hz. Hence hi-fi records and FM radios can reproduce the full range of frequencies required for orchestral music.

There are several other factors which must be considered in the production of high-fidelity music. For example, to produce true-fidelity music the room acoustics, the faithfulness of the original recording, the quality of the record player and the pickup mechanism, of the preamplifier and amplifier, and of the speaker and its enclosure must be considered. We shall find more about room acoustics at the end of this chapter, but simply point out here that the room wherein the music is recorded and that in which it is reproduced should not have excessive reverberation, although some is needed because an acoustically "dead" room is not desirable. A rule of thumb is that in auditoriums and places where sound is recorded and reproduced, a reverberation time of 1 to 2 s is desirable.

The nature of the sound recorded depends upon the marks made by the lateral movement of a stylus on a master record. Approximately 325 complex grooves per inch are cut on a single record. The more complex the groove, the more overtones produced and the greater the quality. In the production of stereo records, the groove has two sides cut in it, one for each of two microphones picking up the sound to produce a binaural effect. Another factor which determines the quality of the sound reproduced on a record is the degree to which noise is eliminated or reduced in intensity. This is done at the time of recording by deliberately *increasing* the recorded signal level for *higher* and *decreasing* the recorded signal for the *lower* frequencies. The lower frequencies carry most of the undesirable noises, such as the sound, or rumble, of the motor.

All the care that one can use in recording will be wasted if it is reproduced by inferior means. The quality of the record player and the pickup also determine the fidelity of the reproduced music. The pickup usually consists of a sapphire or diamond needle which follows the microgroove of the record as it turns with (one hopes) a uniform rate of $33\frac{1}{3}$ rpm. The needle, usually 0.001 in thick, moves within a magnetic field producing an induced electric current which is sent to the preamplifier. We shall see in our study of electricity that when a conductor (the needle) moves back and forth between the poles of a magnet, an electric current can be produced.

Since the current produced by the needle moving in the magnetic field is extremely minute, it must be amplified millions of times before it has enough energy to operate a loudspeaker. This is the function of the preamplifier and the amplifier. Further, the preamplifier preferentially amplifies the energy of the low-frequency notes which were reduced in recording and reduces emphasis on the higher frequencies so that the ratio is as nearly like the original sound as possible.

The output of an amplifier, a fluctuating electrical current of many frequencies and energies, is converted into audible music in an electromagnetic speaker, usually a thin paper cone set into forced vibration by a fluctuating magnetic field actuated by the fluctuating current from the amplifier. The vibrating cone generates sound waves (music). Actually we need more than one speaker to reproduce notes of both high and low frequencies. A minimum of two is needed, a so-called tweeter for high frequencies and a woofer for low frequencies. They may be mounted separately or coaxially.

Finally, a cone speaker in the open is not efficient because sound is produced from the front and the back of the cone leading to interference effects. An attempt is made to totally absorb the waves coming from the back of the speaker by enclosing it in a box called an *infinite baffle.*

22-4 QUALITY AND COMPLEXITY

It is a fact of experience that a tone of a given pitch sounded on the piano is easily distinguished from one of the same pitch sounded, for example,

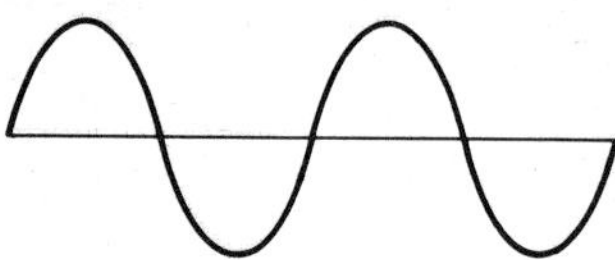

Figure 22-2
When pressure is plotted against time for a sound wave of a single frequency, a simple sine curve is obtained.

on the clarinet. The difference in the two tones is said to be one of *tone quality*. This characteristic of sound is associated with the *complexity* of the sound wave that arrives at the ear.

We have found (Chap. 21) that sound waves are produced by vibrating bodies. In a few cases the body vibrates with a single frequency, but most bodies vibrate in a very complex manner. The sound wave that is sent out from such a vibrating body is a combination of all the frequencies present in the vibration. If we plot a curve of change in pressure (i.e., difference between the pressure in the wave and the normal pressure in the air) against time for the wave sent out by a tuning fork, the curve is similar to that shown in Fig. 22-2. Since the tuning fork vibrates with only a single frequency, the curve is a simple sine curve whose frequency is the same as that of the fork. If a second fork whose frequency is three times as great as the first but with amplitude only half as great is sounded with the first, the two waves combine as shown in Fig. 22-3. The pressure at each point is the algebraic sum of the individual pressures. By adding the ordinates of *a* and *b* at each point, we get the ordinate of the complex wave *c*.

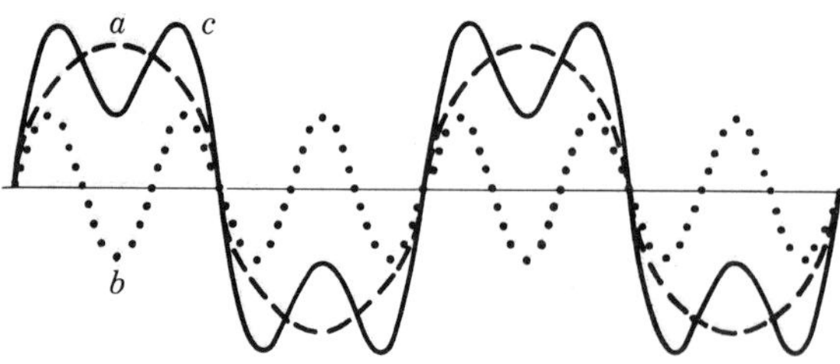

Figure 22-3
Compounding of two simple waves *a* and *b* to form a complex wave *c*.

Every body that vibrates with more than one frequency sends out a complex wave. The complexity, which determines the quality of the sound, is controlled by the number and relative intensity of the harmonics that are present. A "pure" tone (no overtones) may not be so pleasing as the "rich" tone of a violin, which contains 10 or more harmonics. Any complex wave can be resolved into a number of simple waves. The more complex the wave, the greater the number of harmonics that contribute to it.

In Fig. 22-4 are shown the wave forms and harmonic structure of sound produced by different instruments. Such wave forms can be obtained by means of a cathode-ray oscilloscope (Chap. 45). The sound wave is received by a microphone, which converts the pressure changes into electrical impulses. These in turn are amplified and cause motion of a spot on a sensitive screen to make a record of the wave form.

22-5 LOUDNESS AND INTENSITY

The *loudness* of a sound is the magnitude of the auditory sensation produced by the sound. The associated physical quantity, *intensity*, refers to the rate at which sound energy flows through unit area. Intensity may also be expressed in terms of the changes in pressure, since the rate of flow of energy is proportional to the square of the pressure change.

The loudness of sound depends upon both intensity and frequency. For a given frequency an increase in intensity produces an increase in loudness, but the sensitivity of the ear is so different in the various frequency ranges that equal intensities produce far different sensations in the different regions. In Fig. 22-5 is shown a diagram giving the relations between frequency, intensity, and hearing. Intensities below the line indicating

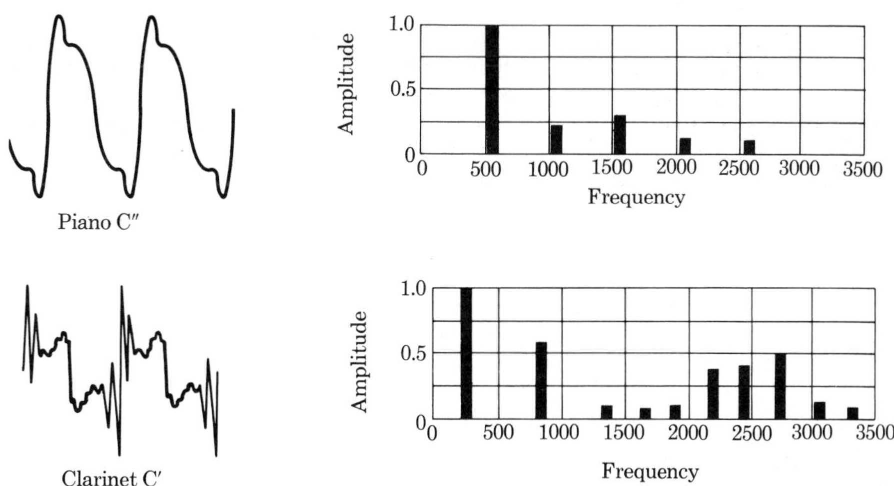

Figure 22-4
Wave forms and frequency distributions in sound waves produced by musical instruments.

the threshold of hearing are insufficient to produce any sensation of hearing. The curve indicates that the normal ear is most sensitive in the frequency range 2,000 to 4,000 Hz, that is, in this range it requires the least energy to cause a sensation of sound. The intensity necessary for hearing in the regions near the high and low limits of audibility is many times as great as that necessary in the region of greatest sensitivity.

The smooth threshold curve represents the

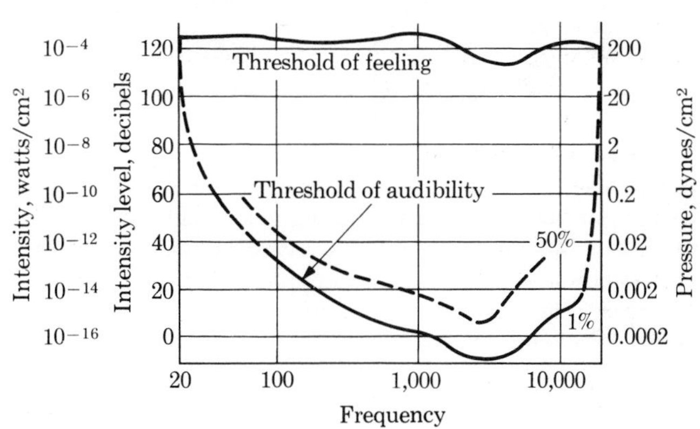

Figure 22-5
Limits of audibility. Only within the region of frequency and intensity enclosed by the curves is the sensation of sound excited.

average of many individuals. For a single person the curve would not be smooth but would show numerous dips and peaks. The threshold curve in Fig. 22-5 is one for the most sensitive ears. Of a large group tested only 1 percent were able to hear sounds of intensity less than this threshold. The broken line indicates thresholds below which 50 percent of the group were unable to hear.

If the intensity becomes too great, the sensation becomes one of feeling or pain rather than of hearing. This is indicated by the upper curve showing the threshold of feeling. Only in the region of intensity and frequency bounded by the two curves is there a sensation of hearing.

In the region of maximum sensitivity the ear is able to hear sounds over an extremely wide range of intensity. At the threshold of audibility the intensity at the ear is almost unbelievably small, about 10^{-16} W/cm^2. At the threshold of feeling the maximum intensity that the ear records as sound is about 10^{-4} W/cm^2. The maximum intensity in this range is thus *a thousand billion times the minimum*. The sensitivity of the ear to pressure changes is very great. At the threshold of audibility the pressure in the wave varies from normal only by about 0.0002 dyn/cm^2 and for the most intense sounds by about 200 dyn/cm^2. For the most intense sounds the pressure change is about a million times as great as for the least intense.

There are always those people around who like to uncover odd statistics. Someone with such a bent has estimated that a force of about 1 dyn is sufficient to lift a mosquito. Therefore, the energy expended in lifting a mosquito to a height of 1 cm is $\mathcal{W} = \mathbf{F} \times \mathbf{s} = 1$ dyn $\times$ 1 cm $= 1$ erg.

The power of 1 W is equal to approximately 10^7 erg/s. Therefore $10^{-10}\,\mu$W, the minimum energy level which can stimulate the ear, equals 10^{-9} erg/s. The ear is so sensitive that it can record the impact of energy in such small amounts that in order for it to accumulate enough energy to lift the mosquito, the energy would have to be received at this rate continually for 1 erg/(10^{-9} erg/s) $= 10^9$ s, or about 30 years.

In a sound wave the particles of air that take part in the vibration move neither far nor fast. Using Eq. (17), Chap. 19, we can calculate the amplitude of the particles at a moderate frequency, say, 2,000 Hz. For the most intense sounds (threshold of feeling) the displacement is only about 5.4×10^{-3} cm, while at the threshold of hearing it is about a millionth of this value. The corresponding maximum speed calculated from simple harmonic motion is about 68 cm/s.

The measurement of loudness is important for practical purposes but is difficult to achieve. There is an approximate law of psychology which states that the magnitude of a sensation is proportional to the logarithm of the intensity,

$$S = k \log \frac{I}{I_0} \tag{1}$$

where S represents the magnitude of the sensation (for sound loudness), I the intensity, and I_0 the intensity at a reference level. This law is called the *Weber-Fechner law*. It does not exactly represent the relationship between loudness and intensity for sound but is a fair approximation for pure tones at most frequencies. For complex tones, however, there is no simple relationship.

In place of the sensation S, we may use *intensity level* α, defined by the equation

Table 1
INTENSITY LEVELS OF CERTAIN SOUNDS

Source	Decibels
Average threshold of hearing	0
Quiet home, average living room	40
Ordinary conversation (3 ft)	70
Street-corner traffic, large city	80
Boiler factory	100
Threshold of pain	120
DC4 takeoff (150 ft)	120
F3D takeoff (J34 engine)	140
J-57-type engine	160–170

$$\alpha = \log \frac{I}{I_0} \qquad (2)$$

There is a gain of one unit in intensity level when the actual *power* of the second sound is 10 times as great as the first. Hence it is now customary to state the differences in the intensity levels of two sounds as the *exponent* of 10, which gives the ratio of the powers. The unit exponent is called the *bel,* for Alexander Graham Bell (1847–1922), whose researches in sound are famous. If one sound has 10 times as much power as a second sound, the difference in their intensity levels is 1 bel. The bel is an inconveniently large unit, and hence the *decibel* (0.1 bel) is the unit that is generally used in practice. A 26 percent change in intensity alters the level by 1 dB (decibel). This is practically the smallest change in intensity level that the ear can ordinarily detect. Under the best laboratory conditions a 10 percent (0.4-dB) change is detectable.

Intensity level is measured from an arbitrarily chosen intensity. An intensity of $10^{-10}\ \mu W/cm^2$ ($10^{-6}\ \mu W/m^2$) is chosen as zero level, or 0 dB. It is roughly the intensity of the threshold of audibility for tones between 500 and 2,500 per second. The intensity levels of various sounds are given in Table 1.

Example A small source of sound radiates acoustic energy uniformly in all directions at a rate of 1.5 W. Find the intensity and the intensity level at a point 25 m from the source if (*a*) there is no absorption and (*b*) if there is 10 percent absorption in the 25-m path.

(*a*) $$I = \frac{P}{A} = \frac{1.5\ \text{W}}{4\pi(25\ \text{m})^2} = 1.9 \times 10^{-4}\ \text{W/m}^2$$

$$\alpha = \log \frac{I}{I_0} = \log \frac{1.9 \times 10^{-4}\ \text{W/m}^2}{10^{-12}\ \text{W/m}^2}$$

$$\alpha = \log (1.9 \times 10^8)$$

$$= \log 190{,}000{,}000 = 8 + \log 1.9$$

$$= 8 + 0.3 = 8.3\ \text{bels}$$

$$= 83\ \text{dB}$$

(*b*) $$I = 0.90 \times 1.9 \times 10^{-4}\ \text{W/m}^2$$

$$= 1.7 \times 10^{-4}\ \text{W/m}^2$$

$$\alpha = \log \frac{1.7 \times 10^{-4}}{10^{-12}} = \log (1.7 \times 10^8)$$

$$= \log 170{,}000{,}000 = 8 + \log 1.7$$

$$= 8 + 0.2 = 8.2\ \text{bels}$$

$$= 82\ \text{dB}$$

22-6 THE EAR

The ear is essentially a device which transmits and magnifies the pressure changes which come to it in sound waves. It consists of three sections: the outer ear, the middle ear, and the inner ear (Fig. 22-6). The outer section consists of the external ear, or pinna, and the canal. The canal is separated from the middle ear by a membrane called the *eardrum T.* In the middle ear are three small bones, the *hammer,* the *anvil,* and the *stirrup,* which transmit the pressure changes to the inner ear. The hammer is attached to the eardrum, the stirrup to a membrane *O* that separates

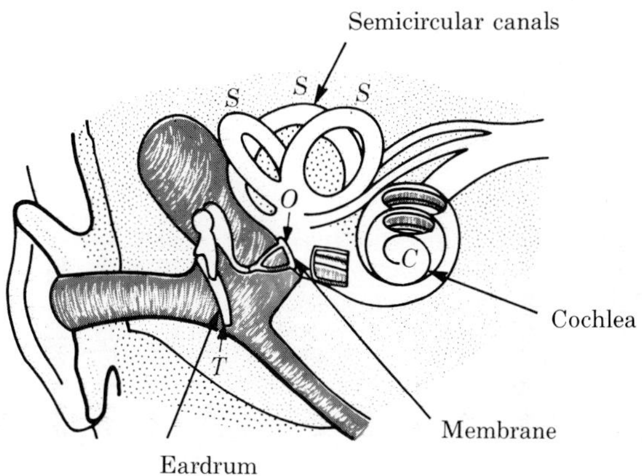

Figure 22-6
Diagram of the ear. The parts of the ear are not drawn to scale.

the middle and inner ear. The three bones make up a system of levers that is so arranged as to increase the force. In addition, the area of the stirrup bearing on the membrane at the inner ear is much smaller than the eardrum. By these means the pressure changes are increased 30 to 60 times. The inner ear is filled with liquid. The semicircular canals S are organs of balance and take no part in hearing. The cochlea C is really the end organ of hearing, where the nerve enters. Here the pressure changes in vibrations excite the sensation of sound.

It is possible for sound vibrations to bypass the outer and middle ear and be conducted through the bone directly to the inner ear. One type of hearing aid is designed to make use of such bone transmission. Others are amplifiers that increase the energy that reaches the eardrum.

22-7 VOICE SOUNDS

Voice sounds are formed by passage of air through the vocal cords, lips, and teeth. As the air stream passes through the vocal cords, they are set in vibration. The cavities of the nose and throat impress resonant characteristics on these vibrations to produce speech sounds. All the vowel sounds and some of the consonant sounds are produced in this manner. Other sounds, called *unvoiced* sounds, for example, f, s, th, sh, t, and k, are produced by passage of air over the teeth and tongue without use of the vocal cords. *Voiced* consonants, such as b, d, g, j, v, and z, are combinations of the two processes.

The various vowel sounds are made by changing the shape of the resonant chamber so that different frequencies are enhanced. Each of the vowel sounds has certain characteristic frequency groups as shown in the frequency chart in Fig. 22-7 and in Table 2. The values given are average values, and there is considerable variation for different individuals and for a single individual at different times. If one of these speech sounds is passed through a sound filter that absorbs frequencies in the neighborhood of one of the characteristic frequencies, the vowel sound is no longer recognizable.

22-8 MUSICAL SCALES

The simplest music consists of a succession of musical tones of the same or different pitches. This constitutes a melody. Any succession of pitches can be chosen, but it is found that the effect is most pleasing if the ratio of the frequencies of succeeding tones is a ratio of small integers. By use of this fact we construct a *musical scale,* and in the melody only frequencies that appear in the scale are used.

Two tones sounded in succession or together constitute an *interval.* The ear recognizes intervals in the musical scale by ratios of frequencies rather

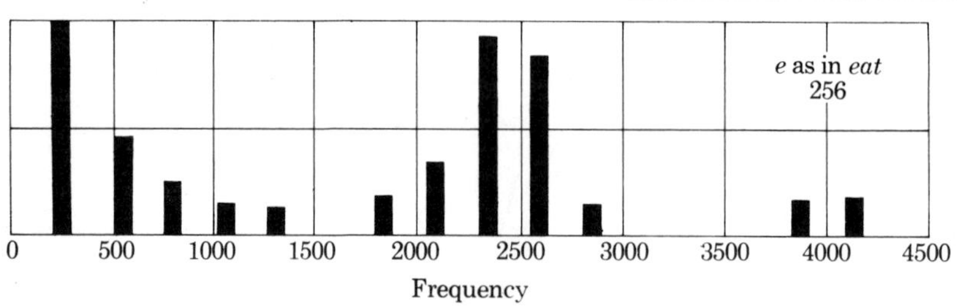

Figure 22-7
Frequency chart for a vowel sound.

Table 2
CHARACTERISTIC FREQUENCIES OF VOWEL SOUNDS

Vowel sound	Low frequency	High frequency
a (tape)	550	2,100
a (father)	825	1,200
e (eat)	375	2,400
e (ten)	550	1,900
i (tip)	450	2,200
o (tone)	500	850
u (pool)	400	800

than by differences in frequency. The interval between frequencies 600 and 1,200 per second is the same as the interval between 150 and 300 per second, since the ratio in each case is $\frac{2}{1}$. This is the simplest ratio other than $\frac{1}{1}$. Other simple ratios used in our common musical scale are $\frac{3}{2}$, $\frac{4}{3}$, $\frac{5}{3}$, $\frac{5}{4}$, $\frac{9}{8}$, and $\frac{15}{8}$.

22-9 A DIATONIC SCALE

A diatonic scale is made up of eight notes which have frequencies such that the intervals between the first note and the others are those just given. The frequency of the first note may be chosen arbitrarily. The others are then fixed. Suppose that middle C on the piano is fixed at 256 Hz. Then, the frequencies and intervals are those given in Table 3.

We note that there are three intervals between successive notes, $\frac{9}{8}$ the major tone, $\frac{10}{9}$ the minor tone, and $\frac{16}{15}$ the semitone. Intervals between the basic frequency and the other frequencies are also given names. Since the ratio $\frac{2}{1}$ occurs at the eighth note, this interval is called the *octave*. Similarly, the ratio $\frac{5}{4}$ is called the major third, $\frac{4}{3}$ the major fourth, $\frac{3}{2}$ the major fifth, and $\frac{5}{3}$ the major sixth.

The ear recognizes as harmonious three notes when the frequencies are proportional to the three numbers, 4, 5, and 6. This combination is called a *major triad*. The major diatonic scale is made up of three major triads as indicated in Table 3.

Example Construct a diatonic musical scale using scientific pitch.

In a diatonic scale three major triads exist:

C E G
F A C′
G B D′

Since the frequency of C in scientific pitch is

Table 3
FREQUENCIES AND INTERVALS IN THE DIATONIC SCALE

	Note							
	C	D	E	F	G	A	B	C′
Frequency, scientific pitch	256	288	320	341	384	427	480	512
Frequency, concert pitch	264	297	330	352	396	440	495	528
Ratio to C	1	$\frac{9}{8}$	$\frac{5}{4}$	$\frac{4}{3}$	$\frac{3}{2}$	$\frac{5}{3}$	$\frac{15}{8}$	2
Ratio to preceding frequency		$\frac{9}{8}$	$\frac{10}{9}$	$\frac{16}{15}$	$\frac{9}{8}$	$\frac{10}{9}$	$\frac{9}{8}$	$\frac{16}{15}$
Triads	4		5		6			
				4		5		6
		(6)			4		5	

256 Hz, and the ratio of C, E, and G is 4, 5, and 6, we can find that 256/4 Hz = 64 Hz, which is a base frequency for that triad. Therefore, E has a frequency of 5 × 64 = 320 Hz and G has a frequency of 6 × 64 = 384 Hz.

Similarly, since we know that C′, an octave higher than C, has a frequency twice C, C′ = 512 Hz, then in the triad F, A, C′, the ratio 4, 5, and 6 shows that 512/6 = 85.3 Hz is the base frequency for that triad. Therefore, F has a frequency of 4 × 85.3 = 341.2 Hz and A has a frequency of 5 × 85.3 = 426.5 Hz.

Finally, in the triad G, B, D′, we know the value of G, 384 Hz. Therefore, 384/4 = 96 Hz is the base frequency and B = 5 × 96 = 480 Hz, and D′ = 6 × 96 = 576 Hz. Since the frequency of D, an octave lower, is one-half the frequency of D′, D = 576/2 Hz and D = 288 Hz.

These computed values compare very well with those appearing in Table 3.

22-10 CONSONANCE AND DISSONANCE

When two notes are sounded together or consecutively, they seem pleasing if the interval is a simple ratio. Such a combination is a *consonance*. All other combinations are *dissonances*. There is no sure test as to whether the ratio of frequencies is "simple" since the decision depends upon the past training of the observer. What seems dissonance to one observer may seem consonance to another.

Dissonances appear when beat frequencies between about 10 and 50 are formed as the two notes are sounded together. The beat frequency may be formed by the fundamental frequencies or by *any pair* of overtones present in the complex tones. Dissonances thus will not be present in pure tones except at low frequencies. Consonance represents the absence of dissonance.

22-11 EQUAL TEMPERED SCALE

The intervals of the diatonic scale are larger than is desirable in much music. However, if we try to split the larger intervals by semitones, it is not possible to find a single frequency that will divide the interval satisfactorily. For example, the semitone above G would not have the same frequency as the semitone below A. In other words, on this scale G♯ would differ from A♭. To overcome this difficulty, a scale of 12 equal intervals or semitones is set up. If x represents the semitone interval, then $(x)^{12} = 2$ or $x = 2^{\frac{1}{12}} = 1.059$, approximately. This scale of even intervals is called the *equal tempered* scale and is that actually used in music at the present time. Bach wrote his compositions for the "well-tempered clavichord" to demonstrate the usefulness of the evenly tempered scale in changing from one key to another. Only a very acute ear will recognize the difference between the evenly tempered and diatonic scales. In Table 4 is a comparison of the intervals.

Table 4
FREQUENCY INTERVALS IN MUSICAL SCALES

	Note												
Interval	C	D♭ C♯	D	E♭ D♯	E	F	G♭ F♯	G	A♭ G♯	A	B♭ A♯	B	C′
Diatonic interval	1.000		1.125	1.200	1.250	1.333		1.500	1.600	1.667		1.895	2.000
Tempered interval	1.000	1.059	1.122	1.189	1.260	1.335	1.414	1.498	1.587	1.682	1.782	1.888	2.000

22-12 SOUND PRODUCTION

Any vibrating body whose frequency is within the audible range will produce sound provided that it can transfer to the medium enough energy to reach the threshold of audibility. Even though this limit is reached, it is frequently necessary to amplify the sound so that it will be readily audible where the listener is stationed. For this purpose sounding boards and loudspeakers may be used, the purpose of each being to increase the intensity of the sound.

When a sounding board is used, the vibrations are transmitted directly to it and force it to vibrate. The combined vibrations are able to impart greater energy to the air than the original vibration alone. If the sounding board is to reproduce the vibrations faithfully, there must be no resonant frequencies, for such resonance will change the quality of sound produced.

The loudspeaker is used to increase the intensity of sound sent out, either by electrical amplification or by resonance. Two general types are used: the direct radiator, such as the cone loudspeaker commonly used in radios, and the horn type. The direct radiator is used more commonly because of its simplicity and the small space required. The horn speaker consists of an electrically or mechanically driven diaphragm coupled to a horn. The air column of the horn produces resonance for a very wide range of frequencies and thus increases the intensity of the sound emitted. The horn loudspeaker is particularly suitable for large-scale public-address systems.

22-13 SOUND DETECTORS

The normal human ear is a remarkably reliable and sensitive detector of sound, but for many purposes mechanical or electrical detectors are of great use. The most common of such detectors is the microphone, in which the pressure variations of the sound wave force a diaphragm to vibrate. This vibration, in turn, is converted into a varying electric current by means of a change of resistance or generation of an emf, which is then commonly transmitted to a loudspeaker. For true reproduction the response of the microphone should be uniform over the whole frequency range. Such an ideal condition is never realized, but a well-designed instrument will approximate this response. Microphones are used when it is necessary to reproduce, record, or amplify sound.

Parabolic reflectors may be used as sound-gathering devices when the intensity of sound is too small to affect the ear or other detectors or where a highly directional effect is desired. The sound is concentrated at the focus of the reflector, and a microphone is placed there as a detector. Such reflectors should be large compared with the wavelength of the sound received, and hence they are not useful for low frequencies.

22-14 LOCATION OF SOUND

Although a single ear can give some information concerning the direction of a source of sound, the use of two ears is necessary if great accuracy is desired. The judgment of direction is due to a difference between the impression received at the two ears, these differences being due to the differences in loudness or in time of arrival. This is sometimes called the *binaural* effect. Certain types of locators exaggerate this effect by placing two listening trumpets several feet apart and connecting one to each ear. The device is then turned until it is perpendicular to the direction of the sound. In this way the accuracy of location is increased.

22-15 REVERBERATION; ACOUSTICS OF AUDITORIUMS

A sound, once started in a room, will persist by repeated reflection from the walls until its intensity is reduced to the point where it is no longer audible. If the walls are good reflectors of sound waves—for example, hard plaster or marble—the sound may continue to be audible for an appreciable time after the original sound stops. The repeated reflection that results in this persistence of sound is called *reverberation.*

In an auditorium or classroom, excessive reverberation may be highly undesirable, for a given speech sound or musical tone will continue to be heard by reverberation while the next sound is being sent forth. The practical remedy is to cover part of the walls with some sound-absorbent material, usually a porous substance like felt, compressed fiberboard, rough plaster, or draperies. The regular motions of the air molecules, which constitute the sound waves, are converted into irregular motions (heat) in the pores of such materials, and consequently less sound energy is reflected.

Suppose that a sound whose intensity is one million times that of the faintest audible sound is produced in a given room. The time it takes this sound to die away to inaudibility is called the *reverberation time* of the room. Some reverberation is desirable, especially in concert halls; otherwise the room sounds too dead. As noted earlier, for a moderate-sized auditorium the reverberation time should be of the order of 1 to 2 s. For a workroom or factory it should, of course, be kept to much smaller values, as sound deadening in such cases results in greater efficiency on the part of the workers, with much less attendant nervous strain.

The approximate reverberation time of a room is given by the expression,

$$T = \frac{(0.049\ \text{s/ft})V}{\Sigma kA} \tag{3}$$

where T is the time in seconds, V is the volume of the room in cubic feet, and ΣkA is the *total absorption* of all the materials in it. The total absorption is computed by multiplying the area A, in square feet, of each kind of material in the room by its *absorption coefficient* k (see Table 5) and adding these products together.

The absorption coefficient is merely the fraction of the sound energy that a given material will absorb at each reflection. For example, an open window has a coefficient of 1, since all the sound that strikes it from within the room would be lost to the room. Marble, on the other hand, is found to have a value of 0.01, which means that it absorbs only 1 percent of the sound energy at each reflection. Equation (3) usually gives satisfactory results except for very large or very small halls, for rooms with very large absorption, or for rooms of peculiar shape.

Table 5
ABSORPTION COEFFICIENTS FOR SOUNDS OF MEDIUM PITCH

Material	Coefficient
Open window	1.00
Ordinary plaster	0.034
Acoustic plaster	0.20–0.30
Carpets	0.15–0.20
Painted wood	0.03
Hair felt, 1 in thick	0.58
Draperies	0.40–0.75
Marble	0.01

By means of Eq. (3) we can compute the amounts of absorbing materials needed to reduce the reverberation time of a given room to a desirable value. The absorbing surfaces may be placed almost anywhere in the room, since the waves are bound to strike them many times in any case. In an auditorium, however, they should not be located too close to the performers.

In addition to providing the optimum amount of reverberation, the designer of an auditorium should make certain that there are no undesirable effects due to regular reflection or focusing of the sound waves. Curved surfaces of large extent should in general be avoided, but large, flat reflecting surfaces behind and at the sides of the performers may serve to send the sound out to the audience more effectively. Dead spots, due to interference of direct and reflected sounds, should be eliminated by proper design of the room.

The acoustic features of the design of an audi-

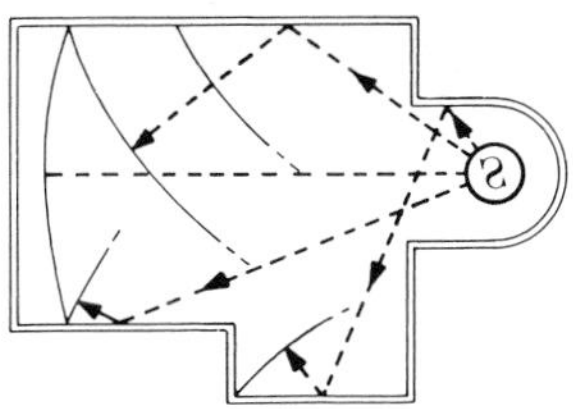

Figure 22-8
Ripple-tank model for an auditorium, showing reflection from the walls.

torium may be investigated before the structure is built by experimenting with a sectional model of the enclosure in a ripple tank (Fig. 22-8). In this way the manner in which waves originating at the stage are reflected can be observed and defects in the design remedied before actual construction is undertaken.

Example A ballroom is 150 ft long, 50 ft wide, and 30 ft high. The walls are to be totally covered with drapes having an absorption coefficient of 0.50. If the floor is made of wood having an absorption coefficient (k) of 0.03 and the ceiling is made of acoustic plaster having an absorption coefficient of 0.20, what is the reverberation time for the ballroom when empty?

The area of the ceiling is 150×50 ft $=$ 7,500 ft^2. Therefore, its $kA = 0.2 \times 7{,}500 =$ 1,500 ft^2. The floor has the same area and its $kA = 0.03 \times 7{,}500$ ft $= 225$ ft^2. The walls have a total area of $(2 \times 50 \times 30)$ ft^2 $+$ $(2 \times 150 \times 30)$ ft^2 or 3,000 ft^2 $+$ 9,000 ft^2 $=$ 12,000 ft^2 and its kA $=$ 6,000 ft^2. The volume of the ballroom is $150 \times 50 \times 30$ ft $=$ 225,000 ft^3. Therefore, from Eq. 3,

$$T = \frac{0.049 \text{ s/ft} \times 225{,}000 \text{ ft}^3}{1500 \text{ ft}^2 + 225 \text{ ft}^2 + 6{,}000 \text{ ft}^2}$$

$$= \frac{4.9 \times 10^{-2} \text{ s/ft} \times 2.25 \times 10^5 \text{ ft}^3}{7.725 \times 10^3}$$

$$= 1.4\bar{3} \text{ s}$$

Therefore, the room will be too dead for good music production, especially when it becomes filled with people. A possible solution would be to have less drapery in the room or to select drapery with a lower absorption coefficient.

SUMMARY

The science of *acoustics* ties together the production and transmission of sound to our sense of hearing.

Sounds differ in *pitch, quality,* and *loudness.*

A *musical tone* is produced by a regular succession of compressions and the following rarefactions.

An *unpitched sound* is produced by an irregular succession of compressions and rarefactions or by a disturbance of such short duration that the ear is unable to distinguish a regular succession.

A *noise* is any undesired sound.

The *pitch* of a sound is associated with the physical characteristic of *frequency* of vibration. The average human ear is sensitive to frequencies over a range of 20 to 20,000 Hz.

High-fidelity reproduction of sound requires the reproduction of the full audio range of frequencies with little distortion. To produce high-fidelity music, the room acoustics, the faithfulness of the original recording, the quality of the record player and pickup mechanism, the preamplifier and amplifier, and the speaker and its enclosure must all be taken into consideration.

The *quality* of a sound depends upon the *complexity* of the wave, i.e., upon the number and relative prominence of the overtones.

The *loudness* of sound is the magnitude of the auditory sensation.

The *intensity* of sound is the energy per unit area per unit time.

Intensity level is the logarithm of the ratio of the intensity of a sound to an arbitrarily chosen intensity. The *bel* and *decibel* are units of intensity level. One bel is the change in intensity level which represents a tenfold ratio of *power,*

$$\alpha = \log \frac{I}{I_0}$$

The human ear is extremely sensitive to sounds and can record them as small as 10^{-16} W/cm^2 and as loud as 10^{-4} W/cm^2, a range where the maximum sound recorded is a thousand billion times the minimum level of recorded sound.

A *musical scale* is a succession of tones which bear a simple relation to each other.

Reverberation is the persistence of sound in an enclosed space, due to repeated reflection of waves. It may be reduced by distributing sound-absorbing material about the enclosure.

Questions

1 As a wave passes from one medium to another, which property of the wave usually stays constant? Explain.

2 Is the following statement correct? Higher-pitched sounds travel more rapidly through air than do lower-pitched sounds. Discuss your answer.

3 As a record on a turntable is slowed down, what happens to the pitch of the recorded sounds? Why?

4 Draw a simple wave and its first harmonic overtone along the same axis, making the amplitude of the latter half as great as that of the fundamental. Combine the two graphically by adding the ordinates of the two curves at a number of different points, remembering that the ordinates must be added algebraically. If the resulting curve is taken to represent a complex sound wave, what feature of the curve reveals the quality of the sound?

5 Draw a simple wave and its first harmonic overtone, making the latter 45° out of phase with the fundamental and the amplitude half that of the fundamental. Combine the two graphically by adding the ordinates at a number of different points. Compare the curve obtained with that of Question 4. Is the wave form the same? Is the complexity the same? Will the two combinations ordinarily be distinguishable?

6 What difference is there between the sound of middle C as it is played by a flute and a violin?

7 Distinguish between ultrasonics, supersonics, and infrasonics, and give an example of each.

8 Not all compressional waves can be called sound waves. In what respects can they differ from sound waves?

9 If two pure tones have the same frequency and amplitude, can the ear distinguish one from the other?

10 Define the terms pitch, intensity, and quality as they apply to musical tones.

11 Does the evenly tempered musical scale have any advantages as compared with the diatonic scale? Explain.

12 How many equal frequency intervals are there between the notes C and G on the evenly tempered scale?

13 What is meant by an octave? About how many octaves are there in the frequency range of human hearing?

14 By what means can the overtones of a musical note be isolated and identified?

15 Three things are necessary for the reproduction of sound, a vibrating object with a frequency in the range of human hearing, an elastic medium, and a receiver. What is the vibrating object in the following instruments: saxophone, piano, trumpet, flute, and drum?

16 Musical instruments in an orchestra are classified by the nature of the vibrating object that creates their sound. What are these general classifications and give examples of each.

17 Why is a megaphone an effective device for projecting your voice? Explain.

18 Carefully identify the differences between music and noise. What two characteristics must a sound have to be classified as a musical tone rather than as a noise?

19 Define a tweeter and a woofer and explain their function in a sound-reproduction system.

20 A noise meter is calibrated in decibels. What does the meter measure?

21 Describe two types of hearing devices used to assist people with different types of hearing problems and the principles upon which they work.

22 Describe carefully the manner in which voice sounds are produced by a human.

23 Two tuning forks of 350 and 370 Hz, respectively, are sounded at the same time. Does this produce consonance, dissonance, or neither? Discuss.

24 Why is the absorption of sound waves an important consideration in architecture? What is the physical meaning of absorption? How is it accomplished? What becomes of the energy of the sound wave absorbed?

25 In an auditorium with wooden seats how does the presence of a large audience affect (*a*) the reverberation and (*b*) the average loudness of the sound?

26 Which of the following actions would be more effective in decreasing the reverberation time in an auditorium in which the audience occupies wooden chairs (*a*) requiring the audience to rise from their seats, (*b*) opening the windows, or (*c*) opening doors to adjoining rooms?

27 One hall is considered ideal for music, another for speaking. Which hall is likely to have the longer period of reverberation?

Problems

1 How many beats would be produced if two tuning forks, one being set for scientific pitch to produce the note G and the other set for concert pitch to play the note G in the same octave, were struck simultaneously?

2 What note is sounded by a siren having a disk with 16 holes and making 20 r/s?

Ans. E, 320 Hz.

3 What is the wavelength in air of sound to which the ear is most sensitive?

4 In a certain concrete grandstand the distance from one riser back to the next is 30 in. If a person claps his hands in front of this grandstand, what is the frequency of the note that comes back to him? *Ans.* 220 per second.

5 An observer hears a whistle whose frequency is 750 Hz. What will be the apparent frequency when he moves toward the whistle at a speed of 60 mi/h? What will be the apparent frequency after he passes the whistle? Assume speed of sound to be 1,100 ft/s.

6 Calculate the amplitude in air under standard conditions of a sound wave whose frequency is 400 per second at (*a*) the threshold of audibility and (*b*) the threshold of feeling.

Ans. 8.6×10^{-9} cm; 2.7×10^{-3} cm.

7 Calculate the amplitude in air under standard conditions of a sound wave whose frequency is 1,200 per second at the intensity level of ordinary speech.

8 If one sound is 5.0 dB higher than another, what is ratio of their intensities? *Ans.* 3.16.

9 Two sounds of the same frequency have intensities of 10^{-16} and 10^{-12} W/cm^2. What is the difference between the intensity levels of these sounds?

10 Two sound waves have intensities of 0.5 and 10 W/m^2, respectively. How many decibels is one louder than the other? How many bels?

Ans. 13 dB; 1.3 bels.

11 At a certain point the power received from one loudspeaker is 100 times as great as that from a second. What is the difference in intensity level between the two sounds at that point?

12 Two sounds have intensities of 100 and 400 μW/cm^2, respectively. How much louder is one than the other? *Ans.* 6.0 dB.

13 For ordinary conversation the intensity level is given as 60 dB. What is the intensity of the wave?

14 A source of sound radiates energy uniformly in a spherical pattern at a rate of 4 W. Calculate the intensity and the intensity level at a point 100 m from the source if (*a*) there is no absorption and (*b*) if 20 percent of the energy is absorbed by the time the wave reaches 100 m.

Ans. 3.19×10^{-5} W/m^2, 75 dB; 2.56×10^{-5} W/m^2, 74 dB.

15 A small source ($f = 200$ per second) radiates uniformly in all directions at a rate of 0.0050 W. If there is no absorption, how far from the source is the sound audible?

16 A pipe organ, pipes closed at one end, will be designed to operate at 25°C. What are the shortest lengths of pipes that can be used to create a major triad in the middle octave of the diatonic scale? Use musical or concert C.

Ans. C = 1.08 ft; E = 0.86 ft; G = 0.72 ft.

17 The first note of a diatonic scale has a frequency of 480 per second. Compute the frequencies of the other seven notes.

18 Compute the frequencies of the major diatonic scale based on E as 320 per second. What notes of this scale are common to the scale based on C as 256? *Ans.* 320, 360, 400, 427, 480, 533, 600, 640 Hz.

19 What is the frequency of middle E on the evenly tempered scale? Use scientific pitch.

20 Construct a diatonic musical scale using concert pitch. *Ans.* 264, 297, 330, 352, 396, 440, 495, and 528 Hz.

21 If the note A is assigned a frequency of 400 Hz on an arbitrarily defined diatonic musical scale, what frequency would the note E have on this scale?

22 A musical recorder, a wind instrument, emits a sound having a frequency of middle C, scientific pitch, when air is blown through it. If the effective length of the resonating air column is 2.2 ft, what is the temperature of the air in the instrument? *Ans.* 20°C.

23 What is the reverberation time of a hall whose volume is $100,\bar{0}00$ ft^3 and whose total absorption is 2,000 ft^2? How many square feet of acoustic wallboard of absorption coefficient 0.60 should be used to cover part of the present walls (ordinary plaster) in order to reduce the reverberation time to 2.0 s?

24 A room having a volume of 400 ft^3 has a total wall area of 175 ft^2 and a floor and ceiling area of 100 ft^2, respectively. If the coefficient of absorption for the wall is 0.030 per ft^2, the ceiling is 0.020 and the floor is 0.050, find the reverberation period of the room.

Ans. 1.63 s.

25 An auditorium is rectangular in shape, 115×75 ft, and 30 ft high. It has plaster walls and ceiling, a wood floor, and 750 seats, each with an equivalent complete absorption area of 0.10 ft^2. (*a*) Find the reverberation time of the empty auditorium. (*b*) What is the reverberation time when the auditorium is filled if each auditor has an equivalent area of 4.0 ft^2?

26 An auditorium has plaster walls and ceiling, a wooden floor, and 500 seats: (*a*) What is the reverberation time when empty? (*b*) What is the reverberation time when the seats are filled? The auditorium is 100 ft long, 50 ft wide, and 30 ft high. K for plaster is 0.034, wood is 0.03, seats are 0.10 per seat, and people are 4.0 per person.

Ans. 11.0 s; 2.75 s.

27 A classroom $10 \times 8 \times 4$ m has a cork-tile floor (absorption coefficient 0.030) and plastered walls. (*a*) What is the reverberation time for this room? (*b*) How many square meters of draperies (absorption coefficient 0.60) must be placed on the walls of the room to make the reverberation time 2.0 s? Note: 0.049 s/ft = 0.164 s/m.

Sir James Chadwick, 1891–
Born in Manchester. Assistant Director of research at the Cavendish Laboratory, Cambridge. In 1935 appointed professor of physics in Liverpool University. In 1935 Chadwick was presented the Nobel Prize for Physics for his discovery of the neutron.

23

Light and Illumination

From his earliest days, man has been aware of that type of radiation appearing in the form of light. However, he has always had considerable uncertainty about its nature. Historical evidence indicates that the study of the behavior of light was a part of the earliest science. Man's interest in this type of radiation has not diminished through the centuries. For example, the recent development of lasers demonstrates that optical research is still a vital part of present-day science as well.

It will be shown in later chapters that many of the techniques developed in the study of light have found application in connection with other forms of radiation which, like light, are electromagnetic in nature (Chap. 43): radio, microwave, infrared, ultraviolet, and x-ray. In fact the methods of optics have been applied to some extent in dealing with streams of particles in electron microscopes, cathode-ray tubes, and the like, so as to give rise to the term *electron optics*. Most of our ideas on the atomic and molecular character of matter derives from the study of spectra.

23-1 NATURE OF LIGHT

The history of the attempts to understand the nature of light provides us with one of the most interesting examples of the gradual development of a scientific theory. Since the earliest times, men have been familiar with some specific properties of light. For example, it was known by Plato in the fourth century B.C. that light would rebound from certain substances, that is, it possessed the property of *reflection*. *Refraction,* the property of light to be bent in passing from one medium to another, was studied by the Greeks in the second century A.D. Further, the *rectilinear* nature of light, that is, that light normally travels in straight lines, was known to the early Greeks who depended, as we do, upon this property for sighting.

These properties, reflection, refraction, and rectilinear propagation, were known before Isaac Newton (1642–1727) applied his genius to the problem in the seventeenth century. By that time the ancient theory that vision resulted from something (visual rays) sent out from the eye had been discarded. In Newton's time there were seen to be two possible explanations for the three known properties of light. The first, held by Newton and Laplace (1749–1827), was that light energy was propagated by particles of matter (corpuscles) emanating from luminous objects, such as a flame, or reflecting from nonluminous objects. The second explanation given in 1664 by the English physicist Robert Hooke (1635–1703) and then published in 1690 by Christian Huygens (1629–1695), a Dutch scientist for whom the theory was named, was that light was a special form of wave disturbance.

The particle (corpuscular) theory of light, also known as the "theory of emission," was developed by Newton along with a mathematical explanation of refraction and reflection which was in agreement with the then-known facts. Many practical examples can be given to show that Newton's theory was mechanically correct concerning these properties. A simple mechanical analog to reflection is the rebounding of balls from a resilient surface which behave as light does when it rebounds from a smooth reflecting surface. An analog to refraction that is commonly given is a ball rolling across two flat surfaces at unequal elevations joined by an incline. The ball will accelerate on the incline due to gravity and will therefore have a greater velocity across the lower surface. If the ball crosses the upper surface at an angle, the angle will change on the incline, considered to be an interface between two media, and the ball will cross the lower surface at a different angle, as light does going from one medium to another where the velocity of the light changes. Further, rectilinear propagation can be illustrated by considering the fact that although a ball thrown horizontally from a tower curves downward due to gravity, the faster the ball goes, the less curvature is noted in its path. Since the particles proposed by Newton were traveling very fast, they would appear to an observer to travel in a straight line.

According to Newton's particle theory, light would have to travel faster in water and in other denser substances than it does in air, in that an "optically denser medium possesses greater attraction." He believed that water attracted these particles much as gravity does.

At about the same time (late seventeenth century), Huygens proposed his wave theory of light, sometimes called the "undulatory theory," which was an equally good explanation of refraction and reflection and accounted somewhat more plausibly than did the particle theory for the fixed speed of light in space. The geometric constructions that Huygens used to support his theory are explained in detail in Chap. 24, so we will defer presenting them until that point in our study. It should be noted carefully, however, that the wave theory required that the speed of light in a *dense* medium be *less* than in air, exactly opposite to Newton's prediction. (It was not until 1850 that Jean Foucault was to provide convincing support for Huygens' wave theory by discovering that the speed of light *was* less in water than in air.)

Huygens' theory was not a complete success, because of the difficulty it had in explaining rectilinear propagation. In fact, Newton and other supporters of the particle theory could not see how waves could travel in straight lines and in their arguments against the wave theory pointed to the bending of sound, which was known to be a wave form, around corners.

As, at that time, the phenomena of interference and diffraction had not been discovered for light, the two theories seemed at once equally good and mutually contradictory.

More than a hundred years elapsed before the discovery of three other properties of light was to apparently solidly establish the wave nature of light. These properties were *interference,* where light waves could pass through each other causing constructive and destructive interference but then continue on as though the interference had never occurred, which would be an impossibility for particles colliding with particles to duplicate; *diffraction,* the slight bending of light around corners, which particles would not do; and *polarization,* the process by which the vibrations of light are confined to a definite plane (described in detail in Chap. 31). Subsequent studies of the speed of light and measurement of spectral wavelengths confirmed this view.

No theory could have seemed more securely established at that point than did the wave theory. But further discoveries threw it again into doubt. Planck's explanation of the distribution of blackbody radiation, Einstein's treatment of the photoelectric effect, and Compton's study of the scattering of x-rays by electrons all required the assumption of a particle nature for electromagnetic radiation, without in any way invalidating the evidence for a wave nature.

Despite the obvious difficulty of imagining a wave which acts like a particle or a particle which acts like a wave, we are nevertheless in the position of having to do so. Sometimes the wave nature of light is invoked, as in describing the propagation of light through optical systems. At other times the particle nature of light is called upon to account for the observed phenomena, as in the case of interchange of energy between light and matter.

Energy in transit in the form of radiant energy can be detected and studied only when it is intercepted by matter and converted into thermal, electric, chemical, or mechanical energy. The energy and wavelength of radiant energy can thus be measured by purely physical means. Other aspects of radiant energy depend upon the presence and response of an observer. *Light* is the aspect of radiant energy of which a human observer is aware through the visual sensations which arise from the stimulation of the retina of the eye.

The wavelength of radiant energy capable of visual detection varies from about 3.9 to 7.6 ten-thousandths of a millimeter. Light wavelengths are usually expressed in angstrom units (Å) or in micrometers (mμ). One angstrom unit is 10^{-8} cm. One micrometer (a millionth of a meter) is 10^{-4} cm. The range of visible radiation is 390 to 760 nm (nanometer) or 3,900 to 7,600 Å.

Because of the very short wavelength, there is little spreading of light around and behind obstacles as is observed with water waves and sound waves. Except for such diffraction effects, to be discussed later, light travels in straight lines in a homogeneous medium. The rectilinear propagation of light is made use of in sighting with a plumb line, in aiming a gun, or in forming a photographic image in a pinhole camera (Fig. 23-1).

Radiation similar to light and having wavelengths between 390 and 100 nm constitutes *ultraviolet* radiation and is detected by photographic means. That in the wavelength range from 760 nm to 1 mm is called *infrared* radiation. The longer wavelengths of infrared radiation are

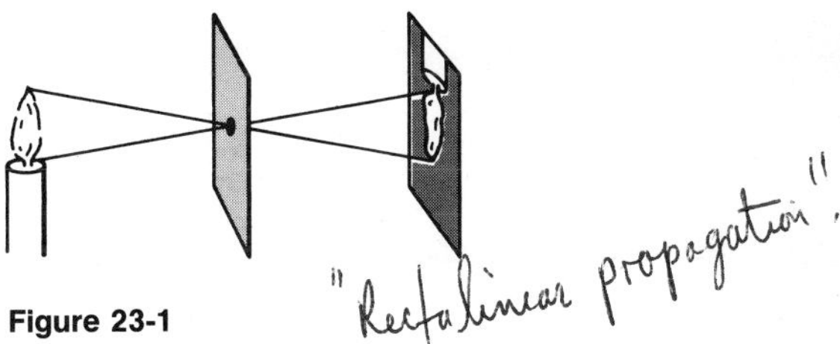

Figure 23-1
Formation of an image in a pinhole camera by the straight-line propagation of light.

most readily detected by their thermal effects. According to the electromagnetic theory of light, originated by Maxwell, all these waves are the same in kind as those which constitute the electromagnetic oscillations of radio waves.

Study of the interaction between matter and energy has added to our ideas about light an important assumption which is the basis of the *quantum theory*. According to this theory energy transfers between light and matter occur only in discrete amounts of energy (quanta), which are proportional to the frequency as given by

$$E = hf \qquad (1)$$

If f is the frequency in vibrations per second and h is Planck's constant, whose value is determined experimentally to be 6.625×10^{-34} J·s, the energy E is in joules (Chap. 45).

23-2 WAVES AND RAYS

The representation of a wave motion was discussed in Chap. 19. Figure 23-2 shows spherical waves spreading from a small source and also the radial lines, called *rays,* drawn to show the direction in which the waves are moving. The rays are merely convenient construction lines that often enable us to discuss the behavior of light more simply than by drawing the waves.

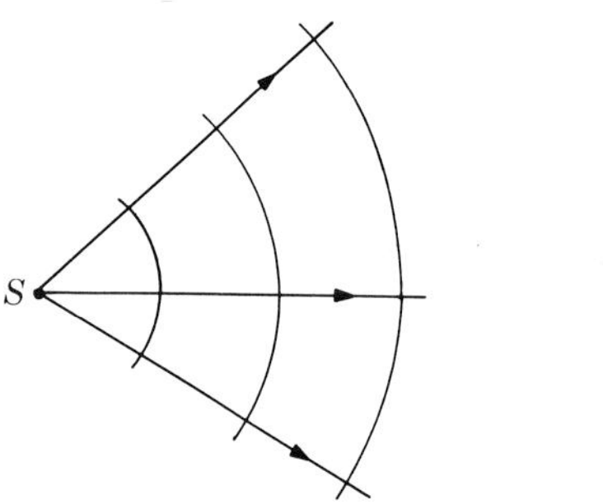

Figure 23-2
Light waves and rays. The concentric arcs represent sections of wave fronts. The straight lines represent rays.

23-3 SOURCES OF LIGHT

Our most important source of light and life-sustaining radiation is the sun. Most artificial sources of light are hot bodies which radiate light but which also emit much infrared radiation. Hence as producers of visible radiation they have a low efficiency. Generally the efficiency of such light sources improves as the operating temperature is increased. The early carbon-filament electric lamp, which supplanted the open-flame gas light, employed an electrically heated filament as the source of radiation. The filament was mounted in an evacuated glass envelope to prevent its oxidation. In a modern electric lamp the use of a tungsten spiral filament in a bulb containing inert gas permits operation at a higher temperature (3000 °C) without excessive evaporation of the filament.

The carbon arc also uses the heating effect of an electric current. Most of the light originates at the crater of the positive carbon, which attains a temperature of 3700 °C. The brilliant white light of a carbon arc is frequently used for searchlights and commercial motion-picture projectors. High-temperature light sources approach sunlight in the whiteness of their radiation. But no artificial sources of light operate at temperatures as high as the 6000 °C measured for the sun's photosphere, the layer of ionized gases in which sunlight originates.

Light sources may be divided into three main categories: thermal, gas discharge, and luminescent. Light from an incandescent solid contains all visible wavelengths, though in varying intensities. Light obtained by maintaining electric current in a gas at low pressure has its intensity concentrated in one or several narrow wavelength bands. A low-pressure mercury-arc lamp has a characteristic bluish light and also emits ultraviolet radiation. Such lamps are used for photographic work, some kinds of industrial illumination, germicidal purposes, and advertising signs, but they are not suited for general indoor illumination. By operating the discharge in a tiny quartz tube containing mercury vapor at high pressure, 50 to 100 atm, the quality of the light is improved, all wavelengths now being present, although some are still emphasized.

A fluorescent lamp consists of a thin-walled glass tube in which an electric current is maintained in mercury vapor at low pressure. The ultraviolet radiation from the glow discharge is absorbed by fluorescent substances affixed to the inner wall. These reemit the radiant energy with a shift of wavelength into the visible range. The color of the light can be adjusted by the choice of fluorescent powders: calcium tungstate for blue, zinc silicate for green, cadmium borate for pink, or mixtures for white.

23-4 LUMINOUS FLUX

Practical measurements of light, called *photometry,* are concerned with three aspects: the luminous intensity of the source, the luminous flux, or flow, of light from the source, and illuminance of a surface.

The energy radiated by a luminous source is distributed among many wavelengths. Only radiant energy in the wavelength interval from 390 to 760 nm produces a visual sensation, and in that

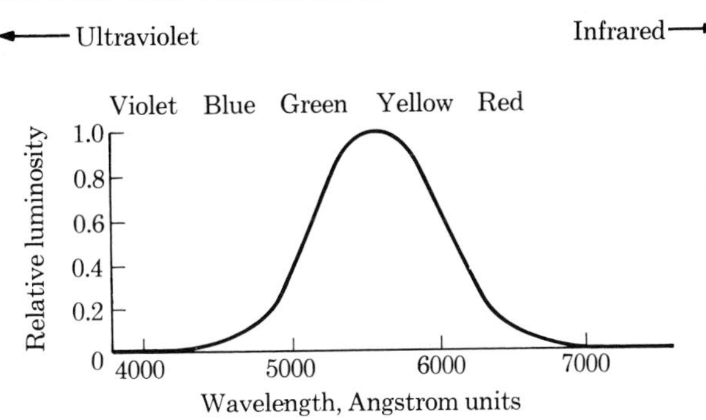

Figure 23-3
Standard luminosity curve.

interval the radiant energy is not all equally effective in stimulating visual sensation. The standard luminosity (visibility) curve of Fig. 23-3 represents the ratio of the power at the wavelength of the eye's greatest sensitivity, required to produce a given brightness sensation, to the power at the chosen wavelength, necessary to produce the same brightness sensation. The maximum ordinate is thus arbitrarily assigned a value of 1. The curve represents the average response of many individuals and is assumed to be the normal response.

We define luminous flux in terms of the curve of Fig. 23-3. We may divide the visible region into many wavelength intervals so short that the response of the eye can be considered the same over any one such interval. For each interval we multiply the radiant energy per unit time in the interval by the corresponding ordinate of the curve and add the products so obtained. This sum is called the *luminous flux*. It represents the part of the total radiant energy per unit time that is effective in producing the sensation of sight. The other photometric quantities are defined in terms of luminous flux. Since the process here described is difficult to carry out, the unit of luminous flux is not defined directly from this procedure but in terms of the flux from a standard source.

23-5 LUMINOUS INTENSITY OF A POINT SOURCE

While no actual source of light is ever confined to a point, many are so small in comparison with the other dimensions considered that they may be regarded as point sources. From such a source light travels out in straight lines. If we consider a solid angle ω, with the source at the apex, the luminous flux included in the angle remains the same at all distances from the source. *The luminous intensity of a point source is defined as the luminous flux per unit solid angle* subtended at the source,

$$I = \frac{F}{\omega} \tag{2}$$

where F is the flux in the solid angle ω.

If a sphere of radius s is described about the apex as center (Fig. 23-4), the solid angle intercepts an area A on the surface of the sphere. The ratio of the intercepted area of the spherical surface to the square of the radius is the measure of the solid angle ω in steradians (or sterads)

$$\omega = \frac{A}{s^2} \tag{3}$$

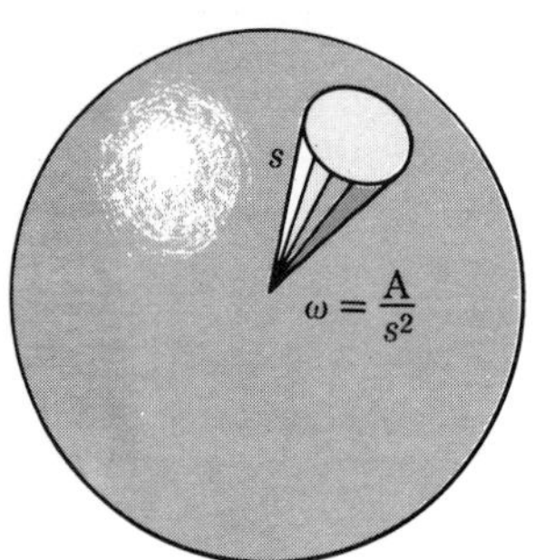

Figure 23-4
Flux from a point source.

In building a set of photometric units, it is convenient to start with luminous intensity rather than the more fundamental luminous flux. The common unit of luminous intensity is the candle. The *candle* was originally defined as the luminous intensity *in a horizontal direction* of the flame of a standard spermaceti candle of specified dimensions burning wax at the rate of 120 grains per hour. Since the flame of a candle is a rather unsatisfactory source, this primary standard has been replaced by others measured in comparison with it. Standardized electric lamps are most commonly used as secondary sources. However, in 1948 a *new international candle* was adopted, defined as one-sixtieth of the luminous intensity of a square centimeter of blackbody radiator (Fig. 23-5) operated at the temperature of freezing platinum, 2046°K. The new unit is about 1.9 percent smaller than the former international candle, a difference which does not affect significantly most photometric ratings.

Most light sources have different luminous intensities in different directions. The average luminous intensity of a source measured in all directions is called its *mean spherical luminous intensity*. The total flux emitted by the source is 4π times the mean spherical luminous intensity since there are 4π sterads about a point.

The unit of luminous flux is defined from the candle. A *lumen* is the luminous flux in a unit solid angle from a point source of *one candle.*

Example A spotlight equipped with a 32-cd bulb concentrates the beam on a vertical area of 125 m² at a distance of 100 m. What is the luminous intensity of the spotlight?

The purpose of the reflector and lens is to concentrate the beam into a small solid angle. Since the surface area of a sphere is $4\pi r^2$, then

$$\omega = \frac{A}{s^2} = \frac{4\pi r^2}{r^2}$$

and the total flux emitted by the bulb is given by

$$F = 4\pi I = 4\pi \times 32 \text{ cd} = 400 \text{ lm}$$

For the beam, the solid angle is

$$\omega = \frac{125 \text{ m}^2}{(100 \text{ m})^2} = 0.0125 \text{ sterad}$$

$$I = \frac{F}{\omega} = \frac{400 \text{ lm}}{0.0125 \text{ sterad}} = 32{,}000 \text{ cd}$$

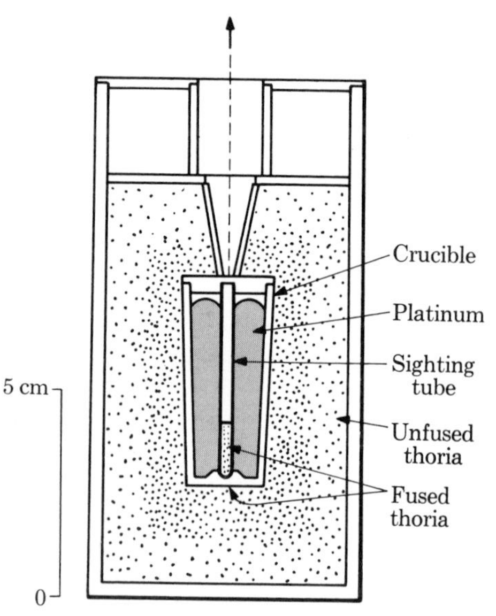

Figure 23-5
The national primary standard of light. The incandescent cavity in an ingot of platinum at 2046°K provides 60 cd/cm².

23-6 ILLUMINANCE

When visible radiation comes to a surface, we say that the surface is illuminated. The measure of the illumination is called illuminance. The *illuminance* of a surface is the *luminous flux per unit area that reaches the surface,*

$$E = \frac{F}{A} \tag{4}$$

The flux F may come from one or many sources; it may come to the area from any direction. The flux used in Eq. (4) is the sum total of the flux from all the sources that irradiate the surface being considered. Among the units of illuminance are the *lumen per square foot* (footcandle) and the *lumen per square meter* (lux).

For a *point source* there is a simple relationship between illuminance E and luminous intensity I. By definition

$$I = \frac{F}{\omega}$$

or

$$F = I\omega \tag{5}$$

When luminous flux from the point source P (Fig. 23-6) falls on the surface around O, the normal to the surface at O makes an angle θ with the direction PO of the flux. The solid angle ω subtended at P by a small area A of the surface is

$$\omega = \frac{A \cos \theta}{s^2}$$

where s is the distance PO from the source to the screen. From Eq. (5),

$$F = I\omega = \frac{IA \cos \theta}{s^2}$$

From Eq. (4),

$$E = \frac{F}{A} = \frac{IA \cos \theta}{s^2 A} = \frac{I}{s^2} \cos \theta \tag{6}$$

For light from a point source the illuminance of a surface varies inversely with the square of the distance from the source and directly with the cosine of the angle between the direction of flow and the normal to the surface. When the surface is perpendicular to the light beam, the angle θ becomes zero and $\cos \theta = 1$. For this special case, Eq. (6) reduces to

$$E = \frac{I}{s^2} \tag{7}$$

For a uniform point source whose luminous intensity is 1 cd, the luminous flux sent out is 4π lm. If this source is taken as the center of a sphere 1 ft in radius, the flux through the area of the sphere (4π ft^2) is 4π lm. The illuminance of this surface is 1 lm/ft^2. This unit is also called the *footcandle* (fc), since it is the illuminance of a surface 1 ft from a point source of 1 cd. Similarly, the lumen per square meter is also called the *metercandle* (mc).

Example A "point-source" unshaded electric lamp of luminous intensity 100 cd is 4.0 m above the top of a table. Find the illuminance of the table (*a*) at a point directly below the lamp and (*b*) at a point 3.0 m from the point directly below the lamp.

Directly below the lamp the light falls normally on the surface. From Eq. (7),

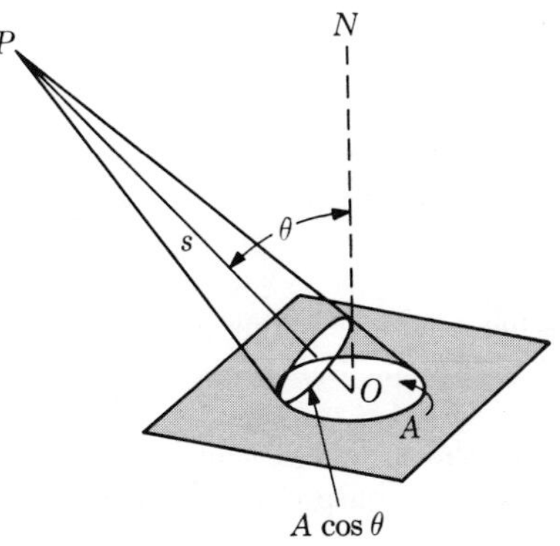

Figure 23-6
Illuminance produced by a point source.

$$E = \frac{I}{s^2} = \frac{100 \text{ cd}}{(4.0 \text{ m})^2} = 6.25 \text{ lm/m}^2$$
$$= 6.25 \text{ mc}$$

At the second point

$$s = \sqrt{(4.0 \text{ m})^2 + (3.0 \text{ m})^2} = 5.0 \text{ m}$$

and $$\cos\theta = \frac{4.0 \text{ m}}{5.0 \text{ m}} = 0.80$$

From Eq. (6),

$$E = \frac{I}{s^2}\cos\theta = \frac{100 \text{ cd}}{(5.0 \text{ m})^2} \times 0.80$$
$$= 3.2 \text{ lm/m}^2$$
$$= 3.2 \text{ mc}$$

Example A small, unshaded electric lamp hangs 6.0 m directly above a table. To what distance should it be lowered to increase the illuminance to 2.25 times its former value?

$$E_2 = 2.25E_1$$

From Eq. (7),

$$\frac{I}{s_2^2} = 2.25\frac{I}{s_1^2}$$
$$s_2^2 = \frac{s_1^2}{2.25} = \frac{(6.0 \text{ m})^2}{2.25}$$
$$s_2 = 4.0 \text{ m}$$

23-7 LIGHTING

In planning the artificial lighting of a room, the type of work to be done there or the use to which the room is to be put is the determining factor. Experience has shown that certain illuminances are desirable for given purposes. Some figures are given in Table 1.

The values listed refer to illuminance that should be maintained in service. The initial illuminance observed for a new installation generally declines, because of deterioration of the source and reflecting surfaces. Dull daylight supplies an illuminance of about 100 fc, while direct sunlight when the sun is at the zenith gives about 9,600 fc.

In addition to having the proper illuminance it is essential to avoid *glare,* or uncomfortable local brightness, such as that caused by a bare electric lamp or by a bright spot of reflected light in the field of vision. Glare may be reduced by equipping lamps with shades or diffusing globes and by avoiding polished surfaces or glossy paper.

Table 1
LEVELS OF ILLUMINATION—GOOD PRESENT-DAY PRACTICE

Place	Illuminance, fc
Baseball outfield, major league	100
Home:	
Living room, general illumination	5
Supplementary illumination, reading	50
Kitchen work counters	50
Library:	
Reading and stack rooms	50
Research study areas	50–100
Machine shop:	
Rough bench and machine work	40
Jewelry and watch manufacturing	300
Schools:	
Auditorium	20
Corridors	5
Classrooms	50–100
Show windows	100–200
Intense interior illuminance to minimize daylight reflections in window glass	1,000

23-8 PHOTOMETERS

A *photometer* is an instrument for comparing the luminous intensities of light sources. A familiar laboratory form of such an instrument usually consists of a long graduated bar, with the two lamps to be compared mounted at or near the ends (Fig. 23-7). A movable, dull-surfaced white screen is placed somewhere between the lamps and moved back and forth until both sides of the screen appear to be equally illuminated. When this condition is attained,

$$E_1 = E_2$$

From Eq. (7),

$$\frac{I_1}{s_1^2} = \frac{I_2}{s_2^2} \tag{8}$$

where I_1 and I_2 are the luminous intensities of the two sources and s_1 and s_2 are their respective distances from the screen. If one source is a standard lamp of known luminous intensity, that of the other may be found by comparison using Eq. (8).

Example A standard 48-cd lamp placed 36 cm from the screen of a photometer produces the same illumination there as a lamp of unknown intensity located 45 cm away. What is the luminous intensity of the latter lamp?

Substitution in Eq. (8) gives

$$\frac{I_1}{48 \text{ cd}} = \left(\frac{45 \text{ cm}}{36 \text{ cm}}\right)^2$$

$$I_1 = 75 \text{ cd}$$

Notice that the distances may be expressed in any unit when substituting in the equation, so long as they are both in the same unit.

In ordinary lighting the value of total flux from a lamp is more significant than is the luminous intensity in a particular direction. Total flux is most simply measured by use of an integrating sphere. The lamp is placed in a large sphere with a white diffusing interior wall. The flux emerging from a small hole in the wall is measured by means of a photometer set at the hole. The flux emerging from the hole is proportional to the total flux emitted by the source within the sphere. In practice, a source of known output is placed in the sphere and a reading made to calibrate the comparison source.

In planning a practical lighting installation for a room, one should take into account not only the direct illumination from all light sources but also the light that is diffused or reflected by the walls and surrounding objects. For this reason it is often very difficult to compute the total illuminance at a given point, but this quantity can be measured by the use of instruments known as *footcandle meters*. The most commonly used type of this instrument makes use of the photoelectric effect (Chap. 44). The light falling on the sensitive surface causes an electric current whose value is proportional to the radiant flux and hence to the illuminance. This current operates an electric meter whose scale is marked directly in footcandles.

This instrument gives objective readings, but it has the disadvantage that the sensitivity curve for the photoelectric cell is not the same as that for the human eye. However, the two curves are somewhat similar, and if close similarity is desired, it can be obtained by the use of a special filter over the photoelectric cell.

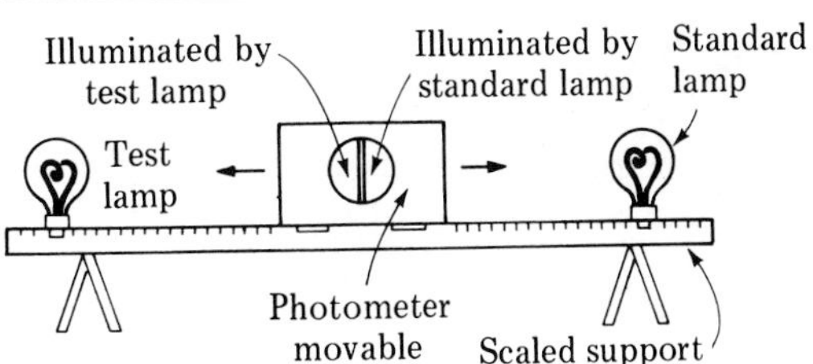

Figure 23-7
Bench-type photometer.

23-9 LUMINANCE OF EXTENDED SOURCES

Since the introduction of diffusing shades, fluorescent lamps, and indirect lighting fixtures, many light sources must be treated as extended sources rather than points. The *luminance* B_θ of a surface element A (Fig. 23-8) in any direction θ is defined as the luminous flux per unit solid angle (I_θ) per unit area of source projected on a plane perpendicular to that direction,

$$B_\theta = \frac{I_\theta}{A \cos \theta} \tag{9}$$

or *luminance is the source intensity per unit projected area of emitting surface.*

A perfectly diffusing surface is one for which luminance is independent of direction of observation. The intensity (lumens per unit solid angle, or candles) of a perfectly diffusing plane source is therefore proportional to the cosine of the angle between the perpendicular to the surface of the source and the direction of observation.

Luminance can be expressed in candles per square centimeter (of projected area) or in candles per square foot (Table 2). A special unit of luminance, the *lambert,* is defined as $1/\pi$ candle per square centimeter.

The new international candle mentioned above is actually a unit of luminance rather than of luminous intensity. That is, the luminance of a complete (blackbody) radiator at 2046°K is defined as 60 new candles per square centimeter.

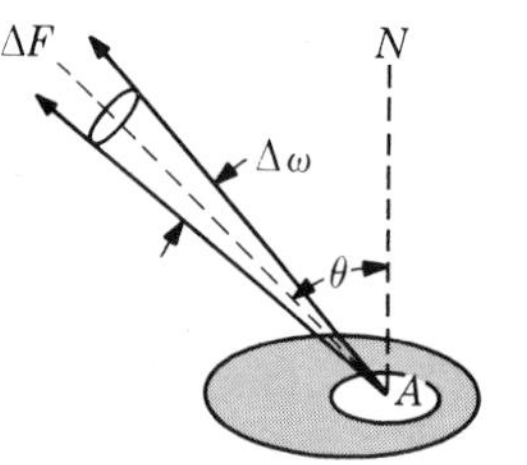

Figure 23-8
The intensity of a surface element A in direction θ is $I_\theta = \Delta F/\Delta\omega$.

Table 2
TYPICAL VALUES OF LUMINANCE

Source	Luminance, cd/m²
Surface of sun	2×10^9
Tungsten-lamp filament at 2700°K	10^7
Standard source (blackbody at 2046°K)	600,000
White paper in sunlight	25,000
Fluorescent lamp	6,000
Clear sky	3,200
White paper in moonlight	0.03

23-10 EFFICIENCY OF LIGHT SOURCES

The term *efficiency* as applied to light sources has the same meaning as it does when applied to machines, namely, the ratio of output to input power. The useful output of a light source is that part of the radiant flux to which the eye is sensi-

Table 3
LUMINOUS EFFICIENCIES

Source	Flux, lumens	Efficiency, lm/W
14-W 15-in fluorescent lamp	490	35
24-W tungsten lamp (vacuum)	260	10
30-W 36-in fluorescent lamp	1,500	50
40-W tungsten lamp (gas-filled coiled-coil filament)	465	12
60-W tungsten lamp (similar)	835	14
100-W tungsten lamp (similar)	1,630	16
100-W 60-in fluorescent lamp	4,400	44
500-W tungsten lamp	9,950	20

tive, i.e., the *luminous* flux. The input power is usually supplied electrically and is measured in watts. Thus the efficiency of a lamp is stated in *lumens per watt,* in contrast to the dimensionless ratio or percentage used in giving the efficiency of a machine.

The efficiencies of practical light sources are small. Not all the electric power is converted into radiant flux; some is lost as heat conducted away from the lamp. Of the radiant flux produced, only a small fraction lies in the region 390 to 760 nm which causes a visual sensation. The efficiency of incandescent sources increases with increase in temperature (Table 3). A blackbody would have its greatest luminous efficiency at a temperature of 6500°K, which is roughly the temperature of the sun's surface. It is a challenge to engineers to develop cool sources of light which may use phenomena such as fluorescence to channel more of the electrical input into luminous flux and thus achieve luminous efficiencies greater than that of a thermal (blackbody) radiator.

23-11 SPEED OF LIGHT

The speed with which light, and in fact all radiant energy, travels through space is one of the most precisely measured quantities in the physical sciences. This speed, moreover, is one of the most important constants in physical theory. Galileo was one of the first to suggest that light takes a finite time to travel between two points and to attempt to measure this time. The immense speed of light calls for the measurement of its passage over great distances or the precise determination of small time intervals, or both.

Römer in 1675 made a measurement of the speed of light over an astronomical distance. He noticed that when the earth was closest to Jupiter (Fig. 23-9) the eclipses of one of the moons of Jupiter occurred about 500 s ahead of the time predicted on the basis of yearly averages and that they were late by the same amount when the earth was farthest from Jupiter. He concluded that the difference of 1,000 s was the time required for light to traverse the diameter of the earth's orbit (186,000,000 mi).

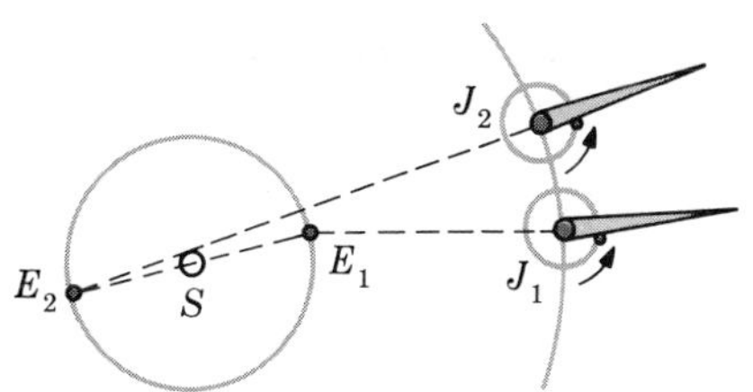

Figure 23-9
Römer's method of measuring the speed of light.

Since the time of Römer several investigators have devised methods for measuring the speed of light. Fizeau in 1848 and Foucault in 1850 developed methods for timing pulses of light. Using Foucault's method, Michelson in 1926–1929 measured the speed of light over an accurately measured terrestrial distance, the 22 mi between Mount Wilson and Mount San Antonio, using an ingenious method of timing. An octagonal mirror M (Fig. 23-10) is mounted on the shaft of a variable speed motor. Light from a source S falls on the mirror M, at an angle of 45°, and is reflected to a distant mirror m, which returns it. With M stationary the reflected ray strikes M_3 at an angle of 45° and is reflected into the telescope T. When mirror M is set in rotation, light returning to it from mirror m will generally

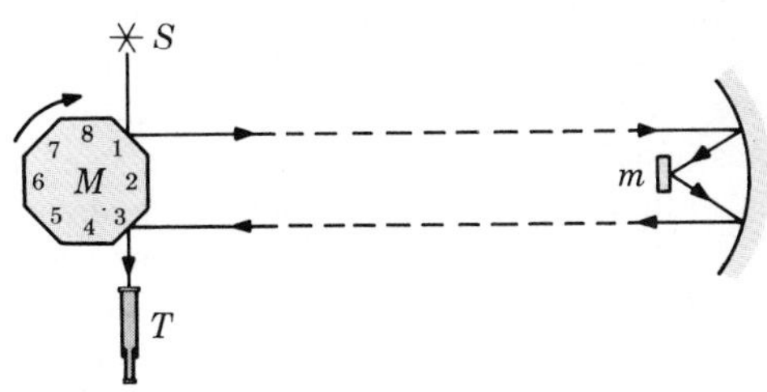

Figure 23-10
Michelson's method of measuring the speed of light.

strike M_3 at an angle different from 45° and hence light will not enter the telescope. When, however, the speed of M is sufficient so that section 2 of the octagonal mirror M is brought into the position formerly occupied by section 3 in the time required for light to go from M_1 to m and back to M_2, light will again enter the telescope. The experiment consists in varying the speed of the motor until light reappears in the telescope. This speed is accurately measured by stroboscopic comparison with a signal of standard frequency. The time for the light to travel a distance of 2 Mm is then one-eighth of the time required for one revolution of M. In these experiments the speed of the motor was about 500 r/s, and the final adjustment was made by noticing the displacement of the beam from the crosshair of the telescope. The currently accepted value for the speed of light in empty space is

$$c = 2.9979 \times 10^8 \text{ m/s}$$

or about 186,000 mi/s.

Recent measurements of the speed of electromagnetic waves have usually taken the form of measurements of the frequency and wavelengths of microwaves in resonant cavities.

Measurements continue to be made of the speed of light because of its central role in many branches of physical theory which involve the principles of relativity. In the Einstein theory, the speed of material particles is limited to the speed of light in a vacuum; and the relation of mass to energy involves the speed of light. These relationships will be discussed later; but mention may be made here of radar measurements of the distances of the moon and planets. A transmitter sends a pulse or message to the planet, and a precise time measurement is made of the arrival of the echo. On the assumption that the speed of light (or of radar waves) is known, the distance to the planet is obtained. Such distances are used, with Kepler's third law (Chap. 9), to obtain a more accurate value for the earth's distance from the sun; this value, in turn, is used as a standard for measuring star distances.

23-12 SPEED OF LIGHT IN DIFFERENT MATERIALS

The speed of light is different in each material, being less in any material medium than in a vacuum. Foucault (1850) and Michelson each measured the speed of light in water by placing tubes of water in the path of the light. In this way the speed of light in water was found to be about three-fourths its speed in air. This was a decisive experiment, for it eliminated temporarily one of two rival theories of light.

It was noted earlier that Newton (1666) had suggested that light might consist of particles or corpuscles shot from a body. He showed that ordinary laws of mechanics could account for the observed characteristics of light, provided that light traveled faster in a dense material (water) than in air. The wave theory developed by Huygens (1678) competed with the corpuscular theory for a hundred years by offering an equally logical explanation of reflection and refraction and a simpler explanation of diffraction. The wave theory predicted that light must travel more slowly in a dense medium (water) than in air. Foucault's direct measurement of the speed of light in water decided the controversy in favor of the wave theory. The corpuscular theory of light thus became dormant.

Later studies of the interaction of light and matter, notably the photoelectric effect, have shown that light behaves as if it were comprised of energy particles (quanta). While we have not returned to Newton's concept of light as made up of mass particles, current theory is forced to view light as possessing both wave and particle characteristics.

In 1934 P. A. Cherenkov (1904–) observed that water and other transparent substances, such as glass, produced a faint blue-white glow now known as the *Cherenkov radiation* when gamma rays passed through them. Cherenkov radiation is described as an electromagnetic shock wave produced when a charged particle passes through a substance at a speed greater than the speed of light in that substance.[1] A spectacular demonstration of Cherenkov radiation occurs when viewing a functioning swimming-pool nuclear reactor in a darkened room. The core of the reactor is seen to be encased by a glow due to the Cherenkov effect. This is caused by the electrons which are leaving the reactor core with a speed very nearly equal to the speed of light in a vacuum. While light travels at approximately 3.00×10^{10} cm/s in a vacuum, (*note: c* is speed of light in a vacuum) the speed of light in water is 2.25×10^{10} cm/s, about 0.7 c, and in glass is 2.2×10^{10} cm/s, hence, the radiation from the core passes through the water more rapidly than light does and the Cherenkov radiation occurs. It has been determined that the angle between the direction of light emission and the initiating particle depends on the velocity of the particle.[2] Because of this property, Cherenkov radiation is used to determine the speed of rapidly moving charged particles by observing the angle at which the blue-white light is emitted as the particles pass through the medium.

SUMMARY

Man's understanding of the nature of light is the result of a long and gradual refinement of this scientific theory. Newton, who supported the *corpuscular* theory of light, and Huygens, who supported the *wave* theory of light, were two of many scientists who contributed to our present comprehension of the *dual* nature of light.

Radiant energy within certain limits of wavelength (3,900 to 7,600 Å) is visible as *light*. Neighboring ranges of wavelengths comprise the ultraviolet and infrared radiations.

Light is transmitted in waves, which can pass not only through some (transparent) materials, such as glass, but also through empty space (vacuum).

Lines drawn in the direction of travel of light waves are called *rays*. In a homogeneous substance the rays are straight lines.

Most sources of light are hot bodies, which simultaneously emit invisible radiations.

Measurement of the luminous intensity of a light source takes account only of the portion of the emission that evokes *visual sensation*.

The *luminous intensity* of a point source is evaluated in terms of the power that arouses the brightness sensation experienced from a standard candle and is expressed in candles.

The *candle* is now defined in terms of the blackbody radiation emitted at the temperature of freezing platinum.

Luminous flux is the quantity of visible radiation passing per unit time. The unit of luminous flux is the *lumen,* which is the flux emitted by a point source of one candle through a solid angle of one sterad.

$$I = \frac{F}{\omega}$$

The total luminous flux emitted by a point source is

$$F = 4\pi I$$

The *illuminance* of a surface is the luminous flux per unit area that reaches the surface.

$$E = \frac{F}{A}$$

Illuminance is expressed in lumens per square foot (footcandles) or lumens per square meter (luxes).

For light from a point source, the illuminance on a surface is given by the inverse-square law.

$$E = \frac{I}{s^2} \cos \theta$$

A *photometer* is an instrument for comparing the luminous intensities of two point sources.

---

[1] David B. Hoisington, "Nucleonics Fundamentals," McGraw-Hill Book Company, New York, 1959.
[2] Samuel Glasstone, "Sourcebook on Atomic Energy," D. Van Nostrand Company, Inc., Princeton, N.J., 1958.

A *footcandle meter* is an instrument which measures a constant fraction (whose value depends upon spectral distribution) of the radiant flux and which is calibrated in units of illuminance.

Luminance is the luminous intensity per unit projected area emitted by an extended source. It is measured in candles per square foot or in candles per square meter.

$$B_\theta = \frac{I_\theta}{A \cos \theta}$$

In a vacuum, the speed of light is about 3.00×10^8 m/s. In any substance the speed is always less than this.

Cherenkov radiation is an electromagnetic shock wave produced when a charged particle passes through a substance at a velocity greater than the velocity of light in that substance.

Questions

1 List several forms of electromagnetic radiation.

2 Name three properties of light that were known before Newton developed his theory. Describe each of these properties.

3 Discuss Newton's theory relative to the speed of light in water and in air.

4 Describe the three properties of light which were discovered after Newton's era which helped to support the wave nature of light.

5 What evidence indicates that different colors (wavelengths) of light travel with the same speed in free space?

6 Describe a theoretical and an experimental verification of the inverse-square law of illumination. Does the photometer experiment verify the inverse-square law? Explain.

7 How is the luminous intensity of a lamp usually specified?

8 How would you determine experimentally the effective luminous intensity of a searchlight?

9 Light from a student lamp falls on a paper on the floor. Mention the factors on which the illuminance of the paper depends.

10 With the aid of a diagram, show that an illuminance of 1 m·c is the same as 1 lm/m^2.

11 Define luminous efficiency. What is the unit for it? Show how the luminous efficiency of a lamp may be measured.

12 Show that the total luminous flux emitted by a source is given by $F = 4\pi I$.

13 Would you expect more, less, or an equal amount of light from one 100-W bulb as compared with two 50-W bulbs? Justify your answer.

14 Sketch a shadow photometer with which you could compare the intensity of a light source with that of a lamp of known luminous intensity, using only metersticks and a white wall. What measurements would be made?

15 When a diffusing globe is placed over a bare electric lamp of high intensity, the total amount of light in the room is decreased slightly, yet eyestrain may be considerably lessened. Explain.

16 What is the significance of the equation $E = hf$? Explain its importance.

17 Show that the illuminance on a given surface distant s from a line source of light (fluorescent lamp) is given by

$$E = \frac{2I}{Ls}$$

where I/L is the intensity per unit length of the lamp in candles per foot. *Hint:* Consider a cylinder of radius s and length L concentric with the lamp. Calculate the illuminance on the cylinder produced by light proceeding radially from the lamp.

18 Derive an equation for calculating the speed of light from the quantities measured in the Michelson method for measuring the speed of light.

19 Michelson assumed that the light coming from one mountain to the other in his experiment to determine the speed of light traveled in a straight line and at a constant velocity. Was he correct in both assumptions? Discuss your answer.

20 A member of a radio studio audience in New York is seated 50 m from the performer, while a radio listener hears him in Cedar Rapids, 1,600 km away. Which auditor hears the performer first?

21 Is it possible to design an optical instrument which will produce an image brighter than the original object as seen by the unaided eye? (Consider a "burning glass.")

22 Describe Cherenkov radiation and a use that is made of it.

Problems

Assume point sources of light, unless otherwise stated.

1 Calculate the frequency of the longest and shortest wavelengths of light that are visible to the eye. Speed of light = 3×10^8 m/s.

2 Convert 3,900 Å and 7,600 Å to feet.
Ans. 1.28×10^{-6} ft; 2.5×10^{-6} ft.

3 A meterstick is standing vertically at the edge of a horizontal table. A point source of light is on the table, 264 cm from the foot of the meterstick. (*a*) How long is the shadow of the meterstick on a vertical wall 582 cm from it? (*b*) The light source is raised vertically 50 cm. What is now the length of the shadow?

4 The flame of a vertical candle is 2.16 cm high. It is 135 cm from a metal sign, circular in shape with radius 15.4 cm, that is suspended at the same distance above the floor as the candle. Calculate the diameters of the parts of the shadow of the sign on a vertical wall 426 cm from the sign. *Ans.* 1.21 m; 1.35 m.

5 It is desired to have an illuminance of 25 mc on a drafting table. What incandescent lamp should be used if it is to be located 2.0 m directly above the table and if two-thirds of the light received on the table is reflected from the walls and ceiling?

6 An automobile headlamp has a 16-cd bulb which floods an area of 200 ft^2 at a distance of 100 ft. Calculate the luminous intensity of the lamp. *Ans.* 10,000 cd.

7 A lamp, considered to be an unshaded point source of light, has a luminous intensity of 120 cd and is 2 m above a table. Find the illuminance directly under the lamp and at a point 30° away from vertical line from the lamp to the table.

8 A darkened room has a tiny aperture in the roof through which sunlight enters when the sun is directly overhead. What is the size of the sun's image on the floor 12.0 ft below the aperture? The sun is 93 million mi away, and its diameter is $86\bar{5}$,000 mi. *Ans.* 13 in.

9 A screen near a light source receives 2.86 W of radiant flux, of wavelength 600 nm. Find the luminous flux if 1 lm is equivalent to 1.47 mW of monochromatic light of wavelength 555 nm.

10 If an unshaded electric lamp is 6.0 m above a table and is lowered to 3.0 m, how much has its illuminance been increased?
Ans. 4.0 times.

11 What is the illuminance on the pavement at a point directly under a street lamp of 800 cd hanging at a height of 10 m?

12 A lamp produces a certain illuminance on a screen situated 85 cm from it. On placing a glass plate between the lamp and the screen, the lamp must be moved 5.0 cm closer to the screen to produce on it the same illuminance as before. What percent of the light is stopped by the glass?
Ans. 11 percent.

13 A pool table measuring 4.0 by 8.0 ft is illuminated by three light bulbs. There are identical bulbs 4.0 ft directly above the midpoint of each end of the table and a bulb rated at 1,965 lm 3.0 ft above the center of the table. To maintain a high degree of sportsmanship, it is agreed that the illuminance on the table at the side pockets shall be no less than 50 ft·c. Find (*a*) the luminous intensity of the bulbs over the ends of the table and (*b*) the illuminance on the table at the corner pocket.

14 If a lamp that provides an illuminance of 8.0 mc on a book is moved 1.5 times as far away, will the illumination then be sufficient for comfortable reading? *Ans.* 3.6 mc.

15 If it is desired to increase the illuminance of an unshaded lamp by five times and the lamp is 3.0 m above a table, to what height should the lamp be lowered?

16 If the light of the full moon is found to produce the same illuminance as a 1.0-cd source does at a distance of 4.0 ft, what is the effective luminous intensity of the moon? (The mean distance to the moon is 239,000 mi.)
Ans. 1.0×10^{17} cd.

17 A table 7.0 ft long is illuminated by two lamps. A 200-cd lamp is 4.0 ft above the left-hand end, and a 150-cd lamp is 3.0 ft above the right-hand end. Find the illuminance of the table 3.0 ft from the left-hand end.

18 An unknown lamp placed 6 m from a photometer screen provides the same illumination as an 80-cd lamp placed 4 m from that same screen. What is the candlepower of the unknown lamp?
Ans. 180 cd.

19 What illuminance will be given on a desk by a 36-cd fluorescent lamp 1.0 ft long placed 18 in above the surface? (For an extended line source, the illumination decreases as the inverse first power of the distance, $E = 2I/Ls$.)

20 A standard 48-cd lamp is placed 30 cm from a photometer screen and produces the same illumination as a 60-cd lamp placed some distance away. How far away is the 60-cd lamp from the screen?
Ans. 33.6 cm.

21 A 500-W tungsten lamp is placed at the center of a spherical photometer that has a diameter of 1.60 m. What is the illuminance at the surface of the sphere?

22 Two point sources of light are placed 4.0 m apart on a photometer. If one is a standard of 50 cd and they produce equal illuminance at a point 2.5 m from the standard source, what is the intensity of the second source?
Ans. 18 cd.

23 Two point sources of 50 and 110 cd are placed 2.0 m apart on an optical bench. At what point between them will the illuminance from each source be the same?

24 In an experiment with a bar photometer it was found that two lamps produced the same illuminance when the screen was 40.0 cm from a standard lamp and 160 cm from a second lamp. The standard lamp was rated at 8.00 cd. Determine (*a*) the luminous intensity of the second lamp, (*b*) the overall luminous efficiency of the second lamp if it used 150 W of power, and (*c*) the illuminance at the screen due to each lamp.
Ans. 128 cd; 16.2 lm/W; 4.66.

25 A sample 60-W lamp produces on a screen 2.00 m away the same illuminance as that produced by a 16-cd standard lamp 1.00 m from the screen. What is the efficiency of the sample lamp?

26 A point source of light of 10 cd is enclosed at the middle of a hollow sphere having a radius of 4.0 m. If an opening of 10 m^2 exists in the sphere, what is the luminous flux through the opening?
Ans. 6.25 lm.

27 A 150-cd lamp is suspended 2.5 m above a sheet of white blotting paper which reflects 75 percent of the light incident upon it. Calculate (*a*) the illuminance of the paper and (*b*) its luminance.

28 Two unshaded lamps of 100 cd each are suspended 5 ft above a pool table and 6 ft apart. Find the illuminance on the tabletop (*a*) directly under one of the lamps and (*b*) at a point midway between the two lamps.
Ans. 22.56 ftc; 5 ftc.

29 A filament in a tungsten lamp emits 6,000 lm. Determine the candlepower of a lamp.

30 A high-intensity lamp causes an illumination of 40 ftc when it falls on a book 4 ft from the lamp. If the print in the book is fine and requires greater illumination to be read, what will be the illumination if the lamp is moved to within 2 ft of the book?
Ans. 160 ftc.

31 How far above a surface should an unshaded 100-W lamp be placed to produce an illumination of 10 ftc. The luminous efficiency of the lamp is 2 cd/W.

32 Find the candlepower of a point source of light that provides 20 mc of illumination on a surface 4 m away.
Ans. 320 cd.

33 The speed of light is to be measured by means of a revolving mirror. If the distance between the rotating mirror and the fixed mirror is 5.00 mi, how fast must the mirror rotate in order that the angle between the incident beam and the reflected beam shall be 3.00°?

34 What minimum speed of rotation is necessary for an eight-sided mirror used in a Michelson experiment for measuring the speed of light if the distance from the rotating mirror to the fixed reflector is 22 mi?
Ans. 5.3×10^2 r/s.

35 In 1935, Pease and Pearson repeated Michelson's experiment on the speed of light. They sent a beam of light through an iron pipe from which the air had been evacuated. Their mirror had 32 faces, and they caused the light to travel back and forth eight times while the mirror rotated from one face to the next. If the pipe were exactly 1 mi long and the speed of light 2.99793×10^8 m/s, what would the minimum rotational speed of the mirror have to be?

36 The speed of light has been measured by means of a toothed wheel rotating at high speed. Assume such a wheel to have 480 teeth that are just as wide as the spaces between them. A beam of light perpendicular to the wheel passes between two teeth and falls normally upon a stationary mirror 500 m away. Compute the minimum speed of rotation which will cause a tooth to intercept the reflected beam. *Ans.* 312 r/s.

Victor Franz Hess, 1883–1964

Born in Waldstein Castle near Peggau, Austria. Director of the Institute for Radium Research, Innsbruck University. Professor of Physics at Fordham University. The 1936 Nobel Prize for Physics was awarded jointly to Hess and Anderson, to the former for his discovery of cosmic radiation.

Carl David Anderson, 1905–

Born in New York City. Professor at the California Institute of Technology. The 1936 Nobel Prize for Physics was awarded jointly to Hess and Anderson, to the latter for his discovery of the positron.

24

Reflection of Light

An object is seen by the light that comes to the eye from the object. If the object is not self-luminous, it is seen only by the light it reflects. The reflection of light makes a room with white walls much lighter than a similar room with black walls. The "high lights" produced by reflection on polished doorknobs and car fenders are so characteristic of convex surfaces that an artist uses them to suggest curved surfaces in a painting. Reflection at the concave surface behind a head lamp sends light where it is needed to make other objects visible, by reflected light.

When light reaches a boundary of the medium in which it is traveling, one or more of three things can happen. Usually some of the light is reflected back into the first medium, and part is transmitted into the second. The part that enters the second medium may be all absorbed, if the medium is opaque, or partly absorbed and partly transmitted if the medium is transparent. For example, if light comes normally (perpendicularly) from air to a surface of ordinary glass, about 4 percent of the light is reflected. Of the 96 percent that enters the glass some is absorbed; the remainder proceeds to the second glass surface, where again about 4 percent is reflected at normal incidence. If the incidence is not normal at either surface, a higher portion is reflected at that surface. The laws of reflection and refraction of light may be derived from the equations of electromagnetism (Chap. 43), but for the purposes of this book simple geometrical constructions may be used.

24-1 HUYGENS' PRINCIPLE

Suppose that waves originate at a point source S (Fig. 24-1) and travel through a uniform medium. The waves will arrive simultaneously at all parts of a surface such as AB that are equidistant from S. Thus the waves at AB are all in the same phase, and hence that surface represents a part of a *wave front*. Christian Huygens (1629–1695) devised a method of construction for finding new wave fronts. His principle (Chap. 36) states that each point on a wave front may be considered as a new source of disturbance sending out wavelets. In Fig. 24-1, wavelets from A, a, b, c, d, etc., are shown. The new wave front is the surface CD tangent to all the wavelets. Huygens' principle, published in 1690, was fruitful in explaining and predicting many properties of light.

The propagation of light can be represented in a diagram by drawing successive wave fronts, located by applying Huygens' principle. However, the construction of a fairly accurate wave-front diagram is usually a tedious task. It is easier and often quite as satisfactory to draw a few *rays*

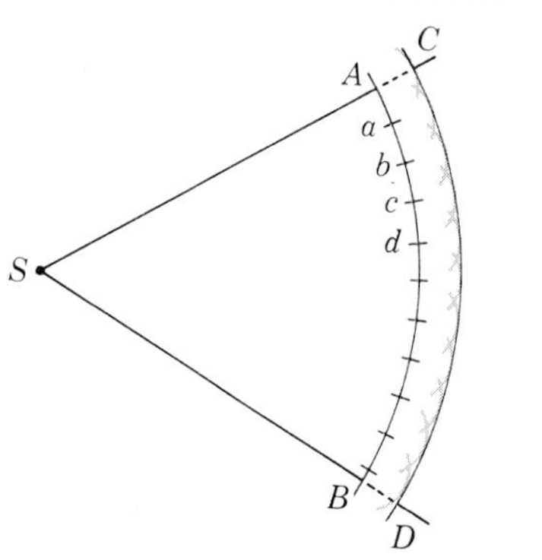

Figure 24-1
Huygens' construction.

representing the direction of propagation of the light wave. In an isotropic medium (a material whose physical properties are the same in all directions) the rays will be straight lines perpendicular to the wave fronts. In Fig. 24-1, the lines *SAC* and *SBD* are rays.

24-2 LAWS OF REFLECTION

It is found by experience that when light, or any wave motion, is reflected from a surface, the reflected ray at any point makes the same angle with the perpendicular, or normal, to the surface as does the incident ray. The angle between the incident ray and the normal to the surface is called the *angle of incidence i;* the angle between the reflected ray and the normal is called the *angle of reflection r* (Fig. 24-2). The laws of reflection may then be stated:

1 *The angle of incidence is equal to the angle of reflection.*
2 *The incident ray, the reflected ray, and the normal to the surface lie in the same plane.*

These laws hold for any incident ray and the corresponding reflected ray.

In Fig. 24-2 a parallel beam of light is incident on the mirror surface *MM'*. A wave front in the incident beam is represented by *AB* at the instant one edge of this front reaches the mirror. At later times this wave front reaches positions A_1B_1, A_2B_2, A_3B_3, and *A'* successively. At this latter instant the wave front would be at *A'D* if the mirror were not present. The presence of the reflecting surface starts a Huygens' wavelet upward from *B*, which spreads outward to *B'* in the time it takes for the upper part of the wave front *AB* to travel to the mirror at *A'*. When the wave front reaches B_1, B_2, etc., wavelets start upward also and will reach B_1', B_2', etc., at the instant the upper edge of the incident wave front reaches *A'*. The line *A'B'* tangent to the arcs drawn from *B*, B_1, etc., indicates the position of the wave front immediately after reflection. The reflection has inverted the wave front.

In the right triangles *BAA'* and *A'B'B* of Fig. 24-2, the sides *AA'* and *BB'* were drawn equal. Thus the triangles are similar. Since angle *B'BA'* is equal to angle *AA'B*, and since angles *r* and *i* are respectively complementary to these angles, it follows that angle *i* equals angle *r*.

It should be noted that this construction is valid only for wave fronts and incident surfaces that may be considered to have infinite extent. This is well approximated in actual practice, where the reflecting surfaces are very large in comparison with the wavelengths of light.

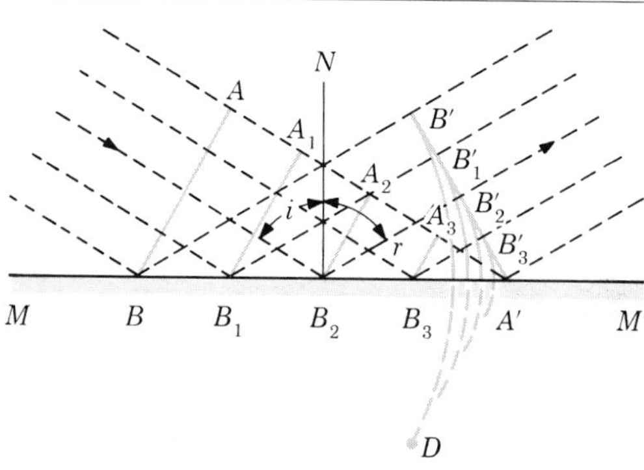

Figure 24-2
Reflection of light by Huygens' principle.

24-3 REGULAR AND DIFFUSE REFLECTION

Sunlight falling on a mirror surface, such as still water or polished glass or metal, is reflected regularly, i.e., the angle of reflection equals the angle of incidence. If an observer is to see the light, his eye must be placed within the reflected beam. When sunlight falls on a piece of white blotting paper, the light is scattered in all directions and the observer may see it from any position near it. These are examples of regular and diffuse reflection. A smooth, plane surface reflects parallel rays falling on it all in the same direction, while a rough surface reflects them diffusely in many directions (Fig. 24-3). At each point on the rough surface the angle of incidence is equal to the angle of reflection, but the normals have many directions.

An observer sees an object only if light comes to his eye from that object. In this process the observer assumes that the light has traveled in straight lines from the source to his eye. If the light has been reflected diffusely by a surface, it comes to the eye in the same manner as it would from a self-luminous surface and one reports that he sees that surface. If, however, the surface reflects regularly, the light comes, or appears to come, to the eye from some point in front of or behind the reflecting surface. In this case one does not see the reflecting surface, but, rather, he sees an *image* located at the point from which the light comes or appears to come. If the light, after reflection, actually passes through the points where the image is located, the image is called a *real* image; if the light only appears to come from the points where the image is located, the image is called a *virtual* image. The eye sees a virtual image as well as it sees a real image. However, a real image will appear on a screen placed at its position while a virtual image will not.

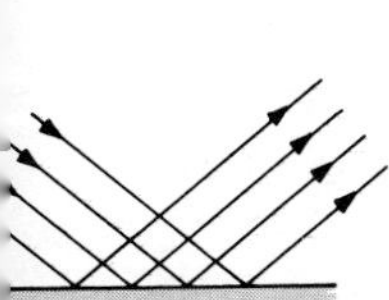

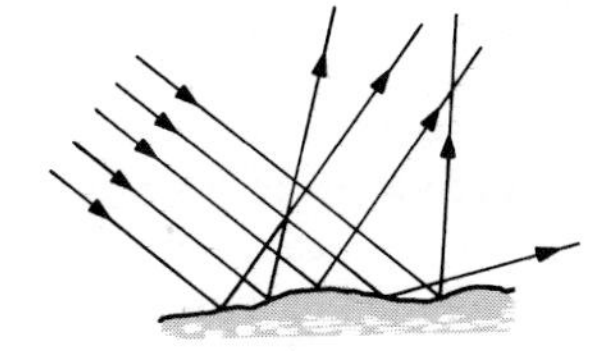

Figure 24-3
Regular and diffuse reflection.

24-4 PLANE MIRRORS

When light from a point source, proceeding in spherical wave fronts, falls on a plane mirror, the wave fronts are reflected with their curvatures reversed. In Fig. 24-4, there is a point source of light above a plane mirror. At the instant considered, points a and b on the wave front have just reached the mirror, while the central part of the wave has been reflected upward a distance mn equal to the distance mn' which it would have progressed below the line amb in the absence of the mirror. The center of curvature of the reflected wave front anb is at point I, whose distance mI below the mirror is equal to the distance mO at which the source O lies above the mirror. The reflected wave front proceeds toward an observer at E as if it came from I. Point I is the virtual image of the real source at O. This image lies below the mirror the same distance the object

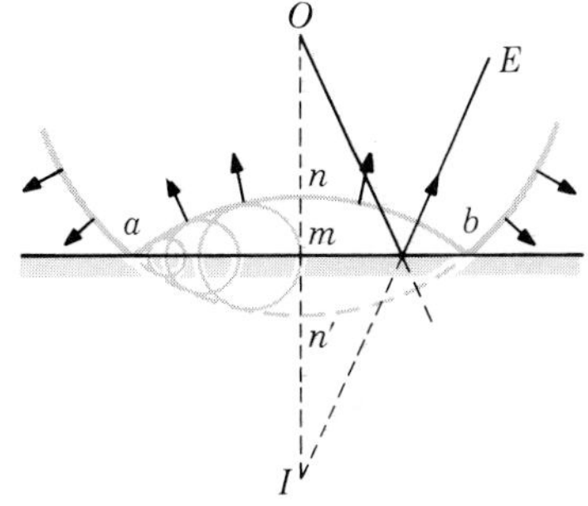

Figure 24-4
Reflection in a plane mirror, wave-front diagram.

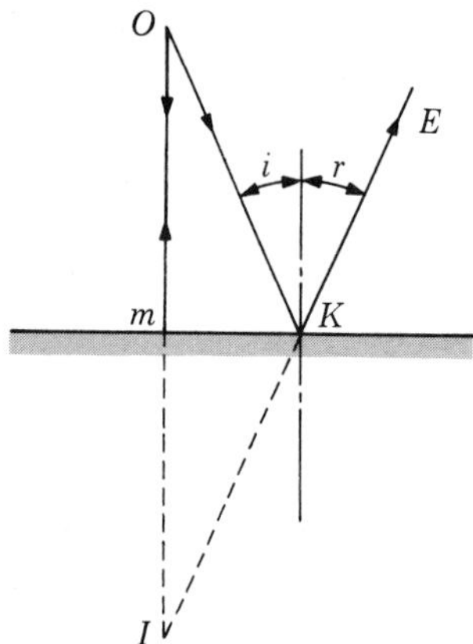

Figure 24-5
Reflection in a plane mirror, ray diagram.

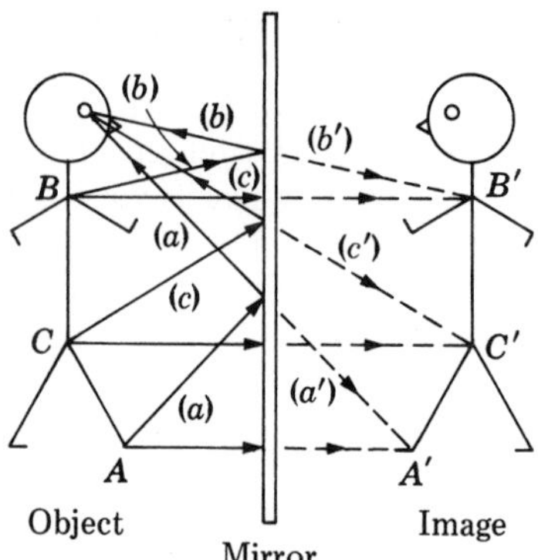

Figure 24-7
Production of an image in a plane mirror.

lies above it. The image is virtual because the light does not actually pass through it.

The position of an image formed by a mirror can be located even more readily in a ray diagram. In Fig. 24-5, a ray OK from the source O, reaching the mirror at K, is reflected along KE, making angle r equal angle i. A ray reaching the mirror along the normal Om makes zero angle of incidence and is reflected back on itself. Lines Om and EK, when extended below the mirror, intersect at I. Rays from O after reflection at the

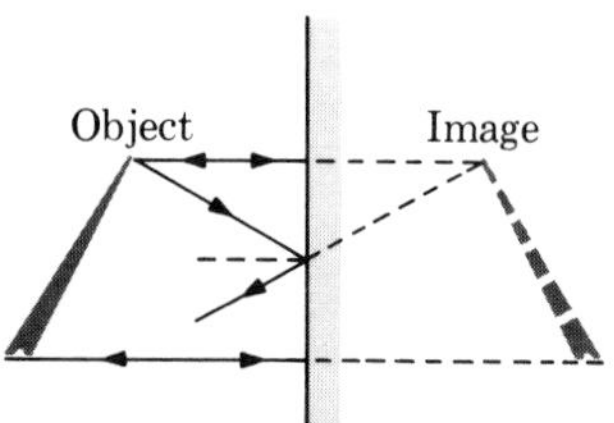

Figure 24-6
Image of an extended object (arrow) formed by a plane mirror.

plane mirror travel as if they originated at I. The point I is the virtual image of the source O. The triangles OmK and ImK are similar, and the image distance mI is equal to the object distance Om.

The image of an extended source or object in a plane mirror is found by taking one point after another and locating its image. The familiar result is that the complete image is the same size as the object and is placed symmetrically with respect to the mirror (Fig. 24-6).

A plane mirror provides an erect, virtual image of the same size as the object, hence, the image is the same distance behind the mirror as the object is in front. The ray diagram illustrating the image formation for a person standing in front of a full-length mirror is shown in Fig. 24-7. Two rays of light from each reference point are selected. For example, the ray of light reflecting from point A (the man's foot) strikes the mirror and rebounds to his eye, ray a. The eye sees the ray as an extension of a through a' to A'. A second ray of light is drawn from A perpendicularly to the mirror which apparently continues on through the mirror until it intersects with a' producing a virtual image at the point A'.

Similarly, rays of light from any point on the

body can be selected to further define the image, such as c and c' and b and b'.

The fact that one sees in a plane mirror a virtual image of an illuminated object is used in many optical illusions. An old stage trick is performed by having a well-illuminated actor or object located out of sight of the audience. A large sheet of plate glass is placed between the audience and the stage. In a darkened room the audience does not know of the glass, and the images seen are highly realistic. By suitable placement of these invisible mirrors very striking "ghosts" are observed, apparently walking up a wall or on a ceiling.

If a small beam of light falls on a mirror at an angle of incidence θ, the angle of reflection is also θ and the angle between the incident and reflected beams is 2θ. If the mirror is turned through an angle $\Delta\theta$, the angle of incidence is changed by $\Delta\theta$ and likewise the angle of reflection is changed by $\Delta\theta$. Hence, the angle between incident and reflected beams is changed by $2\,\Delta\theta$. Thus the reflected beam is changed in direction by twice the angle of rotation of the mirror. This principle is used in such devices as the sextant, galvanometers, and the optical lever.

The image in a plane mirror is reversed, right for left, as will be noticed when one looks at the image of printed material. This image technically is said to be *perverted*. An interesting example of this is the case of two plane mirrors placed at right angles to each other (Fig. 24-8). The location of the second image may be determined by considering the image of the object formed by the first mirror as an object for the second mirror. Hence a double perversion occurs, and the second image appears like the object.

It can be shown through a ray diagram that the image produced in a plane mirror is perverted. Selecting two rays of light from each reference point, one at such an angle that it reflects back to the eye (for purpose of clarity, the rays in Fig. 24-9 are shown reflecting to a spot located between the eyes) and the other ray perpendicular to the mirror as in the discussion of the image formation in a plane mirror presented above. Through this procedure it is seen that the man's hair part, which is on the left, appears on the right of the image.

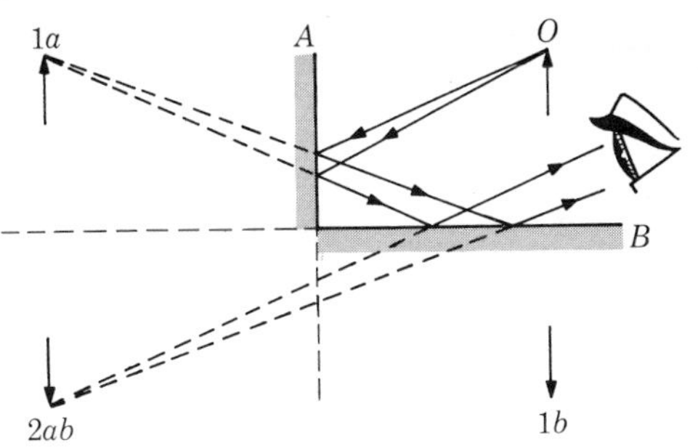

Figure 24-8
Images formed by two plane mirrors placed at right angles to each other.

Example A car is backing up at a rate of 10 km/h. If the driver looks through his rear-view mirror, how fast does a tree which is behind him appear to approach him?

The image of the tree at time t is just as far behind the mirror as the tree is in front of the mirror. However, as the car backs up, the distance

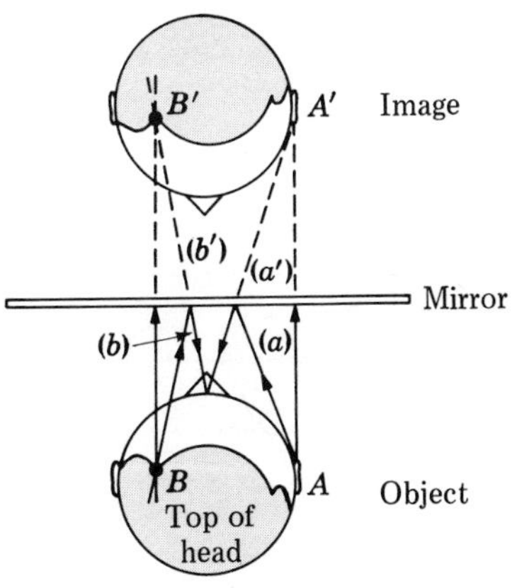

Figure 24-9
Left-to-right perversion of an image. Hair part on the left of object appears on the right of image.

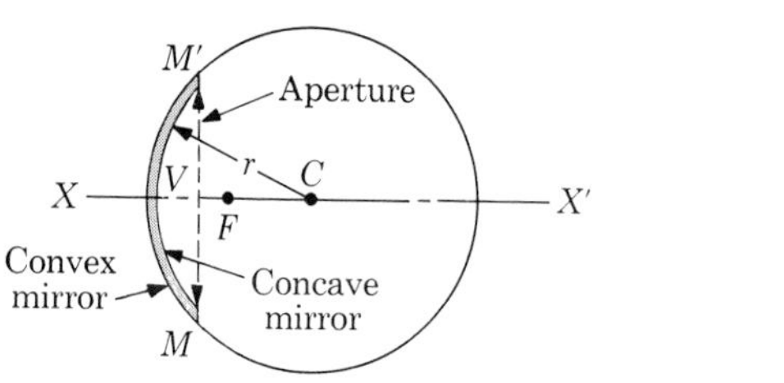

Figure 24-10
Spherical mirrors.

between the tree and the mirror shortens by a distance s. Meanwhile, the image distance becomes shorter by the same distance s. Therefore, as the driver observes the image it becomes closer by $s + s$ or $2s$. Thus, the apparent speed of the tree as viewed through the mirror is $v = 2s/t$, or twice the actual speed of the car. As a result, the tree, as seen through the mirror, is approaching the car at 20 km/h.

24-5 SPHERICAL MIRRORS

If the reflecting surface is curved rather than plane, the same law of reflection holds but the size and position of the image formed are quite different from those of an image formed by a plane mirror.

Curved mirrors are commonly made as portions of spherical surfaces. Spherical mirrors are classified as *concave* or *convex* according to whether the surface reflects light from inside the sphere or from outside (Fig. 24-10). The *center of curvature, C,* of the mirror is the center of the sphere. The *radius of curvature, r,* of the mirror is the radius of the sphere. The radius is conventionally taken as positive for concave mirrors, negative for convex mirrors. A line connecting the middle point or vertex V of the mirror and the center of curvature is called the *principal axis* of the mirror; it is marked XX'. The diameter MM' of the circular outline of the mirror is called the *aperture* of the mirror. Most spherical mirrors used for optical purposes have apertures small compared with their radii of curvature.

Figure 24-11 shows a concave and a convex mirror, with the beam of light directed on each made up of rays parallel to the principal axis. By applying the law of reflection to each ray, it is seen that the bundle of rays parallel to the principal axis will converge approximately through a common point F after reflection from a concave mirror or will diverge after reflection from a convex mirror as though they originated from a common point F behind the mirror. The point F to which rays parallel to the principal axis

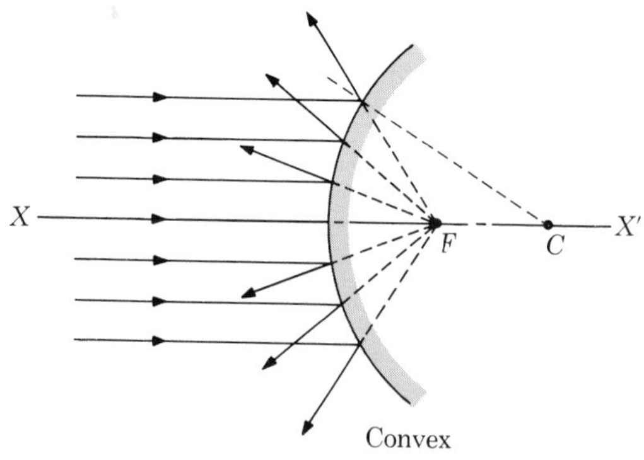

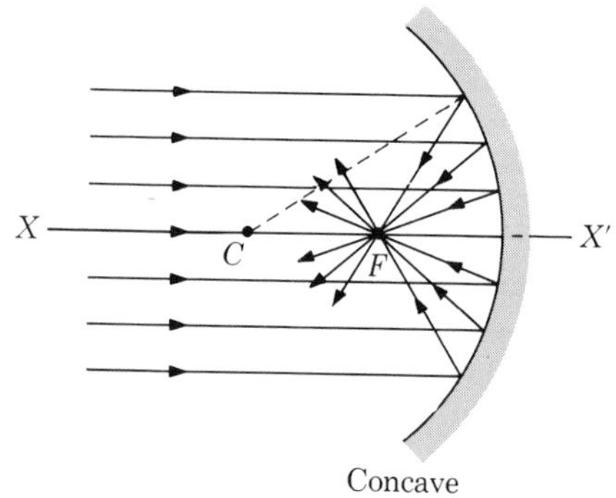

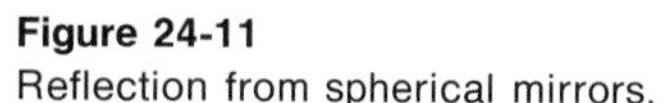
Figure 24-11
Reflection from spherical mirrors.

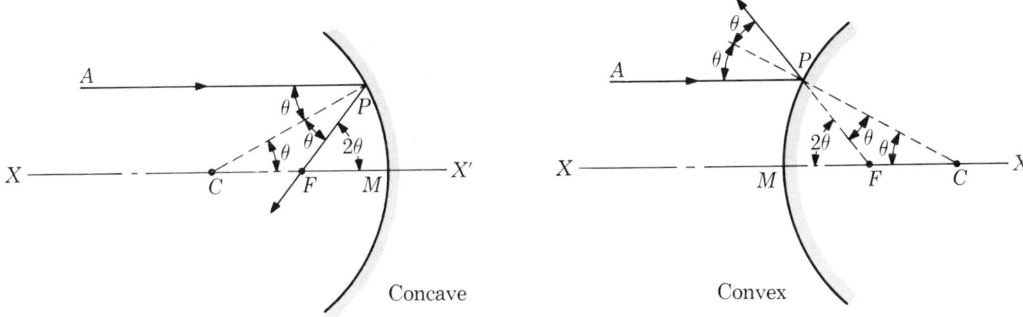

Figure 24-12
Location of the principal focus of a spherical mirror.

converge (or from which they diverge) is called the *principal focus* of the mirror. The distance f of the principal focus from the mirror is called the *focal length*.

The concentration of rays at the principal focus of a concave mirror can be shown experimentally by allowing sunlight to fall on the mirror along XX' and moving a bit of paper or a match along the axis to find the point F where the spot of light is brightest. The principal focus of a convex mirror is a *virtual* focus, because the rays do not actually pass through it but merely appear to do so.

When the aperture of the mirror is small, the principal focus of a spherical mirror lies on the principal axis halfway between the mirror and its center of curvature. This relation can be proved from the law of reflection for either of the mirrors of Fig. 24-12. A ray AP parallel to the principal axis strikes the mirror at P and is reflected along PF. At P the angle of reflection is equal to the angle of incidence, θ. The angle between the reflected ray and the principal axis at F is 2θ, since it is an exterior angle of the triangle PCF. For mirrors whose aperture is small compared with the radius of curvature, the angles θ are small, and the arc PM may be considered a line perpendicular to the axis XX'. Hence

$$\frac{PM}{CM} = \tan\theta \quad \text{and} \quad \frac{PM}{FM} = \tan 2\theta \qquad (1)$$

For small angles we can set the angles (in radians) equal to their tangents, so that

$$\frac{PM}{CM} = \theta = \frac{1}{2}\frac{PM}{FM} \qquad (2)$$

or

$$CM = 2FM$$

showing that the principal focus F lies halfway between the center of curvature and the middle of the mirror surface M when the aperture is small. When the aperture of the mirror is not small, the approximation made here is not valid and the rays parallel to the principal axis do not all come to a single point.

24-6 IMAGES FORMED BY SPHERICAL MIRRORS

Concave mirrors have wide application because of their ability to make rays of light converge to a *focus*. If rays coming from a point S (Fig. 24-13) strike the concave spherical mirror, the reflected rays may be constructed by applying the law of reflection at each point of reflection, the direction of the normal being that of the radius in each case. All rays from S will be found to pass after reflection very nearly through the single point I, which is the real image of the source S. If the

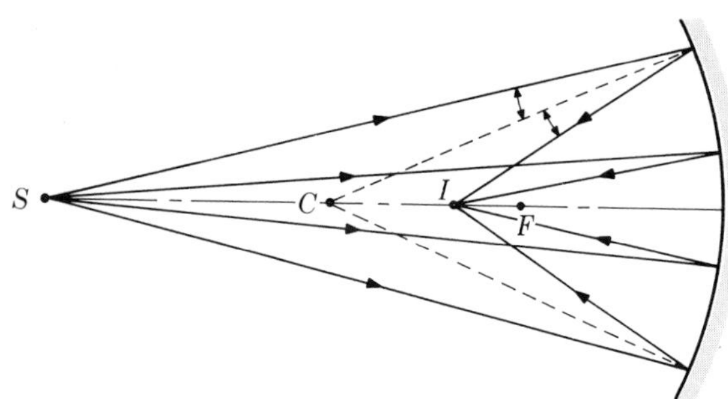

Figure 24-13
The points S and I are conjugate foci.

object is placed at I, the image will be formed at S, for the direction of each ray will merely be reversed from that shown in the figure. Any two points so situated that light from one is concentrated at the other are *conjugate foci*.

Since the image is formed by converging rays that actually pass through it, the image is real. This real image can be formed on a screen and viewed in that way.

The problem of locating the image of an extended source or object can be solved graphically by drawing rays to locate a few pairs of conjugate foci. The graphic method is indicated in Fig. 24-14. From any point O on the object, two rays are drawn to the mirror and their directions after reflection indicated. The point of intersection of these rays after reflection will be the image I (conjugate focus) of the point O on the object from which they originated. Two rays whose directions can be predicted readily are: first, the ray from O, parallel to the principal axis, which after reflection passes through F; and, second, the ray from O, in the direction OC along a radius of the mirror, which after reflection returns along the same line. The intersection of these rays at I locates the image of the head of the arrow. All other rays from O pass through I after reflection. Another pair of rays could be drawn from O' to locate I'.

The distance at which an object is located in front of a concave mirror affects the nature of the image produced. Figure 24-15*a* to *e* illustrates the formation of images where the object distances vary.

Case A The object is placed at a distance greater than twice the focal length in front of a concave mirror. Since C, the center of curvature of the mirror, equals $2F$, the object is placed in front of C (Fig. 24-15*a*). The image formed is real, inverted, and diminished in size.

Case B The object is placed at a distance equal to twice the focal length in front of the lens (Fig. 24-15*b*). The image formed is real, inverted, and the same size as the object.

Case C The object is placed between $2F$ and F (Fig. 24-15*c*). The image formed is real, inverted, and magnified in size.

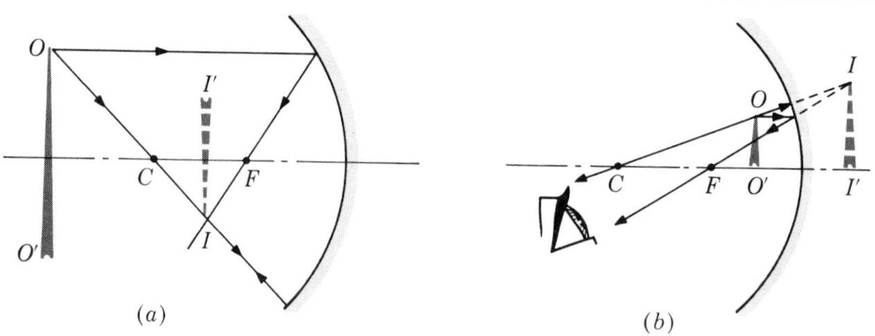

Figure 24-14
Location of the image formed by a concave mirror.

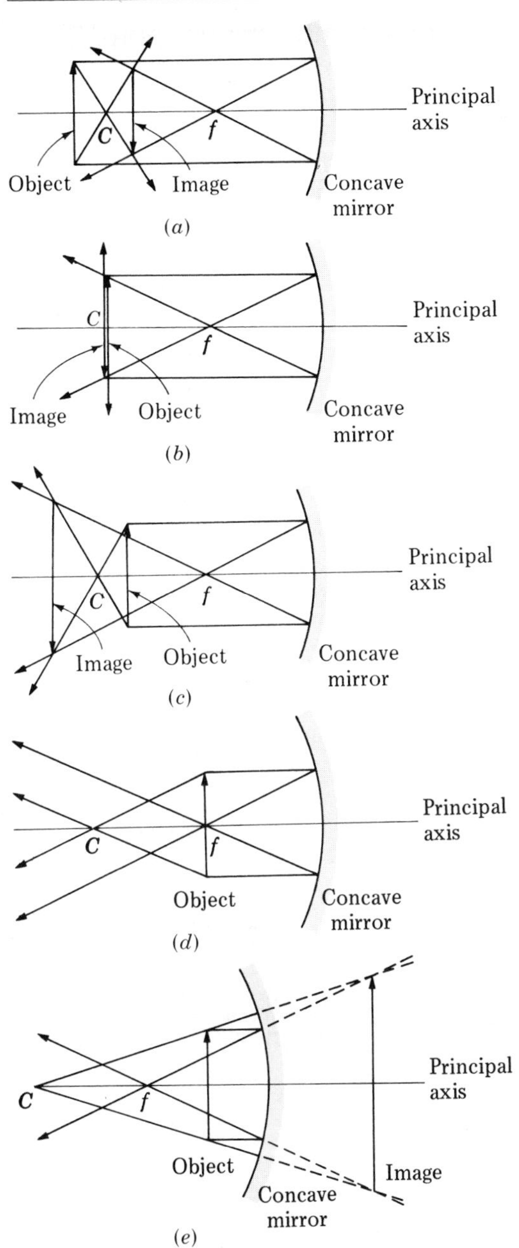

Figure 24-15
(*a*) Case *A*. (*b*) Case *B*. (*c*) Case *C*. (*d*) Case *D*. (*e*) Case *E*.

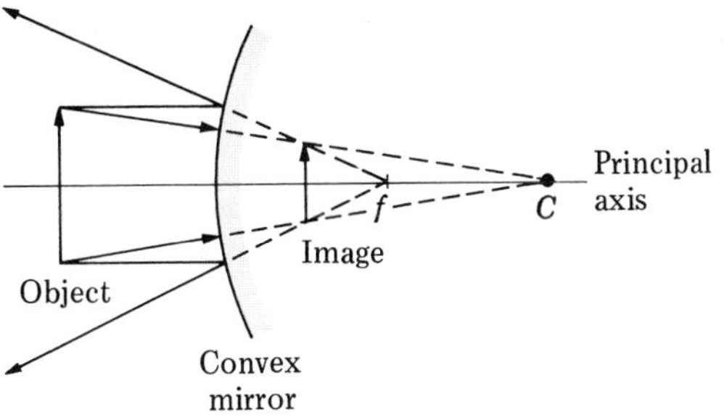

Figure 24-16
Image formed by a convex mirror.

Case D The object is placed at the principal focus F (Fig. 24-15*d*). Since the reflected rays are parallel, they do not intersect and no image is formed.

Case E The object is placed between the principal focus and the mirror (Fig. 24-15*e*). Since the image appears to form on the opposite side of the mirror (an opaque substance) to the object, the image is virtual (imaginary). A virtual image cannot be formed on a screen but can be observed by looking into a mirror. The rays do not pass through the virtual image but arrive at the eye as *if* they had originated at the virtual image. The image formed in this case is also erect and magnified in size.

The image of any real object formed by a convex mirror (Fig. 24-16) is always virtual, erect, and diminished.

24-7 THE MIRROR EQUATION

There is a simple relation between the distance p of the object from the mirror, the distance q of the image from the mirror, and the focal length f. To derive this relationship, consider Fig. 24-17 in which three rays have been drawn from point O on the object to locate the corresponding point I on the image. From one pair of similar triangles (shown shaded), taking $PM = OO'$, we have

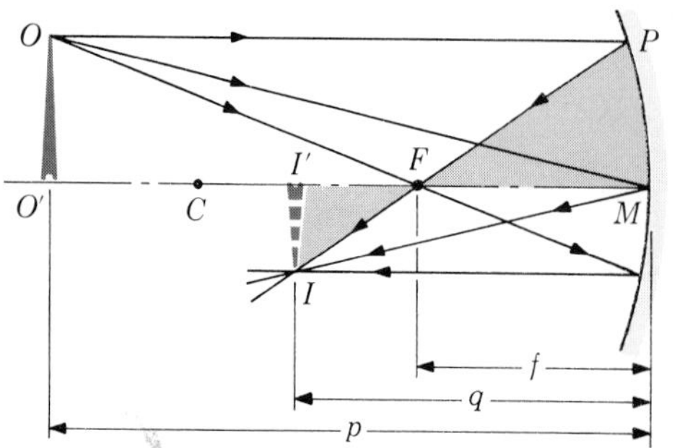

Figure 24-17
Diagram for the derivation of the mirror equation.

$$\frac{II'}{PM} = \frac{I'F}{FM} = \frac{q-f}{f} = \frac{q}{f} - 1 \qquad (3)$$

From another pair of similar triangles, $OO'M$ and $II'M$, it follows that

$$\frac{II'}{OO'} = \frac{I'M}{O'M} = \frac{q}{p} \qquad (4)$$

Since $PM = OO'$, the left-hand members of Eqs. (3) and (4) are equal and

$$\frac{q}{p} = \frac{q}{f} - 1$$

If we divide by q, we have

$$\frac{1}{p} + \frac{1}{q} = \frac{1}{f} \qquad (5)$$

Since the focal distance f equals $r/2$, the mirror equation may be written

$$\frac{1}{p} + \frac{1}{q} = \frac{2}{r} \qquad (6)$$

If any two of the quantities of Eq. (5) or Eq. (6) are known, the third can be calculated. Equations (5) and (6) apply to both concave and convex mirrors when the proper convention regarding signs is followed. The focal length f (or the radius of curvature r) is taken as *positive* for a *concave* (converging) mirror, *negative* for a *convex* (diverging) mirror. The object distance p and the image distance q are taken as *positive* for *real* objects and images and *negative* for *virtual* objects and images formed behind the mirror. Table 1 summarizes these sign conventions.

The mirror equation is always written with positive signs when expressed in algebraic symbols, as in Eqs. (5) and (6). Negative signs are introduced only with the values that are substituted for these symbols, as required by the sign conventions just stated.

Table 1
SIGN CONVENTION FOR MIRRORS

	Object	Image	Focal length
Concave mirror			
Real	+	+	+
Virtual	+	−	+
Convex mirror			
Virtual	+	−	−

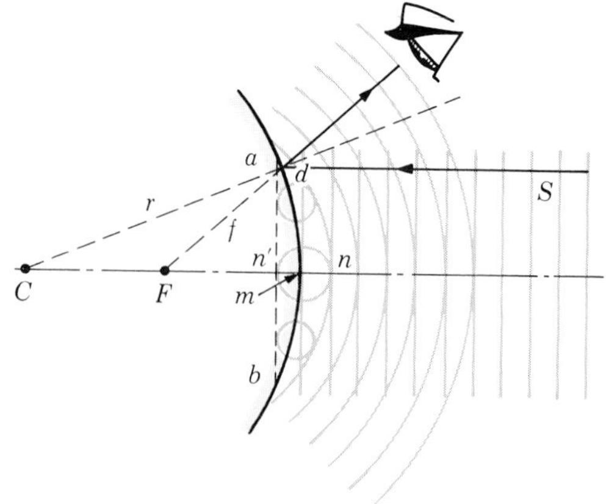

Figure 24-18
Reflection from a convex mirror, represented by a wave-front diagram.

The methods of physical optics may also be used to find the relation between image distance, object distance, and focal length in any mirror problem. As an illustration of this method, consider a parallel beam of light *Sd* impinging upon a convex mirror *amb* in Fig. 24-18. If the mirror were not present, the wave front could continue its motion undisturbed to the dotted position *an′b*. But when the center of the wave front meets the mirror at *m*, it is turned back and by Huygens' principle the reflected wavelet reaches *n* in the same time in which it would have reached *n′* if there were no mirror. By constructing the wavelets from other points on the mirror one obtains the reflected wave front *anb* whose center is at *F*. The light that enters the eye appears to originate at point *F* behind the mirror. Hence *F* is the principal focus of the mirror, since the waves were proceeding in a direction parallel to the principal axis before reflection. The curvature of the wave front as it reaches the mirror is represented by $1/p$ (positive for a convex wave front). The mirror changes this curvature by amount $1/f$ to produce a curvature $-1/q$; since q is positive in our sign convention for light converging to a real image, and a concave wave front was negative curvature, the curvature is given by $-1/q$. Then $1/p - 1/f = -1/q$ which is equivalent to Eq. (5). In the present case $1/p$, the curvature of the incident wave is zero, and $1/q = 1/f$.

Example A candle is held 3.0 cm from a concave mirror whose radius is 24 cm. Where is the image of the candle?

Figure 24-14*b* illustrates the conditions of the problem. From the general mirror equation

$$\frac{1}{p} + \frac{1}{q} = \frac{1}{f} = \frac{2}{r}$$

we have $$\frac{1}{3.0\text{ cm}} + \frac{1}{q} = \frac{1}{12\text{ cm}}$$

$$\frac{1}{q} = \frac{1-4}{12\text{ cm}} = -\frac{3}{12\text{ cm}}$$

$$q = -4.0\text{ cm}$$

The negative sign for q indicates that the image lies behind the mirror and is a virtual image.

24-8 MAGNIFICATION

The linear magnification produced by a mirror is the ratio of image size to object size. From the similar triangles $OO'M$ and $II'M$ in Fig. 24-17, it is apparent that the size of the image II' is to the size of the object OO' as the image distance q is to the object distance p. This will be true for any spherical mirror, without regard to signs.

$$\text{Magnification} = \frac{\text{image height}}{\text{object height}}$$

$$M = \frac{q}{p} \tag{7}$$

Example A man has a concave shaving mirror whose focal length is 40 cm. How far should the mirror be held from his face in order to give an image of twofold magnification?

An erect virtual magnified image is desired. Figure 24-14*b* illustrates the conditions of the problem. The equation

$$M = \frac{q}{p} = 2$$

gives a relation between p and q without regard to sign. But since the image is virtual, p and q have opposite signs, or

$$q = -2p$$

Substitution in the general mirror equation gives

$$\frac{1}{p} + \frac{1}{-2p} = \frac{1}{40\text{ cm}}$$

$$\frac{2-1}{2p} = \frac{1}{40\text{ cm}}$$

$$p = 20\text{ cm}$$

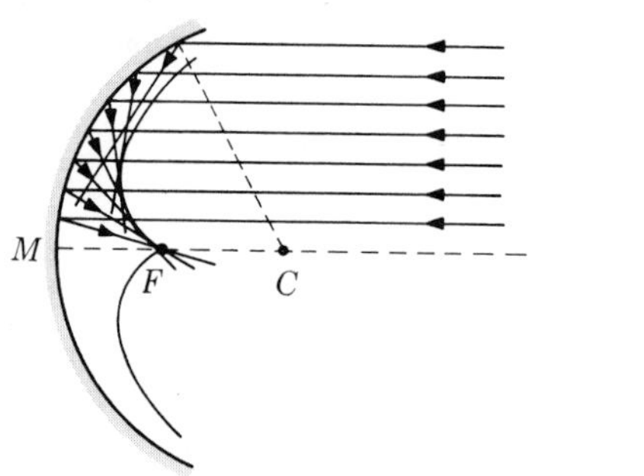

Figure 24-19
Spherical aberration.

24-9 SPHERICAL ABERRATION

The foregoing discussion of spherical mirrors applies only to mirrors whose apertures are small compared with their radii of curvature and for objects on or near the principal axis. Under other conditions the images formed are blurred and imperfect. For example, the extreme rays reflected from a mirror of large aperture (Fig. 24-19) cross the axis closer to the mirror than do the rays which are reflected nearer the center of the mirror surface and pass through the focal point *F*. This imperfection is called *spherical aberration*. The image is drawn out into a surface formed by the intersecting rays; the trace of this surface in the plane of the paper is a curved line called the *caustic* of the mirror. A caustic curve can be observed on the surface of liquid in a glass tumbler when the inner surface of the glass is illuminated obliquely.

It is possible to design the reflecting surface of a mirror of such shape that all rays reaching it from a definite object point will be brought to a common focus. For an object point at a great distance (incident rays parallel) the mirror should be a paraboloid. The most common use is in the mirror of the automobile headlight. When the filament is placed at the focus of the mirror, the rays sent out form a parallel beam (Fig. 24-20). A very slight shift in the position of the filament causes a marked displacement of the beam. The searchlight mirror and the big reflectors of astronomical telescopes are other applications of the parabolic mirror.

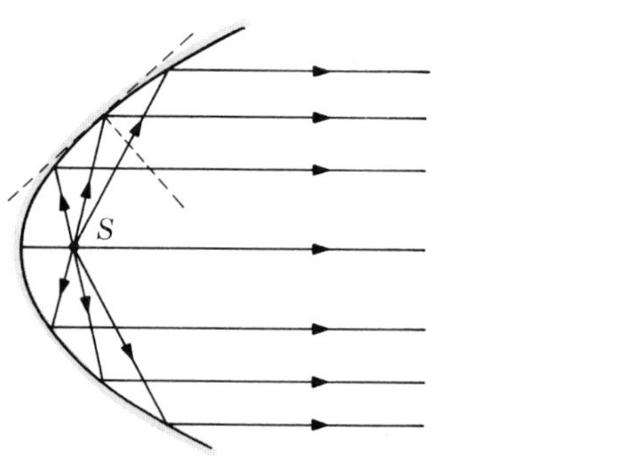

Figure 24-20
Reflection from a parabolic mirror.

SUMMARY

Huygens' principle states that light progresses as a wave, that every point on a wave front acts as a secondary source sending ahead wavelets, and that at any instant the new wave front is the surface tangent to all these wavelets.

When light is regularly reflected, the angle of reflection is equal to the angle of incidence. The reflected ray, the incident ray, and the normal to the surface are in the same plane. These are the *laws of reflection*.

The *image* formed by a plane mirror is the same size as the object, is erect, and is located as far behind the mirror as the object is in front of it.

In a plane mirror the image is *perverted;* i.e., right and left are interchanged with respect to the object.

The image formed by multiple mirrors can be found by considering the image of the first mirror to be the object for the second mirror, and so on.

When a plane reflector is rotated through a given angle, the reflected ray is deviated through twice as large an angle. This is the principle of the optical lever.

The *principal focus* of a spherical mirror is the convergence point for rays parallel to and close to the principal axis of the mirror. It is located halfway between the mirror and its center of curvature.

The *mirror equation* is

$$\frac{1}{p} + \frac{1}{q} = \frac{2}{r} = \frac{1}{f}$$

The mirror equation applies to both concave and convex mirrors. The radius of curvature, r, and focal length f are taken as positive for concave (converging) mirrors and negative for convex (diverging) mirrors. The object distance p and the image distance q are taken as positive for real objects or images and negative for virtual objects or images.

Linear magnification, defined by

$$\text{Magnification} = \frac{\text{image size}}{\text{object size}}$$

is given, for spherical mirrors, by the equation

$$M = \frac{q}{p}$$

This magnification equation holds for both concave and convex mirrors, where the right-hand term stands for the absolute value of q/p, without regard to sign.

Not all rays parallel to the principal axis of a spherical mirror are reflected to a single focus. The rays farther from the axis are reflected to cross the axis nearer to the mirror than those close to the axis. This imperfection is called *spherical aberration*.

Questions

1 Describe the image produced when an object is placed half the focal length in front of a convex mirror.

2 Distinguish between a wave front and a ray. Use a diagram to illustrate your answer.

3 State and show by diagrams the laws of reflection of light.

4 Give some practical illustrations in which diffuse reflection is preferred to regular reflection.

5 A plane wave strikes a plane mirror at an angle of incidence of 30°. By the use of Huygens' wavelets construct the reflected wave front.

6 A carpenter who wishes to saw through a straight board at an angle of 45° places his saw at the correct angle by noting when the reflection of the edge of the board in the saw seems to be exactly perpendicular to the edge itself. Explain by the use of a diagram.

7 How can a real image be distinguished from a virtual image? Can each type of image be projected on a screen? Why?

8 Using Fig. 24-5, prove geometrically that the image point I is the same distance from the mirror as the object point O.

9 If a plane mirror projects an image upon a second plane mirror and then upon a third, will the third image be perverted compared with the original object?

10 Show that for a plane mirror the image moves away from the object twice as fast as the mirror moves from the object.

11 Show that as a person approaches a plane mirror his velocity toward the image in the mirror doubles.

12 Show how to construct the image of an arrow 2 cm long held 4 cm in front of and parallel to a plane mirror. Describe fully the image formed.

13 At what distance should an object be placed in front of a concave mirror so that a real image is formed? Describe this image.

14 At what distance must an object be placed with respect to a concave mirror so that a virtual image is formed? Describe the image formed.

15 Identical twins stand at equal distances on

opposite sides of an opaque wall. What is the minimum size of window which must be cut in the wall so that they can obtain a full view of each other? How must the window be placed? What would the answers to these questions be as the twins move farther away from the wall?

16 The principal focus of a convex mirror is called a virtual focus. Why is this the case?

17 List several uses of concave mirrors; of convex mirrors.

18 If light waves are to converge to a point after reflection from a plane mirror, what must be their form before reflection? Explain by a sketch.

19 Two plane mirrors are placed at an angle of 90°. A ray of light falls on one mirror. What is the direction of the ray after two reflections? How is this principle used in reflectors on highway signs?

20 Prove the fact that when a plane mirror is rotated a beam of light reflected from it will rotate twice as fast as the mirror. Suggest some practical applications of this fact.

21 Show geometrically that the image of an object formed by a plane mirror is as far behind the mirror as the object is in front of the mirror. Use the spherical mirror equation to justify this same fact.

22 Two concave spherical mirrors have equal focal lengths but different apertures. Which mirror will form the hotter image of the sun? Why? Answer the same question for two mirrors of equal aperture but different focal lengths.

23 Does a convex mirror ever form an inverted image? Why? Illustrate by ray diagrams.

24 Draw appropriate ray diagrams to locate the approximate position and size of the image formed by a concave spherical mirror (Fig. 24-14) when the object lies (*a*) beyond C, (*b*) at C, (*c*) between C and F, and (*d*) inside F. Identify the nature of each image, real or virtual.

25 Make a geometrical construction to show that if the plane mirrors of Fig. 24-8 are set at 0° there will be $360/\theta - 1$ images.

26 Construct a graph showing the image distance (ordinates) against object distance (abscissas) for a concave mirror as the object distance is varied from plus infinity to minus infinity. Construct a similar graph for the case of a convex mirror.

27 A distant object is brought toward a concave spherical mirror. Describe the changes in the size of the image as the object distance varies from infinity to zero.

28 Two spherical mirrors of the same size and having focal lengths 20 cm and 10 cm are placed in the path of the parallel rays from a searchlight. Compare the sharpness of the images formed by the two mirrors.

29 A searchlight comprises a light source and a reflecting mirror of radius R. What should be the distance between the source of illumination and the mirror if as much light as possible is to be concentrated into a beam that neither converges nor diverges? What form of mirror is needed for this purpose? What characteristics should the luminous source have?

30 Show by diagram that a head-lamp reflector in the shape of a parabola will project a parallel beam of light when the filament is placed at the focal length of the mirror.

31 A girl stands between two plane mirrors that are parallel to each other, with one mirror in front of her and the other mirror behind her. What images can the girl see?

Problems

1 Sketch and calculate the image produced by an object placed 20 cm in front of a convex mirror whose focal length is 10 cm. What is its magnification?

2 What is the vertical length of the smallest plane mirror in which a man 2 m tall can just see his full height from the top of his head to his feet? *Ans.* 1 m.

3 Two vertical plane mirrors face each other 8.0 m apart. A candle is set 1.5 m from one mirror. An observer in the middle of the room looks into the other mirror and sees two distinct images of the candle. How far are these images from the observer?

4 A man 5 ft 10 in tall stands 4.0 ft from a large vertical plane mirror. (*a*) What is the size of the image of the man formed by the mirror? (*b*) How far from the man is his image? (*c*) What is the shortest length of mirror in which the man can see himself full length? (*d*) What length of mirror would suffice if he were 10.0 ft away?

Ans. 5 ft 10 in; 8.0 ft; 2 ft 11 in; 2 ft 11 in.

5 An observer walks toward a plane mirror at a speed of 3 m/s. With what speed does he approach his image?

6 A person backing a car up sees in the rear-view mirror a person walking toward him at 5 km/h. If the car is moving backward at 10 km/h, how fast does the person as viewed through the mirror approach the car?

Ans. 30 km/h.

7 A concave mirror has a radius of curvature of 12 cm. A small object is placed on the axis of the mirror 9 cm from it. Find the position of the image and the magnification.

8 An electric lamp with a concentrated filament of 300 cd intensity is placed 2.5 ft in front of a plane mirror. What is the illuminance 9.0 ft in front of the lamp? Neglect the radiation from the walls of the room. *Ans.* 5.2 fc.

9 Calculate and show by a ray diagram the image produced by an object placed 15 cm in front of a convex mirror whose focal length is 10 cm. What is its magnification?

10 The distance of comfortable, distinct vision is about 25 cm for the average person. Where should a person hold a plane mirror in order to see himself conveniently? *Ans.* 12.5 cm.

11 An object is located 300 mm in front of a convex spherical mirror whose radius of curvature is 400 mm. (*a*) Determine the location of the image and its relative size. (*b*) Is the image real or virtual? (*c*) Is it erect or inverted?

12 An object is placed 20 cm in front of a concave mirror of radius 60 cm. Where is the image?

Ans. -60 cm.

13 What is the radius of curvature of a shaving mirror that will produce an image twice normal size when a man stands 1.0 m in front of it?

14 What is the focal length of a concave mirror which produces an image five times the size of an object placed 9 in from the mirror?

Ans. 7.5 in.

15 A small candle is placed in front of a convex mirror which produces an image half the size of the flame. An observer then places a wire behind the mirror and shifts it back and forth until there is no parallax between the image as seen in the mirror and the wire projecting above the mirror. The distance of the wire from the mirror in this position is 250 mm. What is the focal length of the mirror?

16 An object is located on the principal axis of a convex mirror 20 in in front of the mirror. The image is 6.7 in behind the mirror. What is the radius of curvature of the mirror?

Ans. -20 in.

17 The moon is approximately 2,160 mi in diameter. What is the size of the image of the moon formed by a concave mirror of 3.0 m radius when the moon is at its nearest distance from the earth, 221,000 mi?

18 In what position in front of a spherical mirror should an object be placed to produce a real image which is magnified three times if the radius of curvature of the mirror is 18 cm?

Ans. 12 cm.

19 In a solar heater, water flows through a glass pipe located above and parallel to a semicylindrical concave reflecting surface. If the reflector has a diameter of 1.0 m, how far above the reflector should the center of the pipe be located? Should the axis of the cylinder be located in an east-west or north-south direction? Is it desirable to rotate the mirror during the day?

20 A dentist holds a concave mirror of radius of curvature 6.0 cm at a distance 2.0 cm from a filling in a tooth. What is the magnification of the image of the filling? *Ans.* 3.0.

21 Find the position, nature, and size of the image of an object 4.0 cm long formed by a concave spherical mirror, if the object is 100 cm from the mirror and the radius of curvature is 40 cm.

22 A convex mirror whose focal length is 15 cm has an object 10 cm tall and 60 cm away. Find the position, nature, and size of the image.

Ans. -12 cm; 2.0 cm.

23 A concave mirror has a radius of curvature of 24 cm. A small lamp bulb is held on the axis 18 cm from the mirror. Find the position and nature of the image and its magnification.

24 Describe the image produced by placing an object 30 cm in front of a convex mirror having a focal length of 10 cm. *Ans.* Virtual; 7.5 cm behind mirror; magnification = $\frac{3}{4}$.

25 The sun has a diameter of 864,100 mi and is distant, on the average, 92,900,000 mi from the earth. What will be the size of its image formed by a concave mirror of 6.0 ft radius?

26 In what position should an object be placed in front of a concave mirror having a focal length of 20 cm so that an erect image which is twice as large as the object is formed?

Ans. $p =$ 10 cm.

27 A concave mirror has a radius of curvature of 20 cm. Locate the image and determine its size when an object 4.0 cm high is 5.0 cm in front of the mirror.

28 A concave and a convex mirror, each of radius 40 cm, face each other at a distance of 60 cm. An object 5.0 mm high is placed midway between the mirrors. Find the position and size of the image formed by successive reflections from the two mirrors (*a*) if the first reflection is at the concave mirror and (*b*) if the first reflection is at the convex mirror.

Ans. (*a*) 60 cm, 10 mm, at convex surface; 10 mm; (*b*) −12 cm, 2.0 mm, 28 cm, 0.78 mm.

29 A concave mirror has a radius of curvature of 50 cm. Find two positions at which an object may be placed in order to give an image four times as large. What are the position and character of the image in each case?

30 A person stands between two large vertical plane mirrors which face each other and are 10 m apart. If he stands 6 m from the mirror that he faces, how far away from him are the first, second, and third images of himself that he sees?

Ans. 12 m; 20 m; 32 m.

31 An object is placed (*a*) 1.5 m in front of one and (*b*) 3.0 m in front of the other of two plane mirrors placed at right angles to each other. Locate three images formed by this arrangement.

32 A concave mirror of radius 80.0 cm and a plane mirror face each other 60.0 cm apart. An object is placed midway between the mirrors. Find the position of the image and the magnification when the first reflection is (*a*) by the plane mirror and (*b*) by the concave mirror.

Ans. 30.0 cm behind plane mirror; 72.0 cm; 0.800; −120 cm; 4.00.

33 A lamp, a point source of light, is placed on an optical bench and a plane mirror is placed 4 ft away from the lamp. An object which is 6 ft away from the lamp and on the opposite side to the mirror receives illumination both directly from the lamp and from the reflected light from the mirror. If the direct illuminance on the object by the lamp is 40 fc, what is the total illuminance on the object?

34 A convex mirror having a radius of 40 cm is placed 50 cm from a concave mirror having a radius of 30 cm. If an object is placed 20 cm from the concave mirror, describe the image produced by the combined mirrors by light which reflects from the concave mirror.

Ans. Real, inverted, 20 cm from convex mirror, magnification = 6.

Clinton Joseph Davisson, 1881–1958

Born in Bloomington, Illinois. Physicist at the Bell Telephone Laboratories. Shared the 1937 Nobel Prize for Physics with G. P. Thomson for their discovery of the interference phenomena arising when crystals are exposed to electron beams.

Sir George Paget Thomson, 1892–

Born in Cambridge, England. Professor at the Imperial College of Science and Technology of the University of London. Shared the 1937 Nobel Prize with Davisson for their discovery of interference phenomena of electrons.

25

Refraction of Light

We have seen that light travels in straight lines in homogeneous media. This is frequently not the case, however, for heterogeneous media, such as air, which vary in density with temperature or the lens of the eye in which the optical properties change from the surface to the center. When light encounters an abrupt change of medium at a boundary surface, some of the light is reflected and some is transmitted. The relative amount of light that is transmitted depends on each of the media and on the angle at which the light strikes the boundary. Of the light that enters the second medium, some is absorbed and some penetrates the material and encounters the far boundary, if any. Here again reflection and transmission take place.

A beam of light in one medium that enters a second medium obliquely will undergo an abrupt change in direction if the speed of the wave in the second medium is different from that in the first. This bending of the light path is called *refraction*. It is because of refraction that we are able to see transparent objects. When the boundary between two media is curved, the light paths are altered in such a way as to cause distortion or change in size of an object seen through the boundary. An understanding of the laws of refraction makes possible the design and application of spectacles, cameras, telescopes, microscopes, and the other optical instruments which have extended our vision.

25-1 REFRACTION

Consider the simplest case of refraction, that of a plane wave meeting a plane surface. In Fig. 25-1, AB represents an advancing plane wave, OA and PB are rays normal to the wave front, and NA is the normal to the surface of the medium at A. The direction of the incident ray OA is defined by the angle of incidence i which it makes with the normal NA. The plane containing the incident ray and the normal to the surface is called the plane of *incidence*. The angle r between the refracted ray AC and the normal is called the angle of *refraction*.

The incident ray, the refracted ray, and the normal to the surface lie in the same plane.

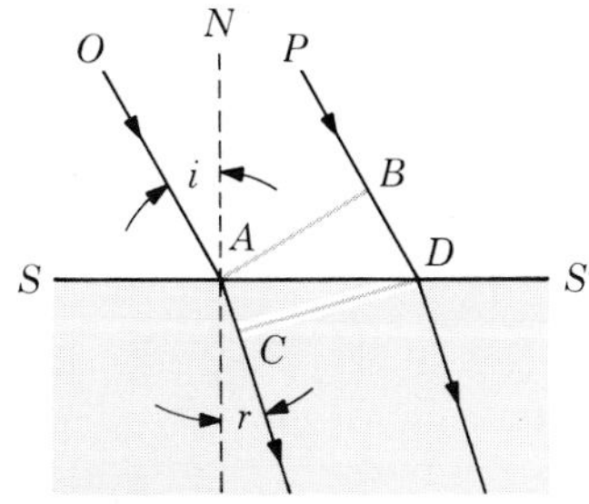

Figure 25-1
Refraction of a plane wave.

Refraction may be explained quite simply on the basis of the wave theory. Let SS' in Fig. 25-2 represent the boundary surface between two optical materials and AB represent a plane wave front in the first medium at the instant one edge of the beam enters the second medium. A Huygens' wavelet starting from A will move a distance AC in the second medium in the same time that the wave moves from B to D in the first medium. As the original wave front AB moves on, secondary wavelets will be started from successive points, such as A' and A'', on the surface SS' to the right of A. When the point B of the original wave front has reached D, all the secondary waves will have spread into the second medium. There will be a new wave front CD in the second medium which is the envelope of all the secondary wavelets. Because the wave travels more slowly in the second medium than in the first, the new wave front CD is not parallel to the original wave front AB. The light has been refracted. The dotted construction lines of Fig. 25-2 show that the wave would have reached the position $C'D$ in the time considered if it had continued at its original speed.

From this consideration of Fig. 25-2 we note that, when a ray of light passes from one material into another in which its speed is less than in the first, the ray is bent toward the normal. If the light travels from the medium of less speed into that of greater speed, the ray will be bent away from the normal. This would represent a condition in which the direction of propagation of the light in Fig. 25-2 is reversed.

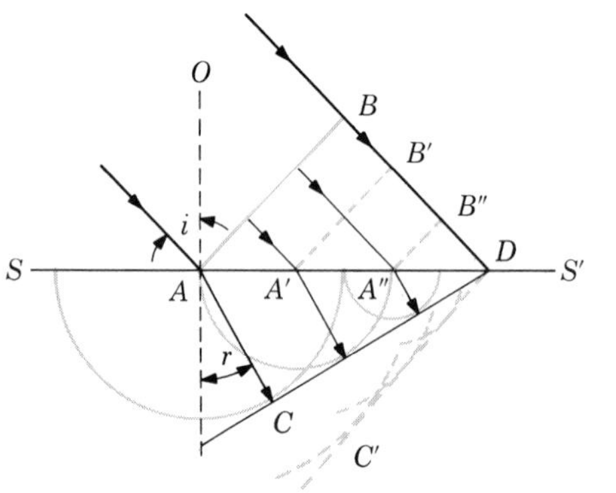

Figure 25-2
Refraction represented by a Huygens' wave diagram.

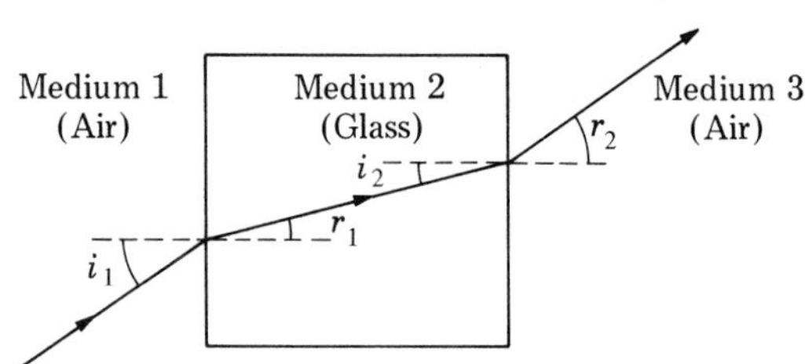

Figure 25-3
Refraction of light through a glass plate.

If a ray of light after leaving one medium and then passing through a second medium leaves that second medium and enters a third medium through which light passes at a different speed than in the second medium, the direction of the ray will once again be changed according to the procedure noted above. For example, a light ray passing from air through a glass plate having parallel faces and then back out into air will undergo two refractions, as shown in Fig. 25-3. Since in this example the medium on both sides of the glass is the same, the ray of light will leave the glass at the same angle that it entered the glass plate on the opposite side, that is, $i_1 = r_2$.

25-2 INDEX OF REFRACTION

Since the change in direction of light as it goes from one medium to another depends upon the speeds of light in the two media, we may use the ratio of the two speeds in expressing relationships in refraction. The *index of refraction* of the second medium relative to the first medium may be defined as the ratio of the speed v_1 in the first medium to the speed v_2 in the second medium,

$$n_r = \frac{v_1}{v_2} \quad (1)$$

If the first medium is empty space, v_1 becomes

the speed c of light in empty space. The special index of refraction for this case is called the *absolute* index of refraction, and we shall represent it by n without the subscript. Thus

$$n = \frac{c}{v} \qquad (2)$$

The relative index is related to the absolute indices of the two mediums, since $v_1 = c/n_1$ and $v_2 = c/n_2$. Then

$$n_r = \frac{v_1}{v_2} = \frac{c/n_1}{c/n_2} = \frac{n_2}{n_1} \qquad (3)$$

The relationship between the angle of incidence i and the angle of refraction r may be expressed in terms of the indices of refraction. In Fig. 25-1 light travels from B to D in the first medium in the same time t that it travels from A to C in the second medium. Thus

$$BD = v_1 t \qquad \text{and} \qquad AC = v_2 t$$

Since DA is perpendicular to AN and BA is perpendicular to AO, $\measuredangle DAB = \measuredangle i$. Likewise $\measuredangle ADC = \measuredangle r$.

$$\frac{BD}{AD} = \sin i$$

$$\frac{AC}{AD} = \sin r$$

By dividing we obtain

$$\frac{BD}{AC} = \frac{\sin i}{\sin r}$$

or

$$\frac{v_1 t}{v_2 t} = \frac{v_1}{v_2} = \frac{\sin i}{\sin r} = \frac{n_2}{n_1} = n_r \qquad (4)$$

This important relationship may be represented by the equation

$$n_1 \sin i = n_2 \sin r \qquad (5)$$

Equation (4) expresses the fact that, for a given wavelength and a given pair of substances, the ratio of the sine of the angle of incidence to the sine of the angle of refraction is a constant, independent of the angle of incidence. This statement was developed by Snell and is known as *Snell's law.*

Table 1
INDICES OF REFRACTION (for wavelength 5,893 Å)

Solids:	
Crown glass	1.517
Barium flint glass	1.568
Barium crown glass	1.574
Light flint glass	1.580
Dense flint glass	1.655
Fluorite	1.434
Calcite (ordinary ray)	1.658
Calcite (extraordinary ray)	1.486
Canada balsam	1.530
Diamond	2.419
Quartz, fused	1.4585
Ice at −8°C	1.31
Lucite	1.50
Liquids:	
Benzene at 20°C	1.501
Carbon disulfide at 20°	1.643
Carbon tetrachloride at 20°	1.461
Ethyl alcohol at 20°	1.354
Water at 0°C	1.334
Water at 20°	1.333
Water at 40°	1.331
Water at 80°	1.323
Gases and vapors at 0°C and 760 mm Hg:	
Dry air	1.000292
Carbon dioxide	1.00045
Ethyl ether	1.00152
Water vapor	1.000250

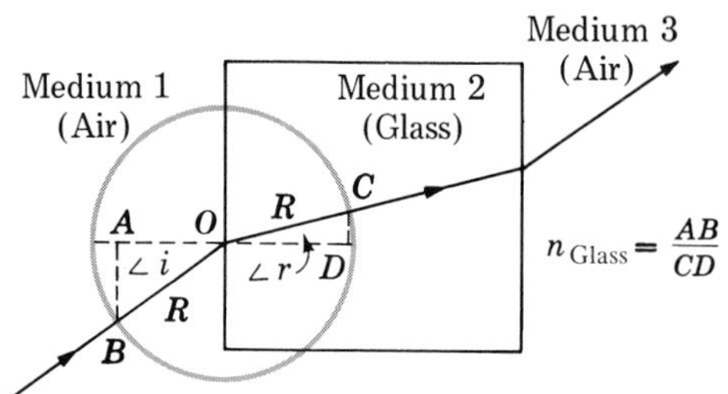

Figure 25-4
A geometric procedure to determine the index of refraction.

A simple geometric exercise will yield a good approximation of the index of refraction of a substance.

Consider a ray of light traveling through a glass plate as in Fig. 25-3. The amount that the ray is refracted depends upon the relative indices of refraction. After having constructed a ray diagram through standard laboratory procedure, a circle is drawn around point O with a radius R, as shown in Fig. 25-4. Then $\sin r = CD/R$ and the $\sin i = AB/R$. Since

$$n_r = \frac{n_2}{n_1} = \frac{\sin i}{\sin r}$$

where medium 1 is air and $n_1 = 1.0$, then,

$$n_2 = \frac{AB/R}{CD/R} = \frac{AB}{CD}$$

Therefore, by directly measuring the length of AB and CD, and relating AB to CD, the index of refraction of the glass can be determined.

The numerical value of an index of refraction is characteristic of the two media, but it depends also on the wavelength of the light. Hence an index of refraction is specified definitely only when the wavelength of the light is stated. Unless otherwise mentioned, an index is usually given for yellow light. The absolute index of refraction for air under standard conditions is 1.0002918 for light having the wavelength of the D line of sodium (5,893 Å). Since the absolute index n for air is so near to unity, it follows that for a solid or a liquid the absolute index and the index relative to air differ only slightly and it is usually not necessary to distinguish between them.

Example Light passes at an angle of incidence of 30° from glass into the air. What is the angle of refraction if the index of refraction of this glass is 1.50?

From Eq. (5),

$$n_1 \sin i = n_2 \sin r$$

Here the first medium is glass and the second is air.

$$\sin r = \frac{n_1}{n_2} \sin i = \frac{1.50}{1.00}(0.500) = 0.750$$

$$r = 49°$$

Example A ray of light in water ($n_w = \frac{4}{3}$) is incident upon a plate of crown glass ($n_g = 1.517$) at an angle of 45°. What is the angle of refraction for the ray in the glass? What is the index of glass with respect to water?

$$n_1 \sin i = n_2 \sin r$$

$$\sin r = \frac{1.333}{1.517}(0.707) = 0.621$$

$$r = 38°$$

and

$$n_r = \frac{n_2}{n_1} = \frac{1.517}{1.333} = 1.138$$

25-3 THE SHALLOWING EFFECT OF REFRACTION

When a spherical wave front passes from one material to another through a plane surface, the form of the wave front is changed and it is in general no longer spherical. A particular case of interest, represented by Fig. 25-5, is that of waves from a point source O in some optically dense medium emerging into air.

When the wave front from O has reached A, the secondary wavelet from P will have reached

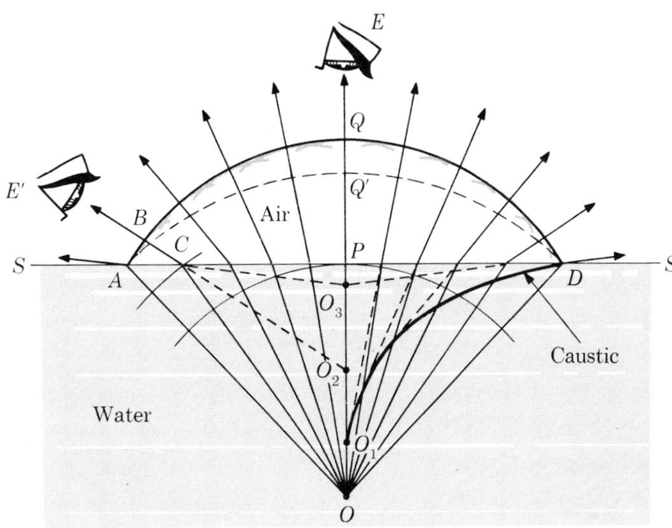

Figure 25-5
The apparent depth of an object under water is less than its depth below the surface.

Q where $PQ = \frac{4}{3}(OA - OP)$, since light has a speed in air four-thirds of that in water. The secondary wavelet from C will have spread to a radius $CB = \frac{4}{3}(OA - OC)$. By drawing similar secondary wavelets from successive points along APD, one can locate the new position of the advancing wave front AQD and show that this wave front is no longer spherical. For an observer at E looking vertically down, the wave front arrives with a radius of curvature O_1P, and the source appears to be at O_1 at a depth three-fourths the actual depth of the object O below the surface of the water. To an observer at E' the object seems to be somewhere on the line CO_2, but since the wave front is not spherical, the image shows astigmatism (Chap. 26) and is not uniquely located. If the object is viewed in a direction normal to the boundary between media, the ratio of the real depth to the apparent depth in the medium in which the object is located is equal to the ratio of the two indices.

$$\frac{\text{Real depth}}{\text{Apparent depth}} = n_r \qquad (6)$$

Example A plate of glass 2.00 cm thick is placed over a dot on a sheet of paper. The dot appears 1.280 cm below the upper surface of the glass when viewed from above through a microscope. What is the index of refraction of the glass plate?

From Eq. (6),

$$n_r = \frac{2.00 \text{ cm}}{1.280 \text{ cm}}$$
$$= 1.57$$

Example (*a*) A fisherman standing on a dock attempts to spear a fish swimming alongside of the dock. Should he aim directly at, above, or below the fish to hit it? (*b*) Where should he aim if the fish is directly under him?

(*a*) Since the fish seems to be farther away than it actually is (Fig. 25-6), the fisherman must aim *below* or closer to himself than the fish appears to be.

(*b*) As the fisherman looks at the fish which is directly below him, there will be no bending effect due to refraction. Therefore, he will aim directly at the fish.

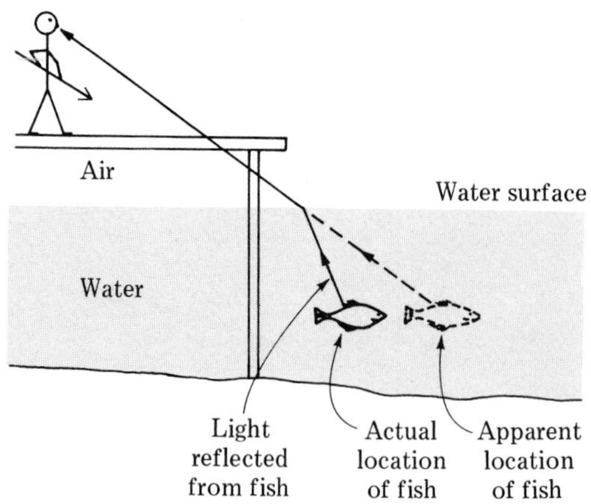

Figure 25-6
Fisherman spearing fish.

25-4 MIRAGES, ATMOSPHERIC REFRACTION

On still, sunny days there may be a layer of hot, expanded air in contact with the heated ground. Light travels faster in the rarer hot air than in the denser cool air above it. Hence light rays entering the warm air obliquely from above will be refracted upward. One may see inverted images of distant objects (Fig. 25-7) suggestive of the reflections in a smooth pool of water. Mirages on a small scale are often observed over concrete highways on still, hot days.

A mirage of another sort called *looming* may occur when atmospheric conditions are reversed

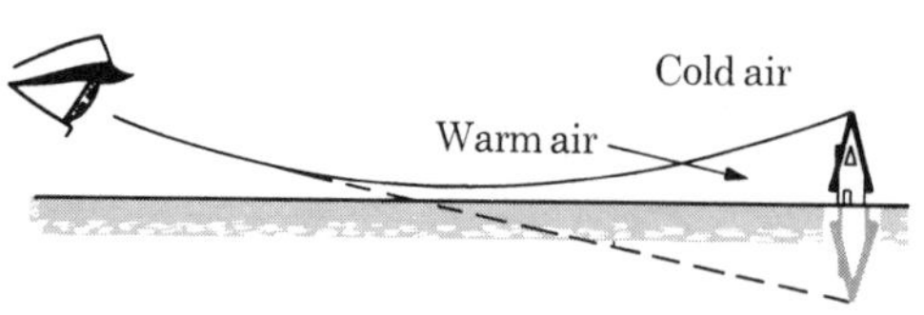

Figure 25-7
A mirage is formed by atmospheric refraction upward.

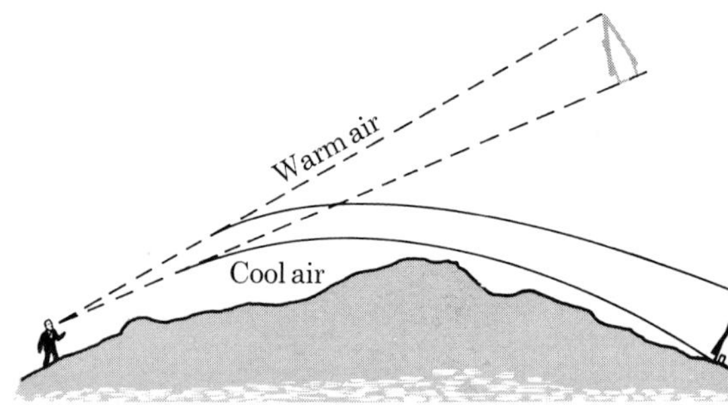

Figure 25-8
Looming is caused by atmospheric refraction downward.

and the lower strata of air are cooler than the upper strata, as would be the case over a snow field or a body of cold water. Rays of light from a distant object are then deviated downward. One may see an image of a ship above the ship itself, or the curvature of the light rays may bring into view objects normally below the horizon (Fig. 25-8).

The change of the refractive index of air with changing temperature is easily observed in a turbulent stream of hot air rising from a stove or radiator, since there is an apparent wavering of objects seen through the nonhomogeneous air. A ray of light entering the earth's atmosphere obliquely is bent toward the normal. Hence we see light from the sun while it is slightly below the horizon; and the sun is flattened in appearance at sunrise and sunset. The positions of the sun and stars always appear to be higher than their actual positions, except when they are directly overhead.

25-5 TOTAL INTERNAL REFLECTION

In speaking of optical materials, it is customary to refer to relative speeds of light in terms of *optical density*. A material of lesser speed is called *optically more dense*.

If we consider the passage of light from an

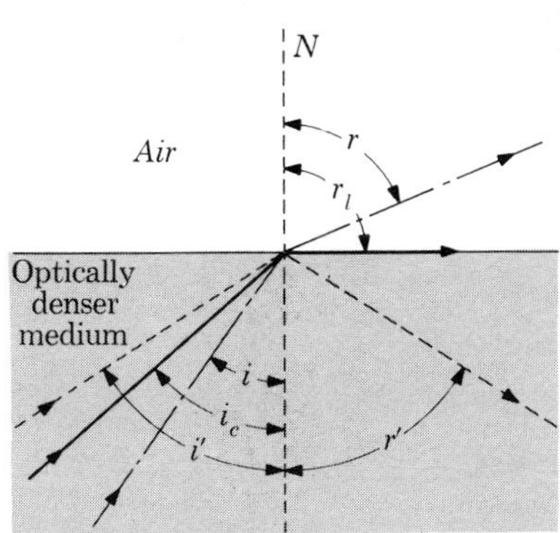

Figure 25-9
Total internal reflection occurs when the angle of incidence exceeds the critical angle.

optically denser medium out into air, as in Fig. 25-9, we observe that as the angle of incidence i is increased the angle of refraction r increases and approaches the limiting value $r_1 = 90°$, beyond which, of course, there could be no light refracted into the air. The limiting angle of incidence in the denser medium, which makes the angle of refraction 90°, is called the *critical angle* of incidence, i_c. From the law of refraction,

$$n_1 \sin i = n_2 \sin 90°$$

$$\sin i_c = \frac{n_2}{n_1} \qquad (7)$$

When the angle of incidence is increased beyond its critical value i_c, we find that the light is totally reflected, making the angle of reflection r' equal to the angle of incidence i'. Total reflection can take place only when the light in the medium of lesser speed is incident on the surface separating it from the medium of greater speed.

Example What is the critical angle between carbon disulfide and air?

$$\sin i_c = \frac{n_2}{n_1} = \frac{1.000}{1.643} = 0.609$$

$$i_c = 37°27'$$

Total reflection is utilized in various optical instruments. A beam of light may be turned through 90° by a 45° right-angle prism of glass having polished faces (Fig. 25-10*a*). Total reflection in the prism of Fig. 25-10*b* inverts the image, and such an inverting prism may be used in binoculars or in a projection lantern to give an upright image of an object that otherwise would appear inverted.

25-6 REFRACTION THROUGH PLANE-PARALLEL PLATES

When a ray of light passes through a layer of transparent material that has plane-parallel surfaces and emerges again into the first medium, the emergent ray is parallel to its original direction but is displaced laterally. This may be seen by applying the laws of refraction at each of the surfaces represented in Fig. 25-11.

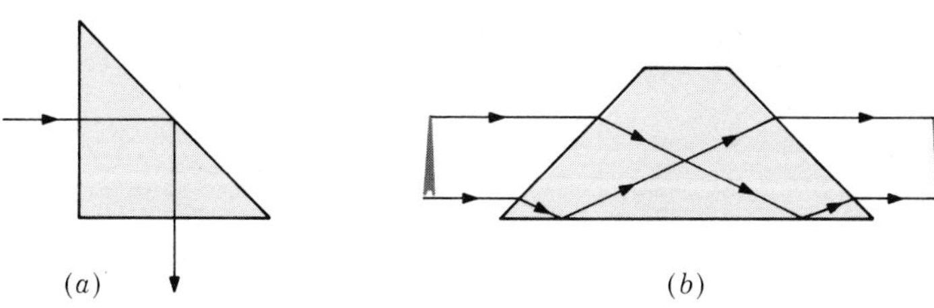

Figure 25-10
Totally reflecting prisms.

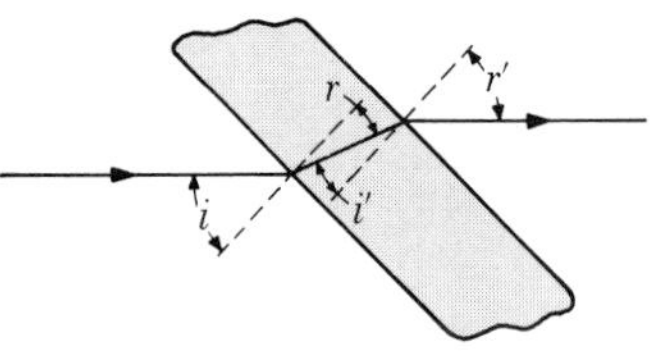

Figure 25-11
Refraction through a plane-parallel glass plate.

25-7 REFRACTION BY A PRISM

When light passes from air through a glass prism, it is bent toward the thicker part of the prism. In Fig. 25-12, A is the refracting angle, and δ is the angle of deviation, measured between the original direction of the incident ray and the direction of the refracted ray. The deviation is found to depend on the prism angle, the index of refraction of the prism material, and the angle of incidence i. Minimum deviation D occurs when the ray passes through the prism symmetrically, making i equal to r'.

From Fig. 25-12*a*, the deviation δ is the sum of the deviations $i - r$ taking place at the first surface and $r' - i'$ at the second surface,

$$\delta = i - r + r' - i' = (i + r') - (r + i')$$

But $r + i' = A$; and for minimum deviation $i = r'$, and hence $r = i'$. Therefore, for minimum deviation

$$D = 2i - A \qquad \text{or} \qquad i = \tfrac{1}{2}(A + D)$$

and

$$r = \frac{A}{2}$$

If we substitute these values into Eq. (5) for the first surface, we have

$$n_1 \sin \tfrac{1}{2}(A + D) = n_2 \sin \tfrac{1}{2}A$$

$$n_2 = n_1 \frac{\sin \frac{1}{2}(A + D)}{\sin \frac{1}{2}A} \qquad (8)$$

Equation (8) provides a precise method of determining the index of refraction of a transparent material in the form of a prism.

Example Light from a sodium lamp when passed through a 60°00′ prism has a minimum angle of deviation of 51°20′. What is the index of refraction of the glass?

From Eq. (8),

$$n = 1.000 \frac{\sin \frac{1}{2}(60°00' + 51°20')}{\sin \frac{1}{2}(60°00')} = \frac{\sin 55°40'}{\sin 30°00'}$$

$$n = \frac{0.8258}{0.5000} = 1.652$$

Equation (8) can be used only when the prism is so adjusted that the ray passes through the

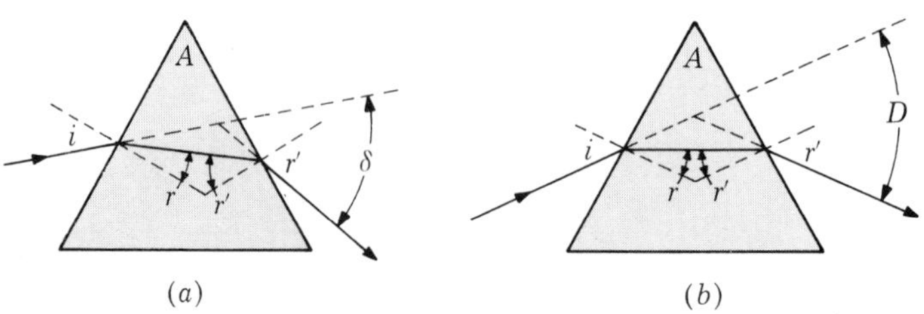

Figure 25-12
Refraction by a prism. The deviation in (*a*) is greater than the minimum deviation D in (*b*). For minimum deviation, $i = r'$.

prism perpendicular to the bisector of the angle of the prism. For any other angle of incidence, it is possible to compute the deviation (now greater than the minimum) by following the ray through the prism and calculating the angles of incidence and refraction at each surface.

25-8 REFRACTION AT A SPHERICAL SURFACE

When light is refracted at a spherical surface the refraction at each point is consistent with the general law of refraction. If the rays from an object point O (Fig. 25-13) make small angles with the axis (line through the center of curvature), the rays meet at an image point I.

From the geometry of triangle COA of Fig. 25-13,

$$i = \theta + \alpha$$

And from CAI,

$$\alpha = r + \phi \qquad \text{or} \qquad r = \alpha - \phi$$

$$\tan\theta = \frac{y}{p + \delta}$$

$$\tan\phi = \frac{y}{q - \delta}$$

$$\sin\alpha = \frac{y}{R}$$

If the angles are small, the angle, its sine, and its tangent are equal and δ is negligible. To this degree of approximation

$$\sin i = i = \frac{y}{p} + \frac{y}{R}$$

and

$$\sin r = r = \frac{y}{R} - \frac{y}{q}$$

Then since $n_1 \sin i = n_2 \sin r$,

$$\frac{n_1 y}{p} + \frac{n_1 y}{R} = \frac{n_2 y}{R} - \frac{n_2 y}{q}$$

$$\frac{n_1}{p} + \frac{n_2}{q} = \frac{(n_2 - n_1)}{R} \qquad (9)$$

Equation (9) is valid only if the angles are small. If the rays diverge from the axis more than allowed by this approximation, the rays do not all pass through a single image point.

Conventions of sign must be used in Eq. (9). We may use the same convention for the signs of p and q as that used for mirrors (Sec. 24-7). It is convenient to call R positive when it is measured from the surface to the center in the direction of the light leaving the surface, negative

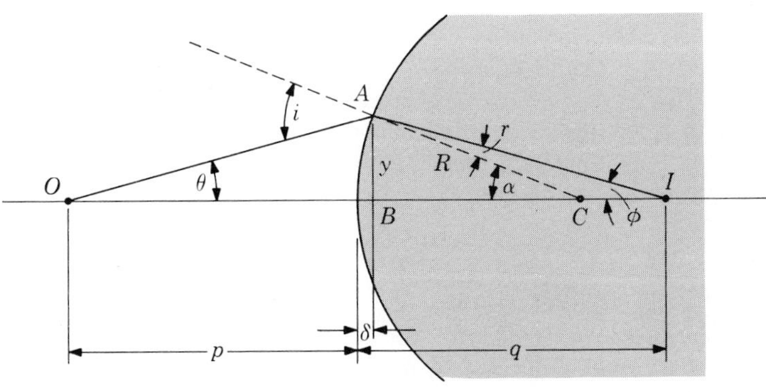

Figure 25-13
Refraction at a spherical surface.

if in the opposite direction. This convention is consistent with that used for the radii of spherical mirrors in Sec. 24-7. It follows from this convention that a refracting surface that is convex to the incident light has a positive R while one that is concave has a negative R.

Example A water tank has a small, thin window that is a section of a sphere of 6.00 cm radius. The convex side of the window is on the outside of the tank. A small light is placed 30.0 cm from the window (*a*) on the outside and (*b*) on the inside of the tank. Find the position of the image in each case.

(*a*)
$$\frac{n_1}{p} + \frac{n_2}{q} = \frac{n_2 - n_1}{R}$$

$$\frac{1.00}{30.0\text{ cm}} + \frac{1.333}{q} = \frac{1.333 - 1.00}{6.00\text{ cm}}$$

$$q = 60.0\text{ cm}$$

The image is real and inside the water.

(*b*)
$$\frac{1.333}{30.0\text{ cm}} + \frac{1.00}{q} = \frac{1.000 - 1.333}{-6.00\text{ cm}} = \frac{0.333}{6.00\text{ cm}}$$

$$q = 90.0\text{ cm}$$

The image is real and outside the water.

SUMMARY

Refraction is the change in direction of a light ray because of change in speed.

The *index of refraction* of a substance relative to the surrounding medium is the ratio of the speed v_1 of light in the first medium to the speed v_2 in the second medium.

$$\frac{v_1}{v_2} = n_r$$

The *absolute* index of refraction is the ratio of the speed of light in empty space to the speed in the medium.

$$\frac{c}{v} = n$$

Snell's law states that

$$n_1 \sin i = n_2 \sin r$$

$$\frac{\sin i}{\sin r} = \frac{n_2}{n_1} = n_r$$

A transparent body appears to be less thick than it really is because of the refraction at its surface. The amount of this shallowing effect depends upon the angle at which it is viewed.

Mirages result from atmospheric refraction.

Total internal reflection may occur when light passes from a medium of less speed to one of greater speed. As the light proceeds in this direction, it is bent away from the normal. The angle of incidence in the denser material for which the angle of refraction is 90° is called the *critical angle*. If the angle of incidence in the denser material is greater than the critical angle, total reflection occurs.

When light passes through a body whose surfaces are plane and parallel, the rays are displaced but not deviated.

When light passes from air through a glass prism, it is bent toward the thicker part of the prism. The amount of deviation depends upon the angle of the prism, the angle of incidence, and the index of refraction of the prism. Minimum deviation occurs when the ray passes through the prism symmetrically, making $i = r'$. For minimum deviation D

$$n_r = \frac{\sin \frac{1}{2}(A + D)}{\sin \frac{1}{2}A}$$

For refraction at a spherical surface, radius R,

$$\frac{n_1}{p} + \frac{n_2}{q} = \frac{n_2 - n_1}{R}$$

Questions

1 A lead glass beaker when immersed in tetrachloroethylene "disappears." Since it actually is still intact, though submerged in the solution, what optical explanation can be given for this optical mystery?

2 Define the terms opaque, transparent, and translucent and give examples of each.

3 State the law of refraction. Explain why a beam of light is refracted when it enters a new medium at an angle.

4 How could one show experimentally that light travels faster in air than in water?

5 A bundle of rays parallel to the principal axis is reflected by a convex mirror in air. If the mirror is placed in water, will the divergence of the rays from the mirror be the same? Illustrate by sketches.

6 Compare and give reasons for the difference between the index of refraction of glass for red light and for violet light.

7 Sketch a diagram and explain how refraction lengthens the daylight period of the day. Would smog affect this process? How?

8 Show how the "shallowing effect" in water appears to a fisherman as he moves away from the shore of a lake.

9 When a piece of thick plate glass is placed in a beam of convergent light, what happens to the point of convergence? Explain by the use of a diagram.

10 Under what conditions will total reflection occur? Give examples of uses made of total reflection.

11 As a ray of light passes obliquely from a medium which is optically less dense to one which is optically more dense, what happens to the speed of the light ray? If the ray is deflected, will it bend toward or away from the normal in this case?

12 Show by a Huygens' wavelet diagram why a beam of light falling perpendicularly on a flat piece of glass is not refracted.

13 Explain the waviness frequently observed over a hot surface. Is this effect similar to the "wet" mirage often seen by a motorist ascending a hill on a hot, dry day?

14 If the air is warmer near the ground than it is at the level of a target, will a marksman aiming a rifle at the bull's-eye tend to hit the target above or below the bull's-eye?

15 If there were no air surrounding the earth, how would the appearance of the sky be affected?

16 A coin is placed in a tin can in such a position that a person cannot see it resting on the bottom. As water is added, the coin comes into view even though neither the observer nor the coin changed positions. By use of a diagram explain what has happened.

17 What causes stars to twinkle? Do stars twinkle as seen by astronauts in space? Explain.

18 Describe the effect of the variation of density with altitude of the earth's atmosphere on the direction of a beam of light from a star. Would you expect this effect to be significant? Explain.

19 What is meant by the minimum deviation of a ray by a prism? For what position of a prism is the deviation minimum?

20 It is impossible for light falling upon a pane of glass in an ordinary window to be totally reflected. Explain why this must be true.

21 Show by the use of a diagram how light can be "piped" through a curved quartz rod with little loss in intensity. What is a practical application of this effect?

22 Describe how a prism spectrometer could be used to measure wavelengths.

23 From a consideration of the critical angle, explain why a diamond examined in a beam of light sparkles more brilliantly than a piece of glass of the same shape.

24 Why is a right-angle prism a better reflector than a plane silvered mirror? Would the reflection be improved by covering with a metallic coating the polished surface at which the reflection takes place?

25 If a prism is set for minimum deviation for yellow light, what change, if any, would be necessary in order to adjust it for minimum deviation for blue light?

26 When one looks down into a glass of water, he cannot see the table through the walls of the tumbler although the fingers where they are in contact with the glass walls may be seen. Explain by the aid of a ray diagram.

27 Trace a beam of light through a crown glass prism that is immersed in (*a*) air, (*b*) water, and (*c*) carbon disulfide.

28 The principle of reversibility is a phrase used in connection with the refraction of a light ray as it enters a medium with different optical density. To what do you suppose this principle refers?
29 Compare the angle of minimum deviation for yellow light for a 60° diamond prism with that for a 60° prism made of ice.
30 A ray of light is incident upon a piece of optical glass at such an angle that the angle between the reflected and refracted beams is 90°. Prove that $n_r = \tan i$ for this case.
31 Show that a light wave that passes through a plate of optical glass with parallel sides of thickness s suffers a sidewise displacement x given by

$$x = \frac{s\,[\sin\,(i - r)]}{\cos r}$$

Problems

1 A ray of light strikes a water surface at an angle of incidence of 40°. What is the angle of refraction in the water? $n_w = 1.33$.
2 The angle of incidence of a ray of light at the surface of water is 40° and the observed angle of refraction is 29°. Compute the index of refraction. *Ans.* 1.33.
3 When a ray of light is incident on the surface of a certain liquid at an angle of incidence of 48°, the angle of refraction within the liquid is 35°. What is the relative index of refraction of the liquid?
4 The velocity of light in a liquid is 0.80 as fast as it is in air. What is the index of refraction of the liquid? *Ans.* 1.25.
5 A beam of light is incident on a surface of water at an angle of 30° with the normal to the surface. The angle of refraction in the water is 22°. What is the speed of the light in the water?
6 A ray of light strikes a water surface at an angle. The angle of refraction in the water is measured to be 22°. What must the angle of incidence of the light ray have been? *Ans.* 30°.
7 A ray of light goes from air into glass ($n = \frac{3}{2}$), making an angle of 60° with the normal before entering the glass. What is the angle of refraction in the glass?
8 Yellow sodium light has a wavelength in air of 589.3 μm. What is the frequency? What is the wavelength in water? $n = 1.33$. *Ans.* 5.09×10^{14} per second.
9 A layer of ice ($n = 1.33$) lies on a glass ($n = 1.50$) plate. A ray of light makes an angle of incidence of 60° on the surface of the ice. Find the angle between this ray and the normal in the glass.
10 A layer of ethyl alcohol having an index of refraction of 1.354 rests on top of some water. A ray of light strikes the alcohol surface from the air at an angle of 30°. Calculate the angle of refraction in the alcohol, the angle of refraction in the water, and the total deviation of the ray from its original direction. *Ans.* 21.7°; 22.7°; 7.3°.
11 A tank of benzene is 1.5 m deep. How deep does it appear when one is looking vertically downward?
12 A ray of light makes an angle of 30° with the normal in glass ($n = \frac{3}{2}$), and passes into a layer of ice ($n = \frac{4}{3}$). What is the sine of the angle the ray makes with the normal in the ice? *Ans.* $\frac{9}{16}$.
13 The water in a swimming pool is 6 m deep. How deep does it appear to a diver looking straight down into it? $n_w = 1.33$.
14 A diver needs a minimum of 4 m of water in a pool to dive off a 15-m tower. From his viewpoint at the top of the tower, the water appears to be 3 m deep. Is it in fact deep enough? $n_w = 1.33$. *Ans.* Yes; 4 m.
15 A coin is placed beneath a rectangular slab of glass 5.0 cm thick. If the index of refraction of the glass is 1.50, how far beneath the upper surface of the glass will the coin appear to be when viewed vertically from above?
16 A tank 308 cm deep is filled with ethyl alcohol ($n = 1.35$). (*a*) How deep will it appear to an observer who looks directly downward? (*b*) A plane mirror is suspended with its face horizontal at a depth of 154 cm. An object is suspended

254 mm above the mirror. Find the apparent depth of the image formed by the mirror. *Ans.* 228 cm; 132 cm.

17 A telescope is pointed at a angle of 40° to the vertical over the edge of a tank full of water so that the edge of the tank limits the field of vision. How high must the tank be to make a small body at the bottom, 25.4 cm from the side nearest the telescope, just visible in the telescope?

18 A diver looks straight down at the water in a swimming pool and sees that the pool is only two-thirds full. How full is it in reality? *Ans.* eight-ninths full.

19 A point source of light is located 275 cm below the surface of a lake. Calculate the area of the surface that transmits all the light that emerges from the surface.

20 A ray of light from a sodium-vapor lamp falls on the surface of a liquid at an angle of 45° and passes into the liquid making an angle of refraction of 30°. (*a*) What is the index of refraction of the liquid with respect to the air? (*b*) What is the speed of yellow light in the liquid? (*c*) What is the value of the critical angle in the liquid? *Ans.* 1.41; $\bar{1}30{,}000$ mi/s; 45°.

21 What is the largest possible angle of refraction for yellow light incident upon a surface of fused quartz?

22 A ray of light strikes a plate glass at an angle of 45°. If the index of refraction of the glass is 1.52, through what angle is the light deviated at the air-glass surface? *Ans.* 17.3°.

23 At what angle of incidence should a ray of light approach the surface of diamond ($n = 2.42$) from within, in order that the emerging ray shall just graze the surface?

24 Red light strikes a 60° prism (apex angle). The beam of red light has an index of refraction of 1.64 in the prism. What is the angle of minimum deviation for this prism? *Ans.* 50°.

25 An aquarium 1.0 m long is filled with water. A beam of light is incident upon one end of the aquarium at an angle of 25°. Neglecting the effect of the glass walls of the aquarium, calculate the lateral displacement of the emergent beam.

26 A hollow prism having a 60° apex angle is made with parallel glass plates and filled with carbon disulfide which has an index of refraction of 1.643. Find the angle of minimum deviation for this prism. *Ans.* 50.8°.

27 A glass ($n = 1.50$) cube is 4.0 cm on a side. A ray of light is incident on one face 1.0 cm from the edge at an angle of incidence of 60° and directed toward the near side. Where and at what angle will the ray emerge from the cube?

28 A ray of light enters the short side of a 30–60–90° prism, reflects from the long side, and passes out the third side with an angle of refraction equal to 50°. Determine the index of refraction of the prism. Was all the light ray reflected internally from the long side? *Ans.* 1.53; yes, critical angle is 40.8°, which was exceeded.

29 A fused-quartz rod is 30.0 mm in diameter. One end is in the form of a hemisphere. A small source of light is placed on the axis of the rod 45 cm from the hemispherical end. Find the position of the image.

30 Solve Prob. 26 with the hollow prism being filled with water. $n_w = 1.33$. *Ans.* 23.2°.

31 A 20.0-cm glass ($n = 1.50$) sphere has half its surface silvered. Sunlight falls on the unsilvered half. Where is the image of the sun formed by the emergent light?

32 The left end of a glass rod ($n = 1.58$) is ground and polished to a convex hemispherical surface of radius 51.6 mm. An object is located on the axis 205 mm to the left of the vertex of the convex surface. Find the position of the image. *Ans.* 251 mm.

33 A tank of ethyl alcohol has a small window that is a section of a sphere of radius 90 cm. The concave surface is on the outside. Find the position of the image of a source that is placed 80 cm from the window (*a*) outside the tank and (*b*) inside the tank.

Enrico Fermi, 1901–1954

Born in Rome. Professor at the University of Rome, and later at the University of Chicago. Received the 1938 Nobel Prize for Physics for his identification of new radioactive elements produced by neutron bombardment and his discovery of the nuclear reactions effected by slow neutrons.

Ernest Orlando Lawrence, 1901–1958

Born in Canton, South Dakota. Director of radiation laboratory, University of California. Awarded the 1939 Nobel Prize for Physics for his invention and development of the cyclotron and for the production of artificial radioactive elements.

26

Thin Lenses

When light passes through an object made of a transparent material, it is, in general, deviated both at entrance and at emergence. Just how much resultant deviation there will be depends upon the shape of the refracting body as well as upon the relative index of refraction.

We have observed that a ray of light is bent toward the thicker part of an optically dense prism. Consider the double prism of Fig. 26-1*a*. Two rays parallel to the common base pass through one of the prisms, and each is bent around the base. They emerge parallel to each other, since the prism has not changed the curvature of the wave front. If a second pair of rays parallel to the first (Fig. 26-1*b*) passes through the other prism, they will also be deviated and each will cross the first pair. However, the four rays do not reach any common point. Such a pair of prisms does not focus a set of rays.

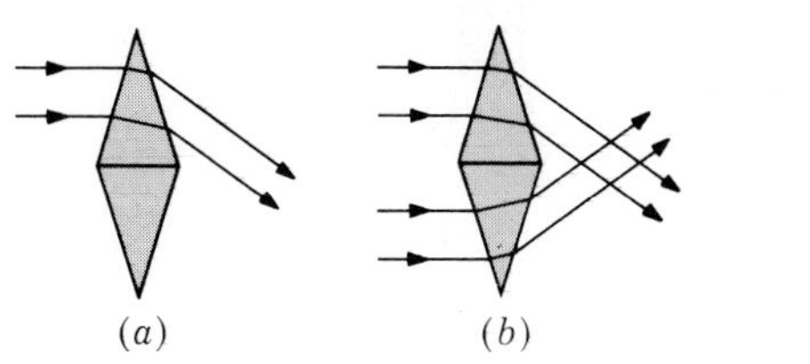

Figure 26-1
Refraction by a double prism.

If, however, we change the shape of the transparent body, we can arrange it so that each of the parallel rays is bent by a different amount and all will intersect at a common point. Such a body with its curved surfaces changes the shape of the wave front. A transparent body with regular curved surfaces that produce changes in the shape of the wave front is called a *lens*.

26-1 LENSES

The curved surfaces of lenses may be of any regular shape, such as spherical, cylindrical, parabolic, or even curves that deviate from these regular surfaces.

It will be seen that spherical surfaces are nearly the correct shape for focusing light which emerges from a point source so that it converges to a point image. Since spherical surfaces are easy to make and the ideal surfaces for particular purposes are frequently difficult to manufacture, most lenses are constructed of one or more pieces of glass with spherical surfaces. Although a single spherical surface does not possess a unique axis, a lens with two such surfaces has only one axis common to both surfaces. The line joining the

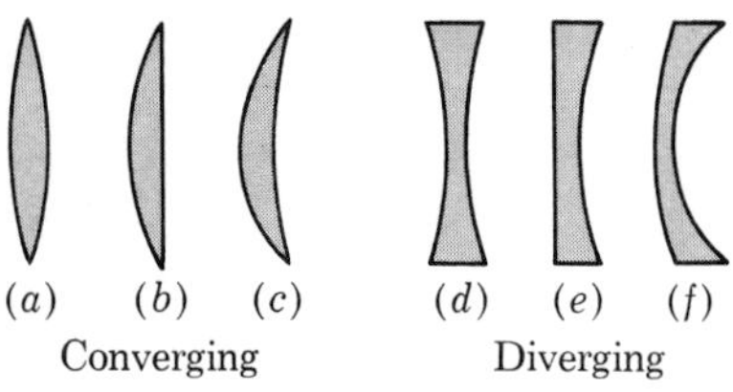

Figure 26-2
Lenses of various forms.

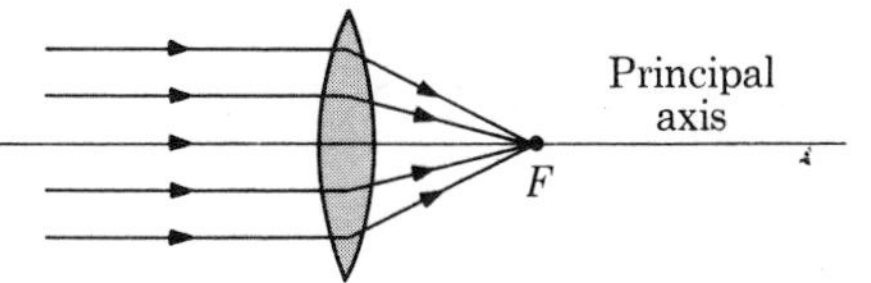

Figure 26-4
Focusing of light by a converging lens.

centers of the two spheres is called the *principal axis* of the lens. Typical forms of spherical lenses are shown in Fig. 26-2. The lenses are named from the two surfaces as double convex (*a*), double concave (*d*), plano-concave (*e*), etc.

A lens produces a change in the curvature of a wave front passing through it. If the lens is made of a material in which the speed of light is less than it is in air, a thin-edged lens retards light passing through the center more than light passing near the edge. Hence a plane wave (Fig. 26-3*a*) will have its central part retarded on passing through a thin-edged lens and will converge toward a point F after emerging. Conversely, a thick-edged lens will render a plane wave divergent (Fig. 26-3*b*).

The propagation of light through lenses and the formation of images may be represented in wave-front diagrams based on Huygens' principle. It is easier, however, to use ray diagrams. Consider a glass lens such as *a* of Fig. 26-2 on which is incident a set of rays from a very distant source on the axis of the lens. These rays will be parallel to the axis. Each ray is bent toward the thicker part of the glass. As they leave the lens, they converge toward a point F (Fig. 26-4). Any lens that will cause a set of parallel rays to converge is called a *converging* lens. The point F to which the *rays parallel to the principal axis* are brought to a focus is called the *principal focus*. The distance from the center of the lens to the principal focus is called the *focal length* of the lens. A thin lens has two principal foci, one on each side of the lens and equally distant from it.

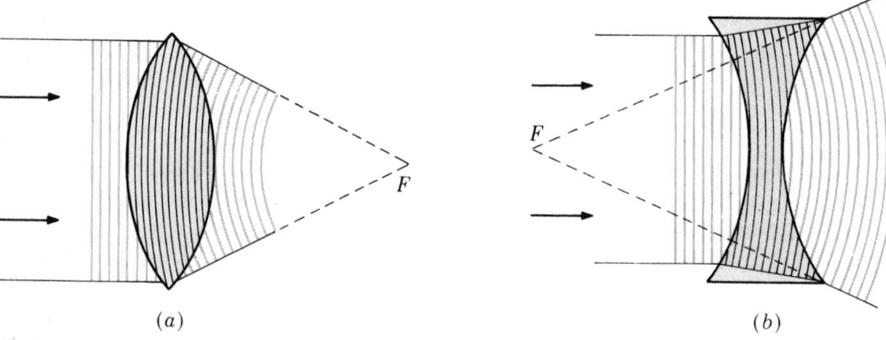

Figure 26-3
Refraction of a plane wave by a lens.

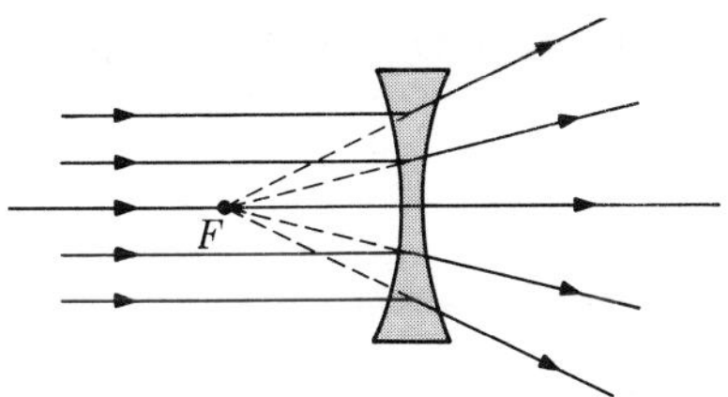

Figure 26-5
The principal focus (virtual) of a diverging lens.

If a lens such as *d* of Fig. 26-2 is used in the same manner, the rays will again be bent toward the thicker part and in this case will diverge as they leave the lens (Fig. 26-5). Any lens in which light travels more slowly than in its surroundings and which is thicker at the edge than at the middle will cause a set of rays parallel to the axis to diverge as they leave the lens and is called a *diverging* lens. The point *F* from which the rays diverge on leaving the lens is the principal focus. Since the light is not actually focused at this point, it is known as a *virtual* focus.

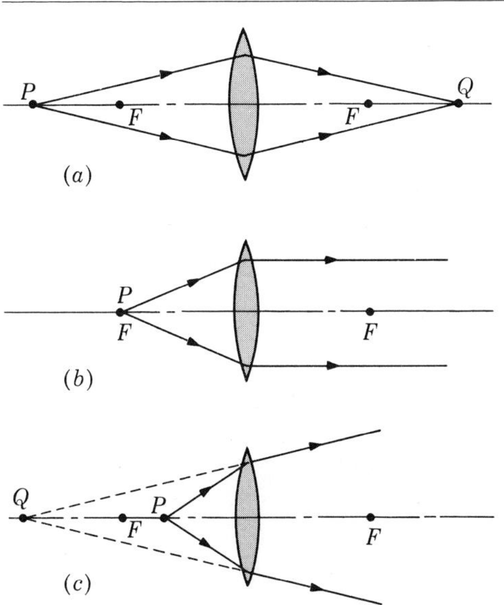

Figure 26-6
Effect of a converging lens on light originating (*a*) beyond the principal focus, (*b*) at the principal focus, and (*c*) between the lens and the principal focus.

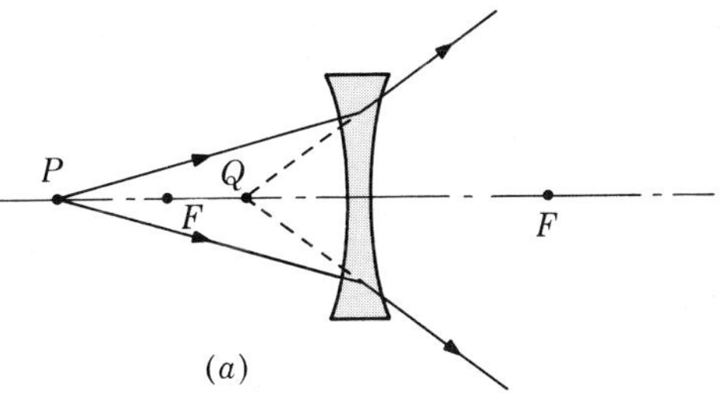

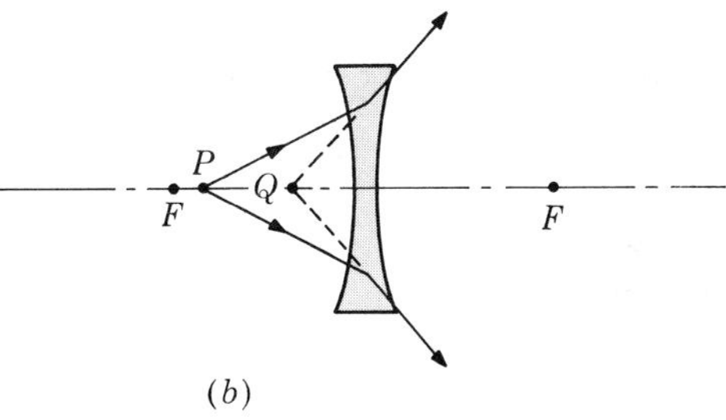

Figure 26-7
Effect of a diverging lens on light originating (*a*) beyond the principal focus and (*b*) closer than the principal focus.

If the source is not very distant from the lens, the rays incident upon the lens are not parallel but diverge as shown in Fig. 26-6. The behavior of the rays leaving a converging lens depends upon the position of the source. If the source is farther from the lens than the principal focus, the rays converge as they leave the lens, as shown in Fig. 26-6*a*; if the source is exactly at the principal focus, the emerging rays will be parallel to the principal axis, as shown in Fig. 26-6*b*. If the source is between the lens and the principal focus, the divergence of the rays is so great that the lens is unable to cause them to converge but merely reduces the divergence. To an observer beyond the lens, the rays appear to come from a point Q rather than from P, as shown in Fig. 26-6*c*. The point Q is a virtual focus.

A diverging lens causes the rays emerging from the lens to diverge more than those which enter. No matter what the position of the real source, the emergent rays diverge from a virtual focus as shown in Fig. 26-7.

26-2 IMAGE FORMATION BY LENSES

When the rays converge after passing through the lens, they pass through the points occupied by the image. The image can be formed on a screen and viewed there. Such an image is called a *real* image. If the rays diverge on leaving the lens, they do not pass through the points occupied by the image and the image cannot be formed on a screen. The image can be seen by looking through the lens. This type of image is called a *virtual* image. Its position is the place from which the diverging rays appear to come. Thus Figs. 26-4 and 26-6*a* represent the formation of real images, while Figs. 26-5, 26-6*c*, and 26-7 represent virtual images. Note that a diverging lens produces only virtual images of real objects, while a converging lens may produce either real or virtual images, depending upon the location of the real object.

26-3 THE THIN-LENS EQUATION

The relationship between object distance, image distance, and focal length of a lens may be found by considering the refraction at each surface. If an object is at O (Fig. 26-8), the first surface of the lens forms an image I_1. The position of this image is found by use of Eq. (9), Chap. 25,

$$\frac{n_1}{p} + \frac{n_2}{q_1} = \frac{n_2 - n_1}{R_1} \tag{1}$$

The image formed by the first surface becomes the object for the second surface. In Fig. 26-8 the light leaving the first surface would reach I_1 if the second surface did not intervene and hence I_1 is a real image for the first surface. If the second surface does intervene, the light does not reach I_1 and hence I_1 becomes a *virtual* object for the second surface. The object distance for the second surface is therefore negative and is given by $-(q_1 - t)$, where t is the thickness of the lens. For the second surface,

$$\frac{n_2}{-(q_1 - t)} + \frac{n_1}{q} = \frac{n_1 - n_2}{R_2} \tag{2}$$

If the lens is *thin*, i.e., if its thickness is small compared with the other dimensions involved, t can be neglected and Eq. (2) reduces to

$$\frac{n_2}{-q_1} + \frac{n_1}{q} = \frac{n_1 - n_2}{R_2} \tag{3}$$

By adding Eqs. (1) and (3) we obtain

$$\frac{n_1}{p} + \frac{n_1}{q} = (n_2 - n_1)\left(\frac{1}{R_1} - \frac{1}{R_2}\right) \tag{4}$$

or

$$\frac{1}{p} + \frac{1}{q} = \left(\frac{n_2}{n_1} - 1\right)\left(\frac{1}{R_1} - \frac{1}{R_2}\right) \tag{5}$$

If the object is at infinity, $1/p = 0$, the image is at the principal focus and $q = f$. Thus

$$\frac{1}{f} = \left(\frac{n_2}{n_1} - 1\right)\left(\frac{1}{R_1} - \frac{1}{R_2}\right) \tag{6}$$

Hence, for *thin* lenses

$$\frac{1}{p} + \frac{1}{q} = \frac{1}{f} \tag{7}$$

Equations (1) and (2) may be applied to any spherical lens to find object and image distances, subject to the restriction that the rays must not make large angles with the axis. This restriction is inherent in Eq. (9), Chap. 25. The rest of these equations apply to thin lenses. Since the lens is thin, the measurement of p, q, and f is to the lens; its thickness is negligible. If the lens is turned around, R_1 and R_2 are interchanged in Eq. (6) and the sign of each is reversed. Thus there is no change in f. For a thin lens the focal length is independent of the order of the surfaces.

Note: In using Eqs. (1) to (7) we must follow a consistent set of conventions of sign. Such a set is listed in Secs. 24-7 and 25-8. These conventions are:

1 Object and image distances are taken positive for real objects and images, negative for virtual objects and images.
2 The radius of curvature R is positive when it is measured from the surface to the center in the direction of the light leaving the surface and negative if in the opposite direction. It follows from this convention and Eq. (6) that, for any thin lens that is thicker in the middle than at the edges, f is positive if the index of refraction n_2 of the lens is greater than that, n_1, of the surrounding medium and negative if the reverse condition is true. The third convention then follows.
3 The focal length f is positive for a converging lens and negative for a diverging lens.

Example A double-convex lens has radii of curvature of 10.00 and 15.00 cm, respectively, for its first and second surfaces and is 0.40 cm thick at the center. The index of refraction of the glass is 1.600. Find the image of an object placed 14.00 cm from the first surface. Do not neglect the thickness of the lens.

Using Eq. (1) for the first surface,

$$\frac{1.000}{14.00\text{ cm}} + \frac{1.600}{q_1} = \frac{1.600 - 1.000}{10.00\text{ cm}}$$

$$\frac{1.600}{q_1} = (0.0600 - 0.07143)\text{ cm}^{-1}$$

$$= -0.01143\text{ cm}^{-1}$$

$$q_1 = \frac{1.600}{-0.01143\text{ cm}^{-1}}$$

$$= -139.9\text{ cm}$$

The image due to the first surface is virtual and to the left of the lens. For the second surface, since the light is still diverging, this image acts as a real object distant from the second surface by 139.9 cm + 0.4 cm = 140.3 cm. Since the wave front arrives at the second surface from within the glass, the index of refraction n_1 for the next equation is now 1.600. The radius of the second surface is negative, according to the conventions stated above. Applying Eq. (1) again, we obtain

$$\frac{1.600}{140.3\text{ cm}} + \frac{1.000}{q_2} = \frac{1.000 - 1.600}{-15.00\text{ cm}}$$

$$q_2 = 35.0\text{ cm}$$

Example Find the image for the previous example, using the thin-lens approximation.

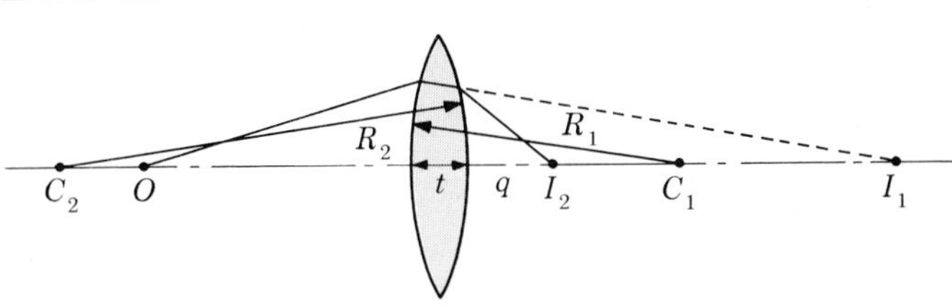

Figure 26-8
Formation of image by refraction at the two surfaces of a thin lens.

First the focal length of the lens must be found. From Eq. (6),

$$\frac{1}{f} = (1.600 - 1.000)\left(\frac{1}{10.00\text{ cm}} - \frac{1}{-15.00\text{ cm}}\right)$$

$$f = 10.00\text{ cm}$$

Applying Eq. (7) and remembering that the object is 14.2 cm from the center of the lens, we obtain

$$\frac{1}{14.2\text{ cm}} + \frac{1}{q} = \frac{1}{10.00\text{ cm}}$$

$$\frac{1}{q} = 1.000\text{ cm}^{-1} - 0.704\text{ cm}^{-1}$$

$$\frac{1}{q} = 0.296\text{ cm}^{-1}$$

$$q = 33.8\text{ cm}$$

The image distance q is measured from the center of the lens, so the image is found to be 33.6 cm from the second surface. This result differs by 1.4 cm from that obtained in the previous example—an error of 4 percent.

Clearly the thin-lens equation cannot be used in accurate optical work; but it is extremely useful for approximate computations.

Example A plano-convex lens (Fig. 26-2*b*) of focal length 12 cm in air is to be made of glass of refractive index 1.50. What should be the radius of curvature of the curved surface?

From Eq. (6),

$$\frac{1}{12\text{ cm}} = \left(\frac{1.50}{1.00} - 1\right)\left(\frac{1}{R_1} - \frac{1}{\infty}\right)$$

$$R_1 = 6.0\text{ cm}$$

Example A converging lens made of glass ($n = 1.66$) has a focal length f_a of 5.0 cm in air. What is the focal length f_w when it is placed in water ($n = 1.33$)?

From Eq. (6)

$$\frac{1}{f_w} = \left(\frac{1.66}{1.33} - 1\right)\left(\frac{1}{R_1} - \frac{1}{R_2}\right)$$

$$\frac{1}{f_a} = \left(\frac{1.66}{1.00} - 1\right)\left(\frac{1}{R_1} - \frac{1}{R_2}\right)$$

If we divide the second of these equations by the first, we obtain

$$\frac{f_w}{f_a} = \frac{0.66}{0.25}$$

and $$f_w = \frac{0.66}{0.25}\,5.0\text{ cm} = 13\text{ cm}$$

Example When an object is placed 20 cm from a certain lens, its virtual image is formed 10 cm from the lens. Determine the focal length and character of the lens.

From Eq. (7),

$$\frac{1}{20\text{ cm}} + \frac{1}{-10\text{ cm}} = \frac{1}{f}$$

$$f = -20\text{ cm}$$

The negative sign shows that the lens is diverging.

26-4 IMAGE DETERMINATION BY MEANS OF RAYS

When an object is placed before a lens, it is possible to determine the position of the image graphically. By drawing at least two rays whose complete paths we know, the image point corresponding to a given object point may be located. Suppose that we have as in Fig. 26-9*a* a converging lens with an object represented by an arrow placed some distance in front of the lens. Let f represent the principal foci on the two sides of the lens. A point on the object, such as the tip of the arrow, may be considered to be the source of any number of rays. Consider the ray which proceeds toward the center of the lens. Since the surface at the point of entrance of this ray is parallel to the surface at the point of emergence, there is no deviation of this ray. We are here treating only *thin* lenses, and hence the displacement of the ray is negligible. Thus for thin lenses the ray through the optical center of the lens is a straight line.

Now consider another ray from the tip of the arrow—one that travels parallel to the principal axis. We saw from Fig. 26-4 that all rays parallel to the principal axis which strike a converging lens pass through the principal focus after emerging. Thus the ray we have drawn from the tip of the arrow will, after refraction by the lens, pass through f. If this line is continued, it will cut the ray through the center of the lens at some point. This is the image point corresponding to the tip of the arrow. The other image points, corresponding to additional points of the arrow, will fall in the plane through I perpendicular to the lens axis. In particular, the image of the foot of the arrow will be on the axis if the foot of the arrow itself is so placed. An inverted real image of the arrow will actually be seen if a card is held in the plane of the object and the lens at the predicted image location in Fig. 26-9 (*a*), (*b*), and (*c*). Inversion takes place also in the sidewise direction so that if the object has any extent in a direction normal to the plane of the figure, right and left will be reversed.

The distance at which an object is located in front of a double-convex lens affects the nature of the image produced. Figure 26-9*a* to *e* illustrates the formation of images when the object distances vary.

Case A The object is placed at a distance greater than twice the focal length in front of the lens (Fig. 26-9*a*). The image formed is real, inverted, and diminished in size.

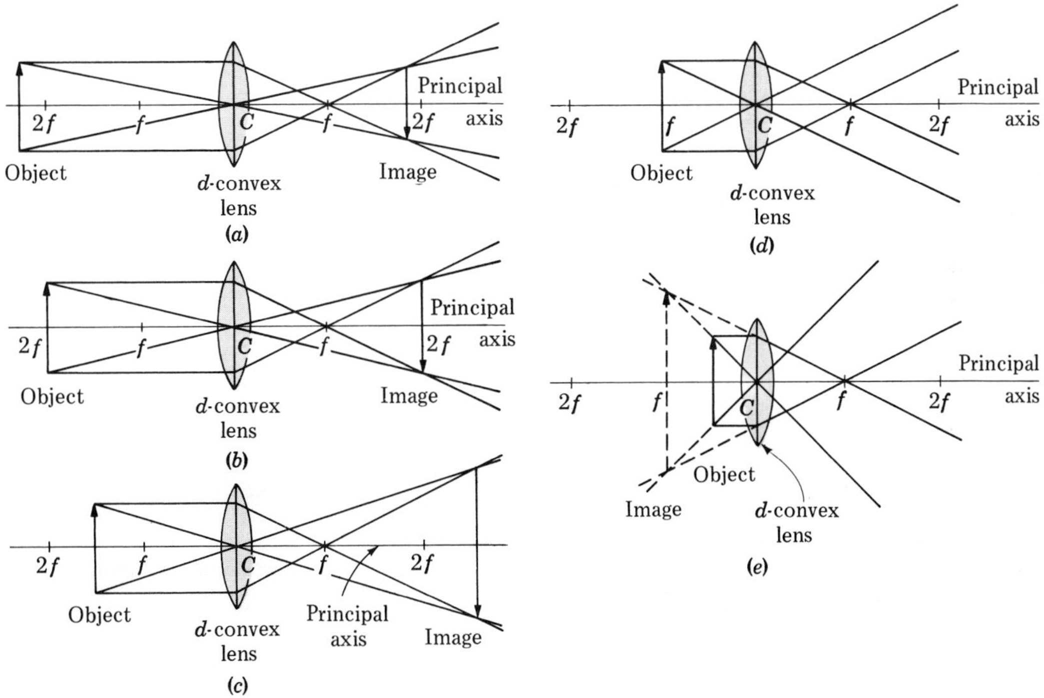

Figure 26-9
(*a*) Case *A*; (*b*) case *B*; (*c*) case *C*; (*d*) case *D*; (*e*) case *E*.

Case B The object is placed at a distance equal to twice the focal length in front of the lens (Fig. 26-9*b*). The image formed is real, inverted, and the same size as the object.
Case C The object is placed between $2f$ and f (Fig. 26-9*c*). The image formed is real, inverted, and magnified.
Case D The object is placed at the principal focus f (Fig. 26-9*d*). Since the refracted rays do not intersect, no image is formed.
Case E The object is placed between the principal focus and the lens (Fig. 26-9*e*). Since the image appears to form on the same side of the lens (a transparent substance) as the object, the image is virtual (imaginary). Also the image is erect and enlarged in size. The reason that the image is virtual, from the point of view of the ray construction, is that the ray through the lens center and the ray passing through f do not intersect on the right-hand side of the lens but diverge instead. However, the rays *appear* to have come from some point located by projecting them back to the left until they cross. This point is the *virtual image* of the tip of the arrow. The virtual image cannot be formed on a screen but may be viewed by looking into the lens from the right.

In a similar way, the formation of a virtual image by a diverging lens is shown in Fig. 26-10. The type of image formed when an object is placed at any distance in front of a double-concave lens always is of the same nature regardless of the distance the object is in front of the lens. The image formed is always virtual, erect, and diminished in size.

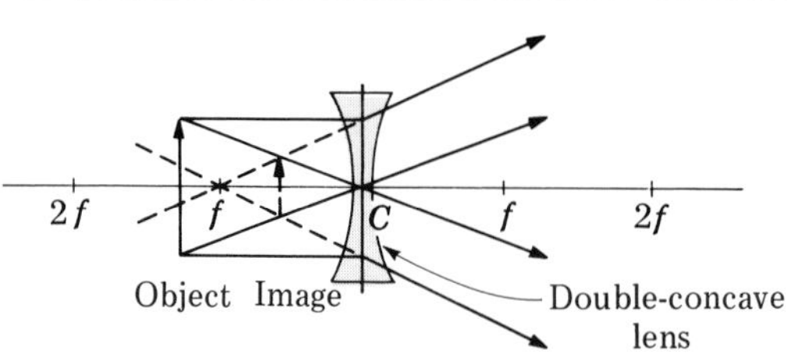

Figure 26-10
Image formed by double-concave lens.

26-5 MAGNIFICATION

In every example of image formation described, we may see from the graphical construction that

$$\frac{\text{Size of image}}{\text{Size of object}} = \frac{\text{distance of image from lens}}{\text{distance of object from lens}}$$

The first ratio is called the *linear magnification*, or simply the magnification. Hence, in symbols

$$M = \frac{q}{p} \tag{8}$$

where p is the distance of the object from the lens and q is that of the image.

Example The lens system of a certain portrait camera may be considered equivalent to a thin converging lens of focal length 10.0 cm. How far behind the lens should the plate be located to receive the image of a person seated 50.0 cm from the lens? How large will the image be in comparison with the object?

Substitution in Eq. (7) gives

$$\frac{1}{50.0\text{ cm}} + \frac{1}{q} = \frac{1}{10.0\text{ cm}} \qquad \text{or} \qquad q = 12.5\text{ cm}$$

From Eq. (8),

$$M = \frac{12.5\text{ cm}}{50.0\text{ cm}} = 0.250$$

The image will be one-fourth as large as the object.

Example Determine the location and character of the image formed when an object is placed 9.0 cm from the lens of the previous example.

Substitution in Eq. (7) gives

$$\frac{1}{9.0\text{ cm}} + \frac{1}{q} = \frac{1}{10.0\text{ cm}}$$

$$q = -90\text{ cm}$$

The negative sign shows that the image lies to the left of the lens and is therefore virtual. It is larger than the object in the ratio

$$M = \frac{90\text{ cm}}{9.0\text{ cm}} = 10$$

26-6 POWER OF A LENS—DIOPTER

The power of a lens is the amount by which it can change the curvature of a wave. Thus the power D of a lens is the reciprocal of its focal length,

$$D = \frac{1}{f} \qquad (9)$$

Hence, the shorter the focal length of a lens, the greater its power. Opticians express the power of a lens in terms of a unit called the *diopter,* the power of a lens that has a focal length of one meter. In using Eq. (9), the focal length must be expressed in meters to give the power in diopters.

Example What is the power (diopters) of a diverging lens whose focal length is -20 cm?

$$f = -20\text{ cm} = -0.20\text{ m}$$

$$\text{Power} = \frac{1}{-0.20\text{ m}} = -5.0\text{ diopters}$$

26-7 LENS COMBINATIONS

Two or more lenses are used in combination in most optical instruments. The location, size, and nature of the final image can be determined by the use of the lens equation or by use of a ray diagram. By either method we find first the image formed by the first lens, use that image as the object of the second lens, and locate the image formed by the second lens. If there are more than two lenses, we continue this process; the object of each lens is the image formed by the preceding lens.

When we have combinations of lenses, we frequently have *virtual objects* for the second and succeeding lenses. For real objects, the rays diverge from each point on the object. Thus the rays entering a lens from a point on a real object are always diverging. When the object of one lens of a combination is the image formed by the preceding lens, the rays entering the second lens may be converging toward a position beyond the lens. The object then is said to be *virtual.* Such a situation is shown in Fig. 26-11. For a virtual object, the object distance is negative.

When lenses are used in combination, each magnifies the image from the preceding lens. Hence the total magnification produced by the combination is the product of the magnifications of the individual lenses,

$$M = M_1 \times M_2 \times \cdots$$

Example Two converging lenses O and E having focal lengths 12 cm and 4 cm, respectively, are placed 39 cm apart on a common principal axis. A small object is placed 18 cm in front of lens O. Find the position, nature, and magnification of the image formed by the combination of lenses.

A conventional ray diagram (Fig. 26-11) is drawn to scale to locate the image A' of object A formed by lens O. Image A' is found to be real. It serves as a real object for lens E. Starting from A', a ray diagram is drawn through lens E, showing that the final image is virtual, enlarged, and inverted (compared with the object A).

The lens equation applied to lens O gives

$$\frac{1}{18\text{ cm}} + \frac{1}{q_o} = \frac{1}{12\text{ cm}}$$

$$q_o = 36\text{ cm}$$

$$\text{Magnification } M_o = \frac{q_o}{p_o} = \frac{36\text{ cm}}{18\text{ cm}} = 2.0$$

Since the image formed by lens O lies 36 cm from O, the object distance for lens E is 39 cm $-$ 36 cm $=$ 3 cm $= p_e$.

If we apply Eq. (7) to lens E, we have

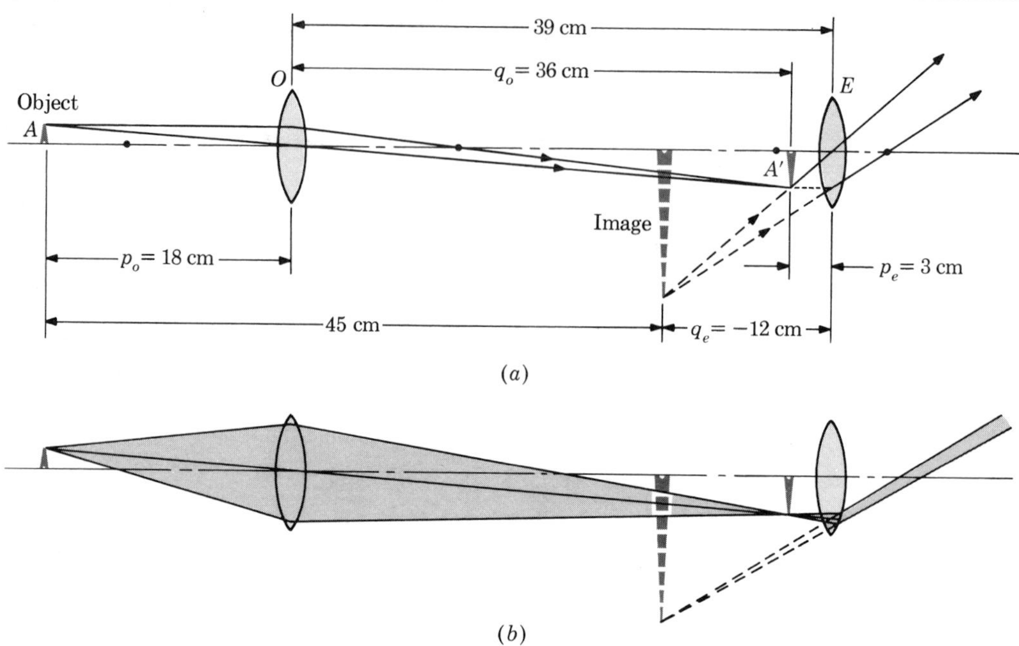

Figure 26-11
An enlarged virtual image formed by successive refractions by two lenses.

$$\frac{1}{3\text{ cm}} + \frac{1}{q_e} = \frac{1}{4\text{ cm}}$$

$$q_e = -12\text{ cm}$$

$$\text{Magnification } M_e = \frac{q_e}{p_e} = \frac{12\text{ cm}}{3\text{ cm}} = 4$$

$$\text{Total magnification } M = M_o M_e = 2 \times 4 = 8$$

The final image is formed 45 cm from the object; it is virtual, inverted, and magnified eight times.

Example A converging lens O of focal length 12 cm and a diverging lens E of focal length −4.0 cm are placed 33 cm apart on a common principal axis. A small object is placed 18 cm in front of lens O. Find the position, nature, and magnification of the image formed by the combination of lenses (Fig. 26-12).

For the first lens,

$$\frac{1}{18\text{ cm}} + \frac{1}{q_o} = \frac{1}{12\text{ cm}}$$

$$q_o = 36\text{ cm}$$

$$M_o = \frac{36\text{ cm}}{18\text{ cm}} = 2$$

The image formed by lens O lies behind lens E and is therefore a virtual object for that lens.

For the second lens,

$$-\frac{1}{3\text{ cm}} + \frac{1}{q_e} = -\frac{1}{4.0\text{ cm}}$$

$$q_e = 12\text{ cm}$$

$$M_e = \frac{12\text{ cm}}{3\text{ cm}} = 4$$

$$\text{Total magnification} = M_o M_e = 2 \times 4 = 8$$

The final image is formed 63 cm from the

object; it is real, inverted, and magnified eight times.

In the ray diagram for this problem (Fig. 26-12) the real image A', which would be formed by lens O alone, falls behind lens E. It is treated as the virtual object for lens E, and the final image is found by drawing a conventional ray diagram for lens E. This is merely a convenient device. It does not, of course, imply that rays return through lens E. Furthermore, the rays that are drawn from the intermediate image A' do not represent continuations of the rays from A. A dotted line is projected back from the head of A' parallel to the axis. This would represent the continuation of a ray that entered the lens from the left parallel to the axis. This ray leaves the lens as though it had come from the principal focus.

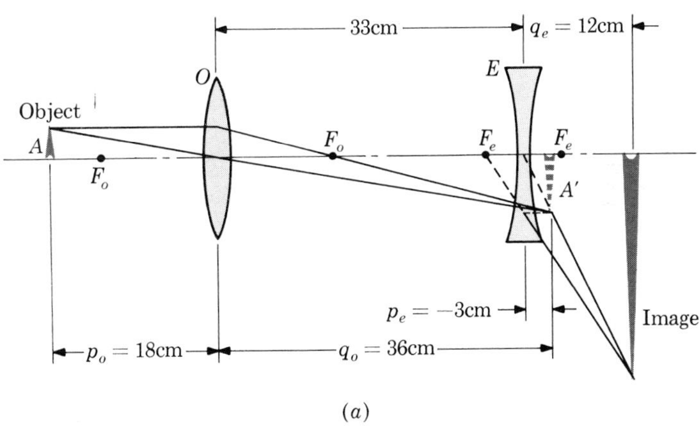

(a)

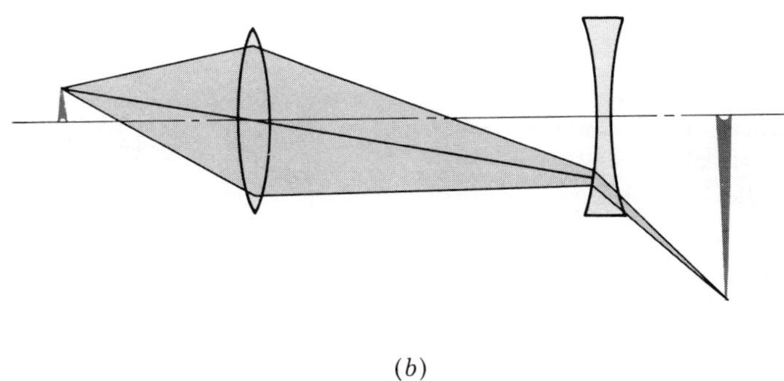

(b)

Figure 26-12
An enlarged real image formed by successive refractions by two lenses.

26-8 THIN LENSES IN CONTACT

When two or more thin lenses are in contact, they may be treated as a single lens by finding the focal length of a single lens that would produce the same refraction as the combination. One method of finding the equivalent focal length is to treat the image of the first lens as the object of the second lens, as with other lens combinations. A somewhat simpler method uses the powers of the lenses. Each lens changes the curvature of the wave by the amount of its power D. The total change in curvature is the sum of the two changes. That is,

$$D = D_1 + D_2 \tag{10}$$

or

$$\frac{1}{f} = \frac{1}{f_1} + \frac{1}{f_2} \tag{11}$$

where f is the focal length of the combination and f_1 and f_2 are the focal lengths of the individual lenses.

Example A double-convex lens has a focal length of 20 cm. When it is placed in direct contact with a double-concave lens, the new combined focal length is 30 cm. What is the focal length of the concave lens?

$$\frac{1}{f} = \frac{1}{f_1} + \frac{1}{f_2}$$

where f is the combined focal length, f_1 the focal length of the convex lens, and f_2 the focal length of the concave lens.

Then, $$\frac{1}{30\text{ cm}} = \frac{1}{20\text{ cm}} + \frac{1}{f_2}$$

and $$\frac{1}{f_2} = \frac{1}{30\text{ cm}} - \frac{1}{20\text{ cm}} = \frac{2}{60\text{ cm}} - \frac{3}{60\text{ cm}}$$

$$= -\frac{1}{60\text{ cm}}$$

$$f_2 = -60\text{ cm}$$

26-9 SPHERICAL ABERRATION

In the derivation of the equations for refraction at spherical surfaces the treatment was limited to rays that make small angles with the axis. Unless the aperture of a lens is very small, there are rays that make angles larger than justified by this approximation. Thus the rays that enter the lens near its edge are brought to a focus closer to the lens than are the central rays. This characteristic of a spherical lens is called *spherical aberration* (see also Sec. 24-9). Its effect may be minimized by using a diaphragm in front of the lens to decrease its effective aperture, a sharper image being then produced with consequent loss of light.

26-10 COMA

Another form of spherical aberration occurs for object points that are laterally displaced from the principal axis. Rays from such points may intersect in the focal plane, not as a sharp focus, but in a comet-shaped figure (hence the term "coma"). This effect can be minimized by using a compound lens having several surfaces, or more simply by an aperture stop that eliminates rays which are not near the principal axis.

26-11 ASTIGMATISM

Astigmatism is the lens defect whereby horizontal and vertical lines in an object are brought to a focus in different planes, as the two lines AB and CD in Fig. 26-13. Astigmatism arises from the lack of symmetry of a lens or a lens system about the line from the center of the lens to an object. In the eye this occurs when the curvature of the cornea is different in different axial planes. But astigmatism occurs even for a perfectly spherical lens whenever the source P is not on the optical

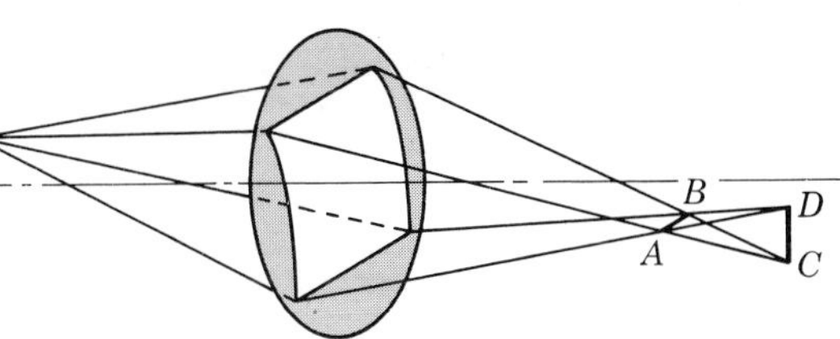

Figure 26-13
Astigmatism.

axis of the lens. It is corrected, along with other defects, by a combination of lenses.

26-12 DISTORTION

Another aberration, known as *distortion*, is caused by the fact that the magnification of an image varies at different parts of the image. The central part of the image may be accurate, but the outer parts are not faithful images of the object. When a diaphragm is used to reduce spherical aberration and is inserted between the lens and the object, the image is distorted so that a square-mesh object appears as the barrel-shaped image shown in Fig. 26-14*a*. If the diaphragm is placed on the opposite side of the lens from the object, the square-mesh object gives an image of pincushion shape, as shown in Fig. 26-14*b*. The effects of distortion may be minimized by using two lenses and placing a diaphragm between them.

A lens that is corrected for such defects as spherical aberration, astigmatism, distortion, and chromatic aberration (Chap. 28) is called an *anastigmatic* lens.

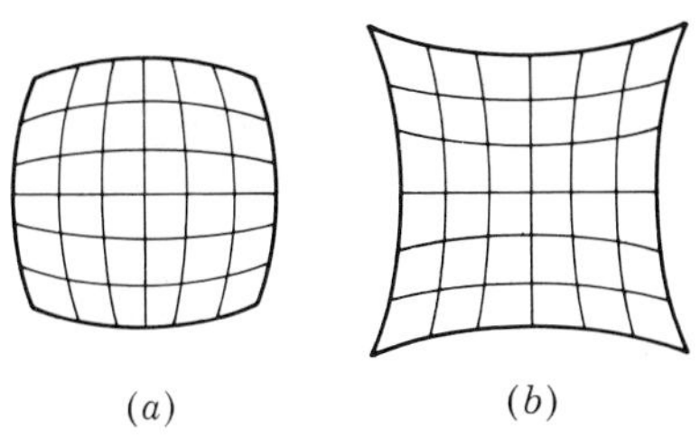

Figure 26-14
"Barrel" and "pincushion" distortion of square-mesh object.

SUMMARY

When rays of light pass through a lens, they are bent toward the thicker part of the lens if the lens has an index of refraction greater than that of the surrounding medium. Rays parallel to the principal axis of the lens pass through a point called the *principal focus*. The distance of the principal focus from the lens is called the *focal length*.

A *real image* is formed if the rays actually pass through the image after refraction. A *virtual image* is formed if the rays only appear to come from the image after refraction.

A *converging lens* makes a set of parallel rays converge after refraction. A *diverging lens* renders a set of parallel rays divergent after refraction.

A converging lens forms *real images* when the real object is farther from the lens than the principal focus, but *virtual images* when the real object is between the lens and the principal focus. A diverging lens forms a virtual image of any real object.

The position of an image can be determined by use of the equation

$$\frac{n_1}{p} + \frac{n_2}{q_1} = \frac{n_2 - n_1}{R_1}$$

for the first surface, followed by the use of

$$\frac{n_2}{-(q_1 - t)} + \frac{n_1}{q_2} = \frac{n_1 - n_2}{R_2}$$

for the second surface.

If the lens is thin, the focal length is given in terms of its radii of curvature R_1 and R_2 and the indices of refraction n_2 of the lens and n_1 of the surrounding medium.

$$\frac{1}{f} = \left(\frac{n_2}{n_1} - 1\right)\left(\frac{1}{R_1} - \frac{1}{R_2}\right)$$

The position, size, and nature of an image can be determined by means of a *ray diagram* or by use of the *lens equation*.

$$\frac{1}{p} + \frac{1}{q} = \frac{1}{f}$$

Conventionally, f is to be taken as positive for a converging lens and negative for a diverging lens; p and q are positive for real objects and images, negative for virtual objects and images.

The *linear magnification* M is the ratio of the size of the image to the size of the object. It is related to the object and image distances by the equation

$$M = \frac{q}{p}$$

The power of a lens is the amount by which it can change the curvature of a wave. It is the reciprocal of the focal length. When the focal length is expressed in meters, the power is in *diopters*.

Lens combinations are treated by using the image of the first lens as the object of the second, the image of the second as the object of the third, and so on, through the whole combination. When the image of one lens is located beyond the next lens, it serves as a *virtual object* for the second lens and the object distance for that lens is negative.

Spherical aberration is the defect of a lens by which rays entering near the edge of the lens are brought to a focus nearer the lens than the rays that enter near the center.

Coma is a form of lateral spherical aberration whereby rays from object points not on the principal axis are focused in a comet-shaped image.

Distortion is caused by variations in the magnification of the outermost portions of an image, resulting in barrel-shaped or pin-cushion-shaped images of a square-mesh object.

Astigmatism is the defect whereby horizontal and vertical lines are brought to a focus at different distances.

Both spherical aberration and astigmatism can be reduced by decreasing the aperture of the lens.

Questions

1 Explain by use of a sketch why light is bent toward the thickest part of an optically dense prism.

2 Describe the image produced by a convex lens if an object is placed $1\frac{1}{2}$ times the focal length in front of the lens.

3 How large does a lens have to be in order to give a complete image of a distant object? Compare this case with the corresponding one for the plane mirror.

4 Trace the paths of a beam of parallel rays which are incident upon a hollow-glass sphere immersed in water.

5 As you look directly at a goldfish through the side of a spherical bowl, does it appear its normal size? What factors enter into determining the answer to this question?

6 Trace a beam of rays parallel to the principal axis of a crown-glass convex lens when the lens is in (*a*) air, (*b*) water, and (*c*) carbon disulfide. Describe the path of these rays if an "air" lens were used in each of these materials.

7 Describe and sketch by a ray diagram the image produced of an object placed between the focal point of a thin, double-convex lens and the lens.

8 Describe and illustrate with ray diagrams three different laboratory methods of finding the focal length of a concave lens.

9 Describe an experimental method for measuring the focal length of a diverging lens, using real images only. Derive the equation to be used.

10 To obtain an image which has the same size

as the object but inverted, where should the object be placed in front of a thin, double-convex lens?

11 If it is desired to project an image of an object on a screen, should a double-concave lens be used? Explain.

12 Draw a graph showing the variation of $1/p + 1/q$ for a converging lens. What is the significance of the x and y intercepts of this curve?

13 Draw an object-distance-vs.-image-distance graph for a diverging lens of focal length 12 cm for object distances varying from plus to minus infinity.

14 What is meant by the power of a lens? In what units can it be expressed?

15 Is the observed focal length of a lens in water longer or shorter than in air? Explain.

16 Does the column of mercury in a clinical thermometer look broader or narrower than it really is? Explain.

17 Two identical convex lenses of focal lengths f are placed a distance $3f$ apart. Indicate whether the combination produces a real or a virtual image when a lamp is placed, relative to the first lens, (*a*) between infinity and $2f$, (*b*) between $2f$ and f, (*c*) at f, and (*d*) between the principal focus and the lens.

18 Which of the following statements is true? If a converging beam of light is incident upon a double-convex lens the image (*a*) is always real, (*b*) is always virtual, and (*c*) may be real or virtual, depending upon the convergence of the beam and the power of the lens.

19 What two factors determine whether a lens is converging or diverging?

20 As a review, list the algebraic sign conventions for each factor in the thin-lens equation.

21 Describe some cases in which it is convenient to use a combination of lenses rather than a single lens.

22 Prove by sketches that the following is a valid procedure to distinguish between converging and diverging spherical lenses. Looking through the lens and moving the lens at right angles to your line of sight causes the image to move in the same direction for a diverging lens and conversely for a converging lens.

23 What type of lens is a tumbler filled with water? Would one expect such a lens to have a large or a small amount of spherical aberration?

Problems

1 Describe the differences between the images formed by (*a*) a convex lens having a focal length of +6 cm which has a 4-cm object placed 8 cm in front of the lens and (*b*) the same object placed the same distance in front of a concave lens having a focal length of −6 cm.

2 A straight-filament lamp is placed 5.0 cm in front of a convex lens of focal length 2.0 cm. How far from the lens will an image of the filament be formed? *Ans.* 3.3 cm.

3 A double-convex lens, both surfaces of which have radii of 20 cm, is made of glass whose index of refraction is 1.50. Find the focal length of the lens.

4 A laboratory spotlight consists of an incandescent lamp mounted on the common principal axis of a concave mirror of focal length 1.5 cm and a convex condensing lens of focal length 8.0 cm. (*a*) What is the distance from lamp filament to the mirror, and from lamp filament to lens when the spotlight is arranged to give a parallel beam? (*b*) What is the usefulness of the mirror? (*c*) Where is the image of the lamp filament formed by the mirror? *Ans.* 3.0 cm; 8.0 cm; coincides with filament.

5 A certain converging lens made of glass of index of refraction 1.52 has a focal length of 10.0 cm. What is the focal length of this lens in water?

6 A converging lens of focal length 10 cm is placed in contact with a diverging lens. If an image of a distant object is formed 30 cm from the lenses, what is the focal length of the diverging lens? *Ans.* −15 cm.

7 A lens is constructed by cementing together the edges of two watch glasses, each of radius of curvature 25.0 cm. The resulting lens is immersed in ethyl alcohol. What is the focal length of this air lens? Is it converging or diverging?

8 At what distance from a converging lens of focal length 18 cm must an object be placed in order that an erect image may be formed twice the size of the object? *Ans.* 9.0 cm.

9 A glass lens ($n = 1.50$) has a concave surface of radius 206 mm and a convex surface of radius 618 mm. A candle flame 12.5 mm high is 126 mm from the lens. Where is the image, and how high is it?

10 Two converging lenses of focal lengths of 12 cm and 10 cm are placed 30 cm apart. If an object AB is 10 cm tall and is placed 60 cm in front of the 12-cm lens, sketch and calculate the final image and the size of the image formed by this combination of lenses.

Ans. Erect; 30 cm to right of 10-cm lens; 5 cm tall; magnification $= \frac{1}{2}$.

11 An object 1.0 cm long is placed 30 cm from a converging lens of focal length 10 cm. Find the position, size, and nature of the image.

12 Two convex lenses each having a focal length of 5 cm are mounted 10 cm apart. Sketch and "calculate" the image produced by an object placed 8 cm in front of the first lens. Describe the final image and the magnification of the lens system. *Ans.* Virtual, inverted, 10 cm behind lens with focal length of 5 cm; magnification = 1.02.

13 An object 40.6 mm high is placed 52.4 cm from a converging lens. An image is formed 25.8 cm from the lens. (*a*) How large is the image? (*b*) Is the image real or virtual? (*c*) What is the focal length of the lens?

14 A fused-quartz lens of index of refraction 1.458 has one convex and one concave surface whose radii are 246 and 408 mm, respectively. An object 15 mm high is placed 165 cm from the lens. Where is the image, and how high is it?

Ans. 770 cm; 7.0 cm.

15 A diverging lens has a focal length of -10 cm. Where is the image when the object is (*a*) 20 cm from the lens? (*b*) 5 cm from the lens? How large is the image in each case if the object is 0.50 cm high?

16 A double-convex lens has one side with a radius of 10 cm and the other side with a radius of 20 cm. What is the focal length of this lens if the index of refraction of the glass is 1.6?

Ans. +11.1 cm.

17 At what distance from a converging lens of focal length 5.24 cm must an object be placed in order that a real image may be formed twice the size of the object?

18 A piece of a meterstick 15.6 mm long stands vertically at the left of a lens and 125 cm from the lens. An observer at the right of the lens sees the image of the stick at a distance of 11.8 cm on the left of the lens. What is the focal length of the lens, and what is the length of the image?

Ans. -13.1 cm; 1.47 mm.

19 A convex lens 25 cm from a straight-filament lamp 5.0 cm high forms an image of the latter on a screen. When the lens is moved 25 cm farther from the lamp, an image is again formed on the screen. Calculate the focal length of the lens, the distance of the screen from the lamp, and the sizes of the two images.

20 A 3-cm-tall object is placed 20 cm in front of a spherical bowl which has a diameter of 10 cm. Determine the position of the image produced if the bowl is filled with water.

Ans. First image = +83 cm from first side, magnification = 0.18; second image = +11.8 cm from second side, magnification = 0.215; total magnification = 0.038.

21 A plano-convex lens ($R = 12.0$ cm, $n = 1.50$) is cemented over an opening in the side of a water tank. A small light source is placed outside the tank 30.0 cm from the opening. Where is the image if (*a*) the curved surface is outside and (*b*) if the plane surface is outside?

22 What minimum distance can there be between an object and its image formed by a converging lens if the image is (*a*) real or (*b*) virtual?

Ans. $4f$; zero.

23 The focal lengths of two lenses are 10 and 20 cm. What is the focal length of the combined lenses when they are placed in contact?

24 Two thin lenses of focal lengths 8.0 and -3.0 cm are placed in contact. What is the focal length of the combination? *Ans.* -4.8 cm.

25 A beam of sunlight falls on a diverging lens

of focal length 10 cm, and 15 cm beyond this is placed a converging lens of 15-cm focal length. Find where a screen should be placed to receive the final image of the sun.

26 Compute the focal length and type of lens which has the following diopters. (*a*) +3 diopters; (*b*) −2 diopters.

Ans. (*a*) 0.33 m, converging; (*b*) 0.50 m, diverging.

27 What would the combined power and equivalent focal length be of the two lenses in Prob. 26 if they were placed in contact?

28 A 12.5-diopter converging lens is to be made from glass of index of refraction 1.500. If each surface is to have the same curvature, what should be the common radius of curvature?

Ans. 8.00 cm.

29 Two lenses are placed in contact. Their powers are +4 and −6 diopters, respectively. What is the power and the focal length of the combination? Is it convergent or divergent?

30 A diverging lens *B* of 20-cm focal length is placed 45 cm to the right of a converging lens *A* of 30-cm focal length. A small object is placed 50 cm to the left of *A*. Find (*a*) the position of the final image and (*b*) the linear magnification. (*c*) Is the image real or virtual? (*d*) Is it erect or inverted? *Ans.* 15 cm to left of *A*; 3.0.

31 A double-convex lens, both surfaces of which have radii of 10 cm, is made of glass whose index of refraction is 1.60. Find the focal length of the lens.

32 An object is placed at the 0.0-cm mark on an optical bench, and a convex lens is placed at the 50.0-cm mark. A real image is formed at the 70.0-cm mark. Without moving the object or the convex lens, a diverging lens is placed at the 60.0-cm mark and a new real image is formed at the 80.0-cm mark. Find the focal length of each lens. *Ans.* −20 cm.

33 Two convex lenses of focal lengths 20 and 30 cm are 10 cm apart. Calculate the position and length of the image of an object 2.0 cm long placed 100 cm in front of the first lens.

34 A magnifying glass having a focal length of 5 cm is used by a person with normal vision (i.e., being able to see objects most distinctly which are 25 cm from his eye). If the magnifier is held close to his eye, what is the best position of the object? *Ans.* 4.17 cm.

35 A converging lens of focal length 10 cm is placed 12 cm from a lighted candle. A diverging lens of focal length −16 cm is placed 36 cm beyond the converging lens. (*a*) Where is the resultant image? (*b*) Is it real or virtual? (*c*) What is the magnification?

Otto Stern, 1888–1969

Born in Sohrau, Germany. Professor at the Carnegie Institute of Technology. Awarded the 1943 Nobel Prize for Physics for his contribution in the development of the molecular-ray method of detecting the magnetic moment of protons.

Isador Isaac Rabi, 1898–

Born in Rymanow, Austria. Professor at Columbia University. Awarded the 1944 Nobel Prize for Physics for his research in the resonance method of recording the magnetic properties of atomic nuclei.

27

The Eye and Optical Instruments

For thousands of years the only optical device was the eye. This optical instrument is truly marvelous, but in the current technological age accessory instruments have been devised to extend the range of man's observations from molecules to galaxies.

Smooth surfaces of water were used for primitive mirrors. Early scientists recognized the reflecting properties of plane and later curved surfaces of metals and glass. Spectacles were introduced during the thirteenth century, and similar lenses were used to make a primitive telescope in the early 1600s. The era of optical instruments may be said to have begun with the invention of a satisfactory telescope by Galileo about 1609.

In the lowest animals, the eye may be merely a collection of pigmented cells capable of distinguishing between light and darkness. In more highly developed forms the eye includes a lens which forms real images and a fine-grained mosaic of receptors which records the pattern of intensities and wavelengths in the image and submits that pattern to the brain for interpretation. The response of the visual system to different intensities and wavelengths of illumination, its ability to distinguish size and position, and the common errors of vision are of such practical importance as to have received extensive study. Spectacle lenses compensate for faults of vision, but physiological remedial measures are as yet little understood. Many optical instruments are designed to extend the usefulness of the human eye by exploiting its advantageous characteristics or by compensating for its shortcomings. Most of them serve to increase the size of the image; but the effect of looking through them may be to change the apparent distance of objects (as does a telescope) rather than to change apparent size (as does a magnifier or a microscope).

27-1 THE HUMAN EYE

The eye (Fig. 27-1) contains a lens, a variable diaphragm (the iris), and a sensitive screen (the retina) on which the cornea and lens form a real, inverted image of objects within the field of vision. The eye is a nearly spherical structure held in a bony cavity of the skull in which it can be rotated to a certain extent in any direction by the complex action of six muscles. The eye has a tough fibrous coat of which about one-sixth, the *cornea,* is transparent, admitting light. Within the eyeball behind the cornea is the opaque, muscular *iris,* which has a central opening, the pupil. This contracts and dilates to control the amount of light admitted. The lens is a transparent bi-

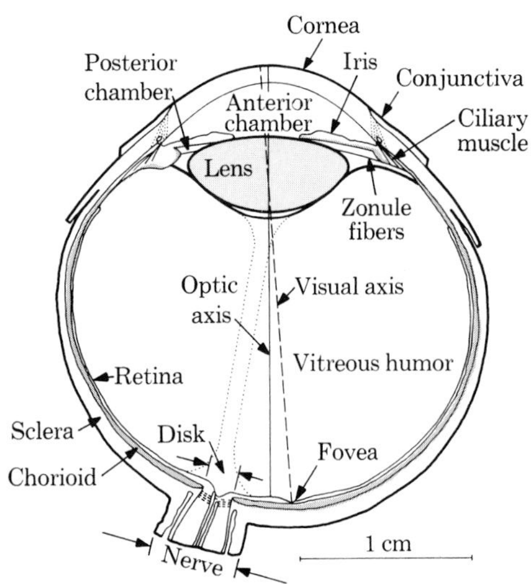

Figure 27-1
The eye.

convex body composed of myriads of microscopic, glassy fibers, which readily slide on one another so that the lens can change its shape. The lens is held just behind the iris by a system of spokelike ligaments (zonule fibers) that are relaxed or tensed by the action of a muscle (the ciliary), producing changes in the curvature of the lens. These changes in the focusing ability of the lens accomplish *accommodation,* the adjustment of the distance of the image for the exact distance of the retina, in accordance with the external distance to the object of regard.

Lining the wall of the eyeball is the *retina,* a sensitive membrane whose stimulation results in the visual sensation. The spaces between the cornea and the zonule (anterior and posterior chambers) are occupied by a salt solution, the aqueous humor, and the interior region between lens and retina is filled by a jellylike vitreous humor. The humors have about the same index of refraction (little higher than that of water). The principal refraction occurs as light enters the curved outer surface of the cornea, and lesser refractions take place as it enters and leaves the lens, whose index is considerably higher than that of the humors.

Many millions of light receptors, the *rods* and *cones,* form one layer of the retina. These are connected to a smaller number of intermediary nerve cells, which in turn are connected to a still smaller number of optic nerve fibers. From all over the inner surface of the retina these converge at one spot, where they pass out through the eyeball wall, forming the optic nerve, which connects the retina and the brain. Here at the head of the optic nerve (disk) is an insensitive region, or "blind spot," since there are no rods or cones there. The existence of the blind spot can easily be verified by closing the left eye and looking intently at the **x** in Fig. 27-2. As the book is moved toward the eye, the square disappears when the page is about 10 in, or 25 cm, from the eye. On moving the page closer the black dot may also be made to disappear. Still closer, the square and dot will reappear in turn. More centrally located than the blind spot is a pit in the retina, the *fovea,* which contains only cones. Here, vision is most acute. It is this portion of the retina, embracing about 1° of the visual field, which is always used when we look directly at an object. Evidently the eyes will have to turn toward each other, so that their axes converge, in order to bring an image of a nearby object into position on the fovea in each eye.

Objects which normally appear brightly colored during the day lose much of their color when seen under bright moonlight. This difference between day and night vision is due to the different functions performed by the rods and the cones in the human eye. The cones, which are color sensitive, need a greater amount of light in order to be stimulated into action than do the rods. The rods, which can only distinguish gray and black and are, therefore, color insensitive, may be stimulated by a minimum amount of light. Day vision, then, functions through the cones while night vision, where there is not enough light to stimulate the cones, functions because of the rods. Further, there is a greater concentration of cones in the center of the retina and a greater concentration of rods in areas of the retina away from the center. Therefore, if you were to try to see

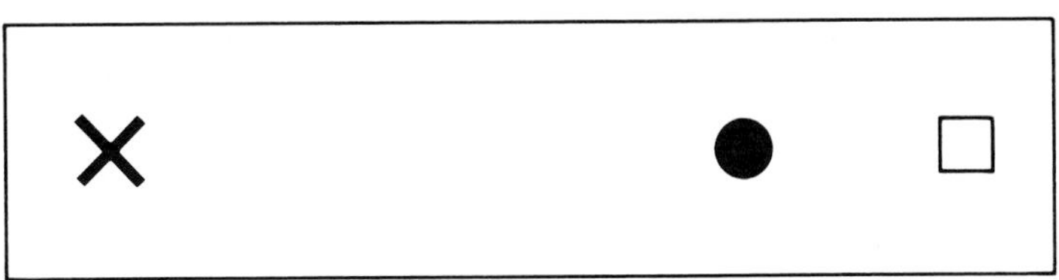

Figure 27-2
Locating the "blind spot" of the eye.

an object in a darkened room, it would be wise not to look directly at the anticipated location but to look either above or below or to one side of that position so that more rods can be called upon to help make the object visible.

27-2 REFRACTIVE ERRORS

An optically normal eye, when the ciliary muscle is entirely relaxed, forms an image of a distant object on the retina (Fig. 27-3*a*). As an object approaches the eye, the ciliary muscle adjusts the curvature of the lens, making it more convex so that the image is held on the retina despite the decreasing object distance. When the object distance decreases to a certain value, called the *near point,* the limit of accommodation is reached and for still smaller object distances the image recedes behind the retina.

Figure 27-3*b* represents a nearsighted (myopic) eye that is relaxed. The rays from a very distant point focus in front of the retina because the eyeball is too long for its lens. The ciliary muscle does not have the power to reduce the curvature of the lens beyond that of the relaxed eye, and so the eye cannot focus on a distant object. Its normal power of accommodation does, however, allow it to focus for very close objects and produce a larger clear image than can a normal eye. To correct this nearsighted vision, a diverging lens is needed to diminish the refraction. The opposite condition, that of a farsighted (hyperopic) eye which is too short, is shown in Fig. 27-3*c*. A converging lens is needed to overcome farsightedness.

Example A certain farsighted person has a minimum distance of distinct vision of 150 cm. He wishes to read type at a distance of 25 cm. What focal-length glasses should he use?

Since the person cannot see clearly objects closer than 150 cm, the lens must form a virtual image at that distance.

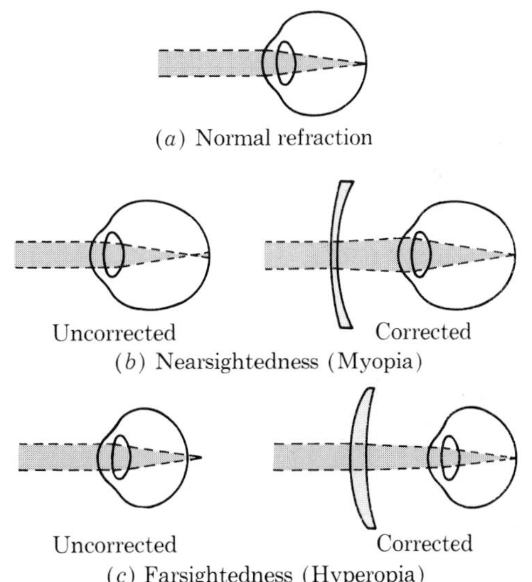

Figure 27-3
Correction of nearsightedness and farsightedness by spectacle lenses.

$$p = 25 \text{ cm}$$

$$q = -150 \text{ cm}$$

$$\frac{1}{p} + \frac{1}{q} = \frac{1}{f}$$

$$\frac{1}{25 \text{ cm}} + \frac{1}{-150 \text{ cm}} = \frac{1}{f}$$

$$f = 30 \text{ cm}$$

Astigmatism, a common defect in human eyes, is a failure to focus all lines of an object-plane in a single image-plane, but it arises from a cause different from that of the astigmatism which is one of the aberrations of spherical lenses or mirrors. Ocular astigmatism is generally due to unequal curvature of the front surface of the cornea, the surface being distorted by a certain amount of cylindrical curvature. A person with this defect will see radial lines (Fig. 27-4) parallel to the axis of the cylindrical curvature of his eye less sharply than other lines. To correct the astigmatism, a cylindrical lens is so arranged that the convergence produced by the eye and spectacle lens together is the same in all meridians. Astigmatism often occurs in combination with nearsightedness or farsightedness, both being neutralized by a single lens incorporating both spherical and cylindrical corrections.

There is a normal dwindling of the power of accommodation so that by the age of about 45 positive lenses are needed for reading. Eyes originally myopic, and formerly corrected by diverging lenses, may in fortunate cases be perfectly adjusted for reading by removing the spectacles. More often a special reading correction, less divergent than for the old spectacles (and perhaps even positive), is needed. Bifocal glasses, for myopics, have the upper area negative for distant viewing, the lower area less strongly negative, or even somewhat positive, for reading. For hyperopic eyes bifocal glasses always have both areas positive, the lower one of greater power (by about 2 diopters).

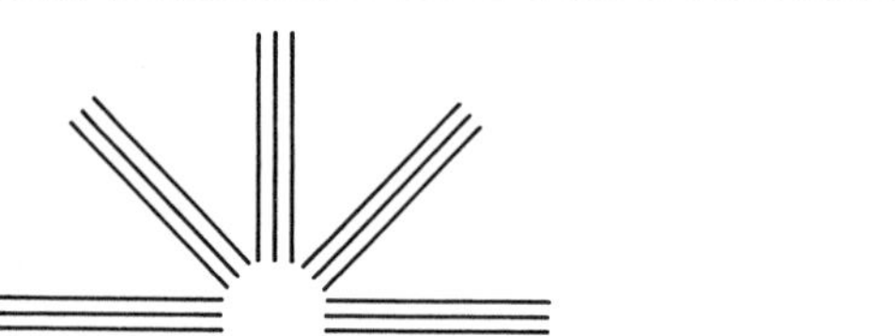

Figure 27-4
Astigmatic dial for locating the meridians of ocular astigmatism.

27-3 SENSITIVITY OF THE EYE

The eye can detect extremely small amounts of luminous energy. A dark-adapted eye can detect light equivalent to that received from a single candle distant 15 to 20 mi (30 km), in which case the retina is receiving only a few quanta of light (Chap. 23). The sensitivity of the eye varies greatly for different wavelengths (Fig. 23-3).

27-4 PERSISTENCE OF VISION

The visual system lags a bit in its response to a stimulus, and the sensation lasts an even greater fraction of a second after the stimulus ceases. This retention of the mental image, referred to as the persistence of vision, prevents the appearance of any "flicker" when a motion-picture film is projected on a screen at the rate of at least 16 screen illuminations per second (24 frames per second or 72 screen illuminations per second are used in commercial motion pictures), even though the screen is completely dark while the film is in motion in the projector. The illusion of movement in a motion picture is due to a fortunate propensity of the visual system to "fill in" the positions of a moving object intermediate between those imaged discretely and successively upon the retina.

27-5 STEREOSCOPIC VISION

The importance of binocular vision in judging position and relative distance may be appreciated by closing one eye and trying to bring two pencil points together when the pencils are held in the hands at arm's length and moved at right angles to the line of sight. When a single nearby object is viewed with both eyes, the axes of the eyes are turned toward each other. Distance is estimated by solution of the triangle whose base is the distance between the two pupils, averaging 64 mm. The amount of convergence enables us to make a crude estimate of the distance of the object; but its size (when known) tells us most about its distance. Far better is our ability to say which of two objects is the nearer. This depends upon the fact that the interpupillary distance subtends a different angle (called the binocular parallax) at each of the objects. The difference in the binocular parallaxes needs to be only a few seconds of arc for the difference in distance to be detectable (Fig. 27-5).

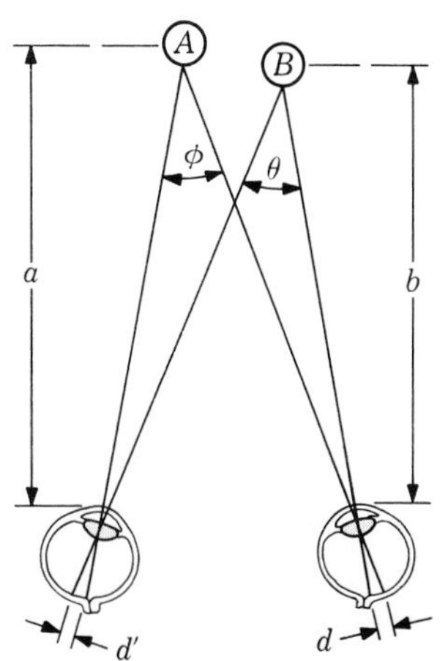

Figure 27-5
Binocular vision: arrangement for the measurement of stereoscopic visual acuity. If *A* and *B* are just discriminable, $\theta - \phi$ is the stereo threshold in angular terms and $d' - d$ is the corresponding retinal disparity. The values of $\theta - \phi$ and $d' - d$ are dependent upon the ratio of *a* to *b*.

For distances greater than about 700 ft, or about 215 m, the inclination of the optic axes is so slight that both eyes see practically the same view and distances are judged by the apparent sizes of familiar objects rather than by stereoscopic vision. Prism binoculars and range finders in effect increase the interpupillary distance and, with an extended base line of known length, permit distances to be calculated by the solution of a triangle in which the base and two angles are known.

27-6 LIMITATIONS OF VISION

The visibility of an object depends on size, contrast, intensity, time, and the adaptation of the eye. A deficiency in one of these factors, within certain limits, may be compensated for by an increase in one or more of the other factors. Thus close machine work or the inspection of small parts may be facilitated by increased illumination and contrasting colors. Illuminations recommended for various visual tasks are listed in Table 1, Chap. 23.

In the retinal mosaic of rods and cones the most sensitive part, the fovea, is about a millimeter in diameter, and its central part, 0.2 mm in diameter, contains only cones. The angular field of most distinct vision is about 1°, which is subtended by a circle of 4.4 mm at a distance 25 cm from the eye. Two fairly wide lines can just be distinguished by the eye when their angular separation is about 1 minute of arc, in which case the centers of the image lines are only a few thousandths of a millimeter apart on the retina. When the smallness or distance of an object exceeds the limitations for direct visibility, the eye requires optical aids, some of which will be briefly described.

27-7 MAGNIFIER

A simple magnifier is a converging lens placed so that the object to be examined is a little nearer to the lens than its principal focus (Fig. 27-6). An enlarged, erect, virtual image of the object is then seen. The image should be at the distance of most distinct vision, which is about 25 cm from the eye, the magnifier being adjusted so that the image falls at this distance.

The *linear magnification* M_l is the ratio of the image size to the object size; i.e.,

$$M_l = \frac{II'}{OO'} = \frac{q}{p} \tag{1}$$

The thin-lens equation

$$\frac{1}{p} + \frac{1}{q} = \frac{1}{f}$$

gives
$$\frac{q}{p} = \frac{q}{f} - 1$$

and if $q = -25$ cm,

$$M_l = \frac{25 \text{ cm}}{f} + 1 \tag{2}$$

where f is in centimeters.

The magnifier in effect enables one to bring the object close to the eye and yet observe it comfortably. When the object is thus brought closer, it subtends a larger angle at the eye than it would at a greater distance.

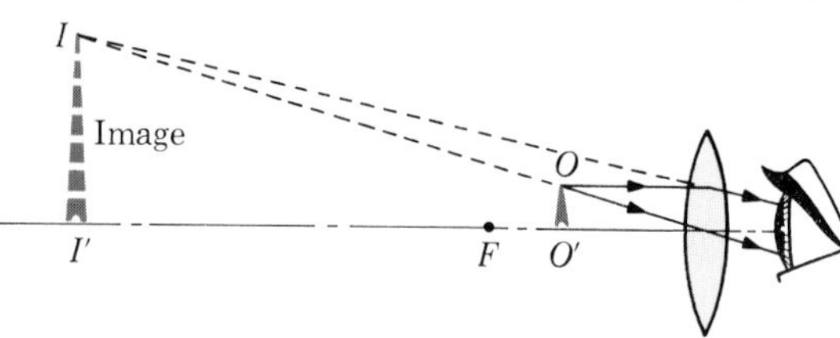

Figure 27-6
A simple magnifier.

Example A converging lens of 5.0-cm focal length is used as a simple magnifier, producing a virtual image 25 cm from the eye. How far from the lens should the object be placed? What is the magnification?

From the lens equation

$$p = \frac{fq}{q-f} = \frac{5.0 \text{ cm } (-25 \text{ cm})}{-25 \text{ cm } -5.0 \text{ cm}} = 4.17 \text{ cm}$$

$$M = \frac{25 \text{ cm}}{4.17 \text{ cm}} = 6.0$$

A magnifier is frequently used as an eyepiece or ocular in an optical instrument in combination with other image-forming lenses. In such a situation experienced users of optical instruments frequently place the object at the principal focus of the eyepiece so that parallel rays enter the eye and the image is at infinity rather than at 25 cm. The visual magnification is then considered to be the ratio of the angle subtended at the eye by the image to the angle which would be subtended by the object in the position of most distinct naked-eye vision. This is equal to the ratio of 25 cm to the focal length of the lens, or

$$M_a = \frac{25 \text{ cm}}{f} \tag{2a}$$

The magnification of a simple magnifier, or other visual instrument, is thus dependent on the habits of the user. The lens of the last example would have a magnification of 5.0 if used so that it produced parallel rays.

27-8 THE MICROSCOPE

Whenever high magnification is desired, the *microscope* is used. It consists of two converging lenses (in practice, lens systems), a so-called *objective* lens of very short focal length and an *eyepiece* of moderate focal length. The objective forms within the tube of the instrument a somewhat enlarged real image of the object. This

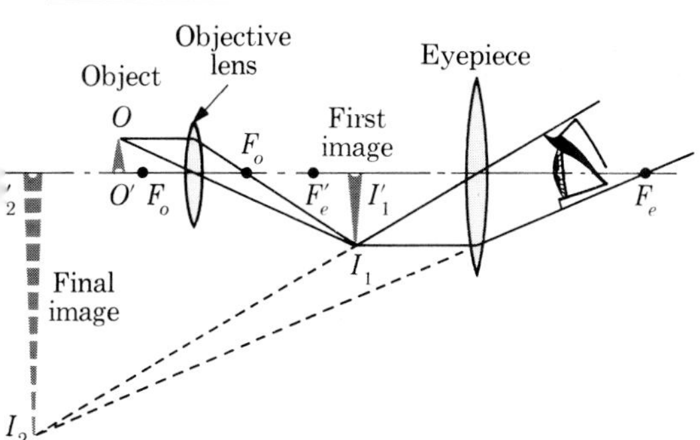

Figure 27-7
Ray diagram for a microscope.

image is then magnified by the eyepiece. The final image seen by the eye is virtual and very much enlarged.

Figure 27-7 shows the ray construction for determining the position and size of the image. The object is placed just beyond the principal focus of the objective lens, and a real image is formed at II'. This image is, of course, not caught on a screen but is merely formed in space. It consists, as does any real image, of the points of intersection of rays coming from the object. This image is examined by means of the eyepiece, which serves here as a simple magnifier. The position of the eyepiece, then, should be such that the real image $I_1I'_1$ lies just within the principal focus F'_e. Hence the final image $I_2I'_2$ is virtual and enlarged, and it is inverted with respect to the object.

The magnification produced by a microscope is the product of the magnification M_e produced by the eyepiece and the magnification M_o produced by the objective lens. Hence, for a final image at the distance of most distinct vision (25 cm)

$$M = M_o M_e = \frac{q_o}{p_o}\left(\frac{25\text{ cm}}{f_e} + 1\right) \qquad (3)$$

where p_o and q_o are the distances of object and first image, respectively, from the objective, and f_e is the focal length of the eyepiece, all distances measured in centimeters. In practice the largest magnification employed is usually about 1,500.

Example A microscope has an objective lens of 10.0-mm focal length and an eyepiece of 25-mm focal length. What is the distance between the lenses, and what is the magnification if the object is in sharp focus when it is 10.5 mm from the objective?

Consider the objective alone:

$$\frac{1}{10.5\text{ mm}} + \frac{1}{q} = \frac{1}{10.0\text{ mm}}$$

$$q = 210\text{ mm}$$

Consider the eyepiece alone, with the virtual image at the distance of most distinct vision (250 mm):

$$\frac{1}{p'} + \frac{1}{-250\text{ mm}} = \frac{1}{25.0\text{ mm}}$$

$$p' = 22.7\text{ mm}$$

$$\begin{aligned}\text{Distance between lenses} &= q + p' \\ &= 210\text{ mm} + 22.7\text{ mm} \\ &= 233\text{ mm} \\ &= 23.3\text{ cm}\end{aligned}$$

Magnification by objective:

$$M_o = \frac{210\text{ mm}}{10.5\text{ mm}} = 20.0$$

Magnification by eyepiece:

$$M_e = \frac{250\text{ mm}}{22.7\text{ mm}} = 11.0$$

Total magnification:

$$M = M_e M_o = 11.0 \times 20.0 = 220$$

As a check, Eq. (3) gives

$$M = \frac{q_o}{p_o}\left(\frac{25\text{ cm}}{f_e} + 1\right) = \frac{21.0\text{ cm}}{1.05\text{ cm}}\left(\frac{25\text{ cm}}{2.5\text{ cm}} + 1\right)$$
$$= 2\bar{2}0$$

27-9 REFRACTING TELESCOPES

The astronomical refracting telescope, like the compound microscope, consists of an objective lens system and an eyepiece. The instruments differ, however, in that the objective of the telescope has a long focal length and a much larger aperture. Light from the distant object enters the objective, and a real image is formed within the tube (Fig. 27-8). The eyepiece, used again as a simple magnifier, leaves the final image inverted.

It should be noted that, whereas the *same* rays might have been traced through both lenses, in Fig. 27-8 two *different* pairs of rays are traced. The first is perhaps the better representation of the physical action of the lenses, but the second procedure is the more practical. In this procedure, the first pair of rays traced through the objective is used to locate the first (real) image. Then from a convenient point on this image two more rays are drawn, one passing through the center of the eyepiece (undeviated), the other entering the eyepiece parallel to its principal axis and refracted to pass through the principal focus F_e. By this artifice the final (virtual) image is located.

A telescope image is usually far smaller than the object being observed; hence linear magnification is not a good index of the value of the telescope. Rather, the comparison to be made is between the size of the retinal image with the telescope and that without the telescope. The ratio of the two retinal image sizes is the same as the ratio of the two visual angles and is called the *angular magnification*.

Angular magnification is defined as the ratio of the angle β subtended at the eye by the image to the angle α which the object subtends at the lens or eye. The angular magnification can be computed from the geometry of Fig. 27-8. Neglecting the length of the telescope, the first image subtends the same angle α at the center of the objective lens as the object does at the observer's naked eye, and similarly the first and second images subtend the same angle β at the optical center of the eyepiece. Hence the angular magnification is given by

$$M_a = \frac{\beta}{\alpha} \doteq \frac{II'/f_e}{II'/f_o} = \frac{f_o}{f_e} \tag{4}$$

Owing to the approximations made, Eq. (4) ap-

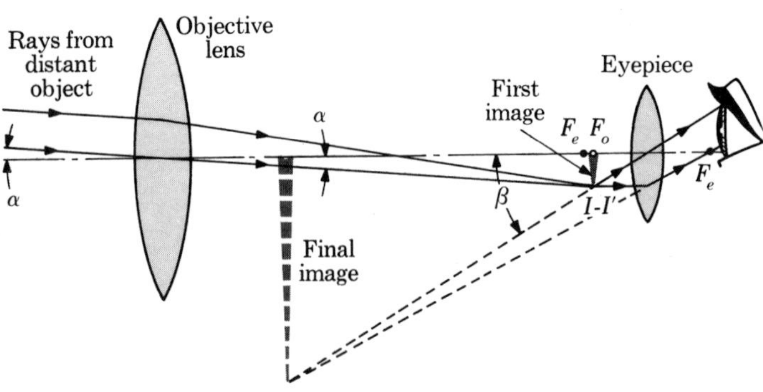

Figure 27-8
Ray diagram for a refracting telescope.

plies only for distant objects. This formula shows that apparently unlimited values of M may be obtained by making f_o very large and f_e very small. Other factors, however, limit the values employed in practice, so that magnifications greater than about 1,000 are rarely used in astronomy.

Which of the features of the telescope make it valuable depends upon whether it is to be used in conjunction with the eye or with a photographic plate. All the cones of light leaving the eyepiece pass through the image (exit pupil) that it forms of the objective aperture. So long as the exit pupil of the instrument (objective diameter/magnification) agrees in size with the pupil of the eye, differences in magnification cannot affect the amount of light per unit area (illuminance) of the retinal image. Any improvement in the visibility of a faint star is then directly attributable to the enlargement of the retinal image. But when photography is employed, the light-gathering ability of the objective lens (or of the mirror of a reflecting telescope such as the 200-in Mount Palomar instrument) becomes of the greatest importance. The amount of light collected by the objective is directly proportional to its area. A lens of 800-mm aperture will gather in $(800/8)^2 = 10{,}000$ times as much light as will the pupil of the eye at night (when it is 8 mm in diameter). Huge telescope lenses and mirrors make possible the photography of invisible stars within reasonable exposure time. The eye, on the contrary, can take only snapshots, no time exposures. But a "night glass" can increase the brightness of a faint star, which is invisible through a field glass of equal power designed for daytime use, only by reason of the fact that the exit pupil of the day glass is too small to fill the eye pupil and that of the night glass is made as large as the nighttime pupil by increasing the aperture of the objective.

Example A reading telescope comprising an objective of 30.0-cm focal length and an eyepiece of 3.0-cm focal length is focused on a scale 2.0 m away. What is the length of the telescope (distance between lenses)? What magnification is produced?

Consider the objective

$$\frac{1}{200\text{ cm}} + \frac{1}{q} = \frac{1}{30.0\text{ cm}}$$

$$q = 35.3\text{ cm}$$

For the eyepiece,

$$\frac{1}{p'} + \frac{1}{-25\text{ cm}} = \frac{1}{3.0\text{ cm}}$$

$$p' = 2.7\text{ cm}$$

Telescope length = 35.3 cm + 2.7 cm = 38.0 cm
Linear magnification by objective:

$$M_o = \frac{35.3\text{ cm}}{200\text{ cm}} = 0.176$$

Linear magnification by eyepiece:

$$M_e = \frac{25\text{ cm}}{2.7\text{ cm}} = 9.4$$

Total linear magnification:

$$M = M_e M_o = (9.4)(0.176) = 1.6$$

The value of the telescope in this application is not indicated by the calculation in this example, however. The image is not only 1.6 times as large as the object, but it is only $\frac{1}{8}$ as far away. The angle subtended at the eye by the image is therefore 12.8 times as great as that subtended by the object at the naked eye. It is this enlarged angle which determines the size of the retinal image, and hence the visual magnification. Note that Eq. (4) is not applicable to a telescope focused on a nearby object, and would lead to an erroneous result if applied here.

A galilean telescope (Fig. 27-9) consists of a converging objective lens, which alone would form a real inverted image $I_1I'_1$ of a distant object practically at the principal focus, and a diverging

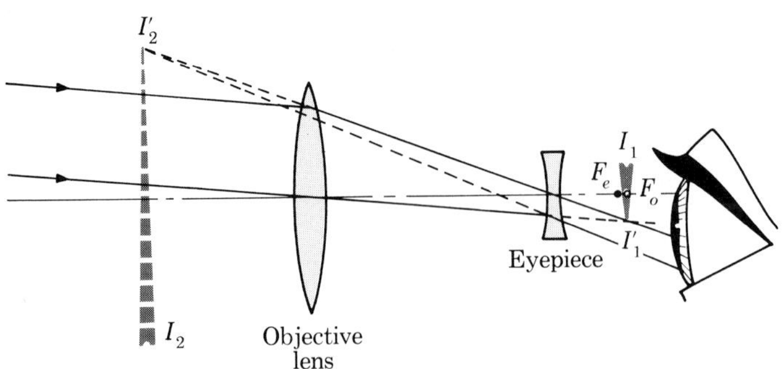

Figure 27-9
A diagram of a galilean telescope.

eyepiece lens. In passing through the concave lens, rays that are converging as they enter are made to diverge as they leave. To an observer the rays appear to come from $I_2I'_2$, the enlarged virtual image. With this design of telescope an erect image is secured. The galilean telescope is also much shorter than the astronomical telescope.

For a distant object the angular magnification is

$$M = \frac{f_o}{f_e} \tag{5}$$

Two galilean telescopes are mounted together in opera glasses. Such glasses are no longer widely used, for their field is very small, and they are limited, in practice, to a magnification of not more than 4. Modern field glasses are prism binoculars, which have several advantages.

27-10 REFLECTING TELESCOPES

Many of the larger astronomical telescopes do not employ lens systems as do refracting telescopes but instead use a parabolic-shaped concave mirror in place of the objective lens (Fig. 27-10). Perhaps the best known reflecting telescope is the 200-in telescope located at Mount Palomar. There are some advantages to using reflecting telescopes in place of refracting telescopes. Since much of the weight of a reflecting telescope is located at the base of the telescope (where the heavy mirror is placed), the telescope has greater stability. Further, one of the major problems encountered in refracting telescopes, chromatic aberration, is eliminated when a concave mirror is used. Because of this, a comparatively large aperture, admitting more light, can be used in the reflecting telescope.

Figure 27-10 shows a typical arrangement of the mirrors in a reflecting telescope. While a variety of mountings are possible, the Cassegrainian mounting is shown. The light entering the telescope from a distant light source is considered to be parallel. When the rays strike the large concave mirror, they reflect in such a manner that they converge at the principal focus f. The rays are intercepted before reaching f by a smaller convex mirror which redirects the rays through a hole in the center of the large concave mirror so that the rays now come to a focus at p. In practice, a camera or some other recording device such as a spectrograph is placed at this point p.

A reflecting telescope has the disadvantage that it is more liable to experience the problems of astigmatism and coma than do refracting telescopes. Because this problem is less serious in

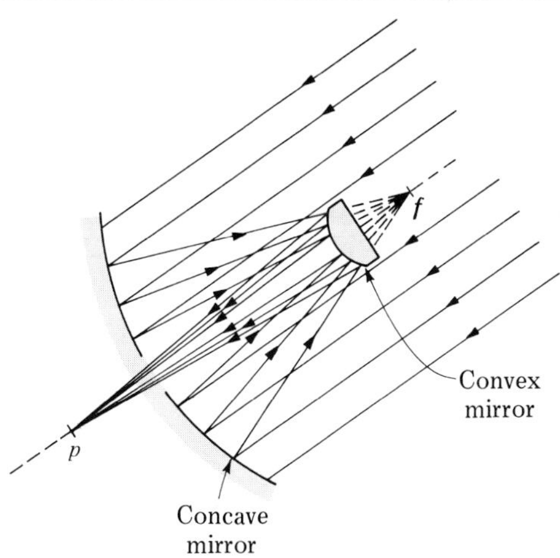

Figure 27-10
A reflecting telescope using a Cassegrainian mounting.

small refracting telescopes, reflecting telescopes are generally used in the larger astronomical telescopes.

27-11 THE PRISM BINOCULAR

The prism binocular (Fig. 27-11) consists of two astronomical telescopes in each of which two totally reflecting right-angle prisms are used to erect the image and shorten the overall length. The two prisms are set with half the hypotenuse face of one in contact with half the hypotenuse face of the other and with the long dimensions of these faces at right angles to each other. The combination inverts the image and exchanges left and right sides. The image formed by the objective lenses and prisms is therefore real and has the same orientation as the object itself, and so too does the final virtual image formed by the eyepiece. With this construction, advantage is taken of the fact that the distance between the objectives can readily be made greater than the distance between the eyes, thus enhancing the stereoscopic effect as an aid to the perception of distances.

27-12 THE PHOTOGRAPHIC CAMERA

A camera consists of a converging lens at one end of a lightproof enclosure and a light-sensitive film or plate at the other end of the enclosure, where it receives the real inverted image formed by the lens. The amount of light that reaches the film depends on the effective area of the lens (usually regulated by an iris diaphragm) and the time of exposure. The lens aperture or diameter of the effective opening is usually given as a fraction of the focal length. Thus an $f/11$ lens means that the diameter of the aperture is one-eleventh the focal length. This notation is convenient for expressing the "speed" of the lens (the relative exposure time required), since the amount of light per unit area reaching the film is proportional to the square of the diameter of the lens opening and inversely proportional to the square of the focal length of the lens.

Example Under certain conditions the correct

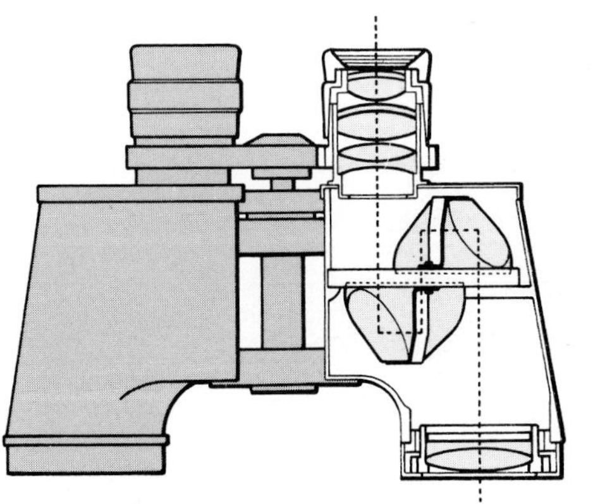

Figure 27-11
Prism binocular.

film exposure time is $\frac{1}{20}$ s, with a lens "speed" of $f/4.5$. What is the correct exposure time when the lens is diaphragmed to $f/6.3$?

The apertures are $A_1 = f/4.5$, $A_2 = f/6.3$.

$$\frac{t_1}{t_2} = \left(\frac{A_2}{A_1}\right)^2 = \left(\frac{4.5}{6.3}\right)^2$$

$$t_1 = \frac{1}{20}\text{ s}$$

$$t_2 = \left(\frac{6.3}{4.5}\right)^2 \left(\frac{1}{20}\text{ s}\right) = \frac{1}{10}\text{ s}$$

When a lens is focused for a certain distance, object points at that distance only are imaged with maximum sharpness. Points at other distances from the lens are imaged as blurred circles, termed *circles of confusion*. The farther a point is from the plane focused on, the greater is the size of the circle of confusion. If the circle of confusion is below a certain size, it appears to the eye as a point and the image appears sharp. The range of distances on the near and far sides of the plane focused upon, within which the details are imaged with acceptable sharpness, is called the *depth of field* and is of particular importance in photography.

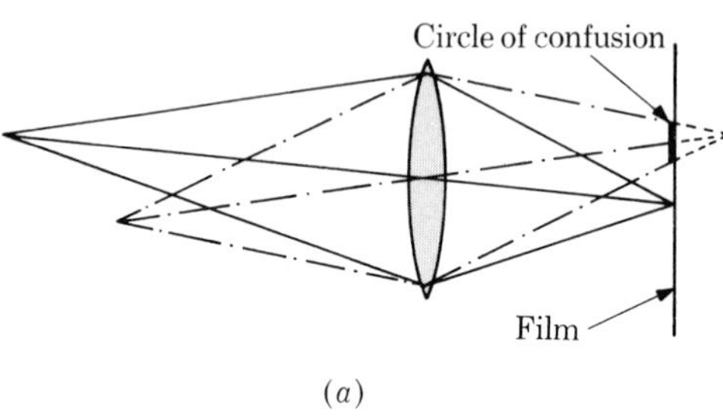

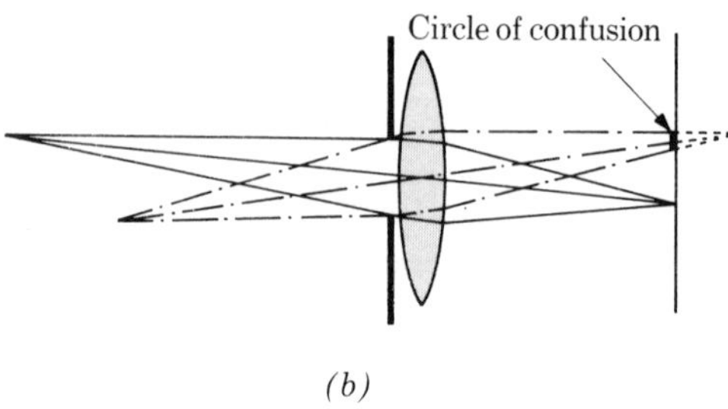

Figure 27-12
The relation of aperture to depth of field.

This situation is indicated in Fig. 27-12. If the diaphragm in front of the lens reduces the effective aperture, the size of the circle of confusion is reduced for points out of the plane on which the lens is focused. Of course the increased depth of field is gained at a cost of decreased illumination of the image on the film. The exposure time must be increased.

Depth of field depends on the relative aperture and the focal length of the lens, the distance focused upon, and the size of the circle of confusion which is acceptable For the same object distance, the depth of field increases for decreasing focal length, one of the advantages of a miniature camera with short-focal-length lens. The depth of field decreases rapidly as the object focused upon approaches the camera. It is therefore important to determine the distances more carefully for near objects than for distant objects. The greatest range of depth is that obtained when the lens is focused for the *hyperfocal distance,* or the shortest distance for which the far limit extends to infinity.

For the most critical definition of sharpness, the circle of confusion should subtend not more than 2 minutes of arc at the eye. If a photographic print is to be viewed at the distance for normal vision, the circle of confusion in the negative should not exceed approximately one two-thousandth the focal length of the camera lens. Many cameras are provided with a depth-of-focus table based upon this criterion.

SUMMARY

The eye contains a lens, a variable diaphragm (iris), and a sensitive screen (retina) on which the lens forms a real, inverted image of objects within the field of vision.

The difference in the eye's ability to discern color in the day and in the night is due to the different functions performed by the *rods* and *cones*—light receptors.

Accommodation, the adjustment of the eye for seeing at different distances, is accomplished by changes in the curvature of the lens.

The conventional distance of most distinct vision is 10 in, or 25 cm.

Nearsightedness is compensated by a diverging (concave) spectacle lens, farsightedness by a converging (convex) spectacle lens.

The linear magnification produced by a lens used as a simple magnifier is $M = 25\text{ cm}/f + 1$, where f is the focal length in centimeters and the lens is adjusted so that the image falls at the distance of most distinct vision, 25 cm.

The microscope consists of a short-focus objective and a longer-focus eyepiece. The linear magnification is given by

$$M = \frac{q_o}{p_o}\left(\frac{25\text{ cm}}{f_e} + 1\right)$$

The astronomical telescope consists of a long-focus objective and an eyepiece. The linear magnification is given by

$$M = \frac{q_o}{p_o}\frac{q_e}{p_e}$$

Angular magnification is the ratio of the angle subtended at the eye by the image to the angle subtended by the object.

The galilean telescope consists of a long-focus objective and a negative eyepiece.

Many large astronomical telescopes use a parabolic-shaped concave mirror as the objective lens.

The f number of a camera lens is the quotient of its focal length and its aperture.

The depth of field of a lens is the range of distances, on the near and far sides of the plane focused upon, within which the object structure is imaged with acceptable sharpness.

Questions

1 Define the terms vitreous humor, aqueous humor, fovea centralis, blind spot, cornea, and retina.

2 What is the primary difference between the method of focusing used by the eye and by a camera? Explain.

3 Compare and contrast the optical arrangements of the human eye and those of a photographic camera.

4 Explain why roses in a garden lose their brilliant colors under moonlight.

5 Name three common eye defects and state the type of spectacle lens that is used to compensate for each.

6 Would a person whose spectacles are corrected for both astigmatism and nearsightedness note the difference if his lenses should become rotated in the frame? For astigmatism only? For nearsightedness only?

7 Explain the optical illusion frequently observed in motion pictures when the wheels of a forward-moving vehicle appear to be stationary or even to be turning backward.

8 Consider that in Fig. 27-6 the object subtends an angle α at the lens or eye and its image subtends a larger angle β. The magnifying power depends on this gain in angle β/α subtended at the eye. On this basis, derive an equation for the magnification of a simple magnifier.

9 Why does a young person who wears glasses not need bifocals?

10 Describe some of the imperfections that occur when a simple microscope is made up of two uncorrected lenses.

11 What determines the magnification produced by a simple magnifier? By a microscope?

12 If a person were looking through a telescope at the full moon, how would the appearance of the moon be changed by covering half the objective lens?

13 Comment on the differences between the astronomical and galilean telescopes and point out where these are useful for certain applications.

14 Cross hairs are sometimes placed in astronomical telescopes for sighting purposes. In what position should they be placed? Explain.

15 Why do astronomical telescopes have objective lenses of such large diameters? Why is this not necessary for the galilean telescopes used in opera glasses?

16 How could the final image of the galilean telescope be photographed?

17 Compare a refracting telescope with a reflecting telescope pointing out the advantages and disadvantages of each.

18 Should one use a lens of long or short focal length for a simple magnifier? For a microscope objective? For a telescope objective?

19 How will changing the aperture of a camera lens from $f/2$ to $f/8$ affect (*a*) the size of the image, (*b*) the illuminance of the image, (*c*) the exposure time, (*d*) the sharpness of the image, and (*e*) the depth of field?

20 Show that the depth of focus of a camera increases as the f number of the stop increases. *Hint:* Show that for a given lens the diameter of the circle of confusion is directly proportional to the diameter of the aperture, and hence inversely proportional to the f number of the stop used.

21 Show that in a camera it is desirable to use a lens of short focal length and a small film size in order to reduce the size of the circle of confusion relative to the picture area and still be able to use low f numbers for high speed.

22 Explain why the use of goggles enables an underwater swimmer to see clearly.

23 Draw to scale a ray diagram for a telescope having an objective of 40-cm and an eyepiece of 5.0-cm focal length, and another ray diagram for a galilean telescope having an objective of 40-cm and an eyepiece of -5.0-cm focal length, when both telescopes are adjusted for viewing a distant object. Mention advantages and disadvantages of each type of telescope.

24 In a copying camera, the image should be of the same size as the object. Prove that this is the case when both object and image are at a distance $2f$ from the lens.

Problems

1 A person has a minimum distance of direct vision of 2.44 m. What kind of lenses and of what focal length are required for spectacles to enable him to read a book at a distance of 0.46 m?

2 A person having hyperopia has a minimum distance of distinct vision of 125 cm rather than the optimum 25 cm. To correct this problem what focal-length glasses should he wear?
Ans. 31.2 cm.

3 Light entering the eye is refracted chiefly at the cornea. Assuming the eye to be 25 mm from cornea to retina and to be filled with a homogeneous medium of refractive index 1.336, calculate (*a*) the radius of curvature of the cornea and (*b*) the length of the retinal image of an object 10 cm long placed 1.0 m from the eye. (Use a wave-front diagram and calculate curvature.)

4 The shortest distance that a person can distinctly see an object is 2 m. What power must a lens have to enable him to see an object which is held 50 cm away? *Ans.* 1.5 diopters.

5 A certain farsighted person can read fairly clearly without spectacles at a distance of 80.0 cm. His glasses have a power of 2.5 diopters. What is his reading distance when he is wearing glasses?

6 A 10× magnifier is one that produces a magnification of 10 times. According to Eq. (2), what is its focal length? How large an image of a flashlight lamp 0.64 cm in diameter will this lens be able to produce on a card held 12.7 cm away?
Ans. 2.8 cm; 2.3 cm.

7 A nearsighted man cannot see clearly objects more than 5.0 m away. Because of an error made by his oculist, he receives spectacles intended for a nearsighted man who cannot see clearly objects more than 10.0 m away. When he is wearing these spectacles, what is the greatest distance at which he can focus clearly?

8 A farsighted man needs a lens with a power of +1.50 diopters. The inner face of the lens is to fit against an eyeball having radius of curvature of 1.20 cm. Glass is available with an index of refraction 1.63. What must be the curvature for the outer face of the lens? The index of refraction of the cornea is 1.33. *Ans.* 2.3 cm.

9 A normal eye has a distance of most distinct vision of about 25 cm. An observer views an object through a convex lens of 4.5-cm focal length, used as a simple magnifier. (*a*) Where should the object be placed for most distinct vision? (*b*) What is the linear magnification thus produced?

10 A linen tester is a simple magnifier for use in counting the number of threads per linear inch in fabrics. The tester shown in Fig. 27-13 consists of a lens of 13-mm focal length, mounted in a folding frame, the lower arm of which is cut out to cover an area of fabric 0.50 in square. What is the approximate magnification when the image is formed at the distance of most distinct vision?

Ans. 20.

Figure 27-13

11 The focal lengths of the objective and the eyepiece of a compound microscope are 0.500 and 2.00 cm, respectively. An object is placed 0.520 cm from the objective, and the instrument is adjusted so that the final image is formed 25.0 cm from the eye. Compute the linear magnification by the microscope.

12 A double-convex lens with a focal length of 6 cm is used as a simple magnifier. If an erect image three times as large is desired, how far should the lens be held from the object?

Ans. 4 cm.

13 A crude microscope is constructed of two spectacle lenses of focal lengths 5.0 and 1.0 cm, spaced 20 cm apart. (*a*) Where must the object be placed to enable the observer to see a distinct image at a distance of 25 cm? (*b*) What is the linear magnification?

14 An objective of a microscope is 10 mm. Assuming it produces an image distance of 160 mm, what is the magnification of the objective and what should be the object distance?

Ans. 15; 10.65 mm.

15 A microscope has an objective lens of focal length 0.90 cm placed 13.0 cm from an eyepiece of 5.00-cm focal length. An insect 0.050 cm long is placed 1.0 cm from the objective. Locate, describe, and find the size of the final image.

16 A compound microscope has as objective and eyepiece, thin lenses of focal lengths 1.0 and 4.0 cm, respectively. An object is placed 1.2 cm from the objective. If the virtual image is formed by the eyepiece at a distance of 25 cm from the eye, what is the magnification produced by the microscope? What is the separation between the lenses?

Ans. 36; 9.4 cm.

17 A certain biological specimen being examined under a microscope cannot be subjected to illumination greater than 10,000 lm/ft^2 without injury. If the apparent illumination in the image must be at least 1 lm/ft^2 for satisfactory viewing, what is the maximum useful magnification?

18 A microscope with an objective of focal length 10 mm and an eyepiece of focal length 50 mm, 20 cm apart, is used to project an image on a screen 1.0 m from the ocular. What is the linear magnification of the image?

Ans. 260×.

19 The Yerkes refracting telescope has an objective of diameter 40 in and a focal length of 65 ft. (*a*) What magnifying power should be used to give an exit pupil of 2.0 mm, matching the entrance pupil of the eye? (*b*) What focal-length eyepiece is needed for this magnifying power? (*c*) Show that if some other magnifying power is used, there is no gain in the total light in the retinal image.

20 A telescope has an objective lens having a focal length of 10 m and a diameter of 100 cm. What is the magnification provided by this telescope if its eyepiece has a focal length of 5 cm?

Ans. 200.

21 What would be the magnifying power of the Yerkes telescope (Prob. 19) when used with an eyepiece of 1.0-in focal length to produce a final image at infinity?

22 A telescope with an angular magnification of 10 is sighted at the sun and focused to give an image at infinity. How far and in what direction must the eyepiece be moved to project a sharp image of the sun on a screen 31 cm behind the eyepiece? The focal length of the objective is 50.0 cm. *Ans.* 0.96 cm.

23 A simple telescope consists of two converging lenses of focal lengths 5.08 and 25.4 cm, respectively. What is (*a*) their distance apart and (*b*) the magnification if the telescope is used to view a scale 3.05 m from the objective, the final image being in the plane of the object?

24 A lens in a camera has a focal length of 10 cm. It is used to take a picture of a person 2 m tall, 4 m away. How tall will the person be on the film? *Ans.* 5.13 cm.

25 A pair of opera glasses has objective lenses of focal length 5.33 cm. (*a*) If the magnifying power of the glasses is 4, what is the focal length of the eyepieces? (*b*) What is the approximate length of the glasses?

26 The objective lens in an opera glass has a focal length of 12.5 cm. The eyepiece has a focal length of −3.18 cm. A distant object is viewed. Calculate (*a*) the magnification and (*b*) the approximate length of the tubes.

Ans. 3.94; 9.3 cm.

27 A miniature camera whose lens has a focal length of 5.08 cm can take a picture 2.54 cm high. How far from a building 36.6 m high should the camera be placed to receive the entire image?

28 The lens of a camera has a focal length of 5.0 cm. If it is desirable to take closeup shots as close as 1.0 m from the lens, what range of adjustment must be provided for the camera?

Ans. 0.27 cm.

29 A certain camera lens has a focal length of 5.0 cm. A second converging lens of focal length 10.0 cm can be mounted immediately in front of this lens to serve as a portrait attachment. Calculate the position of the image formed of an object 20 cm in front of the camera.

30 A camera has a lens with a focal length of 6 in. What is the size of the image of a 4-ft-tall object placed 30 ft away from the camera?

Ans. $\frac{4}{59}$ ft.

31 If a correct camera exposure for a certain scene is $\frac{1}{100}$ s when the diaphragm is set at $f/3.5$, what exposure time is required at $f/12.5$?

32 An enlarging camera is so placed that the lens is 30 cm from the screen on which an image five times the size of the object is to be projected. (*a*) How far is the object from the lens? (*b*) What is the focal length of the lens?

Ans. 6.0 cm; 5.0 cm.

33 Two camera lenses have focal lengths of 7.0 and 5.0 cm, respectively. The first has a free diameter of 0.50 cm. What diameter must the other have in order that they may both have the same exposure time?

34 A slide projector produces an image on a screen 15 m from the focusing lens. If the magnification is 50×, what distance should the slide be placed from the lens and what is the focal length of the lens? *Ans.* 0.30 m; 0.295 m.

35 A reconnaissance plane is equipped with a camera having a lens of 24-in focal length. An observer photographs the ground with the camera properly focused when the plane is 18,000 ft above a river. (*a*) What is the distance between the film and the optical center of the lens? (*b*) If the image of the river on the developed negative is 1.5 in wide, what is the actual width of the river?

36 A lantern slide 3.0 in wide is to be projected onto a screen 30 ft away by means of a lens whose focal length is 8.0 in. How wide should the screen be to receive the whole picture? *Ans.* 11 ft.

37 A projection lantern is to produce a magnification of 50 diameters at a distance of 15 m from the objective. Find the distance of the lens from the lantern slide and the equivalent focal length of the objective lens.

38 A lens system designed to be used for telephoto purpose has a double-convex lens having a focal length of +10 cm which is located 6 cm in front of a concave lens having a focal length of −5.0 cm. Determine the position and the size of the image formed of a distant object.

Ans. +20 cm (opposite side of concave lens from original object); 5×.

Wolfgang Pauli, 1900–1958

Born in Wien. Visiting professor, Institute for Advanced Study, Princeton. Awarded the 1945 Nobel Prize for Physics for his work on the exclusion principle, which deals with regulation of electrons in the outer shell of atoms and molecules.

Percy Williams Bridgman, 1882–1961

Born in Cambridge, Massachusetts. Hollis Professor of Mathematics and Natural Philosophy at Harvard University. Awarded the 1946 Nobel Prize for Physics for his investigation of the physical effect of high pressures.

28

Dispersion; Spectra

In 1666 Isaac Newton discovered that a narrow beam of sunlight on passing through a glass prism could be spread out into an array of colors called a spectrum and that these colors could be recombined to form white light. He concluded that white light is a mixture of colored lights and that there is a wave phenomenon involved, this in spite of the fact that he was the principal champion of a particle theory of light. We now understand that the different colors are due to different wavelengths of light which travel with different speeds in glass. The separation of colors in refraction is called dispersion. Spectra are used for the study of atomic and molecular structure, for chemical analysis, and for the study of the composition of the stars.

28-1 DISPERSION BY A PRISM

Measurements on the refraction of light as it passes from air into glass show that the amount of refraction depends upon the wavelength (Fig. 28-1). Light of all wavelengths is reduced in speed in glass, but blue-producing light, which is refracted the greatest amount, travels slower than red-producing light, which is refracted least. The variation of the index of refraction, n, with the wavelength of light is called the *optical dispersion* ($\Delta n/\Delta\lambda$) of a substance. By measuring values for n for a series of known wavelengths one can determine the dispersion curve characteristic of a given substance.

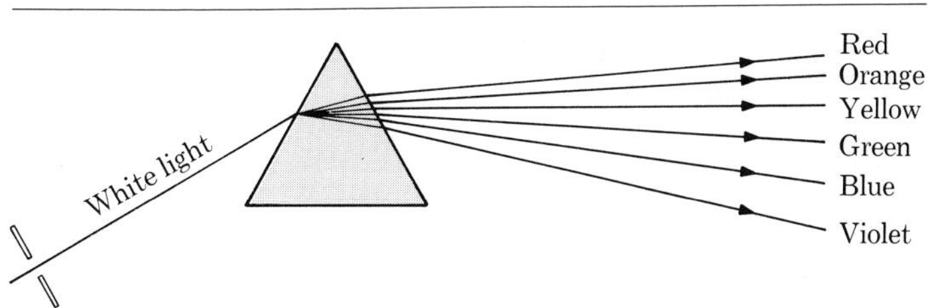

Figure 28-1
Spectrum formed by the passage of light through a glass prism.

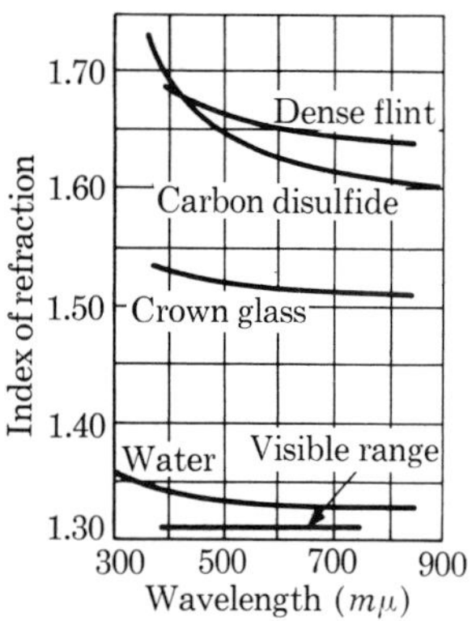

Figure 28-2
Dispersion curves.

The dispersion curves of all transparent substances are of the same general form (Fig. 28-2), the larger values of n being associated with shorter wavelengths, but there are marked differences in the total variation of n throughout the visible spectrum. Substances showing a rapid change in n with wavelength in this region are said to have a high dispersion and usually have also a high index of refraction.

If a substance absorbs light strongly in a narrow-wavelength region, the values of n in that region change rapidly with wavelength.

The amount of dispersion produced by a prism is expressed quantitatively by the angular separation of particular parts of the spectrum. When a ray of white light is incident upon a prism (Fig. 28-3), the light is broken into its components and each component is deviated by a different angle, for example, δ_r for red-producing light and δ_v for violet-producing light. The angular dispersion ψ between the violet and red regions is the difference between the deviations

$$\psi = \delta_v - \delta_r \tag{1a}$$

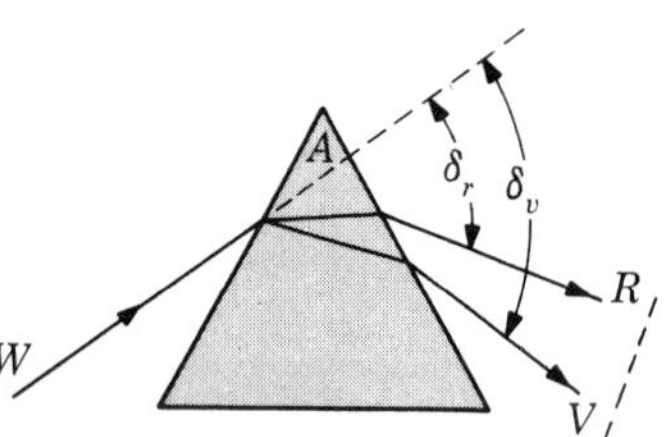

Figure 28-3
Dispersion of light by a prism.

This angular dispersion depends upon the angle of incidence as well as upon the indices of refraction. If the angle of incidence is such as to produce minimum deviation for an intermediate ray, say, yellow, the two deviations will each be near minimum and we may use as approximations the relationships for that condition. Then

$$\psi = D_v - D_r \tag{1}$$

The minimum deviation D of a ray produced by a prism is related to the prism angle A and the index of refraction n by Eq. (8), Chap. 25,

$$n = \frac{\sin \frac{1}{2}(D + A)}{\sin \frac{1}{2}A} \tag{2}$$

Example Find the dispersion from the F line of the spectrum (blue) to the C line (red) produced by a dense flint-glass prism of refracting angle 60°.

Solving Eq. (2) for D,

$$D = 2 \sin^{-1} (n \sin \tfrac{1}{2}A) - A$$

$$\begin{aligned}\psi = D_F - D_C &= 2 \sin^{-1} (n_F \sin \tfrac{1}{2}A) \\ &\quad - 2 \sin^{-1} (n_C \sin \tfrac{1}{2}A) \\ &= 2[\sin^{-1} (1.6691)(0.5000) \\ &\quad - \sin^{-1} (1.6500)(0.5000)] \\ &= 2(56°34' - 55°35') = 1°58'\end{aligned}$$

28-2 CHROMATIC ABERRATION

When light passes through a lens, it is bent around the thicker part of the lens. The short wavelengths are refracted more than the long

Table 1
VARIATION OF THE INDEX OF REFRACTION OF GLASSES WITH WAVELENGTH

Color	Solar spectrum line	Wave-length, Å	Ordi-nary crown	Boro-silicate crown	Medium flint	Dense flint
Violet	*H*	3,969	1.5325	1.5388	1.6625	1.6940
Blue	*F*	4,861	1.5233	1.5297	1.6385	1.6691
Yellow	*D*	5,893	1.5171	1.5243	1.6272	1.6555
Red	*C*	6,563	1.5146	1.5219	1.6224	1.6500

wavelengths. Thus violet rays are focused closer to a converging lens than are red rays (Fig. 28-4). This property of a lens to converge to different foci the rays of light of different wavelengths coming from a single point is called *chromatic aberration*. Its presence in a simple magnifier can be seen by the fringes of color which surround each image. Chromatic aberration would be objectionable in most optical instruments if it were left uncorrected. It was its presence, in telescopes which led Newton to invent the reflecting telescope, in which, of course, all wavelengths are reflected to the same focus. His pioneering work on the dispersion due to a prism, however, laid the foundation for the eventual solution of the problem of chromatic aberration.

Example A double-convex lens made of ordinary crown glass has both its radii of curvature equal to 10.00 cm. Find its focal lengths, in air, for the *F* (blue) line and for the *C* (red) line.

For the *F* line,

$$\frac{1}{f_F} = (n_F - 1)\left(\frac{1}{R_1} - \frac{1}{R_2}\right)$$

$$= (1.5233 - 1)\left(\frac{1}{10.00} - \frac{1}{-10.00}\right)\text{ cm}^{-1}$$

$$= (0.5233)(0.2000)\text{ cm}^{-1} = 0.1047\text{ cm}^{-1}$$

$$f_F = 9.55\text{ cm}$$

For the *C* line,

$$\frac{1}{f_C} = (0.5146)(0.2000)\text{ cm}^{-1} - 0.1029\text{ cm}^{-1}$$

$$f_C = 9.72\text{ cm}$$

Note the difference of 1.7 mm in the focal lengths for the two colors.

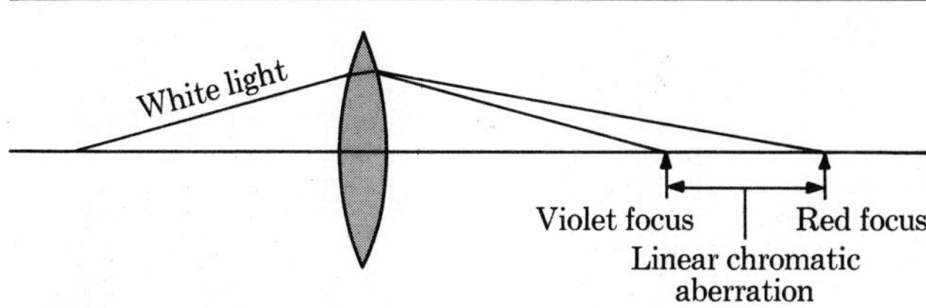

Figure 28-4
Chromatic aberration in a lens.

28-3 THE ACHROMATIC LENS

Chromatic aberration can be largely eliminated by combining lenses made of glasses of different dispersion characteristics so as to obtain a combined lens having practically the same focal length for all wavelengths. Strictly, a two-component lens can be designed to bring to the same focus only two particular wavelengths, but usually close agreement is obtained for all wavelengths between these two.

The correction of chromatic aberration can be accomplished by combining a strong positive lens (short focal length), made of glass with a low dispersion, with a weak negative lens of highly dispersive glass. (See Fig. 28-5.) In this way the dispersive effects can be made to counteract each other, while positive focal effect remains. Each component of the lens has a difference of power (in diopters) for two colors, say, for the F (blue) and the C (red) lines of the spectrum. If these two differences of power are made equal in magnitude and opposite in sign, the net power difference is zero; i.e., the power of the combined lens will be the same for each color.

Example A double-convex crown-glass lens whose radii of curvature are each 10.00 cm (that of the previous problem) is to be combined with a negative lens of dense flint glass, one of whose surfaces is concave with a 10.00-cm radius. Using the thin-lens approximation, find the radius of curvature of the second surface so as to make the entire lens achromatic for F (blue) and C (red) light. What will be the combined focal length for D (yellow) light?

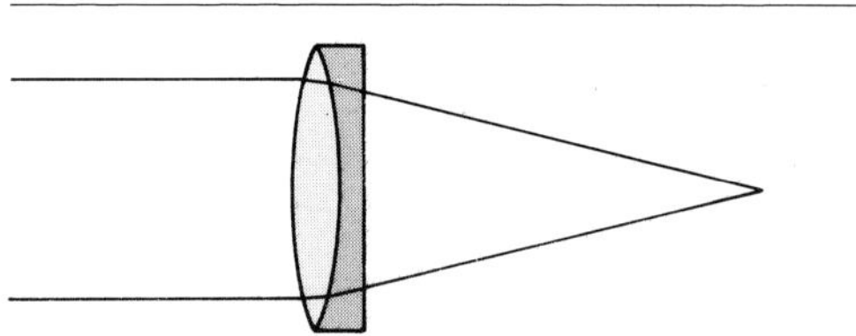

Figure 28-5
An achromatic lens combination.

The difference of the powers of the first lens for F (blue) and C (red) light may be expressed as

$$\frac{1}{f_F} - \frac{1}{f_C} = [(n_F - 1) - (n_C - 1)]\left(\frac{1}{R_1} - \frac{1}{R_2}\right)$$
$$= (1.5233 - 1.5146) \times \left(\frac{1}{10.00 \text{ cm}} - \frac{1}{-10.00 \text{ cm}}\right)$$
$$= (0.0087)(0.2000 \text{ cm}^{-1})$$
$$= 0.00174 \text{ cm}^{-1}$$

For the flint glass the difference of powers is

$$\frac{1}{f'_F} - \frac{1}{f'_C} = (1.6691 - 1.6500) \times \left(\frac{1}{-10.00 \text{ cm}} - \frac{1}{R'_2}\right) \text{cm}^{-1}$$
$$= -0.0191\left(0.1000 + \frac{1}{R'_2}\right) \text{cm}^{-1}$$

Since this difference is to be made equal in magnitude and opposite in sign to that of the first lens,

$$0.00174 \text{ cm}^{-1} = 0.0191\left(0.1000 + \frac{1}{R'_2}\right) \text{cm}^{-1}$$

from which

$$R'_2 = -110 \text{ cm}$$

That is, the second surface is slightly convex.

The power of the crown-glass lens for yellow light is

$$\frac{1}{f_D} = (0.5171)\left(\frac{1}{10.00 \text{ cm}} - \frac{1}{-10.00 \text{ cm}}\right)$$
$$= 0.1034 \text{ cm}^{-1}$$

of energy as its temperature is raised. At comparatively low temperatures most of this energy is in the infrared. As the temperature of the source is raised, more energy is radiated at each wavelength and the maximum in the distribution is shifted toward shorter wavelengths (Fig. 17-9). This change in energy distribution is responsible for the familiar change in the color of a body as it is heated to incandescence and for the indications of high temperature implied by the terms red-hot and white-hot.

There is a definite proportionality between absorption and emission at the same temperature for any wavelength. This is Kirchhoff's law. Thus a good absorber is also a good radiator. A blackbody, which absorbs all radiant energy incident on it, is also the perfect radiator. The total energy radiated at all wavelengths is proportional to the area under the distribution curve, and this was shown by Stefan to vary as the fourth power of the absolute temperature for a perfect radiator (Chap. 17).

28-7 FLUORESCENCE AND PHOSPHORESCENCE

We saw in our discussion of the reflection of light that the color which an object assumes is determined by the color of the light which it selectively reflects. For example, an object which appears red when white light falls upon it does so because it reflects mainly the red portion of the incident white light. Generally, the wavelengths of the reflected light are the same as one or more of the wavelengths found in the incident light. There is a process, however, by which certain substances change the wavelength of the incident light when they reflect it. *Fluorescence,* which literally means "to glow softly," is a process in which a substance absorbs radiant energy and then immediately reemits an appreciable part of it with its wavelengths changed so that they are longer than those absorbed. Often the term fluorescence refers to the changing of ultraviolet to visible light. In the commercial fluorescent lamp the light that is seen

That of the flint-glass lens for yellow light is

$$\frac{1}{f'_D} = (0.6555)\left(\frac{1}{-10.00\text{ cm}} - \frac{1}{-110\text{ cm}}\right)$$

$$= -0.0596\text{ cm}^{-1}$$

For the combined lens,

$$\frac{1}{F_D} = \frac{1}{f_D} + \frac{1}{f'_D} = (0.1034 - 0.0596)\text{ cm}^{-1}$$

$$= 0.0438\text{ cm}^{-1}$$

$$F_D = 22.8\text{ cm}$$

28-4 PRISM SPECTROSCOPE

A spectroscope (Fig. 28-6) is a combination of a prism and achromatic lenses used to segregate the various wavelengths in a beam of light and thus to permit examination of its spectrum. Light from the source to be examined falls on an adjustable narrow slit SS' placed at the principal focus of a convex lens L_1. The rays emerging from the lens are thus collimated, i.e., made parallel. This beam falls on a prism in which it is deviated toward the base and dispersed into rays of different wavelengths. These rays are viewed through a telescope whose objective lens L_2 forms a real image of the slit for each wavelength of light present. These images are side by side, with some overlap, and form a continuous or discontinuous band. This spectrum is magnified by an eyepiece, in which cross hairs or other reference marks may be located. With a narrow slit, a monochromatic image of the slit is formed by every discrete wavelength present. Each image of the slit is called a *spectrum line.*

The collimator and telescope are usually in horizontal tubes arranged to rotate about a common axis perpendicular to the prism table. If an instrument of this sort is provided with a graduated circle and vernier scales on the telescope and prism table, so that angles of deviation can be measured, it is called a *spectrometer*. In a *spectrograph,* a camera is substituted for the telescope so that permanent photographic records can be made. Since photographic plates can be sensitized for wavelengths beyond either end of the visible spectrum, the spectral range of the spectrograph is greater than that of the spectroscope. Since glass is not a good transmitter of radiant energy in the infrared or ultraviolet regions, rock salt or potassium bromide prisms are used in infrared spectrometers and quartz or fluorite prisms are used in ultraviolet spectrometers to give a lower practical limit of about 1,000 Å. In addition the entire optical path may be evacuated to reduce the effects of atmospheric absorption of radiation.

A prism spectrometer is not useful in making primary determinations of wavelengths. Once certain wavelengths have been measured by some other method (grating spectrometer or interferometer), these standards can be used to calibrate a prism spectrometer. The line spectrum from an iron arc is frequently used as a reference, since it comprises many lines conveniently spaced throughout the spectrum.

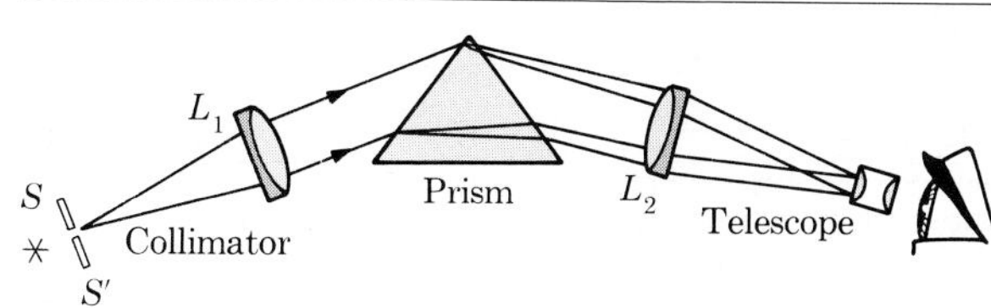

Figure 28-6
Arrangement of the essential parts of a prism spectroscope.

28-5
TYPES OF SPECTRA

When light produced by an incandescent solid, liquid, or gas falls directly on the slit of a spectroscope, an image of the slit is formed for each frequency emitted by the source. Such a spectrum is called an *emission* spectrum. (See the lower four spectra in the optical spectra chart.) If all the visible frequencies are emitted by the source, the images overlap and the spectrum shades continuously from one color to the next as in Fig. 29-5. This type of spectrum is called a *continuous* spectrum. Incandescent solids, liquids, or gases under high pressure are sources of continuous spectra. In the continuous spectrum all wavelengths are present, but the amount of energy radiated at each wavelength depends primarily upon the temperature of the incandescent material and to some extent upon its nature.

When light from a luminous gas or vapor under moderate or low pressure is examined by means of a spectroscope, the spectrum is found to be made up of definitely placed *bright lines*. (See the lower four spectra of the optical spectra chart.) Each *line* is a monochromatic image of the slit through which the light passes. Every gas emits certain definite frequencies that are characteristic of that particular gas. The bright-line spectrum of each provides a convenient and sensitive means of identifying even minute quantities of the substance. In the bright-line spectrum the wavelengths of the lines depend primarily upon the chemical nature of the material, and the number and intensities of the lines depend upon the method by which energy is supplied to the atoms to "excite" them.

A third type of emission spectrum is the type that exhibits a fluted-band structure which on sufficiently high dispersion is found to consist of closely spaced lines arranged in an orderly manner. *Band* spectra are characteristic of molecules, while line spectra are associated with uncombined atoms.

If the light in traveling from the source to the slit of the spectroscope passes through an absorbing medium, some of the frequencies emitted by

4 The wavelength of yellow light from a sodium lamp is 0.00005893 cm. (*a*) How many waves are there in 1.00 cm? (*b*) What is the frequency? (*c*) If the speed of sodium light in water is three-fourths its speed in air, what is the wavelength of sodium light in water?
Ans. 1.70×10^4; 5.08×10^{14} Hz; 4.42×10^{-5} cm.

5 Red light (6,563 Å) enters a medium flint glass. (*a*) What is the frequency of the waves in this glass? (*b*) The speed? (*c*) The wavelength?

6 An equiangular prism has a refractive index of 1.414. Compute its angle of minimum deviation. *Ans.* 30°.

7 Red light strikes a 60° prism (apex angle). The beam of red light has an index of refraction of 1.64 in the prism. What is the angle of minimum deviation for this prism?

8 Compute the angle of minimum deviation for a glass prism whose refracting surfaces form an angle of 60°. Assume that the index of refraction of the glass is 1.6. *Ans.* 46°16′.

9 A beam of white light falls on a plate of medium flint glass, making an angle of incidence of 30°0′. What is the angle of dispersion between the violet and the red rays refracted in the glass?

10 A beam of white light falls upon a 30° prism. Determine the angle of dispersion for red and for blue light if their respective indices of refraction are 1.64 amd 1.65. *Ans.* 0.32°.

11 A prism is made of medium flint glass and has a refracting angle of 60°0′. The prism is adjusted in a spectrometer to give minimum deviation for the yellow wavelengths from a sodium lamp. What deviation will it produce for the violet (3,970 Å) light from a calcium source?

12 Determine the dispersion from the *F* line of the spectrum (blue) to the *C* line (red) produced by a dense flint-glass prism having a refracting angle of 40°. *Ans.* 1°.

13 A converging lens of flint glass has an index of refraction of 1.650 for yellow light and a focal length of 1.250 ft for this light. For a certain red light the focal length is found to be 1.275 ft. What is the index of refraction of the lens for this red light?

14 A converging crown-glass lens has a focal length of 25 cm for the violet (3,969 Å) rays. Find its focal length for red (6,563 Å) rays.
Ans. 26 cm.

15 A plano-convex lens is made of medium flint glass and has a radius of curvature of 30.0 cm. Find the focal length of the lens for light of the *C* and *F* lines of the solar spectrum.

16 A pencil of rays of white light parallel to the principal axis falls upon a double-convex lens of dense flint glass. The lens has radii of curvature each 40.0 cm. Calculate the separation of the focal points for red (6,560 Å) and blue (4,860 Å) rays. *Ans.* 9 mm.

17 A crown-glass plano-convex lens has a radius of curvature of 5.00 in. What radius of curvature has the medium-flint-glass plano-concave lens, which with the first lens makes an achromatic pair for blue- and red-light rays?

18 Two prisms, one of crown glass and the other of flint glass, are arranged so that their apex angles point in opposite directions and are in direct contact with each other. When light having a wavelength of 5,500 Å is passed through the prism, there is no net deviation. If the crown-glass prism has an apex angle of 15°, find the apex angle of the flint-glass prism. (Index of refraction for light of 5,500 Å in crown glass is 1.51; in flint glass, it is 1.625.) Make the assumption that the light passes through the prisms at the angle of minimum deviation. *Ans.* 12.2°.

19 A convex crown-glass lens has a focal length of 10.0 cm for yellow light. What is the focal length of the dense flint lens that will make an achromatic doublet for blue and red light?

20 A prism of crown glass followed by a prism of flint glass can be used to form an achromatic prism (one that deflects a ray of white light without separating it into rays of different colors). A 6° crown-glass prism is placed in contact with an inverted flint-glass prism to form an achromatic prism. What must be the angle of the flint prism to produce deviation without dispersion for the red and blue rays? *Ans.* 3.16°.

21 An 18° crown-glass prism is to be combined with a dense flint-glass prism so that the combination will be achromatic, producing deviation without dispersion for the violet (3,969 Å) and yellow (5,893 Å) rays. What should be the refracting angle of the flint-glass prism?

22 Calculate the value of the angle of a heavy flint prism which produces dispersion without deviation when arranged so as to produce an achromatic prism by being placed in contact with a zinc crown-glass prism having a 10° angle. Index of refraction for D line is 1.514 in crown and 1.879 in flint glass. *Ans.* 5.85°.

23 To get deviation without dispersion for the achromatic prism arrangement in Prob. 22, what must be the angle of the heavy flint prism?

24 (*a*) What angle should a medium flint-glass prism have so that when it is combined with a 10° ordinary crown-glass prism, yellow light is not bent at all? (*b*) What will then be the angular dispersion of red and blue rays?

Ans. −8.2°; 0.045°.

25 Compare the power needed to maintain a carbon filament at 1500°C with that required for its operation at 1200°C, assuming that all power is radiated in blackbody radiation.

26 It is observed that the frequency of the light of a receding star shifts from 5.70×10^{14} Hz to 5.61×10^{14} Hz. How fast is the star moving away from the earth? *Ans.* 5.14×10^{6} m/s.

Sir Edward Victor Appleton, 1892–

Born in Bradford, Yorkshire, Secretary, Department of Scientific and Industrial Research, 1939–1949. Principal and vice-chancellor, University of Edinburgh. Awarded the 1947 Nobel Prize for Physics for his researches in the physics of the atmosphere, particularly for his discovery of the ionized layer called after his name.

Patrick Maynard Stuart Blackett, 1897–

Born in London. Professor at the University of Manchester. Awarded the 1948 Nobel Prize for Physics for his development of the Wilson cloud-chamber method and his discoveries in nuclear physics and cosmic radiation.

OPTICAL SPECTRA

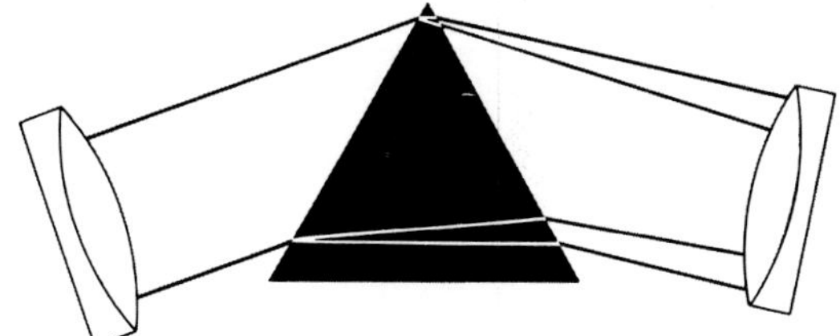

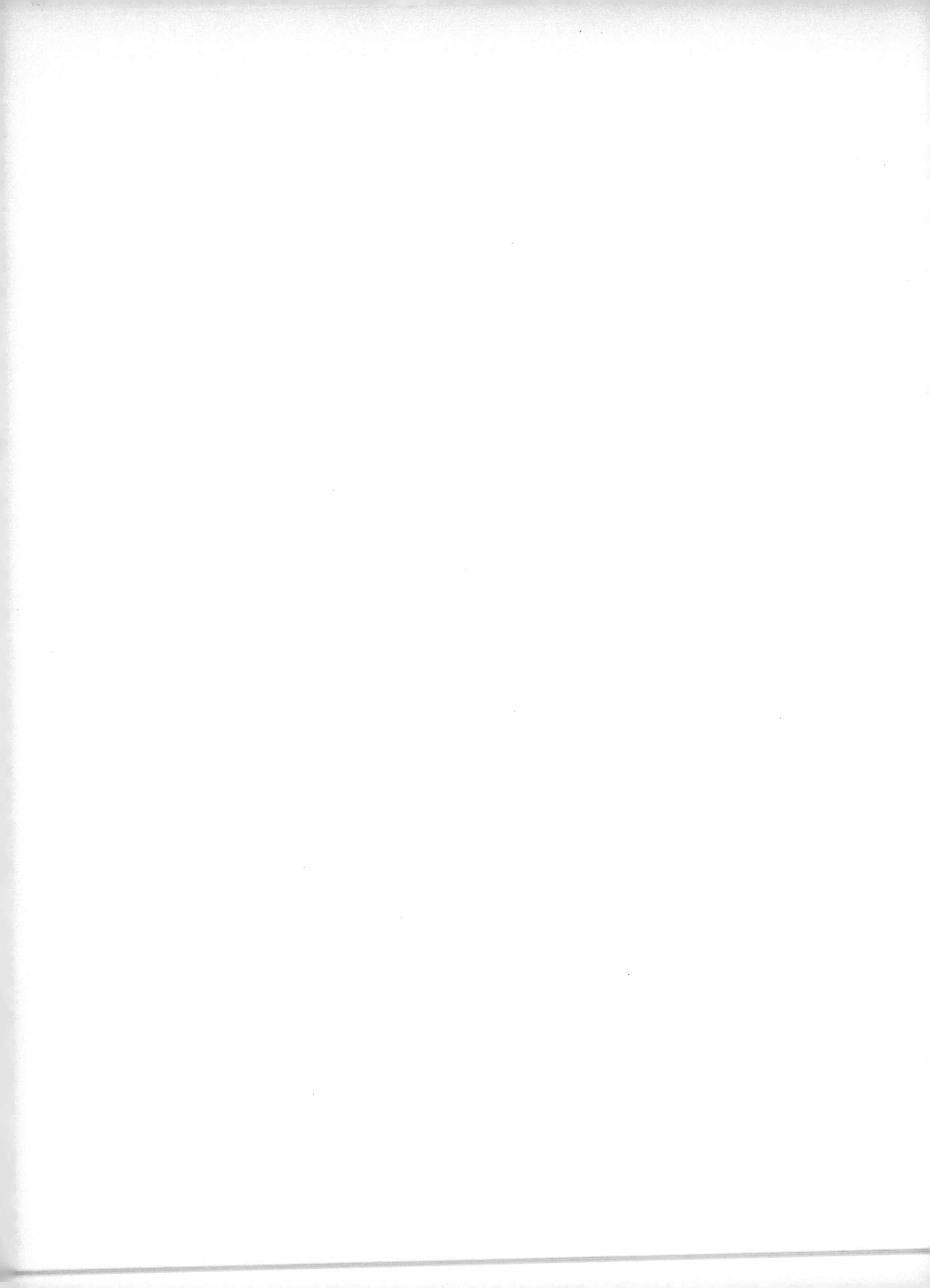

29
Color

Color interests and pleases everyone. The artist, decorator, physiologist, psychologist, and physicist all study color, each with somewhat different emphasis and terminology. A discussion of color is included in this book chiefly to illustrate how far the methods of physics can be used to describe something as subjective as color perception. The logical analysis of colors and their numerical specification have numerous practical applications, as in color photography and printing and in the color coordinating of textiles, leathers, furs, and plastics for next season's fashions.

We shall here relate color to physical stimuli. A given color sensation, however, can be produced by more than one type of stimulus. The science of *colorimetry* seeks to relate the average person's perception of color to the physical light stimulus in such a way as to provide practical graphical and numerical specifications of color.

29-1 COLOR CLASSIFICATION

In the study of spectra we have observed that light from the sun or other source after passing through a narrow slit is dispersed by a prism into a spectrum, each part of which gives a different color sensation. Each part of the spectrum may be named a *hue,* such as red, orange, yellow, green, blue, and violet. Not all hues are observed in the spectrum of sunlight. The purples are notably absent.

The evolution of a graphic method of color classification has been suggested by Deane B. Judd somewhat as follows: Suppose that we have a trunk full of colored papers and that we attempt to classify the colors. After separating them, we note that there are some that lack the quality of hue. Some are white, some are grays, and some are black. Each of these will reflect equally all parts of the sunlight that falls upon it. A white would reflect nearly all, a black none, and the various grays an intermediate amount.

We first separate the papers into two classes: the *grays* (achromatic colors) that lack hue and the *chromatic colors* that have hue. We next group the chromatic colors by hue as red, yellow, etc. We further notice that intermediate groups can be found that form a continuous circle, ranging from red through orange and on through green to blue. Purple completes the circle back to red (Fig. 29-1). This is classification by hue.

The grays are now placed in a series from white through grays of decreasing lightness to black. Among the colors of one hue some of the samples are darker or lighter than others. We find that we can match the lightness of each to one of the grays of the achromatic series. By finding the equivalent gray we classify a color by light-

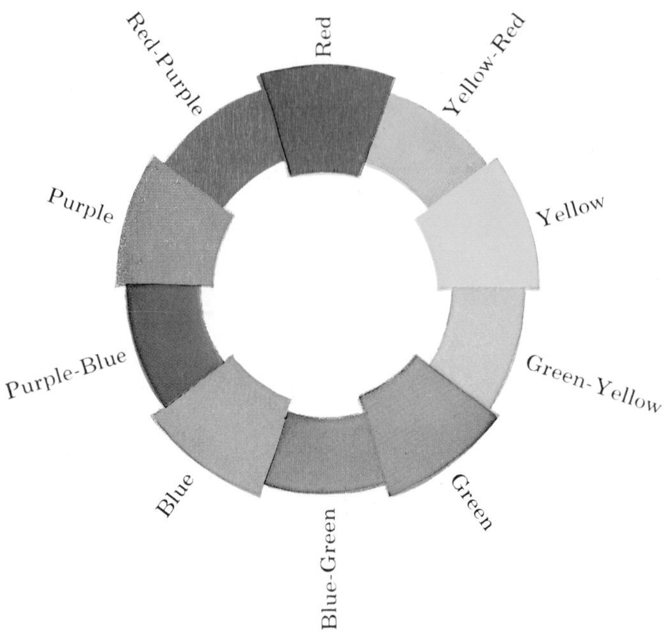

Figure 29-1
Hue circle showing the principal hues.

ness. *Lightness* is the impression of the relative amount of incident light that a surface reflects, irrespective of hue.

We now notice still a third characteristic, *saturation,* which is the degree of difference from a gray of the same lightness. We use such adjectives as vivid or strong to describe high saturation; weak is used to describe low saturation. For the grays, the saturation is zero. For a two-part mixture composed of a chromatic color and a gray, the saturation is increased by increasing the amount of the chromatic color.

These three aspects of color can be represented on a color solid (Fig. 29-2*a*). Hue changes around the circle, as in Fig. 29-1, lightness increases upward, and saturation increases outward from the axis. Many such color solids have been devised.

A single point in the color solid of Fig. 29-2*a* represents one surface color differing from all others in at least one of the three attributes: hue, saturation, or lightness. The color solid can be used to set up an "atlas" system of standard colored surfaces. A page from one such system, the Munsell, is shown in Fig. 29-3. In using such an atlas, the unknown sample is matched visually under standard illumination to one specimen of the atlas and is specified by the designation of that specimen. Such atlas systems have limitations. They are useless for the specification of colored light sources and are applicable only with considerable difficulty to transparent materials. But, in its field the Munsell system has been accepted by the American National Standards Institute.

The color solid (Fig. 29-2*b*) referring to colors of *self-luminous* areas has the dimensions of hue, saturation, and brightness. *Brightness* is the subjective impression of the rate at which light is

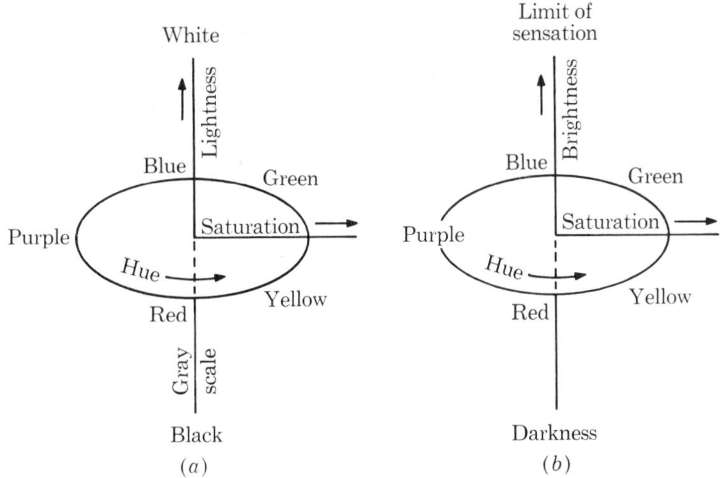

Figure 29-2
Dimensions for the psychological color solid: (*a*) surface colors, with illuminance constant throughout the solid; (*b*) unattached colors, with brightness considered as an aspect of the color itself.

being emitted from unit area toward the observer and ranges from very dim to very bright or dazzling. It is the intensive aspect of the visual sensation.

29-2 COLORIMETRY

We examine colored objects by light transmitted through them or reflected from their surfaces. The color observed depends not only on the characteristics of the object but also on the color of the light illuminating it. Daylight, or a light source approximating daylight, is the usual standard for the illumination of colored surfaces.

The physical basis of the hue and saturation of a sample (called its *chromaticity*) may be completely described by stating the fraction of the incident energy from a standard illuminant that the colored object transmits or reflects at each wavelength. An instrument for measuring this fraction is called a *spectrophotometer*. The instru-

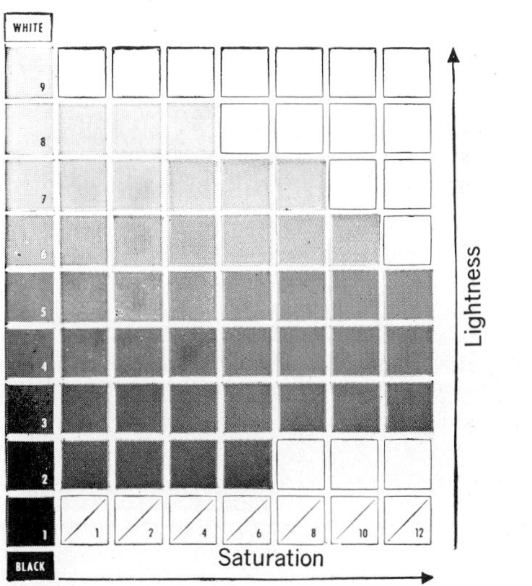

Figure 29-3
Sample page of a Munsell color atlas (with designations of coordinates changed).

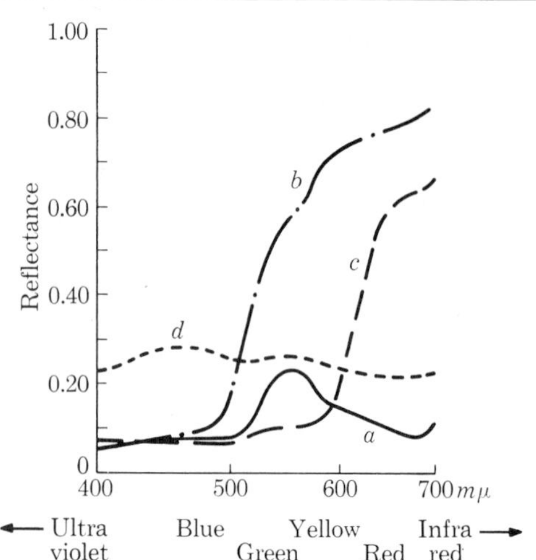

Figure 29-4
Spectrophotometer curves of familiar objects: (*a*) leaf (Boston ivy), (*b*) lemon, (*c*) tomato, (*d*) gray cloth.

ment draws curves representing the spectral distribution similar to those in Fig. 29-4. The spectrophotometric curve could be used as a specification of color stimulus. However, we cannot accurately imagine the appearance of an object from the shape of the curve. Moreover, two objects of the same color may have quite different spectrophotometer curves.

29-3 COLOR VISION

Many theories of color vision have been proposed. None is completely satisfactory. In one, the Young-Helmholtz, or *three-component theory,* it is assumed that in the eye there are three types of retinal cones. All three types are sensitive to nearly the whole of the visible spectrum, but each has a region of maximum sensitivity, one in the red, one in the green, and one in the blue (Fig. 29-5). If one type of receiver alone could be stimulated, the sensation produced would be completely saturated red or green or blue. Actually any stimulus affects all three, and the three effects are integrated in the proportions that each is aroused by the stimulating light. Thus a single sensation is produced. The hue aroused by any wavelength is determined by the curve that is uppermost at that wavelength, with some influence from the relative height of the next uppermost. Blue-green and yellow are positioned in the spectrum by the crossings of two upper curves. Saturation is largely determined by the relative height of the lowest curve, since it determines an amount of desaturating whiteness present in the sensation from even a monochromatic light.

The three-component theory has led to a workable system of color specification and matching. Colors may be synthesized by mixtures of two or more wavelengths. The result can be predicted from Fig. 29-5. If two or more wavelengths reach the retina at the same time, they will produce a *single hue* at some *particular saturation*. Only by chance will the saturation be the same as that of the same hue when it is aroused by a single wavelength.

True purples do not appear in the spectrum. They are not aroused by a single wavelength, but require mixtures of short- and long-wavelength light.

The sensation of white requires *equal* excitation of each of the three sensation-producing factors. There is equal excitation, for example, when the exciting source is daylight, since the areas under the three curves of Fig. 29-5 are all the same. The daylight spectrum is thus an "equal-energy" spectrum. The sensation of white can also be produced by less than the whole spectrum if the combination is such as to give the required equal excitation. Any two pairs of wavelength regions that satisfy this condition and thus produce white are called *complementary* colors. Thus blue and yellow are complementaries (Fig. 29-6). Any wavelength toward one end of the spectrum has a complementary wavelength somewhere in the other end. In Fig. 29-1, complementaries are nearly directly across the color circle.

A midspectral wavelength and one from either

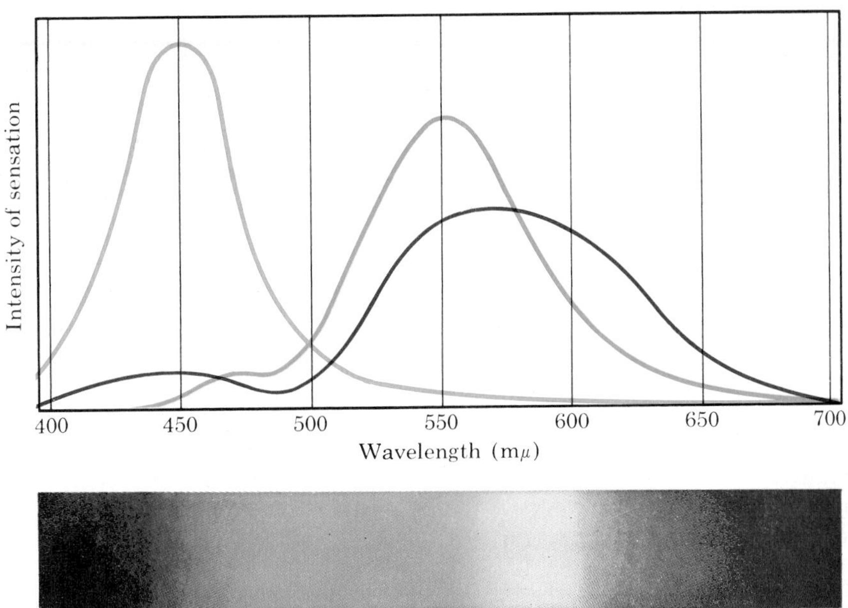

Figure 29-5
Sensation curves of the normal color-vision mechanism of the three-component theory, with the appearance of the solar spectrum (to which the curves pertain) suggested beneath.

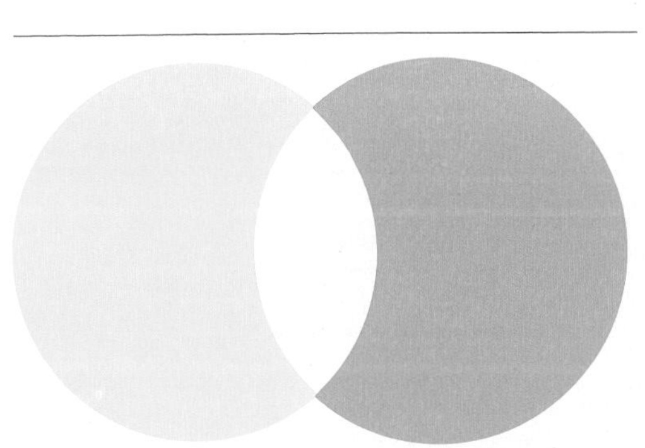

Figure 29-6
Additive mixture of complementary patches of light on a screen.

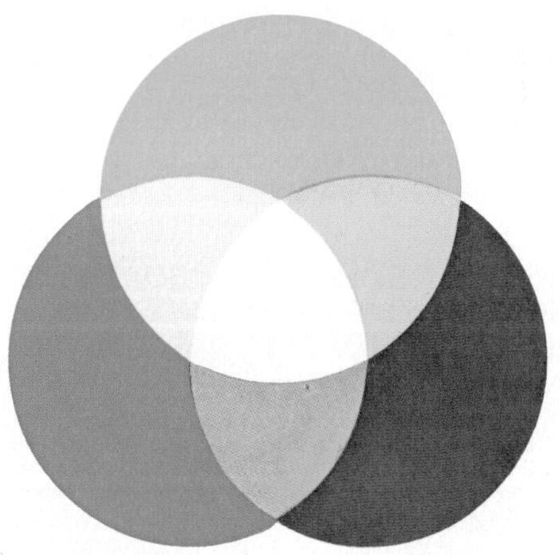

Figure 29-7
Additive mixture of three primary lights in projected patches allowed to overlap on a screen.

end of the spectrum can be mixed in various proportions to produce all the hues in the corresponding half of the spectrum. With a midspectral wavelength and two terminal wavelengths, called three *primary colors,* additive mixtures can be made that will produce any hue of the spectrum or any of the purples. Figure 29-7 represents patches of primary lights falling on a screen and so adjusted in intensity that in the region where the three patches overlap there is equal excitation of the three sensation-producing factors, giving white. Any other mixture of these lights on the screen will produce some hue in some degree of saturation.

29-4 COLOR BY ABSORPTION, ADDITION, AND SUBTRACTION

From the foregoing, we see that we cannot tell by inspection of a colored light or object what the exact physical basis of its color is. A source will ordinarily arouse the sensation of hue if it is confined to a small part of the spectrum. A source may arouse the same hue if it emits a band of wavelengths of considerable width, or several bands, so long as these do not cancel each other's effects by complementation, or even if the source emits energy of all the wavelengths in the spectrum if the energy distribution is substantially different from that of an equal-energy spectrum.

Any appreciable disturbance of the energy distribution in an equal-energy spectrum will create hue and hence chromatic color. If an object is illuminated with daylight and absorbs selectively, it will reflect or transmit an altered spectral distribution of energy and hence will be perceived as colored.

No object reflects only a single wavelength (absorbing the rest), or even a very narrow band of wavelengths. More commonly, a colored object owes its color to a minor subtraction from the daylight falling on it. For example, a yellow pencil can be thought of as a "minus-blue" object. It reflects not merely the yellow but the whole of the spectrum except for the short-wave blue-violet region. The red- and green-producing wavelengths reaching the retina arouse a yellow sensation "additively" in the visual system. This yellowness is added to that aroused by the yellow-producing wavelengths. The pencil is consequently lighter than it could be if it reflected only the yellow-producing energy.

When daylight falls on colored-glass filters, selective absorption "subtracts" some parts of the spectrum and thus affords opportunity for unabsorbed parts to have simple additive effects, resulting in color (Fig. 29-8).

The color of a pigment can be thought of as due to the fact that the pigment subtracts (absorbs) from the light reaching it that light which is supplementary to the light which the pigment reflects. A mixture of yellow and blue *paints* is green, although yellow and blue *lights* are complementary and produce white light when added (Fig. 29-6). The yellow paint (which reflects red, yellow, and green) absorbs blue and violet light; and the blue paint (which reflects violet, blue, and green) absorbs yellow and red. Nothing but green light remains to be reflected by a mixture of the paints (Fig. 29-9, compare with Fig. 29-8). What

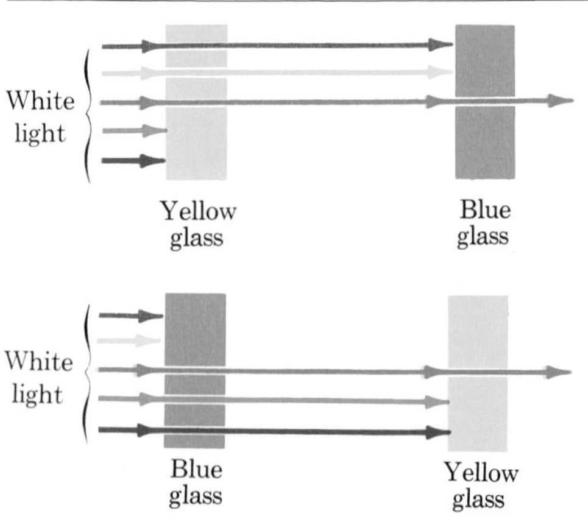

Figure 29-8
Subtractive effects of colored glass upon white light.

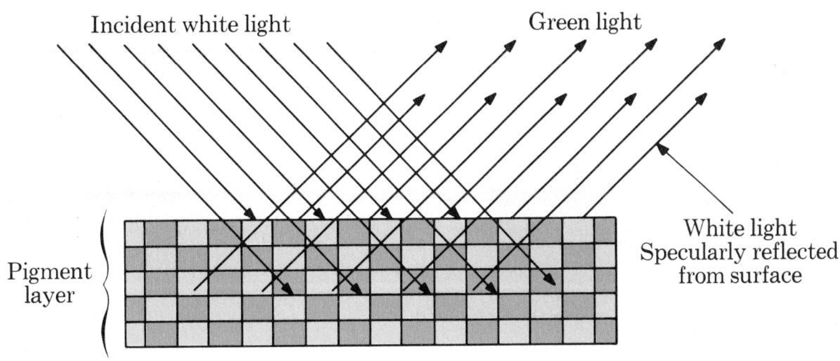

Figure 29-9
Subtractive production of green by a mixture of blue and yellow pigments.

is "additive" or "subtractive" is not the color experience, but the method of production of the spectral distribution of energy in the light reaching the eye. (Fig. 29-10).

29-5 COLOR SPECIFICATION

Colors created by either reflected or transmitted light can be specified in terms of a mixture of lights producing a visual match for the sample. A device (Fig. 29-11) in which a given color is matched by the addition of three primaries is called a *tricolorimeter*. One finds experimentally that by using any three primaries in A, B, and C, in suitable proportions, many colors in S can be matched. But there is *no* set of three primaries that can be added in this way so as to match *all* colors in S. For some, one of the primary beams represented in Fig. 29-11 must be shifted to illuminate the left side of the field, "subtracting" this primary from the right side. If our process is understood to include subtraction, i.e., if negative amounts of the primaries may be used, then it is found that any color can be matched by some mixture in proper proportions of any three primaries whatever.

As an example, let the three primaries be spectrum lights of wavelengths 450, 550, and 620 μm. Illuminate the left side of the field by spectrally homogeneous light of constant radiant flux whose wavelength is successively varied throughout the spectrum, and gauge the relative amounts of the three primaries needed to secure color match with S at each wavelength. The results are given in Fig. 29-12, in which ordinates of curves A, B, and C, at each wavelength, indicate the amounts of the three primaries needed to match a spectrum light of that wavelength. Thus the tristimulus values for light of wavelength 500 μm are

$$A = 12 \qquad B = 55 \qquad C = -30 \tag{1}$$

The system of color specification adopted by the Commission Internationale de L'Éclairage (CIE) expresses color-mixture data in terms of three primaries so chosen that the curves corresponding to A, B, and C in Fig. 29-12 lie everywhere above the axis. This avoids the use of negative values in computations. It requires the use of primaries which lie outside the realm of real colors, but this is not a disadvantage. A color may be specified by the relative amounts of the international primaries (Fig. 29-13) required in a matching mixture. These are called the *tristimulus values* of the color and are designated X, Y,

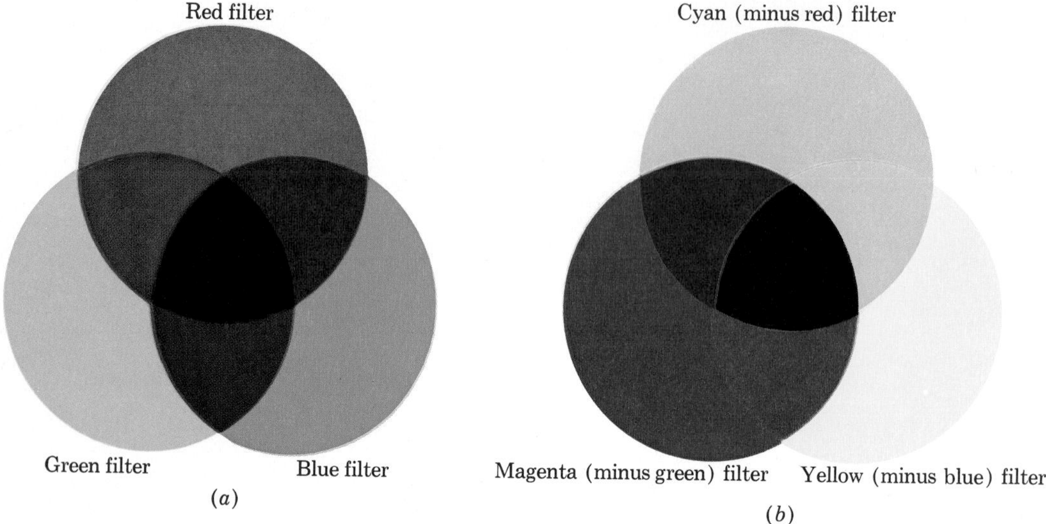

Figure 29-10
Subtractive production of colors by the overlapping of filters in the white-light beam from a single projector. In (*a*) the primaries are the same as those of Fig. 29-7. In (*b*) the primaries are the complementaries of the primaries of (*a*). (This picture should be viewed in strong white light.)

and Z. The tristimulus specifications for a small portion of the spectrum are designated $\bar{x}_\lambda$, $\bar{y}_\lambda$, and $\bar{z}_\lambda$. Thus the CIE tristimulus values for light of wavelength 500 μm are

$$\bar{x}_{500} = 0.00492 \qquad \bar{y}_{500} = 0.32300 \qquad \bar{z}_{500} = 0.27201 \tag{2}$$

29-6 CHROMATICITY DIAGRAM

A three-dimensional diagram would be necessary to specify a color if the three tristimulus values were plotted directly. This inconvenience is avoided by introduction of three related quantities, x, y, and z, known as the *chromaticity coordinates,* defined as follows:

$$x = \frac{X}{X + Y + Z} \qquad y = \frac{Y}{X + Y + Z} \qquad z = \frac{Z}{X + Y + Z} \tag{3}$$

Since $x + y + z = 1$, the values of any two of these coordinates are sufficient to determine the third. In a two-dimensional diagram (Fig. 29-14) the x and y coordinates are plotted at right angles to each other. The value of z for a particular chromaticity point P can be read from the figure by drawing a horizontal or a vertical line from P to the hypotenuse.

Not all points in this triangle can represent colors. Since $\bar{x}$ is never completely zero for any spectrum wavelength (Fig. 29-13), not a single point on the y axis can be a chromaticity point. The same applies to $\bar{y}$, and thus no point on the x axis can be a chromaticity point. But $\bar{z}$ is zero for a large part of the spectrum, so that a large number of chromaticity points lie on the hypotenuse.

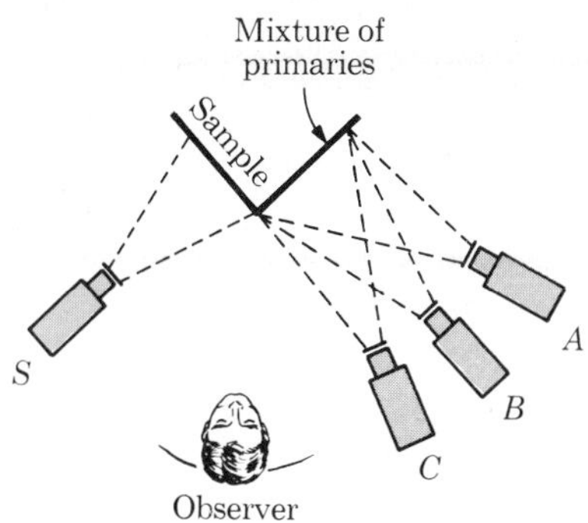

Figure 29-11
Colorimetry: color matching by the additive mixture of three primaries.

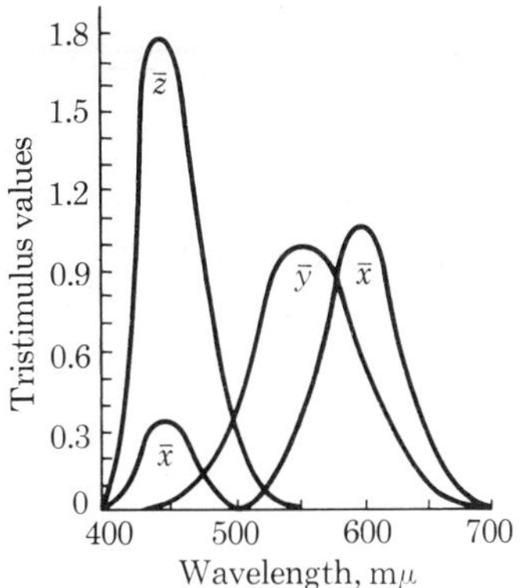

Figure 29-13
Tristimulus values for the spectrum colors. The values of $\bar{x}$, $\bar{y}$, and $\bar{z}$ are the amounts of the three CIE primaries required to color-match a unit amount of energy have the indicated wavelength. *(Adapted from A. C. Hardy, "Handbook of Colorimetry," Massachusetts Institute of Technology,* 1936.)

Spectrum colors The actual spectrum locus is shown in Fig. 29-15. The y values of all possible colors plotted against the x values of those same colors comprise the CIE *chromaticity diagram.*

Again considering light of wavelength 500 μm, and using values from Eq. (2), we find for the chromaticity coordinates

$$x = \frac{0.00492}{0.59993} = 0.00820$$

$$y = \frac{0.32300}{0.59993} = 0.53839 \qquad (4)$$

$$z = \frac{0.27201}{0.59993} = 0.45341$$

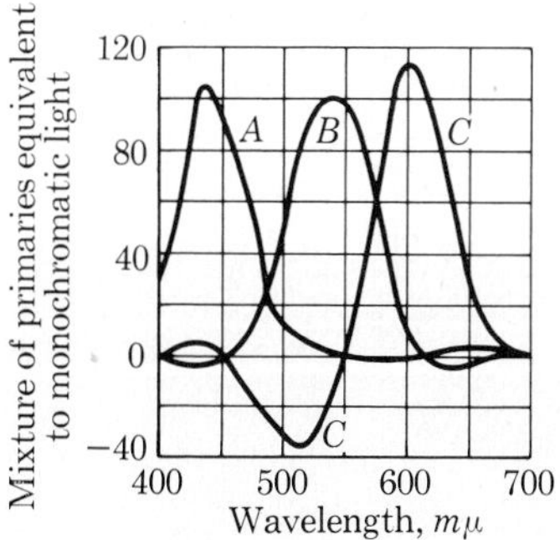

Figure 29-12
Color-mixture data for monochromatic primaries of wavelengths *(A)* 450 mμ, *(B)* 550 mμ, and *(C)* 620 mμ. *(Adapted from A. C. Hardy and F. H. Perrin, "The Principles of Optics," McGraw-Hill Book Company,* 1932.)

Hence this color may be represented in the chromaticity diagram by a point whose coordinates are $x = 0.00820$ and $y = 0.53839$. When this procedure is carried out for each wavelength in the visible spectrum, the curve called the *spectrum locus* (Fig. 29-15) is obtained.

Light beams The tristimulus value Y of a light beam of given spectral distribution is found by multiplying the ordinate of the $\bar{y}$ curve (Fig. 29-13) at each wavelength by the radiant flux of the light beam at that wavelength and summing over the visible spectrum. This integration is usually performed graphically or by an approximate mathematical method. When one has found the X and Z values by a like process, he may calculate the chromaticity coordinates x, y, and z from Eq. (3).

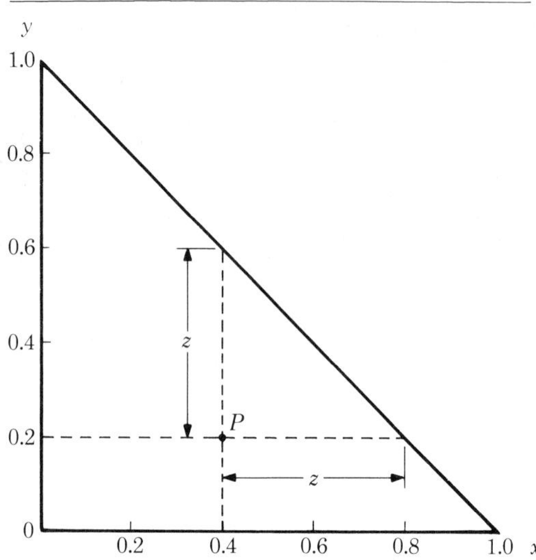

Figure 29-14
Color triangle.

Surface colors The color sensation produced by light reflected from an opaque object depends both on the composition of the incident light and on the reflectivity of the object at each wavelength. The CIE has recommended three standard light sources, produced by incandescent lamps and filters. Illuminant C corresponds closely to average daylight. The chromaticity coordinates of illuminant C (plotted in Fig. 29-15) are $x = 0.3101$, $y = 0.3163$.

Consider now the computation of the chromaticity coordinates of a green paint whose percentage reflection is given as R in Fig. 29-16, when illuminated by illuminant C, whose spectral composition is given by curve C. The product (P) of the relative flux of the original beam at a particular wavelength (ordinate of curve C) and the reflection factor of the paint at that wavelength (ordinate of curve R) gives the relative flux of

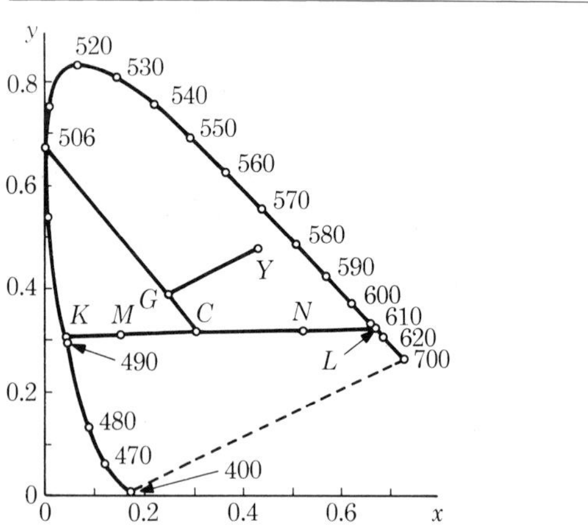

Figure 29-15
Chromaticity diagram. The curved line is the spectrum locus. The green (G) may be regarded as a mixture of illuminant C and a spectrum color having a wavelength of 506 mμ. (*Adapted from A. C. Hardy, "Handbook of Colorimetry," Massachusetts Institute of Technology,* 1936.)

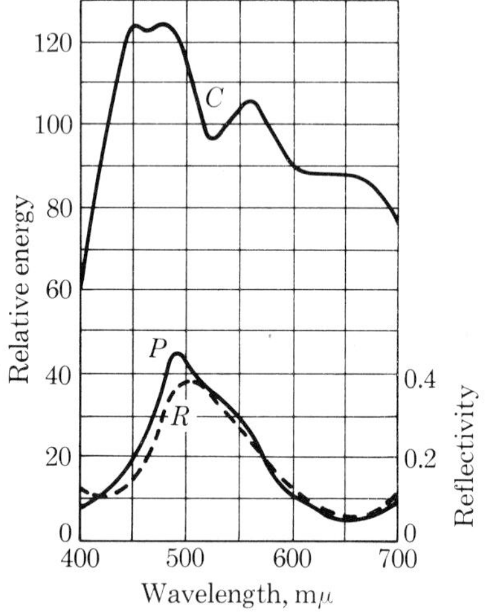

Figure 29-16
(R) Reflectivity (percentage reflection) curve of a green paint. (C) Distribution of energy radiated per unit time by illuminant C. (P) Distribution of energy reflected per unit time from the green paint. (*Adapted from A. C. Hardy, "Handbook of Colorimetry," Massachusetts Institute of Technology,* 1936.)

the light reflected from the paint at that wavelength. The procedure for obtaining the tristimulus values from curve P is now the same as that for computing the tristimulus values for any light beam of known flux distribution. One finds

$$X = 15.5 \qquad Y = 24.2 \qquad Z = 22.6 \tag{5}$$

The corresponding values of the chromaticity coordinates are

$$x = 0.25 \qquad y = 0.39 \tag{6}$$

This point, shown plotted as G in Fig. 29-15, represents the chromaticity of the green paint.

29-7 DOMINANT WAVELENGTH, PURITY, AND LUMINANCE

All colors that may be produced by the additive mixture of any two given colors will be represented by points on the straight line joining the given colors on the chromaticity diagram. Thus all colors obtainable by the additive mixture of the two colors G and Y (Fig. 29-15) are represented on the segment GY.

Since complementary colors are those which, when added, produce white, spectral colors complementary to one another lie at the intersection with the spectrum locus of straight lines passing through the "white point" C. Thus K and L are complementary. Colors M and N, K and N, or M and L are also complementary.

It is evident that all real colors must lie within the area enclosed by the spectrum locus and the dotted line, since every real color can be considered to be a mixture of its spectral components. Furthermore, any color can be considered to be a mixture of illuminant C and spectrum light of a certain wavelength, called the *dominant wavelength* of that color. The fractional distance at which the given color lies along the line joining the white point and the dominant wavelength is called the *purity* of the color. Thus the green paint has a dominant wavelength of 506 μm and a purity of 20 percent.

Before the CIE system and its chromaticity diagram were devised, dominant wavelength and purity were determined directly with *monochromatic colorimeters,* in which a sample was matched by a wavelength and intensity of monochromatic light (mixed with white light) found by trial and error. The specification of a color in terms of dominant wavelength and purity permits the appearance of the color to be visualized more readily than would its specification in terms of chromaticity coordinates, and it is often preferred for this reason. The treatment of purples is "special" in all except atlas systems of color specification, and it is therefore omitted from this chapter.

The luminance (photometric brightness) of a sample is found by evaluating the spectral distribution of the light reflected from it, weighted by the standard luminosity curve of Fig. 23-3. The luminance at any wavelength is the radiance of the reflected light at that wavelength, multiplied by the corresponding ordinate of the luminosity curve. Integration of this product over the entire spectrum gives the luminance of the sample. The necessity of making this integration separately is avoided, because the Y curve of the international primaries was deliberately made to have the same shape as the luminosity curve, by proper choice of the primaries. The same integration which gives the Y tristimulus value therefore also gives the luminance. Thus the green paint for which Y was 24.2 has a luminance of 24.2 percent of that of the standard magnesium oxide. The green paint is then specified by

$$\left.\begin{array}{l}\text{Luminous}\\ \text{reflectance}\end{array} = 24.2\%\right\} \quad \text{or} \quad \left\{\begin{array}{l}\text{Photometric}\\ \text{brightness} =\\ \text{24.2 (relative}\\ \text{to 100 for a}\\ \text{perfect white}\\ \text{paint)}\end{array}\right.$$

$$\left.\begin{array}{l}x = 0.25\\ y = 0.39\end{array}\right\} \quad \text{or} \quad \left\{\begin{array}{l}\text{Dominant}\\ \text{wavelength} = 506\ \mu\text{m}\\ \text{Purity} = 20\%\end{array}\right.$$

SUMMARY

A color sensation is described by *hue, saturation,* and *brightness* (self-luminous area) or *lightness* (non-self-luminous object). These attributes of color depend upon the physical quantities *wavelength, purity,* and *energy.*

Color aspect	Physical determinant
Hue	Assortment of wavelengths and their intensities
Saturation	Colorimetric purity: percentage of monochromatic light in a mixture of monochromatic and white light required to match the sample
Brightness	Proportional, at each wavelength, to the logarithm of the radiant energy per unit time
Lightness	Ratio of the luminance (luminous flux per unit area per unit solid angle) of the object to the luminance of its surroundings, or, in most cases, luminous reflectance

The wavelengths in the visible spectrum extend from about 380 to 760 μm. The chief spectral hues, in order, are violet, blue, green, yellow, orange, and red.

Complementary beams are monochromatic (or polychromatic) pairs which when mixed in the proper proportions produce the sensation of white.

Colors are commonly formed by the *addition* of lights or by the *subtraction* of certain wavelengths from the light source.

Any three beams having wavelengths near the two extremes and the middle of the spectrum are called *primaries.* When added in proper proportions, they produce the sensation of white, or when combined in other proportions (sometimes negative), they can match any color.

Chromaticity coordinates specify a color quantitatively in terms of the relative amounts of three artificial (CIE) primaries necessary to produce a visual equivalent of that color.

A color may be specified also by stating its *dominant wavelength* and its *purity.* These may be obtained graphically from the chromaticity diagram or directly by the use of a monochromatic colorimeter.

References

Bergmans, J.: "Seeing Colours," Philips Technical Library, The Macmillan Company, New York, 1960.

Committee on Colorimetry, Optical Society of America: "The Science of Color," Thomas Y. Crowell Company, New York, 1953.

Judd, Deane B.: "Color in Business, Science, and Industry," John Wiley & Sons, Inc., New York, 1952.

Monographs on Color: Color Chemistry, Color as Light, Color in Use, Research Laboratories of the International Printing Ink Corp., New York, 1935.

Questions

1 All the following appear red to the eye: sunrise, tomato, objects seen through red sunglasses, and a glowing splint. Explain why in each case.

2 What determines the color of an opaque object?

3 Why does a dark-blue suit appear black by candlelight?

4 What color will red printing on the white pages of a book appear when viewed under red, then blue, and then green light? Explain.

5 What is responsible for the color of a nonluminous body? What effect has the character of the illuminant on the resultant color?

6 Explain why a block of ice is transparent, whereas snow is opaque and white.

7 In an earlier discussion in this book it was noted that roses lose their color when observed under moonlight. Why is this so?

8 The sky and the ocean are always seen as a dark and beautiful blue in travel posters advertising the South Sea Islands. What photographic procedure might the camera man use to ensure the predominance of the blue colors?

9 A lamp has a colored lens which transmits chiefly light of wavelengths near 6,300 Å. What wavelength would this light have under water? What color would the lamp appear to a submerged swimmer? Explain.

10 A person who is color-blind can actually see certain colors. Explain why this happens.

11 The manufacture, packaging, and processing of panchromatic photographic film are carried out under the faint illumination of green safelights. Why is this color chosen? Why are red safelights used in handling orthochromatic film?

12 Why is red light often used as a danger signal?

13 Compare the mixing of colored lights with the mixing of pigments.

14 Why does a mixture of blue and yellow paint appear to be green? What would be the apparent color resulting from mixing blue and yellow light? Explain.

15 Is it possible to make white paint from orange paint by adding other color pigments? Explain.

16 What are complementary colors? Explain their production through retinal fatigue. (Look intently at a brightly lighted red object and then immediately look at a white paper.)

17 A diamond will show flashes of light when white light falls upon it. What would happen if red light fell upon a diamond?

18 Stars are classified by their color, i.e., red, yellow, or blue. Which of these have the highest and which the lowest temperature?

19 Interpret the definition of color recommended by the Committee on Colorimetry of the Optical Society of America: "Color consists of the characteristics of light other than spatial and temporal inhomogeneities; light being that aspect of radiant energy of which a human observer is aware through the visual sensations which arise from the stimulation of the retina of the eye."

20 In what different ways is the work of the colorimetrist important?

21 Mention some of the precautions and limitations in specifying colors by reference to a color atlas of standard samples.

22 Is lightness largely independent of brightness or closely related to it? Consider, for example, that a patch of snow in a dim place looks white and a pile of coal out in the sunlight looks black, even though the luminance of the coal may be greater than that of the snow.

Problems

1 By consideration of the chromaticity diagram, what three monochromatic primaries would probably produce the largest possible range of natural object colors when additively mixed?

2 What is the wavelength of monochromatic light complementary to (*a*) the sodium yellow line at 5,890 Å and (*b*) the mercury blue line at 4360 Å? *Ans.* 4,840 Å; 5,700 Å.

3 From the chromaticity diagram, compute the dominant wavelength and purity of the *R*, *G*, and *B* colors of Prob. 3.

4 Three colored glasses *R*, *G*, and *B* produce light having *x* and *y* trichromatic coordinates as follows: *R*, 0.56, 0.34; *G*, 0.30, 0.63; and *B*, 0.20, 0.20. (*a*) On a tracing of the chromaticity diagram, locate these colors. (*b*) Show the location of all colors that can be matched by mixtures of *R* and *G*; *R* and *B*; and *G* and *B*. (*c*) Where are the colors which if mixed with *R* can be matched by a mixture of *G* and *B*?

Ans. 6,080 Å; 5,500 Å; 4,770 Å.

5 The tristimulus values of two pigmented surfaces under illuminant C are

$$X_1 = 16 \qquad Y_1 = 24 \qquad Z_1 = 22$$
$$X_2 = 10 \qquad Y_2 = 20 \qquad Z_2 = 15$$

How do the two surfaces differ with respect to (*a*) dominant wavelength, (*b*) purity, and (*c*) average reflectance?

Hideki Yukawa, 1907–

Born in Tokyo. Professor of physics at Kyoto University and later at Columbia University. Awarded the 1949 Nobel Prize for Physics for his prediction of the existence of mesons, based upon the theory of nuclear forces.

Cecil Frank Powell, 1903–1969

Born at Tonbridge, Kent. Professor of physics in the University of Bristol. Awarded the 1950 Nobel Prize for Physics for his development of a simple method for examining the action of atomic nuclei by photography and for important discoveries concerning the meson.

30

Interference and Diffraction

Thus far we have considered the linear propagation of light, reflection at a boundary between two media, and refraction where the speed of light changes. All these phenomena are explainable on the basis of a wave theory of light. However, if light is to be considered as having a wave nature, interference and diffraction effects must be expected. Among the early experimenters in phenomena of light were Newton and Huygens. These two scientists reached directly opposite conclusions regarding the nature of light from similar observations. Because he failed to observe interference effects, Newton held that light must be particle in nature. Conversely, Huygens believed that the wave theory explained reflection and refraction more satisfactorily and hence upheld that theory in spite of the lack of knowledge of interference phenomena. It was not until nearly a hundred years later that Thomas Young in 1801 performed his famous experiment showing interference in light. The fact that light exhibits interference effects is the best evidence that luminous energy travels in a manner that may be represented by a wave motion. Optical apparatus designed to utilize interference effects permits the measurement of wavelengths or distances with a greater precision than that attainable in almost any other type of physical measurement.

30-1 YOUNG'S DOUBLE-SLIT EXPERIMENT

As pointed out in Chap. 19, whenever two wave trains pass through the same region of a medium, each continues through the medium as though the other were not present. However, at every point in that region the resultant disturbance is the sum of the disturbances created by the individual waves; i.e., the waves interfere.

An arrangement similar to that used by Young to produce an interference pattern is shown in Fig. 30-1. Here S is a narrow slit with a source of monochromatic light behind it, and A and B are two narrow slits parallel to S. Light progresses from S to A and B. A series of wave fronts is shown; the solid arcs represent maxima, the broken arcs minima. In accordance with Huygens' principle, each of the slits A and B may be considered as a new source of disturbance, with wavelets traveling out from each. The wavelets leave A and B in phase with each other, and in the region to the right they reinforce each other at certain places, producing brightness, and annul each other at other places, leaving darkness. Reinforcement occurs where the two waves arrive in phase with each other, for example, where two maxima arrive together, as at the points marked

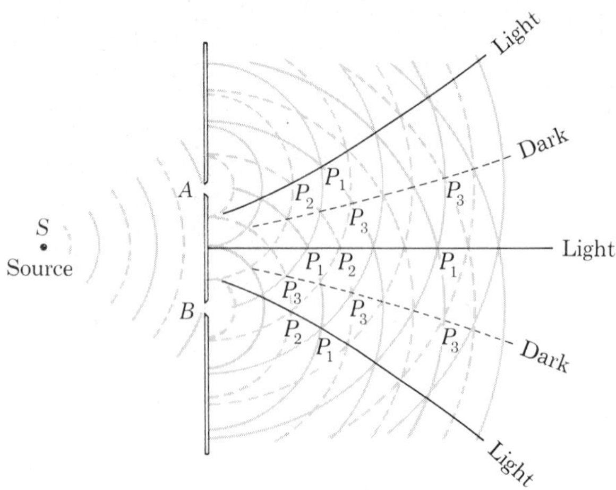

Figure 30-1
Interference of light from two identical slits. Wave-front diagram.

P_1, or where two minima arrive together, as at points marked P_2. For such points the distances traveled by the wavelets from A and B are either equal or differ by a whole number of wavelengths. Such points lie on the heavy solid lines of Fig. 30-1.

Annulment occurs where the wavelets from A and B arrive out of phase with each other by a half wavelength, for example, where a maximum and a minimum arrive together, as at the points marked P_3. For such points the distances traveled by the wavelets from A and B differ by an odd number of half wavelengths. Some such points lie on the dotted lines (marked "dark") of Fig. 30-1.

Within this region of overlapping wavelets we can predict the reinforcement (constructive interference) or annulment (destructive interference) in terms of the path difference from the two slits. There will be constructive interference at every point for which the path difference is any *whole* number of wavelengths ($N\lambda$). There will be destructive interference if the path difference is any *odd* number of half wavelengths [$(2N - 1)\lambda/2$].

Consider a screen that is placed perpendicular to the bisector of AB as shown in Fig. 30-2. The point P_0 is equidistant from A and B, and hence the wavelets reaching P_0 will be in phase, producing constructive interference and a central bright fringe. Moving out from P_0, one reaches a point P_1 that is one-half wavelength farther from A than from B. That is,

$$AP_1 = BP_1 + \tfrac{1}{2}\lambda$$

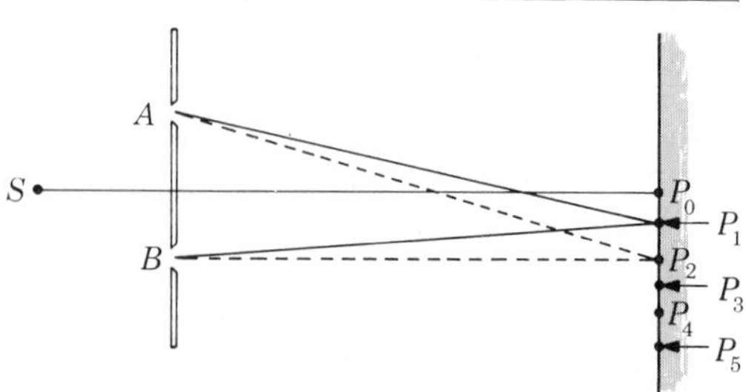

Figure 30-2
Interference of light from two slits. Ray diagram.

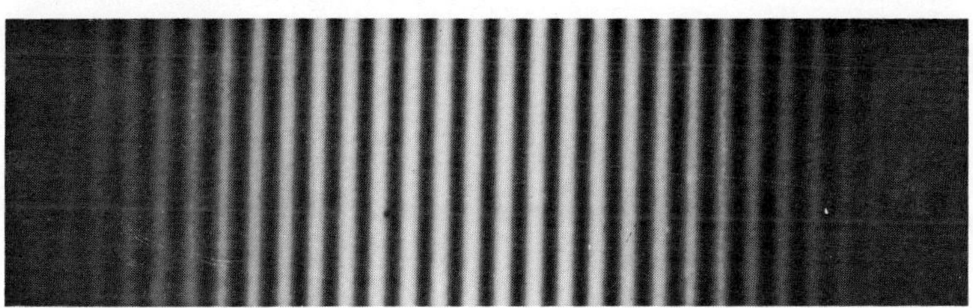

Figure 30-3
Interference fringes produced by light from two identical slits. (*Courtesy of Jenkins and White, "Fundamentals of Optics," 3d ed., McGraw-Hill Book Company,* 1957.)

The wavelets reaching P_1 are out of phase by a half wavelength, producing destructive interference and a dark fringe. Farther out from the central fringe is another point P_2 that is one whole wavelength farther from A than from B, so that

$$AP_2 = BP_2 + \lambda$$

Here there is again constructive interference, since the wavelets are in phase. At P_3, the path difference is $\frac{3}{2}\lambda$; at P_4, 2λ; at P_5, $\frac{5}{2}\lambda$, etc. We observe that there will be alternate bright and dark fringes on either side of the central bright fringe. The general conditions stated above must apply to these fringes. Whenever the path difference is any whole number of wavelengths, including zero, there is constructive interference and a bright fringe. Whenever the path difference is an odd number of half wavelengths, there is destructive interference and a dark fringe.

Bright fringe:

$$\text{Path difference} = N\lambda \qquad N = 0, 1, 2, 3, \ldots$$

Dark fringe:

$$\text{Path difference} = (2N - 1)\frac{\lambda}{2} \qquad N = 1, 2, 3, \ldots$$

In Fig. 30-3, there is shown a picture of interference fringes formed as described here.

The spacing between fringes may be examined with the aid of Fig. 30-4, in which vertical dimensions are exaggerated for clarity. The line SO is drawn perpendicular to the plane containing the slits, and a screen is set perpendicular to SO. Lines AP and BP represent rays from each of the slits reaching the screen at P, distant x from O. By drawing AQ so that AP equals QP the path difference for the two rays may be expressed as $s = BQ$. The dotted line from P to C is perpendicular to AQ, and hence angles BAQ and OCP are equal. Since these angles are small

$$\sin BAQ = \tan OCP \qquad \text{approx}$$

giving

$$\frac{s}{b} = \frac{x}{L} \tag{1}$$

The relations which show that the path difference

$$s = N\lambda \qquad \text{for reinforcement} \tag{2}$$

$$s = (2N - 1)\frac{\lambda}{2} \qquad \text{for annulment} \tag{3}$$

may be combined with Eq. (1) to give for the distance of the fringe from the central bright band

$$x = N\lambda\frac{L}{b} \qquad \text{for bright fringes} \tag{4}$$

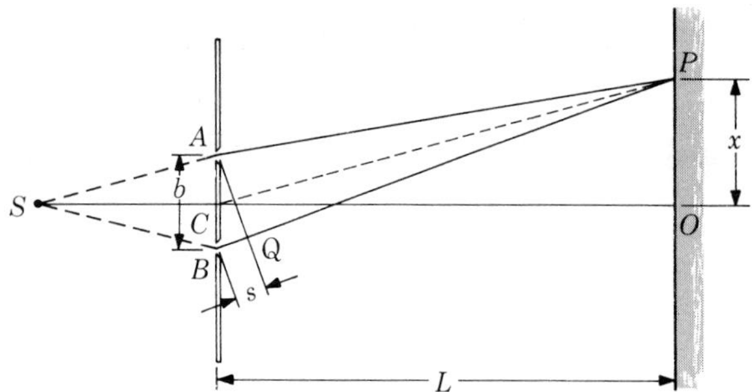

Figure 30-4
Spacing of fringes by light from two slits.

$$x' = (2N - 1)\frac{\lambda}{2}\frac{L}{b} \qquad \text{for dark fringes} \qquad (5)$$

Since the distance of a bright fringe from the central bright fringe is proportional to the number N of the fringe, the fringes must be equally spaced. At the same time the distance is directly proportional to the wavelength, and hence the spacing is greater for long wavelengths (red) than it is for shorter wavelengths (blue).

When white light is used as a source, each wavelength produces its own interference fringes. The fringe pattern will then be colored, the color at each point depending on which wavelengths are reinforced by interference. Such colored designs may be observed by looking at a distant source of white light through a piece of silk or a fine-mesh screen.

Example Yellow light from a sodium-vapor lamp ($\lambda = 5{,}893$ Å) is directed upon two narrow slits 0.100 cm apart. Find the positions of the first dark and first bright fringes on a screen 100 cm away.

$$x' = \frac{(2N-1)(\lambda/2)L}{b}$$

$$= \frac{(2 \times 1 - 1)(5{,}893 \times 10^{-8}\ \text{cm}/2)(100\ \text{cm})}{0.100\ \text{cm}}$$

$$= 0.0295\ \text{cm} \qquad \text{(dark)}$$

$$x = \frac{N\lambda L}{b} = \frac{(1)(5{,}893 \times 10^{-8}\ \text{cm})(100\ \text{cm})}{0.100\ \text{cm}}$$

$$= 0.0589\ \text{cm} \qquad \text{(bright)}$$

Thus the first dark fringe is 0.0295 cm from O (Fig. 30-4), and the first bright fringe is 0.0589 cm from O. The second dark fringe would be 0.0884 cm from the center, etc., the separation of adjacent dark (or bright) fringes being 0.0589 cm.

30-2 COHERENCE

If the slits A and B (Fig. 30-4) are illuminated by two completely independent light sources, no interference fringes are observed. The phase difference between the two beams arriving at P varies with time in a random way. At one instant the wave train from source A may be in phase with that from source B (Fig. 30-5*a*); but in time intervals as short as 10^{-8} s this phase relation may change from reinforcement to cancellation. The same random phase behavior holds for points on screen OP, with the result that the screen is uniformly illuminated.

If, instead, the light waves that travel from A and B to P have a phase difference $\Delta\phi$ that remains constant with time (Fig. 30-5*b*), the two beams are said to be *coherent*. To find the inten-

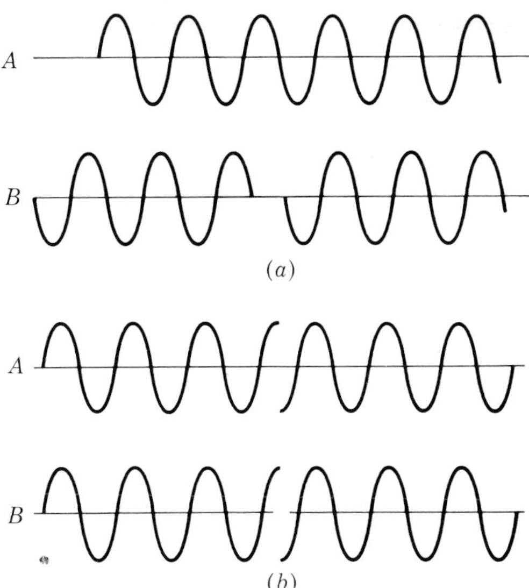

Figure 30-5
(*a*) Incoherent waves; (*b*) waves with a constant phase difference $\Delta\phi$.

sity produced at a point P by coherent light beams, one combines the amplitudes vectorially (taking $\Delta\phi$ into account) and then squares the resulting amplitude to obtain a quantity proportional to luminous intensity. In contrast, for incoherent light beams, one first squares the individual amplitudes to obtain quantities proportional to individual intensities and then adds the individual intensities.

Since the emitting atoms in common light sources do not act cooperatively (coherently), coherent beams for interference experiments have usually been obtained by splitting a light beam from a single source, as in Fig. 30-4. But since 1960, devices have been developed in which atoms in ruby or in a gas mixture are first excited and then triggered to radiate coherently. In such devices, called *optical masers* or *lasers,* the light output is monochromatic, intense, and highly collimated. Practical applications of lasers now being perfected include amplification of weak light signals, the production of sharply defined local heating for medical surgery and industrial cutting, and the transmission of information on modulated laser beams for great distances with little spreading or loss.

30-3 INTERFERENCE IN THIN FILMS

Thin transparent films, such as soap bubbles or oil on water, show streaks that may be accounted for by the principles of interference. In Fig. 30-6, E represents the eye focused on a thin film of thickness t from which is reflected light from a monochromatic source. A ray of light AB incident on the film at angle i will be partly reflected from the front surface along BE and partly refracted into the film along BC. The latter ray will be partly reflected from the back surface of the film to emerge along CPE so that PE is parallel to BE. If these two rays enter the eye, an interference effect can be expected, since they have traveled by different paths to the eye.

To determine the kind of interference produced, we must find the number of waves or fractions of waves by which the paths differ. By drawing PQ perpendicular to the reflected rays, it is evident that the ray reflected within the film travels farther than the ray reflected at the surface by an amount $2a - d$, where a represents the

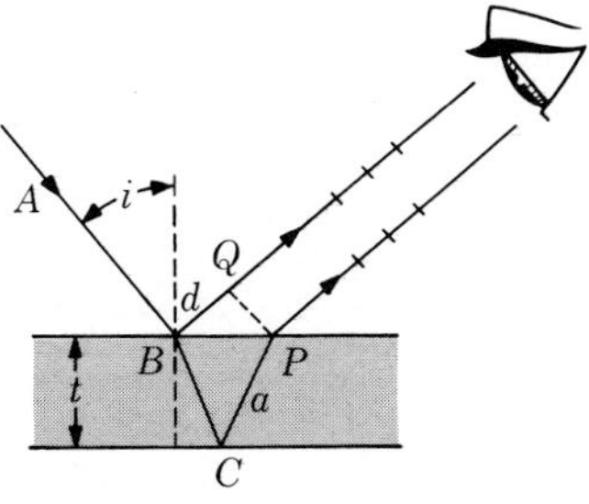

Figure 30-6
Interference in a thin transparent film.

distance BC or CP. Since light travels slower in the film than in air, there are more waves per unit length in the film than in air. Hence the distance $2a$ in the film is equivalent in waves to a distance $2na$ in air, where n is the index of refraction of the material of the film. For normal incidence, $i = 0$, and $a = t$.

It might be expected that an extremely thin film whose thickness was only a small fraction of the wavelength of any visible light would appear bright, since the path difference would be too small to produce destructive interference. Actually such a film appears black; i.e., there is complete destructive interference. This may be observed in a soap film. As it thins just before it breaks, it appears black by reflection. This and other evidence indicate that a phase difference of $\frac{1}{2}\lambda$ or 180° is introduced by the fact that one of the two interfering rays is reflected in air, the other in an optically denser medium. Consequently, the total retardation of ray $BCPE$ with respect to ray BQE is $2na + \lambda/2$. For the special case of normal incidence where $a = t$

$$\text{Retardation} = 2nt + \frac{\lambda}{2} \tag{6}$$

Constructive interference and brightness will occur when the retardation is a whole number of wavelengths; destructive interference and darkness will occur when the retardation is an odd number of half wavelengths.

Example Two rectangular pieces of plate glass are held in contact along one pair of edges while the opposite edges are separated by a thin sheet of paper (Fig. 30-7). When the plates are viewed by the light from a sodium lamp reflected normally, they are seen to be crossed by 17 dark interference fringes. For sodium light $\lambda =$ 5,893 Å. What is the thickness of the paper?

Since the fringes are dark, the retardation must be an odd number of half wavelengths,

$$\text{Retardation} = (2N - 1)\frac{\lambda}{2} = 2t + \frac{\lambda}{2}$$

$$t = \frac{(N-1)\lambda}{2}$$

$$\lambda = 0.00005893 \text{ cm}$$

For the seventeenth dark fringe,

$$N = 17$$

$$t = \frac{(17 - 1)(0.00005893 \text{ cm})}{2} = 0.00047 \text{ cm}$$

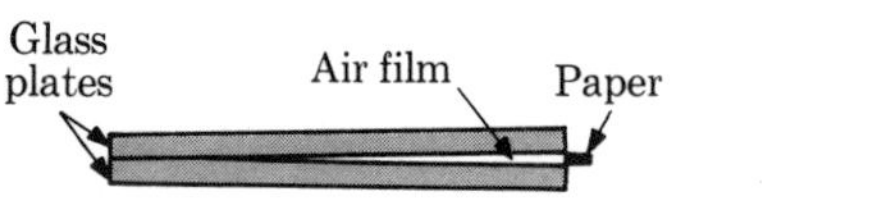

Figure 30-7
Two glass plates separated by a wedge of air.

The interference fringes formed by the film between two surfaces are similar to the contour lines on a topographic map, and they indicate the variations in thickness of the air film separating the two optical parts. The interference patterns (Fig. 30-8) are very useful in routine inspection of the polishing of fine lenses and in the preparation of very flat test plates of glass or metal. By repeated inspection and polishing, these plates may be made optically flat within a tenth of a wavelength of mercury light, or to about 0.000005 cm.

Thin-film interference can be observed when light is reflected from front and back surfaces of any transparent thin film. One of the most common observations is the color of oil films on water or that in bubbles. When the illumination is by monochromatic light, we see light and dark areas dependent upon the thickness of the film. If white light is used, colored areas are observed. The retardation depends upon wavelength. A given thickness will produce destructive interference for one wavelength bank, leaving the complementary color. A film of varying thickness will thus show many colors.

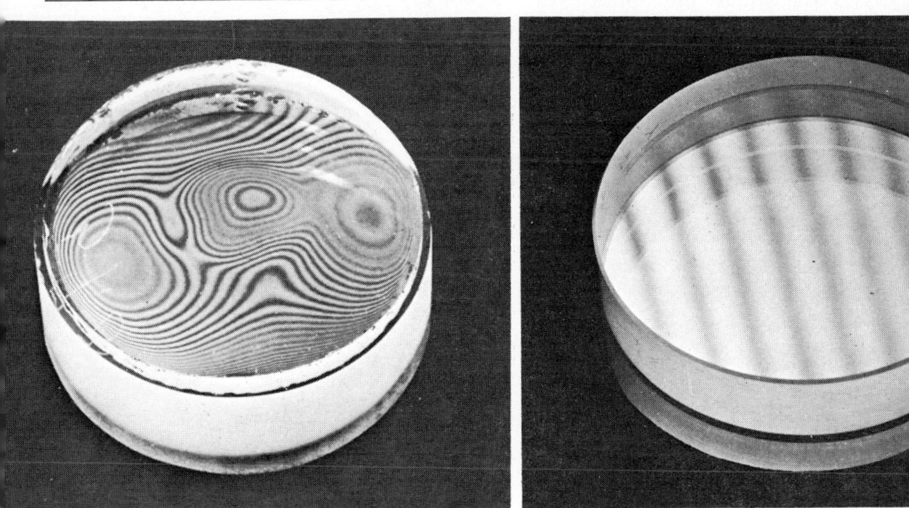

Figure 30-8
Photographs of interference fringes in thin films between two glass plates. Parallel fringes indicate that the surfaces are optically plane. Irregular fringes indicate that the surfaces are not plane.

One example of thin-film interference is that produced in the air film between a convex lens of large radius of curvature and an optically flat plate placed in contact with the lens. When viewed by reflected monochromatic light, there will be destructive interference at the point of contact because of the $\frac{1}{2}\lambda$ change in phase of the ray reflected in the air film at the glass surface. Since the film increases in thickness as the distance from the point of contact increases, the central dark circle is surrounded by alternate bright and dark circular bands. The retardation depends upon the thickness of the film as given in Eq. (6). These interference rings are called *Newton's rings.*

Newton's rings will also be produced by the film between lens and plate when viewed by transmitted light. In this case the central spot is bright because the waves that interfere are those which are transmitted without reflection and those which are twice reflected. Thus the change of phase is a whole wavelength.

30-4 OPTICAL COATINGS

Interference in a thin film may be used to reduce unwanted reflections from the glass of a showcase or a camera lens. Lenses are often coated with thin films of transparent, durable substances such as magnesium fluoride, MgF_2.

Example To produce a minimum reflection of wavelengths near the middle of the visible spectrum (550 μm), how thick a coating of MgF_2 ($n = 1.38$) should be vacuum-coated on a glass surface?

Consider light to be incident at near-normal incidence (Fig. 30-9). We wish to cause destructive interference between rays r_1 and r_2 so that maximum energy passes into the glass. A phase change of $\frac{1}{2}\lambda$ occurs in each ray, for at both the upper and lower surfaces of the MgF_2 film the light is reflected by a medium of greater index of refraction. Since no net change of phase is produced by these two reflections, the optical path difference needed for destructive interference is

$$(N + \tfrac{1}{2}\lambda) = 2nd \qquad N = 0, 1, 2 \text{ for minima}$$

When $N = 0$

$$d = \frac{\frac{1}{2}\lambda}{2n} = \frac{\lambda}{4n} = \frac{550 \text{ m}\mu}{4 \times 1.38} = 100 \text{ m}\mu$$

$$= 1.0 \times 10^{-5} \text{ cm}$$

Light sources used in motion-picture projection radiate intense infrared radiation as well as visible light. The infrared radiation does not contribute to the screen image, but it increases problems of overheating and film distortion. A front-surface-coated projection lamp mirror has been developed (Fig. 30-10) which reflects to the film about 95 percent of the visible light but which allows most of the infrared to escape into the glass, to be absorbed in the glass and in the rear of the housing.

The coatings for such a "cold mirror" consist of alternate layers of materials of high and low refractive index. These layers, of precisely controlled thickness, can separate radiation in different regions of the spectrum. Figure 30-11 shows how the transmission, curve T, of a cold mirror increases sharply above the red end of the visible spectrum, about 700 mμ, while the relative energy reflected, curve R, is large only in the visible part of the spectrum, 400 to 700 mμ.

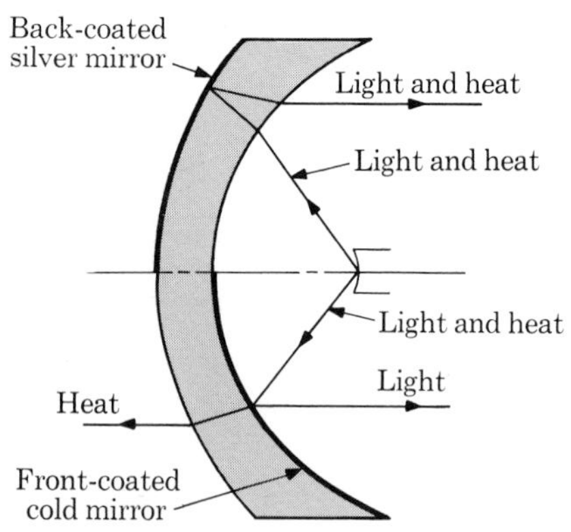

Figure 30-10
Path of rays with conventional silver-back-coated mirror and with a front-surface cold mirror.

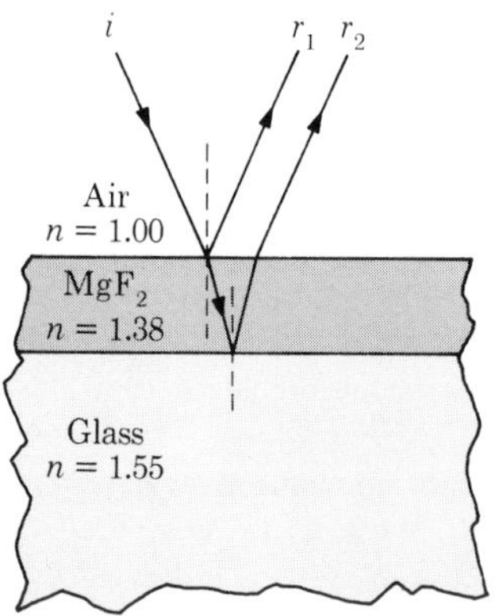

Figure 30-9
Destructive interference in an optical coating minimizes unwanted reflections from glass.

30-5 THE MICHELSON INTERFEROMETER

The interferometer is an instrument that uses interference in the measurement of wavelengths of light in terms of a standard of length or in the measurement of distances in terms of known wavelengths of light. The essential parts of the interferometer devised by Michelson are two plane mirrors M and M' and two glass plates A and B arranged as shown in Fig. 30-12. The plate A is lightly silvered on the back so that of the light which falls on that surface half is reflected and half is transmitted.

A beam of monochromatic light from source S falls on the plate A, where it is divided into two beams. These advance to the mirrors M' and M, return to the plate A, and then proceed to the eye at E. If the mirrors are equidistant from A,

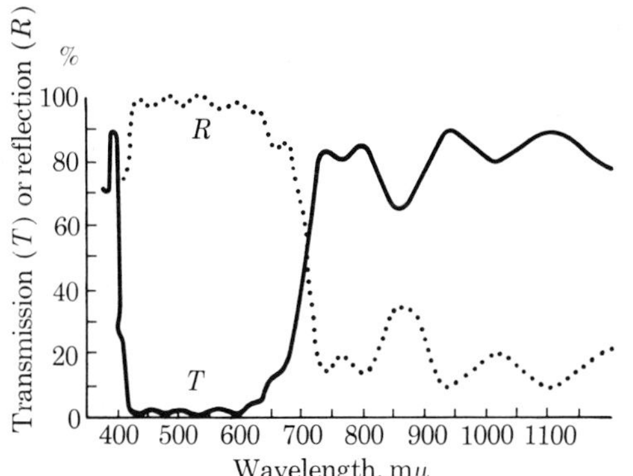

Figure 30-11
Characteristics of coating on a cold mirror: relative transmission *T*; relative reflection *R*.

the two beams will travel optically similar paths, the plate *B* serving only to introduce the same retardation in beam 2 as is introduced in beam 1 by its two passages through *A*.

If the optical paths happen to be equal, the beams 1 and 2 will arrive at *E* in phase and produce a bright field by constructive interference. If the distance *AM* is increased $\frac{1}{4}\lambda$ by moving mirror *M*, the optical path for ray 1 will be lengthened $\frac{1}{2}\lambda$ and the destructive interference of rays 1 and 2 at *E* will give a dark field. Usually mirrors *M* and *M′* are made nearly but not quite perpendicular to give a field crossed by alternate bright and dark interference fringes, which may be counted as they move past a reference mark as *M* is moved. For each fringe that passes the index, the optical path has been changed by one wavelength; i.e., *M* has moved a half wavelength.

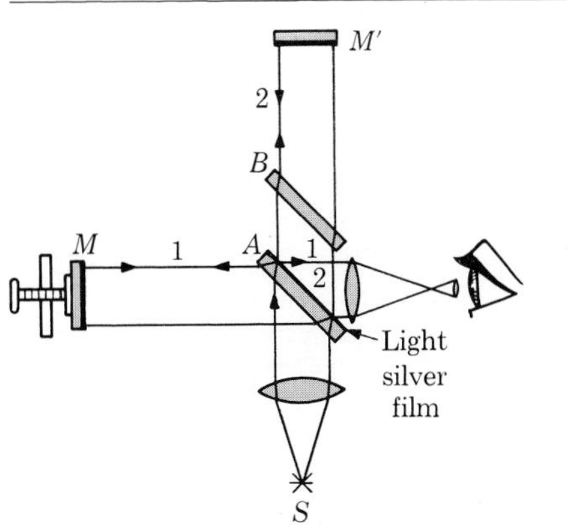

Figure 30-12
Michelson interferometer.

If *N* successive dark fringes are counted as the mirror is moved a distance *D*, then

$$N\frac{\lambda}{2} = D \qquad \text{or} \qquad \lambda = \frac{2D}{N} \tag{7}$$

Michelson measured wavelengths in terms of the standard meter bar. From such measurements a wavelength can be set up as a standard of length that can be reproduced accurately. The standard now in use is the orange line of the spectrum of krypton 86. In terms of this wavelength,

$$1 \text{ m} = 1{,}650{,}763.73\ \lambda_{\text{Kr}}$$

All other wavelengths can be compared with this standard, and from it the standard meter could be reproduced.

Example An interferometer illuminated with red light from cadmium ($\lambda = 6{,}438$ Å) is used to measure the distance between two points. Calculate this distance if 239 fringes pass the reference mark as the mirror is moved from one of the points to the other.

From Eq. (7),

$$D = N\frac{\lambda}{2} = \frac{239(6.438 \times 10^{-5}\text{ cm})}{2} = 0.00769 \text{ cm}$$

30-6 DIFFRACTION

Light travels in straight lines in a uniform medium but commonly changes direction where there is a change of medium or a change of properties of a single medium. Thus we observe

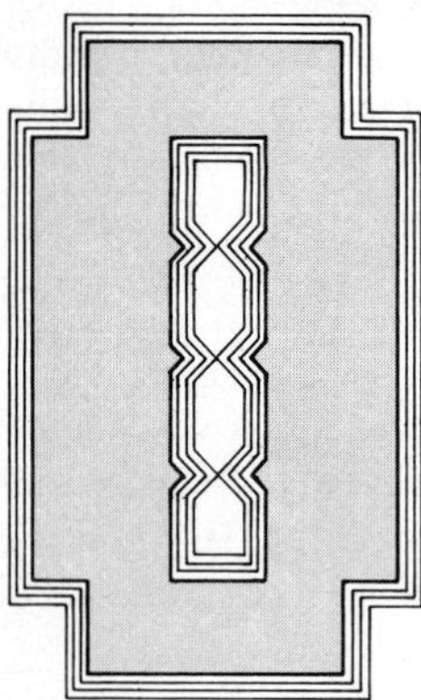

Figure 30-13
Diffraction pattern formed by monochromatic light passing over a razor blade.

reflection and refraction. Careful observation shows that there is also a slight bending around obstacles. The spreading of light into the region behind an obstacle is called *diffraction*. Diffraction occurs in accordance with Huygens' principle and is an interference phenomenon.

Any obstacle introduced into the light beam from a point source will produce diffraction effects under proper conditions. A slit, a wire, a hole, or a straightedge are examples of such obstacles. A straightedge illuminated by a beam of monochromatic light from a point source casts a shadow that is not geometrically sharp. A small amount of light is bent around the edge into the geometrical shadow, and the shadow is bordered by many alternate light and dark bands.

Figure 30-13 shows a diffraction pattern formed by monochromatic light passing over a razor blade. The ability of an optical system to produce clearly defined images is limited by the property of diffraction.

30-7 DIFFRACTION BY A SINGLE SLIT

If the obstacle in the light beam is a single narrow slit, a pattern of fringes will be formed on a screen placed behind the slit. Such an arrangement is shown in Fig. 30-14. If the slit CD is parallel to the wave front, each point in the slit can be considered as a source of Huygens' wavelets, all starting in phase. At every point on the screen these wavelets combine in some manner to produce the effect at that point. At P_0, on the perpendicular bisector of the slit, wavelets from C and D arrive in phase, since these points are equidistant from P_0. Similarly pairs of points in the upper and lower halves of the slit will combine to contribute to the light at P_0. Thus P_0 is bright. At points off the perpendicular bisector, the path lengths from the two edges of the slit are no longer equal. Let P_1 be a point such that $CP_1 = DP_1 + \lambda/2$. The wavelets from C and D will arrive at P_1 a half wave out of phase and hence will annul each other. However, wavelets from all other points in the slit are not annulled by other wavelets, and hence P_1 is still within the bright central band. If P_2 is located so that $CP_2 = DP_2 + 2\lambda/2$, wavelets from C and from D will arrive in phase but $EP_2 = DP_2 + \lambda/2$ and $CP_2 = EP_2 + \lambda/2$. Thus a wavelet from E will annul a wavelet from D, a wavelet from a point just above E will annul a wavelet from a point just above D, and so on, across the two halves of the slit. The wavelets from the upper half of the slit annul those from the lower half, producing darkness at P_2. We may think of the slit as divided into zones whose edges are successively

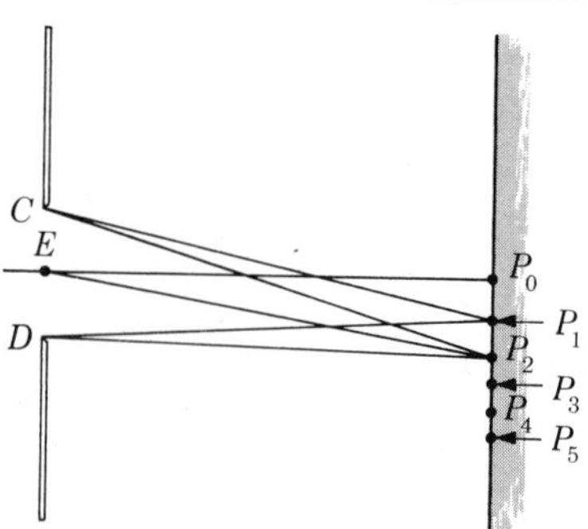

Figure 30-14
Spacing of fringes in a single-slit diffraction pattern.

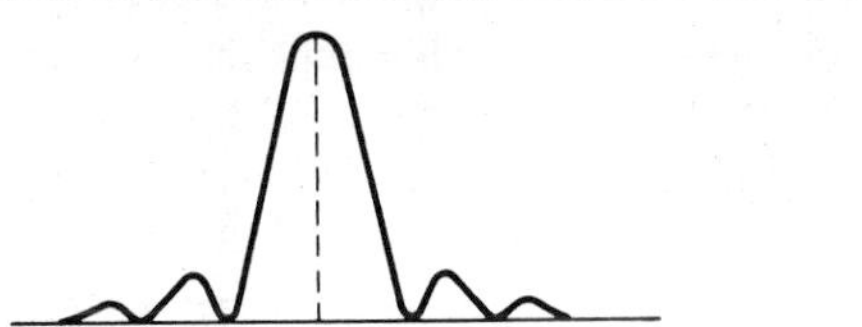

Figure 30-15
Intensity distribution in a single-slit diffraction pattern.

$\frac{1}{2}\lambda$ farther from the point considered. Thus, for the point P_1, the slit makes up only one zone, and P_1 is bright. For P_2, the slit has two zones, and the wavelets from the two zones annul each other. At another point P_3 for which $CP_3 = DP_3 + \frac{3}{2}\lambda$, the slit has three zones. Two of these zones annul, leaving the light from the third, and P_3 is bright but less bright than P_1. Similarly, for P_4, there are four zones and darkness; for P_5, five zones and brightness but again less bright than P_3. The variation of intensity in such a pattern is shown in Fig. 30-15. Here there is a bright and wide central band with narrower and less intense side fringes. The zones are called *Fresnel zones* after the man who originated this method of analysis. Figure 30-16 is a picture of a single-slit diffraction pattern. The picture represents a pattern obtained when the screen is far enough from the slit so that the rays can be considered parallel. Such diffraction is called *Fraunhofer* diffraction.

The conditions that a point shall be bright or dark can be written in terms of the path difference of the wavelets from the two edges of the single slit. We have found that where the path difference is a whole number of wavelengths there is darkness.

Path difference:

$$s = N\lambda \qquad \text{for darkness}$$

$$N = 1, 2, 3, \ldots$$

Where the path difference from the two edges of the slit is an odd number of half wavelengths,

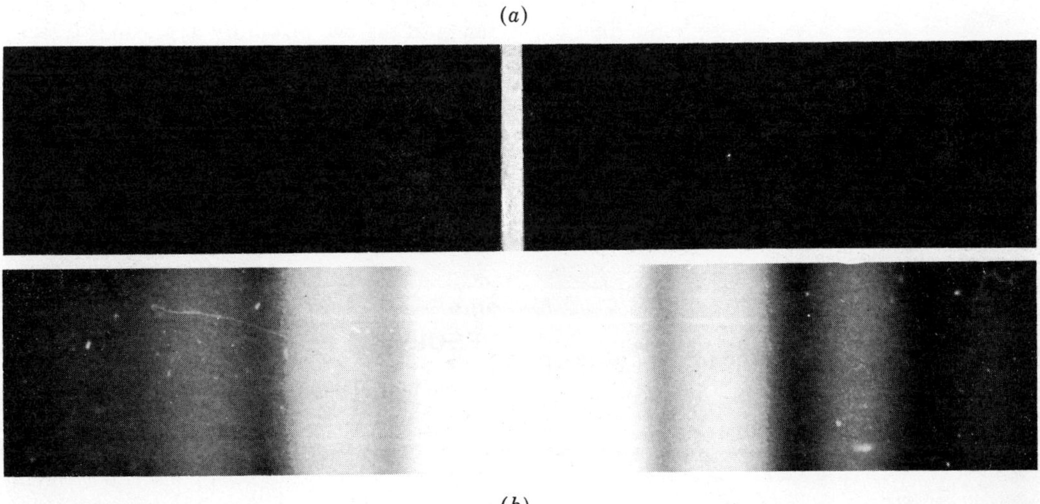

Figure 30-16
A single-slit diffraction pattern. (*a*) Shows the width of the slit illuminated by parallel monochromatic light; (*b*) is an actual photograph of the diffraction pattern. (*Courtesy of Jenkins and White, "Fundamentals of Optics," 3d ed., McGraw-Hill Book Company,* 1957.)

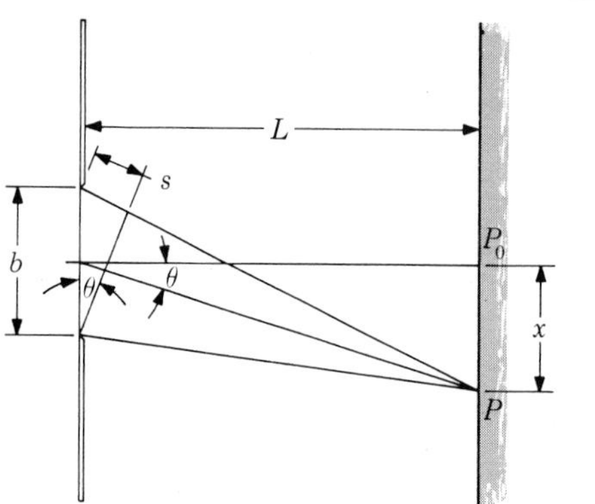

Figure 30-17
Ray diagram for diffraction from a single slit.

there is brightness. We may write this condition.

Path difference:

$$s = (2N + 1)\frac{\lambda}{2} \qquad \text{for brightness}$$

$$N = 1, 2, 3, \ldots$$

The spacing of the fringes can be computed by reference to Fig. 30-17 in terms of the path difference s, the slit width b, and the distance L from the slit to the screen.

$$\sin\theta = \frac{s}{b} \qquad \text{approx}$$

$$\tan\theta = \frac{x}{L}$$

Since the angles are small, the sine and tangent are approximately equal and

$$\frac{s}{b} = \frac{x}{L} \tag{8}$$

To find the distance from the middle of the central bright band to any given fringe, one must insert the appropriate value of the path difference s. When this substitution is made and the equation solved for x, λ appears in the numerator and b in the denominator. Thus the separation of the fringes is greater for longer wavelengths and increases as the width of the slit decreases.

Example A plane wave of monochromatic light of wavelength 5,893 Å passes through a slit 0.500 mm wide and forms a diffraction pattern on a screen 1.00 m away from the slit and parallel to it. Compute the separation of the first dark bands on either side of the central bright band, i.e., the width of the central bright band.

From Eq. (8),

$$\frac{x}{L} = \frac{s}{b}$$

For the first dark band,

$$s = 1 \times \lambda = \lambda$$

$$x = \frac{\lambda L}{b} = \frac{(0.00005893\ \text{cm})(10)}{0.0500\ \text{cm}} = 0.118\ \text{cm}$$

where x is the distance of the first dark band from the middle of the central bright band. The separation of the two first-order dark bands is

$$2x = 2(0.118\ \text{cm}) = 0.236\ \text{cm} = 2.36\ \text{mm}$$

For obstacles of different shape, a similar treatment can be set up. The zones may be different in shape for various obstacles.

30-8 RESOLVING POWER OF A LENS

If a circular opening is illuminated by light from a point source, the diffraction pattern will be a set of circular fringes. If a lens is used to form an image, the image of each point is not a point but a small diffraction disk. The size of the disk depends upon the aperture (diameter) of the lens and the wavelength of the light used. A larger aperture decreases the size of the disk, while a longer wavelength increases it. In the image formed by a lens, two points will appear as separate if their diffraction disks do not overlap by more than the radius of the disk. The *resolving power* of a lens is its ability to separate the images of two points that are close together. The resolving power is directly proportional to the aperture of the lens and inversely proportional to the wavelength of the light. Thus increasing the aperture of a lens increases its resolving power, since it decreases the diameter of the diffraction disk, and thus two images can be closer together without overlapping by as much as the radius of the disk.

30-9 THE DIFFRACTION GRATING

The principles of diffraction and interference find important application in the measurement of wavelength with the optical diffraction grating. A grating for use with transmitted light is a glass plate upon which is ruled a large number of equally spaced opaque lines, usually several thousand per centimeter.

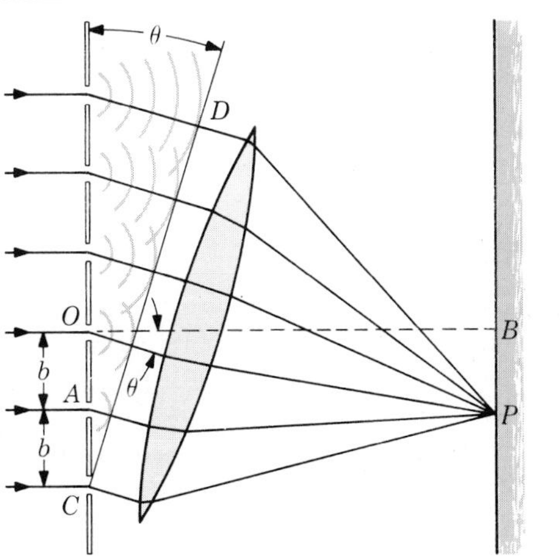

Figure 30-18
Diffraction of light by an optical grating.

A parallel beam of monochromatic light falling on the grating (Fig. 30-18) sends waves in all forward directions from each slit. Along certain definite directions waves from adjacent slits are in phase and reinforce each other. Consider the parallel rays making an angle θ with OB, the normal to the grating, which are brought to focus on the screen at P by an achromatic lens. If the ray AP travels a distance λ farther than ray CP, then waves from A and C will interfere constructively at P. So also will waves from all the slits, for they will differ in phase by a whole number of wavelengths. The wave front CD makes an angle θ with the grating. From the smallest right triangle, the path difference λ is seen to be $CA \sin\theta$. Calling the grating space $CA = b$, the condition for reinforcement in the direction θ may be written

$$b \sin\theta = \lambda \qquad \text{first order} \tag{9}$$

In general, there will be other directions on each side of OB for which waves from adjacent slits differ in phase by 2λ, 3λ, etc., and for which the corresponding bright images P_2, P_3, . . . are called the *second-order, third-order*, etc., images. The grating equation can be written in the more general form

$$b \sin\theta_N = N\lambda \tag{10}$$

where b is the grating space and N is the order of the spectrum.

With white light each diffracted image is dispersed into its component colors, and continuous spectra are produced at P_1, P_2, etc. The dispersion is greater in the higher order spectra. In each order the colors appear in the sequence violet to red with increasing deviation. There is overlapping of the spectra; for example, the red and orange in the second-order spectrum overlap the violet of the third order, since 2λ for the long wavelengths is greater than 3λ for the short wavelengths.

Example The deviation of the second-order diffracted image formed by an optical grating having 5,000 lines per centimeter is 32°. Calculate the wavelength of the light used.

$$b \sin \theta_N = N\lambda$$

$$b = \frac{1}{5{,}000} \text{ cm} = 0.00020 \text{ cm}$$

$$\lambda = \frac{b \sin \theta_N}{N} = \frac{0.00020 \text{ cm} \times 0.53}{2}$$

$$= 0.000053 \text{ cm} = 5{,}\bar{3}00 \text{ Å}$$

SUMMARY

In common with other forms of wave motion, light exhibits the phenomena of *interference*.

In order to observe interference effects with light, the light from a single source is made to traverse different optical paths to introduce a phase difference, and then the beams are reunited. *Constructive* interference occurs whenever the optical path difference $s = N\lambda$; *destructive* interference occurs whenever the path difference $s = (2N - 1)\lambda/2$.

When light is reflected at the boundary of an optically denser medium, it undergoes a phase change of $\frac{1}{2}\lambda$, or 180°.

Newton's rings are the interference fringes formed in the air films between optical surfaces. By their spacing they provide a sensitive means of measuring the curvatures of those surfaces.

The Michelson *interferometer* utilizes the interference of two light beams traveling different paths as the fundamental method of measuring wavelength. If N successive fringes are counted as the mirror is moved a distance D,

$$\lambda = \frac{2D}{N}$$

The spreading of light into the region behind an obstacle is due to *diffraction*. Diffraction occurs in accordance with Huygens' principle and is an interference phenomenon.

The *resolving power* of a lens is the smallest angular distance at which the images of two point sources can be recognized as separate. Owing to diffraction, this minimum angle of resolution varies directly with the wavelength of the light and inversely with the aperture of the lens.

A diffraction grating utilizes the diffraction of light from many closely spaced parallel slits to disperse light into its component wavelengths. Light of wavelength λ deviated through angle θ_N forms a bright image when

$$b \sin \theta_N = N\lambda$$

References

Boys, C. V.: "Soap Bubbles," Doubleday & Company, Inc., Garden City, N.Y., 1959.

Ruechardt, Eduard: "Light, Visible and Invisible," University of Michigan Press, Ann Arbor, Mich., 1958.

Schawlow, Arthur L.: Advances in Optical Masers, *Scientific American,* July 1963, p. 34.

Questions

1 A phonograph record viewed under certain light conditions produces a colored pattern. What are these conditions and why does the color appear?

2 A certain type of jewelry is made from reflection-type diffraction grating. Why would this make attractive jewelry?

3 In double-slit interference how does the spacing of the slits affect the separation of the fringes?

4 Could the mirror glass for a front-surface mirror (Fig. 30-10) be made more cheaply and more heat-resistant than that for a back-coated mirror?

5 When two light waves interfere at some point to produce darkness, what becomes of the energy?

6 Why is interference of the light waves from two different lamps never observed?

7 Derive, in terms of the wavelength, the separation of the slits, and the distance from the slits, an expression for the distance between two adjacent bright fringes in a double-slit interference pattern.

8 Describe in detail what causes soap bubbles and oil films to produce colors when light falls upon them.

9 An interesting experiment provides some clue to the conditions necessary to produce interference in films. A loop which has a thin soap film stretched across it is held on edge vertically and observed as reflected light falls upon it. At first the color appears at the upper edge, with no color at the lower edge. Then the color moves downward and eventually black areas appear in the upper end of the film before it breaks. Explain why this happens.

10 What is the purpose of using coated lenses in optical instruments? Describe the optical properties that this coating must have and how it helps to improve such instruments.

11 How does the origin of the colors observed in a soap bubble in sunlight differ from that of the colors observed when sunlight passes through a glass prism?

12 Show by a sketch what the path difference is between the two reflected rays which form one of the dark Newton's rings.

13 If white light falls upon a Newton's rings apparatus, would the rings be in color or would they be black and white?

14 Why are interference colors not observed in a very thick film—in a piece of glass, for example?

15 Why are Newton's rings circular?

16 The characteristic yellow light from sodium comprises two lines of wavelengths 5,890 and 5.896 Å and is thus not a monochromatic source. How could a Michelson interferometer be used to measure very precisely the wavelength separation of the two sodium yellow lines?

17 If a screen with a small circular opening is placed between a point source of monochromatic light and another screen, a diffraction pattern will be formed on the second screen. For points on the axis of the opening, what will be the shape of the Fresnel zones? Will the number of zones be the same for all points on the axis? How would the intensity of the central spot change as the distance to the second screen increases?

18 Will the angular separation of orders be greater for a grating with many lines per inch or for a grating with fewer lines?

19 Will the angular separation between red and blue rays be greater in the first-order or in the second-order spectrum?

20 Will the image of the slit be more sharply defined in the first-order spectrum or in the third-order spectrum?

21 Contrast the formation of a spectrum by a diffraction grating with the production of the spectrum of the same light source by a glass prism.

22 Imagine a light made up of wavelengths spaced at 100-Å intervals. Make a rough sketch of the appearance of the spectrum viewed in a prism spectroscope and in a grating spectroscope.

23 Describe the procedures needed for the measurement of wavelengths by means of a diffraction grating.

24 What are the differences between a transmission and a reflection diffraction grating? Show these differences by a sketch.

25 What factors govern the choice between a prism and a diffraction grating for use in a spectrograph?

Problems

1 A double slit having a separation of 1 mm has red light (6,200 Å) falling on it. Determine the distance between the central bright band and the fifth bright fringe on one side of an interference pattern formed on a screen 3 m away.

2 What wavelength light would cause interference fringes 20 mm apart when the light is passed through a slit having a separation of 0.5 mm if the screen is 2.0 m away? *Ans.* 5,000 Å.

3 The two slits in a Young's experiment apparatus are 0.200 mm apart. The interference fringes for light of wavelength 6,000 Å are formed on a screen 80.0 cm away. (*a*) How far is the second dark band from the central image? (*b*) How far is the second light band from the central image?

4 Light from a narrow slit passes through two parallel slits 0.20 mm apart. The two central dark interference bands on a screen 100 cm away are 2.95 mm apart. What is the wavelength of the light? *Ans.* 5,900 Å.

5 A Fresnel biprism (Fig. 30-19) receiving light from a very narrow source slit at S_0 produces interference fringes on the screen *OP*. (*a*) Show that for a small prism angle *A*, the light proceeds to the screen from two virtual sources S_1 and S_2 whose separation *y* is given by $y = 2aA\ (N - 1)$, where *n* is the index of refraction of the prism. (*b*) What is the spacing of the fringes of green light ($\lambda = 5{,}000$ Å) on a screen 2.0 m from the biprism for which $n = 1.60$ when the light source is at a distance $a = 20$ cm?

6 A slit 0.500 mm wide causes a diffraction pattern on a screen 300 cm away when illuminated with sodium light (wavelength 5,890 Å). Find the distance from the central image to the first dark band. *Ans.* 3.53 mm.

7 Light from a single slit passing through a double prism of very small angle (Fig. 30-19) proceeds as though it had come from two sources. With such an arrangement, light of wavelength 5,460 Å produces fringes whose separation is 0.200 mm at a distance of 120 cm. What is the distance between the two virtual sources S_1 and S_2?

8 Two slits 0.125 mm apart are illuminated by light of wavelength 4,500 Å. What is the separation of the second bright bands on either side of the central bright band when the screen is 60.0 cm from the plane of the slits? *Ans.* 0.865 cm.

9 Two slits are illuminated by light that consists of two wavelengths. One wavelength is known to be 6,000 Å. On a screen the fourth dark fringe of the pattern for the 6,000-Å wavelength coincides with the fifth light fringe for the other wavelength. What is the unknown wavelength?

10 Two glass plates are arranged to produce an interference pattern by having their one common edge in contact and the opposite edges separated by a thin film of air due to a piece of paper being placed between those edges. If 10 bright fringes are observed when viewed by reflected light having a wavelength of 5,890 Å, what is the thickness of the paper? *Ans.* 2.8×10^{-4} cm.

11 A quartz fiber is placed between the edges of two flat glass plates which are in contact at the other end. The wedge-shaped air film between the plates is viewed by reflected monochromatic light of wavelength 5,890 Å and is found to be crossed by 35 dark interference

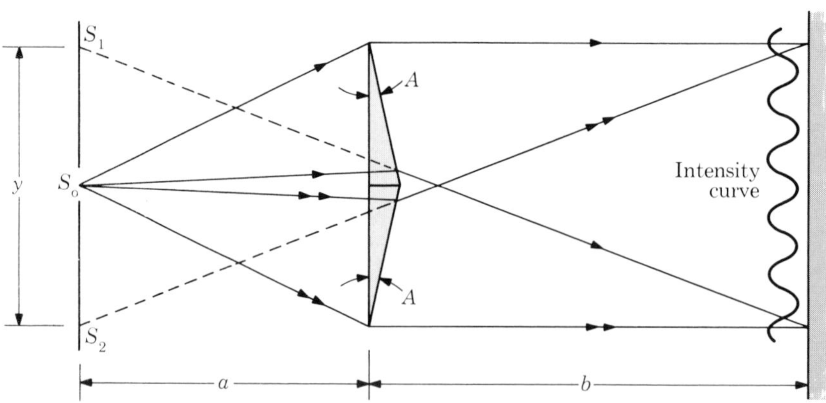

Figure 30-19
A Fresnel biprism.

bands. Calculate the diameter of the quartz fiber.

12 Two glass plates are separated by a thin wedge of air and produce a pattern of interference fringes when light falls from above. If 16 fringes are observed when light having a wavelength of 6,500 Å is used, how thick is the wedge of air? *Ans.* 5.2×10^{-4} cm.

13 Light of wavelength 4,500 Å falls on a sheet of transparent material whose index of refraction is 1.50. What must be the minimum thickness of the sheet in order that there be constructive interference in the reflected beam?

14 A spherical lens whose radius of curvature is 94.2 cm lies on a plane plate glass. When the lens is illuminated by light of wavelength 5,890 Å, a dark spot appears at the point of contact and it is surrounded by alternate dark and bright rings. Find the diameter of the twelfth dark ring; of the twelfth bright ring.

Ans. 4.90 mm; 5.04 mm.

15 Light of wavelength 5,893 Å is reflected at nearly normal incidence from a soap film of index of refraction 1.42. What is the least thickness of the film that will appear (*a*) black and (*b*) light?

16 A soap film has an index of refraction of 1.40. How thick must it be for it to appear black when mercury light, 5,461 Å, falls on it at right angles? *Ans.* 1,950 Å.

17 In a Newton's rings experiment, the radius of the fourth bright ring is 6.4×10^{-2} cm when sodium light having a wavelength of 5,893 Å is used. What is the radius of curvature of the spherical surface?

18 Determine the wavelength of light that would produce the tenth dark ring, if the thickness of the air layer between a lens and a glass plate to form Newton's rings is 3×10^{-4} cm.

Ans. 6×10^{-5} cm.

19 When the movable mirror of a Michelson interferometer is moved 0.100 mm, how many dark fringes pass the reference point if light of wavelength 5,800 Å is used?

20 When the adjustable mirror on a Michelson interferometer is moved in one direction, 400 fringes appear to pass through the field of view when light of 5,000 Å falls on it. Through what distance was the mirror moved?

Ans. 1×10^{-2} mm.

21 An interferometer is illuminated with light of 6,400 Å. How many fringes will pass the reference mark as the mirror is moved a distance of 8×10^{-3} cm?

22 To produce a minimum reflection for light having a wavelength of 7,000 Å, how thick a coating of magnesium fluoride ($n = 1.38$) should be coated on a lens? *Ans.* 1.27×10^{-5} cm.

23 A beam of monochromatic light, $\lambda =$ 6,300 Å, passes through a slit 0.070 cm wide and produces a diffraction pattern on a screen 80.0 cm away. What is the distance from the middle of the pattern to the third dark diffraction band on the side?

24 A diffraction grating having 5,000 lines to the centimeter was used to find the wavelength of a certain color light. The bright image in the second order was formed at an angle of 30° from the bright central image. What was the wavelength of the light? *Ans.* 5,000 Å.

25 A student places a small sodium lamp just in front of the blackboard. Standing 20.0 ft away, he views the light at right angles to the blackboard while holding in front of his eye a transmission grating ruled with 14,500 lines per inch. He has his partner mark on the board the positions of the first-order diffracted images on each side of the lamp. The distance between these marks is found to be 14 ft 2 in. Compute the wavelength of the light.

26 A diffraction grating has slits which are 2.5×10^{-4} cm apart. If the angle of diffraction for the first-order image is 15°, what is the wavelength of the light? *Ans.* 6,480 Å.

27 A beam of monochromatic light is passed through a diffraction grating which has 5,000 lines per centimeter. The angular deviation of the second-order image is 30°. (*a*) What is the wavelength of the light? (*b*) What is the angular deviation of the third-order image?

28 What would be the angle for the second-order and the third-order images for the slit opening and light wavelength used in Prob. 26?

Ans. 31.3°; 51.0°.

29 A grating has a 3.0 × 3.0-in ruled area in which there are 40,000 lines. What is the dispersion of yellow light ($\lambda = 5{,}890$ Å) in the second order?

30 An observer looks through a diffraction grating at a slit illuminated with light whose wave length is 4,000 Å. The second-order spectrum is seen at an angle of 30°. What is the distance between slits and how many lines are there in a centimeter? *Ans.* 1.6×10^{-4} cm; 6,250 lines per centimeter.

31 A yellow line and a blue line of the mercury-arc spectrum have wavelengths of 5,791 and 4,358 Å, respectively. In the spectrum formed by a grating that has 5,000 lines per inch, compute the separation of the two lines in the first-order spectrum and in the third-order spectrum. Compare the angle of diffraction for the yellow line in the third order with that for the blue line in the fourth order.

32 An observer looks through a diffraction grating at a slit illuminated with light whose wavelength is 4,000 Å. The second-order spectrum is seen at an angle of 45°. What is the distance between slits and how many lines are there in a centimeter? *Ans.* 11.3×10^{-5} cm; 8,850 lines per centimeter.

33 The two headlights of an approaching car are 1.0 m apart. What is the maximum distance at which the eye can resolve them? Use Rayleigh's criterion that the lights must have a minimum angular separation for resolution of $\theta_r = \sin^{-1}(1.22\,\lambda/d)$. Assume a pupil diameter $d = 5.0$ mm for the eye, and assume that $\lambda = 5{,}500$ Å.

34 Using Rayleigh's criterion that two points are barely resolvable when they have an angular separation $\theta_r = \sin^{-1}(1.22\,\lambda/d)$, estimate the separation of two points on the moon's surface that can just be resolved by the 200-in telescope at Mount Palomar, assuming $\lambda = 5{,}500$ Å. The distance from earth to moon is 2.4×10^5 mi. *Ans.* 50 m.

Sir John D. Cockcroft, 1897–1967

Born at Todmorden, England. Worked under Lord Rutherford at the Cavendish Laboratory. Director of the Atomic Energy Research Establishment, Harwell. Shared the 1951 Nobel Prize for Physics with Walton for their discovery of the transmutations of atomic nuclei by artificially accelerated particles.

Ernest T. S. Walton, 1903–

Born in Dungarvan, Ireland. Research student of Lord Rutherford. Professor of natural and experimental philosophy, Trinity College, Dublin. Shared the 1951 Nobel Prize with Cockcroft for their production of nuclear transmutations.

31

Polarization of Light

Interference and diffraction effects have given us perhaps our best evidence that light has wave characteristics. But these phenomena leave unanswered the questions of what it is that vibrates, and how. Indeed, with a suitable change of scale, our discussion of reflection, refraction, wavelength, interference, and diffraction is just as true for sound as for light. The fact that light can be polarized, however, its vibrations being confined to a single plane, shows conclusively that its wave properties are transverse, in contrast to the longitudinal waves associated with sound.

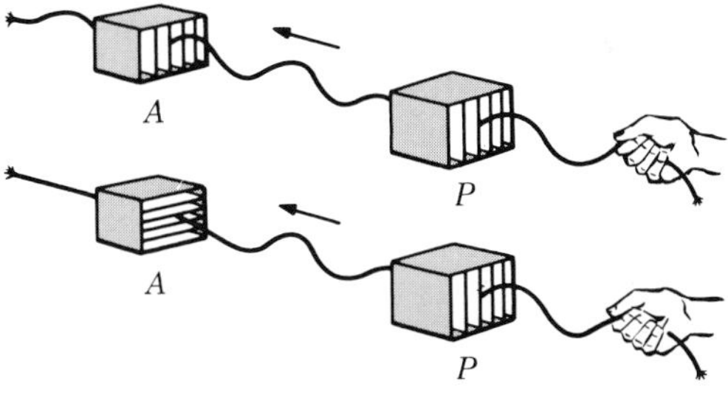

Figure 31-1
Mechanical analog of polarization.

31-1 POLARIZATION

Experiments with transverse waves in a rope (Fig. 31-1) show that a slot P can be used to confine the vibrations to one plane, after which they can be transmitted or obstructed by a second slot A, depending on whether it is placed parallel or perpendicular to the first slot. The slotted frame P is called a *polarizer*. The waves emerging from the polarizer are confined to one plane, and the wave is said to be *plane-polarized*. The slotted frame A is the *analyzer* of the plane-polarized wave. The plane that contains the wave form is called the *plane of polarization*. Polarizability is characteristic of *transverse* waves. If the rope were replaced by a coiled spring, compressional (longitudinal) waves in it would pass through slots regardless of their orientation.

According to the electromagnetic theory of light, a light wave consists of an electric vibration accompanied by a magnetic vibration at right angles to it. In the following discussion the light vibrations refer to the vibrating electric field. A light wave traveling through space can be represented at any point at any instant by a line whose direction and length represent the direction and displacement of the vibrating electric field. Such a line is purely diagrammatic and is called a *light vector*.

The light from ordinary sources is unpolar-

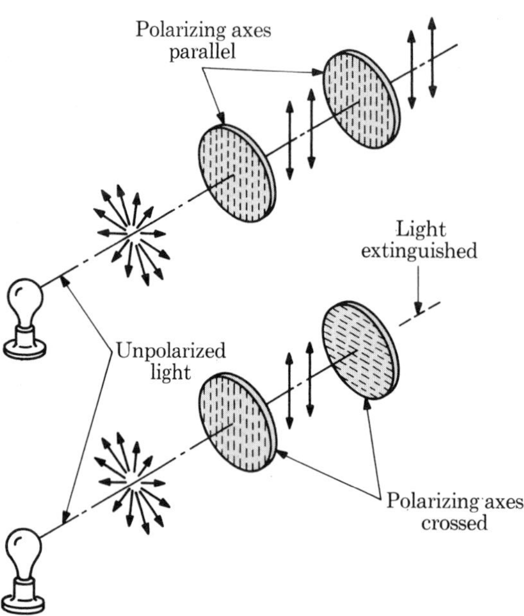

Figure 31-2
Production and detection of polarized light by a polarizer and an analyzer.

ized. By suitable interaction with matter, light can be made plane-polarized. This interaction may be one of reflection from a transparent surface, refraction through a crystal, selective absorption in certain crystals, or scattering by small particles.

Except in the rarest of special conditions, and only momentarily, the unaided eye cannot detect any difference between polarized and unpolarized light. Hence an analyzer must be used to detect the state and plane of polarization of a light beam, as indicated in Fig. 31-2.

31-2 POLARIZATION BY REFLECTION

When a beam of light strikes a piece of glass, it is in part reflected and in part transmitted. When the incident beam is ordinary, unpolarized light it is found that the reflected and refracted rays are each partly plane-polarized. It is observed that, in the reflected beam, the polarized part has vibrations perpendicular to the plane of incidence (parallel to the reflecting surface), while in the refracted beam the polarized part has vibrations parallel to the plane of incidence.

The fact that light may be polarized by reflection can be shown by allowing the light reflected from the glass to fall on a second piece of glass. When the two glass plates have their surfaces parallel, the polarized light reflected from the first glass will be reflected also by the second glass. When, however, the second glass is rotated 90° about the reflected ray as an axis, there will be little or no reflection of the polarized part from the second glass plate. Thus the second glass plate can be used as an analyzer to test the polarization of light.

The degree of polarization of light reflected from a glass plate, or other nonconducting surface, depends upon the angle of incidence. The polarization of the reflected beam is found to be complete when the reflected ray is perpendicular to the refracted ray (Fig. 31-3). The vibrations in the incident ray AO are in all planes. These can be represented by their components perpendicular and parallel to the plane of incidence. These components are shown in Fig. 31-3 by

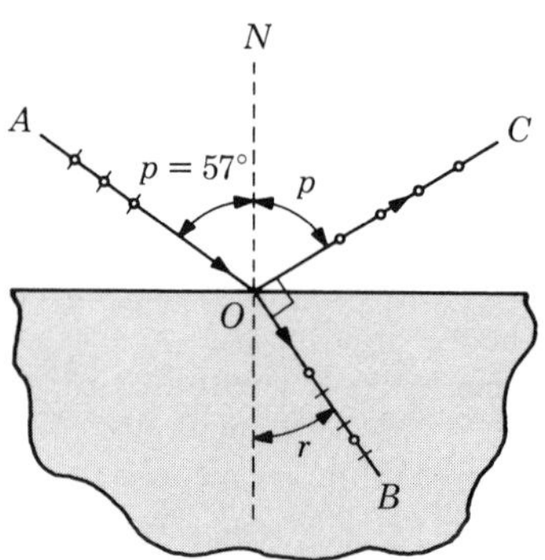

Figure 31-3
Polarization by reflection of light.

small circles and by short lines, respectively. The angle of incidence p, called the *polarizing angle,* for which the polarization of the reflected beam is complete, is related simply to the index of refraction of the reflecting material. In Fig. 31-3 the angle COB is 90°; therefore $p + r = 90°$. Thus $\sin r = \cos p$. Hence

$$n = \frac{\sin i}{\sin r} = \frac{\sin p}{\cos p} = \tan p \qquad (1)$$

This relation between the index of refraction and the polarizing angle is known as *Brewster's law.*

At the polarizing angle (about 57° for glass) none of the vibrations that lie in the plane of incidence is reflected. The reflected beam is then plane-polarized, but of relatively low intensity, since only about 8 percent of the incident beam is reflected at the polarizing angle. A pile of 6 to 12 plates is often used to attain sufficient intensity by combining reflected rays from all the surfaces.

The transmitted beam is not completely plane-polarized at the polarizing angle. When a pile of plates is used, the successive reflections remove more and more of the vibration perpendicular to the plane of incidence from the transmitted beam. If many plates are used, the transmitted beam is also practically plane-polarized in a plane perpendicular to that of the reflected beam.

31-3 INTENSITY IN POLARIZED LIGHT

Consider that a beam of light is passing through a polarizer and an analyzer (Fig. 31-2) and that the analyzer has been rotated by an angle θ from its position of maximum transmission. The amplitude A of the plane-polarized light waves incident on the analyzer may be resolved into the component $A \cos \theta$ which is transmitted and the component $A \sin \theta$ which is reflected or absorbed. Since the intensity is proportional to the square of the amplitude (Chap. 19), the intensity I of the beam transmitted by the analyzer when set at angle θ is given in terms of the maximum intensity I_0 by

$$I = I_0 \cos^2 \theta \qquad (2)$$

a relation called *Malus' law.*

31-4 DOUBLE REFRACTION

If a crystal of Iceland spar (calcite, $CaCO_3$) is laid on a printed page, one observes through it two refracted images of the type shown in Fig. 31-4. This phenomenon was observed as early as 1669

Figure 31-4
Double refraction in calcite compared with single refraction in glass.

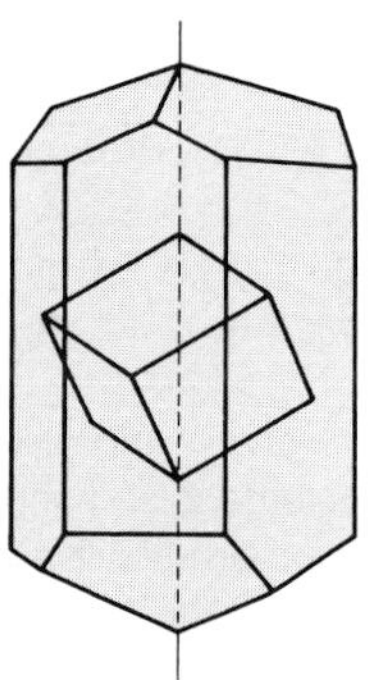

Figure 31-5
The crystal form of calcite.

by Bartholinus, who realized that he had come upon a fundamental question in refraction. It was later observed by Huygens (1690) that the rays which produced the two images were plane-polarized, in mutually perpendicular planes.

The common form of calcite crystal is shown in Fig. 31-5. It has a tendency when struck to cleave obliquely in three definite planes, forming rhomboidal fragments. The external symmetry of the crystal is an indication of a corresponding symmetry in the lattice work of atoms comprising the crystal. This suggests that light energy may be transmitted with different speeds in such a crystal, depending on the orientation of its vibrations with respect to the planes of atoms which make up the crystal. Double refraction is unusual only in the sense that our ordinary experience in optics is with isotropic materials such as air, glass, and water, which do not exhibit this effect.

The line joining the two blunt corners of the equilateral rhombohedron in Fig. 31-5 coincides with the principal crystallographic axis of the calcite. The direction of this line, not the line itself, is called the *optic axis*. In this particular direction light travels with the same speed regardless of the plane of its vibration. Hence in this direction there is no double refraction.

When a parallel beam falls obliquely on a surface of a calcite rhombohedron, the light is split into two parts. Both are refracted. Measurement of the angles of incidence and refraction shows that for every angle of incidence the index

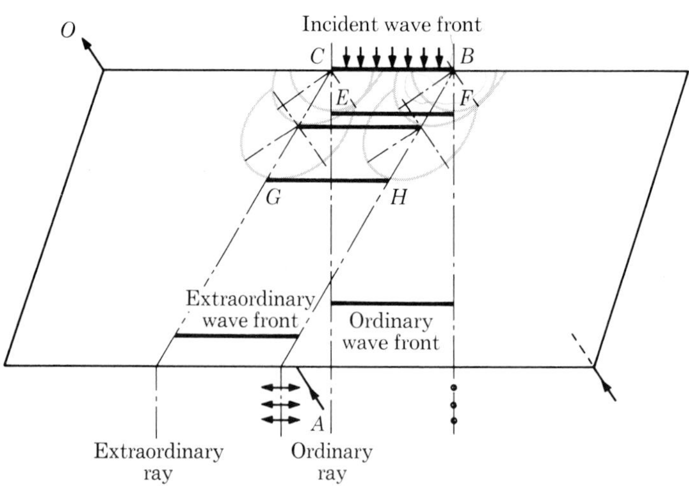

Figure 31-6
Huygens' wave fronts in a doubly refracting crystal.

of refraction for one ray (the ordinary ray) is 1.658 (for sodium yellow light of 5,893 Å). But for the other ray (the extraordinary ray) the refractive index alters with the angle of incidence, varying between 1.486 and 1.658. The one ray is called *ordinary* because its index of refraction is constant, while the other ray is called *extraordinary* because its index of refraction varies with angle of incidence. This shows that the speed of the ordinary ray is less than that of the extraordinary ray except in the special direction along the optic axis when both travel with the same speed.

If a pencil dot is viewed through a calcite cyrstal, the image formed by the ordinary ray will appear closer than that formed by the extraordinary ray. If the obtuse corners of the crystal are ground off and polished to form surfaces perpendicular to the optic axis, a ray entering along the optic axis is neither refracted nor polarized.

Huygens applied his wave construction to explain the ordinary and extraordinary rays. Around points *B* and *C* (Fig. 31-6) are drawn spheres representing the ordinary wave fronts and spheroids representing the extraordinary wave fronts. Each spheroid has its axis of revolution parallel to the optic axis, so that the two surfaces touch at the points that lie in the direction of the crystal axis. The envelope of the spheres is the plane *EF* which determines the ordinary wave front, while the plane *GH* is the envelope of the spheroids which locates the extraordinary wave front. It should be noted that the extraordinary ray is not perpendicular to the wave front.

31-5 THE NICOL PRISM

A beam of light entering a calcite crystal becomes, in effect, divided into two beams. These two beams are completely plane-polarized in directions at right angles to each other. Nicol (1832) used an artifice to separate these rays to give a single beam of plane-polarized light. A calcite rhombohedron (Fig. 31-7) has the ends

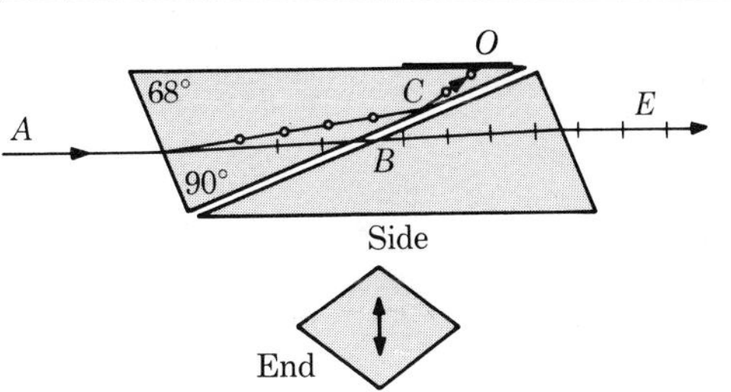

Figure 31-7
A Nicol prism.

polished down so that the acute angle between them and the sides is 68° (instead of the 71° obtained by cleavage). The length is so chosen that the crystal can be sawed in a diagonal plane perpendicular to these two new faces and to the optic axis. The sawed surfaces are polished and are then cemented together with a layer of Canada balsam. The index of refraction of the balsam (1.530) is intermediate between the two indices for calcite, $n_O = 1.658$ and $n_E = 1.486$.

With the crystal angles properly chosen, the extraordinary ray travels through the Nicol prism in the direction *ABE*. The ordinary ray, however, strikes the cement layer at *C* at an angle exceeding its critical angle and is internally reflected toward the side of the crystal, where it may be absorbed in black paint.

The intensity of the polarized light from a Nicol prism or any other polarizing device is always less than half the intensity of the incident beam. One polarized component is removed, and the light also loses intensity by reflection at boundary surfaces and by absorption.

31-6 POLARIZATION BY SELECTIVE ABSORPTION

Certain crystals, known as *dichroic,* produce two internal beams polarized at right angles to each other, and in addition strongly absorb the one beam, while transmitting the other (Fig. 31-8).

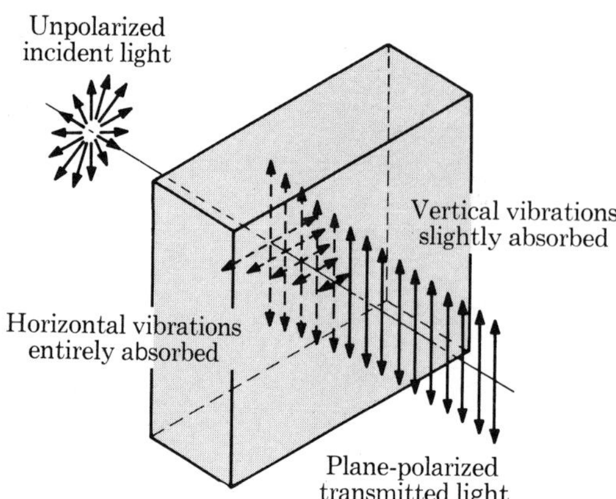

Figure 31-8
Production of polarized light by a dichroic crystal.

Tourmaline has this property. Unfortunately the plane-polarized light transmitted is colored. In 1852, Herapath, an English physician, discovered that dichroic crystals of quinine iodosulfate (herapathite) transmit a beam as plane-polarized light with transmission close to the ideal 50 percent for all wavelengths of visible light (Fig. 31-9). The potential usefulness of this material

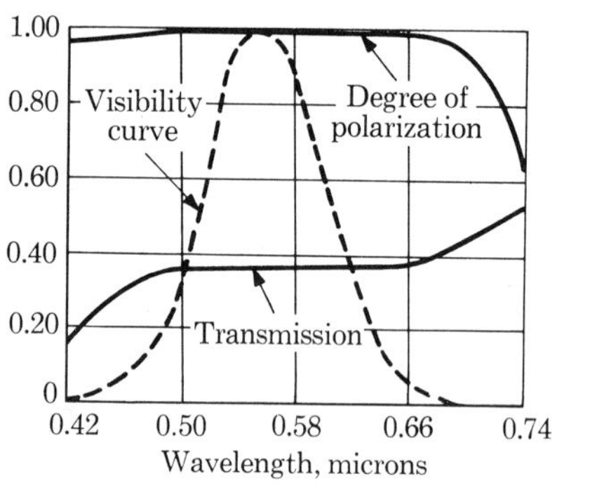

Figure 31-9
The spectral distribution of the relative intensity (lower curve) and the degree of polarization (upper curve) of light transmitted by Polaroid polarizing material.

led to extensive experiments, culminating in the invention by Land (1929) of a practical method for embedding the tiny synthetic crystals (about 10^{11} per square centimeter) in a transparent cellulosic film in uniform alignment. The thin film 0.001 to 0.004 in thick thus acts like a single huge crystal. This Polaroid sheet, sometimes bonded between glass plates, has the advantage of large size, low cost, and a polarizing effectiveness approaching that of a Nicol prism except at the extremities of the spectrum.

More recently Polaroid materials have been prepared by aligning molecules rather than the tiny crystals. They consist of aligned molecules of polymeric iodine in a sheet of polyvinyl alcohol (in H sheet) or aligned molecules of polyvinylene (in K sheet). These molecular polarizers are superior in several respects, such as greater stability and freedom from scattered light.

31-7 POLARIZATION BY SCATTERING

When a strong beam of light is passed through a region containing no fine particles, it is not visible from the side. If, however, the beam is intercepted by fine particles, such as smoke, dust, or colloidal suspensions, the beam is partly scattered and becomes visible. In this *Tyndall effect,* the color and the intensity of the scattered light depend on the size of the particles. Very small particles scatter chiefly blue light, as in the case of cigar smoke or the "blue" sky (Sec. 28-8). As

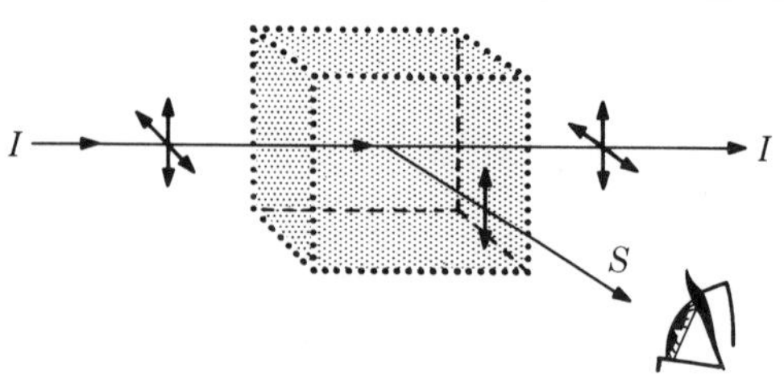

Figure 31-10
Polarization by scattering.

the particles are made larger, the longer wavelengths also are scattered until the scattered light appears white.

Scattered light is partly plane-polarized. The vibrations are perpendicular to the plane determined by the direction of the incident light and the line of sight (Fig. 31-10).

31-8 OPTICAL ROTATION

When two Polaroid light polarizers or two Nicol prisms are held in the line of vision in the crossed position, no light gets through and the field of view is dark. If, now, a crystal of quartz or a tube of sugar solution is placed between the crossed polarizer and analyzer, the light reappears. Monochromatic light can be extinguished by rotation of the analyzer. Materials that have the property of rotating the plane of polarization while transmitting polarized light are called *optically active* substances. The rotation is proportional to the amount of optically active substance in the path, and it provides an accurate way of determining the concentration of, say, sugar in an inactive solvent. *Polarimeters* are instruments for measuring optical rotation.

We are concerned not only with the angle through which the analyzer must be turned but also with the direction in which it is turned. Where it is necessary to turn the analyzer to the *right,* the substance is said to be *dextrorotatory;* where it is necessary to turn the analyzer to the *left,* the substance is *levorotatory*. Most substances do not possess optical activity; for example, water, ethyl alcohol, acetic acid, and chloroform have no effect when placed in a polarimeter. However, the sugars called levulose and dextrose are optically active. In fact, their names indicate the type of optical activity they possess, *levu*lose is levorotatory (left turning) and *dextro*se is dextrorotatory (right turning).

31-9 INTERFERENCE EFFECTS

When light travels through a birefringent material such as calcite, the crystal plate produces a difference in phase between the ordinary and extraordinary rays, because of the difference in the speeds of the two rays. By choosing a suitable thickness of crystal it is possible to retard one ray a half wavelength relative to the other. Nevertheless, interference bands or rings are never observed. One concludes that two rays of light polarized at right angles to each other do not interfere. They can cross one another without disturbance.

To the conditions for interference mentioned in Chap. 30 we must now add that the vibrations

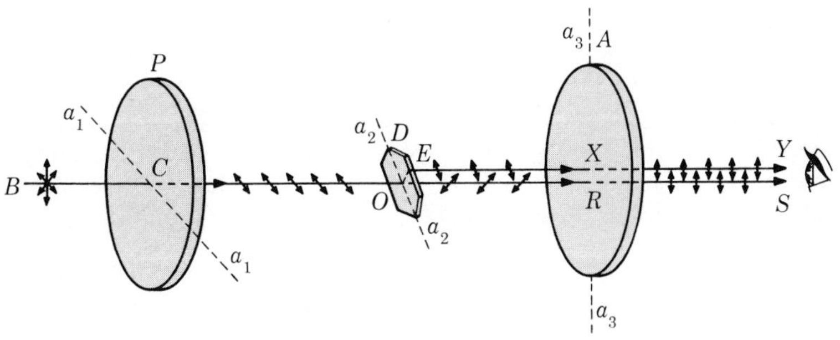

Figure 31-11
Production of color by interference effects in polarized light.

in the interfering rays must be in the same plane. An arrangement for producing such interference is shown in Fig. 31-11. A ray of unpolarized white light *BC* becomes plane-polarized white light after passage through polarizer *P*. It is then allowed to strike a doubly refracting crystal *D* whose optic axis a_2 is inclined at about 45° to the plane of vibration a_1. In the crystal *D* the ray is resolved into an ordinary ray *OR* vibrating perpendicularly to a_2a_2 and an extraordinary ray *EX* vibrating parallel to a_2a_2. There will be a phase difference between the rays emerging from crystal *D*, but they will not interfere, since their vibrations are in mutually perpendicular planes. Finally the analyzer *A* passes only the vertical components of these vibrations, giving two emergent rays *RS* and *XY* whose vibrations are in the same plane.

These rays are now capable of interference. If the relative retardation is an odd number of half wavelengths, the corresponding color is annulled by interference and its complementary color is transmitted.

The arrangement described is useful in the examination of crystals and thin rock sections, for differentiating isotropic and anisotropic materials, for detecting characteristics of structure not otherwise noticed, and for mineral identification.

31-10 APPLICATIONS OF POLARIZED LIGHT

For several centuries polarized light gave valuable evidence about the nature of light, but its production by inconvenient or expensive devices limited its usefulness to the laboratory. The production of polarizing materials in thin sheets, such as the various Polaroid light polarizers, opens new possibilities for utilizing polarized light.

Perhaps the simplest application of Polaroid sheets is the control of the intensity of light. An outer polarizing sheet is fixed in position, and an inner one may be rotated to adjust the amount of light transmitted—from maximum, when the planes are parallel, to zero, when they are perpendicular.

The use of Polaroid glasses for eliminating glare from reflected light makes use of the fact that the polarized component has vibrations parallel to the reflecting surface. The glasses are set so that they will not transmit this component.

In photography it is often desirable to enhance the effect of sky and clouds by eliminating some of the actinic rays from the sky. Since light from the sky is partly polarized by scattering, a suitably oriented polarizing disk in front of the camera lens will serve as a "sky filter," with obvious advantages over the common yellow filter when color film is used.

Three-dimensional pictures may be projected with the two components polarized in directions at right angles to each other. If these double pictures are viewed through Polaroid glasses oriented to allow the right eye to see one picture but the left eye the other, the impression of three dimensions is attained. Similarly aerial pictures may be taken from slightly different angles and when viewed in a manner similar to that of the three-dimensional pictures will give a better perception of depth.

31-11 PHOTOELASTICITY

Certain materials such as glass or transparent Bakelite become doubly refracting under mechanical strain. When the material is placed between crossed polarizer and analyzer, patterns of interference fringes can be observed. These patterns are used in detecting strains in glassware, as in the examination of glass-to-metal seals in electron tubes.

Complex engineering structures may be analyzed by photoelastic studies of transparent models. The regions of greatest strain are those of the closest spacing of fringes (Fig. 31-12). Quantitative measurements can often be obtained from the photoelastic constants of the material and the observed spacing of fringes. The strains in gears or other moving parts are sometimes

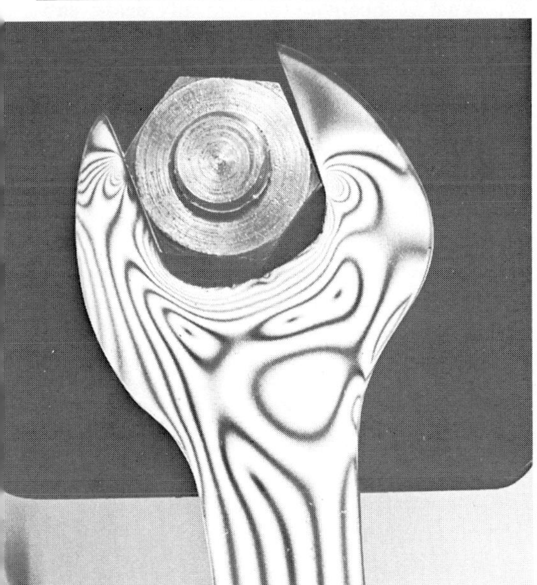

Figure 31-12
Stress lines in a plastic wrench photographed under polarized light. (*Courtesy Instrument Division, Budd Company.*)

studied by chilling the photoelastic model and thus freezing the strains in the sample, which may then be studied at rest.

SUMMARY

Polarization is the process by which the vibrations of a wave motion are confined to a definite pattern.

Polarizability is a characteristic of *transverse* waves.

Light may be plane-polarized by reflection by nonconducting surfaces, by refraction, by transmission through crystals showing double refraction, or by scattering from small particles.

The *angle of polarization* p is that angle at which light reflected from a substance is almost completely plane-polarized. By Brewster's law, $n = \tan p$.

The intensity I of a beam transmitted by a polarizing device upon which plane-polarized light is incident is given by

$$I = I_0 \cos^2 \theta$$

where I_0 is the maximum intensity transmitted and θ is the angle through which the device has been rotated from its position of maximum transmission.

When plane-polarized light passes through an *optically active* material, the plane of polarization is rotated through an angle that depends on the material, the length of the path traversed, and the wavelength of the light.

Certain materials such as glass or transparent Bakelite become doubly refracting under strain. The science of photoelasticity relates these changes in optical properties to the strains producing them and furnishes a method of measuring strains.

References

Frocht, Max M.: "Photoelasticity," vol. 1, 1941, vol. 2, 1948, John Wiley & Sons, Inc., New York.

Halliday, D., and R. Resnick: "Physics for Students of Science and Engineering," 2d ed., pt. II, chap. 46, John Wiley & Sons, Inc., New York, 1960.

Shurcliff, William A.: "Polarized Light: Production and Use," Harvard University Press, Cambridge, Mass., 1962. An advanced treatise, with bibliography.

Strong, John: "Concepts of Classical Optics," W. H. Freeman and Company, San Francisco, 1958. Chapter 6 describes the situation in which the unaided eye can just discern when light from the sky is polarized.

Questions

1 What two hypotheses about light are supported by the fact that light can be polarized?

2 Review the evidence that light is a wave

phenomenon and that the waves are transverse.

3 Explain why it is not possible to polarize a longitudinal wave.

4 Can radio waves be polarized? X-rays? Sound waves? Explain.

5 What are three ways that polarized light can be produced other than by use of a polarizing filter?

6 Why must an analyzer be used to study polarized light?

7 What is the function of the layer of Canada balsam in a Nicol prism?

8 Sunlight bouncing off a surface of water or a road pavement can be reduced in intensity by using sheets of Polaroid material. What clue does this give about the nature of reflected light? How should the Polaroid material be arranged to be effective?

9 Can a polished sheet of aluminum be used to produce polarization by reflection? Explain.

10 Describe a possible method for minimizing automobile headlight glare by using sheet-polarizing material in head lamps and providing the driver with a light-polarizing viewer. How should the polarizing axes be oriented?

11 In the 1950s, theatergoers used Polaroid glasses to see three-dimensional, or 3-D, movies. How would these glasses aid in providing such an illusion?

12 Suggest a method by which polarizing filters could be used to give the illusion of depth in stereoscopic photographs or in motion pictures. Is the method applicable to color pictures?

13 How can a single sheet of Polaroid be used to prove that light from the sky is partially polarized?

14 Explain what happens as light passes through a doubly-refracting material.

15 Under what condition will the degree of polarization be a maximum when light falls on a reflecting surface?

16 Show that when light is incident upon a plane-parallel plate of glass at the angle of polarization for the upper surface, the refracted ray strikes the lower surface at the angle of polarization for that surface.

17 What is meant by an optically active substance? Give examples.

18 Show by a sketch how two glass plates can be arranged to act as a simple polariscope when a beam of natural light falls on them.

19 Explain why and how a polarimeter can be used to show the nature of a sugar solution. On what does the rotation depend?

20 Compare reflection and scattering of light as to their effects on color and on polarization of light.

21 Cite examples of the production of color by each of the following: reflection, refraction, diffraction, interference, selective scattering, and the interference of polarized light.

22 Why do certain transparent materials show interference patterns when under stress?

23 A crumpled piece of clear cellophane will emit beautiful interference patterns when placed between a polarizer and an analyzer. Explain why this happens.

24 Diagram an experimental arrangement for detecting strains by the use of polarized light.

25 Two polarizing sheets are oriented so that when unpolarized light falls on the first sheet, none is transmitted by the second sheet. If a third polarizing sheet is placed between the first two sheets, can light be transmitted through the three sheets?

Problems

1 What is the polarizing angle for dense flint glass having an index of 1.768?

2 The angle of polarization for diamond is 67°34′. Compute the index of refraction of diamond. *Ans.* 2.424.

3 The angle of polarization for rock salt is 57°4′. What is the index of refraction?

4 Find Brewster's angle of incidence for white light on water having an index of refraction of 1.33. *Ans.* 53.1°.

5 A beam of light is incident upon the surface of carbon disulfide of index of refraction 1.64.

The angle of incidence is arranged to give maximum polarization of the reflected light. Compute the angles of polarization and refraction.

6 A certain glass has an index of refraction of 1.65. At what angle of incidence is the light reflected from this glass plane-polarized? *Ans.* 58.8°.

7 At a certain temperature the critical angle of incidence of water for total internal reflection is 48° for a certain wavelength. What are the polarizing angle and the angle of refraction for light incident on the water at the angle that gives maximum polarization of the reflected light?

8 Find Brewster's angle of incidence for microwaves falling on a water surface for which the index of refraction is 9.0. What is Brewster's angle for internal incidence? *Ans.* 83.7°; 6.3°.

9 Calculate the speed of the ordinary ray and also the highest speed of the extraordinary ray in a calcite crystal for the yellow light from a sodium lamp.

10 What are the limits of the polarizing angle for white light (range of 4,000 to 7,000 Å) for crown glass (n: blue = 1.5233; n: red = 1.5146). *Ans.* Blue = 56.71°; red = 56.57°.

11 Photometric measurements show that when light falls on a glass plate at the angle of polarization, 16 percent of the light that has its vibrations perpendicular to the plane of incidence is reflected. From the back surface of the same plate, 16 percent of the remaining light vibrating in that plane is reflected. What is the intensity ratio of vibrations in and at right angles to the plane of incidence in the beam transmitted through two plates?

12 A thin plate of calcite is cut with the optic axis parallel to the plane of the plate. What is the minimum thickness needed to produce destructive interference for sodium light when the plate is placed between crossed Nicol prisms? *Ans.* 1.7×10^{-4} cm.

13 Compute the critical angle for total reflection of the ordinary ray at the layer of Canada balsam in a Nicol prism for $\lambda = 5{,}890$ Å.

14 What thickness of quartz is required for a half-wave plate for yellow light if its indices of refraction are 1.544 and 1.553? *Ans.* 0.0033 cm.

15 Two Nicol prisms have their planes parallel to each other. One of the Nicol prisms is then turned so that its principal plane makes an angle of 40° with the principal plane of the other prism. What percent of the light originally transmitted by the second Nicol prism is now transmitted by it?

16 The two indices of refraction of quartz for yellow light are 1.544 and 1.553. For what thickness will the two rays emerge $1\frac{1}{4}$ wavelengths out of phase? *Ans.* 8.2×10^{-3} cm.

17 The two indices of refraction for yellow light in quartz are 1.544 and 1.553. For what thickness of quartz will the two beams differ in phase by one wavelength when they emerge?

18 What minimum thickness of calcite crystal is needed to introduce a phase difference of (*a*) 45°, (*b*) 90°, and (*c*) 180° between the emergent O and E rays when plane-polarized light is incident normally upon it? *Ans.* 4.28×10^{-4} mm; 8.55×10^{-4} mm; 1.71×10^{-3} mm.

19 A beam of plane-polarized light is incident on a calcite crystal, with its vibrations making an angle θ with the principal section of the crystal. (*a*) Show that the relative amplitudes of O and E beams is given by $A_O/A_E = \tan\theta$. (*b*) If $\theta = 30°$, find the relative amplitudes and intensities of the two beams.

20 A beam of light is sent through three Polaroid disks. The first and third disks are turned to transmit the maximum amount of light. The middle disk has its axis at 60° to the axis of the other two disks. What portion of the incident light is transmitted? Neglect ordinary absorption. *Ans.* 6 percent.

21 The sensitivity of a photographic plate is to be determined by exposing successive portions to light from a standard lamp transmitted through two Nicol prisms. Designating as 1.00 the illuminance on the plate when the prisms are parallel, through what angles should the analyzer Nicol prism be rotated to produce illuminances 0.80, 0.60, 0.20, and 0?

I cannot help thinking . . . that this discovery of magnetic-electricity [induction] is the greatest experimental result ever obtained by an investigator.

J. Tyndall

PART THREE

The Physics of Fields

Edward Mills Purcell, 1912–

Born in Taylorville, Illinois. Studied electrical engineering at Purdue University; attended the Technische Hochschule, Karlsruhe, Germany, and Harvard. Professor of physics, Harvard University. Shared the 1952 Nobel Prize for Physics with Bloch for their studies of nuclear magnetic moments.

Felix Bloch, 1905–

Born in Zurich, Switzerland. Studied with Heisenberg at University of Leipzig. Professor of physics, Stanford University. Shared the 1952 Nobel Prize for Physics with Purcell for their development of new methods to measure nuclear magnetic moments and the discoveries made with the aid of these methods.

32

Electric Charges and Fields

Gasoline trucks dangle chains along the road to prevent static electric charges from accumulating to dangerous proportions. Lightning rods are effective in discharging the electricity from low-hanging clouds and thus minimizing damage caused by lightning. Huge machines are used today for building up high voltages for "atom-smashing" experiments. These are examples of that branch of physics, called *electrostatics,* that deals with the properties of electricity at rest.

While electrostatic studies usually concentrate on electricity at rest, it should be noted that some of the electrostatic properties of electrical charges also exist when the charges are in motion. Important concepts considered in electrostatics include electric charge, electric fields, and electric polarization in dielectric materials. While the measurement of some electrostatic properties requires the transfer of electricity, we are concerned in this chapter with the effects after equilibrium (static) conditions are attained. The term *electrostatic generator* may seem a misnomer when it is applied to a Van de Graaff generator where "static" charges are transported on a moving belt. But here, too, we are usually concerned with characteristics observed after equilibrium has been attained and which would be the same for static charges.

The phenomena of electrostatics have acquired renewed status in this age of "modern" physics because of their importance in the phenomena of atomic structure and the nature of matter. For many centuries some of the basic facts of electrical charges at rest were known, and a few applications were made of these principles. But the current century has brought many more and important devices directly based on electrostatic principles, such as the ultrahigh-speed particle accelerators, air cleaners, electrostatic loudspeakers, and devices for minimizing the effects of static electricity in the paper, printing, and textile industries. Electrical charges are stored in devices called capacitors, temporarily at rest, and then removed and replaced in less than a millionth of a second, thus giving rise to a host of applications in the radio, television, radar, and communications industries.

32-1 ELECTRIFICATION

The ancient Greeks noted the fact that when certain substances were brought into close contact by being rubbed against others, they acquired the ability to attract light objects, such as bits of cork, thread, or paper. The philosopher Thales around 600 B.C. observed that these effects were particu-

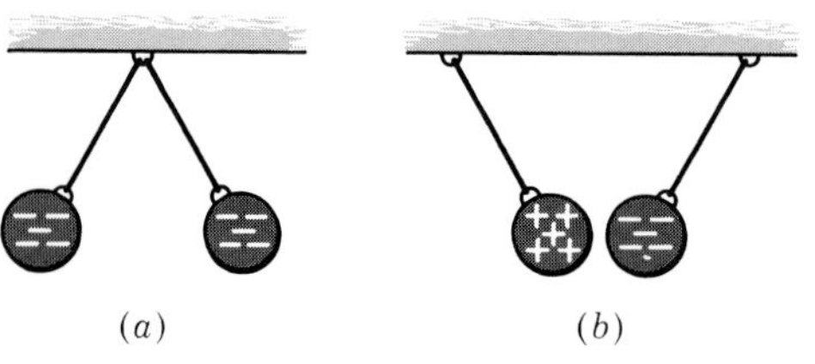

Figure 32-1
(*a*) Like charges repel; (*b*) unlike charges attract.

larly strong in *elektron,* the Greek word for amber. This is the origin of our modern word *electricity.*

The process of rubbing two materials together and then separating them to produce the above effect is now called *electrification,* or charging by contact. The objects are said to be *electrified,* or *charged* with electricity. It should be noted that the rubbing of the two materials does not *create* an electric charge but simply permits a *transfer* of electrons from one material to the other by placing the atoms of each material in intimate contact with each other.

There are two kinds of electrification. For example, when two light hard-rubber balls are electrified by being brushed against fur or flannel and supported by silk strings (Fig. 32-1*a*), they will be observed to repel each other. But a glass ball rubbed with silk will attract either of the rubber balls (Fig. 32-1*b*), although two such glass balls will repel each other. The charge on the glass is evidently unlike that on the rubber. These facts suggest the first law of electrostatics: *Objects that are similarly charged repel each other; bodies with unlike charges attract each other.*

32-2 POSITIVE AND NEGATIVE CHARGES

Benjamin Franklin (1706–1790), a diplomat, printer, and amateur scientist, played a leading role in developing theories on the nature of electricity. Franklin suggested that all bodies contained a single electrical fluid (charge) which might be present in excess or of which there might be a deficiency. In the first case the charge was called positive and in the second case negative. However, Charles DuFay (1698–1739) did not accept this so-called one-fluid, or unitarian, theory but rather a two-fluid, or dualistic, theory. This theory proposed that a neutral body contained equal amounts of two fluids, which were to be considered as opposites; and the electrification of a body consisted in separating the two fluids. In this theory the terms negative and positive were also used arbitrarily. The electricity produced by rubbing a glass rod with his hand DuFay called *vitreous* electricity, and positive. The electricity produced by rubbing wool on resin he called *resinous* electricity and claimed that it was charged negatively.

The present designation of electric charges is based upon the arbitrary classification of the electrification produced in a glass rod that has been in contact with silk as being of *positive* sign. The electrification of a rubber rod after contact with fur is therefore arbitrarily classified as *negative* in sign. Uncharged objects contain equal amounts of positive and negative electricity. When glass and silk are brought into contact, some negative electricity is transferred from the glass to the silk, leaving the glass rod with a net positive charge and the silk with an equal net negative charge. Similarly, hard rubber receives negative electricity from the fur with which it is in contact, causing the rod to be negatively charged and leaving the fur positive. Though a similar explanation could be made by assuming a transfer of positive electricity, it has been shown that in solids only negative electricity is transferred.

These phenomena have justified the present theory that the rubbing of objects does not create electricity but merely changes the electrical neutrality of the substances in contact. Hence it is now accepted that the *law of conservation of charge* applies, and no exceptions have been discovered to this law. This law states simply the fact that the algebraic sum of the electric charge in any closed system remains constant. A spectacular example of this conservation law is the case when a negative electron and a positive electron are caused to collide. In some of these interactions the two particles are transmuted into energy by the annihilation process described later. Their mass is converted into the energy of two very penetrating rays, called gamma rays. Other examples of charge conservation are found in cases of radioactive decay, as discussed in Chap. 49.

32-3 DISPLACEMENT OF CHARGES

The effect of a charged object on one that is uncharged is illustrated in Fig. 32-2. Separation of positive and negative electricity within the uncharged object is produced by the charged rod, which exerts a force of repulsion on the like portion of the charge and an attraction on the unlike. In (*a*) the negatively charged rubber rod causes the adjacent side of the uncharged object to become positively charged, while the opposite side becomes negatively charged. Because the unlike charge is nearer the rod, the force of attraction will exceed that of repulsion and produce a net attraction of the uncharged object by the rod. In (*b*) is shown the case in which a positively electrified glass rod is used. It should be remembered that the separation of the charges described here does not alter the total amounts of positive and negative electricity in the uncharged object. No charge is gained or lost; all that occurs is a shift of negative electricity toward one side of the object, making that side predominantly negative and leaving the other side predominantly positive.

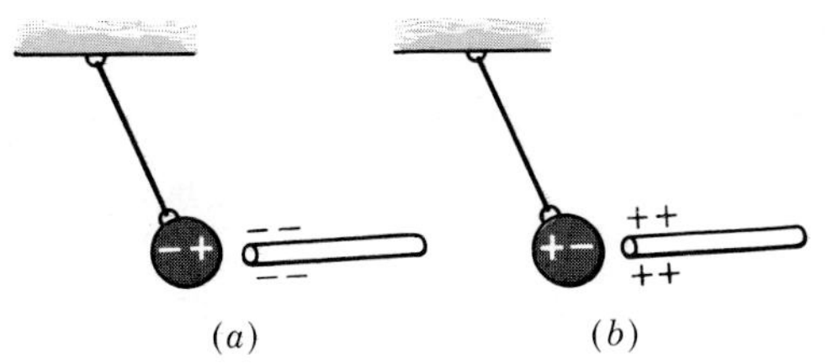

Figure 32-2
A charged body brought near an insulated conductor causes charges in the conductor to separate. This results in an attraction of the conductor by the charge.

32-4 ELECTRON THEORY AND ATOMIC STRUCTURE

According to the theory of atomic structure proposed by Sir Ernest Rutherford and Niels Bohr, all matter is composed of *atoms,* each consisting of a *nucleus,* a small, tightly packed, positively charged mass, and a number of larger, lighter, negatively charged particles called *electrons,* which revolve about the nucleus at tremendous speeds (Fig. 32-3). The centripetal force necessary to draw these electrons into their circular or elliptical paths is supplied by the electrical attraction between them and the nucleus. The nucleus consists of a number of *protons,* each with a single positive charge, and (except for hydrogen) one or more *neutrons,* which have no charge. Thus the positive charge of the nucleus depends upon the number of protons that it contains. This number is called the *atomic number* of the atom. An ordinary, uncharged atom has a number of electrons outside the nucleus equal to the number of protons within the nucleus. Each electron carries a single negative charge of the same magnitude as the positive charge of a proton, so that the attraction between the nucleus of an atom and one of the electrons will depend on the number of protons in the nucleus. An electron at rest has a mass of 9.1083×10^{-31} kg. Since the mass of a proton or a neutron is about 1,836 times that of an electron, the mass of the atom is almost entirely concentrated in the nucleus. The chemical properties of the atom are determined by the number and arrangement of the extranuclear electrons.

This idealized model of atomic structure served a useful step in the historical development of our picture of subatomic phenomena. However, these simple concepts had several fatal weaknesses that were revealed by later developments. These objections could not be eliminated by simple revisions of the atomic model. Revolutionary changes in some of the basic postulates of the "planetary" model of the atom have resulted from the growth of the new quantum mechanics. But for the purposes of the current chapter the symbolic model of the atom pictured

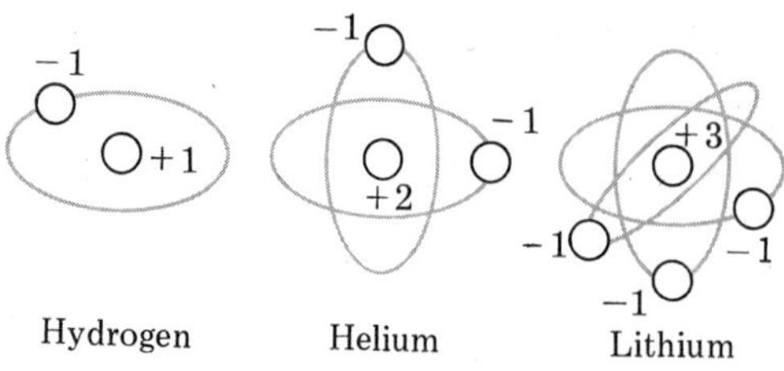

Figure 32-3
Each atom consists of a positively charged nucleus surrounded by electrons. The three simplest atoms, hydrogen, helium, and lithium, are represented diagrammatically. (See the discussion of the fact that this atomic model must not be considered as literal.)

above is quite helpful. For example, this atom model indicates the fact that the outermost electrons may rather easily be removed from the atoms or molecules of a body, thus leaving a net positive charge on this body. The atoms or molecules of the other body that was in contact with the first acquire an excess of electrons, and hence the second body is charged negatively.

32-5 CONDUCTORS AND INSULATORS

Any material body of macroscopic size is made up of an enormous number of atoms. In solids these atoms cling tightly together. Although the atoms in solids vibrate about their normal positions as a result of thermal agitation, their general configuration is not permanently altered by this motion.

Each atom ordinarily contains an equal number of electrons and protons, but, under the influence of many nearby atoms, some electrons lose their association with particular atoms and become *free electrons. The number and freedom of motion of these electrons determine the properties of the material as a conductor of electricity*. An *insulator,* or poor conductor, is a substance that contains few free electrons. The positive charges in atoms are firmly bound and do not participate in ordinary conduction in solids. In Table 1 some common substances are arranged roughly in the descending order of their electrical conductivities.

There is no sharp boundary between materials which are insulators and those which are conductors. All materials can conduct electricity to some extent. In fact there are certain materials, called *semiconductors,* that ordinarily are insulators but become conducting under particular conditions, for example, under the influence of light or increased temperature.

When conductors are to be charged, by contact, for example, they must be mounted on insulating supports. However, an insulator can be charged at one place and retain its charge without having the charge rapidly leak away by conduction. Unlike the case of conductors, when insulators are charged by contact, only the regions near the places of contact become charged.

The reason for describing electrification as occurring through the *transfer of negative electricity* can now be seen. An uncharged object contains a large number of atoms (each of which normally contains an equal number of electrons and protons), but with some electrons temporarily free from atoms. If some of these free electrons are removed, the object is considered to be positively charged, though actually this means that its negative charge is below normal. If extra free electrons are gained by an object, it is said to be negatively charged, since it has more negative charge than is normal. The positive charges in atoms are firmly bound in the nucleus and do not participate in ordinary conduction in solids.

Table 1
ELECTRICAL CONDUCTORS

Good conductors	Poor conductors	Insulators (very poor conductors)
Silver	Tap water	Glass
Copper	Moist earth	Mica
Other metals	Moist wood	Paraffin
Carbon, graphite	Dry wood	Hard rubber
Certain solutions	Leather	Amber

32-6 THE LEAF ELECTROSCOPE

A common device for studying electrostatic phenomena is the *leaf electroscope* (Fig. 32-4). This instrument consists essentially of a strip of very light gold leaf or other thin metal foil, hanging from a contact on a flat metal plate which terminates in a ball at the upper end. This plate is carefully insulated from the metal case, which has glass windows for observation.

When a charged body of either sign is brought near the knob of the electroscope, the leaf diverges from the plate. This is because the charge in the body causes a separation of the charges in the electroscope plate. When the charging body is removed, the charges in the plate flow together and the leaf collapses. To charge the electroscope permanently, one could touch the knob with a charged rod. If the charging rod has a positive charge, some electrons will leave the plate and flow into the charged rod. When the rod is withdrawn, the leaf will remain diverged for a considerable time because of the positive charge on the electroscope. A negatively charged body touching the knob will leave the electroscope negatively charged, since some of the extra electrons on the body will flow into the plate. Note the fact that in this process of *charging by conduction* the sign of the charge left on the electroscope is the same as that of the charging body.

An insulated conductor may be given a permanent charge by *induction*. This process takes no charge from the object which induces the charge in the conductor, as there is no electrical connection between them. The method of charging by induction will be illustrated for the case of the leaf electroscope (Fig. 32-5). When a positively charged glass rod is brought near the electroscope, the electrons are attracted to the knob, leaving the leaf positively charged. The net charge on the leaf and plate is zero, since no charge has been transferred to or from the surroundings. However, the potential of the entire plate and leaf is some positive value, since this conductor is situated near the positively charged glass rod. The second step in the process consists of grounding the plate by touching the knob with a finger.

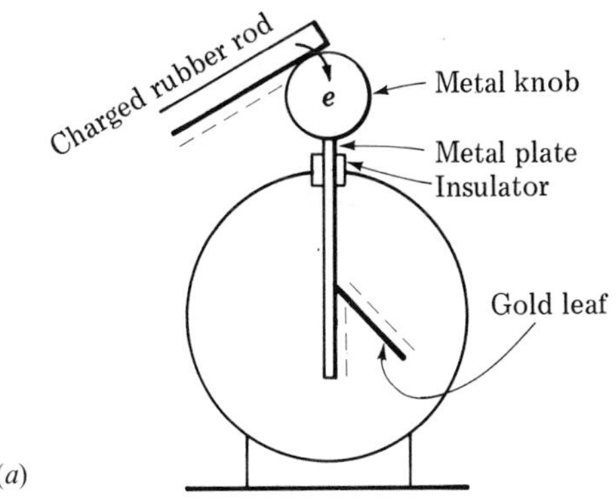

(*a*)

The excess electrons on the rod enter the neutral knob and distribute themselves over the electroscope; when the rod is removed the excess electrons remaining on the electroscope cause the leaf to be repelled by the plate

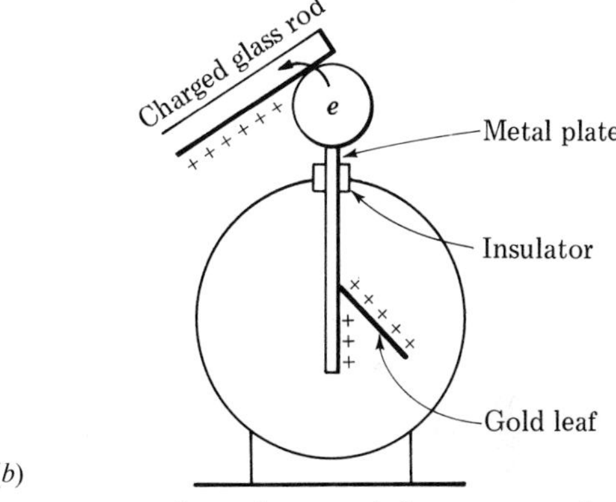

(*b*)

Free electrons leave the neutral electroscope and enter the electron-deficient rod; when the rod is removed, the electroscope is then electron deficient, positively charged, and the leaf and the plate diverge

Figure 32-4
(*a*) Charging an electroscope negatively by conduction. (*b*) Charging a leaf electroscope positively by conduction.

Electrons flow up from the ground through the finger, attracted by the positively charged glass rod. This flow of electrons will cease in an instant when the previous positive potential of the plate has been reduced to zero. This potential must be zero, since the plate and ground have a common

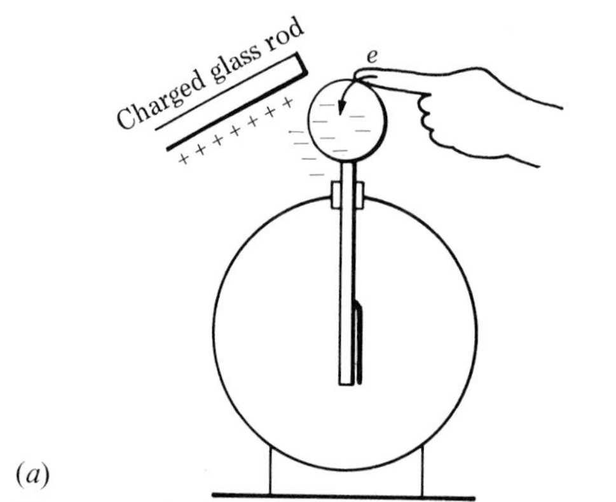

(a)

The approach of a + charged rod attracts free electrons in an electroscope toward that side of the knob; when a finger touches the knob, electrons leave the hand and enter the knob, when the hand is removed, the excess of electrons causes the electroscope to be charged negatively, and the leaf and plate diverge

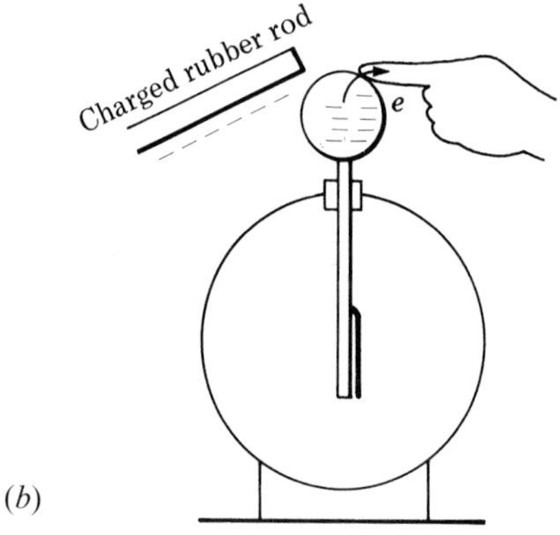

(b)

The approach of a negatively charged rod drives the free electrons to the opposite side of the knob where they escape to the ground through a person's hand; when the hand is removed, the electroscope is deficient of electrons, and is charged positively

Figure 32-5
(a) Charging an electroscope negatively by induction. (b) Charging an electroscope positively by induction.

potential which, by conventional agreement, is taken as zero. The third phase consists of breaking the ground connection by removing the finger from the knob. This does not change the potential or the charge. But when the charged glass rod is removed and the electrons in the knob redistribute themselves over the plate and leaf, the electroscope is left negatively charged and having a negative potential. Note that the sign of the final charge is opposite to that of the charging body, unlike the case of charging by conduction. The procedure by which the electroscope can be charged positively by induction is shown in Fig. 32-5*b*. The source of the energy stored up in the charged electroscope is represented by the additional work done in pulling the positively charged glass rod away from the negative charge on the electroscope.

Electroscopes can be made which have an amazingly high sensitivity, so that noticeable deflections can be observed from exceedingly small charges, for example, those associated with radioactive phenomena.

Specially designed electroscopes can be used to determine the amount of radiation coming from a radioactive source. In 1937 C. C. Lauritsen and T. Lauritsen invented a quartz-fiber electroscope based upon the gold-leaf electroscope but having the fiber as the leaf and a rigid wire as the plate. This electroscope is charged by connecting it for a short time to a battery. When the charged electroscope is exposed to ionizing radiation, ion pairs (+ and −) are formed and the charged fiber and rigid wire collect these ions and lose their charge. Hence the mutual repulsion between the fiber and the wire decreases and the fiber gradually returns to its original position. The more ionizing radiation that falls on the electroscope, the more rapidly it is discharged.

The principle of the quartz-fiber electroscope is used extensively to safeguard those working with radioactive materials. The pocket dosimeter, a pen-shaped electroscope which most of the workers carry, is charged and regularly read by health safety officers to determine the degree of exposure to which each person has been subjected. Since other forms of radiation emitted by radioactive substances, such as gamma rays and neutrons, are extremely injurious to a person's health, these dosimeters are designed to measure exposure to that type of radiation.

Here we have seen an example of a quite simple instrument being used to perform a highly significant and specialized function.

32-7 FORCE BETWEEN POINT CHARGES

The *quantity of electricity*, or *charge Q*, possessed by a body is simply the aggregate of the amount by which the negative charge exceeds or is less than the positive charges in the body. The term *point charge* is used to indicate that the charge is not distributed over a large area but rather is concentrated at a specifically located point.

Charles Augustin de Coulomb was the first investigator to place the law of force between electrostatic charges upon an experimental basis. His relatively rough experiments established in 1784 the law, now known as *Coulomb's law of electrostatics*, that the force F between two point charges Q and Q' varies directly with each charge, inversely with the *square* of the distance s between the charges, and is a function of the nature of the medium surrounding the charges. In symbols, Coulomb's law is

$$F = k\frac{QQ'}{s^2} \qquad (1)$$

where the dimensional factor k is introduced to take care of the units of F, Q, and s and also to provide for the properties of the medium around the charges, insofar as these properties affect the force between the charges. It must be noted that this k is not a dimensionless proportionality constant. The dimensions of k are force $\times$ distance^2/charge^2.

Since F is always a vector quantity, it must be noted that Eq. (1) gives only its magnitude. The direction of the force is always along the line joining Q and Q'.

The analogy between Coulomb's law for electrostatics, as given by Eq. (1), and Newton's law of universal gravitation, as expressed by Eq. (8) in Chap. 3, is quite apparent. Scientists like Einstein have searched for a common basis between gravitational and electrostatic forces, but these efforts have largely been unsuccessful.

32-8 SYSTEMS OF UNITS IN ELECTROSTATICS

Two families of units are useful in the areas of electrostatics: the mks system and the system of cgs electrostatic units (esu). The electrostatic units will first be considered because of their historical significance and their simplicity.

It can be seen from Eq. (1) that there are two new concepts to be defined that have not previously been considered, namely, those represented by the symbols k and Q. It is most convenient in the electrostatic system to select the concept represented by k as the one to be arbitrarily designated as fundamental (like length, mass, and time in mechanics). Then Q can be *defined* from Coulomb's law. From this agreement, k is arbitrarily assigned the value of exactly 1 dyn·cm^2 per unit charge^2 for empty space. The esu of charge, usually called the *statcoulomb* (statC), is defined as a point charge of such a magnitude that it is repelled by a force of one dyne if it is placed one centimeter away from an equal charge in empty space. The size of the statcoulomb makes it convenient for many problems in electrostatics.

Example A charge A of +250 statC is placed on a line between two charges B of +50.0 statC and C of −300 statC. The charge A is 5.00 cm from B and 10.0 cm from C. What is the force on A? See Fig. 32-6.

$$\text{Force on } A \text{ due to } B = F_1 = k\frac{QQ'}{s^2} =$$

$$\frac{(1.00\ \text{dyn}\cdot\text{cm}^2/\text{statC}^2)(250 \times 50.0)\text{statC}^2}{(5.00\ \text{cm})^2}$$

$$= 500\ \text{dyn} \qquad \text{toward } C$$

$$\text{Force on } A \text{ due to } C = F_2 =$$

$$\frac{(1.00\ \text{dyn}\cdot\text{cm}^2/\text{statC}^2)(250 \times 300)\text{statC}^2}{(10.0\ \text{cm})^2}$$

$$= 750\ \text{dyn} \qquad \text{toward } C$$

Since these two component forces are in the

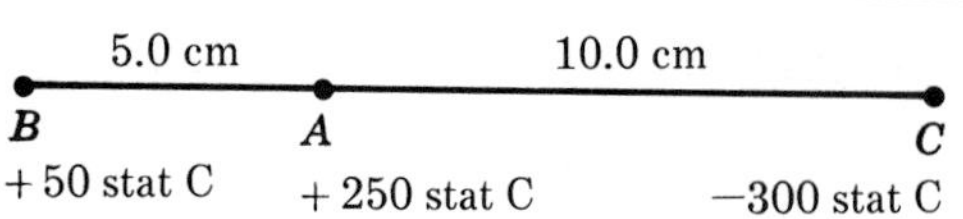

Figure 32-6

same direction, the resultant force is their arithmetical sum

$$F_1 + F_2 = 1{,}250 \text{ dyn} \qquad \text{toward } C$$

In the mks system of units it has been found convenient to make the substitution $k = 1/(4\pi\varepsilon)$ and to write Eq. (1) as

$$F = \frac{1}{4\pi\varepsilon}\frac{QQ'}{s^2} \tag{1a}$$

Here the symbol ε (Greek epsilon) is used to represent the *permittivity* of the medium surrounding the charges. This concept may be considered as a measure of that property of the medium surrounding electric charges which determines the forces between the charges.

Here again we are faced with the alternative of an arbitrary choice of one new concept, ε or Q, which must be accepted (without definition) as fundamental and from which all other concepts in electricity may be defined. International agreement on this selection has not yet been reached. For this book we shall ignore this dilemma and agree to use the concept of charge as defined from the concept of electric current, to be defined later. The mks unit of current is the ampere, and the mks unit of charge, called the *coulomb,* is the charge transferred through any cross section of a conductor in one second by an unvarying current of one ampere.

The coulomb is a convenient unit for use in practical current electricity; for example, it is about the charge that passes through a 100-W electric lamp in 1 s. However the coulomb is an enormous charge in terms of those ordinarily met with in electrostatic phenomena. For this reason the submultiples microcoulomb and micromicrocoulomb are often used ($1\ \mu\text{C} = 10^{-6}$ C and 1 pC $= 10^{-12}$ C). One coulomb is approximately 3×10^9 statC.

When the coulomb is established as the mks unit of charge, it becomes possible to evaluate the value of the permittivity of empty space, for which the symbol ε_0 is used. This value turns out to be approximately $8.85 \times 10^{-12}\ \text{C}^2/\text{N}\cdot\text{m}^2$. The factor $1/(4\pi\varepsilon_0)$ has a value of $8.98740 \times 10^9\ \text{N}\cdot\text{m}^2/\text{C}^2$; this number is usually rounded off to the approximate 9.0×10^9. The permittivity of some commonly used insulators (rubber, mica, glass) ranges from two to ten times that of empty space. The value of ε for air is practically identical with ε_0 for empty space.

Example What is the force between a point charge of $-50\ \mu$C and another of $+25\ \mu$C when they are 50 cm apart, in a vacuum?

$$F = \frac{1}{4\pi\varepsilon_0}\frac{QQ'}{s^2}$$

$$= 9.0 \times 10^9\ \text{N}\cdot\text{m}^2/\text{C}^2 \times \frac{(50 \times 10^{-6}\ \text{C})(25 \times 10^{-6}\ \text{C})}{(0.50\ \text{m})^2}$$

$$= 45\ \text{N}$$

32-9 HINTS TO THE READER

Note the fact that the algebraic signs of the charges in this example have not been included in the computation. The direction of the force is one of attraction along the line joining the charges.

At first glance it might seem that a good unit of charge to choose for the basic quantity of electricity would be the charge of one electron. Since this charge has the value of 1.6020×10^{-19} C, it is obvious that its size and value are not convenient for practical problems.

Certain precautions must be observed in applying Coulomb's law to situations that are not ideal. For example, this law is not valid when the charges are near other conductors, since induced charges may appear on them and such charges will change the force between the given charges. It must again be emphasized that Coulomb's law applies only to *point* charges, or to those which act as if they are concentrated at a point. For example, if charges are uniformly distributed over the surface of a sphere, their effect on other charges outside the sphere is the same as if the charges were concentrated at the center of the sphere.

32-10 ELECTRIC FIELDS

The forces that act between electric charges can be described by visualizing the charges as being situated in an electric field. Such a region has special properties by reason of the presence of the charges. An *electric field* is defined as a region in which there would be a force upon a charge brought into that region.

A close analogy to an electric field is the gravitational field of the earth. A body in the earth's field is acted upon by this field; this force is the weight of the body.

An electric field has two important characteristics: direction and intensity. The *direction of an electric field* at a point is defined as the direction of the force upon a *positive* charge placed at that point. In the case of the earth's gravitational field the direction of the force on a mass near the earth is always directed toward the center of the earth. In the case of the electric field near a positive charge the force on a positive test charge will be directed away from the charge that produces the field, but toward a negative charge.

32-11 ELECTRIC FIELD INTENSITY

The quantitative measure of the strength of an electric field is known as the field intensity. *Electrostatic field intensity* at a point is defined as the force per unit positive test charge at that point.

$$\mathbf{E} = \frac{\mathbf{F}}{+q} \qquad (2)$$

where **E** is the electric field intensity and **F** is the force exerted upon the positive test charge q at the point in question. It is important to observe the fact that electric field intensity is always a *vector* quantity. Component fields must be added vectorially.

In the mks system the unit for E is the newton per coulomb; the esu for E is the dyne per statcoulomb. The electrostatic unit of electric field intensity is a larger unit than the mks unit. The following conversion may be made:

$$\frac{1\text{ N}}{1\text{ C}} = \frac{10^5\text{ dyn}}{3 \times 10^9\text{ statC}} = \frac{1}{3 \times 10^4}\text{ dyn/statC}$$

or 1 esu of electric field intensity is 3×10^4 as large as the corresponding mks unit, since 1 dyn/statC is 3×10^4 N/C.

Example The electric field in the space between the plates of a discharge tube is 3.25×10^4 N/C. What is the force of the electric field on a proton in this field? Compare this force with the weight of the proton, if the mass of the proton is 1.67×10^{-27} kg and its charge is 1.60×10^{-19} C.

$$F = Eq = 3.25 \times 10^4\text{ N/C} \times 1.60 \times 10^{-19}\text{ C} = 5.20 \times 10^{-15}\text{ N}$$

The weight of the proton is given by

$$W = mg = 1.67 \times 10^{-27}\text{ kg} \times 9.80\text{ m/s}^2 = 1.64 \times 10^{-26}\text{ N}$$

Hence

$$\frac{F}{W} = \frac{5.20 \times 10^{-15}\text{ N}}{1.64 \times 10^{-26}\text{ N}} = 3.17 \times 10^{11}$$

This emphasizes the fact that the force caused by an electric field may be much greater than that due to a gravitational field.

32-12 ELECTRIC FIELD INTENSITY NEAR AN ISOLATED POINT CHARGE

A useful expression for the electric field intensity due to an isolated point charge Q may be obtained as follows: Imagine a small positive charge q placed at a distance s from Q. The force between these charges, by Coulomb's law, is

$$F = k\frac{Qq}{s^2} = \frac{1}{4\pi\varepsilon}\frac{QQ'}{s^2}$$

From the definition of electric field strength

$$E = \frac{F}{q} = k\frac{Qq/s^2}{q} = k\frac{Q}{s^2} = \frac{1}{4\pi\varepsilon}\frac{Q}{s^2} \qquad (3)$$

Since electric field intensity is a vector quantity, the resultant of two or more field intensities is obtained by taking the vector sum of the various field intensities. When there are a number of charges near a point, the electric field intensity at that point is obtained by taking the *vector* sum of component intensities.

Example Two charges are 60 cm apart, in air. One charge Q_1 is $+1.67 \times 10^{-7}$ C; the other Q_2 is -1.67×10^{-7} C. What is the electric field intensity at P midway between the charges?

Imagine a test charge $+q$ to be placed at P (Fig. 32-7). The resultant force per unit charge at point P is the vector sum of the component fields E_1 and E_2. Both these fields are in the same direction.

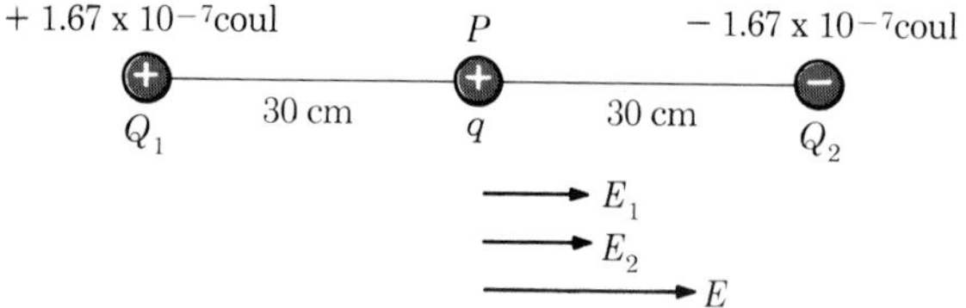

Figure 32-7
Electric field intensity midway between two charges.

$$E = \frac{1}{4\pi\varepsilon}\frac{Q_1}{{s_1}^2} + \frac{1}{4\pi\varepsilon}\frac{Q_2}{{s_2}^2}$$

$$= 9.0 \times 10^9\ \mathrm{N\cdot m/C^2} \times \frac{1.67 \times 10^{-7}\ \mathrm{C}}{(0.30\ \mathrm{m})^2}$$

$$+ 9.0 \times 10^9\ \mathrm{N\cdot m/C^2} \times \frac{1.67 \times 10^{-7}\ \mathrm{C}}{(0.30\ \mathrm{m})^2}$$

$$= 3.33 \times 10^4\ \mathrm{N/C}$$

The resultant field is directed along the line from Q_1 to Q_2.

Example Equal charges of +250 statC are placed in air at the corners of an equilateral triangle. (*a*) What is the electric field intensity produced by one of the charges at a point 18 cm from each corner? (*b*) What is the resultant field intensity at this point (Fig. 32-8)?

$$E = k\frac{Q}{s^2}$$

$$= 1.00\ \mathrm{dyn\cdot cm^2/statC^2} \times \frac{250\ \mathrm{statC}}{(18\ \mathrm{cm})^2}$$

$$= 0.77\ \mathrm{dyn/statC}$$

This field is directed away from the charge, along

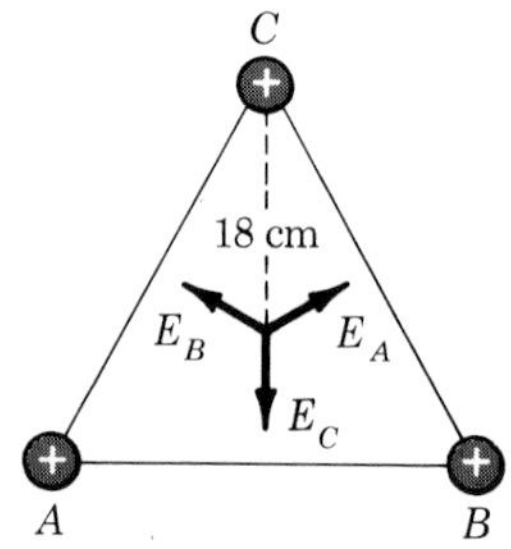

Figure 32-8
Electric field intensity at a point equidistant from three equal charges.

the line joining the center point and the charge in question. From the symmetry of the vector figure it is clear that resultant field at midpoint is zero.

32-13 LINES OF FORCE

The direction of electric fields at various points may be represented graphically by lines of force. A *line of force* in an electric field is a line so drawn that a tangent to it at any point shows the direction of the electric field at that point.

As an example of this definition, consider the isolated positive charge Q placed on sphere A (Fig. 32-9). A small, positive test charge q at b would experience a force of repulsion. Hence the region at b is an electric field. From symmetry it is evident that the direction of the force upon a positive charge anywhere near A would be radially *away from A*. Hence the lines of force are drawn in these directions. If A were the location of an isolated negative charge, the electric field would extend everywhere radially *toward A*.

The diagram of Fig. 32-10 shows a plane section of the electric field near a pair of charges equal in magnitude but of opposite sign. The field at any point is the resultant of the superposition of the two component fields due to the charges at A and B. For example, at point b the direction of the resultant force on a positive charge would be along the vector drawn tangent to the line of force at that point.

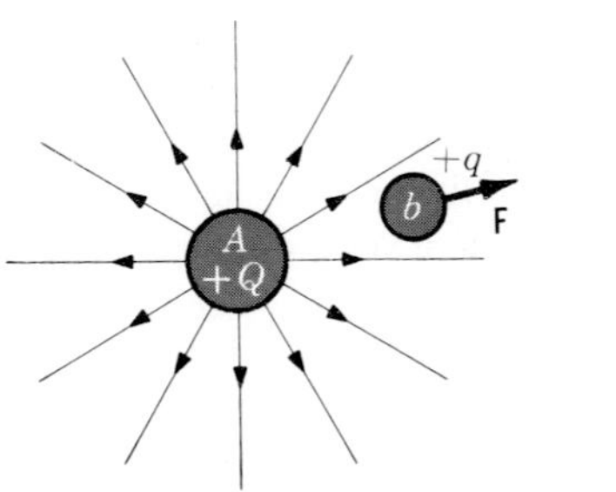

Figure 32-9
Electric field near a charge.

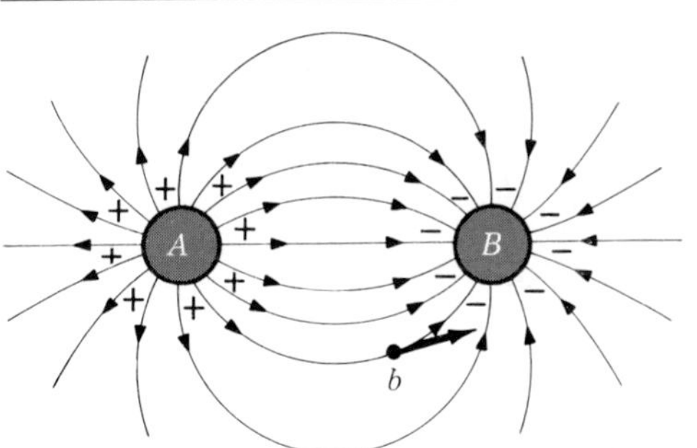

Figure 32-10
Electric field near two equal charges of opposite sign.

32-14 LINES OF FORCE AND ELECTRIC FIELD INTENSITY

It is frequently convenient to represent graphically the intensity as well as the direction of the electric field by means of lines of force (*electric flux*). This is done by the conventional agreement that the intensity of the field at a point in space will be represented by a number of lines of force per unit area through a surface normal to the field in the neighborhood of the point.

We shall define electric *flux density D* as the number of lines of force per unit area crossing the surface at right angles to the direction of the field (Fig. 32-11). Where the field intensity is large, the lines of force are closely spaced, as in the regions near A and B in Fig. 32-10. Where the field is less intense, the lines are more widely spaced, as at b in Fig. 32-10.

Our conventional agreement concerning this electric flux density can be expressed by the relation

$$D = \frac{\psi}{A_n}$$

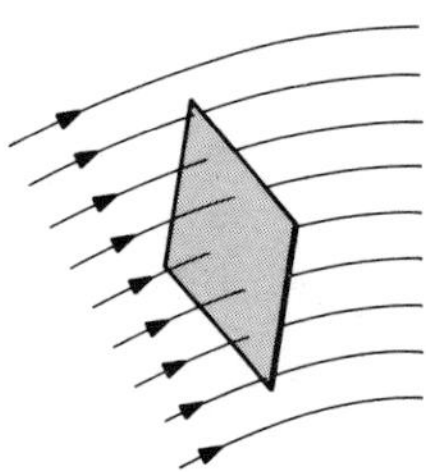

Figure 32-11
Conventional representation of electric field intensity by the number of lines of force passing perpendicularly through a unit area.

where D is the number of lines of force per unit area (flux density), ψ (Greek psi) is the total number of lines of force crossing the surface, and A_n is the area perpendicular to the field at the point considered.

For reasons of convenience in many of the equations in electrostatics it is helpful to introduce into the factor of proportionality between E and D the dimensional constant ε_0, the permittivity of empty space (Sec. 32-10), so that the equation for D in empty space becomes

$$D = \varepsilon E \tag{4}$$

Equation (4) serves as a defining equation to state the number of lines of force per unit area normal to the field that we have selected to represent the electric field graphically.

As an example of the use of Eq. (4), let us consider the electric field in a vacuum around an isolated point charge (Fig. 32-9). The electric field intensity E near this charge is

$$E = \frac{Q}{4\pi\varepsilon_0 r^2}$$

Let us draw an imaginary sphere of radius r around the charge Q. The area of this sphere is $4\pi r^2$, and the flux is everywhere perpendicular to this area. Hence we may write

$$\psi = DA_n = \varepsilon_0 EA_n = \varepsilon_0 \frac{Q}{4\pi\varepsilon_0 r^2} \times 4\pi r^2$$
$$= Q \tag{5}$$

Thus in the choice made in Eq. (4) the total number of lines of force emerging from a charge is just equal to the charge.

From this convention it follows that in the mks system an electric field intensity of 1 N/C in empty space may be represented by 8.85×10^{-12} lines of force normal to a surface of 1 m^2. In the electrostatic system of units an electric field of intensity 1 dyn/statC is conventionally represented by a flux density of 1 line of force per square centimeter of a surface that is perpendicular to the field.

32-15 GAUSS' LAW

A significant relation between the net number of lines of force passing through any closed surface in the outward direction and the net positive charge enclosed within that surface was discovered by Karl Friedrich Gauss (1777–1855). In the preceding section it was shown (for the case of an isolated point charge) that the total number of lines of electric flux emerging from a charge is exactly equal to that charge (in the mks system of units). From Fig. 32-9 it is evident that the same net number of lines of force will pass out of any closed surface of any shape if the surface completely encloses the charge. The generalization of this conclusion is known as *Gauss' law* and may be stated as follows:

The net number of lines of force in an electric field that cross any closed surface in an outward direction is equal (in the mks system) to the net positive charge enclosed within that surface.

In symbols Gauss' law may be stated by the equation previously given as Eq. (5), $\psi = \Sigma Q$.

An advantage of this law is the fact that the net number of lines of force through a surface can be expressed in terms of the electric field intensity at that surface. An alternative use of this law is the computation of the electric field intensity produced by symmetrical charge distributions in terms of the electric flux produced by these charges. For example, we shall consider the case of an isolated hollow spherical charged conductor, as shown in Fig. 32-12. The sphere carries a charge $+Q$. If there are no other charges nearby, symmetry considerations justify the statement that all the charge will be uniformly distributed over the outer surface of the sphere. This can be shown also by considering a gaussian surface drawn just below the outer surface of the sphere. If there were any charges inside the gaussian surface, there would be a flux through this surface and hence an electric field within the metal. Such a field would produce a flow of charge. But this is contrary to our assumption of an electrostatic condition. Hence it follows that there can be no charge and no electric field within the metal and all the charge must be on the outer surface.

The lines of flux from this sphere must be radial. A gaussian spherical surface of radius r is drawn around the sphere, as shown by the dotted circle in Fig. 32-12. The flux lines intersect the gaussian surface perpendicularly. The total flux through this surface is the product of the electric field intensity E, the permittivity of the medium around the sphere ε, and the area of the surface $4\pi r^2$. From Gauss' law

$$\psi = DA_n = \varepsilon EA_n = \varepsilon E \times 4\pi r^2 = Q$$

Hence $$E = \frac{1}{4\pi\varepsilon}\frac{Q}{r^2}$$

This is the same expression we obtained as Eq. (3) for the electric field intensity at a place at distance r from a point charge Q. Hence it follows that the electric field outside a uniformly charged conducting sphere is identical with the field produced by a point charge of the same magnitude at the center of the sphere.

It also follows from Gauss' law that the electric field within the metal and in the hollow space within the inner surface is zero, because there is no net charge situated within these regions. We shall see in the next chapter why this fact is of importance in connection with a study of the work done in moving a test charge within the metal or in the interior hollow space.

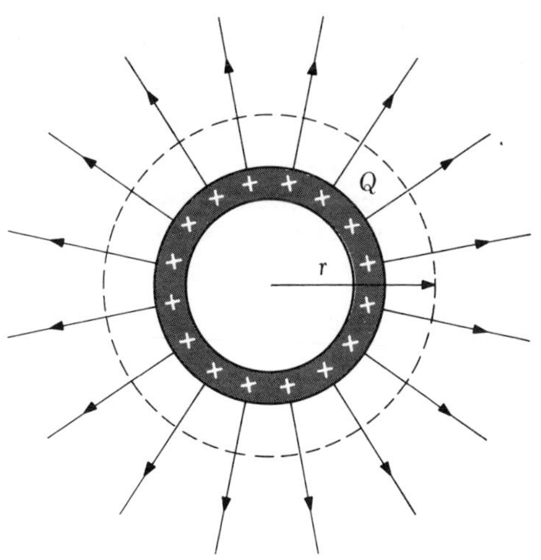

Figure 32-12
Charged sphere, surrounded by a gaussian surface.

32-16 ELECTROSTATICS IN NATURE: LIGHTNING

Man has always been curious about and often awed by the power of lightning. Benjamin Franklin's experiment approximately 200 years ago with a kite in which he showed that lightning consisted of the same type of electricity that could be produced on earth by electrostatic means is possibly the most famous study made of lightning. Although this phenomenon has been the topic of considerable research, scientists have failed to agree upon the cause and the nature of lightning. We know that lightning is a violent example of the tremendous electrostatic charges that can occur in nature, but there are several different explanations given about the manner in which the charges that cause lightning are produced in a cloud. Most of these theories are built upon the premise that because of violent air currents in a cloud, and the interaction of ions and water droplets and ice particles, positive and negative charges are produced in a cloud and are then separated with the positive charges moving upward and the negative charges moving downward.

To understand this generalization, let us first look at the makeup of a thunderhead cloud. In the "mature" stage the cloud top will reach up to 10,000 to 15,000 m (40,000 ft) where the temperature is about −50°C. Within this cloud, which may have its lower layer at a height of 5,000 ft and a temperature of around +20°C, there are strong updrafts reaching a speed of 60 mi/h. The water vapor in the clouds is carried upward, is condensed, and is cooled below its freezing point but does not freeze immediately until it has a "seed," or nucleus, to form upon, and as a result it becomes supercooled. However, once ice starts to form, with large slow-moving ions serving as nuclei, the crystals grow rapidly and fall through the rising air causing a downdraft of cold air. When this cold air reaches the bottom of the cloud, precipitation in the form of rain occurs. The strength of this downdraft is shown by the fact that the rush of cold air reaches to the ground just ahead of the rain, causing the sometimes noticed chill that precedes a rainstorm.

We have not yet considered exactly how the charges are created in a cloud. What has been said up to this point is factual; the following is based upon man's best interpretation of available data. Combining the consistent portion of several theories, we will attempt to develop an overview. In a thunderhead, friction occurs between the water droplets falling through the cloud and the updraft of air currents. This friction is instrumental in producing an electric charge on the falling droplets. Also, in this process the water droplets are broken up. (At this point several theories appear about what charge is created and where the charged particles migrate, i.e., do the positive charges move upward or downward etc.) Let us consider the most popular theory that the lighter "spraylike" parts of the droplets torn off during falling become positively charged and are then carried upward into the cold upper layers of the cloud. The heavier particles become negatively charged and continue to move downward and in so doing acquire more negativity as they grow in size. Some scientists feel that water droplets must be frozen before electric charges will separate to any great degree. This does not seriously affect our theory because it has been shown that ice crystals when falling acquire charges similar to water. The ice in forming will have surface water attached to it which is stripped from the falling ice crystal, the water will then assume a positive charge and migrate upward where it will freeze and start to fall as ice, hence repeating the process. In this theory, the ice will acquire a negative charge as it falls. As a result of these procedures, whether ice or water droplets fall, the upper layer of the cloud becomes positively charged and the lower layer negatively charged. The charge produced in this separation of charges can be very large, with a potential difference of as much as 100 million volts being created between the cloud and the earth.

As a result of this charge separation, it is found that the layers of clouds about 20,000 to 23,000 ft high (temperature −20°C) contain the highest concentration of positive charge and the layers about 10,000 to 13,000 ft high (temperature 0° to −10°C) have the highest concentration of negative charges. The earth has a negative charge but is much less negative than the lower layer of the cloud. When the potential difference is great enough, a discharge occurs in the form of lightning. There are at least three conditions that determine whether such a lightning discharge may occur: the environmental conditions, the strength of the charge on the earth, and the availability of a good conductor on the earth (such as a steeple or perhaps a golfer carrying a golf club).

Careful study has been made of the manner in which a lightning bolt leaves the cloud and moves toward the earth. We have all seen the

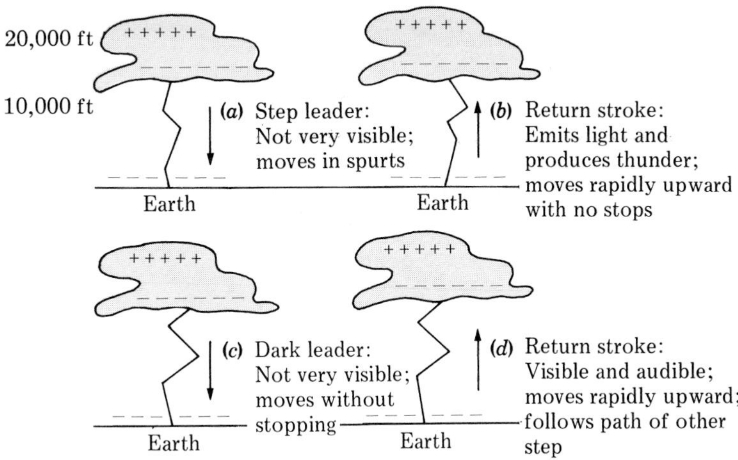

Figure 32-13
Four steps in a lightning charge.

jagged flash of light as lightning occurs, but the thunderbolt that we see is actually a second phase in the lightning discharge. As the lightning process starts, a smaller lightning discharge, called a "step leader," leaves the cloud in a stop-and-go fashion, moving between 150 and 250 ft in 50 μs, and continues to move down to the earth in jagged spurts, taking on the average 0.01 s to reach the ground. When the step leader, which is carrying negative charges, reaches the ground, the charges rapidly escape and this rapid discharge moves with great speed back up the path that has been followed by the step leader, something like that of a fire following a trail of gunpowder back to the barrel. This "return stroke" creates enough heat and light energy to produce a visible lightning bolt and thunder, a result of the rapidly expanding air around the bolt. It has been estimated that the return stroke can produce around 10,000 A. It should be noted that the thunderbolt that we see actually runs backward *up* the path *from* the ground. But the process does not end here. A second discharge leaves the cloud following the path used by the step leader and the return stroke, but this discharge, called the "dark leader," goes all the way to the ground in one rapid stroke and does not go in spurts. This is followed by a second return stroke. The process can be repeated several times in a very short time. One observation of a lightning occurrence showed that more than 40 lightning bolts followed the same path. So lightning can indeed strike the same place twice.

SUMMARY

Electrostatics deals with the phenomena common to electric charges in motion and at rest.

Objects may be *electrified* or *charged* either *positively* or *negatively* by the removal or addition of *electrons*.

Like charges repel; unlike charges attract.

The *leaf electroscope* is a device used to detect the presence of an electric charge. The *quartz-fiber electroscope* is designed to detect radiation emanating from a radioactive source.

Coulomb's law for the force between point charges is expressed by the equation

$$F = k\frac{QQ'}{s^2} = \frac{1}{4\pi\varepsilon}\frac{QQ'}{s^2}$$

In the cgs electrostatic system of units the dimensional factor k is arbitrarily assigned a numerical value of unity for empty space. In the mks family $k = 1/4\pi\varepsilon$, where ε is the *permittivity* of the medium around the charges. For empty space $\varepsilon_0 = 8.85 \times 10^{-12}\ C^2/N \cdot m^2$ and $1/4\pi\varepsilon_0 =$ approximately $9 \times 10^9\ N \cdot m^2/C^2$.

The *coulomb* may be arbitrarily defined, or it may be defined in terms of the ampere. One coulomb is approximately 3.0×10^9 statcoulombs.

The *electrostatic unit charge* (statcoulomb) is one that will act upon a similar charge with a force of 1 dyn when the charges are 1 cm apart, in empty space.

An *electric field* is a region in which a force is exerted upon a charge placed in the field.

The *direction of an electric field* is the direction of the force on a positive charge placed at the point considered.

Electric field intensity is force per unit positive charge.

$$E = \frac{F}{+q}$$

The *electric field intensity* near an isolated point charge is given by

$$E = \frac{kQ}{s^2} = \frac{1}{4\pi\varepsilon}\frac{Q}{s^2}$$

The electrostatic unit of electric field intensity is the dyne per statcoulomb; the mks is the newton per coulomb.

A *line of force* in an electric field is a line so drawn that a tangent to it at any point shows the direction of the electric field at that point.

The total number of lines of electric force is referred to as *electric flux* ψ. *Flux density* D is the flux per unit area normal to the field at the place considered, $D = \psi/A_n$.

By conventional agreement electric field intensity may be represented by the number of lines of flux per unit area normal to the field at the point considered, divided by the permittivity of that region,

$$E = \frac{\psi}{A_n\varepsilon} = \frac{D}{\varepsilon}$$

In terms of this convention *Gauss' law*, applied to the mks system of units, states that the net flux in an electric field that crosses any closed surface, in an outward direction, is equal to the net positive charge enclosed within that surface, or

$$\psi = \Sigma Q$$

Lightning is a violent example of the tremendous electrostatic charges that can occur in nature.

Questions

1 Do the properties of electricity hold only for those situations when electrical charges are at rest? If not, give examples where the properties can be observed where charges are in motion.

2 Define and note the differences between the unitarian and the dualistic theories of the nature of electricity.

3 According to Rutherford and Bohr, what kind of force is necessary to keep atomic electrons in orbit and what provides this force?

4 Describe the fundamental difference between electrical conductors and insulators. Give examples of each other than those listed in the chapter.

5 Explain in detail the techniques used to charge a leaf electroscope by conduction and by induction and the resulting charge in each case.

6 When a charged rod is brought near bits of cork dust, the cork will at first cling to the rod. Very quickly thereafter the bits of cork will fly off from the rod. Explain why this is true.

7 A positively charged rod is brought near a ball suspended by a silk thread. The ball is attracted by the rod. Does this indicate that the ball has a negative charge? Justify your answer. Would an observed force of repulsion be a more conclusive proof of the nature of the charge on the ball? Why?

8 Will a solid metal sphere hold a larger electric charge than a hollow sphere of the same diameter? Where does the charge reside in each case?

9 If a positively charged electroscope collapses at a greater rate when placed in a chamber with a supposedly radioactive substance than it does in air, what can you conclude about the substance? Is it radioactive? If so, what kind of radiation is being emitted?

10 As one gets out of an automobile, he sometimes gets a "shock" when he touches the car. Why is this? Is it likely to happen on a rainy day? Why?

11 A positively charged rod is brought near the terminal of a charged electroscope, and the leaves collapse as the rod approaches the terminal. When the rod is brought still closer (but not touching), the leaves again diverge. What is the sign of the charge on the electroscope? Explain its action.

12 Describe and explain an experiment in which a large and massive body can be set into motion by the use of the charge on a rubber rod. State the reasoning involved.

13 What is the significance of the factor K in the equation for Coulomb's law? Give at least two other examples where a factor with a similar function is used in an algebraic expression of a physical law. Point out any differences that might exist between K and these other factors.

14 In the definition of electric field intensity (force per unit charge), it is assumed that the test charge on which the force acts does not appreciably distort the field. For what charge would this statement be rigorously true?

15 What proportionality constant would have to be introduced into Eq. (2) if the force were given in pounds and the other quantities were in mks units?

16 A map of equipotential lines has places in which the lines are much closer together than in other regions. Is the field strong or weak where the lines are close together? Explain.

17 State some similarities between gravitational and electric fields and the laws of force in these cases.

18 Show by a sketch the lines of force near (*a*) a positive charge, (*b*) a negative charge, and (*c*) a positive charge near an equal negative charge.

19 Sketch the appearance of the lines of force (*a*) between two charged plates, one positive and the other negative, and (*b*) between a small positive charge and a negatively charged plate.

20 Sketch an approximate curve to show how the electric field intensity near an isolated point charge varies as the distance away from the charge is increased.

21 Can the force caused by an electric field be greater than the force due to a gravitational field? Give an example to prove this.

22 As a review, state the precautions which must be taken and the restrictions placed upon the use of Coulomb's law.

23 A pith ball gilded with metallic paint is attracted by a glass rod that has been rubbed with silk. Does this experiment show that the pith ball has a positive, negative, or zero charge? Explain fully by the aid of diagrams.

24 A copper wire is connected to a brass sphere that has a large positive charge. The wire is touched in turn to equal-sized glass, hard-rubber, and aluminum spheres. The spheres are carefully insulated from the ground. Compare the charges that the spheres receive.

25 Three charges of +20, +30, and +40 μC are placed at the corners of an equilateral triangle. At the center of the triangle, toward which charge is the electric field most nearly directed?

26 Assume that the electron of charge e and mass m in the hydrogen atom revolves in a circular orbit of radius r and that the centripetal force is supplied by electrostatic attraction between the electron and the nucleus. Derive the expression for the speed v of the electron.

27 What is the basic premise upon which most theories regarding the nature of lightning are constructed?

28 According to the theories of lightning presented in the chapter, can lightning strike the same place twice? Explain.

Problems

1 In an electric lamp 2.50 C pass through each second. (*a*) How many electrons does this represent? (*b*) How many statcoulombs?

2 A small charged pith ball is suspended 2 cm above a second charged pith ball resting on an insulated surface. If the charge of the lower ball is +20 statC and it has a mass of 0.25 g, what charge must be on the upper ball to lift the lower ball? *Ans.* −49 statC.

3 Two small gilded pith balls carrying charges of +0.00250 and −0.00600 μC are 250 mm apart, in air. Compute the force between them.

4 Two gilded pith balls, spherical in shape and each having a mass of 1 g, are supported from a common point by a thread 1 m long. Determine the magnitude of a positive charge that must be given to each ball to cause them to be separated by 10 cm. *Ans.* 70 statC.

5 A charge of 0.700 μC is placed on a small copper sphere. A similar uncharged sphere is placed near the charged sphere. (*a*) Show clearly why there is a force between these spheres. Is the force one of attraction or repulsion? (*b*) The spheres are placed in contact and then separated until their centers are 60.0 mm apart. Determine the magnitude and sense of the force between the spheres.

6 A solid brass sphere of 5.00-mm radius on which there is a positive charge of 800 statC is put in contact with a hollow brass sphere of the same radius on which is a negative charge of 500 statC. The spheres are then separated so that there is a distance of 200 mm between their centers. What is the force between them? *Ans.* 56.2 dyn.

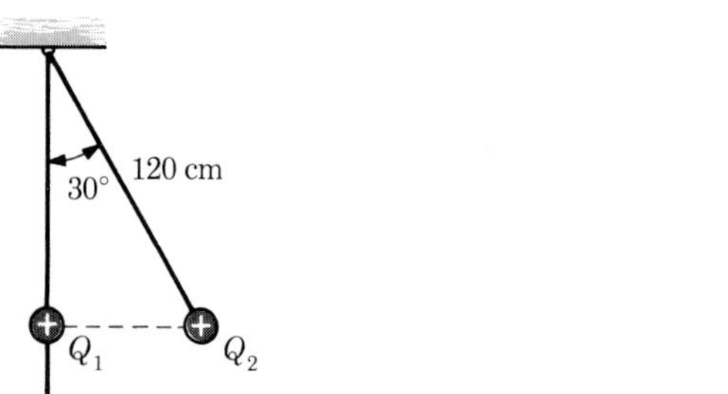

Figure 32-14

7 A charge *A* of +300 statC is placed on a line between two charges *B* of +100 statC and *C* of −400 statC. The charge *A* is 10.0 cm from *B* and 20.0 cm from *C*. What is the force on *A*?

8 Three +100-statC charges are arranged in a straight line, the second charge (*B*) is 20 cm to the right of the first (*A*) and the third (*C*) is 50 cm to the right of *A*. (*a*) What force is exerted by charges *A* and *B* on *C*. (*b*) What force is exerted by *A* and *C* on *B*?

Ans. 15.1 dyn; 13.9 dyn toward *C*.

9 A small charged body, placed 3.0 cm vertically over a charge of +100 statC, has its apparent weight increased by 49 dyn. What are the sign and magnitude of the charge on the body?

10 What is the force between a point charge of −100 μC and another of +50 μC when they are 100 cm apart, in a vacuum? *Ans.* 45 N.

11 Calculate the position of the point in the neighborhood of two point charges of 1.67 μC and −0.600 μC, situated 400 mm apart where a third charge would experience no force.

12 What is the force of the electric field on a proton if the field is set up between the plates of a discharge tube and has a magnitude of 5.00×10^4 N/C? *Ans.* 8.0×10^{-14} N.

13 A small brass sphere *A* has a charge of 0.0250 μC. A similar sphere *B* has a charge of −0.0075 μC. The distance between the centers of the spheres is 75.0 mm. (*a*) What is the force between these charges? (*b*) The spheres are brought into contact and then separated to a distance of 750 mm between their centers. Calculate the force between the spheres for this case.

14 Find the electric field intensity at a point *P* midway between two charges of $+2.0 \times 10^{-7}$ C and -2.0×10^{-7} C if they are 100 cm apart.

Ans. 1.44×10^4 N/C away from the negative charge.

15 Two Ping-Pong balls having equal charges and equal masses of 0.250 g are suspended from a nail by nylon threads 125 cm long. Electrostatic repulsion keeps the balls 200 mm apart. Calculate the charge on each ball.

16 Two small spheres having a mass of 300 mg each and having equal charges are suspended from the same point by silk strings 100 cm long. When the spheres are kept 15.0 cm apart by electrostatic repulsion, what is the charge on each? *Ans.* 70.5 statC.

17 A small, charged pith ball is suspended by a light thread, as shown in Fig. 32-14. The pith ball is charged with a quantity of electricity Q_2 of 0.075 μC. A charge Q_1 of 0.125 μC is held fixed. From these data and the dimensions shown in the figure, calculate the weight of the pith ball.

18 The charge on a proton is $+1.6 \times 10^{-19}$ C and on an electron is -1.6×10^{-19} C. (*a*) What is the attractive force between the nucleus of a hydrogen atom (having no neutrons—protium) and an electron in a circular orbit of 6.0×10^{-7} cm radius. (*b*) How fast must this electron be traveling in revolutions per second to keep from being drawn into the nucleus? The mass of the electron is 9.11×10^{-31} kg. *Ans.* 6.4×10^{-8} N; 5.45×10^{15} r/s.

19 What is the intensity of the electric field which will just support a water droplet having a mass of 10 μg and a charge of 1.0×10^{-7} μC?

20 Two similar small conductors have charges of +25 and −10 statC, respectively. They are placed in contact and then separated until their centers are 8.0 cm apart. What is the force between them at this position? *Ans.* 0.88 dyn.

21 Three points *A*, *B*, and *C* are on the corners of an equilateral triangle. At *A* there is a point charge of +0.100 μC. What is the magnitude of the electric field intensity at a point midway between *B* and *C*, if *BC* is 10.0 cm?

22 The sides of a right triangle are $BC = 3$, $AC = 4$, and $AB = 5$ cm. If +10-statC charges are placed at the corners *B* and *C*, what is the magnitude and direction of the electric intensity at *A*?

Ans. 0.991 dyn/esu; 11.64° below line *AC*.

23 A charge *A* of 400 μC is 12.0 m from a charge *B* of −100 μC. (*a*) What is the strength of the electric field at a point *C* that is 5.00 m from *B* and 13.0 m from *A*? (*b*) How many electric lines of force pass through an area of 1.00 cm² normal to the field at *C*?

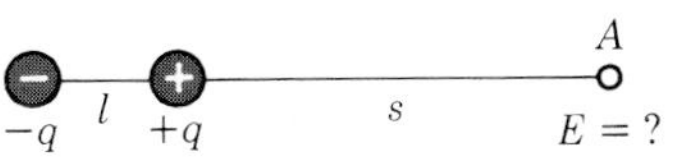

Figure 32-15

24 A rectangle, 10 cm high and 20 cm wide, has +200-statC charges placed on each corner. What is the magnitude and direction of the electric intensity (*a*) at the center of the rectangle and (*b*) at a point midway between the corners of one of the long sides.

Ans. zero; 1.414 dyn/esu, vertical.

25 Two point charges *A* and *B* are 500 mm apart. The charge *A* is +525 statC, and the charge *B* is −275 statC. Calculate the number of lines of force that pass through a circle of 1.25-mm diameter, with its plane normal to the electric field, midway between the charges.

26 Two small, charged metal spheres are placed 20 cm apart in air. If they have +20- and +40-statC charges on them, respectively, what force acts on them? What is the electric intensity halfway between them? At what point between them would the electric intensity be zero?

Ans. 2 dyn, repulsive; 0.8 dyn/esu toward +20-statC charge; 8.3 cm from +20-statC charge.

27 The electric field between two large horizontal parallel plates, 5.00 cm apart, is 2.50×10^5 N/C. A negatively charged oil drop of mass 4.67×10^{-11} g is in equilibrium at a point midway between the plates. (*a*) Which plate is positive? (*b*) Calculate the charge on the drop. (*c*) What is the electrostatic force on the drop if it is 1.0 cm from the lower plate?

28 What is the field strength at a reference point if a force of 40 N is necessary to hold a charge of 200 μC at that point? *Ans.* 2×10^5 N/C.

29 An electric dipole consists of two point charges equal in magnitude and opposite in sign separated by a distance *l* (Fig. 32-15). Show that when *l* is small in comparison with *s*, $E_A = 2kql/s^3$.

Frits Zernike, 1888–1966

Born in Amsterdam. Studied at the University of Amsterdam. Professor of mathematical physics, University of Groningen. Awarded the 1953 Nobel Prize for Physics for his method of phase contrast and especially for his invention of the phase-contrast microscope.

33

Electric Potential

One of the great unifying concepts of physics is that of energy. Earlier in our study we defined energy as that property of a system by virtue of which work may be done. There are numerous ways in which energy in some form is converted into electric energy. For example, a battery transforms chemical energy into electric energy. In other cases electric energy is transformed into some other form of energy. The electric motor, for instance, converts electric energy into mechanical energy.

In electrical phenomena the concept that is of primary importance in all energy transfers is that known as potential difference. In our study of gravitational fields we discussed the concept of the potential energy of a body in such a field and found that certain problems, such as those involving gravitational fields, could be much simplified by the use of energy considerations. The same will be found to be true for the electric field, which is so similar, mathematically, to the gravitational field. For example, work must be done to lift a body above the earth against gravitational force. Similarly, energy exchanges must occur in order to move an electric charge in an electric field. Many difficulties due to the vector nature of electric fields can be avoided by dealing with electric potential energy and electric potential rather than with force and electric field intensity. The use of electric potential extends into the field of current electricity and is basic to an understanding of circuits and all practical electrical devices.

33-1 ENERGY IN AN ELECTRIC FIELD

If a charge is moved between two points in an electric field, work is usually done. To be specific, consider the simple field shown in Fig. 33-1. If a positive test charge q is moved along path a from B to C, over most of the distance the motion will be against a component of the outward force due to the field and work will have to be done by some outside agency to move the charge. It is one of the most important properties of the static electric field that this work is independent of the path taken. Thus the work needed to move q from B to C along path b is exactly the same as that needed along path a. It is also true that if the charge moves back from C to B, work will be done on it by the field, so that either it will accelerate along the path or work will be done on some agent which prevents it from gaining speed. This work will be exactly the same in amount as that needed to move it from B to C. The student will note that this situation is analogous to that of a mass moved in a gravitational

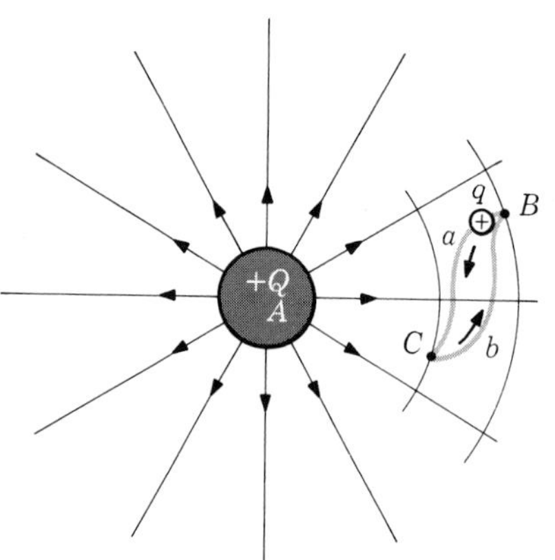

Figure 33-1
Potential difference in an electric field.

field. Since the work done in raising an object to a higher altitude can be recovered when it descends, the work is considered to be stored as potential energy at the higher altitude. Similarly, when work is done by an outside agency to move a test charge between two points in an electric field, potential energy is stored up. For this reason the static electric field is called a *conservative* field, and the law of conservation of energy applies to the movement of charged bodies in such a field. Since the work needed to move a charge between two given points in a steady field is always the same, and since the work is proportional to the charge moved, it is fruitful to use these facts to define a property of the two points in the field, namely, the electrical potential difference between the points.

33-2 POTENTIAL DIFFERENCE

The *potential difference* V between two points is the work done per unit charge when a charge is moved from one point to the other. The defining equation is

$$V = \frac{\mathcal{W}}{q} \qquad (1)$$

Conventionally q is always understood to be a *positive* charge, thus making definite the algebraic sign of the potential difference in any particular case.

The unit of potential difference in the mks system is the volt. One *volt* is defined as the potential difference between points in an electric field such that one joule of work must be done to move a charge of one coulomb between the points considered.

Small potential differences are often expressed in millivolts ($1\ \text{mV} = 10^{-3}\ \text{V}$) or microvolts ($1\ \mu\text{V} = 10^{-6}\ \text{V}$). Large potential differences can be expressed in kilovolts ($1\ \text{kV} = 10^{3}\ \text{V}$) or megavolts ($1\ \text{MV} = 10^{6}\ \text{V}$).

Since the difference of potential between two points is expressed in volts, the potential difference is often referred to as the *voltage* between these points. Thus, if an electric power line has a voltage of 120 V, it follows that 120 J of work have to be expended for each coulomb of electricity which is transferred through any apparatus connected between the two wires.

Example A Van de Graaff generator accelerates electrons so that they have energies equivalent to that attained by falling through a potential difference of 15.0 MV. What is the energy of an electron that has been accelerated in this machine?

$$\mathcal{W} = V \times q = 1.50 \times 10^{7}\ \text{V} \times 1.60 \times 10^{-19}\ \text{C} = 2.40 \times 10^{-12}\ \text{J}$$

In the cgs family of electrostatic units the unit of potential difference is the statvolt. One *statvolt* is the potential difference between points in an electric field when an energy of one erg must be expended to move a charge of one statcoulomb between the points considered.

The relationship between the volt and the statvolt is calculated as follows:

$$1 \text{ statV} = \frac{1 \text{ erg}}{1 \text{ statC}} \times \frac{3 \times 10^9 \text{ statC}}{1 \text{ C}} \times \frac{1 \text{ J}}{10^7 \text{ ergs}} = 300 \text{ J/C}$$

Since 1 J/C is 1 V, it follows that 1 statV = 300 V.

Example Two points in an electric field have a potential difference of 3.0 V. What work is required to move a charge of 5.0 C between these points?

$$V = \frac{\mathcal{W}}{q}$$

or

$$\mathcal{W} = V \times q = 3.0 \text{ V} \times 5.0 \text{ C} = 15 \text{ J}$$

Since both work and charge are scalar quantities, it follows that *potential difference is a scalar quantity*.

33-3 REFERENCE POINT FOR POTENTIAL DIFFERENCE

In our study of gravitational effects on the earth, we noted the fact that there is need for an *arbitrary* choice of the zero of gravitational potential (level). The conventional choice of this zero is usually taken as sea level. We then speak of a body as having a particular height above sea level. At other times we consider the difference of level between two points, neither of which is at sea level. Similarly in electricity we find it convenient to agree upon an arbitrarily selected zero of potential. This zero reference potential is frequently taken as the potential of the earth. For other purposes the zero of potential is considered as the potential of a point greatly distant from all charges (at infinity). When a conducting body is connected to the earth (grounded), there is a flow of charge until the potential difference between the body and the earth is zero. There are often large potential differences between the earth and portions of the atmosphere. Lightning is an example of an electrical discharge that often takes place between the earth and clouds whose potentials differ by billions of volts.

33-4 POTENTIAL AT A POINT

Instead of considering the potential difference between two points, we sometimes wish to consider the potential V_A at a particular point. The *potential at a point* is the difference between the potential of the point and an arbitrarily selected zero of potential. The potential at a point is therefore the work done per unit charge when a charge is moved from a point at zero potential to the point in question. This work equals the increase in the potential energy of the charge. Hence the definition of the potential of a point is equivalent to the statement: The potential at a point is the ratio of the potential energy of a test charge placed at that point to the magnitude of the test charge. In equation form,

$$\text{Potential} = \frac{\text{potential energy}}{\text{charge}} \qquad V = \frac{\mathcal{W}}{q} \qquad (2)$$

It will be noted that Eqs. (1) and (2) are alike and that the respective terms are measured in the same units.

Example A 120-V potential difference is maintained between the ends of a long high-resistance wire. The center of the wire is grounded. What is the potential of the center and each end of the wire?

$$\text{Potential at center} = 0$$

$$\text{Potential at one end} = +60 \text{ V}$$

$$\text{Potential at other end} = -60 \text{ V}$$

33-5 ELECTRIC POTENTIAL NEAR AN ISOLATED POINT CHARGE

In order to obtain an expression for the special but very important case of potential near an isolated uniformly charged sphere or point charge, we consider points A and B at distances r_A and r_B from a positive charge Q and find the work needed to move a small positive test charge q from A to B (Fig. 33-2). We consider the work to be done in a large number of steps, each so small that the field intensity within each step is nearly constant. In the first step the work done $\Delta\mathcal{W}_1$ is

$$\Delta\mathcal{W}_1 = Eq(r_A - r_1) = k\frac{Qq}{r^2}(r_A - r_1)$$

Since the r in the denominator should be neither r_A nor r_1, but an average radius between them, the quantity $r_A r_1$ is used as a good approximation of the proper r^2.

$$\Delta\mathcal{W}_1 = kQq\frac{r_A - r_1}{r_A r_1} = kQq\left(\frac{1}{r_1} - \frac{1}{r_A}\right)$$

In the next step the work done to move q from r_1 to r_2 is

$$\Delta\mathcal{W}_2 = kQq\left(\frac{1}{r_2} - \frac{1}{r_1}\right)$$

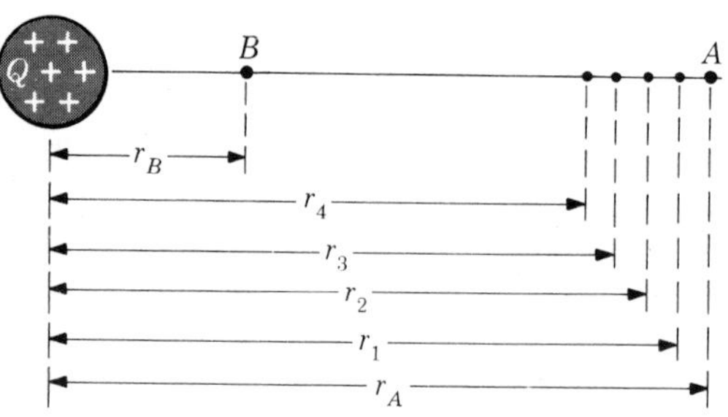

Figure 33-2
Potential near an isolated point charge.

As the charge q moves to r_n and finally to r_B, the total work becomes

$$\begin{aligned}\mathcal{W} &= \Delta\mathcal{W}_1 + \Delta\mathcal{W}_2 + \Delta\mathcal{W}_3 + \cdots \\ &= kQq\left[\left(\frac{1}{r_1} - \frac{1}{r_A}\right) + \left(\frac{1}{r_2} - \frac{1}{r_1}\right)\right. \\ &\quad \left. + \left(\frac{1}{r_3} - \frac{1}{r_2}\right) + \cdots + \left(\frac{1}{r_B} - \frac{1}{r_n}\right)\right]\end{aligned}$$

Removing parentheses and collecting terms, we obtain

$$\mathcal{W} = kQq\left(\frac{1}{r_B} - \frac{1}{r_A}\right)$$

Dividing the work done by the charge moved, we get an expression for the potential difference ΔV between A and B,

$$\Delta V = kQ\left(\frac{1}{r_B} - \frac{1}{r_A}\right) \tag{3}$$

If we now allow the point A to be located at an infinite distance or so far from Q that $1/r_A$ is negligible, the expression gives the work per unit charge to bring a charge from infinity to the point B. If we further assign the value zero to the potential at infinity, the expression gives the *absolute potential* at B. For any point in the field of Q, at distance r, we have

$$V = k\frac{Q}{r} \tag{4}$$

The student will note an exact analogy between the foregoing derivation of Eq. (4) and that given in Chap. 9 for the gravitational potential at distance r in the gravitational field of the earth. The negative sign in that equation has its counterpart in Eq. (4) here if the charge Q is negative, for in this case the force between Q and the test charge is one of attraction, just as is the gravitational force.

In Sec. 32-15 mention was made of the fact that charges uniformly distributed on a sphere produce electrostatic effects outside the sphere the same as if the charges were concentrated at the center of the sphere. Hence the potential at the surface as well as outside of the sphere is given by Eq. (4). Inside the sphere the field intensity is zero everywhere. This fact may be seen by considering a gaussian surface constructed inside the sphere through any point. Since there is no charge within the surface, no lines of force cut the surface and the electric field must be zero. Consequently a test charge moved about inside the sphere would experience no force and no work would be done during the motion. Hence the potential at the surface of a charged metallic sphere, and the potential at any place within the sphere, is given by $V = kQ/r$, where r is the radius of the sphere.

If mks units are used, we substitute for k its equivalent $1/(4\pi\varepsilon_0)$, which gives

$$V = \frac{1}{4\pi\varepsilon_0}\frac{Q}{r} \qquad (5)$$

In the mks system V is in volts, Q in coulombs, and r in meters. The value of $1/4\pi\varepsilon_0$ is $9.0 \times 10^9\ \mathrm{N\cdot m^2/C^2}$.

Example A spherical shell on the top of a small electrostatic generator in air has a charge of 2.0 μC. What is the potential at a point 10 cm from the center of the sphere?

$$V = k\frac{Q}{r} = \frac{1}{4\pi\varepsilon_0}\frac{Q}{r}$$

$$= \frac{(9.0 \times 10^9\ \mathrm{N\cdot m^2/C^2})(2.0 \times 10^{-6}\ \mathrm{C})}{0.10\ \mathrm{m}}$$

$$= 1.8 \times 10^5\ \mathrm{V}$$

The quantities in Eq. (4) are easily handled in esu. In this system k is numerically equal to unity for empty space, Q is in statcoulombs, and r is in centimeters.

If a point such as B in Fig. 33-2 is situated near a number of charges, each charge contributes a share to the resultant potential. The resultant potential of a point in empty space situated near a number of charges is the algebraic sum of all such potentials; in symbols this may be stated

$$V = \sum \frac{1}{4\pi\varepsilon_0}\frac{Q}{r} \qquad (6)$$

Example In Fig. 33-3 is shown a charge A of 8.0 μC situated 1.00 m from a charge B of −2.0 μC. What is the potential at point C located at the midpoint between A and B? What is the potential of point D located 80 cm from A and 20 cm from B? How much work would be required to move a charge of 0.030 μC from D to C?

$$V_C = \sum \frac{1}{4\pi\varepsilon_0}\frac{Q}{r}$$

$$= (9.0 \times 10^9\ \mathrm{N\cdot m^2/C^2}) \times \left(\frac{8.0 \times 10^{-6}\ \mathrm{C}}{0.50\ \mathrm{m}} + \frac{-2.0 \times 10^{-6}\ \mathrm{C}}{0.50\ \mathrm{m}}\right)$$

$$= 1.1 \times 10^5\ \mathrm{V}$$

$$V_D = \sum \frac{1}{4\pi\varepsilon_0}\frac{Q}{r}$$

$$= (9.0 \times 10^9\ \mathrm{N\cdot m^2/C^2}) \times \left(\frac{8.0 \times 10^{-6}\ \mathrm{C}}{0.80\ \mathrm{m}} - \frac{2.0 \times 10^{-6}\ \mathrm{C}}{0.20\ \mathrm{m}}\right)$$

$$= 0$$

$$\mathcal{W}_{DC} = q(V_C - V_D)$$

$$= 0.030 \times 10^{-6}\ \mathrm{C}\,(1.1 \times 10^5\ \mathrm{V} - 0)$$

$$= 3.3 \times 10^{-3}\ \mathrm{J}$$

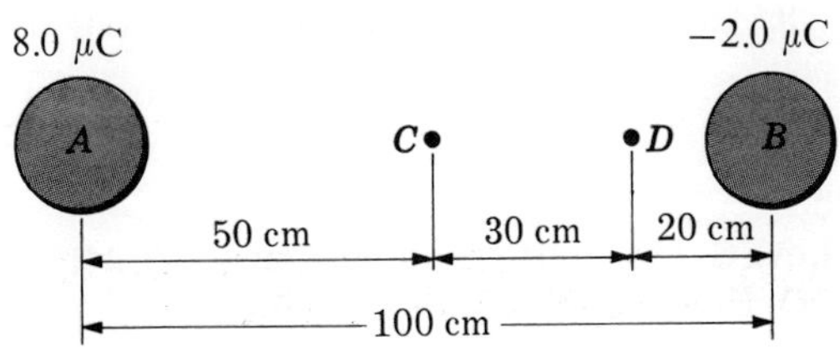

Figure 33-3
Potentials of points near two charges.

Note that, although the potential at D is zero, the field intensity is large (5.6×10^5 N/C). This corresponds in analogy to a steep hill at sea level.

33-6 EQUIPOTENTIAL SURFACES

It is conveniently possible to portray graphically the distribution of the potential at various places in an electric field by means of equipotential lines or surfaces. An *equipotential line* is a line in an electric field so drawn that all points on the line are at the same potential. Such lines are similar to isothermal lines on a weather map drawn to show all points that have the same temperature, or lines on a contour map drawn to show all places that have the same elevation. An *equipotential surface* is a surface that is drawn through points all of which have the same potential.

In general it is easier to locate equipotential lines than it is to measure electric fields directly. As shown in the following sections, once the equipotential surfaces have been mapped, the electric field intensities may be computed.

It is apparent that the surface of a charged conductor, with the charges at rest, must be an equipotential surface. If there were a potential difference between points on the surface, charges would move along the surface and this is contrary to our postulate of an electrostatic condition.

The equipotential surfaces around a point charge are a series of concentric spheres that represent potentials which vary inversely with the distances from the charge. When two electric fields are superimposed, the resulting equipotential surfaces are determined by adding algebraically the individual potentials at the various points.

33-7 LINES OF FORCE PERPENDICULAR TO EQUIPOTENTIAL SURFACES

Equipotential surfaces in an electric field are always perpendicular to the lines of force. This must be true because of the fact that the line of force shows, by definition, the direction of the force upon a test charge and there can be no force normal to this direction. Hence there is no work done in producing a small displacement of a test charge normal to a line of force, and this normal is therefore an equipotential line. Thus, if the equipotential lines can be drawn, the lines of force can be immediately constructed, since they are at all points perpendicular to the equipotential lines which they intersect. For example, in Fig. 33-4 there are shown the lines of force and the equipotential lines in a plane surface containing two charges of equal magnitude and opposite sign.

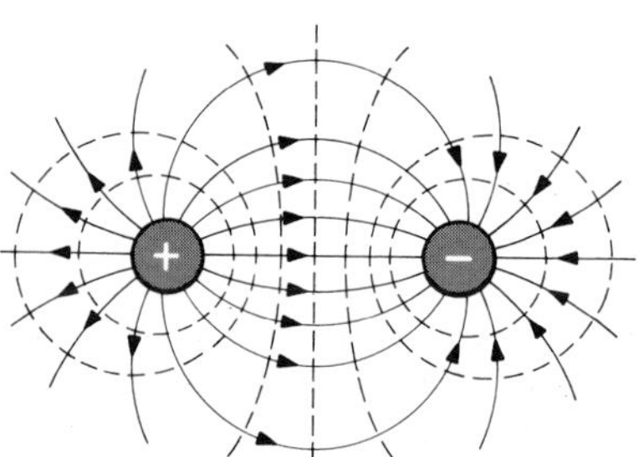

Figure 33-4
Lines of force (solid lines, with arrows) and equipotential lines (shown broken) near equal charges of unlike sign.

33-8 POTENTIAL GRADIENT

The *gradient* of a quantity is the space rate of its increase in the direction in which it increases most rapidly. Thus the gradient of a hillside is the amount of rise per unit horizontal distance, in the steepest direction. The concept of a gradient is important in such cases as temperature gradient, pressure gradient, and density gradient. *Electric potential gradient* is the rate of change of potential with distance along a line of force, $\Delta V/\Delta s$. The mks unit is the volt per meter, and the electrostatic unit is the statvolt per centimeter.

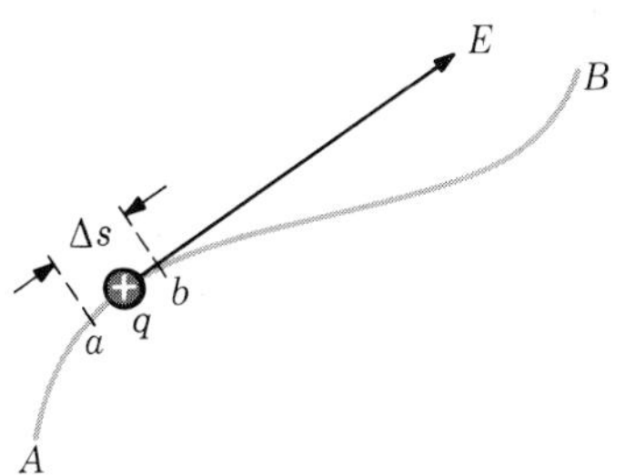

Figure 33-5
Work done in moving a test charge in an electric field.

There is a direct relationship between electric field intensity and potential gradient. Consider two points a and b that are separated an infinitesimal distance Δs on a line of force AB in an electric field (Fig. 33-5). The field is practically constant over the small distance Δs, although the field may vary widely over larger distances. If a test charge $+q$ is moved from a to b, work is done by the field on this charge. The algebraic sign for this work is obtained by the following conventional agreement: We shall consider work done by an outside agent in moving a positive charge *against* the field as positive and work done *by the field* as negative. Thus positive work increases the potential energy of the system and negative work decreases the potential energy of the system. Hence the work $\Delta \mathcal{W}$ done by the field is given by

$$\Delta \mathcal{W} = -F\,\Delta s \tag{7}$$

If we substitute in this equation the value of F from the defining equation for field intensity ($F = Eq$), we obtain

$$\Delta \mathcal{W} = -qE\,\Delta s$$

Upon combining this expression for $\Delta \mathcal{W}$ with the defining equation for potential, it follows that

$$\Delta V = \frac{\Delta \mathcal{W}}{q} = -E\,\Delta s$$

Hence

$$E = -\frac{\Delta V}{\Delta s} \tag{8}$$

Equation (8) states an important relationship: The electric field intensity at a point in an electric field is equal to the negative of the potential gradient of the field at this point.

If the test charge $+q$ is moved in some direction that is not along the line of force, the work done is given by $\Delta \mathcal{W} = -qE \cos\theta\ \Delta s$ where θ is the angle between Δs and the tangent to the line of force at the point considered. Then the potential difference between the ends of Δs is

$$\Delta V = \frac{\Delta \mathcal{W}}{q} = -(E\cos\theta)\,\Delta s$$

and the rate of change of potential with distance in this direction is $\Delta V/\Delta s = -E\cos\theta$. Hence the component of the electric field intensity in any given direction is equal to the negative of the rate of change of potential with distance in that direction,

$$E\cos\theta = -\frac{\Delta V}{\Delta s} \tag{9}$$

Example An electron in an oscilloscope tube is situated midway between two parallel metal plates 0.50 cm apart. One of the plates is maintained at a potential of 60 V above the other. What is the potential gradient between the plates? What is the force on the electron?

On the assumption of ideal conditions, the lines of force are straight and perpendicular to the plates, which are equipotential surfaces. The rate of change of potential with distance along a line of force is

$$\frac{\Delta V}{\Delta s} = \frac{60\ \text{V}}{5.0 \times 10^{-3}\ \text{m}} = 1.2 \times 10^{4}\ \text{V/m}$$

Hence the electric field intensity is 1.2×10^{4} N/C toward the plate at lower potential. The force on the electron is

$$F = Eq = (1.2 \times 10^4\ \mathrm{N/C}) \times (-1.60 \times 10^{-19}\ \mathrm{C})$$

$$= -1.92 \times 10^{-15}\ \mathrm{N}$$

This force is toward the plate of higher potential.

It may be noted that Eq. (8) leads to the statement made in Sec. 33-7 that no work is done in moving a test charge in a direction perpendicular to the direction of the electric field.

33-9 DISTRIBUTION OF CHARGE ON IRREGULAR CONDUCTOR

It has previously been assumed that the charges on a conductor, such as a sphere, were uniformly distributed over the surface. This assumption is not always true, particularly in the case of an irregular body. A familiar illustration is the case of an irregularly shaped conductor, such as that shown in Fig. 33-6. It can be shown, both theoretically and experimentally, that there is a greater density of charge on regions of large curvature and a lower charge density on surfaces of small curvature (large radius of curvature).

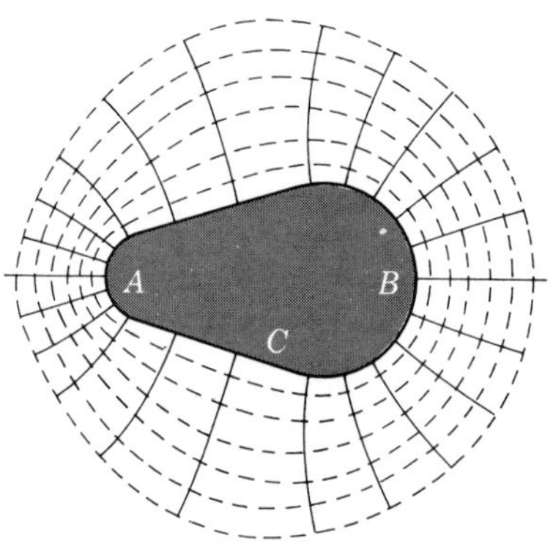

Figure 33-6
Distribution of charges on irregular conductor.

Since the electric field intensity near a point charge is proportional to the charge, it must be true that the electric field will be greatest near the places where the charge density is highest. Lines of force may be sketched in accordance with these facts, by spacing them more closely together where the charge density is larger. Equipotential lines are shown in Fig. 33-6 as dashed lines. They are everywhere normal to the lines of force. The surface of the conductor is an equipotential surface, and hence the lines of force are normal to the surface. It is often true that the potential gradient near a pointed surface that is charged becomes so great that the air in this vicinity becomes ionized and conduction takes place rapidly to reduce the charge at the point. This effect is used widely to reduce the accumulation of static electricity in such devices as lightning rods, airplane structures, moving belts, and printing presses. The electric discharge from such points is known as a *corona* discharge.

It is a fact of considerable importance that all portions of a conductor in an electrostatic field are everywhere at the same potential if the charges are at rest. This must be true, for otherwise there would be a flow of charge from points of higher potential to points of lower potential. Furthermore, all points of a conductor or inside a hollow conductor are at the same potential, regardless of the shape of the conductor when the charges are at rest. If there were a potential difference between the parts of a conductor, there would be a potential gradient and hence an electric field along the conductor. This field would result in a flow of charge, and this would be contrary to our postulate of an electrostatic situation.

Another item of significance in this connection is the fact that there can be no electric field inside a conductor in an electrostatic field. If such a field existed, the free charges within the conductor would continue to move, an electric current being thus produced. Such a current is not possible in electrostatics, as by definition the charges must be at rest in an electrostatic field. It therefore follows that the excess charges on a conductor in an electrostatic field must reside on the outside surface of the conductor. This statement is true regardless of whether the conductor is solid or hollow. In electric shielding, use is made of the fact that external electric fields do not produce an electric field inside a conductor. For example, radio tubes are shielded from external electric fields by placing the tubes inside a conducting "can." A room full of apparatus may be shielded by the use of a wire-mesh cage that encloses the room.

33-10 THE FARADAY ICE-PAIL EXPERIMENTS

Several significant experiments, originally performed by Faraday with a metallic ice pail and an electroscope, are useful in illustrating some of the facts stated above. The pail is connected to the electroscope by a conducting wire, as in Fig. 33-7*a*. When a charged ball, held by an insulating thread, is lowered into the pail, the leaves of the electroscope diverge, showing that they possess an induced charge (Fig. 33-7*b*). No change in the divergence of the leaves is noticed when the charged ball is moved to various places inside the pail. This shows that the number of induced charges inside the pail is just equal to the charge on the ball. Now, if the ball is touched to a wall of the pail, no change in the divergence of the leaves is observed (Fig. 33-7*c*). The ball is found to have lost its charge, and the outside of the pail and the electroscope have gained the charge lost by the ball (Fig. 33-7*d*). When the ball touches the wall, its charge just neutralizes the charges of opposite sign that were on the inside of the pail.

If the ball is again similarly charged and reintroduced into the pail and touched to the pail, it will be found that the pail acquires an additional charge, equal in magnitude and sign to the original charge. This procedure can be continued until the pail is charged to a very high potential, as is described in Sec. 33-11.

33-11 THE VAN DE GRAAFF GENERATOR

Use is often made of high-voltage electrostatic sources for the production of "atom-smashing" devices that are useful in experiments in nuclear physics. One such generator was designed by Robert J. Van de Graaff and further developed with his associate John G. Trump.

A highly simplified schematic diagram of this generator is shown in Fig. 33-8. A large metal sphere *S* is supported by an insulating tube *T*. A wide belt *AB* of some insulating material is motor driven by pulley *C*; at the upper end it passes over the idler pulley *D*. A series of needle points at *F* is kept at a high negative potential (10 to 50 kV) by the source *E*, and a corona discharge sprays electrons onto the belt as it moves upward from *C*. Another series of needle points at *G* transfers some of the electrons from the belt to the sphere *S*, giving it a continually

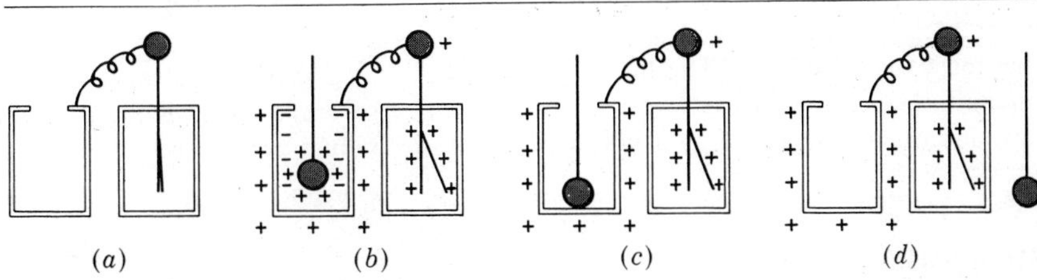

Figure 33-7
Faraday ice-pail experiments.

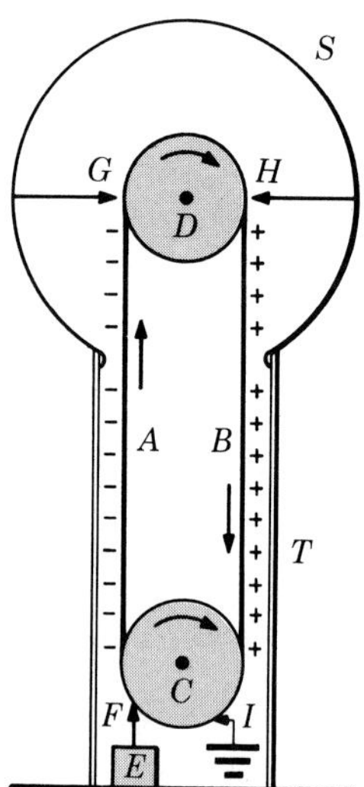

Figure 33-8
Van de Graaff generator.

increasing negative charge. A second set of needle points at H creates a high potential gradient which ionizes the air in the region. Positive ions are attracted to the belt and are carried downward to be neutralized by electrons sprayed onto the belt by the grounded set of needles at I.

Thus there is a steady transfer of negative electricity up the belt to the sphere and a downward flow of positive charges from the sphere to the ground. This results in giving the sphere a high negative potential with respect to the ground. Some large machines of this type have produced potentials up to 10 MV, and commercial models of various sizes are currently available for a variety of uses. The limiting potential is determined by various leakages including the surrounding air and the supporting insulators.

SUMMARY

Potential difference is the work done per unit charge when a charge is moved from one point to another,

$$\Delta V = \frac{\mathcal{W}}{q}$$

The mks unit of potential difference is the volt. One *volt* is the potential difference between points when one joule of work must be done to move one coulomb of charge between the points.

The electrostatic unit of potential difference is the statvolt. One *statvolt* is the potential difference between points when one erg of work must be done to move one statcoulomb of charge between the points.

Potential difference is a scalar quantity.

The potential difference between two points is independent of the path.

The *potential at a point* is the potential difference between the point and an arbitrarily selected zero of potential. This zero is frequently taken as the earth (ground). If the reference zero of potential is chosen as a point very distant from all charges, the potential difference between this distant point (at infinity) and the point in question is called the *absolute potential* of the point. The potential at a point may also be defined as the ratio of the potential energy of a test charge at the point considered to the magnitude of the charge.

$$V = \frac{\mathcal{W}}{q}$$

The potential at a point near an isolated point charge is given by

$$V = k\frac{Q}{r} \qquad \text{or} \qquad V = \frac{1}{4\pi\varepsilon_0}\frac{Q}{r}$$

The potential at a point in the field due to several charges is the algebraic sum of the individual potentials.

$$V = \sum \frac{1}{4\pi\varepsilon_0}\frac{Q}{r}$$

The work done in moving a charge from point B to point A, which are at different potentials, is given by

$$\mathcal{W}_{BA} = q(V_A - V_B)$$

An *equipotential surface* is a surface on which all points have the same potential. The surface is everywhere perpendicular to the lines of force. A charge may be moved anywhere on an equipotential surface without work being performed on or by the field.

Electric potential gradient is the rate of change of potential with distance along a line of force, $\Delta V/\Delta s$. Units are the volt per meter and the statvolt per centimeter.

Electric field intensity is equal to the negative of the potential gradient of the field.

$$E = -\frac{\Delta V}{\Delta s}$$

The *charge density* of charges on an irregular conductor is largest on regions of large curvature (sharp surfaces). The electric fields near such conductors are most intense near the sharp surfaces, thus leading to *corona discharges*.

All portions of a conductor in which the charges are at rest are at the same potential.

When charges are at rest on a conductor there can be no electric field in the conductor.

High-voltage electrostatic generators are made by utilizing the phenomena of electrostatic induction.

Questions

1 Show why the potential difference between two points in an electric field is independent of the path.

2 What factors determine whether a difference of potential exists between two points?

3 State some good analogies between gravitational and electric fields, with reference to field intensity and to potential.

4 Explain clearly why electric potential is not identical with energy. Show why it is not correct to say that potential difference is the work required to move unit charge between the points considered.

5 Is potential difference a vector or a scalar quantity? Prove your answer.

6 Draw a curve to portray the variation of electric potential with distance near an isolated point electric charge.

7 Can two lines of force cross in an electric field? Can two equipotential lines intersect? Why?

8 In working with electrostatics, would one have more success on a dry or on a humid day? Explain.

9 Given a positively charged insulated sphere, how could you charge two other spheres, one positively, the other negatively, without changing the charge on the first sphere? What is the source of the energy represented by the charges acquired by the spheres?

10 Is work required to move a charge on the surface of an isolated charged conductor? Explain.

11 One can connect a conductor to the earth to "ground" it. What is meant by the term ground and why does the earth serve this function?

12 A glass ball and a copper ball are mounted on hard-rubber insulating rod supports. Each is rubbed with a silk cloth. The electrification of each is tested at various places by means of a proof plane and an electroscope. Describe how the electrification is found to vary.

13 Explain how and why a spark discharge occurs between two charged bodies in air.

14 Can a hollow metal sphere of the same diameter hold the same, a smaller, or a larger charge? Explain.

15 In Fig. 33-9 is shown a positively charged sphere A and an insulated cylinder B. The con-

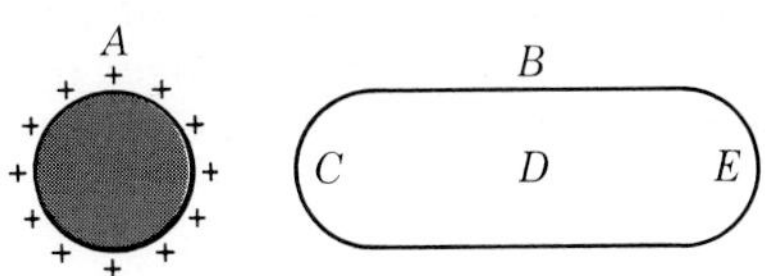

Figure 33-9
Charges and potentials at different places on a conductor near a charged body.

ductor B was uncharged before it was brought near A. (*a*) Compare the charges at C, D, and E. (*b*) Compare the potentials of these points.

16 Can an electric potential exist at a point in a region where the electric field has zero value? Can the potential be zero at a place where the electric field intensity is not zero? Give examples to illustrate your reasoning.

17 Describe why a Van de Graaff generator may be used as a particle accelerator in nuclear physics experiments.

Problems

1 Two points in an electric field have a potential difference of 5.0 V. If 25 J of work is required to move a charge between these points, how large must that charge be?

2 The potential difference between two points in an electric field is 8.0 V. How much work is required to move a charge of 400 μC between these points? *Ans.* 3.2×10^{-3} J.

3 What is the difference of potential between A and B if 125 ergs of work must be done to carry a charge of 6.40 statC from B to A?

4 Two similar charges of 250 statC are situated on small spheres 15.0 cm apart. What is the electric field intensity midway between the spheres? What is the potential at this point?

Ans. zero; 66.7 statV.

5 A charge of 2.75×10^{-8} C is uniformly distributed over a sphere of diameter 150 mm. (*a*) What is the potential at the surface of the sphere? (*b*) Calculate the potential energy of a charge of 1.25×10^{-8} C that is placed 250 mm from the center of the sphere.

6 Calculate the electrostatic field intensity and potential midway between two point charges, one of +500 statC, the other of −200 statC, placed 50.0 cm apart, in air. What work would be required to bring a charge of +23.5 statC to this point from a very distant point?

Ans. 1.12 dyn/statC; 12.0 statV; 282 ergs.

7 How much work is needed to move a charge of 3.00 μC from the earth to a point that is 150 mm from a charge of 35.5 μC and 50.0 mm from a charge of −20.0 μC?

8 If a force of 4.8×10^{-2} N is required to move a 40-μC charge in an electric field between two points 20 cm apart, what potential difference is there between the points?

Ans. 2.4×10^{2} V.

9 Two electrostatic charges, A of 500 statC and B of −8.00 statC, are 260 mm apart. (*a*) What is the potential at a point C which is 240 mm from A and 100 mm from B? (*b*) What work is required to bring a charge of 18.0 statC from infinity to the point C?

10 How much of a potential difference must be maintained between the ends of a long high-resistance wire if the center of the wire is grounded and a +100-V and −100-V potential occurs at opposite ends of the wire?

Ans. 200 V.

11 Point A is 250 mm from a positive charge C of 100 statC and 150 mm from a negative charge D of 300 statC. Point B is 200 mm from C and 300 mm from D. How much work is required to move a charge of 50.0 statC from A to B?

12 Two small metal spheres 25 cm apart have charges of 10 and 20 statC, respectively. Calculate (*a*) the electric field intensity and (*b*) the potential, at a point midway between them.

Ans. 0.064 dyn/statC, toward the 10-statC charge; 2.4 statV.

13 Calculate the potential energy of a system composed of four charges of 0.0125 μC each situated at the corners of a square that is 1.00 m on each side.

14 Compute the work required to bring a charge of 5.0 statC from a point 24 cm from a charge of 60 statC to a point 3.0 cm from it.

Ans. 88 ergs.

15 How much work is required to move a charge of 250 statC a distance of 175 mm against an electrostatic field that increases uniformly from 10.0 to 30.0 dyn/statC?

16 A sphere having a radius of 10 cm is charged in air. If the maximum electric field which can be maintained without producing ionization and permitting the charge to flow is 10^{6} V/cm, what is the maximum potential to which the sphere may be charged? *Ans.* 10^7 V.

17 A charge of 6.75 μC in an electric field is acted upon by a force of 2.50 N. What is the potential gradient at this point?

18 Two drops of water which are identical carry equal potential of 1 V. If these two drops merge into one large drop, what is the potential of the new drop? *Ans.* $V_2 = (\sqrt[3]{4})\ V_1$ volts.

19 A pair of large parallel plates 5.25 mm apart are charged until they have a potential difference of 1.50 kV. What is the electric field intensity between the plates?

20 Two point charges of +24 and −36 statC, respectively, are 50 cm apart in air. What is the electric field intensity and what is the potential at a point 30 cm from the former point and 40 cm from the latter? *Ans.* 0.035 dyn/statC; −0.10 statV.

21 A potential difference of 3,000 V is maintained between two large parallel plates 6.00 cm apart. What are the magnitude and direction of the force exerted on an electron which is 1.00 cm from the positive plate?

22 At what distance from the spherical shell on top of an electrostatic generator in air having a charge of 4.0 μC would the potential be 2.4×10^5 V? *Ans.* 0.12 m.

23 The electric field intensity directed downward between two large horizontal plates 3.00 cm apart is 2.50×10^4 N/C. How much energy is required to move a charge of 6.35 μC (*a*) from the lower to the upper plate, (*b*) 10.0 cm horizontally, and (*c*) 4.30 cm at an angle of 25° above the horizontal?

24 At each corner of a square 20 cm on a side is a small charged body. Going around the square, these charges are +60, −30, +60, and −30 statC. Find (*a*) the field intensity and (*b*) the electrostatic potential at the center of the square. *Ans.* zero; 4.3 statV.

25 A hollow, spherical brass shell has a radius of 18.0 cm and a charge of 0.300 μC. What are the electric field intensity and the potential (*a*) at a point 27.0 cm from the center of the sphere, (*b*) at the surface of the sphere, and (*c*) at a point 9.0 cm from the center of the sphere?

26 Two charges of 10 μC and −4 μC are 2.0 m apart. (*a*) What is the potential at a point midway between these charges? (*b*) What is the potential at a point between the charges and 0.667 m away from the −4-μC charge? (*c*) How much work would be required to move a 0.05-μC charge from the midpoint to the 0.50-m mark? *Ans.* 4.5×10^5 V; 0; 2.25×10^{-2} J.

27 At a certain instant two electrons are at rest and 0.250 mm apart. Because of their electrostatic repulsion they move away from each other. Calculate their relative velocity when they are 1.250 mm apart.

28 In a radio tube containing two flat parallel electrodes 0.02 m apart, a 200-V battery is connected between the plates so as to create a maximum potential difference between the electrodes. What is the potential gradient between the electrodes? What is the force on a electron moving between the electrodes? *Ans.* 1.0×10^4 V/m; 1.6×10^{-15} N.

29 A Van de Graaff generator has a potential of 5.0-MV built up when the charging belt carries negative charge upward at a rate of 1.25 mC/s and positive charge downward at a like rate. What power is required to maintain this flow?

30 If an electron is accelerated in an accelerator possessing a potential difference of 40 MV, what energy will the electron gain? *Ans.* 6.4×10^{-12} J.

Max Born, 1882–

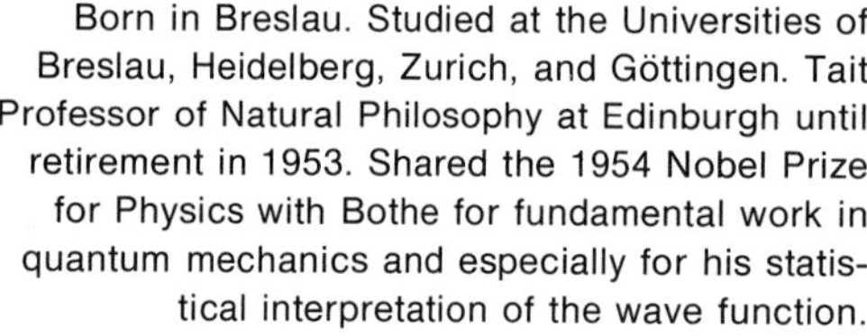

Born in Breslau. Studied at the Universities of Breslau, Heidelberg, Zurich, and Göttingen. Tait Professor of Natural Philosophy at Edinburgh until retirement in 1953. Shared the 1954 Nobel Prize for Physics with Bothe for fundamental work in quantum mechanics and especially for his statistical interpretation of the wave function.

Walter Bothe, 1891–1957

Born in Oranienburg. Studied at the University of Berlin; joined the Physikalisch-Technische Reichsanstalt. Taught at Giessen and Heidelberg. Shared the 1954 Nobel Prize for the method of coincidence and the discoveries which it has made possible.

34

Capacitance

Any device on which electric charge may be stored so as to possess electrical potential energy is called a *capacitor*. (The older term "condenser" is also frequently used.) Isolated charged bodies possess potential energy by virtue of their potential difference with respect to a distant reference potential or, to state this differently, by virtue of the work needed to charge them. Two conductors well insulated from each other may be oppositely charged so as to have a potential difference and will thereby store energy which is readily recovered by allowing charge to flow between them through various electrical devices. If these conductors are of large area and closely spaced, the potential of each is strongly affected by the other so as to reduce the potential difference between them. Hence such a capacitor provides a useful device for the temporary storage of electric energy. Capacitors constructed of two closely spaced conductors are used by the tens of billions in electrical equipment of all kinds, as well as in automobiles and most electrical machinery.

34-1 SIMPLE CAPACITORS

A simple capacitor might consist of two large, parallel metal sheets insulated from each other and their surroundings and placed close together with empty space between them. If a device for moving charge through a potential difference, such as a Van de Graaff generator or a battery, is connected to these plates, electrons are driven onto one plate and removed from the other (Fig. 34-1). The close association of the opposite charges on the two plates has the effect of making the work needed to place additional charge on the plates very much less than would be the case if the plates were not close together. This fact may be understood by considering a sample positive charge q being carried toward the positively charged plate. It experiences a repulsion due to the positive charge on the plate but an almost

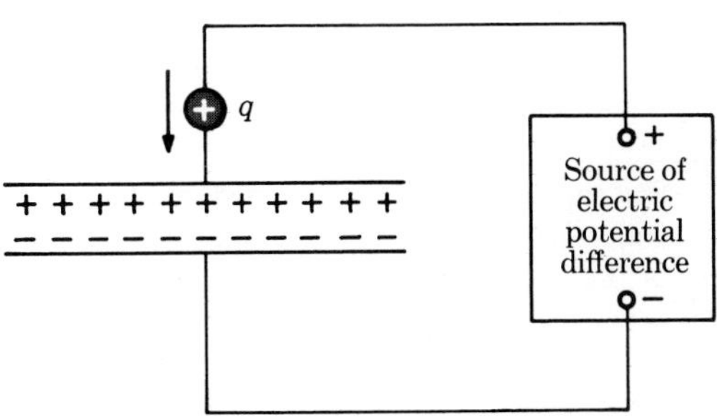

Figure 34-1
A simple capacitor connected to a device for moving charge from one plate to the other.

equal attraction due to the negative charge only slightly farther away. For a given amount of energy available to transfer charge, the amount of charge that can be transferred is increased by the closeness of the plates. Or, for a given charge transferred, the work per unit charge, or the potential difference, is smaller than would be the case if the plates were widely spaced.

The close spacing of most capacitor plates is achieved by separating the conducting sheets with a thin sheet of mica or paper. The insulator contributes more than just insulation, however, as we shall see in Sec. 34-5.

34-2 CAPACITANCE

Since the function of the capacitor is to store a large charge for a given potential difference, or to store a given charge at a low potential difference, the relation of charge and potential difference in a capacitor is used to define capacitance. The *capacitance* of a capacitor is the ratio of the amount of electricity transferred, from one of its plates to the other, to the potential difference produced between the plates.

$$C = \frac{Q}{V} \tag{1}$$

This definition may be extended to include the case of an insulated charged body such as the isolated sphere previously discussed. In such a case the capacitance is the ratio of the charge on the body to its potential.

The capacitance of an insulated conductor or a group of conductors depends upon the geometry of the conductors and the permittivity of the region in which they are situated. It should be clearly understood that this capacitance is not at all analogous to the amount of water that can be held by a vessel, for example. A capacitor is never "full"; if the potential difference is increased, larger charges accumulate on the plates. A fair analogy to a capacitor is an air tank, where an increase in pressure is produced by forcing more air into the tank.

When Eq. (1) is used, one must be careful to note the fact that Q represents the value of the charge on either conductor and not the net charge (zero) of the capacitor. In this chapter we shall consider only cases where the charges are static. In most cases, however, these charges change rapidly with time, often in picoseconds, and frequently these charges vary harmonically with time.

34-3 UNITS OF CAPACITANCE

The unit of capacitance in the mks system is the farad, so named in honor of Michael Faraday. A *farad* is the capacitance of a capacitor which acquires a potential difference of one volt when it receives a charge of one coulomb. The farad is so large a unit that it would take a capacitor of tremendous proportions to have a capacitance of 1 farad; hence, the microfarad (μF, one-millionth of a farad) is the unit most frequently used. In some cases the picofarad (pF) is a convenient unit (1 pF = 10^{-12} F).

Example A capacitor having a capacitance of 3.0 μF is connected to a 50-V battery. What charge will there be in the capacitor?

$$Q = CV = 3.0\ \mu\text{F} \times 50\ \text{V} = 150\ \mu\text{C}$$

34-4 MATERIAL DIELECTRICS

Air is used as a medium between the plates of many types of capacitors, especially those used in high-frequency circuits such as radar. But other capacitors utilize solid or liquid materials such as mica, paraffined paper, or oil. Not only do these materials have good insulating properties, but they also respond to the presence of the electric field between the plates of the capacitor in such a way as to increase the capacitance, often by a large factor. The action is indicated in Fig. 34-2. The molecules of the types of materials used in capacitors, called *dielectrics,* may be considered to have permanent displacement of charges; such molecules are called *polar* molecules. In the absence of an electric field these have a nearly random orientation and are kept in that state by thermal agitation. When the plates of the capacitor are charged, the resulting electric field causes the molecules to align themselves with the field, to an extent determined by the field intensity and the nature of the dielectric. Although this alignment produces no net charge within the body of the dielectric, it results in a sheet of positive charges on one face of the dielectric and a sheet of negative charges on the other. The process very much weakens the electric field within the dielectric as well as without, since the electric field due to these polar charges is opposite in sense to that produced by the charges on the plates. The separation of the sheets of positive and negative charge, already small, is much reduced with the dielectric present. In a typical case 6 units of charge on one plate of a capacitor are matched by 5 units of opposite charge on the adjacent face of the dielectric. The resulting field within and without the capacitor is that due to only 1 unit of charge. Consequently the potential difference between the plates is decreased by a factor of 6, and the capacitance of the device is increased by a factor of 6.

When a charged capacitor is discharged, the alignment of the molecules of the dielectric, called the *polarization* of the dielectric, is destroyed by thermal agitation. In the case of some dielectrics, such as glass, the molecules are such that the polarization may require considerable time to be destroyed. When one discharges such a capacitor by touching a wire to the two plates a spark will be observed as contact is made. If the wire is again touched a minute later, another, although much smaller, spark may be observed. The insulator is said to have dielectric absorption.

In many materials the molecules are not permanently polar, and any response to the electric field in a capacitor takes the form of temporary displacement of charge within the molecules.

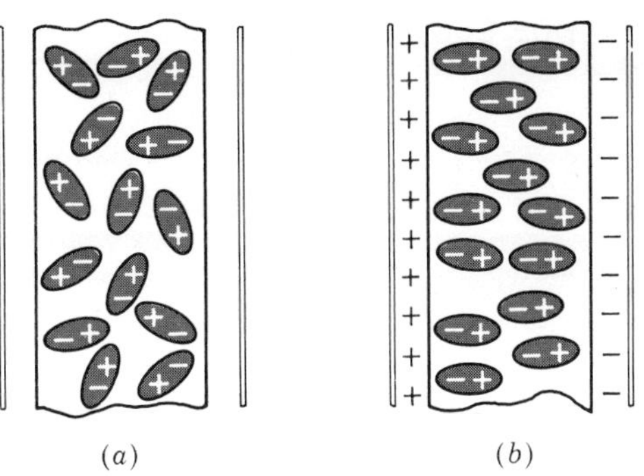

Figure 34-2
Action of the dielectric material in a capacitor: (*a*) uncharged, (*b*) charged.

While the magnitude of the effect is not so great for these materials, the time for disalignment, or *depolarization,* with the removal of the field, is much less than with polar materials.

34-5 PERMITTIVITY

The response of a medium to the presence of an electric field is characterized by a quantity called the *permittivity* of the medium, designated by the symbol ε. A definition of ε might be obtained from Coulomb's law,

$$F = \frac{1}{4\pi\varepsilon}\frac{QQ'}{s^2}$$

The permittivity of empty space is designated by the symbol ε_0. The value of $1/4\pi\varepsilon_0$ is approximately $9 \times 10^9\ \mathrm{N\cdot m^2/C^2}$, and ε_0 is $8.87 \times 10^{-12}\ \mathrm{C^2/N\cdot m^2}$.

For many purposes it is desirable to compare the permittivity of a dielectric with that of empty space, which is accepted as a standard. For this comparison the term relative permittivity will be used in this book. (Other terms sometimes used are dielectric constant, specific inductive capacity, and dielectric coefficient.) We define *relative per-*

Table 1
RELATIVE PERMITTIVITY

Solids:	
Glass	6–10
Mica	5.6–6.6
Paraffined paper	2.1–2.3
Porcelain	6–7
Titanates	15–12,000
Liquids:	
Alcohol	25
Oil	2–2.2
Turpentine	2.2–2.3
Water	80–83
Gases:	
Carbon dioxide	1.00097
Air	1.00060
Hydrogen	1.00026
Water vapor	1.007

mittivity ε_r of a substance as the ratio of its permittivity ε to the permittivity of empty space. In symbols

$$\varepsilon_r = \frac{\varepsilon}{\varepsilon_0}$$

it will be seen that ε_r is a pure number, independent of units. The relative permittivities of some common insulators are given in Table 1.

In Chap. 33 it was pointed out that the potential of a uniformly charged spherical conductor was the same as if all the charge were concentrated at the center of the sphere. This potential is given by

$$V = \frac{1}{4\pi\varepsilon}\frac{Q}{r}$$

Therefore the capacitance of an isolated spherical conductor is given by

$$C = \frac{Q}{V} = 4\pi\varepsilon r \tag{2}$$

Hence the capacitance of a sphere is proportional to its radius.

Example The nearly spherical upper electrode on a small Van de Graaff generator is 10 cm in radius. If the electrode is raised to a potential of 100 kV, what charge does it hold?

$$C = 4\pi\varepsilon r = \frac{0.10\ \text{m}}{9 \times 10^9\ \text{N}\cdot\text{m}^2/\text{C}^2} = 11\ \text{pF}$$

$$Q = CV = (11 \times 10^{-12}\ \text{F}) \times 10^5\ \text{V}$$
$$= 1.1\ \mu\text{C}$$

34-6 CAPACITANCE OF A PARALLEL-PLATE CAPACITOR

In Chap. 32 it was mentioned that in the notation used in this book in the mks system, the total number of lines of force ψ which emerge from a body having charge $+Q$ (or which converge on a body having a charge $-Q$) is numerically equal to Q and hence $\psi = Q$. We shall utilize this fact in the derivation of an expression for the capacitance of a pair of charged plates, separated by a dielectric, as shown in Fig. 34-3. It is seen that virtually all the lines of force are straight lines uniformly spaced between the plates. If the space is filled by a material dielectric, however, many of them terminate on the faces of the dielectric as shown in the "exploded" view in Fig. 34-3*a*. The number of lines which cross the dielectric is consequently less than the number which would cross the space if it were evacuated (see Fig. 34-3*b*). If the plates are large and their separation small, the electric field intensity between them is everywhere uniform. This field is given by

$$E = \frac{\psi}{\varepsilon A} = \frac{Q}{\varepsilon A}$$

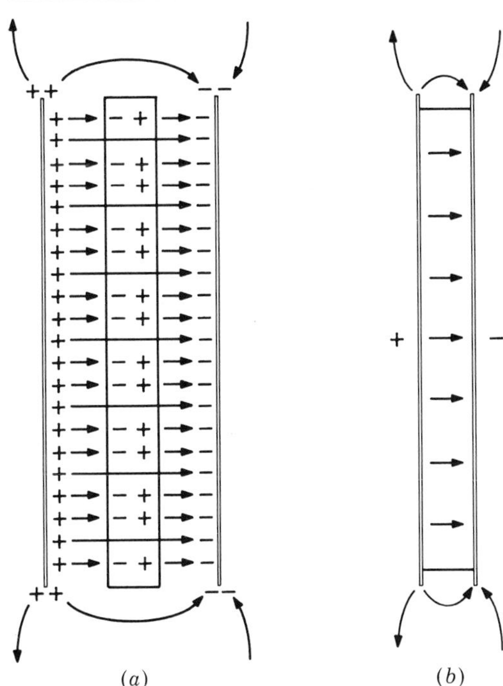

Figure 34-3
Distribution of the electric field of a parallel-plate capacitor: (*a*) exploded view; (*b*) resultant field.

From this equation and the definitions of electric field intensity ($E = F/q$), potential ($V = \mathcal{W}/q$), and work ($\mathcal{W} = F \times s$), it follows that

$$V = \frac{\mathcal{W}}{q} = \frac{F}{q}s = Es = \frac{Q}{\varepsilon A}s$$

where s is the distance between the plates. If we make use of the defining equation for capacitance and substitute the value of V from the equation above, we obtain

$$C = \frac{Q}{V} = \frac{Q}{Qs/\varepsilon A}$$

Hence
$$C = \frac{\varepsilon A}{s} = \frac{\varepsilon_r \varepsilon_0 A}{s} \tag{3}$$

The capacitance is in farads when A is in square meters, s is in meters, and ε ($= \varepsilon_r\varepsilon_0$) is in coulombs^2 per newton-meter^2.

The equation for the capacitance of a parallel-plate capacitor enables one to give an alternative measure of relative permittivity. Consider a given parallel-plate capacitor of capacitance C when it has a dielectric of permittivity ε and a capacitance C_0 when the dielectric is empty space of permittivity ε_0. Then the relative permittivity ε_r is given by

$$\varepsilon_r = \frac{\varepsilon}{\varepsilon_0} = \frac{C}{C_0} \tag{2a}$$

34-7 COMMERCIAL CAPACITORS

In order to increase the capacitance of capacitors so that they will hold sufficiently large quantities of electricity, use is made of a large number of plates, each of large area. An idealized diagram of a multiplate capacitor is shown in Fig. 34-4. In high-grade capacitors mica is used as a dielectric. Inexpensive capacitors of capacitance up to 10 μF are usually made of alternate layers of tin or aluminum foil and waxed paper. These are frequently wound into rolls under pressure and sealed into moisture-resisting metal containers. *Electrolytic capacitors* of large capacitance, up to 1,000 μF, are made by using an insulating layer formed by chemical action directly on the metal plates of the capacitor. The slight space between the layers is filled with an electrolyte in liquid or paste form, this constituting one of the plates.

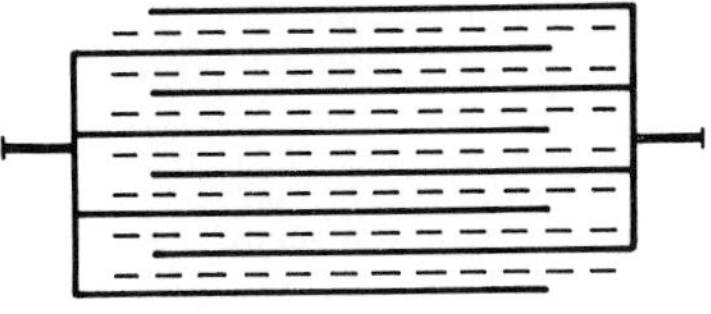

Figure 34-4
A multiplate capacitor.

Table 2
DIELECTRIC STRENGTH
(average values)

Substance	kV/cm to puncture
Air	30
Transformer oil	75
Turpentine	110
Paraffin oil	160
Kerosene	160
Paraffin, solid	250–450
Paraffined paper	300–500
Polystyrene	250–1,500
Mica	300–700
Ebonite	300–1,000
Glass	300–1,600

For a multiple-plate capacitor, such as that shown in Fig. 34-4, the capacitance given by Eq. (3) must be increased by the factor $N - 1$, where N is the total number of plates. It will be seen that this is logical, since N plates must be used to make up $N - 1$ individual capacitors.

Example A parallel-plate capacitor is made of 350 plates, separated by paraffined paper 0.0010 cm thick ($\varepsilon_r = 2.5$). The effective size of each plate is 15 by 30 cm. What is the capacitance of this capacitor?

$$\varepsilon = \varepsilon_r\varepsilon_0 = 2.5 \times 8.87 \times 10^{-12}\ \mathrm{C^2/N\cdot m^2}$$

$$s = 1.0 \times 10^{-5}\ \mathrm{m}$$

$$A = 0.15\ \mathrm{m} \times 0.30\ \mathrm{m} = 0.045\ \mathrm{m^2}$$

$$C = \frac{\varepsilon A}{s}(N - 1)$$

$$= \frac{2.5 \times 8.87 \times 10^{-12}\ \mathrm{C^2/N\cdot m^2}}{1.0 \times 10^{-5}\ \mathrm{m}} \times 0.0045\ \mathrm{m^2} \times (350 - 1)$$

$$= 35 \times 10^{-6}\ \mathrm{F} = 35\ \mu\mathrm{F}$$

34-8 DIELECTRIC STRENGTH

We previously have noted the fact that the quantity of electricity that can be stored in a capacitor is not limited by its capacitance. As more charge is placed on the plates of a capacitor, the potential difference builds up, until finally the dielectric can no longer support the potential gradient. The term *dielectric strength* refers to that property of the insulator which determines the maximum potential gradient which can be applied to the material before its insulating properties are destroyed by a disruptive discharge of electricity through the insulator. Thus dielectric strength is really a measure of the insulating quality of the material. A capacitor is rated to be able safely to withstand only a given potential difference, and it should not be exposed to higher values. Referring back to the analogy of a capacitor and an air tank, the tank will burst when a certain air pressure is reached. The capacitor will "break down" by sparks through the dielectric material when a critical potential gradient is exceeded. This potential gradient frequently is given in tables in units of kilovolts per centimeter of thickness. Approximate values of the dielectric strengths of some insulators are given in Table 2.

34-9 COMBINATIONS OF CAPACITORS

In Fig. 34-5 are shown three capacitors of separate capacitances, C_1, C_2, and C_3, connected in parallel and joined to a device for establishing a potential difference, such as a cell. Obviously,

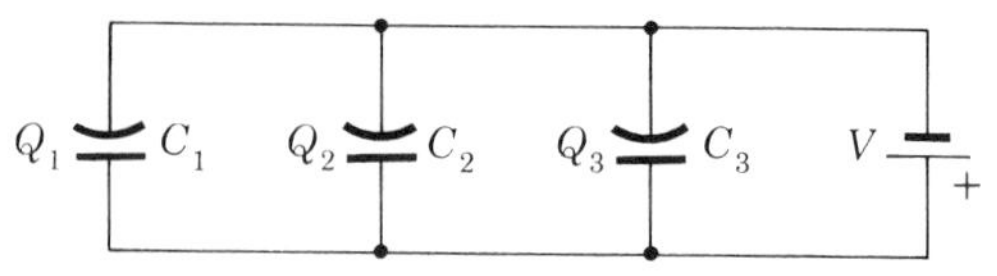

Figure 34-5
Capacitors in parallel.

all the capacitors are charged to the same potential, since they are all connected directly to the same source of potential difference. By definition of capacitance, the charge on each capacitor is

$$Q_1 = C_1V \qquad Q_2 = C_2V \qquad Q_3 = C_3V$$

the potential difference over each being the same. The total charge Q is equal to the sum of the separate charges,

$$Q = Q_1 + Q_2 + Q_3$$

If C represents the joint capacitance, $Q = CV$, and therefore by substitution for Q, Q_1, Q_2, and Q_3, we get

$$CV = C_1V + C_2V + C_3V$$

Dividing both sides of the equation by V, we obtain

$$C = C_1 + C_2 + C_3 \tag{4}$$

That is, *for capacitors connected in parallel, the joint capacitance is the sum of the several capacitances.*

In Fig. 34-6, the capacitors are shown connected in series. Each of these capacitors holds the same quantity of electricity. This follows from the fact that if the capacitors are *initially uncharged* a charge upon one plate always attracts upon the other plate a charge equal in magnitude and opposite in sign. Let V_1, V_2, and V_3 represent the potential differences over the several capacitors and V the potential difference over the whole. For the series connection, if follows that

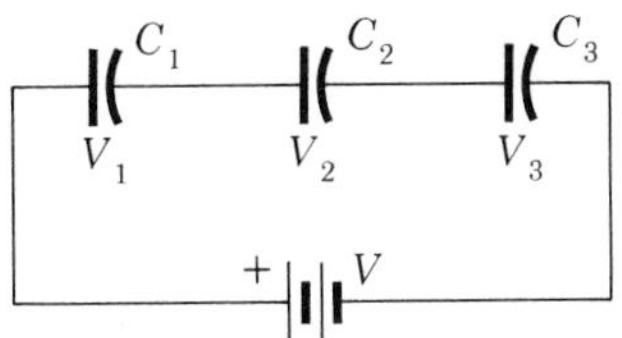

Figure 34-6
Capacitors in series.

$$V = V_1 + V_2 + V_3$$

If we substitute in this equation the various values

$$V = \frac{Q}{C} \qquad V_1 = \frac{Q}{C_1} \qquad V_2 = \frac{Q}{C_2} \qquad V_3 = \frac{Q}{C_3}$$

we have

$$\frac{Q}{C} = \frac{Q}{C_1} + \frac{Q}{C_2} + \frac{Q}{C_3}$$

Dividing both sides of the equation by Q, we obtain

$$\frac{1}{C} = \frac{1}{C_1} + \frac{1}{C_2} + \frac{1}{C_3} \tag{5}$$

That is, for capacitors connected in *series,* the *reciprocal* of the joint capacitance is equal to the sum of the *reciprocals* of the several capacitances.

Example Three capacitors have capacitances 0.50, 0.30, and 0.20 μF. What is their joint capacitance when arranged to give (*a*) a minimum capacitance and (*b*) a maximum capacitance?

In series,

$$\frac{1}{C} = \frac{1}{0.50\ \mu\text{F}} + \frac{1}{0.30\ \mu\text{F}} + \frac{1}{0.20\ \mu\text{F}}$$

$$C = 0.097\ \mu\text{F}$$

In parallel,

$$C = C_1 + C_2 + C_3 = (0.50 + 0.30 + 0.20)\ \mu\text{F}$$

$$= 1.00\ \mu\text{F}$$

A series connection of capacitors is used when it is desired to divide a high voltage among several capacitors, any one of which could not sustain the entire voltage. The voltage per capacitor in such a series connection is *inversely* proportional to the capacitance of the capacitor; i.e., the capacitor of least capacitance has the largest voltage across it. Capacitors are connected in parallel when a large capacitance is desired at a moderate or low potential. Such a combination of capacitors will hold a large charge.

34-10 ENERGY STORED IN A CHARGED CAPACITOR

Consider a process in which we charge an initially uncharged capacitor by transferring successive small charges Δq from one plate to the other. At each step, the work will be $v\,\Delta q$. Since in this process the potential difference starts at zero (uncharged capacitor) and increases uniformly to a final value V (fully charged capacitor), the average potential difference during the process is $\frac{1}{2}(0 + V)$, or $\frac{1}{2}V$. Hence the work done is the total charge Q transferred times the average potential difference,

$$\mathcal{W} = \tfrac{1}{2}VQ \tag{6}$$

Since $Q = CV$, the energy stored in the capacitor may also be written as

$$\mathcal{W} = \tfrac{1}{2}CV^2 = \frac{1}{2}\frac{Q^2}{C} \tag{7}$$

When C is in farads, V in volts, and Q in coulombs, the energy is in joules.

Example A parallel-plate air capacitor has plates 1.50 m² in area, spaced 0.800 mm apart. It is charged to 1,200-V potential difference. What energy does it contain? What energy would it contain if it were filled with a dielectric of relative permittivity 3.0 and then charged? What energy would it contain if it were first charged as an air capacitor and then filled with this dielectric?

$$C_1 = \frac{\varepsilon_r\varepsilon_0 A}{s}$$

$$= \frac{1\,(8.87 \times 10^{-12}\ \mathrm{C^2/N \cdot m^2})(1.50\ \mathrm{m^2})}{8.00 \times 10^{-4}\ \mathrm{m}}$$

$$= 1.66 \times 10^{-8}\ \mathrm{F}$$

$$\mathcal{W}_1 = \tfrac{1}{2}CV_2$$

$$= \tfrac{1}{2}(1.66 \times 10^{-8}\ \mathrm{F})(1.20 \times 10^3\ \mathrm{V})$$

$$= 1.20 \times 10^{-2}\ \mathrm{J}$$

$$C_2 = \frac{\varepsilon_r\varepsilon_0 A}{s}$$

$$= \frac{3\,(8.87 \times 10^{-12}\ \mathrm{C^2/N \cdot m^2})(1.50\ \mathrm{m^2})}{8.00 \times 10^{-4}\ \mathrm{m}}$$

$$= 4.99 \times 10^{-8}\ \mathrm{F}$$

$$\mathcal{W}_2 = \tfrac{1}{2}CV^2$$

$$= \tfrac{1}{2}(4.99 \times 10^{-8}\ \mathrm{F})(1.20 \times 10^3\ \mathrm{V})$$

$$= 3.59 \times 10^{-2}\ \mathrm{J}$$

If the capacitor is charged first and then filled with the material dielectric, the charge remains constant and the potential difference changes from 1,200 to 400 V.

$$Q = C_1V_1 = (1.66 \times 10^{-8}\ \mathrm{F}) \times (1.20 \times 10^3\ \mathrm{V})$$

$$= 20.0\ \mu\mathrm{C}$$

$$\mathcal{W}_3 = \frac{1}{2}\frac{Q^2}{C_2} = \frac{1}{2}\frac{(2.00 \times 10^{-5}\ \mathrm{C})^2}{4.99 \times 10^{-8}\ \mathrm{F}}$$

$$= 4.00 \times 10^{-3}\ \mathrm{J}$$

What happens to the energy that is lost when the dielectric is inserted? The dielectric is attracted into the space, and work is done on the agent, restraining it from accelerating. This energy would be restored to the capacitor if the dielectric were pulled out.

SUMMARY

A *capacitor* is a device in which electricity temporarily may be stored.

The *capacitance* of a capacitor is defined by

$$\text{Capacitance} = \frac{\text{charge}}{\text{potential}} \qquad C = \frac{Q}{V}$$

The *farad* is the capacitance of a capacitor that acquires a charge of one coulomb when the potential difference between its plates is one volt ($1\ \mu F = 10^{-6}$ F; 1 pF $= 10^{-12}$ F).

The *permittivity* ε of a dielectric is defined from Coulomb's law for the force between charges. For empty space

$$\varepsilon_0 = 8.87 \times 10^{-12}\ C^2/N\cdot m^2$$

The *relative permittivity* ε_r is the ratio $\varepsilon/\varepsilon_0$ and is dimensionless.

A *parallel-plate capacitor* has a capacitance given by

$$C = \frac{\varepsilon A}{s} = \frac{\varepsilon_r \varepsilon_0 A}{s}$$

The *dielectric strength* of a material is the minimum potential gradient that will cause a disruptive discharge.

When capacitors are connected in *parallel,*

$$V = V_1 = V_2 = V_3 = \cdots$$

$$Q = Q_1 + Q_2 + Q_3 + \cdots$$

$$C = C_1 + C_2 + C_3 + \cdots$$

For capacitors in *series,*

$$V = V_1 + V_2 + V_3 + \cdots$$

$$Q = Q_1 = Q_2 = Q_3 = \cdots$$

$$\frac{1}{C} = \frac{1}{C_1} + \frac{1}{C_2} + \frac{1}{C_3} + \cdots$$

A charged capacitor contains energy as indicated by

$$\mathcal{W} = \tfrac{1}{2}QV = \tfrac{1}{2}CV^2 = \frac{1}{2}\frac{Q^2}{C}$$

Questions

1 State the relationships between the total and individual voltages, charges, and capacitances for capacitors connected in parallel; in series. When is each arrangement used?

2 To achieve very large capacitances, what three variables must be considered in designing the capacitor?

3 Explain why in the design of capacitors the binding posts and surface insulators are made of materials which have exceedingly high resistivities. Is this equally true of resistance boxes?

4 A fairly good analogy to a capacitor is a steel tank into which air is pumped. Show how the phenomena in these two cases are analogous.

5 In automobile-ignition circuits a capacitor is placed in parallel with the breaker points. Show how this greatly reduces the sparking at these points. What is the cause of the sparking?

6 An experimental parallel-plate air capacitor is charged to a certain voltage. It is then immersed in oil. What happens to the capacitance? What happens to the charge and to the potential difference?

7 A gold-leaf electroscope is fitted with a pair of plates instead of the usual knob. The plates are separated by a thin insulator. A 90-V battery is connected to the plates. What happens to the gold leaf? With the battery still connected, the plates are separated to a considerable distance. Explain what happens to the gold leaf. Explain what would happen if the battery were disconnected before the plates are separated.

8 Show how the relative permittivity of a material may be measured by the use of an experimental parallel-plate capacitor.

9 The potential difference across a given capacitor is varied and the charge is noted for each value of the potential difference. (*a*) Plot a curve of the charge as a function of the potential difference. (*b*) Plot a curve of capacitance as a function of the potential difference.

10 The relative permittivity of a material is frequently measured in the laboratory by measuring the capacitance of a parallel-plate capacitor with the material as a dielectric and measuring the capacitance with the same plates separated by the same thickness of air and then dividing the first capacitance by the second. Show why this procedure is justified.

11 Show that the electric field intensity between the plates of a parallel-plate capacitor is given by $\varepsilon = V/s$, where V is the potential difference between the plates and s is the thickness of the insulator.

12 The relative permittivity of distilled water is very high, about 80. Why is such water never used for the dielectric in capacitors? Why is mica used so widely when porcelain and glass are much better insulators?

13 Two unlike capacitors of different potentials and charges are placed in parallel. What happens to their potential differences? How are their charges redistributed?

14 What is the capacitance of a short-circuited capacitor?

15 Suggest some differences in design which might be found in a 1-μF, 120-V capacitor and a 1-μF, 600-V capacitor of equal quality. Also between an inexpensive capacitor and a high-grade type with the same electrical characteristics.

16 Three capacitors rated at 100 V and 1, 2, and 3 μF are connected in series. If the group is joined to a 300-V circuit, which capacitor is likely to puncture first? Answer the same question for a parallel connection of these capacitors. Explain.

17 Plot rough graphs to show the growth and decay of *current* in the charge and discharge of capacitors.

18 The dielectric properties of many insulators used in capacitors are greatly affected by temperature. Explain why capacitors designed for dc or low-frequency circuits cannot be used in high-frequency circuits. Would this be true of air capacitors?

19 Discuss the following characteristics of dielectrics in their relationships to use in capacitors: dielectric constant, dielectric strength, dielectric loss (electric hysteresis), insulation resistance, temperature coefficients, and tensile and compressive properties.

20 What type of energy is stored up in a capacitor? in a storage cell? in an inductive circuit? Why are capacitors not used in place of secondary batteries for the storage of electric energy?

21 A capacitor is separately charged to a series of voltages, and the energy thus stored is determined for each value of the voltage. Sketch the energy-vs.-voltage curve which would be obtained from such an experiment.

Problems

1 Two parallel plates are arranged as a capacitor having air as the dielectric and being 0.5 cm apart. If a capacitance of 20×10^{-12} F exists when a 200-V source is connected across the plates, what is the charge on the capacitor?

2 If the air dielectric is replaced by mica of the same thickness (0.5 cm) in the capacitor in Prob. 1, what additional charge does the capacitor acquire? *Ans.* 12×10^{-11} F; 2×10^{-8} C.

3 Calculate the area in square miles of a parallel-plate air capacitor having a capacitance of 1.0 F when the plates are separated by 1.0 mm.

4 A certain capacitor having a capacitance of 2.00 μF is charged to a difference of potential of 100 V. What is the charge on this capacitor? *Ans.* 200 μC.

5 Calculate the capacitance of a capacitor made of 21 circular plates separated by sheets of mica 0.518 mm thick. The diameter of each plate is 125 mm and the relative permittivity of the mica is 7.18.

6 A 0.500-μF capacitor is placed in parallel with a 0.750-μF capacitor and the group is joined to a 110-V dc source. What charge is taken from the source? What are the charges on each capacitor? *Ans.* 137 μC; 55.0 μC; 82.5 μC.

7 Two capacitors of 3.0 and 5.0 μF, respectively, are connected in series and a 110-V difference in potential is applied to the combination. What is the potential difference across the 3.0-μF capacitor? What is the energy in the 5.0-μF capacitor?

8 Three capacitors made of paper have capacitances of 0.15, 0.20, and 0.40 μF. If they are arranged in parallel and are charged to a potential difference of 200 V, (*a*) what is the charge on each capacitor, (*b*) what is the total capacitance, and (*c*) what is the total charge?

Ans. 30 μC, 40 μC, 80 μC; 0.75 μF; 1.5×10^2 μC.

9 A potential difference of 75.4 V is applied to a combination consisting of a 1.25-μF capacitor and a 0.572-μF capacitor, connected in series. (*a*) What is the charge on each capacitor? (*b*) What is the potential difference across the 1.25-μF capacitor?

10 Three capacitors of capacitance 2.0, 3.0, and 4.0 μF are connected in series across a 1,300-V line. What is the drop in potential across each capacitor? *Ans.* 6$\bar{0}$0 V; 4$\bar{0}$0 V; 3$\bar{0}$0 V.

11 A 2.0-μF capacitor is charged to a potential difference of 100 V, momentarily connected to an uncharged 8.0-μF capacitor and then removed. Determine the potential and charge of the 8.0-μF capacitor.

12 Six $\frac{1}{2}$-μF capacitors are connected first in series and then in parallel. What are the respective joint capacitances of the combinations? What charge appears on each capacitor in each case when the group is connected to a 600-V battery? *Ans.* 83 pF; 3.0 μF; 50 μC; 3$\bar{0}$0 μC.

13 Two capacitors of capacitance 2.0 and 4.0 μF are connected in parallel. This group is connected in series with a 3.0-μF capacitor across an 800-V line. Find the potential difference across each capacitor.

14 Three capacitors of capacitance 0.100, 0.200, and 0.500 μF, respectively, are connected in parallel and the group then connected in series with another group of 0.100, 0.200, and 0.500 μF, respectively, connected in series. Calculate the total capacitance. *Ans.* 54.7 pF.

15 Two capacitors of 3.0 and 5.0 μF are connected in series, and a 110-V difference in potential is applied to the combination. What is the potential difference across the 3.0-μF capacitor? What is the energy in the 5.0-μF capacitor?

16 A capacitor of capacitance 3.00 μF is charged to a potential of 150 V. If the voltage were changed to 500 V in 5.25 ms, what average current would there be? *Ans.* 0.20 A.

17 A 10-μF capacitor charged to a potential difference of 1,200 V is connected terminal to terminal to an uncharged 20-μF capacitor. What is the resulting potential difference?

18 A 5.0-μF capacitor is charged to a potential difference of 800 V and discharged through a conductor. How much energy is given to the conductor during the discharge? *Ans.* 1.6 J.

19 Three 1.00-μF capacitors are charged to potentials of 100, 200, and 300 V. The capacitors are then connected in series and joined to a 450-V battery. What is the total energy of the system before and after the series connection is made? How do you account for the difference?

20 Three capacitors of capacitance 2.0, 3.0, and 6.0 μF, respectively, are charged by a 60-V battery. Find the energy of the stored charge when the capacitors are connected (*a*) in parallel and (*b*) in series. *Ans.* 0.020 J; 0.0018 J.

21 A 10-μF capacitor charged to a potential difference of 1,000 V is connected terminal to terminal to an uncharged 40-μF capacitor. What is the resulting potential difference?

22 A 2.0-μF capacitor is connected in series with a resistor of 5,$\bar{0}$00 Ω across a 1,$\bar{0}$00-V line. What is the current at the instant the switch is closed? If the current remained at this value, how long would it take to charge the capacitor fully? *Ans.* 0.30 A; 1.0 × 10^{-2} s.

23 A 10-μF capacitor *A* is charged to a potential difference of 500 V. A 15-μF capacitor *B* is charged to a potential difference of 1,000 V. The positive terminal of *A* is now connected to the negative terminal of *B* and the negative terminal of *A* to the positive terminal of *B*. What is the resulting potential difference?

24 A 4.0-μF capacitor is connected in series with a 2,$\bar{5}$00-Ω resistor across a 500-V line. What is the current in the resistor at the instant (*a*) the switch is closed, (*b*) the capacitor is half charged, and (*c*) the capacitor is fully charged? *Ans.* 0.20 A; 0.10 A; zero.

Polykarp Kusch, 1911–

Born in Blankenberg, Germany; lived in United States from 1912. Studied at Case Institute of Technology, University of Illinois, University of Minnesota. Conducted research at Westinghouse Electric Corporation, Bell Telephone Laboratories, and Columbia University. Professor of physics, Columbia University. Shared the 1955 Nobel Prize for Physics with Lamb for his precise determination of the magnetic moment of the electron.

Willis E. Lamb, Jr., 1913–

Born in Los Angeles. Studied at the University of California at Berkeley. Taught at Columbia and Stanford Universities. Professor of physics at Yale University. Shared the 1955 Nobel Prize for Physics with Kusch for his discoveries concerning the structure of the spectrum of hydrogen.

35

Electric Current

We live in an age of electricity. Homes and factories are lighted by electricity; communication by telegraph, telephone, radio and television depends upon the use of electricity; and the industrial applications of electricity extend from the delicate instruments of measurement and control to giant electric furnaces and powerful motors. Electricity is a useful servant of man—a practical means of transforming energy to the form in which it serves his particular need. The effects of electricity both at rest and in motion are well known, and the means to produce these effects are readily available.

This chapter is designed to give a preliminary (and hence necessarily superficial) preview of the major sources and effects of electric currents. It is helpful to survey the field broadly before beginning more detailed studies of the individual portions.

We have discussed at length the properties of charges and charged bodies, the forces between them, their electric fields, and their attributes as systems possessing electric potential energy. Now that this groundwork has been laid, we are prepared to study charge in motion, the understanding and development of which has brought about the electrical marvels with which we are familiar. In this chapter we shall be concerned with steady currents, the conditions for their production, and a few of their effects.

35-1 POTENTIAL DIFFERENCE IN CONDUCTORS

It has been shown in the foregoing chapters that, if a distribution of charge is created on a conductor in such a way that a potential difference exists between two points on the conductor, an electric field exists in that region. This field tends to cause a flow of positive charge in the direction of the field or a flow of negative charge in the opposite direction. Work is done on the charge by the field as this flow takes place. In the case of an isolated conductor the work is done at the expense of the potential energy stored in the original charge distribution; when the potential difference across the conductor is removed by the flow of charge, the field disappears and the flow ceases.

Let us suppose that a constant potential difference between two points on a conductor can be maintained in spite of the resulting flow of charge. Then the flow of charge will continue between the points at a constant rate. A time rate of flow of charge is called an *electric current*. Such currents constitute one of the most widely used means of transmitting energy in the modern world. In order to maintain the potential difference and current, charge must be continuously supplied to the place of higher potential and removed from the place of lower potential along some path other than that taken by the current already mentioned. Energy must also be supplied by some outside agency to effect the transfer of charge and to maintain the potential difference. Hence, in order to maintain a constant current, a complete conducting loop, or *circuit,* must be established as well as a means for converting some other form of energy to electric energy.

35-2 QUANTITY OF ELECTRICITY AND CURRENT

In Sec. 32-8 reference was made to the coulomb as the mks unit of quantity of electricity or charge. Electric current may be related to charge by the definition: *Electric current is the time rate of flow of charge.* In the form of an equation,

$$\text{Current} = \frac{\text{charge}}{\text{time}} \qquad I = \frac{Q}{t} \qquad (1)$$

If the rate of flow of charge is variable, the instantaneous current is given by $I = \lim_{\Delta t \to 0} \Delta Q/\Delta t$.

The mks unit of current, called the ampere, is named in honor of André Marie Ampère (1775–1836), French physicist. The ampere is defined in Chap. 38 in terms of the magnetic effects of electric current. It is equivalent to a rate of flow of charge of one coulomb per second.

35-3 THE NATURE OF ELECTRIC CURRENT

The flow of charge caused by an electric field may, in the case of a gas or liquid, consist of a flow of positive ions in the direction of the field or of negative ions or electrons opposite to that direction, or of both at once. In a metal the flow is known to consist largely of a movement of electrons opposite to the direction of the field. A large number of electrons in a metal are relatively free to move about, with random thermal motions. In the absence of an electric field this motion produces no net flow of charge in any particular direction. Upon the application of an electric field to the metal, however, the velocities of the electrons moving opposite to the field are increased, and those in the direction of the field are decreased, causing a net flow of charge. The electrons are repeatedly deflected or stopped by processes associated with imperfections in the metallic crystals, impurities, and the thermal motions of the atoms. Hence the flow is not an accelerated one but rather a drift or diffusion process. The average drift velocity is low—of the order of 0.1 mm/s in a typical case. However, changes in the electric field which produce changes in the flow rate are propagated with a speed approaching the speed of light.

Some objects are classified as *conductors* of electricity, for example, metals, and others as *nonconductors* or *insulators,* such as glass or rubber. Conductors differ from insulators in the ease with which electrons leave their "parent" atoms and move through the conductor to constitute an electric current. There is a third type of material called *semiconductors* which have few electrons available for conduction. Silicon is an example of a semiconductor. The resistance to electron flow in semiconductors can be significantly affected by adding small amounts of an impurity, such as arsenic or gallium, to the silicon crystal (Sec. 48-9). With certain types of impurities added, additional conduction electrons are supplied by the impurity (arsenic) and the conduction is increased. This is called a *n-type* semiconductor. On the other hand, if we add other types of impurities to the semiconductor, gallium to the silicon, for example, the impurity can borrow electrons from the semiconductor and produce "electron gaps," or "holes." When an electric field is applied across the crystal with this type of impurity, electrons move toward the anode (+ electrode) by successively filling holes. The flow of charge in such a case is described by referring to the movement of the holes, which act like positive charges and move to the cathode (− electrode). This so-called *p-type* semiconductor has its conductive ability increased by the migration from atom to atom of vacancies caused by the removal of electrons. This technique of adding impurities to semiconductors is used in making *transistors,* which will be discussed later in the modern physics section of this book.

We can see, then, that an electric current may consist of a flow of positive charges, or of negative charges, or of both at once (see Fig. 35-1).

To imply that all materials fit into either a conductor, semiconductor, or nonconductor category does not take into account the fact that variations of conductive ability sometimes appear due to changes in environmental conditions. For example, one of the most striking changes occurs when certain objects are cooled to very low temperatures. It had been anticipated that the slowing down of the vibrational motion of the atoms as the object approached absolute zero, 0°K, would cause the resistance to electron flow to decrease. However, it was not anticipated, as Heike Kamerlingh Onnes discovered in 1911, that as they were cooled to very low temperature certain materials would become perfect conductors of electricity—*superconductors,* in which the resistance actually disappears. The causes of superconductivity are actively being studied through a whole new field of physics called "low-temperature physics," or *cryogenics.* Through these studies it has been found that metals which are normally poor conductors (e.g., lead and tin) become the best superconductors. Also objects made up of heavier atoms will less likely become superconductors. In a related study, Walther Meissner, a German physicist, found that metals in a superconducting state repel magnetic fields. When placed between magnetic poles, the superconducting material will cause the magnetic force lines to deflect around it. While it is difficult to maintain objects at such low temperatures, engineers are currently working on developing ways that superconductors can be used. A list of potential uses includes frictionless bearings, tiny switching devices for computers, noiseless amplifiers, and electric motors having great efficiency.[1]

Benjamin Franklin recognized that a current could be thought of as the motion of only one type of charge. Either a flow of positive charge to the right or of negative charge to the left could be called a current to the right. Any current direction is a convention, and the choice is arbitrary. But it is highly desirable that the choice be made so that a current directed toward a specific region, such as a capacitor plate, contributes to an increase of positive charge at that region, for the sake of consistency in setting up equations regarding charge, current, potential, etc.

[1]Charles McCabe and Charles Bauer, "Metals, Atoms and Alloys," Vistas of Science Book, National Science Teachers Association, Washington, D.C., 1964.

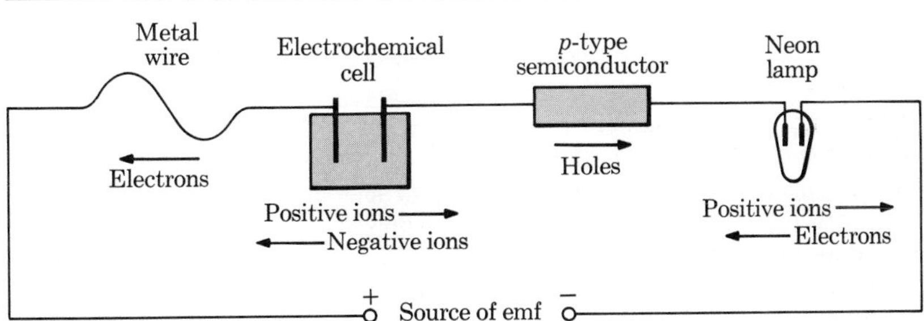

Figure 35-1
The clockwise conventional current in this circuit represents the flow of different kinds of charge carriers in different parts of the circuit.

35-4 HINTS TO THE READER

The choice which has historically been used by most physicists and electrical engineers is that *the current in a conductor shall be designated to have the direction of the net flow of positive charge.* Hence in the cases of metallic conductors and some vacuum tubes, in which electron flow predominates, it is necessary to say that the *conventional current is in one direction and the electron flow is in the other.* Thinking in both terms at once is very desirable for the student, particularly in the study of transistors and semiconductors.

35-5 ELECTROMOTIVE FORCE

In order to maintain an electric current, some agency is required to expend energy in moving the charge around a circuit. With the exception of a few metals near absolute zero, the superconductors, all conductors present some opposition to the flow of charge so that work must be done to maintain a current. An agency capable of causing such a flow by converting other forms of energy to electrical work is called a seat of electromotive force or a *source* of current. It should be clearly understood that a source of current does not manufacture charge but merely moves the charge through a circuit. In most circuits this agency is concentrated in one or a few parts of the circuit. The source must create an electric field in all parts of the circuit to cause the charges to move against the various opposing effects they may encounter. The *electromotive force,* or *emf,* of a source is the energy per unit charge transformed in a reversible process. (The term "electromotive force" is an old term now rooted in the language of physics; its choice was unfortunate, as this quantity is not a force. Hence its abbreviation *emf* will be used hereafter.) In the mks system, emf is measured in volts. An emf causes differences of potential to exist between points in the circuit. Thus there is an intimate relation between emf and potential difference. An emf is associated only with reversible conversions of energy, whereas potential differences exist not only in sources of emf but also in resistors, which convert energy to heat irreversibly. The distinction is sometimes useful and will become clearer as we proceed.

In the simple circuit shown in Fig. 35-2, as charge flows through the circuit, the cell converts chemical energy to electrical energy, giving rise to an emf. The lamp is a resistive conductor called a *resistor;* it converts electric energy to heat, and the work done on the charge by the electric field, as the charge moves through the resistor, is evidenced by the presence of a potential difference between the ends of the resistor. A small amount of the total electric energy converted from chemical energy in the cell also produces heat inside the cell.

Electric circuits are conventionally represented by circuit diagrams employing standard symbols.

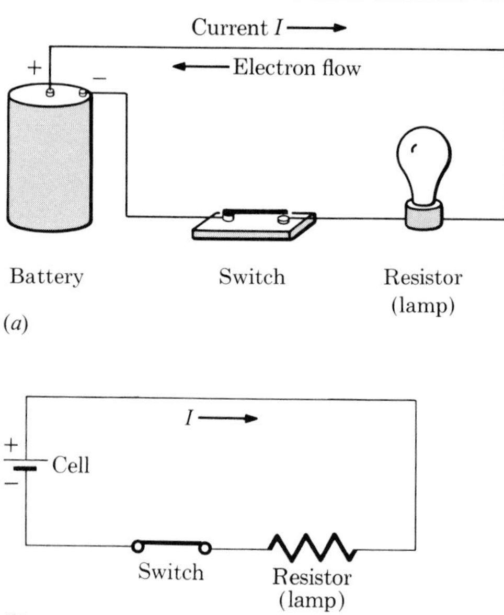

Figure 35-2
(*a*) A simple electric circuit. (*b*) A schematic diagram of the simple circuit of *a*.

The student will become acquainted with these as we proceed. An example is given with Fig. 35-2*b*.

35-6 SOURCES OF EMF

Chemical sources of emf make use of the fact that in many chemical reactions, electrons are liberated at one place and absorbed at another. The cycle is frequently closed in a small region, producing no external electrical effects. But it was discovered by Luigi Galvani (1737–1798) and Alessandro Volta (1745–1827) that a cell could be constructed in which electrons were liberated from a solution onto a zinc plate and absorbed into the solution from a copper one, causing a potential difference to exist between the plates. Such galvanic cells provided the first source of large, continuous currents. Although the use of chemical sources is now largely limited to moving or portable equipment, their development is still actively pursued. The *primary cell,* of which the dry cell is an example, deteriorates as it supplies electrical energy to an external circuit and must be replaced.

The most commonly used galvanic cell is the so-called *dry cell* (Fig. 35-3). The positive electrode of this cell is a carbon rod and the negative terminal is the zinc container for the cell. A layer of paper moistened with ammonium chloride (NH_4Cl) is placed in contact with the zinc, while the space between this and the central carbon rod is filled with manganese dioxide and granulated carbon moistened with ammonium chloride solution. The ammonium chloride is the electrolyte, and in the chemical reaction, hydrogen is liberated at the carbon electrode. The hydrogen reduces the effectiveness of the cell for two reasons: (1) the gas increases the internal resistance of the cell, and (2) the hydrogen produces a reverse emf at the carbon electrode. The latter effect is known as *polarization*. The manganese dioxide acts as a depolarizing agent by reacting with the hydrogen to form water. The cell polarizes when it is used but recovers slowly as the manganese dioxide reacts with the hydrogen. Because of this behavior, the cell should not be used continuously. The voltage of the dry cell is slightly more than 1.5 V.

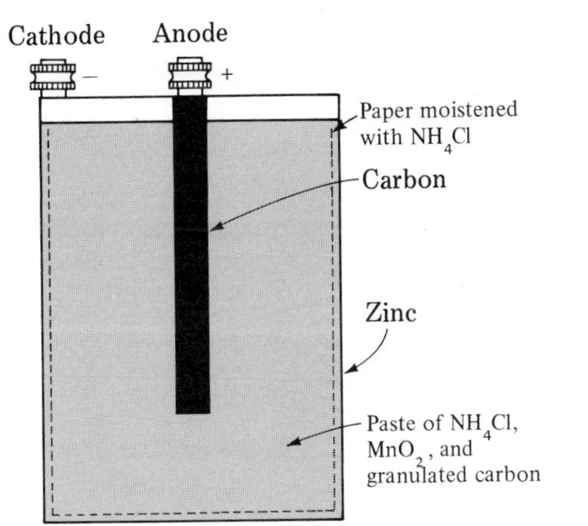

Figure 35-3
A dry cell.

The storage cell differs from the primary cell in that it can be recharged by the use of a reverse current from an outside source. This cell transforms electric energy into chemical energy during the charging process. During discharge, chemical energy is transformed into electric energy, as in the case of the primary cell. The amount of energy that can be stored depends upon the size of the plates. A large cell has exactly the same voltage as a small cell, but the energy available in it when fully charged is much greater than that in the small cell. A relatively new type of electrochemical converter, the *fuel cell,* which is also discussed in Chap. 18, utilizes a continuous supply of hydrogen or hydrocarbon fuel to produce electrical energy. It is now under intensive theoretical and experimental development.

It was discovered independently by Michael Faraday and Joseph Henry that, when a magnetic field changes in intensity, an electric field is set up in that region. This field will produce an electric current in a properly placed conductor. One such arrangement is shown in Fig. 35-4. When the magnetic field through the coil is increased

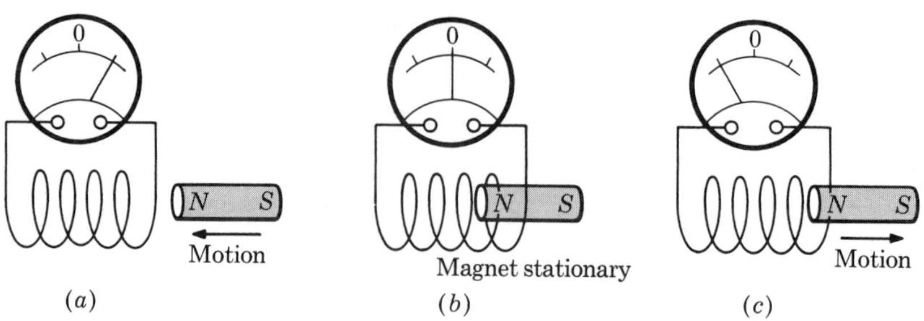

Figure 35-4
Electromagnetic induction. The induced current is indicated by a galvanometer.

in intensity, by moving a magnet or otherwise, an emf is created in one direction in the loop, thus producing a current. When the magnetic field intensity is decreased or reversed, a current in the opposite sense is induced. If the magnetic field is steady, no electric field or current appears. This principle is the basis of innumerable sources of emf, from large power-station generators to tiny microphones.

A commonly used source of electric current in which heat is transformed into electric energy is the thermocouple illustrated in Fig. 35-5. In the diagram there is shown a loop consisting of a piece of iron wire joined to a piece of copper wire. One of the junctions is heated by a flame, causing electrons to flow around the circuit. The flow will continue as long as one junction is at a higher temperature than the other junction. Such a device, consisting of a pair of junctions of dissimilar metals, is called a *thermocouple*. The main commercial use of thermocouples is for the measurement and control of temperature. However, recent development of thermoelectric generators utilizing ceramic and other semiconducting materials promises important application in supplying electric energy.

Figure 35-5
A thermocouple.

If light falls on a clean surface of certain metals, such as potassium or sodium, electrons are emitted by the surface. This phenomenon is called the *photoelectric* effect. If such a metallic surface is made a part of an electric circuit, the electric current in the circuit is controlled by the light.

In the circuit of Fig. 35-6*a* the photoelectric effect is not the principal source of emf in the circuit, but its emf adds to that of the battery to produce a total emf dependent on the character and brightness of the light. If the light is bright, the current will be larger than if the light is dim. This device is known as a *photoelectric cell* and serves as a basis for most of the instruments that are operated or controlled by light, such as television, talking motion pictures, wire or radio transmission of pictures, and many industrial devices for counting, rejecting imperfect pieces, and control. The photoelectric cell is an example of a device that transforms radiant energy into electric energy.

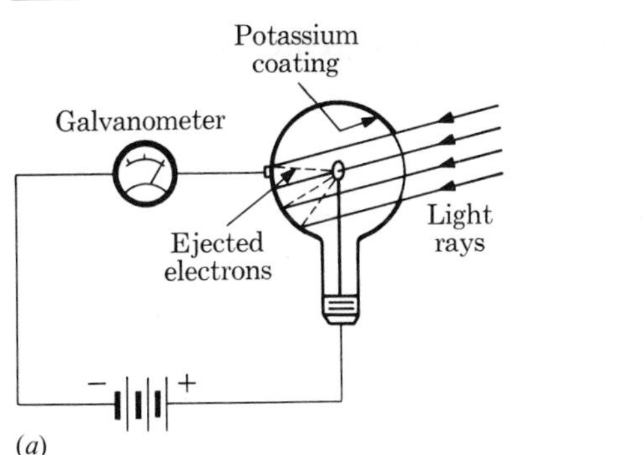

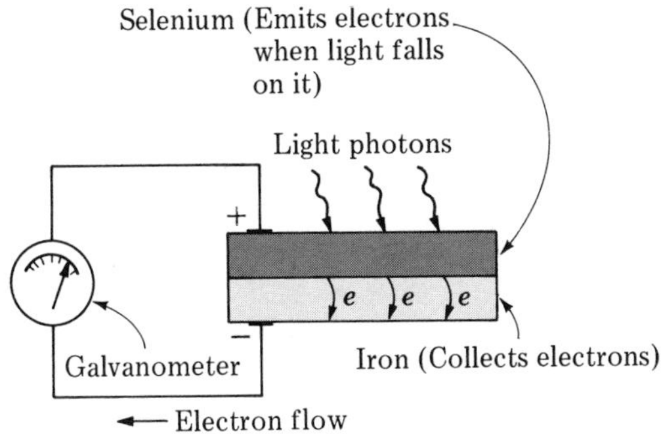

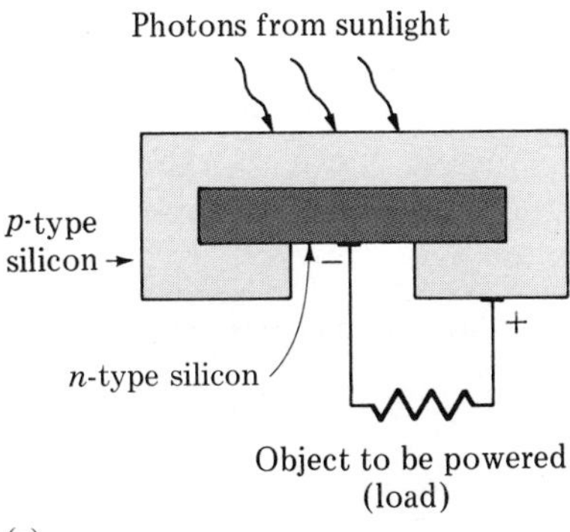

Figure 35-6
(*a*) A photoelectric cell. (*b*) Photoelectric cell consisting of a metal which loses electrons upon being struck by light (photons) and a metal which collects these electrons. (*c*) A photoelectric cell used as a solar battery.

In Figure 35-6*b*, the operating principle of the light meter used in photography is illustrated. The energy of visible light, falling upon a thin layer of selenium, strips electrons from these atoms and deposits them on the iron which readily collects electrons. This builds up an excess of electrons in the iron and a deficiency of electrons in the selenium. If a conductor, a wire, is connected from the iron to the selenium, incorporating an ammeter in the circuit to show the passage of charge, an electric current will be shown. Since the amount of electrical current observed is proportional to the intensity of the light falling on the selenium, the instrument will indicate the amount of light present and can be used as a *light meter*.

A different type of photoelectric cell, popularly called the "solar cell," is used as the principal source of emf for some circuits in space devices. Solar batteries are being developed for the more effective utilization of solar power on earth.

A typical solar battery consists of a silicon crystal "sandwich," the outside of which has had boron, an electron acceptor, added to it as an impurity, making this a *p*-type silicon. The impurity arsenic, an electron donor, has been added to the inside layer of silicon making it an *n*-type silicon. When photons from the sun hit the crystals, ejected electrons move to the *n*-type region and the "holes" move to the *p* type between the *p* and the *n* sections as in Fig. 35-6*c*, electrons will flow which can be used to power a load, e.g., a lamp.

Solar batteries are practically maintenance-free, are long lasting, and are pollutant-free, but are not yet especially efficient. Even though the latter is true, improvements in design have led to the development of solar batteries which can convert up to 15 percent of the energy of the sunlight falling upon them. A square yard of flat land receives about 1,000 W of power from bright sunlight. Covered with present-day solar cells, a square yard would yield a constant 150 W of electrical power, enough to power a television receiver. Silicon solar cells were used to power the radios on space satellites and provided sufficient power to transmit from nine milllion miles in space. The long life of solar batteries is shown by the fact that the satellite Vanguard I, launched March 17, 1958, has been transmitting to earth through the power created by solar cells ever since. The potential uses of solar batteries is almost unlimited. In this age when the world's supply of fossil fuel is diminishing and our concern for atmospheric pollution grows, it becomes ever more important that this source of power be developed.

Another source of electric current has become of importance in such devices as microphones, oscillators, phonograph pickups, and frequency stabilizers. These instruments utilize crystals which, when slight pressures are applied, produce tiny emfs which may be amplified and used. This is known as the *piezoelectric effect*.

Such crystals, for example quartz, tourmaline, and Rochelle salts, when placed between a pair of metal plates and subjected to mechanical stress, i.e., by squeezing or striking, show a separation of electric charge with some regions of the crystal becoming positively charged and others negatively charged. In such crystals, if the stress is alternated so that it at first compresses and then "stretches" the crystals, the polarity is reversed. In such an alternating situation, a sharply defined resonance point may be determined. Because of this, such crystals are used as frequency standards and in controlling the frequencies of radio stations.

In all these sources of electric current some type of energy is used to set the electrons in motion. Chemical, mechanical, thermal, or radiant energy is transformed into electric energy.

35-7 EFFECTS OF ELECTRIC CURRENT

In Fig. 35-7 is shown a lamp L and an electrolytic cell Z that contains two platinum electrodes a and b immersed in a weak acid solution. A magnetic compass C is placed directly over the wire. When a current from left to right is maintained in this apparatus, characteristic effects of the electric current are observed. The filament of wire in the incandescent lamp becomes so hot that it begins to glow. The water in Z presents a very interesting appearance. Bubbles of gas come from the surfaces of the electrodes a and b (twice as much from a as from b). Tests show that hydrogen gas is being given off at a and oxygen at b. Since oxygen and hydrogen are the gases that combine to form water and since the water in Z is disappearing, it is natural to conclude that the water is being divided into its constituents (hydrogen and oxygen) by the action of the electric current.

Compass C, which points north (along the wire in Fig. 35-7) when the switch is open, is deflected when the switch is closed. This indicates that a magnetic effect is produced in the vicinity of an electric current. A phenomenon so simple as the deflection of a compass needle hardly indicates the importance of the magnetic effect of an electric current, but it is this magnetic effect which makes possible the operation of electric motors as devices by means of which electric currents perform mechanical work.

Various other effects of electric currents might be mentioned, but they can be classified as combinations of the three main effects. For example,

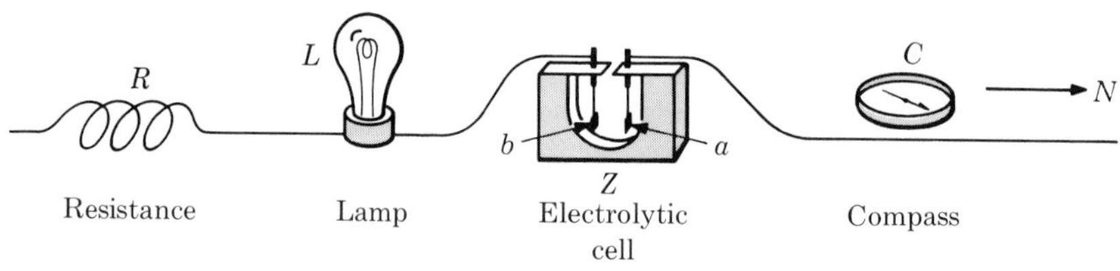

Figure 35-7
A circuit showing three effects of an electric current.

the *optical effect* observed in light sources such as electric lamps, advertising signs, and fluorescent lights is caused chiefly by heating effects. The *physiological effects* that one experiences when he receives an electric "shock" are caused by a combination of heating and chemical effects.

35-8 OHM'S LAW

The various sources of emf described above are all agencies for moving charges through a circuit at the expense of energy. This energy is converted in the circuit from electrical to some other form. In a motor the final form is mechanical energy; in a radio transmitter it is radiant energy. The most common conversion in circuits is from electrical energy to heat. This occurs because the electrons are scattered and stopped repeatedly in their passage through a conductor, yielding some of their energy to add to the thermal vibrations of the atoms of the conductor. Various materials are widely different in their opposition to the flow of electrons, but, except for a few metals at temperatures near absolute zero, all materials absorb energy when a current is maintained in them.

In 1826 Georg Simon Ohm discovered that for metallic conductors there is a substantially constant ratio of the potential difference between the ends of a conductor to the current in the conductor. This constant ratio is called the *resistance* of the conductor. This relationship is called *Ohm's law*. In equation form,

$$\frac{\text{Potential difference}}{\text{Current}} = \text{resistance (a constant)}$$

$$\frac{V}{I} = R \qquad (2)$$

The mks unit of electrical resistance is the ohm Ω, which is the resistance of a conductor such that a potential difference of one volt will maintain a current of one ampere.

Example The difference of potential V_1 between the terminals of an electric heater is 120 V when there is a current I_1 of 8.00 A in the heater. What current will be maintained in the heater if the difference of potential is increased to 180 V?

Ohm's law indicates that the resistance R will remain the same when the potential difference is increased; hence we can write

$$R = \frac{V_1}{I_1} = \frac{120\ \text{V}}{8.00\ \text{A}} = 15.0\ \Omega$$

$$I_2 = \frac{V_2}{R} = \frac{180\ \text{V}}{15.0\ \Omega} = 12.0\ \text{A}$$

Ohm's law has been found to be valid for a wide range of currents in metallic conductors, and hence their resistances are a definite measure of one of their physical properties, provided that other properties (temperature, pressure, tension, etc.) are kept constant. However, Ohm's law is not valid for all conducting media; for example, in some conductors (electronic tubes, arcs, ionic conductors) there is not a direct proportion between V and I, and I may actually decrease as V is increased.

Ohm's law may be applied to an entire circuit or to any part of a metallic circuit, provided that the part does not contain a source of emf. Hence Ohm's law may not be applied to any part of a circuit containing, for example, a cell, a generator, or a motor.

It is important to distinguish carefully between the application of Ohm's law to a complete circuit and to a part of a circuit. When a complete circuit is to be considered, one must take into account all the emfs in the circuit and all the resistances in the circuit. In equation form,

$$\frac{\text{Net emf}}{\text{Current}} = \text{total resistance}$$

$$\frac{\mathcal{E}}{I} = R_t \qquad (3)$$

Whenever only a part of a circuit is to be considered, the potential difference V_1 is the drop in potential across that part and the resistance R_1 is the resistance of that part only,

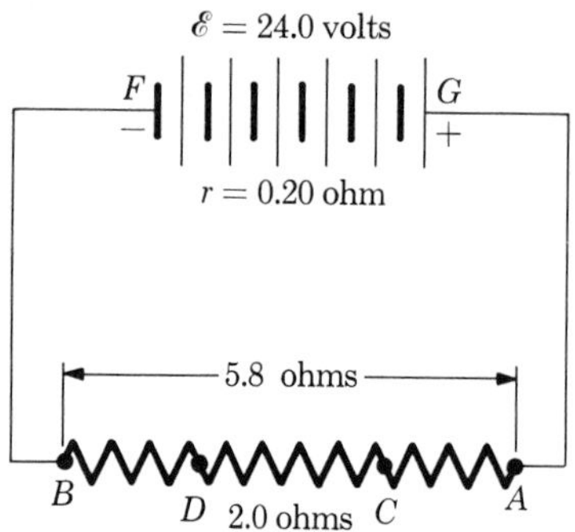

Figure 35-8
Ohm's law applied to a part of a circuit.

$$\frac{V_1}{I_1} = R_1 \qquad (2a)$$

In Fig. 35-8 a simple circuit is shown in which there is both resistance in the resistor AB and internal resistance in the battery. The conductor GA is idealized to have no resistance. All parts of GA are at the same potential, which is the highest potential in the circuit. A potential drop occurs in the conductor AB, in the direction of the conventional current. Conductor BF, again idealized to have zero resistance, is at the lowest potential in the circuit. Considering the entire circuit, Eq. (3) gives

$$I = \frac{\mathcal{E}}{R_t} = \frac{24.0\ \text{V}}{6.0\ \Omega} = 4.0\ \text{A}$$

For the part of the circuit AB, Eq. (2a) gives

$$V_{AB} = IR_{AB} = 4.0\ \text{A} \times 5.8\ \Omega = 23\ \text{V}$$

and for CD

$$V_{CD} = IR_{CD} = 4.0\ \text{A} \times 2.0\ \Omega = 8.0\ \text{V}$$

In the part of the circuit from F to G there is a battery, and Eq. (2a) cannot be applied.

It is important to understand that in a simple circuit, such as that of Fig. 35-8, in which there is only one path for the current, the current is the same in all parts of the circuit. If this were not so, charge would accumulate at certain points, raising the potential of some and lowering the potential of others until the potential differences were such as to equalize the current throughout.

35-9 MEASUREMENT OF EMF, POTENTIAL DIFFERENCE, AND CURRENT

We have discussed emf, potential difference, and current without mentioning how these quantities are measured. This is not yet the place to consider the operation of the instruments used, but it is necessary to understand how they are connected into a circuit. Since the current in a simple circuit is the same in all parts of the circuit, the current-measuring instrument, called an *ammeter,* may be inserted at any place in the circuit and the circuit current will be the reading of the ammeter. A typical connection is shown in Fig. 35-9. One terminal of the ammeter is usually marked with a plus sign. This terminal is connected so that the conventional current will be directed into it. Or, to put it another way, the positive terminal is connected so that when the circuit is traced outward from that terminal it leads to the positive terminal of the principal source of emf. The re-

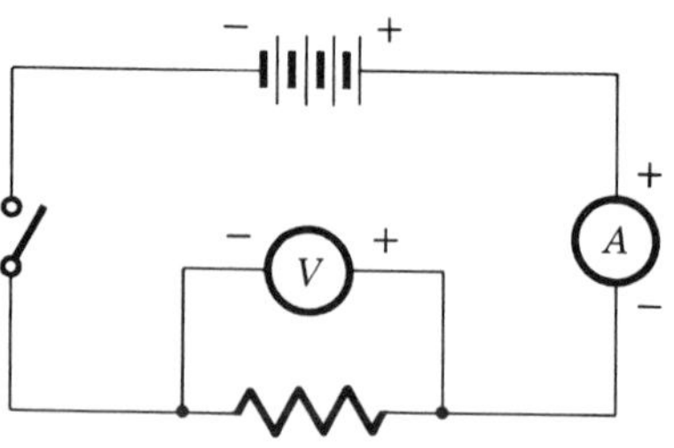

Figure 35-9
Connection of an ammeter and a voltmeter in a simple circuit.

sistance of an ammeter is very low so as to interfere as little as possible with the current that existed in the circuit before the meter was inserted.

The second basic instrument, the *voltmeter,* measures the potential difference between the two points to which the meter is connected. Voltmeters are designed to require negligible currents in proportion to those in the main circuits. In Fig. 35-9 a voltmeter is shown connected to determine the voltage across the resistor. By substituting simultaneous voltmeter and ammeter readings in Eq. (2), an unknown resistance may be measured.

The measurement of emf is less direct than that of current or potential difference, because all sources of emf that deliver current convert some of their own electrical energy into heat within themselves. In the circuit of Fig. 35-10, the effect of internal resistance r in the battery may be found by applying Ohm's law to the entire circuit. The current in the circuit is given by

$$I = \frac{\mathcal{E}}{R_t} \qquad (3a)$$

where $\mathcal{E}$ is the net emf in the circuit and R_t the total resistance. In this case R_t is the sum of the external resistance R and the internal resistance r. Hence

$$I = \frac{\mathcal{E}}{R + r}$$

$$IR = \mathcal{E} - Ir \qquad (3b)$$

A voltmeter placed either across the terminals of the battery or across the ends of the resistor will read this potential difference IR. Hence the terminal potential difference of the battery is dependent on the current. If R is very large and r and I are small, the terminal potential difference is nearly equal to the emf of the battery. If I or r is large, the terminal potential difference drops accordingly by the amount Ir, called the "internal drop" of the battery.

Example What is the terminal potential difference V_B of the battery of Fig. 35-10?

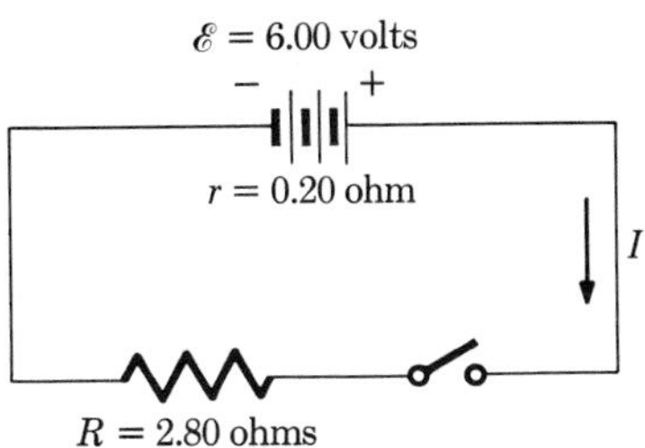

Figure 35-10
The terminal potential difference of a battery.

Applying Eq. (3*b*) to the entire circuit,

$$I = \frac{\mathcal{E}}{R + r} = \frac{6.00\ \text{V}}{3.00\ \Omega} = 2.00\ \text{A}$$

$$V_B = \mathcal{E} - Ir$$

$$= 6.00\ \text{V} - (2.00\ \text{A})(0.20\ \Omega)$$

$$= 5.60\ \text{V}$$

Note the fact that $V_B = IR = (2.00\ \text{A})(2.80\ \Omega) = 5.60\ \text{V}$.

In the case of a storage battery being charged, the emf of the battery is opposite in sign to the current in the circuit. In Fig. 35-11 such a battery is being charged by a generator which has an emf and negligible internal resistance, while the battery has emf and an internal resistance r. Apply-

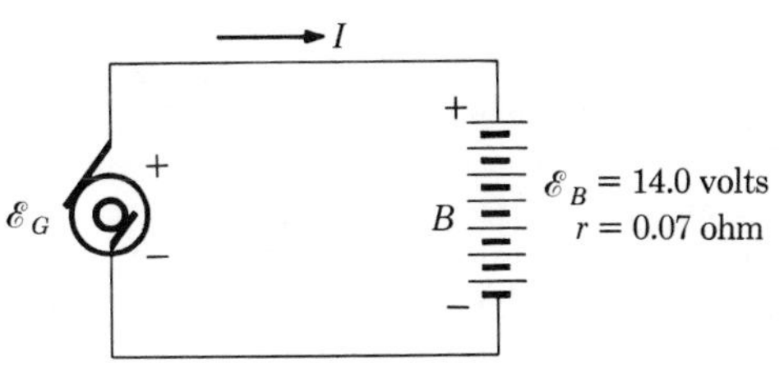

Figure 35-11
A storage battery being charged by a generator.

ing Eq. (3*a*) to the complete circuit,

$$I = \frac{\mathcal{E}_G - \mathcal{E}_B}{r}$$

since $\mathcal{E}_B$ is opposite in sign to $\mathcal{E}_G$ and to the current. Then

$$\mathcal{E}_G = \mathcal{E}_B + Ir$$

Example Find the emf of the generator necessary to maintain a charging current of 20 A in the circuit of Fig. 35-11.

$$\begin{aligned}\mathcal{E}_G &= \mathcal{E}_B + Ir \\ &= 14.0\text{ V} + (20\text{ A})(0.07\ \Omega) \\ &= 14.0\text{ V} + 1.4\text{ V} = 15.4\text{ V}\end{aligned}$$

We may observe from the two examples above the fact that the terminal potential difference of a battery is less than its emf when the battery is discharging and more than its emf when the battery is being charged.

35-10 ELECTRICALLY INDUCED CHEMICAL REACTIONS

The foregoing example of charging a storage cell is one of many cases in which a chemical reaction which would normally proceed in one direction to a condition of lower chemical potential energy is reversed so as to increase the potential energy of the system. A battery is not charged in the sense in which a capacitor is charged, but rather a flow of charge contributes to the removal of material from one set of plates and the deposition of material on another set of plates. In the lead-acid storage cell water is also decomposed into hydrogen and oxygen. This is an example of the process called *electrolysis* (literally, "splitting by electricity") in which a chemical compound is decomposed by an electric current.

35-11 FARADAY'S LAWS OF ELECTROLYSIS

Quantitative measurements made by Faraday (1833) contributed to the understanding of the processes occurring in electrolytic cells and showed a striking relation between the electrolytic behavior and the chemical properties of various substances. Faraday established by experiment the following two laws of electrolysis:

First law. The mass of a substance separated in electrolysis is proportional to the quantity of electricity that passes.

Second law. The mass of a substance deposited is proportional to the chemical equivalent of the ion, i.e., to the atomic mass of the ion divided by its valence.

Faraday's laws may be expressed by the following symbolic statements:

$$m \propto Q \qquad (Q = It)$$

$$m \propto c \qquad \left(c = \frac{\text{atomic mass}}{\text{valence}}\right)$$

whence

$$m = kcQ = zQ = zIt \qquad (z = kc) \tag{4}$$

where k is a proportionality constant, whose value depends only upon the units involved, m is the mass deposited, and z is a constant for a given substance (but different for different substances), which is known as the electrochemical equivalent of the substance under consideration. *The electrochemical equivalent of a substance is the mass deposited per unit charge.* In the mks system it is numerically the number of kilograms deposited in one second by an unvarying current of one ampere.

Example How long will it take to electroplate 3.00 g of silver onto a brass casting by the use of a steady current of 15.0 A? The electrochemical equivalent of silver is 1.1180×10^{-6} kg/C.

$$m = zIt$$

$$3.00 \times 10^{-3}\,\text{kg} = (1.1180 \times 10^{-6}\,\text{kg/C})(15.0\,\text{A})t$$

$$t = 179\,\text{s}$$

35-12 CALCULATIONS OF ELECTROCHEMICAL EQUIVALENTS

From Faraday's second law and the standard value of z for silver, the value of z for any other substance can be calculated if its chemical equivalent is known. From Faraday's second law the following proportion is valid:

$$\frac{\text{Unknown electrochemical equivalent}}{\text{Electrochemical equivalent of silver}} = \frac{\text{chemical equivalent of the substance}}{\text{chemical equivalent of silver}}$$

In symbols

$$\frac{z}{z_{Ag}} = \frac{c}{c_{Ag}}$$

From the standard values for silver,

$$z = (1.1180 \times 10^{-6}\,\text{kg/C})\frac{c}{107.87}$$

$$= \frac{c}{107.87/(1.1180 \times 10^{-6}\,\text{kg/C})}$$

$$= \frac{c}{9.649 \times 10^{7}}\,\text{kg/C} \qquad (5)$$

Example Calculate the electrochemical equivalent of copper.

$$z = \frac{c}{9.649 \times 10^{7}}\,\text{kg/C} = \frac{31.77\,\text{kg}}{9.649 \times 10^{7}\,\text{C}}$$

$$= 3.293 \times 10^{-7}\,\text{kg/C}$$

Table 1
ELECTROCHEMICAL DATA

Element	Atomic mass	Valence	Electrochemical equivalent, kg/C
Aluminum	27.1	3	9.36×10^{-8}
Copper	63.6	2	32.94×10^{-8}
Copper	63.6	1	65.88×10^{-8}
Gold	197.2	3	68.12×10^{-8}
Hydrogen	1.008	1	1.05×10^{-8}
Iron	55.8	3	19.29×10^{-8}
Iron	55.8	2	28.94×10^{-8}
Lead	207.2	2	107.36×10^{-8}
Nickel	58.68	2	30.41×10^{-8}

35-13 THE FARADAY CONSTANT

The mass m deposited by any charge Q is given by the equation $m = zQ$. From Eq. (4) it follows that a charge of 9.65×10^{7} C will deposit a mass of any substance numerically equal to its chemical equivalent. The mass equal to the chemical equivalent expressed in kilograms is called the *kilogram equivalent*. The *Faraday constant* in the mks system is the charge per mass equivalent required to deposit any substance. A recent study of the experimental data gives the value of 9.6487×10^{7} C/kg for the Faraday constant. Some electrochemical data are listed in Table 1.

SUMMARY

A distribution of charge can be maintained on a conductor so that a potential difference exists between various points on the conductor.

A potential difference in a conductor tends to cause a flow of positive charge in the direction of the electric field or a flow of negative charge in the opposite direction.

Energy from an outside source must be supplied to maintain the potential difference in a conductor.

An *electric current* exists when there are charges in motion.

In metallic conductors the current is essentially a stream of electrons forced through the circuit by the source. The direction of the *conventional* current is that of the flow of positive charges; this direction is opposite to that of electron flow.

Electric current and charge are related by

$$\text{Current} = \frac{\text{charge}}{\text{time}} \qquad I = \frac{Q}{t}$$

Materials are generally classified as conductors, semiconductors, or nonconductors.

Certain materials at very low temperature become superconductors in which resistance disappears.

The *ampere* is the unit of electric current.

The *emf of a source* is the energy per unit charge transformed in a reversible process (emf is the potential difference *generated* by a source).

Important sources of emf are electrochemical devices, electromagnetic induction, the thermoelectric effect, the photoelectric effect, and the piezoelectric effect.

The light meter and the solar cell are examples of devices that transform radiant energy into electric energy.

The principal effects of electric current are the production of heat, the production of magnetic fields, electrolysis, and other electrochemical effects, although there are other, less important effects.

Electric *resistance* is the ratio of the potential difference to the current,

$$R = \frac{V}{I} \qquad 1\ \Omega = \frac{1\ \text{V}}{1\ \text{A}}$$

One *ohm* is the resistance of the conductor which requires a potential difference of one volt to maintain a current of one ampere in the conductor.

Ohm's law states that the ratio of the potential difference to the current in a metallic conductor is constant when the physical characteristics of the conductor are unchanged.

Ohm's law may be applied either to an entire circuit or to a part of a circuit containing resistance only, provided that the proper voltages, currents, and resistances are used. For the entire circuit

$$\frac{\mathcal{E}}{I} = R_t$$

For a part of a circuit having resistance only,

$$\frac{V_1}{I_1} = R_1$$

The maximum potential difference generated by a cell or generator is its emf. When such a source is maintaining a current, its terminal potential difference is the same as the voltage across the external circuit and is given by

$$IR = \mathcal{E} - Ir$$

Electrolysis is the chemical action which is connected with the passage of electricity through an electrolyte.

Faraday's laws of electrolysis are as follows:

1 The mass of a substance deposited by an electric current is proportional to the amount of electric charge transferred.

2 For the same quantity of electricity transferred, the masses of different elements deposited are proportional to their atomic masses and inversely proportional to their valences.

$$m = zIt$$

The *electrochemical equivalent* of a substance is the mass per unit charge.

The *Faraday constant* is the charge per mass equivalent required to deposit a material in an electrochemical process. The approximate value of this constant is 9.65×10^7 C/kg. Electrochemical equivalents may be calculated from

$$z = \frac{c}{9.65 \times 10^7} \text{ kg/C}$$

Questions

1 Discuss the statement that an electric current may consist of a flow of positive charges, or of negative charges, or of both at once.
2 What happens when two initially charged conducting bodies are connected by a wire? What determines the direction of the current? How long will it continue?
3 Distinguish between semiconductors and superconductors as to their function and method of operation.
4 Why is the name dry cell given to the common battery used in flashlights and radios?
5 Although the cost of electric energy from dry cells is very high, such sources are widely used. Why is this the case?
6 Arrange the following sources in the order of relative amounts of electric energy which they might ordinarily supply: (*a*) electrostatic generator, (*b*) electromagnetic generator, (*c*) battery, (*d*) thermocouple, (*e*) photoelectric cell, and (*f*) piezoelectric source.
7 One should never hold a lighted match near an open storage battery. Explain why.
8 Does a storage cell store up electricity? Why is it properly called a storage cell?
9 A battery is connected through an ammeter to two plates which dip into a vessel of distilled water. What current is observed? What would happen if a little acid were poured into the water? Explain. What would be noticed at the plates?
10 Is it possible to tell from the rate of evolution of gas at the terminals of a pair of electrodes in an acidic water cell which is the positive terminal? Explain.
11 Describe what would happen if pure platinum plates were used for the electrodes instead of copper and silver plates in an electrolysis apparatus.
12 A pair of platinum plates is inserted into a solution of copper sulfate and connected to a battery. Describe the electrochemical actions which take place. Will this process continue indefinitely? Why?
13 Define electrochemical equivalent. In what units is it usually expressed?
14 A person standing on damp earth sometimes gets a shock by touching electric apparatus. Why is this? Would one get a similar shock if he were standing on a dry floor? Why? Why is it more dangerous to touch a 500-V line than a 110-V line? Why is it dangerous to have an electric switch within reach of a bathtub?
15 The direction of current in Fig. 35-7 is reversed. What effect will this have on each of the devices in the circuit?
16 The oceans contain incredibly large quantities of such materials as gold, silver, and many other valuable substances. Explain why the precious metals are not obtained from seawater by electrolysis when such substances as chlorine, caustic soda, and magnesium are so made in large quantities.
17 Do bends in a wire affect the value of the current in a dc circuit? Why?
18 In a research laboratory a sign reading "Danger: Ten thousand ohms" was placed upon some apparatus. Comment upon the scientific appropriateness of this sign.
19 Plot a curve to show how the reading of the voltmeter in Fig. 35-12 varies as the slider is moved from A to C. What is the significance of the slope of this curve?

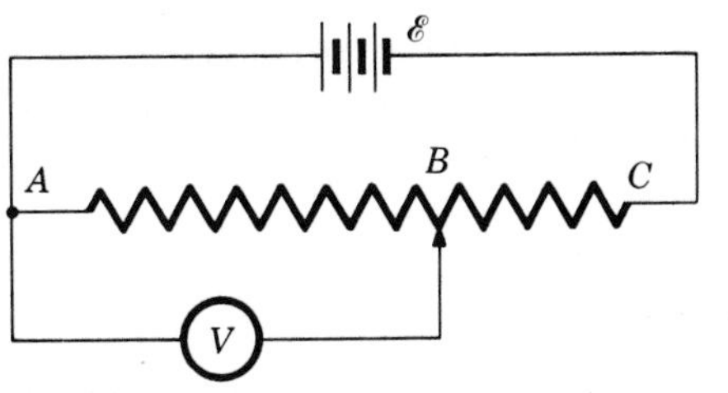

Figure 35-12
Potential drop along a slide wire.

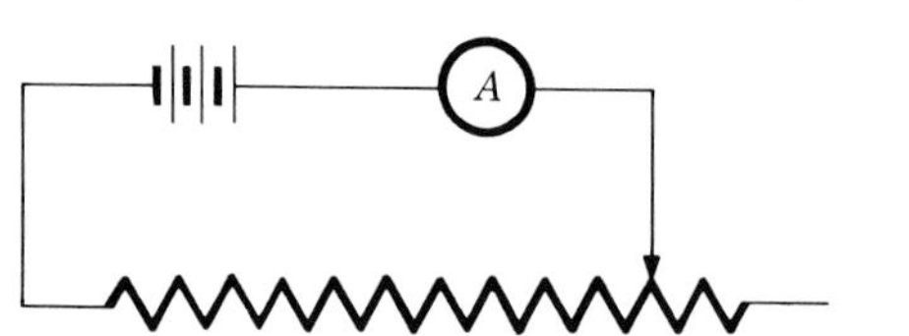

Figure 35-13
Variation of current with resistance.

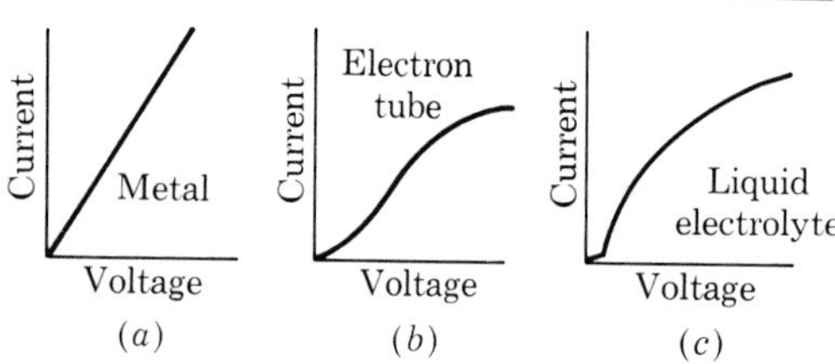

Figure 35-14
Variation of current with voltage.

20 Plot a curve to show the variation of the ammeter reading in Fig. 35-13 as a function of the rheostat resistance. Interpret the shape and intercepts of this curve.

21 What limitations are there on the use of solar cells to provide more of the energy needed by man?

22 The curves in Fig. 35-14 show the variation of current as a function of voltage for three typical classes of conductors. (*a*) How does the resistance change with current in each case? (*b*) Discuss the validity of Ohm's law for each of these cases.

23 Derive the numerical relationship between (*a*) the ampere-hour and the coulomb and (*b*) the ampere-hour and the farad.

Problems

Unless otherwise stated, in all the following problems the electrochemical equivalent of silver may be taken as known, and atomic data may be assumed as known, but other values should be worked out as a part of the problem. For silver, $z = 0.00111800$ g/C; atomic mass = 107.9; valence = 1.

1 A steady current of 10 A is maintained in a metal conductor for 2 min. What charge in coulombs is transferred through it in that time?

2 The current in a common electric heater is 5.0 A. What quantity of electricity flows through the heater in 8.0 min? *Ans.* 2,400 C.

3 A current of 12.3 A is steadily maintained for 2.67 h. What charge flows (*a*) in mks units and (*b*) in cgs electrostatic units?

4 A charge of 3,600 C passes through an electric lamp in 3.0 min. What is the current in the lamp? *Ans.* 20 A.

5 A steady current of 4.00 A is maintained for 10.0 min in a solution of silver nitrate. Find how much silver is deposited on the cathode.

6 What current flows through an electric iron having a resistance of 11 Ω when connected across a line having 110 V? *Ans.* 10 A.

7 How much hydrogen is liberated each day in an acidulated-water cell in which a current of 30.0 A is maintained?

8 A 12.5-V battery operates a 14.2-Ω flashlight intermittently for a cumulative time of 65.3 min. (*a*) What is the average current? (*b*) What quantity of electricity is furnished by the battery? *Ans.* 880 mA; 3.45×10^3 C.

9 A current of 2.00 A is maintained in two coulometers, in series, one of silver, the other an "unknown" metal of atomic mass 55.0. In 2.00 h, 2.73 g is deposited from the unknown. How much silver is deposited, and what is the valence of the other metal?

10 What is the internal resistance of a dry cell which has an emf of 1.55 V if it has a short-circuit current of 22 A? *Ans.* 0.070 A.

11 A 6.24-V battery is connected for 3.33 h to a rheostat and a current of 147 mA is noted. (*a*) What is the resistance of the rheostat? (*b*) What charge is taken from the battery?

12 A steady current of 10 A maintained for 1 h deposits 12.2 g of zinc at the cathode. What is the equivalent mass of zinc? *Ans.* 32.7.

13 A charge of 15.0 C is sent through an electric lamp when the difference of potential is 120 V. What energy is expended?

14 A charge of 600 C flows through a rheostat in which a steady current is maintained for 120 s. What is the current? *Ans.* 5.00 A.

15 A battery which has a terminal voltage of 6.00 V is connected in series with a rheostat and an ammeter which reads 5.00 A. Neglecting the resistance of the ammeter, what must be the resistance of the rheostat?

16 What is the electrochemical equivalent of copper? Copper has an atomic mass of 63.54 and a valence of 2. *Ans.* 3.29×10^{-4} g/C.

17 In a certain rheostat there is a current of 0.45 A when the difference of potential between the terminals is 60 V. What is the resistance of the rheostat?

18 What constant current would be needed for a copper cathode to accumulate 3.06 g of copper in 30 min. The electrochemical equivalent of copper is 3.29×10^{-4} g/C. *Ans.* 5.16 A.

19 A certain wire used in electric heaters has a resistance of 1.75 Ω/ft. How much wire is needed to make a heating element for a toaster which takes 8.25 A from a 115-V line?

20 How long will it take a current of 50 A to produce 32 g of oxygen by the electrolysis of water? Oxygen has an atomic mass of 16 and a valence of 2. *Ans.* 7,720 s.

21 A simple series circuit consists of a cell, an ammeter, and a rheostat of resistance R. The ammeter reads 5 A. When an additional resistance of 2 Ω is added, the ammeter reading drops to 4 A. Determine the resistance R of the rheostat.

22 What is the equivalent mass of a substance if 80 g of it are produced per hour by a current of 20 A? *Ans.* 107.2 g.

23 A fan motor is designed to operate at a current of 3.50 A at a potential difference of 115 V. It is desired to use this motor in another city, where the line voltage is 125 V. How large a resistor must be placed in series with the motor to maintain the rated current?

24 A battery having a voltage of 3.00 V is connected through a rheostat to a uniform wire 100 cm long. The wire has a resistance of 2.00 Ω. What must the resistance of the rheostat be in order that the voltage per millimeter of the wire shall be exactly 1 mV? *Ans.* 4.00 Ω.

25 One electron has a charge of 1.60×10^{-19} C. How many electrons flow each day through an electric lamp in which there is a current of 1.25 A?

26 Measurements show that there are about 10^{22} electrons per milliliter that take part in the conduction of electricity through a wire. Calculate the average speed of the electrons that maintain a current of 2.50 A in the wire leading to an electric lamp if the wire has a diameter of 0.645 mm. *Ans.* 4.8 mm/s.

27 An electron (charge 1.60×10^{-19} C) moves with a speed of one one-hundredth the speed of light (3.0×10^{10} cm/s) in a photoelectric tube. If the plate which collects the photoelectrons is 5.0 cm from the emitting surface, how long does it take for an electron to travel this distance? How many such electrons per second arrive at the collecting plate when there is a current of 9.6×10^{-12} A?

William Shockley, 1910–
(*left*)
Born in London. Studied at California Institute of Technology and Massachusetts Institute of Technology. Member, technical staff, Bell Telephone Laboratories; later director of Shockley Semiconductor Laboratory of Beckman Instruments, Inc.

Walter H. Brattain, 1902–
(*center*)
Born in Amoy, China. Studied at Whitman College, University of Oregon, University of Minnesota. Member of technical staff, Bell Telephone Laboratories.

John Bardeen, 1908–
(*right*)
Born in Madison, Wisconsin. Studied at the University of Wisconsin and Princeton University. Professor of electrical engineering and physics at the University of Illinois. Shared the 1956 Nobel Prize for Physics with Shockley and Brattain for their discovery of the transistor effect.

36

Direct-current Circuits

The application of the principles of electricity to practical problems has called for the design of a vast number of electric circuit arrangements. Usually these are combinations of a very few fundamental circuit elements. Hence a familiarity with a relatively few circuit "building blocks" is basic to the understanding of complex circuits. In this chapter are developed the most important ideas concerning dc circuits which carry steady currents. In later chapters changing, or transient, currents and alternating currents are dealt with.

36-1 RESISTORS IN SERIES

Suppose that a box contains three coils of wire whose resistances are R_1, R_2, and R_3 which are connected in series as shown in Fig. 36-1. If one were asked to determine the resistance of whatever is inside the box without opening it, he might place it in the circuit shown and measure the current I in the box and the voltage V across it. He would then write

$$R = \frac{V}{I} \tag{1}$$

where R is the resistance of the part of the circuit inside the box.

Let us now determine the relation of R, the combined resistance, to the individual resistances R_1, R_2, and R_3. The current in each of the resistors is I, since the current is not divided in the box. The voltages across the individual resistors are

$$V_1 = IR_1 \qquad V_2 = IR_2 \qquad V_3 = IR_3$$

The sum of these three voltages must be equal to V, the voltage across the box; thus

$$\begin{aligned} V &= V_1 + V_2 + V_3 \\ &= IR_1 + IR_2 + IR_3 \\ &= I(R_1 + R_2 + R_3) \end{aligned} \tag{2}$$

$$R_1 + R_2 + R_3 = \frac{V}{I}$$

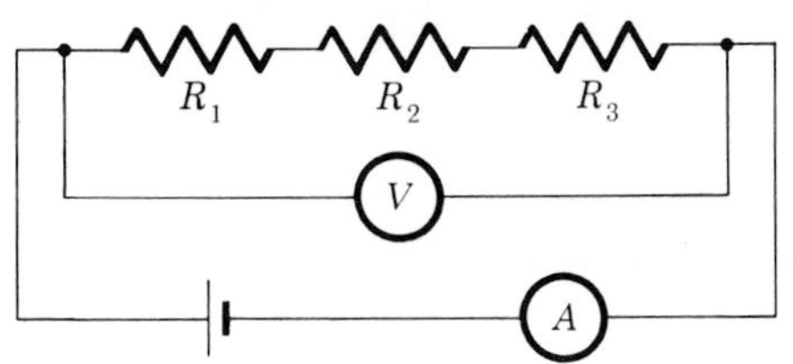

Figure 36-1
Resistors in series.

but this is identical with $R = V/I$, so that

$$R = R_1 + R_2 + R_3 \qquad (3)$$

The following facts may therefore be noted for series connection of resistors:

1 The *current* in all parts of a series circuit *is the same.*
2 The *voltage* across a group of resistors connected in series is equal to the *sum of the voltages* across the individual resistors.
3 The *total resistance* of a group of conductors connected in series is equal to the *sum of the individual resistances.*

Example The resistances of four rheostats are 10.0, 4.0, 6.0, and 5.0 Ω. These rheostats are connected in series to a battery, which produces a potential difference of 75 V across its terminals. Find the current in each rheostat and the voltage across each.

The total resistance is

$$R = (10 + 4 + 6 + 5)\ \Omega = 25\ \Omega$$

so that $$I = \frac{V}{R} = \frac{75\ \text{V}}{25\ \Omega} = 3.0\ \text{A}$$

The voltage across each rheostat is the product of its resistance and the current. Thus

$$V_1 = (3.0\ \text{A})(10\ \Omega) = 30\ \text{V}$$
$$V_2 = (3.0\ \text{A})(4.0\ \Omega) = 12\ \text{V}$$
$$V_3 = (3.0\ \text{A})(6.0\ \Omega) = 18\ \text{V}$$
$$V_4 = (3.0\ \text{A})(5.0\ \Omega) = 15\ \text{V}$$

36-2 RESISTORS IN PARALLEL

Suppose that a box contains a group of three resistors in parallel, of resistances R_1, R_2, and R_3, as shown in Fig. 36-2. The resistance of the combination will be $R = V/I$, where V is the voltage

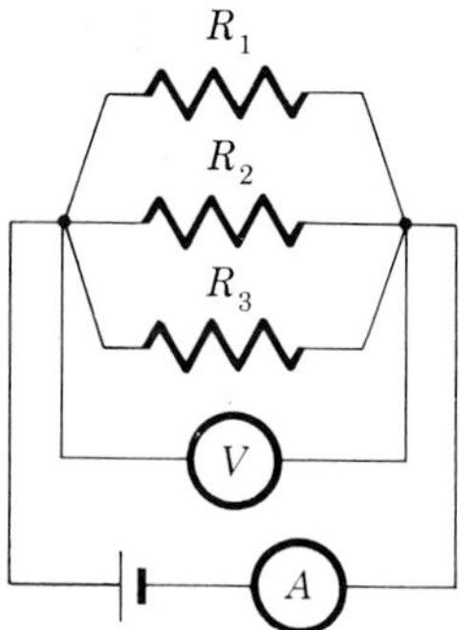

Figure 36-2
Resistors in parallel.

across the terminals of the box and I is the total current in it. Since the voltage across each of the resistors is V, the voltage across the terminals of the box, the currents in the individual resistors are

$$I_1 = \frac{V}{R_1} \qquad I_2 = \frac{V}{R_2} \qquad I_3 = \frac{V}{R_3}$$

The sum of these three currents must be the total current I, so that

$$I = I_1 + I_2 + I_3$$

or $$I = \frac{V}{R_1} + \frac{V}{R_2} + \frac{V}{R_3}$$

This can be written

$$I = V\left(\frac{1}{R_1} + \frac{1}{R_2} + \frac{1}{R_3}\right)$$

or $$\frac{I}{V} = \frac{1}{R_1} + \frac{1}{R_2} + \frac{1}{R_3}$$

Since $V/I = R$, we know that $I/V = 1/R$, so that

$$\frac{1}{R} = \frac{1}{R_1} + \frac{1}{R_2} + \frac{1}{R_3} \qquad (4)$$

For parallel connection of resistors, the following conditions obtain:

1 The currents in the various resistors are different and are inversely proportional to the resistances. The total current is the sum of the separate currents.
2 The voltage across each resistor of a parallel combination is the same as the voltage across any other resistor. Moreover, the voltage across each separate resistor is identical with the voltage across the *whole group* considered as a unit.

This statement may become clearer from a consideration of the fact that the terminals of each resistor are connected to a common point; i.e., each conductor has its beginning at a common potential and its end at another (different) common potential. Hence the potential differences across all conductors in parallel must be identical.

This fact provides the basis of the best method we have for calculating the currents in the separate branches of a parallel group of resistors. Consequently, its importance is emphasized, because this type of problem is one of the most common in elementary electricity.
3 The reciprocal of the total resistance of a number of resistors connected in parallel is equal to the sum of the reciprocals of the separate resistances.

Example The values of three resistances are 10, 4.0, and 6.0 Ω. What will be their combined resistance when connected in parallel?

$$\frac{1}{R} = \frac{1}{R_1} + \frac{1}{R_2} + \frac{1}{R_3}$$

$$= \frac{1}{10\ \Omega} + \frac{1}{4.0\ \Omega} + \frac{1}{6.0\ \Omega}$$

$$= (0.10 + 0.25 + 0.17)/\Omega = 0.52/\Omega$$

$$R = \frac{1}{0.52}\ \Omega = 1.9\ \Omega$$

Note that the resistance of the combination is smaller than any one of the individual resistances.

Note It is important to note that connecting additional resistors in series *increases* the total resistance, while connecting additional resistors in parallel *decreases* the total resistance. For example, in an ordinary house installation, when we "turn on" more lamps, we are inserting additional resistors *in parallel*. We thus *reduce* the total resistance of the house circuit, and hence (since the voltage is constant) we *increase* the current in the mains.

Example Determine the current in each of the resistors in Fig. 36-3.

For the parallel group, Eq. (4) gives

$$\frac{1}{R_{BC}} = \frac{1}{6.0\ \Omega} + \frac{1}{9.0\ \Omega} + \frac{1}{18.0\ \Omega}$$

$$= \frac{6.0}{18.0\ \Omega}$$

$$R_{BC} = \frac{18.0}{6.0}\ \Omega = 3.0\ \Omega$$

$$R_{AC} = R_{AB} + R_{BC} = 4.0\ \Omega + 3.0\ \Omega$$

$$= 7.0\ \Omega$$

The current I_t in the circuit is obtained from

$$I_t = \frac{V_{AC}}{R_{AC}} = \frac{35\ \text{V}}{7.0\ \Omega} = 5.0\ \text{A} \qquad (a)$$

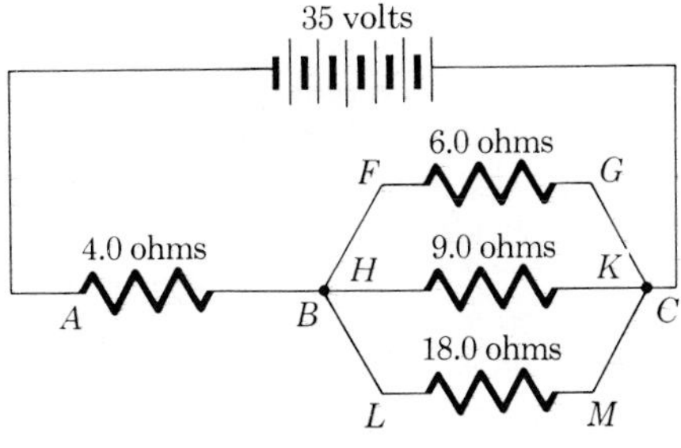

Figure 36-3
Currents in a divided circuit.

This total current is the same as that in the 4-Ω resistor.

The voltage across the parallel group V_{BC} is equal to the resistance of the entire parallel group multiplied by the current in the entire parallel group, i.e.,

$$V_{BC} = R_{CB}I_{BC} = R_{BC}I_t$$
$$= 3.0\ \Omega \times 5.0\ \text{A} = 15\ \text{V} \qquad (b)$$

Note that Ohm's law was applied in Eq. (*a*) to the total current, total voltage, and total resistance, while in Eq. (*b*) the law was applied consistently to the current, voltage, and resistance, *all of the part BC*.

Since the voltage across *BC* is identical with that across *FG*, *HK*, and *LM*, the currents in each of these resistors may be calculated, namely,

$$I_{FG} = \frac{V_{FG}}{R_{FG}} = \frac{15\ \text{V}}{6.0\ \Omega} = 2.5\ \text{A}$$

$$I_{HK} = \frac{V_{HK}}{R_{HK}} = \frac{15\ \text{V}}{9.0\ \Omega} = 1.7\ \text{A}$$

$$I_{LM} = \frac{V_{LM}}{R_{LM}} = \frac{15\ \text{V}}{18.0\ \Omega} = 0.83\ \text{A}$$

Note that

$$I_t = (2.5 + 1.7 + 0.8)\ \text{A} = 5.0\ \text{A}$$

36-3 CELLS IN SERIES

A group of cells may be connected together in *series* or in *parallel* or in a *series-parallel* combination. Such a grouping of cells is known as a *battery,* although this word is often loosely used to refer to a single cell.

Cells are said to be connected in series when they are joined end to end so that the same quantity of electricity must flow through each cell. In the ordinary series connection of cells the positive terminal of one cell is connected to the negative terminal of the next, the positive of the second to the negative of the third, etc., the negative of the first and the positive of the last being joined to the ends of the external resistor (see Fig. 36-4).

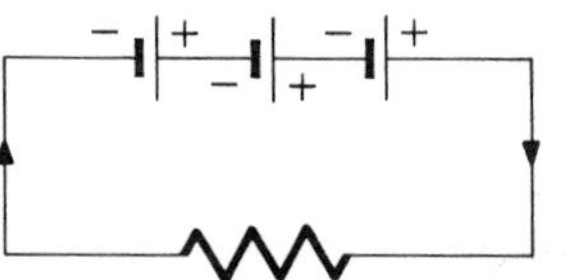

Figure 36-4
Cells in series.

For a series arrangement of cells, the following statements apply:

1 The emf of the battery is equal to the sum of the emfs of the various cells.
2 The current in each cell is the same and is identical with the current in the entire series arrangement.
3 The total internal resistance is equal to the sum of the individual internal resistances.

Example If each of the cells in Fig. 36-4 has an emf of 2 V and an internal resistance of 0.4 Ω, what will be the current in the middle cell when the battery is connected to an 18.8-Ω external resistor?

The emf of the battery is (2 + 2 + 2) V = 6 V. The internal resistance of the battery is (0.4 + 0.4 + 0.4) Ω = 1.2 Ω.

$$\text{Total current} = \frac{\text{total emf}}{\text{total resistance}}$$
$$= \frac{6\ \text{V}}{(18.8 + 1.2)\ \Omega} = 0.3\ \text{A}$$

Since the current in each of the cells is the same and is identical with the current in the entire series arrangement, it follows that the current in the middle cell is also 0.3 A.

36-4 CELLS IN PARALLEL

Cells are connected in parallel when the current is divided between the various cells. In the normal parallel connection of cells all the positive

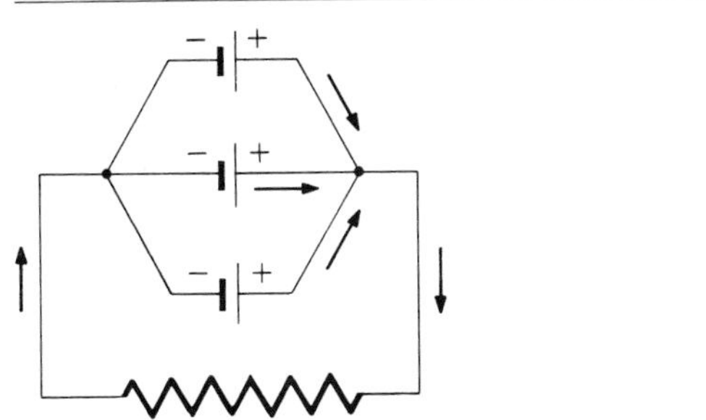

Figure 36-5
Cells in parallel.

poles are connected together and all the negative poles are connected together. Connection is made to an external resistor from the positive and negative terminals at any point along the wires connecting the various cells (see Fig. 36-5).

For a parallel arrangement of *identical* cells, the following statements are true:

1 The emf of the battery is the same as the emf of a single cell.
2 The reciprocal of the total internal resistance is equal to the sum of the reciprocals of the resistances of the individual cells.
3 The current in the external circuit is divided equally among the cells.

Example Compare the currents maintained in a 3.0-Ω resistor by *each cell* of the following arrangements: (*a*) a single cell, (*b*) three cells in series, and (*c*) three cells in parallel. Each cell has an emf of 2.0 V and a negligible internal resistance.

The emf of the first arrangement is 2.0 V; of the second, 6.0 V; and of the third, 2.0 V. Hence, the total current in each case is given by

$$I_{\text{total}} = \frac{\mathcal{E}_{\text{total}}}{R_{\text{total}}}$$

$$I_a = \frac{2.0\ \text{V}}{3.0\ \Omega} = 0.67\ \text{A}$$

$$I_b = \frac{6.0\ \text{V}}{3.0\ \Omega} = 2.0\ \text{A}$$

$$I_c = \frac{2.0\ \text{V}}{3.0\ \Omega} = 0.67\ \text{A}$$

Note that the *total* current in case *a* is the same as that in case *c*. Since there is only one cell in arrangement *a*, the current in that cell is the same as the current in the entire circuit, namely, 0.67 A. Since the cells in case *b* are all joined in series and are in series with the external resistor, the currents in the various cells must be identical and equal to the total current, namely, 2.0 A. In case *c*, the total current is divided equally among three cells; hence the current in any single cell is (0.67/3) A = 0.22 A.

36-5 POTENTIAL CHANGES AROUND A COMPLETE CIRCUIT

If one branch of a parallel set contains a source of emf, the foregoing methods for analyzing the circuit do not apply; i.e., there is no resistance equivalent to the combination of a resistance and an emf. In such cases and even for certain combinations of resistance alone the circuit analysis requires the use of Kirchhoff's rules, which are discussed in the next section. But first it is desirable to consider the relationship of potential values around a complete circuit. In the circuit of Fig. 36-5 we may trace an element of charge from *A* to *B* to *C*, etc., back to *A*, considering rises and falls of potential on the way.

First encountered is the emf $\mathcal{E}_1$ directed from *A* toward *B*. An element of charge passing through a source of emf in the direction of the emf is given energy by the source and is raised in potential by the amount of the emf. Hence in going from the negative terminal *A* to the positive terminal *B* the potential rises by $\mathcal{E}_1$. In going from *B* to *C* the moving charge loses energy to heat and falls through the potential difference IR_1. It falls similarly in the resistor R_2 through the potential difference IR_2. On passing through the source of emf $\mathcal{E}_2$ the element of charge goes opposite to the direction of the emf and gives up energy, which is stored or transformed to some form other than heat. In losing energy the moving

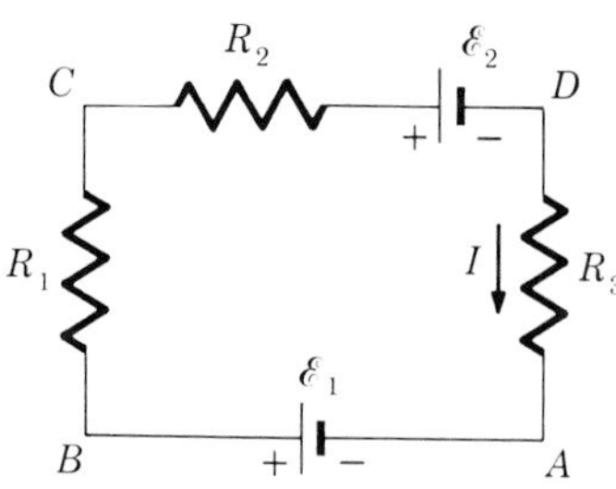

Figure 36-6
A series circuit.

charge falls in potential by amount $\mathcal{E}_2$. Finally the charge element experiences a fall in potential by IR_3 as it completes the circuit. By the law of conservation of energy, the total of the gains of potential plus the total of the losses of potential (treated as negative) must be zero for the complete path from A around to A again. That is,

$$\mathcal{E}_1 - IR_1 - IR_2 - \mathcal{E}_2 - IR_3 = 0$$

Although we followed the circuit of Fig. 36-6 in the direction of the current, an equivalent equation would have resulted had we traced it the other way. Then one would encounter a rise of potential when going through a resistor in the direction opposite to the current. The series of potential changes for this case would be

$$+IR_3 + \mathcal{E}_2 + IR_2 + IR_1 - \mathcal{E}_1 = 0$$

which is equivalent to the previous equation.

36-6 KIRCHHOFF'S RULES

The two principles stated below are so simple as to seem obvious; yet they provide a method for analyzing *any* circuit carrying either transient or steady-state currents, including alternating currents. The first rule is based on the principle of conservation of charge, namely, that charge is neither created nor destroyed but only moved from place to place. The second rule is based on the principle of conservation of energy: energy is neither created nor destroyed but only transformed from one form to another.

Kirchhoff's first rule states that at any point in a circuit, the sum of the currents directed toward the point minus the sum of the currents directed away from the point is equal to zero.

Kirchhoff's second rule states that around any closed path, the algebraic sum of all the changes of potential is zero.

The determination of unknown quantities for a complex circuit calls for the writing of as many independent equations as there are unknowns in the circuit. These unknowns may be voltages, currents, resistances, or other circuit quantities which will be discussed later. Any relationships which provide independent equations are useful. One equation of a set is independent of the others if it is not algebraically equivalent to any of the others or to any combination of them. It is not always easy to tell whether an equation is equivalent to a combination of other equations; but for reasonably simple circuits certain rules may be followed, to avoid difficulty.

Consider the circuit of Fig. 36-7, in which the emfs and the resistances are known but none of the current values, nor their directions, are known. The first step in the application of Kirchhoff's rules to this circuit is to identify with letters the five unknown currents I_1 through I_5. Arbitrary current directions are assigned and indicated by arrows on the diagram (Fig. 36-8). They need not agree with the actual current di-

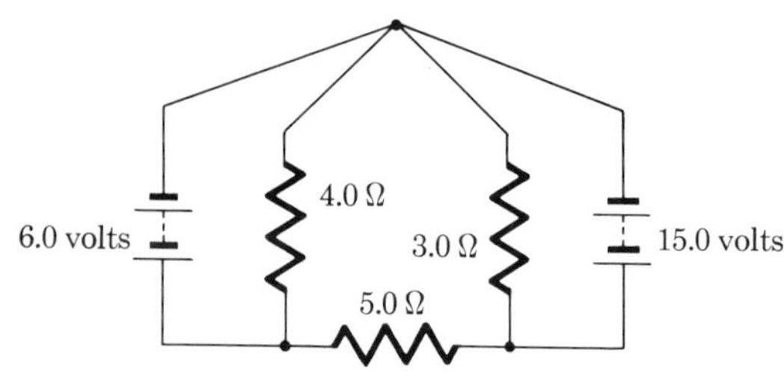

Figure 36-7
A circuit for analysis by Kirchhoff's laws.

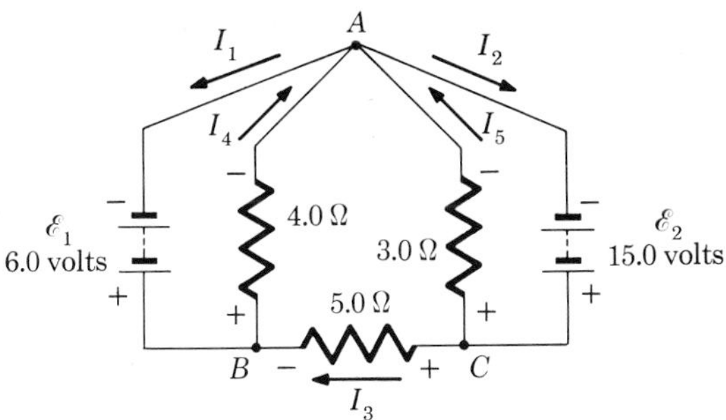

Figure 36-8
The circuit of Fig. 36-7 labeled for the application of Kirchhoff's laws.

rections, for if a current is opposite to the assumed direction, its value will be found to be negative. Since the current in a resistor is from the end at higher potential to that at lower, it is helpful to label with a plus sign the end of each resistor at which the current is assumed to enter and label with a minus sign the end at which the current leaves. Next the positive and negative terminals of all sources of emf are labeled.

In Fig. 36-8 there are three junctions of interest, A, B, and C, at which three or more currents meet. Adding all currents which enter a junction and subtracting those which leave the junction, we write equations from Kirchhoff's first rule for *all but one* of the junctions. (If an equation were written for the final junction, it would be equivalent to the sum of the other equations, and hence not independent of them.) For the circuit shown in Fig. 36-8 the following equations result:

$$\text{For } A: \quad -I_1 - I_2 + I_4 + I_5 = 0 \qquad \text{(i)}$$

$$\text{For } B: \quad +I_1 + I_3 - I_4 = 0 \qquad \text{(ii)}$$

There are a number of different closed paths which could be traced on the circuit, each of which would result in a different equation representing Kirchhoff's second rule. Choosing at random would involve the risk of obtaining equations which were not independent. However, for a circuit of this kind, which can be drawn without crossing or overlapping conductors, there is a simple rule which may be followed to avoid difficulty. The circuit elements on the drawing bound three areas of the surface which do not overlap. Tracing around the boundary of each region will result in an equation which is independent of the others. The first region may be traced from A to B through the 4-Ω resistor and back to A through $\mathcal{E}_1$. The second region may be traced from A to B to C to A through the three resistors, while the third circuit may be traced from A through $\mathcal{E}_2$ to C and back through the 3-Ω resistor. Note that the paths may be traced in either direction, regardless of the assumed current directions. The equations for the rises and falls of potential which result from the above paths are

$$+(4.0\ \Omega)\,I_4 - 6.0\ \text{V} = 0 \qquad \text{(iii)}$$

$$+(4.0\ \Omega)\,I_4 + (5.0\ \Omega)\,I_3 - (3.0\ \Omega)\,I_5 = 0 \qquad \text{(iv)}$$

$$+15.0\ \text{V} - (3.0\ \Omega)\,I_5 = 0 \qquad \text{(v)}$$

These five equations (i) to (v) may be solved simultaneously for the five currents I_1 to I_5. The results are $I_1 = -0.30$ A, $I_2 = 6.8$ A, $I_3 = 1.8$ A, $I_4 = 1.5$ A, and $I_5 = 5.0$ A. It should be noted that the value for I_1 is negative, indicating that this current direction is opposite to that assumed, and the current is opposite to the emf of the battery.

If the sources of emf have appreciable internal resistance, this can be represented by placing resistor symbols next to the sources and including their Ir drops in the equations. *Any* of the electrical quantities in the circuit may be unknown, up to the number of equations in the set.

36-7 FACTORS UPON WHICH THE RESISTANCE OF A CONDUCTOR DEPENDS

Georg Simon Ohm, who formulated the law that bears his name, also reported the fact that the resistance of a conductor *varies directly with its length, inversely with its cross-sectional area, and depends upon the material* of which it is made.

From the study of resistors in series, one would expect that the resistance of a piece of uniform wire is directly proportional to its length, since it can be thought of as a series of small pieces of wire whose total resistance is the sum of the resistances of the individual pieces.

Consider a wire 1 ft in length and having a cross-sectional area of 0.3 in^2. By thinking of this as equivalent to three wires (1 ft in length) each having cross-sectional area of 0.1 in^2 *connected in parallel,* we may infer that

$$\frac{1}{R} = \frac{1}{R_1} + \frac{1}{R_2} + \frac{1}{R_3}$$

or since $R_1 = R_2 = R_3$,

$$\frac{1}{R} = \frac{3}{R_1} \qquad \text{and} \qquad R_1 = 3R$$

showing that the resistance of one of the small wires is three times as great as that of the large wire. This suggests (but does not prove) that the resistance of a wire is inversely proportional to the cross section, a fact that was verified experimentally by Ohm.

Using $R \propto l$ and $R \propto 1/A$, as indicated at the beginning of this section, we can write $R \propto l/A$, where l is the length and A the cross-sectional area of a uniform conductor. This relation can be written in the form of an equation

$$R = \rho \frac{l}{A} \tag{5}$$

where ρ is a quantity, characteristic of the material of the conductor, called the *resistivity* of the substance. (The term *specific resistance* is sometimes used instead of resistivity.)

From Eq. (5),

$$\rho = R\frac{A}{l} \tag{6}$$

If A and l are given values of unity, it is seen that ρ is *numerically equal to the resistance of a conductor having unit cross section and unit length.*

If R is in ohms, A in square centimeters, and l in centimeters, then ρ is in ohm-(centimeters)^2/centimeter or simply ohm-centimeters. This unit is somewhat more convenient than the mks unit, the ohm-meter.

Example The resistance of a copper wire 2,500 cm long and 0.090 cm in diameter is 0.67 ohm at 20°C. What is the resistivity of copper at this temperature?

From Eq. (6)

$$\rho = R\frac{A}{l} = \frac{0.67\ \Omega}{2{,}500\ \text{cm}} \frac{\pi\,(0.090\ \text{cm})^2}{4}$$

$$= 1.7 \times 10^{-6}\ \Omega\cdot\text{cm}$$

The unit of resistivity in the British engineering system of units differs from that just given in that different units of length and area are employed. The unit of area is the *circular mil,* the area of a circle 1 mil (0.001 in) in diameter, and the unit of length is the foot. Since the areas of two circles are proportional to the squares of their diameters, the area of a circle in circular mils is equal to the square of its diameter in mils. In this system of units the resistivity of a substance is numerically equal to the resistance of a sample of that substance 1 ft long and 1 circular mil in area, and is expressed in ohm-circular mils per foot.

The abbreviation CM is often used for circular mils. This should not be confused with the abbreviation used for centimeters (cm). We will use the more standard cmil.

Example Find the resistance of 100 ft of copper wire whose diameter is 0.024 in and whose resistivity is 10.3 Ω·cmils/ft.

$$d = 0.024\ \text{in} = 24\ \text{mils}$$

$$A = d^2 = 24^2\ \text{cmils}$$

$$R = \rho\frac{l}{A} = \frac{(10.3\ \Omega\cdot\text{cmils/ft})(100\ \text{ft})}{24^2\ \text{cmils}}$$

$$= 1.8\ \Omega$$

36-8 CONDUCTANCE AND CONDUCTIVITY

Since the reciprocal of resistance $1/R$ occurs so often in parallel circuits, it is frequently convenient to designate this concept as the *conductance* of the resistor. The symbol used for conductance is G, and the unit is the *mho*. In a parallel circuit the total conductance is given by $G = G_1 + G_2 + G_3$. Less often the reciprocal of resistivity $1/\rho$ is used, and this concept is called the *conductivity* of the material. The symbol for conductivity is σ and the unit is $1/\Omega\cdot\text{cm}$, or mho/cm.

36-9 CHANGE OF RESISTANCE WITH TEMPERATURE

The electric resistance of all substances is found to change more or less with changes of temperature. Three types of change are observed. The resistance may increase with increasing temperature. This is true of all pure metals and most alloys. The resistance may decrease with increase of temperature. This is true of a semiconductor like carbon and of glass and many electrolytes. The resistance may be independent of temperature. This is approximately true of many special alloys, such as manganin (Cu 0.84, Ni 0.12, Mn 0.04).

Experiments have shown that, for a moderate temperature range, the change of resistance with temperature of metallic conductors can be represented by the equation

$$R_t = R_0 + R_0\alpha t = R_0(1 + \alpha t) \qquad (7)$$

where R_t is the resistance at temperature t, R_0 is the resistance at 0°C, and α is a quantity characteristic of the substance and known as the temperature coefficient of resistance. The defining equation for α is obtained by solving Eq. (7), giving

$$\alpha = \frac{R_t - R_0}{R_0 t} \qquad (8)$$

The temperature coefficient of resistance is defined as the change in resistance per unit resistance per degree rise in temperature, based upon the resistance at 0°C.

Although Eq. (7) is only approximate, it can be used over medium ranges of temperature for all but very precise work.

Since $R_t - R_0$ and R_0 have the same units, their units will cancel in the fraction of Eq. (8). Hence, the unit of α depends only upon the unit of t. For instance, for copper $\alpha = 0.004/\text{C}°$, but only $\frac{5}{9} \times 0.004/\text{F}°$.

Example A silver wire has a resistance of 1.25 Ω at 0°C and a temperature coefficient of resistance of 0.00375 per C°. To what temperature must the wire be raised to double the resistance?

From Eq. (7),

$$t = \frac{R_t - R_0}{R_0\alpha}$$

$$= \frac{(2.50 - 1.25)\ \Omega}{1.25\ \Omega \times 0.00375/\text{C}°} = 266\,°\text{C}$$

Note It should be clearly understood that R_0 in the above equations ordinarily refers to the resistance at 0°C and not to the resistance at any other temperature. A value of α based upon the resistance at room temperature, for example, is appreciably different from the value based upon 0°C. This may be made clearer by a graphic analysis of the variation of resistance with temperature.

In Fig. 36-9, the resistance R_t of a conductor at any temperature t is plotted. For a pure metal, this curve gives a linear relation (approximately). Note the fact that the curve does not pass through the origin; i.e., at 0°C the resistance is not zero. Hence we cannot say that $R \propto t$. The slope of the curve $\Delta R/\Delta t$ is a constant. Since

$$\alpha = \frac{\Delta R/\Delta t}{R_0} = \frac{\text{slope}}{R_0}$$

it is clear that the value of α depends upon the base temperature chosen for R_0. In computations

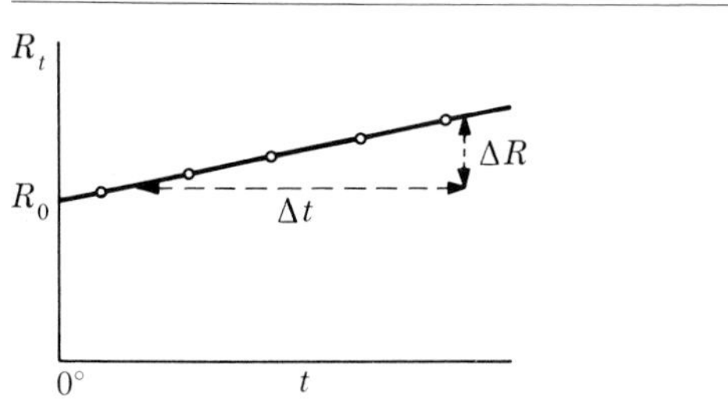

Figure 36-9
Variation of resistance with temperature.

involving temperature variation of resistance, the value of R_0 must first be obtained by using Eq. (7).

Example A tungsten filament has a resistance of 133 Ω at 150°C. If $\alpha = 0.00450/\mathrm{C}^\circ$, what is the resistance of the filament at 500°C?

From Eq. (7),

$$R_0 = \frac{R_t}{1 + \alpha t} = \frac{133\ \Omega}{1 + (0.00450/\mathrm{C}^\circ) \times 150^\circ\mathrm{C}}$$

$$= 79.4\ \Omega$$

$$R_{500} = R_0\,(1 + \alpha t_{500})$$

$$= 79.4\ \Omega\,[1 + (0.00450/\mathrm{C}^\circ) \times 500^\circ\mathrm{C}]$$

$$= 258\ \Omega$$

Since it is the resistivity factor that changes with temperature, Eqs. (7) and (8) may be written with ρ in place of R,

$$\rho_t = \rho_0\,(1 + \alpha t) \tag{9}$$

The resistivities and temperature coefficients of resistivity of some materials are given in Table 1.

It is interesting to note the fact that the value of α for all pure metals is roughly the same, namely, $\frac{1}{273}$, or about 0.004, per Celsius degree. Observe that this value is the same as the coefficient of expansion of an ideal gas. It suggests that, at the absolute zero of temperature (−273°C), a conductor would have zero resistance; i.e., a current once started would continue indefinitely without the expenditure of any energy to keep it going.

The variation of resistance with temperature in the low-temperature region is not so simple in fact. As was noted in Sec. 35-3, experiments in this range led to the discovery in 1911 by the Dutch physicist Heike Kamerlingh Onnes that some metals lose all their resistance at temperatures of a few degrees above the absolute zero. Intensive experimentation on and theoretical study of this phenomenon of superconductivity are now being carried out in many laboratories, both because of its theoretical importance and because of the expectation of many important applications.

36-10 SEMICONDUCTORS

In recent years important discoveries have been made concerning the peculiar properties of certain semiconductors. These materials show an increase of conductivity as the temperature is raised. Prominent examples are silicon, germanium, and a variety of oxides and sulfides. Other materials exhibit a large resistance variation with applied potential difference so that Ohm's law in its original form does not apply. These materials find practical application in circuit elements called varistors.

When a semiconductor, such as copper oxide, is placed in contact with a metal, say, copper, charges accumulate at the interface and interfere with the passage of electricity. These *barrier layers* decrease when a voltage in one direction is applied but increase when the potential is in the opposite sense. Hence such a device may be used as a *rectifier* to produce a unidirectional current from an alternating current. These varistors are finding wide application in modern electrical devices, particularly in the communication industry.

Table 1
RESISTIVITIES AND TEMPERATURE COEFFICIENTS

Material	ρ (at 20°C), $\mu\Omega \cdot$cm	ρ (at 20°C), $\Omega \cdot$cmils/ft	Temperature coefficient of resistance (based upon resistance at 0°C), per C°
Copper, commercial	1.72	10.5	0.00393
Silver	1.63	9.85	0.00377
Aluminum	2.83	17.1	0.00393
Iron, annealed	9.5	57.4	0.0052
Tungsten (wolfram)	5.5	33.2	0.0045
German silver (Cu, Zn, Ni)	20–33	122–201	0.0004
Manganin	44	266	0.00000
Carbon, arc lamp	3,500		−0.0003
Paraffin	3×10^{24}		

The wide range in the resistivities of materials, extending from good conductors, through semiconductors, to the best insulators, is illustrated in Fig. 36-10.

The most important application of semiconductors to date is the *transistor,* which in its various forms can *rectify* (or change alternating to direct current), can *amplify* (produce large current variations in one circuit in response to small ones in a different circuit), can *control* one circuit in response to another, and can *produce oscillations.* All these functions have also been performed by vacuum tubes. The transistor is discussed further in the section on Modern Physics.

36-11 THE MEASUREMENT OF RESISTANCE

As we have seen, the value of an unknown resistance may be determined by placing the resistor in a circuit and simultaneously measuring the current in the resistor and the potential difference between its ends. This is known as the voltmeter-ammeter method of resistance measurement. The resulting resistance values are subject to the errors of both instruments. Another method in wide use compares the current in a known resistance with that in the known and an unknown resistor in

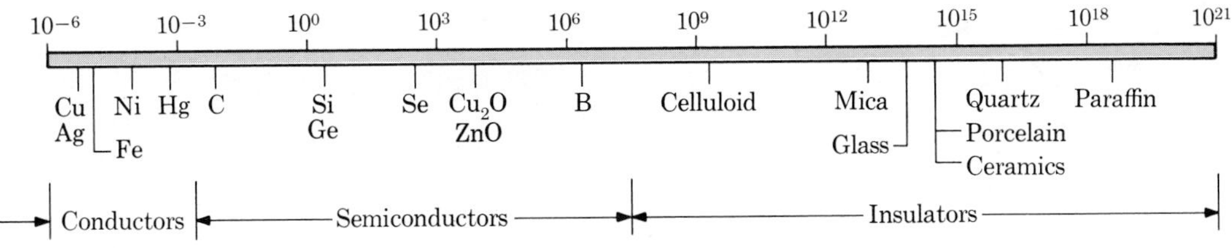

Figure 36-10
Resistivities (ohm-centimeters) vary over the enormous range of about 10^{25}.

series. The same battery is used for both measurements. This is the basis of the ohmmeter, a self-contained instrument which carries its own dry cells and reads directly in ohms. The ohmmeter method involves the errors of only one instrument.

36-12 THE WHEATSTONE BRIDGE

One of the most convenient, precise, and widely used instruments for measuring resistance is the Wheatstone bridge. It consists of a battery, a galvanometer, and a network of four resistors, three of which are known and the fourth to be determined. In Fig. 36-11, the conventional diagram of a Wheatstone bridge is illustrated. The combination is called a *bridge* because the galvanometer circuit is bridged across the two parallel branches *MAN* and *MBN*. In general, the current divides unequally between the two branches. By adjusting the values of the resistances the current in the galvanometer is made zero, as indicated by zero deflection. The bridge is then said to be balanced; i.e., the points *A* and *B* are at the *same potential*. When the bridge is balanced, the fall in potential from *M* to *A* is the same as that from *M* to *B* and similarly the potential difference between *B* and *N* is identical with that between *A* and *N*.

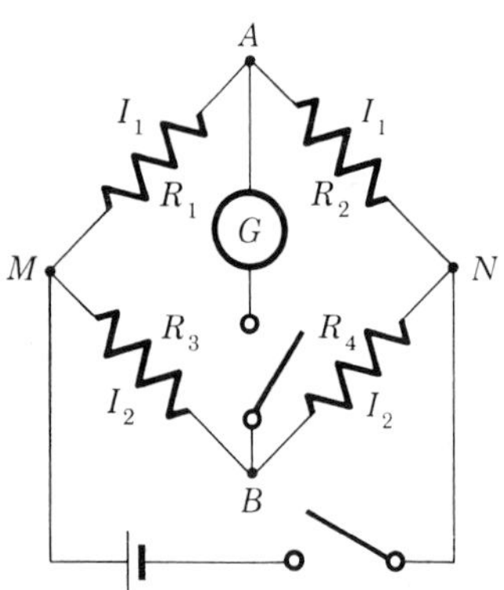

Figure 36-11
Conventional diagram of a Wheatstone bridge.

For the balanced bridge, let the current in the resistors R_1 and R_2 be called I_1 and the current in the resistors R_3 and R_4 be called I_2.

Since the fall of potential over *MA* equals the fall over *MB*,

$$R_1 I_1 = R_3 I_2 \tag{10}$$

Similarly

$$R_2 I_1 = R_4 I_2 \tag{11}$$

If we divide Eq. (11) by Eq. (10), we obtain

$$\frac{R_2}{R_1} = \frac{R_4}{R_3} \tag{12}$$

or

$$R_2 = R_1 \frac{R_4}{R_3} \tag{13}$$

Hence, if any three resistances are known, the fourth resistance can be computed.

From Eq. (13) it will be noted that it is not necessary to know the values of R_4 and R_3, but merely their ratio, in order to determine R_2. Commercial bridges are built with a known adjustable resistance that corresponds to R_1 and ratio coils that can be adjusted at will to give ratios that are convenient powers of 10, usually from 10^{-3} up to 10^3. In some instruments both battery and galvanometer are built into the same box as the resistors, with binding posts for connection to the unknown resistor.

Example A typical commercial Wheatstone bridge has the resistance R_1 variable from 1.00 to 9,999 Ω and the ratio R_4/R_3 variable from 0.00100 to 1,000. What are the maximum and minimum values of unknown resistances that can be measured by this bridge? Are these values feasible in practice? Why?

$$R_2 = R_1 \frac{R_4}{R_3} = 9{,}999 \times 1{,}000$$

$$= 9{,}99\bar{9}{,}000\ \Omega$$

or $$R_2 = 1.00 \times 0.00100 = 0.00100\ \Omega$$

Neither of the extreme values in this example is feasible in practice. Insulation leakage resistance limits the accuracy of the higher range, and contact- and lead-wire resistances introduce serious errors in the lower range. In practice the bridge is most suitable for measuring resistances in the range from 1 to 100,000 Ω.

Although the Wheatstone bridge is widely used with manual adjustment for the zero, or *null*, galvanometer reading, it is increasingly used with electronic and mechanical arrangements to make it self-balancing, the measurements then being recorded on a paper chart. In this form the Wheatstone bridge is the basis for many measurement and control devices used in industry. Any quantity, such as temperature, humidity, strain, displacement, liquid level in a tank, etc., which can be made to produce change in the value of a resistance can be measured with the Wheatstone bridge.

36-13 RESISTANCE THERMOMETER

Because of the accuracy and ease with which resistance measurements may be made and the well-known manner in which resistance varies with temperature, it is common to use this variation to indicate changes in temperature. Devices for this purpose are called *resistance thermometers*. Platinum wire is frequently used in these instruments, because it does not react with chemicals and has a high melting temperature. A coil of fine platinum wire contained in a porcelain tube is placed in the region whose temperature is to be measured. The resistance of the coil is measured by a Wheatstone bridge. Often these instruments are made to record temperatures on a chart, and they are also frequently linked with devices to control temperatures at predetermined values. They can be made to measure temperatures with great precision from exceedingly low temperatures to about 1200°C.

36-14 THE POTENTIOMETER PRINCIPLE

The potentiometer, as the name implies, is a device for measuring potential differences. The essential principle of the potentiometer is the balancing of one voltage against another in parallel with it. In the diagram of Fig. 36-12*a* a branched circuit is shown in which there is a current, because of the potential difference along the slide-wire *AC*, to which it is connected. In Fig. 36-12*b* a cell has been introduced into the lower branch. Depending upon the voltage of this cell, the current in the lower branch may now be in either direction, as indicated by the arrows. As a very special case, the current in the lower branch may be zero when the emf of the cell just

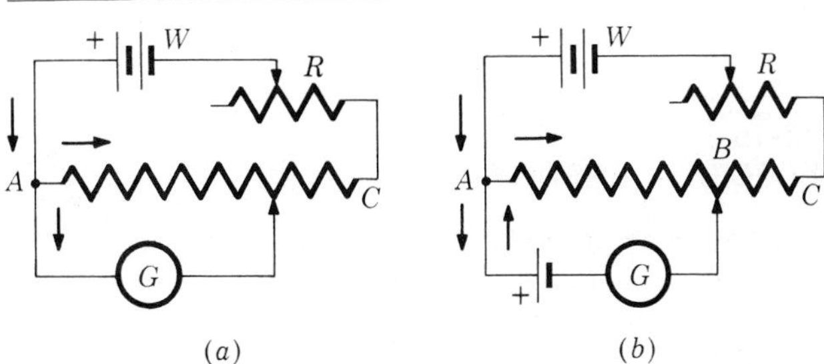

Figure 36-12
(*a*) A potential divider; (*b*) a simple slide-wire potentiometer.

equals the potential difference between A and B and the positive terminal of the cell is connected to the same end of the slide wire as the positive terminal of the working battery W. In the actual potentiometer an unknown emf, here represented by the lower cell, is balanced against a potential drop along a calibrated slide wire.

A great advantage of the potentiometer is that at the moment of balance there is no current in the source of emf under test. Hence the lead wires do not carry any current, so that errors due to line drop or contact resistances do not occur. But even more important is the fact that the true emf is obtained and not just the terminal potential difference, which may differ from the true emf by the drop of potential over the internal resistance of the source in accordance with the equation $V = \mathcal{E} - Ir$, where V is the terminal voltage of the source of emf $\mathcal{E}$ and internal resistance r when there is a current I in the source. At the moment of balance $I = 0$, and hence $V = \mathcal{E}$, regardless of the value of the internal resistance of the cell.

By the use of comparatively simple additional apparatus the potentiometer may be utilized to make measurements, not only of emf but also of current, resistance, and power. Such devices are therefore widely used in industrial processes and laboratories of all kinds. Any quantity that can be made to produce or control an emf or potential difference may be measured with the potentiometer. Temperature, stress, radiation, pH, frequency, angular velocity, and numerous other quantities are both measured and controlled by the use of potentiometers.

36-15 ELECTRIC ENERGY

The flow of electricity in a wire or other conductor always produces heat. Electric soldering, electric welding, electric heating, and electric lighting provided by arcs or incandescent lamps are among the important processes that utilize the heating effect of an electric current. With suitable devices the energy of an electric current may be utilized to produce mechanical work, chemical change, or radiation.

The definition of the potential difference between two points in an electric circuit is the work per unit charge expended in transporting the electricity from the one point to the other. In equation form

$$\text{Potential difference} = \frac{\text{work}}{\text{charge}} \qquad V = \frac{\Delta \mathcal{W}}{\Delta q} \tag{14}$$

Equation (14) may be written in the form

$$\Delta \mathcal{W} = v\,\Delta q = vi\,\Delta t \tag{15}$$

The lowercase letters are conventionally used to represent *instantaneous* values. When v and i are constant, the energy equation may be written in the form

$$\mathcal{W} = VIt \tag{16}$$

In the mks system V is in volts, I in amperes, t in seconds, and $\mathcal{W}$ in joules. Equations (14) to (16) are applicable to all cases in which there are conversions of electric energy. These equations are valid for both sources and sinks, provided that the V term includes all the potential differences in the arrangement being considered. If we consider the case of a resistor and write $V = IR$, the energy equation can be written in the form

$$\mathcal{W} = I^2Rt \tag{17}$$

Example A 60-Ω electric lamp is left connected to a 240-V line for 3.00 min. How much energy is taken from the line?

$$I = \frac{V}{R} = \frac{240\ \text{V}}{60\ \Omega} = 4.0\ \text{A}$$

$$\mathcal{W} = I^2Rt = (4.0\ \text{A})^2 \times 60\ \Omega \times 180\ \text{s}$$

$$= 1.72 \times 10^4\ \text{J}$$

36-16 HEATING EFFECT OF ELECTRIC CURRENT

It is a fact of everyday experience that a conductor in which there is an electric current is thereby heated. In some cases, such as the electric iron and toaster, this heating is desirable. In many other cases, particularly in electric machinery such as dynamos and transformers, the heating is most undesirable. Not only does this heat represent an expensive loss of energy, but it necessitates careful design of the apparatus to get rid of the heat.

In the heating devices the wire in which the useful heat is produced is called the heating element. It is often embedded in a refractory material, which keeps it in place and retards its oxidation. If the heating element is exposed to air, it should be made of metal that does not oxidize readily. Nickel-chromium alloys (such as Nichrome) have been developed for this purpose.

36-17 MECHANICAL EQUIVALENT OF HEAT

From the principle of the conservation of energy it follows that whenever electric energy is expended and heat is evolved, the quantity of heat produced is always strictly proportional to the energy expended. This statement is a form of the first law of thermodynamics (Chap. 18).

Energy is expressed in Eq. (17) in terms of the joule, which is basically a mechanical unit. Energy in the form of heat is often measured in terms of the calorie. Experiments are necessary to establish the relation between the joule and the calorie or between any unit of mechanical energy and heat. These experiments have demonstrated the fact that there is a direct proportion between the expenditure of mechanical energy $\mathcal{W}$ and the heat Q developed. This fact is represented by the equation

$$\mathcal{W} = JQ \tag{18}$$

where J (after Joule) is the proportionality factor called the *mechanical equivalent of heat*. Relationships for the conversion of heat to mechanical energy as given in Chap. 18 include: 1 cal = 4.19 J.

36-18 JOULE'S LAW

By combining Eqs. (17) and (18) a useful form of the equation for heat developed in an electric circuit is obtained, namely,

$$Q = \frac{\mathcal{W}}{J} = \frac{I^2Rt}{J} \tag{19}$$

Example How many calories are developed in 1.0 min in an electric heater which draws 5.0 A when connected to a 110-V line?

From Eq. (19),

$$Q = \frac{I^2Rt}{J} = \frac{VIt}{J} = \frac{(110\text{ V})(5.0\text{ A})(60\text{ s})}{4.19\text{ J/cal}}$$

$$= 7.9 \times 10^3\text{ cal}$$

The facts represented by Eq. (19) are sometimes referred to as Joule's law for the heating effect of the electric current. From this equation it is evident that the heat developed in an electric conductor varies directly with:

1 The square of the current (if R and t are constant)
2 The resistance of the conductor (if I and t are constant)
3 The time (if I and R are constant)

Example In a typical experiment performed to measure the mechanical (electrical) equivalent of heat the following data were obtained: resistance of the coil, 55 Ω; applied voltage, 110 V; mass of water, 153 g; mass of calorimeter, 60 g; specific heat of calorimeter, 0.10 cal/(g)(C°); time of run, 1.25 min; initial temperature of water, 10.0°C; final temperature, 35.0°C. Find the value of J.

$$I = \frac{V}{R} = \frac{110\ \text{V}}{55.0\ \Omega} = 2.00\ \text{A}$$

$$J = \frac{I^2Rt}{Q} = \frac{I^2Rt}{(M_w c_w + M_c c_c)(t_f - t_i)}$$

$$= \frac{(2.00\ \text{A})^2(55.0\ \Omega)(75\ \text{s})}{[153\ \text{g} \times 1.00\ \text{cal/(g)(C°)} + 60\ \text{g} \times 0.10\ \text{cal/(g)(C°)}](35.0°\text{C} - 10.0°\text{C})}$$

$$= 4.15\ \text{J/cal}$$

36-19 TRANSFORMATIONS OF ENERGY

The relations between work and energy are the same whether we are dealing with electricity, heat, or mechanics. The production and use of electric energy involve a series of transformations of energy. Radiation from the sun plays a part in providing potential energy for a hydroelectric plant or the coal for a steam-generating plant. In the latter the chemical energy of coal is converted into heat in the furnace, from heat to work by the steam engine, and from work to electric energy by the generator driven by the steam engine. The energy of the electric current may be converted into work by an electric motor, into heat by an electric range, into light by a lamp. It may be used to effect chemical change in charging a storage battery or in electroplating. The expression $\mathcal{W} = VIt$ (as applied to dc circuits) represents the electric energy used in any of these cases.

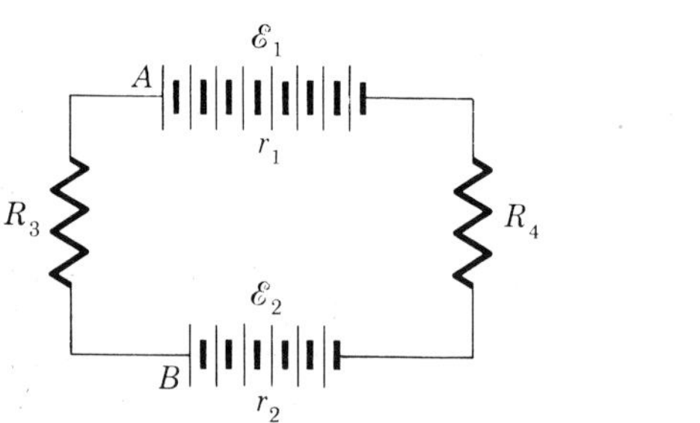

Figure 36-13
Sources and sinks of electric energy.

Consider the circuit shown in Fig. 36-13. Energy is supplied to the circuit by the battery A. It converts chemical energy into electric energy in an amount

$$\mathcal{W}_1 = I\mathcal{E}_1 t$$

where $\mathcal{E}_1$ is the emf of the battery A. Because of internal drop, the terminal potential difference V_1 of battery A will be less than $\mathcal{E}_1$, and there will be a loss of energy I^2r_1t because of heating within the battery itself. Hence the energy $\mathcal{W}_2$ delivered to the external circuit is

$$\mathcal{W}_2 = IV_1t = I(\mathcal{E}_1 - Ir_1)\,t = I\mathcal{E}_1t - I^2r_1t$$

Battery B is inserted into the circuit in such manner as to oppose the emf of battery A. Therefore battery B takes energy out of the circuit. Any device that introduces a counter emf in a circuit, for example, a storage battery being charged or an electric motor, receives energy from the circuit.

The energy supplied to battery B in this circuit is

$$\mathcal{W}_3 = IV_2t = I(\mathcal{E}_2 + Ir_2)\,t = I\mathcal{E}_2t + I^2r_2t$$

Note that V_2, the terminal potential difference of battery B, is greater than $\mathcal{E}_2$, since there is a reverse current in this battery. Of the energy supplied to battery B, a portion, I^2r_2t, is expended as heat within the battery because of its internal resistance. The remainder, $I\mathcal{E}_2t$, is converted into chemical energy in charging the battery.

The product of the current, counter emf, and time in any device represents the energy that can be taken from the circuit by that device for purposes other than heating. Since no such machine is 100 percent efficient, the actual conversion of energy is less than this maximum.

In this circuit the battery A transforms chemical energy into electric energy in amount $I\mathcal{E}_1 t$. Of this energy, $I\mathcal{E}_2 t$ is available for reconversion into chemical energy at battery B. Energy $I^2 r_1 t + I^2 r_2 t$ is converted into heat inside the batteries, and $I^2 R_3 t + I^2 R_4 t$ is converted into heat by the external resistors.

Example In the circuit of Fig. 36-13, $\mathcal{E}_1 = 27$ V; $\mathcal{E}_2 = 6.0$ V; $r_1 = 0.60\ \Omega$; $r_2 = 0.40\ \Omega$; $R_3 = 2.0\ \Omega$; and $R_4 = 4.0\ \Omega$. Calculate the following: the total energy $\mathcal{W}_1$ supplied to the circuit, the energy $\mathcal{W}_2$ delivered to the external circuit, the energy $\mathcal{W}_3$ supplied to battery B, the energy $\mathcal{W}_4$ converted by B into chemical energy, the energy $\mathcal{W}_5$ which is converted into heat in the circuit, each in 3.5 min.

$$\mathcal{E} = \mathcal{E}_1 - \mathcal{E}_2 = 27\text{ V} - 6.0\text{ V} = 21\text{ V}$$

$$\begin{aligned} R &= r_1 + r_2 + R_3 + R_4 \\ &= (0.60 + 0.40 + 2.0 + 4.0)\ \Omega \\ &= 7.0\ \Omega \end{aligned}$$

$$I = \frac{\mathcal{E}}{R} = \frac{21\text{ V}}{7.0\ \Omega} = 3.0\text{ A}$$

$$t = 3.5\text{ min} = 210\text{ s}$$

$$\begin{aligned} \mathcal{W}_1 &= I\mathcal{E}_1 t = 3.0\text{ A} \times 27\text{ V} \times 210\text{ s} \\ &= 1\bar{7},000\text{ J} \end{aligned}$$

$$\begin{aligned} \mathcal{W}_2 &= I\mathcal{E}_1 t - I^2 r_1 t = 1\bar{7},000\text{ J} - (3.0\text{ A})^2 \\ &\qquad \times 0.60\ \Omega \times 210\text{ s} \\ &= 1\bar{6},000\text{ J} \end{aligned}$$

$$\begin{aligned} \mathcal{W}_3 &= I\mathcal{E}_2 t + I^2 r_2 t = 3.0\text{ A} \times 6.0\text{ V} \\ &\qquad \times 210\text{ s} + (3.0\text{ A})^2 \times 0.40\ \Omega \\ &\qquad \times 210\text{ s} \\ &= 4,\bar{6}00\text{ J} \end{aligned}$$

$$\begin{aligned} \mathcal{W}_4 &= I\mathcal{E}_2 t = 3.0\text{ A} \times 6.0\text{ V} \times 210\text{ s} \\ &= 3,\bar{8}00\text{ J} \end{aligned}$$

$$\begin{aligned} \mathcal{W}_5 &= I^2 R t = (3.0\text{ A})^2 \times 7.0\ \Omega \times 210\text{ s} \\ &= 1\bar{3},000\text{ J} \end{aligned}$$

36-20 POWER AND ENERGY

Since power P is the rate of doing work or the rate of use of energy, it may always be obtained by dividing the energy $\mathcal{W}$ by the time t which is taken to use or to generate the energy. Symbolically (for dc circuits)

$$\bar{P} = \frac{\mathcal{W}}{t} = \frac{VIt}{t} = VI \tag{20}$$

In mks units, $\bar{P}$ is the average power in joules per second, i.e., in watts, if V is given in volts and I in amperes. Thus the power in watts used by any electric device, such as a calorimeter, is found by multiplying the ammeter reading by the voltmeter reading. If the electric power is entirely used in producing heat in a resistor R, then from Eq. (17)

$$\bar{P} = \frac{\mathcal{W}}{t} = \frac{I^2 R t}{t} = I^2 R \tag{21}$$

(Note that the symbol $\mathcal{W}$ in these equations represents *energy*, in joules. It does not stand for watt, which is a unit of power.)

Example An electric furnace, operating at 120 V, requires 3.0 hp. Calculate the current and the resistance.

$$\begin{aligned} P &= 3.0\text{ hp} \times 746\text{ W/hp} \\ &= 2,\bar{2}00\text{ W} \\ &= VI \\ 2,200\text{ W} &= 120\text{ V} \times I \\ I &= 18\text{ A} \\ R &= \frac{V}{I} = \frac{120\text{ V}}{18\text{ A}} = 6.7\ \Omega \end{aligned}$$

If the power is not constant the instantaneous value p is the time rate of change of the energy at the given instant. In this case Eqs. (20) and (21) can be rewritten in the forms

$$p = vi = i^2R = \frac{v^2}{R} \tag{22}$$

where the lowercase letters are used to indicate *instantaneous* values.

36-21 UNITS AND COST OF ELECTRIC ENERGY

A very practical aspect of the use of any electric device is the cost of its operation. It should be noted that the thing for which the consumer pays the utility company is *energy and not power.*

$$\text{Energy} = \text{power} \times \text{time}$$

A power of 1 W used for 1 s requires 1 J of energy. This is a rather small unit for general use. The most common unit is the *kilowatthour* (kWh), which is the energy consumed when 1 kW of power is used for 1 h. One kilowatthour is equal to 3.6×10^6 J.

The cost of electric energy is given by the equation

$$\text{Cost} = P \times t \times (\text{unit cost})$$
$$= \frac{VIt \times (\text{cost/kWh})}{1{,}000\ \text{W/kW}}$$

when V is expressed in volts, I in amperes, and t in hours.

Example What is the cost of operating for 24 h a lamp requiring 1.0 A on a 100-V line if the cost of electric energy is \$0.050/kWh?

$$\text{Cost} = \frac{(100\ \text{V})(1.0\ \text{A})(24\ \text{h})(\$0.05/\text{kWh})}{1{,}000\ \text{W/kW}}$$
$$= \$0.12$$

36-22 MEASUREMENT OF POWER BY VOLTMETER-AMMETER METHOD

A simple and widely used method for measuring power (in dc circuits) is to measure the current in the circuit by an ammeter and the voltage with a voltmeter. The power is simply the product of these two readings. (For precise measurements certain corrections must be made, owing to the effect one instrument has on the reading of the other.)

The wattmeter, which measures power directly, will be discussed later.

SUMMARY

For resistors in series

$$I = I_1 = I_2 = I_3 = \cdots$$
$$V = V_1 + V_2 + V_3 + \cdots$$
$$R = R_1 + R_2 + R_3 + \cdots$$

When resistors are connected in parallel,

$$I = I_1 + I_2 + I_3 + \cdots$$
$$V = V_1 = V_2 = V_3 = \cdots$$
$$\frac{1}{R} = \frac{1}{R_1} + \frac{1}{R_2} + \frac{1}{R_3} + \cdots$$

When cells are connected in series,

$$\mathcal{E} = \mathcal{E}_1 + \mathcal{E}_2 + \mathcal{E}_3 + \cdots$$

and $$r = r_1 + r_2 + r_3 + \cdots$$

For identical cells in parallel

$$\mathcal{E} = \mathcal{E}_1 = \mathcal{E}_2 = \mathcal{E}_3 = \cdots \qquad \text{and} \qquad r = \frac{r_1}{N}$$

Kirchhoff's first rule states that at any point in a circuit the sum of the currents directed toward the point minus the sum of the currents directed away from the point is equal to zero.

Kirchhoff's second rule states that around any closed path the algebraic sum of all the changes of potential is zero.

The resistance of a conductor

1 Varies directly with the length
2 Varies inversely with the area of cross section
3 Depends upon the nature of the material and the temperature

The resistance of a wire is given by

$$R = \rho \frac{l}{A}$$

The defining equation for *resistivity* is

$$\rho = R \frac{A}{l}$$

Resistivity is measured in ohm-centimeters in the metric system and in ohm-circular mils per foot in the British system. (The cross-sectional area of a wire in circular mils is found by squaring its diameter expressed in mils.)

Conductance, the reciprocal of resistance, $G = 1/R$, is measured in mhos.

Conductivity, the reciprocal of resistivity, may be measured in $1/\Omega \cdot \text{cm}$ or mho/cm.

The resistivity, and hence the resistance of a conductor, varies with temperature. For many materials this variation may be expressed, over a moderate temperature range, by

$$\rho_t = \rho_0 (1 + \alpha t) \qquad \text{and} \qquad R_t = R_0 (1 + \alpha t)$$

The *temperature coefficient of resistance* is defined by

$$\alpha = \frac{R_t - R_0}{R_0 t}$$

For all pure metals, α is roughly $\frac{1}{273}$ per C°. Near absolute zero many metals have negligible resistance.

Semiconductors show an increase of conductivity with temperature rise. Major uses of semiconductors are the *semiconductor diode,* for rectifying current, and the *transistor,* for replacing vacuum tubes.

A *Wheatstone bridge* is a device for the comparison of resistances. The working equation is

$$R_2 = R_1 \frac{R_4}{R_3}$$

The basic *potentiometer* principle is the balancing of one voltage against another. An advantage of the potentiometer over a voltmeter is the fact that the potentiometer takes no current from the source being measured and hence gives a reading of terminal potential difference equal to the emf.

Electric energy is expressed by

$$\Delta \mathcal{W} = v \, \Delta q$$

For constant voltage and current

$$\mathcal{W} = VIt$$

In the case of energy supplied to a resistor, *Joule's law* states that

$$\mathcal{W} = I^2 Rt$$

The *mechanical equivalent of heat* is defined by

$$J = \frac{\mathcal{W}}{Q}$$

One calorie = 4.19 J.

The heat developed by an electric current is given by

$$Q = \frac{\mathcal{W}}{J} = \frac{I^2 Rt}{J}$$

Electric energy supplied by a source of emf is $\mathcal{E}It$; that which is expended as heat is I^2Rt; and that going into useful mechanical work or chemical energy is $\mathcal{E}'It$, where $\mathcal{E}'$ is the counter emf.

Electric *power* is given by

$$\bar{P} = \frac{\mathcal{W}}{t} = VI$$

At any instant the power is

$$p = \frac{\Delta w}{\Delta t} = vi$$

For a resistor

$$P = I^2R = \frac{V^2}{R}$$

The power is in watts when the current is in amperes, the potential difference in volts, and the resistance in ohms; 1 hp = 746 W.

Electric energy is often measured in kilowatt-hours. One *kilowatthour* is the energy consumed when a power of one kilowatt is used steadily for one hour.

Electric power may be measured by a voltmeter-ammeter method (in dc circuits only).

Questions

1 As electrons flow through a resistor connected to a cell, the electric energy decreases. The energy increases as the electrons go through the cell. Show the source and sink of this energy.

2 A battery is connected to a long resistance wire. One terminal of a voltmeter is connected to one end of the wire. The other terminal of the voltmeter is connected to a sliding contact, which is then moved along the wire. Describe the way in which the voltmeter reading varies as the contact is placed at different points on the wire.

3 When a dry cell is connected to an external resistor, what happens to the terminal voltage of the cell? Why? Describe how this effect differs in a fresh cell of low-internal resistance and an old cell of very high internal resistance.

4 Explain why a good-quality voltmeter connected to a cell does not read the correct emf of the cell. What characteristics should a voltmeter have so that its reading will closely approximate the emf of the cell? Describe how one might use a voltmeter of known resistance to measure the internal resistance of a cell of known emf.

5 Some automobiles have an ammeter with a center zero mounted on the instrument panel. Show how and why the readings of this meter change when the car is (*a*) moving slowly, (*b*) moving rapidly, (*c*) standing still (with the engine idling), and (*d*) standing still with the lights turned on.

6 Suppose a voltmeter with a center zero to be connected across a storage battery. Describe how the pointer on the meter changes for the following cases: (*a*) battery connected to the voltmeter only; (*b*) battery connected to a high-resistance rheostat; (*c*) battery connected to a low-resistance rheostat; (*d*) battery short-circuited; and (*e*) battery being charged.

7 What is the effect of the internal resistance of the battery used with a Wheatstone bridge on the precision of the measurements? How is the precision altered by using similar galvanometers of different resistances? Is contact resistance important in (*a*) the battery circuit, (*b*) the galvanometer circuit, (*c*) the circuit containing the known resistance, and (*d*) the circuit containing the unknown resistance?

8 Strings of Christmas-tree lights are sometimes made of miniature lamps connected in series. For an eight-lamp 120-V set, what is the voltage across each lamp? If one lamp were removed, what would happen? The voltage across the empty socket becomes equal to the line voltage. Why is this?

9 A Wheatstone bridge is balanced in the usual manner. The galvanometer and the working battery are then interchanged. Describe how the galvanometer will behave when the switches are closed.

10 In using a slide-wire Wheatstone bridge it is desirable to adjust the known resistance so that the final balance is obtained when the contact on the slide wire is near the center of the scale. Why is this desirable?

11 A piece of copper wire is cut into 10 equal parts. These parts are connected in parallel. How will the joint resistance of the parallel combination compare with the original resistance of the single wire?

12 As one turns on more lamps in an ordinary household circuit, what happens to the current in the first lamp? to the line current? to the line voltage? Why is it not customary to connect household electric lamps in series?

13 Classify and trace the various sources of energy transformed into electric energy, and show that ultimately all the energy came from the sun.

14 Describe a way in which a frozen water line could be thawed out by electricity. Is this feasible for a long underground pipe? Explain.

15 A resistor forms part of a series circuit. How is the resistance of the circuit affected if a second resistor is connected (*a*) in series with the first and (*b*) in parallel with the first?

16 Resistors *A*, *B*, and *C* are connected in series to a battery. To find the current in resistor *A*, is it satisfactory to divide the terminal voltage of the battery by the resistance of *A*? Why? Answer the same question for these resistors connected in parallel.

17 Arrange the following energy units in order of increasing magnitude: kilowatthour, foot-pound, erg, and joule.

18 Which process has the greater efficiency: the conversion of chemical energy of coal into electric energy in a power plant or the conversion of electric energy into radiant energy in the coils of a dc electric stove?

19 Four 1.5-V dry cells connected in series have a total emf of 6.0 V. Could they be substituted satisfactorily for the 6.0-V lead storage battery in an automobile? Explain.

20 If in connecting two identical cells in parallel, one of the cells is connected by error with polarity reversed from the normal direction, how will this affect the total internal resistance? What other effect may be observed?

21 What is it that is dangerous to the human body: current, voltage, or some combination of these and other factors?

22 A 115-V 25-W lamp and a 115-V 150-W lamp are connected in parallel to a 115-V source of power. What is the effect on the brightness of each lamp if the connections are changed so that the lamps are now connected in series? Assume that the resistances of the lamps remain constant.

23 An external resistor is to be connected to one or more identical cells each having an internal resistance equal to the external resistance. Which arrangement furnishes the largest current: (*a*) one cell, (*b*) 10 cells in series; or (*c*) 10 cells in parallel?

24 In the voltmeter-ammeter method for the measurement of power, the voltmeter may be connected either in parallel with the device studied or in parallel with the device plus the ammeter. Draw the wiring diagram for these two arrangements. Discuss the relative errors introduced by the current in the voltmeter in the first case and the potential difference across the ammeter in the second case.

Problems

1 A dry cell of internal resistance 0.0624 Ω when short-circuited will furnish a current of 25.0 A. What is its emf?

2 The terminal voltage of a battery is 9.0 V when supplying a current of 4.0 A and 8.5 V when supplying 6.0 A. Find its internal resistance and emf. *Ans.* 0.25 Ω; 10.0 V.

3 Two lamps need 10.0 V and 2.5 A each in order to operate at a desired brilliancy. They are to be connected in series across a 120-V line. What is the resistance of the rheostat which must be placed in series with the lamps?

4 Calculate the value of R_x in Fig. 36-14 which

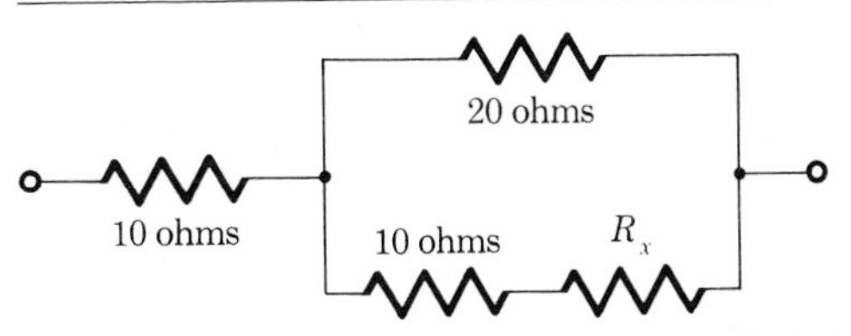

Figure 36-14
Unknown resistor in series-parallel circuit.

makes the total resistance of the circuit also equal to R_x. *Ans.* 22 Ω.

5 A 3.00-V battery of internal resistance 1.00 Ω is connected to a slide wire which is 100 cm long and has a resistance of 5.00 Ω. A voltmeter that takes a negligible current is connected across 60.0 cm of the wire. (*a*) What is the reading of the voltmeter? (*b*) If the voltmeter had a resistance of 150 Ω, what would its reading be?

6 A small generator gives a potential reading of 100 V when a voltmeter is placed across a resistor of 100.0-Ω resistance connected in series with the generator. When the 100.0-Ω resistor is replaced by a 200.0-Ω resistor, the voltmeter reads 105.0 V. Calculate the internal resistance and the emf of the generator. *Ans.* 10.5 Ω; 111 V.

7 A 12.0-V battery is to be charged from a 110-V line. (*a*) If the internal resistance of the battery is 0.500 Ω, how much resistance must be put in series with the battery in order that the charging current shall be 5.00 A? (*b*) What will be the difference in potential between the terminals of the battery?

8 From the curve of Fig. 36-15 determine (*a*) the emf of the cell, (*b*) the short-circuit current, and (*c*) the internal resistance.

Ans. 1.10 V; 0.265 A; 4.2 Ω.

9 The emf of each cell in Fig. 36-16 is 1.50 V, and the internal resistance of each cell is 0.40 Ω. Calculate the (*a*) total current, (*b*) current in the 36-Ω resistor, and (*c*) potential difference between *A* and *B*. (Note the polarity of cell *AB*.)

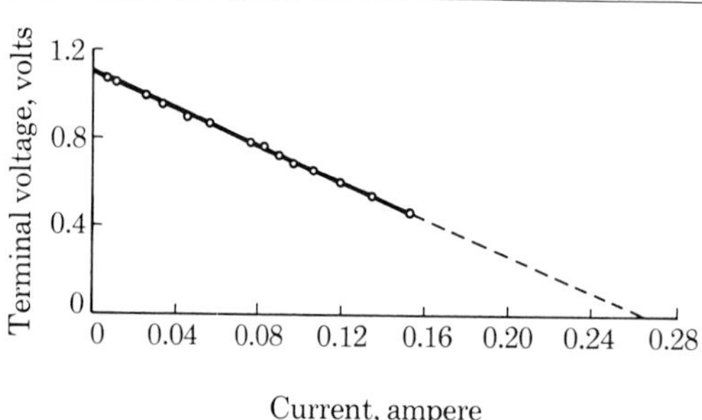

Figure 36-15
Variation of terminal potential difference delivered by a high-resistance cell.

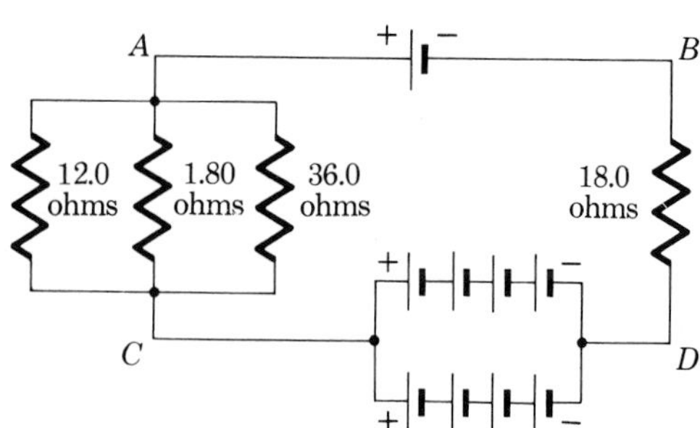

Figure 36-16

10 Two lamps need 50 V and 2.0 A each in order to operate at a desired brilliancy. They are to be connected in series across a 120-V line. What is the resistance of the rheostat which must be placed in series with the lamps?

Ans. 10 Ω.

11 A battery of 12.0-V terminal potential difference is connected to a group of three resistors, joined in series. One resistance is unknown; the others are 3.00 and 1.00 Ω. A voltmeter connected to the 3.00-Ω resistor reads 6.00 V. Determine the value of the unknown resistance.

12 Each cell in Fig. 36-17 has an emf of 2.18 V

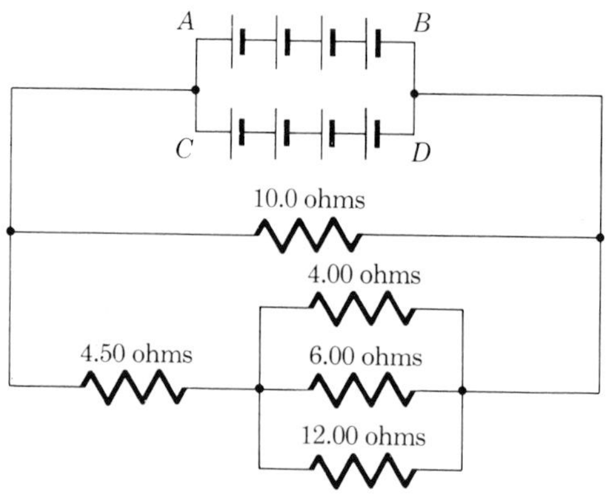

Figure 36-17

and an internal resistance of 0.130 Ω. Calculate (*a*) the current in the 10.0-Ω resistor, (*b*) the current in one of the cells, and (*c*) the terminal potential difference of the battery *AB*.

Ans. 0.818 A; 1.04 A; 8.18 V.

13 A battery of 1.5-Ω internal resistance is connected to two resistors in series of 2.0 and 3.0 Ω. The voltage across the 2.0-Ω resistor is 8.0 V. What is the emf of the battery?

14 Three conductors whose resistances are 20.0, 30.0, and 40.0 Ω, respectively, are connected in parallel. What is the joint resistance of this group?

Ans. 9.25 Ω.

15 A Wheatstone bridge of the slide-wire type uses a 1-m wire of resistance 4.50 Ω for the resistors R_3 and R_4 of Fig. 36-11. In this bridge there is contact resistance where the wire is attached to the junctions *M* and *N*. If the contact resistance is 0.15 Ω at *M* and 0.25 Ω at *N*, what percentage error is introduced when the slider *B* is at (*a*) 50 cm and (*b*) 20 cm? What fact does this problem emphasize?

16 A slide-wire Wheatstone bridge has resistor *MBN* (Fig. 36-11) in the form of a uniform wire 1.00 m long, with the galvanometer contact *B* continuously adjustable along *MBN*. The bridge is balanced with the known resistor at 450 Ω and the slider on the 42.8-cm mark. What is the value of the unknown resistance? *Ans.* 600 Ω.

17 Sketch the outline of a cube. Imagine that each of the 12 edges represents a conductor having a resistance of 1.0 Ω. What is the resistance between corners diagonally opposite each other?

18 Two rheostats of resistances 10.0 and 3.00 Ω are connected in parallel and joined to a battery of negligible internal resistance. There is a current of 0.200 A in the 10.0-Ω resistor. Determine (*a*) the current in the 3.00-Ω resistor and (*b*) the emf of the battery. *Ans.* 0.677 A; 2.00 V.

19 The resistance of a wire 25.4 m long and 0.932 mm in diameter is 0.670 Ω at 20°C. What is its resistivity at this temperature?

20 The Wheatstone bridge shown in Fig. 36-11 is balanced for the following values: $R_1 = 12\ \Omega$, $R_2 = 15\ \Omega$, $R_3 = 60\ \Omega$, $R_4 = 75\ \Omega$, $R_G = 125\ \Omega$, voltage of battery = 2.48 V. Calculate the following: V_{MA}; V_{MB}; V_{AN}; V_{BN}; V_{AB}. *Ans.* 1.10 V; 1.10 V; 1.38 V; 1.38 V; zero.

21 Large electromagnets are sometimes wound with wire of square cross section in order to conserve space. Calculate the width of a square wire that has the same area as a cylindrical wire of area 1.33×10^{-5} cm^2.

22 A cell of internal resistance 0.20 Ω is connected to two coils of resistances 6.00 and 8.00 Ω, joined in parallel. There is a current of 0.200 A in the 8.00-Ω coil. Find the emf of the cell.

Ans. 1.69 V.

23 A pair of resistors of 5.0 and 7.0 Ω, respectively, are connected in parallel. This group is connected in series with another pair in parallel whose resistances are 4.0 and 3.0 Ω. What is the total resistance?

24 A 12-V battery having an internal resistance of 1.0 Ω is connected in series with the following: a 10.0-Ω coil and a parallel group consisting of three branches having resistances of 2.0, 3.0, and 6.0 Ω, respectively. Find (*a*) the current in the 3-Ω coil and (*b*) the potential difference at the terminals of the battery. *Ans.* 0.33 A; 11 V.

25 A simple slide-wire potentiometer is made by using a wire 2.00 m long that has a resistance of 5.25 Ω. The working battery has a terminal voltage of 6.50 V. A Weston cell is connected to make *AB* = 1,018.3 mm (Fig. 36-12*b*), and the rheostat is adjusted to balance the potentiometer. What must the resistance of the rheostat be in order for the potential difference per millimeter of the slide wire to be 1.00 mV?

26 In defining the temperature coefficient of resistance, the resistance at 0°C is usually taken as the reference base. What percentage change would be made in the value of this coefficient for aluminum if the base temperature were taken as 20°C instead of 0°C? *Ans.* 7.4 percent.

27 A storage battery of emf 6.50 V and internal resistance 0.36 Ω is charged for 12 h with a current of 15.0 A. (*a*) How much energy is supplied to the battery? (*b*) What energy is lost because of the heat developed in the battery?

28 Power lines are sometimes provided with electric heating devices to melt the ice that forms during storms. Consider a wire 500 m long, covered with ice 12.5 mm thick. The radius of the wire is 15.4 mm. The ice drops off when one-third of the coating melts. How much energy is expended in melting this ice? *Ans.* 2.08×10^3 cal.

29 A factory uses 55.0 kW of power at 2,200 V supplied from a generator which is some distance from the factory. If the IR drop in the line is 10.6 V, what power is wasted in the line?

30 Two resistors of 6.00 and 2.00 Ω are connected in parallel. This arrangement is connected in series with a 4.00-Ω resistor and a battery having an internal resistance of 0.500 Ω. The current in the 2.00-Ω resistor is 0.800 A. Determine the emf of the battery. *Ans.* 6.40 V.

31 A 15-Ω resistor is connected in series with a 30-Ω resistor. In series with these resistors is a parallel group of two resistors of 60 and 120 Ω. (*a*) Compare the rates at which heat is developed in the 15- and the 30-Ω resistors. (*b*) Compare the rates of heating in the 60- and 120-Ω resistors. (Give quantitative answers, and justify.)

32 Of the energy expended in an electric lamp bulb only about 5 percent is given off as light. Assume that a 75-W bulb costs 25 cents, and electric energy costs 3.0 cents/kWh. (*a*) If the bulb has a life of 1,000 h, what is the cost of operation? (*b*) How much nonluminous heat is developed in the lamp? (*c*) What fraction of the cost of operation is the cost of the bulb?

Ans. \$2.25; 71 kWh; 8.9 percent.

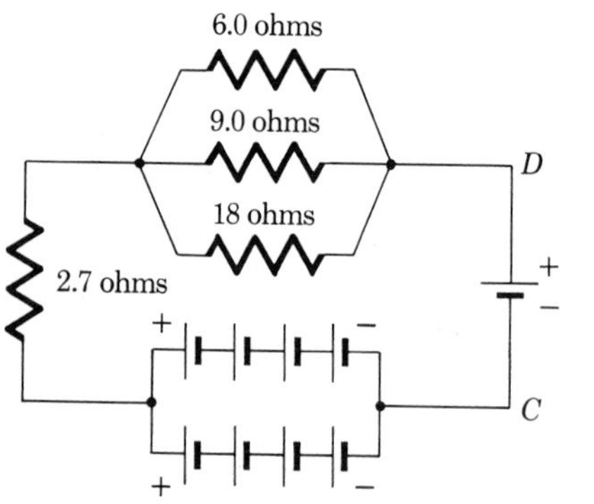

Figure 36-18

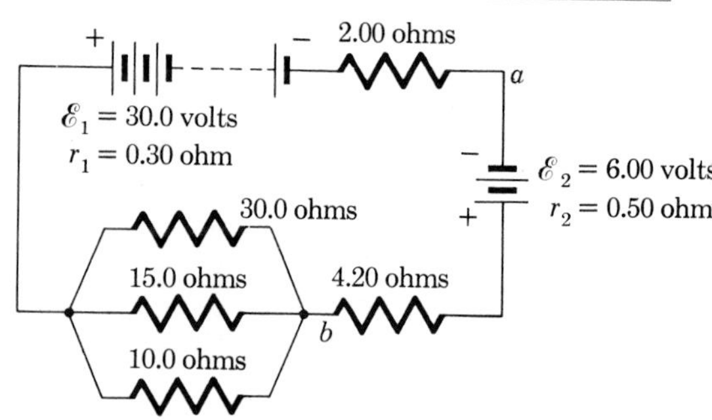

Figure 36-19

33 The emf of each cell (Fig. 36-18) is 1.50 V, and the resistance of each cell is 0.10 Ω. Find (*a*) the current in *CD*, (*b*) the heat produced in the 9.0-Ω resistor in 15.0 min, (*c*) V_{CD}, and (*d*) the electric energy transformed into chemical energy in cell *CD* in 1 h.

34 The voltage between the line wires entering a house is kept nearly constant at 115.0 V. A line leading to the laundry has a resistance of 0.35 Ω. What is the voltage across a lamp in the laundry in which there is a current of 1.5 A? What does this voltage become when a 10-A iron near the lamp is turned on? *Ans.* 114.5 V; 111.0 V.

35 In Fig. 36-7 read the circuit from left to right and insert the following values in place of those given: 3.00 V, 2.00 Ω, 3.00 Ω, 4.00 Ω, 4.50 V. Solve for all currents.

36 For the circuit shown in Fig. 36-19 calculate (*a*) the current in the circuit, (*b*) the potential difference V_{ba}, (*c*) the power supplied to the entire circuit by the source, and (*d*) the electric energy transformed into chemical energy in the 6.0-V battery in 1.00 h. *Ans.* 2.00 A; 150 V; 60.0 W; 12.0 Wh.

37 In the circuit of Fig. 36-11 assume the following values: $R_1 = 1.00\ \Omega$, $R_2 = 2.00\ \Omega$, $R_3 = 3.00\ \Omega$, $R_4 = 400\ \Omega$. The battery emf is 12.0 V and its internal resistance is negligible. Solve for all currents.

38 Two cells have emfs of 2.0 and 4.0 V and internal resistances of 0.10 and 0.20 Ω, respectively. The two are connected in series with each other and with a resistance of 2.7 Ω. Find the difference of potential between the terminals of the 2.0-V cell (*a*) when the two cells are in helping series and (*b*) when they are in opposing series. *Ans.* 1.8 V; 2.1 V.

39 A storage battery consists of 12 lead cells connected in series. Each cell has an emf of 2.1 V and an internal resistance of 0.050 Ω. (*a*) What external voltage must be impressed on the terminals of the battery in order to charge it with a 15-A current? (*b*) What is the potential difference between the terminals of the battery when it is being discharged at a rate of 30 A?

40 A battery is made up of two parallel groups, each of six cells connected in series. Each cell has an emf of 2.0 V and an internal resistance of 0.40 Ω. The battery is connected to an external circuit comprising an 8.8-Ω resistor connected in series with a group of three resistors of 20, 30, and 60 Ω, respectively, connected in parallel. Calculate (*a*) the current in the 20-Ω resistor, (*b*) the current in one cell, and (*c*) the terminal potential difference of the battery. *Ans.* 0.30 A; 0.30 A; 11 V.

41 A battery of emf 6.24 V has an internal resistance of 0.105 Ω. The battery is to be charged for one day with a current of 5.18 A from a source that maintains a potential difference of 12.65 V. (*a*) How much heat is developed in the rheostat that must be included in this circuit? (*b*) How much energy is supplied by the source? (*c*) How much energy is converted into chemical energy?

42 It is desired to illuminate a building with 300 lamps, operating in parallel. The lamps have a hot resistance of 200 Ω each and should operate at 110 V. A battery of storage cells is to be used, each of which has an emf of 2.20 V and an internal resistance of 0.00400 Ω. The cells should discharge continuously at the rate of about 28.0 A. What is the minimum number of cells that can be used and how should they be arranged? *Ans.* 6 rows of 53 cells each.

Chen Ning Yang, 1922–

Born in Hofei, China. Studied at Southwest Associated University, China, and the University of Chicago. Member of the Institute for Advanced Study, Princeton, New Jersey. Shared the 1957 Nobel Prize for Physics with Lee for their overthrow of the principle of the conservation of parity.

Tsung Dao Lee, 1926–

Born in China. Studied at the University of Chicago, the University of California, and the Institute for Advanced Study. Professor of physics, Columbia University. Shared the 1957 Nobel Prize for Physics with Yang for experiments which destroyed the long-accepted principle of the conservation of parity.

37

Magnetic Fields of Electric Currents

Magnetic effects have been recognized for centuries. An understanding of the relationships between magnetic effects and electric current has developed more recently. This relationship is emphasized in the meter-kilogram-second-ampere system of units we are using.

The ancient Greeks observed that "lodestone" (Fe_3O_4) has the ability to attract bits of iron. Long, narrow pieces of magnetized material are found to set themselves in an approximate north-south position, and hence we have the magnetic compass. In both these observations we deal with *magnets*.

The fact that electric currents (electric charges in motion) exert magnetic effects was first observed by the Danish scientist Hans Christian Oersted in 1820. He found that a compass needle is deflected from its normal north-south orientation when a current-bearing conductor is parallel to the compass and above or below it.

Ampère (1820) observed that two current-bearing conductors exert forces on each other and suggested that the magnetic condition of magnets is caused by currents within the body of the magnet. These *amperian currents* may now be identified with the motion of electrons in the atoms of the magnetic material.

37-1 MAGNETIC FORCE ON A MOVING CHARGE; MAGNETIC INDUCTION

When electric charges are at rest, they exert electrostatic forces of attraction or repulsion on each other. When the charges are in motion, they still exert these electrostatic forces, but in addition, magnetic forces appear because of the motion. The force of a current on a magnet is one example of this force.

A *magnetic field* is a region in which there is a force on a moving charge in addition to the Coulomb forces between charges. This force depends upon the charge q, the speed v of the moving charge, the direction of the motion, and some property of the field. This property of the field we call magnetic induction. The *magnetic induction B* is a vector quantity defined from the relation

$$\mathbf{F} = (\mathbf{v} \times \mathbf{B})q \tag{1}$$

That is, the magnitude of B is given by the equation

$$B = \frac{F}{qv \sin\theta} \tag{1a}$$

The direction of B is taken as the direction in which the force F on the moving charge is zero. The angle θ is then the angle between B and the direction of motion of the charge. Then the factor $v \sin \theta$ is the component of the velocity of the charge in a direction perpendicular to B.

If a small compass needle is brought into a magnetic field, it will set itself parallel to the direction here defined. Thus we observe that the orientation of the compass needle is an indication that there is a magnetic field associated with the earth. We may use the compass needle to indicate the positive sense of the magnetic induction. The N end of the compass (the end that seeks the north in the earth's field) points in the positive direction of the magnetic induction.

The direction of the force F on the moving charge is perpendicular to B and to the component $v \sin \theta$ of the motion of the charge.

If the charge q is moving along a conductor, Eq. (1) may be rewritten in terms of the current I and the length l of the conductor on which there is the force F,

$$\mathbf{F} = I(\mathbf{l} \times \mathbf{B}) \tag{2}$$

or

$$B = \frac{F}{Il \sin \theta} \tag{2a}$$

A unit for B may be defined from Eq. (1) or Eq. (2). In the mksa system this unit is the newton per ampere-meter. It is customary to express B in *teslas* T, where $1\ \text{N/A}\cdot\text{m} = 1\ \text{T}$. In the cgs system, where the force is in dynes, the current in abamperes ($1\ \text{abA} = 10\ \text{A}$), and the length in centimeters, the unit of B is the dyne per abampere-centimeter.

Magnetic fields are observed in the neighborhood of magnetized materials and near currents. When the region near a current in a long straight conductor is explored by the small compass needle, it is found that the needle sets itself tangent to some circle with the center at the conductor. A set of such circles is shown in Fig. 37-1. A vector tangent to one of the circles shows the direction of the field. The sense of the magnetic field around a current is given by a right-hand rule: If the conductor is grasped by the right hand with the thumb in the direction of the (conventional) current, the fingers encircle the conductor in the direction of the field.

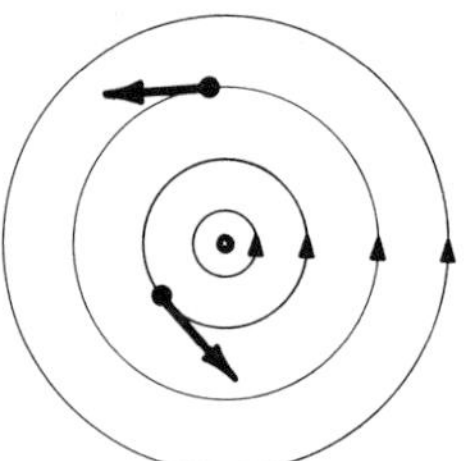

Figure 37-1
Magnetic field around a conductor carrying a current "out" of the plane of the page.

37-2 MAGNETIC FLUX; FLUX DENSITY

In a manner similar to that used for electric fields, the magnetic induction B can be represented by *lines of induction*. A line of induction is a line so drawn that it is everywhere parallel to the direction of the magnetic field. The lines of induction are called *magnetic flux* Φ_m. The selection of the number of lines to represent the magnetic induction quantitatively is arbitrary, and the choice made sets up the system of units to be used. The usual choice is to take the number of lines of induction per unit area passing through a surface perpendicular to the flux as equal to B. In the mksa system a magnetic induction of $1\ \text{N/A}\cdot\text{m}$ is represented by 1 line of induction per square meter. A line thus defined is called a *weber*.

The magnetic induction B is the flux per unit area of the surface perpendicular to B, and hence magnetic induction is also called *flux density*. For a normal surface area ΔA

$$B = \frac{\Delta \Phi_m}{\Delta A} \qquad \text{or} \qquad B\,\Delta A = \Delta \Phi_m \tag{3}$$

In the mks system flux density is measured in webers per square meter. From the way in which the weber is defined, 1 Wb/m^2 = 1 N/A·m, and 1 Wb = 1 N·m/A.

When cgs electromagnetic units are used, the lines of induction are selected so that the number of lines of induction per square centimeter of the area normal to the field is equal to B. The unit of flux thus defined is called the *maxwell*. The corresponding unit of flux density, the maxwell per square centimeter, is called the *gauss:* 1 g = 1 dyn/abA·cm. It follows that 1 Wb/m^2 = 10^4 g.

In a uniform magnetic field the lines of induction are straight lines and uniformly spaced. For this type of field the flux through any plane area A is

$$\Phi_m = BA \cos \alpha \qquad (4)$$

where α is the angle between B and the normal to the surface.

37-3 CURRENT AND FLUX DENSITY

Each element of a current contributes to the magnetic induction at all points in the field about the conductor. This contribution ΔB depends upon the current I in the element, the length Δl of the element, the distance s from the element to the point P considered, and the angle θ between the tangent to the conductor at the element and the line joining the element with P (Fig. 37-2).

$$\Delta B = K \frac{I\, \Delta l \sin \theta}{s^2} \qquad (5)$$

The factor K depends upon the properties of the medium between O and P, as well as upon the units used. The flux density at P is the sum of the contributions from all the elements of the wire, i.e.,

$$B = \sum K \frac{I\, \Delta l \sin \theta}{s^2} \qquad (6)$$

where the summation is taken over the whole length of the conductor. The direction of the flux

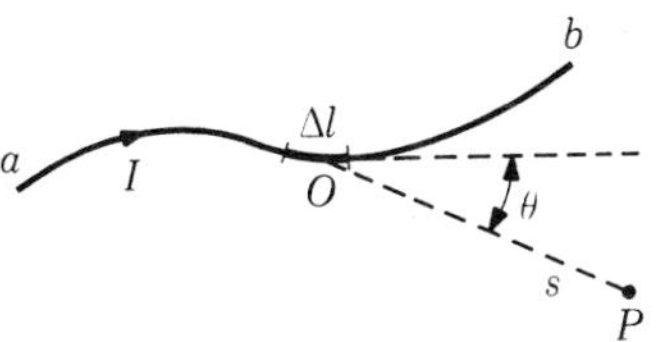

Figure 37-2
Laplace's law for the magnetic flux density near a current.

is at right angles to the plane determined by Δl and the line joining P and Δl.

The factor K may be related to the magnetic properties of the material around a current. Use of this factor K is similar to the way permittivity was used in Coulomb's law to give consideration to the dielectric properties of the medium around a charge. In the mksa system

$$K = \frac{\mu_0}{4\pi} = 10^{-7}\ \text{Wb/A·m}$$

where μ_0 is called the *permeability* of free space (vacuum), or

$$\mu_0 = 4\pi \times 10^{-7}\ \text{Wb/A·m}$$

We cannot check Eq. (5) directly by experiment, for we cannot isolate the effect of one current element from the rest of the circuit. But we can use Eq. (6) to find expressions for B for various current arrangements, and the fact that calculated results are consistent with experimental measurements is a verification of the general validity of Eq. (5).

37-4 MAGNETIC INDUCTION AT THE CENTER OF A CIRCULAR CURRENT (COIL)

One of the most direct applications of Eq. (6) is the case of a current in a circular wire or a coil of a few closely wound turns. When the coil is

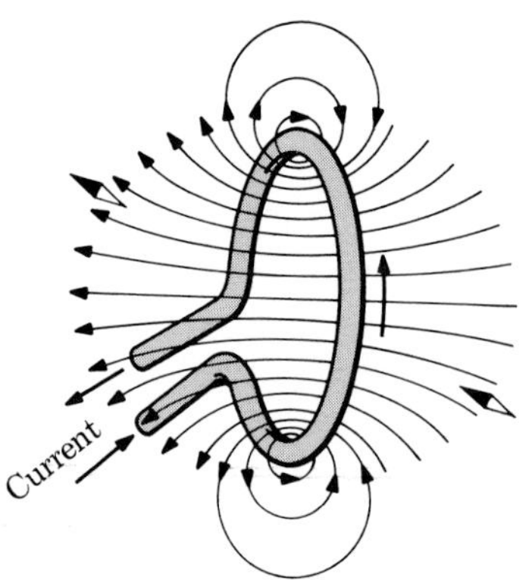

Figure 37-3
Magnetic flux through a circular current.

placed at right angles to the plane of the paper, the field that is produced has something of the appearance shown in Fig. 37-3. It is clear that the field is not at all uniform. However, the value of B at the center of the coil can be obtained from Eq. (6). Since s is everywhere perpendicular to the wire, the angle θ is $90°$ and $\sin\theta = 1$. Also s is constant and equal to the radius r of the coil. From Eq. (6),

$$B = \frac{\mu_0}{4\pi}\sum\frac{I\,\Delta l}{r^2} = \frac{\mu_0 I}{4\pi r^2}\sum\Delta l$$

The summation $\Sigma\,\Delta l$ is simply the length of the circular conductor. For N turns $\Sigma\,\Delta l = 2\pi rN$. Thus the flux density at the center of a circular coil is

$$B = \frac{\mu_0 I}{4\pi r^2}2\pi rN = \frac{\mu_0 NI}{2r} \tag{7}$$

The direction of B is perpendicular to the plane of the coil in the direction given by the right-hand rule. At other points in the plane of the coil, B is perpendicular to the plane, but Eq. (7) gives the value only at the center.

Example There is a current of 30 A in a flat circular coil having 15 closely wound turns, with a radius of 20 cm. What is the flux density at the center of the coil?

$$B = \mu_0\frac{NI}{2r}$$

$$= (4\pi\times10^{-7}\ \text{Wb/A}\cdot\text{m})\frac{15\times30\ \text{A}}{2\times0.20\ \text{m}}$$

$$= 1.4\times10^{-3}\ \text{Wb/m}^2$$

37-5 FLUX DENSITY NEAR A LONG STRAIGHT CURRENT

In Fig. 37-4 an infinitely long, straight wire AB carries a current I. The flux density at a point P distant s from the wire may be found from Eq. (6).

$$\Delta B = \frac{\mu_0}{4\pi}\frac{I\,\Delta l\sin\theta}{r^2}$$

Since α and θ are complementary angles, we may write

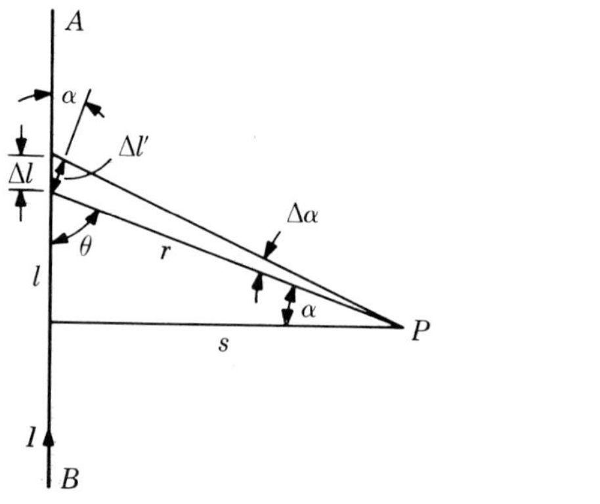

Figure 37-4
Biot-Savart law.

$$\Delta B = \frac{\mu_0}{4\pi} \frac{I \, \Delta l \cos \alpha}{r^2}$$

By construction, $\Delta l' = \Delta l \cos \alpha$ and $\Delta \alpha = \Delta l'/r$.

Hence $\Delta l \cos \alpha = \Delta l' = r \, \Delta \alpha$

Also $r = \dfrac{s}{\cos \alpha}$

When these values are substituted, we obtain

$$\Delta B = \frac{\mu_0}{4\pi} \frac{I \cos \alpha \, \Delta \alpha}{s}$$

The total flux density at P is the sum of the contributions from all the elements, i.e., the summation when α varies from $-90°$ to $+90°$. It is shown in Appendix A-9 that this summation gives the equation

$$B = \frac{\mu_0}{2\pi} \frac{I}{s} \tag{8}$$

This equation is valid only for an infinitely long current, but it applies satisfactorily to shorter wires if the distance from the wire is not too great. Equation (8) is sometimes known as the *law of Biot and Savart*.

The field due to a long, straight conductor is not a uniform field, even at a given distance from the conductor. Although the *magnitude* of the field may be constant, the *direction* changes from point to point and hence the field is not a uniform one.

Example There is a current of 25.0 A in a long, straight wire. What is the flux density at a point 3.00 cm from the wire?

$$B = \mu_0 \frac{I}{2\pi s}$$

$$= (4\pi \times 10^{-7} \text{ Wb/A} \cdot \text{m}) \frac{25.0 \text{ A}}{2\pi \times 0.0300 \text{ m}}$$

$$= 1.67 \times 10^{-4} \text{ Wb/m}^2$$

37-6 FLUX DENSITY ON THE AXIS OF A COIL

Consider the case of a concentrated coil of N turns and radius r, with its plane normal to the plane of the paper (Fig. 37-5). When there is a current I in the coil, there will be a magnetic field at all points nearby. We shall determine the flux density at a point P on the axis of the coil at a distance x from the center O.

From Eq. (5) the contribution of the element Δl to the flux density at P is given by

$$\Delta B = \frac{\mu_0}{4\pi} \frac{I \, \Delta l}{s^2}$$

since s is normal to Δl. This flux density ΔB is perpendicular to s. It may be resolved into a component ΔB_y perpendicular to x and a component ΔB_x parallel to x. When the effect of an entire turn in the coil is considered, it may be seen that each component such as ΔB_y may be paired off against another one that is equal in magnitude but opposite in direction, due to the oppositely directed element of current in a portion of the wire at the other end of the diameter from Δl. But the elements of field ΔB_x are all in the same direction and hence add up to give the resultant B, directed along x. The value of ΔB_x may be obtained from

$$\Delta B_x = \Delta B \sin \phi = \frac{\mu_0}{4\pi} \frac{I \, \Delta l}{s^2} \frac{r}{s}$$

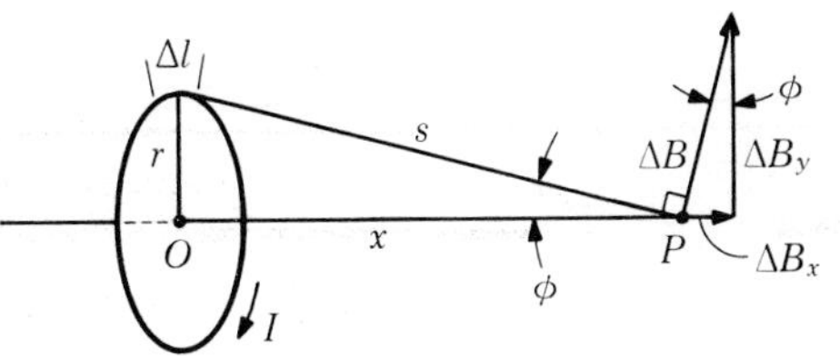

Figure 37-5
Magnetic flux density on the axis of a coil.

$$B = \frac{\mu_0}{4\pi} \frac{Ir}{s^3} \sum \Delta l$$

It is shown in Appendix A-9 that this summation leads to the equation

$$B = \mu_0 \frac{NIr^2}{2s^3}$$

If we substitute $s^2 = r^2 + x^2$, we obtain

$$B = \mu_0 \frac{NIr^2}{2(r^2 + x^2)^{3/2}} \tag{9}$$

It will be seen that Eq. (9) reduces to Eq. (7) when $x = 0$, that is, for a point at the center of the coil.

Example A flat, circular coil of 120 turns has a radius of 18 cm and carries a current of 3.0 A. What are the magnitude and direction of the flux density at a point on the axis of the coil at a distance from the center equal to the radius of the coil?

$$B = \mu_0 \frac{NIr^2}{2(r^2 + x^2)^{3/2}}$$

$$= (4\pi \times 10^{-7}\ \text{Wb/A}\cdot\text{m}) \times \frac{120 \times 3.0\ \text{A} \times (0.18\ \text{m})^2}{2(0.18^2\ \text{m}^2 + 0.18^2\ \text{m}^2)^{3/2}}$$

$$= 4.4 \times 10^{-4}\ \text{Wb/m}^2$$

The direction of the flux density is along the axis of the coil and directed away from the coil when the current in the coil is counterclockwise as viewed by an observer at P (Fig. 37-5).

37-7 FLUX DENSITY ON THE AXIS OF A SOLENOID

Consider the case of a closely wound helical coil of N turns, which has an axial length l and radius r in which there is a current I. In an element of the solenoid of length Δx (Fig. 37-6) there will be $(N/l)\ \Delta x$ turns. In accordance with the equation derived above for the field on the axis of a coil, the field at P due to the element Δx is given by

$$\Delta B = \mu_0 \frac{NIr^2\ \Delta x}{2ls^3}$$

To sum up this expression, it is convenient to express Δx and s in terms of the single variable ϕ. From Fig. 37-6*b* it is seen that

$$\sin \phi = \frac{s\ \Delta\phi}{\Delta x}$$

from which

$$\Delta x = \frac{s\ \Delta\phi}{\sin \phi}$$

Since $\sin \phi$ also equals r/s, we may write the equation for ΔB as

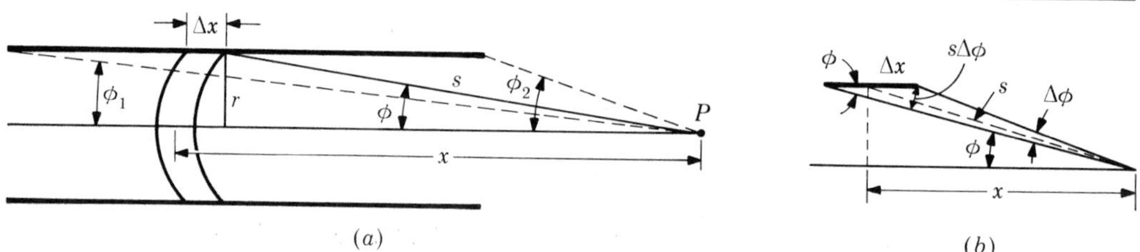

Figure 37-6
Magnetic flux density on the axis of a solenoid.

$$\Delta B = \mu_0 \frac{NI}{2l} \sin \phi \, \Delta\phi$$

The resultant field at P due to the whole solenoid is given by

$$B = \mu_0 \frac{NI}{2l} \sum \sin \phi \, \Delta\phi$$

It is shown in Appendix A-9 that this summation results in the equation

$$B = \mu_0 \frac{NI}{2l} (\cos \phi_1 - \cos \phi_2) \qquad (10)$$

For the center of a long solenoid, when l is large in comparison with r, $\phi_1 = 0$ and $\phi_2 = 180°$. Hence $\cos \phi_1 - \cos \phi_2 = 2$, and the field at the center of the solenoid is given by

$$B = \mu_0 \frac{NI}{l} \qquad (11)$$

Example A closely wound solenoid of 1,250 turns has an axial length of 98 cm and a radius of 1.50 cm. When there is a current of 1.30 amp in the solenoid, what is the flux density on the axis (*a*) at the center and (*b*) at one end of the solenoid?

At the center,

$$B_c = \frac{\mu_0 NI}{l}$$
$$= (4\pi \times 10^{-7} \text{ Wb/A·m}) \times \frac{1{,}250 \times 1.30 \text{ A}}{0.98 \text{ m}}$$
$$= 2.09 \times 10^{-3} \text{ Wb/m}^2$$

At one end,

$$B_e = \mu_0 \frac{NI}{2l} (\cos \phi_1 - \cos \phi_2)$$
$$= (4\pi \times 10^{-7} \text{ Wb/A·m}) \times \frac{1{,}250 \times 1.30 \text{ A}}{2 \times 0.98 \text{ m}} (\cos 0.8° - \cos 90°)$$
$$= 1.04 \times 10^{-3} \text{ Wb/m}^2$$

It will be noticed from this example that, for a solenoid which has a length that is large in comparison with its radius, the flux density at the end is one-half that at the center.

The symbol n is often used for the number of turns per unit axial length of a solenoid ($n = N/l$). When this notation is used, Eq. (11) is written

$$B = \mu_0 nI \qquad (12)$$

In problems dealing with the design of electromagnetic machinery, Eqs. (11) and (12) are highly important. Although it should be clearly understood that these equations apply strictly to an infinitely long solenoid, it is found that reasonably satisfactory results are obtained for shorter solenoids if the radius of the circular turns is small in comparison with the axial length of the helix. The equations are almost exactly valid for a closed solenoid such as the ring-wound *toroid* illustrated in Fig. 37-7, where l is the mean circumference of the toroid, $2\pi R$. Here the field is nearly uniform within the coils and does not flare out anywhere, as it does at the ends of an open solenoid. Such toroids are frequently used for laboratory measurements where a calculable uniform field is desired.

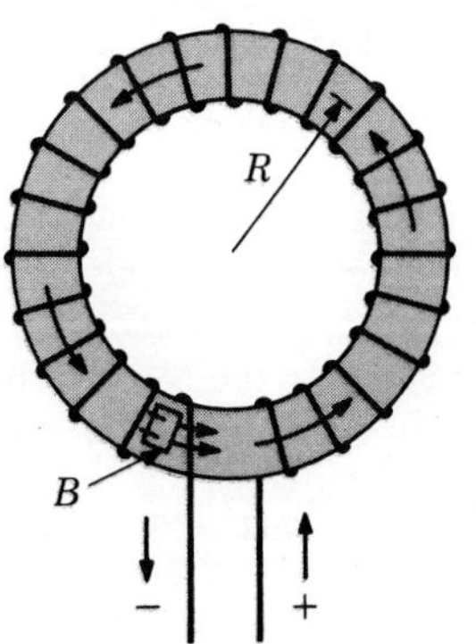

Figure 37-7
A solenoid wound in the form of a toroid.

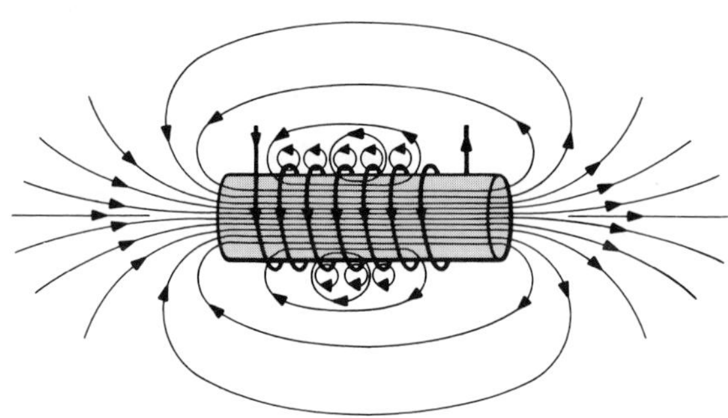

Figure 37-8
Magnetic flux about a solenoid.

Example A toroidal coil has 3,000 turns. The inner and outer diameters are 22 cm and 26 cm, respectively. Calculate the flux density inside the coil when there is a current of 5.0 A.

The mean radius R is 12 cm.

$$B = \frac{\mu_0 NI}{l}$$

$$= (4\pi \times 10^{-7}\ \mathrm{Wb/A\cdot m}) \times \frac{3{,}000 \times 5.0\ \mathrm{A}}{2\pi \times 0.12\ \mathrm{m}}$$

$$= 0.025\ \mathrm{Wb/m^2}$$

It should be noted that the field referred to in the case of the solenoid is the field *within* the solenoidal winding. If the turns are closely packed (not as shown in Fig. 37-7), there will be little flux leakage and the field outside the cross section of the windings will be nearly zero. Students are warned not to confuse the (zero) field at the center of the toroid with the case of the field at the center of a circular coil.

The field caused by a flat *coil* (Fig. 37-3) and that produced by a solenoid (Fig. 37-8) should not be confused. To differentiate, we can think of a coil as a concentrated winding of very short axial length, while a solenoid should have many closely wound turns extending over a considerable axial length.

37-8 AMPÈRE'S LAW

The physical relationship which Eq. (6) expresses as the contribution of current elements $I\,\Delta l$ is also expressed in Ampère's law in terms of the tangential component of B summed over the elements of any closed path,

$$\Sigma\, B_t\,\Delta l = \mu_0 \Sigma I \qquad (13)$$

where ΣI stands for the total current linked by the path chosen.

Example Find the magnetic induction at a distance s from a long, straight wire carrying a current I (Fig. 37-9).

Take as the path to which Ampère's law will be applied a circle concentric with the wire and having radius s (Fig. 37-9). At any element Δl the magnetic induction will be tangential to this circle (as in Fig. 37-1). In evaluating Eq. (13), $\Sigma\,\Delta l$ is the perimeter of the circle, $2\pi s$. Hence

$$B(2\pi s) = \mu_0 I$$

or

$$B = \frac{\mu_0}{2\pi s} I$$

which is in accordance with Eq. (8).

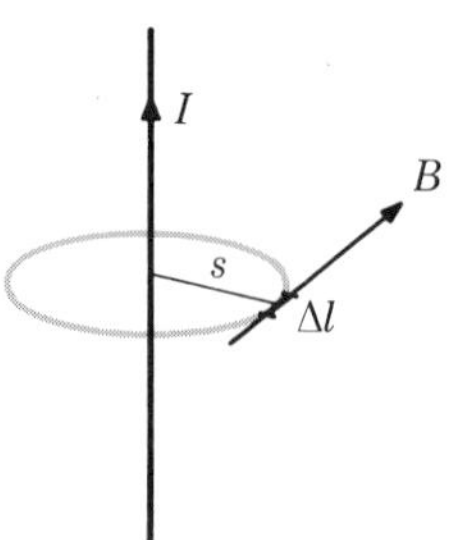

Figure 37-9
Magnetic induction B near a long, straight conductor.

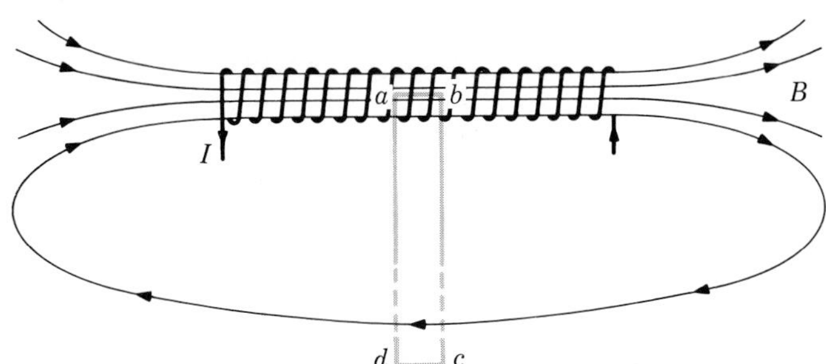

Figure 37-10
Calculation of magnetic induction B near the center of a long solenoid.

Example Find the magnetic induction at the center of a long solenoid having n closely wound turns per unit length (Fig. 37-10).

Ampère's law can be applied to a narrow, very long, rectangular path at the median plane. The left side of Eq. (13) can now be expressed as the sum of four terms corresponding to the four sides of the rectangular path,

$$\sum_a^b B_t \,\Delta l + \sum_b^c B_t \,\Delta l + \sum_c^d B_t \,\Delta l + \sum_d^a B_t \,\Delta l = \mu_0 \Sigma I$$

The second and third summations are zero, for B_t is zero in the median plane, where B is perpendicular to the lines bc and cd. The fourth summation is zero, since $B = 0$ at infinite distance from the solenoid. Hence the previous equation reduces to

$$\sum_a^b B_t \,\Delta l = \mu_0 \Sigma I$$

The term ΣI represents $n(ab)\,I$ current linkages; so

$$B(ab) = \mu_0 nI\,(ab)$$

or

$$B = \mu_0 nI$$

SUMMARY

The forces between magnets are believed to be caused by the forces that moving electric charges in atoms exert on each other.

A *magnetic field* is a region in which there is a force on a moving charge in addition to the Coulomb forces.

Magnetic induction is defined from the force on the moving charge by the equation

$$B = \frac{F}{qv \sin \theta}$$

It is a vector quantity whose direction is the direction of motion of the charge for which the force is zero. The magnetic induction, the normal velocity $v \sin \theta$, and the force are mutually perpendicular.

Magnetic induction is represented by *lines of induction,* parallel to B, called *magnetic flux* Φ, chosen so that the number of lines of induction per unit area of a surface perpendicular to B is equal to B. When B is in newtons per amperemeter and the area is in square meters, the flux is in *webers*. B is also expressed in *teslas*.

The magnetic induction is also called *flux density,* since it is the flux per unit area,

$$B = \frac{\Phi}{A} \qquad \text{or} \qquad \Phi = BA$$

Flux density is measured in webers per square meter.

Magnetic fields are observed in the neighborhood of currents and near magnetized materials.

Each element of a current contributes to the flux density at all points about the conductor by an amount

$$\Delta B = K\frac{I \sin\theta\, \Delta l}{s^2}$$

In the mksa system

$$K = \frac{\mu_0}{4\pi} = 10^{-7}\ \mathrm{Wb/A{\cdot}m}$$

where μ_0 is called the *permeability* of empty space.

The flux density in empty space near a conductor may be found by summing up the contributions of all the elements of the current.

For several arrangements these sums are:

At the center of a circular coil,

$$B = \mu_0\frac{NI}{2r}$$

Near a long, straight current,

$$B = \mu_0\frac{I}{2\pi s}$$

On the axis of a coil,

$$B = \mu_0\frac{NIr^2}{2(r^2 + x^2)^{3/2}}$$

On the axis of a solenoid,

$$B = \mu_0\frac{NI}{2l}(\cos\phi_1 - \cos\phi_2)$$

At the center of a long solenoid,

$$B = \mu_0 nI$$

Questions

1 Is there a magnetic field near an electrostatic charge? Would one expect a magnetic field to exist in the vicinity of a moving charge? Why? Outline experiments which might be used to demonstrate each of these cases.

2 A cable carrying a direct current is buried in a wall that stands in a north-south plane. On the west side of the wall a horizontal compass needle points south instead of north. What are (*a*) the position of the cable and (*b*) the direction of the current in the cable?

3 A coil in the plane of the paper is connected to a battery. The current in the coil is clockwise. (*a*) What is the direction of the flux density near the center of the coil? (*b*) What will be the relative change in flux density if the radius of the coil is doubled and (*c*) if the number of turns is doubled also?

4 Assuming the validity of Ampère's law as an empirical relation, show how Laplace's law could be derived from Ampère's law.

5 From the known relation of the abampere and the ampere deduce the relation between the coulomb and the abcoulomb; between the volt and the abvolt.

6 A "tangent galvanometer" consists of a flat, circular coil of a few turns N and radius r. The coil is placed with its major plane parallel to the horizontal component of the earth's magnetic field H. A small compass needle mounted at the center of the coil is deflected through an angle θ when there is a current I in the coil. Derive the equation for this angle in terms of I and H and the constants of the coil.

7 The small circle in Fig. 37-11 represents a

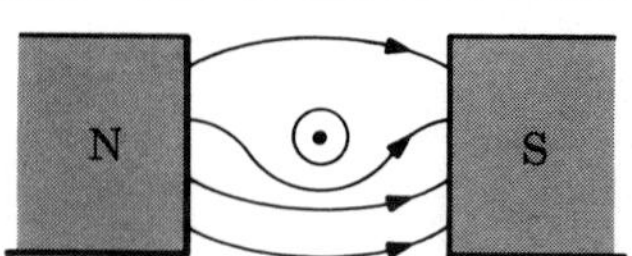

Figure 37-11

current-carrying conductor, perpendicular to the page. The curved lines represent the flux between poles of an electromagnet. What is the direction of the current in the conductor?

8 Sketch a curve to show how the magnetic flux density near a long, straight current varies with the distance away from the current.

9 A piece of flexible wire is wrapped loosely around a strong cylindrical bar magnet. When there is a heavy current in the wire, it entwines itself around the magnet. If the current is reversed, the wire uncoils and winds itself around the magnet in the opposite sense. Explain the reasons for this behavior.

10 A stream of electrons is projected horizontally toward the right. A vertical magnet with the N pole downward is brought near the electron beam. Explain what happens.

11 "Helmholtz coils" consist of two large, flat, circular coils mounted with a common axis and at a distance apart equal to their common radius. (*a*) Show that this arrangement produces a uniform flux density over a small space midway between the coils. (*b*) Derive an equation for this flux density when the coils have the same current in the directions to produce additive effects.

Problems

1 Find the magnetic induction in air at a point 10 cm from a long, straight wire which has a current of 30 A flowing through it.

2 A current of 25 A is maintained in a storage-battery charging outfit for 6.0 h. How many abcoulombs of charge will flow?

Ans. 5.4×10^4 abC.

3 Find the current in conductor *CDEFG* (Fig. 37-12) when the magnetic flux density at *O* is 1.0×10^{-7} Wb/m^2 if the radius *OD* is 4.5 cm and the length of the arc *DEF* is 12.0 cm. What is the direction of the magnetic flux at 0?

4 A current of 18 A in a circular segment of wire (such as *DEF* in Fig. 37-12) produces a magnetic induction at the center of the arc of 4.48×10^{-4} Wb/m^2. The wire makes an angle of

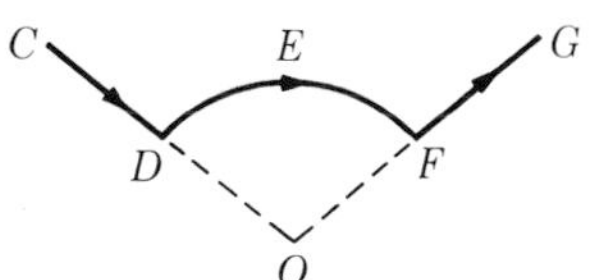

Figure 37-12

3.0 rad at the center *O*. What is its radius of curvature?

Ans. 1.2 cm.

5 What current would have to be maintained in a circular coil of wire of 50 turns and 2.54-cm radius in order to just cancel the effect of the earth's magnetic field at a place where the horizontal component of the earth's field has an induction of 1.86×10^{-5} Wb/m^2? How must the coil be set up?

6 Two long, straight, parallel wires in which there are currents of 2.0 and 3.0 A, respectively, in the same direction, are 10 cm apart. Find the magnitude and the direction of the magnetic field strength at a point halfway between them.

Ans. 4×10^{-6} Wb/m^2.

7 Consider three parts of a circuit: a straight wire of length l_1 and a straight wire of length l_2 lying along the same direction and connected by a semicircular segment of radius *R*. If the same current *I* is maintained in these parts, what is the magnetic induction at the center of the semicircular segment arising from (*a*) the straight segment l_1, (*b*) the straight segment l_2, (*c*) the semicircular segment of radius *R*, and (*d*) the entire circuit?

8 Two long, straight, parallel wires in which there are currents of 5.0 and 10.0 A in opposite directions are 10 cm apart. Find the magnitude and direction of the magnetic field strength at a point halfway between them.

Ans. 6×10^{-5} Wb/m^2.

9 A flat, circular coil has 30 turns wound closely on a radius of 18.5 cm. A small compass needle is mounted on a vertical axis at the center of the coil. The coil is adjusted until its plane is parallel to the earth's magnetic field, which at that place has a horizontal component of

2.08×10^{-5} Wb/m^2. There is a current of 0.358 A in the coil. Through what angle will the compass needle be deflected?

10 Derive an expression for the magnetic induction B at point P in Fig. 37-13 due to current I in the path shown. What is the direction of B?

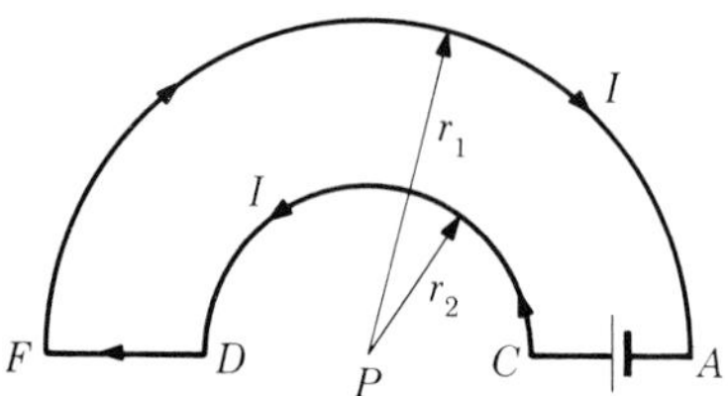

Figure 37-13

Ans. $B = (\mu_0 I/4)[(1/r_2) - (1/r_1)]$, out of page.

11 A very long, straight wire of radius a carries a current producing a magnetic induction of 1.0×10^{-5} Wb/m^2 at the surface of the wire. What is the value of the flux density at (*a*) a point $2a$ from the center of the wire, (*b*) at a point $a/2$ from the center of the wire, and (*c*) at the center of the wire?

12 A toroid has a closely wound coil with 30 turns per inch of length. What is the magnetic induction in the coil for a current of 5.25 A? Show the direction of this field.

Ans. 7.8×10^{-7} Wb/m^2.

13 An upward current of 7.2 A is maintained in a long, straight wire in a place where the horizontal component of the earth's magnetic induction is 2.38×10^{-5} Wb/m^2. Calculate the resultant magnetic induction at a point 8.0 cm from the wire (*a*) north of the wire and (*b*) west of the wire.

14 A beam of electrons is bent in a circle of radius of 4 cm under the influence of a field of magnetic induction of 5.0×10^{-3} Wb/m^2. What is the velocity of the electrons?

Ans. 3.5×10^7 m/s.

15 Two long, parallel wires carry currents perpendicular to the plane of the paper through points A and B. The currents are each 10 A, in the same direction. Consider an equilateral triangle ABC, 25 cm on a side. Determine the magnetic induction at point C.

16 A long solenoid has 35 turns per centimeter of length, and there is a current of 8.00 A in it. A short, closely wound coil of 25 turns and 6.0 cm radius is wound over the center of the solenoid. A current of 12 A in this coil is in the opposite sense to that in the solenoid. What is the magnetic induction at the center of the system?

Ans. 3.20×10^{-2} Wb/m^2.

17 Find the magnetic induction produced by a 5.0-A current in each of the following cases: (*a*) 2.0 cm from a long, straight wire, (*b*) at the center of a 40-turn coil of 2.0 cm diameter, and (*c*) anywhere inside an infinitely long solenoid of 2.0 cm diameter and with 50 turns per centimeter.

18 Calculate the period of rotation of a free electron moving in a plane perpendicular to the magnetic induction of the earth, 0.60×10^{-4} Wb/m^2.

Ans. 0.60 μs.

19 A particle of mass 0.020 gm travels in a straight horizontal line between two horizontal charged plates, between which there is a electric field of 2.0 V/cm upward. A magnetic induction of 2.0 Wb/m^2 is directed at right angles to the electric field, into the paper. If the particle bears a charge of 1.0×10^{-6} C, what is its speed?

20 In a suitable vacuum system, from a mixture of positive ions at a (Fig. 37-14*a*), it is desired to collect at b those which have a common momentum mv. (*a*) Show that a uniform magnetic

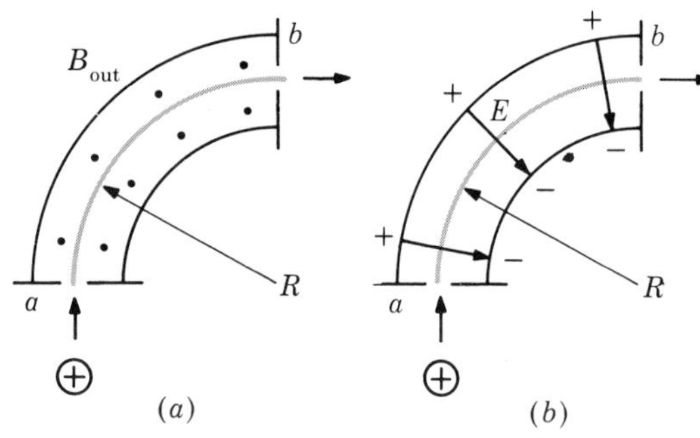

Figure 37-14

induction B directed out of the page will control the ions so that those which travel the circular are ab have momentum given by $mv = BqR$. (*b*) What happens to ions which enter at a with larger or smaller values of momentum? (*c*) Show that a radial electric field E (Fig. 37-14*b*) can be used to select ions which have the same kinetic energy $E_k = EqR/2$.

21 In a cloud-chamber photograph an electron path is bent into a circle of 12-cm radius by a magnetic induction of 0.0080 Wb/m^2 directed perpendicular to the plane of this path. (*a*) Calculate the energy of the electron. (*b*) Calculate the energy of an electron whose path radius is 20 cm in the same magnetic field.

Pavel A. Cherenkov, 1904–
(*left*)
Graduated from Voronezh University. Worked under Sergei Vavilov on luminescence of solutions of uranyl salts subjected to gamma radiation. Shared the 1958 Nobel Prize for Physics with Tamm and Frank for the discovery and interpretation of the Cherenkov effect.

Ilya M. Frank, 1908–
(*center*)
Graduated from Moscow University; worked at the Leningrad Institute of Physics. Member of the Physics Institute of the Soviet Academy of Sciences since 1934 and professor at Moscow University since 1944. His work is principally in optics and nuclear physics. Shared the 1958 Nobel Prize for Physics with Cherenkov and Tamm.

Igor Y. Tamm, 1895–
(*right*)
Member of the Soviet Academy of Sciences and of its Physics Institute. Lecturer at Moscow University. Suggested the quantum theory of the dispension of light in solids, and in 1934, a theory of nuclear forces which influenced modern mesonic theory of nuclear forces. Shared the 1958 Nobel Prize with Frank and Cherenkov.

38

Forces and Torques in a Magnetic Field

We have used, in Chap. 37, the magnetic force on a moving charge as a means of defining magnetic induction B. If the moving charges are in a current-carrying conductor placed in a magnetic field, the conductor will experience a force which is a side thrust perpendicular to the directions of both current I and magnetic induction B. If currents are maintained in two parallel wires, we may regard the second as conducting charges through the magnetic field produced by the first (Sec. 37.5). The second wire will experience a magnetic force. A reacting force is exerted on the first wire. The mutual attraction of two parallel currents is used as the experimental basis for defining the ampere, one of the fundamental units in the mksa system.

Another consequence of the force experienced by a current-carrying conductor in a magnetic field is that a coil suspended in a magnetic field experiences a torque. In many electrical instruments the torque experienced by a sensitively mounted coil is used as an accurate measure of current. In an electric motor the torque acting on the windings of the armature rotates the armature and provides an important means of converting electrical energy into mechanical energy.

38-1 FORCE ON A MOVING CHARGE OR ON A CURRENT IN A MAGNETIC FIELD

From the definition of magnetic induction (Sec. 37-1) a charge q moving in a magnetic induction B experiences a force given by

$$\mathbf{F} = q(\mathbf{v} \times \mathbf{B}) \tag{1}$$

The magnitude of this force is

$$F = qvB \sin\theta \tag{1a}$$

The force is mutually perpendicular to B and to the component $v \sin\theta$ of the velocity of the charge normal to the field. If the fingers of the right hand are curled in the direction in which the vector $v \sin\theta$ must be rotated toward the vector B, the thumb will point in the direction of the force on a positive charge.

If a charge moves at right angles to a uniform field, its path is circular, since the force is constant in magnitude and always perpendicular to the motion. The magnetic force is the centripetal force.

$$Bqv = \frac{mv^2}{r} \tag{2}$$

When the charges are moving along a conductor in a magnetic field, the force ΔF on any element Δl of the conductor is, from Eq. (2*a*) of Chap. 37,

$$\Delta F = BI\,\Delta l \sin\theta \tag{3}$$

where θ is the angle between the conductor and the flux. The resultant force on the conductor is the vector sum of the contributions of the elements. For a straight conductor of length l in a uniform field the sum becomes

$$F = BIl \sin\theta \tag{4}$$

In Fig. 38-1, a magnetic field of flux density B is shown directed horizontally toward the right. A current I is directed vertically downward in this field. From the rule given above the force F on this conductor must be normal to the page and directed out from the paper.

Viewed from above, the conductor and field of Fig. 38-1 appear as in Fig. 38-2. Conventionally the symbol ⊙ is used to represent a current toward the reader, while a current away from the reader is indicated ⊕. Above the conductor in Fig. 38-2, the clockwise magnetic field due to the electric current is in the same direction as the externally applied field. The magnetic field above the wire is therefore strengthened. Below the wire, the field due to the current is opposite in direction to the external field. The field below the wire is therefore weakened. It is found by experiment that a current in a magnetic field experiences a force directed from the strong part of the field toward the weak part of the field. It is seen that the result is a force downward on the wire. The force is at right angles both to the current and to the field.

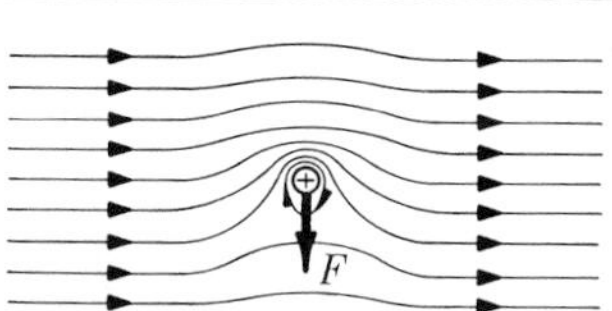

Figure 38-2
Force on a current in a magnetic flux.

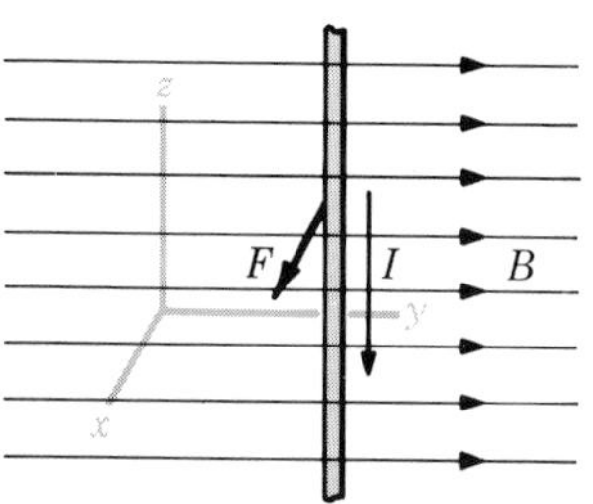

Figure 38-1
Force on a current normal to a magnetic flux.

In any situation involving an electric current in a magnetic field, the direction of the side push on the conductor may be predicted by analyzing the fields, as has been done for the case shown in Fig. 38-2.

It should be emphasized that the B in Eqs. (3) and (4) represents the flux density of the field in which the current is immersed, and not the field near the conductor due to the current. In other words, Eq. (4) gives a measure of the force which a current experiences when placed in a magnetic field set up by some other means.

It will be noted that the force is a maximum when the current is normal to the field and zero when the current is parallel to the field.

Example In Fig. 38-3 is shown a current of 25 A in a wire 30 cm long and at an angle of 60° to a magnetic field of flux density 8.0×10^{-4} Wb/m². What are the magnitude and direction of the force on this wire?

$$F = BIl\sin\theta = (8.0 \times 10^{-4}\ \text{Wb/m}^2) \times 25\ \text{A} \times 0.30\ \text{m} \times 0.87$$

$$= 5.2 \times 10^{-3}\ \text{N}$$

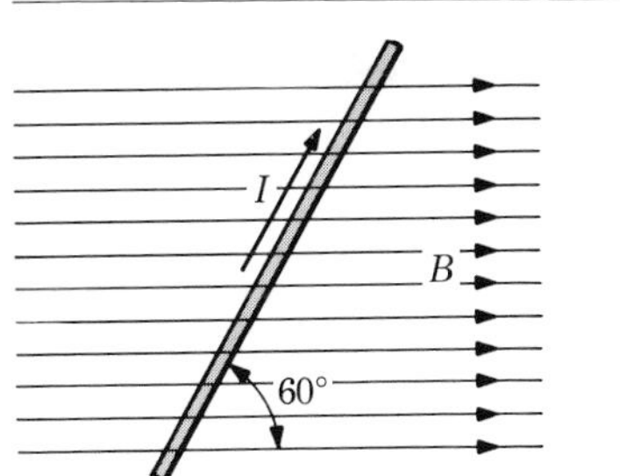

Figure 38-3
Force on a straight current not at right angle with a uniform induction *B*.

The direction of the force can most safely be determined by visualizing as in Fig. 38-2 the resultant field caused by the field due to the current and the field in which the wire is placed. Because of the field due to the current, the field above the wire will be strengthened, while that below the wire will be weakened. Hence the conductor is pushed away from the observer, from the stronger and toward the weaker field.

38-2 FORCES BETWEEN CURRENTS

When parallel current-carrying conductors are adjacent, each exerts a force on the other. We may think of one of the currents, with its accompanying field, as being situated in the field caused by the other current. By Eq. (4) this will result in a force between the current-carrying conductors. The two circular fields combine in the manner shown in Fig. 38-4. In terms of the flux picture, we visualize the effects as if they were caused by the lines acting like stretched rubber bands, with tension in the direction of their lengths. Hence currents in the same direction, as in Fig. 38-4*a*, produce an attractive force. We may visualize this in another manner by noting that the field between the wires is much weaker than the field outside the wires, and hence they tend to move away from the strong field and toward the weaker one. Unlike the rubber-band analogy, the lines are visualized as having a repulsive force in a direction perpendicular to their lengths. This assists us in seeing why currents in opposite directions, as in Fig. 38-4*b*, produce a force of repulsion. Since the two currents in this case combine to produce a strong field between the wires and weak fields outside the wires, it follows that the wires are repelled from each other, since they tend to move from the stronger toward the weaker fields.

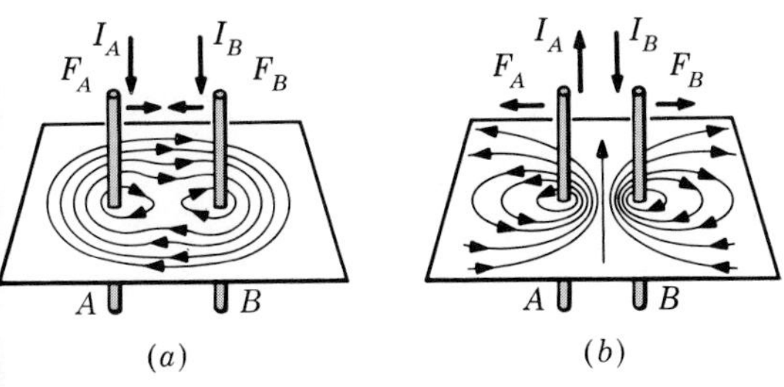

Figure 38-4
Resultant fields and forces between parallel current-carrying conductors.

A measure of the forces between parallel current-carrying conductors may be obtained from Eq. (4) and the expression for the field near a long, straight conductor. From Eq. (8) of Chap. 37, the field of current I_A in wire A is given by

$$B = \mu_0 \frac{I_A}{2\pi s}$$

From Eq. (4), the force on current I_B in wire B is therefore

$$F = BIl = \mu_0 \frac{I_A}{2\pi s} \times I_B l = \mu_0 \frac{I_A I_B l}{2\pi s} \tag{5}$$

where s is the distance from A to B and l is the length of wire B.

The relation expressed by Eq. (5) may be used to define the mks unit of current: The *ampere* is a current that when it is maintained in two infinitely long conductors, parallel to each other and 1 m apart in a vacuum, will cause a force on each conductor of exactly 2×10^{-7} newton per meter length. The numbers chosen in this definition are such as to make the unit thus defined essentially the same as the ampere defined by other methods.

Example Two straight parallel wires each 90 cm long are 1.0 mm apart. There are currents of 5.0 A in opposite directions in the wires. What are the magnitude and sense of the force between these currents?

$$F = \mu_0 \frac{I_A I_B l}{2\pi s} = \frac{4\pi \times 10^{-7}\ \text{Wb/A}\cdot\text{m}}{2\pi \times 0.0010\ \text{m}} \times 5.0\ \text{A} \times 5.0\ \text{A} \times 0.90\ \text{m}$$

$$= 4.5 \times 10^{-3}\ \text{N}$$

38-3 TORQUE ON A CURRENT-CARRYING COIL IN A MAGNETIC FIELD

A current-carrying coil in a magnetic field will experience a torque tending to rotate the coil whenever the plane of the coil is not perpendicular to the direction of the field (Fig. 38-5*a*). From a consideration of the resultant field that is produced (Fig. 38-5*b*), it will be observed that the wire *GH* is urged upward, while wire *CD* is forced downward. These forces, equal in magnitude but opposite in direction, give rise to a couple, the net torque being the product of one of the forces and the perpendicular distance between their lines of action. By Eq. (4), the magnitude of the force is

$$F = BIb \sin 90° = BIb$$

and hence the magnitude of the torque is

$$L = BIab$$

But *ab* is the *area A* of the coil, and hence

$$L = IAB \tag{6}$$

When the plane of the coil makes an angle α with the field, the torque is

$$L = IAB \cos \alpha \tag{7}$$

In general, a coil will tend to rotate in a field until its plane is normal to the field. This is the position where the coil links the maximum flux. It may therefore be stated that a coil freely suspended in a magnetic field will align itself so as to include a maximum number of lines of induction through the cross section of the coil. Equa-

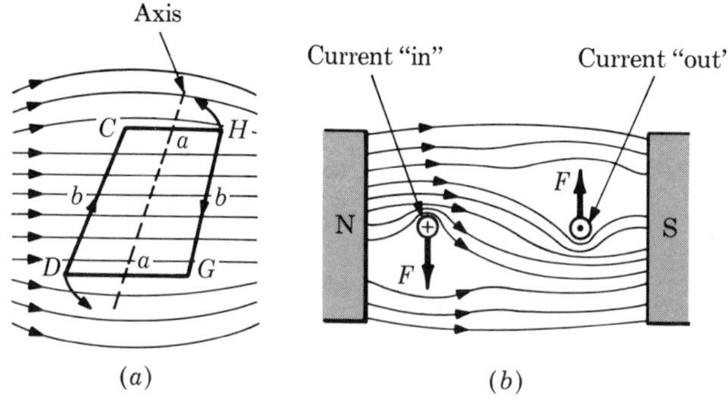

Figure 38-5
A current-carrying coil placed in a magnetic flux experiences a torque.

tion (7) is a basic equation for electric motors. If the coil has more than one turn, the torque will be increased in proportion to the number of turns N. The general equation is

$$L = NIAB \cos \alpha \tag{8}$$

The torque is expressed in meter-newtons when B is in newtons per ampere-meter (webers per square meter), I is in amperes, and A is in square meters. Although Eq. (8) has been derived for the case of a rectangular coil, it can be shown that the same equation applies to a coil of any shape.

Example A rectangular coil 30 cm long and 10 cm wide is mounted in a uniform field of flux density 8.0×10^{-4} N/A·m. There is a current of 20 A in the coil, which has 15 turns. When the plane of the coil makes an angle of 40° with the direction of the field, what is the torque tending to rotate the coil?

$$L = NIAB \cos \alpha = 8.0 \times 10^{-4} \text{ N/A·m} \times 15 \times 20 \text{ A} \times (0.30 \times 0.10) \text{ m}^2 \times 0.77$$

$$= 0.0055 \text{ m·N}$$

38-4 GALVANOMETER

The basic electric instrument is the galvanometer, a device with which very small electric currents can be detected and measured. The d'Arsonval or permanent-magnet moving-coil type of galvanometer is shown in Figs. 38-6 and 38-7. In Fig. 38-6a the coil C is suspended between the poles N and S of a U-shaped magnet by means of a light metallic ribbon. Connections are made to the coil at the terminals marked t. The cylinder of soft iron B and the pole faces N and S are skillfully shaped so as to produce a *radial* magnetic field in the air gap. This has the virtue of giving a field that is constant in magnitude and always parallel to the plane of the coil as the coil rotates. These conditions are necessary if the instrument is to have a scale with uniform graduations. The mirror M is used to indicate the position of the coil, either by reflecting a beam of light onto a scale or by producing an image of a scale to be viewed through a low-power telescope.

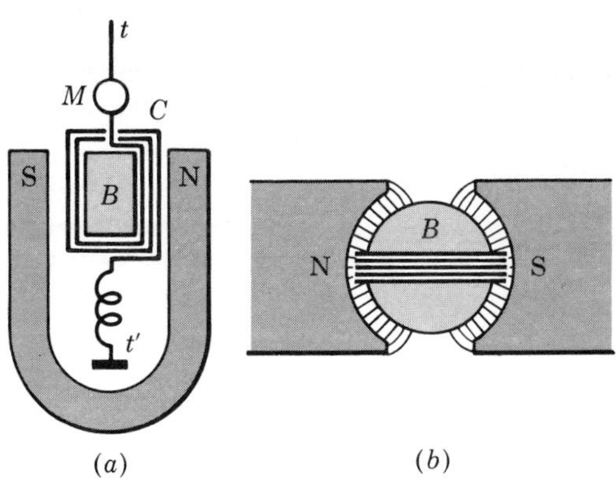

Figure 38-6
Permanent-magnet moving-coil type of galvanometer.

When a current is set up in a coil which is between the poles of a magnet, the coil is acted upon by a torque, which tends to turn it until its plane is perpendicular to the line joining the poles. If a current is set up in the coil (as viewed from above) in Fig. 38-6b, the coil will turn toward a position at right angles to the position

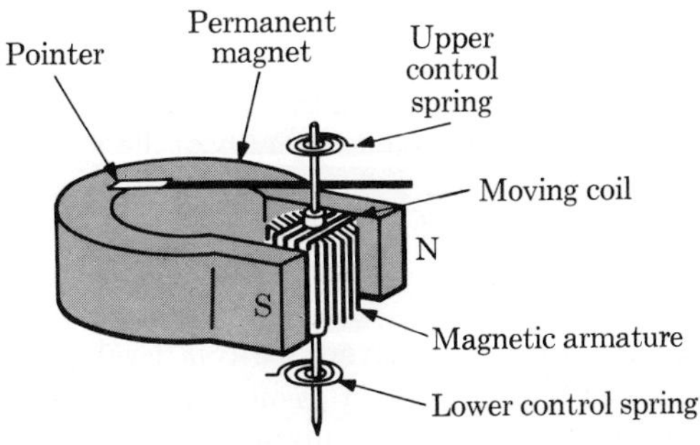

Figure 38-7
Construction of a portable-type galvanometer.

shown. In turning, however, it must twist the metallic ribbon that supports it; hence it turns to the position in which the torque exerted on it by the magnet is just neutralized by the reaction of the twisted ribbon.

The torque exerted on the coil by the magnet is proportional to the current in the coil [Eq. (8)], and the torque of reaction of the ribbon is proportional to the angle through which it is twisted. Since these torques are equal in magnitude and opposite in sense when the coil reaches the equilibrium position, the angle through which the coil turns is proportional to the current in it; that is, $\theta \propto I$, where θ is the angular deflection of the coil. This condition is realized only in well-designed instruments, in which case the galvanometer scale is properly made with equally spaced divisions. In practice, readings are made on a linear scale, since the deflection s read on the scale is proportional to the angle of deflection θ (when θ is small).

38-5 DAMPING

If the current in a galvanometer circuit is interrupted, the coil will ordinarily swing back beyond the zero position and then vibrate with progressively decreasing amplitude through the zero point. This reduction of motion is called *damping*. Similarly, if a current is suddenly established in a galvanometer, the coil will ordinarily swing beyond its final equilibrium position and vibrate several times back and forth through this position before finally coming to rest. Since it is tedious to wait for this gradual dying away of the motion of the coil, artificial means are usually provided to bring the suspended system quickly to its final position.

The most common method of producing rapid damping consists in having the coil develop induced currents because of its motion in the magnetic field. Such induced currents tend to oppose the motion of the coil in the field (Chap. 40). In one of the most commonly used laboratory wall-type galvanometers, these induced currents are developed in a rectangular loop of fairly heavy copper wire which is attached side by side with the movable coil. Since the resistance of this single turn of wire is low, the induced current caused by the motion of the coil is comparatively high and hence there is a suitable countertorque, which tends to reduce the swinging of the coil about its equilibrium position.

In many ammeters and voltmeters, damping is accomplished by winding the movable coil on a light frame of aluminum. The currents induced in this frame are quite effective in producing satisfactory damping. By skillful design the pointer is caused to reach its equilibrium position very quickly with no noticeable oscillation.

For some purposes an external damping resistor is placed either in series or in parallel with the galvanometer coil. This resistance is adjusted to a critical value for a given galvanometer and circuit so as to produce a very slight underdamping. If the resistance is such as to cause an overdamping, the coil creeps to its final position with annoying slowness.

38-6 GALVANOMETER SENSITIVITY

The deflection caused by a given current depends upon the design of the instrument. This characteristic is known as the sensitivity of the galvanometer. There are numerous ways of expressing galvanometer sensitivity, each involving a statement of the electrical conditions necessary to produce a standard deflection. This standard deflection in galvanometers having attached scales is assumed to be one scale division. In galvanometers not equipped with scales (for example, galvanometers read with auxiliary telescope and scale) *the standard deflection is assumed to be* 1 *mm on a scale at a distance of* 1 *m*. Some of the most frequently used methods of expressing galvanometer sensitivity follow.

As previously stated, the galvanometer deflection s is proportional to the current I,

$$s \propto I \qquad \text{or} \qquad I = ks \qquad k = \frac{I}{s} \tag{9}$$

where k, the current per standard unit deflection, is called the *current sensitivity* of the galvanometer.

For sensitive galvanometers which are read with telescope and scale, the current sensitivity k is expressed in microamperes per millimeter deflection on a scale 1 m from the mirror. The current sensitivity is numerically equal to the current in *microamperes* (millionths of an ampere, commonly abbreviated μA) required to cause a 1-mm deflection of the image on a scale 1 m distant. For the highly sensitive types of d'Arsonval galvanometers, k is about 0.00001 μA/mm, or 10^{-11} A/mm. Other expressions of galvanometer sensitivity are derived from the current sensitivity. It should be carefully noted that the term sensitivity is a technical word meaning the reciprocal of sensitiveness. A sensitive galvanometer has a low sensitivity k.

Example A galvanometer of the type shown in Fig. 38-6 has a current sensitivity of 0.002 μ A/mm. What current is necessary to produce a deflection of 20 cm on a scale 1 m distant?

$I = ks$, where s is in millimeters (on a scale 1 m away), so that

$$I = (0.002\ \mu\text{A/mm})(200\ \text{mm}) = 0.4\ \mu\text{A}$$

On a scale twice as far away, the deflection would be twice as great.

Example If the moving coil of the galvanometer of the example above has a resistance of 25 Ω, what is the potential difference across its terminals when the deflection is 20 cm?

$$V = IR = (0.4\ \mu\text{A})(25\ \Omega) = 10\ \mu\text{V}$$

Example A current of 2.0×10^{-4} A causes a deflection of 10 divisions on the scale of a portable-type galvanometer. What is its current sensitivity?

$$k = \frac{I}{s} = \frac{0.00020\ \text{A}}{10\ \text{divisions}} = \frac{2\bar{0}0\ \mu\text{A}}{10\ \text{divisions}}$$
$$= 20\ \mu\text{A/division}$$

Example What current will cause a full-scale deflection (100 divisions) of a portable galvanometer for which $k = 20\ \mu$A per division?

$$I = ks = (20\ \mu\text{A/division})(100\ \text{divisions})$$
$$= 2{,}\bar{0}00\ \mu\text{A} = 0.0020\ \text{A}$$

A galvanometer is often used merely to indicate the presence and direction of a current or its absence. In fact the most common use of a galvanometer is as a null-indicating instrument, i.e., as an instrument which indicates when a current is reduced to zero (Chap. 36). For this purpose the galvanometer scale need not be calibrated in terms of current.

38-7 BALLISTIC GALVANOMETER

For some purposes (such as measuring *quantity* of electricity rather than *current*) galvanometers are designed with the moving coil having a large moment of inertia and a long period. This enables a small quantity of electricity to be discharged through the galvanometer in a time that is small in comparison with the period of the galvanometer. Under these circumstances a deflection is obtained that is proportional to the charge. This follows from the fact that the angular momentum $I\omega$ (Sec. 7-8) imparted to the coil is given by

$$\Sigma I\,\Delta\omega = \Sigma L\,\Delta t = K\Sigma i\,\Delta t$$

whence
$$I\omega = KQ \tag{10}$$

where I is the moment of inertia of the coil, ω is the angular speed imparted to it, L is the torque produced by the current i, K is a constant, and Q is the total charge sent through the coil. As a result of its angular momentum the coil deflects until its kinetic energy of rotation is transferred into the potential energy of the twisted suspension. The deflection s thus produced is therefore proportional to the charge, or

$$Q = k_b s \tag{11}$$

The constant k_b is known as the *ballistic constant* of the galvanometer. Ballistic galvanometers are useful for the comparison of capacitors (Chap. 34).

38-8 VOLTMETERS

The voltage across the usual galvanometer is comparatively low. If higher voltages are to be measured, it is necessary to insert a resistor of high resistance *in series* with the moving coil of the instrument. Most of the potential drop will then occur across the multiplying resistor. By properly choosing this resistance any desired voltage may be measured.

Consider, for instance, the galvanometer mentioned in the last two examples above. The voltage that is required to produce full-scale deflection is 10 mV. In order to use this galvanometer as a voltmeter registering to 10 V, it is necessary only to increase the resistance until a potential difference of 10 V is just sufficient to produce in the galvanometer and resistor a current of 0.0020 A, or enough for a full-scale deflection. Hence

$$R = \frac{V}{I} = \frac{10\ \text{V}}{0.0020\ \text{A}} = 5,\bar{0}00\ \Omega$$

so that the resistance of the meter (5.0 Ω) must be increased by the addition of a series resistance R_m of 4,995 Ω, as in the diagram of Fig. 38-8. The scale of the instrument should be labeled 0-10 V, so that each division represents 0.1 V. If a potential difference of 5 V is applied to the terminals of this instrument, the current is

$$I = \frac{V}{R} = \frac{5\ \text{V}}{5,000\ \Omega} = 0.001\ \text{A}$$

Since 0.002 A is the full-scale current, the deflection will be just half scale, or 50 divisions, indicating 5 V on the 0–10-V scale. It should be noticed that the resistance of the voltmeter is $R = R_m + R_g$, where R_m is the series resistance and R_g is that of the galvanometer.

Example What series resistance should be used with the galvanometer just discussed in order to employ it as a voltmeter of range 0 to 200 V?

$$R = \frac{V}{I} = \frac{200\ \text{V}}{0.002\ \text{A}} = 100,000\ \Omega$$

total resistance, obtained by making R_m = 99,995 Ω. Each division on this instrument will represent 2 V, and its scale will be labeled 0–200 V.

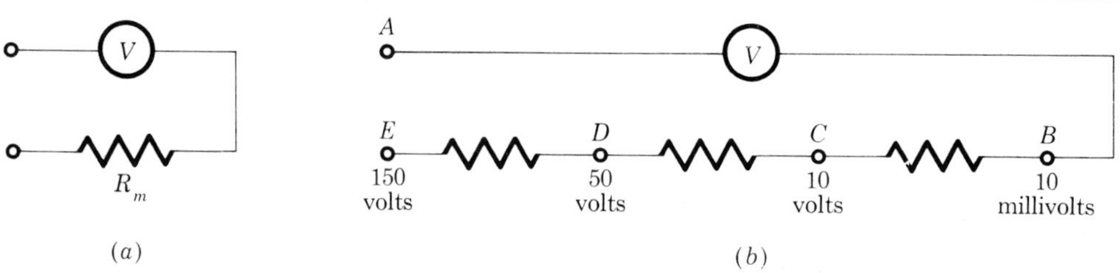

Figure 38-8
Use of resistors for voltmeter multipliers. In (*a*) the voltmeter has a single range; (*b*) illustrates a multirange instrument. Connections at *A* and *B* give a 10-mV range, *A* and *C* a 10-V range, *A* and *D* a 50-V range, and *A* and *E* a range of 150 V.

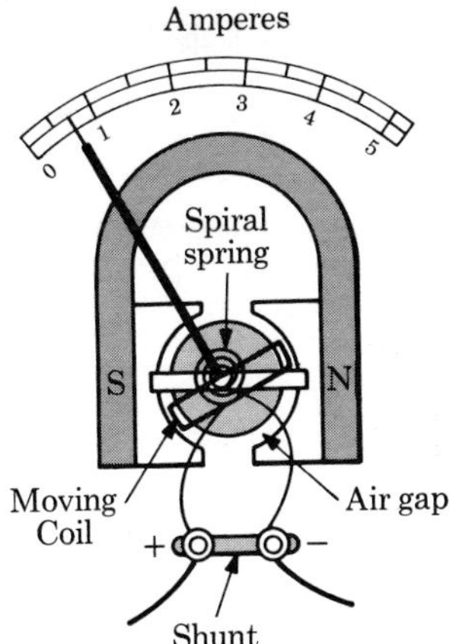

Figure 38-9
The coil of an ammeter is shunted by a low resistance.

38-9 AMMETERS

In order to convert the galvanometer described above into an ammeter for measurements up to 5.0 amp, it is necessary to connect a low resistance, called a shunt, across its terminals, as in Fig. 38-9. In order to be deflected full scale, the galvanometer must carry just 0.0020 A, hence the shunt S must carry the remainder of the 5.0-A current, or 4.998 A.

The potential difference across the meter is

$$V = IR = (0.0020\ \text{A})(5.0\ \Omega) = 0.010\ \text{V}$$

which must be the same as that across S, thus

$$R_s = \frac{V}{I_s} = \frac{0.010\ \text{V}}{4.998\ \text{A}} = 0.0020\ \Omega$$

This resistance is so small that a short piece of heavy strip or wire might be used for S in this case.

In practice, since it is very difficult to make the resistance R_s exactly a certain value when it is to be very low, one commonly obtains a shunt whose resistance is slightly larger than is needed, inserts a comparatively large resistance r in series with the coil (Fig. 38-10*a*), and then adjusts the value of the resistance r to make the meter operate as desired.

A galvanometer may be converted into an ammeter of several different ranges through the use of a number of removable shunts, or the shunts may be self-contained by the use of a circuit such as that shown in Fig. 38-10*b*. Connection is made to the + terminal and to one of the three terminals marked "high," "medium," and "low," respectively. The advantage of this circuit is that the shunt connections are permanently made, eliminating the error due to the variation of contact resistance when a removable shunt is used.

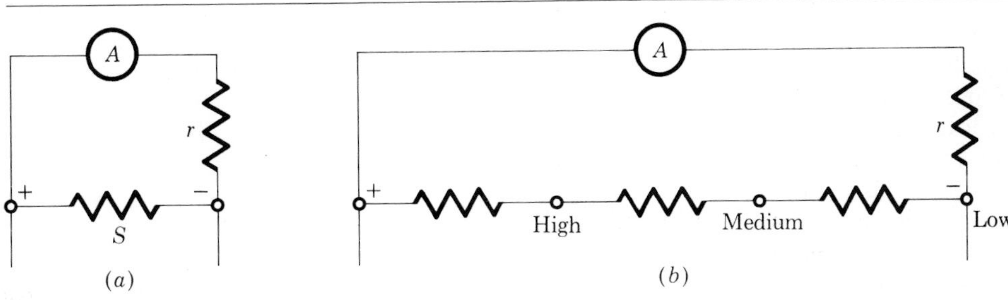

Figure 38-10
Ammeter circuits.

38-10 EFFECTS OF METERS IN THE CIRCUIT

When an ammeter is inserted into a circuit, the current to be measured is changed by the introduction of the ammeter. It is essential that the change in current thus caused shall be a very small fraction of the current itself; i.e., the resistance of the ammeter must be a small fraction of the total resistance of the circuit.

Similarly, when a voltmeter is connected between points whose potential difference is to be measured, the potential difference is changed by the presence of the voltmeter. When the voltmeter is placed in parallel with a portion of the circuit, the resistance is reduced; hence the potential difference across that part of the circuit is decreased, and the total current is increased. The voltmeter introduces two errors: changing the current in the circuit and reducing the potential difference that is to be measured. In order that these errors be small, it is essential that the resistance of the voltmeter be very large in comparison with the resistance across which it is connected. This will ensure also that the current in the voltmeter be small in comparison with that in the main circuit.

38-11 THE WATTMETER

The wattmeter, as the name suggests, is an instrument for measuring power. It consists essentially of two coils at right angles, one fixed and one movable. The fixed coil is made of heavy wire of low resistance and is connected in series with the load. The movable coil is made of small wire and is connected in series with a *multiplier* of high resistance; this coil is connected in parallel with the load. By analogy with the methods of connecting ammeters and voltmeters, these two coils are called the *current* and *voltage* coils, respectively. The torque acting on the movable coil is proportional to both the current in the fixed coil and the voltage across the potential coil. Hence

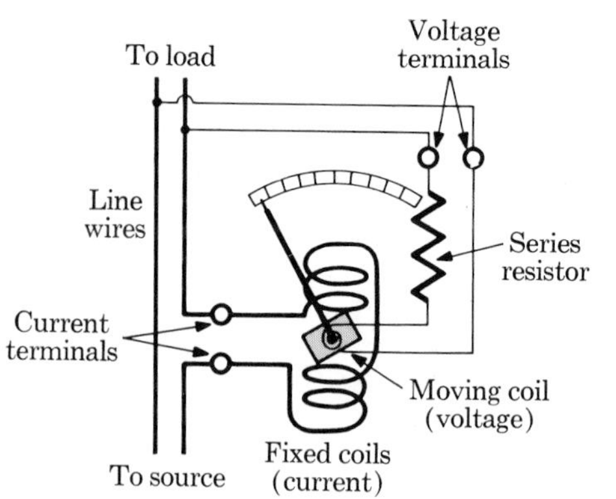

Figure 38-11
Idealized diagram of a wattmeter.

the resultant indication of the meter is proportional to the product of the current and the voltage, i.e., to the power (for direct current). A typical wattmeter is illustrated in Fig. 38-11.

38-12 THE WATTHOUR METER

The watthour meter is a device for the measurement of electric energy. This is the type of meter which is so commonly found in houses for indicating the amount of energy which has been furnished by the electric utility company. The electrodynamometer type of meter (Fig. 38-12) is basically a special type of motor. It is designed so that the armature A of the motor revolves at a speed that is proportional to the power used. This is accomplished by having the field coils F and F' of the motor connected in *series* with the load, thus producing a magnetic field that is proportional to the *current* in the load. The armature is connected in parallel with the load, through a multiplying resistor R. Hence its magnetic field is proportional to the voltage of the load. The resultant torque, being separately proportional to each of these fields, is therefore dependent upon their product and hence to the product of current and voltage, or the power. Suitable pointers,

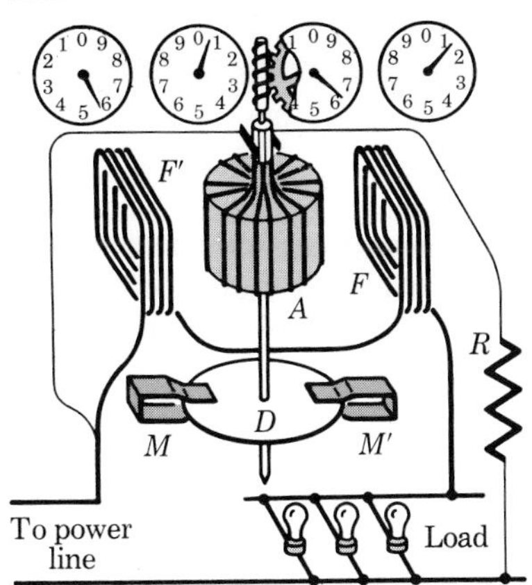

Figure 38-12
A watthour meter.

geared to turn with the armature, indicate on their respective dials the number of watthours (more often kilowatthours) used in the load. The aluminum disk D connected to the armature spindle rotates in the field of the magnets M and M'. The induced eddy currents thereby generated (see Sec. 40-15) act as a brake on the armature and provide a mechanism for adjusting the rate of the rotation when the magnets are moved toward or away from the disk in servicing the device.

38-13 THE INDUCTION TYPE OF WATTHOUR METER

The electrodynamometer type of watthour meter is ordinarily used on dc circuits. For ac use, the instrument is modified, in accordance with the schematic diagram of Fig. 38-13. This device is essentially a single-phase induction motor. The coil P is the voltage coil, in parallel with the load. The current coils S and S' are in series with the load. The rotor is an aluminum disk D that is caused to turn by the rotating magnetic field established by the combination of the currents in the voltage and current windings. The disk rotates between the poles of the permanent magnets MM. Eddy currents induced in the disk produce a drag which is the mechanical load for the motor action of the rotor. Through ingenious design, the speed of the rotor at each instant is proportional to the power. To record energy, the rotor shaft is geared to a set of indicating dials, as in Fig. 38-12.

It is emphasized that all watthour meters measure energy, not power.

SUMMARY

The force on a current-carrying conductor in a magnetic field is given by

$$\mathbf{F} = I(\mathbf{l} \times \mathbf{B})$$

The magnitude of the torque on a coil in a magnetic field is given by

$$L = NABI \sin \theta$$

The force of a current I_1 on a parallel current I_2 is given by

$$F = \mu_0 \frac{I_1 I_2 l}{2\pi s}$$

The ampere is a current that when it is maintained in two infinitely long conductors, parallel

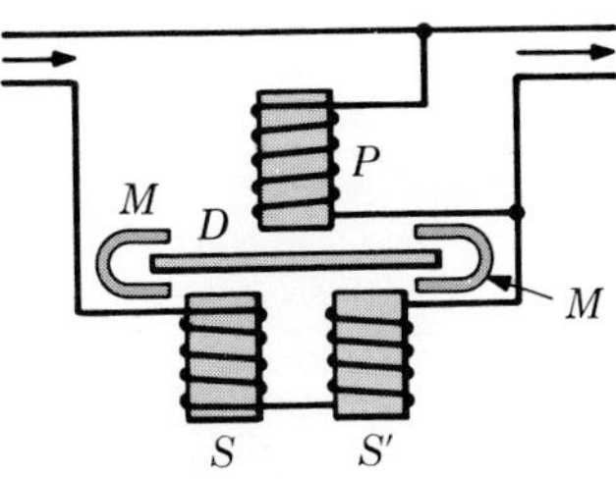

Figure 38-13
Induction type of watthour meter.

to each other and 1 m apart in free space, will cause a force on each conductor of 2×10^{-7} N/m length.

The *galvanometer* is the basic instrument for the detection or measurement of currents and related quantities. The d'Arsonval galvanometer consists of a permanent magnet, a movable coil, and an indicating device.

Damping is the progressive dying away of a motion until the equilibrium position is reached. Electromagnetic damping is used in most electric meters.

The *current sensitivity* of a galvanometer is the current per unit deflection.

$$k = \frac{I}{s}$$

A *ballistic galvanometer* gives deflections that are proportional to quantity of electricity.

$$Q = k_b s$$

A galvanometer may be converted into a voltmeter by the use of a series-resistance multiplier.

A galvanometer may be converted into an ammeter by the use of a low-resistance shunt in parallel with the meter.

The range of a voltmeter may be increased by the use of suitable series resistors or multipliers.

The range of an ammeter may be increased by placing low-resistance shunts in parallel with the ammeter.

A watthour meter is used for the measurement of electric energy. It is a motor whose speed depends upon the power utilized, and its indications integrate power × time = energy.

Questions

1 What sort of path will electrically charged particles follow if they are initially moving at right angles to a uniform flux density in a vacuum? Will such particles experience an energy change during this motion? Explain briefly.

2 Why are magnetic fields used in many machines designed to accelerate charged particles to high energy?

3 Imagine a coil mounted on universal joints (gimbals) which are nearly frictionless. What would happen to such a current-carrying coil placed at random in a magnetic field? Give reasons.

4 A circular loop of wire hangs by a thread in a vertical plane. An electric current is maintained in the loop in a counterclockwise direction as seen from the front face *A*. To what direction (north, south, east, or west) will the front face of the coil turn when it is free to rotate in the earth's magnetic field?

5 Explain what would happen to the configuration of a loosely wound loop of flexible wire when a current is maintained in the loop.

6 A loosely wound helix made of stiff wire is mounted vertically, with the lower end just touching a dish of mercury. When a current from a battery is started in the coil, the wire executes an oscillatory motion, with the lower end jumping out of and into the mercury. Explain the reasons for this behavior. Would the apparatus behave similarly if an ac source were used instead of the battery?

7 Suppose a perfect galvanometer could be built in which the angular deflection would be directly proportional to the current. If such an instrument were used with a lamp and straight scale, plot a rough curve to show how the scale readings would vary with the current.

8 State some ways in which the sensitivity of a galvanometer may be increased. Describe the limitations of each of these methods; that is, why may the instrument not be made infinitely sensitive by each change?

9 In what respects are the actions of a galvanometer and a motor similar?

10 Does the use of an external damping resistor change the current sensitivity of a galvanometer? the voltage sensitivity? Why is the use of such a resistor a convenience?

11 A tap key is often placed in parallel with a galvanometer and manipulated to bring the coil quickly to rest when the current is interrupted. Describe the technique used, and show why this happens. Is such a circuit underdamped, overdamped, or critically damped?

12 A "hot-wire" ammeter uses the expansion of a wire caused by the heating effect of the current to be measured. Show why this type of ammeter can be used for the measurement of both alternating and direct currents. Is this also true for d'Arsonval-type meters? Explain.

13 Give the logical reasoning to show why the deflections of a well-designed dc voltmeter are directly proportional to the voltage at the terminals of the instrument.

14 Explain why voltaic and thermoelectric effects are objectionable in galvanometer circuits. How may they be minimized?

15 State some of the reasons why it is a disadvantage to use a galvanometer which has too high a sensitivity for the purpose in question.

16 Most ammeter binding posts are made of heavy bare metal, whereas voltmeter terminals are usually much lighter and well insulated. Explain why this is desirable.

17 An ammeter and a voltmeter of suitable ranges are to be used to measure the current and voltage of an electric lamp. If a mistake were made and the meters interchanged, what would happen?

18 Some types of fuses used to protect electric meters have resistances of several ohms. Is this objectionable (*a*) in voltmeter circuits and (*b*) in ammeter circuits? Why?

19 What essential differences are there between the common types of galvanometers and ammeters? Between ammeters and voltmeters? Is it desirable for an ammeter to have a high resistance or a low one? Should a voltmeter have a high resistance or a low one? Why?

20 Make wiring diagrams to show two ways in which the meters could be connected in the ammeter-voltmeter method widely used for the measurement of resistance. Explain when each arrangement should be used.

21 A wattmeter is connected to the following arrangements of lamps inserted in a 120-V power circuit: (*a*) one 60-W lamp, (*b*) two of these lamps in parallel, (*c*) two of these lamps in series, (*d*) a pair of lamps in parallel joined in series, and (*e*) a pair of lamps in parallel joined in series to another pair in parallel. What will the reading of the wattmeter be in each case?

22 An electron of charge e and mass m moves with a velocity v normal to a uniform magnetic field of strength H. Because of the force on the electron at right angles to H and v, it will be accelerated and move with uniform circular motion in a circle of radius r. Derive the equation which gives r in terms of H, e, m, and v. Solve this equation for e/m, and discuss the significance of this ratio.

23 Plot a curve to show the variation of the torque on a coil in a magnetic field as the plane of the coil is rotated with respect to the field.

24 Two long, straight, insulated wires are suspended vertically. The wires are connected in series and a heavy current from a battery is maintained in them. What happens to these wires? The battery is replaced by an ac source. Explain what happens in this case.

25 Two current-carrying coils are placed at a distance from each other, with their centers on a straight line. This line is perpendicular to one of the coils. What must be the position of the second coil and the relative direction of the current in order that the coils may attract each other? In what position will there be no force?

Problems

1 A current of 30 A flows through a 50-cm-long wire and at an angle of 45° to a magnetic field having a flux density of 7×10^{-4} Wb/m^2. Find the magnitude and direction of the force acting on this wire.

2 In Fig. 38-14, A, B, and C are long, thin wires arranged so that their centers lie at the corners of an isosceles triangle. Wire A carries a current of 9.0 A into the plane of the paper, wire B carries 9.0 A out of the paper, and wire C carries 3.0 A out of the paper. Find the magnitude and direction of the force per unit length exerted on wire C, due to the currents in wires A and B. The wires maintain their relative positions.

Ans. 2.0×10^{-5} N/m.

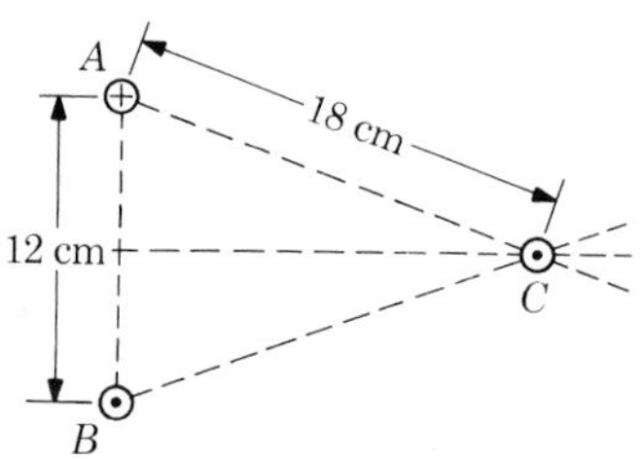

Figure 38-14

3 Two long, straight wires 750 mm apart deliver a steady current of 50.0 A to a motor. (*a*) Find the flux density produced at one wire by the current in the other wire. (*b*) Find the force exerted on a 300-cm length of the second wire.

4 A wire 125 cm long and carrying a current of 30 A is placed at an angle of 30° with a uniform magnetic field of induction 4.0×10^{-4} Wb/m². What force is exerted by the field on the current? By the aid of a figure show the directions of the field, current, and force.

Ans. 7.5×10^{-3} N.

5 A cable 14.0 ft in length runs along the top of a railway car 4.5 ft below the trolley wire. When there is a current of 800 A in the trolley wire and 90 A in the cable, what is the force between them?

6 What is the current sensitivity of a galvanometer that is deflected 20.0 cm on a scale 250 cm distant by a current of 3.00×10^{-5} A?

Ans. 0.375 μA/mm.

7 A conductor 500 mm long makes an angle of 60° with a uniform magnetic field of induction 5.00×10^{-4} Wb/m². What force is exerted on the conductor when it carries a current of 18.0 A? Make a sketch to show the direction of the force.

8 Two long, parallel wires are 20 cm apart in air and carry currents of 10 A and 5 A. What force acts on each meter of wire if the currents are (*a*) in the same direction and (*b*) in opposite directions.

Ans. 5.0×10^{-5} N, attractive;
5.0×10^{-5} N, repulsive.

9 A conductor 80 cm long carrying a current of 20 A is perpendicular to a uniform magnetic field of 5.00×10^{-4} Wb/m². What is the magnitude of the force on the conductor? if the field is directed east and the current is directed upward, what is the direction of the force?

10 A 25-turn rectangular coil 12 × 15 cm is placed with its plane parallel to a magnetic field of strength 4×10^{-3} Wb/m². For a current of 400 mA in the coil, what is the torque when the 12-cm side is (*a*) parallel and (*b*) perpendicular to the field? *Ans.* 7.2×10^{-4} m·N.

11 A galvanometer has a rectangular coil which is 1.80 × 4.50 cm. It is suspended to move through a magnetic field that has a magnitude 2.5×10^{-2} Wb/m². The coil has 90 turns. What maximum torque acts on the coil when it carries a current of 150 μA?

12 In Fig. 38-15 the solenoid consists of 1,200 turns and measures 35 cm in length by 2.3 cm in diameter. It carries a current of 1.5 A in the sense indicated. The windings are separated so that a

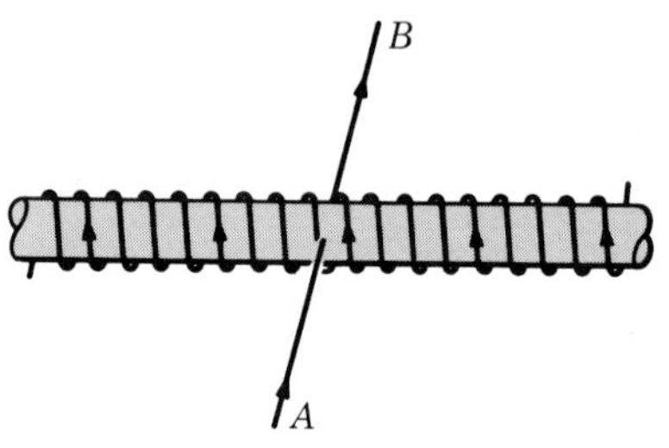

Figure 38-15

straight wire carrying a current of 18 A from *A* to *B* is inserted, to pass through the center of the solenoid, at right angles to its axis. Find the force on the portion of straight wire which is inside the solenoid. *Ans.* 2.7×10^{-3} N.

13 The one-turn rectangular coil of Fig. 38-16 carries a current of 10 A in the sense shown, "in" at *a*. The width *ab* of the coil is 10 cm, the length 20 cm. Its plane makes an angle $\phi = 30°$ with the magnetic flux density $B = 0.25$ Wb/m². (*a*) Find

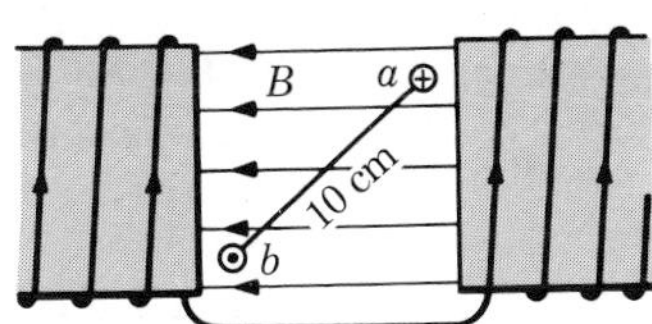

Figure 38-16

the force on the conductor *a* of length 20 cm. (*b*) Find the torque required to hold the coil in the position shown.

14 What is the megohm sensitivity of the galvanometer of Prob. 6? *Ans.* 2.67 mΩ.

15 A rectangular loop of wire in which there is a current of 5.00 A is placed in a uniform field of induction 2.14×10^{-3} Wb/m^2. The loop is 30.0 cm long and 15.0 cm wide. (*a*) How much torque must be applied to keep the loop from rotating? (*b*) How much torque would be required if the loop were circular and its area were 450 cm^2?

16 An ammeter has a resistance of 0.0090 Ω and reads up to 10 A. What resistance shunt is needed to make full-scale deflection of the meter correspond to 100 A? *Ans.* 0.0010 Ω.

17 A galvanometer has a current sensitivity of 0.375 μA/mm. (*a*) What deflection is observed on a scale 250 cm from the mirror when the galvanometer current is 3.00×10^{-5} A?

18 What part of the total current will there be in an instrument of resistance 0.60 Ω when a 0.20-Ω shunt is connected across its terminals? *Ans.* 25 percent.

19 A 50.0-mV meter has a resistance of 5.00 Ω. A multiplier has been inserted to produce a voltmeter of range 3.00 V. How can the multiplier be modified so that the new meter will have a range of 15.0 V?

20 A portable galvanometer is given a full-scale deflection by a current of 1.00 mA. If the resistance of the meter is 7.0 Ω, what series resistance must be used with it to measure voltages up to 50 V? *Ans.* 5×10^4 Ω.

21 A voltmeter of range 120 V and resistance $9,6\bar{0}0$ Ω is placed in series with another voltmeter of range 150 V and resistance $1,5\bar{0}0$ Ω. What will each meter read when they are connected to a 125-V battery?

22 A certain meter gives a full-scale deflection for a potential difference of 50.0 mV across its terminals. The resistance of the instrument is 0.400 Ω. (*a*) How could it be converted into an ammeter with a range of 25 A? (*b*) How could it be converted into a voltmeter with a range of 125 V? *Ans.* 0.00201 Ω; 999.6 Ω.

23 A dry cell, an adjustable resistor, and a voltmeter are connected in series. With the resistor set at zero resistance, the voltmeter reads 1.480 V. When the resistor is adjusted to 450 Ω, the voltmeter reads 0.372 V. What is the resistance of the voltmeter?

24 A millivoltmeter with a resistance of 0.800 Ω has a range of 24 mV. How could it be converted into (*a*) an ammeter with a range of 30 A and (*b*) a voltmeter with a range of 12 V? *Ans.* 800 μΩ; 399 Ω.

25 A millivoltmeter indicates a full-scale deflection for a current of 100 mA. When this meter is provided with a shunt of resistance 0.01727 Ω, the combination produces a milliammeter of range of 300 mA. What is the resistance of the millivoltmeter?

26 A milliammeter has a resistance of 5.00 Ω, and shows a full-scale deflection when there is a current of 10.0 mA. (*a*) What resistor should be connected in series with the milliammeter in order that a full-scale deflection may correspond to 150 V? (*b*) What is the resistance of a shunt that can be connected across the terminals of the milliammeter in order that a full-scale deflection may correspond to 10 A? *Ans.* 1.50×10^4 Ω; 0.00500 Ω.

27 An ammeter is inserted in a circuit in which the current is maintained at 500 mA. When a shunt of 0.0111 Ω is placed across the ammeter, its reading drops to 50 mA. What is the resistance of the meter?

28 An ammeter of range 1.50 A and resistance 0.033 Ω is connected in series with an ammeter of range 1.00 A and resistance 0.050 Ω. How much will the current change when both of these meters are inserted into a circuit of resistance of

2.50 Ω? How will the reading of the meters compare? *Ans.* 3.1 percent.

29 Some vacuum-tube voltmeters have been designed that have resistances of 1 mΩ/V of range. What current is there in such an instrument when it is giving full-scale deflection? What is the great advantage of this instrument?

30 An unknown resistor is to be measured by the voltmeter-ammeter method. A 3,000-Ω voltmeter is connected in parallel with the unknown resistor, and a 0.250-Ω ammeter is placed in series with this combination. The ammeter reads 0.0350 A, and the voltmeter reads 8.50 V. Calculate the value of the unknown resistance. (Do not neglect the errors introduced by the use of the meters.) *Ans.* 264 Ω.

31 A moving-coil galvanometer has a resistance of 2.5 Ω and gives full-scale deflection for a potential difference of 50 mV. If the galvanometer is converted into an ammeter with full-scale deflection at 5.0 A, what is the current in the coil when the ammeter reads 4.0 A?

32 A bifilar wire has a total length of 12 m. The wires are separated only by the thickness of their insulation, a total distance of 3.6 mm. When there is a current of 25 A in the wires, what is the magnitude and sense of the force between them? *Ans.* 4.2×10^4 dyn.

Owen Chamberlain, 1920–

Born in San Francisco. Studied at Dartmouth College, the University of California, and the University of Chicago. Served in the Manhattan Project at Berkeley and Los Alamos. Since 1948 Professor at the University of California at Berkeley. Awarded the 1959 Nobel Prize for Physics with Segre for their discovery of the antiproton.

Emilio Gino Segre, 1905–

Born in Rome. First doctoral student of Professor Fermi. Director of Physics Laboratory at University of Palermo; since 1938 Professor of Physics at the University of California at Berkeley. Awarded the 1959 Nobel Prize for Physics with Chamberlain for their discovery of the antiproton.

39

Magnetism in Matter

We have considered the magnetic fields of induction produced in free space by currents in circuits of different geometrical shapes. If the current is surrounded with a material, the magnetic flux in this medium is found to be different from the flux in free space. We shall now consider a way of investigating and classifying magnetic properties, a theory which explains magnetic properties in terms of electric current, the picture of a magnetic circuit analogous to an electric circuit, and the application of these ideas to the familiar bar magnet and the magnetic compass.

39-1 MAGNETIC FIELD STRENGTH (INTENSITY)

It is desirable to define a new magnetic vector H, known as magnetic field strength (also called intensity). We shall define *magnetic field strength* as the ratio of the magnetic flux per unit area B_0 in empty space to the permeability μ_0 of the space,

$$H = \frac{B_0}{\mu_0} \qquad (1)$$

The mks unit of B_0 (given in Chap. 37) is the weber per square meter, $\mu_0 = 4\pi \times 10^{-7}$ Wb/A·m, and H is in amperes per meter. When the magnetic field is produced by a current, it is customary to express H in ampere-turns per meter. (See example below.)

An inspection of the equations for the flux density B near various arrangements of current-bearing conductors, as given in Chap. 37, will show that equations for H can be written for each of these cases by dividing the equation for B by μ_0, the permeability of free space. In all cases the value of H is found to be a proportionality constant times NI/l.

The magnetic field strength H depends on the geometry of the circuit and the current, but not on the medium.

Example There is a current of 6.0 A in a closely wound solenoid of 200 turns and axial length 50 cm. What is the magnetic field strength at the center of the solenoid?

$$H = \frac{NI}{l} = \frac{200 \text{ turns} \times 6.0 \text{ A}}{0.50 \text{ m}}$$

$$= 2,\bar{4}00 \text{ A-turns/m}$$

39-2 MAGNETIC PERMEABILITY

Measurements for calculating magnetic permeability μ are often made by using a Rowland ring, a toroid around which a short section of secondary coil has been wound and connected to a ballistic galvanometer (Fig. 39-1). An ammeter A measures the current I supplied by a battery to the primary coil. When this current is reversed by switch S, there is a change in the magnetic flux BA that threads both coils. A momentary current is induced in the secondary coil (Chap. 40), resulting in a deflection D by the galvanometer,

$$D \propto \frac{BAN_s}{R} \tag{2}$$

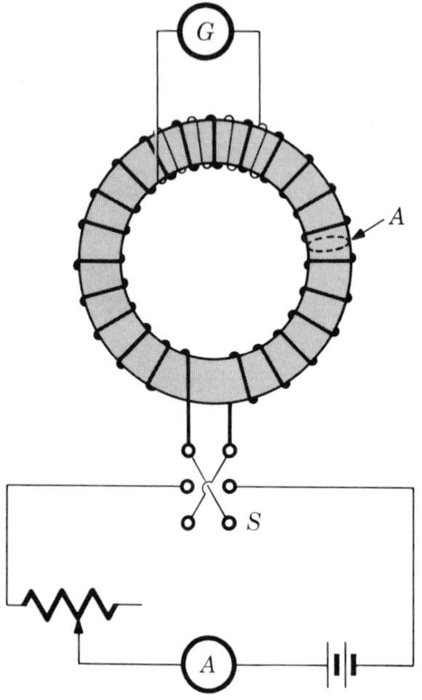

Figure 39-1
Determination of permeability μ with Rowland ring and ballistic galvanometer.

Table 1
RELATIVE PERMEABILITIES

Diamagnetic substances	
Antimony	0.99995
Argon	0.99999
Bismuth	0.99982
Copper	0.99999
Silver	0.99997
Water	0.99999
Paramagnetic substances	
Air	1.0000004
Dysprosium oxide (Dy_2O_3)	1.0225
Liquid oxygen	1.0040
Platinum	1.0003
Ferromagnetic substances	
Magnetic iron	200
Nickel	100
Permalloy (0.785 Ni, 0.215 Fe)	8,000
Mumetal (0.75 Ni, 0.02 Cr, 0.05 Cu, 0.18 Fe)	20,000
Cu-Zn ferrite	1,500

where $BA = \Phi$ is the change in flux, N_s is the number of turns in the secondary coil, and R is the total resistance of the secondary circuit. The magnetic flux corresponding to a current I in the primary coil of N_p turns is

$$BA = \mu \frac{N_p IA}{l} \tag{3}$$

The flux change, and hence the galvanometer deflection, is proportional to μ. The *relative permeability* μ/μ_0 of a material can be measured as the ratio of the galvanometer deflection using that material in the Rowland ring to the deflection obtained with nothing in the core,

$$\mu_r = \frac{\mu}{\mu_0} = \frac{B}{B_0} = \frac{D}{D_0} \qquad (4)$$

Some values of relative permeability measured at a flux density of 0.002 Wb/m^2 are listed in Table 1.

The mks unit of permeability is the same as that for μ_0, the permeability of a vacuum, namely, the weber per ampere-turn meter. From the definition of the weber, the unit of μ may also be expressed in newtons per ampere^2.

39-3 TYPES OF MAGNETIC SUBSTANCES

It is found that materials may be classified, by their relative permeability, into three groups having different magnetic properties, as suggested in Table 1. These are:

1 *Diamagnetic* substances, for which μ_r is very slightly less than unity. A tiny rod of diamagnetic material suspended in the nonuniform field between the poles of a strong electromagnet will align itself at right angles to the field.
2 *Paramagnetic* substances, for which μ_r is very slightly greater than unity. A rod of paramagnetic material tends to align parallel to an external magnetic field.
3 *Ferromagnetic* substances, for which μ_r may range from 10^2 to 10^5 and depends on the magnitude of the external magnetizing field as well as on the past magnetic, mechanical, and thermal history of the sample. In this group are the transition elements iron, cobalt, and nickel, some alloys of these elements, and a few alloys containing none of the "magnetic" elements.

In casual observation one usually notices only the force which a magnetic field exerts on ferromagnetic materials; the force exerted on a paramagnetic or diamagnetic material is so much smaller as to escape detection. Hence we often speak of materials which are not ferromagnetic as being "nonmagnetic," for example, brass in contrast to steel. But, in the theory to be described, all matter possesses magnetic properties.

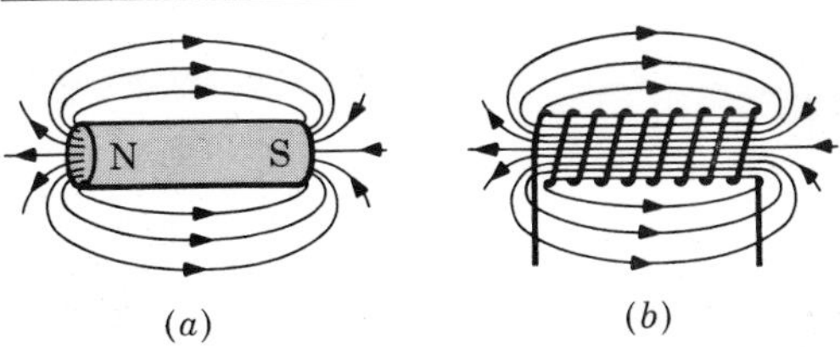

Figure 39-2
Lines of induction B for (*a*) a magnetized rod and (*b*) a solenoid of similar shape.

39-4 ATOMIC THEORY OF MAGNETISM

The magnetic field in the neighborhood of a bar magnet may be visualized by tracing the imaginary lines of B using a tiny compass needle or by dusting iron filings on a horizontal sheet of paper resting on the magnet to form a field pattern. If a current-carrying solenoid is substituted for the permanent magnet, there is a striking similarity in the field patterns (Fig. 39-2).

Progress in understanding magnetism has followed a course of identifying all magnetism closely with electricity. A significant first step was Ampère's "circular-current" theory of magnetism (1820). In 1852, Weber suggested that each atom is a permanent magnet capable of orientation. In 1890, Ewing pointed out that there must be strong torques exerted among the atomic magnets to produce the observed phenomena. Bohr's model (1913) of the shell structure of electrons in atoms and the formulation of quantum mechanics led Heisenberg and others from 1928 to the present to develop an experimentally satisfactory theory of magnetism, unattainable by means of classical physics alone.

Consider a cross section (Fig. 39-3) of a magnetized specimen such as the core of a Rowland ring. The elementary magnets (dipoles) may be regarded as little current loops. These amperian currents cancel each other in the interior of the specimen, but at the edge they are equivalent to the large, single-current loop i. The

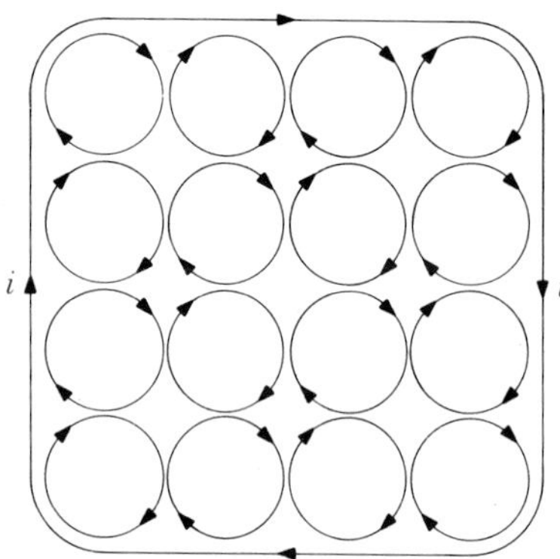

Figure 39-3
Amperian currents in a slice of a Rowland ring are equivalent to a current i around the boundary.

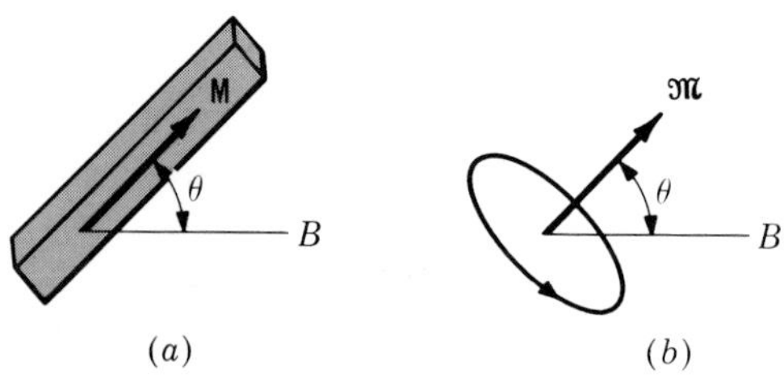

Figure 39-5
Magnetic moment $\mathfrak{M}$ of a magnet and a small coil. Torque $\mathbf{L} = \mathfrak{M} \times \mathbf{B}$.

magnetic induction in the core of the toroid is made up of two contributions: (1) that due to the actual conduction currents in the copper windings and (2) that due to the amperian currents that represent the aligned magnetic dipoles of the core material (Fig. 39-4). For a material of large relative permeability, the contribution of the amperian currents may be thousands of times greater than that of the conduction current.

The torque L on a small current loop (Fig. 39-5b) was found in Sec. 38-3 to be $L = NIAB \sin\theta = NIA\mu_0 H \sin\theta$. We now define the *magnetic moment* $\mathfrak{M}$ as the maximum torque the loop experiences (when $\sin\theta = 1$) per unit magnetic induction B. The magnetic moment is represented by a vector whose direction is perpendicular to the coil in the sense given by the right-hand rule. The magnitude of $\mathfrak{M}$ is proportional to the product of the number of turns, the

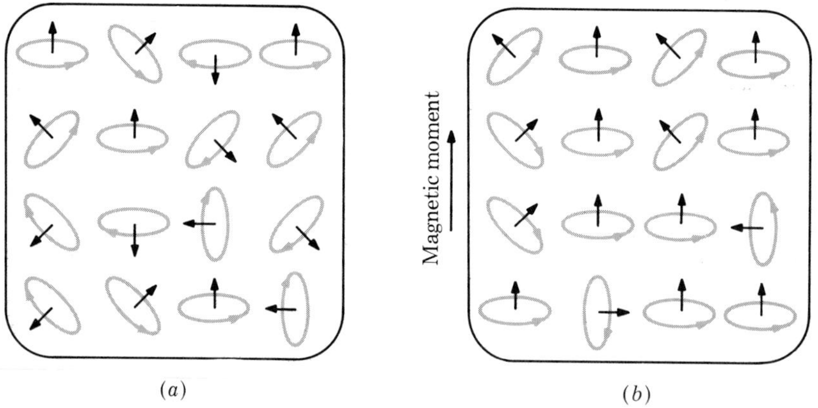

Figure 39-4
Representation of orientations of elementary magnetic moments in (a) unmagnetized and (b) magnetized material. Orbits represent individual atoms for a fluid but represent domains for a ferromagnetic solid.

current in each, and the area of the loop, $\mathfrak{M} = \mu_0 NIA$. The magnetic moment of a small magnet (Fig. 39-5*a*) is equal to the magnetic moment of a small current loop that would experience the same maximum torque ($\sin\theta = 1$) when placed in the same magnetic field. Then, for either the small magnet or the current loop, the torque is expressible as

$$\mathbf{L} = \mathfrak{M} \times \mathbf{B}$$

or in magnitude

$$L = \mathfrak{M}B \sin\theta \qquad (5)$$

The *magnetization* M of a substance is the magnetic moment per unit volume. The magnetic susceptibility X per unit volume is defined as the magnetization produced per unit magnetic field intensity, $x = M/H$. Frequently, the susceptibility is related to unit mass or to a mole of the substance. When we seek to relate our picture of the moments of amperian currents (Fig. 39-4) to atoms, it is convenient to define an atomic susceptibility as $\chi_A = \chi A/\rho$, where A is the atomic mass and ρ the density.

A graph of atomic susceptibilities as related to atomic number (Fig. 39-6) summarizes important relations which a satisfactory atomic theory of magnetism must include. The rare-earth metals are strongly paramagnetic. Most metallic conductors of electricity are weakly paramagnetic. (This implies that the conducting electrons are paramagnetic.) The rare gases are diamagnetic.

39-5 DIAMAGNETISM

In 1905 P. Langevin explained the origin of diamagnetism. Consider first a model of an atom with a single electron of charge e and mass m traveling with speed v in a circular orbit of radius r. This "current" produces a magnetic moment $\mathfrak{M}$ equal to the product of the current and the area of the orbit,

$$\mathfrak{M} = \frac{ve}{2\pi r}(\pi r^2) = \frac{ver}{2} \qquad (6)$$

When an induction B is applied, the electron experiences a force $-e(\mathbf{v} \times \mathbf{B})$ at right angles to its direction of motion. With r constant, the effect of applying a magnetic field is to increase or decrease the angular speed of the electron, depending on its direction of circulation. This increases or decreases its orbital magnetic moment. Now consider an atom with two or more electrons, with the orbits so oriented that there is no net magnetic moment. Such atoms will show diamagnetic behavior. For when an external field is applied, the orbital magnetic moments of the electrons circulating in opposite directions in each atom will no longer cancel. A magnetic moment will be induced whose direction is opposite to B. A rod of such material when placed in the nonuniform field between the poles of a strong electromagnet will align itself at right angles to the field.

39-6 PARAMAGNETISM

The electrons of a substance produce magnetic fields in two ways: An electron revolving in an orbit about the nucleus of an atom is equivalent to an amperian current loop producing a magnetic moment. The electron also has a magnetic moment which may be interpreted as due to the spin of the electron about an axis through its center. In an atom containing many electrons it may happen that their orbits and spins are so oriented as not to cancel completely but to give the atom a net magnetic moment. The magnetic-moment vector of such atoms will experience a torque tending to align it with an external magnetic field. A bar made up of such atoms will show the paramagnetic characteristic of aligning with an external field. But since the aligning of individual atoms is interfered with by the collisions of atoms in thermal vibration, paramagnetism is temperature-dependent.

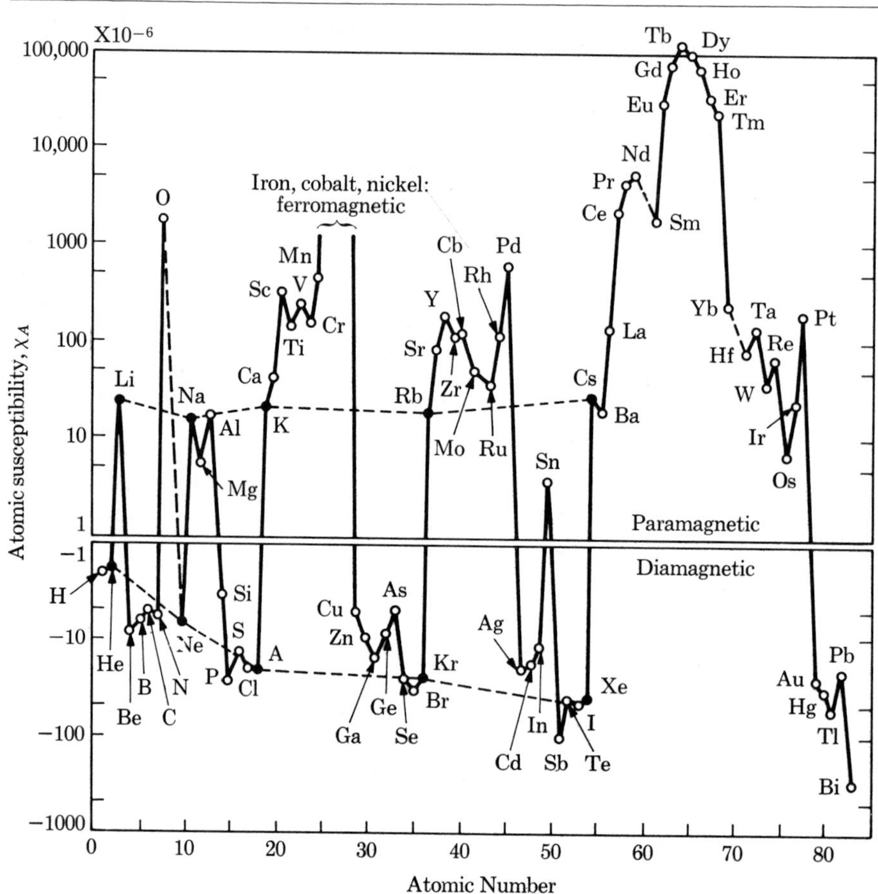

Figure 39-6
Atomic susceptibilities of elements, about 20°C. Dotted lines connect alkali metals (paramagnetic) and rare gases (diamagnetic). (*Courtesy of Dr. R. M. Bozorth.*)

The data of Fig. 39-6 as interpreted by the shell model of an atom and by quantum theory suggest that strong paramagnetism occurs when inner electron shells are incomplete and therefore have a resultant moment that is large compared with the spin of conduction electrons or the diamagnetic moment of closed shells. Incomplete shells occur notably in the iron group, the rare-earth group of elements, the platinum group, and the palladium group, all of which show strong paramagnetism. Weak paramagnetism occurs when loosely bound electrons in an outer shell become conduction electrons in a metal or valence electrons in a compound. The spins of some of these electrons can be influenced by an external field, in a way explainable by quantum mechanics, to exhibit weak paramagnetism that is practically independent of temperature.

39-7 FERROMAGNETISM

The fact that iron vapor and iron ions in solution exhibit only paramagnetism suggests that the ferromagnetism of solid iron is a property, not of individual atoms, but of the crystals that make up a ferromagnetic material. In such materials, an interaction called *exchange coupling* (a quantum effect) couples the magnetic moments of adjacent atoms in rigid parallel alignment. Quantum physics successfully predicts from electron configurations that the only elements for which this coupling will occur are Fe, Co, Ni, Gd, and Dy—just those elements for which ferromagnetism is observed.

Typical magnetization curves, obtainable with a Rowland ring, are shown in Fig. 39-7. The magnetic induction B in the iron core is made up of B_0, due to the current in the copper winding, plus B_M, the magnetic induction due to the iron, which is proportional to the magnetization of the iron. Often B_M is much greater than B_0. As the external magnetizing field is increased, saturation is observed as B_M reaches its maximum value, corresponding to complete alignment of atomic moments (Fig. 39-4) in the iron.

In the *domain theory* of ferromagnetism, one pictures the specimen as made up of domains, microcrystal regions within which there is practically perfect alignment of atomic moments. But the vectors representing the moments in various domains are not parallel at low values of B_0. When an external magnetizing field is increased, the domains which are near parallel orientation with this field increase in size, displacing their boundaries at the expense of other domains. At first this boundary displacement is reversible, but in stronger fields irreversible displacements occur. In still stronger fields rotation of the magnetic moment within domains takes place, and saturation is reached when this is completed.

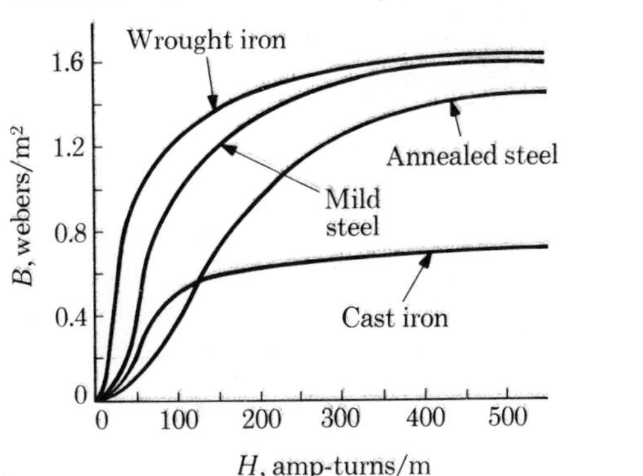

Figure 39-7
Typical B-H curves.

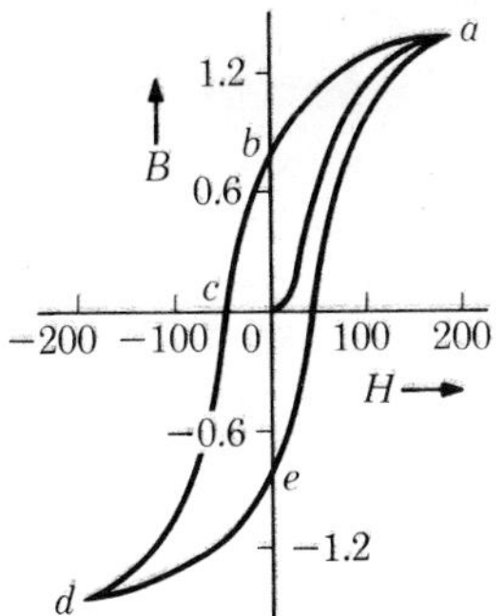

Figure 39-8
Initial magnetization curve and hysteresis loop for a ferromagnetic specimen.

The initial magnetization curve is not retraced as the magnetizing current in the toroid is first increased, then decreased (Fig. 39-8). The lack of retraceability is called *hysteresis,* and it is associated with energy needed to reorient the domains. When the magnetizing field is made zero, the specimen retains some "permanent" magnetism. The area of a hysteresis loop is a measure of the energy dissipated, as heat, in taking a specimen of unit volume through a magnetic cycle.

Because of the hysteresis loss, the magnitude of the area of the hysteresis loop is a factor of great importance in the design of electric ma-

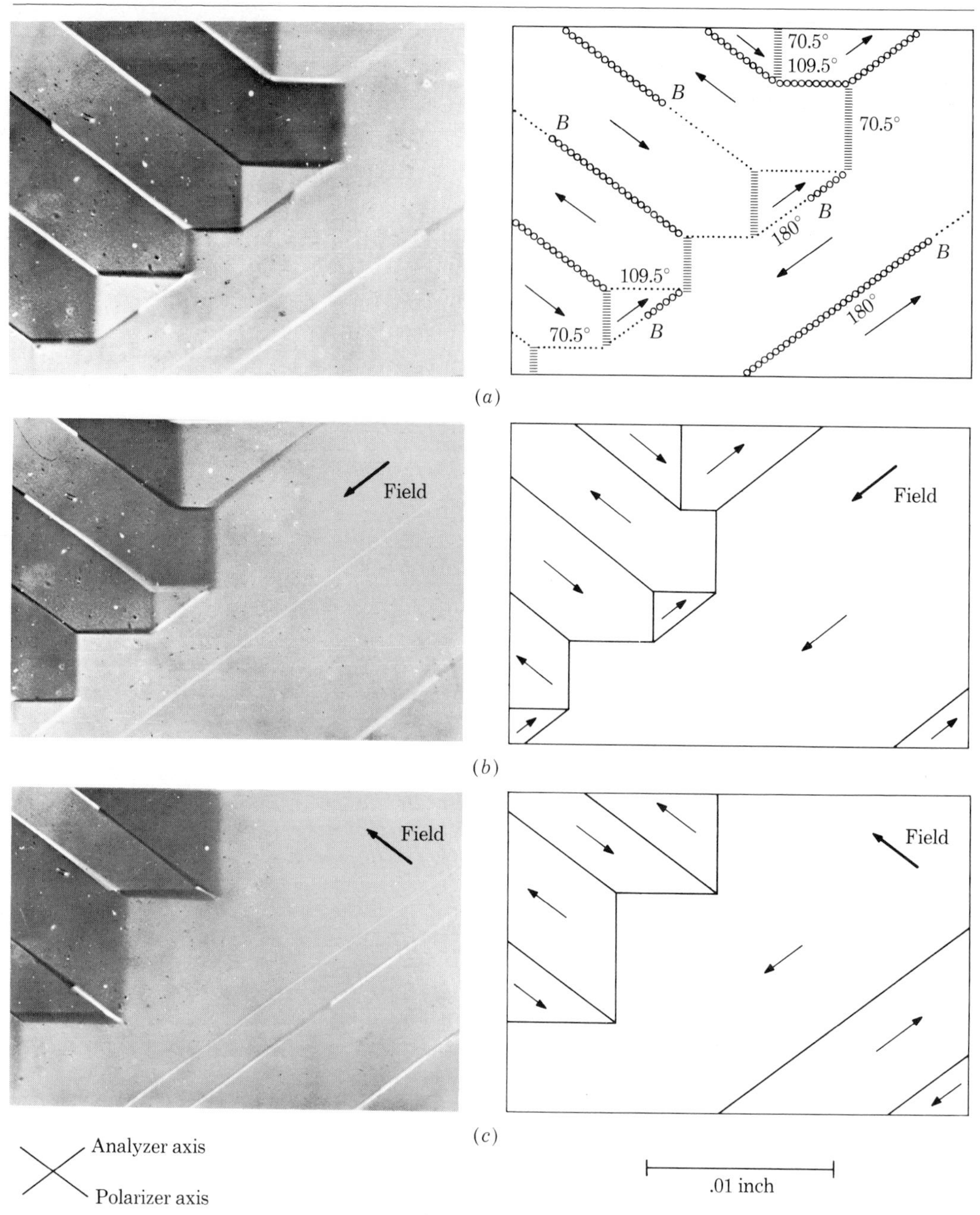

Figure 39-9
Domains in single-crystal yttrium garnet. (*Courtesy of Dr. J. F. Dillon, from "The Smithsonian Institution Report for 1960," pp. 385–404.*)

chinery, particularly for ac use. In ac circuits the iron is continuously being carried through complete cycles of magnetization at the rate, commonly, of 60 per second. The loss of energy by hysteresis is expensive and arrangements have to be made to dissipate the heat evolved.

Soft iron is used in many parts of electric machinery because it has a relatively low hysteresis loss, together with a high permeability. For permanent magnets, a material is desired which has a high residual magnetism; here the large hysteresis loop is immaterial. In transformers certain steels are used which have large hysteresis loops, the designer being willing to accept a comparatively high iron loss in order to obtain the maximum values of flux density that are possible with these materials.

If the temperature of a ferromagnetic specimen is raised above a critical value, called the *Curie temperature,* the coupling of atoms within domains is disrupted by thermal agitation and ferromagnetism disappears. Above its Curie temperature the specimen is paramagnetic. This is additional evidence that ferromagnetism is a property of the interaction of paramagnetic atoms in a crystal lattice. For iron, the Curie temperature is about 770°C.

There is striking visible and audible evidence for the existence of magnetic domains. When a colloidal suspension of finely powdered iron oxide (Fe_2O_3) is deposited on a polished single crystal of iron, the powder is attracted by intense local fields at the domain boundaries and it outlines these boundaries. If the magnetization is changed, the movement of the domains is visible with a microscope. The size of domains varies greatly, from about 10^{-2} to 10^{-6} cm^3 (containing some 10^{21} to 10^{17} molecules).

Photographs may be made of domains in garnet, which is magnetic and which is transparent to visible light. Because polarized light (Chap. 31) passing through it interacts with the magnetization, we are able, with a suitable optical system, to see directly the magnetization distribution within a crystal. In Fig. 39-9*a*, the crystal is demagnetized. The magnetization prefers to lie along certain crystal directions, the body diagonals or {111} axes of the cubic crystal cell. In the demagnetized state each of these directions is about equally populated with domains. Those domains parallel to the polarizer or analyzer axis appear dark, relative to those which make a considerable angle with either of those axes. The domain walls represent transition layers at the center of which the magnetization is directed either up or down, depending on whether they are right-handed or left-handed walls. In this photograph these two kinds of walls appear as light or dark. Since the walls do not always go straight through the crystal, they appear to have a thickness here, though their actual thickness is below the resolution of the optical system. Note that we can easily distinguish several kinds of walls, 180°, 109.5°, and 70.5°.

Figure 39-9*b* and *c* shows what happens when small magnetic fields are applied parallel to two of the easy directions. Favored domains grow at the expense of others. Note that favored domains include those with a component along the field, and not merely those whose magnetization is parallel to the field.

Audible evidence in favor of the domain theory is given by the Barkhausen effect. The secondary coil of a Rowland ring may be connected to an audio amplifier and speaker. Then, when the magnetizing current is slowly increased, a crackling sound is heard in the speaker. As domain boundaries change, the changes in B_M suggested by the magnified portion of the *B-H* curve (Fig. 39-10) induce momentary currents in the secondary coil and these surges are heard as noise from the speaker.

39-8 MAGNETIC POLES AND DIPOLES

A uniformly magnetized bar of iron attracts iron filings in small regions about the ends. These regions where lines of induction enter and leave the magnet (Fig. 39-2*a*) are called *magnetic poles.* When a uniformly magnetized bar is placed at an angle θ with a uniform magnetic induction B (Fig. 39-11), a torque is required to hold it in this

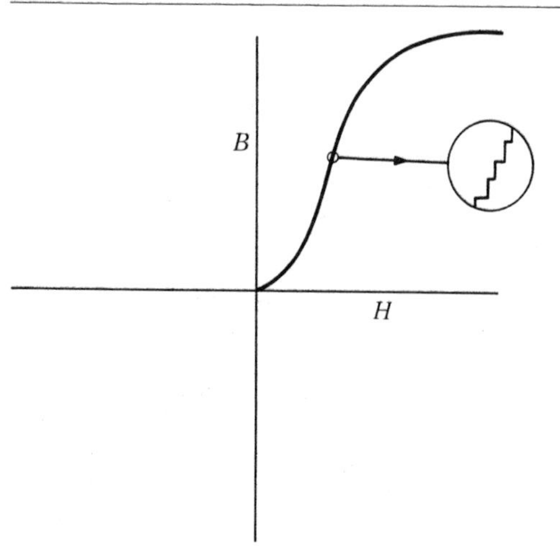

Figure 39-10
Barkhausen effect: Response of domains to uniformly increasing *H* is suggested by magnified portion of *B-H* curve.

position. This torque may be considered as due to forces *F* acting on the poles *m*, which are a distance *l* apart,

$$L = Fl \sin \theta \tag{7}$$

For the equivalent current loop (Fig. 39-5*b*), magnetic moment $\mathfrak{M} = IA$, and the torque on the loop may be written

$$L = \mathfrak{M}B \sin \theta \tag{8}$$

If the loop with current *I* experiences the same torque as the bar when both are oriented as in Fig. 39-5, combining Eqs. (7) and (8) gives

$$\mathfrak{M} = \frac{Fl}{B} \tag{9}$$

The force F/B on the poles per field of unit induction is called the *pole strength m*. The magnetic moment of a bar magnet may be written

$$\mathfrak{M} = ml \tag{10}$$

Although generally the location of the poles of a magnet cannot be expressed precisely, the magnetic moment *ml* can be determined precisely in terms of torque per unit field.

Magnetic poles are "fictitious," and our description of electricity and magnetism using the mks system of units has made no basic use of "poles." However, the historic concept of poles is still useful. The force between poles of strengths m_1 and m_2 a distance *r* apart in space is given by Coulomb's law,

$$F = \frac{\mu_0}{4\pi} \frac{m_1 m_2}{r^2} \tag{11}$$

where *m* is in ampere-meters and μ_0 in webers per ampere-meter. This force is one of repulsion between like poles, attraction between unlike poles. Equation (11) is of limited application: it applies to long, thin magnets with well-separated poles. In the region near a pole of one magnet m_1, the field lines are radial; the force on another "isolated" pole m_2 may be thought of as the product of the strength of m_2 and the induction *B* due to m_1. From Eq. (11),

$$B = \frac{\mu_0}{4\pi} \frac{m_1}{r^2} \tag{12}$$

To find the total number of lines of flux Φ radiating from an N pole of strength *m*, consider the flux through a spherical surface of radius *r* centered at the pole. This flux is the product of the

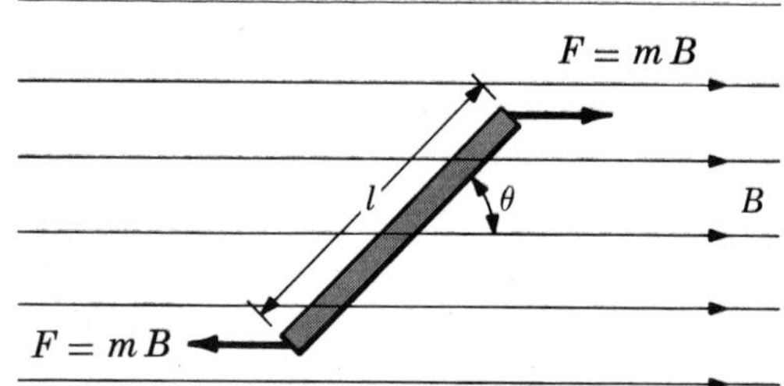

Figure 39-11
Torque on a bar magnet = $mlB \sin \theta$.

spherical area ($4\pi r^2$) and the magnetic induction:

$$\Phi = (4\pi r^2)\frac{\mu_0 m}{4\pi r^2} = \mu_0 m \qquad (13)$$

Since fields, and particularly the earth's field, are often expressed in terms of H, as defined in Sec. 39-1, we may note that the force on a magnetic pole can be expressed

$$F = \mu_0 Hm \qquad (14)$$

Atoms can be ionized, and positive and negative charges can be isolated. But no one has isolated a magnetic pole. When a bar magnet is broken, each fragment exhibits an N pole and an S pole. If we interpret magnetism in terms of amperian current loops, it is apparent that in each elementary magnet (Fig. 39-4), the N face and the S face are inseparable aspects of the same thing. So, while plus and minus charges are basic in electricity, it is the *dipole* rather than the pole which is basic in magnetism. Also, since isolated magnetic poles do not exist, the net flux Φ through any closed surface must be zero.

39-9 MAGNETIC RESONANCE

Magnetic dipoles associated with atomic nuclei are smaller than those associated with electronic motions by a factor of 10^{-2} or 10^{-3} and Rowland-ring techniques are too insensitive to detect nuclear magnetism. But in 1946 Edward Purcell at Harvard and Felix Bloch at Stanford announced independently that they had found a way to "tune in" on the magnetic fields of spinning nuclei.

A proton is considered to spin around an axis (Fig. 39-12), as do the other elementary particles of the atom. Since the proton carries an electric charge, its spin produces a magnetic field; the proton is a tiny magnet. If an external magnetic field is applied to a spinning proton, the axis of spin precesses about the field much as a spinning top or gyroscope (Chap. 7) precesses in a gravitational field. Increasing the strength of the gravitational or magnetic field causes the top or proton, not to fall over, but merely to precess faster.

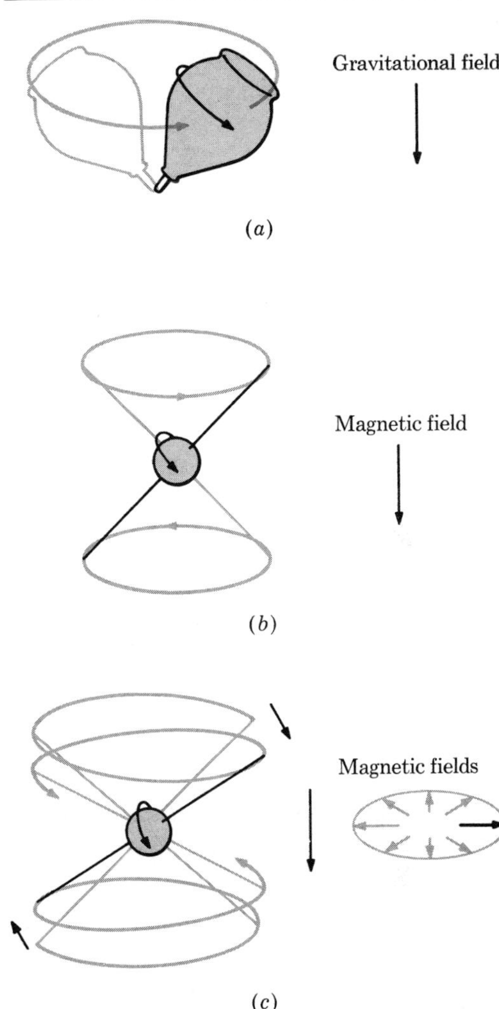

Figure 39-12
Gyroscopic precession of (*a*) a spinning top and (*b*, *c*) a spinning proton.

Purcell and Bloch designed apparatus (Fig. 39-13) in which the spinning particles were subjected to a second magnetic field at right angles

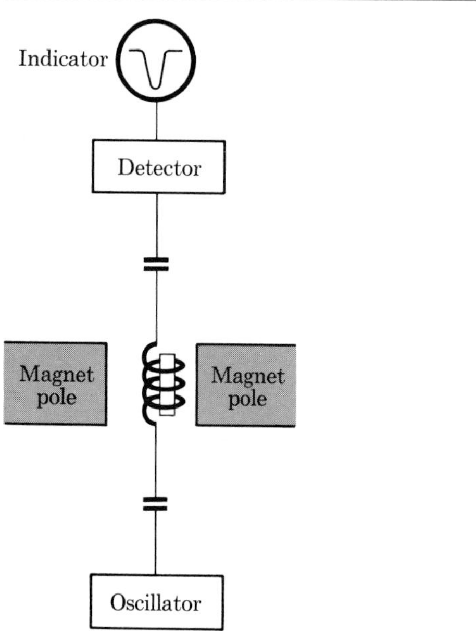

Figure 39-13
Detection of nuclear magnetic resonance.

to the main field. Theory suggested that if the second field (produced by an alternating current in a coil) could be made to rotate around the first field, when the frequency of rotation coincided with the proton's rate of precession (i.e., at resonance) energy would be communicated to the proton, causing it to tip over. The experimenters found that resonance occurred with protons probed with radio waves in the range of a few megacycles per second. Absorption of energy as the protons flipped over was indicated (Fig. 39-13) by a sudden dip in the strength of signal reaching a detector. Nuclear spectroscopy based on the magnetic probe reveals many things about atomic nuclei, atoms, and molecules. Also, an instrument called the *proton precessional magnetometer* permits measurement of the earth's magnetic field by measuring its effect on prealigned protons. An accuracy approaching 1 part in 10^7 makes this magnetometer valuable in prospecting for mineral deposits.

39-10 THE MAGNETIC CIRCUIT

Every line of magnetic flux forms a closed path. In any device, some of the flux may pass through an air portion of the circuit, other parts of the flux may go through iron or other materials. For example, in a dynamo (Fig. 39-14), the flux passes through the pole pieces, the air gaps near the rotor, then through the rotor, and back through the frame to the pole piece. Frequently the flux is divided, a part going through one portion of the device and other parts through the different materials of the apparatus.

The flux in a magnetic circuit can frequently be calculated under ideal conditions. For example, consider the case of a toroid (Fig. 39-7). The use of the defining equation for Φ, with some rearrangement of terms, gives

$$\Phi = BA = \mu HA = \mu A \frac{NI}{l} \tag{15}$$

whence
$$\Phi = \frac{NI}{l/\mu A} = \frac{NI}{\mathcal{R}} \tag{16}$$

It will be observed that this equation is somewhat analogous to Ohm's law for a metallic circuit. Hence the product NI is called the *magnetomotive force* (mmf), and the $l/\mu A$ term is called the *reluctance* of the magnetic circuit. The symbol $\mathcal{R}$ is ordinarily used for magnetic reluctance, in order to avoid confusion with R for resistance. The mks unit of mmf is the ampere-turn. The unit for $\mathcal{R}$ is the ampere-turn per weber.

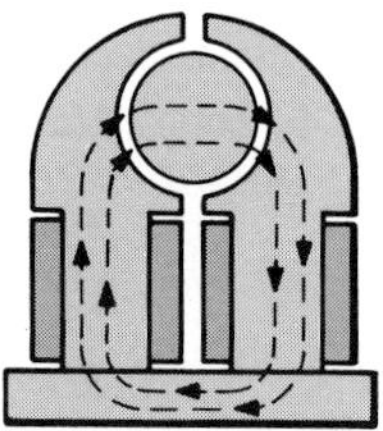

Figure 39-14
A magnetic circuit.

Example A circular ring of iron (toroid) has a cross-sectional area of 5.00 cm², an average diameter of 30.0 cm, and is wound with a coil of 1,000 turns. A current of 3.00 A in the coil magnetizes the iron so that its relative permeability is 250. What is the flux?

$$l = \pi D = \pi \times 0.300 \text{ m} = 0.942 \text{ m}$$

$$\mu = \mu_r \mu_0 = 250 \times 4\pi \times 10^{-7}$$
$$= \pi \times 10^{-4} \text{ Wb/A-turn-m}$$

$$\Phi = \frac{NI}{l/\mu A}$$
$$= \frac{1{,}000 \text{ turns} \times 3.00 \text{ A}}{0.942 \text{ m}/[\pi \times 10^{-4} \text{ (Wb/A-turn-m)} \times 5.00 \times 10^{-4} \text{ m}^2]}$$
$$= 5.00 \times 10^{-4} \text{ Wb}$$

39-11 RELUCTANCES IN SERIES AND IN PARALLEL

In electric circuits, resistances may be connected either in series or in parallel or in series-parallel combinations. In magnetic circuits, reluctances may be connected in similar ways. Two pieces of iron joined end to end constitute two reluctances in series. If these pieces were placed so that the flux could divide between them, they would constitute a case of reluctances in parallel. For the series case the total reluctance is given by

$$\mathcal{R}_t = \mathcal{R}_1 + \mathcal{R}_2 + \mathcal{R}_3 + \cdots \qquad (17)$$

The joint reluctance of a parallel circuit is given by

$$\frac{1}{\mathcal{R}_t} = \frac{1}{\mathcal{R}_1} + \frac{1}{\mathcal{R}_2} + \frac{1}{\mathcal{R}_3} + \cdots \qquad (18)$$

Example If an air gap 1.00 mm wide were cut across the iron ring in the preceding example, what number of ampere-turns would be necessary to maintain the same flux?

Here we have a case of two reluctances in series. Hence

$$\mathcal{R}_t = \mathcal{R}_1 + \mathcal{R}_2 = \frac{l_1}{\mu_1 A_1} + \frac{l_2}{\mu_2 A_2}$$
$$= \frac{0.941}{\pi \times 10^{-4} \times 5.00 \times 10^{-4}} \text{ A-turns/Wb}$$
$$+ \frac{1.00 \times 10^{-3}}{4\pi \times 10^{-7} \times 5.00 \times 10^{-4}} \text{ A-turns/Wb}$$
$$= 7.75 \times 10^6 \text{ A-turns/Wb}$$

$$NI = \Phi \times \mathcal{R}_t = 5.00 \times 10^{-4} \text{ Wb} \times 7.75 \times 10^6 \text{ A-turns/Wb}$$
$$= 3{,}8\bar{7}0 \text{ A-turns}$$

As compared with the 3,000 A-turns in the preceding example, it will be noted that the addition of even a very short air gap greatly increases the mmf necessary to maintain a given flux.

39-12 TERRESTRIAL MAGNETISM

Many centuries ago it was observed that a compass needle aligns itself in a north-south position. About 1600, William Gilbert, physician to Queen Elizabeth I, published results of his experiments which indicated that the earth acts as a great magnet and gave the first satisfactory evidence for the existence of terrestrial magnetism. The observed magnetism of the earth can be roughly portrayed as if it were due to a huge bar magnet within the earth, with its axis displaced about 17° from the earth's axis and considerably shorter than the earth's diameter. The two magnetic poles of the earth are located in northern Canada and in Antarctica, both at considerable distances from the geographical poles.

On his first voyage to America, Columbus observed that a compass needle does not point directly north and that it does not everywhere point in the same direction. This variation of the compass from the true north is called *magnetic declination*. Lines drawn on a map through places

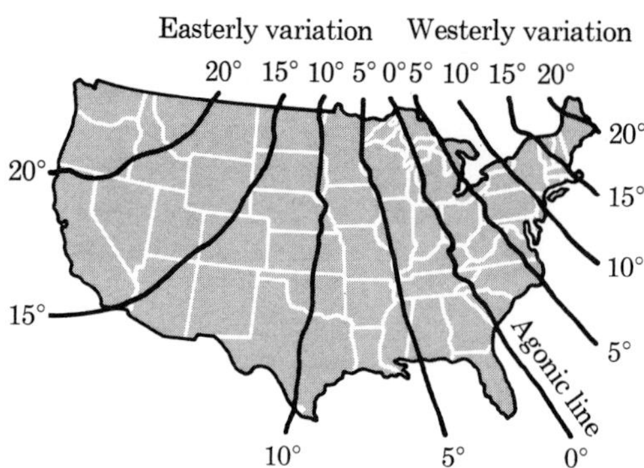

Figure 39-15
Isogonic chart of the United States.

that have the same declination are called *isogonic* lines. On the map in Fig. 39-15 is shown a series of isogonic lines. The line drawn through places that have zero declination is known as an *agonic* line. In the United States places east of the agonic line have *west declination,* and those west of this line have *east declination*. The navigator who uses a magnetic compass must continually make allowance for this *variation* of the compass.

The actual direction of the magnetic field of the earth at most places is not horizontal, and only the horizontal component of the field is effective in the indications of the ordinary compass. If a magnetized needle is mounted on a horizontal axis through its center of gravity, the needle will dip from the horizontal. In the vicinity of New York City the needle would come to rest dipping at an angle of about 72° below the horizontal. Such a needle is known as a *dip needle.*

The strength and direction of the earth's magnetic field not only vary from place to place but also vary in time. There is a very small and periodic daily variation, an even smaller annual variation, but a very material, though erratic, secular variation, or long-time change. For example, at London, England, the declination in 1580 was 11°E, in 1655 it was zero, in 1810 it became 24°W, and in 1940 was about 8°W. Large and very erratic variations occur at certain times. These are known as magnetic "storms." They do not necessarily occur simultaneously with meteorological storms but are probably related to variations in electric currents in the earth's atmosphere. There is correlation in time between magnetic storms and the occurrence of sunspots, and it is generally agreed that the two are intimately related.

The facts of terrestrial magnetism are so complex and the observed data are comparatively so meager and contradictory that theories as to the origin and nature of the earth's magnetism are not at present on a firm basis.

SUMMARY

Magnetic field strength (*intensity*) is the ratio of the magnetic flux density in free space to the permeability of the space.

$$H = \frac{B}{\mu_0}$$

The mks unit of H is the ampere-turn per meter.

Magnetic permeability μ is defined by

$$\text{Permeability} = \frac{\text{magnetic flux density}}{\text{magnetic field strength}}$$

$$\mu = \frac{B}{H}$$

Relative permeability is the ratio of the permeability of a substance to the permeability of empty space.

$$\mu_r = \frac{\mu}{\mu_0}$$

Relative permeability is a pure number. Its value for air and many "nonmagnetic" substances is practically equal to 1.

Substances are classified according to their relative permeabilities as follows: diamagnetic, μ_r less than 1; paramagnetic, μ_r greater than 1; and ferromagnetic, μ_r much greater than 1 and dependent upon H.

In the atomic theory of magnetism, electrons are considered to have magnetic moments associated with both their orbital motion and their spin. If the net magnetic moment of the electrons in an atom is zero, the atom is *diamagnetic*. If electron moments combine to give the atom a net magnetic moment, the atom is *paramagnetic*. For certain electron configurations paramagnetic atoms align in microcrystal domains to produce *ferromagnetism*.

Above its *Curie temperature* the alignment of atoms in a ferromagnetic substance is broken by thermal agitation, and the substance shows only paramagnetic properties.

Hysteresis is the lagging of the magnetization of a magnetic material behind the magnetizing force. The area within a hysteresis loop is a measure of the energy per unit volume lost during a cycle of magnetization.

Coulomb's law for magnetic poles is

$$F = \frac{\mu_0}{4\pi} \frac{m_1 m_2}{r^2}$$

Magnetomotive force in magnetic circuits is analogous to emf in electric circuits.

$$\text{mmf} = NI$$

The mks unit of mmf is the ampere-turn.

Magnetic *reluctance* is the ratio of mmf to magnetic flux

$$\mathcal{R} = \frac{\text{mmf}}{\Phi}$$

The mks unit of $\mathcal{R}$ is the ampere-turn per weber.

The reluctance of a magnetic path is given by

$$\mathcal{R} = \Sigma \frac{l}{\mu A}$$

The law of magnetic circuit is

$$\Phi = \frac{\text{mmf}}{\mathcal{R}}$$

The laws of reluctances in series and in parallel correspond closely with those for resistances in series and in parallel.

In series,

$$\mathcal{R}_t = \mathcal{R}_1 + \mathcal{R}_2 + \mathcal{R}_3 + \cdots$$

In parallel,

$$\frac{1}{\mathcal{R}_t} = \frac{1}{\mathcal{R}_1} + \frac{1}{\mathcal{R}_2} + \frac{1}{\mathcal{R}_3} + \cdots$$

The earth acts as a great magnet. A magnetic compass indicates the direction of the magnetic north; this differs from the geographic north by an angle known as the *declination* (or variation). A *dip needle* shows the angle between the horizontal and the direction of the earth's magnetic field.

References

Bates, L. F.: "Modern Magnetism," 4th ed., Cambridge University Press, New York, 1961.

Bitter, Francis: "Magnets, The Education of a Physicist," William Heinemann, Ltd., London, 1959.

Elsasser, Walter M.: The Earth as a Dynamo, *Scientific American,* May 1958, p. 44.

Gilbert, William: "On the Magnet," Derek J. Price (ed.), Basic Books, Inc., New York, 1958.

Hogan, C. Lester: Ferrites, *Scientific American,* June 1960, p. 92.

Pake, George E.: Magnetic Resonance, *Scientific American,* August 1958, p. 58.

Stoner, Edmund C.: "Magnetism," 4th ed., Methuen & Co., Ltd., London, 1948.

Questions

1 Explain clearly why iron filings line up in a magnetic field. Would copper filings be equally satisfactory?

2 Explain why an iron core is pulled into a current-carrying solenoid.
3 A circular iron ring (toroid) may be magnetized in various ways. What would be the nature of the magnetic field external to the ring in each case?
4 What is the effect of iron beams and pipes in a laboratory on the assumed standard value of the earth's magnetic field for the particular locality?
5 Describe a simple demonstration by which one might illustrate roughly the molecular theory of magnetism.
6 Why are soft-iron pole pieces ("keepers") placed over the ends of horseshoe magnets during storage?
7 Explain what happens to the force when the permeability of the medium is increased: between two poles; between a pole and a current; between two currents.
8 Describe and explain what happens when a bar magnet is repeatedly broken into smaller pieces and the pieces rolled in iron filings.
9 State some of the properties of a material which should be selected for the core of a large lifting magnet, such as that used in steel mills.
10 Describe an experiment to demonstrate the difference between diamagnetism, paramagnetism, and ferromagnetism.
11 Is magnetomotive force a force? Does it have the same dimensions as work? Compare its dimensions with those of electric current.
12 On the basis of modern theory about the electrical nature of atoms, suggest a possible explanation for the magnetic nature of certain materials. Why are other materials nonmagnetic?
13 After a piece of wrought iron has been magnetized until it approaches magnetic saturation, the field strength is doubled. What happens to the induction?
14 An unmagnetized steel rod and an exactly similar magnetized bar are available, but no other apparatus is to be used. How could one show which of the bars is the permanent magnet?
15 Explain clearly why the nature and size of the hysteresis loop are so important to the designer of ac machinery. Describe the loops for various types of commonly used ferromagnetic materials.
16 One method of demagnetizing a watch is to place it in a coil carrying an alternating current and then gradually reduce the current to zero. Explain.
17 Compare the similarities and dissimilarities of magnetic permeability and electric resistivity in the equations

$$\mathcal{R} = \frac{l}{\mu A} \qquad \text{and} \qquad R = \rho \frac{l}{A}$$

18 By multiplying the units of the two coordinates in the hysteresis loop shown in Fig. 39-8, show that the product BH does give the units of energy per volume.
19 Why are gyrocompasses and radio compasses now so commonly used in navigation, especially in the polar regions, instead of the magnetic compass?

Problems

1 What is the magnetic field strength at the center of a closely wound solenoid of 500 turns and axial length 100 cm when a current of 5.0 A is passing through?
2 A toroidal coil has 3,000 turns. The inner and outer diameters are 22 and 26 cm, respectively. Calculate the flux density inside the coil when there is a current of 5.0 A.

Ans. 0.025 Wb/m^2.

3 A specimen of annealed iron has a permeability of 6.75×10^{-3} N/A^2 when it is in a magnetic field of strength 152 A-turns/m. What is the magnetic induction in this iron?
4 What is the permeability of a certain steel in which a magnetic field strength of 543 A-turns/m produces an induction of 1.13 N/A·m? *Ans.* 2.08×10^{-3} Wb/A·m.
5 In a dynamo the magnetic lines of force pass normally from the poles into an iron armature (see Fig. 39-14). Assume that a certain armature is in the form of a uniform cylinder 25 cm long with a radius of 7.0 cm. The magnetic field in the

air gap has a strength, constant in magnitude, of 1,190 A-turns/m. How many lines of induction enter the armature?

6 A cylindrical iron bar 150 cm long and diameter 6.48 cm is placed with its long axis parallel to a uniform magnetic field of strength 18.2 A-turns/m. The permeability of the iron under these circumstances is 4.75×10^{-4} N/A^2. What is the magnetic flux in the iron?

Ans. 2.86×10^{-5} Wb.

7 A circular iron ring (toroid) has a cross-sectional area of 10.0 cm^2 and an average diameter of 50.0 cm. If it has a flux of 8.0×10^{-4} Wb when a current of 5.00 A in the coil has magnetized the iron so that its relative permeability is 250, how many turns are in the coil?

8 A Rowland ring consists of a core of magnetic material around which is wound a primary coil (toroid) connected through a reversing switch to a battery. Around the toroid is wound a secondary coil connected to a ballistic galvanometer.

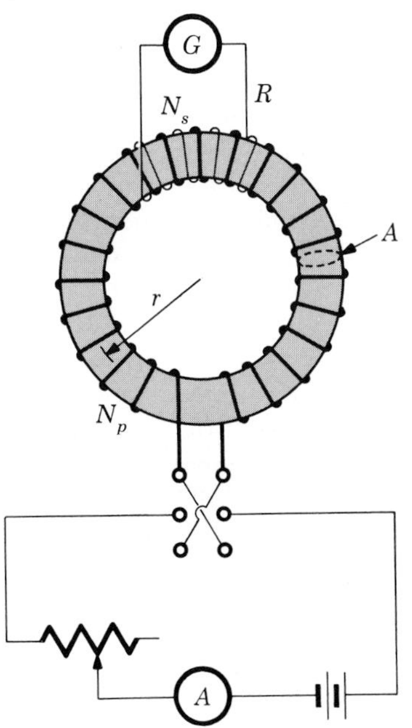

Figure 39-16

Primary coil	Secondary circuit
Number of turns N_p = 200	Total resistance R = 25 Ω
Mean radius r = 5.0 cm	Number of turns N_s = 15
Cross-section area A = 2.0 cm^2	Galvanometer constant $k = 1.5 \times 10^{-8}$ C/division

When a current of 0.10 A is reversed in the primary, a galvanometer deflection of 25.6 divisions is observed. From this observation and the data given in the table above, calculate the relative permeability μ_r of the core (Fig. 39-16).

Ans. 20

9 The iron core of an electromagnet in a field of 150 A-turns/m has a flux density of 1.07 Wb/m^2. What is its relative permeability?

10 The hysteresis loss in a certain piece of iron is 12,000 ergs/cm$^3\cdot$cycle. How many kilowatt-hours per day is used in the core of a transformer operating at 60 cycles/s and having a volume of 25 l? *Ans.* 43 kWh

11 A toroid is 100 cm long and has a cross-sectional area of 30.0 cm^2. It is wound with a coil of 800 turns of wire, and there is a current of 2.50 A in it. The iron core has a relative permeability under the given conditions of 300. Calculate the magnetic field strength in the coil, the total flux, and the flux density.

12 An iron ring has an average diameter of 126 mm and a cross-sectional area of 5.84 cm^2. A toroid of 600 turns wrapped on the iron carries a current of 15.8 A. If the relative permeability is 412, find the (*a*) mmf, (*b*) reluctance of the magnetic circuit, and (*c*) total flux.

Ans. 9.50×10^3 A-turns; 1.32×10^6 A-turns/Wb; 7.15×10^{-3} Wb.

13 A circular ring of iron has a cross section of 6.0 cm^2 and a mean radius of 8.0 cm. The ring is wound with 400 turns of wire. Find the total magnetic flux in the ring when there is a current of 10 A in the coil and the relative permeability of the iron is 300.

14 A solenoid is 50 cm long, has a cross section area of 10 cm^2, an iron core (relative permeability

500), and 300 turns through which a current of 1.5 A flows. Find (*a*) the field intensity, (*b*) the flux density, and (*c*) the flux in the iron core.

Ans. 225 A-turns/m; 0.141 Wb/m^2; 1.41×10^{-4} Wb.

15 A circular iron ring has a mean diameter of 20.0 cm and a sectional area of 30.0 cm^2. It is wound with a coil of 1,000 turns. The ring contains an air gap 1.00 mm long. If the iron has a relative permeability of 200, what current should there be in the coil to produce a flux of 1.00×10^{-3} Wb?

16 With how many turns of wire should a circular ring of 125 cm length be wound in order that a current of 4.18 A may produce a field intensity of 864 A-turns/m within the toroid? What is the flux density in the ring if it is filled with iron of relative permeability 250?

Ans. 258 turns; 0.272 Wb/m^2.

17 A circular ring, wound with 300 turns of wire on an iron core of 146 mm mean diameter, carries a current of 4.15 A. If the relative permeability is 500, calculate (*a*) the magnetic field intensity produced by the current in the toroid and (*b*) the flux density.

18 An electromagnet is wound with a coil of 1,000 turns through which a current of 1.2 A flows and has a total flux of 1.5×10^{-3} Wb. Find the reluctance of this magnetic circuit.

Ans. 8×10^5 A-turns/Wb.

19 A magnetic circuit of three parts, in series, consists of a wrought-iron portion 50 cm long and 120 cm^2 area, relative permeability 1,000; a cast-iron portion 40 cm long and 220 cm^2 in sectional area, relative permeability 200; and an air gap 1.5 mm long and 300 cm^2 in sectional area. Allowing 10 percent for magnetic leakage, determine how many ampere-turns are required to produce a flux of 1.6×10^{-2} Wb in the circuit.

20 A coil of 600 turns is wound uniformly on an iron ring whose mean diameter is 15.0 cm and whose cross section is 5.0 cm^2. The relative permeability of the iron is 500. Calculate the magnetic flux and the flux density within the ring when a current of 15 A is maintained in the coil.

Ans. 6.1×10^{-3} Wb; 12 Wb/m^2.

21 A circular iron ring of relative permeability 400 has an average diameter of 30.0 cm and a sectional area of 120 cm^2. A transverse cut, 1.00 mm long, is made at one place in the iron. If the ring is wound uniformly with a solenoid of 250 turns, what current must there be in order to produce a flux of 1.00×10^{-2} Wb in the core?

22 There is a current of 500 mA in a solenoid 100 cm long, 10.0 cm^2 in cross section, and having 600 turns. Calculate the field strength at the center of the solenoid, the mmf of the solenoid, the reluctance of the region within the solenoid, and the flux.

Ans. 300 A-turns/m; 300 A-turns; 8.0×10^8 A-turns/Wb; 3.76×10^{-7} Wb.

23 A magnetic circuit is made in the form of a rectangle with a vertical bar connecting the midpoints of the horizontal pieces. The central bar is 40 cm long and 200 cm^2 in cross section. The outer rectangle is 100 cm wide, 40 cm high, and 100 cm^2 in cross section. The relative permeability of the iron is 300. A coil of 400 turns is wound on the central bar. (*a*) What is the flux density in the central bar when the current in the coil is 20 A? (*b*) If a gap 1.5 mm wide is cut transversely in the central vertical bar, what would be the flux density in the gap?

24 A toroidal iron ring with a mean radius of 10 cm and cross section of 10 cm^2 is magnetized by a coil of 100 turns in which there is a 5.0-A current. Find the mmf and the magnetic flux in the ring if the relative permeability for iron under these conditions is 1,000.

Ans. 500 A-turns; 1.0×10^{-3} Wb.

25 A magnet is made of iron in the shape of a toroid of mean radius 30 cm, area of cross section 80 cm^2, and relative permeability 400. A bar 50 cm long, of cross section 160 cm^2 and relative permeability 200, is connected across the circle along a diameter. What mmf produced by a coil on the crossbar is required to produce a flux density of 1.2 Wb/m^2 in the straight bar?

26 If a gap 2.0 cm wide is cut in the straight bar of Prob. 25, what will be the flux density in the gap for an mmf of 1,200 A-turns?

Ans. 6.0×10^{-2} Wb/m^2.

27 At a point in central Pennsylvania the earth

has a field strength of 6.5 A-turns/m, and the angle of dip is 70°. What are the horizontal and vertical components of the field strength at this place?

28 At a certain point the horizontal component of the earth's magnetic field has a strength of 2.8 A-turns/m and the vertical component a strength of 4.0 A-turns/m. (*a*) What is the angle of dip at this point? (*b*) Calculate the total strength of the earth's field at this place.

Ans. 55°; 4.88 A-turns/m.

29 The earth's magnetic field in a laboratory is 6.0 A-turns/m, and the angle of dip is 60°. Find the magnetic flux passing through (*a*) the floor, (*b*) an east-west wall, and (*c*) a north-south wall, each of which has an area of 108 ft^2.

Donald Arthur Glaser, 1926–
(*left*)

Born in Cleveland. Studied and taught at Case Institute of Technology and California Institute of Technology. Began career of full-time teaching and research at University of Michigan. In 1959 became Professor of Physics at the University of California at Berkeley. Awarded the 1960 Nobel Prize for Physics for his invention of the bubble chamber.

Robert Hofstadter, 1915–
(*center*)

Born in New York. Studied at the College of the City of New York and at Princeton University. In 1950 at Stanford University initiated a program on scattering of energetic electrons from the linear accelerator. Shared with Mössbauer the 1961 Nobel Prize for Physics for his pioneering studies of electron scattering in atomic nuclei and for his consequent discoveries concerning the structure of the nucleus.

Rudolf-Ludwig Mössbauer, 1929–
(*right*)

Born in Munich. Studied at Technical University in Munich. While at Max Planck Institute in Heidelberg he provided experimental evidence of recoilless nuclear resonance absorption. Appointed Professor of Physics at the Case Institute of Technology in 1961. Shared the 1961 Nobel Prize for Physics with Hofstadter for his researches concerning the resonance absorption of radiation and his discovery in this connection of the effect which bears his name.

40

Induced EMFs

In the years following the discovery by Oersted that a magnetic field is associated with an electric current, it was suspected that a reverse effect exists, making it possible to cause an electric current by means of a magnetic field. In 1831 Michael Faraday in England and independently Joseph Henry in the United States observed that an emf is set up in a conductor when it moves across a magnetic field. This basic discovery is behind the machines by which we are able to convert large amounts of energy from mechanical to electric energy and to reverse the conversion process from electric to mechanical energy. Generators, motors, transformers, and numerous other devices make use of induced emfs.

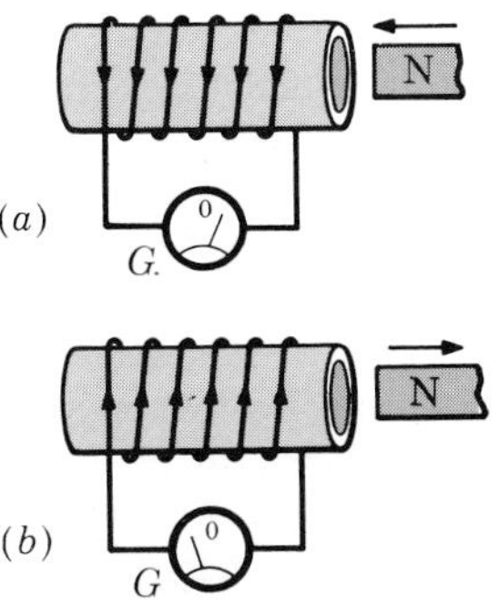

Figure 40-1
Magnet and coil to show induced currents.

40-1 INDUCED EMFs AND CURRENTS

Figure 40-1 represents a coil of wire connected to a sensitive galvanometer G. If the N pole of a bar magnet is thrust into the coil, the galvanometer will deflect, indicating a momentary current in the coil in the direction specified by the arrows in Fig. 40-1*a*. This current is called an *induced current,* and the process of generating the emf is known as *electromagnetic induction*. As long as the bar magnet remains at rest within the coil, no current is induced. If, however, the magnet is quickly removed from the coil, the galvanometer will indicate a current in the direction (arrows, Fig. 40-1*b*) opposite to that at first observed. Faraday found that an emf is induced when there is *any change of magnetic flux linked by the conductor*.

An emf may also be induced in a coil by the change in the magnetic field associated with a change in current in a nearby circuit. For example, in Fig. 40-2 is shown a coil M connected to a battery through switch S. A second coil N connected to a galvanometer is nearby. When the

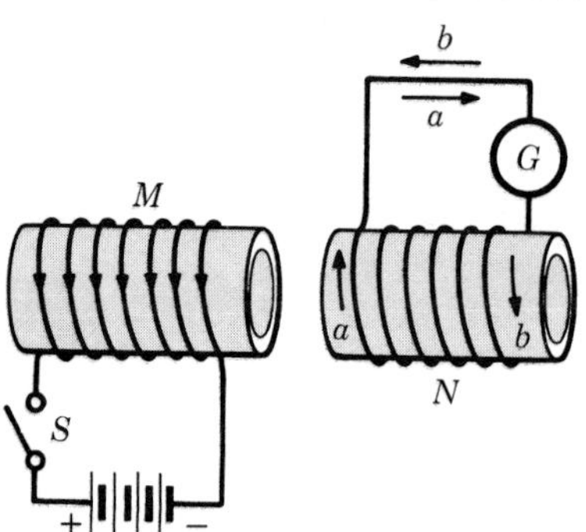

Figure 40-2
Current induced by the interaction of two circuits.

switch S is closed, producing a current in the coil M in the direction shown, a momentary current is induced in coil N in a direction (arrow a) opposite to that in M. If S is now opened, a momentary current will appear in N, having the direction of arrow b. In each case there is a current in N only while the current in M is *changing*. A steady current in M accompanied by a motion of M relative to N is also found to induce a current in N. We observe that, in all cases in which a current is induced in N, the magnetic flux through N is also changing.

40-2 FARADAY'S LAW OF INDUCED EMF

The value of the emf induced in a circuit is found to depend only upon the number of turns N in the circuit and the time rate of change of the magnetic flux linked with the circuit. The average emf $\bar{\mathcal{E}}$ is given by

$$\bar{\mathcal{E}} = -N\frac{\Delta\Phi}{\Delta t} \qquad (1)$$

In Sec. 37-2 it was shown that the unit of magnetic flux Φ is the newton-meter per ampere. Hence we see that the units in the quantities in Eq. (1) are newton-meters/ampere-second = joules/coulomb = volts. If $\Delta\Phi/\Delta t$ is expressed in maxwells per second, Eq. (1) must have a factor 10^{-8} on the right-hand side to give the emf in volts.

The negative sign is often introduced in Eq. (1) to express the fact that the induced emf is in such a direction as to oppose (by its magnetic action) the change that produced it, as explained below.

Equation (1) gives the average induced emf no matter how the emf may vary during the change. Frequently we are interested in instantaneous values of the emf. Then we must use the instantaneous time rate of change of flux. The instantaneous emf e is given by

$$e = -N \lim_{\Delta t \to 0} \frac{\Delta\Phi}{\Delta t} \qquad (2)$$

In Eq. (2) we have used the small e to represent an instantaneous value of the emf. We shall follow this notation in this and the next chapters, where instantaneous values are of particular interest. We shall also refer to instantaneous current by the small i.

Example A coil of 600 turns is threaded by a flux of 8.0×10^{-5} Wb. If the flux is reduced to 3.0×10^{-5} Wb in 0.015 s, what is the average induced emf?

From Eq. (1)

$$\bar{\mathcal{E}} = -N\frac{\Delta\Phi}{\Delta t} = -600\frac{(8.0 - 3.0) \times 10^{-5}\ \text{Wb}}{0.015\ \text{s}}$$

$$= -2.0\ \text{V}$$

40-3 EMF INDUCED BY MOTION

Whenever a charge q moves in a magnetic field, the charge experiences a force F the magnitude of which [Eq. (1a), Chap. 37] is given by

$$F = qvB \sin\theta = \frac{q}{t} lB \sin\theta \qquad (3)$$

In vector form the force is given by

$$F = q\mathbf{v} \times \mathbf{B} = \frac{q}{t}\mathbf{l} \times \mathbf{B} \qquad (3a)$$

The force on the charge $+q$ is at right angles to v and B. In the example illustrated in Fig. 40-3 the force on a positive charge would be upward. A moving charge constitutes a current. The force on a charge in motion is that on the equivalent conventional current (Chap. 37). The vectors representing v, B, and F are mutually perpendicular. If the charge is free to respond to this force, it will move in the direction of F.

An electric conductor, such as a copper wire, has free electrons in it. Consider a wire moving across a magnetic field (Fig. 40-3). The component $B \sin \theta$ perpendicular to the velocity will exert a force on charges in the wire along the direction of the wire. *Positive* charges in the wire would experience a force directed toward b; *electrons* experience a force in the opposite direction, and the free electrons accumulate at a, leaving a deficiency of electrons at b.

Equation (3) gives $F/q = vB \sin \theta$. Thus an electric field is set up in the conductor directed from a toward b, with a magnitude $E = F/q = vB \sin \theta$. The emf e induced in the wire of length l is

$$e = \frac{W}{q} = \frac{Fl}{q} = \frac{Eql}{q} = lvB \sin \theta \qquad (4)$$

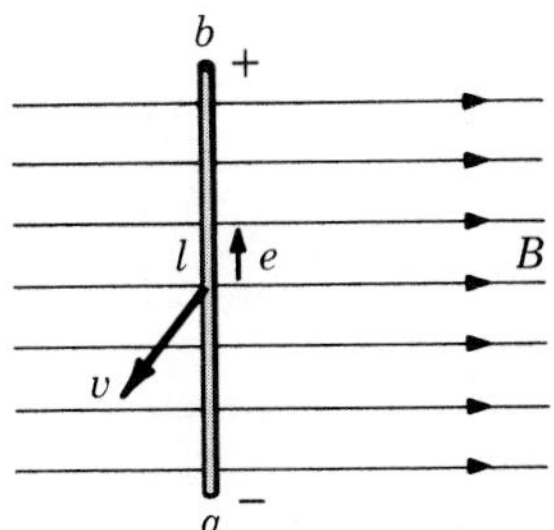

Figure 40-3
Flux density B, motion v, and induced emf e when a conductor moves in a uniform magnetic field.

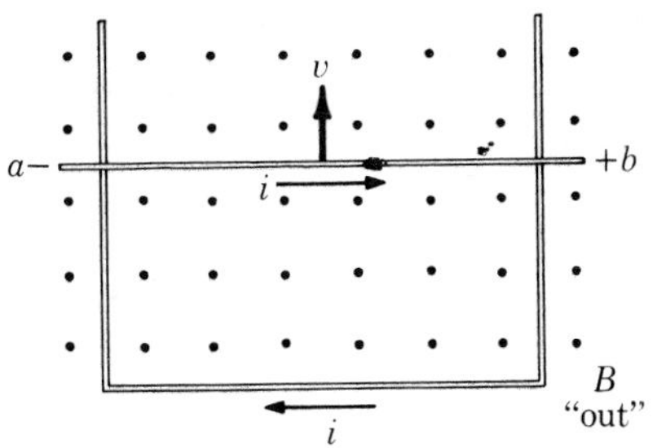

Figure 40-4
Current resulting from the emf induced in conductor ab moving across a uniform flux density B (out of page).

When B is expressed in webers per square meter, l in meters, and v in meters per second, the emf is in joules per coulomb, or volts.

The emf exists whether or not there is a complete circuit for current. If the moving conductor slides along stationary conducting rails (Fig. 40-4), a current will be established in the sense shown.

Example The flux density B in a region between the pole faces of an electromagnet is 0.76 Wb/m^2, horizontally toward the right. Find the emf induced in a straight conductor 10 cm long, perpendicular to B ($\theta = 90°$), when it moves downward at an angle α of 60° with respect to B with a speed of 1.0 m/sec.

$$\text{Induced emf } e = lvB \sin \theta \sin \alpha$$

$$e = (0.75 \text{ Wb/m}^2)(0.10 \text{ m})(1.0 \text{ m/s}) \sin 90° \times \sin 60°$$

$$= 0.066 \text{ V} = 66 \text{ mV}$$

40-4 LENZ'S LAW

Lenz's law states that, whenever an emf is induced, the induced current is in such a direction as to oppose (by its magnetic action) the change inducing the current.

Lenz's law is a particular example of the principle of conservation of energy. An induced current can produce heat to do chemical or mechanical work. The energy must come from the work done in inducing the current. When induction is due to the motion of a magnet or a coil, work is done; therefore the motion must be resisted by a force. This opposing force comes from the action of the magnetic field of the induced current. When a change in current in a primary coil induces an emf in a neighboring secondary coil, the current in the secondary will be in such a direction as to require the expenditure of additional energy in the primary to maintain the current.

Example A horizontal circular coil of wire is lowered toward a bar magnet held vertically with its N pole uppermost (Fig. 40-5). What is the direction of emf induced in the coil?

As the coil is lowered from position *a* to position *b*, the upward flux through it increases. By Lenz's law, the flux due to the induced emf must be downward. The right-hand rule predicts that the induced current will be clockwise, as viewed from above.

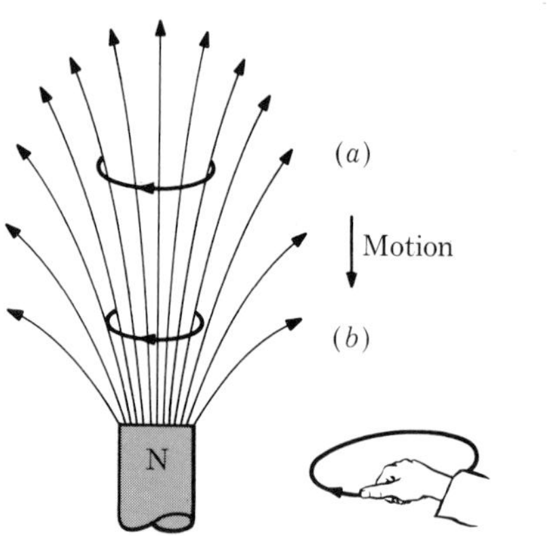

Figure 40-5
Emf induced in a coil moving from *a* to *b* in a field of nonuniform flux density.

Whenever a straight wire, such as *ab* in Fig. 40-4, is drawn across a magnetic field, an emf is induced in the conductor. There will be an induced current in the wire if it is made a part of a closed circuit as indicated in the figure. In accordance with Lenz's law, the direction of induced emf opposes the motion of the conductor.

Therefore the direction of the induced current is such as to add to the field ahead of the current and weaken the field behind the current, since the force on the current is directed from the strong part of the field to the weaker portion of the field. Hence the direction of the induced current depends upon the direction of the field and that of the motion. These three directions are mutually at right angles to each other.

Example A horizontal straight wire 10 m long extending east and west is falling with a speed of 5.0 m/s, at right angles to the horizontal component of the earth's magnetic field, 0.30×10^{-4} Wb/m^2. (*a*) What is the instantaneous value of the emf induced in the wire? (*b*) What is the direction of the emf? (*c*) Which end of the wire is at the higher electrical potential?

(*a*) From $e = lvB \sin \theta$,

$$e = (10 \text{ m})(5.0 \text{ m/s})(0.30 \times 10^{-4} \text{ Wb/m}^2)$$
$$= 0.0015 \text{ V}$$

(*b*) From Lenz's law, the magnetic force on the wire must be upward. But the force on a conductor is from the region of strong magnetic field toward weak. To produce a strong field under the wire, the field due to the induced current should be in the same direction as the earth's field, north. The right-hand rule predicts an emf (or current) from west to east to produce such a magnetic field.

(*c*) Under the action of an emf directed from west to east, positive charge will migrate eastward. The eastern end will be at the higher potential. (If an external circuit were connected to this generator, the positive current in the external circuit would be from east to west, or from high potential to low.)

40-5 AVERAGE EMF IN A ROTATING COIL

An important case of induced emf is that in which a coil rotates at a uniform angular velocity in a uniform magnetic field. In Fig. 40-6*b* is shown a magnetic field that is parallel to the plane of the paper. The axis of rotation *O* is perpendicular to the plane of the paper. As the coil rotates in the field, it "cuts" flux at a variable rate, since only the motion perpendicular to the field is effective in producing an induced emf. When the coil is at position *GH*, the plane of the coil is perpendicular to the field and the conductors *C* and *D* are moving parallel to the field; hence there is zero induced emf. When the plane of the coil is parallel to the field at position *JK*, the conductors are moving normal to the field and the induced emf is a maximum.

If the coil rotates through an angle of 90° from the position *GH* to *JK*, the change of flux in the coil will be *BA*, all of the flux linked by the area of the coil. Hence the *average* emf is

$$\overline{\mathcal{E}} = -N\frac{\Delta\phi}{\Delta t} = -\frac{NBA}{t}$$

where *A* is the area of the coil and *t* is the time required for a quarter revolution. It must be clearly noted that this average emf is *not* one-half the maximum emf, since the emf does not vary linearly.

Example A rectangular coil of 300 turns has a length of 25.0 cm and a width of 15.0 cm. The coil rotates with a constant angular speed of 1,800 r/min in a uniform field of induction 0.365 Wb/m². (*a*) What average emf is induced in a quarter revolution after the plane of the coil is perpendicular to the field? (*b*) What is the average emf for a rotation of 180° from the zero position? (*c*) What is the average emf for a full rotation?

$$t = \frac{1}{4}\left(\frac{1 \text{ min} \times 60 \text{ s}}{1{,}800 \text{ r} \times 1 \text{ min}}\right) = \frac{1}{120} \text{ s}$$

$$(a)\ \overline{\mathcal{E}} = \frac{NBA}{t}$$

$$= \frac{300 \times 0.365 \text{ Wb/m}^2 \times (0.250 \times 0.150) \text{ m}^2}{\frac{1}{120}}$$

$$= 493 \text{ V}$$

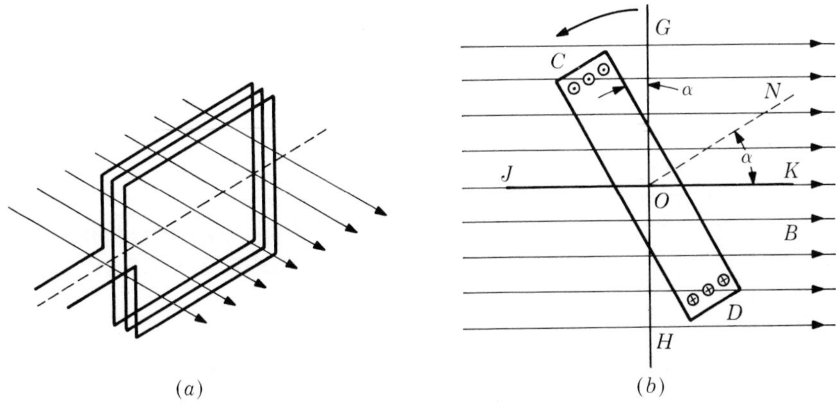

Figure 40-6
Coil rotating in a uniform magnetic field.

(*b*) When the coil rotates through 180° from the zero position, the conductors cut all the flux twice in twice the time of the 90° rotation. Hence the average emf for 180° is identical with that for 90°.

(*c*) For a 360° rotation the emf in one 180° movement is just equal in magnitude and opposite in sense to the emf induced in the next 180° movement. Hence the *average* emf for each revolution is zero.

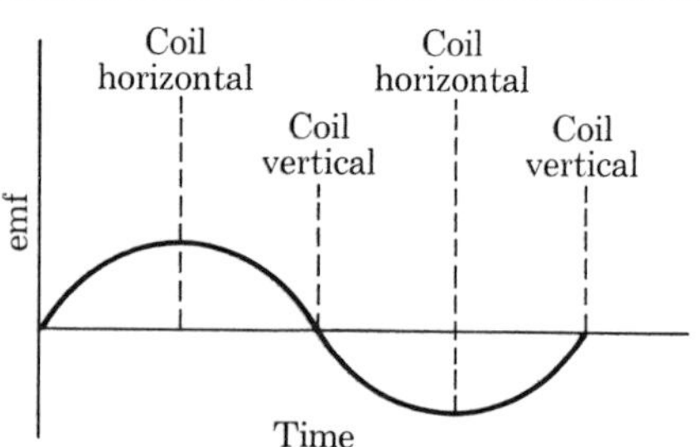

Figure 40-7
Variation of emf with time in a single coil turning in a uniform magnetic field.

40-6 INSTANTANEOUS EMF INDUCED IN A ROTATING COIL

Consider a coil (Fig. 40-6) rotating counterclockwise at constant angular speed ω in a uniform magnetic field. At a certain instant the coil has moved through an angle α past the plane GH (where the induced emf is zero). This angle α is identical with the angle between the induction B and a normal to the coil. At the given instant the flux actually linking the coil is $BA \cos \alpha$. The instantaneous generated emf e is given by Eq. (2),

$$e = \lim_{\Delta t \to 0} - N\frac{\Delta\Phi}{\Delta t}$$

It is shown in Appendix A-9 that this results in an expression for e given by

$$e = NBA\omega \sin \alpha \qquad (5)$$

Equation (5) gives the instantaneous emf at each position α of the plane of the coil after it passes the position of zero emf where the plane GH is normal to the field. The emf varies sinusoidally, reaching its maximum value when $\alpha = 90°$ (coil horizontal) and zero value for $\alpha = 0°$ or $180°$ (coil vertical) as shown in Fig. 40-7). The emf *alternates,* reversing direction twice in each rotation of the coil.

One complete rotation of the coil produces one *cycle* of the emf, causing one cycle of current in any circuit connected across the terminals. The number of cycles per second is called the *frequency*.

As we see from Eq. (5), the maximum value e_m of the emf of the simple generator depends upon the number of turns of the coil, the area of the coil, the flux density of the field, and the angular speed of rotation

$$e_m = NBA\omega \qquad (6)$$

The instantaneous emf e may be expressed by the relation

$$e = e_m \sin \alpha \qquad (7)$$

If B is in webers per square meter, A in square meters, and ω in radians per second, the emf is in volts.

40-7 THE ALTERNATING-CURRENT GENERATOR

The simplest possible generator is a single coil of wire rotating in a uniform magnetic field. The emf induced in such a case is an alternating emf, and hence such a generator is referred to as an *ac generator*. The coil in which the emf is induced is called the *armature*. A high voltage may be obtained in an ac generator by having the coil

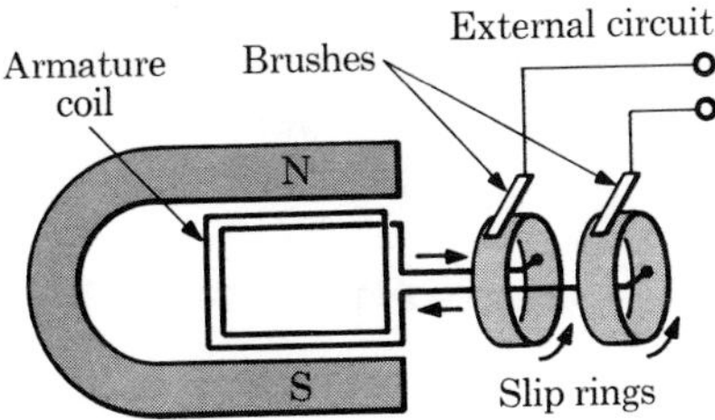

Figure 40-8
Slip rings on an ac generator.

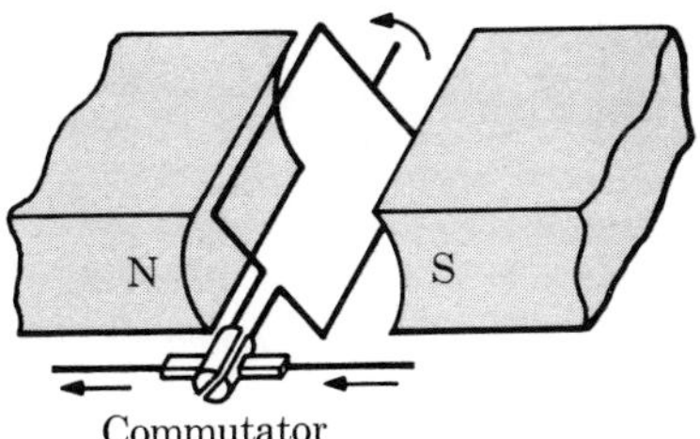

Figure 40-9
A simple generator with a commutator produces a one-directional current in the external line.

wound on an iron core, the flux linked by the coil being thus increased, and also by having a large number of turns in series for each coil. Where the coil rotates, the ends of the coils are connected to circular rings called *collecting rings* or *slip rings*. Carbon (graphite) brushes bearing on these rings make connection to the outside circuit (Fig. 40-8). The basic elements of an ac generator are (1) a *field magnet,* (2) the *armature,* and (3) the *slip rings* and *brushes*. In many ac generators the armature is made stationary and the field magnet is caused to rotate.

40-8 THE DC GENERATOR

An ideal generator can never have a one-directional current in the coil itself, but it is possible to have a one-directional current in the outside circuit by reversing the connections to the outside circuit at the same instant as the emf changes direction in the coil. The change in connections is accomplished by means of a *commutator* (Fig. 40-9). This device is simply a split ring, each side being connected to the respective end of the coil. Brushes, usually of graphite, bear against the commutator as it turns with the coil. The position of the brushes is so adjusted that they slip from one commutator segment to the other at the instant the emf changes direction in the rotating coil. In the external line there is a one-directional voltage, which varies as shown in Fig. 40-10. The curve is similar to a sine curve, with the second half inverted. To produce a steady, one-directional current, many armature coils are used rather than a single coil. These are usually wound in slots distributed evenly around a laminated soft-iron cylinder. With this arrangement of the coils there must be many commutator segments as well as many coils. The connections are so arranged that emfs add at any instant. The average of the instantaneous emfs during a revolution is the emf of the generator.

40-9 THE MAGNETIC FIELD

The magnetic field of a generator is usually produced by an electromagnet, but for a few special uses a permanent magnet system is used. If a permanent magnet is used for the field, the ma-

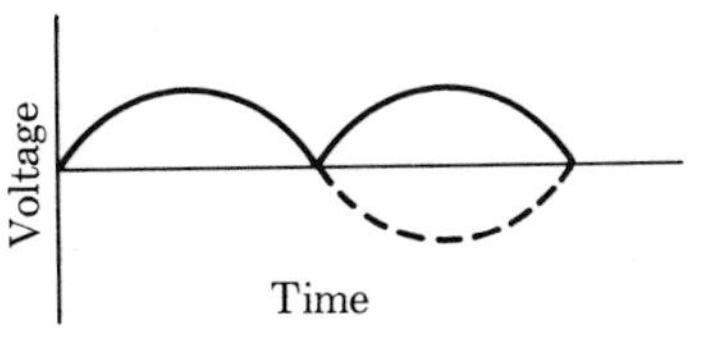

Figure 40-10
Variation of voltage with time in the external line of a simple generator with a commutator.

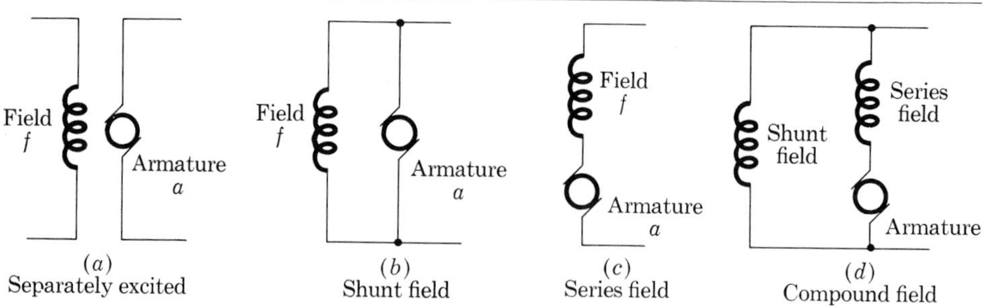

Figure 40-11
Diagrams of the connections of field and armature in dc generators.

chine is commonly called a *magneto*. When the field is supplied by an electromagnet, several connections are possible.

Four common possibilities are illustrated in Fig. 40-11. On the assumption that the speed of the generator remains constant, the emf of the separately excited generator will remain constant as the load increases, while the terminal potential difference decreases by the amount of the I_aR_a drop in the armature. In the shunt generator, the TPD decreases more rapidly as the load increases, since the lowered TPD decreases the field and thus the emf. In the series generator, an increase in load increases the field and hence the emf. By use of a compound field it is possible to keep the TPD almost constant as the load increases.

Whenever there is a shunt field, there are three currents involved in the generator: the armature current I_a, the field current I_f, and the line current I_l. In this generator, the source of current is the armature, and hence

$$I_a = I_f + I_l \tag{8}$$

The field coils act as a resistor for steady currents, while the armature must be treated as a part of a circuit that has a source of energy. For the field

$$V = I_f R_f$$

For the armature

$$V = \mathcal{E} - I_a R_a \tag{9}$$

The power $I_a\mathcal{E}$ converted from mechanical to electric power in the armature supplies the power $I_a{}^2R_a$ used in heating the armature, the power $I_f{}^2R_f$ supplied to the field windings, and the power I_lV delivered to the line.

$$I_a\mathcal{E} = I_a{}^2R_a + I_f{}^2R_f + I_lV \tag{10}$$

Example A shunt generator has a terminal potential difference of 120 V when it delivers 1.80 kW to the line. Resistance of the field coils is 240 Ω, and that of the armature is 0.400 Ω. Find the emf of the generator and the efficiency.

$$P = I_lV$$

$$I_l = \frac{P}{V} = \frac{1{,}8\bar{0}0\text{ W}}{120\text{ V}} = 15.0\text{ A}$$

$$I_f = \frac{V}{R_f} = \frac{120\text{ V}}{240\ \Omega} = 0.500\text{ A}$$

$$I_a = I_f + I_l = 0.50\text{ A} + 15.0\text{ A}$$
$$= 15.5\text{ A}$$

$$V = \mathcal{E} - I_aR_a$$

$$\mathcal{E} = V + I_aR_a = 120\text{ V} + 15.5\text{ A} \times 0.400\ \Omega = 126\text{ V}$$

Input

$$P_i = I_a\mathcal{E} = 15.5\text{ A} \times 126\text{ V} = 1{,}9\bar{5}0\text{ W}$$

Output

$$P_o = I_l V = 1{,}8\bar{0}0 \text{ W}$$

$$\text{Eff} = \frac{P_o}{P_i} = \frac{1{,}8\bar{0}0 \text{ W}}{1{,}9\bar{5}0 \text{ W}} = 0.923 = 92.3\%$$

40-10 THE MOTOR EFFECT

In Sec. 38-3 it was shown that a current-carrying coil in a magnetic field is acted upon by a torque whose magnitude is given by

$$L = NBIA \cos \alpha \tag{11}$$

This torque tends to rotate the coil from its original position until the plane of the coil is perpendicular to the field, when the torque becomes zero. If the coil turns beyond this point, there will be a torque to return it to this position unless the direction of the current is changed. For the coil to continue to rotate, there must be a commutator to reverse the direction of the current at the proper instant.

Example An armature coil of 25 turns is 60 cm long and 20 cm wide. The armature current is 100 A, and the effective flux density in the air gap is 0.30 Wb/m². What is the torque when the plane of the coil is parallel to the field?

$$A = (0.60 \text{ m})(0.20 \text{ m}) = 0.12 \text{ m}^2$$

$$L = NBIA \cos \alpha = 25 \times 0.30 \text{ Wb/m}^2 \times 100 \text{ A} \times 0.12 \text{ m}^2 \times 1.0 = 90 \text{ m} \cdot \text{N}$$

40-11 DC MOTORS

The side push that a current-bearing conductor experiences in a magnetic field is the basis for the operation of the common dc motor. In construction the motor is similar to the generator, having a magnetic field, a commutator, and an armature. When a current is maintained in the armature coils, the force on the conductors produces a torque tending to rotate the armature. The amount of this torque depends upon the current, the flux density, the diameter of the coil, and the number and length of the active conductors on the armature. The commutator is used to reverse the current in each coil at the proper instant to produce a continuous torque.

40-12 COUNTER EMF IN A MOTOR

We have observed that there is an emf induced in a coil that is rotating in a magnetic field. Hence there must be an induced emf in the armature of a dc motor. This emf is opposite in direction to the voltage that is impressed upon the motor from the power source. The induced emf is known as a *counter emf.* The presence of the counter emf serves to reduce the current that would otherwise be produced in the armature. When the motor armature revolves faster, the counter emf is greater and the difference between the impressed voltage and the counter emf is therefore smaller. This difference determines the current in the armature, so that there is a larger current in a motor when it is running slowly than when it is running fast. The current is much larger when the motor is starting than when it is operating at normal speed. For this reason adjustable starting resistors in series with the motor are frequently used to minimize the danger of excessive current while starting.

Since the emf $\mathcal{E}$ induced in the armature of the motor opposes the impressed voltage V, we may apply to the armature the equation for a part of a circuit that includes a sink of energy,

$$V = \mathcal{E} + I_a R_a \tag{12}$$

For a series motor, the relationship of Eq. (12) applies to the motor as a whole. For the motor that has a shunt field, it applies only to the branch including the armature.

Example A dc series motor operates at 120 V and has a resistance of 0.300 Ω. When the motor is running at rated speed, the armature current is 12.0 A. What is the counter emf in the armature?

$$V = \mathcal{E} + IR$$

$$\mathcal{E} = V - IR = 120\ \text{V} - 12.0\ \text{A} \times 0.300\ \Omega$$

$$= 116\ \text{V}$$

The increase in current with decrease in speed makes a motor somewhat self-regulating. An increase in load causes the motor to slow down, thus causing an increase in current. Since the torque is proportional to the current, an automatic increase in torque accompanies an increase in load.

The power supplied to a motor is the product of the line current and the impressed voltage ($I_l V$). A part of this power $I_f^2 R_f$ is used to heat the field coils, a part $I_a^2 R_a$ is used to heat the armature, and the remainder $I_a \mathcal{E}$ is available for operating the motor and thus represents the useful power

$$I_l V = I_f^2 R_f + I_a^2 R_a + I_a \mathcal{E} \tag{13}$$

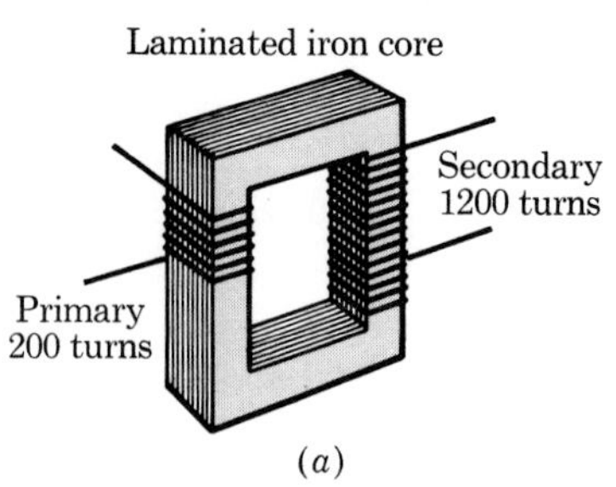

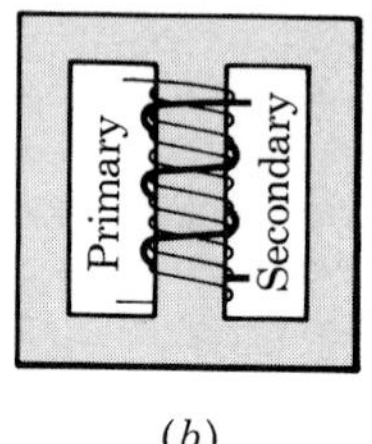

Figure 40-12
Transformers: (*a*) core type, (*b*) shell type. Transformer (*a*) is a step-up and (*b*) is a step-down transformer.

40-13 THE TRANSFORMER

It was previously shown that a change in the current in one of two neighboring coils causes an emf to be generated in the other coil. The induced emf and, therefore, the induced current can be greatly increased by winding the two coils on a closed, laminated iron core, as in Fig. 40-12. This combination of two coils and an iron core is one type of transformer. Suppose that an alternating current is maintained in the primary coil of the transformer. This current is constantly changing; hence the magnetic flux in the iron core also varies periodically, thereby producing an alternating emf in the secondary coil.

Since the rate of change of flux is nearly the same in the primary and secondary coils, it follows that the induced emfs in the coils are directly proportional to the respective numbers of turns in the coils. Symbolically

$$\frac{\mathcal{E}_p}{N_p} = \frac{\mathcal{E}_s}{N_s} \tag{14}$$

where $\mathcal{E}$ is used for induced emf and N for the number of turns. In practice the voltage impressed on the primary from the outside source is somewhat greater than $\mathcal{E}_p$, the induced emf in the primary, and the secondary voltage at the terminals is slightly less than $\mathcal{E}_s$, the emf induced in the secondary.

The efficiency of a commercial transformer operating under favorable conditions is very high, only a few percent less than ideal. Hence the power input to the primary is nearly equal to the power output from the secondary.

$$V_p I_p = V_s I_s \qquad \text{approximately}$$

or $$\frac{I_p}{I_s} = \frac{V_s}{V_p} \quad \text{approximately} \qquad (15)$$

The currents are thus seen to be inversely proportional to the respective voltages (approximately).

40-14 DISTRIBUTION OF ELECTRIC ENERGY

Whenever electric energy is to be used at any considerable distance from the generator, an ac system is used, because the energy can then be distributed without excessive loss, whereas, if a dc system were used, the losses in transmission would be very great.

In an ac system the voltage may be increased or decreased by means of transformers. The terminal voltage at the generator may be, for example, 12,000 V. By means of a transformer the voltage may be increased to 66,000 V or more in the transmission line. At the other end of the line "step-down" transformers reduce the voltage to a value that can safely be used. In a dc system these changes in voltage cannot readily be made.

One might ask why all this increase and decrease in voltage is needed. Why not use a generator that will produce just the needed voltage, say, 115 V? The answer lies in the amount of energy lost in transmission. In dc circuits, and in the ideal case in ac transmission lines, the power delivered is $P = VI$, where V is the (effective) voltage and I the current. (It will be shown later that in ac circuits $P = VI$ only in special cases.) If a transformer is used to increase the available voltage, the current will be decreased.

Suppose that a 10-kW generator is to supply energy through a transmission line whose resistance is 10 Ω. If the generator furnishes 20 A at 500 V,

$$P = VI = (500 \text{ V})(20 \text{ A})$$

$$= 10{,}000 \text{ W}$$

The heating loss in the line is

$$I^2R = (20 \text{ A})^2(10\ \Omega) = 4{,}000 \text{ W}$$

or 40 percent of the original power. If a transformer is used to step up the voltage to 5,000 V, the current will be only 2 A and the loss

$$I^2R = (2 \text{ A})^2(10\ \Omega) = 40 \text{ W}$$

or 0.4 percent of the original power.

A second transformer can be used to reduce the voltage at the other end of the line to whatever value is desired. With 2 or 3 percent loss in each of the transformers, the overall efficiency of the system may be increased from 60 to 95 percent by the use of transformers. Thus alternating current, through the use of transformers producing very high voltages, makes it possible to furnish electric power over transmission lines many miles in length.

Example A step-down transformer at the end of a transmission line reduces the voltage from 2,400 V to 120 V. The power output is 9.0 kW, and the overall efficiency of the transformer is 92 percent. The primary ("high-tension") winding has 4,000 turns. How many turns has the secondary, or "low-tension," coil? What is the power input? What is the current in each of the two coils?

$$P_s = 9.0 \text{ kW} = 9{,}\bar{0}00 \text{ W}$$

$$\frac{\mathcal{E}_p}{\mathcal{E}_s} = \frac{N_p}{N_s}$$

In Eq. (14), the emfs are the induced emfs. If the efficiency is high, the error introduced by using the terminal voltages is not excessive. Then

$$\frac{2{,}400 \text{ V}}{120 \text{ V}} = \frac{4{,}000 \text{ turns}}{N_s}$$

Hence $$N_s = 200 \text{ turns}$$

$$\text{Eff} = \frac{P_s}{P_p}$$

$$0.92 = \frac{9{,}\bar{0}00 \text{ W}}{P_p}$$

Therefore $P_p = 9,\bar{8}00 \text{ W}$

$$I_p = \frac{9,\bar{8}00 \text{ W}}{2,400 \text{ V}} = 4.1 \text{ A}$$

$$I_s = \frac{9,\bar{0}00 \text{ W}}{120 \text{ V}} = 75 \text{ A}$$

To assist in insulating the coils and to dissipate the heat in a power transformer, the core and coils are submerged in oil contained in a metal housing. Sometimes in the larger sizes this oil is circulated through the transformer and cooled by various means.

The losses in a transformer are of two types. The *copper losses* are caused by the heat developed from the I^2R power consumption on account of the resistance of the wire in the coils. In practice these losses are minimized by the use of wire of large size, although there is an obvious economic limit to the size of wire that should be used. Other losses are due to the hysteresis and eddy-current losses in the iron core. These are known as *iron losses*.

40-15 EDDY CURRENTS

When a large block of a conducting substance is moved through a nonuniform magnetic field or when in any manner there is a change in the magnetic flux through the conductor, induced currents will exist in eddies through the solid mass. These are referred to as *eddy currents*. This effect was discovered by the physicist Foucault (1819–1868), and the currents are sometimes referred to as "Foucault currents." In Fig. 40-13 is shown a metal disk that is being moved to the right across a nonuniform magnetic field assumed to be directed toward the reader. The right-hand rule shows the direction of the currents in the eddies in the disk. These currents tend to set up magnetic fields, which, reacting with the field which gave rise to them, will oppose the motion of the disk through the field. The disk acts as though it were embedded in a very viscous medium. Eddy currents produce heat and hence consume energy. They are frequently a source of considerable difficulty to the electrical designer.

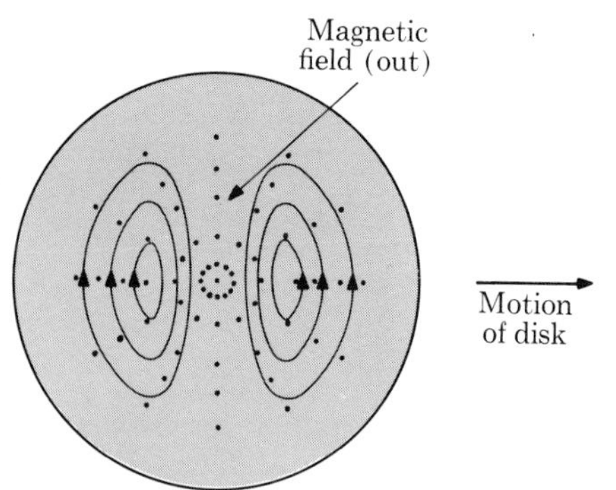

Figure 40-13
Eddy currents in a disk moving in a nonuniform magnetic field.

Eddy currents may be minimized by *laminating* the block of metal, i.e., by making it up out of many very thin sheets or laminations. The planes of these laminations must be arranged parallel to the magnetic flux, so that they cut across the eddy-current lines. This increases the resistance of the current paths and thus reduces the eddy currents. Many ac machines, such as transformers and dynamos, have their magnetic parts made up of laminated sections.

SUMMARY

Most commercial sources of electric energy involve the generation of an emf by *electromagnetic induction*.

An emf is induced whenever there is a change of magnetic flux linking an electric circuit, or motion of a conductor through a magnetic field.

Lenz's law states that the induced emf is always in such a direction as to oppose the change which gives rise to it. The direction of a current induced in a moving conductor is such as to build up the magnetic field ahead of the conductor and weaken the field behind the conductor.

The average value of the induced emf is given by

$$\bar{\mathcal{E}} = -N\frac{\Delta\Phi}{\Delta t}$$

The instantaneous emf is given by

$$e = \lim_{\Delta t \to 0} -N\frac{\Delta\Phi}{\Delta t}$$

The instantaneous emf induced in a straight wire moving through a uniform magnetic field of induction B with a velocity v at an angle θ with the field is given by

$$e = lvB \sin\theta$$

The instantaneous emf in a rotating coil is given by

$$e = NBA\omega \sin\alpha$$

where α is the angle between the plane of the coil at the instant considered and the plane normal to the field.

The maximum emf in a rotating coil is given by

$$e_m = NBA\omega$$

and the instantaneous emf can be expressed by

$$e = e_m \sin\alpha$$

In the mks system of units e is in volts when B is in webers per square meter, A is in square meters, and ω is in radians per second.

Generators transform mechanical energy into electric energy, while *motors* transform electric energy into mechanical energy.

An elementary generator consists of a coil of wire rotating in a uniform magnetic field. The emf induced in an ideal generator is alternating in character.

A dc generator differs from an ac generator in that the dc generator has a *commutator* to make it deliver a unidirectional current to the external circuit.

The terminal potential difference of a dc generator is given by

$$V = \mathcal{E} - I_a R_a$$

Of the *power* generated, part is used in heating the field coils and armature of a dc generator, and the remainder is delivered to the line.

$$I_a\mathcal{E} = I_a^2R_a + I_f^2R_f + I_lV$$

The *torque* on an armature coil of a motor is given by

$$L = NBIA \cos\alpha$$

The *counter emf* generated in a dc motor armature nearly equals the voltage impressed upon it. This counter emf reduces the current which there would otherwise be in the armature, in accordance with the equation

$$V = \mathcal{E} + I_a R_a$$

The power delivered to a dc motor is I_lV. The power available to operate the motor after the heating losses in the armature and field coils is $I_a\mathcal{E}$.

$$I_lV = I_f^2R_f + I_a^2R_a + I_a\mathcal{E}$$

The basic equations for an ideal transformer are

$$\frac{\mathcal{E}_p}{\mathcal{E}_s} = \frac{N_p}{N_s}$$

$$\frac{I_s}{I_p} = \frac{N_p}{N_s} \qquad \text{approximately}$$

Electric energy is transmitted long distances at high voltage and low current in order to minimize the I^2R heating loss in the line.

Eddy (Foucault) currents are the currents induced in relatively large bodies of conducting material when they are linked with variable magnetic fluxes.

Questions

1 What is the source of the energy of the induced current when a bar magnet is inserted into a coil of wire?

2 Discuss the emf induced in a coil as a bar magnet is passed through a coil and withdrawn from the other side.

3 Two identical hoops, one of copper, the other of aluminum, are similarly rotated in a magnetic field. Explain the reasons for the different torques required.

4 A circular coil lies on a horizontal table. Discuss the emf and current induced by cutting the vertical component of the earth's field as the coil is slid along the table.

5 A bar magnet falls with uniform speed through a coil of wire with its plane horizontal. The long axis of the magnet is normal to the plane of the coil. As the N pole of the magnet approaches the coil, in what sense is the emf induced? What is the direction of the force which acts upon the wire? Describe the way in which the induced emf in the coil varies with time, as the magnet drops through the coil.

6 A metal clothesline stretched east and west falls to the ground. In what direction is the emf induced by cutting the earth's field? Illustrate by a diagram.

7 The coils wound on the spools of resistance boxes are often made of bifilar wires, i.e., a wire doubled back upon itself. Make a sketch of such a resistance spool. Explain why this arrangement gives coils which are practically noninductive.

8 Explain how a primary coil could be wound noninductively so that it would have no effect upon the secondary coil.

9 Derive the general equation for induced emf, beginning with the law for the force on a current-carrying conductor in a magnetic field.

10 Two secondary coils are identical in every respect except that one is wound with copper wire and the other with nickel-silver wire. Compare the induced emfs and the induced currents when these coils are used with the same primary coil.

11 Explain clearly why Lenz's law must follow from the principle of conservation of energy.

12 Sometimes students believe that Ohm's law does not apply to an inductive circuit. Show that the law is applicable when it is properly interpreted.

13 A bar magnet which is resting in the center of a coil is quickly withdrawn from the coil. Discuss the situation in detail with respect to Lenz's law and Oersted's relationship.

14 Explain what happens to the electrons in a straight wire when the wire is moved through a magnetic field. Give an example to show the direction of the induced emf and the induced current. Indicate the end of the conductor which has the higher potential.

15 Upon what factor does the magnitude of an induced emf depend? Discuss.

16 Since there is no electrical connection between the primary and secondary of a transformer, what is the mechanism for the transfer of energy between these circuits? Even a well-designed transformer becomes warm in use. State reasons why this is true.

17 What are some of the advantages of using iron for the core of a 60-cycle transformer? Why do ultrahigh-frequency transformers utilize air cores?

18 What happens to the motion of a copper pendulum bob when it swings between the poles of a strong electromagnet? What becomes of the kinetic energy of the pendulum? Describe some devices in which eddy currents are of value. Mention some in which such currents are objectionable. State reasons in each case.

19 A flat aluminum disk is rotated between the poles of a U-shaped magnet, with the axis of rotation parallel to the magnetic field. Assign letters A, B, C, and D to the edges of the disk at points 90° apart and letter O to the center of the disk. Between which two lettered points is the induced emf a maximum? Make a sketch to show the paths of the induced currents.

20 Is an electric generator a "generator of electricity"? Where is the electricity before it is "generated"? What does such a machine "generate"?

21 A rural electric line supplies power to two industries by means of step-down transformers. Both plants use the same voltage, but one plant takes much more power than the other. What differences in design are there in the transformers supplying these industries? State the reasons for these differences.

22 Draw a curve showing the variation of the terminal potential difference of a separately excited generator as the load increases.

23 A railway electric locomotive uses its dc motor as a generator for braking purposes on long downgrades. Show how this may be done and what happens to the energy.

24 Usually an automobile has a generator and a starting motor as separate units. How do they differ and why? Would it be possible to make one machine that would serve both purposes?

25 A dc shunt motor rotates in a clockwise direction when power is supplied to it with the current in a certain direction. How must it be rotated as a generator so that the current will be in the same direction?

26 Describe an experiment that one might make to measure the input and output power of a dc motor.

27 From a consideration of the connections shown in Fig. 40-11, suggest characteristics of the way that the torque produced by a motor of each kind would change as the load is increased.

Problems

1 A wire "cuts" a flux of 0.030 Wb in 0.020 s. What average emf is induced in the wire?

2 A circular coil of wire of 3,000 turns and diameter 6.0 cm is situated in a magnetic field so that the major plane of the coil is normal to the field. If the flux density in the coil changes uniformly from 0.5 to 1.7 Wb/m^2 in 3.14 min, what emf is induced in the coil? *Ans.* 54 mV.

3 The N pole of a magnet is brought down toward a flat coil of 20 turns lying on a table. If the flux from the N pole passing through the coil changes from 1.42×10^{-3} Wb to 8.66×10^{-3} Wb in 0.215 s, what is the magnitude of the induced emf? Is it clockwise or counterclockwise as you look down at the coil?

4 A flat, circular coil has 150 turns, a radius of 12 cm, and a resistance of 0.85 Ω. The coil is placed with its plane normal to the earth's magnetic field, where B is 7.5×10^{-5} Wb/m^2. When the coil is rotated through 90° in $\frac{1}{3}$ s, what charge circulates in the coil? *Ans.* 6.0×10^{-4} C.

5 A circular coil of 2,500 turns and radius 125 mm is situated in a magnetic field that is perpendicular to the plane of the coil. The flux density changes from 0.0142 Wb/m^2 to 0.0678 Wb/m^2 in 2.33 s. What average emf is induced in the coil?

6 In one of the coils of a generator armature, 100 wires pass across a pole near which there is a magnetic flux density of 1.5 Wb/m^2. The pole face has an area of 1,000 cm^2. If the armature moves across the pole in $\frac{1}{120}$ min, what is the average emf induced in the coil by the "cutting" of this flux? *Ans.* 30 V.

7 An east and west wire falls to the ground with a speed of 25 cm/s. If the horizontal component of the earth's magnetic flux density is 2.0×10^{-5} Wb/m^2, find the emf induced in 100 m of the wire.

8 A train is traveling north with a speed of 20 m/s. The distance between the rails is 1.3 m, and the vertical component of the earth's flux density is 4.0×10^{-5} Wb/m^2. (*a*) Find the potential difference between the rails. (*b*) If the leakage resistance between the rails is 100 Ω, what is the retarding force on the train? *Ans.* 1.6 mV; 1.3×10^{-6} N.

9 A straight wire 254 cm long at right angles to a magnetic field of flux density 0.124 Wb/m^2 is moved across the field with a speed of 50.0 cm/s. If the path of the wire makes an angle of 50° with the field, what is the emf between the ends of the wire? Show the direction of the emf by means of a sketch.

10 In an experiment to determine the flux density of the region between the pole pieces of a magnet a 30-turn coil of 50-Ω resistance and area 2.0 cm^2 is placed with its plane perpendicular to

the field and then quickly withdrawn from the field. The charge flowing in the operation is found to be 5.0×10^{-4} C. Determine the flux density.
Ans. 0.42 Wb/m^2.

11 A straight wire 75.0 cm long is perpendicular to a field of flux density 0.172 Wb/m^2. The wire is moved across the field at a speed of 2.00 m/s, the direction of motion making an angle of 40° with the field. What emf is generated? Sketch the apparatus, and show the direction of the induced emf.

12 A coil of wire of 10 turns, each enclosing an area of 900 cm^2, is in a plane perpendicular to a magnetic field of flux density 5.0×10^{-5} Wb/m^2. The coil is turned to a position parallel to the field in 0.50 s. Find the average emf induced. *Ans.* 9.0×10^{-5} V.

13 A rectangular coil of 30 turns, 20 by 40 cm, is rotated at 25 r/s about an axis perpendicular to a uniform magnetic field. The field has a flux density of 0.046 Wb/m^2. If the coil starts from the position where the induced emf is zero, calculate (*a*) the average emf induced in the first quarter turn and (*b*) the average emf for the following half turn.

14 A closed rectangular coil of wire 20×50 cm rotates at a uniform speed of 5.0 r/s about an axis perpendicular to a uniform magnetic field of 8×10^{-2} Wb/m^2. The loop has a resistance of 0.20 Ω. (*a*) What is the maximum magnetic flux included by the rotating coil? (*b*) What is the maximum emf developed in the coil? (*c*) What is the average current in the coil during 1 r?
Ans. 8×10^{-3} Wb; 0.25 V; zero.

15 A rectangular coil having 150 turns, 300×200 mm, is rotating at 1,2000 r/min about an axis normal to a uniform field of induction 5.24×10^{-3} Wb/m^2. Find the instantaneous value of the induced emf when the plane of the coil is (*a*) perpendicular to the field and (*b*) 30° from this position.

16 A metal spoke in a wheel is 80 cm long. If the wheel makes 30 r/min in a plane perpendicular to the earth's magnetic field where the flux density is 15.0×10^{-5} Wb/m^2, find the difference in potential between the axle and the rim of the wheel. Which part is at the higher potential when the wheel rotates clockwise as seen by one looking in the positive direction of the field?
Ans. 5.0×10^{-5} V; rim.

17 A rectangular coil of 30 turns, 200×400 mm, is rotated at 24 r/s about an axis perpendicular to a magnetic field of induction 0.108 Wb/m^2. Starting from the position of the coil where the induced emf is zero, calculate (*a*) the average emf induced in the next half turn and (*b*) the instantaneous value of the emf when the coil is 58° from the starting point.

18 The wire spoke of a wheel is 45 cm long. The wheel turns through 90° in 0.050 s. If this motion is perpendicular to a magnetic flux density of 3.5×10^{-5} Wb/m^2, what average emf is induced in the wire? *Ans.* $1\bar{1}0$ μV.

19 A Ferris wheel is 100 ft in diameter. Its axis of rotation is on a north-south line. (*a*) If the horizontal component of the earth's flux density is 2.00×10^{-5} Wb/m^2 and the wheel is rotating at 2.00 r/min, what is the potential difference existing between the axle and the end of one spoke? (*b*) Which is at the higher potential?

20 A copper disk of radius 10.0 cm rotates 5.0 r/s with its axis parallel to a magnetic field of 0.50 Wb/m^2. The rotation is clockwise, as viewed in the direction of the field. (*a*) Determine the difference in potential developed between the center and the edge of the disk. (*b*) Which is at the higher potential? *Ans.* 0.079 V; edge.

21 A toroid (ring) of circumference 10 m and cross section 0.0010 m^2 is magnetized by a coil of 1,000 turns carrying a current of 1.0 A. The

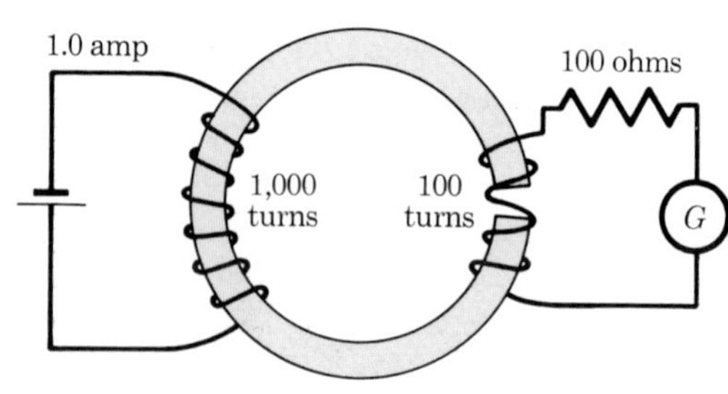

Figure 40-14

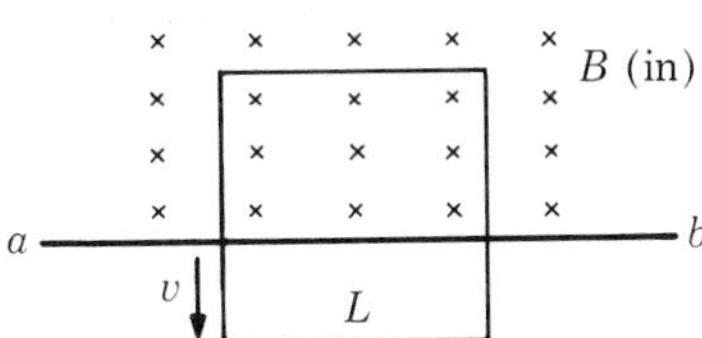

Figure 40-15

ring is broken and its ends separated by a distance of 1.0 mm. A second coil with 100 turns is wound around the gap as shown in Fig. 40-14 and connected in a ballistic galvanometer circuit of total resistance 100 Ω. If the permeability of the ring is assumed to remain constant and equal to 7.43×10^{-3} Wb/A-turn·m, how much charge passes through the galvanometer when the ring is suddenly pushed together, closing the gap? Neglect the emf induced in the galvanometer coil, as well as any fringing of the magnetic field.

22 A magnetic field of uniform flux density B is established in a horizontal direction above a certain horizontal line **ab.** Below the line, $B = 0$. If a square wire frame (Fig. 40-15) of side L, mass m, and resistance R is placed part inside and part outside the field with the sides perpendicular to the field, at what velocity of fall will there be no acceleration? *Ans.* $v = mgr/BL^2$.

23 A 110-V potential difference is applied to the primary of a step-up transformer that delivers 2.0 A from its secondary. There are 25 times as many turns in one winding as in the other. Find (*a*) the voltage of the secondary and (*b*) the current in the primary. Assume no losses in the transformer.

24 A transformer is designed to operate at 15 kW and $\frac{2,200}{122}$ V. Neglecting all losses, what is the turn ratio? the primary and secondary currents? *Ans.* $\frac{18}{1}$; 6.8 A; 120 A.

25 The primary and secondary coils of a transformer have 500 and 2,500 turns, respectively. (*a*) If the primary is connected to a 110-V ac line, what will be the voltage across the secondary? (*b*) If the secondary were connected to the 110-V line, what voltage would be developed in the smaller coil? (Assume ideal conditions.)

26 A radio transformer has 690 turns of wire in the primary coil, which is connected to a 120-V ac source. The secondary coil supplies 6.3 V to the tube filaments. How many secondary turns are required? (Allow 5 percent more for losses.) *Ans.* 38.

27 The overall efficiency of a transformer is 90 percent. The transformer is rated for an output of 12.5 kW. The primary voltage is 1,100 V, and the ratio of primary to secondary turns is 5:1. The iron losses at full load are 700 W. The primary coils have a resistance of 1.82 Ω. (*a*) How much power is lost because of the resistance of the primary coils? (*b*) What is the resistance of the secondary coils?

28 Find the power loss in a transmission line whose resistance is 1.5 Ω, if 50 kW are delivered by the line (*a*) at 50,000 V and (*b*) at 5,000 V. *Ans.* 1.5 W; 150 W.

29 A shunt generator has a field resistance of 240 Ω and an armature resistance of 1.50 Ω. The generator delivers 3.00 kW to the external line at 120 V. Find the emf of the generator and its efficiency.

30 A certain amount of power is to be sent over each of two transmission lines to a distant point. The first line operates at 220 V, the second at 11,000 V. What must be the relative diameters of the line wires if the line loss is to be identical in the two cases? (Carefully justify the reasons for the answer.) *Ans.* $\frac{50}{1}$.

31 A certain shunt generator has a field resistance of 240 Ω and an armature resistance of 0.600 Ω. The machine will overheat if the armature loss is greater than 240 W. What is the maximum armature current? If the emf under these conditions is 150 V, what is the power output? What is the efficiency?

32 A series motor connected across a 110-V line has an armature current of 20 A and develops a counter emf of 104 V when running at normal speed. What current would there be in the armature at the instant the switch were closed if no starting box were used? *Ans.* 3.7×10^2 A.

33 A series motor has a counter emf of 115 V when the current is 10.0 A and the applied voltage is 120 V. What will be the current in this motor at the instant when the switch is closed?

34 The voltage impressed on a series motor is 115.0 V; the counter emf is 112.0 V; the current is 6.00 A. What current would there be if the motor were stopped altogether?

Ans. 2.3×10^2 A.

35 A shunt motor has a field resistance of 240 Ω and an armature resistance of 1.50 Ω. When it is connected to a 110-V line, an ammeter in the line reads 5.00 A. Find the counter emf, the power delivered, and the efficiency of the motor.

Lev Davidovich Landau, 1908–1968

Born in Baku. Graduated from Physical Department of Leningrad University. Head of Theoretical Department, Physico-Technical Institute at Kharkov, 1932–1937. In 1937 became head of the Theoretical Department of the Institute for Physical Problems of the Academy of Sciences of the U.S.S.R. in Moscow simultaneously teaching in the Kharkov and Moscow State Universities. Awarded the 1962 Nobel Prize for Physics for his pioneering theories for condensed matter, especially liquid helium.

41

Inductance; Capacitance; Transients

In our study of electricity we have considered first electric charges at rest. In this study, charges often moved, but in our experiments, we waited until the motion had stopped, then observed the static effects.

In the chapter on Electric Current we studied the effects of charges in steady motion. Again the motions of charges were not always steady, but we waited the necessary time for a steady state to be established and then observed the steady-state effects. In our study of induced emfs we found new effects that appear during the time when some change is taking place in the circuit.

We are now ready to examine the events that occur during the time, usually very short, in which the current changes from its initial zero value at the instant a switch is closed to the final steady state. We shall examine special circuits to find the manner in which these changes take place and the time required for the changes. We shall study inductive and capacitive circuits.

41-1 SELF-INDUCTANCE

We have observed in Chap. 40 that there is an emf induced in a coil whenever there is a change in the magnetic flux threading the coil. In the examples there considered the source of the flux was outside the coil in which the change occurred. But even when there is a change in the current in the coil itself, the flux threading the coil changes and an emf is induced in the coil. Such an emf, induced by changes of current in the coil itself, is called an *emf of self-induction*. In accordance with Lenz's law, the emf of self-induction is always in such a direction as to oppose the change that caused it.

Consider a toroidal coil, such as that shown in Fig. 41-1, which has N turns, an average circumference l, and in which there is a steady current I. The flux density within the toroid is given by Eq. (11), Chap. 37, as $B = \mu_0 NI/l$,

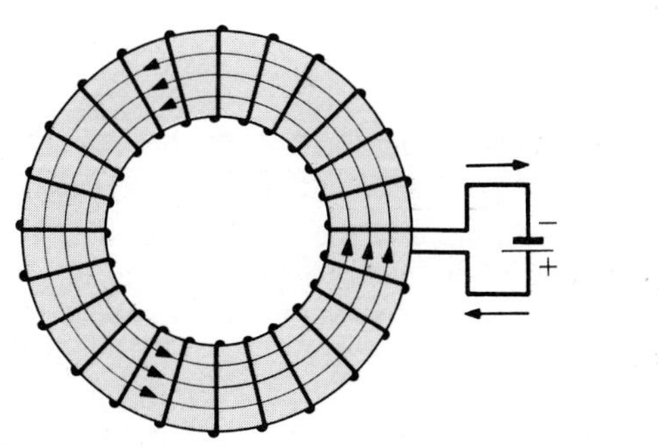

Figure 41-1
Toroidal coil with air core.

and the flux threading the coil is $\Phi = BA = (\mu_0 NI/l)A$. If the current in the coil is changed by an amount ΔI in time Δt, there will be an induced emf whose average value during the time is

$$\overline{\mathcal{E}} = -N\frac{\Delta\Phi}{\Delta t} = -N\left(\mu_0 \frac{N}{l} A\right)\frac{\Delta I}{\Delta t} \qquad (1)$$

The induced emf is proportional to the rate of change of current. We may express the instantaneous induced emf e_L in terms of the instantaneous rate of change of current, using small e and i to represent instantaneous values.

$$e_L = -L\frac{\Delta i}{\Delta t} \qquad (2)$$

or

$$L = -\frac{e_L}{\Delta i/\Delta t} \qquad (3)$$

The factor of proportionality L used here depends upon the physical characteristics of the coil and the units in which e, i, and t are expressed. This factor is called the *self-inductance* of the coil and is defined as the ratio of the emf of self-induction to the rate of change of current in the coil.

By the use of Eq. (1) self-inductance may be related to the flux. For the coil of Fig. 41-1, the flux Φ links the N turns of the coil when there is a current i in the coil. The product $N\Phi$ represents the *flux linkage*. By comparison of Eq. (1) and Eq. (2) we see that

$$L = N\left(\mu_0 \frac{N}{l} A\right) = N\left(\mu_0 \frac{N}{l} A \frac{i}{i}\right) = \frac{N\Phi}{i} \qquad (4)$$

The self-inductance of the coil is thus equal to the number of flux linkages per unit current.

The mks unit of self-inductance, called the *henry*, is the self-inductance of a coil in which an emf of one volt is induced when the current is changing at the rate of one ampere per second.

We may think of the self-inductance of a circuit as that property of the circuit which causes it to oppose any change in the current in the circuit. The role of inductance is therefore analogous to that of inertia in mechanics, since inertia is the property of matter which causes it to oppose any change in its velocity.

It should be noted that even a straight wire has inductance, since a current in the wire contributes magnetic flux, linking any circuit of which it is a part. The inductance of even the shortest and straightest wire is a considerable factor in the performance of high-frequency circuits.

Example A circuit in which there is a current of 5 A is changed so that the current falls to zero in 0.1 s. If an average emf of 200 V is induced, what is the self-inductance of the circuit?

$$\overline{\mathcal{E}} = L\frac{\Delta i}{\Delta t}$$

$$200\text{ V} = L\frac{5\text{ A}}{0.1\text{ s}}$$

$$L = 4\text{ H}$$

41-2 MUTUAL INDUCTANCE

When there is a current in one of two adjacent circuits, some of the magnetic flux associated with the first, or primary, circuit A links the nearby, or secondary, circuit B (Fig. 41-2). If the current changes in A, there results a change of flux in B. Hence an emf is induced in the secondary circuit. This phenomenon is called *mutual induction*. Mutual inductance is the property of a pair of circuits by virtue of which any change of current in one of the circuits induces an emf in the other circuit. It is evident that the value of this emf will depend upon the rate of change of the current in the primary coil and the geometrical constants of the two circuits. *Mutual inductance* M may therefore be defined as the ratio of the emf induced in the secondary to the time rate of change of current in the primary,

$$e_s = -M\frac{\Delta i_p}{\Delta t} \qquad \text{or} \qquad M = \frac{-e_s}{\Delta i_p/\Delta t} \qquad (5)$$

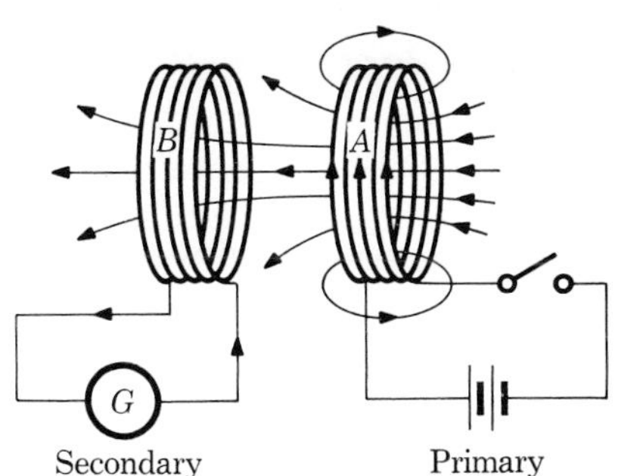

Figure 41-2
An emf is induced in the secondary coil when the current changes in the primary coil.

The negative sign is used to indicate the fact that the induced emf is in such a direction as to oppose the change of current.

The mutual inductance is expressed in henrys when e_s is in volts and $\Delta i_p/\Delta t$ is in amperes per second. The unit of mutual inductance, the henry, is the mutual inductance of that pair of circuits in which a rate of change of current of one ampere per second in the primary causes an induced emf of one volt in the secondary. Mutual inductance is of importance to the electrical designer in calculations involving transformer coils, dynamo apparatus, and other electric machinery.

Example A pair of adjacent coils has a mutual inductance of 1.5 H. If the current in the primary changes from 0 to 20 A in 0.050 s, what is the average induced emf in the secondary? If the secondary has 800 turns, what is the change of flux in it?

$$\bar{\mathcal{E}}_s = M\frac{\Delta i_p}{\Delta t} = 1.5\ \text{H}\frac{20\ \text{A}}{0.050\ \text{s}}$$

$$= 6\bar{0}0\ \text{V}$$

$$\bar{\mathcal{E}}_s = N_s\frac{\Delta\Phi}{\Delta t}$$

$$6\bar{0}0\ \text{V} = 800\frac{\Delta\Phi}{0.050\ \text{s}}$$

$$\Delta\Phi = 0.038\ \text{Wb}$$

41-3 GROWTH AND DECAY OF CURRENT IN AN INDUCTIVE CIRCUIT

Consider the circuit of Fig. 41-3 which includes a source of terminal potential difference V and negligible internal resistance, a resistor R of negligible inductance, and an inductor of inductance L and negligible resistance. When the switch S is thrown to position 1, the current is zero at the instant the switch is closed. At that instant the current starts to rise so that there is an induced emf $e_L = -L(\Delta i/\Delta t)$. The induced emf opposes the rise in current; i.e., when $\Delta i/\Delta t$ is positive, e_L is negative, a *counter* emf. We shall represent the counter emf v_L as

$$v_L = L\left(\lim_{\Delta t\to 0}\frac{\Delta i}{\Delta t}\right)$$

As the current rises, the rate of change of current is not constant, but at each instant the sum of the iR drop in potential and the instantaneous counter emf v_L must equal the potential difference V of the source.

$$V = iR + v_L \tag{6}$$

The solution of Eq. (6) for the current i is a calculus problem and is given in Appendix A-9.

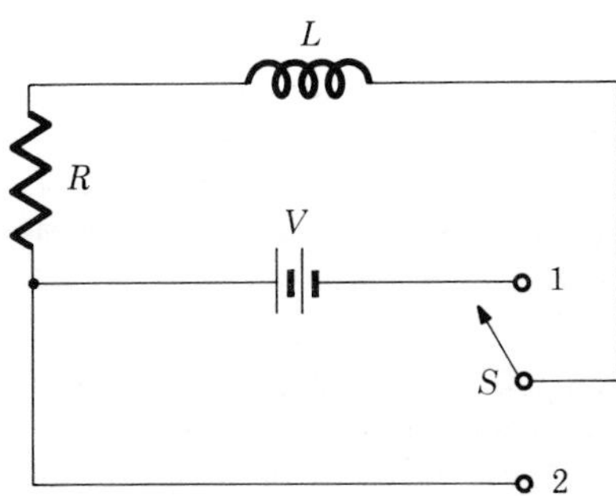

Figure 41-3
An inductive circuit. Switch S can be used (1) to apply a potential V to the coil or (2) to "short" the coil.

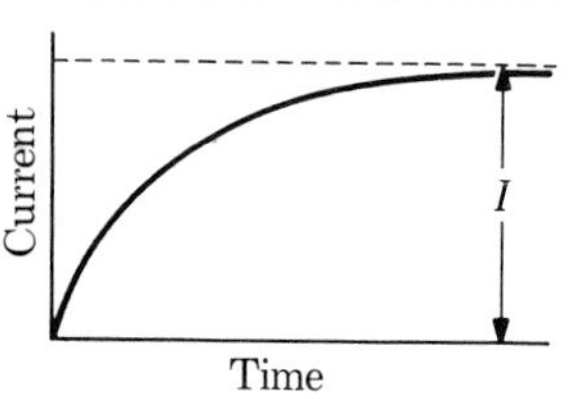

Figure 41-4
Rise of current after a steady potential is applied to an inductive circuit.

This solution is an exponential equation,

$$i = \frac{V}{R}(1 - e^{-(R/L)t}) \tag{7}$$

A curve showing the variation of Eq. (7) is given in Fig. 41-4. At the instant $t = 0$, $i = 0$. As the current increases, it approaches the final steady current V/R. Whether the growth is fast or slow depends upon the relative values of L and R. The exponent of the logarithmic base e must be a pure number. Thus L/R has the dimensions of time and is called the *time constant* of the circuit. In Eq. (7) let $t = L/R$. Then $i = (V/R)(1 - e^{-1}) = (V/R)(1 - 1/e)$. The time constant is the time required for the current to rise to $1 - 1/e$ of the final steady value. This is a fraction $1 - 1/2.72 = 0.63$ approximately. Alternatively the time constant may be described as the time required for the current to rise to the final steady value if it continued to rise at the initial rate. If the inductance of the circuit is low, the current rises very rapidly and very soon approaches close to the steady value.

Example A coil of resistance 3.0 Ω and inductance 0.25 H is connected to a battery of negligible internal resistance and terminal potential difference 60 V. What is the induced emf at the instant the current has risen to one-fourth its final steady value? At what rate is the current changing at this instant? What is the time constant for this circuit?

The steady current is given by

$$I = \frac{V}{R} = \frac{60 \text{ V}}{3.0 \ \Omega} = 20 \text{ A}$$

The instantaneous current is $20 \text{ A}/4 = 5.0 \text{ A}$,

$$V = iR + v_L$$

$$v_L = V - iR = 60 \text{ V} - (5.0 \text{ A} \times 3.0 \ \Omega)$$

$$= 45 \text{ V}$$

$$\frac{\Delta i}{\Delta t} = \frac{v_L}{L} = \frac{45 \text{ V}}{0.25 \text{ H}} = 180 \text{ A/s}$$

The time constant is

$$\frac{L}{R} = \frac{0.25 \text{ H}}{3.0 \ \Omega} = 0.083 \text{ s}$$

If, after the current has reached the steady value, the switch S of Fig. 41-3 is thrown quickly from position 1 to position 2, the current will die down exponentially. In the new circuit $V = 0$, and during the decay

$$0 = iR + v_L \tag{8}$$

In the decay the current decreases, and the induced emf is in such a direction as to resist this change. The induced emf is thus a forward emf. The current direction remains the same as before, but v_L reverses. The variation of iR and v_L during the rise and decay is shown in Fig. 41-5. Note that during the rise the ordinates of the two curves add at all times to V; during the decay they add to zero.

41-4 ENERGY IN INDUCTIVE CIRCUITS

During the time when the current is being established in an inductive circuit, energy must be expended against the counter emf of self-induction in creating a magnetic field. At each

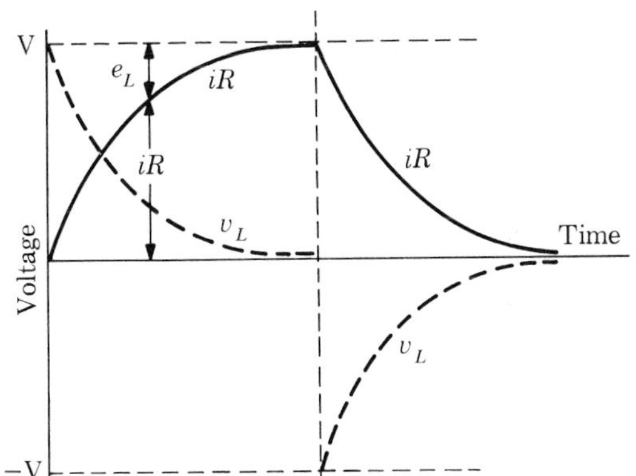

Figure 41-5
Voltage variations in an inductive circuit as the current rises and decays.

instant during this time the power expended in this purpose is

$$p = v_L i$$
$$\Delta \mathcal{W} = p\, \Delta t = v_L i\, \Delta t$$

Since $v_L = L(\Delta i/\Delta t)$,

$$\Delta \mathcal{W} = L \frac{\Delta i}{\Delta t} i\, \Delta t = Li\, \Delta i$$
$$\mathcal{W} = \Sigma Li\, \Delta i$$

In the Appendix this summation is carried out to show that

$$\mathcal{W} = \tfrac{1}{2}LI^2 \tag{9}$$

where I is the final current.

The energy stored in the magnetic field is recoverable when the magnetic field decreases. During the rise in current illustrated in Fig. 41-5, the energy of the magnetic field is built up to a final value given by Eq. (9). This energy is used in maintaining the current during the decay. In Eq. (9) the energy is in joules if the inductance is in henrys and the current is in amperes.

41-5 GROWTH AND DECAY OF CHARGE IN A CAPACITOR

When a steady voltage is introduced into a circuit containing only capacitance and resistance (Fig. 41-6), the charge flows into the capacitor rapidly at first. As the potential difference q/C between the plates of the capacitor rises, the rate of growth of charge decreases until the transfer of electricity stops entirely when the voltage of the capacitor becomes equal to the potential V of the charging source. At every instant during the rise

$$V = iR + \frac{q}{C} \tag{10}$$

where i is the instantaneous current and q is the instantaneous charge. The form of Eq. (10) is the same as that of Eq. (6), and the resulting equation for the growth of charge is similar to that for the growth of current in the inductive circuit. The charge at any time t is given by

$$q = CV(1 - e^{-t/RC}) \tag{11}$$

The rate of flow of charge, the current, is high initially and decreases exponentially,

$$i = \frac{V}{R} e^{-t/RC} \tag{12}$$

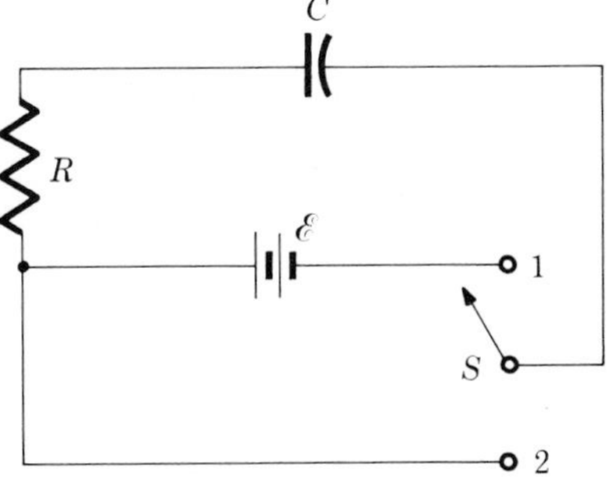

Figure 41-6
A capacitive circuit.

When the switch (Fig. 41-6) is thrown quickly to position 2, V becomes zero and the capacitor discharges exponentially. During the discharge the current reverses direction, and hence iR is negative, as shown in Fig. 41-7.

In the charging of the capacitor energy was expended and stored in amount $\frac{1}{2}CV^2$ [Eq. (7), Chap. 34]. In the discharge this stored energy is dissipated as heat. If the resistance is sufficiently high so that all the energy has been dissipated by the time the capacitor is completely discharged, the simple exponential curve represents the manner of decay. If the resistance is so low that the energy is not all dissipated in this time, the remaining energy will be stored in the magnetic field of the connecting wires and will maintain the current of self-inductance. This energy is then transferred to the capacitor, and it becomes charged oppositely from the original charge. The discharge becomes oscillatory. The charge and discharge continue until all the energy has been dissipated by the resistance.

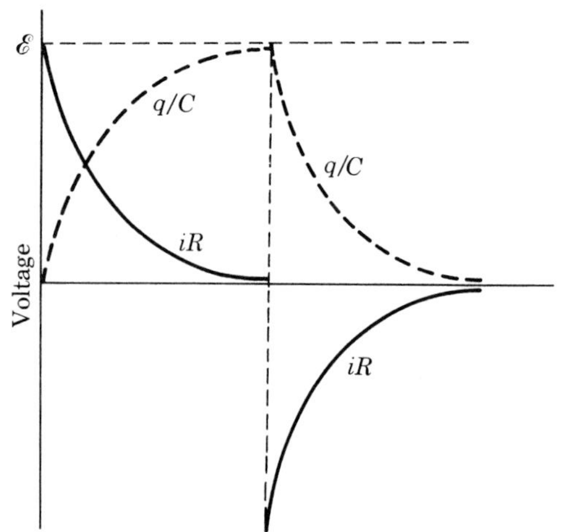

Figure 41-7
Potentials in a capacitive circuit as the current rises and decays.

SUMMARY

Self-inductance is that property of an electric circuit which causes the circuit to oppose any change of current in the circuit. Self-inductance is defined by the equation

$$L = \frac{-e_L}{\Delta i/\Delta t}$$

Mutual inductance is the property of a pair of circuits whereby a change of current in one circuit causes an induced emf in the other circuit. Mutual inductance is defined by the equation

$$M = \frac{-e_s}{\Delta i_p/\Delta t}$$

The unit of inductance is the *henry*, the inductance such that there will be an induced emf of one volt when the current is changing at the rate of one ampere per second.

In an inductive circuit the current rises exponentially from zero to a final steady current. The instantaneous current i is given by

$$i = \frac{V}{R}(1 - e^{-(R/L)t})$$

The rate at which the current rises depends upon the relative values of L and R. The ratio L/R is called the *time constant* of the inductive circuit.

Energy is stored in the magnetic field of an inductor. This energy is returned to the circuit as the magnetic field decreases.

$$\mathcal{W} = \tfrac{1}{2}LI^2$$

The charge of a capacitor rises exponentially,

$$q = CV(1 - e^{-t/RC})$$

During the charging of the capacitor the current changes exponentially,

$$i = \frac{V}{R}e^{-t/RC}$$

Questions

1 Discuss the meaning of inductance as applied to a straight conductor.
2 The self-inductance of an air-core coil is independent of the value of the current in the coil. This is not true for an iron-core coil. Explain.
3 What will be the effect on the self- and mutual-inductance of solenoids when iron cores are inserted? State reasons. Mention some of the factors upon which the mutual inductance of a pair of adjacent coils depends, indicating the manner in which the mutual inductance varies with such factors.
4 Discuss the effect of inductance upon the transmission of a signal through a conductor. Might a circuit be designed so that a signal introduced into the circuit would reach selected points at predetermined times?
5 A person can close an electric circuit through his body to a 30-V battery and a coil of high inductance without receiving a noticeable shock. A severe shock may be noticed when he releases one of the wires, thus opening the circuit. Explain why this is true.
6 Show that the henry is one weber per ampere.
7 Since there is no electrical connection between the two coils of a mutual inductor, how is energy transferred from one coil to the other?
8 Explain why in the design of capacitors the binding posts and surface insulators are made of materials which have exceedingly high resistivities. Is this equally true of resistance boxes?
9 A fairly good analogy to a capacitor is an automobile tire into which air is pumped. Show how the phenomena in these two cases are analogous.
10 An air capacitor having parallel plates is charged to a certain potential difference. The capacitor is then immersed in transformer oil. What happens to the capacitance of the system? to the charge? to the potential difference?
11 A gold-leaf electroscope is fitted with a pair of plates instead of the usual knob. The plates are separated by a thin insulator. A 90-V battery is connected to the plates. What happens to the gold leaf? With the battery still connected, the plates are separated to a considerable distance. Explain what happens to the gold leaf. Explain what would happen if the battery were disconnected before the plates are separated.
12 In automobile ignition circuits a capacitor is placed in parallel with the breaker points. Show how this greatly reduces the sparking at these points. What is the cause of the sparking?
13 Show that the electric field intensity between the plates of a parallel-plate capacitor is given by $E = V/s$, where V is the potential difference between the plates and s is the thickness of the insulator.
14 The dielectric constant of distilled water is very high, about 80. Why is such water never used for the dielectric in capacitors? Why is mica used so widely when porcelain and glass are much better insulators?
15 Two unlike capacitors of different potential charges are placed in parallel. What happens to their potential differences? How are their charges redistributed?
16 Plot rough graphs to show the growth and decay of current in the charge and discharge of capacitors.
17 The dielectric properties of many insulators used in capacitors are greatly affected by temperature. Explain why capacitors designed for dc or low-frequency circuits cannot be used in high-frequency circuits. Would this be true of air capacitors?
18 What type of energy is stored up in a capacitor? In a storage cell? In an inductive circuit? Why are capacitors not used in place of secondary batteries?

Problems

1 What is the self-inductance of a circuit in which there is induced an emf of 100 V when the current in the circuit changes uniformly from 1.0 to 5.0 A in 0.30 s?
2 An inductive coil has a resistance of 9.50 Ω. A battery of 115 V is suddenly connected to the coil. After the current has risen to 10.0 A, it is

changing at the rate of 160 A/s. What is the self-inductance of the coil? *Ans.* 125 mH.

3 An air-core solenoid is 50 cm long and has 120 turns per centimeter and a diameter of 12.0 cm. What is the self-inductance of the solenoid?

4 The solenoid of Prob. 3 has a resistance of 500 Ω. What is the initial rate of change of current when the solenoid is connected to a source of 80 V? *Ans.* 79 A/s.

5 The current in a coil of 325 turns is changed from zero to 6.32 A, thereby producing a flux of 8.46×10^{-4} Wb. What is the self-inductance of this coil?

6 An impressed emf of 50 V at the instant of closing the circuit causes the current in a coil A to increase at the rate of 20 A/s. Find the self-inductance of the coil. At the same instant the changing flux from A through a nearby coil B induces an emf of 100 V in B. What is the mutual inductance of the coils? *Ans.* 2.5 H; 5.0 H.

7 A 60-V potential difference is suddenly applied to a coil of inductance 60 mH and resistance 180 Ω. What is the current after 5.0×10^{-4} s and at what rate is it rising?

8 Two coils A and B are placed near each other. A current change of 6.00 A in A produces a flux change of 12×10^{-4} Wb in B, which has 2,000 turns. What is the mutual inductance of the system? *Ans.* 400 mH.

9 A steady potential difference of 36.5 V is applied to a coil of 1,000 turns, 12.8 Ω resistance, and self-inductance 460 mH. (*a*) At what rate is the current changing at the instant the current has reached a value of 2.00 A? (*b*) At what rate is the flux changing at this instant? (*c*) How much energy is stored in the circuit after a time of 10.0 min?

10 A steady potential difference of 120 V is applied to a coil of 30 Ω resistance and 600 mH inductance. (*a*) What is the final steady current? (*b*) What is the time constant of the circuit? (*c*) At what rate is the current rising at the instant the current is 3.0 A? (*d*) At the instant the current is 3.0 A, at what rate is the source supplying energy? (*e*) At what rate is energy being converted into heat? (*f*) At what rate is energy being stored in the magnetic field?

Ans. 4.0 A; 0.020 s; 50 A/s; 90 W; 270 W; 4.8 W.

11 A coil of wire has a self-inductance of 0.0020 H in air, and carries a current of 0.10 A. What is the change in the magnetic energy associated with the coil when it is immersed in liquid oxygen and still carries a current of 0.10 A? For air, $u_r - 1 = 4.0 \times 10^{-3}$.

12 Two circuits have a mutual inductance of 2.25 H. When the current in the primary changes from 0 to 18 A in 7.5 ms, what emf is induced in the secondary? What energy is stored as a result of mutual induction? *Ans.* 5.4 kV; 365 J.

13 A toroid 350 cm long is wound on an air core of cross-sectional area 21.4 cm². The coil of 1,800 turns carries a current of 4.56 A. (*a*) What is the magnetic flux in the coil? (*b*) What is the self-inductance of the system? (*c*) If the current decreases at a rate of 2.25 A/s, what is the emf induced?

14 In the circuit of Fig. 41-3, $V = 120$ V, $R = 30$ Ω, and $L = 600$ mH. After the current has reached a final steady value with S at position 1, the switch is quickly thrown to position 2. (*a*) What is the current 0.002 s after the switch is changed? (*b*) At what rate is the current changing at the time of *a*? *Ans.* 3.62 A; 181 A/s.

15 In the circuit of Fig. 41-3, $V = 80.0$ V, $R = 20.0$ Ω, and $L = 50.0$ mH. With the switch thrown to 1, the current reaches a steady value. The switch is then thrown quickly to position 2. (*a*) After what time will the current be $1/e$ of the steady value? (*b*) How fast is the current changing at that instant?

16 A coil of 20.0 Ω resistance is connected to a 100-V dc line. At the instant when the current has grown to 3.00 A, it is increasing at the rate of 80.0 A·s. What is the inductance of the coil? What energy is stored in it when the induced emf becomes zero? *Ans.* 500 mH; 6.25 J.

17 In the circuit of Fig. 41-6, $V = 120$ V, $R = 600$ Ω, and $C = 12.0$ μF. (*a*) What is the initial current when the switch is closed to position 1? (*b*) What is the current at $t = 7.2 \times 10^{-3}$ s?

18 A toroid, Fig. 41-8, made of a paramagnetic

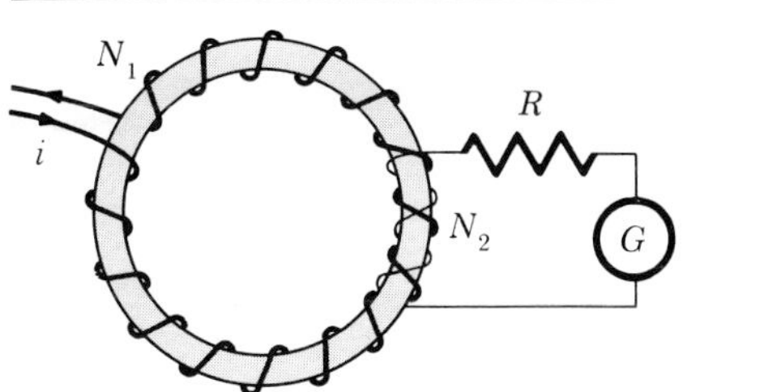

Figure 41-8

substance, is wound uniformly with N_1 turns of wire carrying a current i. The mean circumferential length of the ring is l; its cross section is A. When the magnetizing current is cut off, a charge q is sent through the galvanometer connected to a secondary of N_2 turns. The resistance of the galvanometer is R. In terms of i, N_1, N_2, q, l, R, and A, calculate (*a*) the relative permeability of the substance and (*b*) the self-inductance of the toroid. *Ans.* $\mu = qRl/N_1N_2Ai_1$; $L = N_1{}^2A\mu/l$.

19 A storage battery that maintains a terminal potential difference of 1.250 V is connected to a 21.4-μF capacitor through a 268-Ω rheostat. (*a*) Calculate the initial current in the circuit. (*b*) What is the current after the capacitor is charged to three-fourths its maximum value? (*c*) How much energy is stored in the capacitor after 1 h?

20 A 2.0-μF capacitor is connected in series with a resistor of 5,$\bar{0}$00 Ω across a 1,$\bar{5}$00-V line. What is the current at the instant the switch is closed? If the current remained at this value how long would it take to charge the capacitor fully? *Ans.* 0.30 A; 1.0×10^{-2} s.

21 A 4.0-μF capacitor is connected in series with a 2,$\bar{5}$00-Ω resistor across a 500-V line. What is the current in the resistor at the instant (*a*) the switch is closed? (*b*) the capacitor is half charged? (*c*) the capacitor is fully charged?

22 A 25-μF capacitor is connected in series with a $\frac{1}{2}$-MΩ resistor and a 120-V storage battery. What are the charge and potential difference in the capacitor 6.0 s after the circuit is closed? What is the current at this instant? *Ans.* 1.1×10^{-3} C; 46 V; 0.15 mA.

23 Two capacitors of 3.0 and 5.0 μF, respectively, are connected in series and a 110-V difference in potential is applied to the combination. What is the potential difference across the 3.0-μF capacitor? What is the energy in the 5.0-μF capacitor?

Eugene Wigner, 1902–

Born in Budapest. Studied at the Technical University in Berlin. Taught at Göttingen, Princeton, and Wisconsin. Prominent in obtaining government support for studies of nuclear reactions. In 1938 became Professor of Mathematical Physics at Princeton University. Shared the 1963 Nobel Prize for Physics for his many important contributions in the field of atomic physics.

Maria Goeppert Mayer, 1906–

Born in Kattowitz, Poland. American citizen since 1933. Associated with The Johns Hopkins University, Columbia University, Sarah Lawrence College, and the Argonne National Laboratory. Professor of Physics at the University of Chicago; since 1960 at the University of California at San Diego. Shared the 1963 Nobel prize for Physics for adding to man's knowledge of the structure of the atomic nucleus.

J. Hans D. Jensen, 1905–

Born in Germany. Received his Ph.D. from the University of Hamburg and served on faculty there and at the Institute of Technology in Hanover before becoming Professor at the University of Heidelberg in 1949. Shared the 1963 Nobel Prize for Physics for adding to man's knowledge of the structure of the atomic nucleus.

42

Alternating-current Series Circuits

In the preceding two chapters we have seen that an *alternating* emf is induced in a coil rotating in a magnetic field and that, by means of a transformer with no moving parts, an alternating voltage can be changed conveniently and efficiently to either higher or lower amplitudes. Chiefly for these reasons, about 99 percent of the electrical energy that is generated in the United States is distributed in ac circuits. Where direct current is required, a vacuum tube or solid-state rectifier may be used or an ac motor may be used to run a dc generator.

For electrical power networks in the United States the usual frequency is 60 cycles/s. In aircraft and guided rockets 400 cycles/s is frequently used. In telephone circuits ac frequencies in the audio range, chiefly 100 to 2,000 cycles/s, are transmitted directly or are superposed on carrier waves in the radio frequency range (0.1 to 100 megacycle/s) for long-distance transmission.

In the many electrical-communications devices and power circuits that use alternating current, inductances and capacitances of the circuit elements are just as significant as the resistances. In Chap. 41, we examined the transient effects associated with inductance and capacitance. In this chapter, we shall examine the steady-state conditions in simple ac circuits whose elements are all in series. We shall be concerned especially with the relationships that determine the current, the phase relations of current and potential difference, and the power.

42-1 NOMENCLATURE IN AC CIRCUITS

In Sec. 41-6, the form of voltage curve generated in an ideal alternator was described. The instantaneous values of the emf e are related to the maximum values e_m as shown by curve a of Fig. 42-1 and given by the equation

$$e = e_m \sin \theta \qquad (1)$$

where θ is the angle between the given position of the coil and its position when the generated emf is zero. If this sinusoidal emf is applied to a resistor, the current at each instant is

$$i = \frac{e}{R} = \frac{e_m}{R} \sin \theta = i_m \sin \theta \qquad (2)$$

The current then varies in exactly the same manner as the voltage. For a resistor the voltage is in phase with the current. This relationship is shown in Fig. 42-1.

Example The voltage in an ac circuit varies

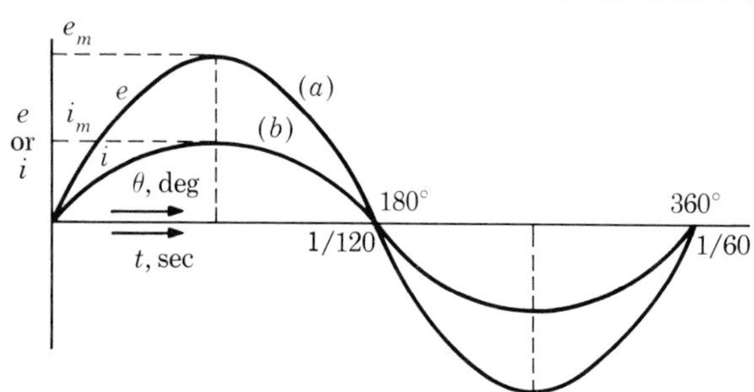

Figure 42-1
Instantaneous values of voltages and currents in an ac circuit.

harmonically with time with a maximum of 170 V. What is the instantaneous voltage when it has reached 45° in its cycle?

$$e = e_m \sin \theta = 170 \text{ V} \times 0.71 = 120 \text{ V}$$

In the common 60-cycle ac circuit, there are 60 complete cycles each second; i.e., the time interval of 1 *cycle* is $\frac{1}{60}$ s. It should be noted that this corresponds to a reversal of the *direction* of the current every $\frac{1}{120}$ s (since the direction reverses twice during each cycle).

Radio broadcast frequencies are of the order of 1 million cycles/s. A few modern ultrahigh-frequency devices use frequencies of the order of 3×10^{10} cycles/s and higher.

42-2 EFFECTIVE VALUES OF CURRENT AND VOLTAGE

Suppose that in a rheostat of resistance R there is an alternating current whose maximum value is 1.0 A. Certainly the rate at which heat is developed in the resistor is not so great as if a steady direct current of 1.0 A were maintained in it.

Remembering that the rate at which heat is developed by a current is proportional to the square of its value ($P = I^2R$), one can see that the average rate of production of heat by a varying current is proportional to the *average value* of the *square* of the current. The square root of this quantity is called the *effective,* or root-mean-square (rms), current. The effective value of a current is equal to the magnitude of a steady direct current that would produce the same heating effect. The value ordinarily given for an alternating current is its effective, or rms, value.

A curve of current squared for one cycle of a sinusoidal current is shown in Fig. 42-2. The value of the average of the square of the current can be obtained by calculating the area of one lobe of this curve and dividing this area by the width of the lobe. In order to compute the lobe area, we sum up all the infinitesimal areas $i^2\,\Delta\theta$, such as that shown in the shaded rectangle. It is shown in the Appendix that from this summation we can obtain

$$\Sigma i^2\,\Delta\theta = \Sigma i_m{}^2 \sin^2\theta\,\Delta\theta = \frac{\pi}{2}\,i_m{}^2$$

The average ordinate is obtained by dividing this lobe area by π, the width of the lobe; thus the average ordinate is $\frac{1}{2}i_m{}^2$. Hence the square of the steady current I which would produce the

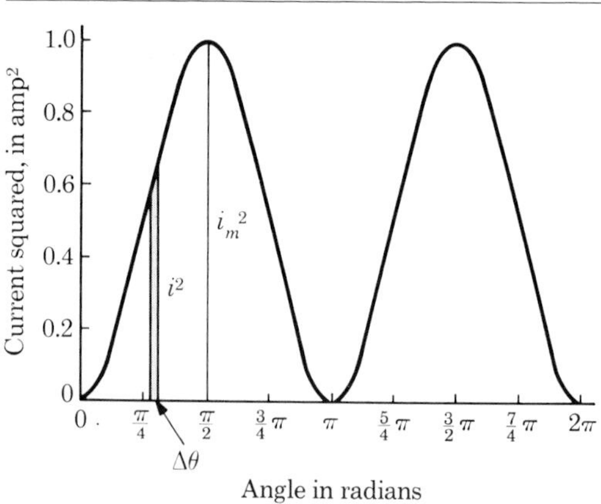

Figure 42-2
Effective value of alternating current in the square root of the mean-square current.

same heating effect as the alternating current is given by

$$I^2 = \tfrac{1}{2} i_m{}^2$$

The square root of this value is the root-mean-square, or effective, current, namely,

$$I = \frac{i_m}{\sqrt{2}} = 0.707 i_m \qquad (3)$$

Similarly, the effective value $\mathcal{E}$ of an alternating emf is defined as its rms value. If the emf varies sinusoidally,

$$\mathcal{E} = 0.707 e_m \qquad (4)$$

Example What is the maximum value of a 6.0-A alternating current?

$$I = 0.707 i_m = 6.0 \text{ A}$$

so that $$i_m = \frac{6.0}{0.707} \text{ A} = 8.5 \text{ A}$$

It is standard practice to use capital letters, such as I and V, for effective values in ac circuits. The lowercase letters, such as i and v, are used to indicate instantaneous values. Unless otherwise clearly stated, effective values are understood when ac quantities are mentioned. For example, a "voltage of 25 V" refers to an effective value of 25 V. These are the values indicated by ordinary meters.

42-3 PHASE RELATIONS OF CURRENT AND VOLTAGE IN AC CIRCUITS

In an ac circuit containing only pure resistance (i.e., no inductance or capacitance) the instantaneous values of current and voltage are always *in phase*. This means that they both are zero at the same instant and both pass through their maximum values at the same instant and always have similar time relationships. This is illustrated in Fig. 42-1 and is as demanded by Eqs. (1) and (2).

It is rare for a circuit to include only pure resistance. Most circuits also contain inductance and frequently capacitance. In such circuits there is a time interval, or *phase difference*, between the maximum value of the applied voltage and the maximum value of the current. This phase difference is usually given in terms of an angle ϕ and is expressed in degrees.

Consider a circuit which contains only pure inductance, that is, no resistance or capacitance. The induced emf e_L in such a circuit is given by

$$e_L = -L \frac{\Delta i}{\Delta t} \qquad (5)$$

where L is the inductance and $\Delta i / \Delta t$ is the time rate of change of current. The negative sign is used to show conventionally that the induced emf is in such a direction as to oppose the change of current. Therefore, to maintain the flow, there must be an impressed voltage

$$v_L = -e_L$$

and hence $$v_L = L \frac{\Delta i}{\Delta t}$$

It is shown in the Appendix that this results in the equation

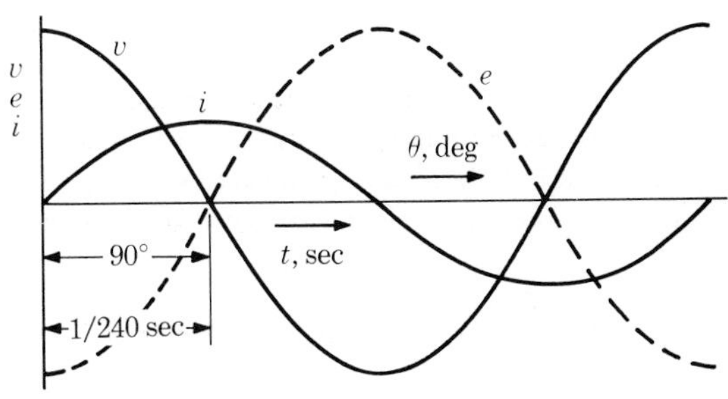

Figure 42-3
In a circuit having pure inductance the current lags the voltage by 90°. Curve *v*, applied voltage; curve *i*, current; curve *e*, induced emf.

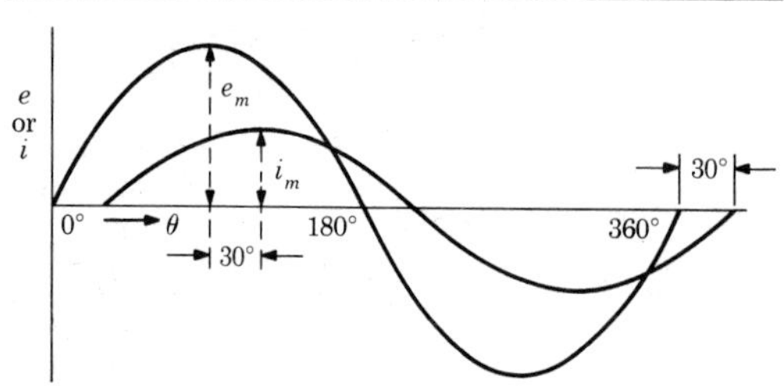

Figure 42-4
Curves showing the current lagging the voltage by 30° in a circuit having both resistance and inductance.

$$v_L = 2\pi f L i_m \cos\theta \qquad (6)$$

From Eq. (6) we see that the inductive voltage is represented by a cosine curve when the current is represented by a sine curve (Fig. 42-3). Therefore the inductive voltage v_L is 90° ahead of the current in phase.

It is impossible to have an inductor with only pure inductance, since the windings must contain some resistance. In such cases the angle of lag is less than 90°. Hence the phase angle for circuits containing resistance and inductance may vary from 0° for a circuit with resistance only to 90° for a circuit with inductance only. An example of an angle of lag of 30° is shown in Fig. 42-4.

When an alternating voltage is impressed upon a capacitor, there are surges of charge into and out of the plates of the capacitor. This results in an alternating current in the circuit outside the capacitor, even though the plates of the capacitor are separated by a perfect insulator. From the defining equation for capacitance it follows that

$$v_C = \frac{q}{C} = \frac{\Sigma i\,\Delta t}{C} = \frac{1}{C}\Sigma i_m \sin 2\pi f t\,\Delta t \qquad (7)$$

In the Appendix it is shown that this summation produces the equation

$$v_C = -\frac{i_m}{2\pi f C}\cos 2\pi f t = -\frac{i_m}{2\pi f C}\cos\theta \qquad (8)$$

The cosine factor of Eq. (8) indicates that the capacitive voltage is 90° out of phase with the current, and the negative sign shows that it is behind the current in phase. In circuits containing both capacitance and resistance the phase angle may vary from −90° for a pure capacitance to 0° for a circuit containing resistance only.

The most general series circuit is one containing resistance, inductance, and capacitance. These may be present in all proportions so that the current may either lag or lead the voltage by 90° or less. This phase relation is expressed by the equations

$$e = e_m \sin\theta \qquad i = i_m \sin(\theta - \phi) \qquad (9)$$

The phase angle ϕ is considered positive when the current lags the voltage and negative when the current leads the voltage; hence the sign of ϕ must always be introduced properly in Eq. (9).

Example In an ac circuit with capacitance predominating, the current leads the voltage by 30°. The effective value of the current is 100 A. What is the instantaneous current when the voltage passes through its zero value?

$$i_m = \frac{I}{0.707} = \frac{100\text{ A}}{0.707} = 141\text{ A}$$

$$i = i_m \sin(\theta - \phi)$$

$$= 141\text{ A} \times \sin[0° - (-30°)] = 70.5\text{ A}$$

42-4 INDUCTIVE REACTANCE

The *inductive reactance* X_L of an inductor is defined as the ratio of the effective value of the inductive voltage V_L to the effective value of the current I,

$$X_L = \frac{V_L}{I} \qquad (10)$$

It was shown in Eq. (6) that

$$v_L = 2\pi f L i_m \cos\theta$$

Hence the effective value of V_L is given by

$$V_L = \frac{(v_L)_m}{\sqrt{2}} = \frac{2\pi f L i_m}{\sqrt{2}} = 2\pi f L I$$

since $I = \dfrac{i_m}{\sqrt{2}}$

Therefore $X_L = \dfrac{V_L}{I} = \dfrac{2\pi f L I}{I} = 2\pi f L \qquad (11)$

This inductive reactance $X_L = 2\pi f L$ is expressed in ohms when f is in cycles per second and L is in henrys.

The portion of the applied voltage which must be used because of this induced emf is called the *reactance drop in potential.* It is equal to the product of the current and the inductive reactance and may be written as

$$V_L = IX_L$$

In our discussion of ac circuits we shall consider series circuits only. In such circuits the current is the same in all parts. Therefore we refer phases to the common current phase. Thus the various voltages are ahead or behind the current in phase.

The conventional wiring diagram for a circuit containing resistance and inductance is shown in Fig. 42-5. The resistance portion is shown diagrammatically as physically separated from the inductor; in practice this is rarely the case.

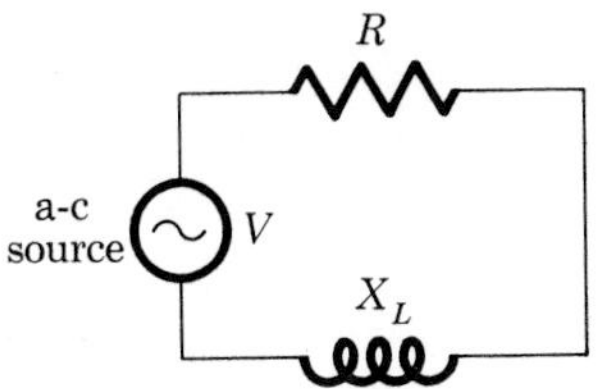

Figure 42-5
Circuit containing resistance and inductance.

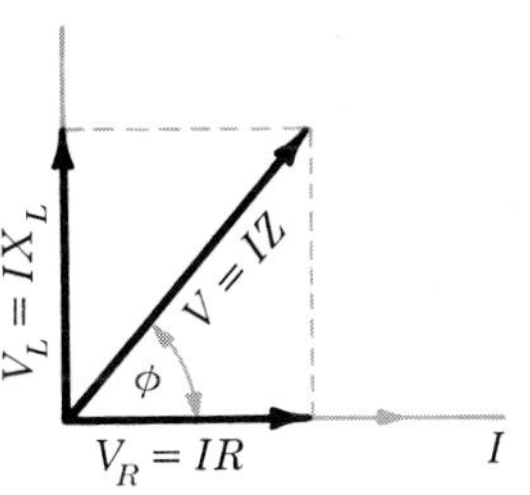

Figure 42-6
Reactive and resistance components of the applied voltage in an inductive circuit.

The voltage and phase relations in a circuit containing a pure resistor and a pure inductor in series are shown in Fig. 42-6. Such a figure is known as a *phase diagram.* The resistance drop (IR) is in phase with the current and is laid off along the positive direction of the x axis. The reactance drop (IX_L) leads the current by 90° and is laid off along the y axis. The impressed voltage V is the vector sum of IR and IX_L. The angle ϕ is the phase angle.

42-5 IMPEDANCE

The joint effect of resistance and reactance in an ac circuit is known as impedance; it is designated by the symbol Z. *Impedance* is defined as the ratio of the effective voltage to the effective current. The defining equation is

$$Z = \frac{V}{I} \qquad (12)$$

Impedance is measured in ohms, since V is in volts and I in amperes.

For a circuit containing only resistance and inductance it is clear from Fig. 42-6 that

$$Z^2 = R^2 + X_L^2 \qquad (13)$$

The current in an ac circuit containing resistance and inductance is given by

$$I = \frac{V}{Z} = \frac{V}{\sqrt{R^2 + X_L{}^2}} \tag{14}$$

An application of this equation offers a very direct method for measuring X_L and hence L by Eq. (11). The current in an inductive circuit is measured when a known voltage is impressed. The resistance R is then measured and, if the frequency is known, every factor in Eqs. (11) and (14) is known except L.

Example In an inductive circuit a coil having a resistance of 10 Ω and an inductance of 2.0 H is connected to a 120-V 60-cycle ac source. What is the current?

$$Z^2 = R^2 + X_L{}^2 = 10^2 + (2\pi \times 60 \times 2.0)^2$$

$$Z = 760\ \Omega$$

$$I = \frac{V}{Z} = \frac{120\ \text{V}}{760\ \Omega} = 0.16\ \text{A}$$

42-6 CAPACITIVE REACTANCE

The *capacitive reactance* X_C of a capacitor is defined as the ratio of the effective value of the capacitive drop in potential V_C to the effective value of the current I,

$$X_C = \frac{V_C}{I} \tag{15}$$

From Eq. (8) the instantaneous value of the capacitive voltage is $v_C = (-i_m/2\pi fC)\cos\theta$, and the effective value is $V_C = I/2\pi fC$. Therefore

$$X_C = \frac{V_C}{I} = \frac{I/2\pi fC}{I} = \frac{1}{2\pi fC} \tag{16}$$

This capacitive reactance $X_C = 1/2\pi fC$ is in ohms when f is in cycles per second and C is in farads.

The conventional wiring diagram for a circuit containing resistance and capacitance in series is shown in Fig. 42-7. The corresponding phase diagram is shown in Fig. 42-8. It is apparent from Fig. 42-8 that the impedance of a circuit containing only resistance and capacitive reactance is given by the equation

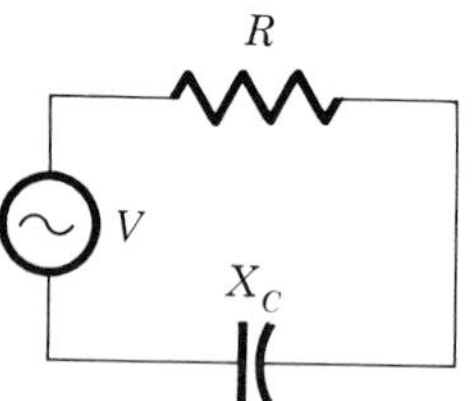

Figure 42-7
Circuit containing resistor and capacitor in series.

$$Z^2 = R^2 + X_C{}^2 \tag{17}$$

The current in a circuit containing resistance and capacitive reactance is expressed by the equation

$$I = \frac{V}{Z} = \frac{V}{\sqrt{R^2 + X_C{}^2}} \tag{18}$$

Example In the circuit of Fig. 42-7, the values are as follows: $C = 30\ \mu\text{F}$, $V = 120$ V, and

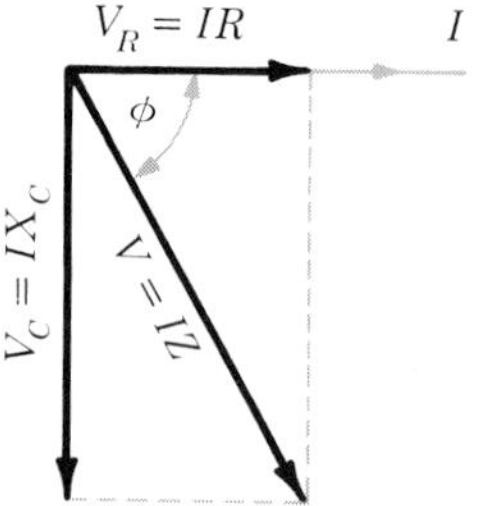

Figure 42-8
Phase diagram for circuit containing resistor and capacitor in series.

$R = 25\ \Omega$. What is the current? What is the phase angle?

$$X_C = \frac{1}{2\pi fC} = \frac{1}{2\pi \times 60 \times 30 \times 10^{-6}} = 88\ \Omega$$

$$Z = \sqrt{R^2 + X_C{}^2} = \sqrt{25^2 + 88^2} = 91\ \Omega$$

$$I = \frac{V}{Z} = \frac{120\ \text{V}}{91\ \Omega} = 1.3\ \text{A}$$

$$\phi = \arctan\frac{-X_C}{R} = \arctan\frac{-88\ \Omega}{25\ \Omega}$$

$$= -74.2°$$

42-7 SERIES CIRCUITS CONTAINING RESISTANCE, INDUCTANCE, AND CAPACITANCE

We shall first consider an ideal case of a series circuit (Fig. 42-9) containing pure resistance (without inductive or capacitive effects), pure inductance (without resistance or capacitive effects), and pure capacitance (without resistance or inductive effects). The voltages in this circuit are shown in the "clock" (phase) diagram of Fig. 42-10. In Fig. 42-10*a*, the voltage V_R across the resistor is laid off in phase with the current. The voltage V_L across the inductive coil is 90° ahead of the current. The voltage V_C across the capacitor lags the current by 90°. Hence the net reactive voltage is $V_L - V_C$ and is designated V_X in Fig. 42-10*b*. From

$$V_X = IX = IX_L - IX_C = I(X_L - X_C)$$

it follows that the net reactance X of the series circuit is $X = X_L - X_C$. The impedance Z is given by

$$Z = \sqrt{R^2 + X^2} = \sqrt{R^2 + (X_L - X_C)^2}$$

$$= \sqrt{R^2 + \left(2\pi fL - \frac{1}{2\pi fC}\right)^2} \qquad (19)$$

Hence

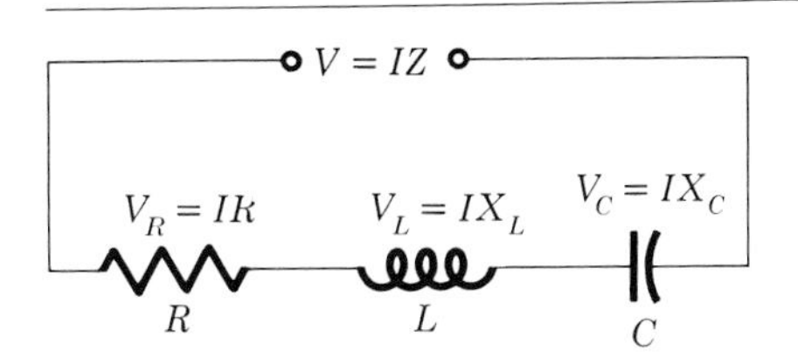

Figure 42-9
A series circuit containing pure resistance, pure inductance, and pure capacitance.

$$I = \frac{V}{Z} = \frac{V}{\sqrt{R^2 + \left(2\pi fL - \frac{1}{2\pi fC}\right)^2}} \qquad (20)$$

The ideal case represented by Fig. 42-9 never exists in practice. The resistor usually contains more or less inductance; the inductor necessarily includes some resistance; the capacitor may have sufficient losses to offer appreciable equivalent resistance. However, the phase diagram of the voltages in such a circuit may be reduced to one like that in Fig. 42-10, where in this case V_R represents the voltage across all the resistance in the circuit, V_L the total inductive voltage of the circuit, and V_C the total capacitive voltage. As

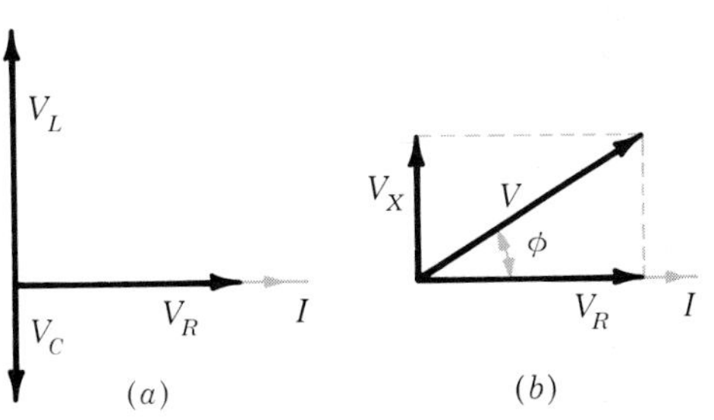

Figure 42-10
Phase relations of voltages in a series circuit having resistance, inductance, and capacitance. In this circuit the inductive reactance is shown predominant over the capacitive reactance.

before, V_X represents the net reactive voltage and V the voltage of the entire circuit.

42-8 THE CHOKE COIL

A coil made up of a large number of turns of heavy copper wire has little resistance in a dc circuit, but, because of its high inductance, it offers a large impedance in an ac circuit. By fitting the coil with a variable iron core, the impedance may easily be varied within wide limits. Such a device is useful for many purposes instead of a rheostat, such as for dimming lights, and possesses the important advantage over a rheostat in that the I^2R heating loss is much less. In other words, the coil acts as a "choke" without requiring a large expenditure of energy in the form of heat, as is the case of an ordinary resistor.

42-9 POWER IN AC CIRCUITS

In dc circuits the power is given by $P = VI$. In such circuits both the current and voltage are assumed to be steady and in phase. In ac circuits this is not the case. During half of the cycle, energy is supplied to the reactive component of the circuit, but this energy is returned to the source during the other half of the cycle. Hence no power is required to maintain the current in the part of the circuit which is purely reactive. All the power is used in the resistance portion of the circuit. From Fig. 42-10,

$$P = IV_R = I(IR) = I^2R$$

Since $V_R = V\cos\phi$, the power equation ($P = IV_R$) can be written

$$P = VI\cos\phi \qquad (21)$$

where P is the average power in watts when V is the effective value of the voltage and I is the effective value of the current. The angle ϕ is the angle of lag of the current behind the voltage. The quantity $\cos\phi$ is called the *power factor* of the circuit. Note that the power factor can vary anywhere from zero for a purely reactive circuit to unity for a pure resistance.

From Eq. (21) and Fig. 42-10*b* it may be seen that

$$\cos\phi = \frac{IR}{IZ} = \frac{R}{Z} \qquad (22)$$

Example (*a*) Find the current in a circuit consisting of a coil and capacitor in series, if the applied voltage is 110 V, 60 cycles/s; the inductance of the coil is 0.80 H; the resistance of the coil is 50.0 Ω; and the capacitance of the capacitor is 8.0 μF. (*b*) Find the power used in the circuit.

$$X_L = 2\pi fL = 2\pi(60)(0.80)\ \Omega = 3\bar{0}0\ \Omega$$

$$X_C = \frac{1}{2\pi fC} = \frac{1}{2\pi \times 60 \times 8.0 \times 10^{-6}}\ \Omega$$

$$= 3\bar{3}0\ \Omega$$

$$Z = \sqrt{R^2 + (X_L - X_C)^2}$$

$$= \sqrt{(50)^2 + (300 - 330)^2}\ \Omega$$

$$= \sqrt{50^2 + (-30)^2}\ \Omega = 58\ \Omega$$

$$I = \frac{V}{Z} = \frac{110\text{ V}}{58\ \Omega} = 1.9\text{ A}$$

$$\cos\phi = \frac{R}{Z} = \frac{50\ \Omega}{58\ \Omega} = 0.86$$

$$P = VI\cos\phi = 110\text{ V} \times 1.9\text{ A} \times 0.86$$

$$= 1\bar{8}0\text{ W}$$

Note also that

$$P = I^2R = (1.9\text{ A})^2 \times 50\ \Omega = 1\bar{8}0\text{ W}$$

There are power losses in inductors other than the usual I^2R copper losses. Such losses are due to eddy currents and hysteresis in the core of the inductor, especially if the coil is wound on an iron core. Hence the equivalent resistance of such a coil may be considerably larger than the ordinary ("ohmic") resistance as measured by the use of direct current. The equivalent resistance of an inductor may be obtained from the relation

$$R_{\text{equiv}} = \frac{P}{I^2} \tag{23}$$

where P is the power loss and I is the current in the coil.

There may be losses in a capacitor because of inadequate insulation and dielectric hysteresis. In such cases the capacitor losses may be considered as equal to those in an equivalent resistance given by Eq. (23). In using this equation, it must be remembered that the equivalent resistance does not include any other series resistance that may be in the circuit.

42-10 ELECTRIC RESONANCE

It is apparent from an inspection of the net reactance term $(2\pi fL - 1/2\pi fC)$ in Eq. (20) that it is possible for the inductive reactance to be just equal in magnitude and opposite in sense to the capacitive reactance. In this important special case, since $X_L - X_C = 0$, the current and the voltage in the circuit are in phase. In this case the current is a maximum given by $I = V/R$, and the power factor is unity. The current is limited only by resistance, just as if no inductance or capacitance were present.

In a series circuit at resonance the voltage across the inductor and that of the capacitor may each be very large, with only a moderate applied voltage. The inductive voltage is equal in magnitude and opposite in sense to the capacitive voltage. Under these circumstances the current and the applied voltage are in phase with each other. Such a state of adjustment in a series ac circuit when $X_L = X_C$ and the current is a maximum for a given applied potential difference is known as *electric resonance*.

It should be noted that a given circuit with constant values of inductance and capacitance can be in resonance only for a particular frequency such that $2\pi fL = 1/2\pi fC$. A circuit of given L and C, therefore, has a natural resonant frequency. If the frequency of the impressed voltage equals this resonant frequency the current in the circuit is much larger than for higher or lower frequencies.

42-11 ELECTRIC OSCILLATIONS

In the study of mechanical vibration (Chap. 11), it was found that oscillations can be set up in a body if certain conditions are present. The body must have inertia, a distortion must produce a restoring force, and the friction must not be too great. A massive object suspended in air by a spring meets these conditions.

In an electric circuit, analogous conditions are necessary for electric oscillations. Just as inertia opposes change in mechanical motion, inductance opposes change in the flow of electrons. The building up of charge on the plates of a capacitor causes a restoring force on the electrons in the circuit. Resistance causes electric energy to be changed into heat, just as friction changes mechanical energy to heat. To produce electric oscillations, it is necessary to have inductance, capacitance, and not too much resistance. As the frequency of mechanical vibrations depends upon the inertia (mass) and the restoring force (force constant), so the frequency of electric oscillations depends upon inductance and capacitance.

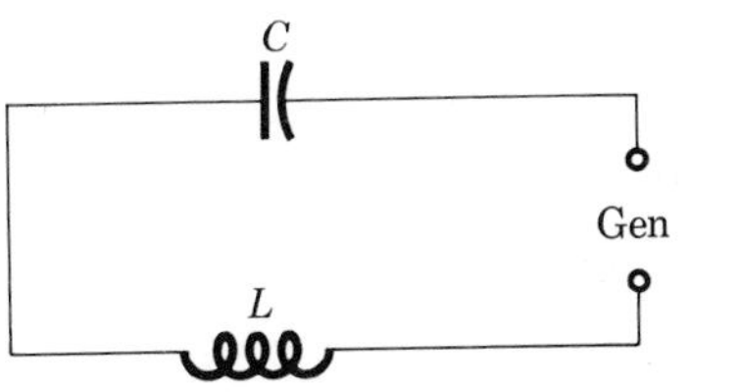

Figure 42-11
Circuit for production of electric oscillations.

In the circuit of Fig. 42-11 a capacitor of capacitance C and a coil of inductance L are connected in series with a sphere gap G. The sphere gap has a high resistance until a spark jumps across but low resistance after the spark jumps. If the voltage across G is gradually increased, the charge on the capacitor will increase. When the voltage across G becomes high enough, a spark will jump and the capacitor will then discharge. The current does not cease at zero when the capacitor is completely discharged but continues, charging the capacitor in the opposite direction. It then discharges again, the current reversing in the circuit. The current oscillates until all the energy stored in the capacitor has been converted into heat by the resistance of the circuit. However, if the resistance is high, all the energy is used in the first surge and hence there are no oscillations.

The frequency f of the oscillation is determined by the values of L and C and is the frequency for which the impedance of the circuit is the least, i.e., the frequency for which the net reactance is zero. From the relation

$$X = 2\pi fL - \frac{1}{2\pi fC} = 0$$

it follows that

$$f = \frac{1}{2\pi\sqrt{LC}} \tag{24}$$

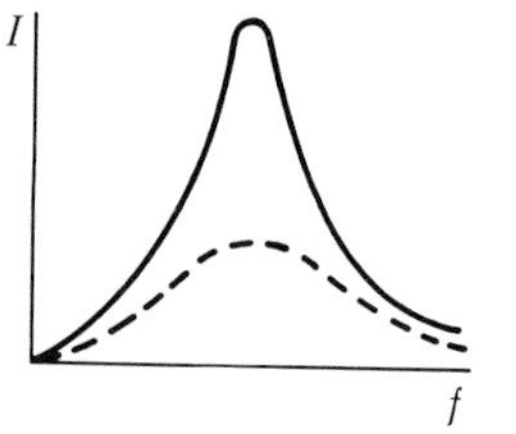

Figure 42-12
Resonance in a series circuit. Small resistance, solid line; larger resistance, dashed line.

where L is the inductance in henrys and C is the capacitance in farads.

Figure 42-12 shows how the current varies with frequency in an oscillating circuit at and near resonance conditions. For a very small range of frequencies, the current is rather large, but outside this region the current is small. This response over a very limited range of frequencies makes possible the tuning of a radio circuit. The incoming wave has a fixed frequency, and the circuit is tuned so that its natural frequency is the same as that of the incoming wave. The tuning is usually done by adjusting the value of the capacitance.

The value of the current in a resonant circuit depends also upon the resistance of the circuit (Fig. 42-12). If the resistance is small, the current curve has a higher and more distinct peak; such a circuit may be more sharply tuned. For a larger circuit resistance, the currents are smaller, and the resonant value is not so sharply marked.

42-12 ELECTROMAGNETIC WAVES

A knowledge of electromagnetic wave theory is so important to an understanding of modern physics that the next chapter has been devoted to a discussion of these waves. However, since electromagnetic waves are produced by electric oscillations, it is appropriate that we consider them briefly in that context here. In an oscillating electric circuit the varying alternating current produces correspondingly varying magnetic and electric fields. These fields are at right angles to each other and travel out into space with the speed of light, approximately $18\bar{6},000$ mi/s, or 3.00×10^8 m/s. This radiation of energy into space is facilitated by connecting the circuit to a suitable antenna that is comparable in size with the wavelength of the radiation. These electromagnetic waves were investigated in 1887 by Hertz and are sometimes called Hertzian waves. Hertz used apparatus similar to that shown in Fig. 42-11, one set as a "sending" station and a similar set as a "receiving" station. He produced "standing" electromagnetic waves and was able to measure the distance between adjacent nodes and thus determine the wavelength. Since the frequency of the oscillating circuit was known, he was able to measure the speed of the electromagnetic waves from the wave equation $v = f\lambda$. His measurements showed that this speed was the same as that for the speed of light.

The Italian scientist Marconi (1874–1937) applied the discoveries of Hertz to the field of electrical communication. His researches began the radio age that has been so fruitful in the communication industry (see Fig. 42-13).

A remarkable contribution to electromagnetic theory was made by the British mathematical physicist Maxwell. In 1864 he had shown that an oscillating electric circuit should be a source of electromagnetic waves that should travel with a speed related to the magnetic permeability and electric permittivity of the transmitting medium. This speed is given by the equation

$$v = \sqrt{\frac{1}{\mu\varepsilon}} \tag{25}$$

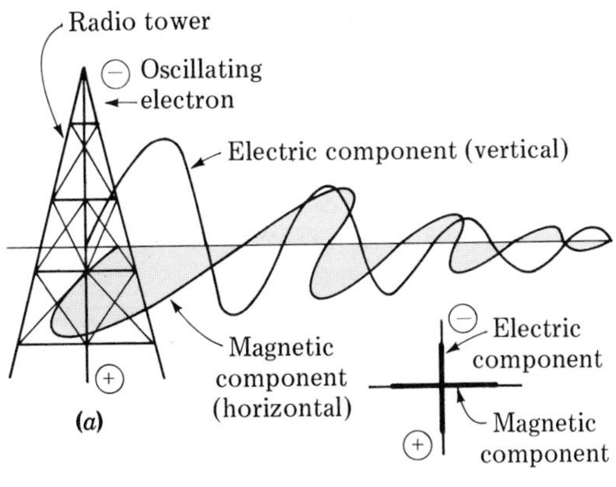

Figure 42-13
The formation of an electric and magnetic wave at right angles to each other by an oscillating electron.

When the values of μ and ε for free space are inserted in Eq. (25), the value of v comes out equal to the accepted value for the speed of light in empty space.

SUMMARY

Alternating currents are used very much more than direct currents because of the economy of transmission made possible by the ease with which ac voltages may be stepped up or stepped down.

The effective or rms value of a sinusoidal current is

$$I = 0.707 \times i_m$$

In ac circuits the current and voltage are

1 In phase for noninductive resistors
2 90° out of phase for pure inductances or pure capacitances

The current *lags* the voltage in *inductive* circuits by angles ranging from zero to 90°, depending upon the relative values of the resistance and inductive reactance.

The current *leads* the voltage in *capacitive* circuits by angles ranging from zero to 90°, depending upon the relative values of the resistance and capacitive reactance.

In circuits containing resistance, inductance, and capacitance, the phase relation between current and voltage is given by

$$e = e_m \sin\theta \qquad i = i_m \sin(\theta - \phi)$$

Inductive reactance is defined by the ratio

$$X_L = \frac{V_L}{I} = 2\pi f L$$

Capacitive reactance is defined by the ratio

$$X_C = \frac{V_C}{I} = \frac{1}{2\pi fC}$$

Impedance is defined by the ratio

$$Z = \frac{V}{I} = \sqrt{R^2 + X^2}$$

The impedance of a series circuit containing R, L, and C is given by

$$Z = \sqrt{R^2 + \left(2\pi fL - \frac{1}{2\pi fC}\right)^2}$$

The power in an ac circuit is

$$P = VI \cos \phi$$

Power factor is the cosine of the angle by which the current is out of phase with the voltage, $\cos \phi = R/Z$.

The equivalent resistance due to losses in a capacitor or inductor is given by

$$R_{\text{equiv}} = \frac{P}{I^2}$$

When $X_L = X_C$, the circuit is said to be in *electrical resonance* and the current is a maximum for a given voltage and resistance.

The frequency of the oscillations in a resonant circuit is determined by L and C.

$$f = \frac{1}{2\pi \sqrt{LC}}$$

An oscillating electric circuit may be connected to an antenna to radiate electromagnetic waves that travel with the speed

$$v = \sqrt{\frac{1}{\mu \varepsilon}}$$

The speed of electromagnetic waves is the same as the speed of light, approximately $18\bar{6}{,}000$ mi/s, or 3.00×10^8 m/s.

References

Josephson, Matthew: The Invention of the Electric Light, *Scientific American,* November 1959, p. 99.

Sharlin, Harold I.: From Faraday to the Dynamo, *Scientific American,* May 1961, p. 107.

Winch, Ralph P.: "Electricity and Magnetism," 2d ed., chaps. 5–8, Prentice-Hall, Inc., Englewood Cliffs, N.J., 1963.

Questions

1 Prove by means of vectors the statement $e = e_m \sin \theta$, which holds for the ideal generator. Justify the reasoning.

2 Sketch and describe an analogy between an ac generator and circuit and an hydraulic "circuit."

3 In some ac ammeters the deflections are nearly proportional to the square of the current (Fig. 42-14). What are some of the desirable and undesirable features of such instrument scales?

4 What is the average value of an alternating current of 10 A for one complete cycle?

5 In considering the voltage in an ac circuit required to puncture a capacitor, should one be concerned with the effective, maximum, or average values? Explain.

6 A long section of wire is available. Compare the inductive reactance of the wire under the following conditions: (*a*) long, straight wire; (*b*) wire doubled back upon itself at the center; (*c*) wire wound as a long solenoid of small radius; (*d*) wire wound as a short coil of large radius; and (*e*) iron core inserted in coil *c*.

7 As noted in question 3, in some ac ammeters the deflections are nearly proportional to the square of the current. Plot a curve to show the deflection-vs.-current relations in such an instrument.

8 A circuit contains a fixed resistor in series with a variable capacitor. Plot a curve to show the variation of the impedance with the capacitive

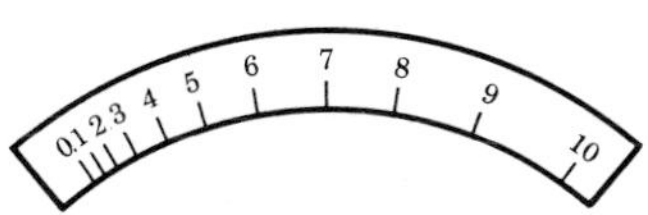

Figure 42-14
Scale of an instrument whose deflection is proportional to the square of the quantity being measured.

reactance as the capacitance is varied from zero to an infinitely large value.

9 An air-cored solenoid is connected to an ac source. Then an iron core is brought from a distance and inserted into the solenoid. Plot a curve to show the variation of the angle of phase lag as a function of the distance from the center of the solenoid to the iron core.

10 A noninductive resistor, an iron-cored solenoid, and a good-quality capacitor are in series in an ac circuit. The current in the circuit and the voltage across each part are measured by ac meters. The applied voltage is then raised and the readings repeated. The process is continued for several increasing voltages. Explain what values would be expected for the V/I ratios for each part of the circuit.

11 Compare the equivalent resistance (due to iron losses) of a reactor in a 60-cycle and a 500-cycle circuit.

12 A choke coil placed in series with an electric lamp in an ac circuit causes the lamp to become dimmed. Why is this? A variable capacitor added in series in this circuit may be adjusted until the lamp glows with normal brilliance. Explain why this is possible.

13 The set of three curves in Fig. 42-15*a* represents conditions in an ac series circuit which has resistance only. Which curve may represent the voltage? The current? The power?

14 Describe the differences between the currents that exist in the wires leading to a capacitor when these wires are connected to (*a*) a dc source and (*b*) an ac source.

15 The set of three curves in Fig. 42-15*b* represents conditions in an ac series circuit which has inductance but negligible resistance. Which curve represents the voltage? The current? The power? Does the current lag or lead the voltage in phase?

16 Compare the power in watts with the apparent power in volt-amperes for the following apparatus in an ac circuit: (*a*) electric lamp, (*b*) choke coil, and (*c*) capacitor.

17 How would one expect the equivalent resistance of a 500-V, 1-μF mica capacitor to compare with a 500-V, 1-μF paper capacitor?

18 Show clearly how it is much more nearly possible to have "wattless" current in ac capacitor circuits than it is in ac circuits containing choke coils.

19 A resistor, capacitor, and inductor are connected in series to a 120-V ac generator. Sketch a curve to show how the current in the circuit

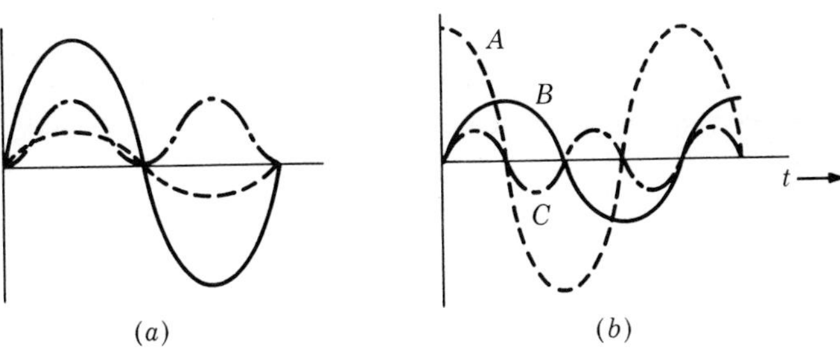

Figure 42-15

varies with the frequency of the generator, the voltage being kept constant.

20 A familiar type of ac motor has three similar coils wound on an iron frame so that the coils make an angle of 120° with each other. Each of these coils is energized from one of the phases in a three-phase supply line, the voltages of which differ in phase by 120°. Show trigonometrically why these coils produce a rotating magnetic field.

Problems

In each of the following problems in which it is appropriate, draw a phase diagram, roughly to scale, to represent the voltages involved. Current and voltage mean effective current and effective voltage, unless otherwise stated.

1 The voltage in an ac circuit varies harmonically with time with a maximum of 180 V. At what angle in its cycle will instantaneous voltage be 120 V?

2 Find the maximum value of an 8.0-A alternating current. *Ans.* 11.3 A.

3 A 60-cycle ac circuit has a voltage of 120 V and a current of 6.00 A (effective values). (*a*) What are the maximum values of these quantities? (*b*) What is the instantaneous value of the voltage $\frac{1}{720}$ s after the voltage has zero value?

4 A capacitor is frequently placed across a 110-V line to reduce the noise in radios. What is the smallest voltage rating such a capacitor should have? *Ans.* 155 V.

5 Two generators are connected in series. One develops an emf of 225 V, the other an emf of 120 V. If the first generator is 90° in phase ahead of the second, what is the emf of the two generators in series?

6 A solenoid has an emf of 43.8 V induced when the current is changed by 12.5 A in 0.100 s. What is the inductive reactance of this solenoid in a 60-cycle circuit? *Ans.* 132 Ω.

7 An inductor has a resistance of 15.00 Ω. On a 120-V 60-cycle line this inductor takes a current of 2.50 A. What is the self-inductance of the coil?

8 A coil connected to 120-V dc mains takes a power of 432 W. When this coil is connected to ac mains of the same voltage, a current of 2.5 A and a power of 281 W are observed. What are the actual and equivalent resistances of this coil? *Ans.* 33 Ω; 45 Ω.

9 A choke coil has a resistance of 4.00 Ω and a self-inductance of 2,390 μH. It is connected to a source of 500-cycle 110-V alternating emf. Calculate the reactance, impedance, and current of the circuit.

10 A coil has an inductance of 478 μH. (*a*) What is its reactance in a 1,000-cycle ac circuit? (*b*) If connected in series with a resistor of 4.00 Ω, what current would there in a 1,000-cycle, 110-V line. *Ans.* 3.00 Ω; 22.0 A.

11 A coil of wire has a resistance of 30 Ω and an inductance of 0.10 H. (*a*) What is its inductive reactance X_L in a 60-cycle circuit? (*b*) Its impedance Z? (*c*) What current will there be if the coil is connected to a dc source of 120 V? (*d*) To a 60-cycle ac source of 120 V?

12 A current of 2.50 A is observed in a 120-V 60-cycle circuit which consists of a "pure" resistor and a "pure" inductor in series. The voltages across the resistor and the inductor are found to be identical. Calculate the value of the resistance and the inductance. *Ans.* 34.0 Ω; 90.5 mH.

13 An ac series circuit consists of an inductor that has a resistance of 41.5 Ω and an inductive reactance of 112 Ω, connected through an 80.2-Ω rheostat to a 125-V 500-cycle power source. (*a*) Calculate the current in the circuit. (*b*) What is the potential difference across the rheostat, the resistance component of the inductor, the inductive component of the inductor, and the inductor? (*c*) Find the inductance of the inductor.

14 What is the reactance of a 2.00-μF capacitor on a 110-V 60-cycle line? What is the current? *Ans.* 1,3$\bar{2}$0 Ω; 0.0833 A.

15 The resistance in a certain 220-V 60-cycle ac series circuit is 82.4 Ω, and the capacitive reactance is 60.0 Ω. (*a*) What is the total impedance? (*b*) Calculate the current in the circuit. (*c*) What is the capacitance?

16 (*a*) What is the reactance of a 3.00-μF capacitor in a 60.0-cycle ac circuit? (*b*) What is the impedance of this capacitor in series with a resistance of 300 Ω? (*c*) What current would there be if this capacitor and resistor were connected to a 1,200-V line? *Ans.* 883 Ω; 933 Ω; 1.29 A.

17 A resistor of 4.00 Ω, an inductive coil of negligible resistance and inductance 2.39 mH, and a good-quality 30.0-μF capacitor are connected in series to a source of 500-cycle 110-V alternating emf. Calculate the reactance of each part of the circuit and the current in the line.

18 A pure inductor and a pure resistor are connected in series in an ac circuit. A voltmeter reads 30 V when connected across the inductor and 40 V when connected across the resistor. What will it read when connected across both? *Ans.* 50 V.

19 A coil is connected across 220-V 60-cycle mains. The current in the coil is 4.0 A, and the power delivered is 324 W. Find the resistance and the inductance of the coil.

20 A 120-Ω rheostat and a good-quality 15.2-μF capacitor are connected in series and joined to a 60-cycle 600-V line. What is the voltage across each? *Ans.* 341 V; 495 V.

21 A 250-μF capacitor has an equivalent resistance of 12.5 Ω. What power does it take from a 60-cycle 120-V line? What is the power factor?

22 A 25.0-μF capacitor having an equivalent resistance of 3.00 Ω is connected in series with a 50.0-Ω resistor. When a 500-cycle 300-V potential difference is impressed on the circuit, what is the voltage across the capacitor? Across the resistor? *Ans.* 70.6 V; 276 V.

23 There is a current of 0.600 A in a 60-cycle ac circuit, which consists of a 40.0-Ω lamp, a choke coil of 50.0-Ω resistance, and a capacitor having a negligible resistance, connected in series. The voltage across the coil is 50.0 V, and that across the capacitor is 170 V. Find the total voltage of the circuit, the power, and the power factor.

24 A choke coil takes a current of 5.00 A from a 120-V ac source. A wattmeter connected to the coil reads 450 W. What is the (*a*) apparent power, (*b*) real power, and (*c*) power factor? *Ans.* 600 V·A; 450 W; 0.75.

25 A capacitor and a coil are connected in series across a 520-V 60-cycle line. The capacitor has a capacitive reactance of 240 Ω. The coil has a resistance of 60.0 Ω and an inductive reactance of 320 Ω. Find (*a*) the current in the circuit, (*b*) the power factor, (*c*) the power delivered, and (*d*) the capacitance of the capacitor. Is the current lagging or leading?

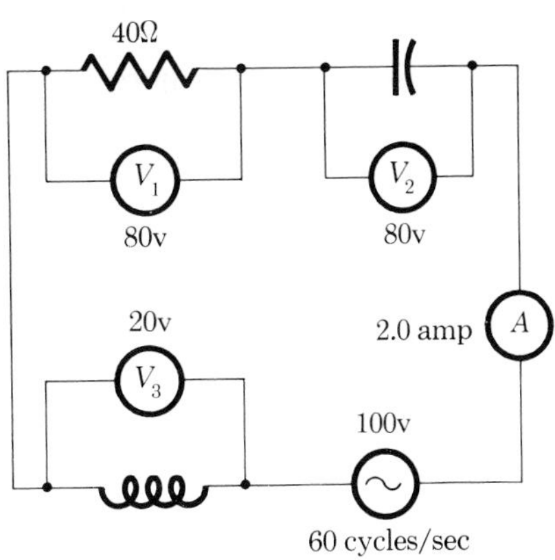

Figure 42-16

26 A fluorescent lamp unit connected to a 110-V ac line takes 1.20 A and requires 110 W power. What is the power factor? *Ans.* 0.833.

27 An ac series circuit includes the following: a 220-V 60-cycle source; a 52.6-Ω resistor; an inductor of negligible resistance and inductance of 204 mH; and a 14.7-μF capacitor. Calculate (*a*) the inductive reactance, (*b*) the capacitive reactance, (*c*) the impedance, (*d*) the current, and (*e*) the phase angle.

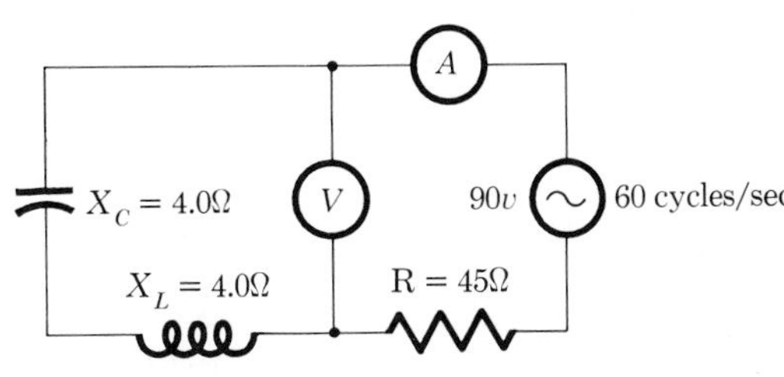

Figure 42-17

28 In the circuit of Fig. 42-16 the readings of the ac meters are indicated. Calculate the power supplied to this circuit, using two different expressions. Show that your two answers agree.
Ans. 160 W.

29 What current and potential difference will be indicated by meters in the positions shown in Fig. 42-17.

30 A bank of lamps operates at a current of 12 A and a voltage of 120 V. What power is taken from (*a*) dc mains and (*b*) ac mains?
Ans. 1.44 kW; 1.44 kW.

31 A coil of negligible resistance and inductance 80 mH, a 7.0-Ω resistor, and a 25-μF capacitor are connected in series across 110-V ac supply mains of variable frequency. For what frequency is the current a maximum? What is that maximum value? When the current has this value, what is the voltage across the inductor? Across the resistor? Across the capacitor?

32 A 0.10-H coil (resistance, 100 Ω) and a 10-μF capacitor are connected in series across a 110-V ac line. Find the current and the power if the frequency is (*a*) 60 Hz and (*b*) 25 Hz.
Ans. 0.44 A; 20 W; 0.18 A; 3.2 W.

33 What is the range of wavelengths associated with the usual commercial radio-broadcasting frequencies that vary from 550 to 1,575 kHz? What is the wavelength associated with a 60-cycle ac source?

34 It is desired to construct a broadcasting station that will transmit at a wavelength of 300 m. A capacitor of 2.40 μF is available. What is the inductance of the coil that must be used for a resonant circuit?
Ans. 10.5×10^{-12} H.

35 An oscillating circuit consists of a 2.18-μF capacitor and a coil of 12.5-mH inductance. (*a*) What is the resonant frequency of this circuit? (*b*) What is the wavelength of the radiation that is broadcast?

Alexandr Mikhailovich Prokhorov, 1916–
(*left*)
Graduated from Leningrad University in 1939. Served in the Red Army in World War II. Joined the Lebedev Physics Institute in 1950, where he and Basov developed molecular generators. Elected corresponding member of the Soviet Academy of Sciences in 1960. Shared the Lenin Prize in 1959 and the 1964 Nobel Prize for Physics for work in quantum electronics.

Charles Hard Townes, 1915–
(*center*)
Born in Greenville, S.C. Earned academic degrees at Furman, Duke, and California Institute of Technology. Versatile research scientist and academic administrator at Bell Telephone Laboratories 1939–1947, Columbia University 1948–1955, and Massachusetts Institute of Technology, where since 1961 he has been Provost. His interests include microwave spectroscopy, atomic time standards, masers, radioastronomy, and phonon masers. Awarded half the 1964 Nobel Prize for Physics for his research in quantum electronics.

Nikolai Gennadievich Basov, 1922–
(*right*)
Graduated from Moscow Engineering and Physics Institute in 1950, earned the degree of Doctor of Physical-Mathematical Sciences in 1957, and was elected corresponding member of the Soviet Academy of Sciences in 1962. Since 1958 Basov has been deputy director of the Lebedev Institute. Basov's work with Prokhorov on the oscillators and amplifiers used to produce laser beams earned for them both the 1959 Lenin Prize and the 1964 Nobel Prize for Physics.

43 Electromagnetic Waves

In all the waves that we have discussed so far we considered the motion of particles of a material medium. As the wave passes through the medium we may describe the displacement, or velocity, or in some cases pressure changes, that are observed in the medium. We now wish to consider a very important type of wave for which no material medium is required. This type of wave is an *electromagnetic* wave, in which the properties that vary are electric and magnetic fields. We have already observed that electric and magnetic fields may exist in empty space.

The development of our ideas of electromagnetic waves followed a series of observations that we have previously studied. The first of these is the observation that electric fields exist in the region around electric charges. Second, Oersted (1820) observed that there is a magnetic effect (and hence a magnetic field) associated with moving charges. The third is the discovery of induced emfs (Faraday, 1831) associated with changing magnetic fields.

James Clerk Maxwell studied these observations mathematically and in 1864 set forth a theory that accelerated charges cause waves to travel out, waves that consist of variation of electric and magnetic fields. Heinrich Hertz carried out an experimental study and first detected such waves in 1886. The waves produced in Hertz's experiment were of wavelength greater than 1 m, but they exhibited properties previously associated with light: linear propagation, reflection, refraction, diffraction, and polarization.

Maxwell's theory of electromagnetic radiation stands with Newton's laws of motion and the laws of thermodynamics as masterpieces of intellectual achievement. For many persons, the structure of physics seemed almost complete at the close of the nineteenth century. But, beginning in 1900, developments took place which indicated that Maxwell's theory does not predict accurately all aspects of electromagnetic radiation, especially at high frequencies. These developments led to the quantum theory of radiation.

43-1 OSCILLATIONS, DIPOLE FIELDS

We have considered the oscillations that may occur in a circuit containing capacitance and inductance. Consider a circuit such as that of Fig. 43-1. If the capacitor is charged to a potential difference V, energy $\frac{1}{2}CV^2$ is stored in the electric field. When the switch is closed, the capacitor discharges and oscillations are set up in the circuit. The frequency of oscillation is determined by the capacitance and inductance of the circuit, $f = 1/(2\pi\sqrt{LC})$. At low frequency, if no energy were dissipated in the oscillation, the amplitude

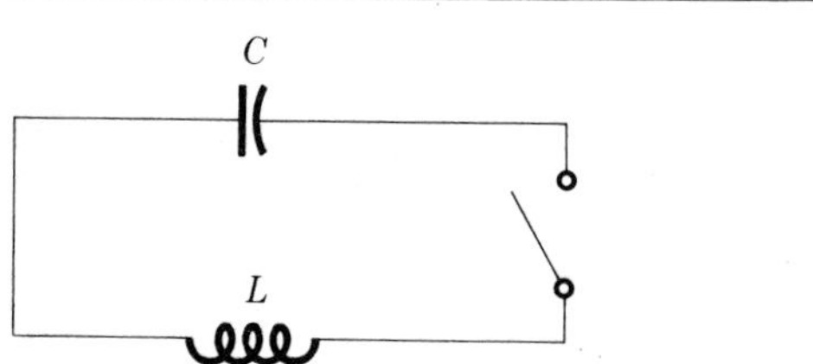

Figure 43-1
An oscillation circuit.

of the oscillation would remain constant and the energy would go from the electric field to the magnetic field and back. When energy is dissipated in the circuit, the amplitude decreases with time and we have damped oscillations. When energy is dissipated so rapidly that energy is all used before the capacitor is recharged, there is no oscillation.

A second process by which energy is lost by an oscillating circuit is by radiation of electromagnetic waves. At low frequency the rate of radiation is very low. As the frequency rises, the rate of radiation increases rapidly, approximately as the fourth power of the frequency. The rate of radiation is also related to the size of the radiator, increasing sharply as the size approaches the wavelength in space of the radiated wave.

Let us consider a dipole antenna, shown schematically in Fig. 43-2. It consists of two straight, equal conductors with an oscillator between. At one instant during the oscillation one of the conductors has its maximum positive charge, the other maximum negative (Fig. 43-2*a*). The two conductors may be considered a capacitor, and the electric field in the plane of the paper is suggested by the lines of force shown. As the oscillation continues, the capacitors lose their charges and a current exists during the discharge. At the instant when the charge on each conductor becomes zero, the current is maximum and directed downward. The electric field due to the charges is now zero near the oscillator, but there is a magnetic field perpendicular to the plane of the paper in the directions shown in Fig. 43-2*b*. As the oscillation continues, the lower conductor becomes charged positively, thus setting up an electric field which rises to a maximum as the magnetic field disappears, as shown in Fig. 43-2*c*, and which then dies down to zero with the magnetic field rising to a maximum, as in Fig. 43-2*d*. In the region near the dipole we have oscillating electric and magnetic fields that are mutually at right angles to each other and out of phase by 90°.

Both electric and magnetic fields are set up at larger distances, but the changes that take place near the dipole during the oscillation do not reach

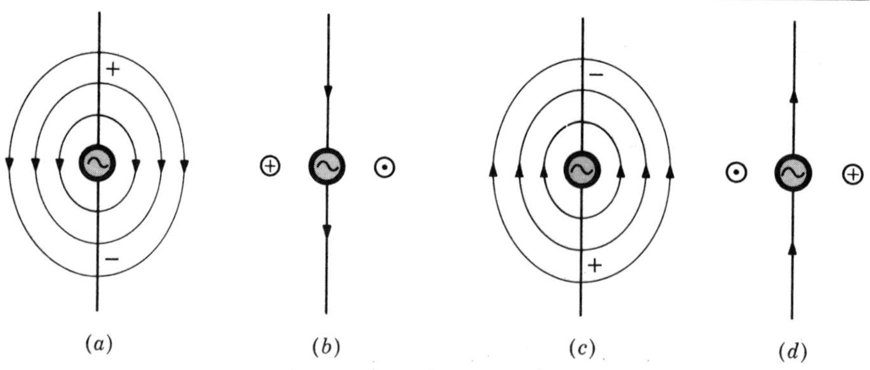

Figure 43-2
Oscillator with dipole radiator.

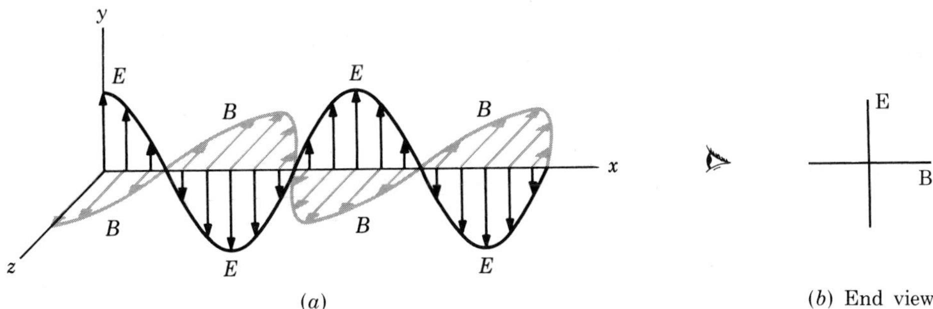

Figure 43-3
Electric and magnetic components of an electromagnetic wave traveling in the x direction.

distant points instantaneously. A finite time is required for the change to travel out from the source. The periodic variations of E and B traveling out from the source constitute the electromagnetic wave that carries energy from the source.

We observe that the electric and magnetic fields are at right angles to each other and also at right angles to the direction of propagation. In Fig. 43-3 is shown the relation between B, E, and the direction of propagation. We have noted that near the radiator B and E are out of phase. We shall see that this phase difference does not continue as the wave travels but that after a few wavelengths the electric and magnetic components are in phase. In the diagram the waves are shown in phase.

43-2 MAXWELL'S EQUATIONS

We have expressed several relations between E and B and the properties of the surrounding space. We shall refer to four of these relations. The first we expressed as Gauss' law, which states that the sum of the outward-directed normal components of E over any closed surface is proportional to the sum of the charges inside the surface,

$$\varepsilon_0 \Sigma E_n \, \Delta S = \Sigma Q \qquad (1)$$

Here the electric flux begins on positive charges and ends on negative charges.

For the magnetic field the equation corresponding to Eq. (1) is

$$\frac{1}{\mu_0} \Sigma B \, \Delta S = 0 \qquad (2)$$

In the magnetic field there is nothing corresponding to the isolated electric charges, and the magnetic lines are closed loops. Hence, a line that emerges through the surface must return into the surface.

Faraday's law of induction, expressed for a single loop, is

$$\mathcal{E} = -\frac{\Delta\Phi}{\Delta t} \qquad (3)$$

The emf $\mathcal{E}$ is work per unit charge done in taking a charge around a closed loop. This emf may be expressed as the sum of the products of force per unit charge E and the element of distance Δl around the closed loop.

$$\Sigma E \, \Delta l = -\frac{\Delta\Phi}{\Delta t} \qquad (4)$$

Ampère's law relates the tangential component of the magnetic induction to the current. For any closed loop the sum of the products $B_T \Delta l$ is proportional to the current enclosed by the loop,

$$\frac{1}{\mu_0} \Sigma B_T \Delta l = \Sigma I \qquad (5)$$

Each of these equations should be expressed in limiting form, but we have here avoided the calculus notation.

Maxwell studied these relationships and the conclusions that can be reached from them. He extended Eq. (5) by noting that there may be effects due to changing electric fields as well as those due to amperian currents. That is, he stated that a magnetic field exists, not only in the region near an electric current but also wherever there is a changing electric field. We have already seen that an electric field exists wherever there is a changing magnetic field, as expressed in Eq. (4). Then, in place of Eq. (5) we have

$$\frac{1}{\mu_0} \Sigma B_T \Delta l = \Sigma I + \varepsilon_0 \frac{\Delta E}{\Delta t} \qquad (6)$$

The foregoing equations are one form of Maxwell's equations. From these equations he showed that the space and time variation of the electric and magnetic fields is that characteristic of waves.

When the mathematical expressions for the electric and magnetic components are expanded, it is found that there are terms in each expression that are in phase and others that are out of phase by 90°, that is, sine and cosine terms. These terms also involve the distance from the source. The terms which are out of phase by 90° decrease as the second or third power of the distance, while those which are in phase decrease only by the first power. Thus, near the source, the out-of-phase components predominate, while at greater distances the electric and magnetic components are in phase.

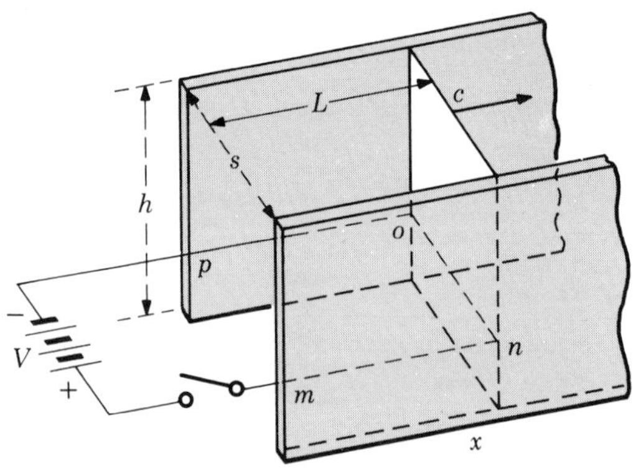

Figure 43-4
Propagation of an electromagnetic wave between parallel plates.

43-3 SPEED OF ELECTROMAGNETIC WAVES

We have seen that the speed of a wave depends upon the properties of the medium through which it travels. The dependence of the speed of electromagnetic waves on the properties of the region appeared directly in Maxwell's derivation. Since that work is beyond the scope of our discussion, we shall rationalize the relationship by a simpler discussion.[1]

Assume a parallel-plate capacitor of infinite extent in the x direction and of height h in the y direction and spaced a distance s apart. If the switch is suddenly closed at time $t = 0$, we assume that a steep-front voltage wave travels with speed c down the capacitor in the x direction. At time Δt the wave front will have traveled a distance $L = c\,\Delta t$, and at this time charge $\Delta q = C'V$ will have flowed onto the charged part of the plates, where C' is the capacitance of the charged part of the capacitor. Then

[1] Adapted from a derivation suggested by F. G. Werner and D. R. Brill, *American Journal of Physics,* **28:**126 (1960).

$$\Delta q = C'V = \frac{\epsilon_0 A}{s} V = \epsilon_0 \frac{hL}{s} V$$

$$= \epsilon_0 \frac{hc\,\Delta t}{s} V \qquad (7)$$

The rate of flow of charge, the current I, is

$$I = \frac{\Delta q}{\Delta t} = \epsilon_0 \frac{hc}{s} V \qquad (8)$$

This current will set up a magnetic field in the space between the plates. This magnetic field is due to the two sheets of current on the two plates, one to the right, the other to the left. Let us compare this situation with that in a solenoid, where we have currents in opposite directions on opposite sides of the coil. The field of induction inside the solenoid is $B = \mu_0 nI$, where n is the number of turns per unit length. Then nI is the current per unit length of the solenoid. If we apply this same reasoning to our capacitor, the current per unit length is I/h and $B = \mu_0 I/h$. The magnetic field is directed upward at right angles to the horizontal electric field.

$$B = \mu_0 \frac{I}{h} = \mu_0 \frac{\epsilon_0 c}{s} V \qquad (9)$$

The flux Φ through a circuit in the plates bounded by *mnop* will be

$$\Phi = BA = \mu_0 \frac{\epsilon_0 c}{s} Vsc\,\Delta t = \mu_0\epsilon_0 c^2\, V\,\Delta t \quad (10)$$

The rate of change of flux $\Delta\Phi/\Delta t$ is a counter emf equal to V. The fact that this counter emf must be at every instant equal to the applied voltage indicates that the speed of advance along the plates is controlled by the speed of advance of the wave. Thus

$$\frac{\Delta \Phi}{\Delta t} = V = \mu_0\epsilon_0 c^2\, V \qquad (11)$$

or

$$1 = \mu_0\epsilon_0 c^2$$

and

$$c^2 = \frac{1}{\mu_0\epsilon_0}$$

or

$$c = \frac{1}{\sqrt{\mu_0\epsilon_0}} \qquad (12)$$

In Eqs. (10) to (12) the height h of the assumed plates and the distance s between them disappear. We can then think of the plates as spreading until our region is just empty space.

Example What is the speed of an electromagnetic wave in empty space?

For empty space,

$$\mu_0 = 4\pi \times 10^{-7}\ \text{N/A}^2$$

$$\epsilon_0 = 8.85 \times 10^{-12}\ \text{C}^2/\text{N}\cdot\text{m}^2$$

$$c = \frac{1}{\sqrt{\mu_0\epsilon_0}}$$

$$= \frac{1}{\sqrt{4\pi \times 10^{-7} \times 8.85 \times 10^{-12}\ \text{C}^2/\text{A}^2\cdot\text{m}^2}}$$

$$= 3.00 \times 10^8\ \text{m/s}$$

Note to the reader: *Energy travels out in empty space in the electromagnetic wave with speed c and does not return to the source.* The wave is, as it were, "broken off" and travels out until it is deflected or absorbed. *The discovery by Maxwell that electromagnetic waves of electrical origin travel with the speed of light in empty space was one of the great unifying discoveries and a great triumph of theoretical physics.*

43-4 THE ELECTROMAGNETIC SPECTRUM

With an appropriate source electromagnetic waves of any frequency may be produced. All such waves travel with the same speed in empty space. There are major differences in the way the waves of various frequency ranges are produced and the methods by which they are studied. In Fig. 43-5 we show the frequency and wavelength

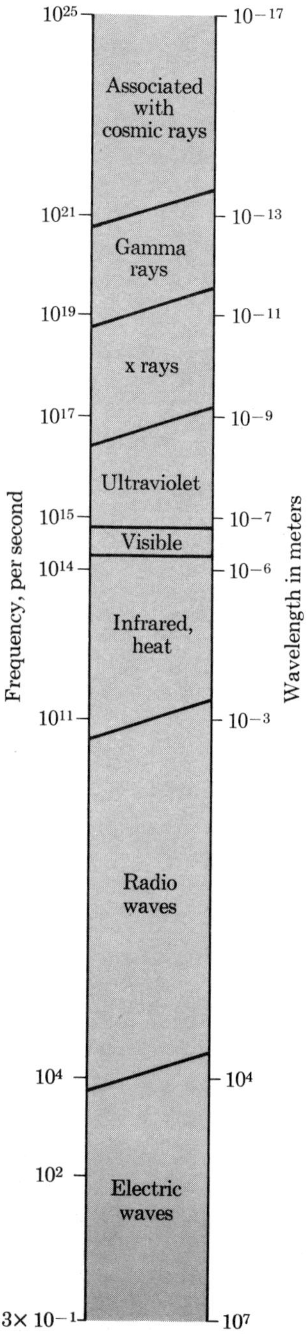

ranges of various parts of the electromagnetic spectrum. Not all parts are drawn to scale.

In the low-frequency–long-wavelength region little energy is radiated unless the source equipment is very large. As the frequencies are increased, the radiating (or receiving) devices become smaller and smaller until finally a practical lower limit of machined parts is reached in the microwave section. Higher frequencies of the electromagnetic spectrum are produced by molecules and by atoms. The visible region is that range to which the human eye is sensitive. In the visible, ultraviolet, and x-ray regions, the radiation is due to processes that occur within atoms; gamma rays are due to transitions within atomic nuclei.

In the electric-, radio-, and microwave regions the radiation (and reception) may be considered continuous, but in the higher-frequency regions the energy is emitted in quanta, discrete bundles of energy proportional to the frequency. These quanta correspond in energy to finite energy transitions within the atoms.

43-5 STATIONARY ELECTROMAGNETIC WAVES

In a traveling electromagnetic wave we may represent the electric and magnetic components of the wave at a considerable distance from the source by the usual wave expression

$$E = E_m \sin \frac{2\pi}{\lambda}(x - ct) \tag{13}$$

and

$$B = B_m \sin \frac{2\pi}{\lambda}(x - ct) \tag{14}$$

Equations (13) and (14) represent the variation

Figure 43-5
Electromagnetic spectrum; frequencies on a logarithmic scale. Slanted lines are used to indicate that waves of various regions overlap.

in space and time of the electric and magnetic fields as the wave travels through the medium.

If the wave is reflected back upon itself, we have the two waves traveling in opposite directions and a stationary wave may be set up. For a perfectly conducting reflector the electric field is always zero at the conducting surface. Therefore, for such a reflector, the boundary is fixed, and the reflection is one with a 180° phase change. Hence there is a node of the stationary electric wave at the surface. For the magnetic component the conducting surface is a free boundary, and there is an antinode at the surface. If this stationary wave is explored with two probes, one that responds to changes in the electric field, the other to changes in the magnetic field, it will be found that the electric nodes are coincident with the magnetic antinodes, and vice versa.

Stationary electromagnetic waves may be produced in a manner analogous to other stationary waves.

SUMMARY

Electromagnetic waves are periodic changes in electric and magnetic fields that travel out through space. The electric and magnetic field components are at right angles to each other and at right angles to the direction of travel. At some distance from the source the two components are in phase with each other.

The *speed* of electromagnetic waves in empty space is related to the permittivity and permeability,

$$c = \frac{1}{\sqrt{\mu_0 \epsilon_0}}$$
$$= 3.00 \times 10^8 \text{ m/s}$$

Maxwell worked out a theory of electromagnetic radiation embodied in *Maxwell's equations.* Hertz confirmed the conclusions of Maxwell's work by producing waves, as predicted by the theory.

With suitable sources, electromagnetic waves may be produced with any of a very wide range of frequencies, from nearly zero up to the order of 10^{25} per second.

Questions

1 How are E and B related in a plane electromagnetic wave?

2 Describe the nature and relationship of the electrical and magnetic forces associated with an electromagnetic wave.

3 Show why electric oscillations are obtained when a capacitor is discharged through an inductor of low resistance. Discuss briefly the energy transformations that occur in such a circuit.

4 Explain why it is true that an oscillating circuit is a source of radiant energy transported by electromagnetic waves. For efficient radiation what kind of relation should there be between the wave and the radiator?

5 What is meant by "distributed" capacitance in referring to a transmission line? Describe the flow of electricity in a long transmission line when a constant potential difference is suddenly connected across one end of the line.

6 Discuss how the current and charge are distributed along a transmission line when an ac source is impressed at one end of the line.

7 Equation (12) applies to electromagnetic waves in empty space. What implications could be drawn with regard to waves in air or other gases?

8 Having determined upon the wavelength to be radiated by an oscillating circuit, how might one decide the characteristics of the component parts of the circuit?

9 In Eq. (12) how can c always have the same value if μ_0 is determined arbitrarily and ϵ_0 is evaluated by experiment?

Problems

1 What is the range in centimeters of the wavelength of electromagnetic waves?

2 What is the wavelength of the electromagnetic waves that are radiated from an oscillating circuit that consists of a 2.18-μF capacitor and an inductor of 12.5 mH inductance? Discuss a suitable arrangement of the radiator.

Ans. 3.1×10^{-5} m.

3 A broadcasting station transmits radio waves of wavelength 300 m. What is the inductance of an inductor that could be connected with a 2.40-μF capacitor to be in resonance with these waves?

4 At a point 10 km east of a vertical radio-transmitting antenna, the peak electric field is 10^{-3} V/m. (*a*) Draw a sketch to show the direction of the magnetic induction B at that point when E is "up." (*b*) Find the magnitude of B.

Ans. 0.047 Wb/m^2.

5 The solar constant, a measurement of the amount of radiation that reaches the earth's atmosphere from the sun, has been calculated to be 1.36×10^3 W/m^2. Find the magnitude of E and B for the electromagnetic waves emitted from the sun at a point near the earth's surface.

6 A radio antenna radiates power of 10 kW. Find the amplitude of E and B in the wave at a distance 10 km from the antenna. Assume that the power is uniform over a hemisphere with the antenna at its center.

Ans. 0.11 V/m, 2.8×10^{-4} A/m.

Richard B. Feynman, 1918–
(*left*)
Born in New York. Professor of Physics at California Institute of Technology. Shared the 1965 Nobel Prize for Physics with Schwinger and Tomonaga for their research in electrodynamics that contributed to the understanding of elementary particles in high-energy physics.

Julian S. Schwinger, 1918–
(*center*)
Born in New York. Professor of Physics at Harvard University. Shared the 1965 Nobel Prize for Physics with Feynman and Tomonaga for their fundamental work in quantum electrodynamics.

Sin-Itiro Tomonaga, 1906–
(*right*)
Physics professor at the Tokyo University of Education. Shared the 1965 Nobel Prize for Physics with Feynman and Schwinger for their fundamental work in quantum electrodynamics.

44

Conduction in Gases; Electronics

In the modern development of physics one of the important ideas is that of the electron. The existence of an elementary quantity of electricity was suggested by the experiments on electrolysis conducted by Faraday, which indicated that each ion taking part in electrolysis has a fixed charge, the smallest being that on the monovalent ion. Stoney (1874) attempted to determine the magnitude of this charge by using the measured mass of silver deposited by a coulomb and Avogadro's number N, the number of atoms in a gram atom. From these data he computed the charge per atom. To this unit of charge he gave the name *electron*. The value obtained by Stoney was inaccurate because his value of N was in error.

In 1870, Sir William Crookes discovered the phenomenon of cathode rays and later (1879) suggested that the rays consist of streams of negatively charged particles. J. J. Thomson confirmed this hypothesis, showed that all these particles are alike, and named the particles *electrons*.

The subject of electronics is the study of the motion of electrons and of those phenomena in which electrons have a basic part. The engineering applications of electronics are especially important in the fields of communication and automatic control.

44-1 CONDUCTION OF ELECTRICITY IN SOLIDS AND IN LIQUIDS

In Chap. 35 conduction of electricity in solids and in liquids was discussed. There it was stated that in a conducting solid, conduction consists of the drift of electrons that have been temporarily detached from the parent atoms. On the other hand, conduction in liquid electrolytes is ionic in nature. Ions, produced by dissociation of molecules, drift through the solution when a potential difference is maintained. Whereas in solid conduction a single kind of charged particle, the negative electron, moves in the process, in electrolytic conduction both positively and negatively charged particles take part in the motion, the positive particles moving in one direction while the negative move in the opposite. Moreover, the particles moving in electrolytic conduction are of atomic or molecular mass, consisting of charged atoms or groups of atoms, while in solids the moving particles have the mass of the electron, much smaller than that of the smallest atom.

44-2 CONDUCTION OF ELECTRICITY IN GASES AT ATMOSPHERIC PRESSURE

A third type of conduction occurs in gases. This type of conduction is similar to liquid conduction in that both positive and negative ions move in the process, but it differs in the very important respect that very few of the ions exist before the beginning of the conduction process. Most of the ions are produced as a result of collisions between moving particles and molecules of the gas. Also, the ions are of both atomic and electronic nature.

Under normal conditions a gas is a very poor conductor of electricity. There are very few ions present to take part in the conduction. If a low voltage is applied to the specimen of gas, each ion moves toward the appropriate terminal. In this motion the ions collide frequently with molecules of the gas. In these collisions further ionization rarely takes place, because the ion colliding with a molecule seldom has enough energy to remove an electron from the molecule. As the potential difference applied to the gas is increased, each ion will acquire more energy, on the average, between collisions. When the voltage is great enough so that an ion acquires between collisions sufficient energy to ionize the atom or molecule that it strikes, two or more new ions are produced, one being the electron knocked off the atom and the other being the atom less its electron. Thus the number of ions builds up very rapidly, and a disruptive discharge, or spark, occurs. This process of cumulative ionization is called *ionization by collision*.

Ionization by collision is dependent upon the interaction of individual ions and atoms or molecules. Whether or not an atom is ionized will depend upon the energy acquired by the ion and also upon the energy necessary to ionize the atom struck. Thus the potential required to produce a disruptive discharge will depend upon the kind of gas, since the energy necessary to ionize the molecules varies from gas to gas. The potential V through which an electron must fall in order to have enough energy to ionize the atom that it hits is called the *ionizing potential* of that atom and is characteristic of the atom. The energy acquired is Ve, where e is the charge of the electron.

The potential necessary to produce a disruptive discharge is also dependent upon the distance between the atoms or molecules, since that distance must be great enough that the fall in potential is equal to the ionization potential of the gas. That is, the potential required for the disruptive discharge depends upon the pressure. When the pressure is high, the ions will move only small distances between collisions, since the molecules are close together. A very high potential gradient is then required to produce the discharge. The voltage gradient necessary to produce a spark between fairly large terminals in air at atmospheric pressure is about 30,000 V/cm. If the pressure is reduced, the average distance between molecules is greater and hence, on the average, the ion moves farther between collisions. Hence a smaller potential gradient is required for the ion to acquire sufficient energy between collisions in order to ionize the molecule that it strikes.

44-3 DISCHARGE OF ELECTRICITY THROUGH GASES AT LOW PRESSURE

An interesting experiment on phenomena of electric discharges through gases at reduced pressures may be performed by the use of the apparatus shown in Fig. 44-1. A glass tube about 3 ft long is connected to a vacuum pump. Cylindrical aluminum electrodes sealed into the ends of the tube are attached to the terminals of a source of high voltage, such as an induction coil. When the gas in the tube is at atmospheric pressure, the sparks will pass across the short air gap between the terminals of the induction coil. As the gas

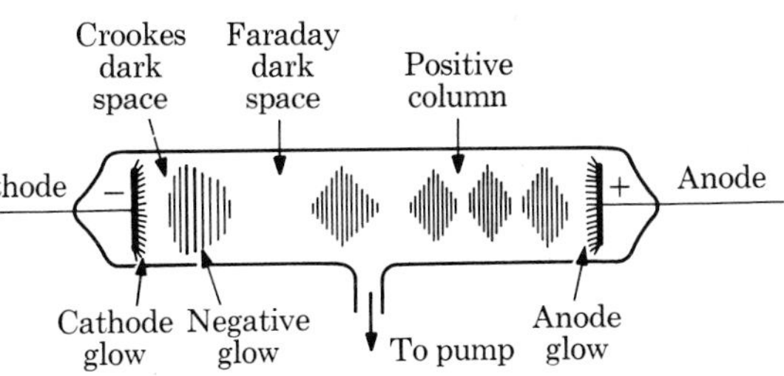

Figure 44-1
Discharge of electricity through a gas at reduced pressure.

pressure in the tube is continuously reduced, the discharge begins to pass through the long tube, in preference to the shorter path between the terminals of the coil in air at atmospheric pressure. The gas in the tube emits light of a color characteristic of the particular gas used. When air is used, the first discharge to appear consists of long, sparklike streamers emitting bluish-violet light. As the pressure is reduced further, a glow appears on the cathode, or negative terminal. At still lower pressure, this glow moves away from the cathode, and a pinkish glow appears throughout most of the tube. The appearance of the tube at a pressure of a few tenths of a millimeter of mercury is shown schematically in Fig. 44-1. Each of the electrodes is covered by a velvety glow, known, respectively, as the cathode and anode glow. A comparatively dark space near the cathode is called the *Crookes dark space*. Near it is a short region of light known as the *negative glow*, this being followed by another darker portion designated the *Faraday dark space*. The major portion of the tube is filled with a striated series of bright and dark regions called the *positive column*.

If the pressure of the gas in the tube is lowered below the optimum value for the type of discharge just described, it will be observed that the Crookes dark space becomes larger, finally filling the whole tube. At this stage most of the ions traverse the whole length of the tube without colliding with a molecule of the gas. Consequently, there are few ions produced, and the discharge current decreases. The voltage required to maintain the discharge rises rapidly until finally, at very high vacuum, the discharge becomes nearly impossible.

44-4 CATHODE RAYS

At intermediate pressure, about 0.001 mm Hg, many of the positive ions will traverse nearly the whole length of the tube without collision and will strike the cathode with enough energy to cause it to emit a considerable number of electrons. These electrons stream away from the negatively charged cathode in directions at right angles to the surface. Such electrons are called *cathode rays*. Historically the study of their properties has yielded very rich returns.

Some of the properties of cathode rays can be demonstrated rather easily. A few of these properties are listed below.

1 *Cathode rays travel in straight lines perpendicular to the surface of the cathode.* If a discharge tube contains a metal obstruction in the path of the cathode rays, the shadow cast by the obstruction is sharp, indicating that the rays travel in straight lines (Fig. 44-2). If the surface of the cathode is made concave, the rays are focused at the center of curvature of the cathode (Fig. 44-3),

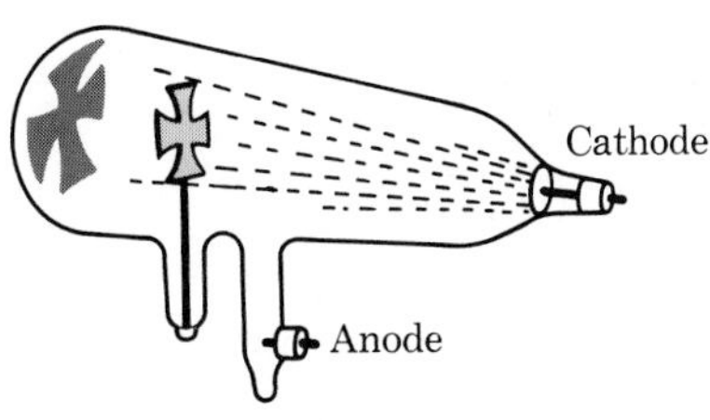

Figure 44-2
Cathode rays travel in straight lines.

which shows that they travel at right angles to the surface.

2 *Cathode rays have kinetic energy.* When an intense beam of cathode rays is allowed to fall on a target, the target is heated by the impact of the rays (Fig. 44-3). Also, if the target is movable, it can be set into motion by the impact of the particles.

3 *Cathode rays can produce fluorescence.* If a beam of cathode rays is allowed to fall on a suitable material, light is emitted. The color is characteristic of the fluorescing material. If the walls of the tube are ordinary soft glass, they will fluoresce with an apple-green color. Some materials will continue to emit light for a short time after the beam has been discontinued.

4 *Cathode rays can be deflected by a magnetic field.* If cathode rays are confined in a small beam and a magnet is brought near, the pencil of rays is deflected in the direction in which moving negative charges would be deflected.

5 *Cathode rays are deflected by an electrostatic field.* If the narrow pencil of rays is passed between two plates, one of which is positive and the other negative, the electrons will be deflected toward the positive plate.

6 *Cathode rays can produce x-rays.* If the energy of the cathode rays is sufficiently high, very penetrating radiation is emitted when they strike a target. These penetrating rays are called *x-rays*. Their properties will be discussed in a later chapter.

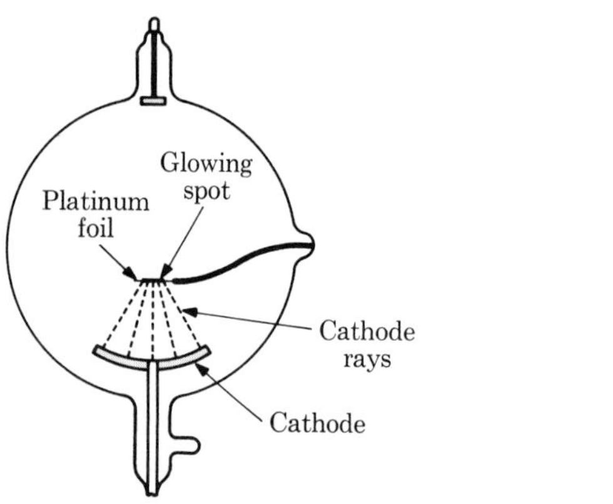

Figure 44-3
Heating effect of cathode rays.

44-5 ELECTRONS

J. J. Thomson studied the properties of cathode rays by use of the deflection in electric and magnetic fields. In Fig. 44-4 is shown a diagram of apparatus similar to that used by Thomson. The cathode rays are accelerated as they move toward the anode. Some of them pass through the opening in the anode, and a narrow pencil proceeds into the region beyond the anode. There the particles pass between two plates spaced a distance s apart and differing in potential by V. Each particle will experience a force as it moves between the plates and it will be deflected downward,

$$F = Ee \tag{1}$$

where E is the strength of the electric field and e is the charge of the particle. But $E = -\Delta V/\Delta s$. If the field between the plates is uniform, the magnitude of the field strength will be $E = V/s$ and

$$F = \frac{V}{s} e \tag{2}$$

If V is in volts, s in meters, and e in coulombs, Eq. (2) gives the force in newtons. Since the force is constant and perpendicular to the initial path, the path of the particle between the plates is parabolic like that of a projectile.

If in place of the electric field a magnetic field is used, directed out of the paper toward the reader, the particles again experience a force, this time perpendicular to the direction of motion of the particles and the particles will move in a circular path while in the field. The force on a current is

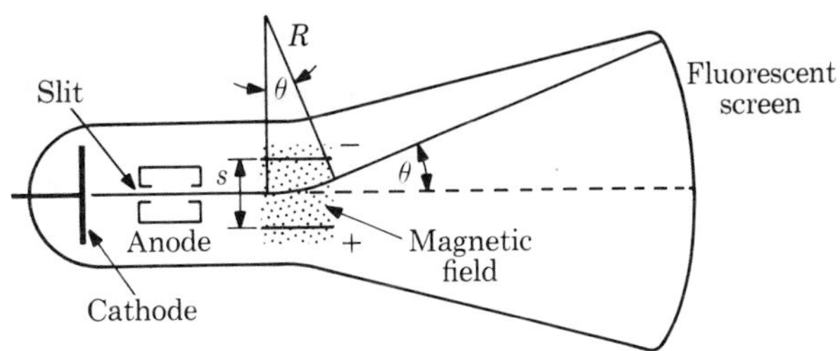

Figure 44-4
Diagram of apparatus used to measure e/m for the electron.

$$F_1 = BIl$$

If in the stream of electrons there are n electrons per unit length, each moving with a speed v, the current is $I = nev$ and

$$F_1 = Bnevl$$

But there are nl electrons in the field. Thus the centripetal force F on a single electron is $F = F_1/nl$, and

$$F = Bev = \frac{mv^2}{R} \tag{3}$$

$$v = \frac{BeR}{m} \tag{4}$$

If both electric and magnetic fields are present at the same time, the magnetic field deflects the particles upward, while the electric field deflects them downward. Thus one can adjust the field so that the resultant force is zero and there will be no deflection of the beam as shown by no change in position of the spot on the fluorescent screen. Then, from Eqs. (2) and (3)

$$\frac{V}{s}e = Bev \tag{5}$$

When we substitute for v from Eq. (4), we obtain

$$\frac{V}{s}e = Be\frac{BeR}{m} \tag{6}$$

$$\frac{e}{m} = \frac{V}{sB^2R} \tag{7}$$

Thus from measurements of V and B Thomson was able to determine the ratio of charge to mass for the electron and found it to be much greater than the corresponding ratio for the hydrogen ion. Later, more accurate measurements showed its value to be about 1,837 times that of the hydrogen ion. The present most probable value of e/m for the electron is 1.75890×10^{11} C/kg.

44-6 CHARGE OF THE ELECTRON

Thomson's experiment enabled him to determine e/m but not to determine independently either e or m for the electron. Townsend, Thomson, and Millikan carried out experiments designed to measure the charge e. Townsend and Thomson worked with clouds of water droplets which when charged were supported by electric fields between two parallel horizontal plates. These experiments were not very accurate, partly because of evaporation of the water droplets. Millikan, in his famous oil-drop experiment, modified and improved the procedure, minimizing evaporation by the use of oil and observing individual small droplets for comparatively long times. His appa-

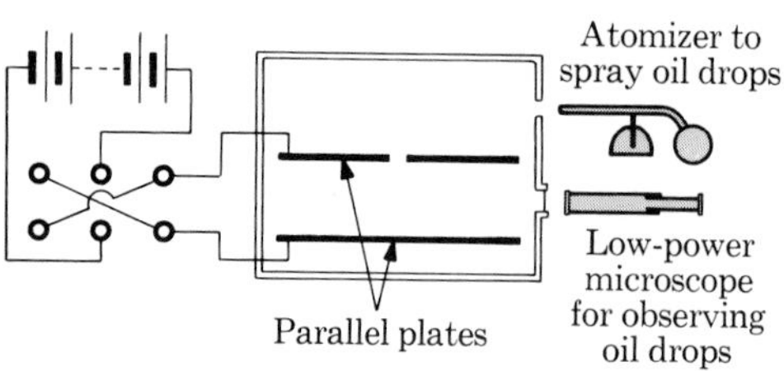

Figure 44-5
Apparatus for an oil-drop experiment. Ionization in the region between the plates may be produced by a radioactive material or an x-ray tube, not shown.

ratus is shown schematically in Fig. 44-5. The size of the droplets could be determined from the rate of fall when the plates were uncharged. By adjusting the potential difference between the plates, the droplet could be held stationary or caused to rise or fall in the field of view. When the droplet is stationary, its weight is balanced by the force of the field

$$\frac{V}{s}q = mg \tag{8}$$

where V is the potential difference of the plates, $q\,(= ne)$ is the charge on the droplet, s is the distance between the plates, and m is the mass of the droplet. By using many droplets of different sizes and various charges, it was found that all charges were whole-number multiples of a smallest charge e. At present the most probable value for e, which may be determined by various methods, is $e = 1.60206 \times 10^{-19}$ C. The electronic charge e is a natural unit of electric charge.

From the values of e and e/m, one can compute the mass of the electron,

$$m = \frac{e}{e/m} = \frac{1.60206 \times 10^{-19}\ \text{C}}{1.75890 \times 10^{11}\ \text{C/kg}}$$
$$= 9.1083 \times 10^{-31}\ \text{kg}$$

44-7 CONDUCTION OF ELECTRICITY IN A VACUUM

In all conduction of electricity there is a transfer of some kind of charged particle. Hence in a perfect vacuum there can be no conduction at all. At high vacuum, the number of ionic carriers is so small that the current is negligible even at high voltages. Few ions are produced by collision because of the rarity of such occurrences. Hence if there is to be appreciable conduction in a region of high vacuum, ions must be introduced by some process.

44-8 ELECTRON EMISSION

Electrons may be obtained from atoms in all states of matter: solid, liquid, and gas. However, we shall here be concerned with four methods of liberating electrons from metal surfaces.

In *field emission,* electrons are stripped from a metal surface by a high electrostatic field, of the order of millions of volts per centimeter. In the field-emission microscope (Fig. 44-6), the surface structure of a pointed metal emitter may be made visible in an image produced on a fluorescent screen by electrons diverging from the tip. In a field ion microscope, Dr. E. W. Müller has used the emission of positive ions from a needle-like specimen to obtain images in which individual atoms can be distinguished.

In *secondary emission,* a metal is bombarded with high-speed particles which cause the ejection of electrons from the surface. Depending on the type and energy of incident particle, generally the arrival of each particle causes the emission of two or more electrons. This effect is utilized in the photoelectric multiplier tube (Fig. 44-7), to amplify weak light signals.

It was shown earlier that in *photoelectric emission,* light or other radiation of suitable frequency falling on a surface causes emission of electrons. A photoelectric tube may consist of a cathode

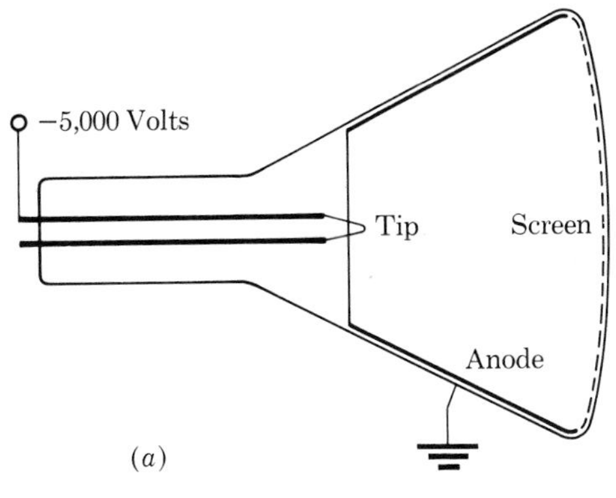

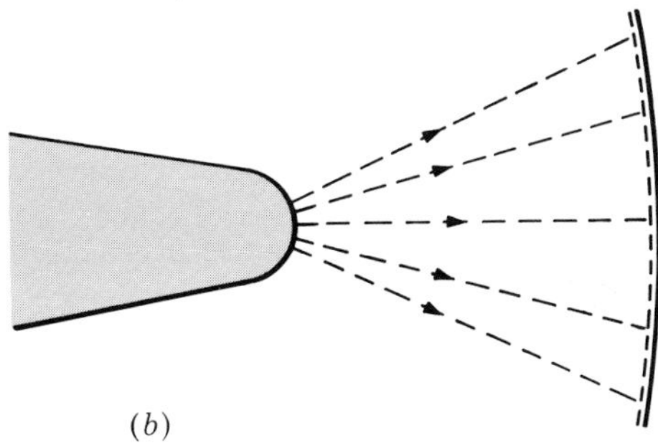

Figure 44-6
Field-emission microscope. (*a*) Cross section of tube; (*b*) paths of electrons leaving rounded tip of needle, here enormously enlarged and moved close to the screen.

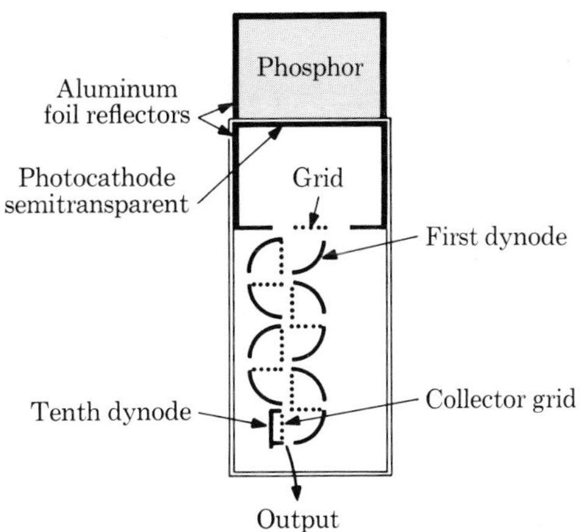

Figure 44-7
Photomultiplier in a counter for alpha particles. An individual alpha particle produces a scintillation in the phosphor, a clear crystal of naphthalene. This light causes ejection of an electron from the photocathode surface. This electron is accelerated to the first dynode, where it starts a cascade of secondary emission.

having a photosensitive surface and an anode suitably arranged in a glass bulb, which is either evacuated or filled with an inert gas at low pressure. Common light-sensitive surfaces are potassium, cesium, sodium, rubidium, and certain oxides. The photoelectric current from the cathode is proportional to the intensity of the incident radiation. We have seen that there are numerous applications of the photoelectric effect in photoelectric exposure meters, in the reproduction of sound from motion-picture films, and in light-operated relays and signaling devices. This topic will be considered again in the Modern Physics section.

In *thermionic emission,* electrons are "boiled" from a filament maintained at a high temperature. This effect was discovered by Thomas A. Edison during his development of light bulbs and was applied by J. A. Fleming, Lee De Forest, and others in a variety of electron tubes for the generation, amplification, and detection of signals in the form of electrical oscillations.

44-9 THERMIONIC VACUUM TUBE; THE DIODE

The simplest electron tube is of the type used in Edison's experiment (Fig. 44-8). It is a two-element tube, or diode, with a heated filament F and a plate P. The diode conducts a current

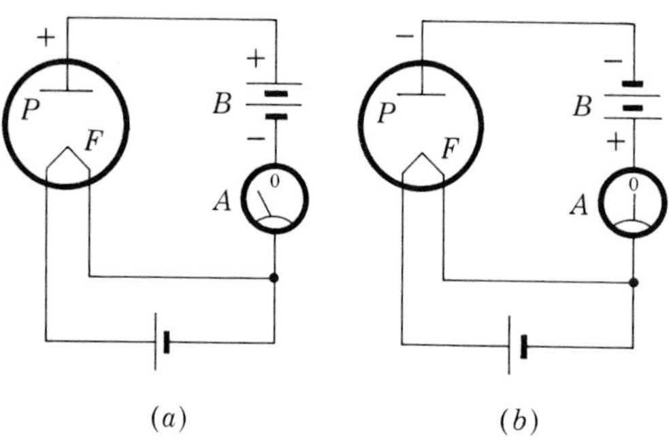

Figure 44-8
Thermionic emission in a two-element tube, discovered by Edison: (*a*) current from plate to filament; (*b*) no current in tube.

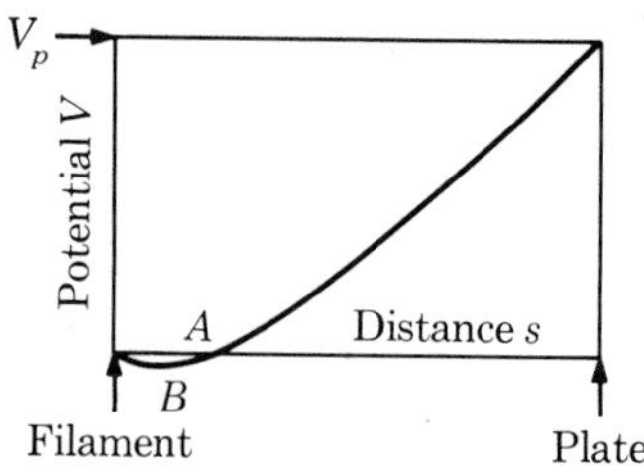

Figure 44-10
Variation of potential between filament and plate.

when the plate is positive with respect to the filament, but the current is zero when the plate is negative. The electric field is far too small to cause field emission from the plate, but electrons are liberated from the filament through thermionic emission. Because the vacuum tube does not obey Ohm's law, it is called a *nonlinear* circuit element.

When the filament is heated by an electric current, electrons are emitted. If the plate is made positive with respect to the filament, electrons will be attracted to the plate and there will be a current in the tube. If, however, the plate is made negative with respect to the filament, the electrons will be repelled and there will be no current. The diode thus acts as a *valve,* permitting flow in one direction but not in the other. If it is connected in an ac line, the diode acts as a rectifier; there is a current during the half cycle in which the plate is positive.

If the plate is positive with respect to the filament, electrons will flow across but not all the electrons that come out of the filament reach the plate, because of the space charge (Fig. 44-9). Figure 44-10 shows a graph of potential against distance across the tube. Because of the space charge, the potentials at points out to A are below the potential of the filament. An electron will reach the plate only if it has sufficient speed as

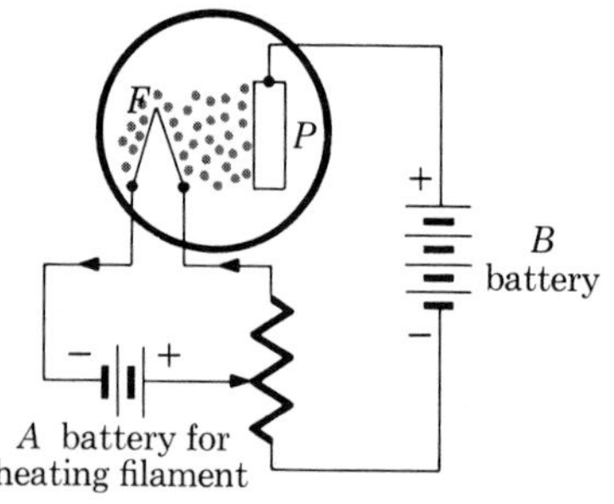

Figure 44-9
Space charge in a diode.

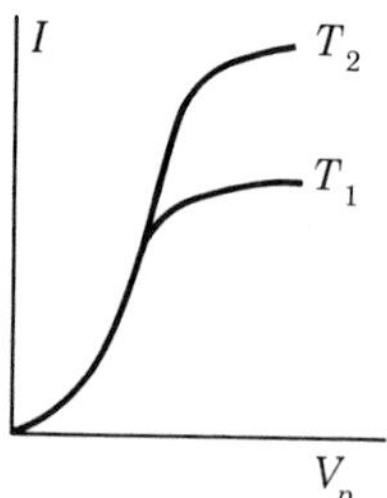

Figure 44-11
Plate current as a function of plate potential.

it leaves the filament to reach B, the point of lowest potential, before it is stopped. If the difference of potential V_p between filament and plate is increased, the potential at B rises and more electrons will be able to reach the plate. The current depends upon V_p, as is shown in the graph of Fig. 44-11. At a temperature T_1 the current increases as the potential is raised until nearly all the electrons emitted by the filament reach the plate. Further increase of V_p produces no change. *Saturation* has been reached. At a higher temperature T_2 more electrons are emitted; hence the saturation current is higher.

If the plate potential is kept constant while the filament temperature is increased, the current increases at first but reaches saturation because of the increase in the electron cloud around the filament.

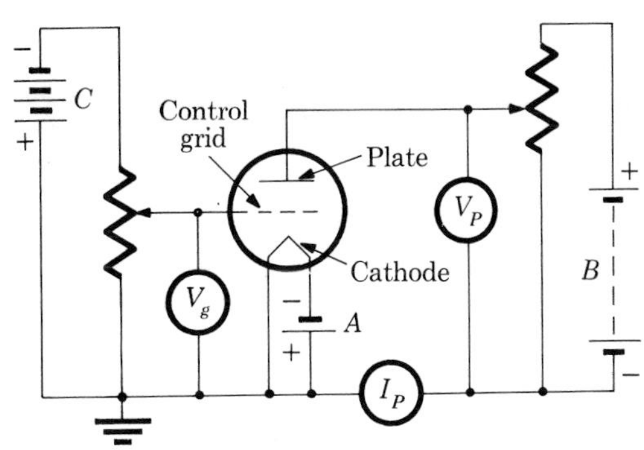

Figure 44-12
Circuit for determining characteristics of a triode.

44-10 THE TRIODE

If a third element, the *grid,* is inserted into the tube near the filament, it can be used as a control for the tube current. Such a tube is called a *triode,* or three-element tube. The grid usually consists of a helix, or spiral, of fine wire so that the electrons may freely pass through it. Small variations of the grid potential will cause large changes in the plate current, much larger than those caused by similar changes in the plate potential. If the grid is kept negative with respect to the filament, electrons will not be attracted to the grid itself and there will be no grid current.

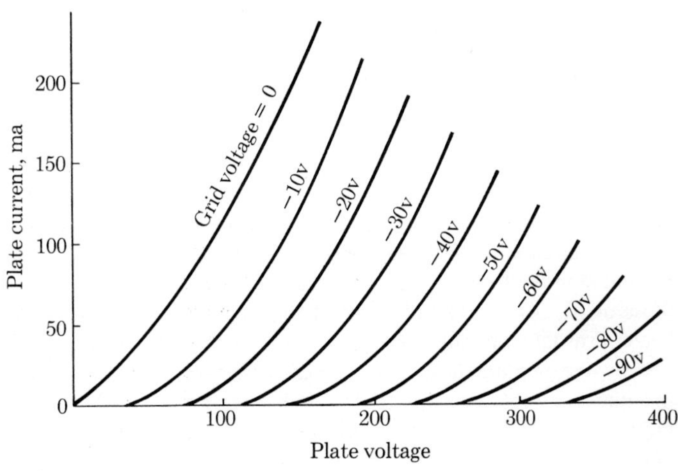

Figure 44-13
Plate-current characteristics for a particular triode.

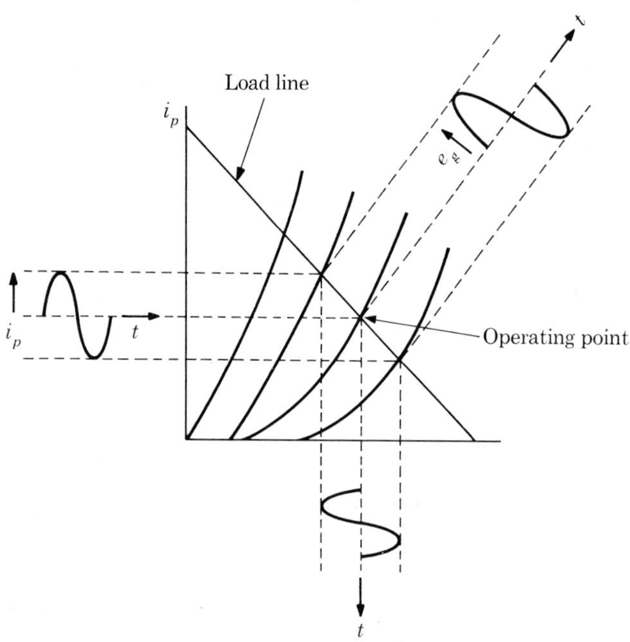

Figure 44-14
Operating characteristics of a vacuum-tube amplifier. The load line passes through corresponding values of plate voltage e_p and plate current i_p for a given load resistor and plate-voltage supply. At the operating point grid voltage e_g equals the negative applied grid bias voltage. Variations of grid voltage produce variations of plate current and plate voltage.

Using the circuit shown in Fig. 44-12, one may study the dependence of plate current on cathode temperature, plate voltage, and grid voltage. The relations cannot be expressed conveniently in algebraic equations; so the information is usually presented in the form of characteristic curves like those of Fig. 44-13. The filament of the cathode heater is usually operated at a fixed voltage specified by the manufacturer; the curves then give the plate current for various grid voltages and plate voltages relative to the cathode. It will be noted that changes in grid voltage have much more effect on plate current than do changes in plate voltage.

A typical variation of plate current with grid potential V_g is shown in Fig. 44-14. A part of the curve is practically a straight line. If the grid voltage varies about a value in this region, the fluctuations of the plate current will have the same shape as the variations of grid voltage. The tube will *amplify* the disturbance without distorting it.

The triode also acts as a *detector,* or partial rectifier, if the grid voltage is adjusted to the bend of the curve. With this adjustment an increase in grid voltage above the average produces considerable increase in plate current, but a decrease in grid voltage causes little change in plate current. The plate current fluctuates in response to the grid signal, but the fluctuations are largely on one side of the steady current.

In Fig. 44-15 is shown a simple receiving cir-

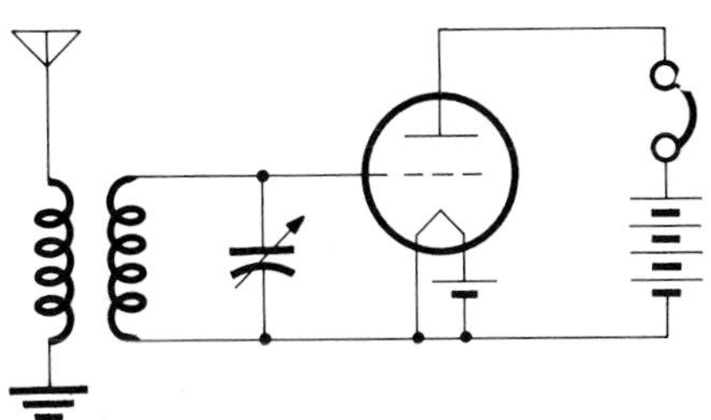

Figure 44-15
Simple receiving circuit.

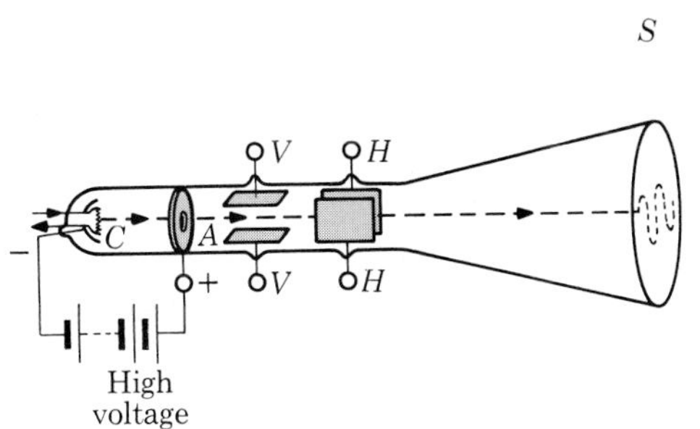

Figure 44-16
Cathode-ray tube.

cuit. When the waves strike the antenna, they set up oscillations in the circuit, which is tuned to the frequency of the waves. This causes the potential of the grid to vary, and the tube, acting as a detector, produces a variation in current according to the amplitude of the signal. This causes the earphones to emit sound.

In radio circuits, triode electron tubes are used to produce high-frequency oscillations, to act as detectors or rectifiers, and to act as amplifiers. Tubes of different characteristics are used for each of these purposes.

In many tubes the filament merely acts as a heater of a sleeve that covers it and is insulated from it. The sleeve, or *cathode,* is the element that emits electrons.

For many purposes tubes are constructed with more than three active elements. They are named from the number of active elements, as *tetrode, pentode,* etc.

44-11 CATHODE-RAY TUBES

The fact that cathode rays are really electrons traveling at high speed away from the cathode is made use of in instruments such as television tubes and cathode-ray oscilloscopes. The oscilloscope tube was first developed by the physicist Braun in 1897. The essentials of a modern tube are shown in Fig. 44-16. Cathode rays from the thermionic cathode C pass through a small hole in a control grid and are accelerated by the voltage applied to the anode A. The electrons then pass between two pairs of parallel plates to which various test voltages may be applied. One pair of plates, V, is horizontal; the other, H, is vertical. The electric fields of these plates therefore cause deflections, either vertically or horizontally in proportion to the values of the test voltages applied. The spot of light on the fluorescent screen S at the large end of the tube can travel around with extraordinary rapidity, because of the very small mass of the electrons and their high speeds. Hence these tubes can be used to follow transient variations of voltage, which are of entirely too high frequency for any mechanical device to follow.

44-12 TELEVISION

Of the many things that the development of electronics has provided, possibly the most striking contributions have been made in the field of communication and especially in that form of visual communication known as television. The production of television pictures and their transmission, along with accompanying audio signals, is an involved and technical process. However, related material discussed in the earlier portions of this book enable us to present at least a superficial explanation of the nature of television here.

Television pictures are sent basically in the

same manner that newspaper pictures have been transmitted since 1907. Black-and-white pictures actually consist of a large number of black dots appearing on a white background. The density of these dots and their arrangement determines what the picture looks like. In the process of transmitting pictures, a scanning device passes over the original photograph and transmits impulses received by it as it moves across the light and dark portions of the picture. The first commercially successful method of transmitting pictures by wire was developed in 1925 by Ives and Horton, who used a photoelectric cell to receive reflected light from a picture. The variation of the light was interpreted by the photocell circuitry in terms of varying electric currents. These electric signals were transmitted by some communication channel (in earlier days by wire and later by radio) to a receiver which then reconverted the signals back to light rays, which retraced the picture on a photographic film. A similar technique is still in use today. The quality of the reproduction depends upon the size of the spot of light used in scanning and the number of lines scanned per picture in the reproduction process. The smaller the area scanned with each beam of light and the greater the number of lines used to scan each picture, the finer the quality.

Since in television an attempt is made to reproduce a constantly changing scene, the scanning process must be very rapid. Whereas the process for newspaper photos, or single-picture transmission, requires up to 20 min of scanning, in television each scene is scanned from 24 to 30 times per second. Due to the persistence of human vision, an observer at the receiving end of the transmission where the scan is retraced on a "picture" tube does not see the point-by-point illumination of the screen but sees the picture as a whole. This is the same process that occurs in moving pictures, where a succession of still pictures, seen at the rate of 24 per second, appear as continuous action.

In television this rapid scanning process is made possible mainly through a specially developed electronic tube called an *orthicon tube,* which is the heart of a television camera. Figure 44-17 is a sketch of the principal parts of an orthicon tube. Light (photons) from the object enters the lens of the camera and is directed on to a screen, which contains millions of small light-sensitive globules of silver. When struck by photons, these globules emit electrons in a manner like tiny photocells. The number of electrons emitted in a given time is proportional to the intensity of the illumination, the brighter portions

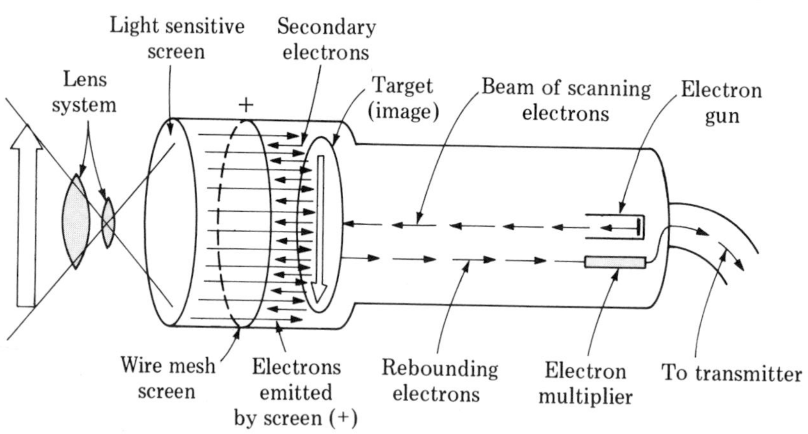

Figure 44-17
Parts of an orthicon tube.

of the scene producing more electrons. These electrons are then attracted to a target where an image forms and more electrons are released by secondary emission. The secondary electrons produced on the target are attracted to a positively charged, fine-mesh screen placed very close to and in front of the target. This leaves the target with a positive charge, concentrated in those areas where secondary electrons were released to the screen. The amount of charge on the target is proportional to the brightness of the portion of the image being viewed. A beam of electrons coming from an electron gun at the opposite end of the tube moves across the target from left to right scanning every point and every line on the target, normally 30 times a second. That is, in every second, 30 pictures are transmitted by the orthicon tube. Each scan traces 525 lines on the target. As the scanning beam explores the target, it releases just enough electrons to each target area to counteract the positive charge residing there due to the loss of electrons to the screen. Then the beam rebounds from the target, lacking electrons in those areas where they were given up to the screen. The rebounding beam, called the *video signal,* varies exactly as the light that originally entered the tube, but the beam is now an "electrical picture." The video signal beam then flows to an electron multiplier where it is amplified and sent to the transmitter where the beam is sent out in much the same manner as are radio waves, but with higher frequencies and with the carrier wave being modulated with the video current from the orthicon tube.

When this signal is picked up by a television antenna and carried to a receiver, the signal is rectified into a fluctuating direct current. The receiver is a cathode-ray tube with a fluorescent screen on one end and an electron gun at the other end. A reverse action to what occurred in the orthicon tube takes place here; that is, each change in electrical current must be converted into a change in illumination. The electron beam is moved across the screen synchronously with the motion of the electron beam in the orthicon tube. The sweep action is controlled by employing electromagnetic deflection and focusing. Each electron that hits the screen causes the screen to fluoresce, and the parts of the beam which contain the most electrons appear brightest on the screen. The signal from the transmitter travels so rapidly that for all intents and purposes one can say that the scene being viewed on the television screen is occurring simultaneously with the original live "happening."

44-13 COLOR TELEVISION

The wonders of television became even more amazing with the perfection of color television. Color systems are based upon the fact that all colors can be produced by combinations or mixtures of the three primary colors, blue, green, and red. One of the basic tasks of a color-television system is to separate the image into the three basic colors. Once this is accomplished, the electrical signals corresponding to these colors are combined, amplified, and then transmitted.

In the color camera there are three separate orthicon tubes, each one designed to respond to one of the three colors, blue, green, and red. Light from the object enters the camera through the lens system and then by a system of mirrors and half-mirrors, as shown in Fig. 44-18, the beam of light is broken up into three parts and directed to the three orthicon tubes. Colored filters are placed in the path of these three beams so that the red portions of the picture go to one tube, the green portions to another tube, and the blue to the third tube. In each orthicon tube the same process occurs as it did in the orthicon tubes used in black-and-white television. In fact the color-television receiver is capable of receiving that type of transmission, since all three orthicon tubes can react to black-and-white images.

The chief difference between a color-television receiver and a black-and-white-television receiver is that there are three electron guns in the cathode-ray tube used for color reception. Each of these guns emits a beam of electrons toward a phosphor plate which contains thousands of color-coded phosphorescent dots, generally arranged in a triangle with a red, a green, and a blue dot. Before the electron beams reach the phosphor

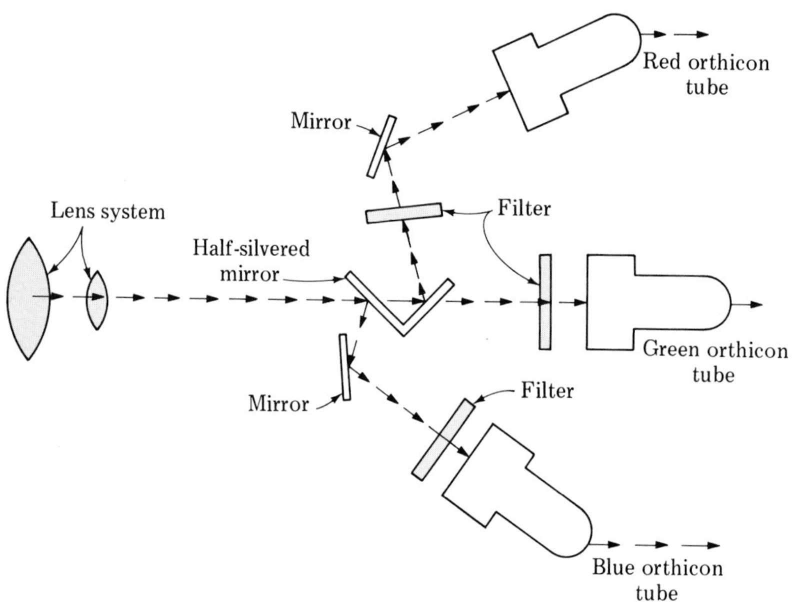

Figure 44-18
The separation of color in a color-TV camera.

plate, they encounter a plate containing thousands of tiny holes, called an aperture plate, so arranged that electrons reach each of the groups of colored dots. The stream of electrons from each gun is focused so that each hits its own color dot and is diverted away from the other color dots in that group. As with the case of black-and-white television, the greater the number of electrons in the beam, the brighter the light and the more intense the color. As in noncolor television, the electron guns scan the phosphor plate rapidly to give the effect of continuous motion. One of the great problems in color-TV transmission has been that of mixing colors in the proper ratio to get "true" color. This technique has now been perfected to a suitable degree, and color is reproduced with some fidelity in present-day color-television receivers. To provide the more detailed picture required in color transmission and to avoid interfering with other stations, color television is transmitted on the ultrahigh-frequency (UHF) wavelengths.

44-14 ELECTRON MICROSCOPE

In our study of optics we found that light can be bent and focused by passing it through lenses. Because of this property, lenses can be used in optical devices such as telescopes and microscopes to produce enlarged images of objects. In the electron microscope, high-speed electrons are used instead of rays of light. The development of the electron microscope was based on the observation that when electrons are passed through a magnetic field, their paths are deflected. By adjusting the strength of the magnetic field, the beam of electrons may be focused, much as a lens focuses light. Further, if a magnetic field is created by an electromagnet in the form of a coil and electrons are passed through it, we can determine the focal length of this "magnetic lens" according to the rules of geometric optics.

This behavior of electrons in magnetic fields has been used to construct electron microscopes to help remove limitations placed upon researchers who were working with small particles. In optical microscopes particles having a diameter of less than 2,000 Å, or 2×10^{-5} cm, or lines on an object which are closer together than that distance cannot be resolved and the image blurs. Electron beams of high energy (60 kV or more) have very short wavelengths (a point to be discussed in detail in Chap. 48) of approximately 5 Å, or 5×10^{-8} cm, which if they could be used, would provide the needed resolution to work with such small objects. However, electrons cannot be used in optical microscopes because they cannot pass through glass. With the discovery that electrons can be focused by electromagnetic fields, the need for lenses was removed and the possibility of using electrons in a microscope became a reality. With electron microscopes the resolving power is many times that of an optical microscope. At present, electron microscopes have a resolving power of about 10 Å which is over 200 times finer than the limit of resolution of the optical microscopes. Resolution to this degree permits observations in the range of molecular dimensions. Also, while the best optical microscopes have a magnification of 2,000 times, the electron microscopes have a magnification of 100,000 times. Figure 44-19 depicts the striking similarities between an optical microscope and an electron microscope.

The electron gun emits a fine stream of electrons. These diverge as they move downward

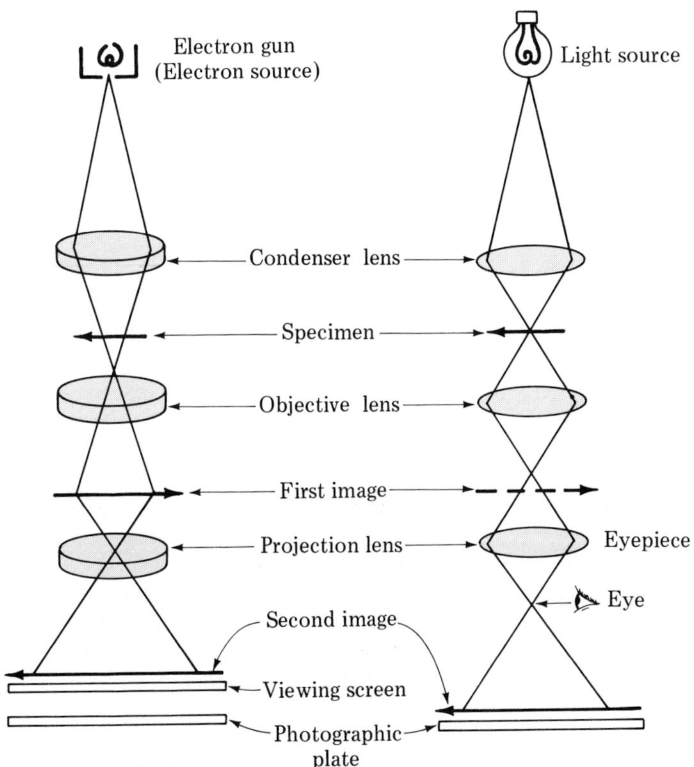

Figure 44-19
Comparison of an electron microscope (left) with an optical microscope (right).

until they pass through the area of the first magnetic field produced by the condensor which causes them to converge and brings them to a focus. At this point the specimen is placed. Where the specimen is dense, many electrons are absorbed; and as the electron beam continues, it beams an image of the material it has traversed. A second magnetic field, the objective, then causes the beam of electrons to diverge, spreading it out so that the image is magnified. A third magnetic field, the projection coil, focuses a small section of the magnified beam on a viewing screen, usually made of a fluorescent material, which is then viewed through an eyepiece. It is commonly the practice to take photographs of the image instead of looking directly at it. This has the added advantage of permitting a permanent record to be made and of further magnification through the process of enlarging the photograph. It is possible to enlarge these photographs 4 to 6 times without much loss of detail, giving pictures 500,000 or more times as large as the specimen.

As was noted earlier in our study of waves, a wave is affected only by an object which is several times larger than its wavelength. For example, sound waves are neither blocked nor seriously affected by objects in their paths which have diameters of a few inches. On the other hand, visible light cannot "see" objects that are smaller than ten-thousandth of an inch. Although electrons and hard x-rays have approximately the same wavelength, electrons are more affected by the solid media they pass through than x-rays and are therefore more effective in fine microscopy. However, x-rays provide a greater penetration than electrons and can serve to make observations which are impossible with electrons by the use of x-ray microscopes.

There are still problems to be solved before electron microscopes reach their maximum potential. A greater resolution than 10 Å is theoretically possible, but problems of construction have prevented reaching this goal. The expense of these electron microscopes is another handicap to large-scale research with them. The price range currently is from $15,000 to $100,000. Also, a problem that may never be solved is that live specimens cannot survive in an electron beam, so study must be limited to dead specimens. Yet, the electron microscope has already yielded great amounts of information about viruses and other small bits of matter and is truly an important tool for the microscopists.

44-15 MICROWAVES

In our study of electromagnetic waves it was shown that the wavelength of these waves can vary greatly (Fig. 43-5). For example, gamma rays have wavelengths of only 10^{-13} m, whereas radio waves have wavelengths from 10^{-3} to 10^{4} m and electric waves can be as long as 10^{7} m. The longest electromagnetic wave has a wavelength 100 billion billion times that of the shortest electromagnetic wave. In the electromagnetic spectrum between radio waves and the shorter infrared waves lies a narrow band of waves having a wavelength range of 1 to 100 cm, called *microwaves*. These microwaves are of special interest in that they have many important applications.

The development of microwaves came about more slowly than that of radio waves, with not much success being realized until the invention of continuous-wave generators, such as klystron or magnetron tubes, which are capable of producing oscillations of a single tunable frequency. These were accompanied by the development of microwave guides, hollow metallic tubes which can confine and guide the microwaves. A typical microwave transmitter is similar to a radio transmitter, consisting of a microwave generator, which in the form of a klystron tube (Fig. 44-20) uses two cavity resonators, one called a "buncher" and the other the "catcher," to produce alternating voltages on the collector and producing waves about a centimeter long, plus a power amplifier and other necessary circuitry.

Microwaves are of value for many reasons. First, they are useful in the field of microwave optics because they have similar properties to light waves and are effective in that their wave-

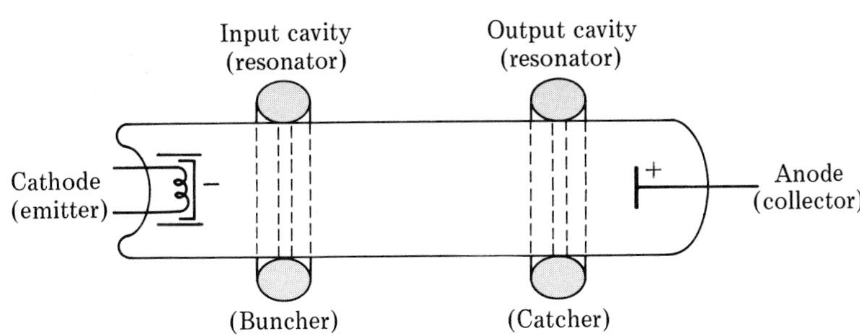

Figure 44-20
Klystron tube.

lengths are comparable to or smaller than the objects which they fall upon, permitting good resolution. They, like light waves, possess the properties of rectilinear propagation, reflection, refraction, and polarization. They usually are monochromatic in frequency and coherent in phase and can be easily manipulated by frequency modulation and amplification. Since microwaves travel in straight lines in a homogeneous medium, they can be used in communication and in direction-finding devices and can easily be reflected and focused. In radio communication, microwaves provide more "space" in useful frequencies. For example, microwaves provide about 100 times more frequency space than the present combined frequency range of radio broadcasting, communication, and television. Another important property of microwaves is that they have the ability to amplify the natural vibrations of matter when they penetrate it.

Possibly the most important application of microwaves is *radar*. Radar is the abbreviated name for *ra*dio *d*etection *a*nd *r*anging. We can trace its inception back to 1926, when Breit and Tuve observed the reflection of short-wave radio signals back from the ionized upper layer of the earth's atmosphere. The first demonstration of locating aircraft by radar was given on February 26, 1935. This technique was developed extensively during World War II in the early 1940s. Radar is used to detect the presence and the range of an object, such as an aircraft, by transmitting a pulsed microwave and receiving the "echo," or reflected wave, of the same pulse from the reflecting object. Since radar waves travel at the speed of light, the length of time between transmission and reflection can be observed and from this the range quickly determined. By the use of a directive antenna, the location as well as the range of the object can be determined with great accuracy.

As radar was being perfected, other microwave applications were being developed as byproducts. For example, microwaves are now used as carrier waves in relay links for multichannel transmission of television, telegraph, and telephone. Earlier in this book we discussed atomic clocks, which are based upon the vibration of cesium or ammonia atoms. These clocks use microwave resonance interactions to amplify their natural vibrations. The *maser* developed in 1955 is another application of microwaves; in fact, the name maser is an acronym for *m*icrowave *a*mplification by *s*timulated *e*mission of *r*adiation. One of the most important applications of the maser is in the field of radio astronomy.

By the use of a microwave generator it is possible to set up interference patterns which can be accurately measured. From such interference patterns, the wavelength of the microwave can be determined. It is a common laboratory exercise to determine the wavelength of such waves

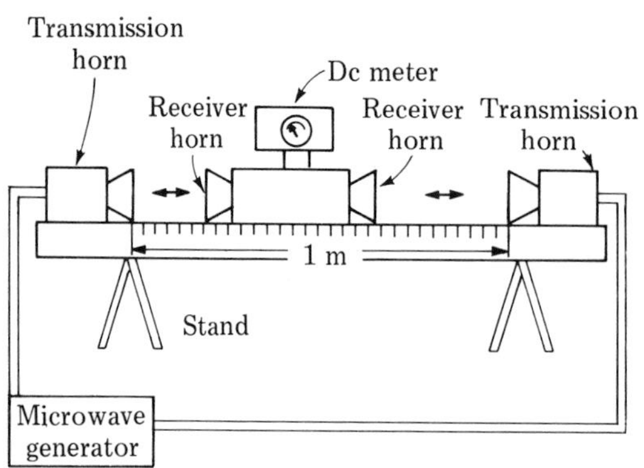

Figure 44-21
Microwave apparatus to determine wavelengths.

using apparatus based upon that used by K. D. Froome in 1954 to determine the speed of electromagnetic waves. Most commercial apparatus designed for this purpose have microwave generators, each consisting of a klystron tube, which produce waves with a frequency of 8,500 to 9,600 megahertz/s providing a wavelength range of 3.15 to 3.53 cm. By applying the data obtained in an experiment into the equation $c = f\lambda$, a third variable can be found if two are known. Figure 44-21 depicts the apparatus used by Froome in his work. He used a wavelength of about 1 cm and a frequency of 24,000 mHz/s to determine the speed of the wave. By moving the receiving horns back and forth, the maximum and minimum interference points were located and the wavelength could be measured accurately.

Microwaves are now being used in the home. Microwave ovens which can bake a potato in 4 min, a $3\frac{1}{2}$-lb roast in 25 min, and a cake in 7 min are, like many other home products, based on research and development for industrial and military uses. These ovens use a magnetron vacuum tube, similar in operation to the klystron, to generate microwaves. How do the microwaves cook food? In an ordinary oven food is cooked either by heat convection, when something is baked, or by radiation, when something is broiled. In both cases, the outside of the food absorbs the heat energy and the food itself conducts the heat inside. The outside gets hot first and cooks faster, hence meat and cakes get brown on the outside. In microwave cooking, the food is heated throughout and all at the same time by molecular agitation. Since molecular activity depends upon the amount of heat energy present in an object, the microwaves are increasing heat energy in food when they penetrate it and, hence, cause it to be cooked. One of the problems with this type of cooking, however, is that since the meat cooks evenly throughout, the outside will be the same color as the inside of the meat, and if you wait for the outside to turn brown, the inside will be the same degree of brown (burnt).

A similar process is used in diathermy machines which direct microwaves to various parts of the body where they cause a uniform heating of the body through increased internal molecular vibration. The heat produced in this manner has many applications, such as in treating arthritis and muscular strains and sprains.

44-16 POSITIVE RAYS; MASS SPECTROGRAPH

The atomic masses of elements historically have been measured by conventional chemical methods, using comparatively large amounts of the material under test. Developments in electronics have provided scientists with apparatus far more precise and capable of making measurements on individual atoms and ions. One such device is known as the *mass spectrograph,* a well-chosen designation, since it enables measurements to be made of atomic masses and to separate a mixture of ions into a veritable spectrum of atoms having different masses.

A form of mass spectrograph designed by Dempster (1936) is shown schematically in Fig. 44-22. This apparatus is based upon the fact that a beam of ions moving through electric and magnetic fields suffers a deflection that depends upon

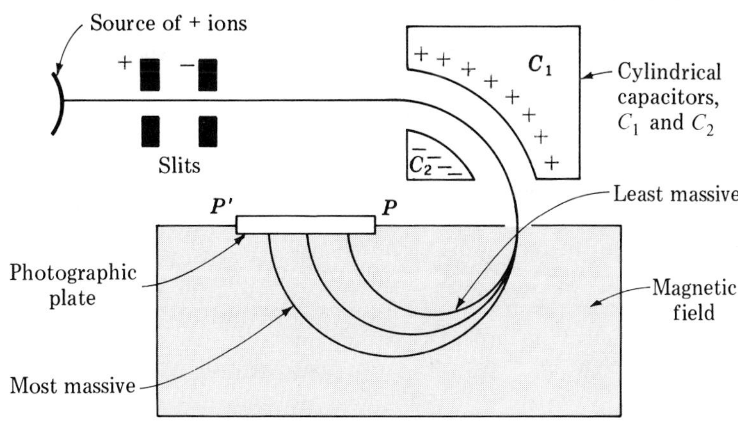

Figure 44-22
A Dempster-type mass spectrograph.

the charges and masses of the ions. Hence ions of various masses are deflected differently. Positive ions from a source near the slits are made to pass through a radial electric field set up by the cylindrical capacitor plates C_1 and C_2. The ions are thereby bent into paths that are arcs of circles. Those which emerge from the slits have a common velocity and enter a magnetic field that is at right angles to the electric field. The magnetic field causes the ions of different masses to be separated so that they fall at different places on the photographic plate PP'. The blackening of the plate gives clear indication of the masses of the respective ions. A typical mass spectrogram is shown in Fig. 44-23.

44-17 ISOTOPES

Studies of positive rays first showed that not all atoms of a given element have identical atomic masses. Although the various atoms of a given element have exactly the same chemical properties, it was shown by mass spectrographic studies that such atoms frequently have different atomic masses. Elements which have the same chemical properties (atomic number, or position in the periodic table) but different masses are known as *isotopes*. Chlorine, atomic mass 35.46, is a mixture of isotopes of masses 35 and 37. The case of hydrogen is an extreme example. Its atomic mass is 1.00813, and it is a mixture of ordinary hydrogen of mass 1 and "heavy hydrogen" of mass 2. This isotope (unlike other cases) has been given the special name of *deuterium*. A third isotope of mass 3 has also been identified. It has been named *tritium*. Various elements are composed of quite different numbers of isotopes, ranging from 1 for many elements to 10 for tin.

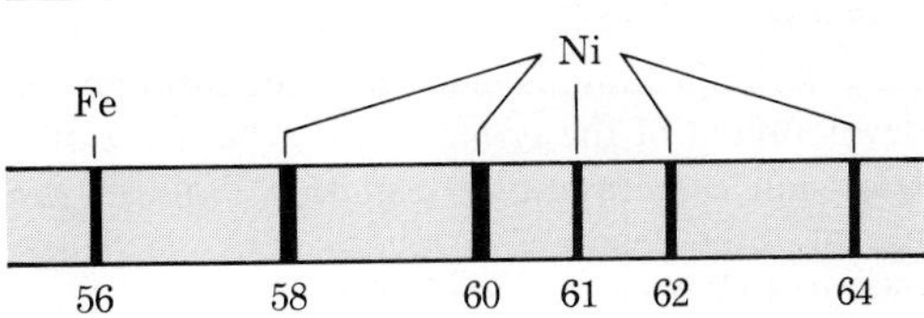

Figure 44-23
A mass spectrogram showing isotopes of nickel and iron.

SUMMARY

All electrical conduction consists of motion of some charged particles. In metals the carriers are electrons; in electrolytes they are positive and negative ions formed by dissociation; in gases they are positive and negative ions produced by collisions between ions and molecules of the gas.

Gases under standard conditions are very poor conductors but become rather good conductors at reduced pressure.

Cathode rays are streams of electrons. They emerge normally from the surface of the cathode, travel in straight lines, may be easily deflected by magnetic or electric fields, cause fluorescence when they strike many minerals, and produce heating, chemical, and physiological effects upon matter.

Electrons are basic particles of electricity. All charges are multiples of the electronic charge. The ratio of charge to mass of the electron can be determined by means of deflection in magnetic and electric fields. The charge was first accurately determined by the Millikan oil-drop experiment.

In order to have conduction in a vacuum, charges must be introduced. This is accomplished by *thermionic emission, photoelectric emission, secondary emission,* or *field emission.*

Two element tubes, or *diodes,* act as rectifiers in ac circuits.

Three-element tubes, or *triodes,* may be used as amplifiers, oscillators, or detectors.

Cathode-ray oscilloscopes are used to study rapidly varying voltages and other transient or cyclic phenomena. As such they are useful in radar and television.

Television, an example of the use of electronics for communication, is made possible through the development of the *orthicon tube,* the heart of a television camera. *Color television* requires the use of three separate orthicon tubes, each one designed to respond to one of three colors—blue, green, and red.

The *electron* microscope, which may have a resolution more than 200 times finer than that of light microscopes, is based on the property of electrons which permits them to be bent and brought to a focus by a magnetic field.

Microwaves are electromagnetic waves having wavelengths from 1 to 100 cm. They have many important applications in communication, direction finding (radar), atomic clocks (maser), and heating.

Positive rays are streams of positive ions. They are used in the mass spectrograph for the identification and measurement of atomic masses and the study of isotopes.

Questions

1 The development of the triode provided an important advantage over the diode. Explain what this advantage was.

2 Prepare a list of electronic devices based upon one or more of the properties of electrons presented in Sec. 44-4.

3 Would a spark discharge begin at the same potential difference for all gases at a given pressure?

4 How might the voltage required to produce lightning discharges between clouds vary from clouds near the earth as compared with those very high in the atmosphere?

5 List and describe four methods of liberating electrons from metal surfaces.

6 Could there ever be a disruptive discharge of electricity between the earth and the moon? Why?

7 What potential with respect to the cathode is maintained on the grid of a triode? Why?

8 State and describe briefly the three main functions that are performed by thermionic vacuum tubes in modern radio receivers.

9 In using a three-element tube as an amplifier, it is desirable to arrange the circuit so that the linear portion of the characteristic curve is used. Show why this is desirable and what happens when other portions of the curve are used.

10 Why is it true that a thermionic tube should produce a louder radio signal than a crystal detector?

11 A small paddle wheel is balanced on a pair of horizontal parallel bars located between a cathode and an anode in an evacuated tube. When a discharge occurs between the cathode and the anode the paddle wheel begins to turn. What important properties of electrons does this demonstration show?

12 Explain why it is expected that the wireless transmission of power over long distances is commercially impractical. Why, then, is it possible to utilize receiving sets at great distances from the transmitter?

13 Radio reception is frequently very seriously interfered with when an electric shaver is being used near the radio receiver. Show why this is to be expected. How may this be minimized?

14 What properties determine whether a television transmission and reception is of high quality?

15 "Persistence of vision" is necessary to see action on a television picture tube. What does the phrase mean?

16 Define carefully the nature and importance of a magnetic lens. In what type of instrument is it used?

17 A sign over the doorway to a laboratory where microwaves are being used cautions people wearing pacemakers not to enter. Why do you suppose this rule is established?

18 Show clearly why the mass spectrograph gives data on the atomic masses of individual ions, whereas conventional chemical methods yield results only on average atomic masses.

19 A circuit consists of a battery, a rheostat, and a photoelectric cell, connected in series. The voltage across the rheostat is found to change with the illumination on the photoelectric cell. Explain why this is true.

20 What actually cooks the meat in a microwave oven? Explain the process.

Problems

Whenever needed in the following problems, take the electronic charge to be 1.60×10^{-19} C, the electronic mass 9.11×10^{-31} kg, and the mass of an atom of unit atomic mass 1.66×10^{-27} kg. An alpha particle has a mass of four atomic mass units and a positive charge of two electronic units.

1 In 1 h how many electrons pass a point in a wire in which there is a current of 3.00 A?

2 An electronvolt is a unit of energy. Express this unit in joules. What is the speed of an electron accelerated by a potential difference of 1.00 V? *Ans.* 1.60×10^{-19} J; 5.93×10^{5} m/s.

3 An electron is accelerated by a potential difference of 12 V. Find the energy acquired by the electron and its speed.

4 What energy is acquired by an electron in falling through a potential difference of 50.0 V? What is its speed after this acceleration if it starts from rest? *Ans.* 8.00×10^{-18} J; 4.19×10^{6} m/s.

5 The potential difference between the filament and the plate of a diode is 120 V. What speed does an electron acquire in moving from the cathode to the anode of this tube?

6 A beam of cathode rays equivalent to a current of 10.0 mA impinges on a thin sheet of metal with a speed of 5.00×10^{9} cm/s. (*a*) How many particles strike the metal sheet per second? (*b*) If their speed is halved in passing through the metal sheet, how much heat do they develop per second? *Ans.* 6.25×10^{16}; 12.8 cal.

7 A stream of electrons from the cathode of a triode equivalent to a current of 1.36 mA strikes a thin metallic grid with a speed of 8.64×10^{6} m/s. The grid reduces the speed of the electrons to one-third the original value. At what rate is heat developed in the grid?

8 What speed must an electron have for its path to be a circle of radius 1.00 cm if it is projected normal to a magnetic field where $B = 0.0020$ Wb/m^2? *Ans.* 3.5×10^{6} m/s.

9 A stream of particles of atomic mass 4.0 and double electronic charge is accelerated by a potential difference of 100 V and projected midway between two parallel plates 1.00 cm apart and 3.00 cm long. What is the deflection of the beam on a screen 15.0 cm beyond the plates when a potential difference of 20.0 V is maintained between the plates?

10 Calculate the period of revolution of a free electron moving in a plane perpendicular to the magnetic field of the earth with a speed of 2.65×10^6 m/s at a place where the magnetic induction is 0.76×10^{-4} Wb/m^2. *Ans.* 0.47 μs.

11 An alpha particle is constrained to move in a circular path of radius 354 mm by a uniform magnetic field of induction of 0.085 Wb/m^2. (*a*) What is the speed of the particle? (*b*) What work is done by the field on the particle?

12 A proton moving with a speed of 1.15×10^5 m/s enters at an angle of 40° a uniform magnetic field of induction 0.283 Wb/m^2. Calculate the acceleration of the proton.

Ans. 2.02×10^{12} m/s^2.

13 The electron in a hydrogen atom revolves in an approximately circular orbit of radius 5.28×10^{-9} cm. (*a*) If a hydrogen atom is situated in a uniform magnetic field of induction 11.2 Wb/m^2 perpendicular to the orbit, what is the centripetal magnetic force? (*b*) Compare this force with the electrostatic force of attraction by the nucleus.

14 A stream of electrons is accelerated by a potential difference of 50 V and proceeds into a uniform magnetic field where $B = 0.0080$ Wb/m^2. What is the radius of the path in the magnetic field? *Ans.* 3.0 mm.

15 Calculate the nature and magnitude of the path described by an alpha particle which is projected with a speed of 4.25×10^2 m/s at right angles to a uniform magnetic field of induction 0.382 Wb/m^2.

16 A radar transmitter generates waves which are 50 cm in wavelength. Determine the frequency of the signal. *Ans.* 6.0×10^8 Hz.

17 A radio station broadcasting on an assigned frequency sends out waves 600 m long. What is its assigned frequency?

18 What electric field would just support a water droplet 1.0×10^{-4} cm in diameter, carrying one electronic charge? *Ans.* 3.2×10^{-4} N/C.

19 An oil droplet whose mass is 2.5×10^{-11} g and which carries two electronic charges is between two horizontal plates 2.0 cm apart. Assume that the droplet is entirely supported by electric forces. What must be the potential difference between the plates to support the droplet?

20 Two isotopes of copper have mass numbers of 63 and 65. If the positive ions are accelerated by a potential difference of 25 V and deflected 180° by a magnetic field ($B = 0.0200$ Wb/m^2), what will be the separation of the lines on the photographic plate? Assume that each ion carries a single electronic charge. *Ans.* 1.0 cm.

21 Positive ions each carrying a single electronic charge are accelerated in a Dempster-type mass spectrograph by a potential difference of 30 V. They leave the slit moving normal to a magnetic field in which $B = 0.018$ Wb/m^2. They are deflected 180° in the field and strike a photographic plate. One group of ions is known to be sodium, atomic mass 23; the others are unknown. On the plate the second and third lines are, respectively, 1.0 and 1.8 cm beyond the sodium line. Find the atomic mass of each of the unknowns.

The most incomprehensible thing about the universe is that it is comprehensible.

Albert Einstein

PART FOUR

Modern Physics

Alfred Kastler, 1902–

Professor at L'Ecole Normale Supérieure in Paris. Awarded the Nobel Prize for Physics in 1966 for his discovery and development of optical methods for studying Hertzian resonances in atoms.

Hans Albrecht Bethe, 1906–

Born in Strasbourg, Germany. Professor of Physics at Cornell University. Awarded 1967 Nobel Prize for Physics for his contributions to the theory of nuclear reaction, and especially his discoveries concerning the energy production of stars.

45

Relativity

Relativity and quantum mechanics are the two great theories of twentieth century physics. They are regarded as fundamental in the understanding of atomic and nuclear phenomena. Hence, although both theories are mathematically complex and cannot be developed here in detail, in this and following chapters we shall examine the observations which inspired the development of these theories and some of the important consequences of each.

The "special" theory of relativity predicts that the mechanics of particles with speeds approaching the speed of light is somewhat different from Newton's mechanics. The concepts of space and time are shown to be interrelated. A particle speed greater than the speed of light in empty space is shown to be impossible. Mass and energy are shown to be interconvertible. (The "general" theory of relativity is Einstein's theory of gravitation and will not be discussed here).

45-1 THE "ETHER"

The wave properties of light demonstrated during the first part of the nineteenth century were explained in Maxwell's brilliant electromagnetic theory of radiation. It was difficult for scientists of the nineteenth century, as for us, to conceive of a wave motion apart from a material medium to transmit its vibrations. So they invented a medium called the *ether* for the propagation of light. This suggested two consequences worthy of experimental check:

1 Light waves should travel with a definite speed ($c = 3 \times 10^8$ m/s in free space) with respect to ether itself. Then the apparent speed of light relative to a body moving through the ether should differ from c and should depend on the speed of the body.
2 By measurements on light waves, it should be possible to determine an "absolute velocity" of the earth or any other body.

An experiment to detect motion of the earth relative to the ether would require very sensitive apparatus. The speed of the earth in its orbit is about 3×10^4 m/s, or only about 10^{-4} times the speed of the light signals that would be used in the experiment.

45-2 MICHELSON'S EXPERIMENT

With the problem in mind of measuring the earth's speed through the ether, Michelson devised an interferometer (Fig. 45-1), an instrument

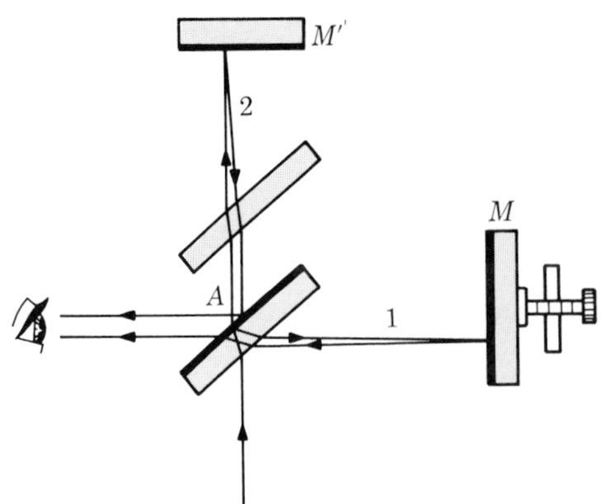

Figure 45-1
Michelson interferometer.

in which interference patterns produced by a divided light beam are used to reveal differences in the optical paths (see Sec. 30-5).

Assume that the earth travels through stationary ether with a speed v and that light has a speed c in the ether. Consider a Michelson interferometer arranged so that one of its two equal arms is parallel to the earth's velocity v (Fig. 45-2). Then the times required for the light beams to travel the distances AMA and $AM'A$ will be unequal. The speed of a beam traveling from A to M is $c - v$ relative to the interferometer. On the return from M to A, the speed of the beam relative to the interferometer is $c + v$. The time for the round trip AMA is thus

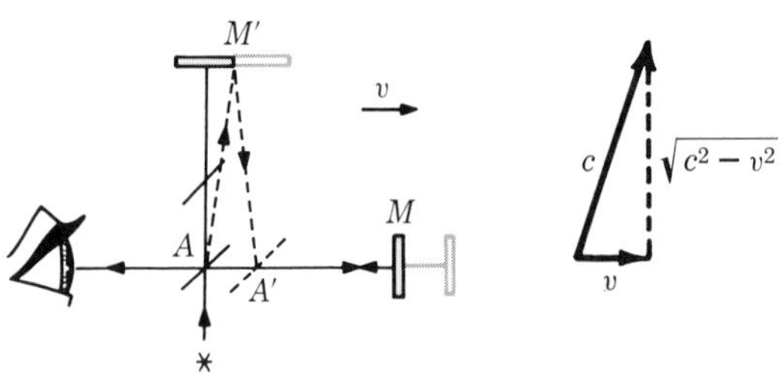

Figure 45-2
Light paths in moving interferometer and velocity vector diagram.

$$t_1 = \frac{s}{c - v} + \frac{s}{c + v} = \frac{2sc}{c^2 - v^2} \tag{1}$$

Since v is small compared with c, we may use the binomial theorem to obtain the approximation

$$t_1 = \frac{2s/c}{1 - v^2/c^2} = \frac{2s}{c}\left(1 + \frac{v^2}{c^2} + \cdots\right) \tag{2}$$

A wave front leaving A toward mirror M' will be returned, according to Huygens' principle, but only after A has moved to a new position A' (Fig. 45-2). The component of the velocity of the light in the direction perpendicular to the motion of the interferometer is $\sqrt{c^2 - v^2}$. The time for the round trip $AM'A$ is

$$t_2 = \frac{2s}{\sqrt{c^2 - v^2}} = \frac{2s/c}{\sqrt{1 - v^2/c^2}} = \frac{2s}{c}\left(1 + \frac{1}{2}\frac{v^2}{c^2} + \cdots\right) \tag{3}$$

Waves which are in phase when they reach A from the monochromatic source will differ in phase when they return to A after reflection, because of the time difference

$$\Delta t = t_1 - t_2 = \frac{sv^2}{c^3} \tag{4}$$

If the interferometer is rotated 90°, paths 1 and 2 will have their roles interchanged and the total retardation will be $2sv^2/c^3$. The number of fringes passing the reference mark should be

$$N = \frac{\text{path difference}}{\text{wavelength}} = \frac{c\,\Delta t}{\lambda} = \frac{c2sv^2}{\lambda c^3} = \frac{2sv^2}{c^2\lambda} \tag{5}$$

To estimate the magnitude of fringe shift to be expected, we may assume that the earth's velocity through the ether is the same as its orbital velocity, about 30 km/s. By using multiple reflections, Michelson and Morley attained an effective path s of about 10 m. For light of wavelength 5,000 Å we should then estimate a maximum fringe shift of

$$N = \frac{2 \times 10 \text{ m } (3 \times 10^4 \text{ m/s})^2}{(3 \times 10^8 \text{ m/s})^2 (5.0 \times 10^{-7} \text{ m})} = 0.4 \text{ fringe} \quad (6)$$

A fringe shift of this amount is readily detectable with the apparatus. It should then be possible to measure the fringe shift and from it compute the velocity of the earth relative to the ether, i.e., the absolute velocity of the earth.

Surprisingly, Michelson and Morley found *no* fringe shift when the interferometer was rotated in a pool of mercury. Measurements were made during day and night, at various locations in Europe and America and at various seasons of the year, to assure that at least some measurements would be made when the earth had a large velocity relative to the ether. Always null results were obtained, as if the interferometer were at rest ($v = 0$) relative to the medium in which light is transmitted. It appeared that optical experiments cannot detect motion of the earth relative to the ether.

Michelson and Morley reported their results in 1887. No subsequent experimental evidence contradicts them. Some lingering doubts were laid to rest in a review article published in the *Reviews of Modern Physics* (pages 167–178) in 1955.

Several attempted explanations for the apparent impossibility of measuring the earth's absolute motion failed to gain acceptance when they did violence to established theory, disagreed with known astronomical data, or introduced too many special hypotheses.

45-3 POSTULATES OF THE SPECIAL THEORY OF RELATIVITY

In a paper entitled On the Electrodynamics of Moving Bodies, published in 1905, Einstein introduced a theory that explained the negative result of the Michelson-Morley experiment. This new theory resulted in a revolution in our concepts of space and time. Einstein interpreted the negative result of the Michelson-Morley experiment as indicating that only *relative* velocities can be measured. Accordingly, the laws of physics must be independent of the velocity of the particular reference system of coordinates used to state them; otherwise it would be possible to ascribe some absolute meaning to different velocities.

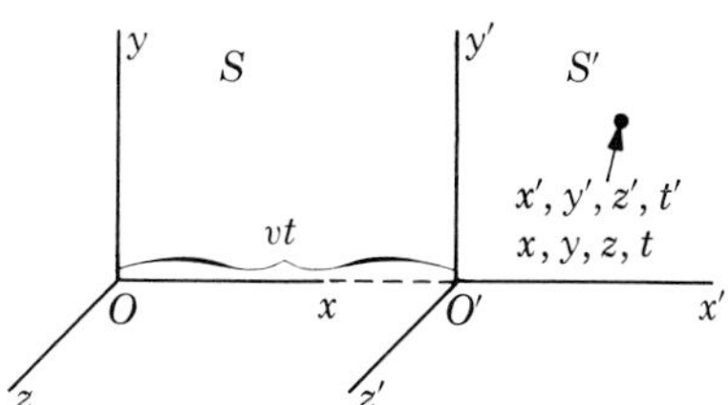

Figure 45-3
Reference system S' moves with constant velocity v in x direction relative to reference system S.

The special or restricted theory of relativity of 1905 dealt with reference systems moving at a constant velocity with respect to each other (Fig. 45-3). Einstein based his theory on two postulates:

1 The laws of physical phenomena are the same when stated in terms of either of two reference systems moving at constant velocity relative to each other.
2 The velocity of light in empty space is the same for all observers and is independent of the velocity of the light source relative to the observer.

Since this theory involved no reference to an absolute reference system, Einstein believed the concept of the ether to be superfluous and therefore rejected it.

These two postulates require us to give up our intuitive conception of time, which Newton described as "absolute, true, and mathematical time (which) of itself and by its own nature, flows uniformly on, without regard to anything external." Indeed, we now find that, if two events take place at different locations, it is not always possible to say which of the two precedes the other or that they occur simultaneously.

45-4 LORENTZ TRANSFORMATION OF COORDINATES

Suppose that person A in a laboratory assigns to every event he observes a position (x, y, z) relative to a particular origin (O, Fig. 45-3) fixed in his laboratory and time t as indicated by a clock in his laboratory. Now let person B move through A's laboratory with speed v in A's positive x direction. Let person B measure positions relative to an origin (O', Fig. 45-3) moving with him and measure times with a clock (just like A's clock) also moving with him. Then to each event B will assign a position (x', y', z') and a time t'. Assume that the clocks are synchronized to read $t = t' = 0$ when the $x'y'z'$ axes momentarily coincide with the xyz axes. The relations which connect the distance and time intervals between two events as measured from the two inertial reference frames are the equations in Table 1a.

Consider a light signal that starts at the origin O of Fig. 45-3 at time $t = 0$ and moves in the positive x direction. In time t_1 it travels a distance x. Since its speed in the laboratory is c, then $x_1/t_1 = c$. An observer in the moving system S' will observe the light signal to travel a distance x_1' in time t_1', where x_1' and t_1' are related to x_1' and t_1' through the classical transformation equations. The moving observer will observe the speed of the light signal to be

$$c' = \frac{x_1'}{t_1'}$$

but $$\frac{x_1'}{t_1'} = \frac{x_1 - vt_1}{t_1'} = \frac{x_1}{t_1} - v = c - v$$

or $$c' = c - v$$

This result is incompatible with the second postulate. Thus, a new set of transformation equations must be developed.

Table 1*a*
CLASSICAL TRANSFORMATION

$x' = x - vt$	(7)
$y' = y$	(8)
$z' = z$	(9)
$t' = t$	(10)

The transformation equations presented in Table 1b preserve the constancy of the speed of light.

These transformation equations were developed by Voigt (1887) and Lorentz (1904) in exploring the ether hypothesis. Einstein adopted these equations for his theory since they satisfied the postulates.

This set of equations is referred to as the *Lorentz transformation*. Note that if $v \ll c$, the Lorentz transformation reduces to the classical transformation.

Example Show that relativistic Eqs. (11) to (14) predict that the speed of light will be the same in each coordinate system of Fig. 45-3.

Suppose that a light signal starts from origin O of Fig. 45-3 at time $t = 0$ and moves in the x direction. It will arrive at a point $x = X$ at the time $t = X/c$, since its speed through the laboratory is c. A second observer, moving with the reference frame S', will say that the light signal arrives at the point

$$x' = \frac{X - v(X/c)}{\sqrt{1 - v^2/c^2}}$$

at the time $$t' = \frac{X/c - (v/c^2)X}{\sqrt{1 - v^2/c^2}}$$

Table 1*b*
RELATIVISTIC TRANSFORMATION

$x' = \dfrac{x - vt}{\sqrt{1 - v^2/c^2}}$	(11)
$y' = y$	(12)
$z' = z$	(13)
$t' = \dfrac{t - vx/c^2}{\sqrt{1 - v^2/c^2}}$	(14)

He will compute its speed u' in the primed (moving) coordinate system as

$$u' = \frac{x'}{t'} = \frac{X - (v/c)\,X}{X/c - (v/c^2)\,X} = c$$

In a mathematical sense, the principle of relativity is that the equations of physical phenomena must be invariant in form under Lorentz transformations. The basic physical assumption of relativity is that no mechanical or electromagnetic influence can transport energy from one point to another with a speed exceeding the speed of light.

Events which happen at the same place at different times, as measured in one reference system, may be seen from another system to happen at different places as well. Also, a difference in spatial position with respect to one system may correspond to a difference in both space and time with respect to another. In the four dimensions of "space-time," space differences can be converted partly into time differences, and vice versa.

45-5 ADDITION OF VELOCITIES

Consider two observers, A at rest in reference system S (Fig. 45-3) and B at rest in reference system S'. System S' moves with constant velocity v relative to S. Suppose that an object flies past these two observers in the x direction. Observer B measures the speed of the object relative to himself as u'. To express u' in terms of the coordinates of observer A at rest in the laboratory reference system S, we have

$$u' = \frac{\Delta x'}{\Delta t'} = \frac{\Delta\left(\dfrac{x - vt}{\sqrt{1 - v^2/c^2}}\right)}{\Delta\left(\dfrac{t - vx/c^2}{\sqrt{1 - v^2/c^2}}\right)}$$

$$= \frac{\Delta x - v\,\Delta t}{\Delta t - (v/c^2)\,\Delta x} = \frac{u - v}{1 - uv/c^2} \qquad (15)$$

where in the last step we have divided numerator and denominator by Δt and let $\Delta x/\Delta t = u$, the speed of the object in the laboratory S. Thus

$$u' = \frac{u - v}{1 - uv/c^2} \qquad \text{or} \qquad u = \frac{u' + v}{1 + vu'/c^2} \qquad (16)$$

Example A rocket moves above the laboratory with a speed $v = 0.90c$. A "flying saucer" passes the rocket with a speed $u' = 0.90c$. Find the speed of the flying saucer as determined at the laboratory.

Classical mechanics would predict this speed as the speed relative to the rocket plus the speed of the rocket relative to the laboratory, or $1.80c$. But the relativistic theorem for addition of velocities [Eq. (16)] gives

$$u = \frac{0.90c + 0.90c}{1 + \dfrac{0.90c \times 0.90c}{c^2}} = \frac{1.80c}{1 + 0.81} = 0.9945c$$

No moving body has a speed greater than c, the speed of light.

45-6 LORENTZ CONTRACTION

Consider a meterstick traveling along the positive x direction with speed v. At time t, its ends are at the points x_1 and x_2, respectively. An observer in a moving reference frame, moving with the same velocity as the meterstick, would record the ends of the meterstick to be at the points x'_1 and x'_2, respectively, where

$$x'_1 = \frac{x_1 - vt}{\sqrt{1 - v^2/c^2}}$$

$$x'_2 = \frac{x_2 - vt}{\sqrt{1 - v^2/c^2}}$$

The length of the stick in the moving frame is

$$x'_2 - x'_1 = \frac{x_2 - x_1}{\sqrt{1 - v^2/c^2}}$$

The length of the stick in the moving frame is 1 m since it is at rest in that frame. To a person in the frame that is not moving, the length is $x_2 - x_1$. Since $\sqrt{1 - v^2/c^2}$ is always less than 1, $x_2 - x_1$ is less than $x_2' - x_1'$ by the factor $\sqrt{1 - v^2/c^2}$. Thus, if L_0 is the length of an object measured by an observer at rest with respect to the object, and L is the length of the object measured by an observer moving with the speed v in the direction in which the length is measured, then

$$L = L_0\sqrt{1 - \frac{v^2}{c^2}}$$

This effect, called the *Lorentz-Fitzgerald contraction,* was proposed many years before the special theory of relativity to explain the results of the Michelson-Morley experiment.

Example A spaceship is moving with the speed of 10 percent that of light. What is the fractional change in length due to the Lorentz contraction?

$$\frac{L}{L_0} = \sqrt{1 - (0.1)^2}$$

$$\frac{L_0 - L}{L_0} = 1 - \frac{L}{L_0} = 0.005 = 0.5\%$$

45-7 VARIATION OF MASS

We have previously written the equation of motion for a body as $F = ma$, and also in terms of momentum as

$$F = \frac{mv_2 - mv_1}{t} = \frac{\Delta(mv)}{\Delta t} \tag{17}$$

In relativity theory, the conservation of momentum remains a basic principle of mechanics. It turns out that in order to have the total momentum of an isolated system remain constant, the momentum of a particle must be defined as

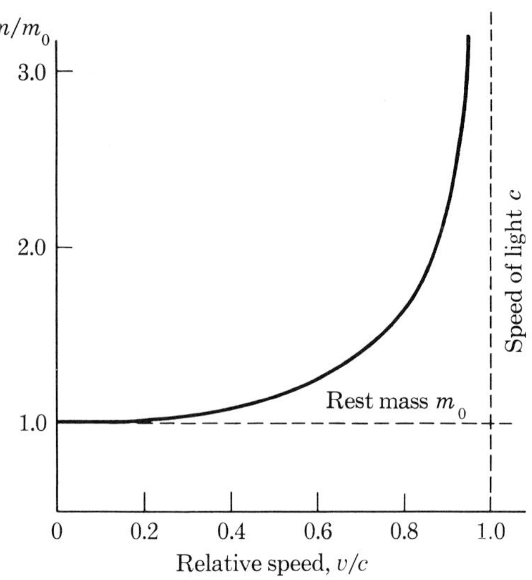

Figure 45-4
The mass of a particle increases with its speed.

$$\text{Momentum} = mv = \frac{m_0 v}{\sqrt{1 - v^2/c^2}} \tag{18}$$

That is, it is necessary to assume that a particle which has a mass m_0 when at rest has a mass

$$m = \frac{m_0}{\sqrt{1 - v^2/c^2}} \tag{19}$$

when it is moving with speed v. The quantity m_0 is called the *rest mass.*

Variation of mass with speed (Fig. 45-4) can readily be observed for small particles, such as electrons, which can be accelerated to enormously large speeds.

Example An electron at rest has a mass of 9.11×10^{-31} kg. At what speed would the mass of the electron be doubled?

We want $m = 2m_0$. From Eq. (19)

$$1 - \frac{v^2}{c^2} = \frac{m_0{}^2}{m^2} = \frac{1}{4}$$

$$\frac{v^2}{c^2} = \frac{3}{4} \qquad v = 0.86c = 2.6 \times 10^8 \text{ m/s}$$

45-8 KINETIC ENERGY

In classical physics, force may be defined as the time rate of change of momentum. This definition of force is retained in relativistic mechanics, but the mass must be recognized as a variable. The kinetic energy of an object can be calculated as the energy required to accelerate it from rest to its final speed v. The result is

$$E_k = \frac{m_0c^2}{\sqrt{1 - v^2/c^2}} - m_0c^2$$

$$E_k = mc^2 - m_0c^2 \qquad (20)$$

Equation (20) says that as a particle is speeded up its energy is increased and that the increase in energy $(m - m_0)c^2$ is proportional to the increase in mass. This result suggests that we identify c^2 times the relativistic mass of the particle with the total energy of the particle

$$\text{Total energy} = mc^2 = E_k + m_0c^2 \qquad (21)$$

The total energy of a free particle consists of its rest energy (the energy m_0c^2 it has at rest) plus its kinetic energy (the energy E_k it has because of its motion).

Example Consider an electron which has been accelerated from rest through a potential difference of 500 kV. Find (*a*) its kinetic energy, (*b*) its rest energy, (*c*) its total energy, (*d*) its mass, and (*e*) its speed.

(*a*) $E_k = e\,\Delta V = 500 \text{ keV}$

(*b*) $m_0c^2 = (9.108 \times 10^{-31} \text{ kg}) \times (2.998 \times 10^8 \text{ m/s})^2$

$= 8.186 \times 10^{-14} \text{ J} = 511 \text{ keV}$

(*c*) Total energy $= (500 + 511)$ keV $= 1{,}011$ keV

(*d*) Since $\dfrac{m}{m_0} = \dfrac{mc^2}{m_0c^2} = \dfrac{\text{total energy}}{\text{rest energy}} = \dfrac{1{,}011 \text{ keV}}{511 \text{ keV}}$

$= 1.98$, the mass of the electron is $m = 1.98m_0$

$= 18.1 \times 10^{-31}$ kg.

(*e*) From the relation

$$m = 1.98m_0 = m_0\left(1 - \frac{v^2}{c^2}\right)^{-1/2}$$

$$1 - \frac{v^2}{c^2} = \frac{1}{(1.98)^2}$$

Hence $v/c = [1 - 1/(1.98)^2]^{1/2} = 0.863$, and the electron speed $v = 0.863c = 2.59 \times 10^8$ m/s.

Table 2
DATA ON ELECTRONS

v/c	*Energy, eV*	m/m_0
0	0	1.00000
0.0100	25.54	1.00005
0.0200	102.2	1.00020
0.0500	638.5	1.00125
0.100	2,575	1.00504
0.200	10,530	1.02062
0.500	79,030	1.1547
0.600	127,700	1.2500
0.700	204,300	1.4002
0.800	340,500	1.6666
0.900	661,000	2.2941
0.990	3,110,000	7.0888

45-9 TIME DILATATION: THE CLOCK PARADOX

Consider now the effect of relative motion on a clock. Two events occur at a point in coordinate system S': one at time t_1', the other at a later time t_2'. To an observer in S these events take place at different points in space, (x_1, y, z) and (x_2, y, z), as well as at different times, such that $x_2 - x_1 = v(t_2 - t_1)$. From the Lorentz transformations

$$t_2 - t_1 = \frac{t_2' - t_1'}{\sqrt{1 - v^2/c^2}}$$

Thus the sequence in time of the two events is the same, but Δt appears longer for the observer in S than for the observer in S'. This is interpreted as meaning that a moving clock appears to run at a slower rate than does an identical clock at rest, in the ratio $\sqrt{1 - v^2/c^2}/1$.

The imminence of space travel has revived interest in the "clock paradox," or "twin paradox." One of two identical twins leaves his brother on earth and voyages at high speed into distant space. On his return, he finds that his brother has grown much older than he, because of time dilatation in the spaceship. Superficially, this is a paradox, for it challenges "common sense." Also, it seems to contradict the assertion of special relativity that in describing physical events all observers are equivalent; none has a preferred or absolute reference system. The aging, or clock, effect seems to provide a way of distinguishing among observers. But relativity asserts the equivalence of observers in *inertial* systems, and since one of the twins accelerated at the start of his space trip and again when he altered course to return, he did not view his brother from the same inertial system before and after the trip. So there is no paradox.

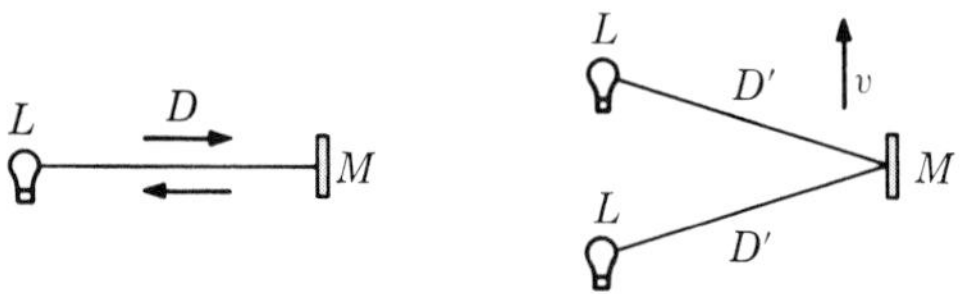

Figure 45-5
Clock paradox.

The intriguing question remains: Did the stay-at-home brother grow older faster? Yes. In his 1905 paper On the Electrodynamics of Moving Bodies, Einstein wrote,

> If at the points A and B there are two stationary clocks which, viewed by a stationary observer, are synchronous, and if the clock at A is moved with the velocity v along the line AB to B, then on its arrival at B the two clocks no longer synchronize, but the clock moved from A to B lags behind the other which has remained at B by $tv^2/2c^2$ (up to magnitudes of fourth and higher orders) t being the time required for the journey from A to B.
>
> It is at once apparent that this result still holds if the clock moves from A to B in any polygonal line, and *also when the points A and B coincide.*

Bergmann has suggested the following elucidation of the clock effect (Fig. 45-5): Observer A arranges for periodic light signals to go from lamp L to mirror M and back (a kind of optical clock). Light travels a distance $2D$ for each LML circuit. Observer B is moving with constant speed v at right angles to the line LM. For him, the same light signal travels the larger distance $2D'$. If observer B set up a similar experiment in his coordinate system S', his light signals would complete their round trips in shorter times than noted by observer A. The discrepancies arise because the two observers do not agree on which of two distant events (completion of the nth round trip by either light signal) takes place first.

Now let observer B suddenly reverse his velocity ($v\downarrow$). He is now in a different Lorentz frame. (He accelerated.) His notions of simultaneity have changed. Observer A sees B coming toward him, with B's light signals arriving slower than his own. When they meet, A's signals have completed a larger number of LML circuits than have B's signals. Observer A has aged more than B.

Example What will be the difference in the rates of two identical clocks, one of which is on a spaceship moving at 300 mi/s relative to the other?

$$v = 300 \text{ mi/s} = 5.25 \times 10^5 \text{ m/s}$$

$$c = 3.0 \times 10^8 \text{ m/s}$$

$$\text{Relative change in rate} = \sqrt{1 - \frac{v^2}{c^2}}$$

$$= \sqrt{1 - \frac{5.25 \times 10^5}{3.0 \times 10^8}}$$

$$= \sqrt{1 - 3.06 \times 10^{-6}}$$

$$= 0.002\ \%, \text{ approx}$$

Experimental detection of time dilatation was achieved by Ives and Stilwell (1938) on viewing the spectral lines of hydrogen atoms which were given a high speed directed away from the spectroscope. An arrangement was used to distinguish relativity effects from Doppler effects. Light from the atoms fell on the spectrograph slit directly, and also after reflection in a mirror set at some distance and normal to the velocity of the atoms. Owing to the Doppler effect, each spectrum line was split into two frequencies. Then light from hydrogen atoms at rest was viewed with the same spectrograph. This gave lines slightly displaced, in frequency, from the middle of the Doppler pairs, in an amount predicted by relativity.

Measurements of the lifetimes of mesons have been used to check relativity predictions. The mean life of μ mesons (about 2×10^{-6} s) has been found to depend on their *speed* roughly in the way predicted by relativity.

In 1971 NASA supported some interesting experiments to test the theory of relativity by measuring the relativistic effects of motion on time. In these experiments an astronomer, J. C. Hafele, and a physicist, R. E. Keating, circled the earth twice, once from east to west in two days and then from west to east in the same amount of time in a commercial jet airliner carrying with them two atomic clocks capable of measuring time to a billionth (10^{-9}) of a second. Upon completing the circumnavigation of the earth, the plan was to compare the "on-board" clocks with a master atomic clock which had not been taken on the trips.

The theory to be tested was that a clock flying eastward at 600 mi/h, added to the speed of the earth's spin, should record less time than a clock on the ground, the einsteinian "clock paradox." It was proposed that if the traveling clocks ran about 100 billionths of a second behind the master clock, it would lend further support to Einstein's theory of relativity. Conversely, it was anticipated that on the west-to-east flight the clocks should gain time.

The results were consistent with Einstein's theory. The eastward-bound clocks in effect "ran slow" by 40 billionths of a second. That is, time contracted for the airborne clocks. Traveling in a westerly direction, the clocks "ran fast" by 275 billionths of a second and time was expanded. These experiments showed that time apparently does vary according to the motion of the clocks which measure it as Einstein predicted.

SUMMARY

The result of the Michelson-Morley experiment is interpreted as showing that only relative velocities can be measured.

Einstein's relativity theory of 1905 assumes that (1) physical phenomena are described by the same laws in either of two reference systems moving at constant velocity relative to each other and (2) all observers measure the same velocity for light in free space, regardless of the velocity of the light source relative to the observer.

An object moving with velocity u' relative to an axis which itself has a velocity v relative to an observer will appear to the observer to have velocity

$$u = \frac{u' + v}{1 + vu'/c^2}$$

The *Lorentz-Fitzgerald contraction* was proposed many years before the special theory of relativity to explain the results of the Michelson-Morley experiment.

A particle which has mass m_0 when at rest has a mass

$$m = \frac{m_0}{\sqrt{1 - v^2/c^2}}$$

when it is moving with speed v.

The total energy of a free particle consists of its rest energy plus its kinetic energy

$$\text{Total energy} = mc^2 = m_0c^2 + E_k$$

References

Coleman, James A.: "Relativity for the Layman," Signet Books, New American Library, Inc., New York, 1969.

Dingle, Herbert: "The Special Theory of Relativity," Chemical Publishing Company, Inc., New York, 1941.

Einstein, Albert: "The Meaning of Relativity," 5th ed., Princeton University Press, Princeton, N.J., 1955.

French, A. P.: "Special Relativity," W. W. Norton & Company, Inc., New York, 1968.

Lieber, Lillian R.: "The Einstein Theory of Relativity," Farrar & Rinehart, Inc., New York, 1945.

Lindsay, Robert Bruce, and Henry Margenau: "Foundations of Physics," chaps. 7 and 8, John Wiley & Sons, Inc., New York, 1936.

Richtmyer, F. K., E. H. Kennard, and T. Lauritsen: "Introduction to Modern Physics," McGraw-Hill Book Company, Inc., New York, 1955. Especially chaps. 3, 4, 5, and 10.

Russell, Bertrand: "The ABC of Relativity," Mentor Books, New American Library, Inc., New York, 1969.

Questions

1 Would it be possible for an electron beam to move across the screen of a cathode-ray tube at a speed faster than the speed of light without contradicting special relativity?

2 Derive the relation between the total energy E and the momentum p of a particle: $E = \sqrt{m_0{}^2c^4 + p^2c^2}$.

3 Justify the following useful relations involving the kinetic energy E_k of a particle:

a $$E_k = \left[\frac{1}{\sqrt{1 - v^2/c^2}} - 1\right] m_0c^2$$

b $$\frac{v}{c} = \left[1 - \frac{1}{1 + (E_k/m_0c^2)^2}\right]^{1/2}$$

c $$1 + \frac{E_k}{m_0c^2} = \frac{1}{1 - v^2/c^2}$$

4 Assume that the speed of light is only 30 m/s. Describe how everyday events would appear to us.

5 Assume two events occur simultaneously at the points x_1 and x_2, respectively. Show that an observer moving with speed v would see the events separated by a time interval

$$t_2' - t_1' = \frac{(x_1 - x_2)\, v/c^2}{1 - v^2/c^2}$$

Problems

1 What would be the mass of an electron traveling half as fast as light?

2 What energy would be required to change the speed of an electron from rest to $0.8c$?

Ans. 5.5×10^{-14} J.

3 Assume that you are moving with a speed $0.80c$ past a man who picks up a watch and then sets it down. If you observe that he held the watch for 10 s, how long does he think he held it? [*Hint:* Use the time transformation of Eq. (14). You want to find $\Delta t = t_2 - t_1$ when you know $\Delta t_2' = t' - t_1'$.]

4 If your spaceship is flying with a speed of $0.50c$ relative to the earth and another spaceship passes you with a speed of $0.50c$ relative to you, how fast is it going relative to the earth?

Ans. $0.80c$.

5 Find the contraction in length of a 1.0-mi-long train when traveling at 80 mi/h.

6 For a particle of rest mass m_0 and speed v, which is the largest: $\frac{1}{2}m_0v^2$, $\frac{1}{2}mv^2$, or its kinetic energy?

Ans. $E_k > \frac{1}{2}mv^2 > \frac{1}{2}m_0v^2$.

7 The earth receives radiant energy from the sun at an average rate of 2.00 cal/(cm^2)(min) or 1.35×10^3 W/m^2. How much of the sun's mass reaches the earth in 1 year? Is the earth getting heavier as a result?

Luis W. Alvarez, 1911–

Born in San Francisco, California. Professor of Physics at the University of California, Berkeley. Awarded the Nobel Prize for Physics for 1968 for his decisive contributions in the early 1960s to the physics of subatomic particles and to the techniques of their detection.

Murray Gell-Mann, 1929–

Born in New York. The Robert A. Millikan Professor at California Institute of Technology. Awarded the 1969 Nobel Prize for Physics for his theoretical contributions to the study of particle physics.

46

Photons and Quanta

Beginning about 1900, developments took place which indicated that classical electromagnetic theory (Chap. 43) does not predict accurately all aspects of electromagnetic radiation and absorption of energy. These developments led to the quantum theory, which along with relativity (Chap. 45) set physics upon a new course.

We shall trace briefly the quantum hypothesis from its origin in explaining blackbody radiation, through its confirming success in explaining the photoelectric effect and Compton effect and its striking but limited success in the Bohr model of the atom (Chap. 47), to its merging with other hypotheses in wave mechanics (Chap. 48).

46-1 BLACKBODY RADIATION

Electromagnetic radiation that exists inside an enclosure or cavity whose walls are at some uniform temperature is called blackbody radiation. Its properties may be investigated if we make a tiny hole to allow a small amount of radiation (not enough to disturb appreciably the thermal equilibrium inside) to emerge from the cavity. The cavity with the tiny hole is designated a *blackbody* source of radiation.

As we have already seen (Sec. 17-10), Max Planck succeeded in 1901 in explaining the distribution of energy in a blackbody spectrum (Fig. 17-9) by assuming that the microoscillators comprising the wall of the cavity can have only energies which are whole-number multiples of $h\nu$, where ν is the frequency and h is called Planck's constant. The microoscillators do not radiate continuously (as expected from classical theory), but only in quanta of energy, emitted when an oscillator changes from one of its quantized energy states to a lower one (Fig. 46-1). The quantum (or photon) radiated has energy proportional to the frequency

$$E = h\nu \qquad (1)$$

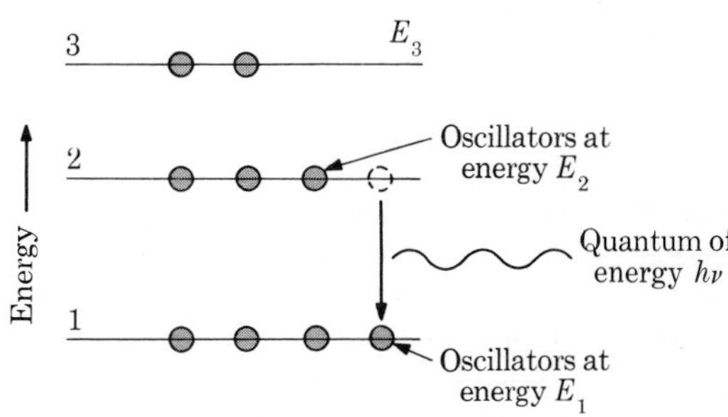

Figure 46-1
A three-level energy-level diagram showing Planck's hypothesis.

A numerical value for Planck's constant can be calculated from measurements on blackbody radiation and confirmed from experiments on the photoelectric effect, the spectrum of hydrogen, and the Compton scattering of x-rays. All yield a consistent result for the value of Planck's constant, $h = 6.6252 \times 10^{-34}$ J·s.

46-2 THE PHOTON

Einstein suggested (1905) that not only is radiation emitted and absorbed in whole numbers of energy quanta, as proposed by Planck, but also the electromagnetic energy is propagated through space in definite quanta or photons, moving with the speed of light. This extension of the quantum theory implies a modification of the wave theory of light in favor of a particle theory.

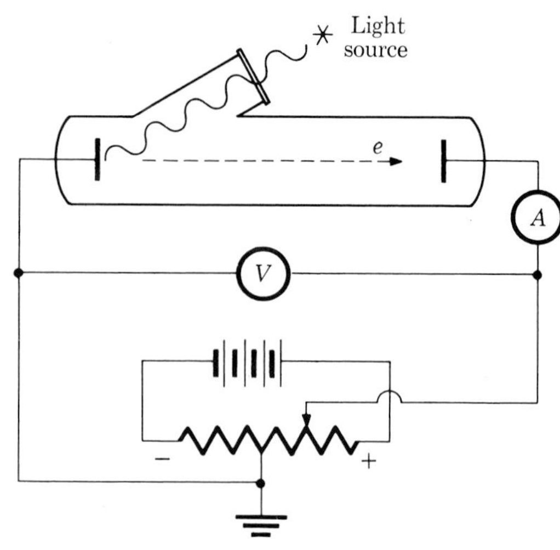

Figure 46-2
Apparatus for measuring photoelectric current and the energy of photoelectrons.

46-3 PHOTOELECTRIC EFFECTS

Einstein used Planck's idea of interchange of energy in quanta to explain the emission of electrons from the surface of a metal receiving electromagnetic radiation (Sec. 44-8). With apparatus suggested by Fig. 46-2, the number of electrons emitted per second can be determined by measuring the photoelectric current. The energy distribution of the electrons can be determined by applying a retarding (negative) potential to the collector and increasing it gradually until the stopping potential V_s is found for which no electrons reach the collector.

The chief features of photoemission are: (1) There is no detectable time lag (greater than 10^{-9} s) between irradiation of an emitter and ejection of photoelectrons. (2) The number of electrons ejected per second is proportional to the intensity of radiation, at a given frequency. (3) The photoelectrons have energies ranging from zero up to a definite maximum, which is proportional to the frequency of the radiation and independent of its intensity. (4) For each material there is a threshold frequency ν_0 below which no photoelectrons are emitted.

These characteristics of photoemission cannot be explained by Maxwell's classical theory of electromagnetic radiation. In 1905 Einstein tried the assumption that light of frequency ν can give energy to the electrons in the metal only in quanta of energy $h\nu$. Either an electron absorbs one of these quanta, or it does not. If it is given energy $h\nu$, an electron may use an amount of energy w in escaping from the metal, where it has negative potential energy, into the vacuum, where it has zero potential energy. The quantity w is called the *work function* of the surface. The maximum kinetic energy which the electron can have when it leaves the surface is therefore

$$E_{k,\,\max} = h\nu - w \tag{2}$$

This is called *Einstein's photoelectric equation*. It explains the linear relationship $E_k = a\nu + b$ shown in Fig. 46-3. The slope a measured from the graph agrees with the value of Planck's constant h; the negative intercept b is identified with

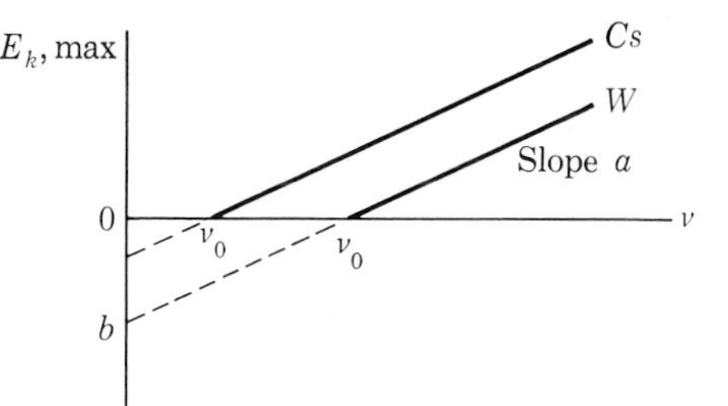

Figure 46-3
Dependence of maximum energy of photoelectrons on frequency, for two different metals.

the work function w of the metal. The intercept on the frequency axis is the minimum frequency of light that will liberate electrons from the particular metal. At this threshold frequency ν_0, the photon delivers just enough energy to enable the electron to get out of the metal, with $E_k = 0$,

$$h\nu_0 = w \tag{3}$$

From Eq. (2) $E_{k,\,\max}$ is independent of the intensity of illumination, in agreement with experiment. At a given frequency, increasing the intensity of illumination merely increases the number of photons reaching the metal per second. This increases the number of electrons that can be emitted per second, but not their energy.

The photoelectric effect gives strong support to Planck's hypothesis that light of frequency ν can be emitted or absorbed only in packets of energy $h\nu$. The citation which accompanied the award of the Nobel prize to Einstein stated that it was for "his attainments in mathematical physics and especially for his discovery of the law of the photoelectric effect."

46-4 THE CONTINUOUS X-RAY SPECTRUM

Another confirmation of the photon idea, and an independent way of calculating the value of Planck's constant h, is provided by continuous x-ray spectra. X-rays are electromagnetic waves of very short wavelength, about 10^{-8} to 10^{-11} m. In an x-ray tube (Fig. 46-4), an electric current heats a tungsten filament (cathode) so that it emits electrons. A potential difference of several thousand volts between cathode and target accelerates the electrons. The fast-moving electrons are quickly decelerated when they strike the metal target. Most of their energy is converted into heat by collisions among atoms of the target. But as the electrons are decelerated, they are expected to radiate, according to classical electromagnetic theory. The radiation is emitted in all directions. One finds a continuous distribution of frequencies up to a certain maximum. This maximum frequency depends on the potential difference at which the tube is operated: $\nu_{\max}/V =$ a constant for a wide range of voltages (Fig. 46-5).

This high-frequency limit in the continuous x-ray spectrum is difficult to explain classically. It is easily clarified by the photon hypothesis. An electron may suffer numerous decelerations as it encounters various atoms in the target. Each time, a photon is emitted, whose energy $h\nu$ is equal to the decrease in kinetic energy, ΔE_k, of the electron. Clearly the highest-frequency photon that can be produced is that which results from the complete conversion of the electron's kinetic en-

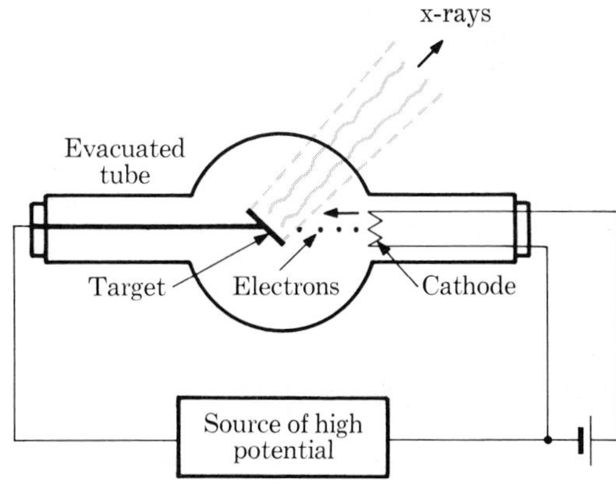

Figure 46-4
An x-ray tube.

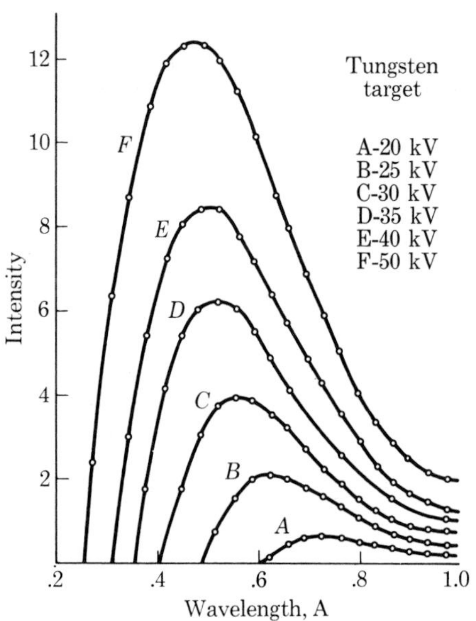

Figure 46-5
Continuous x-ray spectra, showing relationship between minimum wavelength emitted and operating voltage.

ergy into a single photon. Since electrons arrive at the target with energy Ve,

$$h\nu_{\max} = Ve \tag{4}$$

From this law of Duane and Hunt, e/h may be determined from the sharp cutoff of the x-ray-intensity-vs.-frequency curve at $\nu_{\max}$. There is good agreement with the value of h determined in other ways.

46-5 THE COMPTON EFFECT

Another, even more direct confirmation of the photon hypothesis came about 1923 in A. H. Compton's explanation of properties of scattered x-rays. Compton allowed a beam of monochromatic x-rays (Fig. 46-6) to fall on a scattering material C and then examined the scattered radiation with an x-ray spectrometer. Scattered radiation having the same wavelength as the incident beam was found, as expected from the electromagnetic wave theory. But in addition Compton

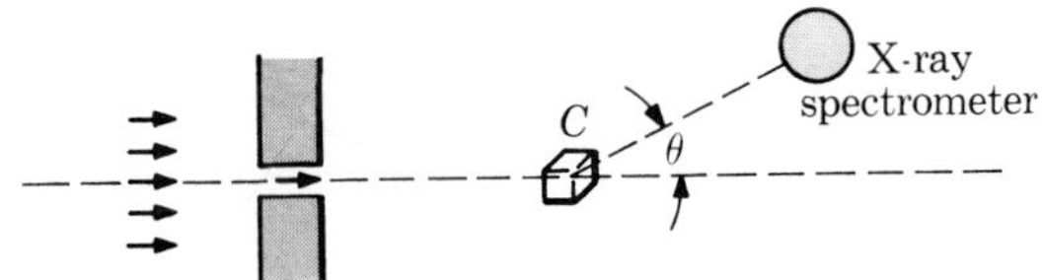

Figure 46-6
Apparatus for observing Compton scattering of x-rays.

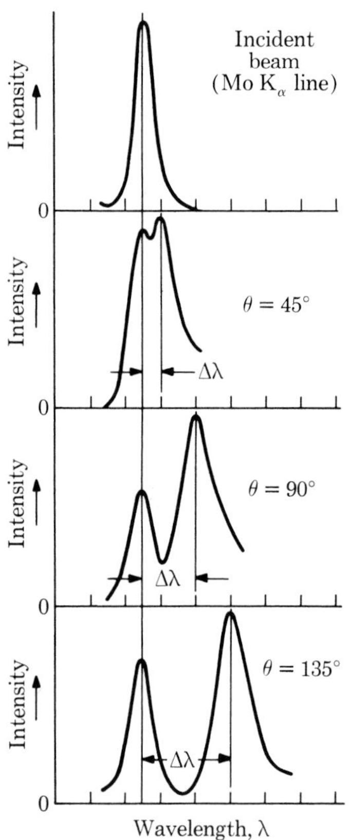

Figure 46-7
Wavelength shift due to Compton scattering.

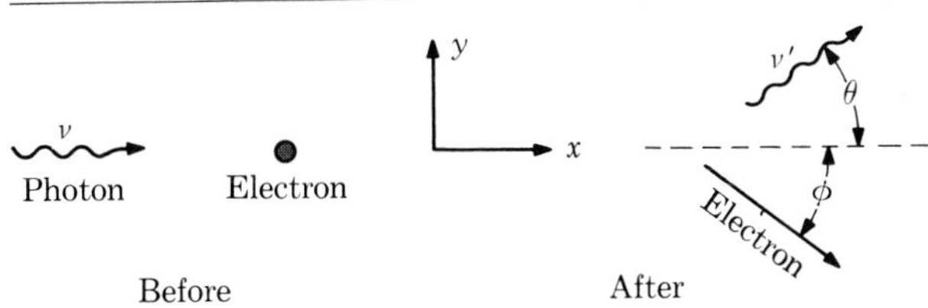

Figure 46-8
Compton scattering of a photon.

observed a scattered wavelength λ' greater than that of the original beam (Fig. 46-7). The shift in wavelength, $\Delta\lambda = \lambda' - \lambda$, was found to become larger as the scattering angle θ was increased. This scattering with an increase in wavelength is called the *Compton effect* and cannot be explained by the classical wave theory, which suggests no way of introducing a frequency (or wavelength) change in scattering.

The photon picture of electromagnetic radiation provides a straightforward explanation of the Compton effect. We picture a photon as having zero rest mass; its total energy is its kinetic energy. Since the energy of a photon is $E = h\nu$, we can calculate its mass from $E = mc^2$,

$$mc^2 = h\nu \qquad \text{or} \qquad m = \frac{h\nu}{c^2}$$

The momentum of a photon is

$$p = mc = \frac{h\nu}{c} \tag{5}$$

We now interpret the Compton scattering as the elastic collision of a photon with an electron of the scattering material which is free and at rest before the collision (Fig. 46-8). To satisfy conservation of momentum,

$$X \text{ components:} \quad \frac{h\nu'}{c}\cos\theta + p\cos\phi = \frac{h\nu}{c}$$

$$Y \text{ components:} \quad \frac{h\nu'}{c}\sin\theta - p\sin\phi = 0$$

where p is the momentum of the electron after the collision and ν' is the frequency of the scattered photon. Since we assume an elastic collision, the final kinetic energy must be equal to the initial kinetic energy

$$h\nu = h\nu' + E_k$$

where E_k is the final kinetic energy of the electron.

If we write the momentum equations as

$$pc\cos\phi = h\nu - h\nu\cos\theta$$

$$pc\sin\phi = h\nu'\sin\theta$$

square these equations, and add, we get

$$p^2c^2 = (h\nu)^2 - 2 \times h\nu \times h\nu'\cos\theta + (h\nu')^2 \tag{6}$$

By combining two equations we get a useful relationship between the total energy and the momentum of a particle,

$$E = m_0c^2/\sqrt{1 - \frac{v^2}{c^2}}$$

and

$$p = m_0v/\sqrt{1 - \frac{v^2}{c^2}}$$

give

$$E^2 = p^2c^2 + m_0{}^2c^4 \tag{7}$$

Hence for the recoil electron

$$p^2c^2 = (E_k)^2 + 2(E_k)m_0c^2$$

From the energy equation $E_k = h\nu - h\nu'$,

$$p^2c^2 = (h\nu - h\nu')^2 + 2m_0c(h\nu - h\nu') \quad (8)$$

By substitution of Eq. (8) in Eq. (7) we find

$$m_0c^2h\nu - m_0c^2h\nu' = h\nu \times h\nu'(1 - \cos\theta)$$

On dividing by h^2c^2 and substituting

$$\frac{1}{\lambda} = \frac{\nu}{c}$$

we finally obtain

$$\lambda' = \lambda + \frac{h}{m_0c}(1 - \cos\theta) \quad (9)$$

which fits the experimentally measured wavelength shift as a function of scattering angle θ.

That part of the radiation which is scattered without wavelength change (Fig. 46-7) is accounted for by assuming that these photons were scattered by electrons strongly bound in atoms. The recoiling mass is that of the whole atom, the photons experience negligible loss of energy in collision with the heavy atom, and hence the wavelength of these photons will be very close to that of the incident photons.

SUMMARY

Planck's theory of radiation is based on the assumption that a body emits and absorbs radiant energy only in multiples of a quantum of energy whose value is proportional to the frequency

$$E = h\nu$$

Einstein's explanation of photoelectric emission predicts that the maximum kinetic energy a photoelectron will acquire is equal to the energy $h\nu$ of the incident photon minus the energy w required for escape.

$$E_{k,\,\max} = \tfrac{1}{2}mv^2 = h\nu - w$$

The maximum frequency emitted in a continuous x-ray spectrum occurs when the entire energy of a bombarding electron Ve is converted into a single photon.

$$h\nu_{\max} = Ve$$

Compton's interpretation of the scattering of x-rays confirms Planck's quantum hypothesis and assigns to a photon the momentum $h\nu/c$. The increase in wavelength observed in the scattered radiation is given by

$$\lambda' - \lambda = \frac{h}{m_0c}(1 - \cos\theta)$$

Problems

1 The mass of a proton at rest is 1.67×10^{-27} kg. What is its mass when it is moving with a speed 78 percent that of light?

2 What are the speed and mass of an electron whose kinetic energy is (*a*) 5,000 eV and (*b*) 1.0 MeV? (1 eV $= 1.6 \times 10^{-19}$ J.)
Ans. $0.14c$, $1.0098m_0$; $0.941c$, $2.96m_0$.

3 The threshold frequency for potassium is 3.0×10^{14} per second. What is the work function of potassium?

4 Calculate a value for Planck's constant h, using the intercept of curve C in Fig. 46-5 and the known value for the charge of an electron.
Ans. 6.4×10^{-34} J·s.

5 A 10-kW radio transmitter operates at a frequency of 1.40 MHz. (*a*) How many photons does it emit every second? (*b*) How many photons are received per second by a receiver tuned to this frequency and located at a place where the signal strength is 5.0×10^{-6} W?

6 In a student experiment on the photoelectric effect, when the photosensitive surface was irradiated with light of frequency 8.0×10^{14} per second, the stopping potential for emitted electrons was -0.18 V. For light of frequency 5.5×10^{14} per second, V_s was -1.25 V. Calculate a value for Planck's constant h from these data. What is the percentage error?
Ans. $h = 6.85 \times 10^{-34}$ J·s; 4 percent.

7 Light falls upon a nickel surface which has a work function of 8.02×10^{-19} J. Will photoelectrons be emitted?

8 A 1-cd source of yellow light ($\lambda = 5{,}900$ Å) gives a flow of luminous energy of about 0.50 erg/(cm²)(s) at a distance of 1.0 m from the source. A human eye when seeing a star of sixth magnitude is receiving light equivalent to that from a 10^{-8}-cd source at 1.0 m distance. If the diameter of the pupil is 3.0 mm, how many quanta enter per second? *Ans.* $1,\bar{1}00$.

9 Radiation of wavelength $2{,}536 \times 10^{-10}$ m from a mercury arc is used to eject photoelectrons from silver. The potential difference needed to bring the photoelectrons to rest is 1.10 V. What is the work function of silver?

10 How many photons of red light of wavelength 7×10^{-7} m constitute 1 J of energy?

Ans. 3.6×10^{18} photons.

11 For a certain photocell it is found that a potential difference of −1.6 V between collector and emitter reduces the photocurrent to zero when light of wavelength 3,000 Å falls on the emitter. What is the threshold wavelength for the emitter surface? Photons associated with light of wavelength 12,400 Å have an energy of 1.0 eV.

12 Calculate the energy of a photon whose wavelength is (*a*) the size of an atom (10^{-10} m), and (*b*) the size of a nucleus (10^{-15} m).

Ans. 1.99×10^{-15} J; 1.99×10^{-10} J.

13 The work function of a metal surface is 3.2 eV. Calculate the threshold wavelength for this surface.

14 The surface of a certain metal is illuminated by light of wavelength 4,000 Å. The photoelectrons emitted are all stopped by a negative potential of 0.80 V. What is the maximum kinetic energy of an electron emitted? What is the longest wavelength that will cause photoelectrons to be emitted at this surface?

Ans. 1.28×10^{-19} J; $5,\bar{3}90$ Å.

15 Photoelectrons are emitted when ultraviolet light having a wavelength of 2,400 Å falls upon a metal surface (nickel). If the work function of nickel is 5.01 eV (1 eV = 1.6×10^{-19} J), what potential difference must be used to stop the most rapidly moving photoelectrons emitted by the nickel?

16 Radiation of wavelength 3.750 Å falls upon a surface whose work function is 2.0 eV. Compute the maximum speed of the emerging photoelectrons. *Ans.* 6.8×10^5 m/s.

17 When the surface of potassium is illuminated with yellow light (5,890 Å), electrons are liberated which require 0.36 V to bring them to rest. When the surface is illuminated with ultraviolet light (2,537 Å), the stopping potential for the photoelectrons is 3.14 V. Derive values for (*a*) Planck's constant h, (*b*) the work function w_0 for potassium, and (*c*) the long-wavelength limit of the photoelectric effect for potassium.

18 An x-ray beam has a wavelength of 0.20 Å. Find (*a*) the mass and (*b*) the momentum of its photons.

Ans. 1.1×10^{-32} kg; 3.3×10^{-23} kg·m/s.

19 A 250-keV photon is scattered by a free electron (Compton effect). The kinetic energy of the recoil electron is 200 keV. What is the wavelength of the scattered photon?

20 The longest wavelength that will cause photoelectrons to be emitted from a sodium surface is 5,830 Å. What energy is necessary to take the electron through the surface? If the surface is illuminated by light of wavelength 4,500 Å, what is the maximum speed of the photoelectrons emitted? *Ans.* 3.42×10^{-5} J; 4.70×10^5 m/s.

21 An x-ray photon of wavelength 0.20 Å is scattered at an angle of 90° with its original direction after colliding with an electron initially at rest. Find the Compton shift in wavelength and the wavelength of the scattered photon.

Louis Néel, 1904–

Born in Lyon, France. Director of the National Center of Scientific Research at Grenoble. Awarded the 1970 Nobel Prize for Physics for fundamental work on antiferromagnetism and ferrimagnetism which have led to important applications in solid state physics.

Hannes Alfvén, 1908–

Born in Sweden, professor of physics at the Institute of Technology in Stockholm and the University of California at La Jolla. Awarded the Nobel Prize for Physics for 1970 for fundamental work in magnetohydrodynamics with fruitful applications in different parts of plasma physics.

47

Atomic Physics

During the nineteenth century, the golden age of classical physics, science advanced at a much more rapid pace than during any earlier period. The avalanche in the development of science led to specialization: physics, astronomy, chemistry, engineering, etc., became recognized as separate disciplines. At the same time the greatest achievements in physics were in the correlation of apparently diverse phenomena.

During the twentieth century, physics has developed at an even more accelerated pace, and entirely new fields, unsuspected in 1890, are of concern to physicists today. Also, there has been a subtle change in the approach to nature. The earlier emphasis on mechanical models has given way to theories which are much more mathematical (often statistical) and abstract. Interest has extended to submicroscopic phenomena not directly observable. Finally, an earlier trend toward specialization may have become reversed in the joining of chemistry and physics and the gradual merging with biochemistry and biophysics. The present great interest in atomic and nuclear physics arises both from the basic nature of the problems studied and from the technical applications, which have worldwide importance.

47-1 DEVELOPMENT OF THE ATOMIC VIEWPOINT

Evidence of the existence of atoms came first from the study of simple chemical reactions during the latter half of the eighteenth century. By about 1800 it had been demonstrated that:

1 A particular compound always comprises the same elements chemically united in the same proportions by mass.
2 When two elements A and B combine as constituents of more than one compound, the masses of B which unite with a fixed mass of A (and vice versa) are related to each other in the ratios of whole numbers, which are usually small.

A simple kinetic theory of gases, based on the atomic viewpoint, gave a convincing interpretation of the pressure exerted by a gas, as we have seen in Chap. 14. Through this theory, measurements of latent heat and surface tension gave an estimate of the diameters of molecules, of the order of 10^{-8} cm. The theory gained further credibility by predicting the relation between the viscosity and density of a gas and the coefficient of diffusion of a streaming gas. The development of kinetic theory made it evident that heat is associated with the kinetic energy of molecules. It appeared likely that the propagation of sound is also a manifestation of molecular motion. But the atomic theory so far discussed offered no explanation of the nature of electricity and magnetism or of the nature of light.

Among the many important discoveries made during the nineteenth century, some apparently gave answers to long-standing controversies; others introduced new phenomena. Interference, diffraction, and polarization phenomena of light were observed in the early years of the century and led to acceptance of the wave theory of light. The work of Faraday in electrolysis helped confirm the atomic nature of matter and supported ideas of electricity as discrete particles. Faraday and Henry (1831) discovered electromagnetic induction, providing the basis for the electrical industry. Joule (1847) demonstrated the identity of mechanical energy and heat. Maxwell, with the background of Faraday's experiments, worked out a theory of electromagnetic radiation. This theory predicted that an oscillating charge would radiate energy in waves. The prediction was confirmed by Hertz (1887), who produced and detected electromagnetic waves with an oscillating electric circuit.

By 1890, the system of "natural philosophy" looked fairly satisfactory and complete. Not all known phenomena were fully explained, but each had a background that looked promising. In fact some scientists believed that all major discoveries had been made and that future research would be mere refinement of measurement.

Suddenly physicists were confronted with a remarkable series of discoveries. During the decade between 1895 and 1905:

Röntgen discovered x-rays (1895).
Becquerel discovered radioactivity (1896).
Thomson identified electrons (1897).
Planck stated the basic postulate of quantum theory (1900).
Einstein formulated the basic viewpoint of relativity (1905).

These discoveries, in focusing attention on the structure of atoms, led to the profound change of viewpoint in physics implied by the contrasting terms *classical physics* and *modern physics*.

47-2 THE DISCOVERY OF X-RAYS

At the time of Röntgen's discovery, the most active field of research in physics was that of electric discharge in gases at low pressure. Wilhelm Röntgen, professor of physics at Würzburg, studying the conduction of electricity in gas at low pressure covered the discharge tube with black paper. With the apparatus (Fig. 47-1) in a darkened room, he observed that a fluorescent screen nearby lighted up. Röntgen soon found that the agency that caused the fluorescence originated at the end of the tube where the cathode rays struck the glass wall. He recognized immediately the importance of his discovery and proceeded at once to study the properties of the new radiation, which he called *x-rays*.

Among the observations that were included in Röntgen's first report were the following:

1 Many substances fluoresce under the action of x-rays.

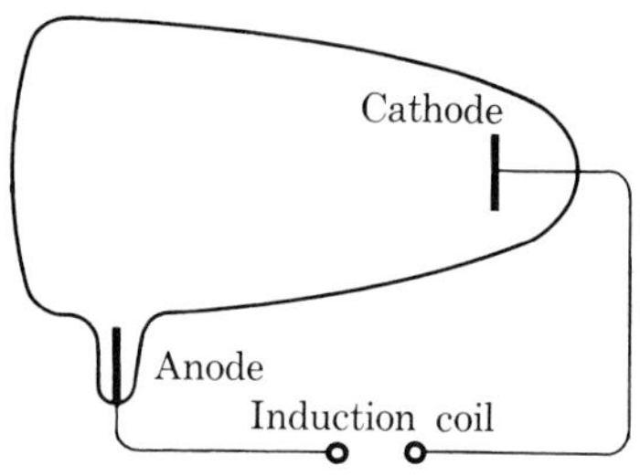

Figure 47-1
Diagram of a Röntgen tube.

2 Photographic plates are affected by x-rays.
3 All substances are more or less transparent to x-rays. Books and thin sheets of wood are very transparent; aluminum weakens the effect but does not destroy it. Lead glass is quite opaque, but other glass of the same thickness is much more transparent. Dark shadows of the bones of the hand are apparent. The opacity of a substance is found to depend not only on its density but also on some other property.
4 X-rays travel in straight lines. Röntgen failed to find reflection or refraction of x-rays, because his apparatus was not sufficiently sensitive. Later both phenomena were observed.
5 X-rays are not deflected by a magnetic field.
6 X-rays discharge electrified bodies whether charged positively or negatively.
7 X-rays are generated when cathode rays strike any solid body. Röntgen found that a heavier element such as platinum is a more efficient generator of x-rays than a lighter element such as aluminum. He designed a tube using a concave cathode to concentrate the cathode rays on a platinum target set at an angle of 45°. This design became almost standard until the introduction of the high-vacuum Coolidge tube about 1913.

Within three months after the announcement of the discovery of x-rays they were being used for medical purposes in the hospitals of Vienna.

47-3 DIFFRACTION OF X-RAYS

In the years following the discovery of x-rays it was generally assumed that they are electromagnetic waves of very short wavelength. Since no means of deflecting them was known, most of the experiments depended upon absorption. It was found that when x-rays fall on a particular substance, x-rays are reemitted. These are fluorescent radiation. Barkla (1908), by the use of the rather crude absorption method, was able to show that some of the x-rays emitted by the secondary material were scattered without change in quality but that others were characteristic of the fluorescing material and that this characteristic radiation consisted of at least two types, one more penetrating than the other. He called these K and L radiations.

In 1912, von Laue with Friedrich and Knipping succeeded in producing diffraction of x-rays. A narrow pencil of x-rays was passed through a zinc sulfide crystal and onto a photographic plate. The central spot on the plate was surrounded by regularly spaced spots. This was the first *direct* evidence of the wave nature of x-rays. Also it showed that atoms are regularly spaced in the crystal. A modern diffraction pattern is shown in Fig. 47-2.

Following the experiment of von Laue, W. H. and W. L. Bragg (1912), using the cleavage face of a crystal, obtained a type of reflection, really diffraction, as in the von Laue experiment. W. L. Bragg predicted the conditions under which it is possible to diffract x-ray beams from a crystal. He considered a family of parallel planes, represented by the dashed lines of Fig. 47-3*a*, passing through the atoms of a crystal. Other sets of planes with different interplanar spacing d could also be defined.

In Fig. 47-3*b* a plane wave is represented falling on one of the planes. It is customary to specify the direction of an x-ray by giving the glancing angle θ between the ray and the plane on which it is incident. (In optics one usually specifies instead the angle of incidence between the ray and the normal to the surface.) Neighboring rays in the plane of Fig. 47-3*b* will reinforce if the path difference ΔP between adjacent rays is an integral number of wavelengths,

$$\Delta P = ae - bd = x(\cos\phi - \cos\theta) = n\lambda \quad (1)$$

where $n = 0, 1, 2, \ldots$

For $n = 0$, this gives $\phi = \theta$, and the plane of atoms *abc* reflects the incident wave like a mirror, for any grazing angle θ. For n different from zero, ϕ does not equal θ, but the diffracted beam can still be regarded as being "reflected" from a

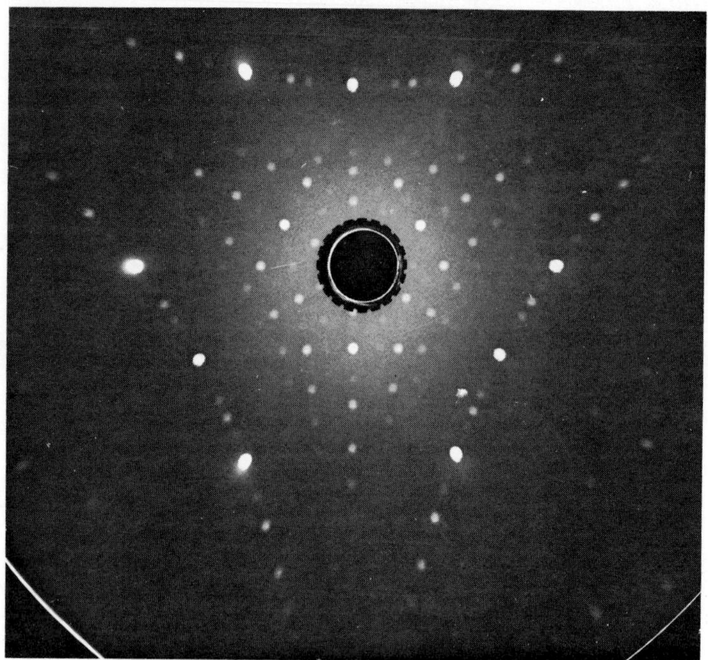

Figure 47-2
Diffraction pattern (white spots) formed by x-rays passing through a tungsten crystal indicates triangular arrangement of atomic layers. Bright spots represent reflections from symmetrical layers of atoms. (*Courtesy RCA Laboratories.*)

different set of planes from those shown in Fig. 47-3*b*, with different spacing d. However, if a beam of x-rays is to be reflected in the direction θ from a whole set of atomic planes (Fig. 47-3*c*), the rays from separate planes must reinforce each other. The path difference *abc* must be an integral number of wavelengths,

$$n\lambda = 2d \sin \theta \qquad (2)$$

This relation is called *Bragg's law*.

X-ray diffraction is a valuable method of studying the arrangement of atoms in a crystal. Quantitative measurements require that the wavelength of the x-rays be known. One might start by determining the dimensions of the fundamental repetitive unit in the crystal, called the *unit cell* (Fig. 47-4), from the observed symmetry of the crystal, the atomic masses, density, and Avogadro's number. Then x-ray diffraction measurements on NaCl in a Bragg spectrometer can be used to determine the wavelength of the x-ray beam. This in turn can be used to determine the structure of other crystalline solids.

Example Find the length a of a unit cell for NaCl (Fig. 47-4).

There are $2N$ ions or diffracting centers in a mole; hence the volume associated with each ion is $V = M/2\rho N_A$. The distance between ions in the cubic structure is then $d = (M/2\rho N_A)^{1/3}$. For NaCl the molecular mass $M = 58.45$, and density

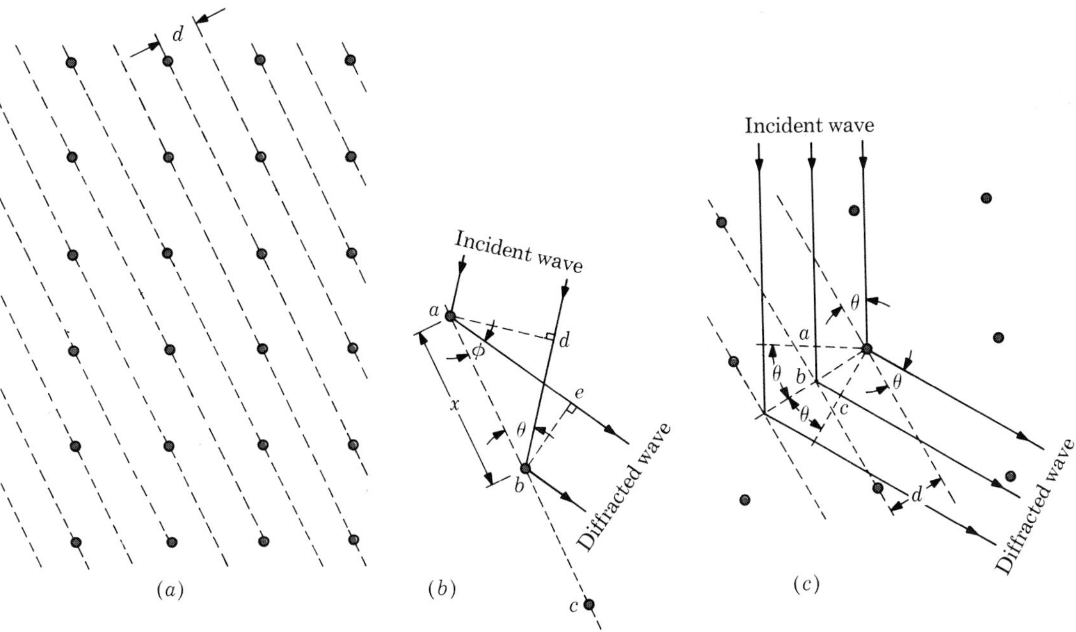

Figure 47-3
Basis for Bragg diffraction of x-rays from a cubic crystal. (*a*) A family of atom-rich planes with interplanar spacing *d*; (*b*) a plane wave incident on one of planes shown in (*a*) at grazing angle θ; (*c*) constructive interference of waves from adjacent atom planes.

$\rho = 2.164$ gm/cm^3. Substituting for Avogadro's number $N_A = 6.0247 \times 10^{23}$ atoms per mole gives $d = 2.814 \times 10^{-8}$ cm. The edge of a unit cell (Fig. 47-4) has a length twice the distance between adjacent ions; so $a = 2d = 5.628$ Å.

Example When a monochromatic x-ray beam falls on the surface of an NaCl crystal, it is found that the first-order Bragg reflection occurs at a grazing angle of 8°30′. What is the wavelength of the x-ray?

In the preceding example it was found that successive atomic planes of an NaCl crystal have a spacing of 2.81 Å. From Eq. (2), with $n = 1$,

$$\lambda = 2(2.81 \text{ Å}) \sin 8.5° = 0.830 \text{ Å}$$

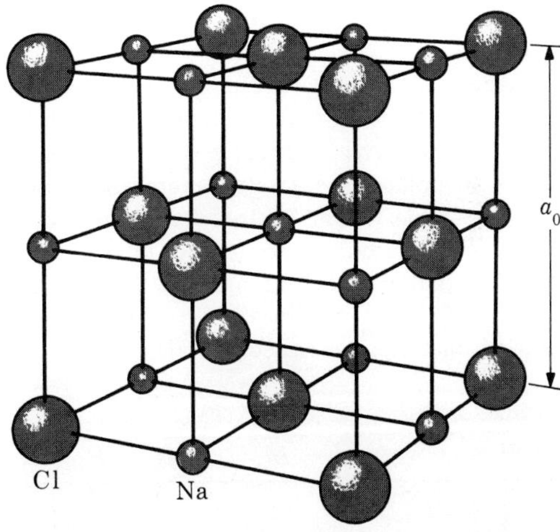

Figure 47-4
Model of Na^+ and Cl^- ions arranged to form a unit cell of NaCl crystal. The edge of the cubical unit is 5.627 Å long.

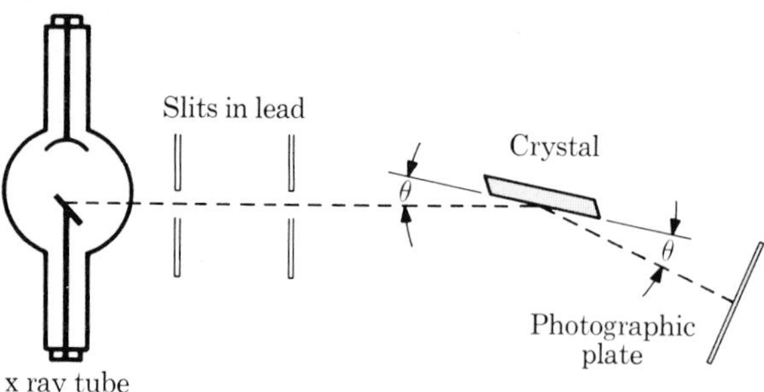

Figure 47-5
A Bragg crystal x-ray spectrometer.

By the use of a crystal spectrometer to measure x-ray wavelengths the Braggs found in the x-ray beam rays that are characteristic of the material (Fig. 47-5). These are superimposed on a general x radiation (Sec. 46-7).

47-4 THE HYDROGEN SPECTRUM

An electric discharge in a gas at low pressure produces a glow of light whose wavelengths are characteristic of the particular gas (Chap. 45). It was early suspected that the light emitted by a gas is a clue to the structure of its individual atoms or molecules. Interest centered on hydrogen, the simplest atom. By 1885 Balmer proposed an empirical formula to account for the four lines that he was able to see and measure, the so-called H_α, H_β, H_γ, and H_δ lines (Fig. 47-6). Balmer showed that the measured wavelengths fit the

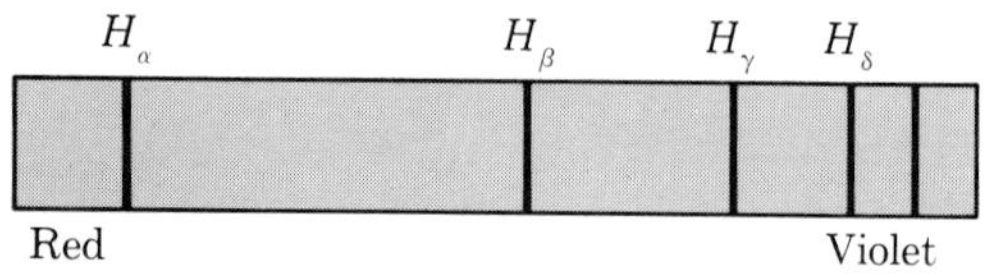

Figure 47-6
Balmer-series lines of hydrogen.

following formula:

$$\lambda = 3{,}645.6 \frac{n^2}{n^2 - 4} \quad \text{Å} \qquad n = 3, 4, 5, \ldots \tag{3}$$

This equation was rewritten by Rydberg in the form

$$\frac{1}{\lambda} = R\left(\frac{1}{2^2} - \frac{1}{n^2}\right) \qquad n = 3, 4, 5, \ldots \qquad R = 1.097 \times 10^{-3}\,\text{Å}^{-1} \tag{4}$$

This Rydberg formula allows one to compute the wavelengths from differences between terms which turn out to be naturally connected with the process of radiation. The possibility of varying the constant term in Eq. (4) suggests other series. When the $1/2^2$ of Eq. (4) is replaced by $1/1^2$, the equation fits the Lyman series photographed in the ultraviolet. When $1/2^2$ is replaced by $1/3^2$, the equation fits the lines of the Paschen series in the infrared (Fig. 47-7).

While these formulas represented a remarkable achievement in discovering relationships among a large number of measured wavelengths, the equations were obtained by empirically fitting measured wavelength values. A derivation of the equations was not achieved until 1913, in terms of an atomic model and assumptions made by Niels Bohr.

47-5 THE NUCLEAR ATOM

Speculation as to the structure of atoms began soon after their existence was accepted. Prout (1815) suggested that "all elements are made up of atoms of hydrogen." Since many of the elements were found to have atomic weights that were not exact multiples of that of hydrogen, the suggestion was not very seriously considered. After the discovery of radioactivity and the elec-

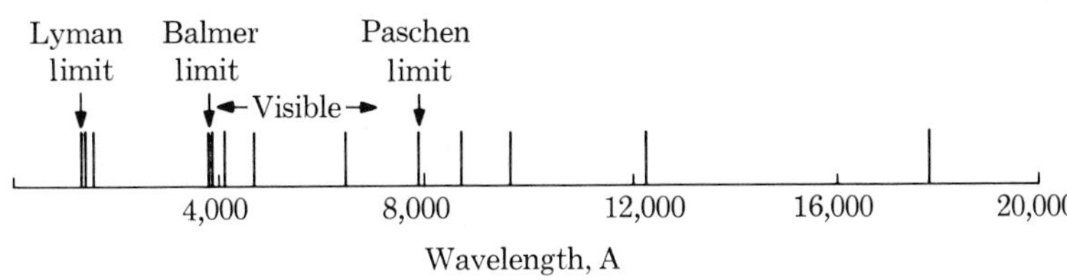

Figure 47-7
Some series of spectral lines in the spectrum of hydrogen.

tron, interest in atomic structure became greater. The electrical nature of matter required that there be equal quantities of positive and negative electricity in the uncharged atom. J. J. Thomson suggested an atom model consisting of a sphere throughout which there is uniform distribution of positive electricity with electrons embedded within the sphere in such positions as to be stable. There was no evidence to support or to contradict this, and the model was retained until experimental evidence inconsistent with it was found.

In 1911, H. Geiger and E. Marsden, in association with Ernest Rutherford, reported the results of a most illuminating experiment. Working with a fluorescent material which flashed when struck by alpha particles, they were able to show that these positively charged radioactive particles when directed at a thin sheet of solid gold could pass through it with ease. However, the most important fact that they observed was that a few of the alpha particles were deflected from their straight path and that some of these bounced back from the gold foil. Rutherford later noted that this "was about as credible as if you had fired a 15-inch shell at a piece of tissue paper and it came back and hit you."

Rutherford was unable to explain these large angles of scattering on the basis of the Thomson atom model but could explain the large deflection on the assumption that *the positive charge is concentrated in a very small region of the atom, less than* 10^{-12} *cm in diameter*. The negative charge must therefore surround a positive nucleus.

Figure 47-8 is a diagram of the interpretation of the results of this experiment. Path *a* is the most commonly followed path in that these particles passed through the large areas of empty space of the gold atoms and were not deflected. Those particles which followed path *b* were fewer in number than those in path *a* and were deflected due to near misses with the positively charged nucleus. Still fewer particles followed path *c*, that is, they were deflected at an angle greater than 90° due to a direct hit on the massive nucleus. These observations are of special significance in that they set the stage for the development of the Bohr atom model in 1913 in which the quantum theory was applied to the nuclear atom.

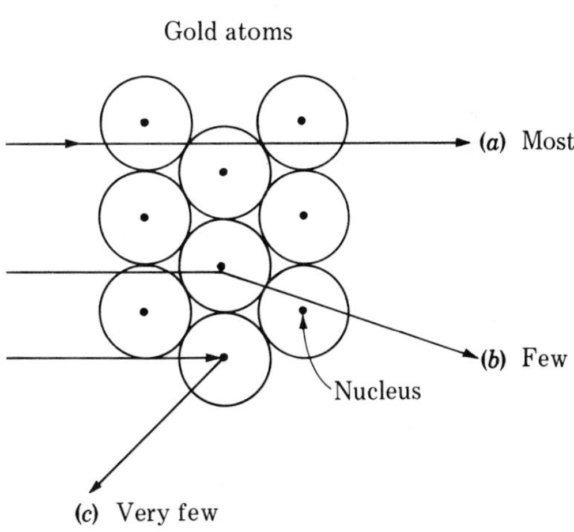

Figure 47-8
Results of Rutherford's gold foil experiment.

47-6 THE BOHR ATOM MODEL

Niels Bohr (1913) combined several earlier developments to set up a specific picture of atomic structure. He assumed a nuclear atom, with electrons moving in circular paths about the nucleus. This picture he applied to the simplest of all atoms, hydrogen.

He assumed that the atom obeys the laws of mechanics and electrostatics. He applied to the atom the quantum ideas of Planck. In applying the quantum idea, he assumed that the electrons are restricted to certain particular orbits: those for which the angular momentum of the electron is an integral multiple of $h/2\pi$. A set of such orbits is shown in Figs. 47-9 and 47-10. He further assumed that there is no radiation as long as an electron remains in one orbit but that energy is absorbed by the atom when the electron is moved from an orbit of lower energy to one of higher energy and emitted when the electron moves in the reverse direction. Since there are only discrete orbits, this restriction requires that energy be absorbed or emitted in quanta.

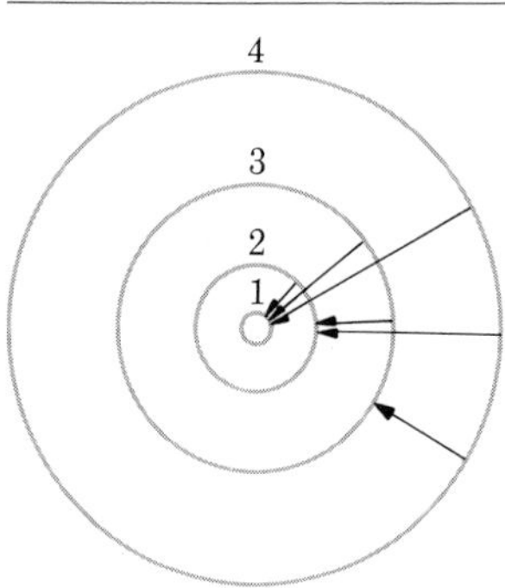

Figure 47-9
Diagram of a few of the possible orbits in the Bohr atom model for hydrogen. The arrows indicate transitions that produce lines in the spectrum series of hydrogen. Those ending at orbit 2 correspond to the red and blue lines of the spectrum of hydrogen produced in the ordinary Geissler tube (see Chap. 28).

The calculation of the frequencies of radiation emitted or absorbed may be outlined as follows. The centripetal force needed to hold the electron in its circular orbit is supplied by the electrostatic (Coulomb) force of attraction toward the nucleus

$$\frac{mv^2}{r} = \frac{Ze^2}{(4\pi\varepsilon_0)r^2} \text{ for hydrogen, } Z = 1 \tag{5}$$

Since only those orbits are permitted for which

$$mvr = \frac{nh}{2\pi} \tag{6}$$

the speed of the electron may be written

$$v = \frac{nh}{2\pi mr} \tag{7}$$

From Eq. (5), the kinetic energy of the electron is

$$\frac{1}{2}mv^2 = \frac{Ze^2}{(4\pi\varepsilon_0)2r}$$

Elimination of v from the last two equations gives for the radius of an orbit

$$r = \frac{n^2h^2(4\pi\varepsilon_0)}{4\pi^2mZe^2} \tag{8}$$

The potential energy of the electron at distance r from the nuclear charge Ze is

$$\frac{Ze}{(4\pi\varepsilon_0)r}(-e) = -\frac{Ze^2}{(4\pi\varepsilon_0)r} \tag{9}$$

The total energy (sum of kinetic and potential) is

$$\mathcal{W} = \frac{Ze^2}{(4\pi\varepsilon_0)2r} + \left[-\frac{Ze^2}{(4\pi\varepsilon_0)r}\right]$$

$$\mathcal{W} = -\frac{Ze^2}{(4\pi\varepsilon_0)2r} = -\frac{2\pi^2mZ^2e^4}{(4\pi\varepsilon_0)^2h^2}\frac{1}{n^2} \tag{10}$$

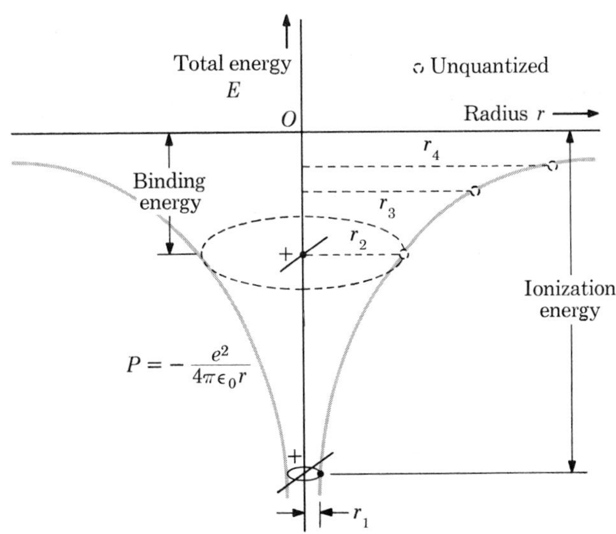

Figure 47-10
Bohr's model of the hydrogen atom.

The assumption that radiation is emitted only for an orbital transition and that the energy of the photon is given by

$$h\nu = \mathcal{W}_i - \mathcal{W}_f \tag{11}$$

leads to an expression for the frequency of a spectrum line, when Eq. (10) is substituted for $\mathcal{W}$,

$$\nu = \frac{2\pi^2 m Z^2 e^4}{(4\pi\varepsilon_0)^2 h^3}\left(\frac{1}{n_f^{\,2}} - \frac{1}{n_i^{\,2}}\right)$$

$$= cR\left(\frac{1}{n_2^{\,2}} - \frac{1}{n_1^{\,2}}\right) \tag{12}$$

To compare the predictions of the Bohr theory with the value of the Rydberg constant previously determined empirically, we may note that since $1/\lambda = \nu/c$, the constant R of Eq. (4) should be derivable from the atomic constants which appear in Eq. (12), that is,

$$R = \frac{2\pi^2 m(1)^2 e^4}{(4\pi\varepsilon_0)^2 h^3 c}$$

The value so calculated for R agrees fully with the empirical value. The Bohr model and assumptions predict the experimentally determined wavelengths in the various series of the spectrum of hydrogen.

Example Calculate the wavelength of the first line in the visible (Balmer) series of the hydrogen spectrum.

$$\frac{1}{\lambda} = R\left(\frac{1}{n_2^{\,2}} - \frac{1}{n_1^{\,2}}\right)$$

$$= 1.097 \times 10^{-3}\,\text{Å}^{-1}\left(\frac{1}{2^2} - \frac{1}{3^2}\right)$$

$$= 1.52 \times 10^{-4}\,\text{Å}^{-1}$$

$$\lambda = 6{,}560\ \text{Å}$$

Example Calculate the radius of the first Bohr orbit for hydrogen, and the speed of the electron in the first orbit.

$$r = n^2 \frac{h^2(4\pi\varepsilon_0)}{4\pi^2 me^2} = (1)^2$$

$$\times \frac{(6.63 \times 10^{-34}\ \text{J}\cdot\text{s})^2}{4\pi^2 \times 9.108 \times 10^{-31}\ \text{kg}\ (1.60 \times 10^{-19}\ \text{C})^2}$$

$$\times \frac{1}{8.99 \times 10^9\ \text{N}\cdot\text{m}^2/\text{C}^2}$$

$$r = 0.53 \times 10^{-10}\ \text{m}$$

$$v = \frac{1}{n}\frac{2\pi e^2}{h(4\pi\varepsilon_0)} = \frac{2\pi(1.60 \times 10^{-19}\ \text{C})^2}{1 \times 6.63 \times 10^{-34}\ \text{J}\cdot\text{s}}$$

$$\times 8.99 \times 10^9\ \text{N}\cdot\text{m}^2/\text{C}^2$$

$$= 2.2 \times 10^6\ \text{m/s}$$

Note that $v/c = 0.0073$ so that the relativity correction for increase in mass of the electron is slight. From $m = m_0/\sqrt{1 - v^2/c^2}$, $m = 1.000027\ m_0$.

The Bohr atom model gave a basis of attack on all line spectra. It was not possible to compute completely the spectra of other more complicated atoms, but approximations were made that yielded satisfactory results. It was soon found that the picture was too simple to explain all the facts; however, certain fundamentals have remained. The nuclear atom with outside electrons is firmly established. The concrete picture of the orbits is not retained, but the concept of *energy levels* for the electrons in each atom remains. An atom is excited when an electron is moved from a lower (inner) energy level to a higher (outer) level, and a definite amount of energy is required to accomplish the change. Radiation accompanies the reverse transition. The energy is radiated in quanta whose size depends upon the energy change. All energy changes between atoms and radiation, whether absorption or emission, occur in quanta. Thus we have a sort of particle characteristic in these interchanges. The packets of energy are called *photons*. Each photon has energy $h\nu$.

47-7 CHARACTERISTIC X-RAYS

H. G. J. Moseley (1913) studied the characteristic x-rays (Sec. 47-3) systematically for many target elements. He identified them with Barkla's K and L radiation. Moseley also found a simple relation between the frequency of a given line for successive elements and the atomic number of the element. This relationship is expressed graphically in Fig. 47-11*a*. The equation of any one line on such a Moseley diagram is given, to a good approximation, by

$$\nu = C(Z - a)^2 \qquad (13)$$

In his study of x-ray frequencies, Moseley clarified the role of the atomic number. The periodicity of chemical properties displayed in the periodic table (Chap. 48) of the elements was originally discovered by arranging the elements in order of increasing atomic mass. Moseley had to rearrange the order of nickel and cobalt, assigning the lower atomic number to the element of higher atomic mass, to obtain a straight line graph of $\sqrt{\nu}$ against Z. Also, a gap had to be left at $Z = 43$, showing the existence of an element then unknown. The nuclear atom model developed by Rutherford and Bohr makes it clear that the atomic number Z (rather than atomic mass) determines the behavior of the element in chemical combinations and in the radiation of energy.

The close relationship between Moseley's findings and the Bohr theory can be shown by writing Eq. (13) in the form

$$\nu_{K\alpha} = cR(Z - 1)^2\left(\frac{1}{1^2} - \frac{1}{2^2}\right) \qquad (14)$$

The K_α line is emitted when an electron goes from the orbit of quantum number $n = 2$ to the orbit of quantum number $n = 1$. The factor $Z - 1$ replaces the Z of Bohr's frequency equation because the electron that goes from orbit $n = 2$ to orbit $n = 1$ is "screened" from the total

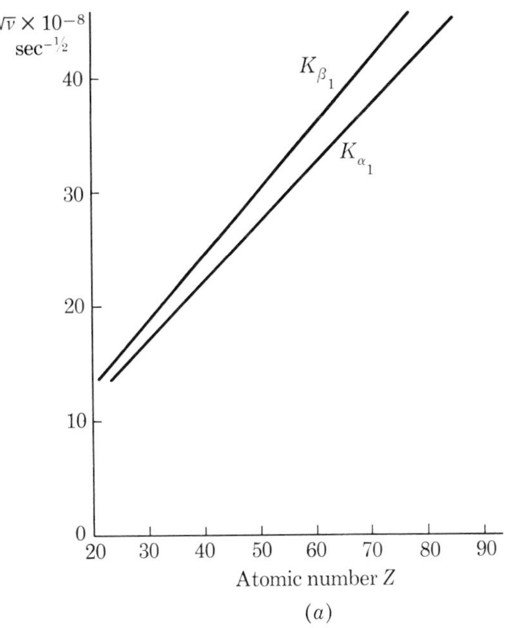

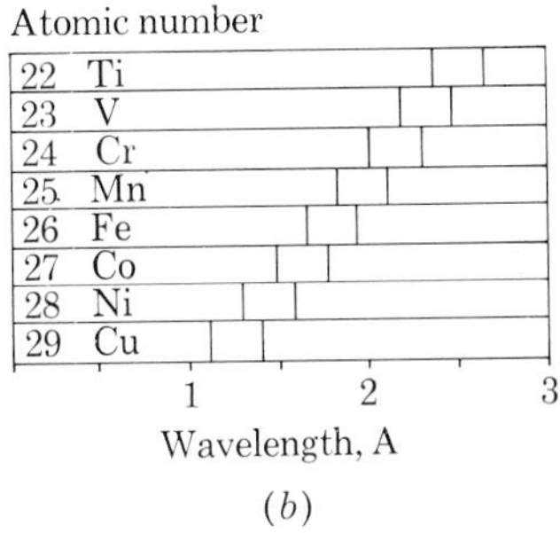

Figure 47-11
(*a*) Moseley diagram in which the square root of the frequency is plotted against the atomic number of the element emitting two lines of the *K* series. (*b*) Characteristic *K* spectra of several elements. The wavelengths are seen to decrease systematically with increasing atomic number.

positive charge *Z* of the nucleus by the negative charge of one electron.

An x-ray energy-level diagram (Fig. 47-12) suggests the origin of the principal lines in the characteristic spectrum of an element in terms of changes in its electron configuration. The zero energy level is taken as that of the normal state of the neutral atom (in contrast to the convention used in Sec. 47-6). A requirement of quantum mechanics known as *Pauli's exclusion principle* (Chap. 48) limits the number of electrons which can be in one shell or energy level: 2 for the lowest, or *K*, level, 8 for the *L* level, 18 for the *M* level, etc. Hence *K* radiation can occur from the target material of an x-ray tube only if the cathode ray impinging on that material has enough energy to remove an electron from the *K* shell, creating a vacancy into which an electron from another level may transfer.

Example Estimate the minimum voltage needed to excite *K* radiation from an x-ray tube with a tungsten target, and calculate the wavelength of the K_α line.

From the Bohr model and Moseley's equation, the energy to remove an electron from the *K* shell of an atom of atomic number *Z* is

$$E_{\text{ionization}} = 13.6 \text{ eV}(Z - 1)^2$$

For tungsten, $Z = 74$, and the energy input required is $(13.6 \text{ eV})(73)^2 = 72{,}800$ eV. So the anode must be at least 72,800 V above the cathode potential.

The first line in the *K* series of tungsten has a wavelength given by

$$\frac{hc}{\lambda_{K\alpha}} = 13.6 \text{ eV}(Z - 1)^2\left(\frac{1}{1^2} - \frac{1}{2^2}\right)$$

$$\lambda_{K\alpha} = \frac{6.625 \times 10^{-34} \text{ J}\cdot\text{s}\,(3.0 \times 10^8 \text{ m/s})}{(13.6 \times 1.60 \times 10^{-19} \text{ J})(73)^2(1 - \frac{1}{4})}$$

$$= 0.21 \times 10^{-10} \text{ m} = 0.21 \text{ Å}$$

47-8 THE CORRESPONDENCE PRINCIPLE

Bohr stated that the predictions of quantum mechanics should merge with classical physics at high quantum numbers. He called this the corre-

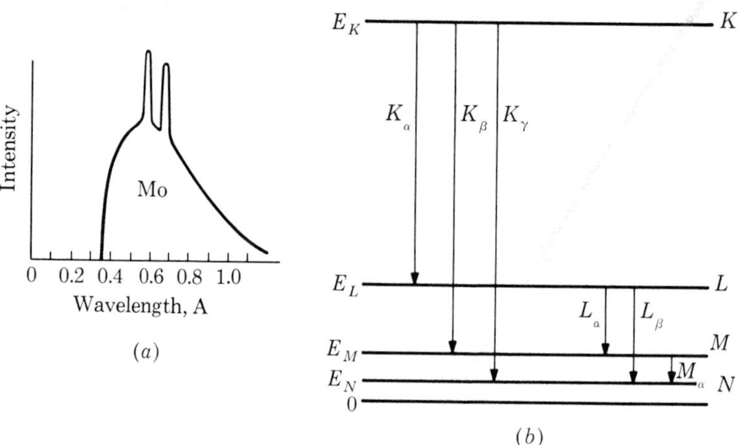

Figure 47-12
Characteristic x-ray spectrum: (*a*) molybdenum target with $V = 35$ kV; (*b*) simplified energy-level diagram.

spondence principle. Bohr's theory of the hydrogen atom shows such agreement with classical physics in the limit when n becomes so large (many thousands) that we may think of the atom as expanded to macroscopic size.

The frequency of orbital motion of the electron in an atom of hydrogen is

$$\nu_{\text{orbital}} = \frac{\omega}{2\pi} = \frac{1}{2\pi}\frac{v}{r} = \frac{1}{2\pi}\left[\frac{e^2}{(4\pi\varepsilon_0)mr^3}\right]^{1/2}$$

$$= \frac{me^4}{4\varepsilon_0^2h^3n^3}$$

From Maxwell's (classical) electromagnetic theory, we expect the electron orbit to act like an antenna, radiating electromagnetic energy at this frequency, and possibly at its harmonics. But the quantum theory applied to Bohr's atom model predicts for the radiation frequency

$$\nu_{\text{Bohr}} = \frac{me^4}{8\varepsilon_0^2h^3}\left(\frac{1}{n_f^2} - \frac{1}{n_i^2}\right)$$

We may write

$$\frac{1}{n_f^2} - \frac{1}{n_i^2} = \frac{n_i^2 - n_f^2}{n_i^2n_f^2} = \frac{(n_i - n_f)(n_i + n_f)}{n_i^2n_f^2}$$

If n_i and n_f are both large compared with unity and are close together, then

$$\frac{1}{n_f^2} - \frac{1}{n_i^2} \approx \frac{2n\,\Delta n}{n^4} = \frac{2\,\Delta n}{n^3}$$

where $\Delta n = n_i - n_f$ and $n \approx n_i \approx n_f$. Then we have

$$\nu_{\text{Bohr}} \approx \frac{me^4}{4\varepsilon_0^2h^3n^3}\Delta n$$

Hence, for $\Delta n = 1$ we get for large orbits the radiation classical theory predicts. For $\Delta n > 1$ we get harmonics. It is apparent that there is a transition region between macroscopic and microscopic physics where the laws of classical and quantum physics overlap.

SUMMARY

The discovery of *x-rays* by Röntgen in 1895 opened a new era of advance in physics. X-rays are of the same nature as light but of very short wavelength.

The discovery of *diffraction of x-rays* by von Laue (1912) led to the Bragg crystal spectrometer, used for measurement of x-ray wavelengths, and to the use of x-rays for determining crystal structure.

Experiments on the scattering of α particles at large angles led Rutherford (1911) to propose a *nuclear atom,* a positive nucleus surrounded by electrons.

Bohr (1913) applied the quantum theory to the nuclear atom to devise an atom model that was successfully used in computing wavelengths in the spectrum of hydrogen.

The emission of the characteristic x-ray lines of heavy elements can be explained in terms of energy levels and an extension of the Bohr atom model. Because heavy atoms have larger nuclear charge and smaller orbits for inner electrons, the large energy changes associated with emission of x-rays produce photons roughly 10^4 times as energetic as the photons of visible light.

According to the correspondence principle, predictions of quantum physics merge with classical physics for very large quantum numbers.

References

Blackwood, O. H., W. C. Kelly, and R. M. Bell: "General Physics," John Wiley & Sons, Inc., New York, 1963. Chapters 41, 42.

Geiger, H., and E. Marsden: The Laws of Deflexion of α Particles through Large Angles, *Phil. Mag.*, **25**(6):604–623 (1913).

Harrison, G. R., R. C. Lord, and J. R. Loofbourow: "Practical Spectroscopy," Prentice-Hall, Inc., Englewood Cliffs, N.J., 1948.

Holton, G., and D. H. D. Roller: "Foundations of Modern Physical Science," chap. 34, Addison-Wesley Publishing Company, Inc., Reading, Mass., 1958.

Pauli, W. (ed.): "Niels Bohr and the Development of Physics: Essays Dedicated to Niels Bohr on his Seventieth Birthday," McGraw-Hill Book Company, Inc., New York, 1955.

Rutherford, E.: The Scattering of α and β Particles by Matter and the Structure of the Atom, *Phil. Mag.*, **21**(6):669–688 (1911).

Weidner, R. T., and R. L. Sells: "Elementary Modern Physics," chaps. 5–7, Allyn and Bacon, Inc., Boston, 1960.

Questions

1 Why do x-rays penetrate material which is opaque to light?

2 When x-ray pictures of internal organs are made, one is frequently asked to drink a preparation containing heavy atoms. Why is this helpful in making the picture?

3 How do x-rays ionize gases and produce chemical changes in organic substances?

4 If transfers of light energy always occur in quanta, why do we not perceive a discontinuous structure in light that comes to the eye?

5 Why are relativity changes seldom used in studying the mechanics of everyday objects?

6 State some of the observed facts that lead to the idea of the equivalence of matter and energy.

7 In a rock-salt crystal the atoms are at corners of cubes. What various planes of atoms might be used in the reflection of x-rays? How would the grating space depend upon the choice of plane?

8 How can x-rays be used in determining the structure of crystals? Would they be of any use in studying a substance that has no crystalline structure?

9 List four types of experimental evidence for the statement: Atoms have characteristic energy levels.

10 Justify the following statement: The observed fact that the lines in the spectrum of hydrogen are very sharp is strong evidence that all electrons have exactly the same charge, e.

11 In a certain hydrogen atom, the electron is in the energy level $n = 4$. If this electron loses energy by radiation in going to level $n = 1$, what is the maximum number of photons it can emit? The minimum number?

12 The minimum wavelength in angstrom units of the x-radiation produced when electrons are accelerated by a potential difference V is given by $\lambda = 12{,}345/V$. Justify the expression on the basis of the quantum equation.

13 (*a*) By using the Rydberg formula, show that the Balmer-series limit (shortest wavelength) should be 3,647 Å. (*b*) The wavelength of the first line in the Lyman series is 1,215 Å. Show from these wavelengths that the ionization energy of hydrogen is 13.6 eV.

Problems

In the following problems, use the electronic charge as 1.60×10^{-19} C and the electronic mass at rest as 9.11×10^{-31} kg.

1 What would be the wavelength of the radiation emitted if an electron that had been accelerated by a potential difference of 300 V had all its kinetic energy transformed into a single quantum?

2 What would be the wavelength of the quantum of radiant energy emitted if an electron were transmuted into radiation and went into one quantum? *Ans.* 0.0201 Å.

3 If the grating space of a calcite crystal is 3.03 Å, find the wavelength of x-rays that are reflected at an angle of 15.0° in the first order.

4 In crystals of rock salt the atoms are set at the corners of cubes, 2.81 Å on a side. Compute the angle in the first and the second order at which x-rays of wavelength 0.721 Å would be reflected. *Ans.* 7.4°; 14.8°.

5 Compute the ratio of the gravitational attraction to the electrostatic (Coulomb) attraction between electron and proton separated by 0.53 Å in a hydrogen atom. Does this justify neglect of gravitational forces in the Bohr theory?

6 Find the frequency of revolution f of the electron circling about a nucleus of charge Ze in terms of the quantum number n.

Ans. $f = 4\pi^2 mZ^2e^4/n^3h^3\ (4\pi\varepsilon_0)^2$.

7 Construct an energy-level diagram for hydrogen. (*a*) Compute the energies of the first six states, in electronvolts. (*b*) Calculate the wavelengths for the first two lines and the series limit for the Lyman, Balmer, and Paschen series.

8 Assume that a free electron having kinetic energy of 2.42×10^{-13} erg unites with an H^+ atom, goes to the lowest ($n = 1$) level, and gives up its energy in radiating a single photon. Find the frequency of the radiation.

Ans. 3.65×10^{15}/s.

9 (*a*) Discuss the possibility of observing the Balmer series in absorption. (*b*) If a tube containing atomic hydrogen is irradiated by an ultraviolet source at a wavelength of 974 Å, what state would be excited? What wavelengths might be emitted? (*c*) Through what potential difference would a beam of electrons have to be passed to excite this same state by collision? (*d*) At what temperature would the average energy of the hydrogen atoms correspond to the energy of this state?

10 Calculate the energy necessary to remove an electron in a hydrogen atom if it is in the $n = 3$ state. *Ans.* 1.5 eV.

11 The bright yellow line in the helium spectrum has a wavelength of 5,876 Å. Compute the difference in energy between the two levels responsible for this line.

12 The red and blue lines of the hydrogen spectrum have wavelengths of 6,563 and 4,861 Å, respectively. Compute the difference in energy between the corresponding energy levels.

Ans. 3.01×10^{-19} J.

13 What is the highest state to which a hydrogen atom initially in its lowest state can be excited by a photon of 12.2 eV energy?

14 An x-ray tube has an anode-to-cathode potential difference of 50 kV. Find the minimum wavelength of the continuous x-ray spectrum emitted. *Ans.* 0.248 Å.

15 From the Bohr theory, calculate the ionization potential of helium.

16 The characteristic K_α x-rays obtained when molybdenum is used as a target in an x-ray tube have a wavelength of 0.71×10^{-10} m. Calculate the voltage on the x-ray tube just sufficient to produce this radiation. What is the energy difference in electron volts between the K and L electron energy levels with the molybdenum atom?
Ans. 22,$\bar{8}$00 eV; 17,$\bar{0}$00 eV.

17 Calculate the wavelength of the first line in the K series of x-rays for silver ($Z = 47$). Could x-rays of this wavelength eject an electron from the K shell of a silver atom?

18 From the wavelength, 1,655 Å, of the K_α line of Ni, use Moseley's law to calculate the wavelength of the K_α line of Ag. *Ans.* 0.57 Å.

19 The spacing between successive atomic planes in a NaCl crystal is 2.82 Å. When an x-ray beam falls on this surface, the second-order Bragg "reflection" occurs at a grazing angle of 17°22′. What is the wavelength of the x-rays?

48

Particles and Waves

The success of the Bohr theory in the study of atomic spectra led to an examination of the relationship of particles and waves. Planck's derivation of the law for the energy distribution of blackbody radiation (1900) first showed the particle (quantum) aspect of electromagnetic radiation. Einstein strikingly established this viewpoint with his explanation of the photoelectric emission of electrons from solids (1905). Photons were endowed with momentum ($h\nu/c$) in Compton's explanation (1924) of the scattering of x-rays accompanied by an increase in their wavelengths.

Light has many characteristics that are essentially properties of waves. However, in those phenomena which involve transfer of energy, explanations require particle properties. Thus there is a dual picture of light: wave and particle.

48-1 MATTER WAVES

If light has a dual wave and particle nature, might not all particles also have a wave nature? Among those who considered this question was Louis de Broglie (1925). A photon has energy $E = h\nu$. It can be assigned a mass from the relation $E = mc^2$. Equating these energies, one can express the wavelength of the photon in terms of the mass

$$mc^2 = h\nu = \text{-}$$

$$\lambda = \frac{h}{mc}$$

The product mc represents the
photon. Similarly, for a partic
a wavelength as h divided by

$$\lambda = \frac{h}{mv} = \frac{h}{p}$$

where the symbol p is used to
tum mv. On this view a singl
can be represented by a little

Example Calculate the
1,000-kg automobile traveling
30 m/s.

The wavelength is given by

$$\lambda = \frac{h}{mv} = \frac{6.63 \times 10^{-34}\ \text{J} \cdot \text{s}}{1{,}000\ \text{kg} \times 30\ \text{m/s}}$$

This wavelength is too small
Thus, the wave nature of matte
in everyday experience.

Example Calculate the w
electron traveling with a speed

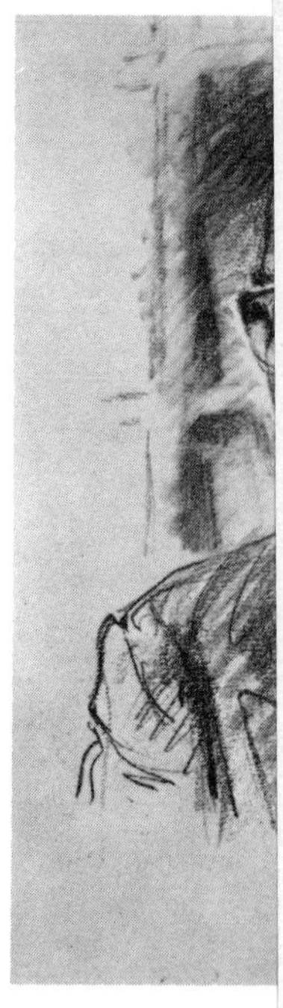

Born in Budapes
at the Imperial (
ogy, London. A
Physics for his di

The wavelength is given by

$$\lambda = \frac{h}{mv} = \frac{6.63 \times 10^{-34}\ \text{J} \cdot \text{s}}{9.11 \times 10^{-31}\ \text{kg} \times 10^{6}\ \text{m/s}}$$

$$= 7.28 \times 10^{-10}\ \text{m}$$

This wavelength is about the size of an atom. Thus, the wave nature of matter is important in atomic phenomena.

In 1926, shortly after de Broglie proposed his theory of matter waves, Erwin Schrödinger, an Austrian physicist, developed a wave equation, the solution of which described the wave amplitude ψ of the matter wave as well as the allowed energy values the particle can have. Schrödinger solved this equation for the case of the hydrogen atom and showed that the allowed energy values of the electron were precisely those predicted by the Bohr theory.

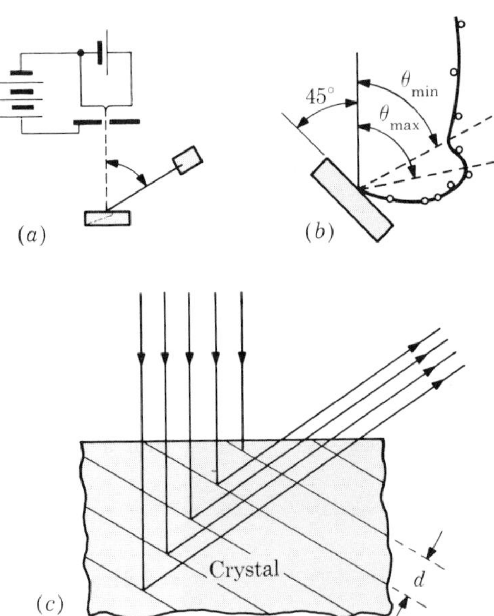

Figure 48-1
(*a*) Davisson and Germer apparatus; (*b*) angular distribution of secondary electrons; (*c*) interpretation in terms of Bragg reflection of electrons (refraction of rays has been omitted).

But what is the physical significance of the wave amplitude? For an electromagnetic wave, the amplitude describes the propagating electric and magnetic fields. For the vibrating string, the amplitude describes the displacement of the string from its equilibrium position. In 1927 the German physicist Max Born interpreted the wave amplitude $\psi(x)$ as being related to the probability of finding the particle at a certain point in space. He stated that the square of the magnitude of $\psi(x)$ is the probability per unit volume that the particle is at the point x.

In the de Broglie hypothesis about wave-particle duality, an electromagnetic wave is the de Broglie wave for a photon and proceeds with speed c. The de Broglie waves for electrons, protons, neutrons, etc., are not electromagnetic waves but "matter waves," which travel with the speed of the particle.

In this chapter we shall discuss a verification of de Broglie's wave hypothesis, something about how the amplitude of a de Broglie wave at various points in space may be calculated, and finally the application of wave mechanics to the conduction of electricity in solids.

48-2 ELECTRON DIFFRACTION

Since the de Broglie equation predicts that 100-eV electrons should have wavelengths of about 1 Å, it seems that the wave nature of matter might be tested in the same way as the wave nature of x-rays was first tested. A beam of electrons of appropriate energy might be directed onto a crystalline solid (Fig. 48-1*a*). The atoms of the crystal form a three-dimensional array of diffracting centers for the de Broglie wave guiding the electrons. There should be strong diffraction of electrons in certain directions, just as for the Bragg diffraction of x-rays.

Experiments started by C. J. Davisson and L. H. Germer for other purposes provided a successful test of the de Broglie hypothesis. They directed a beam of 54-eV electrons on a crystal of nickel (Fig. 48-1*b*). The emergent beam

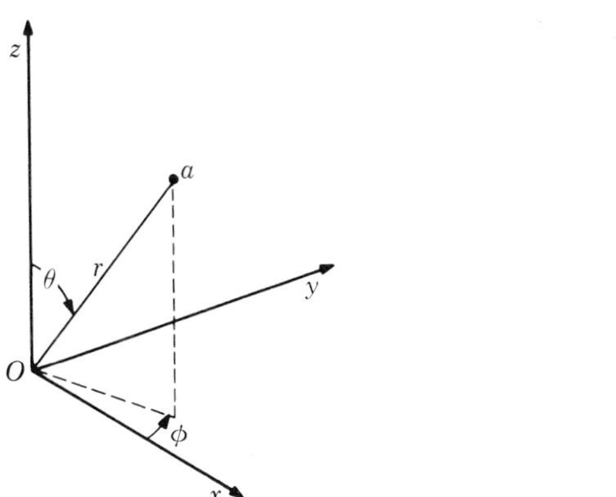

Figure 48-5
Rectangular and spherical coordinates.

the hydrogen atom is found to have one or more solutions described by a second quantum number, the *orbital quantum number l*. This quantum number takes on only the values $l = 0, 1, 2, 3, \ldots, n-1$. It is related to the angular momentum of the atom in that state.

For each value of l, the third part of the wave equation (dependent on coordinate ϕ) is found to have one or more solutions designated by the *magnetic quantum number* m_l. This takes on only integral values for $-l$ to $+l$,

$$m_l = -l, -(l-1), -(l-2), \ldots, -1, 0, 1, 2, \ldots, (l-1), l$$

No solution of the wave equation for the hydrogen atom exists for any other values of n, l, and m_l.

The Bohr model and the wave mechanics so far discussed explain the simplicity of the line spectrum of hydrogen and the appearance of the lines in series of related frequencies. The spectra of atoms of higher atomic number (with more electrons) show another kind of regularity. These spectra may contain multiplet lines—lines that appear as doublets (pairs), triplets (groups of three), etc. This suggests that another principle of quantization is involved. In 1925 Goudsmit and Uhlenbeck proposed that every electron has an intrinsic angular momentum of magnitude $\frac{1}{2}(h/2\pi)$ (Fig. 48-6). The electron may be visualized as a "spherical" mass with negative charge spinning around an axis of rotation. The spinning charge acts like a tiny current, and hence the spinning electron possesses a magnetic moment. When an external magnetic field is applied to atoms of a gas which are radiating, the slight differences in energy levels associated with electron spin are revealed in the spectrum lines.

48-7 THE PAULI EXCLUSION PRINCIPLE

In 1925 W. Pauli suggested that a complete description of the atom must include a unique description of each electron in the atom. No two electrons in an atom may have identical values for a set of four quantum numbers.

To see how this rule operates, consider the number of electrons permitted in the first orbital

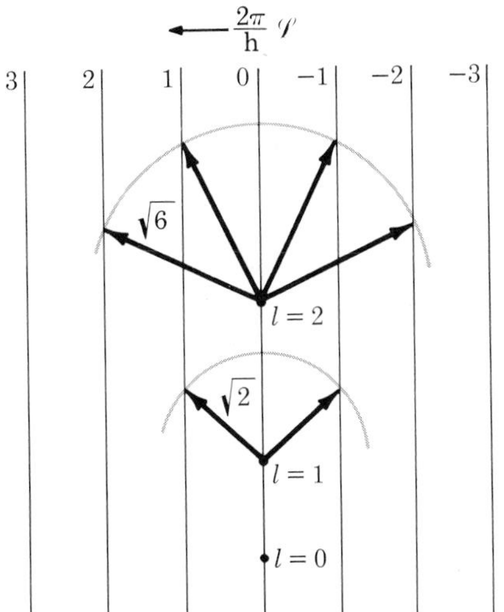

Figure 48-6
Possible orientations of angular-momentum vectors.

group (or shell) for which $n = 1$. When n is 1, $l = 0$ and $m_l = 0$. But the spin quantum number m_s may be $+\frac{1}{2}$ or $-\frac{1}{2}$. So in this first group there may be two electrons, distinguished only by having their spins in opposite directions. This assigning of quantum numbers to electrons in many-electron atoms is further shown in Table 1.

In Table 1, we replace the term "orbit" by "group" or "shell" (determined by n). This emphasizes the three-dimensional nature of the atom. The shells are often named the $K, L, M, \ldots, Q$ shells, corresponding to $n = 1, 2, 3, \ldots, 7$. Within a shell, electrons with a common value of l form a subshell. These are designated s, p, d, or f subshells, according to whether l has the value 0, 1, 2, or 3.

48-8 BUILDING THE PERIODIC TABLE OF ELEMENTS

When the elements are arranged in order of increasing atomic number, a periodicity in their chemical properties is apparent, as shown by Mendeleeff. The structure of the periodic table is in agreement with the ideas of filled shells and subshells as predicted by the Pauli principle. We may "build up" an atom by putting each electron in the shell of lowest energy until the quota of permitted states is filled. Any additional electrons must be put in other shells, as shown in Table 2. The electron configuration of an atom is described by an abbreviated notation. For example, $3p^2$ means that there are two electrons in the $n = 3$, $l = 1$ subshell.

The quantum numbers we are using were originated for the case of one electron. It is remarkable that by assigning occupied states in terms of these numbers we get an accurate description of many of the properties of complex atoms. Evidently the various electrons in a complex atom must disturb each other's orbits very little.

One sort of disturbance, called screening, has been mentioned (Sec. 47-7). An outer electron is in a weak electric field because inner electrons

Table 1
NUMBERS OF ELECTRONS IN GROUPS (OR SHELLS) AS DETERMINED BY PAULI'S EXCLUSION PRINCIPLE

Orbital group	n	l	m	s	No. electrons in subgroup	No. electrons in completed group
1	1	0	0	$+\frac{1}{2}$	$2s$	2
	1	0	0	$-\frac{1}{2}$		
2	1	0	0	$+\frac{1}{2}$	$2s$	8
	1	0	0	$-\frac{1}{2}$		
	2	1	-1	$+\frac{1}{2}$	$6p$	
	2	1	-1	$-\frac{1}{2}$		
	2	1	0	$+\frac{1}{2}$		
	2	1	0	$-\frac{1}{2}$		
	2	1	1	$+\frac{1}{2}$		
	2	1	1	$-\frac{1}{2}$		
3	3	0	0	$+\frac{1}{2}$	$2s$	18
	3	0	0	$-\frac{1}{2}$		
					$6p$	
					$10d$	

Table 2
ELECTRONIC STRUCTURE OF ATOMS*

Principal quantum number, n				1	2		3			4	
Orbital quantum number, l				0	0	1	0	1	2	0	1
Designation of state				1s	2s	2p	3s	3p	3d	4s	4p
Z	**Element**		V_i, **volts**								
1	H	Hydrogen	13.60	1							
2	He	Helium	24.58	2							
3	Li	Lithium	5.39		1						
4	Be	Beryllium	9.32		2						
5	B	Boron	8.30		2	1					
6	C	Carbon	11.26	Helium core	2	2					
7	N	Nitrogen	14.54		2	3					
8	O	Oxygen	13.61		2	4					
9	F	Fluorine	17.42		2	5					
10	Ne	Neon	21.56		2	6					
11	Na	Sodium	5.14				1				
12	Mg	Magnesium	7.64				2				
13	Al	Aluminum	5.98				2	1			
14	Si	Silicon	8.15		Neon core		2	2			
15	P	Phosphorus	10.55				2	3			
16	S	Sulfur	10.36				2	4			
17	Cl	Chlorine	13.01				2	5			
18	A	Argon	15.76				2	6			
					Argon core						

*For additional information, see C. E. Moore, Atomic Energy Levels, vol. 2, *National Bureau of Standards Circular* 467, 1952.

screen it from the positive charge of the nucleus. Hence states in which the electron has some probability of being found very near the nucleus will have lower energy (greater binding) than those states in which the electron tends to stay outside the screening inner electrons. Of the solutions to the wave equation for a given n, those with lower values of l will tend to penetrate the cloud of screening electrons most. Hence, for atoms containing more than one electron, penetration causes the energy of an orbit to depend on l, as well as on n. (In terms of the Bohr picture, energy depends on the shape of the orbit as well as on its size.)

Electrical measurements correlate well with electron configurations. The ionization energy $V_i e$ is the work needed to remove the least tightly bound electron from an atom. The variation of ionization energy E_i ($E_i = eV_i$) shown in Fig. 48-7 and the variation of ionization potential V_i shown in Table 2 for different values of atomic number Z suggest that certain electron configurations have relatively great stability. The first is for helium, where the $n = 1$ shell has its quota of two electrons. The sharp drop to the binding energy for lithium is attributed to the fact that the third electron must be added to the $n = 2$ shell and is therefore farther from the nucleus. For the elements after lithium, there is a trend toward increasing binding energy until another maximum is reached at neon, when the $n = 2$ shell is filled. Like He, Ne is an inert gas. This variation in binding energy is repeated several times in the periodic table, each time giving a maximum binding at an inert gas, followed by a minimum for the succeeding alkali metal. The

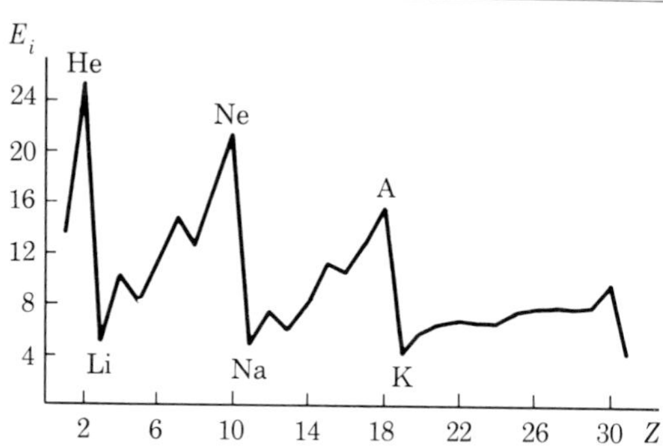

Figure 48-7
Variation of ionization energy (E_i in electronvolts) with atomic number Z, suggesting greater stability of certain electron configurations.

size of atoms also oscillates from shell to shell, about a value approximately 1 Å for the radius. In each shell, the alkali metal has the largest radius.

48-9 CONDUCTION IN METALS

Now that we have a reasonably satisfactory atomic theory, based on the nuclear atom model and quantum mechanics, we may use and extend this theory as a guide in acquiring knowledge in two attractive branches of physics: nuclear physics (Chap. 49) and physics of the solid state. Atomic theory implies that under ordinary conditions of temperature and pressure the nuclei of atoms will never get very close to one another. The chemical combination of atoms should therefore be explainable entirely through the exchange or sharing of electrons. From similar considerations one might expect to be able to describe crystal lattices and the mechanical, thermal, and electrical properties of solids. A practical difficulty is the complexity of the computations. The success of quantum mechanics in explaining important properties of solids is based upon the wave concept of an electron, an extension of the concept of atomic energy levels to an entire crystal, an extension of Pauli's exclusion principle to the electrons in a crystal, and consideration of the effect of the positive ions in the lattice on the multiplicity of electron energy levels and on the motion of "free" electrons.

A classical theory proposed by Drude and Lorentz soon after the discovery of the electron assumed, as have later theories, that some of the electrons are free to travel throughout the whole volume of a crystalline material. It was assumed that in a "good" metal there is about one free electron per atom and that the number of conduction electrons is independent of temperature. These electrons dart in all directions with the high speeds of thermal agitation. But if an electric field is applied, the "electron gas" experiences a relatively slow drift, superposed on the random thermal motions. The electron drift is the electric current. The transfer of any increase in the energy of random motion in a particular direction constitutes thermal conduction. To make quantitative predictions, it is necessary to make some assumptions about the distribution of electron speeds. Theories have differed in these assumptions.

The classical theory assumed that the electron speeds followed the same distribution law as Maxwell and Boltzmann had used for molecular speeds in developing a successful kinetic theory of gases (Chap. 14). Among a large number N of electrons, the fractional number N_v/N having speed v is given by

$$\frac{N_v}{N} = \frac{4}{\sqrt{\pi}}\left(\frac{m}{2kT}\right)^{3/2} v^2 e^{-mv^2/2kT} \tag{6}$$

If we plot this expression against v, the area under the curve between v_1 and v_2 equals the fraction of all the electrons whose speeds lie between v_1 and v_2. Since kinetic energy depends upon the speed squared, the average kinetic energy depends on the average of the squares of the speeds. The square root of this average is called the root-mean-square (rms) speed. The distribution curve becomes flatter and the maximum shifts

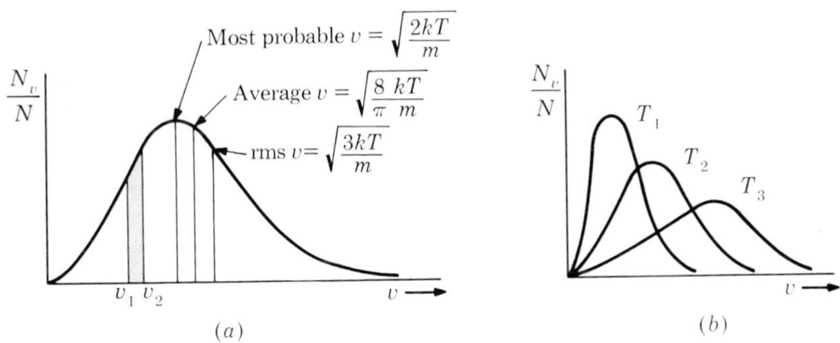

Figure 48-8
(*a*) Maxwell-Boltzmann speed distribution; (*b*) distribution function at three different temperatures, $T_3 > T_2 > T_1$.

toward the higher speeds as the temperature increases (Fig. 48-8).

The classical theory gives rough predictions of the electrical and thermal conductivities of metals. It is in accordance with the experimental observation that the best conductors of electricity are also the best conductors of heat. Wiedemann and Franz (1850) showed that the electrical-conductivity-to-thermal-conductivity ratio is a constant for metals. The classical theory, using known values for e and k, predicts that the ratio is 2.56×10^{-8} W·Ω/K°. This checks well with values measured for platinum and other pure metals.

But the classical theory has significant shortcomings. It predicts that the free electrons should contribute $\frac{3}{2}R$ to the specific heat of a crystal. This considerable electronic specific heat is not observed experimentally. Also, the theory is unable to explain the enormous range of electrical conductivity for different materials (Sec. 36-10). Further, the theory suggests that, since the free electrons have magnetic moments, even a weak magnetic field should produce a large paramagnetic magnetization (magnetic moment per unit volume). It does not. Finally, the classical theory has difficulty in predicting the sign of the Hall coefficient.

48-10 THE LASER

One of the most important and interesting applications of the quantum theory was in the development of the laser. The term *laser*, as noted earlier, is an acronym for *l*ight *a*mplification by the *s*timulated *e*mission of *r*adiation. The laser is a device that produces an intense beam of coherent monochromatic and unidirectional light. There are many types of lasers; we will restrict our discussion to the ruby laser developed in 1960 by T. H. Maimon of the Hughes Aircraft Company.

One of the basic ideas that allowed the development of the laser was proposed in 1917 by Einstein. He introduced the concept of stimulated emission of radiation. We know that an electron in an excited state will spontaneously emit a photon and fall to a lower energy level. This is called *spontaneous emission.* In stimulated emission, however, a photon can stimulate the electron in the excited state to emit a photon identical to the original one and fall to a lower level. The frequency of the photon causing the emission must be equal to the difference in the electron's energy levels divided by Planck's constant. The change in the electron's energy due to

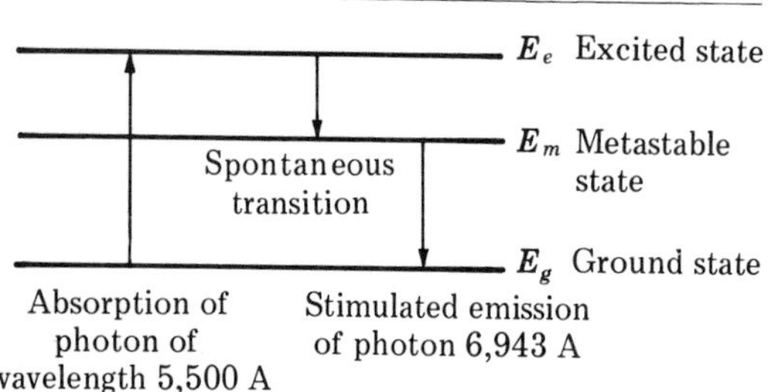

Figure 48-9
Stimulated emission in a laser.

the transition will equal the energy of the photon created. The net result is two identical photons. These photons travel in phase in the same direction, with the same frequency. Each of these photons can stimulate the production of another identical photon from another electron in an excited state, and so on. A chain reaction of photon production is achieved and, hence, light amplification, if more than half the photons are able to stimulate further emission. On the other hand, if most of the electrons are in the lower energy state to begin with, they will absorb the photons and jump to an excited state. The net result is a depletion of the number of photons rather than their production. Thus, there must be a larger population of electrons in the higher level than the lower level in order to achieve light amplification. Under normal equilibrium conditions, however, the lower energy levels are more populated than the higher levels. The reverse of this is called population inversion. We thus see that to cause light amplification we must create a population inversion. This can be caused in a ruby crystal.

A ruby crystal is primarily made up of aluminum oxide with a few chromium atoms in place of aluminum atoms. The electrons in the chromium atoms are responsible for the light amplification.

The electrons in the ground state of the chromium atoms are stimulated by light of wavelength 5,500 Å to jump to an excited state. The approximate length of time an electron will remain in this state is about 10^{-8} s. Many of these electrons jump down to a level intermediate between the ground level and the upper level. The average time the electrons will remain in this intermediate level is about 3×10^{-3} s, a very long time for electrons to remain in an excited state. These long-lived states are called *metastable states*. Thus, for this period of time we have caused a population inversion. An electron will return to the ground level and produce a photon of wavelength 6,943 Å. This photon will stimulate another electron to produce another photon. Hence, the chain reaction has begun.

To increase the production of photons and to produce a unidirectional beam, the photons are reflected back and forth from the ends of the crystal which are partially silvered. About 1 percent of the photons striking the end of the crystal emerge. This constitutes the beam of red light. The continual traversals of the crystal will produce more and more photons and therefore a large amplification. Only those photons that are traveling perpendicular to the ends of the crystal will be reflected back and forth and cause the production of more photons. Hence, the beam is unidirectional. Maimon's laser produced a pulse of radiation frequency 6,943 Å (red light) with a peak power of 10^4 W, for a beam measuring less than a square centimeter in cross section. The beam only spreads 5 feet in moving 1 mile. With the proper optical instrumentation it is possible to project on the moon a spot of light only 2 mi in diameter.

Because of the fact that the laser beam is unidirectional and intense, it is useful in a wide variety of applications. In medicine it has been used to destroy tumors as well as repair detached retinas. A sharply focused laser beam can vaporize rapidly any substance on which it is incident. Thus, minute holes in hard solids such as diamond can be produced. The laser shows promise in the field of communications. The laser beam has the ability to carry much more information than any existing communication system. Thus, the overcrowded communication facilities can be relieved.

Many types of lasers have been developed

since 1960. Lasers that produce continuous beams rather than pulses have been developed. An example of this is the helium-neon laser. This type of laser is commonly used in undergraduate physics experiments.

48-11 THE HALL EFFECT

When a current-carrying conductor is placed at right angles to a magnetic field, an electric potential gradient is developed in the conductor in a direction transverse to the magnetic field. This transport phenomenon, called the Hall effect, can be detected in an arrangement suggested by Fig. 48-10. When a magnetic field is applied at right angles to the conducting strip, the moving conduction electrons are deflected to one side. A charge builds up along the edge of the strip. Additional electrons approaching this edge are deflected back. A balanced state is quickly attained in which the deflecting force of the magnetic field is exactly balanced by the electrostatic force due to the charges along the edge. The Hall electric field can be detected by measuring the potential difference across the conducting strip at right angles to the current direction and to the magnetic field.

An electron moving with drift velocity v experiences a force due to the magnetic field, $\mathbf{F}_B = e(\mathbf{v} \times \mathbf{B})$. This is balanced if the Hall field has a strength $\mathbf{E}_H = -\mathbf{F}/e = -(\mathbf{v} \times \mathbf{B})$. Expressed in terms of the electric current density, $\mathbf{j} = ne\mathbf{v}$, the Hall field is

$$\mathbf{E}_H = \frac{1}{ne}(\mathbf{B} \times \mathbf{j}) \tag{7}$$

In the configuration of Fig. 48-10, where **j** and **B** are at right angles, we can say that the Hall voltage is proportional to the magnetic induction, where the constant of proportionality, called the Hall coefficient, is

$$R_H = \frac{1}{ne} \tag{8}$$

Thus measurement of the Hall effect should give valuable information on n, the number of charge carriers per unit volume.

Example For copper at room temperature the resistivity ρ is $1.72 \times 10^{-8}\ \Omega\cdot\text{m}$ and the Hall coefficient R_H is found to be $-5.5 \times 10^{-11}\ \text{m}^3/\text{C}$. (*a*) Find n, the number of conducting electrons per unit volume. (*b*) What is the average time between collisions, called the *relaxation time?*

(*a*) From $R_H = 1/ne$

$$-5.5 \times 10^{-11}\ \text{m}^3/\text{C} = \frac{1}{n(-1.6 \times 10^{-19}\ \text{C})}$$

$$n = 1.1 \times 10^{29}/\text{m}^3 = 1.1 \times 10^{23}/\text{cm}^3$$

(*b*) If an electron starts from rest and travels for a time τ before making a collision, the velocity it acquires in the electric field is

$$v = a\tau = \frac{Ee}{m}\tau$$

Ohm's law may be expressed in terms of the electric field intensity **E** and current density **j** as $\mathbf{E} = \mathbf{j}\rho$. Since current density **j** equals $ne\mathbf{v}$

$$\mathbf{j} = \frac{n\mathbf{E}e^2}{m}\tau$$

and

$$\rho = \frac{m}{ne^2\tau}$$

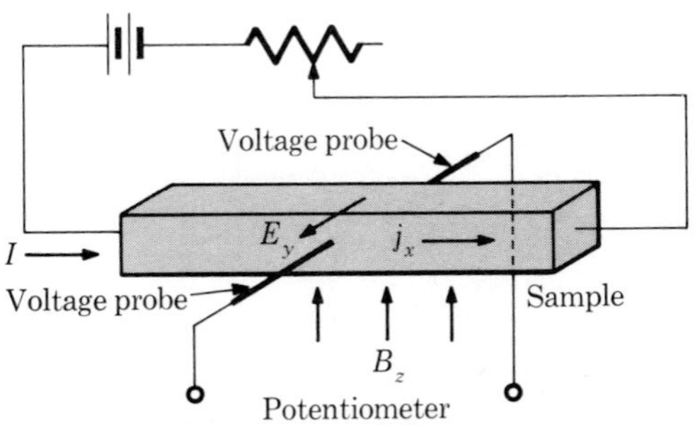

Figure 48-10
Experimental measurement of the Hall coefficient.

Then $\tau = \dfrac{m}{ne^2\rho}$

$$= 9.1 \times 10^{-31}\ \text{kg}/[1.1 \times 10^{29}/\text{m}^3 \times (1.6 \times 10^{-19}\ \text{C})^2\ 1.72 \times 10^{-8}\ \Omega\cdot\text{m}]$$

$$= 2.0 \times 10^{-14}\ \text{s}$$

According to the "electron-gas" model of conduction in metals, we expect the Hall coefficient to have the same sign for all metals. This is not found to be true experimentally, showing the inadequacy of the classical theory of conduction, and suggesting that positive carriers (as well as electrons) may be involved.

48-12 FREE-ELECTRON QUANTUM THEORY OF CONDUCTION

Fermi introduced a radically different description of the free electrons in a metal. He incorporated the exclusion principle, assuming that the "free" electrons in a metal are quantized and that no two can be in precisely the same state.

Momenta are quantized; only two electrons (having opposite spins) can have a given momentum. As the temperature is lowered, electrons settle down by quantized steps to lower momentum values. But as a consequence of the exclusion principle, some electrons will remain at momentum values considerably above zero; i.e., they will have appreciable energy, even at absolute zero temperature. When the temperature rises, only the electrons of highest momentum can accept thermal energy and move to still higher momentum values.

The Fermi distribution law is expressed by

$$\frac{N_v}{N} = \frac{8\pi m^3}{h^3}\,\frac{v^2}{e^{(mv^2/2-E_m)/kT}+1} \qquad (9)$$

where E_m is the maximum energy an electron can have at 0°K. In Fig. 48-11, the progressive rounding of the curve as temperature increases

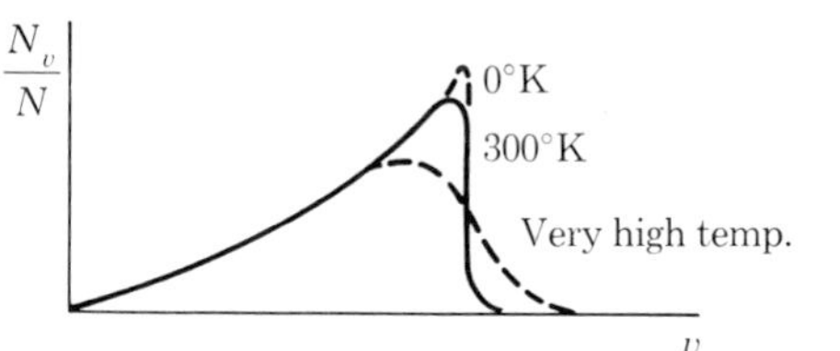

Figure 48-11
Fermi distribution of speeds at various temperatures.

represents the shift of some electrons to higher energies. The Fermi distribution curve should be compared with the Maxwell distribution.

The Fermi theory successfully accounts for the slight participation of electrons in specific heats. In Fig. 48-12, the Fermi distribution of energy is plotted. At 0°K all energy states are occupied up to a certain maximum (Fig. 48-12*a*). At a higher temperature some electrons in upper levels have been able to accept energy and move to still higher levels (Fig. 48-12*b*). But owing to quantum restrictions, relatively few electrons have participated in the temperature rise. The Fermi theory predicts that electrons in a conductor should contribute roughly 1 percent of the amount predicted

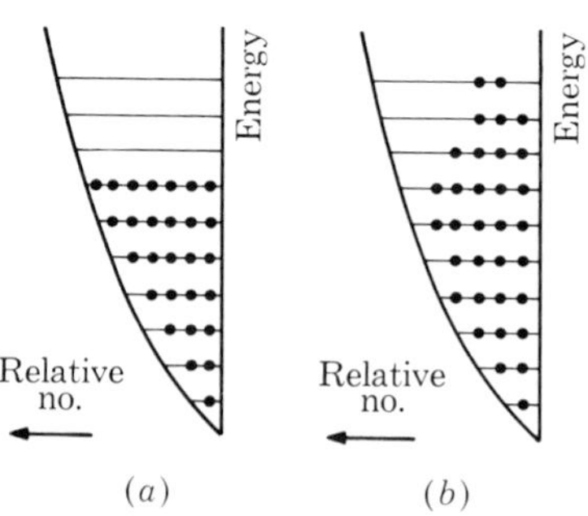

Figure 48-12
Fermi distribution of energies, showing (*a*) all levels filled up to a maximum, at 0°K, and (*b*) some electrons promoted to higher energy levels at a high temperature.

by the Maxwell theory, in agreement with experiments in calorimetry.

The fact that all energy levels, up to a certain maximum, are filled means that for every electron traveling to the right in a metal there is another electron traveling to the left. Thus all electrical conduction in the metal must be due to the relatively few electrons near the top of the distribution (Fig. 48-12*b*) which can be excited easily to an unoccupied quantum level. One concludes that electricity must be conducted by only a small fraction of the free electrons (rather than by all, as assumed in classical theory). In turn, this implies that an electron must be able to travel long distances without being bumped by ions in the crystal lattice. The free-electron quantum theory, like the classical theory, is unable to account for the distinction between conductors and insulators.

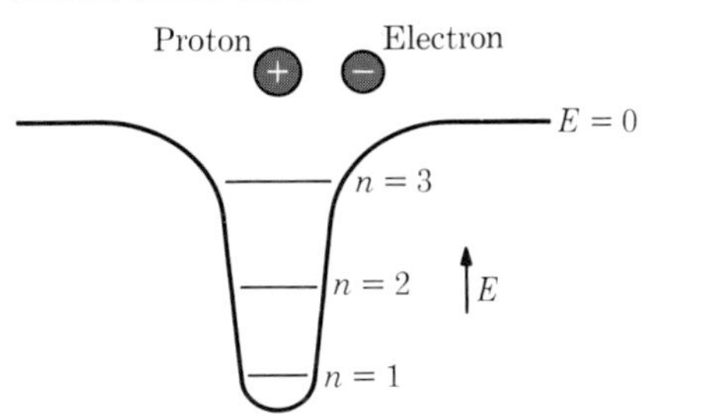

Figure 48-13
Some energy levels for an electron in an isolated H atom.

48-13 BAND THEORY OF CONDUCTION

In the modern band theory of the electronic structure of solids, the effects of the lattice ions on the free electrons are considered to explain the occurrence of conductors, insulators, and semiconductors. The Coulomb attraction of the nuclei causes the potential energy of the electrons to increase and decrease in a periodic way throughout the crystal. The electron moves in a periodic field. Within a potential well such as that of an isolated hydrogen atom, the wave function of an electron is a standing wave (Sec. 48-5). The electron can have only certain discrete energies, as suggested in Fig. 48-13.

Consider what happens to the wave function if an electron is near *two* protons, as is the case in a hydrogen molecular ion, H_2^+. Solution of the Schrödinger equation for this situation of two potential wells about 1 Å apart (Fig. 48-14) yields results unsuspected from classical theory: (1) Each level of Fig. 48-13 ($n = 1, 2$, etc.) now splits into two energy levels in each of the wells of Fig. 48-14. (2) The de Broglie wave penetrates the barrier separating the wells, the decreasing amplitude, as suggested in Fig. 48-4, predicting that an electron placed in, say, level 2 of well A may at some later time appear in level 2 of well B, without "going over the top." The fact that the wave function which describes the probable location of the electron has a nonzero amplitude in the barrier region between A and B is interpreted as meaning that the electron tunnels through the classically forbidden region between A and B and occupies an energy level in A half the time and an energy level in B half the time. This sharing of an electron by two nuclei is an example of a covalent bond, one of the types of chemical bonds which join atoms to form molecules.

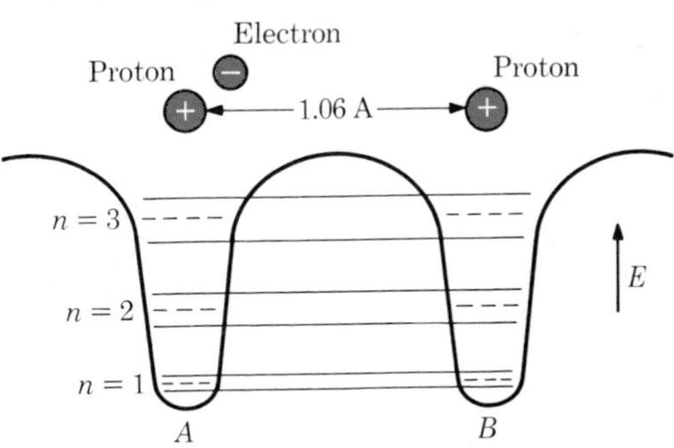

Figure 48-14
For an electron in H_2^+, the energy levels are doubled, and there is some probability that the electron will tunnel from one well to the other.

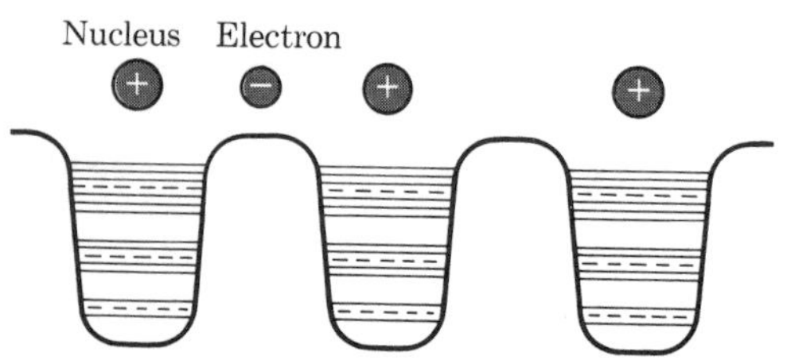

Figure 48-15
Variation of potential energy of an electron in a "one-dimensional" crystal lattice.

We next extend our model to consider a line of atoms regularly spaced in a one-dimensional crystal lattice. The passage of the de Broglie wave representing an electron is treated mathematically by methods similar to those used in investigating the passage of light waves or x-rays through a similar lattice. A result of the periodicity of the electric field in which an electron finds itself is that the sharp energy levels characteristic of an isolated atom (Fig. 48-13) are replaced by *bands,* each comprising many closely spaced energy levels, as crudely represented by Fig. 48-15. The bands are wider at higher electron energies. As long as an electron has energy corresponding to that of one of the energy levels within a band, the electron can travel (tunnel) from one potential well to another. But when an electron acquires energy which falls within one of the gaps or forbidden bands of Fig. 48-15, tunneling does not occur. For these wavelengths, reflection occurs at the "wall" of the potential well in such phase as to cancel the wave that would otherwise extend through the wall.

Our theory still needs to be extended from consideration of a hypothetical single electron in a crystal to the enormous number of electrons in an actual conductor. In copper, for example, there are 29 electrons per neutral atom. We may estimate the number of electrons in 1 cm^3 as $AN_A\rho/m = 29(6.02 \times 10^{23}$ atoms per mole) $(8.89\ g/cm^3)/(63.54\ g/mol) = 2.0 \times 10^{24}/cm^3$. Section 48-12 suggests that not all these will be conduction electrons. But to fit them into the energy-band theory, we use these considerations: (1) the Pauli exclusion principle limits the number of electrons in one energy state to *two* (with different spins); (2) each major energy level for an isolated atom, similar to Fig. 48-13, becomes

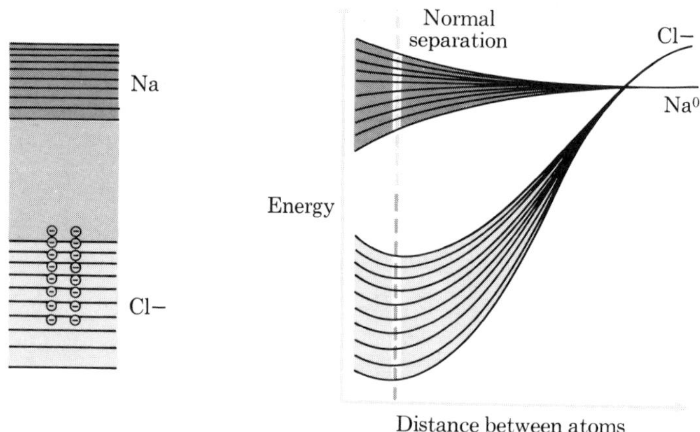

Figure 48-16
Energy bands for a molecule (NaCl) as a function of distance between atoms. The dashed line marks the normal distance existing between the Na and Cl atoms in a crystal.

a band with levels for each atom in the crystal; and (3) electrons normally fill the lowest-lying energy levels first.

The properties of conductors, insulators, and semiconductors can now be interpreted in terms of the conduction bands. If the highest energy band containing electrons is full and is appreciably separated from other bands, the material is an insulator. To produce a current in such a material, electrons have to be advanced across an energy gap large compared with thermal energy kT. In a conductor, however, the highest band containing electrons is not full. Even a small external electric field can produce an unbalanced momentum distribution (a current) by promoting electrons to energy states of small excitation. Semiconductors are an intermediate case in which the highest occupied band is full but the energy jump to the next band is comparable with kT. Increase in temperature would be expected to lower the resistance of a semiconductor.

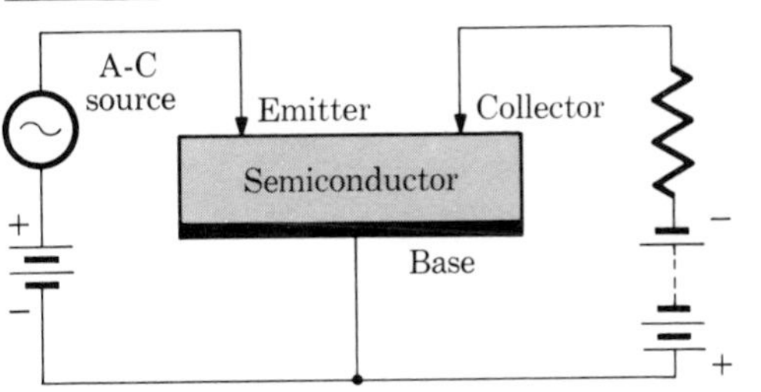

Figure 48-17
Transistor circuit.

48-14 TRANSISTORS

The properties of semiconductors can be used to construct nonlinear circuit elements which have certain advantages over the electron tubes discussed in Chap. 44. In the 1950s, Bardeen, Brattain, and Shockley, at the Bell Telephone Laboratories, did pioneer work in growing crystals of semiconducting materials of extraordinary purity and then modifying them with minute traces of another element to obtain desirable conduction properties.

An example of the semiconductor is germanium. Pure germanium has few electrons available for conduction and hence is a poor conductor. When minute traces of an impurity, for example, antimony, which is capable of supplying conduction electrons, are added to the germanium, the conduction is slightly increased. With this type of impurity the negative conduction electrons are free to migrate within the crystal, and hence this is called *n*-type germanium. If, instead, a different impurity is added, for example, gallium, which is able to borrow electrons from the germanium, gaps, or "holes," appear in the structure that are electron deficiencies and therefore equivalent to positive charges. The holes may migrate through the crystal and furnish the equivalent of a moving positive charge. This is called *p*-type germanium. Both types of germanium are used in making transistors. Other semiconductors are also used.

In one form, the point-contact type, the transistor consists of a small block of semiconducting material, if germanium is usually *n*-type, with three electrodes properly placed (Fig. 48-17). One electrode, the base, makes close contact with the semiconductor; the other two are fine points that make light contact. One point contact is called the emitter, the other the collector. The emitter is connected to the base through a low-voltage battery in such a manner that the emitter is positive. The collector is connected to the base by a higher-voltage battery so that the collector is negative. For the proper values of emitter and collector currents a small change in emitter current causes a large change in collector current. The transistor acts primarily as a current amplifier, whereas the electron tube is primarily a voltage amplifier.

A second kind of transistor is the junction type (Fig. 48-18). It consists of three sections, the two end sections of the same type, *n* or *p*, and the central section of the other, *p* or *n*. They would be labeled *n-p-n* or *p-n-p*. Suitable connections are shown in Fig. 48-18 for a *p-n-p* type. For the

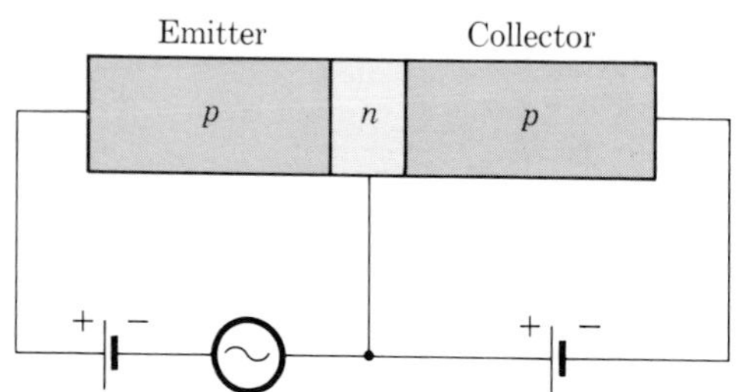

Figure 48-18
Circuit for a junction-type transistor.

n-p-n type the emitter and collector voltages are negative and positive, respectively. Junction transistors are more stable as amplifiers than point-contact transistors, and give higher power gain and less noise.

Certain advantages of the transistor over an electron tube are quite apparent. The transistor is much smaller than the tube, and thus the circuit can be more compact. No energy is necessary to heat a filament; thus there is a saving in space and weight by elimination of the heater source. The elimination of the filament also tends to increase stability in the circuit and to permit longer life.

SUMMARY

The probable location of a particle is represented by the square of the amplitude of its matter wave, suggested by de Broglie (1925) as having a wavelength related to the momentum of the particle: $\lambda = h/mv$.

The amplitude of the matter wave was observed by Born to be related to the probability of finding a particle at a certain point in space.

Wave properties of the electron were verified in diffraction experiments by Davisson and Germer and by G. P. Thomson.

Heisenberg showed through the *uncertainty principle* that in quantum physics there is an inherent limitation on our ability to make measurements. The determinism of classical physics does not exist in quantum physics.

The energy levels of the H atom and its observed spectrum can be predicted from solutions of the wave equation for the bound electron, without requiring the mechanical model of the Bohr atom or Bohr's arbitrary assumption of quantized angular momentum.

Extended to many-electron atoms, wave mechanics and the Pauli exclusion principle explain the periodic table of the elements in terms of the shell configuration of electrons in atoms.

Extended to the electrons of an entire crystal, wave mechanics can explain electrical and thermal properties satisfactorily when account is taken of the influence of the Coulomb fields of the periodic array of positive ions in producing bands of electron energies.

The *laser,* an application of the quantum theory, is based on the concept of *stimulated emission of radiation* proposed by Einstein in 1917.

References

Feather, Norman: "An Introduction to the Physics of Vibrations and Waves," Edinburgh University Press, Edinburgh, 1961.

Kittel, Charles: "Introduction to Solid State Physics," 2d ed., John Wiley & Sons, Inc., New York, 1956.

Pake, George E.: Magnetic Resonance, *Scientific American,* **199**:2, 58–66 (1958).

Waldron, R. A.: "Waves and Oscillations," D. Van Nostrand Company, Inc., Princeton, N.J., 1964.

Questions

1 Assume that Planck's constant is 1 J·s. Describe how everyday events would appear to us.

2 An electron and a proton have the same de Broglie wavelength. Which particle is moving faster?

3 Why are wave properties of particles normally observed only when we study very small particles?

4 If both matter and radiation have both particle and wave properties, how can we decide which property to use in describing a physical phenomenon?

5 Can you reconcile the statement that the *wavelength* of an electron is given by $\lambda = h/p$ with the appearance of momentum p in the formula, which suggests that the electron is a *particle?*

6 In applying Bragg's equation to the data of Fig. 48-1, how could Davisson and Germer be sure that $n = 1$ and that they were observing first-order diffraction?

7 What factors must be present for stimulated emission of radiation to occur in a substance?

8 What is the cause of electrical resistance?

9 Which would you expect to change more for a given temperature change, the resistance of a pure metal or the resistance of a semiconductor? Explain.

10 A standing wave may be regarded as the superposition of two traveling waves (Chap. 20). How can the behavior of a particle confined between two rigid walls (Fig. 48-2) be viewed in terms of wave superposition to explain the motion of the particle?

Problems

1 Calculate the wavelength of a 2-g bullet traveling at 1000 m/s.

2 What is the least uncertainty in the position of the bullet in Prob. 1 if there is an uncertainty in its speed of 2 percent? *Ans.* 2.64×10^{-33} m.

3 Find the de Broglie wavelength for a beam of electrons whose kinetic energy is 100 eV.

4 What is the wavelength associated with an electron whose speed is half the speed of light? *Ans.* 0.04 Å.

5 An electron is accelerated by a potential difference of 1.25 kV. What is the wavelength associated with the moving electron? Neglect any change in mass.

6 (*a*) What is the de Broglie wavelength of an electron with an energy of 45 eV? (*b*) Of a 3.6-MeV alpha particle? (*c*) Of a grain of sand of mass 1.6 mg blown by the wind at 25 mi/h? *Ans.* 1.8 Å; 6.8×10^{-16} m; 3.7×10^{-35} m.

7 A hydrogen atom is in equilibrium with other hydrogen atoms in a container at 20°C. What is the de Broglie wavelength of the atom?

8 Calculate the wavelength of (*a*) an electron whose kinetic energy is 20 eV, (*b*) a 10-MeV neutron, (*c*) a 5.0-μg oil drop that is moving at 2.0 cm/s, and (*d*) a 2-ton truck that is traveling at 30 mi/h. *Ans.* 2.7 Å; 9.05×10^{-15} m; 6.6×10^{-25} m; 2.7×10^{-38} m.

9 In a crystal successive layers of nuclei are spaced at 1.15 Å. At what angle with the crystal surface will first-order Bragg reflection occur for neutrons of kinetic energy 0.025 eV?

10 The smallest detail that can be separated by a microscope (its resolving power) is about equal to the wavelength used. If one hopes to "see" individual atoms (diameter about 1.0 Å), (*a*) what minimum energy of electrons is needed in an electron microscope? (*b*) If a light microscope is used, what minimum energy of photons is needed? (*c*) Which device seems more feasible? Explain. *Ans.* 159 eV; 12,400 eV.

11 What is the kinetic energy of the lowest state of a neutron confined within a distance of 2.0×10^{-15} m?

12 Imagine an electron to be restricted by electrical forces to move between two rigid "walls" (Fig. 48-2) separated by 0.60×10^{-9} m, which is about three atomic diameters. What would be the quantized energy values for the three lowest energy states? *Ans.* $E_1 = 1.04$ eV; $E_2 = 4.2$ eV; $E_3 = 9.4$ eV.

13 A 2.0-μg dust particle moves with a speed 1.0×10^{-6} m/s between walls 0.10 mm apart (Fig. 48-2). (*a*) What quantum number describes this motion? (*b*) Could it be demonstrated experimentally that this motion is quantized?

14 What is the maximum number of electrons in a subshell for which $n = 6$ and $l = 2$? *Ans.* $n = 6$, 10 electrons.

15 A potential hole (Fig. 48-4) has a diameter of roughly 10 Å. Estimate the minimum depth it must have to confine an electron.

16 The Hall coefficient for Na at 20°C is -2.5×10^{-10} m³/C, and its resistivity is 4.3×10^{-8} Ω·m. Find (*a*) the number of conduction electrons per unit volume and (*b*) the relaxation time. *Ans.* 2.5×10^{22}/cm³; 10^{-14} s.

J. Robert Schrieffer (*left*), professor of physics, University of Pennsylvania; **Leon N. Cooper** (*right*), professor of engineering and physics, University of Illinois; **John Bardeen** (*below*), professor of physics, Brown University. Shared the 1972 Nobel Prize for Physics for their jointly developed theory of superconductivity.

49

Nuclear Physics

The Rutherford-Bohr model of an atom as a nucleus surrounded by moving electrons introduced the idea of a nucleus. Experimental evidence about the nature of the nucleus started with the discovery, in 1896, of radioactivity, the spontaneous disintegration of unstable nuclei. The first artificial transmutation of one element into another was achieved by Rutherford in 1919; this date may be taken as the start of the study of nuclear processes of such great importance today.

In the theory of the nucleus there is no counterpart of the simple, easily visualized mechanical model employed in the Bohr theory of the atom. But the concept of energy levels, so useful in studying the atom, is carried over to the description of the nucleus. Nuclear spectroscopy deals with the identification of these energy levels.

49-1 THE DISCOVERY OF RADIOACTIVITY

The discovery of radioactivity was a direct result of the discovery of x-rays. Since cathode rays in striking a target produce x-rays and also produce fluorescence on striking certain materials, the question of a connection between fluorescence and x-rays was raised. In 1896, Becquerel tested this connection, using a fluorescent salt of uranium. He found that even when the specimen was not irradiated it emitted a radiation that penetrated dark paper, thin foils, and other substances. Becquerel shortly discovered that this radiation was characteristic of uranium and that the physical or chemical state of the specimen did not affect the radiation. Other materials, such as thorium, were soon found to show the same properties as uranium. Pierre and Marie Curie discovered two new elements, radium and polonium, that are *radioactive,* radium being much more active than uranium.

It was discovered that the rays from these materials would affect photographic plates and ionize gases as well as produce fluorescence. As the properties were studied, evidence developed that there are three kinds of rays. One is deflected slightly in a magnetic field in the direction in which a moving positive charge would be deflected and is called an α ray. A second, the β ray, is deflected in the magnetic field much more than the α ray, and in the opposite direction. The third, the γ ray, is not deflected in the field. Later experiments showed α rays to be positively charged helium atoms, β rays to be electrons, and γ rays to be electromagnetic radiation of very short wavelength.

49-2 EARLY EXPERIMENTS IN RADIOACTIVITY

The early experiments in the study of radioactivity were almost exclusively in the problems of the nature and properties of the rays themselves. We have noted that the α and β rays can be deflected by magnetic and electric fields, all affect a photographic plate or a fluorescent screen, all are more or less penetrating, and all produce ionization. These properties were further investigated and used in studying the rays.

It was found that α rays are stopped rather easily; a sheet of paper is sufficient. In air at normal atmospheric pressure, the range is a relatively few centimeters. The ranges of the α particles are quite different for the various radioactive substances.

The penetrating power of β rays is much greater than that of α rays. The range may be as great as several meters in air or more than a millimeter of aluminum.

The γ rays are much more penetrating than the β rays. They are little absorbed in air and pass through several inches of lead.

The nature of the rays was determined by a series of experiments. The β rays were shown to be negatively charged by their deflection in magnetic and electric fields. When the ratio of charge to mass was measured (Chap. 44), it was found to be the same as that for cathode rays. Thus the β rays were identified as very-high-speed electrons.

The properties of α rays were also studied by deflection in magnetic and electric fields. By combining the two effects it was possible to compute the speed of the α particles and also the ratio of charge to mass. This latter quantity was found to be about twice that for an ionized hydrogen atom. Rutherford and Geiger measured the charge of the α particle by means of two experiments. The α particles from a known mass of a radioactive material fell on a metal plate for a certain time, giving it their total charge. The plate was connected to an electrometer to measure the charge. The number of particles in that time was determined by means of a "counter" (Fig. 49-1). A metal cylinder C has a wire W insulated from the cylinder along the axis. The cylinder is filled with gas at low pressure, and a potential slightly less than that required for a discharge is maintained between cylinder and wire. A thin window allows α particles to enter the chamber. Each particle ionizes the gas, producing a rush of charge indicated by a pulse to the counter. Thus one can count the number of particles. This device is known as a *Geiger counter*. It has been greatly refined and can be used to "count" particles or radiation that enters the chamber and ionizes the gas inside.

From the measurement of charge and number of particles Rutherford and Geiger determined the charge of the α particle. The value was found

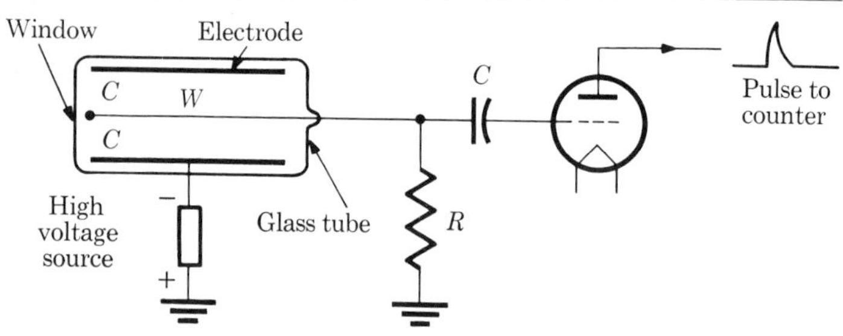

Figure 49-1
Ionization-chamber particle counter.

to be about twice that for the hydrogen atom, and therefore α particle has a mass four times that of hydrogen. This indicates a doubly charged helium atom.

The fact that α particles are charged helium atoms was confirmed by a direct experiment. The α particles were allowed to bombard a glass tube whose walls were thin enough for the particles to penetrate. An electric discharge showed no evidence of a helium spectrum in the tube before the bombardment, but after a considerable time there was a well-defined helium spectrum.

When the electronic structure of an atom acquires some extra energy, the atom almost always gets rid of this extra energy very quickly, returning to its normal state in roughly 10^{-8} s. This is accomplished by emitting one or more photons, or an electron if there is enough extra energy. Many nuclei, however, can exist for long periods of time in an unstable state, i.e., in a state from which the nucleus can and eventually will decay into a stable state. A nucleus may go to a state of lower energy by emitting an α particle (α radioactivity), an electron or positron (β radioactivity), or a photon (γ radioactivity).

Most "natural radioactivity" is found among the very heavy elements ($A > 210$), which tend to be unstable to α decay. These nuclei decay so slowly that there are still some of them left from the time of formation of the elements. Radioactive isotopes not found in nature can be prepared in nuclear reactions.

49-3 IONIZATION OF GASES IN RADIOACTIVITY

When a high-speed α or β particle moves through a gas, there are frequent collisions between the particles and the molecules of the gas. In these collisions ionization is produced. The number of ions produced by the particle is a measure of the energy of the particle. Thus an ionization chamber could be used for such measurements. From the ionization current in the chamber and the number of particles per second, the number of ions produced per particle can be determined. The α particle is much more efficient in ionizing than the β particle and the β particle much better than the γ ray. The ionization is in the ratios of 10,000 to 100 to 1.

The paths of ionizing particles can be photographed for study, with the aid of a cloud chamber or a bubble chamber (Sec. 16-11). In the cloud chamber massive α particles are not easily deflected by collisions, and many ions are formed. Thus the tracks of α particles are mainly heavy and straight, with an occasional fork near the end of the path, where a near direct hit on an atom causes both atom and α particles to show tracks (Fig. 49-2).

On the other hand, β particles produce much less ionization, and hence lighter tracks. Having small mass, they are easily deflected and follow very crooked paths as they are deflected by one collision after another.

49-4 STATISTICAL LAW OF RADIOACTIVE DECAY

The activity of a radioactive sample (the number of disintegrations per second) decreases with time. Each radioactive isotope has its own characteristic rate of decrease. In Fig. 49-3 is the decay curve for a radioisotope which decreases in activity by 50 percent every 4 h. The curve suggests that the decay is a logarithmic function of time. If the logarithm of activity is plotted vs. time, a straight line results, showing that the activity is an exponential (or logarithmic) function of time.

We can derive such an exponential decay law if we assume that: (1) disintegrations occur at random, (2) a large number of radioactive atoms are present, so that statistical methods apply, and (3) the number ΔN of atoms which disintegrate in a time Δt is proportional to the number N of atoms present. Then the decrease in the number of parent atoms is given by

$$-\Delta N = \lambda N \,\Delta t \qquad (1)$$

Figure 49-2
Photograph of α-particle tracks in a cloud chamber. (*Courtesy of P. M. S. Blackett and Proceedings of the Royal Society.*)

where λ is called the *disintegration constant*. It represents the fractional number of radioactive atoms decaying per unit time

$$\lambda = \frac{-\Delta N}{N\,\Delta t}$$

One can express the number N of radioactive atoms remaining after time t in terms of the number N_0 originally present at time $t = 0$.

$$N = N_0 e^{-\lambda t} \qquad (2)$$

The time at which the number of radioactive atoms remaining is just one-half the original number is called the *half-life, T.* Its relation to the disintegration constant is apparent from

$$\frac{N}{N_0} = \frac{1}{2} = e^{-\lambda T}$$

$$T = \frac{\ln 2}{\lambda} = \frac{0.693}{\lambda} \qquad (3)$$

The *average* lifetime T_a of a single atom may be calculated by summing the lives of all the atoms and dividing by the original number of atoms. The result is

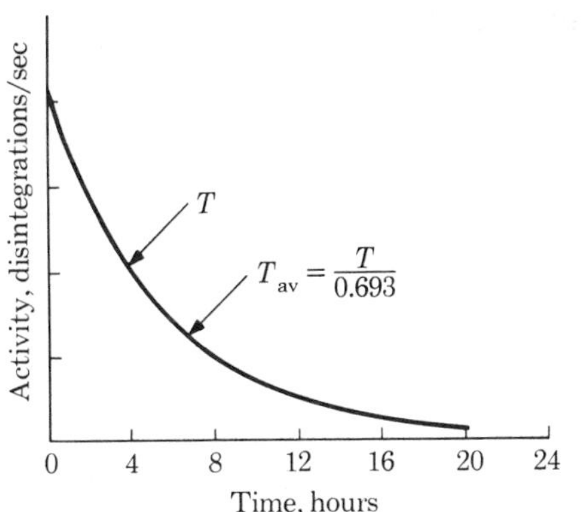

Figure 49-3
Decay of a radioisotope with a 4-h half-life.

$$T_a = \frac{1}{\lambda} \tag{4}$$

The average lifetime is the reciprocal of the disintegration constant.

The unit of activity, the *curie,* is defined as that quantity of any radioactive material which gives 3.70×10^{10} disintegrations per second. Since the curie (Ci) is a relatively large unit, the *millicurie* (1 mCi = 0.001 curie) and the *microcurie* (1 μCi = 10^{-6} Ci) are widely used. A counter near a radioactive source detects a certain fraction of the particles emitted, and the counting rate is proportional to the activity of the source.

When the elements in a radioactive series are allowed to accumulate and to come into equilibrium, the number $N_1\lambda_1$ of atoms of any one isotope which decay per unit time is equal to the number $N_2\lambda_2$ of atoms of the next isotope which decay per unit time, or

$$N_1\lambda_1 = N_2\lambda_2 = N_3\lambda_3 \cdots \text{(equilibrium)} \tag{5}$$

Example The half-life of radon is 3.80 days. After how many days will only one-sixteenth of a radon sample remain?

Solution A After 3.80 d, one-half the original sample remains. In the next 3.80 d, one-half of this decays so that after 7.60 d one-fourth the original amount remains. After 11.4 d, one-eighth the original amount remains, and after four half-lives (15.2 d), one-sixteenth the original amount remains.

Solution B

$$\lambda = \frac{\ln 2}{T} = \frac{0.693}{3.80 \text{ d}} = 0.182/\text{d}$$

We want $\quad N/N_0 = \frac{1}{16} = e^{-\lambda t} = e^{-0.182t}$

$$-0.182t = \ln \tfrac{1}{16} = -2.77$$

so that $\quad t = 15.2$ d

Example Find the activity of a 1.0-g sample of radium ($_{88}Ra^{226}$) whose half-life is 1,620 years.

The decay constant is

$$\lambda = \frac{0.693}{1{,}620 \text{ years}} = 4.2810^{-4}/\text{year}$$
$$= 1.36 \times 10^{-11}/\text{s}$$

One gram of Ra^{226} contains a number of atoms equal to

$$N = \frac{6.025 \times 10^{23} \text{ atoms per mole} \times 1.0 \text{ g}}{226 \text{ grams per mole}}$$
$$= 2.7 \times 10^{21} \text{ atoms}$$

Hence activity $= \lambda N$

$$= (1.36 \times 10^{-11}/\text{s})(2.7 \times 10^{21} \text{ atoms})$$
$$= 3.6 \times 10^{10} \text{ disintegrations per second}$$
$$= 0.97 \text{ Ci}$$

(The activity calculated is due to the Ra^{226} isotope only. An actual sample of radium would show additional activity from the radioactive decay products of Ra^{226}.)

Example It is found that 0.338 μg of radium

is in equilibrium with 1.00 g of U^{238}. Find the half-life of U^{238}.

$$\lambda_1 N_1 = \lambda_2 N_2$$

$$\lambda_1 \times \frac{6.025 \times 10^{23}}{238}$$

$$= \frac{0.693}{1{,}620 \text{ years}} \frac{0.338 \times 10^{-6}}{226} 6.025 \times 10^{23}$$

$$T_1 = \frac{0.693}{\lambda_1} = \frac{226}{238} \times \frac{1{,}620 \text{ years}}{0.338 \times 10^{-6}}$$

$$= 4.55 \times 10^9 \text{ years}$$

49-5 GAMMA DECAY

A nucleus in an excited state (${}_z{}^*X^A$) may go to a state of lower energy by emitting the difference in energy as a photon,

$$_z{}^*X^A \rightarrow {}_zX^A + h\nu \qquad (6)$$

The γ decay does not cause a change in the atomic number or the mass number of the nucleus. The half-lives for γ decay are seldom very long.

Study of γ radiation gives important information about the initial and final states of the nucleus undergoing a γ transition. Like the spectra of atoms, the γ spectra of nuclei are found to consist of sharp lines, showing that the nucleus has discrete energy levels. The observed energies of emitted photons give consistent results for the nuclear energy levels,

$$h\nu = E_i - E_f \qquad (7)$$

The electromagnetic-wave nature of γ radiation is demonstrated experimentally by diffraction. This is feasible only for those γ rays of relatively low energy, because ruled gratings or crystals with effective spacings about equal to very short γ wavelengths are not available.

49-6 ALPHA DECAY

When an α particle is ejected from the nucleus, the original nucleus loses two protons and two neutrons. Its mass number decreases by four units, while its atomic number Z decreases by two. Alpha decay thus causes transmutation of the parent chemical element into a different chemical element,

$$_ZX^A \rightarrow {}_{Z-2}Y^{A-4} + {}_2He^4 + Q \quad \text{(energy)} \qquad (8)$$

Now α decay occurs spontaneously, without any external forces, and it provides kinetic energy ($E_{k,\alpha}$) for the ejected α particle as well as some kinetic energy ($E_{k,d}$) for the recoil "daughter" nucleus. Hence α decay cannot occur unless the total rest mass decreases. The decrease in rest energy is equal to the kinetic energy released, called the *disintegration energy* Q

$$Q = E_{k,d} + E_{k,\alpha} = (m_p - m_d - m_\alpha)c^2 \qquad (9)$$

To predict whether a nucleus will undergo α decay, we may compare its rest mass with the sum of the masses of the product nuclei. Actually we can use the masses of atoms instead of those of the nuclei. The same number of electrons are associated with the initial and final nuclei, so the electron masses cancel in the calculation of Q. From Eq. (9)

$$Q = (m_X - m_Y - m_{He})c^2 \qquad (10)$$

Example Find the Q value for the disintegration ${}_{60}Nd^{144} \rightarrow {}_2He^4 + {}_{58}Ce^{140}$.

From tables of isotope masses

$$\begin{aligned} {}_2He^4 &= 4.00387 \\ {}_{58}Ce^{140} &= 139.94977 \\ &\quad 143.95364 \end{aligned} \qquad \begin{aligned} {}_{60}Nd^{144} &= 143.95556 \\ \text{Products} &= 143.95364 \\ m &= 0.00192 \\ Q = mc^2 &= 1.79 \text{ Mev} \end{aligned}$$

Example In α decay, what fraction of the

disintegration energy appears as kinetic energy of the α particle?

Conservation of energy and conservation of momentum in α decay require

$$Q = E_{k,d} + E_{k,\alpha} = \tfrac{1}{2}m_d v_d^2 + \tfrac{1}{2}m_\alpha v_\alpha^2$$

$$m_\alpha v_\alpha = m_d v_d$$

From the momentum equation, $v_d = (m_\alpha/m_d)v_\alpha$. Substituting this in the energy equation, we have

$$Q = \tfrac{1}{2}m_d \frac{m_\alpha^2}{m_d^2} v_\alpha^2 + \tfrac{1}{2}m_\alpha v_\alpha^2$$

or

$$Q = \tfrac{1}{2}m_\alpha v_\alpha^2 \left(\frac{m_\alpha}{m_d} + 1\right) \qquad (11)$$

or

$$E_{k,\alpha} = \frac{Q}{1 + m_\alpha/m_d}$$

If A is the mass number of the parent nucleus, then $m_\alpha/m_d \approx 4/(A-4)$ and

$$E_{k,\alpha} \approx \frac{A-4}{A}Q \qquad (12)$$

Thus, for large A, the α particle gets most, but not quite all, of the disintegration energy.

An interesting feature of α decay called the tunnel effect may be illustrated by data for a particular case. One can perform an experiment similar to the Rutherford-Geiger-Marsden scattering experiment (Chap. 46), using a thin foil of $_{92}U^{238}$ to scatter the 7.68-MeV α particles from $_{84}Po^{214}$ (also called Ra C′). One finds that the Rutherford scattering law is obeyed. Evidently the α particles from Po^{214} do not have sufficient energy to get over the Coulomb barrier; they are scattered away from the U^{238} nucleus. This is suggested in Fig. 49-4, which shows the potential-energy curve of an α particle near a U^{238} nucleus and a Po^{214} α particle being turned away by the potential barrier. Contrast this with the following fact: U^{238} itself is an α emitter, emitting α particles whose kinetic energy is only 4.20 MeV. We have

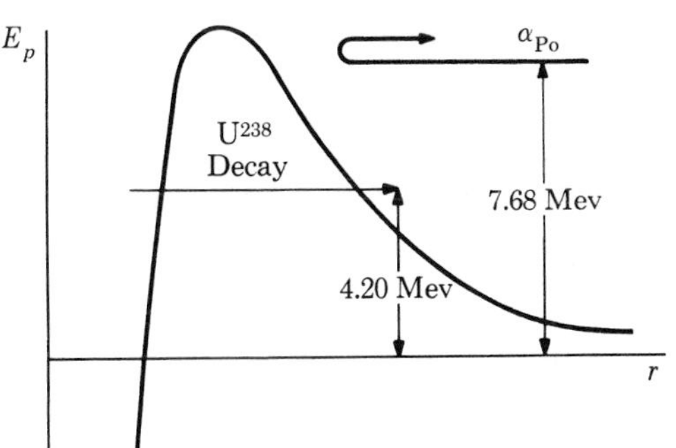

Figure 49-4
Coulomb barrier: Scattering of a high-energy particle and tunneling of a low-energy particle.

a paradoxical situation: The lower-energy U^{238} α particle can cross a barrier which the higher-energy Po^{214} α particles appear unable to cross. An explanation on the basis of classical physics is impossible.

The wave nature of the α particle must be taken into account. When we use wave mechanics to describe an α particle in the nucleus, we find that a little of the wave function will "leak"

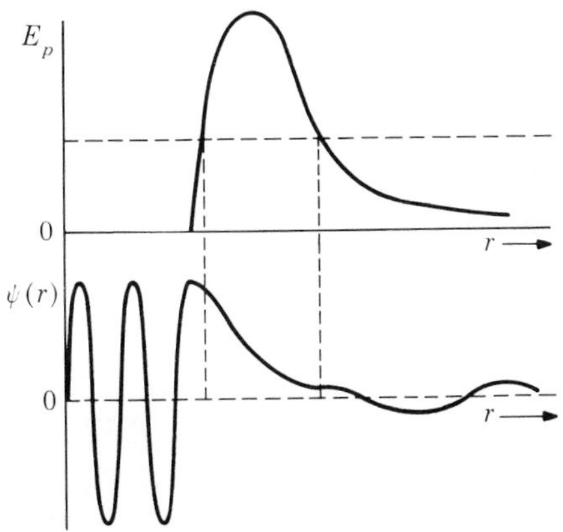

Figure 49-5
Wave-mechanical description of tunnel effect.

through the barrier so that there is a small probability that the particle may be found outside (Fig. 49-5). According to wave mechanics, if the α particle has enough energy to be outside, then there is some probability that it will be found there. This probability is very small for U^{238} and accounts, roughly, for the U^{238} half-life of 4.5 billion years. The tunnel effect works in either direction, so some of the Po^{214} α particles used in the scattering experiment must have penetrated the nucleus, but the fraction which succeeded was negligible. The probability of tunneling depends strongly on the height and width of the potential barrier.

49-7 BETA DECAY

The β particles emitted from a radioactive source are shown by deflection experiments to be high-energy electrons. There are good reasons to believe that these electrons do not exist in the nucleus but are created by a rearrangement of the nucleus into a state of lower energy. Any excess of energy over that required to provide one electron rest mass ($m_e c^2$) appears as kinetic energy of the emitted electron.

Two different types of β decay occur: β^- decay, in which an electron is emitted from the nucleus, and β^+ decay, in which a positron is emitted. If the nucleus consists of neutrons and protons only and if electric charge is conserved, then, upon emission of an electron, a neutron must be converted to a proton, $\Delta Z = +1$. Similarly, positron emission involves the conversion of a proton to a neutron, $\Delta Z = -1$.

$$_{Z}X^{A} \rightarrow {}_{Z+1}Y^{A} + {}_{-1}e^{0} + Q \qquad (13)$$

$$_{Z}X^{A} \rightarrow {}_{Z+1}Y^{A} + {}_{+1}e^{0} + Q \qquad (14)$$

For β^- decay to occur, the mass of the decaying nucleus must be greater than the mass of the product nucleus plus the mass of an electron. An atom which is heavier than the atom with Z one unit greater but with the same A will decay into that atom by β^- emission.

The condition for β^+ decay is slightly more complicated,

$$Q = (m_X - m_Y - 2m_e)c^2 \qquad (15)$$

where m_X and m_Y are the masses of the initial and final atoms, respectively, and m_e is the rest mass of an electron. An atom is β^+-unstable if it is more than two electron masses heavier than the atom with the same A and one less Z.

There is still a third β-decay process whose overall result is the same as β^+ decay. A nucleus may absorb one of its orbital electrons. This process is called *K capture,* since the electrons in the nearest ($n = 1$) shell are most likely to be absorbed. The energy rule is the same as that for β^- decay: If the resulting atom is lighter than the original atom, it is unstable to K capture.

The changes resulting from various nuclear processes are often represented in a proton-neutron diagram (Fig. 49-6) in which each nucleus is plotted in terms of the number (Z) of its protons vs. the number ($A - Z$) of its neutrons. It is a result of the processes we have just discussed that no two adjacent isobars (nuclei with same mass number) can both be stable. The heavier will β-decay into the lighter.

The energies of electrons and positrons from β decay have been determined with various types

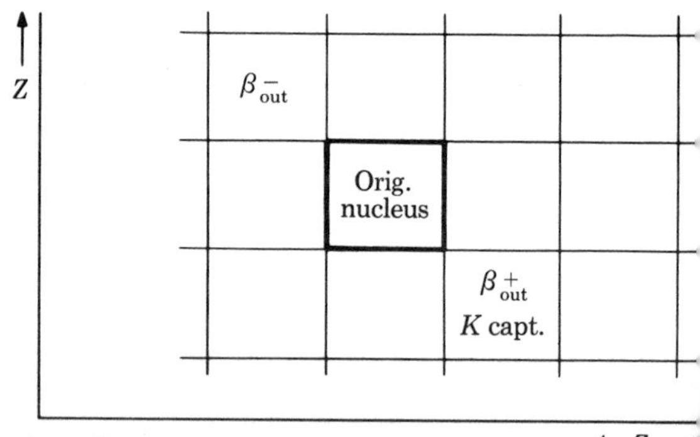

Figure 49-6
A proton-neutron diagram.

of β-ray spectrometers. In principle, they measure the momentum of an electron by finding the curvature of its path in a known magnetic field. It is found that electrons in a given type of α decay may have any energy up to the calculated energy release Q (Fig. 49-7). Here is a difficulty with the hypothesis that β decay consists in the emission of an electron (or positron) and the conversion of a neutron to a proton (or proton to a neutron). For the nuclear change is from one state of definite energy to another state of definite energy. Yet the electrons emitted carry varying amounts of energy, up to the maximum available. There is another difficulty. Consider the β decay of a nucleus containing an even number of nucleons. Its angular-momentum quantum number is an integer, since there is an even number of spin-$\frac{1}{2}$ particles present. If a single electron is now created, there will be an odd number of spin-$\frac{1}{2}$ particles and the total angular-momentum quantum number will be half an odd integer. But a spontaneous change in angular momentum is not possible.

To remove these difficulties, we assume that, along with the electron, another particle, also of spin-$\frac{1}{2}$, is created and emitted. This particle is called the *neutrino*. Since it shares the disintegration energy Q with the electron, the continuous energy distribution observed for the β particles (Fig. 49-7) can be explained. The neutron is assumed to have zero rest mass; so the only change needed in our previous equations is to replace E_k by $E_k + E_{k,\text{neutrino}}$. The neutrino participates only in β reactions. Since it has no rest mass, it travels with the speed of light. It is postulated to have spin $\frac{1}{2}$ and to obey Pauli's exclusion principle. The neutrino has no electric charge, and it is difficult to detect! This remarkable particle has been assumed as necessary by physicists since about 1934. Its existence was first experimentally demonstrated in 1956, by detection of γ rays produced in a planned sequence of events initiated by the neutrino.

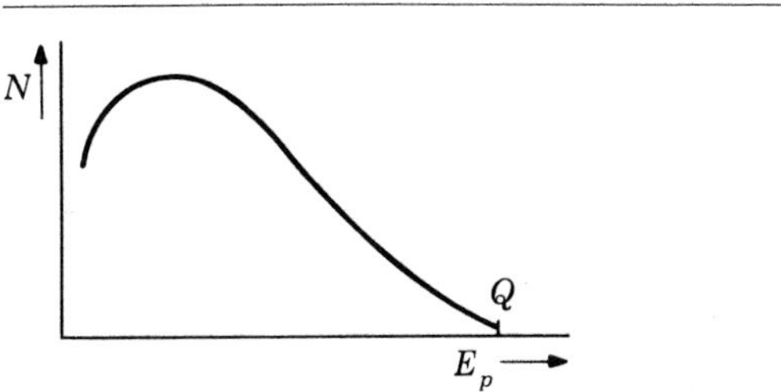

Figure 49-7
A continuous β spectrum.

49-8 NATURAL RADIOACTIVE SERIES

In experiments which followed the discovery of radioactivity, quite a number of substances were found to show activity. It was found that certain of these substances are associated with each other in series, the successive members being formed by the disintegration of the preceding member, until a stable nucleus is reached.

One can predict that there should exist four separate decay chains or radioactive series. A nucleus belongs to one of four classes, depending on whether its mass number A has the form $4n$, $4n + 1$, $4n + 2$, or $4n + 3$, where n is an integer. Radioactive decay of a nucleus in one of these will result in the formation of daughter nuclei in the same class. This follows, since there is no change in mass number in β or in γ decay, while in α decay, $\Delta A = 4$. The four radioactive series are represented in Fig. 49-8. Each bears the name of its longest-lived element. The neptunium series is not observed naturally, because ${}_{93}\text{Np}^{237}$ ($T = 2.2 \times 10^6$ year) has almost completely decayed since the formation of the elements (about 5×10^9 years ago).

The decay schemes of these four series end with stable isotopes of lead. A few radioactive isotopes which do not belong to the heavy-element chains are found in nature: ${}_1\text{H}^3$, ${}_6\text{C}^{14}$, ${}_{19}\text{K}^{40}$, ${}_{37}\text{Rb}^{87}$, ${}_{49}\text{In}^{115}$, ${}_{57}\text{La}^{138}$, ${}_{62}\text{Sm}^{147}$, ${}_{71}\text{Lu}^{176}$, and ${}_{75}\text{Re}^{187}$.

When the elements in a radioactive series are allowed to accumulate, a steady state will be reached (if the parent atom has a long half-life)

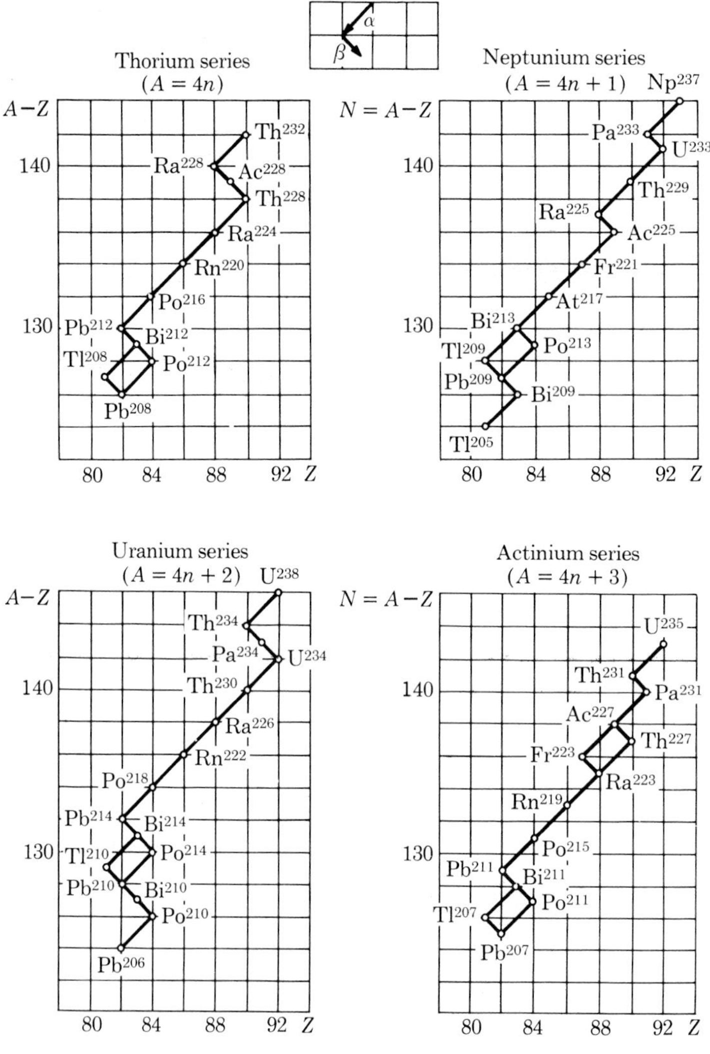

Figure 49-8
Decay schemes of the four families of natural radioactivity.

in which the number $N_1\lambda_1$ of atoms of one isotope which decay per unit time is equal to the number $N_2\lambda_2$ of atoms of the next isotope which decay per unit time, or

$$N_1\lambda_1 = N_2\lambda_2 = N_3\lambda_3 = \cdots \quad \text{(equilibrium)} \tag{16}$$

This equilibrium equation is often used to calculate λ for an isotope whose half-life is too large or too small to make a particle-counting experiment convenient.

49-9 ARTIFICIAL DISINTEGRATION

In natural radioactivity the rate of disintegration is independent of the physical condition of the active material. Attempts to retard or hasten the process produce no result.

In scattering experiments that led to the nuclear-atom theory, the scattering atoms were all very much heavier than the α particles, so that the motion of the scattering nucleus was negligible. If light atoms are used, both α particle and scattering nucleus move after the collision in such a manner that there are conservation of energy and conservation of momentum in the collision. Such collisions cause the forked tracks occasionally seen in cloud-chamber pictures (Figs. 49-2 and 49-9). Rutherford (1918) bombarded air in a cloud chamber with α particles. He found that some of the tracks produced after a collision were those of protons (hydrogen nuclei) of unusually long range. Such a track is shown in Fig. 49-9. The location of the collision is shown by the arrows. The track of the proton slopes downward to the right. If these protons were due to collisions with hydrogen atoms in the air, for the α particles used the range of the protons would be about 28 cm, whereas the observed range was about 40 cm. Furthermore, the protons were found to take almost any direction from the point of collision, whereas hydrogen nuclei must have forward directions only. The long-range particles were found only when nitrogen was present and increased in number when pure nitrogen was used. It was thus shown that the proton was produced when the α particle collided with a nitrogen nucleus. Furthermore, there was no evidence of a recoil α particle after the collision. Thus the α particle strikes and becomes part of the nitrogen nucleus, a proton being emitted by the resulting nucleus. We thus have the first artificial *transmutation* (change from one element to another). In this process the mass and the charge must stay essentially the same. We can write an equation to represent the transition

Figure 49-9
Cloud-chamber photograph of the first artificial disintegration, discovered by Rutherford. The proton track slopes downward to the left, and the recoil nucleus moves upward and to the right. Origin of these particles indicated by the arrows. *(After Blackett and Lees.)*

$$_7N^{14} + {}_2He^4 \rightarrow {}_9F^{18} \rightarrow {}_8O^{17} + {}_1H^1 \quad (17)$$

The subscripts represent atomic number (charge), while the superscripts represent atomic mass.

49-10 PRODUCTION OF HIGH-ENERGY PARTICLES

After the first discovery of artificial disintegration, a number of other substances were found to show the same type of disintegration upon being struck by α particles. Disintegration will occur only if the α particle enters the nucleus. Since both particle and nucleus are positive, the α particle is repelled, and hence must have high energy to reach the nucleus. The higher the charge, the more energy is required for the α particle; hence the higher-atomic-number elements are not readily disintegrated by natural α particles. Protons, having less charge than α particles, could penetrate nuclei more readily if given high energy. Cockcroft and Walton (1930) were the first to produce protons with energy high enough to cause disintegration. They used transformers and vacuum-tube rectifiers in series to produce a potential of 700 kV. The energy gained by the charged particle is Vq. Particle energies are frequently expressed in terms of *electronvolts,* an electronvolt being the energy gained by an electron in falling through a potential difference of one volt. A larger unit, a million electronvolts (MeV), is commonly used as an energy unit. Particles of about 1 MeV have been produced by the method of Cockcroft and Walton. They used 0.15-MeV protons to produce disintegration of lithium.

A second method used an alternating potential of a few thousand volts to accelerate the particles successively. Several cylinders are arranged in series with gaps between them. Alternate cylinders are connected to one side of the line. If the length of each cylinder is so made that the particles traverse it in a half cycle, the particle is accelerated at each gap and acquires high speed.

One of the most important methods for production of high-speed ions is the *cyclotron,* invented by E. O. Lawrence (1930). The cyclotron (Fig. 49-10) consists of a chamber in which are two semicircular, hollow metal boxes insulated from each other. The chamber is placed between the poles of a very strong electromagnet. An ion introduced near the center will travel in a semicircular path within one box. As it crosses from one box to the other, it is accelerated so that in the second box it moves in a larger circular path. The radius of the path may be obtained by equating the centripetal force mv^2/r to the centrally directed force Bqv exerted by the magnetic field B on the particle of charge q,

$$r = \frac{mv}{Bq} \qquad (18)$$

The time for the ion to travel a semicircle is

$$\frac{T}{2} = \frac{\pi r}{v} = \frac{\pi m}{Bq} \qquad (19)$$

which is independent of both the speed v of the ion and of the radius r of its path. A high-frequency alternating potential can be used to accelerate the ions each time they cross from one D-shaped chamber to another. The ions continue on a kind of spiral path, gaining kinetic energy on each crossing but remaining in phase with the alternating voltage. After many trips around the D's the high-energy ions come out of a window.

By introducing various gases into the chamber to be ionized, many different kinds of particles can be produced. Those most generally used are protons (hydrogen nuclei), deuterons (heavy hydrogen nuclei), and α particles (helium nuclei). Large cyclotrons may produce 10-MeV protons, 20-MeV deuterons, and 40-MeV α particles.

A proton with energy 4.7 MeV has a speed one-tenth that of light and experiences an appreciable relativity increase in mass. The increase of mass with speed causes higher-energy ions to lag in phase with respect to the alternating voltage. This relativistic effect places a practical limit on the ion energies attainable with a conventional cyclotron.

The frequency-modulated cyclotron, or synchrocyclotron, starts with a frequency $Bq/2\pi m_0$

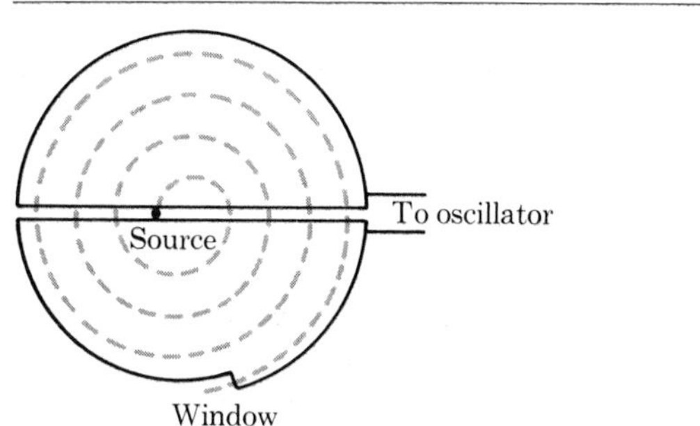

Figure 49-10
Diagram of a cyclotron.

[see Eq. (19)] and then lowers the frequency rapidly as the ion increases in speed and in mass. This cycle is repeated hundreds of times per second to produce pulses of ions with energies as high as 500 MeV.

A source of high-speed electrons is the *betatron*. It has a chamber within a magnetic field into which electrons are introduced. The magnet is energized by alternating current and serves to accelerate the electrons during a part of a cycle and to provide an increasing field to keep them in their circular paths as the speed increases. Electrons have been accelerated to energies of the order of 350 MeV. Because of the loss of energy by radiation the practical limit is probably of the order of 1,000 MeV (1 BeV). The very-high-energy electrons have speeds greater than 0.9999 times that of light, and their mass is several hundred times the rest mass.

The *synchrotron* differs from the betatron in that the acceleration is achieved by a high-frequency field at a gap as in the cyclotron, but it has the rising field of the betatron. Proton-synchrotrons provide very-high-energy particles, of the order of 1 to 6 BeV.

49-11 DISINTEGRATION BY ACCELERATED PARTICLES

By the use of accelerated particles many disintegrations can be produced. The particle energies attainable are sufficient for the particles to penetrate to the nucleus of most atoms and produce disintegration. The following equations represent examples of the many reactions that can occur:

Proton:

$$_{4}\mathrm{Be}^{9} + {}_{1}\mathrm{H}^{1} \rightarrow {}_{3}\mathrm{Li}^{6} + {}_{2}\mathrm{He}^{4} \qquad (20)$$

Deuteron:

$$_{13}\mathrm{Al}^{27} + {}_{1}\mathrm{H}^{2} \rightarrow {}_{12}\mathrm{Mg}^{25} + {}_{2}\mathrm{He}^{4} \qquad (21)$$

α particle:

$$_{13}\mathrm{Al}^{27} + {}_{2}\mathrm{He}^{4} \rightarrow {}_{14}\mathrm{Si}^{30} + {}_{1}\mathrm{H}^{1} \qquad (22)$$

Many of the products of these artificial disintegrations are radioactive. These *artificially radioactive* materials behave much as the natural radioactive materials do. Each emits its characteristic particle (not always an α or β particle); each has a characteristic half-life period. The following equations represent one such change,

$$_{48}\mathrm{Pd}^{108} + {}_{2}\mathrm{He}^{4} \rightarrow {}_{47}\mathrm{Ag}^{111} + {}_{1}\mathrm{H}^{1}$$
$$_{47}\mathrm{Ag}^{111} \rightarrow {}_{48}\mathrm{Cd}^{111} + {}_{-1}e^{0} \qquad (23)$$
$$T = 7.5 \text{ days}$$

where T is the half-life period.

49-12 MASS-ENERGY TRANSITIONS

In the nuclear reactions just discussed there are many changes that involve an increase or decrease in total mass. In considering the energy relations involved in the changes, we must study the change in mass as well as the energies of the moving particles. Whenever there is a change in mass, energy $E = \Delta mc^2$ must be absorbed or emitted. Masses of the atoms are usually expressed in *atomic mass units* (amu). An amu is $\frac{1}{12}$ the mass of $_6\mathrm{C}^{12}$ or 1 amu = 1.660×10^{-24} g. A change of mass of 1 amu involves an energy of

$$E = \Delta mc^2 = 1.660 \times 10^{-24}\ \mathrm{g} \times (2.998 \times 10^{10}\ \mathrm{cm/s})^2$$
$$= 14.94 \times 10^{-4}\ \mathrm{erg} = 14.94 \times 10^{-11}\ \mathrm{J}$$

$$1\ \mathrm{Mev} = 10^6\ \mathrm{V} \times 1.602 \times 10^{-19}\ \mathrm{C} = 1.602 \times 10^{-13}\ \mathrm{J}$$

$$1\ \mathrm{amu} = \frac{14.94 \times 10^{-11}}{1.602 \times 10^{-13}}\ \mathrm{MeV}$$
$$= 931\ \mathrm{MeV}$$

Consider the transition involved in Rutherford's first disintegration experiment,

$$_{7}N^{14} + {_2He^4} \rightarrow {_9F^{18}} \rightarrow {_8O^{17}} + {_1H^1} \qquad (24)$$

The mass of the initial α particle and nitrogen is

$$\underset{(m_N)}{14.00753 \text{ amu}} + \underset{(m_{He})}{4.00389 \text{ amu}} = 18.01142 \text{ amu}$$

The mass of the final products, oxygen and hydrogen nucleus, is

$$\underset{(_8O^{17})}{17.00450 \text{ amu}} + \underset{(_1H^1)}{1.00813 \text{ amu}} = 18.01263 \text{ amu}$$

In the process there is a gain in mass of 0.00121 amu. This gain in mass must come at the expense of energy

$$E = 0.00121 \times 931 \text{ MeV} = 1.12 \text{ MeV}$$

The incident α particle must supply this energy plus the energy of the proton and recoil atom.

In other disintegrations, there may be a decrease in mass, with consequent increase in energy. This energy may appear as kinetic energy of the product particles or as radiant energy ($h\nu$) in the form of γ rays. For example, the bombardment of lithium by protons may produce a γ ray,

$$_{3}Li^7 + {_1H^1} \rightarrow {_4Be^8} + h\nu$$

49-13 THE DISCOVERY OF THE NEUTRON

Bothe and Becker (1930), on bombarding light elements such as beryllium with α particles, found a very penetrating "radiation" that is not deflected by magnetic or electric fields. In the light of previous experience this was assumed to be γ radiation. Measurement of the energy of the ray gave a value in the neighborhood of 10 MeV, which agreed satisfactorily with the mass change in the assumed reaction,

$$_{4}Be^9 + {_2He^4} \rightarrow {_6C^{13}} + h\nu(?) \qquad (25)$$

Curie and Joliot (1932) found that this radiation ($h\nu$) produced much more ionization if it passed through paraffin or other material containing hydrogen but that the ionization was not increased by other substances. These results indicated that the ionization was produced by protons knocked out of the paraffin. If the protons observed were produced by γ rays, the γ-ray photon would be required to have energy of the order of 50 MeV rather than the 10 MeV available from the assumed reaction. Further experiments showed recoil atoms of nitrogen that would require still higher energy for the assumed photon.

Chadwick (1932) showed that these difficulties were removed if it was assumed that the penetrating "radiation" was an uncharged particle of mass about that of the proton. This particle is called a *neutron*. The reaction would then be

$$_{4}Be^9 + {_2He^4} \rightarrow {_6C^{12}} + {_0n^1} \qquad (26)$$

The penetrating ability of the neutron and its failure to show tracks in a cloud chamber are explained by its lack of charge. It will have no reaction with atoms without direct collision. On the other hand, charged particles will react with nuclei when there is a close approach because of the electric forces. Thus these charged particles owe to their electric charge the ionization they produce in matter and the resistance they encounter as they pass through.

Neutrons make ideal particles for bombardment of nuclei. Since they carry no charge, they are not repelled and hence may penetrate the nucleus even when their energy is comparatively low. Examples of transformation by neutron bombardment follow:

$$_{1}H^1 + {_0n^1} \rightarrow {_1H^2} + h\nu$$

$$_{79}Au^{197} + {_0n^1} \rightarrow {_{79}Au^{198}} + h\nu$$

$$_{79}Au^{198} \rightarrow {_{80}Hg^{198}} + {_{-1}e^0}$$

$$T = 2.7 \text{ d}$$

49-14 CHARACTERISTICS OF NUCLEAR REACTIONS

We may consider a general notation for a nuclear reaction in which a particle or nucleus x strikes a nucleus X and results in the emission of particle y, leaving a nucleus Y,

$$x + X \rightarrow y + Y$$

The notation is often abbreviated as $X(x, y)Y$, where the first symbol stands for the struck nucleus, the symbols in parentheses stand for the incoming and outgoing particles, respectively, and the symbol following the parentheses represents the residual nucleus. Thus the reaction associated with Chadwick's discovery of the neutron is written as $Be^9(\alpha, n)C^{12}$.

A reaction which is useful in the detection and counting of slow neutrons is one in which a neutron enters and an alpha particle (a readily detectable ionizing particle) leaves,

$${}_0n^1 + {}_5B^{10} \rightarrow {}_3Li^7 + {}_2He^4 \qquad \text{or} \qquad B^{10}(n,\alpha)Li^7$$

An important reaction initiated by photons is the photodisintegration of the deuteron into a neutron and a proton,

$$h\nu + {}_1H^2 \rightarrow {}_1H^1 + {}_0n^1 \qquad \text{or} \qquad H^2(\gamma,n)H^1$$

or $$H^2(\gamma,p)n$$

A reaction which is useful as a source of high-energy photons is the radiative capture of protons by Li^7,

$${}_1H^1 + {}_3Li^7 \rightarrow {}_4Be^8 + h\nu \qquad \text{or} \qquad Li^7(p,\gamma)Be^8$$

This reaction is followed by the spontaneous disintegration of the Be^8 nucleus into two alpha particles.

In reactions in which electrons or positrons are emitted from an unstable nucleus it appears superficially that energy is not conserved. It is believed that in these reactions an additional particle, the *neutrino,* is emitted, with energy sufficient to balance the energy-conservation equation. The properties postulated for the neutrino (negligible mass and no charge) make its detection extremely difficult. But by 1956 the existence of the neutrino was experimentally demonstrated by detection of γ rays produced in a sequence of events initiated by the neutrino.

In general, there is more than one possible outcome when a given particle x collides with a given nucleus X. Not enough is now known about nuclear forces and nuclear structure to predict the probability of each one of the possible reactions. A few generalizations can be made. In reactions in which nucleons and heavier particles are emitted the sum of mass plus energy remains constant. The number of nucleons (total number of protons and neutrons) also appears to remain constant. In most reactions the atomic number Z of the struck nucleus is changed by at most two units and the mass number A by at most three or four. This limitation does not apply when the bombarding particles have high energies (above 200 MeV), and it does not apply to fission reactions.

49-15 COSMIC RAYS

About 1900 it was established that there is always some ionization of the air. It was assumed that this ionization is caused by radiation from the crust of the earth. Gockel (1910) as the result of observations on balloon flights found the ionization to be greater at high altitude than at the surface, indicating that the source of the ionization came from outside the earth.

Many experiments have been carried out to study the origin and nature of the *cosmic rays* that cause the ionization. Absorption methods, cloud chambers, Geiger counters, and photographic plates have all been used in the studies. The early experiments showed extremely high penetration by cosmic rays, reaching far underground and to the bottoms of deep lakes. Millikan and his collaborators, beginning in 1923, early interpreted the phenomenon as very-short-wave electromagnetic radiation. Later experiments showed variation with latitude that indicated deflection of the cosmic rays in the magnetic field of the earth. This effect would be present only if there are charged particles. Identification of the particles is complicated by the fact that several kinds of secondary particles or rays are produced in the atmosphere by the primary cosmic rays. Most of these particles are very high speed protons, many of the rest are alpha particles with a few being nuclei of the heavier elements. These particles possess energies of billions of electronvolts. The most energetic particles carry energies of greater than 10^{18} eV.

The origin of these rays is not well understood. The sun was once considered as the primary source of cosmic rays. However, there are several arguments which have been presented in disagreement with this hypothesis. First, the majority of the particles ejected by the sun have energies far less than most of the cosmic rays. Also, most cosmic rays approach the earth in equal numbers from all directions. They do not come principally from the direction of the sun.

Many astronomers believe that the majority of cosmic rays originate in exploding stars. These particles originating in this manner are shot out into space and are accelerated by fluctuating magnetic fields in interstellar space to very rapid speeds. Some of these particles travel through great distances in space over a long period of time before eventually reaching the earth.

49-16 THE POSITRON

In 1932, C. D. Anderson was studying cosmic-ray phenomena by means of a cloud chamber placed between the poles of an electromagnet. He found pairs of tracks that, from the ionization produced, were identified as those of electrons, but the tracks were bent in opposite directions by the field, seeming to indicate opposite charges. To determine the direction of travel of the particles, Anderson inserted a sheet of lead across the chamber so that the track would be less curved on the initial side before the particle was slowed down by the lead. By this device he obtained tracks (Fig. 49-11) that were those of positively charged electrons. The curvature, ionization, and range of the particle were inconsistent with a proton track but agreed with the assumption of a positive electron, or *positron*.

Positrons are not at all rare in nature, occurring in cosmic-ray phenomena near the surface of the earth almost as often as negative electrons. Also, they appear frequently as products of artificial disintegration. However, they do not exist long as free positrons, since they combine with negative electrons. The normal life of a free positron is a very small fraction of a second. When the electron and positron combine, they disappear and a photon is produced. Conversely, a photon often disappears and a positron-electron

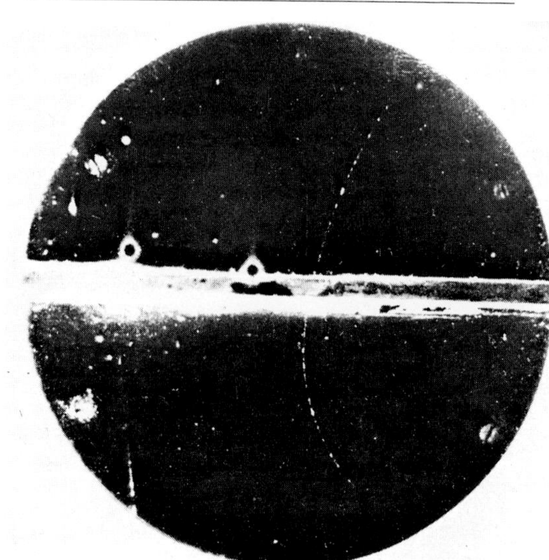

Figure 49-11
Cloud-chamber photograph used to identify the positron. (*Courtesy of C. D. Anderson.*)

pair appears simultaneously. We thus have a direct conversion from particle to wave.

$$_{+1}e^0 + _{-1}e^0 \leftrightarrows h\nu$$

49-17 THE NUCLEUS

A nucleus has been pictured as made up of protons and neutrons. The charge of the nucleus is the sum of the charges of the protons; the mass of the nucleus is approximately the sum of the masses of protons and neutrons. This makes a very simple explanation of isotopes: the different isotopes of a given substance differ only in the number of neutrons in the nucleus.

While the nuclear mass is approximately the sum of the masses of protons and neutrons, it is found that for any permanently stable nucleus the mass is less than this sum. This fact is due to the conversion of some of the mass into *energy of binding*. For example, the combination of a proton and a neutron to form a deuteron results in the release of energy,

$$\begin{aligned}\Delta m &= m_D - m_p - m_n \\ &= 2.014186 - 1.007593 - 1.008982 \\ &= 0.002389 \text{ amu} \\ &= 2.225 \text{ MeV}\end{aligned}$$

since 1 amu is equivalent to 931 MeV.

The forces that hold the particles of the nucleus together must be considered as a distinct type of force, *nuclear force*. Neither gravitational nor electric forces fit the observed conditions. The nuclear forces exist only over very short distances, of the order of 10^{-13} cm.

The major kinds of force in the universe appear to be, in order of decreasing strength:

1 Strong interactions: the forces between nucleons
2 Electromagnetic interactions: those between charged particles and electric and magnetic fields, for example, the Coulomb repulsion between two protons
3 Weak interactions: force acting on all particles, tending to transform elementary particles to electrons and neutrinos as final products
4 Gravitational interactions: those described by Newton's law of universal gravitation

An atom is stable because of the Coulomb force of attraction which binds the electrons to the nucleus. Within the nucleus, however, the Coulomb forces exerted by the protons are forces of repulsion which tend to make the nucleus unstable. The emission of α particles from nuclei and nuclear fission are evidence of this. Somehow the repulsive Coulomb forces within a nucleus must be counterbalanced by strong attractive forces, different from electrical and gravitational forces. The nature of these nuclear forces is only partly understood, but we shall discuss some of the facts which are known about them.

An important and distinctive property of nuclear forces is their short range. The nuclear force between two nucleons becomes negligible if they are separated by more than about 1.4×10^{-15} m. In contrast, gravitational and electrical forces have no upper limit on the distances over which they may act.

A second property of nuclear forces may be deduced from a graph of the binding energy per nucleon, E_B/A, against the number of nucleons A (Fig. 49-12). Except for the lightest nuclei, E_B/A is approximately constant, about 8 MeV per nucleon. Thus the total binding energy increases approximately in proportion to the number of nucleons in the nucleus: $E_B \propto A$. (The relation for a Coulomb force would be $E_B \propto A^2$.) This relation implies that a given nucleon is bound, not to every other nucleon present, but only to its nearest neighbors. Then the addition of more nucleons increases the total binding energy only by an amount proportional to the number of nucleons added; E_B/A does not change appreciably.

Present evidence indicates that the nuclear force between two protons is the same as the force between two neutrons and that these may

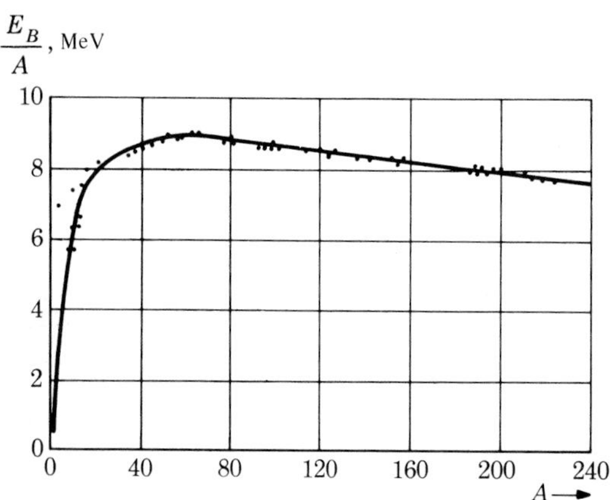

Figure 49-12
Binding energy per nucleon as a function of mass number A.

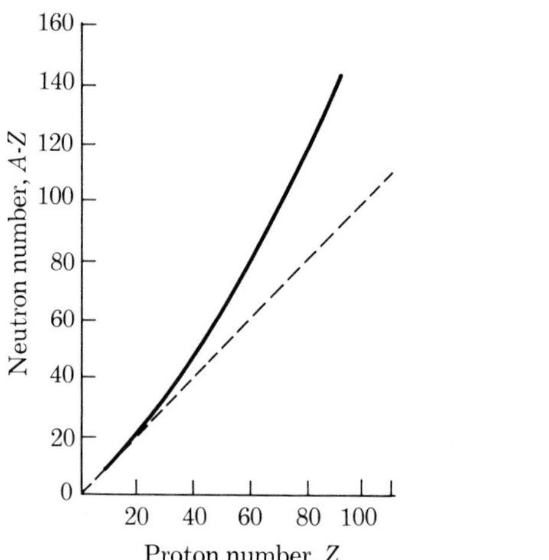

Figure 49-13
Neutron-proton plot for stable nuclei.

be equal to the force between neutron and proton.

The last property of nuclear forces which we shall mention is pairing. The stable nuclei usually have even numbers of protons and of neutrons (Table 1). Only the four light elements ${}_1H^2$, ${}_3Li^6$, ${}_5B^{10}$, and ${}_7N^{14}$ have odd numbers of both neutrons and protons, and for these elements the numbers of neutrons and protons are equal.

When a plot of neutron number vs. proton number is made for all nuclei (Fig. 49-13), one observes a gradual increase in the neutron-proton ratio with increasing Z. This is explained by the fact that the Coulomb (repulsion) force between protons increases more rapidly as the number of protons in the nucleus increases than does the effect of the nuclear force between protons. This difference in the behavior of the Coulomb *pp* force and the nuclear *pp* force accounts for the gradual decrease in E_B/A from about 8.8 MeV for A near 50 to approximately 7.6 MeV for $A = 240$ (Fig. 49-12).

Table 1
EVIDENCE FOR PAIRING

Neutron number (A-Z)	Proton number Z: Even	Proton number Z: Odd
Even	160	52
Odd	56	4

49-18 ELEMENTARY PARTICLES

A few decades ago, matter was believed to consist only of protons, neutrons, and electrons—with the occasional appearance of a neutrino. Today there are said to be over forty "elementary" particles. Although momentarily experiments in the detection of elementary particles seem to have outrun theory, Weisskopf (see References) suggests that the complexity shown in Table 2 results from the following three practices:

1 Each antiparticle of a given particle has been called a new particle. The antiproton (negative

Table 2
MASSES AND DECAY PROPERTIES OF THE ELEMENTARY PARTICLES*

Particle	Symbol	Spin	Mass equivalent, *MeV*	Decay modes	Mean life, *s*
Photon	γ	1	0		Stable
Leptons	ν	$\frac{1}{2}$	0		Stable
	$\bar{\nu}$	$\frac{1}{2}$	0		Stable
	$e^{\pm}$	$\frac{1}{2}$	0.511		Stable
	$\mu^{\pm}$	$\frac{1}{2}$	105.7	$\mu^{\pm} \rightarrow e^{\pm} + \nu + \bar{\nu}$	2.2×10^{-6}
Mesons	$\pi^{\pm}$	0	139.6	$\pi^{+} \rightarrow \mu^{+} + \nu$; $\pi^{-} \rightarrow \mu^{-} + \bar{\nu}$	2.6×10^{-8}
	π°	0	135.0	$\pi^{\circ} \rightarrow 2\gamma$	$\approx 10^{-16}$
	$K^{\pm}$	0	493.9	$K^{\pm} \rightarrow$ $\pi^{\pm} + \pi^{+} + \pi^{-}$; $\pi^{\pm} + 2\pi^{\circ}$; $\pi^{\pm} + \pi^{\circ}$; $\mu^{\pm} + \nu$ or $\bar{\nu}$; etc.	1.2×10^{-8}
	K°	0	497.8		
	$\bar{K}^{\circ}$	0	497.8		
	K°_{1}	0	497.8	$K^{\circ}_{1} \rightarrow$ $\pi^{+} + \pi^{-}$; $2\pi^{\circ}$	10^{-10}
	K°_{2}	0	497.8	$K^{\circ}_{2} \rightarrow$ $\pi^{+} + \mu^{-} + \bar{\nu}$; $\pi^{-} + \mu^{+} + \nu$; $\pi^{+} + e^{-} + \bar{\nu}$; $\pi^{-} + e^{+} + \nu$; $\pi^{+} + \pi^{-} + \pi^{\circ}$	$\approx 10^{-7}$
Baryons	p	$\frac{1}{2}$	938.2		Stable
	n	$\frac{1}{2}$	939.5	$n \rightarrow p + e^{-} + \nu$	10^{3}
	Λ°	$\frac{1}{2}$	1,115.4	$\Lambda^{\circ} \rightarrow$ $\pi^{-} + p$; $\pi^{\circ} + n$	2.5×10^{-10}
	Σ^{+}	$\frac{1}{2}$	1,189.4	$\Sigma^{+} \rightarrow$ $\pi^{+} + n$; $\pi^{\circ} + p$	0.8×10^{-10}
	Σ^{-}	$\frac{1}{2}$	1,196.0	$\Sigma^{-} \rightarrow \pi^{-} + n$	1.6×10^{-10}
	Σ°	$\frac{1}{2}$	1,191.5	$\Sigma^{\circ} \rightarrow \Lambda^{\circ} + \gamma$	$\approx 10^{-20}$
	Ξ^{-}	?	1,318.4	$\Xi^{-} \rightarrow \Lambda^{\circ} + \pi^{-}$	1.3×10^{-10}
	Ξ°	?	1,311.0	$\Xi^{\circ} \rightarrow \Lambda^{\circ} + \pi^{\circ}$	$\approx 10^{-10}$

*From Glossary of Terms Frequently Used in High Energy Physics, American Institute of Physics, 1961.

proton) was produced in high-energy accelerators called proton synchrotrons in 1956. Shortly afterward the antineutron was observed. Astronomers have speculated about the existence of antimatter comprising positrons (antielectrons), antiprotons, and antineutrons. We could halve the number of "elementary" particles by classifying each related pair as one particle.

2 Each excited state has been called a new particle. If we followed this practice with atoms, the number of "different" atoms would now be in the tens of thousands.

3 Entities such as the light quantum have been called particles. This is largely a matter of choice.

Weisskopf has suggested that the physics of particles and their interactions be expressed in terms of field theory. He considers that there are two elementary particles: the baryon and the lepton. These, however, appear in different states. He has further suggested that it may be possible to express the little understood weak interactions in terms of a field whose sources reside in baryons and leptons and that the boson may be its field quantum. The full meaning of the particles listed in Table 2 still remains to be clarified. Why are there so many? Could they be different aspects of several or of even one really fundamental particle?

49-19 FISSION

In 1934, Fermi and his collaborators attempted to produce elements beyond the normal limit at uranium. In bombardment of the lighter elements by slow neutrons the element after the capture is usually transformed by electron emission into the element of next higher atomic number. Therefore, one might expect that a similar bombardment of uranium ($Z = 92$) would produce a new element (93). This reaction has been produced with *neptunium* (93) as the resulting product. Neptunium also disintegrates by emitting a β particle to produce *plutonium* (94). Plutonium is a rather stable material having a half-life period of 30,000 years. From 1944 to 1950 four other new elements were produced in the cyclotron: americium (95), curium (96), berkelium (97), and californium (98). More recently elements einsteinium (99), fermium (100), mendelevium (101), and nobelium (102) have been reported.

In 1939, Hahn and Strassmann found one of the products of neutron bombardment of uranium to be a radioactive barium ${}_{56}\mathrm{Ba}^{139}$. There must then be another fragment such as ${}_{36}\mathrm{Kr}$ associated with the barium fragment to make the charges equal. Neir separated the isotopes of uranium in a mass spectrograph and found that ${}_{92}\mathrm{U}^{235}$ is the one that undergoes the splitting process called *fission*. Fission is a new type of radioactive process, the first that produced particles more massive than α particles.

In the process of fission of uranium there is a decrease in total mass, and therefore there is a corresponding gain in energy. Such a reaction then is a possible source of energy. This energy is controllable, since the process can be started at will.

Among the products of fission one finds one to three neutrons. These neutrons are faster than the ones used to start the fission, but if they strike the uranium nuclei, they can cause fission. Since the fission produces the starting particles and releases energy, the reaction can perpetuate itself, provided that there is enough uranium present so that the neutrons produced will hit other uranium nuclei. Thus a *chain reaction* can be set up. The smallest amount of material in which such a chain reaction can be set up is called the *critical mass*. If the reaction is set up in a sample of fissionable material of greater than the critical mass, the reaction will accelerate and continue as long as the material is together. Thus, in an uncontrolled chain reaction an explosion will result. If suitable materials are introduced to absorb and slow down the neutrons produced by the fission process, a controlled reaction can be established with energy produced at a predicted rate. Such a *reactor* may be used as a source of energy, since the energy released may be converted into heat.

In the fission process there is a decrease in mass of only about one-tenth of 1 percent. Proton and neutron retain their identities, the change in mass resulting from a rearrangement of these particles in the nucleus. Only in transitory particles such as the positron and the meson is there annihilation of particles, with conversion of the whole mass into energy. Such a process is not a net source of energy, since at least an equivalent amount of energy is required to produce the transitory particle.

The process of fission occurs when a uranium, plutonium, or other suitable nucleus captures a neutron. The process will proceed as long as neutrons are present and will accelerate if the number of neutrons is increased. The production of an appreciable amount of nuclear energy depends upon having a mechanism that will produce a sufficiently rapid process.

49-20 NUCLEAR REACTOR

A nuclear reactor is a device for utilizing a chain reaction for any of several purposes: to produce power, to supply neutrons, to induce nuclear reactions, to prepare radioisotopes, or to make fissionable material from certain "fertile" materials. Typical components of a reactor are: the *fissionable fuel* (U or Pu), the *moderator* (graphite or D_2O to slow down the fission-producing neutrons), the *control rods* (usually Cd strips, whose insertion captures neutrons and slows the fission rate), and the *coolant* (water, air, hydrogen, or liquid metal such as Na).

In power reactors the coolant, through a heat exchanger, may furnish steam to operate a conventional turbine and electrical generator. Breeder reactors make new nuclear fuel from fertile substances which cannot themselves sustain a chain reaction but which can be converted into fissionable material. One possible breeding reaction is represented by the following equation

$$_{92}U^{236} + {}_0n^1 \rightarrow {}_{92}U^{239} \xrightarrow[23\text{ min}]{\beta} {}_{93}Np^{239} \xrightarrow[2.3\text{ days}]{\beta} {}_{94}Pu^{239}$$

49-21 FUSION

Nuclear energy can also be released by fusion of small nuclei into larger nuclei if in this process there is a decrease in mass. In such a process the two positively charged nuclei must come into contact, even though there are strong electrical forces of repulsion. This requires that the particles be moving with high speeds. With artificial accelerating apparatus a few nuclei are given very high speeds. Only occasionally will such a particle strike another nucleus before it has lost too much of its energy to make contact. Thus the process is extremely inefficient, and more energy must be supplied to initiate the fusion process than is realized from the reaction.

The only known way in which a fusion process can be carried out to evolve energy is by a thermonuclear reaction at extremely high temperature. The speed of the nuclei is the speed of their thermal motion. Even at the temperature of the interior of the stars, of the order of 2×10^7 °C, the speed of the nuclei is far less than that of the particles produced in the accelerators, but in collisions they are not slowed down, because all the particles are moving with the same high speeds. Hence a very small percentage of the nuclei will fuse to cause the nuclear reaction. In the sun only about 1 percent of the hydrogen is thus transformed into helium in a billion years. This small change maintains the high temperature only because the radiation produced diffuses very slowly to the outside, and there is a great difference in temperature between the center and the outside. On a small body such as the earth, the rate of loss of energy would be so rapid that the temperature necessary to continue such a nuclear reaction could not be maintained.

If thermonuclear reactions are initiated on the earth, they must be very rapid, because the energy is lost so quickly that slow reactions cannot continue. The reaction rate depends upon the charges of the nuclei, being greatest for small charges. Hydrogen, having the smallest charge, should have the highest rate. There are three isotopes of hydrogen, ${}_1H^1$, ${}_1H^2$, and ${}_1H^3$. Of these, the most abundant, ${}_1H^1$, has a very slow rate of reaction, but the other two have a rather high rate and evolve more energy.

$${}_1H^3 + {}_1H^2 \rightarrow {}_2He^4 + {}_0n^1 + 17.6\text{ MeV}$$

Thus a hydrogen-helium reaction is possible if the proper conditions are provided. There must be an extremely high temperature and sufficiently high density of the reacting components so that collisions are probable. It is possible to realize the high temperature in an explosion of uranium or plutonium. The magnitude of the hydrogen reaction depends exactly on the amount of reacting material built into the bomb. Such an explosion does not extend to the general atmosphere, because of the extremely small concentration of reacting materials and the rapid loss of energy by radiation.

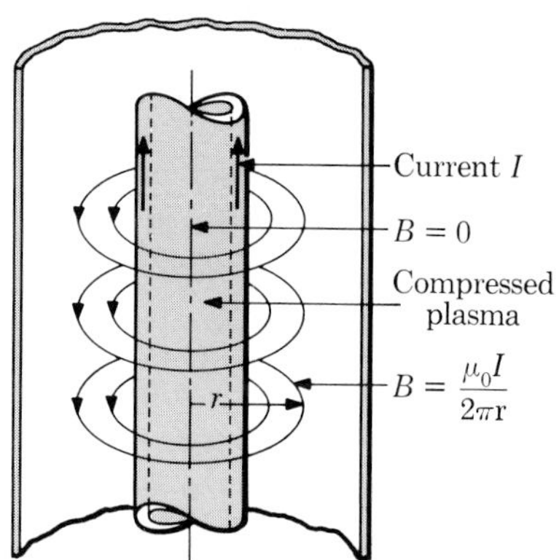

Figure 49-14
The "pinch effect" in a plasma.

49-22 CONTROLLED FUSION

The necessary condition for a thermonuclear process is the attainment of high particle energies for a time interval long enough to bring about kinetic equilibrium. Energy must be supplied initially to attain the "star temperature" needed for thermal fusion. At the same time the reactants must be confined. Ordinary walls will not suffice, for they would vaporize under bombardment of the high-energy particles, and these would be quickly cooled below their fusion temperature. These problems of heating and confinement must be solved in any controlled-fusion reactor.

The choice of fuel for a controlled-fusion reactor is made on the basis of availability and the probability of attaining with it the necessary high temperature. One would prefer elements of low atomic number because of the low Coulomb barrier to be overcome in the fusion reaction. Possible fusion reactions of about equal probability are

$${}_1H^2 + {}_1H^2 \rightarrow {}_2He^3 + {}_0n^1 \qquad \text{releases 3.25 MeV}$$

$${}_1H^2 + {}_1H^2 \rightarrow {}_1H^3 + {}_1H^1 \qquad \text{releases 4.0 MeV}$$

Initial heating first strips the electrons from the atoms to produce a "fourth state of matter," a fully ionized gas, or *plasma*. Further heating of the plasma is done by adding electric energy, in part by using the resistance of the plasma to produce familiar ohmic heating.

The ions reacting at temperatures about 10^8 °K obviously cannot be confined by material walls. But suitably designed magnetic fields can provide a sort of "magnetic bottle." Ions gyrating in a region of varying magnetic induction B will experience a force which results in a plasma drift away from regions of strong B. In the "pinch effect" a cylindrical current (10^6 amp) contracts because of electromagnetic forces (parallel currents attract each other). The plasma inside is thus compressed (Fig. 49-14), producing very high temperatures. Although the simple pinch is unstable, toroidal pinch devices with stabilizing axial magnetic fields have produced thermonuclear temperatures with a confinement time of about 1 ms. So far, however, the power required for these devices has exceeded the useful power gained from fusion.

In the fusion process the energy released goes to both the light charged particles and to the neutrons, as kinetic energy. The neutrons carry energy to the reactor walls. But with magnetic confinement one might be able to couple the energy of the charged particles directly to exterior electric circuits. This scheme would eliminate the heat exchanger of low thermodynamic efficiency conventionally used in fission reactors.

SUMMARY

Radioactive materials emit α particles, β particles, and γ rays.

The α particles are positively charged helium nuclei, are deflected by magnetic and electric fields, show small penetration, and produce great ionization.

The β particles are electrons, are deflected by magnetic and electric fields, show much greater penetration than α particles, and produce less ionization.

The γ rays are highly penetrating electromagnetic radiation, are not deflected by magnetic or electric fields, and produce little ionization.

The *Geiger counter* is a device that enables one to count radioactive radiations or other radiations that will ionize the gas of the counter tube.

The Wilson *cloud chamber* and the Glaser *bubble chamber* are used to photograph tracks of individual particles or rays.

Radioactive materials follow an exponential-decay law. The *half-life* period of a radioactive material is the time required for half its atoms to disintegrate.

Natural radioactivity occurs in four separate series, terminating with stable isotopes of lead.

Artificial disintegration, discovered by Rutherford in 1919, is produced when particles are caused to enter the nucleus.

In artificial disintegration there may be a decrease in mass, resulting in evolution of energy, or an increase in mass, which requires an input of energy.

Neutrons are uncharged particles whose mass is nearly the same as that of the proton.

Cosmic rays enter the atmosphere from outside the earth. The primary rays are positively charged particles, probably protons and other atomic nuclei.

The *positron* is a positively charged particle whose mass is equal to that of an electron.

The *nucleus* is composed of protons and neutrons some of whose mass is converted into energy of binding.

Present theory distinguishes four fundamental kinds of force or interaction: strong, electromagnetic, weak, and gravitational.

Fission is the splitting of a nucleus into two or more large fragments. When this process results in a decrease in mass, it can be a source of energy.

A *chain reaction* can be set up if the reaction produces the particles that produced the reaction originally. The chain reaction cannot be maintained if the mass of fissionable material is less than a *critical mass*.

Nuclear energy can be released by the *fusion* of light nuclei into larger nuclei, with consequent decrease in mass. Such reactions are the probable source of stellar radiation.

References

Beyer, Robert T. (ed.): "Foundations of Nuclear Physics," Dover Publications, New York, 1949. Contains reprints of original papers by Chadwick, Cockcroft and Walton, Rutherford, Fermi, Yukawa, and others.

Burbidge, Geoffrey, and Fred Hoyle: Anti-Matter, *Scientific American,* **198**(4):34–39 (1958).

Eilson, R. R., and R. Littauer, "Accelerators, Machines of Nuclear Physics," Doubleday and Company, Inc., Garden City, N.Y., 1960.

Frisch, D. H., and A. M. Thorndike: "Elementary Particles," D. Van Nostrand Company, Inc., Princeton, N.J., 1964.

Glasstone, Samuel: "Sourcebook on Atomic Energy," 2d ed., D. Van Nostrand Company, Inc., Princeton, N.J., 1958.

Hill, R. D.: "Tracking Down Particles," W. A. Benjamin, Inc., New York, 1963.

———: Elementary Particles, *Physics Teacher,* **2**(1):22–27 (1964).

Hoisington, David B., "Nucleonics Fundamentals," McGraw-Hill Book Company, New York, 1959.

Jordan, Walter H., Nuclear Energy: Benefits Versus Risks, *Physics Today,* **23**(6):32–38 (1970).

Murphy, Glenn, "Elements of Nuclear Engineering," John Wiley & Sons, Inc., New York, 1961.

Weisskopf, V. F.: The Place of Elementary Particle Research in the Development of Modern Physics, *Physics Today,* **16**(6):26–32 (1963).

Questions

1 Explain how gamma rays are produced and what their important properties are.

2 Do α, β, and γ rays come from the same element? Show why we ordinarily find all three in many radioactive experiments.

3 Describe some of the methods for the study and measurement of radioactive rays. Discuss the relative sensitivity of these methods. How does the actual mass of these rays compare with the masses used in ordinary chemical experiments?

4 Distinguish between and give examples of natural radioactivity and artificial radioactivity.

5 In considering the stability of a nucleus three factors must be observed: (*a*) the proton-neutron ratio, (*b*) the mass of the atom, and (*c*) the even-odd relationship of the number of protons to neutrons. Explain why each is a determinant.

6 Suggest several methods by which one could estimate the age of a uranium-bearing rock. How does the age of the earth estimated in this manner (not less than 1,600 million years) compare with independent estimates based on geological evidence?

7 Why was the discovery of the neutron more difficult to achieve than was the discovery of electrons and protons?

8 Compare the methods and achievements of modern "atom smashers" of nuclear physics with those of the ancient alchemists. In what respects are their objectives similar, and how are they unlike?

9 Describe how particles can be accelerated in a cyclotron, a synchrotron, and in an accelerator studied in an earlier chapter, the Van de Graaff generator.

10 Why is it possible to produce higher-energy α particles than protons with a given cyclotron? Is this true with other types of accelerators?

11 What limits the energy that can be given to a particle in a cyclotron?

12 A luminous-dial watch examined through a magnifying lens in a darkened room will be seen to give off tiny flashes of light. Explain the nature of these flashes.

13 Why is it more difficult to produce atomic disintegration by bombarding heavy atoms than by bombarding light atoms?

14 For a time the theory that all elements were made up of combinations of hydrogen atoms was accepted. What facts led to the abandonment of this hypothesis?

15 α particles shot vertically upward are deflected by the earth's magnetic field in which direction?

16 State the number of protons and neutrons in each of the following nuclei: ${}_3Li^6$, ${}_5Be^{10}$, ${}_6C^{13}$, ${}_{16}S^{36}$, and ${}_{72}Hf^{180}$.

17 Explain how the radioactive isotope ${}_{88}Ra^{226}$ can be used to make a luminous watch dial.

18 If a lithium nucleus of atomic mass 7 and atomic number 3 captures a proton and fission results in the disintegration of the nucleus into two equal parts, what is the nature of the materials thus produced?

19 When there is an increase in mass in a nuclear change, what is the source of the added mass?

20 Write an expression for the decay of ${}_{86}Rn^{222}$ into poloninum (Po) by the emission of an alpha particle.

21 When a photon disappears in producing an electron and a positron, is the energy of the photon equivalent to the mass of the particles produced? If not, what accounts for the difference?

22 What are some of the difficulties involved in the commercial utilization of nuclear energy for industrial purposes?

23 Compare the fusion of hydrogen to the fission of uranium-235. Compare the energies released in these reactions.

24 A certain reactor is designed so that there is one chance in two of a neutron leaking out; there is one chance in two of a nonescaping neutron's being absorbed in nonfission capture;

and the fuel gives three free neutrons per fission. Can this reactor be made to go critical?

25 In what form is the energy released in fusion reactions made available to a star?

26 Show that if relativity effects are neglected, the kinetic energy of an ion describing an arc of radius r in the magnetic field B of a cyclotron is given by $(qBr)^2/2m$.

27 What factors determine the optimum ion density in a continuously operating fusion reactor?

28 A fusion reactor loses some energy by electromagnetic radiation, especially in the x-ray region when ions are suddenly decelerated in collision. To minimize this loss, what requirements should one observe in regard to ion density, purity of the gas, and atomic number Z of any impurities?

29 The problem of relativistic increase in ion mass (decrease in its frequency) can be met in another way than by frequency modulation. Show that it should be possible to operate a cyclotron at fixed frequency if the pole faces are shaped so that the magnetic induction varies with radius in a particular way.

30 When nitrogen, atomic mass 14 and atomic number 7, is bombarded with neutrons, the collisions result in disintegrations in which α particles are produced. Write the symbolic equation representing this transmutation.

Problems

1 What product is left after an alpha particle is emitted by ${}_{88}Ra^{226}$?

2 Alpha particles are allowed to bombard some aluminum (${}_{13}Al^{27}$) causing the release of protons from each atom. What new product is produced? *Ans.* ${}_{14}Si^{90}$.

3 An α particle ejected from polonium has a speed of 1.60×10^9 cm/s. What is its energy expressed in (*a*) ergs and (*b*) electronvolts?

4 Express the kinetic energy of a β ray with a speed of 0.95 that of light in terms of electronvolts. *Ans.* 1.12 MeV.

5 An α particle ejected from polonium has a speed of 1.60×10^9 cm/s. What is the radius of curvature of its path in a uniform magnetic field of 1.20 Wb/m^2?

6 How much energy is released when 1 g of matter is converted into energy? *Ans.* 9×10^{13} J.

7 Compute the length of the second and third cylinders of a linear accelerator if the ions are protons starting from rest, each accelerating potential is 10,000 V, and the first cylinder is 2.00 cm long.

8 Radium disintegrates at the rate of approximately 0.045 percent per year. How many α particles are emitted per gram in one day? Would this computation be valid for a short-life substance? *Ans.* 3.3×10^{15}.

9 Calculate the mass of Au^{198} ($T = 2.7$ d) in a source of 1.0 mCi.

10 If 5 mg of Po^{210} ($T = 140$ days) are allowed to decay for 1.0 year, what is the activity of the sample at the end of that time? *Ans.* 1.35×10^{11} disintegrations per second.

11 A radioactive element of atomic mass 218 and atomic number 84 disintegrates with the emission of an α particle. Find the new atomic mass and atomic number.

12 A radioactive element has a half-life of 20 d. How much of the element would be left after 80 d if the original amount was 4 g? *Ans.* 0.250 g.

13 The activity of a radioactive sample decreases from 0.010 to 0.003 Ci in 60 d. What is the half-life of the material?

14 A sample of radioactive sodium (${}_{11}Na^{24}$, $T = 14.8$ h) is assayed at 95 mCi. It is administered to a patient 48 h later. What is the activity at that time? *Ans.* 10 mCi.

15 What is the volume of 1.0 mCi of radon, ${}_{86}Rn^{222}$ ($T = 3.82$ days), at 0°C and 1 atm pressure?

16 Calculate the loss of mass resulting when ${}_{6}C^{12}$ is formed and what the binding energy of ${}_{6}C^{12}$ is. *Ans.* 0.0988; 92.0 MeV.

17 The half-life of thorium (${}_{90}Th^{234}$) is 24.1 d. How many days after a sample of thorium has been purified will it take for 90 percent to change to protactinium (${}_{91}Pa^{234}$)?

18 Imagine that a free neutron gives off an

electron and changes into a proton. Calculate the energy Q which is consumed or liberated in this process. What does your answer suggest about the stability of free neutrons?

Ans. $Q = 0.79$ MeV.

19 The oscillator of a 44-in-diameter cyclotron operates at a wavelength of 26 m. (*a*) What magnetic field is required for resonance when deuterons are used? (*b*) What is the energy of the emerging deuteron beam?

20 When a deuterium atom and a tritium atom combine to produce an alpha particle and a neutron, how much energy is released?

Ans. 17.6 MeV.

21 If the decrease in mass in a fission process is 0.10 percent, how much energy could be obtained from the fission of 1.0 lb of material?

22 When neutrons are produced by bombarding deuterons with deuterons, the reaction is represented by

$$_1H^2 + {}_1H^2 \rightarrow {}_2He^3 + {}_0n^1 + Q$$

The neutrons produced in this reaction will have at least how much energy? *Ans.* 15 MeV plus the kinetic energy of the bombarding deuteron.

23 The energy released per fission of U^{235} is 200 MeV. (*a*) How many fissions occur per second in a reactor releasing 1,000 kW of fission power? (*b*) What mass of U^{235} is consumed per hour?

24 In the following reaction, 3.945 MeV of energy is released: $_3Li^6 + {}_1H^1 \rightarrow {}_2He^4 + {}_2He^3$. What is the mass of the lighter isotope of helium?

Ans. 3.0169 amu.

25 It has been suggested that after a star has converted all its hydrogen into helium, it contracts, producing higher internal pressures, temperatures, and densities until it can convert helium to carbon in triple collisions, $\alpha + \alpha + \alpha \rightarrow C^{12}$. Estimate the energy released in this reaction.

26 In fusion of hydrogen isotopes into helium the decrease in mass is of the order of 0.70 percent. How much energy could be produced by the use of 1.00 lb of hydrogen?

Ans. 2.85×10^{14}.

Appendix

A-1 GREEK ALPHABET

A α	Alpha (a)	N ν	Nu (n)
B β	Beta (b)	Ξ ξ	Xi (x)
Γ γ	Gamma (g)	O o	Omicron (o)
Δ δ or ∂	Delta (d)	Π π	Pi (p)
E ε	Epsilon (e)	P ρ	Rho (r)
Z ζ	Zeta (z)	Σ σ or ς	Sigma (s)
H η	Eta (h)	T τ	Tau (t)
Θ θ	Theta (th)	Υ υ	Upsilon (u)
I ι	Iota (i)	Φ φ or ϕ	Phi (ph)
K κ	Kappa (k)	X χ	Chi (ch)
Λ λ	Lambda (l)	Ψ ψ	Psi (ps)
M μ	Mu (m)	Ω ω	Omega (o)

A-2 SOLUTION OF PHYSICAL PROBLEMS

The ability to solve problems is a mark of an effective and efficient scientist or engineer. Through practice in the solution of problems commensurate with one's knowledge, one attains ability and confidence in independent thinking.

In problem solving, the following systematic approach is highly recommended. First, read the statement of the problem carefully, and decide exactly what is required. Then:

1 Draw a suitable diagram, and list the data given.
2 Identify the type of problem, and write physical principles which seem relevant to its solution. These may be expressed concisely as algebraic equations.
3 Determine whether or not the data given are adequate. If not, decide what is missing and how to get it. This may involve consulting a table, making a reasonable assumption, or drawing upon your general knowledge for such information as the value of g, the acceleration due to gravity, 32 ft/s^2.
4 Decide whether in the particular problem it is easier to substitute numerical values immediately or first to carry out an algebraic solution. Some quantities may cancel.
5 Substitute numerical data in the equations obtained from physical principles. Include the units for each quantity, making sure that they are all in the same system in any one problem.
6 Compute the numerical value of the unknown, preferably with the aid of a slide rule. Determine the units in which the answer is expressed. Examine the reasonableness of the an-

swer. Can it be obtained by an alternative method to check the result?

The student is referred to the numerous solved examples in the text for a demonstration of the form of solution recommended for physics problems.

An orderly procedure aids clear thinking, helps to avoid errors, and usually saves time. Most important, it enables a student to analyze and eventually solve those more complex problems whose solution is not immediately or intuitively apparent.

A-3 SIGNIFICANT FIGURES IN MEASUREMENTS AND COMPUTATIONS

Uncertainty in measurements The word *accuracy* has various shades of meaning, depending on the circumstances under which it is used. It is commonly used to denote the reliability of the indications of a measuring instrument.

As applied to the final result of a measurement, the accuracy is expressed by stating the *uncertainty* of the numerical result, i.e., the estimated maximum amount by which the result may differ from the "true" or accepted value.

Rules for computation with experimental data There is always a pronounced and persistent tendency on the part of beginners to retain too many figures in a computation. This not only involves too much arithmetic labor but, worse still, leads to a fictitiously precise result.

The following rules are recommended and will save much time that would otherwise be spent in calculation; furthermore, their careful use will result in properly indicated accuracies:

1 In recording the result of a measurement or a calculation, *one, and only one, doubtful digit is retained.*

2 In addition and subtraction, do not carry the operations beyond the first column that contains a doubtful figure.

3 In multiplication and division, carry the result to the same number of significant figures which there are in that quantity entering into the calculation which has the *least* number of significant figures.

4 In dropping figures that are not significant, the last figure retained should be unchanged if the first figure dropped is less than 5. It should be increased by 1 if the first figure dropped is greater than 5. If the first figure dropped is 5, the preceding digit should be unchanged if it is an even number but increased by 1 if it is an odd number. For example, $3.4\bar{5}5$ becomes 3.46; $3.4\bar{8}5$ becomes 3.48; $6.79\bar{0}1$ becomes 6.790.

Significant figures The accuracy of a physical measurement is properly indicated by the number of figures used in expressing the numerical measure. Conventionally, only those figures which are reasonably trustworthy are retained. These are called *significant figures.*

In recording certain numbers, the location of the decimal point requires zeros to be added to the significant figures. When this requirement leaves doubt as to which figures are significant, one may *overscore the last significant figure.* This overscored figure is the first digit whose value is doubtful.

Examples

Length of page

= 22.7 cm (three significant figures)

Thickness of page

= 0.011 cm (two significant figures)

Distance to sun

= $9\bar{3}$,000,000 mi (two significant figures)

Speed of light

= $299{,}7\bar{9}0$ km/s (five significant figures)

If each of these numbers is expressed in terms of powers of 10, there is no doubt as to the number of significant figures, for only the significant figures are then retained. Thus

$$\text{Length of page} = 2.27 \times 10^1 \text{ cm}$$

$$\text{Thickness of page} = 1.1 \times 10^{-2} \text{ cm}$$

$$\text{Distance to sun} = 9.3 \times 10^7 \text{ mi}$$

$$\text{Speed of light} = 2.9979 \times 10^5 \text{ km/s}$$

There are some numbers which, by their definition, may be taken to have an unlimited number of significant figures, for example, the factors 2 and π in the relation

$$\text{Circumference} = 2\pi \text{ (radius)}$$

In calculations there is frequently need to use data that have been recorded without a clear indication of the number of significant figures. For example, a textbook problem may refer to a "2-lb weight," or in a cooperative experiment a student may announce that he has measured a certain distance as "5 ft." In such cases the values with the appropriate number of significant figures should be written from what is known or assumed about the way in which the measurements were made. If the distance referred to were measured with an ordinary tape measure, it might

Table 1
ESTIMATES OF ACCURACY AND PRECISION IN MEASURING PHYSICAL QUANTITIES

Physical quantity	Device	Magnitude	Uncertainty, parts per million: Accuracy	Uncertainty, parts per million: Precision
Length	Meter bar	1 m		0.03
	Gauge block	0.1 m	0.1	0.01
	Geodetic tape	50 m	0.3	0.10
Mass	Cylinder	1 kg		0.005
	Cylinder	1 g	1	0.03
	Cylinder	20 kg	0.5	0.1
Temperature	Triple-point cell	273.16°K		0.3
	Gas thermometer	90.18°K	100	20
	Optical pyrometer	3000°K	1,300	300
Resistance	Resistor	1 Ω	5	0.1
	Resistor	0.001 Ω	7	1
	Resistor	1,000 Ω	7	1
Voltage	Standard cell	1 V	7	0.1
	Volt box and standard cell	1,000 V	25	10
Power, dc	Standard cell, resistor	1 W	11	1.5
60-cycle	Wattmeter	10–1,000 W	100	50
X band	Microcalorimeter	0.01 W	1,000	100

Table 2
PERMITTIVITY AND PERMEABILITY FOR FREE SPACE

	ϵ_0		μ_0	
Unit system	**Magnitude**	**Units**	**Magnitude**	**Units**
Mks				
Unrationalized	$\frac{1}{9 \times 10^9}$	Co/V·m or $Co^2/N \cdot m^2$	10^{-7}	Wb/A·turn·m or N/A^2
Rationalized	$\frac{1}{4\pi \times 9 \times 10^9}$	or Fa/m	$4\pi \times 10^{-7}$	or He/m
Cgs				
Electrostatic	1	$\frac{statCo^2}{dyn \cdot cm^2}$	$\frac{1}{9 \times 10^{20}}$	$\frac{1}{\sqrt{\epsilon_0 \mu_0}} = c$
Electromagnetic	$\frac{1}{9 \times 10^{20}}$	$\frac{1}{\sqrt{\epsilon_0 \mu_0}} = c$	1	$\frac{\text{Unit pole}^2}{\text{Dyn} \cdot \text{cm}^2}$

appropriately be written as 5.0 ft. If it were carefully measured with a steel scale to the nearest tenth of an inch, the distance might be recorded as 5.00 ft. In academic problem work a good rule to follow is to retain three figures unless there is reason to decide otherwise.

A systematic use of the rules given above relating to significant figures results in two advantages: (1) time is saved by carrying out calculations only to that number of figures which the data justify, and (2) intelligent recording of data is encouraged by noting always the least accurate of a number of measurements needed for a given determination. Attention can then be concentrated on improving the least accurate measurement, or, if this is not possible, other measurements need be taken only to an accuracy commensurate with it.

A-4 MATHEMATICAL FORMULAS

Quadratic formula If $ax^2 + bx + c = 0$, then

$$x = \frac{-b \pm \sqrt{b^2 - 4ac}}{2a}$$

Pythagorean theorem

$$x^2 + y^2 = r^2$$

Logarithms

1 Definition of e,

$$e = \lim_{n \to \infty} \left(1 + \frac{1}{n}\right)^n = 2.7182818 \ldots$$

2 Natural logarithm, base e,

$$y = \ln x \qquad \text{if } x = e^y$$

Common logarithm, base 10,

$$y = \log x \qquad \text{if } x = 10^y$$

The natural and common logarithms are related by

$$\ln x = 2.303 \log x \qquad \log x = 0.434 \ln x$$

Numerical constants

$$\pi = 3.14159 \qquad \sqrt{2} = 1.414$$

$$e = 2.71828 \qquad \sqrt{3} = 1.732$$

$$\sin 30° = \cos 60° = \tfrac{1}{2} = 0.500$$

$$\sin 60° = \cos 30° = \frac{\sqrt{3}}{2} = 0.866$$

$$\sin 45° = \cos 45° = \frac{\sqrt{2}}{2} = 0.707$$

For small θ (in radians), $\sin\theta \approx \theta$, $\tan\theta \approx \theta$.

Related angle Any trigonometric function of a given angle is numerically equal (+ or −) to the same function of the related angle. For example, the related angle of 160° is 20°; the related angle of 460° is 80°; and the related angle of 570° is 30°. Therefore, $\sin 258° = \sin\ (180° + 78°) = -\sin 78° = -\cos 12°$.

A-5 TRIGONOMETRIC FUNCTIONS

REDUCTION FORMULAS FOR TRIGONOMETRY

90°	$\sin(90° + \theta) = +\cos\theta$	$\sin(90° - \theta) = +\cos\theta$
	$\cos(90° + \theta) = -\sin\theta$	$\cos(90° - \theta) = +\sin\theta$
	$\tan(90° + \theta) = -\cot\theta$	$\tan(90° - \theta) = +\cot\theta$
180°	$\sin(180° + \theta) = -\sin\theta$	$\sin(180° - \theta) = +\sin\theta$
	$\cos(180° + \theta) = -\cos\theta$	$\cos(180° - \theta) = -\cos\theta$
	$\tan(180° + \theta) = +\tan\theta$	$\tan(180° - \theta) = -\tan\theta$
270°	$\sin(270° + \theta) = -\cos\theta$	$\sin(270° - \theta) = -\cos\theta$
	$\cos(270° + \theta) = +\sin\theta$	$\cos(270° - \theta) = -\sin\theta$
	$\tan(270° + \theta) = -\cot\theta$	$\tan(270° - \theta) = +\cot\theta$
360°	$\sin(360° + \theta) = +\sin\theta$	$\sin(360° - \theta) = -\sin\theta$
	$\cos(360° + \theta) = +\cos\theta$	$\cos(360° - \theta) = +\cos\theta$
	$\tan(360° + \theta) = +\tan\theta$	$\tan(360° - \theta) = -\tan\theta$

SIGNS OF THE TRIGONOMETRIC FUNCTIONS BY QUADRANTS

Quadrant	sin	cos	tan	ctn	sec	csc
I (0° to 90°)	+	+	+	+	+	+
II (90° to 180°)	+	−	−	−	−	+
III (180° to 270°)	−	−	+	+	−	−
IV (270° to 360°)	−	+	−	−	+	−

A-6 CONVERSION FACTORS

Angle, plane

$$1 \text{ rad} = 57.3°$$
$$1 \text{ r} = 2\pi \text{ rad} = 360°$$
$$1° = 60' = 3{,}600''$$

Angle, solid

$$1 \text{ sphere} = 4\pi \text{ sterad} = 12.57 \text{ sterad}$$

Length

$$1 \text{ m} = 10^{-3} \text{ km} = 39.37 \text{ in} = 3.281 \text{ ft}$$
$$= 6.214 \times 10^{-4} \text{ mi}$$
$$1 \text{ Å} = 10^{-10} \text{ m}$$
$$1 \text{ mil} = 10^{-3} \text{ in}$$
$$1\ \mu\text{m} = 10^{3} \text{ nm} = 10^{-6} \text{ m}$$
$$1 \text{ statute mi} = 1{,}609 \text{ m} = 5{,}280 \text{ ft}$$
$$1 \text{ nautical mi} = 1{,}852 \text{ m} = 1.1508 \text{ statute mi}$$
$$= 6076.10 \text{ ft}$$

Area

$$1 \text{ m}^2 = 10^4 \text{ cm}^2 = 10.76 \text{ ft}^2 = 1550 \text{ in}^2$$
$$1 \text{ mi}^2 = 27{,}878{,}400 \text{ ft}^2 = 640 \text{ acres}$$

Volume

$$1 \text{ m}^3 = 10^6 \text{ cm}^3 = 35.31 \text{ ft}^3$$
$$= 6.102 \times 10^4 \text{ in}^3$$
$$1 \text{ US fluid gal} = 4 \text{ qt} = 8 \text{ pt}$$
$$= 128 \text{ fluid oz} = 231 \text{ in}^3$$
$$1 \text{ British imperial gal} = \text{volume of 10 lb}$$
$$\text{of water at } 62° \text{ F} = 277.42 \text{ in}^3$$
$$1 \text{ li} = \text{volume of 1 kg of water}$$
$$\text{at its maximum density} = 1000.208 \text{ cm}^3$$

Mass

$$1 \text{ kg} = 1000 \text{ g} = 6.852 \times 10^{-2} \text{ slug}$$
$$= 6.024 \times 10^{26} \text{ amu}$$
$$= 35.27\ \textit{oz} = 2.205\ \textit{lb}$$
$$= 1.102 \times 10^{-3}\ \textit{ton}$$

(The quantities in italics are not mass units but are often used as such. Thus "1 kg = 2.205 *lb*" means that a kilogram is a *mass* that *weighs* 2.205 lb. This equivalence is approximate. It is dependent on the value of g and is meaningful only for terrestrial measurements.)

Density

$$1 \text{ kg/m}^3 = 1.940 \times 10^{-3} \text{ slug/ft}^3 = 0.001 \text{ g/cm}^3$$
$$= 6.243 \times 10^{-2}\ \textit{lb}/\text{ft}^3$$
$$= 3.613 \times 10^{-5}\ \textit{lb}/\text{in}^3$$

Time

$$1\ s = 1.667 \times 10^{-2} \text{ min} = 2.778 \times 10^{-4} \text{ h}$$
$$= 1.157 \times 10^{-5} \text{ d}$$
$$= 3.169 \times 10^{-8} \text{ year}$$

Frequency

$$1 \text{ Hz} = 1 \text{ cycle per second}$$

Speed

$$1 \text{ m/s} = 3.281 \text{ ft/s} = 3.6 \text{ km/h}$$
$$= 2.237 \text{ mi/h} = 1.944 \text{ knots}$$
$$1 \text{ knot} = 1 \text{ nautical mi/h}$$

Force

$$1 \text{ N} = 10^5 \text{ dyn} = 0.2248 \text{ lb} = 7.233 \text{ pdl}$$
$$= 102.0\ g \text{ force} = 0.1020\ kg \text{ force}$$

(Quantities in italics are not force units but are often used as such. Thus "1 gram force =

980.7 dynes" means that a gram *mass* experiences a *force* of 980.7 dyn in the earth's gravitational field.)

Pressure

$$1\ \text{N}\cdot\text{m}^2 = 9.869 \times 10^{-6}\ \text{atm} = 10\ \text{dyn/cm}^2$$
$$= 4.015 \times 10^{-3}\ \text{in of water}$$
$$= 7.501 \times 10^{-4}\ \text{cm Hg}$$
$$= 1.450 \times 10^{-4}\ \text{lb/in}^2$$
$$= 2.089 \times 10^{-2}\ \text{lb/ft}^2$$
$$1\ \text{bar} = 10^6\ \text{dyn/cm}^2$$

Energy, work, heat

$$1\ \text{J} = 9.481 \times 10^{-4}\ \text{Btu} = 10^7\ \text{ergs}$$
$$= 0.7376\ \text{ft}\cdot\text{lb} = 3.725 \times 10^{-7}\ \text{hp}\cdot\text{h}$$
$$= 0.2399\ \text{cal} = 2.778 \times 10^{-7}\ \text{kWh}$$
$$= 6.242 \times 10^{18}\ \text{eV}$$
$$= 1.113 \times 10^{-17}\ \text{kg} = 6.705 \times 10^{9}\ \text{amu}$$

(The last two entries come from the relativistic mass-energy equivalence formula, $E = mc^2$.)

$$1\ \text{J} = 1\ \text{N}\cdot\text{m} = 1\ \text{W}\cdot\text{s}$$
$$1\ \text{erg} = 1\ \text{dyn}\cdot\text{cm}$$

Power

$$1\ \text{W} = 3.413\ \text{Btu/h} = 44.25\ \text{ft}\cdot\text{lb/min}$$
$$= 0.7376\ \text{ft}\cdot\text{lb/s}$$
$$= 1.341 \times 10^{-3}\ \text{hp} = 0.2389\ \text{cal/s}$$
$$= 0.001\ \text{kW}$$

A-7 PHYSICAL DATA FOR THE SOLAR SYSTEM

Body	Mean distance from sun, au	Mass, times earth's mass	Diameter, mi	Gravitational force at solid surface, g's	Intensity of sunlight, relation to earth	Length of day	Length of year
Sun		329,000	864,000	*			
Mercury	0.39	0.05	3,100	0.3	6.7	88 d	88 d
Venus	0.72	0.82	7,500	0.91	1.9	30 ?	225 d
Earth	1	1	7,920	1	1	24 h	365 d
Mars	1.52	0.11	4,150	0.38	0.43	24.6 h	1.9 yr
Jupiter	5.2	317	87,000	2.64†	0.037	10 h	12 yr
Saturn	9.5	95	71,500	1.17†	0.011	10 h	29 yr
Uranus	19.2	15	32,000	0.92†	0.0027	11 h	84 yr
Neptune	30	17	31,000	1.44†	0.0011	16 h	165 yr
Pluto	79	0.8	?	0.46	0.0006	?	248 yr
Moon	1.0	0.012	2,160	0.17	1	27 d	

*Has no solid surface.

†Location of solid surface not known (far below dense atmospheric gases).

Electric charge

$$1 \text{ C} = 0.1 \text{ abC}$$
$$= 2.778 \times 10^{-4} \text{ Ah}$$
$$= 1.036 \times 10^{-5} \text{ faraday}$$
$$= 2.998 \times 10^{9} \text{ statC}$$
$$1 \text{ electronic charge} = 1.602 \times 10^{-19} \text{ C}$$

A-8 METRIC SYSTEMS OF ELECTRIC UNITS

Historically, a cgs *electrostatic* system of units (esu) was defined from Coulomb's law, $F \propto qq'/s^2$, for the force between two point charges. The constant of proportionality was arbitrarily assigned the value unity for free space. The statcoulomb was defined as a charge of such magnitude that it would repel a like charge one centimeter distant in free space with a force of one dyne. Other electrostatic units were derived from the statcoulomb. For example, the unit of current, the statampere, is one statcoulomb per second.

In the early study of magnetism, a cgs *electromagnetic* system of units (emu) was defined on the fiction of isolated magnetic poles which obeyed a Coulomb law of force, $F \propto mm'/s^2$. The constant of proportionality was arbitrarily assigned the value unity for free space. The unit pole was defined as one of such magnitude that it would repel a like pole one centimeter distant in free space with a force of one dyne.

A *practical* system of units was based on a definition of the ampere from Faraday's laws of electrolysis. The ohm was defined as the resistance of a column of mercury of specified dimensions. The unit of potential difference, the volt, was derived from Ohm's law: $V = IR$.

The *meter-kilogram-second* (mks) system of units was devised chiefly to lessen the confusion that results from the intermingled use of the practical system and the electrostatic and electromagnetic units of the cgs system. The mks system was proposed by Prof. Giovanni L. T. C. Giorgi in 1903. The International Committee on Weights and Measures adopted the mks system in 1935.

In mechanics, the mks system introduces only one new unit, the *newton*. From

$$f = ma$$

mks: $$1 \text{ N} = (1 \text{ kg})(1 \text{ m/s}^2)$$

cgs: $$1 \text{ dyn} = (1 \text{ g})(1 \text{ cm/s}^2)$$
$$1 \text{ N} = (1{,}000 \text{ g})(100 \text{ cm/s}^2)$$
$$= 10^5 \text{ dyn}$$

The mks unit of work is the newton-meter which is equivalent to 10^7 dyn·cm = 10^7 ergs = 1 J. The mks unit of power is the newton-meter per second, which is the same as the joule per second or the *watt*, the practical unit of power.

In electricity and magnetism, the mks system employs the well-known practical electrical units, such as the ampere, volt, ohm, and watt; it eliminates the powers of 10 that occur in the relations between the cgs units. The mks unit of charge, the coulomb, is one which in free space would repel an equal charge at a distance of 1 m with a force of 9×10^9 N. Use of this value in Coulomb's law, $F = qq'/\varepsilon_0 s^2$, shows that (in unrationalized units) the permittivity of free space is not unity, but $\varepsilon_0 = 1/(9 \times 10^9)\ \text{C}^2/\text{N}\cdot\text{m}^2$. Similarly, the value for the permeability μ_0 of free space in unrationalized mks units is $10^{-7}\ \text{N/A}^2$.

The mks system based on Coulomb's law in the forms just described leaves an inconvenient factor 4π in the magnetic equations, and even in some electric equations factors 2π or 4π appear where their presence is not immediately suggested by any geometrical considerations. The so-called *rationalized* mks system assigns values to both ε_0 and μ_0 for free space which differ from those stated above in the ratio of 4π to 1. As a consequence, expressions in which ε_0 and μ_0 appear have different values when stated in terms of the rationalized and the unrationalized units; the two forms of the mks system are otherwise the same. The values of ε_0 and μ_0 are given in Table 2 for each of the four unit systems considered.

Table 3
ELECTRIC AND MAGNETIC QUANTITIES IN DIFFERENT METRIC SYSTEMS

Quantity	Symbol	Rationalized mks	Unrationalized cgs electrostatic	Unrationalized cgs electromagnetic
Length	l	1 m	100 cm	100 cm
Mass	m	1 kg	1,000 g	1,000 g
Time	t	1 s	1 s	1 s
Force	F	1 N	10^5 dyn	10^5 dyn
Torque	L	1 m · N	10^7 cm · dyn	10^7 cm · dyn
Energy	E	1 J	10^7 ergs	10^7 ergs
Power	P	1 W	10^7 ergs/s	10^7 ergs/s
Charge	Q	1 C	3×10^9 statC	10^{-1} abC
Current	I	1 A	3×10^9 statA	10^{-1} abA
Potential difference; emf	V; $\mathcal{E}$	1 V	$\frac{1}{300}$ statV	10^8 abV
Electric field intensity	E	1 V/m	$\frac{1}{3} \times 10^{-4}$ statV/cm	10^6 abV/cm
Resistance	R	1 Ω	$\frac{1}{9 \times 10^{11}}$ statΩ	10^9 abΩ
Resistivity	ρ	1 Ω · m	$\frac{1}{9 \times 10^9}$ statΩ · cm	10^{11} abΩ · cm
Permittivity, free space	ε_0	$\frac{1}{36\pi \times 10^9}$ $C^2/N \cdot m^2$	$\frac{\text{statC}^2}{\text{dyn} \cdot \text{cm}^2}$	
Permeability, free space	μ_0	$4\pi \times 10^{-7}$ N/A^2		$\frac{(\text{Unit pole})^2}{\text{dyn} \cdot \text{cm}^2}$
Capacitance	C	1 F	9×10^{11} statF	10^{-9} abF
Inductance	L or M	1 He	$\frac{1}{9 \times 10^{11}}$ statHe	10^9 abHe
Magnetic flux	Φ	1 Wb (or V · s)		10^8 Mxw
Magnetic flux density	B	1 Wb/m^2, or T		10^4 Ga
Magnetic field intensity (magnetizing force)	H	1 A · turn/m		$4\pi \times 10^{-3}$ Oe
Pole strength	m	1 N · m^2/Wb		$\frac{10^9}{4\pi}$ emu

A comparison of expressions for capacitance in unrationalized and rationalized mks units will indicate the easier association for factors 2π and 4π with geometry in the rationalized system. In the rationalized system no π's appear in "plane" formulas, 2π appears in "cylindrical" formulas, and 4π appears in "spherical" formulas.

The advantage claimed for the rationalized mks system is not that it contains fewer π's, but that they appear in the "right places."

Capacitance of:	Unrationalized units	Rationalized units
Plane capacitor	$C = \dfrac{\varepsilon A}{4\pi s}$	$C = \dfrac{\varepsilon A}{s}$
Coaxial cylinders	$C = \dfrac{\varepsilon L}{2 \ln (b/a)}$	$C = \dfrac{2\pi\varepsilon L}{2 \ln (b/a)}$
Concentric spheres	$C = \dfrac{\varepsilon ab}{b - a}$	$C = \dfrac{4\pi\varepsilon ab}{b - a}$

In Table 3, some of the commonly used units of the various systems are listed with equal signs implied along any one row. Electromagnetic units for electrical quantities are given the prefix *ab-* (absolute) and the name of the practical unit. Electrostatic units are indicated by the prefix *stat-* and the name of the practical unit. Nameless units are given the family designation: esu or emu.

The speed of electromagnetic radiation in free space is taken as 3×10^8 m/s.

A-9 DERIVATIONS

Moment of inertia of a solid cylinder Since the moment of inertia of a body is defined as the sum of the products mr^2 for all particles of the body, we can compute this quantity by carrying out the summation $I = \Sigma mr^2$.

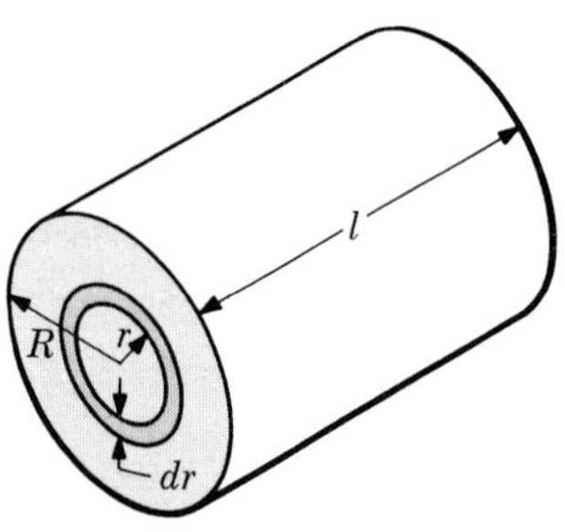

Figure A-1
Moment of inertia of a cylinder.

Consider the cylinder of Fig. A-1. We shall select as an element of mass dm all particles that are at a distance r from the axis. This represents a shell of radius r, length l, and thickness dr. The volume dV of the shell is

$$dV = 2\pi rl\, dr$$

and the mass dm is $\rho\, dV$, where ρ is the mass per unit volume,

$$dm = 2\pi\rho lr\, dr$$

and $$dI = r^2\, dm = 2\pi\rho lr^3\, dr$$

Integration between the limits of 0 and R gives

$$I = \int_0^R 2\pi\rho lr^3\, dr = 2\pi\rho l\left[\frac{r^4}{4}\right]_0^R = 2\pi\rho l\frac{R^4}{4}$$
$$= \tfrac{1}{2}\pi\rho lR^4$$

But the mass m of the cylinder is

$$m = \pi R^2 l\rho$$

and $$I = \tfrac{1}{2}(\pi R^2 l\rho)R^2 = \tfrac{1}{2}mR^2$$

Moments of inertia of other regular bodies, calculated in the same way as that of the cylinder in the example, are given in Table 1 of Chap. 6.

Burnout velocity (see Chap. 9) In differential notation Eq. (9-15) is

$$\int_{v_0}^{V_b} dv = -\bar{v}_e \int_{m_0}^{m_b} \frac{dm}{m} - \bar{g}\int_0^t dt$$

Integration gives

$$v_b = \bar{v}_e \ln \frac{m_0}{m_b} - \bar{g}t$$

Flux density near a long straight current In Fig. A-2 an infinitely long straight wire AB carries a current I. The flux density at a point P distant

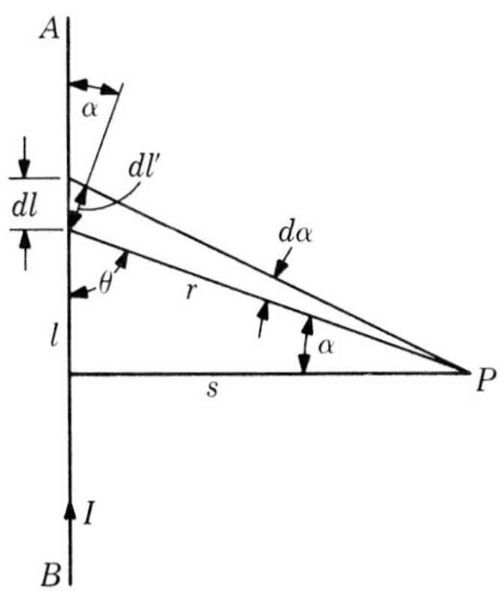

Figure A-2
Biot-Savart law.

s from the wire may be found from

$$dB = \frac{\mu_0}{4\pi} \frac{I\, dl \sin\theta}{r^2}$$

Since α and θ are complementary angles, we may write

$$dB = \frac{\mu_0}{4\pi} \frac{I\, dl \cos\alpha}{r^2}$$

By construction, $dl' = dl \cos\alpha$, and $d\alpha = dl'/r$. Hence

$$dl \cos\alpha = dl' = r\, d\alpha$$

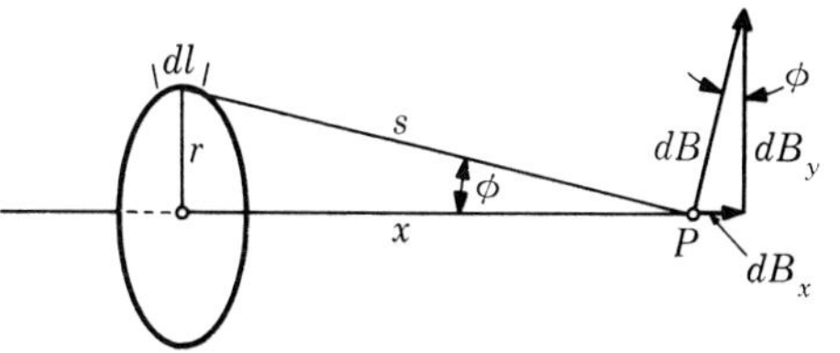

Figure A-3
Magnetic flux density on the axis of a coil.

Also $$r = \frac{s}{\cos\alpha}$$

When these values are substituted, we obtain

$$dB = \frac{\mu_0}{4\pi} \frac{I \cos\alpha\, d\alpha}{s}$$

The total flux density at P is the sum of the contributions from all the elements, that is, the integral when α varies from $-90°$ to $+90°$.

$$B = \frac{\mu_0}{4\pi} \frac{I}{s} \int_{-90°}^{90°} \cos\alpha\, d\alpha = \frac{\mu_0}{4\pi} \frac{I}{s} \Big[\sin\alpha\Big]_{-90°}^{90°}$$
$$= \frac{\mu_0}{2\pi} \frac{I}{s}$$

This is Eq. (5) of Chap. 20.

Flux density on the axis of a coil The value of dB_x (Fig. A-3) may be obtained from

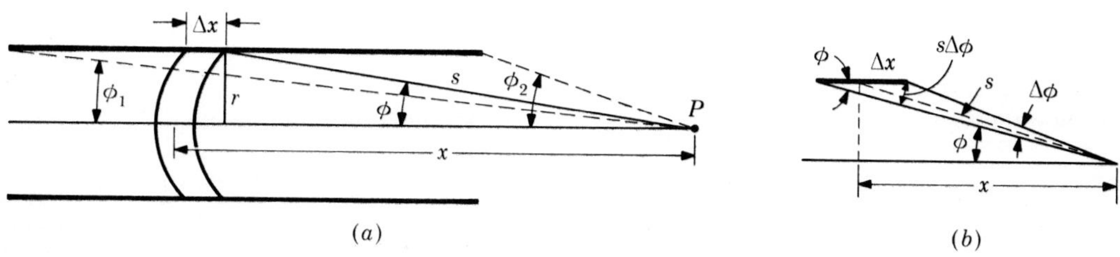

Figure A-4
Magnetic flux density on the axis of a solenoid.

$$dB_x = dB \sin \phi = \frac{\mu_0}{4\pi} \frac{I\,dl}{s^2} \frac{r}{s}$$

$$B = \frac{\mu_0}{4\pi} \frac{Ir}{s^3} \int_0^{2\pi rN} dl$$

$$= \mu_0 \frac{NIr^2}{2s^3}$$

If we substitute $s^2 = r^2 + x^2$, we obtain Eq. (9) of Chap. 20,

$$B = \mu_0 \frac{NIr^2}{2(r^2 + x^2)^{3/2}}$$

Flux density on the axis of a solenoid Consider the case of a closely wound helical coil of N turns, which has an axial length l and radius r and in which there is a current I. In an element of the solenoid of length dx (Fig. A-4) there will be $(N/l)\,dx$ turns. In accordance with the equation derived above for the field on the axis of a coil, the field at P due to the element dx is given by

$$dB = \mu_0 \frac{NIr^2\,dx}{2ls^3}$$

To integrate this expression, it is convenient to express dx and s in terms of the single variable ϕ. From Fig. A-4b it is seen that

$$\sin \phi = \frac{s\,d\phi}{dx}$$

from which $$dx = \frac{s\,d\phi}{\sin \phi}$$

Since $\sin \phi$ also equals r/s, we may write the equation for dB as

$$dB = \mu_0 \frac{NI}{2l} \sin \phi\, d\phi$$

The resultant field at P due to the whole solenoid is given by

$$B = \mu_0 \frac{NI}{2l} \int_{\phi_1}^{\phi_2} \sin \phi\, d\phi$$

$$= \mu_0 \frac{NI}{2l} (\cos \phi_1 - \cos \phi_2)$$

This is Eq. (10) of Chap. 20.

Growth of current in an inductive circuit Consider a circuit in which there is a source of negligible internal resistance and potential difference V, a resistor of resistance R, and an inductor of inductance L. At every instant after the switch is closed, the sum of the iR drop in potential and the counter emf due to inductance is equal to the source potential V.

$$V = iR + L\frac{di}{dt}$$

or $$V - iR = L\frac{di}{dt}$$

and $$\frac{V}{R} - i = \frac{L}{R}\frac{di}{dt}$$

Separating the variables i and t, we obtain

$$-\frac{R}{L}\,dt = \frac{di}{i - V/R}$$

At time $t = 0$, $i = 0$. Let i be the current at time t. Then the limits of integration are given by

$$-\frac{R}{L}\int_0^t dt = \int_0^i \frac{di}{i - V/R}$$

By integrating we obtain

$$-\frac{R}{L}t = \left[\ln\left(i - \frac{V}{R}\right)\right]_0^t = \ln\frac{i - V/R}{-V/R}$$

In exponential form this becomes

$$e^{-(R/L)t} = \frac{(V/R) - i}{V/R}$$

Then $$i = \frac{V}{R}(1 - e^{-(R/L)t})$$

Instantaneous emf induced in a rotating coil Consider a coil (Fig. 40-6) rotating counterclockwise at a constant angular speed ω in a uniform magnetic field B. At the instant considered

$$e = -N\frac{d\Phi}{dt} = -NBA\frac{d\cos\alpha}{dt}$$

$$= NBA\sin\alpha\frac{d\alpha}{dt} = NBA\,\omega\sin\alpha$$

Energy of self-induction When a current is established in an inductive circuit, energy must be expended against the emf of self-induction in creating a magnetic field in the circuit. This energy is given by

$$W = \int_0^t ei\,dt$$

Substituting $e = L(di/dt)$, the equation becomes

$$W = \int_0^t L\frac{di}{dt}i\,dt = \int_0^I Li\,di = \left[\tfrac{1}{2}Li^2\right]_0^I = \tfrac{1}{2}LI^2$$

Effective values of current and voltage The effective value of a current is equal to the magnitude of a steady direct current that would produce the same heating effect. The value ordinarily given for an alternating current is its effective, or rms, value.

A curve of current squared for one cycle of a sinusoidal current is shown in Fig. A-5. The value of the average of the square of the current can be obtained by calculating the area of one lobe of this curve and dividing this area by the width of the lobe. In order to compute the lobe area, we sum up all the infinitesimal areas $i^2\,d\theta$, such as that shown in the shaded rectangle, and obtain

$$\int_0^\pi i^2\,d\theta = \int_0^\pi i_m{}^2\sin^2\theta\,d\theta$$

$$= i_m{}^2\left[\frac{\theta}{2} - \frac{\sin 2\theta}{4}\right]_0^\pi = \frac{\pi}{2}i_m{}^2$$

The average ordinate is obtained by dividing this lobe area by π, the width of the lobe; thus the average ordinate is $\frac{1}{2}i_m{}^2$. Hence the square of the steady current I which would produce the same heating effect as the alternating current is given by

$$I^2 = \tfrac{1}{2}i_m{}^2$$

The square root of this value is the root-mean-square or effective current, namely,

$$I = \frac{i_m}{\sqrt{2}} = 0.707i_m$$

Similarly, the effective value E of an alternating emf is defined as its rms value. If the emf varies sinusoidally

$$E = 0.707e_m$$

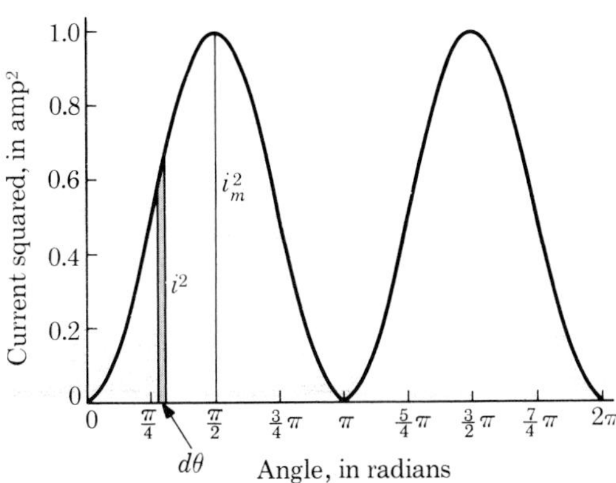

Figure A-5
Effective value of alternating current is the square root of the mean-square current.

Phase relations in ac series circuits In a series circuit that has resistance, inductance, and capacitance, there are three parts to the instantaneous voltage: $v_R = iR$, $e_L = L(di/dt)$, and $e_C = q/C$. The applied potential v at each instant is the sum of the three,

$$v = iR + L\frac{di}{dt} + \frac{q}{C}$$

If the current is sinusoidal,

$$i = i_m \sin\theta = i_m \sin 2\pi ft$$

$$v = i_m R \sin\theta + L\frac{d(i_m \sin 2\pi ft)}{dt} + \frac{1}{C}\int i_m \sin 2\pi ft\, dt$$

$$= i_m R \sin\theta + 2\pi fLi_m \cos(2\pi ft) - \frac{i_m}{2\pi fC}\cos 2\pi ft$$

$$= i_m R \sin\theta + 2\pi fLi_m \cos\theta - \frac{i_m}{2\pi fC}\cos\theta$$

Thus we see that the resistance component of the voltage is in phase with the current, the inductive component is 90° ahead of the current in phase, and the capacitive voltage is 90° behind the current in phase.

Since the component voltages are out of phase, we may represent their maximum (or effective) values by a phase diagram (Fig. A-6). The resultant voltage v_m is

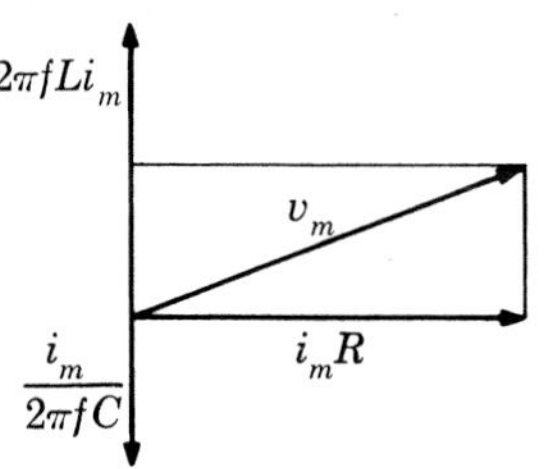

Figure A-6
Phase diagram for a series circuit.

$$v_m = i_m\sqrt{R^2 + \left(2\pi fL - \frac{1}{2\pi fC}\right)^2}$$

or for effective voltages

$$V = I\sqrt{R^2 + \left(2\pi fL - \frac{1}{2\pi fC}\right)^2}$$

since $\quad V = \frac{v_m}{\sqrt{2}} \quad$ and $\quad I = \frac{i_m}{\sqrt{2}}$

A-10 SELECTED REFERENCE TABLES

Table 4
VALUES OF IMPORTANT PHYSICAL CONSTANTS†

Acceleration due to gravity (sea level, 45°)	$g = 9.80665\ m/s^2$
Density of water, maximum	$pH_2O = 0.999972 \times 10^3\ kg/m^3$
Density of mercury	$pHg = 13.5950 \times 10^3\ kg/m^3$
Standard atmospheric pressure	$p_0 = 760.00$ mm Hg
	$= 1{,}013.246$ mb
Gravitational constant	$G = 6.670 \times 10^{-11}\ N \cdot m^2/kg^2$

(Continued)

Table 4 (Continued)

Volume of ideal gas (0°C, 1 atm)	$V_0 = 22.4207\ m^3/kg \cdot mol$
Avogadro's number	$N_0 = 6.02486 \times 10^{26}/kg \cdot mol$
Ideal gas constant	$R_0 = 8316.96\ J/(kg \cdot mol)(K°)$
Boltzmann's constant	$k = R_0/N_0$
	$= 1.38042 \times 10^{-23}\ J/K°$
	$= 8.6164\ eV/K°$
Speed of light (free space)	$c = 2.997930 \times 10^8\ m/s$
Electronic charge (magnitude)	$e = 1.60206 \times 10^{-19}\ C$
Charge-to-mass ratio of electron	$e/m = 1.75890 \times 10^{11}\ C/kg$
Faraday's constant	$F = N_0 e$
	$= 9.65219 \times 10^7\ C/kg \cdot mol$
Planck's constant	$h = 6.62517 \times 10^{-34}\ J \cdot s$
Atomic mass unit ($\frac{1}{12}$ mass of ${}_6C^{12}$)	1 amu $= 1.660 \times 10^{-27}$ kg
Masses of particles at rest	
Neutron:	1.008982 amu
	1.67470×10^{-27} kg
Proton:	1.007593 amu
	1.67239×10^{-27} kg
Hydrogen atom:	1.008142 amu
Electron:	5.4876×10^{-4} amu
	9.1083×10^{-31} kg
Deuterium atom:	2.014735 amu
Mass-energy conversion factors	
1 amu =	931.14 MeV
1 gm =	5.610×10^{26} MeV
1 electron mass =	0.51098 MeV
Stefan-Boltzmann constant	$\sigma = 5.6687 \times 10^{-8}\ J/(m^2)(s)(°K)^4$
Ice point	$T_0 = 273.16°K$

†As experimental techniques improve, "best values" of the physical constants are recomputed by statistical methods. See, for example, E. R. Cohen, J. W. M. DuMond, T. W. Layton, and J. S. Rollett, Analysis of Variance of the 1952 Data on the Atomic Constants and a New Adjustment, 1955, *Reviews of Modern Physics,* **27**:363–380 (1955). The values listed in the table (except those for G, g, and ρ) are taken from the article cited and have been expressed to an appropriate number of significant figures. Uncertainties in these constants are discussed in the original paper. The physical scale is used for all constants involving atomic masses.

Table 5
SATURATED WATER VAPOR

Showing pressure P and density ρ of aqueous vapor saturated at temperature t; or showing boiling point t of water and density ρ of steam corresponding to a pressure P

t, °C	P, mm Hg	ρ, g/cm^3($\times 10^{-6}$)
−10	2.0	2.2
−5	3.0	3.3
0	4.6	4.9
5	6.5	6.8
10	9.2	9.4
12	10.5	10.7
14	12.0	12.1
16	13.6	13.6
18	15.5	15.6
20	17.6	17.3
22	19.8	19.4
24	22.4	21.8
26	25.2	24.4
30	31.8	30.4
40	55.1	51.1
50	92.3	83.2
60	149.2	130.5
70	233.5	198.4
80	355.1	298.8
90	525.8	424.1
95	634.0	505
96	657.7	523
97	682.1	541
98	707.3	560
99.0	733.3	579
100.0	760.0	598
101	787.5	618
102	815.9	639
110	1,074.5	827
120	1,489	1,122
200	11,650	7,840

Table 6
NATURAL SINES AND COSINES

NATURAL SINES

Angle	0.0	0.1	0.2	0.3	0.4	0.5	0.6	0.7	0.8	0.9	Complement difference		
0°	0.0000	0017	0035	0052	0070	0087	0105	0122	0140	0157	0175	89°	
1	0175	0192	0209	0227	0244	0262	0279	0297	0314	0332	0349	88	
2	0349	0366	0384	0401	0419	0436	0454	0471	0488	0506	0523	87	
3	0523	0541	0558	0576	0593	0610	0628	0645	0663	0680	0698	86	
4	0698	0715	0732	0750	0767	0785	0802	0819	0837	0854	0872	85	
5	0.0872	0889	0906	0924	0941	0958	0976	0993	1011	1028	1045	84	
6	1045	1063	1080	1097	1115	1132	1149	1167	1184	1201	1219	83	
7	1219	1236	1253	1271	1288	1305	1323	1340	1357	1374	1392	82	
8	1392	1409	1426	1444	1461	1478	1495	1513	1530	1547	1564	81	
9	1564	1582	1599	1616	1633	1650	1668	1685	1702	1719	1736	80	
10	0.1736	1754	1771	1788	1805	1822	1840	1857	1874	1891	1908	79	
11	1908	1925	1942	1959	1977	1994	2011	2028	2045	2062	2079	78	
12	2079	2096	2113	2130	2147	2164	2181	2198	2215	2233	2250	77	17
13	2250	2267	2284	2300	2317	2334	2351	2368	2385	2402	2419	76	
14	2419	2436	2453	2470	2487	2504	2521	2538	2554	2571	2588	75	
15	0.2588	2605	2622	2639	2656	2672	2689	2706	2723	2740	2756	74	
16	2756	2773	2790	2807	2823	2840	2857	2874	2890	2907	2924	73	
17	2924	2940	2957	2974	2990	3007	3024	3040	3057	3074	3090	72	
18	3090	3107	3123	3140	3156	3173	3190	3206	3223	3239	3256	71	
19	3256	3273	3289	3305	3322	3338	3355	3371	3387	3404	3420	70	
20	0.3420	3437	3453	3469	3486	3502	3518	3535	3551	3567	3584	69	
21	3584	3600	3616	3633	3649	3665	3681	3697	3714	3730	3746	68	
22	3746	3762	3778	3795	3811	3827	3843	3859	3875	3891	3907	67	
23	3907	3923	3939	3955	3971	3987	4003	4019	4035	4051	4067	66	16
24	4067	4083	4099	4115	4131	4147	4163	4179	4195	4210	4226	65	
25	0.4226	4242	4258	4274	4289	4305	4321	4337	4352	4368	4384	64	
26	4384	4399	4415	4431	4446	4462	4478	4493	4509	4524	4540	63	
27	4540	4555	4571	4586	4602	4617	4633	4648	4664	4679	4695	62	
28	4695	4710	4726	4741	4756	4772	4787	4802	4818	4833	4848	61	
29	4848	4863	4879	4894	4909	4924	4939	4955	4970	4985	5000	60	
30	0.5000	5015	5030	5045	5060	5075	5090	5105	5120	5135	5150	59	15
31	5150	5165	5180	5195	5210	5225	5240	5255	5270	5284	5299	58	
32	5299	5314	5329	5344	5358	5373	5388	5402	5417	5432	5446	57	
33	5446	5461	5476	5490	5505	5519	5534	5548	5563	5577	5592	56	
34	5592	5606	5621	5635	5650	5664	5678	5693	5707	5721	5736	55	
35	0.5736	5750	5764	5779	5793	5807	5821	5835	5850	5864	5878	54	
36	5878	5892	5906	5920	5934	5948	5962	5976	5990	6004	6018	53	14
37	6018	6032	6046	6060	6074	6088	6101	6115	6129	6143	6157	52	
38	6157	6170	6184	6198	6211	6225	6239	6252	6266	6280	6293	51	
39	6293	6307	6320	6334	6347	6361	6374	6388	6401	6414	6428	50	
40	0.6428	6441	6455	6468	6481	6494	6508	6521	6534	6547	6561	49	
41	6561	6574	6587	6600	6613	6626	6639	6652	6665	6678	6691	48	13
42	6691	6704	6717	6730	6743	6756	6769	6782	6794	6807	6820	47	
43	6820	6833	6845	6858	6871	6884	6896	6909	6921	6934	6947	46	
44°	6947	6959	6972	6984	6997	7009	7022	7034	7046	7059	7071	45°	
Complement		**0.9**	**0.8**	**0.7**	**0.6**	**0.5**	**0.4**	**0.3**	**0.2**	**0.1**	**0.0**	**Angle**	

NATURAL COSINES

Table 6 (Continued)

NATURAL SINES

Angle	0.0	0.1	0.2	0.3	0.4	0.5	0.6	0.7	0.8	0.9	Complement difference		
45°	0.7071	7083	7096	7108	7120	7133	7145	7157	7169	7181	7193	44°	
46	7193	7206	7218	7230	7242	7254	7266	7278	7290	7302	7314	43	12
47	7314	7325	7337	7349	7361	7373	7385	7396	7408	7420	7431	42	
48	7431	7443	7455	7466	7478	7490	7501	7513	7524	7536	7547	41	
49	7547	7559	7570	7581	7593	7604	7615	7627	7638	7649	7660	40	
50	0.7660	7672	7683	7694	7705	7716	7727	7738	7749	7760	7771	39	
51	7771	7782	7793	7804	7815	7826	7837	7848	7859	7869	7880	38	11
52	7880	7891	7902	7912	7923	7934	7944	7955	7965	7976	7986	37	
53	7986	7997	8007	8018	8028	8039	8049	8059	8070	8080	8090	36	
54	8090	8100	8111	8121	8131	8141	8151	8161	8171	8181	8192	35	
55	0.8192	8202	8211	8221	8231	8241	8251	8261	8271	8281	8290	34	10
56	8290	8300	8310	8320	8329	8339	8348	8358	8368	8377	8387	33	
57	8387	8396	8406	8415	8425	8434	8443	8453	8462	8471	8480	32	
58	8480	8490	8499	8508	8517	8526	8536	8545	8554	8563	8572	31	
59	8572	8281	8590	8599	8607	8616	8625	8634	8643	8652	8660	30	9
60	0.8660	8669	8678	8686	8695	8704	8712	8721	8729	8738	8746	29	
61	8746	8755	8763	8771	8780	8788	8796	8805	8813	8821	8829	28	
62	8829	8838	8846	8854	8862	8870	8878	8886	8894	8902	8910	27	8
63	8910	8918	8926	8934	8942	8949	8957	8965	8973	8980	8988	26	
64	8988	8996	9003	9011	9018	9026	9033	9041	9048	9056	9063	25	
65	0.9063	9070	9078	9085	9092	9100	9107	9114	9121	9128	9135	24	
66	9135	9143	9150	9157	9164	9171	9178	9184	9191	9198	9205	23	7
67	9205	9212	9219	9225	9232	9239	9245	9252	9259	9265	9272	22	
68	9272	9278	9285	9291	9298	9304	9311	9317	9323	9330	9336	21	
69	9336	9342	9348	9354	9361	9367	9373	9379	9385	9391	9397	20	6
70	0.9397	9403	9409	9415	9421	9426	9432	9438	9444	9449	9455	19	
71	9455	9461	9466	9472	9478	9483	9489	9494	9500	9505	9511	18	
72	9511	9516	9521	9527	9532	9537	9542	9548	9553	9558	9563	17	
73	9563	9568	9573	9578	9583	9588	9593	9598	9603	9608	9613	16	5
74	9613	9617	9622	9627	9632	9636	9641	9646	9650	9655	9659	15	
75	0.9659	9664	9668	9673	9677	9681	9686	9690	9694	9699	9703	14	
76	9703	9707	9711	9715	9720	9724	9728	9732	9736	9740	9744	13	4
77	9744	9748	9751	9755	9759	9763	9767	9770	9774	9778	9781	12	
78	9781	9785	9789	9792	9796	9799	9803	9806	9810	9813	9816	11	
79	9816	9820	9823	9826	9829	9833	9836	9839	9842	9845	9848	10	
80	0.9848	9851	9854	9857	9860	9863	9866	9869	9871	9874	9877	9	3
81	9877	9880	9882	9885	9888	9890	9893	9895	9898	9900	9903	8	
82	9903	9905	9907	9910	9912	9914	9917	9919	9921	9923	9925	7	
83	9925	9928	9930	9932	9934	9936	9938	9940	9942	9943	9945	6	2
84	9945	9947	9949	9951	9952	9954	9956	9957	9959	9960	9962	5	
85	0.9962	9963	9965	9966	9968	9969	9971	9972	9973	9974	9976	4	
86	9976	9977	9978	9979	9980	9981	9982	9983	9984	9985	9986	3	1
87	9986	9987	9988	9989	9990	9990	9991	9992	9993	9993	9994	2	
88	9994	9995	9995	9996	9996	9997	9997	9997	9998	9998	9998	1	
89°	9998	9999	9999	9999	9999	1.0000	1.0000	1.0000	1.0000	1.0000	1.0000	0°	0
Complement		0.9	0.8	0.7	0.6	0.5	0.4	0.3	0.2	0.1	0.0	Angle	

NATURAL COSINES

Table 7
NATURAL TANGENTS AND COTANGENTS

NATURAL TANGENTS

Angle	0.0	0.1	0.2	0.3	0.4	0.5	0.6	0.7	0.8	0.9	Complement difference		
0°	0.0000	0017	0035	0052	0070	0087	0105	0122	0140	0157	0175	89°	
1	0175	0192	0209	0227	0244	0262	0279	0297	0314	0332	0349	88	
2	0349	0367	0384	0402	0419	0437	0454	0472	0489	0507	0524	87	
3	0524	0542	0559	0577	0594	0612	0629	0647	0664	0682	0699	86	
4	0699	0717	0734	0752	0769	0787	0805	0822	0840	0857	0875	85	
5	0.0875	0892	0910	0928	0945	0963	0981	0998	1016	1033	1051	84	
6	1051	1069	1086	1104	1122	1139	1157	1175	1192	1210	1228	83	
7	1228	1246	1263	1281	1299	1317	1334	1352	1370	1388	1405	82	
8	1405	1423	1441	1459	1477	1495	1512	1530	1548	1566	1584	81	
9	1584	1602	1620	1638	1655	1673	1691	1709	1727	1745	1763	80	
10	0.1763	1781	1799	1817	1835	1853	1871	1890	1908	1926	1944	79	18
11	1944	1962	1980	1998	2016	2035	2053	2071	2089	2107	2126	78	
12	2126	2144	2162	2180	2199	2217	2235	2254	2272	2290	2309	77	
13	2309	2327	2345	2364	2382	2401	2419	2438	2456	2475	2493	76	
14	2493	2512	2530	2549	2568	2586	2605	2623	2642	2661	2679	75	
15	0.2679	2698	2717	2736	2754	2774	2792	2811	2830	2849	2867	74	
16	2867	2886	2905	2924	2943	2962	2981	3000	3019	3038	3057	73	19
17	3057	3076	3096	3115	3134	3153	3172	3191	3211	3230	3249	72	
18	3249	3269	3288	3307	3327	3346	3365	3385	3404	3424	3443	71	
19	3443	3463	3482	3502	3522	3541	3561	3581	3600	3620	3640	70	
20	0.3640	3659	3679	3699	3719	3739	3759	3779	3799	3819	3839	69	
21	3839	3859	3879	3899	3919	3939	3959	3979	4000	4020	4040	68	20
22	4040	4061	4081	4101	4122	4142	4163	4183	4204	4224	4245	67	
23	4245	4265	4286	4307	4327	4348	4369	4390	4411	4431	4452	66	
24	4452	4473	4494	4515	4536	4557	4578	4599	4621	4642	4662	65	21
25	0.4663	4684	4706	4727	4748	4770	4791	4813	4834	4856	4877	64	
26	4877	4899	4921	4942	4964	4986	5008	5029	5051	5073	5095	63	
27	5095	5117	5139	5161	5184	5206	5228	5250	5272	5295	5317	62	22
28	5317	5340	5362	5384	5407	5430	5452	5475	5498	5520	5543	61	
29	5543	5566	5589	5612	5635	5658	5681	5704	5727	5750	5774	60	23
30	0.5774	5797	5820	5844	5867	5890	5914	5938	5961	5985	6009	59	
31	6009	6032	6056	6080	6104	6128	6152	6176	6200	6224	6249	58	24
32	6249	6273	6297	6322	6346	6371	6395	6420	6445	6469	6494	57	
33	6494	6519	6544	6569	6594	6619	6644	6669	6694	6720	6745	56	25
34	6745	6771	6796	6822	6847	6873	6899	6924	6950	6976	7002	55	
35	0.7002	7028	7054	7080	7107	7133	7159	7186	7212	7239	7265	54	26
36	7265	7292	7319	7346	7373	7400	7427	7454	7481	7508	7536	53	27
37	7536	7563	7590	7618	7646	7673	7701	7729	7757	7785	7813	52	28
38	7813	7841	7869	7898	7926	7954	7983	8012	8040	8069	8098	51	28
39	8098	8127	8156	8185	8214	8243	8273	8302	8332	8361	8391	50	29
40	0.8391	8421	8451	8481	8511	8541	8571	8601	8632	8662	8693	49	30
41	8693	8724	8754	8785	8816	8847	8878	8910	8941	8972	9004	48	31
42	9004	9036	9067	9099	9131	9163	9195	9228	9260	9293	9325	47	32
43	9325	9358	9391	9424	9557	9490	9523	9556	9590	9623	9567	46	33
44°	9657	9691	9725	9759	9793	9827	9861	9896	9930	9965	1.0000	45°	34
Complement		**0.9**	**0.8**	**0.7**	**0.6**	**0.5**	**0.4**	**0.3**	**0.2**	**0.1**	**0.0**	**Angle**	

NATURAL COTANGENTS

Table 7 (Continued)

NATURAL TANGENTS

Angle	0.0	0.1	0.2	0.3	0.4	0.5	0.6	0.7	0.8	0.9	Comp.
45°	1.0000	1.0035	1.0070	1.0105	1.0141	1.0176	1.0212	1.0247	1.0283	1.0319	44°
46	1.0355	1.0392	1.0428	1.0464	1.0501	1.0538	1.0575	1.0612	1.0649	1.0686	43
47	1.0724	1.0761	1.0799	1.0837	1.0875	1.0913	1.0951	1.0990	1.1028	1.1067	42
48	1.1106	1.1145	1.1184	1.1224	1.1263	1.1303	1.1343	1.1383	1.1423	1.1463	41
49	1.1504	1.1544	1.1585	1.1626	1.1667	1.1708	1.1750	1.1792	1.1833	1.1875	40
50	1.1918	1.1960	1.2002	1.2045	1.2088	1.2131	1.2174	1.2218	1.2261	1.2305	39
51	1.2349	1.2393	1.2437	1.2482	1.2527	1.2572	1.2617	1.2662	1.2708	1.2753	38
52	1.2799	1.2846	1.2892	1.2938	1.2985	1.3032	1.3079	1.3127	1.3175	1.3222	37
53	1.3270	1.3319	1.3367	1.3416	1.3465	1.3514	1.3564	1.3613	1.3663	1.3713	36
54	1.3764	1.3814	1.3865	1.3916	1.3968	1.4019	1.4071	1.4124	1.4176	1.4229	35
55	1.4281	1.4335	1.4388	1.4442	1.4496	1.4550	1.4605	1.4659	1.4715	1.4770	34
56	1.4826	1.4882	1.4938	1.4994	1.5051	1.5108	1.5166	1.5224	1.5282	1.5340	33
57	1.5399	1.5458	1.5517	1.5577	1.5637	1.5697	1.5757	1.5818	1.5880	1.5941	32
58	1.6003	1.6066	1.6128	1.6191	1.6255	1.6319	1.6383	1.6447	1.6512	1.6577	31
59	1.6643	1.6709	1.6775	1.6842	1.6909	1.6977	1.7045	1.7113	1.7182	1.7251	30
60	1.7321	1.7391	1.7461	1.7532	1.7603	1.7675	1.7747	1.7820	1.7893	1.7966	29
61	1.8040	1.8115	1.8190	1.8265	1.8341	1.8418	1.8495	1.8572	1.8650	1.8728	28
62	1.8807	1.8887	1.8967	1.9047	1.9128	1.9210	1.9292	1.9375	1.9458	1.9542	27
63	1.9626	1.9711	1.9797	1.9883	1.9970	2.0057	2.0145	2.0233	2.0323	2.0413	26
64	2.0503	2.0594	2.0686	2.0778	2.0872	2.0965	2.1060	2.1155	2.1251	2.1348	25
65	2.145	2.154	2.164	2.174	2.184	2.194	2.204	2.215	2.225	2.236	24
66	2.246	2.257	2.257	2.278	2.289	2.300	2.311	2.322	2.333	2.344	23
67	2.356	2.367	2.379	2.391	2.402	2.414	2.426	2.438	2.450	2.463	22
68	2.475	2.488	2.500	2.513	2.526	2.539	2.552	2.565	2.578	2.592	21
69	2.605	2.619	2.633	2.646	2.660	2.675	2.689	2.703	2.718	2.733	20
70	2.747	2.762	2.778	2.793	2.808	2.824	2.840	2.856	2.872	2.888	19
71	2.904	2.921	2.937	2.954	2.971	2.989	3.006	3.024	3.042	3.060	18
72	3.078	3.096	3.115	3.133	3.152	3.172	3.191	3.211	3.230	3.250	17
73	3.271	3.291	3.312	3.333	3.354	3.376	3.398	3.420	3.442	3.465	16
74	3.487	3.511	3.534	3.558	3.582	3.606	3.630	3.655	3.681	3.700	15
75	3.732	3.758	3.785	3.812	3.839	3.867	3.895	3.923	3.952	3.981	14
76	4.011	4.041	4.071	4.102	4.134	4.165	4.198	4.230	4.264	4.297	13
77	4.331	4.366	4.402	4.437	4.474	4.511	4.548	4.586	4.625	4.665	12
78	4.705	4.745	4.787	4.829	4.872	4.915	4.959	5.005	5.050	5.097	11
79	5.145	5.193	5.242	5.292	5.343	5.396	5.449	5.503	5.558	5.614	10
80	5.67	5.73	5.79	5.85	5.91	5.98	6.04	6.11	6.17	6.24	9
81	6.31	6.39	6.46	6.54	6.61	6.69	6.77	6.86	6.94	7.03	8
82	7.12	7.21	7.30	7.40	7.49	7.60	7.70	7.81	7.92	8.03	7
83	8.14	8.26	8.39	8.51	8.64	8.78	8.92	9.06	9.21	9.36	6
84	9.51	9.68	9.84	10.0	10.2	10.4	10.6	10.8	11.0	11.2	5
85	11.4	11.7	11.9	12.2	12.4	12.7	13.0	13.3	13.6	14.0	4
86	14.3	14.7	15.1	15.5	15.9	16.3	16.8	17.3	17.9	18.5	3
87	19.1	19.7	20.4	21.2	22.0	22.9	23.9	24.9	26.0	27.3	2
88	28.6	30.1	31.8	33.7	35.8	38.2	40.9	44.1	47.7	52.1	1
89°	57.	64.	72.	95.	115.	143.	191.	286.	286.	573.	0°
Angle	**1.0**	**0.9**	**0.8**	**0.7**	**0.6**	**0.5**	**0.4**	**0.3**	**0.2**	**0.1**	

NATURAL COTANGENTS

Table 8
THE RANGE OF DISTANCES
IN THE UNIVERSE

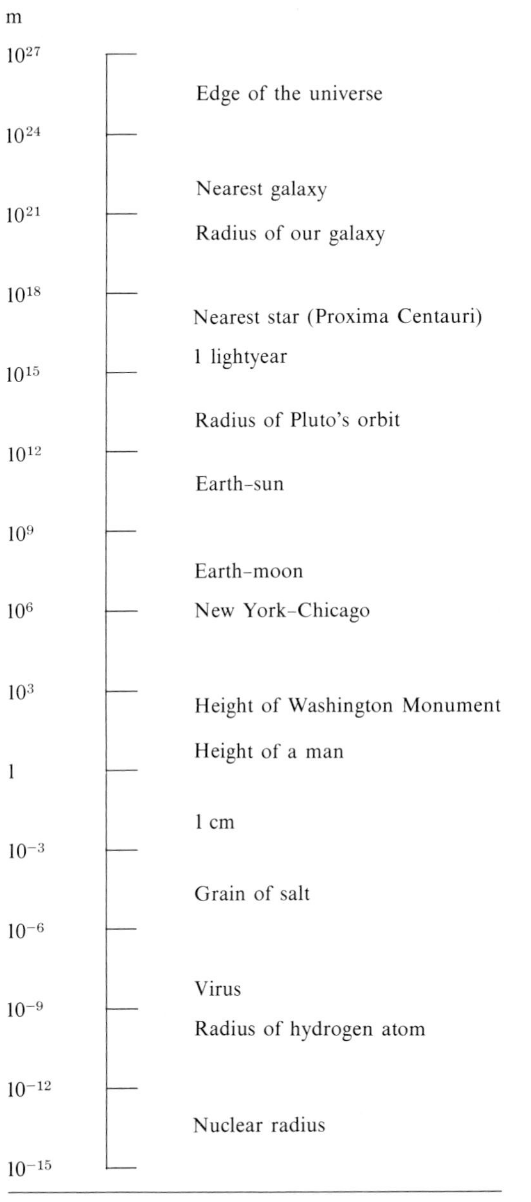

m	
10^{27}	
	Edge of the universe
10^{24}	
	Nearest galaxy
10^{21}	
	Radius of our galaxy
10^{18}	
	Nearest star (Proxima Centauri)
	1 lightyear
10^{15}	
	Radius of Pluto's orbit
10^{12}	
	Earth–sun
10^{9}	
	Earth–moon
10^{6}	New York–Chicago
10^{3}	
	Height of Washington Monument
	Height of a man
1	
	1 cm
10^{-3}	
	Grain of salt
10^{-6}	
	Virus
10^{-9}	
	Radius of hydrogen atom
10^{-12}	
	Nuclear radius
10^{-15}	

Table 9
THE RANGE OF TIME INTERVALS
IN THE UNIVERSE

sec	
10^{18}	Age of universe
	Age of earth
10^{15}	
	Earliest men
10^{12}	Age of pyramids
10^{9}	Lifetime of a man
10^{6}	1 year = 3.156×10^{7} s 1 d = 8.64×10^{4} s
10^{3}	Light travels from sun to earth
1	Interval between heartbeats
10^{-3}	Period of a sound wave
10^{-6}	Period of a radio wave
10^{-9}	Light travels 1 ft
10^{-12}	Period of a molecular vibration
10^{-15}	Period of an atomic vibration
10^{-18}	Light travels an atomic diameter
10^{-21}	Period of a nuclear vibration
	Light travels a nuclear diameter
10^{-24}	

INDEX OF NOBEL LAUREATES IN PHYSICS

Names of physicists who have received the Nobel Prize for Physics and whose portraits appear in this book are listed alphabetically below. The date of the award appears in parentheses.

Subject Index